Handbook of Photonics for Biomedical Science

Series in Medical Physics and Biomedical Engineering

Series Editors: John G Webster, E Russell Ritenour, Slavik Tabakov, and Kwan-Hoong Ng

Other recent books in the series:

Handbook of Anatomical Models for Radiation Dosimetry
Xie George Xu and Keith F. Eckerman (Eds)

Fundamentals of MRI: An Interactive Learning Approach
Elizabeth Berry and Andrew J Bulpitt

Handbook of Optical Sensing of Glucose in Biological Fluids and Tissues
Valery V Tuchin (Ed)

Intelligent and Adaptive Systems in Medicine
Oliver C L Haas and Keith J Burnham

A Introduction to Radiation Protection in Medicine
Jamie V Trapp and Tomas Kron (Eds)

A Practical Approach to Medical Image Processing
Elizabeth Berry

Biomolecular Action of Ionizing Radiation
Shirley Lehnert

An Introduction to Rehabilitation Engineering
R A Cooper, H Ohnabe, and D A Hobson

The Physics of Modern Brachytherapy for Oncology
D Baltas, N Zamboglou, and L Sakelliou

Electrical Impedance Tomography
D Holder (Ed)

Contemporary IMRT
S Webb

The Physical Measurement of Bone
C M Langton and C F Njeh (Eds)

Therapeutic Applications of Monte Carlo Calculations in Nuclear Medicine
H Zaidi and G Sgouros (Eds)

Minimally Invasive Medical Technology
J G Webster (Ed)

Intensity-Modulated Radiation Therapy
S Webb

Series in Medical Physics and Biomedical Engineering

Handbook of Photonics for Biomedical Science

Edited by

Valery V. Tuchin

Saratov State University and

Institute of Precise Mechanics and Control of RAS

Russia

CRC Press is an imprint of the
Taylor & Francis Group, an **informa** business
A TAYLOR & FRANCIS BOOK

CRC Press
Taylor & Francis Group
6000 Broken Sound Parkway NW, Suite 300
Boca Raton, FL 33487-2742

First issued in paperback 2019

ISBN-13: 978-1-4398-0628-9 (hbk)
ISBN-13: 978-0-367-38407-4 (pbk)

Library of Congress Cataloging-in-Publication Data

Handbook of photonics for biomedical science / editor, Valery V. Tuchin.
p. cm. -- (Series in medical physics and biomedical engineering)
Includes bibliographical references and index.
ISBN 978-1-4398-0628-9
1. Imaging systems in medicine--Handbooks, manuals, etc. 2. Phototherapy--Handbooks, manuals, etc. 3. Photobiology--Handbooks, manuals, etc. 4. Photonics--Handbooks, manuals, etc. 5. Nanophotonics--Handbooks, manuals, etc. I. Tuchin, V. V. (Valerii Viktorovich) II. Title. III. Series.

R857.O6H366 2010
616.07'54--dc22 2009038011

Visit the Taylor & Francis Web site at
http://www.taylorandfrancis.com

and the CRC Press Web site at
http://www.crcpress.com

Contents

Preface

Biophotonics as a science and technology is developing very rapidly, therefore to discuss problems and perspectives of this multidisciplinary field, new specific conferences were recently established [1–8]. These meetings focused on the applications of photonics to clinical and investigative medicine, some of them serving as interactive "roadmapping" workshops for leaders in research, industry, and government on the future of biophotonics, others are participant-driven conferences or advanced schools with strong student participation, covering the broad spectrum of biophotonics research and development and applications.

In addition to a specialized *SPIE Journal of Biomedical Optics* [9], two new journals: *Journal of Biophotonics* [10] and *Journal of Innovative Optical Health Sciences* [11] were recently created. These journals serve as international forums for the publication of the latest developments in all areas of photonics for biology and medicine. Many well-recognized and new journals are showing interest in this field of research and applications by publication of special issues or sections on biophotonics [12–15]. There is also a great interest of industry in biomedical photonics [16, 17], which induces educational organizations to open new advanced courses on biophotonics.

At present more than 30 books published by different publishers, which describe methods and techniques of optical biomedical diagnostics and therapy, are available [18–53]. Many of them are devoted to specific topics of biomedical optics such as tissue optics, confocal, two-photon and other nonlinear microscopies, optical coherence tomography, lasers in medicine, optical biomedical diagnostics, applied laser medicine, coherent-domain optical methods in biomedical diagnostics, optical clearing of tissues and blood, optical polarization in biomedical applications, optical glucose sensing, photoacoustic imaging and spectroscopy, etc. Some of them cover a broad range of biomedical photonics problems, including books edited by V.V. Tuchin [33] and Tuan Vo-Dinh [35].

Evidently, a comprehensive Handbook describing advanced biophotonics methods and techniques intensively developed in recent years and written by world-recognized experts in the field may play a significant role in the further development of this prospective technology. We believe that this Handbook is the next step in presenting contemporary advances in biophotonics and will allow researchers, engineers, and medical doctors to be acquainted with major recent published results on biophotonics science and technology, which are dispersed in numerous journals of physical, chemical, biophysical, and biomedical profiles. Thus, this book summarizes and analyzes achievements, new trends, and perspectives of photonics in the application to biomedicine and is very useful.

The most important features of the present book are the following: different advanced photonics methods and techniques effectively used in biomedical science and clinical studies are described on the basis of solid physical background; the presence of a special chapter describing one of the most important methods of mathematical modeling of light transport and interaction with cells – Finite-Difference Time-Domain (FDTD); the discussion of recent optical and terahertz spectroscopy and imaging methods for biomedical diagnostics, based on the interaction of coherent, polarized, and acoustically modulated radiation with tissues and cells; the presentation of novel modalities of photon ballistic (confocal microscopy, OCT), multidimensional fluorescence, Raman, CARS, and other nonlinear spectroscopic microscopies providing molecular-level cell and tissue imaging; the description of key photonic technologies for therapy and surgery, including PDT, low-intensity laser and photothermal therapies; the discussion of nanoparticle photonic technologies and their use for cancer treatment and human organism protection from UV radiation; and advanced spectroscopy and imaging of a variety of normal and pathological tissues (embryonic tissue, eye tissue, skin, brain, gastric tissues, and many others).

The book is for research workers, practitioners, and professionals in the field of biophotonics, and could be used by scientists or professionals in other disciplines, such as laser physics and technology, fiber optics, spectroscopy, materials science, biology, and medicine. Advanced students (MS and PhD), as well as undergraduate students specializing in biomedical physics and engineering, biomedical optics and biophotonics, and medical science may use this book as a comprehensive tutorial helpful in preparation of their research work and matriculation.

This book represents a valuable contribution by well-known experts in the field of biomedical optics and biophotonics with their particular interest in a variety of "hot" biophotonic problems. The contributors are drawn from Canada, China, Cyprus, Denmark, Finland, Germany, Greece, Italy, Ireland, Russia, Scotland, Sweden, UK, Ukraine, and the USA. I greatly appreciate the cooperation and contributions of all authors in the book, who have done great work on preparation of their chapters.

It should be mentioned that this book presents results of international collaborations and exchanges of ideas among all research groups participating in the book project. This book project was supported by many international grants, which are credited in the particular chapters. Here, I would like to mention only one, PHOTONICS4LIFE, which is a consortium of a well-balanced pan-European dimension, self-sufficient in human resources and top-technology showing the proper mass to span across the value chain from photonic components to applications and from fundamental to applied research, while progressing on a single but broad theme: "Biophotonics" [17]. The inclusion of the members of this consortium in this book is of great significance, encompassing eight chapters.

I would like to thank all those authors and publishers who freely granted permissions to reproduce their copyrighted works. I am grateful to Dr. John Navas, Senior Editor, Physics, of Taylor & Francis/CRC Press, for his idea to publish such a book, valuable suggestions, and help on preparation of the manuscript, and to Professor Vladimir L. Derbov, Saratov State University, for preparation of the camera-ready manuscript and help in the technical editing of the book.

I greatly appreciate the cooperation, contributions, and support of all my colleagues from the Optics and Biophotonics Chair and Research-Educational Institute of Optics and Biophotonics of the Physics Department of Saratov State University and the Institute of Precise Mechanics and Control of the Russian Academy of Science.

Last, but not least, I express my gratitude to my family, especially to my wife Natalia and grandchildren Dasha, Zhenya, Stepa, and Serafim for their indispensable support, understanding, and patience during my writing and editing of this book.

References

[1] B. Wilson, V. Tuchin, and S. Tanev (eds.), *Advances in Biophotonics, NATO Science Series I. Life and Behavioural Sciences*, **369**, IOS Press, Amsterdam (2005).

[2] J. Popp, W. Drexler, V.V. Tuchin, and D.L. Matthews (eds.) *Biophotonics: Photonic Solutions for Better Health Care*, Proc. SPIE, **6991**, SPIE Press, Bellingham, WA (2008).

[3] International Congress on Biophotonics, ICOB, The NSF Center for Biophotonics Science & Technology (CBST) of UC Davis, February 3–7, 2008, Sacramento, California, USA, http://cbst.ucdavis.edu/news/first-congress-great-success

[4] 4th International Graduate Summer School on Biophotonics, June 6–13, 2009, Ven, Sweden, www.biop.dk/biophotonics09

[5] II International Symposium Topical Problems of Biophotonics, July 19–24, 2009, Nizhny Novgorod-Samara, Russia, www.biophotonics.sci-nnov.ru

[6] Biophotonics week, series of scientific events, September 22–30, 2009, Quebec, Canada, http://www.biophotonicsworld.org/biophotonicsweek

[7] The second Biophotonic Imaging Graduate Summer School (BIGSS), August 24–28, 2009 Ballyvaughan, Co. Clare, Ireland, http://research1.rcsi.ie/BIP/Events2/Summer_School_-_BIGSS/index.asp

[8] XIII International School for Junior Scientists and Students on Optics, Laser Physics & Biophotonics, 21–24 September, 2009, Saratov, Russia, http://optics.sgu.ru/SFM/

[9] *Journal of Biomedical Optics*, http://spie.org/x866.xml

[10] *Journal of Biophotonics*, www.biophotonics-journal.org

[11] *Journal of Innovative Optical Health Sciences*, http://www.worldscinet.com/jiohs

[12] W.R. Chen, V.V. Tuchin, Q. Luo, and S.L. Jacques (eds.), "Special issue on biophotonics," *J. X-Ray Sci. and Technol.* **10**(3–4), 139–243 (2002).

[13] P. French and A.I. Ferguson (eds.), "Special issue on biophotonics," *J. Phys. D: Appl. Phys.* **36**(14), R207–R258; 1655–1757 (2003).

[14] S. Tanev, B.C. Wilson, V.V. Tuchin, and D. Matthews (eds.), "Special issue on biophotonics," *Adv. Opt. Technol.* **2008**. Article ID 134215(2008); http://www.hindawi.com/journals/aot/csi.html

[15] V.V. Tuchin, R. Drezek, S. Nie, and V.P. Zharov (eds.), "Special section on nanophotonics for diagnostics, protection and treatment of cancer and inflammatory diseases," *J. Biomed. Opt.* **14** (2), 020901; 021001–021017 (2009).

[16] Biophotonics and Life Sciences at the World of Photonics Congress, June 14–19, 2009, May 23–26, 2011, Munich, Germany http://www.world-of-photonics.net

[17] http://www.photonics4life.eu

[18] B. Chance (ed.), *Photon Migration in Tissue,* Plenum Press, New York (1989).

[19] A.V. Priezzhev, V.V. Tuchin, and L.P. Shubochkin, *Laser Diagnostics in Biology and Medicine,* Nauka, Moscow (1989).

[20] A.J. Welch and M.C.J. van Gemert (eds.), *Tissue Optics*, Academic Press, New York (1992).

[21] K. Frank and M. Kessler (eds.), *Quantitative Spectroscopy in Tissue*, pmi Verlag, Frankfurt am Main (1992).

[22] B.W. Henderson and T.J. Dougherty (eds.), *Photodynamic Therapy: Basic Principles and Clinical Applications*, Marcel-Dekker, New York (1992).

[23] G. Müller, B. Chance, R. Alfano et al. (eds.), *Medical Optical Tomography: Functional Imaging and Monitoring*, vol. IS11, SPIE Press, Bellingham, WA (1993).

[24] A. Katzir, *Lasers and Optical Fibers in Medicine*, Academic Press, San Diego (1993).

[25] D.H. Sliney and S.L. Trokel, *Medical Lasers and Their Safe Use*, Academic Press, New York (1993).

[26] G. Müller and A. Roggan (eds.), *Laser–Induced Interstitial Thermotherapy*, **PM25**, SPIE Press, Bellingham, WA (1995).

[27] H. Niemz, *Laser-Tissue Interactions. Fundamentals and Applications*, Springer-Verlag, Berlin (1996).

[28] V.V. Tuchin, *Lasers and Fiber Optics in Biomedical Science*, Saratov Univ. Press, Saratov (1998).

[29] V.V. Tuchin, *Tissue Optics: Light Scattering Methods and Instruments for Medical Diagnosis*, SPIE Tutorial Texts in Optical Engineering **TT38**, Bellingham, WA, USA (2000).

[30] A. Diaspro (ed.), *Confocal and Two-Photon Microscopy: Foundations, Applications, and Advances,* Wiley-Liss, New York (2002).

[31] B.E. Bouma and G.J. Tearney (eds.), *Handbook of Optical Coherence Tomography*, Marcel-Dekker, New York (2002).

[32] D.R. Vij and K. Mahesh (eds.), *Lasers in Medicine*, Kluwer Academic Publishers, Boston, Dordrecht, and London (2002).

[33] V.V. Tuchin (ed.), *Handbook of Optical Biomedical Diagnostics*, **PM107**, SPIE Press, Bellingham, WA (2002).

[34] H.-P. Berlien and G.J. Müller (eds.), *Applied Laser Medicine*, Springer-Verlag, Berlin (2003).

[35] Tuan Vo-Dinh (ed.), *Biomedical Photonics Handbook*, CRC Press, Boca Raton (2003).

[36] P. Prasad, *Introduction to Biophotonics*, Wiley-Interscience, Hoboken, NJ (2003).

[37] V.V. Tuchin (ed.), *Coherent-Domain Optical Methods: Biomedical Diagnostics, Environmental and Material Science*, Kluwer Academic Publishers, Boston, **1 & 2** (2004).

[38] B.R. Masters and T.P.C. So, *Handbook of Multiphoton Excitation Microscopy and Other Nonlinear Microscopies*, Oxford University Press, New York (2004).

[39] V.V. Tuchin, *Optical Clearing of Tissues and Blood*, **PM 154**, SPIE Press, Bellingham, WA (2006).

[40] A. Kishen and A. Asundi (eds.), *Photonics in Dentistry: Series of Biomaterials and Bioengineering*, Imperial College Press, London (2006).

[41] J. Popp and M. Strehle (eds.), *Biophotonics: Visions for Better Health Care*, Wiley-VCH Verlag GmbH & Co. KGaA, Weinheim (2006).

[42] V.V. Tuchin, L.V. Wang, and D.A. Zimnyakov, *Optical Polarization in Biomedical Applications*, Springer-Verlag, New York (2006).

[43] V.V. Tuchin, *Tissue Optics: Light Scattering Methods and Instruments for Medical Diagnosis*, 2nd, **PM 166**, SPIE Press, Bellingham, WA (2007).

[44] R. Splinter and B.A. Hooper, *An Introduction to Biomedical Optics*, Taylor and Francis Publishers, N.Y., London (2007).

[45] L.V. Wang and H.-I. Wu, *Biomedical Optics: Principles and Imaging*, Wiley-Interscience, Hoboken, NJ (2007).

[46] W. Drexler and J.G. Fujimoto (eds.), *Optical Coherence Tomography: Technology and Applications*, Springer, Berlin (2008).

[47] W. Bock, I. Gannot, and S. Tanev, *Optical Waveguide Sensing and Imaging*, *NATO SPS Series B: Physics and Biophysics*, Springer, Dordrecht (2008).

[48] V.V. Tuchin (ed.), *Handbook of Optical Sensing of Glucose in Biological Fluids and Tissues*, CRC Press, Taylor & Francis Group, London (2009).

[49] V.V. Tuchin, *Dictionary of Biomedical Optics*, SPIE Press, Bellingham,WA (2009).

[50] G. Ahluwalia, *Light-Based Systems for Cosmetic Application,* William Andrew, Norwich (2009).

[51] E. Baron, *Light-Based Therapies for Skin of Color*, Springer, NY (2009).

[52] L. Wang (ed.), *Photoacoustic Imaging and Spectroscopy*, CRC Press, Taylor & Francis Group, London (2009).

[53] K.-E. Peiponen, R. Myllylä, and A.V. Priezzhev, *Optical Measurement Techniques, Innovations for Industry and the Life Science*, Springer-Verlag, Berlin, Heidelberg (2009).

Valery V. Tuchin
Research-Educational Institute of Optics and Biophotonics
Saratov State University
Institute of Precise Mechanics and Control of RAS
Saratov, 410012 Russia
e-mail: tuchin@sgu.ru; tuchinvv@mail.ru

The Editor

Professor Valery V. Tuchin holds the Optics and Biophotonics Chair and is a Director of Research-Educational Institute of Optics and Biophotonics at Saratov State University, Head of Laboratory on Laser Diagnostics of Technical and Living Systems, Inst. of Precise Mechanics and Control, RAS. His research interests include biophotonics, biomedical optics and laser medicine, physics of optical and laser measurements. He has authored more than 300 peer-reviewed papers and books, including his latest, *Tissue Optics. Light Scattering Methods and Instrumentation for Medical Diagnosis* (PM166, SPIE Press, second edition, 2007) and *Handbook of Optical Sensing of Glucose in Biological Fluids and Tissues* (CRC Press, Taylor & Francis Group, London, 2009). He has been awarded Honored Science Worker of the Russian Federation and SPIE Fellow, and is a Vice-President of Russian Photobiology Society. In 2007 he was awarded the SPIE Educator Award.

List of Contributors

Garif G. Akchurin
Saratov State University,
Saratov, Russia

Lin An
Division of Biomedical Engineering, School of Medicine, Oregon Health & Science University,
3303 SW Bond Avenue, Portland, Oregon 97239, USA

O.V. Angelsky
Chernivtsi National University, Correlation Optics Department,
2 Kotsyubinskoho Str., Chernivtsi 58012, Ukraine

Egidijus Auksorius
Imperial College London,
London SW7 2AZ, United Kingdom

Pieter De Beule
Imperial College London,
London SW7 2AZ, United Kingdom

Evgenia Bousi
Department of Electrical and Computer Engineering, University of Cyprus,
Nicosia, 1678, Cyprus

Thomas Bruns
Hochschule Aalen, Institut für Angewandte Forschung,
73430 Aalen, Germany

Paul Campbell
Department of Electronic Engineering and Physics, University of Dundee,
Dundee, Scotland

Aaron C.-H. Chen
Wellman Center for Photomedicine, Massachusetts General Hospital,
Boston, MA;
Boston University School of Medicine, Graduate Medical Sciences,
Boston, MA

Wei R. Chen
Biomedical Engineering Program, Department of Engineering and Physics, College of Mathematics and Science, University of Central Oklahoma,
Edmond, OK

Riccardo Cicchi
European Laboratory for Nonlinear Spectroscopy (LENS) and Department of Physics, University of Florence, Sesto Fiorentino, 50019, Italy

Alex Darrell
Institute for Computer Science, Foundation for Research and Technology-Hellas, Heraklion 71110, Greece

Maxim E. Darvin
Center of Experimental and Applied Cutaneous Physiology (CCP), Department of Dermatology, Universitätsmedizin Charité Berlin, Germany

Dan Davis
Imperial College London, London SW7 2AZ, United Kingdom

Kishan Dholakia
SUPA, School of Physics & Astronomy, University of St. Andrews, St. Andrews, Scotland

Mary E. Dickinson
Department of Molecular Physiology and Biophysics, Baylor College of Medicine, One Baylor Plaza, Houston, TX, USA 77030

Christopher Dunsby
Imperial College London, London SW7 2AZ, United Kingdom

Lev A. Dykman
Institute of Biochemistry and Physiology of Plants and Microorganisms, 410049, Saratov, Russia

Daniel S. Elson
Imperial College London, London SW7 2AZ, United Kingdom

Paul French
Imperial College London, London SW7 2AZ, United Kingdom

Neil Galletly
Imperial College London, London SW7 2AZ, United Kingdom

G. Gelikonov
Institute of Applied Physics RAS, 46 Ulyanov Str., 603950 Nizhny Novgorod, Russia

V. Gelikonov
Institute of Applied Physics RAS, 46 Ulyanov Str.,
603950 Nizhny Novgorod, Russia

Mohamad G. Ghosn
University of Houston, Houston, TX 77204, USA

Nirmalya Ghosh
IISER Kolkata, Mohanpur Campus, PO: BCKV Campus Main Office, Mohanpur
741252, West Bengal, India

N. Gladkova
Nizhny Novgorod State Medical Academy 10/1, Minin and Pozharsky Sq.,
603005 Nizhny Novgorod, Russia

David Grant
Imperial College London,
London SW7 2AZ, United Kingdom

Frank J. Gunn-Moore
School of Biology, University of St. Andrews,
St. Andrews, Scotland

Michael R. Hamblin
Wellman Center for Photomedicine, Massachusetts General Hospital,
Boston MA
Department of Dermatology, Harvard Medical School,
Boston MA
Harvard-MIT Division of Health Sciences and Technology,
Cambridge MA

Song Hu
Optical Imaging Laboratory, Department of Biomedical Engineering,
Washington University in St. Louis,
St. Louis, Missouri, 63130-4899, USA

Ying-Ying Huang
Wellman Center for Photomedicine, Massachusetts General Hospital,
Boston MA
Department of Dermatology, Harvard Medical School,
Boston MA
Aesthetic Plastic Laser Center, Guangxi Medical University,
Nanning, Guangxi, P.R. China

Yali Jia
Division of Biomedical Engineering, School of Medicine, Oregon Health & Science University,
3303 SW Bond Avenue, Portland, Oregon 97239, USA

Alexander Karabutov
International Laser Center, Moscow State University,
Moscow, 119991, Russia

Andreas Kartakoullis
Department of Electrical and Computer Engineering, University of Cyprus,
Nicosia, 1678, Cyprus

Gordon Kennedy
Imperial College London,
London SW7 2AZ, United Kingdom

Nikolai G. Khlebtsov
Institute of Biochemistry and Physiology of Plants and Microorganisms,
410049, Saratov, Russia
Saratov State University,
410012, Saratov, Russia

Tatiana Khokhlova
Faculty of Physics, Moscow State University,
Moscow, 119991, Russia

M. Kirillin
Institute of Applied Physics RAS, 46 Ulyanov Str.,
603950 Nizhny Novgorod, Russia

Christoph Krafft
Institute of Photonic Technology,
07745 Jena, Germany

Sunil Kumar
Imperial College London,
London SW7 2AZ, United Kingdom

Juergen Lademann
Center of Experimental and Applied Cutaneous Physiology (CCP), Department of Dermatology,
Universitätsmedizin Charité Berlin, Germany

Peter M.P. Lanigan
Imperial College London,
London SW7 2AZ, United Kingdom

Kirill V. Larin
University of Houston, Houston,
TX 77204, USA
Institute of Optics and Biophotonics, Saratov State University,
Saratov, 410012, Russia

Irina V. Larina
Department of Molecular Physiology and Biophysics, Baylor College of Medicine, One Baylor Plaza, Houston, TX, USA 77030

Martin J. Leahy
University of Limerick, Ireland

Ivo Lenzetti
Ospedale Misericordia e Dolce, Unità Operativa Oculistica, Azienda USL 4, Prato, Italy

Xiaosong Li
Department of Oncology, the First Affiliated Hospital of Chinese PLA General Hospital, Beijing, China

Xingde Li
Department of Biomedical Engineering, Johns Hopkins University, Baltimore, MD, 21205 USA

Hong Liu
Center for Bioengineering and School of Electrical and Computer Engineering, University of Oklahoma, Norman, OK

Irina L. Maksimova
Saratov State University, Saratov, Russia

Michael J. Mandella
Stanford University, Stanford, CA

Hugh Manning
Imperial College London, London SW7 2AZ, United Kingdom

Konstantin Maslov
Optical Imaging Laboratory, Department of Biomedical Engineering, Washington University in St. Louis, St. Louis, Missouri, 63130-4899, USA

Galina N. Maslyakova
Saratov Medical State University, Saratov, Russia

Paolo Matteini
Istituto di Fisica Applicata "Nello Carrara," Consiglio Nazionale delle Ricerche, Sesto Fiorentino, Italy

Anthony Magee
Imperial College London, London SW7 2AZ, United Kingdom

Ewan McGhee
Imperial College London,
London SW7 2AZ, United Kingdom

James McGinty
Imperial College London,
London SW7 2AZ, United Kingdom

Luca Menabuoni
Ospedale Misericordia e Dolce, Unità Operativa Oculistica,
Azienda USL 4, Prato, Italy

Ian Munro
Imperial College London,
London SW7 2AZ, United Kingdom

Risto Myllylä
Optoelectronics and Measurement Techniques Laboratory, Department of Electrical and Information Engineering, Faculty of Technology, University of Oulu and Infotech Oulu,
Oulu, 90014, Finland

Mark F. Naylor
Department of Dermatology, University of Oklahoma College of Medicine at Tulsa,
Tulsa, OK

Maxim Nazarov
Moscow State University, Russia

Mark Neil
Imperial College London,
London SW7 2AZ, United Kingdom

Gert E. Nilsson
Linköping University, Sweden

Robert E. Nordquist
Wound Healing of Oklahoma, Inc.,
14 NE 48th Street, Oklahoma City, OK

Dylan Owen
Imperial College London,
London SW7 2AZ, United Kingdom

Francesco Pavone
European Laboratory for Nonlinear Spectroscopy (LENS) and Department of Physics,
University of Florence,
Sesto Fiorentino, 50019, Italy

Ivan Pelivanov
Faculty of Physics, Moscow State University,
Moscow, 119991, Russia

Lev T. Perelman
Biomedical Imaging and Spectroscopy Laboratory,
Beth Israel Deaconess Medical Center, Harvard University,
Boston, Massachusetts 02215 USA

A.P. Peresunko
Bukovinian State Medical University,
2 Teatralnaya Sq., Chernivtsi, 58000, Ukraine

Roberto Pini
Istituto di Fisica Applicata "Nello Carrara," Consiglio Nazionale delle Ricerche,
Sesto Fiorentino, Italy

V.P. Pishak
Bukovinian State Medical University,
2 Teatralnaya Sq., Chernivtsi, 58000, Ukraine

Costas Pitris
Department of Electrical and Computer Engineering, University of Cyprus,
Nicosia, 1678, Cyprus

James Pond
Lumerical Solutions, Vancouver, BC, Canada

Alexey P. Popov
Optoelectronics and Measurement Techniques Laboratory, Department of Electrical and Information Engineering, Faculty of Technology, University of Oulu and Infotech Oulu,
Oulu, 90014, Finland
International Laser Center, M.V. Lomonosov Moscow State University,
Moscow, 119991, Russia

Jürgen Popp
Institute of Photonic Technology,
07745 Jena, Germany
Institute of Physical Chemistry, University Jena,
07743 Jena, Germany

Alexander V. Priezzhev
Physics Department and International Laser Center, M.V. Lomonosov Moscow State University,
Moscow, 119991, Russia

Le Qiu
Biomedical Imaging and Spectroscopy Laboratory,
Beth Israel Deaconess Medical Center, Harvard University,
Boston, Massachusetts 02215 USA

Fulvio Ratto
Istituto di Fisica Applicata "Nello Carrara," Consiglio Nazionale delle Ricerche,
Sesto Fiorentino, Italy

Jorge Ripoll
Institute for Electronic Structure and Laser, Foundation for Research and Technology-Hellas,
Heraklion 71110, Greece

Francesca Rossi
Istituto di Fisica Applicata "Nello Carrara," Consiglio Nazionale delle Ricerche,
Sesto Fiorentino, Italy

Leonardo Sacconi
European Laboratory for Nonlinear Spectroscopy (LENS) and Department of Physics, University of Florence,
Sesto Fiorentino, 50019, Italy

Ana Sarasa-Renedo
Institute for Electronic Structure and Laser, Foundation for Research and Technology-Hellas,
Heraklion 71110, Greece

Herbert Schneckenburger
Hochschule Aalen, Institut für Angewandte Forschung,
73430 Aalen, Germany,
Institut für Lasertechnologien in der Medizin und Messtechnik an der Universität Ulm,
Helmholtzstr. 12, 89081 Ulm, Germany

A. Sergeev
Institute of Applied Physics RAS, 46 Ulyanov Str.,
603950 Nizhny Novgorod, Russia

N. Shakhova
Institute of Applied Physics RAS, 46 Ulyanov Str.,
603950 Nizhny Novgorod, Russia

Alexander Shkurinov
Moscow State University, Russia

Gordon Stamp
Imperial College London,
London SW7 2AZ, United Kingdom

David J. Stevenson
SUPA, School of Physics & Astronomy, School of Biology, University of St. Andrews,
St. Andrews, Scotland

Wenbo Sun
Science Systems and Applications, Inc., USA

Clifford Talbot
Imperial College London,
London SW7 2AZ, United Kingdom

Luis De Taboada
PhotoThera Inc, Carlsbad, CA

Stoyan Tanev
Integrative Innovation Management Unit, Department of Industrial and Civil Engineering, University of Southern Denmark,
Niels Bohrs Alle 1, DK-5230 Odense M, Denmark

Georgy S. Terentyuk
Saratov State University, The First Veterinary Clinic,
Saratov, Russia

Bebhinn Treanor
Imperial College London,
London SW7 2AZ, United Kingdom

Valery V. Tuchin
Research-Educational Institute of Optics and Biophotonics, Saratov State University,
Saratov, 410012, Russia,
Institute of Precise Mechanics and Control of RAS,
Saratov 410028, Russia

A. G. Ushenko
Chernivtsi National University, Correlation Optics Department,
2 Kotsyubinskoho Str., Chernivtsi 58012, Ukraine

Yu. A. Ushenko
Chernivtsi National University, Correlation Optics Department,
2 Kotsyubinskoho Str., Chernivtsi 58012, Ukraine

Alex Vitkin
Ontario Cancer Institute / Department of Medical Biophysics University of Toronto,
Toronto, Ontario, Canada

Lihong V. Wang
Optical Imaging Laboratory, Department of Biomedical Engineering,
Washington University in St. Louis,
St. Louis, Missouri, 63130-4899, USA

Ruikang K. Wang
Division of Biomedical Engineering, School of Medicine, Oregon Health & Science University,
3303 SW Bond Avenue, Portland, Oregon 97239, USA

Thomas D. Wang
University of Michigan, Ann Arbor, MI

Michael Wagner
Hochschule Aalen, Institut für Angewandte Forschung,
73430 Aalen, Germany

Petra Weber
Hochschule Aalen, Institut für Angewandte Forschung,
73430 Aalen, Germany

Brian C. Wilson
Department of Medical Biophysics, University of Toronto/Ontario Cancer Institute,
Toronto, ON, Canada

Michael Wood
Ontario Cancer Institute / Department of Medical Biophysics University of Toronto,
Toronto, Ontario, Canada

Yicong Wu
Department of Biomedical Engineering, Johns Hopkins University,
Baltimore, MD, 21205 USA

E. Zagaynova
Nizhny Novgorod State Medical Academy 10/1, Minin and Pozharsky Sq.,
603005 Nizhny Novgorod, Russia

X.-C. Zhang
Center for Terahertz Research, Rensselaer Polytechnic Institute,
Troy, USA

1

FDTD Simulation of Light Interaction with Cells for Diagnostics and Imaging in Nanobiophotonics

Stoyan Tanev

Integrative Innovation Management Unit, Department of Industrial and Civil Engineering, University of Southern Denmark, Niels Bohrs Alle 1, DK-5230 Odense M, Denmark [1]

Wenbo Sun

Science Systems and Applications, Inc., USA

James Pond

Lumerical Solutions, Vancouver, BC, Canada

Valery V. Tuchin

Research-Educational Institute of Optics and Biophotonics, Saratov State University, Saratov, 410012, Russia, Institute of Precise Mechanics and Control of RAS, Saratov 410028, Russia

This chapter describes the mathematical formulation of the Finite-Difference Time-Domain (FDTD) approach and provides examples of its applications to biomedical photonics problems. The applications focus on two different configurations – light scattering from single biological cells and Optical Phase Contrast Microscope (OPCM) imaging of cells containing gold nanoparticles. The validation of the FDTD approach for the simulation of OPCM imaging opens a new application area with a significant research potential – the design and modeling of advanced nanobioimaging instrumentation.

Key words: Finite-Difference Time-Domain (FDTD) method, light scattering, biological cell, gold nanoparticle, Optical Phase Contrast Microscope (OPCM) imaging, optical clearing effect, image contrast enhancement, nanobiophotonics

[1]Formerly with the Technology Innovation Management Program in the Department of Systems and Computer Engineering, Faculty of Engineering and Design, Carleton University, Ottawa, ON, Canada.

1.1 Introduction

The development of noninvasive optical methods for biomedical diagnostics requires a fundamental understanding of how light scatters from normal and pathological structures within biological tissue. It is important to understand the nature of the light scattering mechanisms from microbiological structures and how sensitive the light scattering parameters are to the dynamic pathological changes of these structures. It is equally important to quantitatively relate these changes to corresponding variations of the measured light scattering parameters. Unfortunately, the biological origins of the differences in the light scattering patterns from normal and pathological (for example, precancerous and cancerous) cells and tissues are not fully understood. The major difficulty comes from the fact that most of the advanced optical biodiagnostics techniques have a resolution comparable to the dimensions of the cellular and subcellular light scattering structures [1–2]. For example, conventional Optical Coherence Tomography (OCT) techniques do not provide a resolution at the subcellular level. Although there are already examples of ultrahigh resolution OCT capabilities based on the application of ultrashort pulsed lasers [3–4], it will take time before such systems become easily commercially available. This makes the interpretation of images generated by typical OCT systems difficult and in many cases inefficient. Confocal microscopy provides subcellular resolution and allows for diagnostics of nuclear and subnuclear level cell morphological features. However, there are problems requiring a careful and not trivial interpretation of the images [5]. In its typical single photon form, confocal fluorescence microscopy involves an optical excitation of tissue leading to fluorescence that occurs along the exciting cone of the focused light beam, thus increasing the chances of photo-bleaching of a large area and making the interpretation of the image difficult.

In many nanobiophotonics diagnostics and imaging research studies, optical software simulation and modeling tools provide the only means to a deeper understanding, or any understanding at all, of the underlying physical and biochemical processes. The tools and methods for the numerical modeling of light scattering from single or multiple biological cells are of particular interest since they could provide information about the fundamental light-cell interaction phenomena that is highly relevant for the practical interpretation of cell images by pathologists. The computational modeling of light interaction with cells is usually approached from a single particle electromagnetic wave scattering perspective, which could be characterized by two specific features. First, this is the fact that the wavelength of light is larger than or comparable to the size of the scattering subcellular structures. Second, this is the fact that biological cells have irregular shapes and inhomogeneous refractive index distributions, which makes it impossible to use analytical modeling approaches. Both features necessitate the use of numerical modeling approaches derived from rigorous electromagnetic theory such as: the method of separation of variables, the finite element method, the method of lines, the point matching method, the method of moments, the discrete dipole approximation method, the null-field (extended boundary condition) method, the T-matrix electromagnetic scattering approach, the surface Green's function electromagnetic scattering approach, and the finite-difference time domain (FDTD) method [6].

The FDTD simulation and modeling of the light interaction with single and multiple, normal and pathological biological cells and subcellular structures has attracted the attention of researchers since 1996 [7–25]. The FDTD approach was first adopted as a better alternative of Mie theory [26–27] allowing for the modeling of irregular cell shapes and inhomogeneous distributions of complex refractive index values. The emerging relevance of nanobiophotonics imaging research has established the FDTD method as one of the powerful tools for studying the nature of light-cell interactions. One could identify number of research directions based on the FDTD approach. The first one focuses on studying the lateral light scattering patterns for the early detection of pathological changes in cancerous cells such as increased nuclear size and degrees of nuclear pleomorphism and nuclear-to-cytoplasmic ratios [7–17]. The second research direction explores the application of

FDTD-based approaches for time-resolved diffused optical tomography studies [18–20]. A third direction is the application of the FDTD method to the modeling of advanced cell imaging techniques within the context of a specific biodiagnostics device scenario [21–25]. An emerging research direction consists in the extension of the FDTD approach to account for optical nanotherapeutic effects.

The present chapter will provide a number of examples illustrating the application of the FDTD approach in situations associated with the first two research directions. It is organized as follows. Section two provides a detailed summary of the formulation of the FDTD method including the basic numerical scheme, near-to-far field transformation, advanced boundary conditions, and specifics related to its application to the modeling of light scattering and optical phase contrast microscope (OPCM) imaging experiments. Examples of FDTD modeling of light scattering from and OPCM imaging of single biological cells are given in sections 3 and 4. The last section summarizes the conclusions.

1.2 Formulation of the FDTD Method

1.2.1 The basic FDTD numerical scheme

The finite-difference time domain (FDTD) technique is an explicit numerical method for solving Maxwell's equations. It was invented by Yee in 1966 [28]. The advances of the various FDTD approaches and applications have been periodically reviewed by Taflove et al. [29]. The explicit finite-difference approximation of Maxwell's equations in space and time will be briefly summarized following Taflove et al. [29] and Sun et al. [30–32].

In a source-free absorptive dielectric medium Maxwell's equations have the form:

$$\nabla \times \vec{E} = -\mu_0 \frac{\partial \vec{H}}{\partial t}, \tag{1.1}$$

$$\nabla \times \vec{H} = \varepsilon_0 \varepsilon \frac{\partial \vec{E}}{\partial t}, \tag{1.2}$$

where $\vec{E}$ and $\vec{H}$ are the vectors of the electric and magnetic fields, respectively, μ_0 is the vacuum permeability and $\varepsilon_0\varepsilon$ is the permittivity of the medium. Assuming a harmonic $[\propto \exp(-i\omega t)]$ time dependence of the electric and magnetic fields and a complex value of the relative permittivity $\varepsilon = \varepsilon_r + i\varepsilon_i$ transforms Eq. (1.2) as follows:

$$\nabla \times \vec{H} = \varepsilon_0 \varepsilon \frac{\partial \vec{E}}{\partial t} \Leftrightarrow \nabla \times \vec{H} = \omega \varepsilon_0 \varepsilon_i \vec{E} + \varepsilon_0 \varepsilon_r \frac{\partial \vec{E}}{\partial t} \Leftrightarrow \frac{\partial(\exp(\tau t)\vec{E})}{\partial t} = \frac{\exp(\tau t)}{\varepsilon_0 \varepsilon_r} \nabla \times \vec{H}, \tag{1.3}$$

where $\tau = \omega\varepsilon_r/\varepsilon_i$ and ω is the angular frequency of the light. The continuous coordinates (x,y,z) are replaced by discrete spatial and temporal points: $x_i = i\Delta s$, $y_j = j\Delta s$, $z_k = k\Delta s$, $t_n = n\Delta t$, where $i = 0,1,2,\cdots,I$, $j = 0,1,2,\cdots,J$, $k = 0,1,2,\cdots,K$, $n = 0,1,2,\cdots,N$. Δs and Δt denote the cubic cell size and time increment, respectively. Using central difference approximations for the temporal derivatives over the time interval $[n\Delta t,(n+1)\Delta t]$ leads to

$$\vec{E}^{n+1} = \exp(-\tau\Delta t)\vec{E}^{n} + \exp(-\tau\Delta t/2)\frac{\Delta t}{\varepsilon_0\varepsilon_r}\nabla \times \vec{H}^{n+1/2}, \tag{1.4}$$

where the electric and the magnetic fields are calculated at alternating half-time steps. The discretization of Eq. (1.1) over the time interval $[(n-1/2)\Delta t,(n+1/2)\Delta t]$ (one-half time step earlier

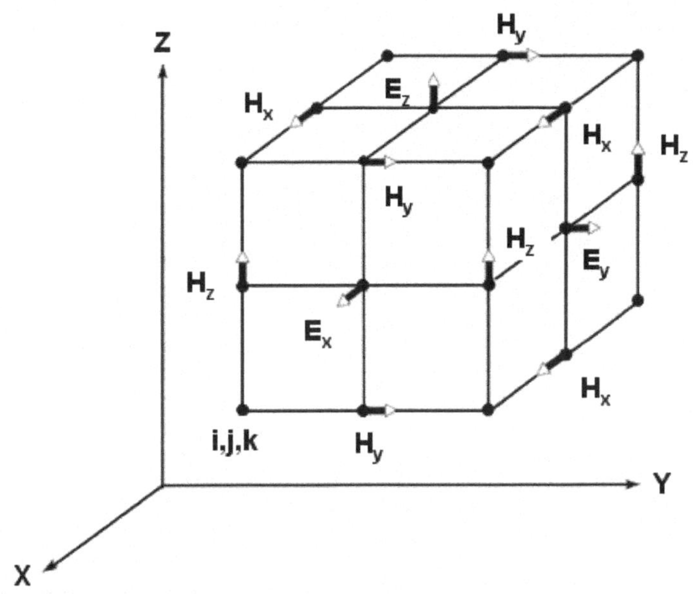

FIGURE 1.1: Positions of the electric- and the magnetic-field components in the elementary (Yee) cubic cell of the FDTD lattice.

than the electric field) ensures second-order accuracy of the numerical scheme. In a Cartesian coordinate system the numerical equations for the x components of the electric and magnetic fields take the form

$$\begin{aligned} &H_x^{n+1/2}(i,j+1/2,k+1/2) = H_x^{n-1/2}(i,j+1/2,k+1/2) + \tfrac{\Delta t}{\mu_0 \Delta s} \\ &\times [E_y^n(i,j+1/2,k+1) - E_y^n(i,j+1/2,k) + E_z^n(i,j,k+1/2) - E_z^n(i,j+1,k+1/2)], \end{aligned} \tag{1.5}$$

$$\begin{aligned} &E_x^{n+1}(i+1/2,j,k) = \exp[-\tfrac{\varepsilon_i(i+1/2,j,k)}{\varepsilon_r(i+1/2,j,k)}\omega\Delta t]E_x^n(i+1/2,j,k) + \\ &\exp[-\tfrac{\varepsilon_i(i+1/2,j,k)}{\varepsilon_r(i+1/2,j,k)}\omega\Delta t/2]\tfrac{\Delta t}{\varepsilon_0\varepsilon_r(i+1/2,j,k)\Delta s} \\ &\times [H_y^{n+1/2}(i+1/2,j,k-1/2) - H_y^{n+1/2}(i+1/2,j,k+1/2) \\ &+ H_z^{n+1/2}(i+1/2,j+1/2,k) - H_z^{n+1/2}(i+1/2,j-1/2,k)], \end{aligned} \tag{1.6}$$

where E_x, E_y, E_z and H_x, H_y, H_z denote the electric and magnetic field components, respectively. The numerical stability of the FDTD scheme is ensured through the Courant-Friedrichs-Levy condition [29]: $c\Delta t \leq \left(1/\Delta x^2 + 1/\Delta y^2 + 1/\Delta z^2\right)^{-1/2}$, where c is the speed of light in the host medium and Δx, Δy, Δy are the spatial steps in the x, y and z direction, respectively. In our case $\Delta x = \Delta y = \Delta z = \Delta s$ and $\Delta t = \Delta s/2c$. The positions of the magnetic and electric field components in a FDTD cubic cell are shown in Figure 1.1.

1.2.2 Numerical excitation of the input wave

We use the so-called total-field/scattered-field formulation [29, 32] to excite the input magnetic and electric fields and simulate a linearly polarized plane wave propagating in a finite region of a homogeneous absorptive dielectric medium. In this formulation, a closed surface is defined inside of the computational domain. Based on the equivalence theorem [29], the input wave excitation within

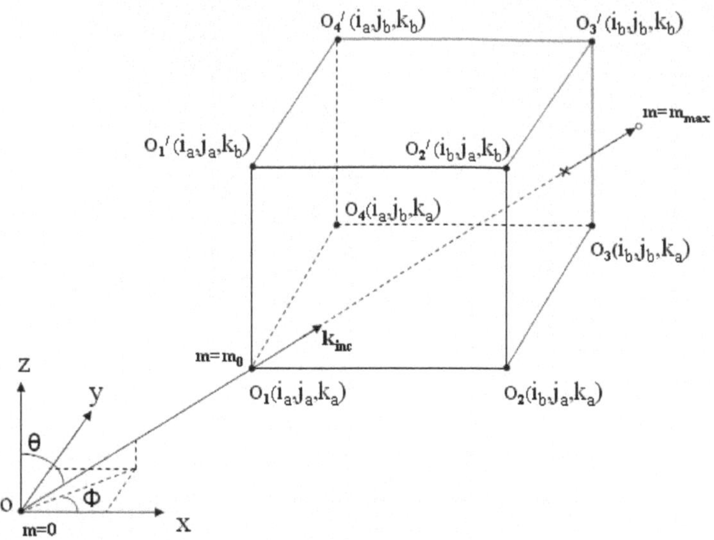

FIGURE 1.2: Example of a closed rectangular surface separating the total fields and scattered fields. The graph also shows the configuration of the one-dimensional auxiliary FDTD grid that is used to calculate the input excitation fields [see the paragraph below Eq. (1.16)].

a given spatial domain can be replaced by the equivalent electric and magnetic currents located at the closed surface enclosing that domain. If there is a scatterer inside the closed surface, the interior fields will be the total fields (incident plus scattered) and the exterior fields are just the scattered fields. An example of geometrical configuration of such closed surface (in this case rectangular) is shown in Figure 1.2.

On the closed surface the electric and magnetic field incident sources are added as follows:

$$\vec{H} \Leftarrow \vec{H} - \frac{\Delta t}{\mu_0 \Delta s}(\vec{E}_{inc} \times \vec{n}), \tag{1.7}$$

$$\vec{E} \Leftarrow \vec{E} - \frac{\Delta t}{\varepsilon_0 \varepsilon \Delta s}(\vec{n} \times \vec{H}_{inc}), \tag{1.8}$$

where $\vec{E}_{inc}$ and $\vec{H}_{inc}$are the incident fields and $\vec{n}$ is the inward normal vector of the closed surface [29]. Eqs. (1.5ab) take different forms at the different interfaces of the closed rectangular surface $O_1O_2O_3O_4O_1^{'}O_2^{'}O_3^{'}O_4^{'}$ shown in Figure 1.2.

At the interface $(i = i_a, ..., i_b; j = j_a - 1/2; k = k_a + 1/2, ..., k_b - 1/2)$:

$$H_x^{n+1/2}(i, j_a - 1/2, k) = \{H_x^{n+1/2}(i, j_a - 1/2, k)\}_{(1a)} + \frac{\Delta t}{\mu_0 \Delta s} E_{z,inc}^n(i, j_a, k); \tag{1.9}$$

at the interface $(i = i_a, ..., i_b; j = j_b + 1/2; k = k_a + 1/2, ..., k_b - 1/2)$:

$$H_x^{n+1/2}(i, j_b + 1/2, k) = \{H_x^{n+1/2}(i, j_b + 1/2, k)\}_{(1a)} - \frac{\Delta t}{\mu_0 \Delta s} E_{z,inc}^n(i, j_b, k); \tag{1.10}$$

at the interface $(i = i_a, ..., i_b; j = j_a + 1/2, ..., j_b - 1/2; k = k_a - 1/2)$:

$$H_x^{n+1/2}(i, j, k_a - 1/2) = \{H_x^{n+1/2}(i, j, k_a - 1/2)\}_{(1a)} - \frac{\Delta t}{\mu_0 \Delta s} E_{y,inc}^n(i, j, k_a); \tag{1.11}$$

at the interface $(i = i_a, ..., i_b; j = j_a + 1/2, ..., j_b - 1/2; k = k_b + 1/2)$:

$$H_x^{n+1/2}(i,j,k_b+1/2) = \{H_x^{n+1/2}(i,j,k_b+1/2)\}_{(1a)} + \frac{\Delta t}{\mu_0 \Delta s} E_{y,inc}^n(i,j,k_b); \tag{1.12}$$

at the interface $(i = i_a + 1/2, ..., i_b - 1/2; j = j_a; k = k_a, ..., k_b)$:

$$\begin{aligned} E_x^{n+1}(i,j_a,k) &= \{E_x^{n+1}(i,j_a,k)\}_{(1b)} \\ &- \exp\left[-\tfrac{\varepsilon_i(i,j_a,k)}{\varepsilon_r(i,j_a,k)}\omega\Delta t/2\right] \tfrac{\Delta t}{\varepsilon_0\varepsilon_r(i,j_a,k)\Delta s} H_{z,inc}^{n+1/2}(i,j_a-1/2,k); \end{aligned} \tag{1.13}$$

at the interface $(i = i_a + 1/2, ..., i_b - 1/2; j = j_b; k = k_a, ..., k_b)$:

$$\begin{aligned} E_x^{n+1}(i,j_b,k) &= \{E_x^{n+1}(i,j_b,k)\}_{(1b)} \\ &+ \exp\left[-\tfrac{\varepsilon_i(i,j_b,k)}{\varepsilon_r(i,j_b,k)}\omega\Delta t/2\right] \tfrac{\Delta t}{\varepsilon_0\varepsilon_r(i,j_b,k)\Delta s} H_{z,inc}^{n+1/2}(i,j_b+1/2,k); \end{aligned} \tag{1.14}$$

at the interface $(i = i_a + 1/2, ..., i_b - 1/2; j = j_a, ..., j_b; k = k_a)$:

$$\begin{aligned} E_x^{n+1}(i,j,k_a) &= \{E_x^{n+1}(i,j,k_a)\}_{(1b)} \\ &+ \exp\left[-\tfrac{\varepsilon_i(i,j,k_a)}{\varepsilon_r(i,j,k_a)}\omega\Delta t/2\right] \tfrac{\Delta t}{\varepsilon_0\varepsilon_r(i,j,k_a)\Delta s} H_{y,inc}^{n+1/2}(i,j,k_a-1/2); \end{aligned} \tag{1.15}$$

at the face $(i = i_a + 1/2, ..., i_b - 1/2; j = j_a, ..., j_b; k = k_b)$:

$$\begin{aligned} E_x^{n+1}(i,j,k_b) &= \{E_x^{n+1}(i,j,k_b)\}_{(1b)} \\ &- \exp\left[-\tfrac{\varepsilon_i(i,j,k_b)}{\varepsilon_r(i,j,k_b)}\omega\Delta t/2\right] \tfrac{\Delta t}{\varepsilon_0\varepsilon_r(i,j,k_b)\Delta s} H_{y,inc}^{n+1/2}(i,j,k_b+1/2). \end{aligned} \tag{1.16}$$

The incident fields $E_{y,inc}^n$, $E_{z,inc}^n$ and $H_{y,inc}^{n+1/2}$, $H_{z,inc}^{n+1/2}$ in Eqs. (1.9)–(1.12) and (1.13)–(1.16) are calculated by means of a linear interpolation of the fields on an auxiliary one-dimensional linear FDTD grid. This auxiliary numerical scheme presimulates the propagation of an incident plane wave along a line starting at the origin of the 3D grid $m = 0$, passing through the closest corner of the closed rectangular surface located at $m = m_0$, and stretching in the incident wave direction to a maximum position $m = m_{\max}$, as shown in Figure 1.2. The incident wave vector k_{inc} is oriented with a zenith angle θ and an azimuth angle ϕ. $m_{\max}$ is chosen to be half of the total number of simulation time steps for the incident wave propagation in the entire absorptive dielectric medium. Since it is impossible to use a transmitting boundary condition for the truncation of the one-dimensional spatial domain, the selected $m_{\max}$ value needs to ensure that no numerical reflection occurs at the forward end of the one-dimensional grid before the 3D FDTD simulation ends. A Gaussian-pulse hard wave source [29] is positioned at the $m = 2$ grid point in the form

$$E_{inc}^n(m=2) = \exp\left[-\left(\frac{t}{30\Delta t} - 5\right)^2\right]. \tag{1.17}$$

By using the hard wave source rather than a soft one at $m = 2$, the field at the grid points $m = 0$ and 1 will not affect the field at the grid points $m > 2$. Therefore, there is no need of boundary conditions at this end of the auxiliary one-dimensional FDTD grid.

Assuming that a plane wave is incident from the coordinate origin to the closed rectangular surface between the total- and scattered-fields as shown in Figure 1.2, the one-dimensional FDTD grid equations become

$$H_{inc}^{n+1/2}(m+1/2) = H_{inc}^{n-1/2}(m+1/2) + \frac{\Delta t}{\mu_0 \Delta s \left[\frac{v_p(0,0)}{v_p(\theta,\phi)}\right]}[E_{inc}^n(m) - E_{inc}^n(m+1)], \tag{1.18}$$

$$E_{inc}^{n+1}(m) = \exp\left[-\frac{\varepsilon_i(m)}{\varepsilon_r(m)}\omega\Delta t\right] E_{inc}^{n}(m)$$
$$+\exp\left[-\frac{\varepsilon_i(m)}{\varepsilon_r(m)}\omega\Delta t\right] \frac{\Delta t}{\varepsilon_0\varepsilon_r(m)\Delta s\left[\frac{v_p(0,0)}{v_p(\theta,\phi)}\right]} \left[H_{inc}^{n+1/2}(m-1/2) - H_{inc}^{n+1/2}(m+1/2)\right], \tag{1.19}$$

where $\varepsilon_i(m)$ and $\varepsilon_r(m)$ denote the imaginary and real relative permittivity of the host medium at position m, respectively. The equalization factor $[v_p(0,0)/v_p(\theta,\phi)] \leq 1$ is the numerical phase velocity ratio in the 3D FDTD grid [29].

1.2.3 Uni-axial perfectly matched layer absorbing boundary conditions

The FDTD numerical scheme presented here uses the Uni-axial Perfectly Matched Layer (UPML) suggested by Sacks et al. [33] to truncate the absorptive host medium in the FDTD computational domain. The UPML approach is based on the physical introduction of absorbing anisotropic, perfectly matched medium layers at all sides of the rectangular computational domain. The anisotropic medium of each of these layers is uni-axial and is composed of both electric permittivity and magnetic permeability tensors. To match a UPML layer along a planar boundary to a lossy isotropic half-space characterized by permittivity ε and conductivityσ, the time-harmonic Maxwell's equations can be written in forms [34–35].

$$\nabla \times \vec{H}(x,y,z) = (i\omega\varepsilon_0\varepsilon + \sigma)\bar{\bar{s}}\vec{E}(x,y,z), \tag{1.20}$$

$$\nabla \times \vec{E}(x,y,z) = -i\omega\mu_0\bar{\bar{s}}\vec{H}(x,y,z). \tag{1.21}$$

The diagonal tensor $\bar{\bar{s}}$ is defined as follows

$$\bar{\bar{s}} = \begin{bmatrix} s_x^{-1} & 0 & 0 \\ 0 & s_x & 0 \\ 0 & 0 & s_x \end{bmatrix} \begin{bmatrix} s_y & 0 & 0 \\ 0 & s_y^{-1} & 0 \\ 0 & 0 & s_y \end{bmatrix} \begin{bmatrix} s_z & 0 & 0 \\ 0 & s_z & 0 \\ 0 & 0 & s_z^{-1} \end{bmatrix} = \begin{bmatrix} s_y s_z s_x^{-1} & 0 & 0 \\ 0 & s_x s_z s_y^{-1} & 0 \\ 0 & 0 & s_x s_y s_z^{-1} \end{bmatrix}, \tag{1.22}$$

where $s_x = \kappa_x + \frac{\sigma_x}{i\omega\varepsilon_0}$, $s_y = \kappa_y + \frac{\sigma_y}{i\omega\varepsilon_0}$, and $s_z = \kappa_z + \frac{\sigma_z}{i\omega\varepsilon_0}$.

The UPML parameters (κ_x, σ_x), (κ_y, σ_y), and (κ_z, σ_z) are independent on the medium permittivity ε and conductivity σ, and are assigned to the FDTD grid in the UPML as follows: i) in the two absorbing layers at both ends of the computational domain in the x-direction, $\sigma_y = \sigma_z = 0$ and $\kappa_y = \kappa_z = 1$; in the two layers at both ends of the y-direction, $\sigma_x = \sigma_z = 0$ and $\kappa_x = \kappa_z = 1$; in the two layers at both ends of the $z-$direction, $\sigma_y = \sigma_x = 0$ and $\kappa_y = \kappa_x = 1$; ii) at the x and y overlapping dihedral corners, $\sigma_z = 0$ and $\kappa_z = 1$; at the z and x overlapping dihedral corners, $\sigma_y = 0$ and $\kappa_y = 1$; iii) at all overlapping trihedral corners, the complete general tensor in Eq. (1.22) is used. To reduce the numerical reflection from the UPML, several profiles have been suggested for incrementally increasing the values of (κ_x, σ_x), (κ_y, σ_y) and (κ_z, σ_z). Here we use a polynomial grading of the UPML material parameters [34–35]. For example,

$$\kappa_x(x) = 1 + (x/d)^m(\kappa_{x,\max} - 1), \tag{1.23}$$

$$\sigma_x(x) = (x/d)^m \sigma_{x,\max}, \tag{1.24}$$

where x is the depth in the UPML and d is the UPML thickness in this direction. The parameter m is a real number [35] between 2 and 4. $\kappa_{x,\max}$ and $\sigma_{x,\max}$ denote the maximum κ_x and σ_x at the outmost layer of the UPML. For example, considering an x-directed plane wave impinging at an angle θ upon a PEC-backed UPML with the polynomial grading material properties, the reflection factor can be derived as [35]

$$R(\theta) = \exp[-\frac{2\cos\theta}{\varepsilon_0 c}\int_0^d \sigma(x)dx] = \exp\left[-\frac{2\sigma_{x,\max} d\cos\theta}{\varepsilon_0 c(m+1)}\right]. \tag{1.25}$$

Therefore, if $R(0)$ is the reflection factor at normal incidence, $\sigma_{x,\max}$ can be defined as

$$\sigma_{x,\max} = -\frac{(m+1)\ln[R(0)]}{2d/(\varepsilon_0 c)}. \tag{1.26}$$

Typically, the values of $R(0)$ are in the range between 10^{-12} to 10^{-5} and $\kappa_{x,\max}$ is a real number between 1 to 30.

The UPML equations modify the FDTD numerical scheme presented by Eqs. (1.5) and (1.6). The modified UPML FDTD numerical scheme is then applied to the entire computational domain by considering the UPMLs as materials in a way no different than any other material in the FDTD grid. However, this is not computationally efficient. The usual approach is to apply the modified scheme only to the boundary layers in order to reduce the memory and CPU time requirements. In the non-UPML region, the unmodified FDTD formulation (Eqs. (1.5), (1.6)) is used. The derivation of the modified UPML FDTD numerical scheme is not trivial at all. To explicitly obtain the updating equations for the magnetic field in the UPML, an auxiliary vector field variable $\vec{B}$ is introduced as follows [35]

$$\begin{gathered} B_x(x,y,z) = \mu_0(\tfrac{s_z}{s_x})H_x(x,y,z), \quad B_y(x,y,z) = \mu_0(\tfrac{s_x}{s_y})H_y(x,y,z), \\ B_z(x,y,z) = \mu_0(\tfrac{s_y}{s_z})H_z(x,y,z). \end{gathered} \tag{1.27}$$

Then Eq. (1.21) can be expressed as

$$\begin{bmatrix} \frac{\partial E_y(x,y,z)}{\partial z} - \frac{\partial E_z(x,y,z)}{\partial y} \\ \frac{\partial E_z(x,y,z)}{\partial x} - \frac{\partial E_x(x,y,z)}{\partial z} \\ \frac{\partial E_x(x,y,z)}{\partial y} - \frac{\partial E_y(x,y,z)}{\partial x} \end{bmatrix} = i\omega \begin{bmatrix} s_y & 0 & 0 \\ 0 & s_z & 0 \\ 0 & 0 & s_x \end{bmatrix} \begin{bmatrix} B_x(x,y,z) \\ B_y(x,y,z) \\ B_z(x,y,z) \end{bmatrix}. \tag{1.28}$$

On the other hand, inserting the definitions of s_x, s_y and s_z into Eqs. (1.27) leads to

$$(i\omega\kappa_x + \frac{\sigma_x}{\varepsilon_0})B_x(x,y,z) = (i\omega\kappa_z + \frac{\sigma_z}{\varepsilon_0})\mu_0 H_x(x,y,z), \tag{1.29}$$

$$(i\omega\kappa_y + \frac{\sigma_y}{\varepsilon_0})B_y(x,y,z) = (i\omega\kappa_x + \frac{\sigma_x}{\varepsilon_0})\mu_0 H_y(x,y,z), \tag{1.30}$$

$$(i\omega\kappa_z + \frac{\sigma_z}{\varepsilon_0})B_z(x,y,z) = (i\omega\kappa_y + \frac{\sigma_y}{\varepsilon_0})\mu_0 H_z(x,y,z). \tag{1.31}$$

Now applying the inverse Fourier transform by using the identity $i\omega f(\omega) \to \partial f(t)/\partial t$ to Eqs. (1.28) and (1.29)–(1.31) gives the equivalent time-domain differential equations, respectively

$$\begin{gathered} \begin{bmatrix} \frac{\partial E_y(x,y,z,t)}{\partial z} - \frac{\partial E_z(x,y,z,t)}{\partial y} \\ \frac{\partial E_z(x,y,z,t)}{\partial x} - \frac{\partial E_x(x,y,z,t)}{\partial z} \\ \frac{\partial E_x(x,y,z,t)}{\partial y} - \frac{\partial E_y(x,y,z,t)}{\partial x} \end{bmatrix} = \frac{\partial}{\partial t} \begin{bmatrix} \kappa_y & 0 & 0 \\ 0 & \kappa_z & 0 \\ 0 & 0 & \kappa_x \end{bmatrix} \begin{bmatrix} B_x(x,y,z,t) \\ B_y(x,y,z,t) \\ B_z(x,y,z,t) \end{bmatrix} \\ + \frac{1}{\varepsilon_0} \begin{bmatrix} \sigma_y & 0 & 0 \\ 0 & \sigma_z & 0 \\ 0 & 0 & \sigma_x \end{bmatrix} \begin{bmatrix} B_x(x,y,z,t) \\ B_y(x,y,z,t) \\ B_z(x,y,z,t) \end{bmatrix}, \end{gathered} \tag{1.32}$$

$$\kappa_x \frac{\partial B_x(x,y,z,t)}{\partial t} + \frac{\sigma_x}{\varepsilon_0} B_x(x,y,z,t) = \mu\kappa_z \frac{\partial H_x(x,y,z,t)}{\partial t} + \mu\frac{\sigma_z}{\varepsilon_0} H_x(x,y,z,t), \tag{1.33}$$

$$\kappa_y \frac{\partial B_y(x,y,z,t)}{\partial t} + \frac{\sigma_y}{\varepsilon_0} B_y(x,y,z,t) = \mu \kappa_x \frac{\partial H_y(x,y,z,t)}{\partial t} + \mu \frac{\sigma_x}{\varepsilon_0} H_y(x,y,z,t), \tag{1.34}$$

$$\kappa_z \frac{\partial B_z(x,y,z,t)}{\partial t} + \frac{\sigma_z}{\varepsilon_0} B_z(x,y,z,t) = \mu \kappa_y \frac{\partial H_z(x,y,z,t)}{\partial t} + \mu \frac{\sigma_y}{\varepsilon_0} H_z(x,y,z,t). \tag{1.35}$$

After discretizing Eqs. (1.32) and (1.33)–(1.35), we can get the explicit FDTD formulations for the magnetic field components in the UPML [29, 32]:

$$\begin{aligned} B_x^{n+1/2}(i,j+1/2,k+1/2) &= \left(\tfrac{2\varepsilon_0\kappa_y - \sigma_y\Delta t}{2\varepsilon_0\kappa_y + \sigma_y\Delta t}\right) B_x^{n-1/2}(i,j+1/2,k+1/2) \\ &+ \left(\tfrac{2\varepsilon_0\Delta t/\Delta s}{2\varepsilon_0\kappa_y + \sigma_y\Delta t}\right) [E_y^n(i,j+1/2,k+1) \\ &- E_y^n(i,j+1/2,k) + E_z^n(i,j,k+1/2) - E_z^n(i,j+1,k+1/2)], \end{aligned} \tag{1.36}$$

$$\begin{aligned} H_x^{n+1/2}(i,j+1/2,k+1/2) &= \left(\tfrac{2\varepsilon_0\kappa_z - \sigma_z\Delta t}{2\varepsilon_0\kappa_z + \sigma_z\Delta t}\right) H_x^{n-1/2}(i,j+1/2,k+1/2) \\ &+ \left(\tfrac{1/\mu}{2\varepsilon_0\kappa_z + \sigma_z\Delta t}\right) [(2\varepsilon_0\kappa_x + \sigma_x\Delta t) B_x^{n+1/2}(i,j+1/2,k+1/2) \\ &- (2\varepsilon_0\kappa_x - \sigma_x\Delta t) B_x^{n-1/2}(i,j+1/2,k+1/2)]. \end{aligned} \tag{1.37}$$

Similarly, for electric field in the UPML, two auxiliary field variables $\vec{P}$ and $\vec{Q}$ are introduced as follows [32, 35]

$$P_x(x,y,z) = \left(\frac{s_y s_z}{s_x}\right) E_x(x,y,z), \tag{1.38}$$

$$P_y(x,y,z) = \left(\frac{s_x s_z}{s_y}\right) E_y(x,y,z), \tag{1.39}$$

$$P_z(x,y,z) = \left(\frac{s_x s_y}{s_z}\right) E_z(x,y,z), \tag{1.40}$$

$$Q_x(x,y,z) = \left(\frac{1}{s_y}\right) P_x(x,y,z), \tag{1.41}$$

$$Q_y(x,y,z) = \left(\frac{1}{s_z}\right) P_y(x,y,z), \tag{1.42}$$

$$Q_z(x,y,z) = \left(\frac{1}{s_x}\right) P_z(x,y,z). \tag{1.43}$$

Inserting Eqs. (1.38)–(1.40) into Eq. (1.20), simply following the steps in deriving Eq. (1.36), leads to the updating equations for the $\vec{P}$ components:

$$\begin{aligned} P_x^{n+1}(i+1/2,j,k) &= \left(\tfrac{2\varepsilon_0\varepsilon - \sigma\Delta t}{2\varepsilon_0\varepsilon + \sigma\Delta t}\right) P_x^n(i+1/2,j,k) + \left(\tfrac{2\Delta t/\Delta s}{2\varepsilon_0\varepsilon + \sigma\Delta t}\right) \\ &\times [H_y^{n+1/2}(i+1/2,j,k-1/2) - H_y^{n+1/2}(i+1/2,j,k+1/2) \\ &+ H_z^{n+1/2}(i+1/2,j+1/2,k) - H_z^{n+1/2}(i+1/2,j-1/2,k)]. \end{aligned} \tag{1.44}$$

From Eqs. (1.41)–(1.43), in an identical way to the derivation of Eq. (1.37), leads to the updating equations for the $\vec{Q}$ components:

$$\begin{aligned} Q_x^{n+1}(i+1/2,j,k) &= \left(\tfrac{2\varepsilon_0\kappa_y - \sigma_y\Delta t}{2\varepsilon_0\kappa_y + \sigma_y\Delta t}\right) Q_x^n(i+1/2,j,k) + \left(\tfrac{2\varepsilon_0}{2\varepsilon_0\kappa_y + \sigma_y\Delta t}\right) \\ &\times [P_x^{n+1}(i+1/2,j,k) - P_x^n(i+1/2,j,k)]. \end{aligned} \tag{1.45}$$

Inserting Eqs. (1.38)–(1.40) into Eqs. (1.41)–(1.43) and also following the procedure in deriving Eq. (1.37), leads to the electric field components in the UPML:

$$E_x^{n+1}(i+1/2,j,k) = \left(\frac{2\varepsilon_0\kappa_z-\sigma_z\Delta t}{2\varepsilon_0\kappa_z+\sigma_z\Delta t}\right)E_x^n(i+1/2,j,k) + \left(\frac{1}{2\varepsilon_0\kappa_z+\sigma_z\Delta t}\right)$$
$$\times[(2\varepsilon_0\kappa_x+\sigma_x\Delta t)Q_x^{n+1}(i+1/2,j,k) - (2\varepsilon_0\kappa_x-\sigma_x\Delta t)Q_x^n(i+1/2,j,k)].$$

1.2.4 FDTD formulation of the light scattering properties from single cells

The calculation of the light scattering and extinction cross sections by cells in free space requires the far-field approximation for the electromagnetic fields [29, 36]. The far-field approach has been also used [37] to study scattering and absorption by spherical particles in an absorptive host medium. However, when the host medium is absorptive, the scattering and extinction rates depend on the distance from the cell. Recently, the single-scattering properties of a sphere in an absorptive medium have been derived using the electromagnetic fields on the surface of the scattering object based on Mie theory [38–39]. Here we derive the absorption and extinction rates for an arbitrarily-shaped object in an absorptive medium using the internal electric field [32]. The absorption and extinction rates calculated in this way depend on the size, shape, and optical properties of the scattering object and the surrounding medium, but do not depend on the distance from it. The single particle scattering approach is perfectly applicable to studying the light scattering properties from single biological cells.

Amplitude scattering matrix

For electromagnetic waves with time dependence $\exp(-i\omega t)$ propagating in a charge-free dielectric medium, we can write Maxwell's equations in the frequency domain as follows

$$\nabla\times\vec{D}=0, \quad \nabla\times\vec{H}=0, \quad \nabla\times\vec{E}\rightarrow=i\omega\mu_0\vec{H}, \quad \nabla\times\vec{H}=-i\omega\vec{D}. \tag{1.46}$$

The material properties of the host medium are defined by the background permittivity ε_h and the electric displacement vector is defined as

$$\vec{D}=\varepsilon_0\varepsilon_h\vec{E}+\vec{P}=\varepsilon_0\varepsilon\vec{E}, \tag{1.47}$$

where here $\vec{P}$ is the polarization vector. Given Eq. (1.47), the first and last equations in Eqs. (1.46) lead to

$$\nabla\cdot\vec{E}=-\frac{1}{\varepsilon_0\varepsilon_h}\nabla\cdot\vec{P}, \tag{1.48}$$

$$\nabla\times\vec{H}=-i\omega(\varepsilon_0\varepsilon_h\vec{E}+\vec{P}). \tag{1.49}$$

Combining the third equation in Eqs. (1.46) and Eqs. (1.48), (1.49) yields a source-dependent form of the electromagnetic wave equation

$$(\nabla+k_h^2)\vec{E}=-\frac{1}{\varepsilon_0\varepsilon_h}[k_h^2\vec{P}+\nabla(\nabla\cdot\vec{P})], \tag{1.50}$$

where $k_h=\omega\sqrt{\mu_0\varepsilon_0\varepsilon_h}$ is the complex wave number in the host medium. Using the unit dyad $\overline{\overline{II}}=\vec{x}\vec{x}+\vec{y}\vec{y}+\vec{z}\vec{z}$ (where $\vec{x}$, $\vec{y}$, and $\vec{z}$ are unit vectors in the x, y, and z direction, respectively), we can rewrite Eq. (1.50) in the form

$$(\nabla+k_h^2)\vec{E}=-\frac{1}{\varepsilon_0\varepsilon_h}(k_h^2\overline{\overline{II}}+\nabla\nabla)\cdot\vec{P}. \tag{1.51}$$

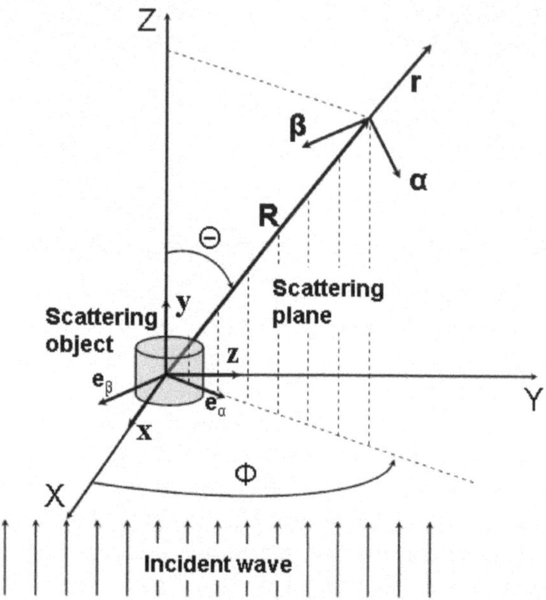

FIGURE 1.3: Incident and scattering wave configurations. The incident wave is propagating in the Z-direction. The unit vectors corresponding to the three coordinate axes are: x,y,z. The scattering direction is defined by the vector $\vec{R}$ with a unit vector $\vec{r} = \left|\vec{R}\right| / \vec{R}$. The z coordinate axis and the vector $\vec{R}$ define the scattering plane. The unit vector $\vec{\alpha}$ is in the scattering plane and is perpendicular to $\vec{R}$ and $\vec{r}$. The unit vector $\vec{\beta}$ is perpendicular to the scattering plane and $\vec{\beta} \times \vec{\alpha} = \vec{r}$. The unit vectors $\vec{e}_\alpha$ and $\vec{e}_\beta$ are in the $X-Y$ plane and are, respectively, parallel and perpendicular to the scattering plane. All vectors in the figure are in bold.

Eq. (1.47) leads to $\vec{P} = \varepsilon_0(\varepsilon - \varepsilon_h)\vec{E}$ which means that $\vec{P}$ is nonzero only in the region inside the cell. The general solution of Eq. (1.51) is given by a volume integral equation [38]:

$$\vec{E}(\vec{R}) = \vec{E}_0(\vec{R}) + \int_V G(\vec{R},\xi)(k_h^2\overline{\overline{II}} + \nabla_\xi \nabla_\xi) \times (\vec{P}/(\varepsilon_0\varepsilon_h))d^3\xi, \tag{1.52}$$

where $\vec{E}_0(\vec{R})$ can be any mathematical solution of $(\nabla^2 + k_h^2)\vec{E} = 0$ but, in practice, the only nontrivial solution here is the incident field in the host medium. The integration volume v is the region inside the particle and $G(\vec{R},\xi)$ is the 3D Green function in the host medium:

$$G(\vec{R},\vec{\xi}) = \frac{\exp(ik_h|\vec{R} - \vec{\xi}|)}{4\pi|\vec{R} - \vec{\xi}|}. \tag{1.53}$$

The scattered field in the far-field region can then be derived from Eq. (1.52) [36]:

$$\vec{E}_s(\vec{R})\Big|_{k_h R \to \infty} = \frac{k_h^2 \exp(ik_h R)}{4\pi R} \int_V [\frac{\varepsilon(\vec{\xi})}{\varepsilon_h} - 1]\left\{\vec{E}(\vec{\xi}) - \vec{r}[\vec{r}\cdot\vec{E}(\vec{\xi})]\right\} \exp(-ik_h\vec{r}.\vec{\xi})d^3\vec{\xi}. \tag{1.54}$$

To calculate the amplitude scattering matrix elements, the incident and the scattered fields are decomposed into their components parallel and perpendicular to the scattering plane (Figure 1.3).

The incident field is decomposed in two components along the unit vectors $\vec{e}_\alpha$ and $\vec{e}_\beta$ both laying in the $X-Y$ plane and defined as parallel and perpendicular to the scattering plane, respectively:

$$\vec{E}_0 = \vec{e}_\alpha E_{0,\alpha} + \vec{e}_\beta E_{0,\beta}. \tag{1.55}$$

$E_{0,\alpha}$ and $E_{0,\beta}$ are related to the x-polarized and y-polarized incident fields used in the FDTD simulation with

$$\begin{pmatrix} E_{0,\alpha} \\ E_{0,\beta} \end{pmatrix} = \begin{bmatrix} \vec{\beta}\cdot\vec{x} & -\vec{\beta}\cdot\vec{y} \\ \vec{\beta}\cdot\vec{y} & \vec{\beta}\cdot\vec{x} \end{bmatrix} \begin{pmatrix} E_{0,y} \\ E_{0,x} \end{pmatrix}. \tag{1.56}$$

The scattered field is decomposed in two components along the unit vectors $\vec{\alpha}$ and $\vec{\beta}$:

$$\vec{E}_s(\vec{R}) = \vec{\alpha} E_{s,\alpha}(\vec{R}) + \vec{\beta} E_{s,\beta}(\vec{R}). \tag{1.57}$$

It is important to note that the incident and scattered fields are specified relative to different sets of basis vectors. The relationship between the incident and scattered fields can be conveniently written in the following matrix form

$$\begin{pmatrix} E_{s,\alpha}(\vec{R}) \\ E_{s,\beta}(\vec{R}) \end{pmatrix} = \frac{\exp(ik_h R)}{-ik_h R} \begin{bmatrix} S_2 & S_3 \\ S_4 & S_1 \end{bmatrix} \begin{pmatrix} E_{0,\alpha} \\ E_{0,\beta} \end{pmatrix}, \tag{1.58}$$

where S_1, S_2, S_3, and S_4 are the elements of the amplitude scattering matrix and, in general, depend on the scattering angle θ and the azimuth angle ϕ.The combination of Eqs. (1.54) and (1.58) leads to the following expressions for amplitude scattering matrix:

$$\overline{\overline{S}} = \begin{bmatrix} S_2 & S_3 \\ S_4 & S_1 \end{bmatrix} = \begin{bmatrix} F_{\alpha,y} & F_{\alpha,x} \\ F_{\beta,y} & F_{\beta,x} \end{bmatrix} \begin{bmatrix} \vec{\beta}\cdot\vec{x} & \vec{\beta}\cdot\vec{y} \\ -\vec{\beta}\cdot\vec{y} & \vec{\beta}\cdot\vec{x} \end{bmatrix}. \tag{1.59}$$

The quantities $F_{\alpha,x}$, $F_{\beta,x}$ and $F_{\alpha,y}$, $F_{\beta,y}$ are calculated for x- and y-polarized incident light, respectively, as follows [36–38]:
for x-polarized incidence:

$$\begin{pmatrix} F_{\alpha,x} \\ F_{\beta,x} \end{pmatrix} = \frac{ik_h^3}{4\pi} \int_V \left[1 - \frac{\varepsilon(\vec{\xi})}{\varepsilon_h}\right] \begin{pmatrix} \vec{\alpha}\cdot\vec{E}(\vec{\xi}) \\ \vec{\beta}\cdot\vec{E}(\vec{\xi}) \end{pmatrix} \exp(-ik_h \vec{r}\cdot\vec{\xi}) d^3\vec{\xi}, \tag{1.60}$$

for y-polarized incidence:

$$\begin{pmatrix} F_{\alpha,y} \\ F_{\beta,y} \end{pmatrix} = \frac{ik_h^3}{4\pi} \int_V \left[1 - \frac{\varepsilon(\vec{\xi})}{\varepsilon_h}\right] \begin{pmatrix} \vec{\alpha}\cdot\vec{E}(\vec{\xi}) \\ \vec{\beta}\cdot\vec{E}(\vec{\xi}) \end{pmatrix} \exp(-ik_h \vec{r}\cdot\vec{\xi}) d^3\vec{\xi}, \tag{1.61}$$

where $k_h = \omega\sqrt{\mu_0 \varepsilon_0 \varepsilon_h}$ and ε_h is the complex relative permittivity of the host medium. When $\varepsilon_h = 1$, Eqs. (1.60), (1.61) will degenerate into a formulation for light scattering by cells in free space.

Eq. (1.58) is now fully defined and can be rewritten in a vectorial form:

$$\vec{E}_s = \overline{\overline{S}} \cdot \vec{E}_0, \tag{1.62}$$

where $S_k = S_k(\theta,\phi)$, $k = 1,2,3,4$. In actual experiments the measured optical signal is proportional to quadratic field combinations [40]. Therefore, to describe the monochromatic transverse wave one introduces four Stokes parameters, which in the case of the scattered wave take the form [37]

$$I_s = \left\langle E_{s,\alpha} E_{s,\alpha}^* + E_{s,\beta} E_{s,\beta}^* \right\rangle,$$

$$Q_s = \left\langle E_{s,\alpha} E_{s,\alpha}^* - E_{s,\beta} E_{s,\beta}^* \right\rangle,$$

$$U_s = \left\langle E_{s,\alpha} E^*_{s,\beta} + E_{s,\beta} E^*_{s,\alpha} \right\rangle, \tag{1.63}$$

$$V_s = \left\langle E_{s,\alpha} E^*_{s,\beta} - E_{s,\beta} E^*_{s,\alpha} \right\rangle.$$

The relation between the incident and the scattered Stokes parameters is given by the Mueller scattering matrix (or simply the scattering matrix) which is defined as follows

$$\begin{pmatrix} I_s \\ Q_s \\ U_s \\ V_s \end{pmatrix} = \frac{1}{k_h^2 R^2} \begin{pmatrix} P_{11} & P_{12} & P_{13} & P_{14} \\ P_{21} & P_{22} & P_{23} & P_{24} \\ P_{31} & P_{32} & P_{33} & P_{34} \\ P_{41} & P_{42} & P_{43} & P_{44} \end{pmatrix} \begin{pmatrix} I_i \\ Q_i \\ U_i \\ V_i \end{pmatrix}, \tag{1.64}$$

where

$$P_{11} = \frac{1}{2}\left(|S_1|^2 + |S_2|^2 + |S_3|^2 + |S_4|^2\right), \quad P_{12} = \frac{1}{2}\left(|S_2|^2 - |S_1|^2 + |S_4|^2 - |S_3|^2\right),$$

$$P_{13} = \mathrm{Re}(S_2 S_3^* + S_1 S_4^*), \quad P_{14} = \mathrm{Im}(S_2 S_3^* - S_1 S_4^*),$$

$$P_{21} = \frac{1}{2}\left(|S_2|^2 - |S_1|^2 - |S_4|^2 + |S_3|^2\right), \quad P_{22} = \frac{1}{2}\left(|S_2|^2 + |S_1|^2 - |S_4|^2 - |S_3|^2\right),$$

and

$$P_{23} = \mathrm{Re}(S_2 S_3^* - S_1 S_4^*), \quad P_{24} = \mathrm{Im}(S_2 S_3^* + S_1 S_4^*),$$

$$P_{31} = \mathrm{Re}(S_2 S_4^* + S_1 S_3^*), \quad P_{32} = \mathrm{Re}(S_2 S_4^* - S_1 S_3^*),$$

$$P_{33} = \mathrm{Re}(S_1 S_2^* + S_3 S_4^*), \quad P_{34} = \mathrm{Im}(S_2 S_1^* + S_4 S_3^*),$$

$$P_{41} = \mathrm{Im}(S_2 S_4^* + S_1 S_3^*), \quad P_{42} = \mathrm{Im}(S_4 S_2^* - S_1 S_3^*),$$

$$P_{43} = \mathrm{Im}(S_1 S_2^* - S_3 S_4^*), \quad P_{44} = \mathrm{Re}(S_1 S_2^* - S_3 S_4^*).$$

The P matrix elements contain the full information about the scattering event. In non absorptive media the elements of the Mueller matrix [Eq. (1.64)] can be used to define the scattering cross-section and anisotropy. The scattering cross-section σ_s is defined as the geometrical cross-section of a scattering object that would produce an amount of light scattering equal to the total observed scattered power in all directions. It can be calculated by the integration of the scattered intensity over all directions. It can be expressed by the elements of the scattering matrix P and the Stokes parameters (I_0, Q_0, U_0, V_0) of the incident light as follows [40]

$$\sigma_s = \frac{1}{k_h^2 I_0} \int_{4\pi} [I_0 P_{11} + Q_0 P_{12} + U_0 P_{13} + V_0 P_{14}]\, d\Omega. \tag{1.65}$$

In the case of a spherically symmetrical scattering object and nonpolarized light, the relationship (1.65) is reduced to the usual integral of the indicatrix with respect to the scattering angle:

$$\sigma_s = \frac{2\pi}{k_h^2} \int_0^{\pi} P_{11}(\theta) \sin(\theta) d\theta. \tag{1.66}$$

The anisotropy parameter g is defined as follows [40]

$$g = <\cos(\theta)> = \frac{2\pi}{k_h^2\sigma_s}\int_0^{\pi}\cos(\theta)P_{11}(\theta)\sin(\theta)d\theta. \tag{1.67}$$

A positive (negative) value of g corresponds to a forward (backward) dominated scattering. The isotropic scattering case corresponds to $g = 0$.

In an absorptive medium, the elements of the Mueller scattering matrix [Eq. (1.64)] depend on the radial distance from the scattering object and cannot be directly related to the scattering cross-section as given above by Eq. (1.65). In this case the different elements of the matrix are used individually in the analysis of the scattering phenomena. In practice, their values are normalized by the total scattered rate around the object in the radiation zone, which can be derived from the integral of P_{11} for all scattering angles.

Absorption, scattering, and extinction efficiencies

The flow of energy and the direction of the electromagnetic wave propagation are represented by the Poynting vector:

$$\vec{s} = \vec{E} \times \vec{H}^*, \tag{1.68}$$

where the asterisk denotes the complex conjugate. To derive the absorption and extinction rates of a particle embedded in an absorptive medium, we can rewrite the last equation in Eq. (1.46) as

$$\nabla\times\vec{H} = -i\omega\varepsilon_0(\varepsilon_r + i\varepsilon_i)\vec{E}. \tag{1.69}$$

Combining the third equation in Eqs. (1.46) with Eq. (1.69) leads to

$$\begin{aligned}\nabla\cdot s &= \nabla\cdot(\vec{E}\times\vec{H}^*) = \vec{H}^*\cdot(\nabla\times\vec{E}) - \vec{E}\cdot(\nabla\times\vec{H}^*)\\ &= i\omega(\mu_0\vec{H}\cdot\vec{H}^* - \varepsilon_0\varepsilon_r\vec{E}\cdot\vec{E}^*) - \omega\varepsilon_0\varepsilon_i\vec{E}\cdot\vec{E}^*\end{aligned} \tag{1.70}$$

For the sake of convenience in the following presentation, we will define the real and imaginary parts of the relative permittivity of the scattering object as ε_{tr} and ε_{ti}, and those for the host medium as ε_{hr} and ε_{hi}, respectively. The rate of energy absorbed by the object is

$$\begin{aligned}w_a &= -\tfrac{1}{2}\,\mathrm{Re}\left[\oint_S \vec{n}\cdot\vec{s}(\vec{\xi})d^2\vec{\xi}\right] = -\tfrac{1}{2}\,\mathrm{Re}\left[\int_V \nabla\cdot\vec{s}(\vec{\xi})d^3\vec{\xi}\right]\\ &= \varepsilon_0\tfrac{\omega}{2}\int_V \varepsilon_{ti}(\vec{\xi})\vec{E}(\vec{\xi})\cdot\vec{E}^*(\vec{\xi})d^3\vec{\xi},\end{aligned} \tag{1.71}$$

where $\vec{n}$ denotes the outward-pointing unit vector normal to the surface of the object. The surface and volume integrals are defined by the volume of the scattering object. When electromagnetic waves are incident on an object, the electric and magnetic field vectors $\vec{E}$ and $\vec{H}$ can be taken as sums of the incident and scattered fields. Therefore, the scattered field vectors can be written as

$$\vec{E}_s = \vec{E} - \vec{E}_i, \tag{1.72}$$

$$\vec{H}_s = \vec{H} - \vec{H}_i, \tag{1.73}$$

where $\vec{E}_i$ and $\vec{H}_i$ denote the incident electric and magnetic field vector, respectively. Therefore, the rate of energy scattered by the object can be expressed as

$$w_s = \frac{1}{2}\,\mathrm{Re}\left[\oint_S \vec{n}\cdot(\vec{E}_s\times\vec{H}_s^*)d^2\vec{\xi}\right] = \frac{1}{2}\,\mathrm{Re}\left\{\oint_S \vec{n}\cdot\left[(\vec{E}-\vec{E}_i)\times(\vec{H}^*-\vec{H}_i^*)\right]d^2\vec{\xi}\right\}. \tag{1.74}$$

Because both absorption and scattering remove energy from the incident waves, the extinction rate of the energy can be defined as

$$\begin{aligned} w_e &= w_s + w_a = \\ &= \tfrac{1}{2}\,\mathrm{Re}\{\oint_S \vec{n}\cdot[(\vec{E}-\vec{E}_i)\times(\vec{H}^*-\vec{H}_i^*)]d^2\vec{\xi}\} - \tfrac{1}{2}\,\mathrm{Re}[\oint_S \vec{n}\cdot(\vec{E}\times\vec{H}^*)d^2\vec{\xi}] \\ &= \tfrac{1}{2}\,\mathrm{Re}[\oint_S \vec{n}\cdot(\vec{E}_i\times\vec{H}_i^* - \vec{E}_i\times\vec{H}^* - \vec{E}\times\vec{H}_i^*)d^2\vec{\xi}] \\ &= \tfrac{1}{2}\,\mathrm{Re}[\int_V \tilde{N}\times(E_i\times H_i^* - E_i\times H^* - E\times H_i^*)d^3\xi]. \end{aligned} \tag{1.75}$$

Using Eqs. (1.70) and (1.75), similar to the derivation of Eqs. (1.71), we can obtain

$$\begin{aligned} w_e = w_a + w_s = &\ \varepsilon_0 \tfrac{\omega}{2} \int_V [\varepsilon_{ti}(\vec{\xi}) + \varepsilon_{hi}(\vec{\xi})]\ \mathrm{Re}[\vec{E}_i(\vec{\xi})\cdot\vec{E}^*(\vec{\xi})]d^3\vec{\xi} \\ &-\varepsilon_0 \tfrac{\omega}{2} \int_V [\varepsilon_{tr}(\vec{\xi}) - \varepsilon_{hr}(\vec{\xi})]\ \mathrm{Im}[\vec{E}_i(\vec{\xi})\cdot\vec{E}^*(\vec{\xi})]d^3\vec{\xi} \\ &-\varepsilon_0 \tfrac{\omega}{2} \int_V \varepsilon_{hi}(\vec{\xi})[\vec{E}_i(\vec{\xi})\cdot\vec{E}_i^*(\xi)]d^3\vec{\xi}. \end{aligned} \tag{1.76}$$

Assuming the rate of energy incident on a particle of arbitrary shape is f, then the absorption, scattering, and extinction efficiencies are $Q_a = w_a/f$, $Q_s = (w_e - w_a)/f$ and $Q_e = w_e/f$, respectively. Consequently, the single scattering albedo is $\tilde{\omega} = Q_s/Q_e$.

In an absorptive medium, the rate of energy incident on the object depends on the position and intensity of the wave source, the optical properties of the host medium, and the object's size and shape. For spherical objects, if the intensity of the incident light at the center of the computational domain is I_0, the rate of energy incident on a spherical scatterer centered at the center of the 3D computational domain is

$$f = \frac{2\pi a^2}{\eta^2} I_0[1 + (\eta - 1)e^{\eta}], \tag{1.77}$$

where a is the radius of the spherical object, $\eta = 4\pi a n_{hi}/\lambda_0$, $I_0 = \frac{1}{2}\left(\frac{n_{hr}}{c\mu_0}\right)|E_0|^2$ is the intensity of the incident light at the center of the computational domain, λ_0 is the incident wavelength in free space, n_{hr} and n_{hi} are the real and imaginary refractive index of the host medium, respectively. $|E_0|$ is the amplitude of the incident electric field at the center of the 3D computational domain. For nonspherical object, the rate of energy incident on the object is calculated numerically.

1.2.5 FDTD formulation of optical phase contrast microscopic (OPCM) imaging

The 3D FDTD formulation provided here is based on a modified version of the total-field/scattered-field (TFSF) formulation that was described earlier [29, 32]. It could be more appropriately called total-field/reflected-field (TFRF) formulation. The 3D TFRF formulation uses a TFSF region that contains the biological cell and extends beyond the limits of the simulation domain (Figure 1.4). The extension of the transverse dimension of the input field beyond the limits of the computational domain through the UPML boundaries would lead to distortions of its ideal plane wave shape and eventually distort the simulation results. To avoid these distortions one must use Bloch periodic boundary conditions (Figure 1.4) in the lateral x- and y-directions, which are perpendicular to the direction of propagation z [29].

Bloch boundary conditions are periodic boundary conditions that take into account the phase effects due to the tilting of the input plane waves incoming at periodic structures, i.e., what we are actually modeling is a periodic row of biological cells. The near scattered fields, however, are calculated in the transverse planes located in the close proximity to the cell where the coupling effect due to waves scattered from adjacent cells is negligible. This effect can be further minimized or completely removed by controlling the lateral dimension of computational domain by using a large enough period of the periodic cell structure. The larger is this period, the smaller is the coupling effect. In the 3D TFRF formulation the location in the computational domain corresponding to the

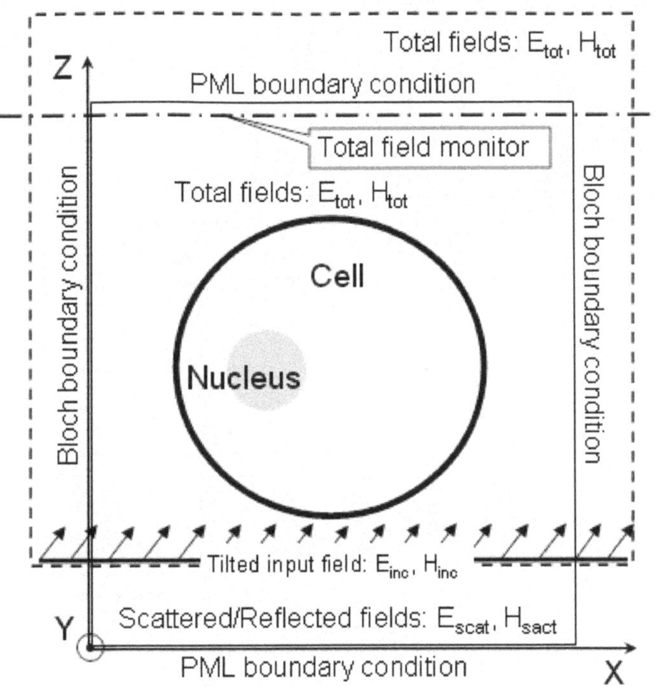

FIGURE 1.4: Schematic representation of the modified 3D FDTD Total Field/Scattered Field formulation.

forward scattered light is positioned within the total field region (Figure 1.4). The OPCM simulation model requires the explicit availability of the forward scattered transverse distribution of the fields. The phase of the scattered field accumulated by a plane wave propagating through a biological cell will be used in the FDTD model of the OPCM that will be described in the next section.

FDTD OPCM principle

Phase contrast microscopy is utilized to produce high-contrast images of transparent specimens such as microorganisms, thin tissue slices, living cells and subcellular components such as nuclei and organelles. It translates small phase variations into corresponding changes in amplitude visualized as differences in image contrast. A standard phase contrast microscope design is shown in Figure 1.5a, where an image with a strong contrast ratio is created by coherently interfering a reference (R) with a diffracted beam (D) from the specimen.

Relative to the reference beam, the diffracted beam has lower amplitude and is retarded in phase by approximately $\pi/2$ through interaction with the specimen. The main feature in the design of the phase contrast microscope is the spatial separation of the R beam from D wave front emerging from the specimen. In addition, the amplitude of the R beam light must be reduced and the phase advanced or retarded by another $\pm\pi/2$ in order to maximize the differences in the intensity between the specimen and the background in the image plane. The mechanism for generating relative phase retardation has two steps: i) the D beam is being retarded in phase by a quarter wavelength (i.e., $\pi/2$) at the specimen, and ii) the R beam is advanced (or retarded) in phase by a phase plate positioned in or very near the objective rear focal plane (Figure 1.5a). This two-step process is enabled by a specially designed annular diaphragm – the annulus. The condenser annulus, which is placed in

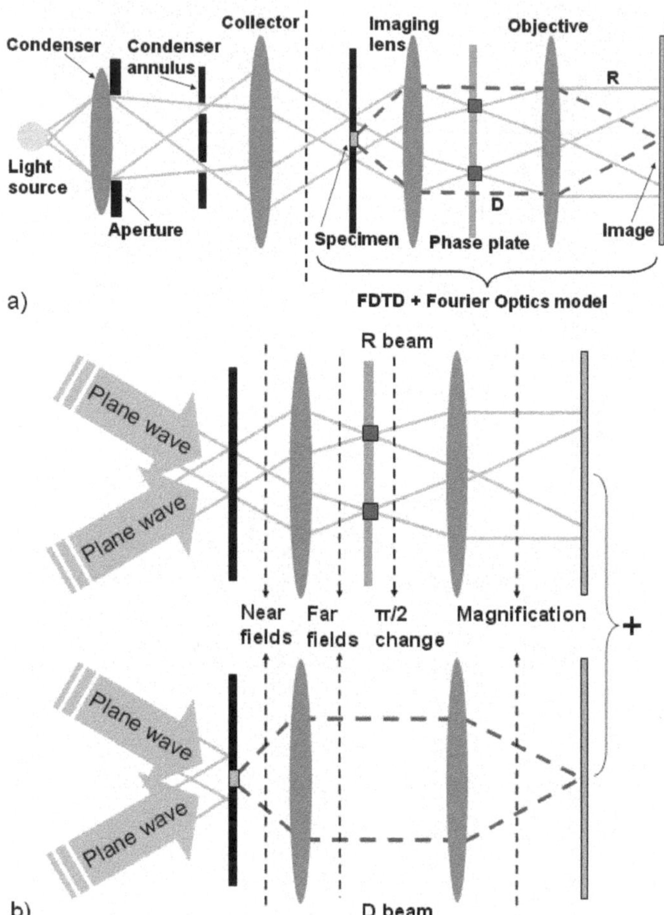

FIGURE 1.5: a) Schematic representation of an OPCM. b) 2D visual representation of the FDTD OPCM model using incoherent illumination by two planes waves at a polar angle of 30 deg. For each of the two plane waves the propagation of light is modeled as a combination of two parallel wave phenomena: i) propagation of the reference (R) beam without the cell, and ii) propagation of the diffracted (D) beam due to the cell.

the condenser front focal plane, is matched in diameter and optically conjugated to the phase plate residing in the objective rear focal plane. The resulting image, where the total phase difference is translated by interference at the image plane into an amplitude variation, can have a high contrast ratio, particularly if both beams have the same amplitude.

Figure 1.5a illustrates the part of the microscope that will become the subject of FDTD modeling combined with Fourier optics. Figure 1.5b provides a visual representation illustrating the major steps in the FDTD OPCM model. The phase contrast microscope uses incoherent annular illumination that could be approximately modeled by adding up the results of eight different simulation using ideal input plane waves incident at a given polar angle (30 deg), an azimuthal angle (0, 90, 180 or 270 deg), and a specific light polarization (parallel or perpendicular to the plane of the graph). Every single FDTD simulation provides the near field components in a transverse monitoring plane located right behind the cell (Figure 1.4).

The far field transformations use the calculated near fields right behind the cell and return the three complex components of the electromagnetic fields far enough from the location of the near fields, i.e., in the far field [29]: $E_r(u_x,u_y)$, $E_\theta(u_x,u_y)$ and $E_\phi(u_x,u_y)$, where r, θ and ϕ refer to the spherical coordinate system shown in Figure 1.3, and the variables u_x and u_y are the x and y components of the unit direction vector $\vec{u}$. The unit direction vector $\vec{u}$ is related to the angular variables θ and ϕ:

$$u_x = \sin(\theta)\cos(\phi), \quad u_y = \sin(\theta)\sin(\phi), \quad u_z = \cos(\theta), \quad u_x^2 + u_y^2 + u_z^2 = 1. \tag{1.78}$$

The in-plane wave vectors for each plane wave are given by $k_x = ku_x$ and $k_y = ku_y$, where $k = 2\pi/\lambda$. What is important here is to note that in this case the near-to-far field transformation must take into account the Bloch periodic boundary conditions in the lateral dimension and is calculated only at angles that correspond to diffracted orders of the periodic structure such as defined by the Bragg conditions. This is done by calculating the direction cosines of all the diffracted orders that meet the Bragg condition, and interpolating the previously far-field distributions onto those specific directions. The near-to-far field projection therefore provides the electric field amplitude and phase corresponding to each diffracted order. The zeroth order, i.e., the light travelling through the scattering object without any deviation in angle, is the reference beam and the phase contrast microscope is designed to provide a phase delay to this order.

The amplitudes and the phases of the calculated far-field components can now be used to do Fourier optics with both the scattered and reference beams. We can assume an ideal optical lens system that could be characterized by a given magnification factor. This simple model could be easily extended to include the numerical equivalent of the two lenses together with an additional model to take into account any aberrational effects. The magnification was implemented by modifying the angle of light propagation, i.e., by multiplying the transverse components of the direction cosines, u_x and u_y by the inverse value of the desired magnification factor M: $U_x = u_x/M$ and $U_y = u_y/M$. In any other circumstances the modification of the direction cosines would lead to complications because of the vectorial nature of the $\vec{E}$ field. In our case, however, working in spherical coordinates (E_r, E_θ, and E_ϕ) leads to the advantage that the vectorial components do not change when u_x and u_y are modified because they are part of a local coordinate system that is tied to the values u_x and u_y. The factor M is applied to the far fields before the interference of the diffracted (D) and reference (R) beams at the image plane.

It was also possible to apply the effect of a numerical aperture NA, which clips any light that has too steep an angle and would not be collected by the lens system. This means that all the light with $U_x^2 + U_y^2 > (NA)^2$ is being clipped. The effect of the aperture is defined by applying the last inequality to the corrected aperture angles θ' and ϕ':

$$\sin(\phi') = U_y/U_{xy}, \quad \cos(\phi') = U_x/U_{xy}, \quad \cos(\theta') = U_z, \quad \sin(\theta') = U_{xy}, \tag{1.79}$$

where $U_{xy} = sqrt(U_x^2 + U_y^2)$, $U_z = sqrt(1 - U_{xy}^2)$ and the "sqrt" labels a square root mathematical operation. The magnified field components will then have the following form:

Diffracted (D) beam:

$$\begin{aligned} E_{x-D}(k_x,k_y) &= -E_{\phi-D}\sin(\phi') + E_{\theta-D}\cos(\phi')\cos(\theta'), \\ E_{y-D}(k_x,k_y) &= E_{\phi-D}\cos(\phi') + E_{\theta-D}\sin(\phi')\cos(\theta'), \end{aligned} \tag{1.80}$$

$$E_{z-D}(k_x,k_y) = -E_{\theta-D}\sin(\theta').$$

Reference (R) beam:

$$E_{x-R}(k_x,k_y) = -E_{\phi-R}\sin(\phi') + E_{\theta-R}\cos(\phi')\cos(\theta'),$$

$$E_{y-R}(k_x,k_y) = E_{\phi-R}\cos(\phi') + E_{\theta-R}\sin(\phi')\cos(\theta'), \tag{1.81}$$

$$E_{z-R}(k_x,k_y) = -E_{\theta-R}\sin(\theta'),$$

where $E_{\theta-R}$ and $E_{\phi-R}$ are the far field components of the reference beam.

The fields given above are then used to calculate back the Fourier inverse transform of the far field transformed fields leading to the distribution of the scattered and the references beams in the image plane:

Diffracted (D) beam:

$$\begin{aligned} E_{x-D} &= \text{sum}\left(E_{x-D}(k_x,k_y)\exp(ik_xx+ik_yy+ik_zz)\right), \\ E_{y-D} &= \text{sum}\left(E_{y-D}(k_x,k_y)\exp(ik_xx+ik_yy+ik_zz)\right), \\ E_{z-D} &= \text{sum}\left(E_{z-D}(k_x,k_y)\exp(ik_xx+ik_yy+ik_zz)\right). \end{aligned} \tag{1.82}$$

Reference (R) beam:

$$\begin{aligned} E_{x-R} &= \text{sum}\left(E_{x-R}(k_x,k_y)\exp(ik_xx+ik_yy+ik_zz)\right), \\ E_{y-R} &= \text{sum}\left(E_{y-R}(k_x,k_y)\exp(ik_xx+ik_yy+ik_zz)\right), \\ E_{z-R} &= \text{sum}\left(E_{z-R}(k_x,k_y)\exp(ik_xx+ik_yy+ik_zz)\right), \end{aligned} \tag{1.83}$$

where the summation is over all angles.

The OPCM images at the image plane are calculated by adding up the scattered and the reference beam at any desired phase offset Ψ:

$$\begin{aligned} I = {} & \text{abs}(E_{x-D} + aE_{x-R}\exp(i\Psi))^2 + \text{abs}(E_{y-D} + aE_{y-R}\exp(i\Psi))^2 \\ & + \text{abs}(E_{z-D} + aE_{z-R}\exp(i\Psi))^2. \end{aligned} \tag{1.84}$$

The coefficient a and the phase Ψ are simulation parameters corresponding to the ability of the OPCM to adjust the relative amplitudes and the phase difference between the two beams.

1.3 FDTD Simulation Results of Light Scattering Patterns from Single Cells

1.3.1 Validation of the simulation results

To simulate the light scattering and absorption by single biological cells, we use a C++ computer program that is based on the FDTD formulation described in section 1.2.4 [17, 32]. A Mie scattering program was also used to provide exact results for the validation of the FDTD simulations. Details about the specific Mie scattering formulation can be found in [27] and [42].

Figure 1.6 shows some preliminary FDTD simulation results for the phase function of a simple spherical biological cell containing only a cytoplasm and a nucleus both embedded in a nonabsorptive extracellular medium. The phase function represents the light scattering intensity as a function of the scattering angle in a plane passing through the center of the cell. It is defined by the angular dependence of the P_{11} element of the Mueller scattering matrix normalized by the total scattered rate

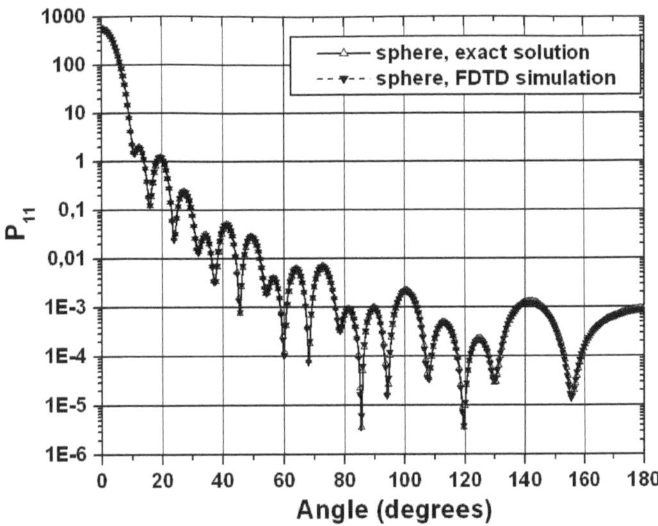

FIGURE 1.6: Normalized light scattering intensity (P_{11} element of the Mueller scattering matrix) distribution with scattering angle – comparison of the FDTD simulation results with exact analytical (Mie theory) results. There is a very good agreement between exact analytical and the numerical solutions with a relative error of approximately 5%.

– the integral of P_{11} in all possible directions around the cell. The cell membrane is not taken into account. The nucleus has a size parameter $2\pi R_c/\lambda_0 = 7.2498$, where R_c is the radius of the nucleus and λ_0 is the incident wavelength in free space. The refractive index of the nucleus is 1.4, of the cytoplasm 1.37 and of the extracellular material 1.35. The size parameter of the whole cell is 19.3328. The FDTD cell size is $\Delta s = \lambda_0/30$ and the UPML parameters are $\kappa_{\max} = 1$ and $R(0) = 10^{-8}$ [see Eqs. (1.12) and (1.13)]. The number of mesh points in all three directions is the same: 209. The number of simulation time steps is 10700. These preliminary results were compared with the exact solutions provided by Mie theory. The relative error of the FDTD simulation results is ~5%. Some absolute values of cell parameters can be derived as follows. If $\lambda_0 = 0.9\mu m$, the FDTD cell size $\Delta s = \lambda_0/30 = 0.03\mu m$, the nucleus' radius $R_c = 7.2498\lambda_0/2\pi = 1.0385\mu m$ and the cytoplasm radius $R_c = 19.3328\lambda_0/2\pi = 2.7692\mu m$. These dimensions are more typical for relatively small cells or bacteria that have no nucleus.

The extinction efficiency Q_e, absorption efficiency Q_a and anisotropy factor g for the cell associated with Figure 1.6 with the parameters given above are listed in Table 1.1. Other P matrix elements are shown in Figure 1.7. To complete the validation of the UPML FDTD scheme we compared the numerical and exact solutions for the light scattering patterns from biological cells embedded in an absorptive extracellular medium.

Accounting for this absorption effect requires a special attention since not all types of boundary conditions can handle absorptive materials touching the boundary of the simulation domains. One of the advantages of the UPML boundary conditions considered here is that they can handle that [17]. To study the effect of the absorption of the extracellular medium we assume that the refractive index of the extracellular medium has a real and an imaginary part: $n + ik$. Figure 1.8 demonstrates the good agreement (~5% relative error) between the FDTD and the exact (Mie theory) results for the normalized light scattering patterns in the case when the refractive index in the extracellular medium is $1.35 + i0.05$.

TABLE 1.1: Comparison between FDTD simulation results and analytical solutions provided by Mie theory for the extinction efficiency Q_e, scattering efficiency Q_s, absorption efficiency Q_a, and anisotropy factor g of a cell with a cytoplasm and a nucleus in nonabsorptive extracellular medium

Results	Q_e	Q_s	Q_a	g
Mie theory	0.359057	0.359057	0.0	0.993889
FDTD	0.358963	0.358963	0.0	0.993914
(FDTD-Mie)/Mie	0.026 %	0.026 %	–	0.0025 %

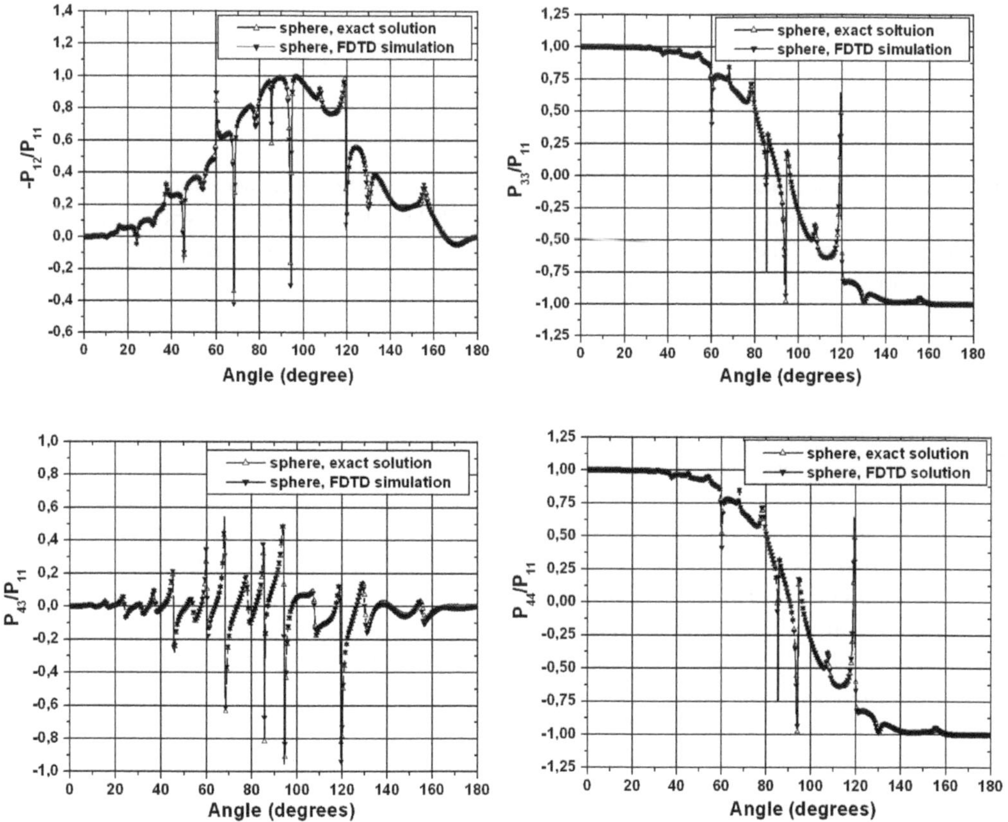

FIGURE 1.7: Angular distributions of the normalized scattering matrix elements P_{12}, P_{33}, P_{43}, and P_{44} calculated by the FDTD method. Cell parameters are the same as for Figure 1.6. There is a very good agreement between exact (Mie theory) and numerical results with a relative error of the FDTD results of approximately 5%.

The extinction efficiency Q_e, scattering efficiency Q_s, absorption efficiency Q_a, and anisotropy factor (g) for a cell in an absorptive medium (the same as in Figure 1.8) are listed in Table 1.2.

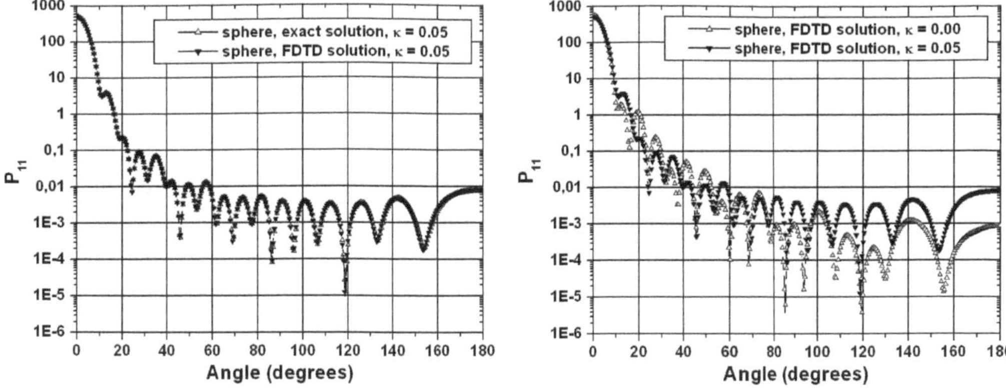

FIGURE 1.8: Normalized light scattering intensity distribution with scattering angle in the case of absorptive extracellular medium. Cell parameters are the same as for Figure 1.6 and Figure 1.7, except a nonzero imaginary part of the refractive index of the extracellular medium.

TABLE 1.2: Comparison between FDTD simulation results and the analytical solutions provided by Mie theory for the extinction efficiency Q_e, scattering efficiency Q_s, absorption efficiency Q_a and anisotropy factor g of a cell with a cytoplasm and a nucleus in a nonzero imaginary part of the refractive index of the extracellular medium $k = 0.05$ and cell parameters as described above in this section.

Results	Q_e	Q_s	Q_a	g
Mie theory	0.672909	0.672909	0.0	0.992408
FDTD	0.676872	0.676872	0.0	0.992440
(FDTD-Mie)/Mie	0.589 %	0.589 %	–	0.0032 %

1.3.2 Effect of extracellular medium absorption on the light scattering patterns

This section describes the FDTD simulation results for the effect of absorption in the extracellular medium on the light scattering patterns from a single cell [17]. We consider two different cell geometries: the one considered in the previous section and another one with a shape of a spheroid (both cytoplasm and nucleus) that can be described by the surface function

$$\frac{x^2}{a^2} + \frac{y^2}{b^2} + \frac{z^2}{c^2} = 1, \tag{1.85}$$

where a, b, and c are the half axes of the spheroid with $a = 2b = 2c$.

Figure 1.9 illustrates the light scattering configuration in the case of a cell with a spheroid shape. The light is incident in the x-direction along the long axis of the spheroid. The size parameters of the cytoplasm are defined by $2\pi R_a/\lambda_0 = 40$ and $2\pi R_{b,c}/\lambda_0 = 20$, where $R_a = a/2$ and $R_{b,c} = b/2 = c/2$. The size parameters of the nucleus are defined by $2\pi r_a/\lambda_0 = 20$ and $2\pi r_{b,c}/\lambda_0 = 10$, where r_a and r_b are the large and small radii of the nucleus, respectively. The cell refractive indices are: cytoplasm: 1.37; nucleus: 1.40, extracellular medium: 1.35, and 1.35+0.05i. The FDTD cell size is $\Delta z = \lambda/20$. The number of mesh point are N_x = 147, N_y = 147, and N_z = 275. The number of simulation time steps is 14000. Figure 1.10 shows: i) the effect of cell size on the light scattering patterns, and ii) the effect of the absorption of the extracellular medium on the phase function of spheroid cells [16–17].

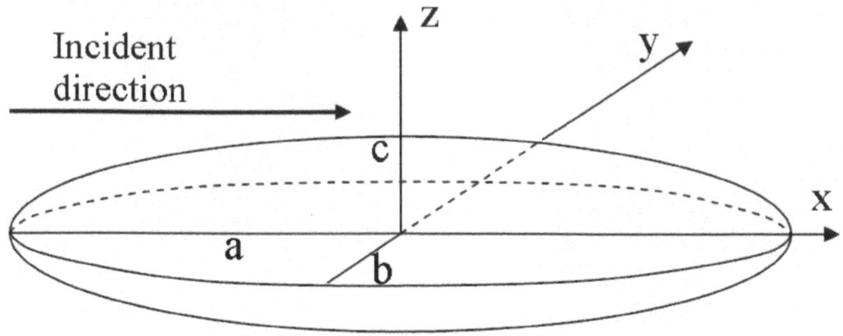

FIGURE 1.9: Coordinate system and geometry of a cell with the shape of a spheroid with half axes a, b, and c, where $a = 2b = 2c$. The light is incident in the positive x-direction, along the long axis a.

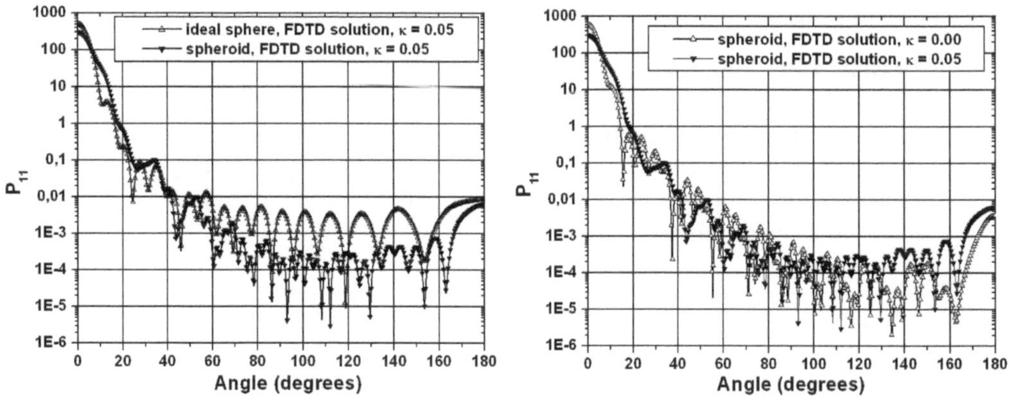

FIGURE 1.10: Normalized light scattering intensity distribution with scattering angle for two different cases: absorptive and nonabsorptive extracellular medium for spherical (left) and spheroid (right) cells. The values of the imaginary part of the refractive index of the extracellular medium are indicated in the insight.

A more detailed analysis of the two graphs shown in Figure 1.10 leads to some interesting findings. First, absorption in the extracellular material of spherical cells (Figure 1.10, right graph) increases the intensity of the light scattering up to one order in the angle range between 90° (transverse scattering) and $\Psi = 180°$(backward scattering). The same light scattering feature was also found in the case of spheroid cells. Second, the influence of absorption in the extracellular material is relatively more pronounced in the case of spherical as compared to spheroid cells (Figure 1.10, left graph). Third, in the case of exact backward scattering and absorptive extracellular material, the light scattering intensity is approximately equal for both cell shapes. However, this is not true for nonabsorptive cell surroundings. This last finding could be highly relevant for OCT and onfocal imaging systems, especially for studies in the wavelength ranges within the hemoglobin, water, and lipids bands.

These findings show that the analysis of light scattering from isolated biological cells should necessarily account for the absorption effect of the surrounding medium. It could be particularly relevant in the case of optical immersion techniques using intra-tissue administration of appropriate chemical agents with absorptive optical properties [43–45]; however, this relevance has not been studied before. It should be pointed out that whenever there is a matching of the refractive indices of a light scatterer and the background material, the scattering coefficient goes to zero and it is only the absorption in the scatterer or in the background material that will be responsible for the light beam extinction. The results presented here provide some good preliminary insights about the light scattering role of absorption in the background material; however, it needs to be further studied.

1.4 FDTD Simulation Results of OPCM Nanobioimaging

1.4.1 Cell structure

This section describes the 3D FDTD modeling results of OPCM imaging of single biological cells in a number of different scenarios. The results are based on the FDTD OPCM model described in subsection 1.2.5 [46]. The optical magnification factor $M = 10$ and the numerical aperture $NA = 0.8$. The cell is modeled as a dielectric sphere with a realistic radius $R_c = 5$ μm (Figure 1.4). The cell membrane thickness is $d = 20$ nm, which corresponds to effective (numerical) thickness of approximately 10 nm. The cell nucleus is also spherical with a radius $R_n = 1.5$ μm centered at a position that is 2.0 μm shifted from the cell center in a direction perpendicular to the direction of light propagation. The refractive index of the cytoplasm is $n_{\text{cyto}} = 1.36$, of the nucleus $n_{\text{nuc}} = 1.4$, of the membrane $n_{\text{mem}} = 1.47$ and of the extracellular material $n_{\text{ext}} = 1.33$ (no Refractive Index Matching – no RIM) or 1.36 (RIM).

1.4.2 Optical clearing effect

The RIM between the cytoplasm and the extracellular medium leads to the optical clearing of the cell image. The optical clearing effect leads to the increased light transmission through microbiological objects due to the matching of the refractive indices of some of their components to that of the extracellular medium [42–45]. If a biological object is homogenous, matching its refractive index value by externally controlling the refractive index of the host medium will make it optically invisible. If the biological object contains a localized inhomogeneity with a refractive index different from the rest of the object, matching the refractive index of the object with that of the external material will make the image of the object disappear and sharply enhance the optical contrast of the inhomogeneity. In the case of biological cells the refractive index of the extracellular fluid can be externally controlled by the administration of an appropriate chemical agent [42–45].

Figure 1.11 shows the cross-sections of two cell images for different values of the phase offset Ψ between the reference and diffracted beam of the OPCM: 180° and 90°. The images illustrate the nature of the optical clearing effect and the value of its potential application for the early detection of cancerous cells by a careful examination of their nucleus size, eccentricity, morphology and chromatin texture (refractive index fluctuations) [14]. At no RIM conditions in both cases ($\Psi = 180°$ and $\Psi = 90°$), the image of the nucleus is represented by a dip in the cell image. At RIM conditions the image contrast of the cell is drastically reduced to zero levels and it is only the image of the nucleus that remains sharply visible. The image of the nucleus is represented by a nice peak associated with the 3-dimensional optical phase accumulation corresponding to its perfectly spherical shape and homogeneous refractive index distribution. A finer analysis of the two graphs shown in Figure 1.11 will show that the diameter of the nucleus (the full width at the half-height of the nucleus image contrast peak) depends on the phase delay Ψ. At $\Psi = 180°$ and no RIM conditions the diameter of the nucleus is estimated at value of $\sim$2.3 μm as compared to RIM conditions where its value is 3.3

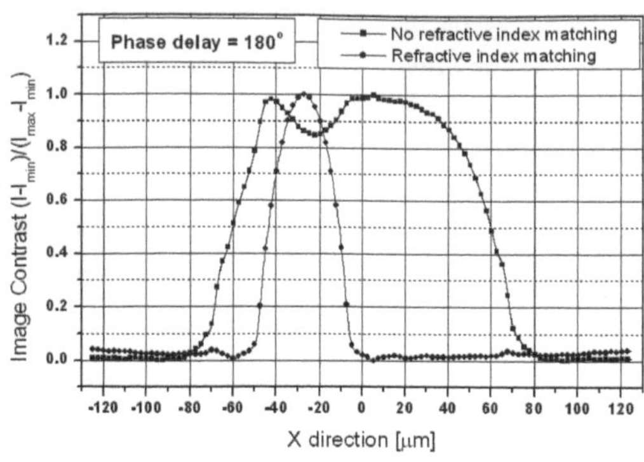

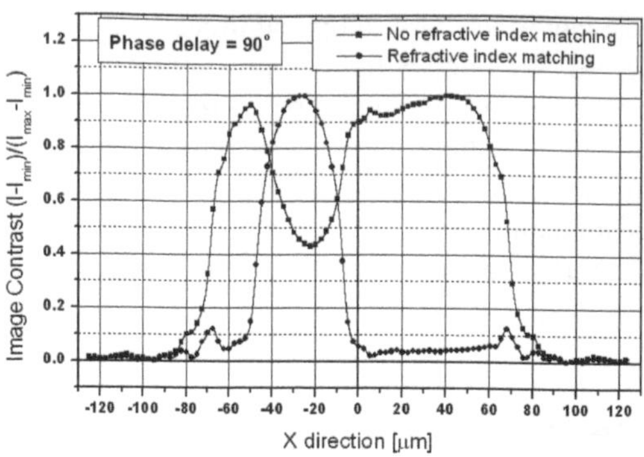

FIGURE 1.11: Cross-sections of FDTD-generated OPCM images of a single cell illustrating the optical clearing effect for different values of the phase offset Ψ between the reference and diffracted beam of the OPCM: 180°(on the left-hand side) and 90° (on the right-hand side). Matching the refractive index value of the extracellular material with that of the cytoplasm enhances the optical contrast and leads to a finer view of the morphological structure of the nucleus.

μm (the estimation accounts for the optical magnification factor $\times 10$ of the system). At $\Psi = 90^\circ$ and no RIM conditions the diameter of the nucleus is estimated at a value of $\sim$3.05 μm as compared to RIM conditions where its value is 3.75 μm (the cell model used in the FDTD simulations has a nucleus with a diameter 3.0 μm). This shows that the OPCM should be preliminary set up at a given optimum phase delay and the OPCM images should be used for relative measurements only after a proper calibration. The analysis of the graphs, however, shows an unprecedented opportunity for using the optical clearing effect for the analysis of any pathological changes in the eccentricity and the chromatin texture of cell nuclei within the context of OPCM cytometry configurations. This new opportunity is associated with the fact that at RIM the cell image is practically transformed into a much finer image of the nucleus.

1.4.3 The cell imaging effect of gold nanoparticles

Optical properties of gold nanoparticles (NPs)

Gold NPs have the ability to resonantly scatter visible and near infrared light. The scattering ability is due to the excitation of Surface Plasmon Resonances (SPR). It is extremely sensitive to their size, shape, and aggregation state offering a great potential for optical cellular imaging and detection labeling studies [47–51]. Our FDTD approach [21, 24] uses the dispersion model for gold derived from the experimental data provided by Johnson and Christy [52] where the total, complex-valued permittivity is given as:

$$\varepsilon(\omega) = \varepsilon_{\text{real}} + \varepsilon_L(\omega) + \varepsilon_P(\omega). \tag{1.86}$$

Each of the three contributions to the permittivity arises from a different material model. The first term represents the contribution due to the basic, background permittivity. The second and third terms represent Lorentz and plasma contributions:

$$\varepsilon_L(\omega) = \varepsilon_{\text{Lorentz}}\omega_0^2/(\omega_0^2 - 2i\delta_0\omega - \omega^2), \quad \varepsilon_P(\omega) = \omega_P^2/(i\omega\nu_C + \omega^2), \tag{1.87}$$

where all material constants are summarized in Table 1.3.

We have modeled both the resonant and nonresonant cases. The ability to model these two different cases, together with the optical clearing effect, provides the opportunity to numerically study

TABLE 1.3: Optical material constants of gold [52]

Background permittivity	Lorentz dispersion	Plasma dispersion
$\varepsilon_{real} = 7.077$	$\varepsilon_{Lorentz} = 2.323$ $\omega_0 = 4.635 \times 10^{15} Hz$ $\delta_0 = 9.267 \times 10^{14} Hz$	$\omega_P = 1.391 \times 10^{16} Hz$ $\nu_C = 1.411 \times 10^{7} Hz$

the possibility for imaging the uptake of clusters of NPs – a scenario that needs to be further studied [21]. We have also used the FDTD technique to calculate the scattering and absorption cross-sections over a 400–900 nm wavelength range for a single 50-nm diameter gold NP immersed in a material having the properties of the cytoplasm ($n_{cyto} = 1.36$) and resolution $dx = dy = dz = 10$ nm. The scattering cross-section is defined as

$$\sigma_{scat} = P_{scat}(\omega)/I_{inc}(\omega), \tag{1.88}$$

where P_{scat} is the total scattered power and I_{inc} is the intensity of the incident light in W/m^2. It was calculated by applying the total-field/scattered-field FDTD formulation described in subsection 1.2.2. We have used the GUI features of the FDTD Solutions software to create 12 field power monitoring planes around the nanoparticle in the form of a box: 6 in the total field region and 6 in the scattered field region. The total scattered power was calculated by summing up the power flowing outward through 6 scattered field power monitors located in the scattered field region.

The absorption cross-section is similarly defined as

$$\sigma_{abs} = P_{abs}(\omega)/I_{inc}(\omega), \tag{1.89}$$

where P_{abs} is the total power absorbed by the particle. The power absorbed by the particle is calculated by calculating the net power flowing inward through the 6 total field power monitors located in the total field region.

The extinction cross-section is the sum of the absorption and scattering cross-sections

$$\sigma_{ext} = \sigma_{scat} + \sigma_{abs}. \tag{1.90}$$

Figure 1.12a shows that the extinction cross-section has a maximum of 3.89 at around 543.0 nm corresponding to one of the radiation wavelengths of He-Ne lasers. Here we also present the results for $\lambda = 676.4$ nm (a Krypton laser wavelength), which corresponds to the nonresonant case (extinction cross-section value 0.322, ~12 times smaller than 3.89). The FDTD results are compared with the theoretical curve calculated by Mie theory. The slight discrepancy between the theoretical and FDTD results for the extinction cross-section is due to the finite mesh size. The consistency of the results could be visibly improved by reducing the mesh size.

OPCM images of gold nanoparticle clusters in single cells

The OPCM cell images are the result of simulations using nonuniform meshing where the number of mesh points in space is automatically calculated to ensure a higher number of mesh points in materials with higher values of the refractive index [53]. Figure 1.12b visualizes the schematic positioning of a cluster of 42 nanoparticles in the cytoplasm that was used to produce the simulation results presented in this section. The cell center is located in the middle ($x = y = z = 0$) of the computational domain with dimensions 15 μm × 12 μm × 15 μm (Figure 1.12b). The center of the nucleus is located at $x = -2$ μm, $y = z = 0$ μm. The cluster of gold nanoparticles is located at $x = 2$ μm, $y = z = 0$ μm. The realistic cell dimensions (including both cell radius and membrane)

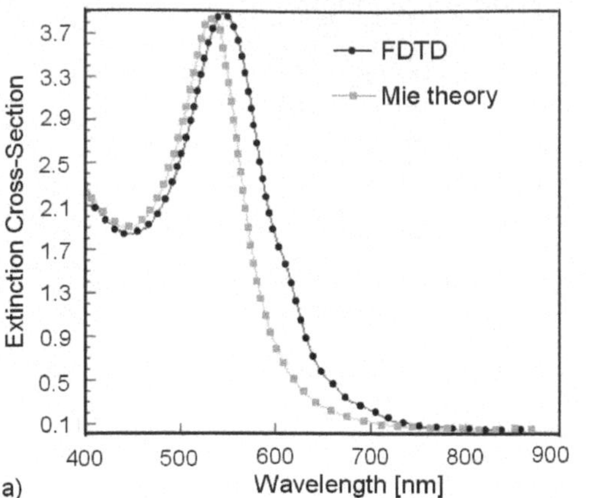

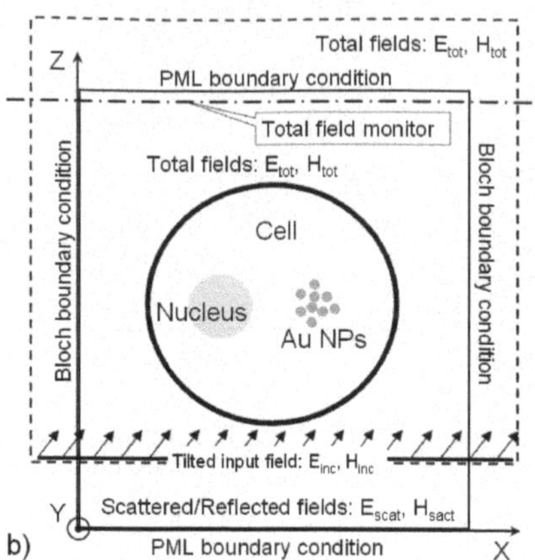

FIGURE 1.12: a) Extinction cross-section of a 50-nm gold nanoparticle immersed in material having the optical properties of the cytoplasm $n = 1.36$ (left). The gold optical properties are described by Eqs. (1.86), (1.87) with the parameters given in Table 1.3. b) Positioning of a cluster of gold nanoparticles randomly distributed within a spatial sphere at a cell location symmetrically opposite to the center of the nucleus.

require a very fine numerical resolution making the simulations computationally intensive. The numerical resolution of the nanoparticles was hard-coded to $dx = dy = dz = 10$ nm to make sure that their numerically manifested optical resonant properties will be the same as the ones shown in Figure 1.12a. This lead to additional requirements for the CPU time and memory (~120 Gbs RAM) requiring high performance computing resources. The time step used during the simulation was defined by means of the Courant stability limit: $c\Delta t = 0.99 \times \left(1/\Delta x^2 + 1/\Delta y^2 + 1/\Delta z^2\right)^{-1/2}$.

Based on the fact that RIM enhances significantly the imaging of the cell components, we have used the FDTD OPCM model to create the OPCM images of the cell at optical immersion conditions including the cluster of 42 gold NPs (Figure 1.13) and for different values of the phase offset Ψ between the reference beam and the scattered beam [assuming $a = 1$, see Eq. (1.62)]. The two graphs in Figure 1.14 compare the geometrical cross-sections ($y = 0\ \mu$m) of the three OPCM images shown at the bottom of Figure 1.13. The right-hand side graph shows the relevant half of the image where the gold NP cluster is located.

At resonance the optical contrast of the gold NP peak is ~2.24 times larger than the one at no resonance and 24.79 times larger than the background optical contrast corresponding to the case when there are no nanoparticles. The enhanced imaging of the gold NP cluster at resonant conditions is clearly demonstrated. It, however, needs to be further studied as a function of particular phase offset Ψ between the reference beam and the scattered beam.

Figure 1.15 visualizes the optical contrast due to the gold nanoparticle cluster as a function of the phase offsets between the reference (R) and the diffracted (D) beams of the optical phase microscope [21]. It shows that the enhancement of the optical contrast due to the nanoparticle resonance changes significantly from minimum of 0.0 ($\Psi = 0°$) to a maximum of 3.60 ($\Psi = -150°$). This finding should be taken into account in real life OPCM imaging experiments.

OPCM images of gold nanoparticles randomly distributed on the nucleus of a cell

Figure 1.16 visualizes the positioning of a cluster of 42 Gold NPs on the surface of the cell nucleus. This was the NP configuration used to produce the FDTD-based simulation results presented in this section [24].

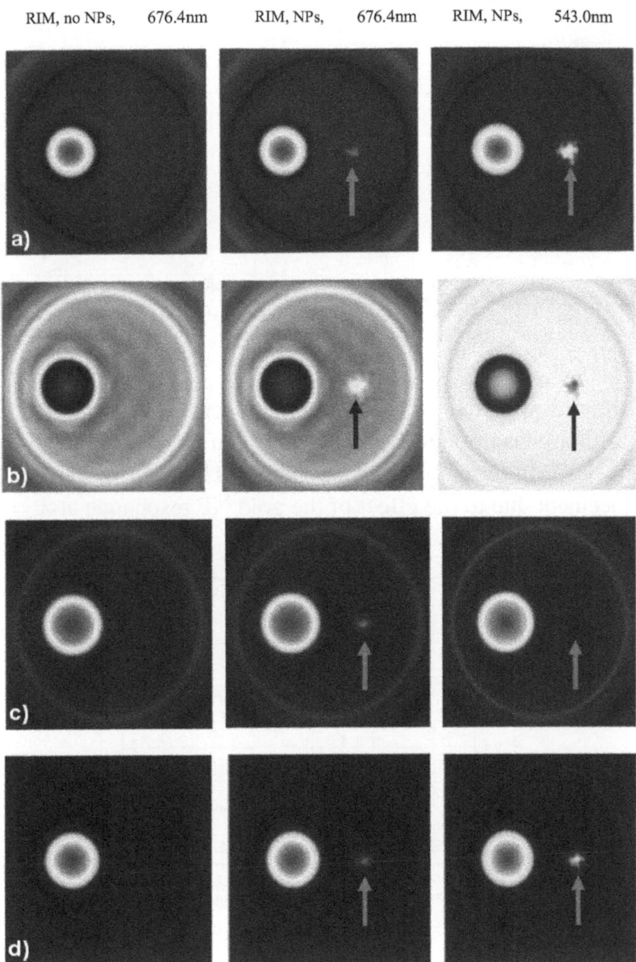

FIGURE 1.13: (See in color after page 572.) OPCM images of a single cell for different values (a: -150°, b: -90°, c: $+90^\circ$, d: $+180^\circ$) of the phase offset Ψ between the reference and diffracted beam of the OPCM at RIM (optical immersion) conditions including a cluster of 42 gold NPs located in a position symmetrically opposite to the nucleus. The arrows indicate the position of the cluster. The left column corresponds to a cell without NPs. The other columns correspond to a cell with NPs at resonant (right) and nonresonant (middle) conditions.

It should be pointed out that the optical wave phenomena involved in the simulation scenario considered here are fundamentally different from the ones considered in the previous section where the gold nanoparticles are randomly distributed within the homogeneous material of cytoplasm and their presence is manifested by means of their own absorption and scattering properties. In the present case the NPs are located at the interface of the nucleus and the cytoplasm, which is characterized by a relatively large refractive index difference ($\Delta n = 0.04$) and which is, therefore, expected to largely dominate and modify the visual effect of the NPs.

A close examination of the OPCM images in Figure 1.17 provides an illustration of this fact. The OPCM cell images are for different values of the phase offset Ψ (a: -90°, b: -30°, c: $+30^\circ$, d: $+90^\circ$) at optical immersion and both resonant and nonresonant conditions. The analysis of Figures 1.17 and 1.18 leads to a number of interesting findings.

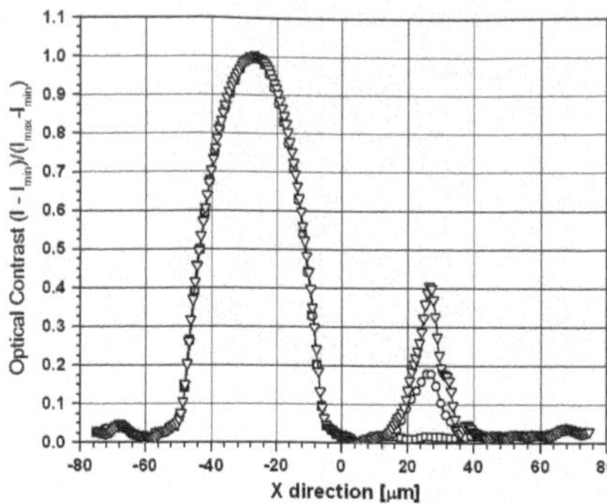

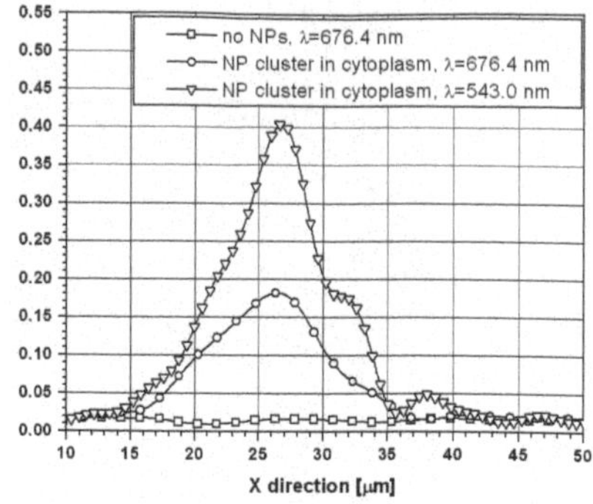

FIGURE 1.14: Comparison of the geometrical cross-sections at $y = 0$ μm of the three OPCM images corresponding to a phase offset $\Psi = 180°$ between the reference and the diffracted beam (bottom row in Figure 1.13) in terms of optical contrast. The right-hand side graph illustrates the optical contrast enhancement due to the effect of the gold NP resonance at $\lambda = 543.0$ nm.

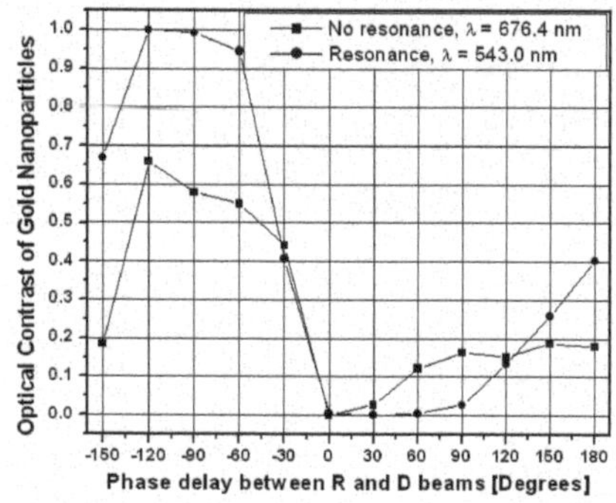

FIGURE 1.15: Optical contrast due to the gold NP cluster as a function of the phase offsets between the reference (R) and the diffracted (D) beams of the optical phase microscope.

First, the images of the cell without the Gold NPs are hardly distinguishable from the images including the Gold NPs at no resonance conditions ($\lambda = 676.4$ nm). Second, the visual effect of the Gold NP presence at resonant conditions ($\lambda = 543.0$ nm) depends significantly on the phase offset Ψ. Third, the presence of the Gold NPs at resonant conditions can be identified by a specific fragmentation of the image of the nucleus for specific values of the offset Ψ (see the right-hand side graphs in Figures. 1.17c and 1.18). This last finding indicates the importance of the ability to adjust the offset Ψ between the reference and diffracted beam of the OPCM in practical circumstances. It is expected to be of relevance for the development and calibration of similar optical diagnostics techniques by medical photonics researchers and clinical pathologists.

1.5 Conclusion

In this chapter we provided a detailed summary of the mathematical formulation of the FDTD method for application in medical biophotonics problems. We have then applied the FDTD approach

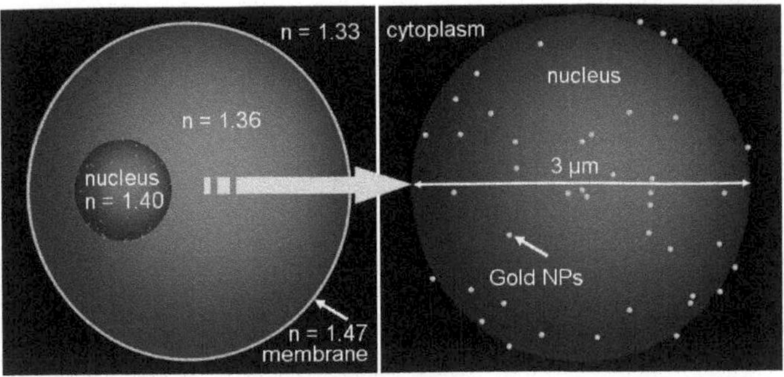

FIGURE 1.16: A cluster of 42 Gold NPs randomly distributed on the surface of the cell nucleus. The NP size on right-hand graph is slightly exaggerated.

to three different modeling scenarios: i) light scattering from single cells, ii) OPCM imaging of realistic size cells, and iii) OPCM imaging of gold NPs in single cells. We have demonstrated the FDTD ability to model OPCM microscopic imaging by, first, reproducing the effect of optical immersion on the OPCM images of a realistic size cell containing a cytoplasm, a nucleus, and a membrane. Second, the model was applied to include the presence of a cluster of gold NPs in the cytoplasm at optical immersion conditions as well as the enhancing imaging effect of the optical resonance of the nanoparticles. Third, we have studied the imaging effect of gold NPs randomly distributed on the surface of the cell nucleus. The results do not allow analyzing the scaling of the NP imaging effect as a function of the number of the NPs. However, the validation of the model provides a basis for future research on OPCM nanobioimaging including the effects of NP cluster size, NP size and number, as well as average distance between the NPs. Another future extension of this research will be to study the capability of the model to provide valuable insights for the application of gold NPs in optical nanotherapeutics.

We believe that the shift from the modeling of the light scattering properties of single cells to the construction of OPCM images of cells containing gold NPs represents a major step forward in extending the application of the FDTD approach to biomedical photonics. It opens a new application area with a significant research potential – the design and modeling of advanced nanobioimaging instrumentation.

Acknowledgment

ST and JP acknowledge the use of the computing resources of WestGrid (Western Canada Research Grid) – a $50-million project to operate grid-enabled high performance computing and collaboration infrastructure at institutions across Canada. VVT was supported by grants: 224014 PHOTONICS4LIFE-FP7-ICT-2007-2, RF President's 208.2008.2, RF Ministry of Science and Education 2.1.1/4989 and 2.2.1.1/2950, and RFBR-09-02-90487_Ukr_f_a, and the Special Program of RF "Scientific and Pedagogical Personnel of Innovative Russia," Governmental contract 02.740.11. 0484.

We are all thankful to Dr. Vladimir P. Zharov for a preliminary discussion on the relevance of the problem of nanoparticle clustering in single cells.

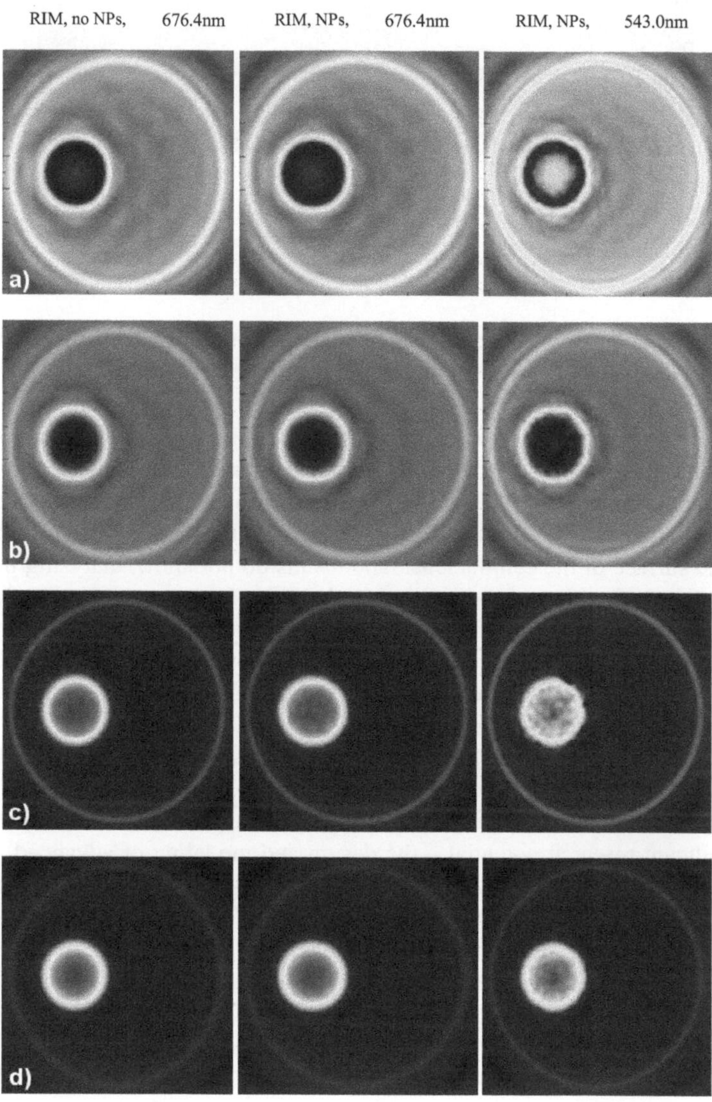

FIGURE 1.17: OPCM images of the cell for different values of the phase offset Ψ (a: -90°, b: -30°, c: $+30^\circ$, d: $+90^\circ$) between the reference and diffracted beam of the OPCM at optical immersion, i.e., refractive index matching, conditions without NPs (left) and including 42 Gold NPs (middle – at no resonance, right – at resonance) randomly located at the surface of the cell nucleus.

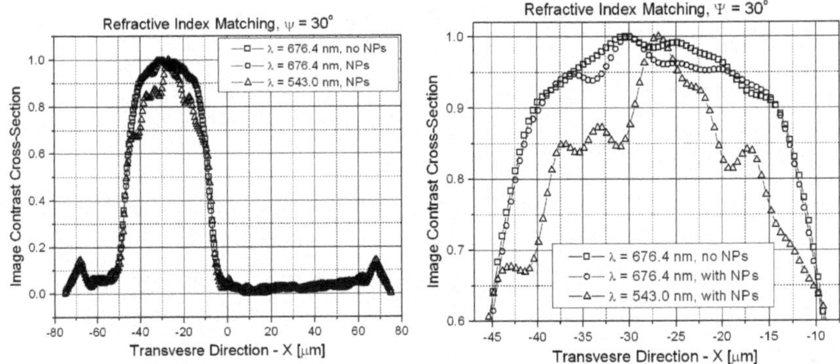

FIGURE 1.18: Cross-sections of the images shown in Figure 1.17c (corresponding to phase offset $\Psi = +30^{\circ}$) expressed in terms of optical contrast. The right-hand side graph provides in finer details an enlarged portion of the left-hand side one. The specific fragmentation of the nucleus' image is due to the presence of the Gold NPs at resonant condition.

References

[1] P.N. Prasad, "Bioimaging: principles and techniques," *Introduction to Biophotonics*, Ch. 7, John Wiley & Sons, New Jersey, 203–249 (2003).

[2] V.V. Tuchin, "Introduction to optical biomedical diagnostics," *Handbook of Optical Biomedical Diagnostics*, Ch. 1, V.V. Tuchin (ed.), SPIE Press, Bellingham, Washington, 1–25 (2002).

[3] J.G. Fujimoto and M.E. Brezinski, "Optical coherence tomography imaging," Ch. 13 in *Biomedical Photonics Handbook*, Tuan Vo-Dinh (ed.), CRC Press, Boca Raton, Florida, 13.1–13.29 (2003).

[4] P.-L. Hsiung, T.H. Ko, S. Bourquin, A.D. Aguirre, P.R. Herz, and J. Fujimoto, "Laser advances benefit optical coherence tomography," *Biophotonics International* **10**, 36–40 (2003).

[5] T. Wilson, "Confocal microscopy," Ch. 10 in *Biomedical Photonics Handbook*, Tuan Vo-Dinh (ed.), CRC Press, Boca Raton, Florida, 10.1–10.18 (2003).

[6] F.M. Kahnert, "Numerical methods in electromagnetic scattering theory," *J. Quant. Spectr. & Radiat. Trans.* **73,** 775–824 (2003).

[7] A. Dunn, C. Smithpeter, A.J. Welch, and R. Richards-Kortum, "Light scattering from cells," *OSA Technical Digest – Biomedical Optical Spectroscopy and Diagnostics*, Optical Society of America, Washington, 50–52 (1996).

[8] A. Dunn and R. Richards-Kortum, "Three-dimensional computation of light scattering from cells," *IEEE J. Selec. Top. Quant. Electr.* **2**, 898–905 (1996).

[9] A. Dunn, *Light Scattering Properties of Cells*, PhD Dissertation, Biomedical Engineering, University of Texas at Austin, Austin TX (1997): http://www.nmr.mgh.harvard.edu/%7Eadunn/papers/dissertation/index.html

[10] A. Dunn, C. Smithpeter, A.J. Welch, and R. Richards-Kortum, "Finite-difference time-domain simulation of light scattering from single cells," *J. Biomed. Opt.* **2**, 262–266 (1997).

[11] R. Drezek, A. Dunn, and R. Richards-Kortum, "Light scattering from cells: finite-difference time-domain simulations and goniometric measurements," *Appl. Opt.* **38**, 3651–3661 (1999); R. Drezek, A. Dunn, and R. Richards-Kortum, "A pulsed finite-difference time-domain (FDTD) method for calculating light scattering from biological cells over broad wavelength ranges," *Opt. Exp.* **6**, 147–157 (2000).

[12] A.V. Myakov, A.M. Sergeev, L.S. Dolin, and R. Richards-Kortum, "Finite-difference time-domain simulation of light scattering from multiple cells," *Technical Digest — Conference on Lasers and Electro-Optics* (*CLEO*), Optical Society of America, Washington, 81–82 (1999).

[13] R. Drezek, M. Guillaud, T. Collier, I. Boiko, A. Malpica, C. Macaulay, M. Follen, and R. Richards-Kortum, "Light scattering from cervical cells throughout neoplastic progression: influence of nuclear morphology, DNA content, and chromatin texture," *J. Biomed. Opt.* **8**, 7–16 (2003).

[14] D. Arifler, M. Guillaud, A. Carraro, A. Malpica, M. Follen, and R. Richards-Kortum, "Light scattering from normal and dysplastic cervical cells at different epithelial depths: finite-difference time domain modeling with a perfectly matched layer boundary condition," *J. Biomed. Opt.* **8**, 484–494 (2003).

[15] S. Tanev, W. Sun, R. Zhang, and A. Ridsdale, "The FDTD approach applied to light scattering from single biological cells," *Proc. SPIE* **5474**, 162–168 (2004).

[16] S. Tanev, W. Sun, R. Zhang, and A. Ridsdale, "Simulation tools solve light-scattering problems from biological cells," *Laser Focus World*, 67–70 (January 2004).

[17] S. Tanev, W. Sun, N. Loeb, P. Paddon, and V.V. Tuchin, "The finite-difference time-domain method in the biosciences: modelling of light scattering by biological cells in absorptive and controlled extra-cellular media," in *Advances in Biophotonics*, B.C. Wilson, V.V. Tuchin, and S. Tanev (eds.), NATO Science Series I, **369**, IOS Press, Amsterdam, 45–78 (2005).

[18] T. Tanifuji, N. Chiba, and M. Hijikata, "FDTD (finite difference time domain) analysis of optical pulse responses in biological tissues for spectroscopic diffused optical tomography," *CLEO/Pacific RIM, The 4th Pacific RIM Conference on Lasers and Electro-Optics* **1**, Optical Society of America, Washington I-376–377 (2001).

[19] T. Tanifuji and M. Hijikata, "Finite difference time domain (FDTD) analysis of optical pulse responses in biological tissues for spectroscopic diffused optical tomography," *IEEE Trans. Med. Imag.* **21**, 181–184 (2002).

[20] K. Ichitsubo and T. Tanifuji, "Time-resolved noninvasive optical parameters determination in three-dimensional biological tissue using finite difference time domain analysis with nonuniform grids for diffusion equations," *Proc. of 27th Annual International Conference of the IEEE Engineering in Medicine and Biology Society*, IEEE-EMBS, Washington, 3133–3136 (2006); doi:10.1109/IEMBS.2005.1617139

[21] S. Tanev, J. Pond, P. Paddon, and V.V. Tuchin, "A new 3D simulation method for the construction of optical phase contrast images of gold nanoparticle clusters in biological cells," *Adv. Opt. Techn.* **2008**, Article ID 727418, 9 pages, doi:10.1155/2008/727418 (2008): http://www.hindawi.com/GetArticle.aspx?doi=10.1155/2008/727418

[22] S. Tanev, V. Tuchin, and P. Paddon, "Cell membrane and gold nanoparticles effects on optical immersion experiments with non-cancerous and cancerous cells: FDTD modeling," *J. Biomed. Opt.* **11**, 064037-1–6 (2006).

[23] S. Tanev, V.V. Tuchin, and P. Paddon, "Light scattering effects of gold nanoparticles in cells: FDTD modeling," *Laser Phys. Lett.* **3**, 594–598 (2006).

[24] S. Tanev, V.V. Tuchin, and J. Pond "Simulation and modeling of optical phase contrast microscope cellular nanobioimaging," *Proc. SPIE* **7027**, 16-1–8 (2008).

[25] X. Li, A. Taflove, and V. Backman, "Recent progress in exact and reduced-order modeling of light-scattering properties of complex structures," *IEEE J. Selec. Top. Quant. Electr.* **11**, 759–765 (2005).

[26] G. Mie, "Beigrade zur optic truber medien, speziell kolloidaler metallsungen," *Ann. Phys.* **25**, 377–455 (1908).

[27] W. Sun, N.G. Loeb, and Q. Fu, "Light scattering by coated sphere immersed in absorbing medium: a comparison between the FDTD and analytic solutions," *J. Quant. Spectr. & Radiat. Trans.* **83**, 483–492 (2004).

[28] K.S. Yee, "Numerical solution of initial boundary value problems involving Maxwell's equation in isotropic media," *IEEE Trans. Anten. Propag.* **AP-14**, 302–307 (1966).

[29] A. Taflove and S.C. Hagness (eds.), *Computational Electrodynamics: The Finite-Difference Time-Domain Method*, 3rd ed., Artech House, Norwood, MA (2005).

[30] W. Sun, Q. Fu, and Z. Chen, "Finite-difference time-domain solution of light scattering by dielectric particles with a perfectly matched layer absorbing boundary condition," *Appl. Opt.* **38**, 3141–3151 (1999).

[31] W. Sun and Q. Fu, "Finite-difference time-domain solution of light scattering by dielectric particles with large complex refractive indices," *Appl. Opt.* **39**, 5569–5578 (2000).

[32] W. Sun, N. G. Loeb, and Q. Fu, "Finite-difference time-domain solution of light scattering and absorption by particles in an absorbing medium," *Appl. Opt.* **41**, 5728–5743 (2002).

[33] Z.S. Sacks, D.M. Kingsland, R. Lee, and J.F. Lee, "A perfectly matched anisotropic absorber for use as an absorbing boundary condition," *IEEE Trans. Anten. Propag.* **43**, 1460–1463 (1995).

[34] S.D. Gedney, "An anisotropic perfectly matched layer absorbing media for the truncation of FDTD lattices," *IEEE Trans. Anten. Propag.* **44**, 1630–1639 (1996).

[35] S.D. Gedney and A. Taflove, "Perfectly matched absorbing boundary conditions," in *Computational Electrodynamics: The Finite-Difference Time Domain Method*, A. Taflove and S. Hagness (eds.), Artech House, Boston, 285–348 (2000).

[36] P. Yang and K.N. Liou, "Finite-difference time domain method for light scattering by small ice crystals in three-dimensional space," *J. Opt. Soc. Am. A* **13**, 2072–2085 (1996).

[37] C.F. Bohren and D.R. Huffman, *Absorption and Scattering of Light by Small Particles*, John Wiley & Sons, New York (1998).

[38] G.H. Goedecke and S.G. O'Brien, "Scattering by irregular inhomogeneous particles via the digitized Green's function algorithm," *Appl. Opt.* **27**, 2431–2438 (1988).

[39] Q. Fu and W. Sun, "Mie theory for light scattering by a spherical particle in an absorbing medium," *Appl. Opt.* **40**, 1354–1361 (2001).

[40] N.G. Khlebtsov, I.L. Maksimova, V.V. Tuchin, and L.V. Wang, "Introduction to light scattering by biological objects," in *Handbook of Optical Biomedical Diagnostics*, Valery Tuchin (ed.), SPIE Press, Bellingham, WA, 31–167 (2002).

[41] W.C. Mundy, J.A. Roux, and A.M. Smith, "Mie scattering by spheres in an absorbing medium," *J. Opt. Soc. Am.* **64**, 1593–1597 (1974).

[42] Q. Fu and W. Sun, "Mie theory for light scattering by a spherical particle in an absorbing medium," *Appl. Opt.* **40**, 1354–1361 (2001).

[43] R. Barer, K.F.A Ross, and S. Tkaczyk, "Refractometry of living cells," *Nature* **171**, 720–724 (1953).

[44] B.A. Fikhman, *Microbiological Refractometry*, Medicine, Moscow (1967).

[45] V.V. Tuchin, *Optical Clearing of Tissues and Blood*, **PM 154**, SPIE Press, Bellingham, WA (2006).

[46] The simulations were performed by the FDTD Solutions^{TM}software developed by Lumerical Solutions Inc., Vancouver, BC, Canada: www.lumerical.com.

[47] K. Sokolov, M. Follen, J. Aaron, I. Pavlova, A. Malpica, R. Lotan, and R. Richards-Kortum, "Real time vital imaging of pre-cancer using anti-EGFR antibodies conjugated to gold nanoparticles," *Cancer Res.* **63**, 1999–2004 (2003): http://cancerres.aacrjournals.org/cgi/content/full/63/9/1999#B9

[48] I.H. El-Sayed, X. Huang, and M.A. El-Sayed, "Surface plasmon resonance scattering and absorption of anti-EGFR antibody conjugated gold nanoparticles in cancer diagnostics: applications in oral cancer," *Nano Lett.* **5**, 829–834 (2005).

[49] N.G. Khlebtsov, A.G. Melnikov, L.A. Dykman, and V.A. Bogatyrev, "Optical properties and biomedical applications of nanostructures based on gold and silver bioconjugates," in *Photopolarimetry in Remote Sensing*, G. Videen, Ya.S. Yatskiv, and M.I. Mishchenko (eds.), NATO Science Series, II. Mathematics, Physics, and Chemistry, **161**, Kluwer, Dordrecht, 265–308 (2004).

[50] V.P. Zharov, J.-W. Kim, D.T. Curiel, and M. Everts, "Self-assembling nanoclusters in living systems: application for integrated photothermal nanodiagnostics and nanotherapy," *Nanomed. Nanotech., Biol., Med.* **1**, 326–345 (2005).

[51] V.P. Zharov, K.E. Mercer, E.N. Galitovskaya, and M.S. Smeltzer, "Photothermal nanotherapeutics and nanodiagnostics for selective killing of bacteria targeted with gold nanoparticles," *Biophys. J.* **90**, 619–627 (2006).

[52] P.B. Johnson and R.W. Christy, "Optical constants of the noble metals," *Phys. Rev. B* **6**, 4370–4379 (1972).

[53] Nonuniform meshing is a standard feature of the FDTD Solutions^{TM} software.

2

Plasmonic Nanoparticles: Fabrication, Optical Properties, and Biomedical Applications

Nikolai G. Khlebtsov
Institute of Biochemistry and Physiology of Plants and Microorganisms, 410049, Saratov, Russia; Saratov State University, 410012, Saratov, Russia

Lev A. Dykman
Institute of Biochemistry and Physiology of Plants and Microorganisms, 410049, Saratov, Russia

Nanoparticle plasmonics is a rapidly emerging research field that deals with the fabrication and optical characterization of noble metal nanoparticles of various size, shape, structure, and tunable plasmon resonances over VIS-NIR spectral band. The recent advances in design, experimental realization, and biomedical application of such nanostructures are based on combination of robust synthesis or nanofabrication techniques with electromagnetic simulations and surface functionalization by biospecific molecular probes. The chapter presents a brief overview of these topics.

Key words: metal nanoparticles, plasmon resonance, light absorption and scattering, bioconjugates, biosensorics, photothermal therapy, bioimaging

2.1 Introduction

The metal (mainly gold) nanoparticles (NPs) [1] have attracted significant interest as a novel platform for nanobiotechnology and biomedicine [2] because of convenient surface bioconjugation with molecular probes and remarkable optical properties related with the localized plasmon resonance (PR) [3]. Recently published examples include applications of NPs to genomics [4], biosensorics [5], immunoassays [6], clinical chemistry [7], detection and control of microorganisms [8], cancer cell photothermolysis [9, 10], targeted delivery of drugs or other substances [11], and optical imaging and monitoring of biological cells and tissues by exploiting resonance scattering [12], optical coherence tomography [13], two-photon luminescence [14], or photoacoustic [15] techniques. One of the outstanding examples is rapid development of methamaterials with optical negative refractive index [16].

Simultaneous advances in *making, measuring*, and *modelling* plasmonic NPs (designated as the 3M's principle [17]) have led to a perfect publication storm in discoveries and potential applications of plasmon-resonant NPs bioconjugates. At present, one can point to several trends in the recent development of physical [18] and biomedical [19] plasmonics. First, the rapid development of the technology of gold [20, 21] and silver [22] NPs synthesis for the last 10–15 years has provided wide possibilities for researchers, beginning from the commonly known colloidal gold nanospheres, nanorods, or silica/gold nanoshells, and ending by exotic structures such as nanorice, nanostars, or nanocages [19]. Furthermore, the development electron-beam-lithography [23] or "nanosphere lithography" [24] technologies allows for fabrication of various nanostructures with fully controlled spatial parameters. Second, the computer simulations of the electromagnetic properties have been expanded from simple and small single-particle scatterers to complex single-particle or multiparticle structures [25, 26]. Third, the traditional characterization methods such as TEM and absorption spectroscopy of colloids have been greatly improved by taking advantage of single-particle technologies. In particular, these techniques are based on the resonance scattering dark-filed spectroscopy [27], the absorption of individual particles [28] (the spatial-modulation spectroscopy [29]) or their surrounding (the photothermal heterodyne imaging method) [30, 31], and on the near-filed optical microscopy [32]. Finally, the development of the surface functionalization protocols [33–35] have stimulated numerous biomedical applications of plasmon-resonant bioconjugates with various probing molecules including DNA oligonucleotides, poly- and monoclonal antibodies (Ab), streptavidin, etc.

Here, we present a short discussion of the above trends. Keeping in mind the limitations of the chapter volume and an enormous volume of the literature data, we avoid any detailed discussion and citation. For the most updated information, the readers are referred to several recent reviews discussing the particle fabrication, functionalization, and optical characterization [1, 18–22, 25, 26, 29, 33–46] as well as analytical and biomedical applications [1, 5, 7–11, 47–58].

2.2 Chemical Wet Synthesis and Functionalization of Plasmon-Resonant NPs

2.2.1 Nanosphere colloids

Perhaps, colloidal gold (CG, Figure 2.1a) is the first nanomaterial, which history goes back to 5th or 4th century B.C. in Egypt and China (see, e.g., reviews [1, 20]). The actual scientific study of CG was pioneered by Faraday [59], Zsigmondy [60], Mie [61], and Svedberg [62]. At present, the most popular colloidal synthesis protocols involve the reduction of $HAuCl_4$ (tetrachloroauric acid) by various reducing agents. In 1973, Frens [63] published very convenient citrate method that produces relatively monodisperse particles with the controlled average equivolume diameter from 10 to 60 nm. Fabrication of smaller particles can be achieved by using sodium borohydride, a mixture of borohydride with EDTA [64] (5 nm) or sodium or potassium thiocyanate [65] (1–2 nm). For particles larger than 20 nm, the citrate method gives elongated particles. To improve the particle shape of large particles, several approaches were suggested [1, 66]. The relevant references and discussion can be found in reviews [1, 20]. Recent advances and challenges in the silver particle fabrication have been considered in a review [22] and paper [67].

2.2.2 Metal nanorods

PR of gold nanorods (NRs) can be easily tuned over 650–1500 nm by simple variation in NRs length or aspect ratio (Figure 2.1b). Historically, the first fabrication of gold nanorods was performed by using the hard-template method (electrochemical deposition in a porous membrane) [68].

Perhaps, Wang and coworkers [69] were the first who introduced a combination of a surfactant additives and an electrochemical oxidation/reduction procedure to prepare colloidal gold nanorods with the average thickness of about 10 nm and the aspect ratio ranging from 1.5 to 11. At present, the most popular is the seed-mediated CTAB-assisted protocol suggested by Murphy [70] and El-Sayed [71] groups. A summary of recent advantages in gold NRs synthesis can be found in reviews [72–74] and papers [75, 76]. Gold nanorods can be used as templates for deposition of a thin silver layer [166]. Such bimetal particles can conveniently be tuned from NIR to VIS through the increase in silver layer.

2.2.3 Metal nanoshells

Silica (core)/gold (shell) particles (Figure 2.1c) present an example of NPs which PR can be tuned over 600–1500 nm by variation in the particle structure, i.e., in their ratio shell thickness/core radius. In 1998, Halas and coworkers [78] developed a two-step protocol involving fabrication of silica nanospheres followed by their amination, attachment of fine (2–4 nm) gold seeds, and formation of a complete gold shell by reducing of $HAuCl_4$ on seeds. A summary of ten-year development of this approach is given in Ref. [79]. Phonthammachai et al. suggested a one-step modification of the Halas method [80]. Fabrication of silver and gold-silver alloy nanoshells (NSs) on silica cores has been described in several publications [81, 82]. By using silica/gold or gold/silica/gold NSs as precursors, Ye et al. [83] developed a versatile method to fabricate hollow gold nanobowls and complex nanobowls (with a gold core) based on an ion milling and a vapor HF etching technique. Such a NPs being assembled on a substrate can be considered as promise SERS platform for biomolecular detection. Xia et al. introduced a new protocol based on galvanic replacement of silver ions by reducing gold atoms [84]. This approach allows production of a variety of particle shapes and structures, beginning from silver cubes (polyol process [85]) and ending by nanocages [86]. The formation of such particles is accompanied by red-shifting of PR from 400 to 1000 nm.

2.2.4 Other particles and nanoparticles assemblies

To date, one can find a lot of published protocols for fabrication of NPs with various shapes and structures. In particular, the reducing of $HAuCl_4$ on 10–15 nm gold seed leads to the formation of gold nanostars [87], while the use of multiple twinned seeds produced gold bipyramids [88]. A combination of ethylene glycol with PDDA cationic surfactant was shown to be an effective polyol route for the controllable synthesis of high-quality gold octahedral [89].

Apart from fabrication of isolated NPs, it is also very important to assemble NPs into 1D-, 2D-, and 3D-structures. The ultimate goal of this approach is to explore the collective physical properties of the assemblies, which are different from those of isolated NPs. To this end, various strategies (solvent evaporation, electrostatic attraction, hydrogen bonding, DNA-driven assembly, and cross-linking induced by biospecific interaction like antigen-antibody and so on) have been developed to form NP assemblies and utilize them in the fabrication of nanostructured devices [33]. Because of volume limitations, we have to restrict ourselves by indicating the above (and other not cited!) recent reviews. In any case, this chapter should be considered as a starting point for further reading. In conclusion, we show a gallery of some particles mentioned in this section (Figure 2.1).

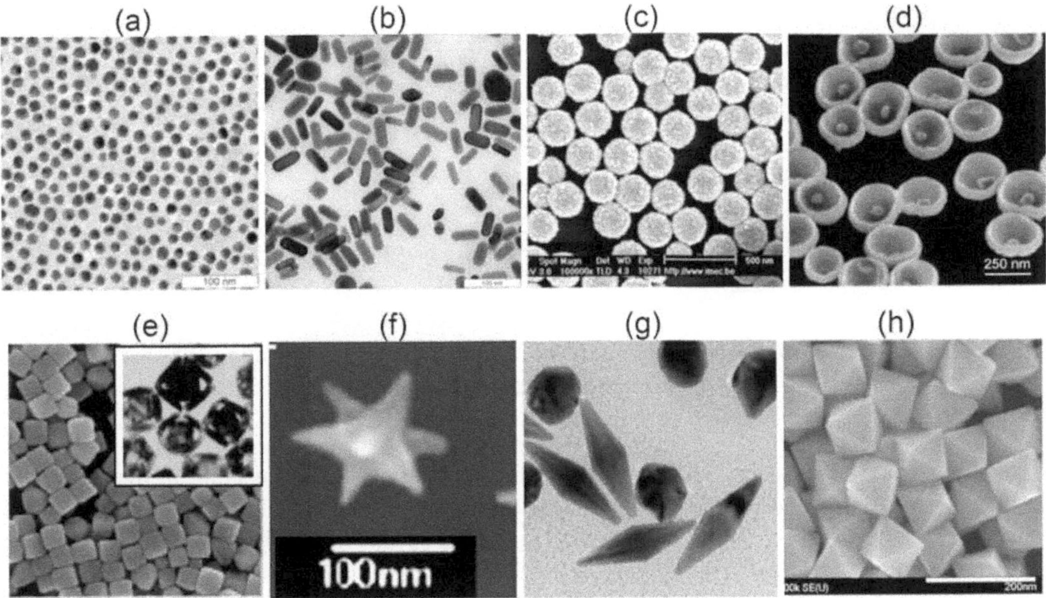

FIGURE 2.1: TEM images of 15-nm colloidal gold (a), 15 × 50-nm gold nanorods (b), 160(core)/17(thickness)-nm silica/gold nanoshells (c, SEM), 250-nm Au nanobowls with 55 Au seed inside (d), silver cubes and gold nanocages (insert) (e), nanostars (f), bipyramids (g), and octahedral (h). Images (d–h) were adapted from Refs. [83], [86], [87], [88], and [89], respectively.

2.3 Optical Properties

2.3.1 Basic physical principles

Collective excitations of conductive electrons in metals are called "plasmons" [3]. Depending on the boundary conditions, it is commonly accepted to distinguish bulk plasmons (3D plasma), surface propagating plasmons or surface plasmon polaritons (2D films), and surface localized plasmons (nanoparticles) (Figure 2.2). A quantum of bulk plasmons $h\omega/(2\pi)_p$ is about 10 eV for noble metals. Because of the longitudinal nature of these plasmons, they cannot be excited by visible light. The surface plasmon polaritons propagate along metal surfaces in a waveguide-like fashion. Below, we shall consider only the nanoparticle localized plasmons. In this case, the electric component of an external optical field exerts a force on the conductive electrons and displaces them from their equilibrium positions to create uncompensated charges at the nanoparticle surface (Figure 2.2c). As the main effect producing the restoring force is the polarization of the particle surface, these oscillations are called "surface" plasmons that have a well-defined resonance frequency.

The most common physical approach to linear NP plasmonics is based on the linear local approximation

$$\mathbf{D}(\mathbf{r},\omega) = \varepsilon(\mathbf{r},\omega)\mathbf{E}(\mathbf{r},\omega), \tag{2.1}$$

where the electric displacement $\mathbf{D}(\mathbf{r},\omega)$ depends on the electric field $\mathbf{E}(\mathbf{r},\omega)$ and on the dielectric function $\varepsilon(\mathbf{r},\omega)$ only at the same position $\mathbf{r}$. Then, the liner optical response can be found by solving the Maxwell equations with appropriate boundary conditions. Actually, in the local approximation, the indicating of position vector in dielectric functions $\varepsilon(\mathbf{r},\omega)$ may be omitted.

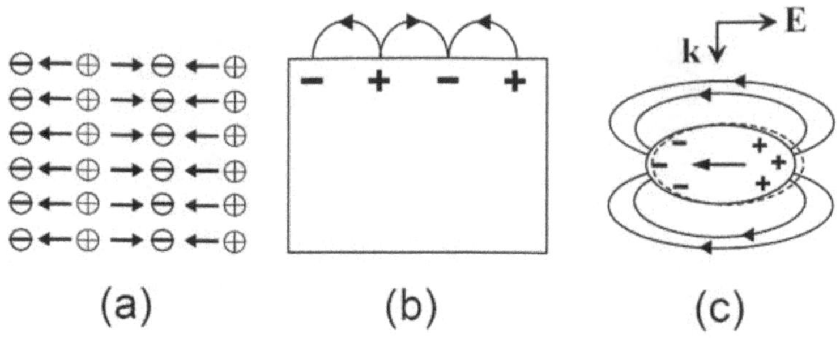

FIGURE 2.2: Schematic representation of the bulk (a), surface propagating (b), and surface localized (c) plasmons. The dashed line shows the electron cloud displacement.

As the particle size is decreased to the value comparable with the electron mean free path ($a \sim l_e$), deviations of the phenomenological dielectric function $\varepsilon(\omega, a)$ of the particle from the bulk values $\varepsilon(\omega) = \varepsilon(\omega, a \gg l_e)$ can be expected. A general recipe for the inclusion of macroscopic tabulated data and size effects to the size-dependent dielectric function consists in the following [3, 19]. Let $\varepsilon_b(\omega)$ be the macroscopic dielectric function, which can be found in the literature from measurements with massive samples [90]. Then, the size-dependent dielectric function of a particle is

$$\varepsilon(\omega, a) = \varepsilon_b(\omega) + \Delta\varepsilon(\omega, a), \tag{2.2}$$

where the correction $\Delta\varepsilon(\omega, a)$ takes into account the contribution of size-dependent scattering of electrons to the Drude part of the dielectric function described by the expression

$$\Delta\varepsilon(\omega, a) = \varepsilon_b^{Drude}(\omega) - \varepsilon_p^{Drude}(\omega, a) = \frac{\omega_p^2}{\omega(\omega + i\gamma_b)} - \frac{\omega_{p,a}^2}{\omega(\omega + i\gamma_p)} \tag{2.3}$$

Here, $\gamma_b = \tau_b^{-1}$ is the volume decay constant; τ_b is the electron free path time in a massive metal; $\omega_{p,a}$ is the plasma frequency for a particle of diameter a (we assume below that $\omega_{p,a} \simeq \omega_p$);

$$\gamma_p = \tau_p^{-1} = \gamma_b + \gamma_s = \gamma_b + A_s \frac{v_F}{L_{\text{eff}}} \tag{2.4}$$

is the size-dependent decay constant equal to the inverse electron mean transit time $\gamma_p = \tau_p^{-1}$ in a particle; L_{eff} is the effective electron mean free path; γ_s is the size-dependent contribution to the decay constant; and A_s is a dimensionless parameter determined by the details of scattering of electrons by the particle surface [91, 92] (which is often simply set equal to 1).

To solve the Maxwell equation, various analytical or numerical methods can be used. The most popular and simple analytical method is the dipole (electrostatic) approximation [93], developed in classic works by Rayleigh [94], Mie [61], and Gans [95] (actually, the Gans solution for ellipsoid had earlier been done by Rayleigh [96]). Among the many numerical methods available, the discrete dipole approximation (DDA) [97], the finite element and finite difference time domain methods (FEM and FDTDM) [98], the multiple multipole method (MMP) [99], and the T-matrix method [100] are most popular. For details, relevant references, and discussion of the advantages and drawbacks of various methods, the readers are referred to recent reviews [18, 19, 26, 27], and a book [100].

Understanding of the interaction of intense laser pulses with PR NPs requires going beyond linear response description. For instance, the second and third harmonic generation is a widely known example [18]. Over the past decade, many transient transmission experiments [101] have been carried out to elucidate the time-dependent physics of nonlinear responses. These experiments involve an

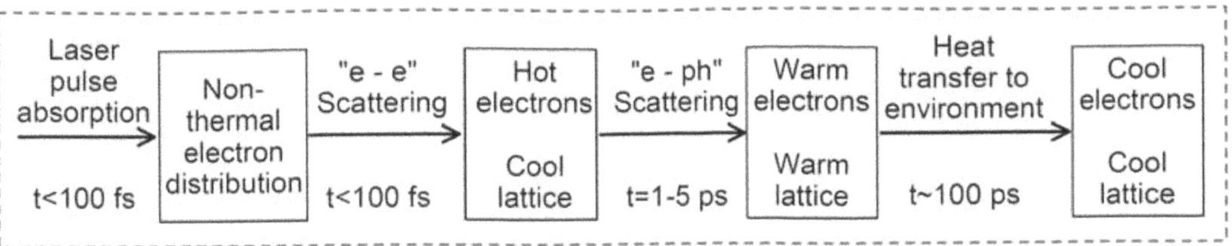

FIGURE 2.3: Schematic of a transient NP response to absorption of a laser pulse followed by the electron-electron (e-e) and electron-phonon (e-ph) scattering and the heat exchange with environment. Adapted from [18].

intense femto-second pump laser pulse followed by measuring the transmittance of a weaker probe laser beam. Figure 2.3 shows a flowchart of the processes [18] accompanied the nonlinear particle response after irradiation with pump pulse. These experiments lead to better understanding the size-dependent dynamical effects (such as electron relaxation) and the temperature-dependent models of the particle dielectric function [18].

Although the local approximation and frequency-dependent dielectric function gives reasonable agreement between theory and experiment, the nonlocal effects can play an important role at small distances in the nanometer region [102]. As the first-principle quantum-mechanical calculations of the optical response [103] are only available for relatively simple systems (hundreds of noble metal atoms) and are insufficient for real nanoparticles, several compromise approximations have been developed to describe quantum-sized and nonlocal effects [102, 104]. In particular, a specular reflection model [102] has been applied to describe the optical properties of small gold and silver spheres, bispheres, and nanoshells. It was shown that the nonlocal effects produce a significant blue shift of PR and its broadening in comparison with the commonly accepted local approach.

By contrast to local relationship 2.1, the nonlocal displacement response is described by integral convolution

$$\mathbf{D}(\mathbf{r},\omega) = \int d\mathbf{r}'\varepsilon(\mathbf{r}',\mathbf{r},\omega)\mathbf{E}(\mathbf{r}',\omega). \tag{2.5}$$

For homogeneous media, $\varepsilon(\mathbf{r}',\mathbf{r},\omega) = \varepsilon(|\mathbf{r}'-\mathbf{r}|,\omega)$, and after Fourier transformation we have the following momentum-space ($\mathbf{q}$) representation

$$\mathbf{D}(q,\omega) = \varepsilon(q,\omega)\mathbf{E}(q,\omega). \tag{2.6}$$

The local approximation means a transition to the $q \to 0$ limit $\varepsilon(q,\omega) \to \varepsilon^{loc}(\omega)$. An approximate recipe to account for the nonlocal effects can be written as a modification of the above Eq. (2.2) [102]

$$\varepsilon(q,\omega) = \varepsilon(\omega) + \Delta\varepsilon(q,\omega) \equiv \varepsilon(\omega) + \varepsilon^{nl}(q,\omega) - \varepsilon^{Drude}(\omega), \tag{2.7}$$

where $\varepsilon(\omega)$ is the experimental (local) tabulated dielectric function, $\varepsilon^{Drude}(\omega)$ is the Drude function, and $\varepsilon^{nl}(q,\omega)$ is the nonlocal valence-electrons response as introduced by Lindhard and Mermin [105]. With the nonlocal dielectric function in hand, further calculations can be made in the same fashion as with the local approximation. Several illustrative examples are presented in the above cited paper [102].

In some cases, an analytical approximation of the Drude type should be used instead of tabulated bulk $\varepsilon_b(\lambda)$ data. Such a situation appears, for example, in the FDTDM method where one needs to calculate an integral convolution of the Eq. (2.5) type. This convolution could be calculated analytically at each step of FDTDM only by using a few simple models for the spectral dependence

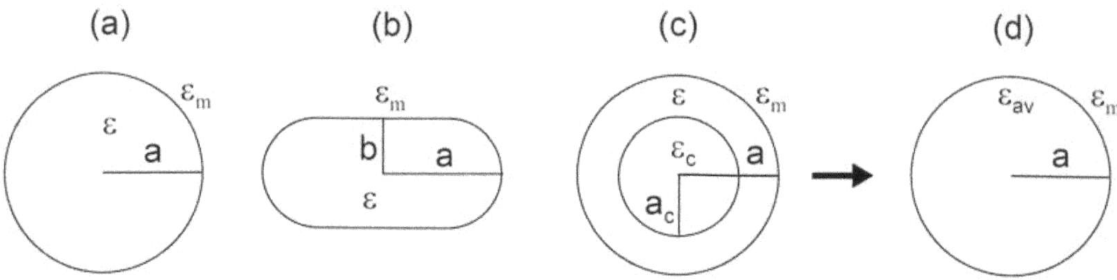

FIGURE 2.4: Three examples of small dipole-scattering metal particles: a nanosphere of a radius a, a nanorod with semiaxes a and b (the s-cylinder model [74]), and a core-shell particle with the corresponding radii a_c and a. The dielectric functions of metal, surrounding medium, and the dielectric core are ε, ε_m, and ε_c, respectively. The core-shell particle is replaced with a homogeneous sphere of a dielectric permittivity ε_{av} according to the "dipole equivalence principle" [107].

of the dielectric function, including the Drude formula. Specifically, in our review [19] we reproduce several sets of interpolation parameters for Johnson and Christy data [106], including those given by Oubre and Nordlander: $\varepsilon_{ib} = 9.5$, ω_p =8.95 eV,$\gamma_b = 0.0691$ (gold) and $\varepsilon_{ib} = 5.0$, ω_p =9.5 eV γ_b =0.0987 eV (silver).

2.3.2 Plasmon resonances

In the general case, the eigenfrequency of a "collective" plasmon oscillator does not coincide with the wave frequency and is determined by many factors, including the concentration and effective mass of conductive electrons, the shape, structure, and size of particles, interaction between particles, and the influence of the environment. However, for an elementary description of the NP optics it is sufficient to use a combination of the usual dipole (Rayleigh) approximation and the Drude theory [93]. In this case, the absorption and scattering of light by a small particle are determined by its electrostatic polarizability α, which can be calculated by using the optical permittivity $\varepsilon(\omega)$ [or $\varepsilon(\lambda)$], where ω is the angular frequency and λ is the wavelength of light in a vacuum.

Let us consider three types of small particles shown in Figure 2.4: a nanosphere of a radius a, a nanorod with semi-axes a and b, and a core-shell particle with the corresponding radii a_c and a. All the particles are assumed to have dipole-like scattering and absorption properties. The core-shell particle is replaced with a homogeneous sphere of a dielectric permittivity ε_{av} according to the dipole equivalence principle [107].

For a small particle of a volume V embedded in a homogeneous dielectric medium with the permittivity ε_m (Figure 2.4), we have the following expressions for the extinction, absorption, and scattering cross-sections

$$C_{ext} = C_{abs} + C_{sca} = \frac{12\pi k}{a^3}\frac{\varepsilon_m \mathrm{Im}(\varepsilon)}{|\varepsilon - \varepsilon_m|^2}|\alpha|^2 + \frac{8\pi}{3}k^4|\alpha|^2 \simeq 4\pi k \mathrm{Im}(\alpha), \tag{2.8}$$

where $k = 2\pi\sqrt{\varepsilon_m}/\lambda$ is the wave number in the surrounding medium, and α is the particle polarizability. Below, we will not distinguish the electrostatic polarizability from the renormalized polarizability [19], which accounts for radiative damping effects. In this approximation, the extinction of a small particle is determined by its absorption $C_{abs} = C_{ext} = 4\pi k \mathrm{Im}(\alpha)$ and scattering contribution can be neglected. For all three types of particles, the polarizability α can be written in the following general form

$$\alpha = \frac{3V}{4\pi}\frac{\varepsilon - \varepsilon_m}{\varepsilon + \varphi\varepsilon_m} = \frac{3V}{4\pi}\frac{\varepsilon_{av} - \varepsilon_m}{\varepsilon_{av} + \varphi\varepsilon_m}, \tag{2.9}$$

where the second expression is written for the metal nanoshell and the parameter φ depends on the particle shape and structure. Specifically, $\varphi = 2$ for spheres, $\varphi = (1/L_a - 1)$ for spheroids (L_a and $L_b = (1 - L_a)/2$ are the geometrical depolarization factors [93]), and

$$\varphi = \frac{1}{2}\left[p_0 + \left(p_0^2 - (\varepsilon_c/\varepsilon_m)\right)^{1/2}\right], \quad p_0 = \frac{\varepsilon_c}{\varepsilon_m}\left(\frac{3}{4f_m} - \frac{1}{2}\right) + \frac{3}{2f_m} - \frac{1}{2} \tag{2.10}$$

for metal nanoshells on a dielectric core [107]. Here, $f_m = 1 - (a_c/a)^3$ is the volume fraction of metal. For thin shells, we have [107]

$$\varphi = \frac{3}{f_m}(1 + \varepsilon_c/2\varepsilon_m). \tag{2.11}$$

One can see from the expressions presented above that the polarizability and optical cross-sections can have a strong resonance under the condition

$$\varepsilon(\omega_{\max} \equiv \omega_0) = \varepsilon(\lambda_{\max}) = -\varphi\varepsilon_m. \tag{2.12}$$

The PR frequency can be estimated from the elementary Drude theory for the permittivity of a bulk metal

$$\varepsilon(\omega) = \varepsilon_{ib} - \frac{\omega_p^2}{\omega(\omega + i\gamma_b)}, \tag{2.13}$$

where ε_{ib} is the contribution of interband electronic transitions; ω_p is the frequency of volume plasma oscillations of free electrons; γ_b is the volume decay constant related to the electron mean free path l_b and the Fermi velocity v_F by the expression $\gamma_b = l_b/v_F$. By combining the above equations, one can obtain the following expressions for the resonance plasmon frequency and wavelength

$$\omega_{\max} \equiv \omega_0 = \frac{\omega_p}{\sqrt{\varepsilon_{ib} + \varphi\varepsilon_m}}, \quad \lambda_{\max} \equiv \lambda_0 = \lambda_p\sqrt{\varepsilon_{ib} + \varphi\varepsilon_m}. \tag{2.14}$$

Here, $\lambda_p = 2\pi c/\omega_p$ is the wavelength of volume oscillations of the metal electron plasma. For gold, $\lambda_p \approx 131$ nm.

Two additional important notes are in order here. First, it follows from Eq. (2.14) that the dipole resonances of gold or silver spheres in water is localized near 520 nm and 400 nm and do not depend on their size. By contrast, the dipole resonance of rods and nanoshells can be easily tuned through variations in their aspect ratio (L_a = 1/3(sphere) $\rightarrow$ 0(needle)) or the ratio shell thickness/core radius (f_m = 1 for a homogeneous sphere and $f_m \rightarrow 0$ for a thin shell). Second, equation (2.14) determines the very first ($n = 1$) dipole resonance of a spherical particle. In addition to the dipole resonance, higher multipoles and corresponding multipole (quadrupole, etc.) resonances can be also excited in larger particles. For each multipole mode the resonance condition exists, which is similar to (2.12) and corresponds to the resonance of the quadrupole, octupole, and so on contributions. For spherical particles, these conditions correspond to the resonance relations for the partial Mie coefficients [93] $\omega_n = \omega_p\left(\varepsilon_{ib} + \varepsilon_m(n+1)/n\right)^{-1/2}$, where n is the mode (resonance) number. With an increase in the sphere size, the multipole frequency decreases [18].

It is important to distinguish two possible scenarios for excitation of multipole resonances. The first case corresponds to a small but nonspherical particle of irregular or uneven shape, when the distribution of induced surface charges is strongly inhomogeneous and does not correspond to the dipole distribution. This inhomogeneous distribution generates high multipoles even in the case when the system size is certainly much smaller than the light wavelength. Typical examples are cubic particles [108] or two contacting spheres [109] where the field distribution near the contact point is so inhomogeneous that multipole expansions converge very slowly or diverge at all. The second scenario of high multipole excitation is realized with increasing the particle size, when the transition from the quasi-stationary to radiative regime is realized, and the contribution of higher spherical harmonics should be taken into account in the Mie series (or another multipole expansion). For example, while the extinction spectrum of a silver 30-nm NP is completely determined by the dipole contribution and has one resonance, the spectrum of a 60-nm sphere exhibits a distinct high-frequency quadrupole peak in addition to the low-frequency dipole peak.

2.3.3 Metal spheres

The optical properties of a small sphere are completely determined by their scattering amplitude matrix [93]. Here we are interested mainly in its extinction, absorption, and scattering cross-sections $C_{ext,abs,sca}$ or efficiencies $Q_{ext,abs,sca} = C_{ext,abs,sca}/S_{geom}$. From an experimental point of view, we shall consider the absorbance (extinction) and the scattering intensity at a given metal concentration c

$$A_{ext} = 0.651 \frac{cl}{\rho} \frac{Q_{ext}}{d}, \quad I_{90}(\lambda, \theta_{sca}) = 0.651 \frac{cl}{\rho d} \left[\frac{16 S_{11}(ka, \theta_{sca})}{3(ka)^2} \right], \tag{2.15}$$

where $S_{11}(ka, \theta)$ is the normalized intensity of scattering at an θ_{sca} angle (the first element of the Mueller scattering matrix [93]); and $k = 2\pi n_m/\lambda$ is the wave number in the surrounding medium (water in our case), ρ and $d = 2a$ are the metal density and the particle equivolume diameter. The expression in square brackets is normalized so that it is equal to the scattering efficiency of small particles. For metal nanoshells, Eq. (2.15) should be slightly modified [110].

Figure 2.5 shows well-known theoretical extinction and scattering spectra calculated by Mie theory for gold and silver colloids at a constant concentration of gold (57 μg / ml, corresponds to complete reduction of 0.01% $HAuCl_4$) and silver (5 μg / ml). Note that for larger particles, the extinction spectra reveal quadrupole peaks and significant contribution of scattering. With an increase in the particle diameter, the spectra become red-shifted and broadened. This effect can be used for fast and convenient particle sizing [111]. The solid line in Figure 2.5c shows the corresponding calibration curve based on long-term set of the literature experimental data (shown by different symbols) [111]

$$d = \begin{cases} 3 + 7.5 \times 10^{-5} X^4, & X < 23 \\ [\sqrt{X - 17} - 1]/0.06, & X \geq 23 \end{cases}, \quad X = \lambda_{\max} - 500. \tag{2.16}$$

In practice, this calibration can be used for particles larger than 5–10 nm. For smaller particles, the resonance shifting is negligible, whereas the resonance broadening become quite evident because of size-limiting effects and surface electron scattering [19]. The extinction spectra width also can be used for sizing of fine particles with diameters less than 5 nm [112].

Figure 2.6a presents the dependences of the maximal extinction and scattering intensity on the diameter of silver and gold particles. For a constant metal concentration, the maximal extinction is achieved for the silver and gold particles of diameters about 25 and 70 nm, respectively. The maximal scattering per unit metal mass can be observed for 40-nm silver and 100-nm gold particles, respectively. Figure 2.6b presents the size dependence of the integral albedo at resonance conditions. Small particles mainly absorb light, whereas large particles mainly scatter it. The contributions of scattering and absorption to the total extinction become equal for 40-nm silver and 80-nm gold particles, respectively.

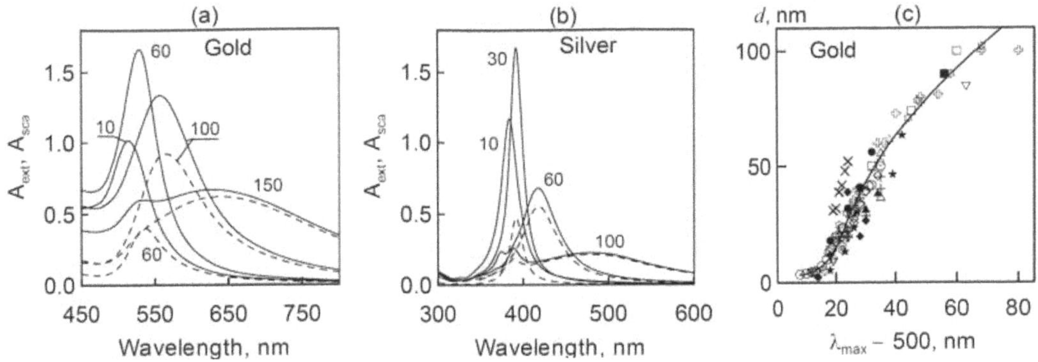

FIGURE 2.5: Extinction (solid lines) and scattering (dashed lines) spectra of gold (a) and silver (b) spheres in water. Numbers near curves correspond to the particle diameter. Panel (c) shows a calibration curve for spectrophotometric determination of the average diameter of CG particles. The symbols present experimental data taken from 14 sources (see References in [111]).

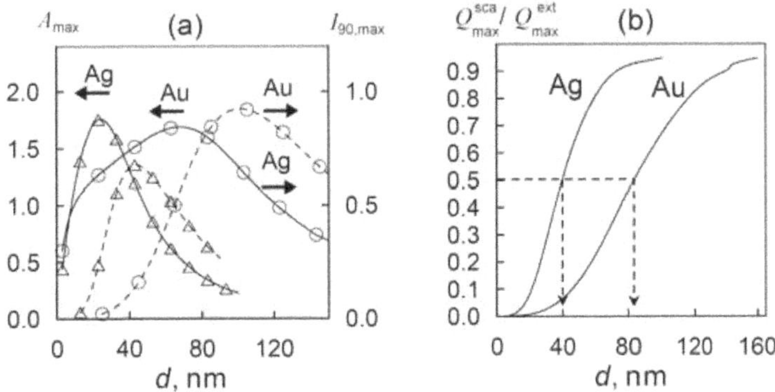

FIGURE 2.6: Dependences of the resonance extinction of suspensions and the intensity of scattering at 90 degrees on the particle diameter at constant weight concentrations of gold (57 μg / ml) and silver (5 μg / ml) (a). Panel (b) shows the ratio of the resonance scattering and extinction efficiencies as a function of particle diameter.

2.3.4 Metal nanorods

2.3.4.1 Extinction and scattering spectra

Figure 2.7 shows the extinction and scattering spectra (in terms of the single-particle efficiencies) of randomly oriented gold (a, c) and silver (b, d) nanorods with the equivolume diameter of 20 nm and the aspect ratio from 1 to 6.

Calculations were carried out by the T-matrix method for cylinders with semispherical ends. We see that the optical properties of rods depend very strongly on the metal nature. First, as the aspect ratio of gold NR is increased, the resonance extinction increases approximately by a factor of five and the Q factor also increases. For silver, and vice versa, the highest Q factor is observed for spheres, whereas the resonance extinction for rods is lower. Second, for the same volume and axial ratio, the extinction and scattering of light by silver rods are considerably more efficient. The resonance scattering efficiencies of silver particles are approximately five times larger than those for gold particles. Third, the relative intensity of the transverse PR of silver particles with the aspect ratio above 2 is noticeably larger than that for gold particles, where this resonance can be simply

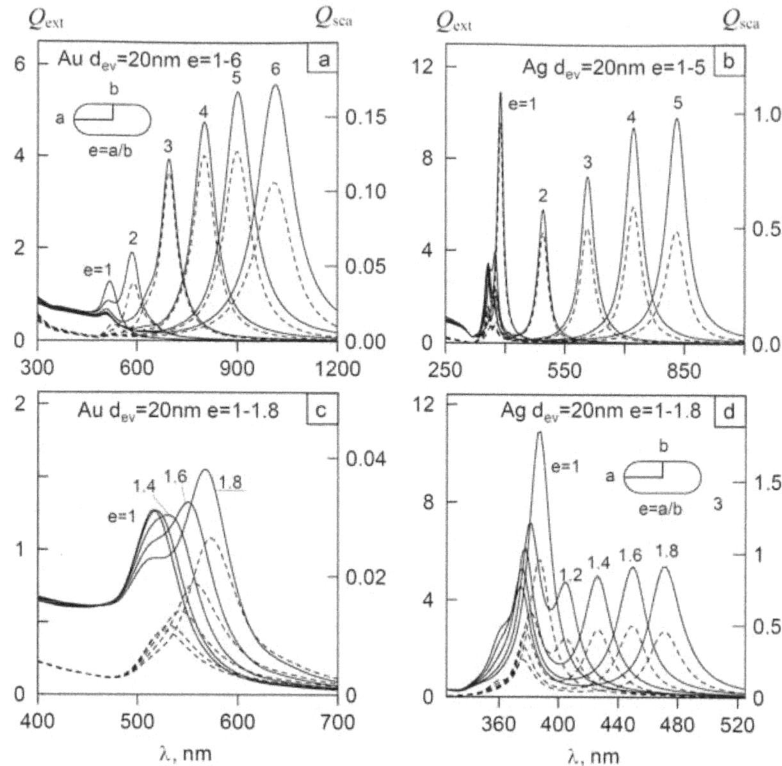

FIGURE 2.7: Extinction (solid curves) and scattering (dashed curves) spectra of randomly oriented gold (a, c) and silver (b, d) s-cylinders with the equivolume diameter 20 nm and the aspect ratio from 1 to 6. Panels (c) and (d) show the transformation of spectra at small deviations of the particle shape from spherical.

neglected. Finally, principal differences are revealed for moderate nonspherical particles (Figures 2.7c, d). The resonance for gold particles shifts to the red and gradually splits into two bands with dominating absorption in the red region. The scattering band shifts to the red and its intensity increases. For silver rods, the situation is quite different. The short-wavelength resonance shifts to the blue, its intensity decreases and it splits into two distinct bands. In this case, the intensity of the long-wavelength extinction band remains approximately constant, it is comparable with the short wavelength band intensity and shifts to the red with increasing nonsphericity. The integrated scattering and absorption spectra approximately reproduce these features.

The position of the longitudinal long-wavelength resonance can be predicted from the axial ratio of particles and, vice versa, the average aspect ratio can be quite accurately estimated from the resonance position. Figure 2.8 presents calibration dependences for measuring the axial ratio of particles from the longitudinal PR position. Along with the T-matrix calculations for s-cylinders of different diameters, we present our and the literature experimental data (for details, see Ref. [19]). In the dipolar limit, the plasmon resonance wavelength is completely determined by the particle shape. However, when the particle diameter is increased, the results of a rigorous solution noticeably differ from the limiting electrostatic curve obtained in the dipole approximation (Figure 2.8). This fact was shown for the first time in Ref. [113] for s-cylinders by the T-matrix method and in [114] for right cylinders by the DDA method. Later, analogous calculations have been extended for larger gold rods [115, 116].

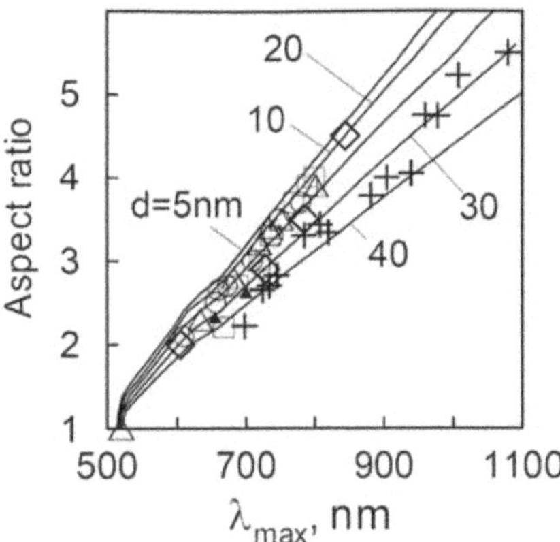

FIGURE 2.8: The calibration plots for determination of the particle aspect ratio through the longitudinal resonance wavelength. Calculations for randomly oriented gold s-cylinders in water by the T-matrix method [113]. The particle diameters are 5 (1), 10 (2), 20 (3), 30 (4), and 40 (5) nm. The symbols show the experimental points (for details, see [74]).

2.3.4.2 Depolarized light scattering

If colloidal nonspherical particles are preferentially oriented by an external field, the suspension exhibits anisotropic properties such as dichroism, birefringence, and orientation-dependent variations in turbidity and in light scattering. Moreover, even for *randomly* oriented particles, there remain some principal differences between light scattering from nanospheres and that from nanorods. Indeed, when randomly oriented nonspherical particles are illuminated by linearly polarized light, the cross-polarized scattering intensity occurs [100, 113], whereas for spheres this quantity equals zero.

The cross-polarized scattered intensity I_{vh} can be characterized by the depolarization ratio $\Delta_{vh} = I_{vh}/I_{vv}$, where the subscripts "$v$" and "$h$" stand for vertical and horizontal polarization with respect to the scattering plane. According to the theory of light scattering by small particles, the maximal value of the depolarization ratio Δ_{vh} cannot exceed 1/3 and 1/8 for dielectric rods and disks with positive values of the real and imaginary parts of dielectric permeability. However, the dielectric limit 1/3 does not hold for plasmon-resonant nanorods whose theoretical depolarization limit equals 3/4 [113].

Quite recently, the first measurements of the depolarized light scattering spectra from suspensions of gold nanorods have been reported for 400–900 nm spectral interval [117]. For separated, highly monodisperse and monomorphic samples, we observed depolarized light scattering spectra with an unprecedented depolarization ratio of about 50% at wavelengths of 600 to 650 nm, below the long-wavelength 780-nm extinction peak (Figure 2.9). These unusual depolarization ratios are between 1/3 and 3/4 theoretical limits established for small dielectric and plasmon-resonant needles, respectively.

To simulate the experimental extinction and depolarization spectra, we introduce a model that included two particle populations: (1) the major nanorod population and (2) a by-product particle population with the weight fraction $0 \leq W_b \leq 0.2$. The optical parameters were calculated for both populations and then were summed with the corresponding weights W_b and $W_{rods} = 1 - W_b$. The nanorod population was modelled by $N = 10 - 20$ fractions of rods possessing a constant thickness $d = 2b$, whereas their aspect ratios were supposed to have normal distribution $\sim \exp\left[-(p/p_{av} - 1)^2/2\sigma^2\right]$. The average value p_{av} and the normalized dispersion σ were obtained from TEM data and were also considered to be fitting parameters for the best agreement between measured and calculated extinction and depolarization spectra. T-matrix calculations with these parameters resulted in excellent

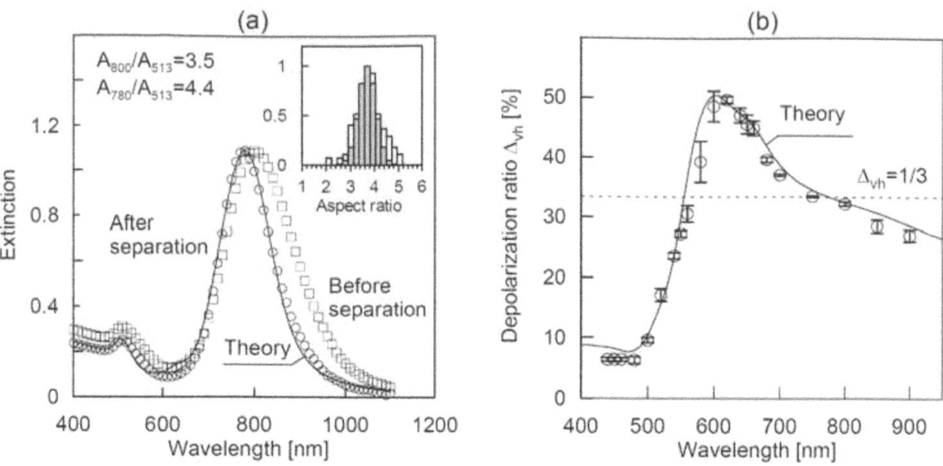

FIGURE 2.9: (a): Extinction spectra of the NR-780 sample before and after separation, together with T-matrix calculations based on best-fitting data ($p_{av} = 3.7$, $\sigma = 0.1$, $W_c = 0.06$). The inset shows a comparison of TEM aspect distribution (light columns) with normal best-fitting distribution (dark columns) for the same sample taken after separation. (b) Experimental and simulated depolarization spectra. The error bars correspond to three independent runs, with analog and photon-counting data included. The dashed line shows the dielectric-needle limit 1/3.

agreement between measured and simulated spectra (Figure 2.9). Thus, we have developed a fitting procedure based on simultaneous consideration of the extinction and depolarization spectra. This method gives an aspect ratio distribution for the rods in solution from which the average value and the standard deviation can be accurately determined in a more convenient and less expensive way than by the traditional TEM analysis.

For larger particles, our T-matrix simulations predict multiple-peak depolarization spectra and unique depolarization ratios exceeding the upper dipolar limit (3/4) because of multipole depolarization contributions (Figure 2.10).

To elucidate the physical origin of strong depolarization [118], let us consider the scattering geometry depicted in Figure 2.11, in which the incident x-polarized light travels along the positive z direction and the scattered light is observed in the plane (x,z) in the x-direction. If the particle symmetry axis is directed along the x, y, or z axis, then no depolarization occurs because of evident symmetry constrains. Moreover, no depolarization occurs either for any particle located in the (x,y) or (x,z) plane. Thus, the maximal depolarization contribution is expected from particles located in the (y,z) plane. For usual dielectric rods, the induced dipoles $\mathbf{d}_b$ and $\mathbf{d}_{a,1}$ oscillate in phase. Accordingly, the deviation of the resultant dipole $\mathbf{d}_1$ from the exciting electric field direction is small. Thus, the depolarized scattering should be weak. However, for metal nanorods, the perpendicular ($\mathbf{d}_b$) and longitudinal ($\mathbf{d}_{a,2}$) dipoles can be excited in opposite phases. In this case, the direction of the resultant dipole $\mathbf{d}_2$ can be close to the z-axis direction, thus causing the appearance of significant depolarization.

Figure 2.11b shows the orientation dependence of the depolarized light scattering calculated for 20×80-nm gold s-cylinder in water. The scattering geometry is depicted in panel (a). The wavelength of 694 nm corresponds to the maximum of the depolarization spectrum calculated at optimal orientation angles $\vartheta = 20^\circ$ and $\varphi = 96^\circ$. It follows from Figure2.11b that there exists two orientation planes ($\vartheta = 20^\circ$ and $\vartheta = 150^\circ$) where gold nanorods can produce an unusual depolarization ratio up to 100% provided that the particle azimuth is close to 90°. Thus, the experimental 50% depolarization maximum is caused by particles located near these optimal orientations.

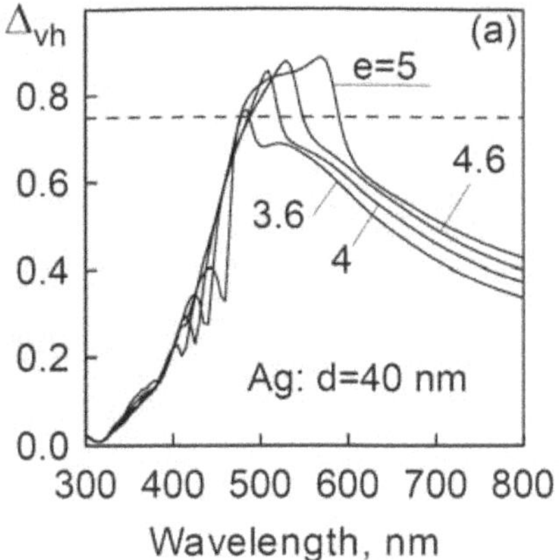

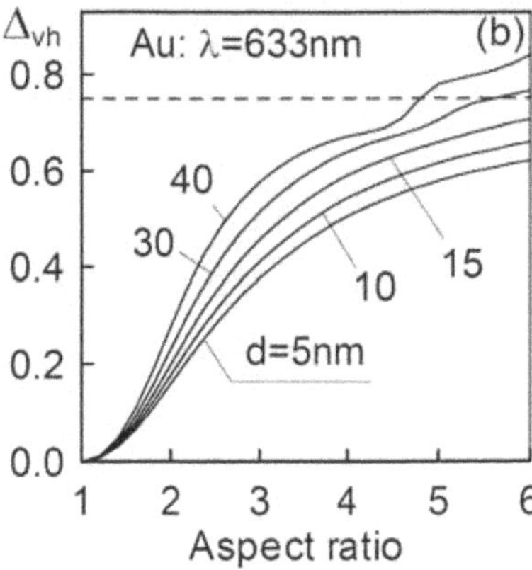

FIGURE 2.10: Spectral dependences of the depolarization ratio for randomly oriented silver s-cylinders of diameter $d = 40$ nm and aspect ratios 3.6-5 (a) and the dependence of the depolarization ratio on the aspect ratio calculated for gold s-cylinders in water at different particle diameters from 5 to 40 nm.

2.3.4.3 Multipole plasmon resonances

Compared to numerous data on the dipole properties of nanorods published in the last years, the studies of multipole resonances are quite limited. The first theoretical studies concerned the quadrupole modes excited in silver and gold spheroids, right circular and s-cylinders, as well as in nanolithographic 2D structures such as silver and gold semispheres and nanoprisms [19]. Recently, the first observations of multipole PRs in silver nanowires on a dielectric substrate [119] and colloidal gold [120] and gold-silver bimetal nanorods [121] were reported.

To elucidate the size and shape dependence of the multipole plasmons, two studies have been performed by using T-matrix formalism [115] and FDTD simulations [116]. Specifically, an extended-precision T-matrix code was developed to simulate the electrodynamic response of gold and silver nanorods whose shape can be modelled by prolate spheroids and cylinders with flat or semispherical ends (s-cylinders). Here, we present a brief summary of the most important results [115].

The multipole contributions to the extinction, scattering, and absorption spectra can be expressed in terms of the corresponding normalized cross-sections (or efficiencies):

$$Q_{ext,sca,abs} = \sum_{l=1}^{N} q^{l}_{ext,sca,abs}. \tag{2.17}$$

For randomly oriented particles, the multipole partial contributions are given by the equation

$$q^{l}_{ext} = (2/k^2 R^2_{ev})\mathrm{Spur}_l(T_{\sigma\sigma}), \tag{2.18}$$

where $\sigma \equiv (l,m,p)$ stands for the T-matrix multi-index and the trace (Spur) is taken over all T-matrix indices except for the multipole order l. For particles at a particular fixed orientation (specified by the angle α between the vector $\mathbf{k}$ and the nanorod axis $\mathbf{a}$), we consider two fundamental cross-sections corresponding to the transverse magnetic (TM) and transverse electric (TE) plane wave configurations, where $\mathbf{E} \in (\mathbf{k},\mathbf{a})$ plane in the TM case and $\mathbf{E} \perp (\mathbf{k},\mathbf{a})$ plane in the TE case (Figure 2.12.).

Figure 2.12 shows the scattering geometry and the extinction, scattering, and absorption spectra of a gold s-cylinder (diameter 40 nm, aspect ratio 6) for the longitudinal, perpendicular (TM polarization), and random orientations with respect to the incident polarized light (a–d). The spectral

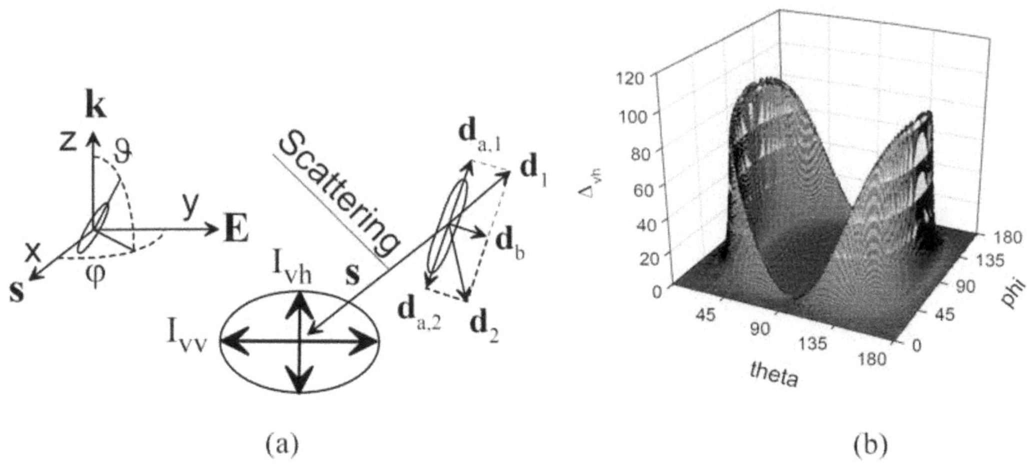

FIGURE 2.11: (a) Scheme to explain the physical origin of enhanced depolarization [118]. The particle is located in the (y,z) plane. If both (perpendicular and longitudinal) dipoles $\mathbf{d}_b$ and $\mathbf{d}_{a,1}$ oscillate in phase, the direction of the resultant dipole $\mathbf{d}_1$ is close to the incident electric field $\mathbf{E}$ and the depolarized scattering intensity is small. However, when the dipoles and $\mathbf{d}_{a,2}$ oscillate out of phase, there appears significant depolarization, as the resultant dipole $\mathbf{d}_{a,2}$ is almost parallel to the z axis. (b) 3D plot of the depolarization ratio as a function of polar, ϑ, and azimuth, φ, angles that define an orientation of the rod symmetry axis (see panel a). Calculations were performed using the extended precision T-matrix code for gold s-cylinder in water. The rods diameter and length are 20 and 80 nm, respectively, the wavelength 694 nm corresponds to the spectral depolarization ratio maximum.

resonances are numbered according to the designation $Q^n_{ext} \equiv Q_{ext}(\lambda_n)$, $n = 1, 2, \ldots$ from far infrared to visible, whereas the symbol "0" designates the shortest wavelength resonance located near the TE mode. Three important remarks can be made on the basis of the plots in Figure 2.12 a–d. First, the maximal red-shifted resonance for random orientations is due to TM dipole ($n = 1$) excitation of particles, whereas the short-wavelength "zero" resonance is excited by TE polarization of the incident light. Second, both scattering and absorption resonances have the same multipole order, are located at the same wavelengths, and give comparable contributions to the extinction for the size and aspect ratio under consideration. Finally, for basic longitudinal or perpendicular orientations, some multipoles are forbidden because of the symmetry constrains. Consider now the case of strongly elongated particles with an aspect ratio $e = 10$ (panels e–g). In this case, the random orientation spectra exhibit a rich multipole structure that includes multipole resonances up to $n = 5$. As in the previous case, at a minimal thickness $d = 20$ nm, the scattering contribution is negligible, and all resonances are caused by the absorption multipoles. Note also that the short-wavelength resonance has negligible amplitude. With an increase in the particle thickness from 20 to 40 and further to 80 nm, all resonances move to the red region, and new scattering and absorption high-order multipoles appear. Again, the scattering contributions dominate for large particles, whereas at moderate thicknesses the scattering contribution is comparable to or less than the absorption contribution, except for the long-wavelength dipole resonance of number 1.

Figure 2.13 shows the extinction spectra $Q^{ext}(\lambda)$ of randomly oriented gold spheroids with a minor axis $d = 80$ nm and an aspect ratio $e = 10$. Besides the usual long-wavelength dipole resonance (not shown), five additional multipole resonances can be identified (the 6th resonance looks like a weak shoulder, but it is clearly seen on the multipole contribution curve, designated q_6). A close inspection of spectra leads to the following conclusions: (1) The parity of multipole contributions

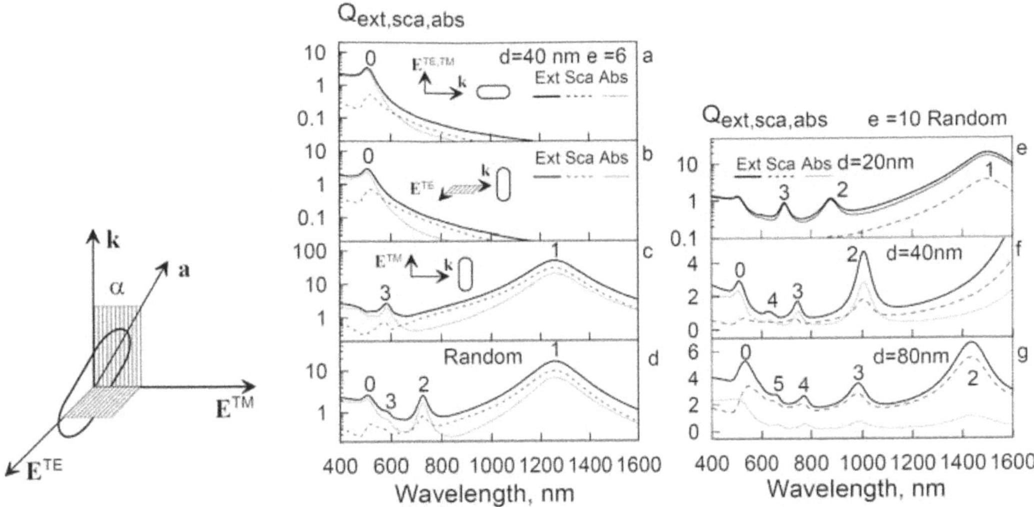

FIGURE 2.12: Two basic TE and TM polarizations of incident light. The orientation of a particle is specified by an angle α between the wave vector **k** and the particle symmetry axis **a**. The panels (a–d) show the extinction, scattering, and absorption spectra of gold s-cylinders for longitudinal, perpendicular TE, perpendicular TM, and random orientations of gold nanorod with respect to the incident polarized light. Panels (e–g) show variation of spectra with an increase in the rod diameter at a constant aspect ratio e =10. The numbers near the curves designate multipole orders according to the T-matrix assignment.

coincides with the parity of the total resonance. (2) For a given spectral resonance number n, the number of partial multipole contributions l is equal to or greater than n. This means that the multipole q_1 does not contribute to the resonances $Q^{2n+1}(n \geq 1)$, the multipole q_2 does not contribute to the resonances $Q^{2n}(n \geq 1)$, and so on. Note that the second statement of the multipole contribution rule is, in general, shape dependent [115]. To elucidate the scaling properties of the multipole resonances, we carried out extensive calculations for various nanorod diameters and aspect ratios. The general course of λ_n *vs* the e plots showed an almost linear shift of the multipole resonances with an increase in aspect ratio

$$\lambda_n = f(e/n) \simeq A_0 + A\frac{e}{n}. \tag{2.19}$$

Figure 2.13b illustrates the scaling law (2.19) for randomly oriented gold spheroids (d =80 nm, $e = 2-20$) in water. It is evident that for resonance numbers $n = 1-8$, all data collapse into single linear functions. Physically, this scaling can be explained in terms of the standing plasmon wave concept and the plasmon wave dispersion law, as introduced for metal nanoantennas [122]. A very simple estimation of the resonance position is given by the equation

$$n\frac{\pi}{L} = \frac{2\pi}{\lambda_n^{eff}} = q^{eff}(\omega_n), \tag{2.20}$$

where $q^{eff}(\omega_n)$ is an effective wave number corresponding to the resonance frequency ω_n. At a constant particle thickness, d, Eq. (2.20) predicts a linear scaling $\lambda_n^{eff} \sim L/n \sim de/n \sim e/n$, in full accord with our finding given by Eq. (2.19).

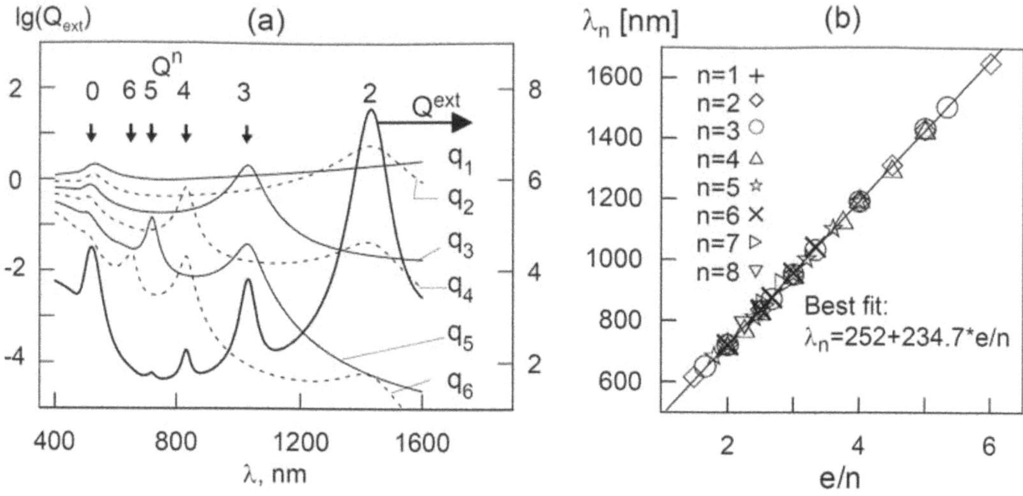

FIGURE 2.13: (a) Extinction spectra $Q^{ext}(\lambda)$ of randomly oriented gold spheroids with a thickness $d = 80$ nm and an aspect ratio $e = 10$. The curves $q_1 - q_6$ show the spectra of multipole contributions. The upper row of numbers $n = 2-6$ designates the total resonances $Q^n = Q_{ext}(\lambda_n)$. The resonance Q^1 is located in the far infrared region and is not shown. (b): Linear scaling of the multipole resonance wavelengths λ_n *vs* the normalized aspect ratio e/n. The regression scaling equation is shown in the plot.

2.3.5 Coupled plasmons

Along with the optics of individual PR particles, the collective behavior of the interacting PR particles is of great interest for nanobiotechnology [123]. The analysis of its features includes the study of various structures, beginning from one-dimensional chains with unusual optical properties [124]. Another example is the optics of two-dimensional arrays [125], in particular, clusters of spherical particles on a substrate and two-dimensional planar ensembles formed by usual gold or polymer-coated spheres [123]. The unusual properties of monolayers of silver nanoparticles in a polymer film [126] and on a glass substrate [127] have been recently discovered. A review of 3D sphere-cluster optics can be found in Refs. 3, 109, and 123.

Apart form direct numerical simulations (e.g., MMP simulation of disk pairs on a substrate [128] and FEM simulations of silver sphere-clusters [129]) or an approximate electrostatic considerations [130], a new concept called "the plasmon hybridization model" [131] has been developed to elucidate basic physics behind plasmon coupling between optically interacting particles. Mulvaney and coworkers [132] reported experimental scattering spectra of gold nanorod dimers arranged in different orientations and separations. The spectra exhibited both red- and blue-shifted surface plasmon resonances, consistent with both the DDA simulations and the plasmon hybridization model. It should be emphasized that for particle separations less than a few nanometers, the plasmon coupling constant did not obey the exponential dependence predicted by the so-called "universal plasmon ruler" (UPR) equation (see, e.g., [133] and references therein). In essence, the UPR means a universal dependence of the couple resonance shift determined only by the separation distance normalized to the characteristic particle size. In particular, the UPR has been confirmed in simulations [128]. Perhaps, deviations from UPR are related to the strong multipole interactions excited at extra small separations.

A more complex assembly of high aspect ratio gold NRs on a substrate has been studied in Ref. [134]. The long axes of NRs were oriented perpendicularly to the substrate. The optical response of such an assembly was governed by a collective plasmonic mode resulting from the strong electromagnetic coupling between the dipolar longitudinal plasmons supported by individual NRs. The spectral position of this coupled plasmon resonance and the associated electromagnetic field distribution in the nanorod array were shown to be strongly dependent on the inter-rod coupling strength.

The unique properties of strongly coupled metal particles are expected to find interesting applications in such fields as the manipulation of light in nanoscale waveguide devices, sensing and nonlinearity enhancement applications, and the subwavelength imaging. As an example, we can refer to recent work by Reinhard and coworkers [135]. These authors extended the conventional gold particle tracking with the capability to probe distances below the diffraction limit using the distance dependent near-field interactions between individual particles. This technology was used for precisely monitoring the interaction between individual gold nanoparticles on the plasma membrane of living HeLa cells.

As coupled plasmonics is an extremely fast growing field, the number of related works exceeds the scope of this review. So, we have to restrict our consideration of collective plasmons by only two instructive examples: (1) the gold and silver interacting bispheres; (2) a monolayer of interacting nanoparticles on a substrate.

2.3.5.1 Metal bispheres

Figure 2.14a,b shows the absorption spectra calculated by the multipole method [109] for two gold particle diameters d =15, 30 (data for 60-nm see in Ref. 109) separated by a variable distance s. When the interparticle separation satisfies the condition $s/d \geq 0.5$, the absorption efficiencies approach the single-particle quantities so that the dipole and multipole calculations give identical results. However, the situation changes dramatically when the relative separation s/d is about several percent. The coupled spectra demonstrate quite evident splitting into two components with pronounced red-shifting of the long-wavelength resonance when the spheres approach each other.

In the case of silver bispheres (Figure 2.14c,d), the resonance light scattering of 60-nm clusters exceeds the resonance absorption so that any comparison between the dipole and the multipole approaches becomes incorrect unless both scattering and absorption are taken into account for the total extinction. That is why we show only the calculated data for silver nanospheres with $d = 15$ nm. At moderate separations ($s/d > 0.05$), the independent-particle spectrum splits into two modes; therefore, the data of Figure 2.14c,d are in great part analogous to those for independent particles. However, at smaller separations $s/d < 0.05$, we observe the appearance of four plasmon resonances related to the quadrupole and the next high-order multipole excitations. The exact multipole approach predicts the well-known enormous theoretical [123] and experimental [136] red-shifting of spectra and their splitting [137] into two modes, whereas the dipole spectra show only a minor red shift.

The dependence of the extinction spectra on the total multipole order N_M is shown in Figure 2.15. Note that N_M means the maximal order of vector spherical harmonics (VSH) that have been retained in the coupled equations rather than the number of multipoles involved in the final calculations of optical characteristics. According to these computer experiments, one has to include extra-high single-particle multipole orders (up to 30–40) into coupled equations to calculate correctly the extinction spectra of 15-nm gold spheres separated by a 0.5–1% relative distance s/d. The extinction spectra were calculated by the exact T-matrix code for randomly oriented gold bispheres in water. It is evident that the convergence problems are related to calculations of the red-shifted resonance peak, which can be reproduced correctly if we retain the VSH with the order of about 30 in the case of 0.15 nm (1%) separation. For smaller separations (e.g., $s = 0.075$ nm, Figure 2.15b), we note the appearance of a quadrupole resonance near 600 nm that can also be reproduced accurately only if we include multipoles of the 40th order into coupled equations.

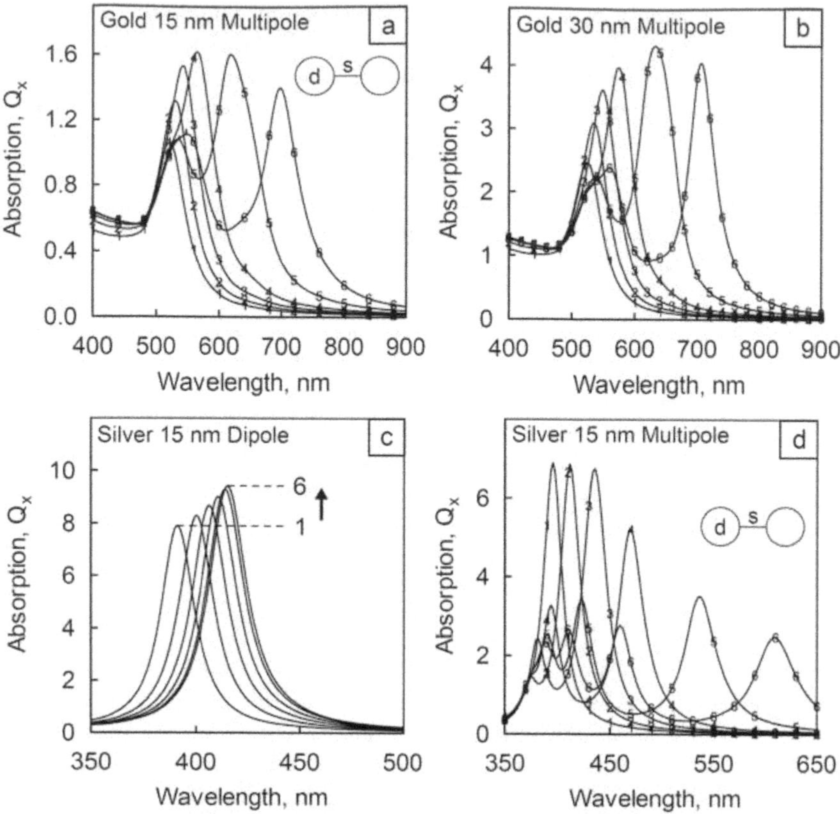

FIGURE 2.14: Absorption spectra at an incident light polarization parallel to the bisphere axis ($x \equiv \|$). Calculations by the exact GMM multipole code (a,b,d) and dipole approximation (c) for gold (a,b) and silver (c,d) particles with diameters $d = 15$ (a,c,d) and 30 nm (b) and the relative interparticle separations $s/d = 0.5$ (1), 0.2 (2), 0.1(3), 0.05 (4), 0.02 (5), 0.01 (6).

The need to retain high multipoles for small spheres, which are themselves well within the dipole approximation, seems to be somewhat counterintuitive. It should be emphasized that the final calculations involve rather small number of multipoles (as a rule, less than 6). However, to find these small-order contributions correctly, one needs to include many more multipoles into coupled equations as given by Eq. (1.44). The physical origin of this unusual electrodynamic coupling was first established by Mackowski [138] for small soot bispheres. He showed that the electric-field intensity can be highly inhomogeneous in the vicinity of minimal separation points between the spheres even if the external field is homogeneous on the scale of bisphere size. Evidently, the same physics holds in our case, as the imaginary part of the dielectric permittivity is the main parameter that determines the spatial electric-field distribution near the contact bisphere point.

2.3.5.2 Monolayers of metal nanoparticles and nanoshells

Recently, Malynych and Chumanov [126] found an unusual behavior of the extinction spectra for a monolayer of interacting silver nanoparticles embedded in a polymer film. Specifically, they showed an intense sharpening of the quadrupole extinction peak resulting from selective suppression of the coupled dipole mode as the interparticle distance becomes smaller. This phenomenon was explained qualitatively by using simple symmetry considerations.

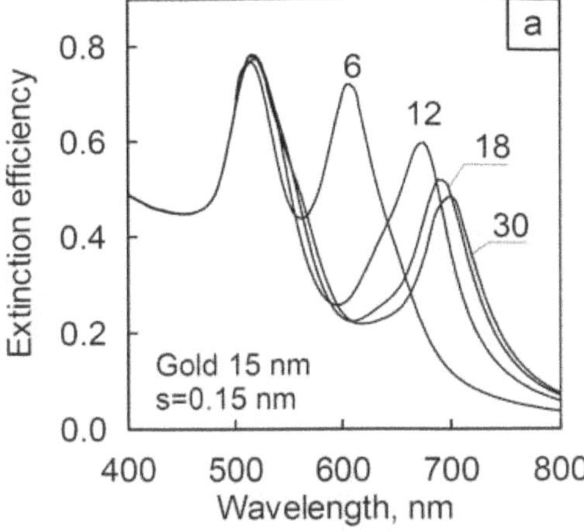

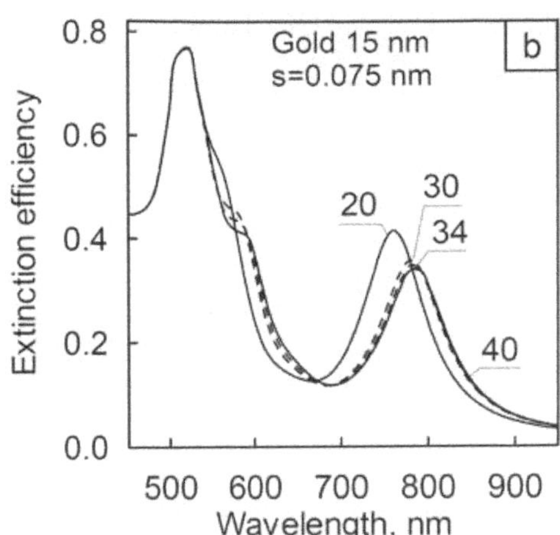

FIGURE 2.15: Extinction spectra of 15-nm randomly oriented gold bispheres in water, calculated by the exact T-matrix method. The sphere diameter is 15 nm, the separation distances between spheres are 0.15 nm (a) and 0.075 nm (b), and the numbers near the curves designate the multipole orders that have been included in the single-particle field expansions of coupled equations.

A comprehensive theoretical analysis of a monolayer consisting of metal or metallodielectric nanoparticles with the dipole and quadrupole single-particle resonances has been done in Ref. [127]. The theoretical models included spherical gold and silver particles and also gold and silver nanoshells on silica and polystyrene cores forming 2D random clusters or square-lattice arrays on a dielectric substrate (glass in water). The parameters of individual particles were chosen so that a quadrupole plasmon resonance could be observed along with the dipole scattering band. By using an exact multipole cluster-on-a-substrate solution, it has been shown that the particle-substrate coupling can be neglected in calculation of the monolayer extinction spectra, at least for the glass-in-water configuration. When the surface particle density in the monolayer was increased, the dipole resonance became suppressed and the spectrum for the cooperative system was determined only by the quadrupole plasmon. The dependence of this effect on the single-particle parameters and on the cluster structure was examined in detail. In particular, the selective suppression of the long-wavelength extinction band was shown to arise from the cooperative suppression of the dipole scattering mode, whereas the short-wavelength absorption spectrum for the monolayer was shown to be little different from the single-particle spectrum.

Here, we provide several illustrative examples. As a simple monolayer model, we used a square lattice with a period $p = d_e(1+s)$, where d_e is the external diameter of particles and s is the relative interparticle distance. Another monolayer model was obtained by randomly filling a square of side L/d_e with a given number of particles N. Then, the relative coordinates of the particles X_i were transformed as $x_i = X_i d_e(1+s)$ where the parameter s controls the minimal interparticle distance. The structure of the resultant monolayer is characterized by the particle number N and the average surface particle density $\rho = NS_{geom}/[L^2(1+s)^2]$.

The interparticle distance is a crucial parameter determining the electrodynamic particle coupling and the cooperative spectral properties of an ensemble. Therefore, we first investigated the influence of the interparticle-distance parameter s on the suppression of the dipole mode. Figure 2.16 shows a comparison of the extinction spectra of lattice and random clusters. All particle and cluster parameters are indicated in the figure caption. It can be seen that in both cases, the spectra for the lattice clusters with an s parameter of 0.2 are much the same as the spectra for the random clusters. Closer agreement between the spectra for the 36-particle clusters can be obtained if the s parameter is 0.35

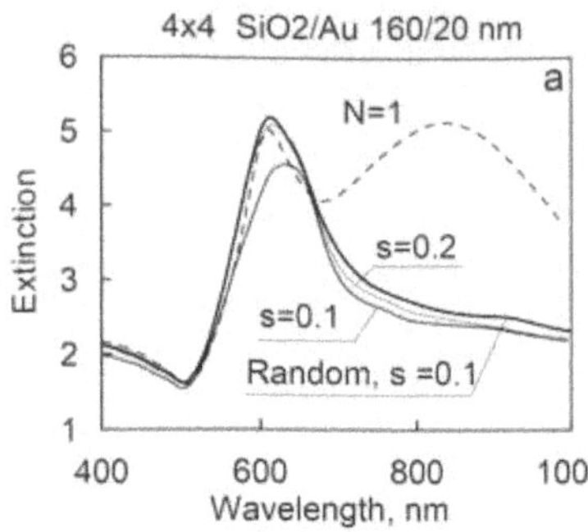

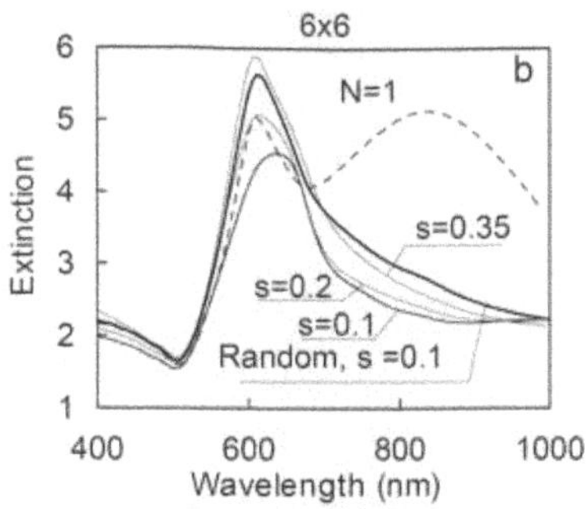

FIGURE 2.16: Comparison of the extinction spectra for lattice (4×4, 6×6) and random clusters made up of SiO_2/Au (160/20 nm) nanoshells with particle numbers of 16 (a) and 36 (b). Also shown are the spectra for isolated particles ($N = 1$). The interparticle-distance parameter is 0.1, 0.2, and 0.35 for the lattice clusters and equals 0.1 for the random clusters. The average particle density for the random clusters is 0.415 (a) and 0.36 (b) for 16 and 36 particles, respectively.

in the lattice case. Beginning with parameter s values about 0.2, there was an effective suppression of the dipole mode, so that the resonance was determined by only the quadrupole mode. A twofold decrease in the s parameter (to as low as 0.1) brought about little change in the system's spectrum. These conclusions are general and depend little on the properties of particles themselves.

To gain an insight into the physical mechanisms responsible for suppression of the coupled dipole mode, one also has to investigate the influence of particle interactions on the cooperative absorption and scattering of light. This question has been studied for several models, including silver and gold spheres as well as silver and gold nanoshells on dielectric cores (polystyrene and silica). In the case of silver spheres, it has been found that the suppression in a monolayer of the dipole extinction band is determined entirely by the decrease in *dipole resonance scattering* that occurs when strongly scattering particles with a dipole and a quadrupole resonance move closer together. The electrodynamic interparticle interaction almost does not change the absorption spectrum, including its fine structure in the short-wavelength region.

Figure 2.17 shows the dependence of the extinction, scattering, and absorption spectra for random clusters of 36 20-nm silver nanoshells on 110-nm polystyrene (PS) spheres. The calculated results were averaged over five independent cluster generations. It can be seen that the transformation of the extinction, scattering, and absorption spectra occurring with increasing particle density in the monolayer is similar to that for silver sphere monolayer [127]. Namely, the clear-cut quadrupole and octupole absorption peaks in the short-wavelength portion of the spectrum almost do not depend on the average particle density in the monolayer. As in the case of solid silver spheres, the disappearance of the dipole extinction band is associated with the suppression of the dominant dipole scattering band.

For experimental studies [127], the silica/gold nanoshell monolayers were fabricated by the deposition of nanoshells on a glass substrate functionalized by silane-thiol cross-linkers. The measured single-particle and monolayer extinction spectra were in reasonable agreement with simulations based on the nanoshell geometrical parameters (scanning electron microscopy data). Finally, the sensitivity of the coupled quadrupole resonance to the dielectric environment was evaluated to find a universal linear relation between the relative shift in the coupled-quadrupole-resonance wavelength and the relative increment in the environment refractive index.

We conclude this section with Figure 2.18, which shows an SEM image of fabricated monolayer, the single-particle extinction spectra, and the coupled spectra of the monolayer. Calculations with fitting size parameters reproduce the spectral positions of both quadrupole and dipole bands. The

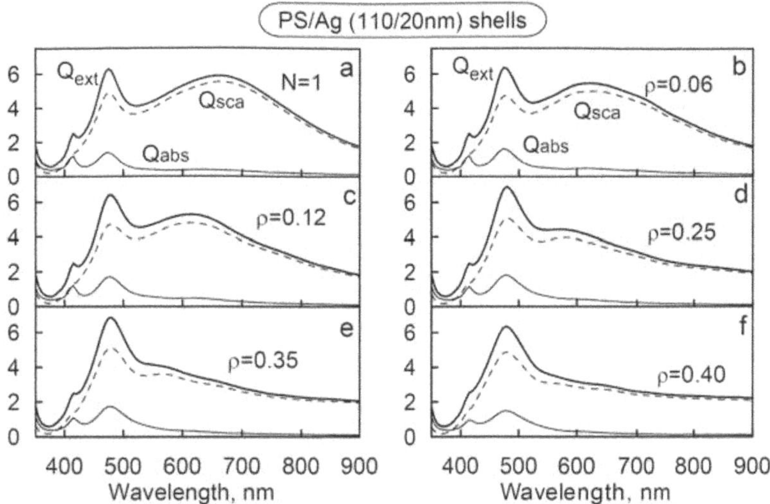

FIGURE 2.17: Extinction, scattering, and absorption spectra for random clusters made up of 36 PS/Ag-type silver nanoshells (core diameter, 110 nm; shell thickness, 20 nm). The minimal-distance parameter s is 0.05, and the average particle density is 0 (a, a single particle), 0.06 (b), 0.12 (c), 0.25 (d), 0.35 (e), and 0.4 (f). The calculated results were averaged over five statistical realizations.

quadrupole extinction peak for a particle suspension is located near 760 nm, and the dipole scattering band lies in the near IR region (about 1100 nm). The theoretical single-particle spectrum reveals an octupole resonance near 650 nm, which is not seen, however, in the experimental plots because of the polydispersity and surface roughness effects. A separate calculation of extinction, absorption, and scattering spectra allow us to attribute the octupole peak to the dominant absorption resonance, the quadrupole peak to both the scattering and the absorption contribution, and the dipole peak to the dominant scattering resonance. The particle interaction in the experimental monolayer brought about a noticeable decrease in the extinction shoulder (800–1100 nm) because of suppression of the scattering resonance. To simulate the monolayer spectrum, we used the fitting average core diameter and gold shell thickness (230 and 15 nm) and five independent calculations for a random array ($N = 36$) with the average surface particle density ρ =0.25 (the s parameter equals 0.05). In general, the model calculations agree well with the measurement. In any case, the suppression of dipole peak is clearly seen in both the experimental and simulated spectra.

2.4 Biomedical Applications

2.4.1 Functionalization of metal nanoparticles

In nanobiotechnology metal nanoparticles are used in combination with recognizing biomacromolecules attached to their surface by means of physical adsorption or coupling through the Au–S dative bond (for example, single-stranded oligonucleotides [139], antibodies [140], peptides [141], carbohydrates [142], etc.). Such nanostructures are called bioconjugates [1, 33, 143], while the attachment of biomacromolecules to the NP surface is often called "functionalization" [19, 33]. Thus, a probe conjugate molecule is used for unique coupling with a target, while a metal core serves as an optical label.

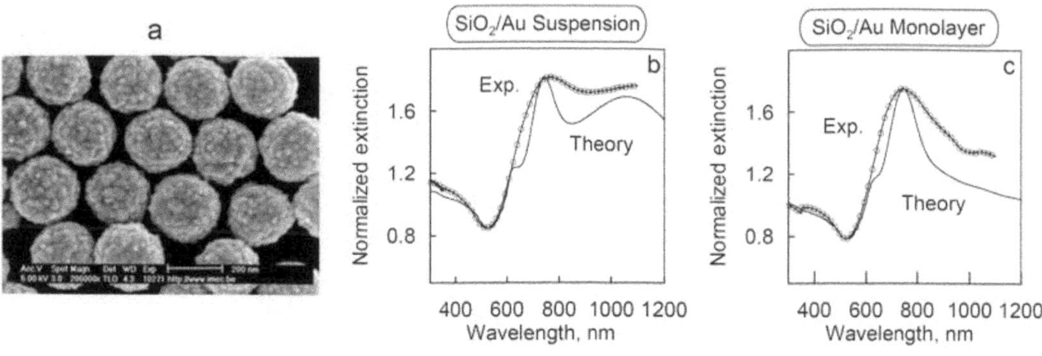

FIGURE 2.18: SEM image of a monolayer portion showing self-assembled silica/gold nanoshells on a silane-functionalized quartz substrate. The average nanoparticle diameter is 245 ± 12 nm, and the average silica core diameter is 200 ± 10 nm (a). Normalized experimental and theoretical extinction spectra for a suspension of nanoshells (b) and for a monolayer on a silane-functionalized quartz substrate in water (c). Details of theoretical fitting can be found elsewhere [127].

Metal hydrosols are typical lyophobic colloids [33, 34] which are thermodynamically unstable and require special stabilization of particles. For instance, citrate CG particles are stabilized due to a negative charge, while the stability of the seed-mediated gold NRs is ensured by CTAB protection. Over last decade, a lot of functionalization protocols have been developed to include various nanoparticles and biomolecules [1, 33, 141]. The most popular technology is based on using the thiolated intermediate linkers such as $HS(CH_2)nR$(R=COOH, OH or SO_3H; n=11-22) [144], or directly the thiolated derivatives of molecular probes such as oligonucleotides, proteins, immunoglobulins, avidin, and even dual functionalization by thiol-oligonucleotides and antibodies [145]. Besides alkanethiols, other ligands containing phosphino, amino, or carboxy groups can also be used as linkers [1]. Quite recently, a new class of so-called "Mix-matrices" has been proposed to facilitate the bioconjugation process [146]. In contrast to the usual CG or NS particles, the functionalization of gold NRs encounters significant difficulties because of presence of CTAB surfactant molecules serving as stabilizers (see, e.g., a relevant discussion and comprehensive citations in Ref. [35]). Nevertheless, even for such a complicated task, several publications have demonstrated successful protocols for gold nanorods [147–150]. Wei and coworkers [56] also provided a useful discussion of five functionalization schemes as applied to gold NRs.

An interesting approach to the enzymatic-driven assembly has been reported recently by the Alivisatos group [151]. It was shown that DNA ligases, which seal single-stranded nicks in double-stranded DNA (dsDNA), also can operate on thiolated DNA-CG conjugates. Specifically, the enzymatic ligation of single CG-DNA conjugates created NP dimer and trimer structures in which the nanoparticles were linked by single-stranded DNA, rather than by double-stranded DNA. This capability was utilized to create a new class of multiparticle building blocks for nanoscale self-assembly.

At present, most biomedical applications are based on gold-particle conjugates. As for silver particles, we can recommend a recent paper [152] where the problem is discussed in general aspects. In addition, in a recent work Huang et al. [153] they successfully designed and synthesized silver nanoparticle biosensors (AgMMUA-IgG) by functionalizing 11.6-nm Ag nanoparticles with a mixed monolayer of 11-mercaptoundecanoic acid (MUA) and 6-mercapto-1-hexanol (1:3 mole ratio). Further, these intermediate particles were additionally functionalized by covalently conjugating IgG of with MUA on the nanoparticle surface via a peptide bond using 1-Ethyl-3-(3-dimethylaminopropyl)-carbodiimide hydrochloride EDC and *N*-hydroxysulfosuccinimide as mediators. Such an example clearly demonstrates that the gold and silver particles can be functionalized with biospecific molecules through common or analogous protocols.

2.4.2 Homogenous and biobarcode assays

There are two variations of the homogeneous assays based on gold-nanoparticle aggregation: (1) aggregation of "bare" particles or particles covered with synthetic polymers that nonspecifically interact with biomacromolecules and (2) biospecific aggregation of conjugates that is initiated by the binding of target molecules in solution or by the binding of target molecules adsorbed on another conjugate. The ability of gold particles interacting with proteins to aggregate with a solution color change served as the basis for development of a method for the colorimetric determination of proteins [154, 155]. As a rule, one uses UV-VIS spectrophotometry of a gold sol (partially stabilized with polymers like Tween-20 or PEG) that can nonspecifically interact with biomacromolecules under acidic condition what is accompanied by change in the color of sol. For instance, the method [155] uses a trypsin conjugate of colloidal gold and consists in recording the interaction of the peptolytic enzyme adsorbed to gold nanoparticles and the protein in solution.

In 1980, Leuvering et al. [156] proposed a new immunoassay method they called the "sol particle immunoassay" (SPIA). A biospecific reaction between target and probe molecules on the colloidal-particle surface results in gold particle aggregation leading to substantial changes in the absorption spectrum, with a noticeable color change from red to blue or gray. It is possible to use the widely available spectrophotometers and colorimeters to assess the reaction outcome. Leuvering et al. [157] also optimized the SPIA by using larger-sized gold particles and monoclonal antibodies to various antigenic sites, to detect. Later, the SPIA was used to determine chorionic gonadotropin in the urine of pregnant women, *Rubella* and *Schistosoma* antigens, affinity constants for various isotypes of monoclonal IgG, immunoglobulins in the blood serum of HIV patients, and cystatin C – an endogenous marker for the glomerular filtration rate. The relevant references can be found in [155]. In 2002, Thanh and Rosenzweig [158] "reinvented" this method for determination of antiprotein A with protein A–colloidal gold conjugates.

In some SPIA experiments, despite the obvious complementarity of the molecular probe-target pair, there were no changes in the solution color. It was suggested that a second protein layer is possibly formed on gold particles without loss of the aggregative stability of the sol [155]. Although the spectral changes induced by the adsorption of biopolymers onto metallic particles are relatively small, Englebienne and coworkers [159] successfully used this approach to develop a quantitative analysis in biomedical applications.

Mirkin et al. [139] proposed a novel SPIA variant, which is based on the colorimetric determination of polynucleotides during their interaction with colloidal-gold particles conjugated through mercaptoalkyls to oligonucleotides of a specific sequence. An alternative variant of DNA-based SPIA without cross-linking interactions was developed by Maeda and coworkers (see, e.g., recent work [160] and references therein). The colorimetric detection of target DNA with using aggregation phenomena was also developed by Baptista [7] and Rothberg [161] groups. The first approach [7] is based on the aggregation of single-stranded DNA conjugates (negative reaction) after the addition of salt, while the red color of a positive sample (particles covered by double-strained DNA after positive hybridization) remains unchanged. This method was recently efficiently applied for diagnostics of tuberculosis by using samples taken from patients after one stage of PCR [7]. The second approach [161] uses the fact that under certain conditions, single-stranded DNA adsorbs to negatively charged gold nanoparticles in a colloid whereas double-stranded DNA does not. Accordingly, these phenomena determine the colloid stability and color.

A single-analyte biobarcode assay, as introduced by Mirkin and coworkers [162], is carried out in a disposable chip through several basic steps [57]. In step 1 the solution with target molecules (TM) sample is mixed with magnetic spheres (MS) conjugated to antibodies and introduced into the separation area of the chip. In step 2 the MS are separated from unreacted molecules by applying a magnetic field and washing. In step 3 gold NPs conjugated to antibodies and double-stranded DNA barcodes are introduced into the chip and mixed with MS. Bound TMs are sandwiched between the MS and gold NPs, while unbound NPs are washed. In step 4 the MS are retained in the chip

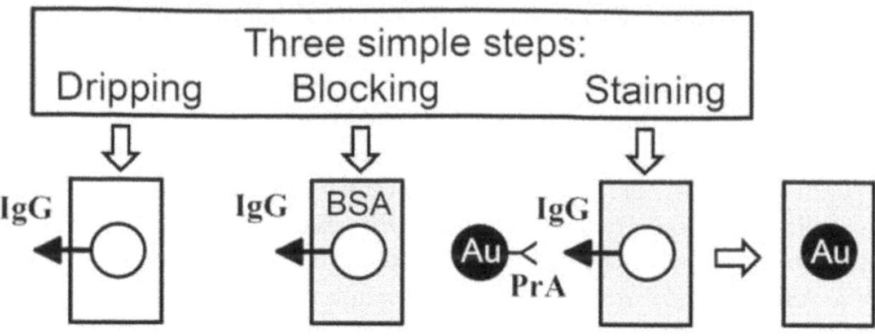

FIGURE 2.19: Basic principles of dot immunoassay include three simple steps: placement of an analyte drop on a membrane strip, blocking of free space with BSA solution, and staining of adsorbed target molecules with biospecific CG conjugates. Here, the target and probing molecules are IgG and protein A, respectively.

and DNA barcodes are dehybridized from them by running water. In step 5 released barcodes are sandwiched between gold NPs and immobilized capture probes. After enhancement with silver and separation from fluidics, the chip is used for an imaging procedure. A multiplexed biobarcode assay is similar to the single analyte assay except for the DNA barcodes are resolved with a two-dimensional array of capture probes before silver enhancement and detection [57].

Along with the extinction-spectrum changes, the particle aggregation is accompanied by the corresponding variation in scattering [163, 164]. This phenomenon has been widely used as an analytical tool for determination of various substances in solution. With some methodical cautions [165], it is possible to use commercial spectrofluorimeters for resonance scattering measurements (see, e.g., [166, 167] and references therein).

2.4.3 Solid-phase assays with nanoparticle markers

A most widespread method for the visualization of biospecific interactions is the solid-phase assay [168], in which the target molecules or antigens are adsorbed on a solid carrier and then revealed with various labels conjugated to recognizing probing molecules, e.g., antibodies (Figure 2.19). Among the diverse solid-phase assay techniques, a special place is occupied by the dot assay, based on the specific staining of a sample drop adsorbed on a membrane [1] or on another substrate. The chief advantage of this method is the possibility of conducting tests without using expensive equipment or means of signal processing. Contrary to, e.g., solid-phase assays using scanning atomic-force microscopy [169], laser-induced scattering around a nanoabsorber (LISNA) [170], or single-particle resonant-light-scattering spectroscopy [171], such tests may be run under home or field conditions. Examples include known tests for early pregnancy and for narcotic substances or toxins in human blood.

In the dot assay, minimal volumes of target-molecule solutions are applied to a substrate as a series of spots, thus making it possible to perform more analyses with the same reagent amounts than, e.g., in the ELISA assay. The sorbent most popular for solid-phase membrane immunoassays is nitrocellulose and modifications thereof [1], although recent years have also seen the use of siliconized matrices [172] and functionalized-glass microarrays [173], which may be constructed in the form of biochips [174].

At the visualization stage, the membrane is placed in a solution of recognizing molecules conjugated to molecular or corpuscular labels. The conjugates interact with the sample only if complementary target molecules have been adsorbed on the membrane. For the detection of biospecific interactions, wide use has been made of such labels as radioactive isotopes, enzymes, and fluo-

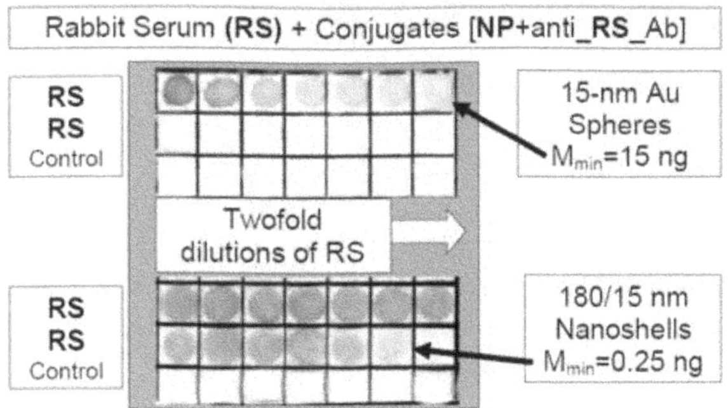

FIGURE 2.20: Dot immunoassays of a normal rabbit serum (RS) by using conjugates of 15-nm colloidal gold particles and silica/gold nanoshells (180/15 nm) with sheep's antirabbit antibodies. The IgG quantity equals $1\mu g$ for the first (upper left) square and is decreased by twofold dilution (left to right). The bottom row of each assay corresponds to a negative control ($10\mu g$ BSA in each square).

rescent dyes. In 1984, several research teams independently proposed using CG as a label in the solid-phase immunoassay [175], in which the intense red color of the marker allows visual determination of the assay results. Information on the principles, technology, and biomedical applications of the immunogold dot assay is detailed in review [1].

In the past few years, the sensitivity of the dot assay using functionalized gold NPs and silver enhancement [6] has been increased down to 1 pg of IgG molecules immobilized on glass or 2.75 ng/ml IgG molecules in solution [176]. There are record-breaking data on the atto- and zeptomolar sensitivity of solid-phase assays [177] (comparable with the sensitivity of resonance scattering [171]) and on the further development of the assay by various means, including the combined use of plasmon-resonant NPs and enzymes [178], peptides [179], fluorescent labels [180], and chitosans [181], as well as the use of gold particles with magnetic cores, showing a sensitivity limit of about 0.14 ng/ml [182].

Despite the obvious progress in further developing the dot assay by using metallic NPs, we believe that the method's potential has not yet been fulfilled – first of all, in respect to improving its reliability and sensitivity yet keeping the simplicity of the original one-step procedure (without silver enhancement or other means). The current state of the technology used for synthesizing NPs with specified geometrical and optical parameters offers a multitude of possibilities for optimization of solid-phase analytical technologies. Recent, more detailed studies [183, 184] showed that the replacement of colloidal gold conjugates by nanoshells improved the analysis sensitivity to 0.2 ng in the case of nanoshells of sizes 180/15 nm (Figure 2.20) and to 0.4 ng for nanoshells of sizes 100/15 and 140/15 nm. An important question arising with the use of various nanoparticles in the dot immunoassay is as follows: What controls the sensitivity limits of the particle-labelled dot immunoassays? By using the developed theory [184], we explained the dependences of the detection threshold, the sensitivity range of the method, the maximum staining, and the probe-amount saturation threshold on the parameters of particles.

Along with dot immunoassay, NP markers can be used in such solid-phase technologies as immunochromatography and multiplexed lateral flow devices (MLFD) based on gold NPs [1, 57]. In particular, MLFD have been successfully applied for simultaneous detection of up to four cardiac markers, parallel detection of up to five threat agents, and serial detection of up to six amplified nucleic acid targets [57].

2.4.4 Functionalized NPs in biomedical sensing and imaging

Because of the sensitivity of PR to the local dielectric environment, plasmonic nanoparticles can act as transducers that convert small changes in the local refractive index into spectral shifts in the extinction and scattering spectra. As a rule, biomolecules have a higher refractive index than the buffer solution. Therefore, when a target molecule binds to the nanoparticles surface, the local refractive index increases and the PR wavelength shifts to the red. Actually, it is the phenomenon that defines the basic principle of all chemical and biosensors utilizing PR shift as an initial input signal. With this principle, molecular binding can be monitored in real time with high sensitivity by using transmission or resonance scattering spectrometry.

The first studies of biopolymer adsorption on the CG nanoparticle surface and related optical effects [185, 186] showed that the biopolymer binding results in small increases in the extinction and scattering maxima (about 10–20%) and in red-shifting of PR wavelength over 1–20 nm range. Because of the local electrodynamic nature of PR shifting, it is clear that the sensing sensitivity should be dependent on the particle size, shape, and structure. For instance, the problem of sensitivity of CG particles to the local refractive index was formulated as a size-dependent optimization problem, for which the solution revealed the existence of optimal diameters about 30–40 nm for silver and 60–80 nm for gold [187]. Furthermore, the first theoretical studies revealed significant differences between such nanostructures as gold spheres, rods, bispheres, and nanoshells [188, 189]. Recent experimental information about the shape-dependent sensitivity of PR to the dielectric environment and relevant references can be found in a review [55] and paper [190].

In contrast to ensemble experiments, the single-particle measurements possess potentially much greater sensitivity. Indeed, the single-particle PR shift is caused by molecules adsorbed on their surface. Typically, the number of adsorbed molecules is less than 100–200, or even smaller. This means that the PR sensitivity of a single PR particle can be about one zeptomole [171]. Recently, Chilkoti and coworkers [191] presented an analytical model for the rational design of a biosensor based on measuring small PR-shifts of individual gold nanoparticles. The model predicts a minimal detection limit of 18 streptavidin molecules for a single-particle sensor, which is in good agreement with experiments and estimates. Moreover, the authors [191] estimated a possible enhancement factor that could potentially provide an 800-fold improvement over the 18-molecule detection limit described in their experimental study. Thus, the ultimate limit of label-free detection of single molecule binding events could be theoretically possible. To conclude this topic, we note that the same idea can be used for a subtle tracking of molecular interaction. For instance, Sönnichsen group [192] demonstrated the ability of probing protein-membrane interaction based on membrane-coated plasmonic particles.

The resonance scattering of PR NPs and their bright colors in dark-filed microscopy determine many emerging applications in biomedical imaging. Depending on their size, shape, composition, degree of aggregation, and nature of the stabilizing shell on their surface, PR nanoparticles can appear red, blue, and other colors and emit bright resonance light scattering of various wavelengths [193]. To give some quantitative estimation and comparison [55], it is sufficient to note that a single 80-nm silver nanosphere scatters 445-nm blue light with a scattering cross-section of 3×10^4 nm^2, a million-fold greater than the fluorescence cross-section of a fluorescein molecule, and a thousand-fold greater than the cross-section of a similarly sized nanosphere filled with fluorescein to the self-quenching limit. Schultz [194] summarized earlier efforts in the labelling of cells with NPs.

To illustrate recent advances in NP-based imaging, we restrict our consideration by only a few examples. In the first experiments, El-Sayed [195, 196] and Halas [197] groups demonstrated SPR imaging of cancer cells by using 35-nm CG, gold NRs and NSs conjugated with antibodies to the molecular cancer markers overexpressed on the cell surface. Sokolov and coworkers used "stellated" particles and the polarization-sensitive technique for visualization of molecular cancer cell markers [12]. Recently, Prasad and coworkers described [198] a multiplex dark-filed microscopy imaging of cancer cells (MiaPaCa) with functionalized Ag and Au nanospheres and gold NRs, that exhibit blue, green, and red scattering colors, respectively.

After publication of the first reports [195, 197], many research groups were focused on studies of NP conjugates in relation to cancer diagnostics and therapy. Conjugates of gold NPs with biospecific antibodies to the cancer cell receptors can be used to fabricate high-performance probes for detection of live cancer cell. For example, Yang et al. [199] showed that the conjugate of Cetuximab with an Au NP-deposited substrate could detect EGFR-high expressed A431 cells related to epithelial cancer with 54-times larger specificity and sensitivity in comparison with EGFR-deficient MCF7 cells. The thiolated PEGs as bifunctional cross-linkers can be used for biospecific functionalization of gold NPs with antibodies to the surface cancer cell receptors. A typical example is recently reported [200] where 15-nm CG NPs were conjugated via carboxy-terminated PEG dithiol with F19 monoclonal antibodies. These humanized murine antibodies are directed against fibroblast activation protein (FAP-α), which is a serine protease produced specifically by activated fibroblasts. Another example [201] demonstrates the use of PEG-SH to obscure gold nanorods against the reticuloendothelial system in the living organism, whereas the biospecific conjugation were performed by covalent attachment of Herceptin (HER) through Nanothinks acid16 (Sigma-Aldrich). Herceptin is a monoclonal antibody that enables molecular recognition of breast cancer cells expressing highly specific tumor-associated antigens. Accordingly, the fabricated conjugates were used for *in vivo* molecular targeting of breast cancer cells.

Although the surface functionalization is typical in biomedical applications of NPs, the "naked" particles themselves may affect biological responses and properties. This topic has been reviewed thoroughly by Bhattacharya and Mukherjee [50].

Besides PR resonance scattering, a photoacoustic technique were recently developed to monitor systemic circulation of single particles [15] or their accumulation and aggregation within cells (e.g., within macrophages in atherosclerotic plaques [202]). The method is based on a generation of acoustic waves upon pulsed heating of particles, which allows the acoustic monitoring of the particle location or movement.

Another approach utilizes the two-photon luminescence (TPL) technique that possesses an obvious advantage over usual dark-field microscopy because of selective excitation of only target particles. This method excludes elastic scattering from nonresonant impurities and it has been successfully applied to visualization of various cells by using CG NPs, NRs, and NSs (see, e.g., [203] and review [19]). The physics behind TPL can be described as follows [19]. The first photon excites an electron through the interband transition from the sp conduction band below the Fermi level to the sp band above the Fermi level and forms a hole in the first band. The second photon excites a d-electron to the lower sp band, where the first photon has formed the hole. This transition is polarization-dependent and it is efficient for nanorods only in the case of longitudinal polarization. The second photon excites a d-electron to the lower sp band, where the first photon has created the hole. After recombination of the electron and hole, a luminescence photon is emitted.

The combination of TPL with confocal microscopy gives the unique possibility to observe and distinguish PR labels on the surface of and inside cells. It was shown recently that the single-particle resonance TPL intensity of nanorods [203] and nanoshells [14] is sufficient to use these labels as contrast agents for the visualization of cancer molecular markers and investigations of intracellular processes. Moreover, it has been previously shown [204] that usual functionalized CG NPs could be used for the TPL visualization of particles on the surface of and inside cancer cells.

Discovered thirty years ago, SERS is now one of the potentially powered techniques to probe single-molecule interactions [205]. Due to its physical nature, SERS is strongly related to plasmonics, which encompasses and profits from the optical enhancement found in PR NPs and clusters. Numerous papers have been published on various aspects of SERS, from physical background to medical applications. As starting points, we recommend a couple of recent papers on analytical [206] and biomedical [207] applications of SERS and SEIRA [208].

2.4.5 Interaction of NPs with living cells and organisms: Cell-uptake, biodistribution, and toxicity aspects

The current developments in nanobiotechnology involve the interaction of PR NPs with living organisms, mostly on the cellular and subcellular levels. Given the breadth of currently arising biomedical applications of NPs, it is of great importance to develop an understanding of the complex processes that govern their cellular uptake and intracellular fate, biodistribution over organs of a living organism, and concerns associated with the toxicology of NPs, which is still in its infancy.

Studies on the interface between the parameters of NPs and cell biology are just beginning and a great deal is unknown about the interaction of NPs with cells and organisms and the potential toxicity that may result. The cellular uptake seems to be the first important step, which may be affected by the particle geometry (size, shape, etc.) and its surface chemistry. In a series of works (see, e.g., [209, 210] and references therein), Chan and coworkers discovered the crucial role of the particle size in the efficiency of cancer cell uptake. Specifically, they showed that gold NPs conjugated with antibodies can regulate the size-dependent processes of binding and activation of membrane receptors, their internalization, and subsequent protein expression. It has been found that 40- and 50-nm particles altered signalling processes most effectively, while other particles within the 2–100 nm size range also demonstrated measurable effects. These finding means that gold NPs can play an active role in mediating the cellular response rather than serve as inert carriers. Note that Zhang et al. [211] recently published a somewhat different observation. Namely, they found that the 20-nm Au NPs coated with TA-terminated PEG5000 showed significantly higher tumor uptake and extravasation from the tumor blood vessels than did the 40- and 80-nm Au NPs. Another approach to the same NR toxicity problem has been recently reported by Parab et al. [212]. In that work, surface modification of CTAB-stabilized gold NRs was carried out by using poly(styrenesulfonate) (PSS) to reduce the toxicity of as-prepared samples because of free CTAB. Electrophoretic measurements confirmed a charge reversal due to PSS coating. Functionalization of NRs with PSS significantly increased the cell viability and showed easy intracellular uptake of the nanorods, which suggests their possible use for different biomedical applications.

The surface chemistry is a second key factor in determining the cellular uptake efficiency. It is generally believed that nanoparticles will remain in the endosome created by the endocytosis process, unless they are microinjected or brought into the cell mechanically. However, it was demonstrated that this endosomal route of cellular uptake can be bypassed by surface modification of the nanoparticles with so-called cell penetrating peptides (CPPs, e.g., CALNN) [213]. Furthermore, successful nuclear targeting was demonstrated using surface modification of NPs with a cocktail of CPPs and a peptide acting as a nuclear localization signal. In the same line, the Chan group [214] used several layer-by-layer polyelectrolyte coatings to tune the mammalian cellular uptake of gold nanorods from very high to very low by manipulating the surface charge and functional groups of polyelectrolytes. Some peptide motif's are known to deliver a "cargo" into a chosen cellular location specifically. Mandal et al. [215] attempted to use this property by modifying gold NPs in order to deliver the particles into osteosarcoma cells through chemical cross-linking with different peptides known to carry protein into cells. Enhanced cellular uptake and anticancer efficiency was also reported for conjugates of gold NPs with a dodecylcysteine surfactant [216].

Some biomolecules are known to be efficient substances for cellular uptake (e.g., transferrin). Thus, a possible way for the design of an efficient nanovector is to attach these molecules to the NP surface. Yang et al described an attractive approach to use transferring-conjugated gold NPs as vehicles for specific uptake and targeted drug-delivery [217].

Gold nanoparticles are also being actively used in DNA vaccination [218] and in the *in vivo* preparation of antibodies specific to complete antigens and various haptens [219]. When serving as an antigen carrier, CG enhances the phagocytic activity of macrophages and exerts an influence on lymphocyte functioning; this fact possibly determines the immunomodulating effect of CG. The most interesting aspect of the manifestation of immunogenic properties by hapten after immobilization on CG is that gold nanoparticles serve as both a carrier and an adjuvant – that is, they present the hapten to T cells in some way. Several authors reported on responses of Kupffer cells [220] and peritoneal macrophages [221] during interaction with gold nanoparticles. Vallhov et al. [222] discussed the influence of dendritic cells on the formation of an immune response upon the administration of an antigen conjugated to gold nanoparticles.

Pharmacokinetic and organ/tissue distribution properties of functionalized NPs are of great interest from a clinical point of view because of potential human cancer treatment. After the initial injection of NPs into an animal, the systemic circulation distributes the NPs towards all organs in the body. Several publications have shown distribution of NPs to multiple animal organs including the spleen, heart, liver, and brain. For instance, Hillyer and Albrecht [223] observed an increased distribution to other organs after oral administration of gold NPs of decreasing size (from 58 to 4 nm) to mice. The highest amount of CG NPs was obtained for smallest particles (4 nm) administered orally. For 13-nm CG NPs, the highest amount of gold was detected in the liver and spleen after intraperitoneal administration [223]. Niidome et al. [224] showed that after an intravenous injection of gold NRs these particles accumulated predominantly in the liver within 30 min. The coating of gold NRs by PEG-thiols resulted in a prolonged circulation.

De Jong et al. [225] performed a kinetic study to evaluate the gold NPs size (10, 50, 100, and 250 nm) effects on the *in vivo* tissue distribution of NPs in the rat. Rats were intravenously injected in the tail vein with gold NPs and the gold particle distribution was measured quantitatively 24 h after injection. The authors showed that the tissue distribution of gold NPs is size-dependent with the smallest 10-nm NPs possessing the most widespread organ distribution. Analogous study for 1.4- and 18-nm gold NPs was reported by Semmler-Behnke et al. [226] for biodistribution in rats and by Zhang et al. [211] for biodistribution of 20-, 40-, and 80-nm Au NPs in nude mice.

Katti and coworkers [227] published detailed *in vitro* analysis and *in vivo* pharmacokinetics studies of gold NPs within the nontoxic phytochemical gum-arabic matrix in pigs to gain insight into the organ-specific localization. Pigs were chosen as excellent animal models because of their similar physiological and anatomical characteristics to those of human beings. Kogan and coworkers reported data on biodistribution of PEG-coated 15×50-nm gold NRs and 130-nm silica/gold NSs in tumor-bearing mice [228]. The kinetics of gold distribution was evaluated for blood, liver, tumor, and muscles over 1-h (NRs) and 24-h (NSs) periods. It was found that the gold NRs were nonspecifically accumulated in tumors with significant contrast in comparison with other organs, while silica/gold NSs demonstrated a more smooth distribution. This study was recently extended in our report [229], where kinetics, biodistribution, and histological examinations were performed to evaluate the particle-size effects on distribution of 15-nm and 50-nm PEG-coated CG particles and 160-nm silica/gold NSs in the rats and rabbits. The above NPs were used as model because of their importance for current biomedical applications such as the photothermal therapy, the optical coherence tomography, and the resonance-scattering imaging. The dynamics of NPs circulation *in vivo* was evaluated after intravenous administration of 15-nm CG NPs to rabbit, and the maximal concentrations of gold were observed 15–30 min after injection. Rats were injected in the tail vein with PEG-coated NPs (about 0.3 mg Au/kg rats). 24 h after injection, the accumulation of gold in different organs and blood was determined by the atomic absorption spectroscopy. In accordance with the published reports, we observed 15-nm particles in all organs with a rather smooth distribution over the liver, spleen, and blood (Figure 2.21). By contrast, the larger NSs were accumulated mainly in the liver and spleen. For rabbits, the biodistribution was similar (72 h after intravenous injection).

Although NPs have been estimated to increase from 2,300 tons produced today to 58,000 tons by 2020 [230], the current knowledge on the NPs toxicity effects is quite limited and also reported results are confusing (see, e.g., citations in Ref. [230]). Despite the great excitement about the potential uses of gold nanoparticles for medical diagnostics, as tracers, and for other biological applications, researchers are increasingly aware that potential NP toxicity must be investigated before any *in vivo* applications of gold nanoparticles can move forward.

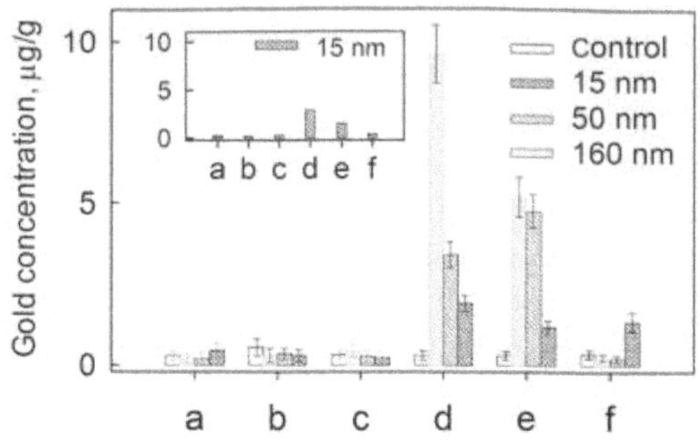

FIGURE 2.21: Au concentrations in rat organs 24 h after intravenous administration of 15-nm, 50-nm, and 160-nm NPs. a – brain, b – kidney, c – lung, d – liver, e – spleen, f – blood. The insert shows the biodistribution of 15-nm gold NPs 72 h after injection in rabbit. The bars show standard errors of the mean.

At present, it would be quite desirable to separate the toxicity effects related to the particles themselves and to their surface coatings. For example, Chan group [214] found that CTAB-coated rods did not produce significant toxicity in media containing serum. This finding suggests that serum proteins bind to nanorods and shield the toxic surface coating. On the other hand, Ray and coworkers [230] reported high toxicity of CTAB-coated NR. A close inspection of their data suggests that the reported toxicity was due to desorption of CTAB molecules. Note, that the reported [230] toxicity of usual CG NPs was very small in accordance with the recent study of biocompatibility of 5-, 15-, and 30-nm CG NPs [231]. The importance of surface chemistry and cell type for interpretation of nanoparticle cytotoxicity studies also has been recently illustrated by Murphy et al. [232].

For a more detailed consideration of the NP toxicity, the readers are referred to recent reviews [53, 233] (general consideration of various NPs) and [48, 232] (gold NPs). Recent review [234] presents a detailed summary of the *in vitro* cytotoxicity data currently available on three classes of NPs: carbon-based NPs (such as fullerenes and nanotubes), metal-based NPs (such as CG NPs, NSs, and NRs), and semiconductor-based NPs (such as quantum dots).

2.4.6 Application of NPs to drug delivery and photothermal therapy

The surface functionalization of NPs and their cellular uptake are closely related to the targeted drug delivery problems. Gold NPs have been considered recently as a promising candidate for delivery of small drug molecules or large biomolecules into their targets [11, 50]. Owing to their inherent properties such as tunable size and low polydispersity, long-term stability and weak toxicity, easy functionalization and large surface-to-volume ratio gold NPs could find promising applications in the drug delivery technologies [235]. Efficient release of these therapeutic agents could be triggered by internal or external stimuli. In the case of cancer therapy, CG NPs also could be used as a platform for developing multifunctional tumor-targeted drug delivery vectors [236]. For example, Patra et al. described recently an efficient application of 5-nm gold Ns conjugated with anti-EGFR antibodies Cetuximab (C225) and Gemcitabine for targeted delivery of Gemcitabine to pancreatic adenocarcinoma [237].

Targeted delivery of conjugated NPs could improve the therapeutic efficiency of known technologies as photodynamic therapy [238] and newly emerged field called plasmonic photothermal therapy (PPTT) [9]. In 2003, nanoshell-based therapy was first demonstrated in tumors grown in mice by Halas and coworkers [239], while Pitsillides et al. [240] and El-Sayed and coworkers [196] published first application of CG NPs and gold NR for PPTT *in vitro*. Basic principles of PPTT are explained in Figure 2.22.

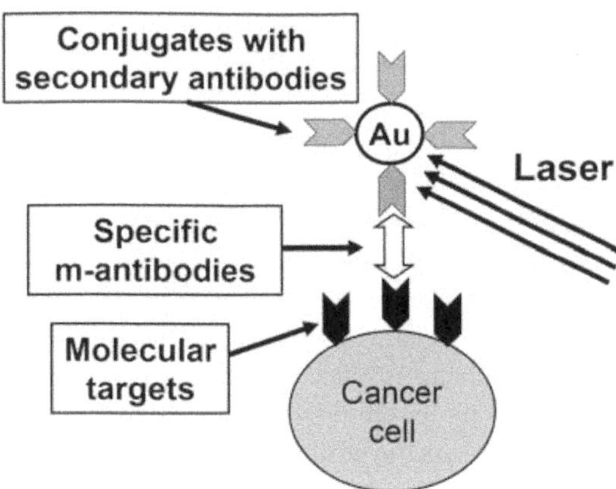

FIGURE 2.22: Scheme of PTT therapy. In step 1, the molecular cancer targets (MCT, e.g. EGFR) are labelled with primary monoclonal specific antibodies. In step 2, these m-antibodies are sandwiched between MCT and gold NPs conjugated with secondary antibodies. In step 3, irradiation with a laser pulse results in drastic heating of NPs and cancer cell death.

As the available hyperthermic techniques for cancer tumor therapy [241] possess low spatial selectivity in heating of the tumors and the surrounding healthy tissues, one of the ways to improve the laser heating spatial selectivity is the labelling of the tumor with PR NPs. By exposing NPs to laser radiation near their plasmon-resonant absorption band, it is possible to produce local heating of labelled cancer cells without harming surrounding healthy tissues. The spectral tuning of nanoparticle resonance to the "therapeutic optical window" (750–1100 nm) and getting the desired ratio between the absorption and scattering efficiencies can be achieved by variations in the particle size, shape, and structure [110].

The first step in the enhancement of the PPTT efficiency is a clear understanding of the relationship between the nanoparticle/cluster parameters (size, shape, particle/cluster structure, etc.) and optical responses. We reported [110] theoretical simulations aimed at finding the optimal single-particle and cluster structures to achieve its maximal absorption that is crucial for PT therapeutic effects. To characterize the optical amplification in laser-induced thermal effects, it is convenient to introduce relevant parameters such as the ratio of the absorption cross-section to the gold mass of a single-particle structure and absorption amplification, defined as the ratio of cluster absorption to the total absorption of noninteracting particles. We have simulated the absorption efficiency of single nanoparticles (gold spheres, rods, and silica/gold nanoshells), linear chains, 2D lattice arrays, 3D random volume clusters, and the random aggregated N-particle ensembles on the outer surface of a larger dielectric sphere, which mimic aggregation of nanosphere bioconjugates on or within cancer cells. The clusters particles are bare or biopolymer-coated gold nanospheres. The light absorption of cluster structures is studied by using the generalized multiparticle Mie solution and the T-matrix method. The gold nanoshells with (silica core diameter)/(gold shell thickness) parameters of (50–100)/(3–8) nm and nanorods with minor/major sizes of (15–20)/(50–70) nm are shown to be more efficient PT labels and sensitizers than the equivolume solid single gold spheres. In the case of nanosphere clusters, the interparticle separations and the short linear-chain fragments are the main structural parameters determining the absorption efficiency and its spectral shifting to the red. Although we have not found a noticeable dependence of absorption amplification on the cluster sphere size, 20–40-nm particles are found to be most effective, in accord with our experimental observations. The long-wavelength absorption efficiency of random clusters increases with the cluster particle number N at small N and reveals a saturation behavior at $N > 20$. Besides the linear optical responses, the second important point is related to the modelling of the heat transfer processes. This question has been addressed in several previous reports (see, e.g., references in [110]) and in a recent work [242].

Among a great variety of possible nanoparticle PPTT platforms, gold nanoshells seem to be a promising candidate because of robust fabrication and functionalization protocols and a significant absorption cross-section. Halas and coworkers [10] recently reviewed the properties of nanoshells that are relevant to their preparation and use in cancer diagnostics and therapy. It was stressed that specific functionalization technologies are necessary for both passive uptake of nanoshells into tumors and for targeting specific cell types. NS-based PPTT in several animal models of human tumors have produced highly promising results, and the corresponding NP dosage information, thermal response, and tumor outcomes for these experiments can be found in Ref. [10].

Although several preclinical reports on the utility of gold NSs for PPTT and OCT applications are available [10], there remain several areas still to be studied more thoroughly. First of all the question is about controlled and localized hyperthermia without significant overheating of both tumor and surrounding normal tissues. Our recent report [243] describes the investigation of heating kinetics, spatial temperature distribution, and morphological alterations in tissues dependent on laser irradiation parameters and nanoparticle concentrations. The experimental results for phantoms and laboratory animals at several temperature regimes and related morphological tissue patterns were also considered. We anticipate that additional research and theoretical models achieving temperature elevations are needed to apply PPTT as generally as possible to tumor remission. There is a need for effective quantification methods for the accumulation of NPs in tumors.

In the future development of PPTT technologies, one needs to address a number of variables, e.g., stability, biocompatibility, and chemical reactions of nanoparticles bioconjugates in physiological environments, blood retention time, tumor extravasation, the fate of the nanoparticles following therapy, etc. [9]. Most of the published *in vivo* studies have been carried out for subcutaneous tumors easily accessible to NIR light applied over the skin. For convenient clinical applications of PPTT one needs to deliver NIR light deep into the tissue by exploiting fiber optic probes or other means.

2.5 Conclusion

In this chapter, we have discussed the chemistry-based bottom-up protocols for fabrication of particles with the desired size, shape, and structure. In the case of silver NPs, however, there remain some problems related with NP stability and the synthetic protocol reproducibility. The geometrical and structural parameters of NPs determine the individual linear optical properties, which can now be characterized at the single particle level. The main effort now should be directed to improve the single-particle imaging techniques and the signal-to-noise ratio. Current development of the electromagnetic simulation tools allows for modelling various complex assemblies of bare or functionalized metal NPs.

For biomedical applications, the surface functionalization of NPs with biospecific molecules is a crucial step that determines the following properties of conjugates and the overall success of applied technologies. Although significant progress has been seen over the last five years, there are evident needs in development of simple and robust protocols.

Recent publications on carbon nanotubes injected into mice and resulting in mesothelioma [10] clearly demonstrate that the toxicity concerns need to be addressed in a serious and systematic way. Every novel application of nanotechnology for the detection, imaging, drug delivery, or therapeutic treatment should take into account all potential toxicities, both those to the patient and those to the environment.

Acknowledgment

This research was supported by grants of RFBR (Nos. 07-04-00301a, 07-04-00302a, 07-02-01434a, 08-02-01074a, 08-02-00399a, 09-02-00496a), RF Program on the Development of High School Potential (Nos. 2.1.1/4989 and 2.2.1.1/2950), and by a grant from the Presidium of RAS Program "The Basic Sciences to Medicine." The authors thank Drs. V. Bogatyrev, B. Khlebtsov, V. Khanadeev, I. Maksimova, G. Terentuk, G. Akchurin, V. Zharov, and V. Tuchin for collaboration work discussed in this chapter.

References

[1] L.A. Dykman and V.A. Bogatyrev, "Gold nanoparticles: preparation, functionalisation and applications in biochemistry and immunochemistry," *Russ. Chem. Rev.* **76**, 181–194 (2007).

[2] S.-Y. Shim, D.-K. Lim, and J.-M. Nam, "Ultrasensitive optical biodiagnostic methods using metallic nanoparticles," *Nanomedicine* **3**, 215–232 (2008).

[3] U. Kreibig and M. Vollmer, *Optical Properties of Metal Clusters.* Springer-Verlag, Berlin (1995).

[4] X. Liu, Q. Dai, L. Austin, J. Coutts, G. Knowles, J. Zou, H. Chen, and Q. Huo, "Homogeneous immunoassay for cancer biomarker detection using gold nanoparticle probes coupled with dynamic light scattering," *J. Am. Chem. Soc.* **130**, 2780–2782 (2008).

[5] M.E. Stewart, Ch.R. Anderton, L.B. Thompson, J. Maria, S.K. Gray, J.A. Rogers, and R.G. Nuzzo, "Nanostructured plasmonic sensors," *Chem. Rev.* **108**, 494–521 (2008).

[6] Sh. Gupta, S. Huda, P.K. Kilpatrick, and O.D. Velev, "Characterization and optimization of gold nanoparticle-based silver-enhanced immunoassays," *Anal. Chem.* **79**, 3810–3820 (2007).

[7] P. Baptista, E. Pereira, P. Eaton, G. Doria, A. Miranda, I. Gomes, P. Quaresma, and R. Franco, "Gold nanoparticles for the development of clinical diagnosis methods," *Anal. Bioanal. Chem.* **391**, 943–950 (2008).

[8] P.G. Luo and F.J. Stutzenberger, "Nanotechnology in the detection and control of microorganisms," *Adv. Appl. Microbiol.* **63**, 145–181 (2008).

[9] X. Huang, P.K. Jain, I.H. El-Sayed, and M.A. El-Sayed, "Plasmonic photothermal therapy (PPTT) using gold nanoparticles," *Lasers. Med. Sci.* **23**, 217–228 (2008).

[10] S. Lal, S.E. Clare, and N.J. Halas, "Nanoshell-enabled photothermal cancer therapy: impending clinical impact," *Acc. Chem. Res.* **41**, 1842–1851 (2008).

[11] G. Han, P. Ghosh, and V.M. Rotello, "Functionalized gold nanoparticles for drug delivery," *Nanomedicine* **2**, 113–123 (2007).

[12] J. Aaron, E. de la Rosa, K. Travis, N. Harrison, J .Burt, M. José-Yakamán, and K. Sokolov, "Polarization microscopy with stellated gold nanoparticles for robust monitoring of molecular assemblies and single biomolecules," *Opt. Express* **16**, 2153–2167 (2008).

[13] A.M. Gobin, M.H. Lee, N.J. Halas, W.D. James, R.A. Drezek, and J.L. West, "Near-infrared resonant nanoshells for combined optical imaging and photothermal cancer therapy," *Nano Lett.* **7**, 1929–1934 (2007).

[14] J. Park, A. Estrada, K. Sharp, K. Sang, J.A. Schwartz, D.K. Smith, Ch. Coleman, J.D. Payne, B.A. Korgel, A.K. Dunn, and J.W. Tunnell, "Two-photon-induced photoluminescence imaging of tumors using near-infrared excited gold nanoshells," *Opt. Express* **16**, 1590–1599 (2008).

[15] V. Zharov, E. Galanzha, E. Shashkov, N. Khlebtsov, and V. Tuchin, "*In vivo* photoacoustic flow cytometry for monitoring circulating cells and contrast agents," *Opt. Lett.* **31**, 3623–3625 (2006).

[16] V.M. Shalaev, "Optical negative-index metamaterials," *Nature Photonics*, **1**, 41–48 (2007).

[17] T.W. Odom and C.L. Nehl, "How gold nanoparticles have stayed in the light: the 3M's principle," *ACS NANO* **2**, 612–616 (2008).

[18] M. Pelton, J. Aizpurua, and G. Bryant, "Metal-nanoparticle plasmonics," *Laser & Photon. Rev.* **2**, 136–159 (2008).

[19] N.G. Khlebtsov, "Optics and biophotonics of nanoparticles with a plasmon resonance," *Quant. Electron.* **38**, 504–529 (2008).

[20] M.C. Daniel and D. Astruc, "Gold nanoparticles: assembly, supramolecular chemistry, quantum–size–related properties, and applications toward biology, catalysis, and nanotechnology," *Chem. Rev.* **104**, 293–346 (2004).

[21] L.A. Dykman, V.A. Bogatyrev, S.Yu. Shchyogolev, and N.G. Khlebtsov, *Gold Nanoparticles: Synthesis, Properties, Biomedical Applications,* Nauka, Moscow (2008) (in Russian).

[22] I. Pastoriza-Santos and L.M. Liz–Marzán, "Colloidal silver nanoplates: state of the art and future challenges," *J. Mater. Chem.* **18**, 1724–1737 (2008).

[23] W. Rechberger, A. Hohenau, A. Leitner, J. R. Krenn, B. Lamprecht, and F.R. Aussenegg, "Optical properties of two interacting gold nanoparticles," *Opt. Commun.* **220**, 137–141 (2003).

[24] C.L. Haynes and R.P. van Duyne, "Nanosphere lithography: a versatile nanofabrication tool for studies of size-dependent nanoparticle optics," *J. Phys. Chem. B* **105**, 5599–5611 (2001).

[25] J. Zhao, A.O. Pinchuk, J.M. McMahon, Sh. Li, L.K. Ausman, A.L. Atkinson, and G.C. Schatz, "Methods for describing the electromagnetic properties of silver and gold nanoparticles," *Acc. Chem. Res.* **41**, 1710–1720 (2008).

[26] 26. V. Myroshnychenko, J. Rodríguez-Fernandez, I. Pastoriza-Santos, A.M. Funston, C. Novo, P. Mulvaney, L.M. Liz-Marzán, and F.J. García de Abajo, "Modelling the optical response of gold nanoparticles," *Chem. Soc. Rev.* **37**, 1792–1805 (2008).

[27] S. Schultz, D.R. Smith, J.J. Mock, and D.A. Schultz, "Single-target molecule detection with nonbleaching multicolor optical immunolabels," *Proc. Natl. Acad. Sci. USA* **97**, 996–1001 (2000).

[28] A. Arbouet, D. Christofilos, N. Del Fatti, F. Vallée, J.R. Huntzinger, L. Arnaud, P. Billaud, and M. Broyer, "Direct measurement of the single-metal-cluster optical absorption," *Phys. Rev. Lett.* **93**, 127401-1–4 (2004).

[29] N. Del Fatti, D. Christofilos, and F. Vallée, "Optical response of a single gold nanoparticles," *Gold Bull.* **41**, 147–158 (2008).

[30] S. Berciaud, L. Cognet, P. Tamarat, and B. Lounis, "Observation of intrinsic size effects in the optical response of individual gold nanoparticles," *Nano Lett.* **5**, 515–518 (2005).

[31] O.L. Muskens, N. Del Fatti, F. Vallée, J.R. Huntzinger, P. Billaud, and M. Broyer, "Single metal nanoparticle absorption spectroscopy and optical characterization," *Appl. Phys. Lett.* **88**, 0634109-1–3 (2006).

[32] T. Klar, M. Perner, S. Grosse, G. von Plessen, W. Spirkl, and J. Feldmann, "Surface-plasmon resonances in single metallic nanoparticles," *Phys. Rev. Lett.* **80**, 4249–4252 (1998).

[33] W.R. Glomm, "Functionalized gold nanoparticles for applications in bionanotechnology," *J. Disp. Sci. Technol.* **26**, 389–414 (2005).

[34] J. Zhou, J. Ralston, R. Sedev, and D.A. Beattie, "Functionalized gold nanoparticles: synthesis, structure and colloid stability," *J. Colloid Interface Sci.* **331**, 251–262 (2008).

[35] K. Mitamura and T. Imae "Functionalization of gold nanorods toward their applications," *Plasmonics* **4**, 23–30 (2009).

[36] M. Hu, J. Chen, Z.-Y. Li, L. Au, G.V. Hartland, X. Li, M. Marquez, and Y. Xia, "Gold nanostructures: engineering their plasmonic properties for biomedical applications," *Chem. Soc. Rev.* **35**, 1084–1094 (2006).

[37] C.L. Nehl and J.H. Hafner, "Shape-dependent plasmon resonances of gold nanoparticles," *J. Mater. Chem.* **8**, 2415–2419 (2008).

[38] K.J. Major, C. De, and S.O. Obare, "Recent advances in the synthesis of plasmonic bimetallic nanoparticles," *Plasmonics* **4**, 61–78 (2009).

[39] A.R. Tao, S. Habas, and P. Yang, "Shape control of colloidal metal nanocrystals," *Small* **4**, 310–325 (2008).

[40] V. Biju, T. Itoh, A. Anas, A. Sujith, and M. Ishikawa, "Semiconductor quantum dots and metal nanoparticles: syntheses, optical properties, and biological applications," *Anal. Bioanal. Chem.* **391**, 2469–2495 (2008).

[41] S.-M. Yang, S.-H. Kim, J.-M. Lim, and G.-R. Yi, "Synthesis and assembly of structured colloidal particles," *J. Mater. Chem.* **18**, 2177–2190 (2008).

[42] M. Tréguer-Delapierre, J. Majimel, S. Mornet, E. Duguet, and S. Ravaine, "Synthesis of non-spherical gold nanoparticles," *Gold Bull.* **41**, 1195–1207 (2008).

[43] T.Y. Olson and J.Z. Zhang, "Structural and optical properties and emerging applications of metal nanomaterials," *J. Mater. Sci. Technol.* **24**, 433–446 (2008).

[44] N. Harris, M.J. Ford, P. Mulvaney, and M.B Cortie, "Tunable infrared absorption by metal nanoparticles: the case for gold rods and shells," *Gold Bull.* **41**, 5–14 (2008).

[45] S. Lal, N.K. Grady, J. Kundu, C.S. Levin, J.B. Lassiter, and N.J. Halas, "Tailoring plasmonic substrates for surface enhanced spectroscopies," *Chem. Soc. Rev.* **37**, 898–911 (2008).

[46] J.Z. Zhang and C. Noguez, "Plasmonic optical properties and applications of metal nanostructures," *Plasmonics* **3**, 127–150 (2008).

[47] M.M.-C. Cheng, G. Cuda, Y. Bunimovich, et al., "Nanotechnologies for biomolecular detection and medical diagnostics," *Curr. Opin. Chem. Biol.* **10,** 11–19, (2006).

[48] C.L. Brown, M.W. Whitehouse, E.R.T. Tiekink, and G.R. Bushell, "Colloidal metallic gold is not bio-inert," *Inflammopharmacology* **16**, 133–137 (2008).

[49] M. Roca and A.J. Haes, "Probing cells with noble metal nanoparticle aggregates," *Nanomedicine* **3**, 555–565 (2008).

[50] R. Bhattacharya and P. Mukherjee, "Biological properties of "naked" metal nanoparticles," *Adv. Drug Deliv. Rev.* **60**, 1289–1306 (2008).

[51] R.A. Sperling, P.R. Gil, F. Zhang, M. Zanella, and W.J. Parak, "Biological applications of gold nanoparticles," *Chem. Soc. Rev.* **37**, 1896–1908 (2008).

[52] K.A. White and N.L. Rosi, "Gold-nanoparticle based assays for the detection of biologically relevant molecules," *Nanomedicine* **3**, 543–553 (2008).

[53] S.T. Stern and S.E. McNeil "Nanotechnology safety concerns revisited," *Toxicolog. Sci.* **101**, 4–21 (2008).

[54] I. Uechi and S. Yamada, "Photochemical and analytical applications of gold nanoparticles and nanorods utilizing surface plasmon resonance," *Anal. Bioanal. Chem.* **391**, 2411–2421 (2008).

[55] J.N. Anker, W.P. Hall, O. Lyandres, N.C. Shah, J. Zhao, and R.P. van Duyne, "Biosensing with plasmonic nanosensors," *Nat. Mater.* **7**, 442–453 (2008).

[56] L. Tong, Q. Wei, A. Wei, and J.-X. Cheng, "Gold nanorods as contrast agents for biological imaging: optical properties, surface conjugation and photothermal effects," *Photochem. Photobiol.* **85**, 21–32 (2009).

[57] R. Wilson, "The use of gold nanoparticles in diagnostics and detection," *Chem. Soc. Rev.* **37**, 2028–2045 (2008).

[58] J. Homola, "Surface plasmon resonance sensors for detection of chemical and biological species," *Chem. Rev.* **108**, 462–493 (2008).

[59] M. Faraday, "Experimental relations of gold (and others metals) to light," *Phil. Trans. Royal. Soc. (Lond.)* **147**, 145–181 (1857).

[60] R. Zsigmondy, "Über wässrige Lösungen metallischen Goldes," *Ann. Chem.* **301**, 29–54 (1898).

[61] G. Mie, "Beiträge zur Optik trüber Medien, speziell kolloidaler Metallösungen," *Ann. Phys.* **25**, 377–445 (1908).

[62] T. Svedberg, *The Formation of Colloids,* Churchill, London (1921).

[63] G. Frens, "Controlled nucleation for the regulation of the particle size in monodisperse gold suspensions," *Nature Phys. Sci.* **241**, 20–22 (1973).

[64] N.G. Khlebtsov, V.A. Bogatyrev, L.A. Dykman, and A.G. Melnikov, "Spectral extinction of colloidal gold and its biospecific conjugates," *J. Colloid Interface Sci.* **180**, 436–445 (1996).

[65] W. Baschong, J.M. Lucocq, and J. Roth, "Thiocyanate gold: Small (2–3 nm) colloidal gold for affinity cytochemical labeling in electron microscopy," *Histochemistry* **83**, 409–411 (1985).

[66] D.V. Goia and E. Matijević, "Tailoring the particle size of monodispersed colloidal gold," *Colloids and Surfaces A.* **146**, 139–152 (1999).

[67] S.T. Gentry, S.J. Fredericks, and R. Krchnavek, "Controlled particle growth of silver sols through the use of hydroquinone as a selective reducing agent," *Langmuir* **25**, 2613–2621 (2009).

[68] C.R. Martin, "Nanomaterials: a membrane based synthetic approach," *Science* **266**, 1961–1966 (1994).

[69] Y.-Y. Yu, S.-S. Chang, C.-L. Lee, and C.R.Ch. Wang, "Gold nanorods: electrochemical synthesis and optical properties," *J. Phys. Chem. B* **101**, 6661–6664 (1997).

[70] N. Jana, L. Gearheart, and C. Murphy, "Wet chemical synthesis of high aspect ratio cylindrical gold nanorods," *J. Phys. Chem. B* **105**, 4065–4067 (2001).

[71] B. Nikoobakht and M.A. El-Sayed, "Preparation and growth mechanism of gold nanorods (NRs) using seed-mediated growth method," *Chem. Mater.* **15**, 1957–1962 (2003).

[72] C.J. Murphy, T.K. Sau, A.M. Gole, C.J. Orendorff, J. Gao, L. Gou, S.E. Hunyadi, and T. Li, "Anisotropic metal nanoparticles: synthesis, assembly, and optical applications," *J. Phys. Chem. B* **109**, 13857–13870 (2005).

[73] J. Pérez-Juste, I. Pastoriza-Santos, L.M. Liz-Marzán, and P. Mulvaney, "Gold nanorods: synthesis, characterization and applications," *Coord. Chem. Rev.* **249**, 1870–1879 (2005).

[74] A.V. Alekseeva, V.A. Bogatyrev, B.N. Khlebtsov, A.G. Melnikov, L.A. Dykman, and N.G. Khlebtsov, "Gold nanorods: synthesis and optical properties," *Colloid. J.* **68**, 661–678 (2006).

[75] W. Ni, X. Kou, Z. Yang, and J. Wang, "Tailoring longitudinal surface plasmon wavelengths, scattering and absorption cross sections of gold nanorods," *ACS NANO* **2**, 677–686 (2008).

[76] B. Pietrobon, M. McEachran, and V. Kitaev, "Synthesis of size-controlled faceted pentagonal silver nanorods with tunable plasmonic properties and self-assembly of these nanorods," *ACS NANO* **3**, 21–26 (2009).

[77] M. Liu and P. Guyot-Sionnest, "Synthesis and optical characterization of Au/Ag core/shell nanorods," *J. Phys. Chem. B* **108**, 5882–5888 (2004).

[78] S. Oldenburg, R.D. Averitt, S. Westcott, and N.J. Halas, "Nanoengineering of optical resonances," *Chem. Phys. Lett.* **288**, 243–247 (1998).

[79] B.E. Brinson, J.B. Lassiter, C.S. Levin, R. Bardhan, N. Mirin, and N.J. Halas, "Nanoshells made easy: improving Au layer growth on nanoparticle surfaces," *Langmuir* **24**, 14166–14171 (2008).

[80] N. Phonthammachai, J.C.Y. Kah, G. Jun, C.J.R. Sheppard, M.C. Olivo, S.G. Mhaisalkar, and T.J. White, "Synthesis of contiguous silica-gold core-shell structures: critical parameters and processes," *Langmuir* **24**, 5109–5112 (2008).

[81] J.-H. Kim, W.W. Bryan, and T.R. Lee, "Preparation, characterization, and optical properties of gold, silver, and gold-silver alloy nanoshells having silica cores," *Langmuir* **24**, 11147–11152 (2008).

[82] D. Chen, H.-Y. Liu, J.-S. Liu, X.-L. Ren, X.-W. Meng, W. Wu, and F.-Q. Tang, "A general method for synthesis continuous silver nanoshells on dielectric colloids," *Thin Solid Films* **516**, 6371–6376 (2008).

[83] J. Ye, P. Van Dorpe, W. Van Roy, G. Borghs, and G. Maes, "Fabrication, characterization, and optical properties of gold nanobowl submonolayer structures," *Langmuir* **25**, 1822–1827 (2009).

[84] Y. Sun and Y. Xia, "Alloying and dealloying processes involved in the preparation of metal nanoshells through a galvanic replacement reaction," *Nano Lett.* **3**, 1569–1572 (2003).

[85] P.-Y. Silvert, R. Herrera-Urbina, N. Duvauchelle, V. Vijayakrishnan, and T.K. Elhsissen, "Preparation of colloidal silver dispersions by the polyol process. Part 1 – Synthesis and characterization," *J. Mater. Chem.* **6**, 573–577 (1996).

[86] J. Chen, J.M. McLellan, A. Siekkinen, Y. Xiong, Z.-Y. Li, and Y. Xia, "Facile synthesis of gold-silver nanocages with controllable pores on the surface," *J. Am. Chem. Soc.* **128**, 14776–14777 (2006).

[87] C.L. Nehl, H. Liao, and J.H. Hafner, "Optical properties of star–shaped gold nanoparticles," *Nano Lett.* **6**, 683–688 (2006).

[88] M. Liu and P. Guyot-Sionnest, "Mechanism of silver(I)-assisted growth of gold nanorods and bipyramids," *J. Phys. Chem. B* **109**, 22192–22200 (2005).

[89] C. Li, K.L. Shuford, M. Chen, E.J. Lee, and S.O. Cho, "A facile polyol route to uniform gold octahedra with tailorable size and their optical properties," *ACS NANO*, **2**, 1760–1769 (2008).

[90] E.D. Palik (ed.), *Handbook of Optical Constants of Solids*, Academic Press, New York, Parts I, II, III, 1985, 1991, 1998.

[91] E.A. Coronado and G.C. Schatz, "Surface plasmon broadening for arbitrary shape nanoparticles: a geometrical probability approach," *J. Chem. Phys.* **119**, 3926–3934 (2003).

[92] A. Moroz, "Electron mean free path in a spherical shell geometry," *J. Phys. Chem. C* **112**, 10641–10652 (2008).

[93] C.F. Bohren and D.R. Huffman, *Absorption and Scattering of Light by Small Particles*, John Wiley & Sons, New York (1983).

[94] D.W. Rayleigh, "On the scattering of light by small particles," *Phil. Mag*. **41**, 447–454 (1871).

[95] R. Gans, "Über die Form ultramikroskopischer Goldteilchen," *Ann. Phys.* **37**, 881–900 (1912).

[96] D.W. Rayleigh, "On the incidence of aerial and electric waves upon small obstacles in the from of ellipsoids or elliptic cylinders, and on the passage of electric waves through a circular aperture in a conducting screen," *Phil. Mag.* **44**, 28–52 (1897).

[97] B.T. Draine "The discrete dipole approximation for light scattering by irregular targets," In: *Light Scattering by Nonspherical Particles: Theory, Measurements, and Applications* / Ed. by M.I. Mishchenko, J.W. Hovenier, L.D. Travis. Academic Press, San Diego, pp. 131–145 (2000).

[98] A. Talfove (ed.), *Advances in Computational Electrodynamics: The Finite-Difference-Time-Domain Method,* Artech House, Boston (1998).

[99] C. Hafner, *Post-modern Electromagnetics: Using Iintelligent MaXwell Solvers*, Wiley and Sons, New York (1999).

[100] M.I. Mishchenko, L.D. Travis, and A.A. Lacis, *Scattering, Absorption, and Emission of Light by Small Particles*, University Press, Cambridge (2002).

[101] S. Link and M.A. El-Sayed, "Optical properties and ultrafast dynamics of metallic nanocrystals," *Annu. Rev. Phys. Chem.* **54**, 331–366 (2003).

[102] F.J. García de Abajo, "Nonlocal effects in the plasmons of strongly interacting nanoparticles, dimers, and waveguides," *J. Phys. Chem. C* **112**, 17983–17987 (2008).

[103] G. Weick, *Quantum Dissipation and Decoherence of Collective Excitations in Metallic Nanoparticles*, PhD Thesis, Universite Louis Pasteur, Strasbourg and Universitat Augsburg, 2006.

[104] E. Prodan and P. Nordlander, "Electronic structure and polarizability of metallic nanoshells" *Chem. Phys. Lett.* **352**, 140–146 (2002).

[105] N.D. Mermin, "Lindhard dielectric function in the relaxation-time approximation," *Phys. Rev. B* **1**, 2362–2363 (1970).

[106] P.B. Johnson and R.W. Christy, "Optical constants of noble metals," *Phys. Rev. B* **6**, 4370–4379 (1972).

[107] B.N. Khlebtsov and N.G. Khlebtsov, "Biosensing potential of silica/gold nanoshells: sensitivity of plasmon resonance to the local dielectric environment," *J. Quant. Spectr. Radiat. Transfer* **106**, 154–169 (2007).

[108] C.J. Noguez, "Surface plasmons on metal nanoparticles: the influence of shape and physical environment," *J. Phys. Chem. C* **111**, 3806–3819 (2007).

[109] B.N. Khlebtsov, A.G. Melnikov, V.P. Zharov, and N.G. Khlebtsov, "Absorption and scattering of light by a dimer of metal nanospheres: comparison of dipole and multipole approaches," *Nanotechnology* **17**, 1437–1445 (2006).

[110] B.N. Khlebtsov, V.P. Zharov, A.G. Melnikov, V.V. Tuchin, and N.G. Khlebtsov, "Optical amplification of photothermal therapy with gold nanoparticles and nanoclusters," *Nanotechnology* **17**, 5167–5179 (2006).

[111] N.G. Khlebtsov, "Determination of size and concentration of gold nanoparticles from extinction spectra," *Anal. Chem.* **80**, 6620–6625 (2008).

[112] J. Sancho-Parramon, "Surface plasmon resonance broadening of metallic particles in the quasi-static approximation: a numerical study of size confinement and interparticle interaction effects," *Nanotechnology* **20**, 23570 (2009)..

[113] N.G. Khlebtsov, A.G. Melnikov, V.A. Bogatyrev, L.A. Dykman, A.V. Alekseeva, L.A. Trachuk, and B.N. Khlebtsov, "Can the light scattering depolarization ratio of small particles be greater than 1/3?" *J. Phys. Chem. B* **109**, 13578–13584 (2005).

[114] A. Brioude, X.C. Jiang, and M.P. Pileni, "Optical properties of gold nanorods: DDA simulations supported by experiments," *J. Phys. Chem. B* **109**, 13138–13142 (2005).

[115] B.N. Khlebtsov and N.G. Khlebtsov, "Multipole plasmons in metal nanorods: scaling properties and dependence on the particle size, shape, orientation, and dielectric environment," *J. Phys. Chem. C* **111**, 11516-11527 (2007).

[116] G.W. Bryant, F.J. García de Abajo, and J. Aizpurua, "Mapping the plasmon resonances of metallic nanoantennas," *Nano Lett.* **8**, 631–636 (2008).

[117] B.N. Khlebtsov, V.A. Khanadeev, and N.G. Khlebtsov, "Observation of extra-high depolarized light scattering spectra from gold nanorods," *J. Phys. Chem. C* **112**, 12760–12768 (2008).

[118] N. Calander, I. Gryczynski, and Z. Gryczynski, "Interference of surface plasmon resonances causes enhanced depolarized light scattering from metal nanoparicles," *Chem. Phys. Lett.* **434**, 326–330 (2007).

[119] G. Laurent, N. Felidj, J. Aubard, G. Levi, J.R. Krenn, A. Hohenau, G. Schider, A. Leitner, and F.R. Aussenegg, "Evidence of multipolar excitations in surface enhanced Raman scattering," *J. Chem. Phys.* **122**, 011102-1–4 (2005).

[120] E.K. Payne, K.L. Shuford, S. Park, G.C. Schatz, and C.A. Mirkin, "Multipole plasmon resonances in gold nanorods," *J. Phys. Chem. B* **110**, 2150–2154 (2006).

[121] S. Kim, S.K. Kim, and S. Park, "Bimetallic gold-silver nanorods produce multiple surface plasmon bands," *J. Am. Chem. Soc.* **131**, 8380–8381 (2009).

[122] G. Schider, J.R. Krenn, A. Hohenau, H. Ditlbacher, A. Leitner, F.R. Aussenegg, W.L. Schaich, I. Puscasu, B. Monacelli, and G. Boreman, "Plasmon dispersion relation of Au and Ag nanowires," *Phys. Rev. B* **68**, 155427-1–4 (2003).

[123] N.G. Khlebtsov, A.G. Melnikov, L.A. Dykman, and V.A. Bogatyrev, "Optical properties and biomedical applications of nanostructures based on gold and silver bioconjugates," In: *Photopolarimetry in Remote Sensing*, G. Videen, Ya.S. Yatskiv, and M.I. Mishchenko, Eds., NATO Science Series, II. Mathematics, Physics, and Chemistry, Vol. 161, Kluwer Academic Publishers, Dordrecht, pp. 265–308 (2004).

[124] V.A. Markel, "Divergence of dipole sums and the nature of non-Lorentzian exponentially narrow resonances in one-dimensional periodic arrays of nanospheres," *J. Phys. B: At. Mol. Opt. Phys.* **38**, L115–L121 (2005).

[125] B. Lamprecht, G. Schider, R.T. Lechner, H. Ditlbacher, J.R. Krenn, A. Leitner, and F.R. Aussenegg, "Metal nanoparticle gratings: influence of dipolar particle interaction on the plasmon resonance," *Phys. Rev. Lett.* **84**, 4721–4724 (2000).

[126] S. Malynych and G. Chumanov, "Light-induced coherent interactions between silver nanoparticles in two-dimensional arrays," *J. Am. Chem. Soc.* **125**, 2896–2898 (2003).

[127] B.N. Khlebtsov, V.A. Khanadeyev, J. Ye, D.W. Mackowski, G. Borghs, and N.G. Khlebtsov, "Coupled plasmon resonances in monolayers of metal nanoparticles and nanoshells," *Phys. Rev. B* **77**, 035440-1–14 (2008).

[128] T. Härtling, Y. Alaverdyan, A. Hille, M.T. Wenzel, M. K'all, and L.M. Eng, "Optically controlled interparticle distance tuning and welding of single gold nanoparticle pairs by photochemical metal deposition," *Opt. Express* **16**, 12362–12371 (2008).

[129] M.W. Chen, Y-F. Chau, and D.P. Tsai, "Three-dimensional analysis of scattering field interactions and surface plasmon resonance in coupled silver nanospheres," *Plasmonics* **3**, 157–164 (2008).

[130] T.J. Davis, K.C. Vernon, and D.E. Gmez, "Designing plasmonic systems using optical coupling between nanoparticles," *Phys. Rev. B* **79**, 155423-1–10 (2009).

[131] E. Prodan, C. Radloff, N.J. Halas, and P. Nordlander, "A hybridization model for the plasmon response of complex nanostructures," *Science* **302**, 419–422 (2003).

[132] A.M. Funston, C. Novo, T.J. Davis, and P. Mulvaney, "Plasmon coupling of gold nanorods at short distances and in different geometries," *Nano Lett.* **9**, 1651–1658 (2009).

[133] P.K. Jain and M.A. El-Sayed, "Surface plasmon coupling and its universal size scaling in metal nanostructures of complex geometry: elongated particle pairs and nanosphere trimers," *J. Phys. Chem. C* **112**, 4954–4960 (2008).

[134] G.A. Wurtz, W. Dickson, D. O'Connor, R. Atkinson, W. Hendren, P. Evans, R. Pollard, and A.V. Zayats, "Guided plasmonic modes in nanorod assemblies: strong electromagnetic coupling regime," *Opt. Express* **16**, 7460–7470 (2008).

[135] G. Rong, H. Wang, L.R. Skewis, and B.M. Reinhard, "Resolving sub-diffraction limit encounters in nanoparticle tracking using live cell plasmon coupling microscopy," *Nano Lett.* **8**, 3386–3393 (2008).

[136] K.-H. Su, Q.-H. Wei, X. Zhang, J.J. Mock, D.R. Smith, and S. Schultz, "Interparticle coupling effects on plasmon resonances of nanogold particles," *Nano Lett.* **3**, 1087–1090 (2003).

[137] B. Lamprecht, A. Leitner and F.R. Aussenegg, "SHG studies of plasmon dephasing in nanoparticles," *Appl. Phys. B* **68**, 419–423 (1999).

[138] D. Mackowski, "Electrostatics analysis of radiative absorption by sphere clusters in the Rayleigh limit: application to soot particles," *Appl. Opt.* **34**, 3535–3545 (1995).

[139] C.A. Mirkin, R.L. Letsinger, R.C. Mucic, and J.J. Storhoff, "A DNA-based method for rationally assembling nanoparticles into macroscopic materials," *Nature* **382**, 607–609 (1996).

[140] W. Faulk and G. Taylor, "An immunocolloid method for the electron microscope," *Immunochemistry* **8**, 1081–1083 (1971).

[141] Ž. Krpetić, P. Nativo, F. Porta, and M. Brust, "A multidentate peptide for stabilization and facile bioconjugation of gold nanoparticles," *Bioconjugate Chem.* **20**, 619–624 (2009).

[142] R. Ojeda, J.L. de Paz, A.G. Barrientos, M. Martín-Lomas, and S. Penadés "Preparation of multifunctional glyconanoparticles as a platform for potential carbohydrate-based anticancer vaccines," *Carbohydr. Res.* **342**, 448–459 (2007).

[143] G.T. Hermanson, *Bioconjugate Techniques*, Academic Press, San Diego (1996).

[144] C.R Lowe, "Nanobiotechnology: the fabrication and applications of chemical and biological nanostructures," *Curr. Opin. Struct. Biol.* **10**, 428–434 (2000).

[145] X.-L. Kong, F.-Y. Qiao, H. Qi, and F.-R. Li, "One-step preparation of antibody and oligonucleotide duallabeled gold nanoparticle bio-probes and their properties," *Biotechnol Lett.* **30**, 2071–2077 (2008).

[146] L. Duchesne, D. Gentili, M. Comes-Franchini, and D. G. Fernig, "Robust ligand shells for biological applications of gold nanoparticles," *Langmuir* **24**, 13572–13580 (2008).

[147] H. Liao and J. H. Hafner, "Gold nanorod bioconjugates," *Chem. Mater.* **17**, 4636–4641 (2005).

[148] A.P. Leonov, J. Zheng, J.D. Clogston, S.T. Stern, A.K. Patri, and A. Wei, "Detoxification of gold nanorods by treatment with polystyrenesulfonate," *ACS NANO* **2**, 2481–2488 (2008).

[149] B. Thierry, J. Ng, T. Krieg, and H.J. Griesser, "A robust procedure for the functionalization of gold nanorods and noble metal nanoparticles," *Chem. Commun.* 1724–1726 (2009).

[150] C. Wang and J. Irudayaraj, "Gold nanorod probes for detection of multiple pathogens," *Small*, **4**, 2204–2208 (2008).

[151] S.A. Claridge, A.J. Mastroianni, Y.B. Au, H.W. Liang, C.M. Micheel, J.M.J. Fréchet, and A.P. Alivisatos, "Enzymatic ligation creates discrete multinanoparticle building blocks for self-assembly," *J. Am. Chem. Soc.* **130**, 9598–9605 (2008).

[152] A.M. Schrand, L.K. Braydich-Stolle, J.J. Schlager, L. Dai, and S.M. Hussain, "Can silver nanoparticles be useful as potential biological labels?" *Nanotechnology* **19**, 235104-1–13 (2008).

[153] T. Huang, P.D. Nallathamby, D. Gillet, and X.-H.N. Xu, "Design and synthesis of single-nanoparticle optical biosensors for imaging and characterization of single receptor molecules on single living cells," *Anal. Chem.* **79**, 7708-7718 (2007).

[154] C.M. Stoschek, "Protein assay sensitive at nanogram levels," *Anal. Biochem.* **160**, 301–305 (1987).

[155] L.A. Dykman, V.A. Bogatyrev, B.N. Khlebtsov, and N.G. Khlebtsov, "A protein assay based on colloidal gold conjugates with trypsin," *Anal. Biochem.* **341**, 16–21 (2005).

[156] J.H.W. Leuvering, P.J.H.M. Thal, M. van der Waart, and A.H.W.M. Schuurs, "Sol particle immunoassay (SPIA)," *J. Immunoassay* **1**, 77–91 (1980).

[157] J.H.W. Leuvering, P.J.H.M. Thal, and A.H.W.M. Schuurs, "Optimization of a sandwich sol particle immunoassay for human chorionic gonadotrophin," *J. Immunol. Meth.* **62**, 175–184 (1983).

[158] N.T.K. Thanh and Z. Rosenzweig, "Development of an aggregation-based immunoassay for anti-protein A using gold nanoparticles," *Anal. Chem.* **74**, 1624–1628 (2002).

[159] P. Englebienne, A. van Hoonacker, M. Verhas, and N.G. Khlebtsov, "Advances in high-throughput screening: biomolecular interaction monitoring in real-time with colloidal metal nanoparticles," *Comb. Chem. High Throughput Screen.* **6**, 777–787 (2003).

[160] A. Ogawa and M. Maeda, "Simple and rapid colorimetric detection of cofactors of aptazymes using noncrosslinking gold nanoparticle aggregation," *Bioorganic & Medicinal Chem. Lett.* **18**, 6517–6520 (2008).

[161] H. Li, E. Nelson, A. Pentland, J. Van Buskirk, and L. Rothberg, "Assays based on differential adsorption of single-stranded and double-stranded DNA on unfunctionalized gold nanoparticles in a colloidal suspension," *Plasmonics* **2**, 165–171 (2007).

[162] J.-M. Nam, C.S. Thaxton, and C.A. Mirkin, "Nanoparticle-based bio-bar codes for the ultrasensitive detection of proteins," *Science* **301**, 1884–1886 (2003).

[163] N.G. Khlebtsov, V.A. Bogatyrev, A.G. Melnikov, L.A. Dykman, B.N. Khlebtsov, and Ya.M. Krasnov, "Differential light-scattering spectroscopy: a new approach to studying of colloidal gold nanosensors," *J. Quant. Spectrosc. Radiat. Transfer* **89**, 133–142 (2004).

[164] D. Roll, J. Malicka, I. Gryczynski, Z. Gryczynski, and J. R. Lakowicz, "Metallic colloid wavelength-ratiometric scattering sensors," *Anal. Chem.* **75**, 3440–3445 (2003).

[165] R.F. Pasternack and P.J. Collings, "Resonance light scattering: a new technique for studying chromophore aggregation," *Science* **269**, 935–939 (1995).

[166] X. Liu, H. Yuan, D. Pang, and R. Cai, "Resonance light scattering spectroscopy study of interaction between gold colloid and thiamazole and its analytical application," *Spectrochim. Acta. Part A* **60**, 385–389 (2004).

[167] S.P. Liu, Y.Q. He, Z.F. Liu, L. Kong, and Q.M. Lu, "Resonance Rayleigh scattering spectral method for the determination of raloxifene using gold nanoparticle as a probe," *Anal. Chim. Acta* **598**, 304–311 (2007).

[168] J.E. Butler (ed.), *Immunochemistry of Solid-Phase Immunoassay,* CRC Press, Boca Raton (1991).

[169] N.C. Santos and M.A.R.B. Castanho, "An overview of the biophysical applications of atomic force microscopy," *Biophys. Chem.* **107**, 133–149 (2004).

[170] G.A. Blab, L. Cognet, S. Berciaud, I. Alexandre, D. Husar, J. Remacle, and B. Lounis, "Optical readout of gold nanoparticle-based DNA microarrays without silver enhancement," *Biophys. J.* **90**, L13–L15 (2006).

[171] G. Raschke, S. Kowarik, T. Franzl, C. S´onnichsen, T.A. Klar, J. Feldmann, A. Nichtl, and K. K´urzinger, "Biomolecular recognition based on single gold nanoparticle light scattering," *Nano Lett.* **3**, 935–938 (2003).

[172] T. Furuya, K. Ikemoto, S. Kawauchi, A. Oga, S. Tsunoda, T. Hirano, and K. Sasaki, "A novel technology allowing immunohistochemical staining of a tissue section with 50 different antibodies in a single experiment," *J. Histochem. Cytochem.* **52**, 205–210 (2004).

[173] L. Duan, Y. Wang, S.S. Li, Z. Wan, and J. Zhai, "Rapid and simultaneous detection of human hepatitis B virus and hepatitis C virus antibodies based on a protein chip assay using nanogold immunological amplification and silver staining method," *BMC Infectious Diseases* **5**, http://www.biomedcentral.com/1471-2334/5/53 (2005).

[174] E. Katz and I. Willner, "Integrated nanoparticle–biomolecule hybrid systems: synthesis, properties, and applications," *Angew. Chem. Int. Ed.* **43**, 6042–6108 (2004).

[175] D. Brada and J. Roth "'Golden blot'-detection of polyclonal and monoclonal antibodies bound to antigens on nitrocellulose by protein A – gold complexes," *Anal. Biochem.* **142**, 79–83 (1984); M. Moeremans, G. Daneles, A. van Dijck, et al., "Sensitive visualization of antigen-antibody reactions in dot and blot immune overlay assays with immunogold and immunogold/silver staining," *J. Immunol. Meth.* **74**, 353–360 (1984); B. Surek and E. Latzko, "Visualization of antigenic proteins blotted onto nitrocellulose using the immuno-gold-staining (IGS)-method," *Biochem. Biophys. Res. Commun.* **121**, 284–289 (1984); Y.-H. Hsu, "Immunogold for detection of antigen on nitrocellulose paper," *Anal. Biochem.* **142**, 221–225 (1984).

[176] R.Q. Liang, C.Y. Tan, and R.C. Ruan, "Colorimetric detection of protein microarrays based on nanogold probe coupled with silver enhancement," *J. Immunol. Meth.* **285**, 157–163 (2004).

[177] S.-Y. Hou, H.-K. Chen, H.-C. Cheng, and C.-Y. Huang, "Development of zeptomole and attomolar detection sensitivity of biotin-peptide using a dot-blot goldnanoparticle immunoassay," *Anal. Chem.* **79**, 980–985 (2007).

[178] C. Cao and S.J. Sim, "Signal enhancement of surface plasmon resonance immunoassay using enzyme precipitation-functionalized gold nanoparticles: a femto molar level measurement of anti-glutamic acid decarboxylase antibody," *Biosens. Bioelectron.* **22**, 1874–1880 (2007); S.-M. Han, J.-H. Cho, I.-H. Cho, et al., "Plastic enzyme-linked immunosorbent assays (ELISA)-on-a-chip biosensor for botulinum neurotoxin A," *Anal. Chim. Acta* **587**, 1–8 (2007).

[179] H.C. Koo, Y.H. Park, J. Ahn, W.R. Waters, M.V. Palmer, M.J. Hamilton, G. Barrington, A.A. Mosaad, K.T. Park, W.K.H. Jung, Y. In, S.-N. Cho, S.J. Shin, and W.C. Davis, "Use of rMPB70 protein and ESAT-6 peptide as antigens for comparison of the enzyme-linked immunosorbent, immunochromatographic, and latex bead agglutination assays for serodiagnosis of bovine tuberculosis," *J. Clin. Microbiol.* **43**, 4498–4506 (2005).

[180] Z. Peng, Z. Chen, J. Jiang, X. Zhang, G. Shen, and R. Yu, "A novel immunoassay based on the dissociation of immunocomplex and fluorescence quenching by gold nanoparticles," *Anal. Chim. Acta* **583**, 40–44 (2007).

[181] S.-B. Zhang, Z.-S. Wu, M.-M. Guo, G.-L. Shen, and R.-Q. Yu, "A novel immunoassay strategy based on combination of chitosan and a gold nanoparticle label," *Talanta* **71**, 1530–1535 (2007).

[182] H. Zhang and M.E. Meyerhoff, "Gold-coated magnetic particles for solid-phase immunoassays: enhancing immobilized antibody binding efficiency and analytical performance," *Anal. Chem.* **78**, 609–616 (2006).

[183] B.N. Khlebtsov, L.A. Dykman, V.A. Bogatyrev, V.P. Zharov, and N.G. Khlebtsov, "A solid-phase dot assay using silica/gold nanoshells," *Nanosc. Res. Lett.* **2**, 6–11 (2007).

[184] B.N. Khlebtsov and N.G. Khlebtsov, "Enhanced solid-phase immunoassay using gold nanoshells: effect of nanoparticle optical properties," *Nanotechnology* **19**, 435703 (2008).

[185] M.D. Malinsky, K.L. Kelly, G.C. Schatz, and R.P. Van Duyne, "Chain length dependence and sensing capabilities of the localized surface plasmon resonance of silver nanoparticles chemically modified with alkanethiol self-assembled monolayers," *J. Am. Chem. Soc.* **123**, 1471–1482 (2001).

[186] N.G. Khlebtsov, V.A. Bogatyrev, L.A. Dykman, B.N. Khlebtsov, and P. Englebienne, "A multilayer model for gold nanoparticle bioconjugates: application to study of gelatin and human IgG adsorption using extinction and light scattering spectra and the dynamic light scattering method," *Colloid J.* **65**, 622–635 (2003).

[187] N.G. Khlebtsov, "Optical models for conjugates of gold and silver nanoparticles with biomacromolecules," *J. Quant. Spectrosc. Radiat. Transfer* **89**, 143–152 (2004).

[188] N.G. Khlebtsov, L.A. Trachuk, and A.G. Melnikov, "The effect of the size, shape, and structure of metal nanoparticles on the dependence of their optical properties on the refractive index of a disperse medium," *Opt. Spectrosc.* **98**, 83–90 (2005).

[189] M.M. Miller and A.A. Lazarides, "Sensitivity of metal nanoparticle surface plasmon resonance to the dielectric environment," *Phys. Chem. B* **109**, 21556–21565 (2005).

[190] H. Chen, X. Kou, Z. Yang, W. Ni, and J. Wang, "Shape- and size-dependent refractive index sensitivity of gold nanoparticles," *Langmuir* **24**, 5233–5237 (2008).

[191] G.J. Nusz, A.C. Curry, S.M. Marinakos, A. Wax, and A. Chilkoti, "Rational selection of gold nanorod geometry for label-free plasmonic biosensors," *ACS NANO* **3**, 795–806 (2009).

[192] C.L. Baciu, J. Becker, A. Janshoff, and C. S′onnichsen, "Protein–membrane interaction probed by single plasmonic nanoparticles," *Nano Lett.* **8**, 1724–1728 (2008).

[193] Z. Wang and L. Ma, "Gold nanoparticle probes," *Coord. Chem. Rev.* **253**, 1607–1618 (2009).

[194] D.A. Schultz, "Plasmon resonant particles for biological detection," *Curr. Opin. Biotechnol.* **14**, 13–22 (2003).

[195] I.H. El-Sayed, X. Huang, and M.A. El-Sayed, "Surface plasmon resonance scattering and absorption of anti-EGFR antibody conjugated gold nanoparticles in cancer diagnostics: applications in oral cancer," *Nano Lett.* **5**, 829–834 (2005).

[196] X. Huang, I.H. El-Sayed, W. Qian, and M.A. El-Sayed, "Cancer cell imaging and photothermal therapy in the near-infrared region by using gold nanorods," *J. Am. Chem. Soc.* **128**, 2115–2120 (2006).

[197] C. Loo, L. Hirsch, M. Lee, E. Chang, J. West, N. Halas, and R. Drezek, "Gold nanoshell bioconjugates for molecular imaging in living cells," *Opt Lett.* **30**, 1012–1014 (2005).

[198] R. Hu, K.-T. Yong, I. Roy, H. Ding, S. He, and P.N. Prasad, "Metallic nanostructures as localized plasmon resonance enhanced scattering probes for multiplex dark-field targeted imaging of cancer cells," *J. Phys. Chem. C* **113**, 2676–2684 (2009).

[199] J. Yang, K. Eom, E.-K. Lim, J. Park, Y. Kang, D.S. Yoon, S. Na, E.K. Koh, J.-S. Suh, Y.-M. Huh, T.Y. Kwon, and S. Haam, "*In situ* detection of live cancer cells by using bioprobes based on au nanoparticles," *Langmuir* **24**, 12112–12115 (2008).

[200] W. Eck, G. Craig, A. Sigdel, G. Ritter, L.J. Old, L. Tang, M.F. Brennan, P.J. Allen, and M.D. Mason, "PEGylated gold nanoparticles conjugated to monoclonal f19 antibodies as targeted labeling agents for human pancreatic carcinoma tissue," *ACS NANO* **2**, 2263–2272 (2008).

[201] A.V. Liopo, J.A. Copland, A.A. Oraevsky, and M. Motamedi, "Engineering of heterofunctional gold nanorods for the *in vivo* molecular targeting of breast cancer cells," *Nano Lett.* **9**, 287–291 (2009).

[202] B. Wang, E. Yantsen, T. Larson, A.B. Karpiouk, J.L. Su, K. Sokolov, and S.Y. Emelianov, "Plasmonic intravascular photoacoustic imaging for detection of macrophages in atherosclerotic plaques," *Nano Lett.* **9**, 2212–2217 (2009).

[203] N.J. Durr, T. Larson, D.K. Smith, B.A. Korgel, K. Sokolov, and A. Ben-Yakar, "Two-photon luminescence imaging of cancer cells using molecularly targeted gold nanorods," *Nano Lett.* **7**, 941–945 (2007).

[204] K. Sokolov, M. Follen, J. Aaron, I. Pavlova, A. Malpica, R. Lotan, and R. Richards-Kortum, "Real-time vital optical imaging of precancer using anti-epidermal growth factor receptor antibodies conjugated to gold nanoparticles," *Cancer Res.* **63**, 1999–2004 (2003).

[205] N.P.W. Pieczonka and R.F. Aroca, "Single molecule analysis by surfaced-enhanced Raman scattering," *Chem. Soc. Rev.* **37**, 946–954 (2008).

[206] J. Kneipp, H. Kneipp, and K. Kneipp, "SERS – a single-molecule and nanoscale tool for bioanalytics," *Chem. Soc. Rev.* **37**, 1052–1060 (2008).

[207] X.-M. Qian and S.M. Nie, "Single-molecule and single-nanoparticle SERS: from fundamental mechanisms to biomedical applications," *Chem. Soc. Rev.* **37**, 912–920 (2008).

[208] A.A. Kamnev, L.A. Dykman, P.A. Tarantilis, and M.G. Polissiou, "Spectroimmunochemistry using colloidal gold bioconjugates," *Biosci. Reports* **22**, 541–547 (2002).

[209] B.D. Chithrani, A.A. Ghazani, and W.C.W. Chan, "Determining the size and shape dependence of gold nanoparticle uptake into mammalian cells," *Nano. Lett.* **6**, 662–668 (2006).

[210] W. Jiang, B.Y.S. Kim, J.T. Rutka, and W.C.W. Chan, "Nanoparticle-mediated cellular response is size-dependent," *Nature Nanotechnol.* **3**, 145–150 (2008).

[211] G. Zhang, Z. Yang, W. Lu, R. Zhang, Q. Huang, M. Tian, L. Li, D. Liang, and C. Li, "Influence of anchoring ligands and particle size on the colloidal stability and *in vivo* biodistribution of polyethylene glycol-coated gold nanoparticles in tumor-xenografted mice," *Biomaterials* **30**, 1928–1936 (2009).

[212] H.J. Parab, H.M. Chen, T.-C. Lai, J.H. Huang, P.H. Chen, R.-S. Liu, M. Hiao, C.-H. Chen, D.-P. Tsai, and Y.-K. Hwu, "Biosensing, cytotoxicity, and cellular uptake studies of surface-modified gold nanorods," *J. Phys. Chem. C* **113**, 7574–7578 (2009).

[213] P. Nativo, I.A. Prior, and M. Brust, "Uptake and intracellular fate of surface-modified gold nanoparticles," *ACS NANO* **2**, 1639–1644 (2008).

[214] T.S. Hauck, A.A. Ghazani, and W.C.W. Chan, "Assessing the effect of surface chemistry on gold nanorod uptake, toxicity, and gene expression in mammalian cells," *Small* **4**, 153–159 (2008).

[215] D. Mandal, A. Maran, M.J. Yaszemski, M.E. Bolander, and G. Sarkar, "Cellular uptake of gold nanoparticles directly cross-linked with carrier peptides by osteosarcoma cells," *J. Mater. Sci.: Mater. Med.* **20**, 347–350 (2009).

[216] E.M.S. Azzam and S.M.I. Morsy, "Enhancement of the antitumour activity for the synthesised dodecylcysteine surfactant using gold nanoparticles," *J. Surfact. Deterg.* **11**, 195–199 (2008).

[217] P.H. Yang, X.S. Sun, J.F. Chiu, H.Z. Sun, and Q.Y. He, "Transferrin-mediated gold nanoparticle cellular uptake," *Bioconjugate Chem.* **16**, 494–496 (2005).

[218] J.J. Donnelly, B. Wahren, and M.A. Liu, "DNA vaccines: progress and challenges," *J. Immunol.* **175**, 633–639 (2005).

[219] L.A. Dykman, M.V. Sumaroka, S.A. Staroverov, I.S. Zaitseva, and V.A. Bogatyrev, "Immunogenic properties of colloidal gold," *Biol. Bull.* **31**, 75–79 (2004).

[220] E. Sadauskas, H. Wallin, M. Stoltenberg, U. Vogel, P. Doering, A. Larsen, and G. Danscher, "Kupffer cells are central in the removal of nanoparticles from the organism," *Part. Fibre Toxicol.* **4**, http://www.particleandfibretoxicology.com/content/4/1/10 (2007).

[221] R. Shukla, V. Bansal, M. Chaudhari, A. Basu, R.R. Bhonde, and M. Sastry, "Biocompartibility of gold nanoparticles and their endocytotic fate inside the cellular compartment: a microscopic overview," *Langmuir* **21**, 10644–10654 (2005).

[222] H. Vallhov, J. Qin, S.M. Johansson, N. Ahlborg, M.A. Muhammed, A. Scheynius, and S. Gabrielsson, "The importance of an endotoxin-free environment during the production of nanoparticles used in medical applications," *Nano Lett.* **6**, 1682–1686 (2006).

[223] J.F. Hillyer and R.M. Albrecht, "Gastrointestinal persorption and tissue distribution of differently sized colloidal gold nanoparticles," *J. Pharm. Sci.* **90**, 1927–1936 (2001).

[224] T. Niidome, M. Yamagata, Y. Okamoto, Y. Akiyama, H. Takahishi, T. Kawano, Y. Katayama, and Y. Niidome, "PEG-modified gold nanorods with a stealth character for *in vivo* application," *J. Control. Release* **114**, 343–347 (2006).

[225] W.H. De Jong, W.I. Hagens, P. Krystek, M.C. Burger, A.J.A.M. Sips, and R.E. Geertsma, "Particle size-dependent organ distribution of gold nanoparticles after intravenous administration," *Biomaterials* **29**, 1912–1919 (2008).

[226] M. Semmler-Behnke, W.G. Kreyling, J. Lipka, S. Fertsch, A. Wenk, S. Takenaka, G. Schmid, and W. Brandau, "Biodistribution of 1.4- and 18-nm gold particles in rats," *Small* **4**, 2108–2111 (2008).

[227] V. Kattumuri, K. Katti, S. Bhaskaran, E.J. Boote, S.W. Casteel, G.M. Fent, D.J. Robertson, M. Chandrasekhar, R. Kannan, and K.V. Katti,"Gum arabic as a phytochemical construct for the stabilization of gold nanoparticles: in vivo pharmacokinetics and X-ray-contrast-imaging studies," *Small* **3**, 333–341 (2007).

[228] B. Kogan, N. Andronova, N. Khlebtsov, B. Khlebtsov, V. Rudoy, O. Dement'eva , E. Sedykh, and L. Bannykh, "Pharmacokinetic study of PEGylated plasmon resonant gold nanoparticles in tumor-bearing mice," In: Technical Proceedings of the 2008 NSTI Nanotechnology Conference and Trade Show, NSTI-Nanotech, Nanotechnology 2, 65–68 (2008).

[229] G.S. Terentyuk, G.N. Maslyakova, L.V. Suleymanova, B.N. Khlebtsov, B.Ya. Kogan, G.G. Akchurin, A.V. Shantrocha, I.L. Maksimova, N.G. Khlebtsov, and V.V. Tuchin, "Circulation and distribution of gold nanoparticles and induced alterations of tissue morphology at intravenous particle delivery," *J. Biophotonics* **2**, 292–302 (2009).

[230] S. Wang, W. Lu, O. Tovmachenko, U.S. Rai, H. Yu, and P.C. Ray, "Challenge in understanding size and shape dependent toxicity of gold nanomaterials in human skin keratinocytes," *Chem. Phys. Lett.* **463**, 145–149 (2008).

[231] J.H. Fan, W.I. Hung, W.T. Li, and J.M. Yeh, "Biocompatibility study of gold nanoparticles to human cells," *ICBME 2008 Proc.* **23**, 870–873 (2009).

[232] C.J. Murphy, A.M. Gole, J.W. Stone, P.N. Sisco, A.M. Alkilany, E.C. Goldsmith, and S.C. Baxter, "Gold nanoparticles in biology: beyond toxicity to cellular imaging," *Acc. Chem. Res.* **41**, 1721–1730 (2008).

[233] N.G. Bastús, E. Casals, S. Vázquez-Campos, and V. Puntes, "Reactivity of engineered inorganic nanoparticles and carbon nanostructures in biological media," *Nanotoxicology* **2**, 99–112 (2008).

[234] N. Lewinski, V. Colvin, and R. Drezek, "Cytotoxicity of nanoparticles," *Small*, **4**, 26–49 (2008).

[235] P. Ghosh, G. Han, M. De, C.K. Kim, and V.M. Rotello, "Gold nanoparticles in delivery applications," *Adv. Drug Deliv. Rev.* **60**, 1307–1315 (2008).

[236] G.F. Paciotti, D.G.I. Kingston, and L. Tamarkin, "Colloidal gold nanoparticles: a novel nanoparticle platform for developing multifunctional tumor-targeted drug delivery vectors," *Drug Dev. Res.* **67**, 47–54 (2006).

[237] C.R. Patra, R. Bhattacharya, E. Wang, A. Katarya, J.S. Lau, S. Dutta, M. Muders, S. Wang, S.A. Buhrow, S.L. Safgren, M.J. Yaszemski, J.M. Reid, M.M. Ames, P. Mukherjee, and D. Mukhopadhyay, "Targeted delivery of gemcitabine to pancreatic adenocarcinoma using cetuximab as a targeting agent," *Cancer Res.* **68**, 1970–1978 (2008).

[238] Y. Cheng, A.C. Samia, J.D. Meyers, I. Panagopoulos, B. Fei, and C. Burda, "Highly efficient drug delivery with gold nanoparticle vectors for *in vivo* photodynamic therapy of cancer," *J. Am. Chem. Soc.* **130**, 10643–10647 (2008).

[239] L.R. Hirsch, R.J. Stafford, J.A. Bankson, S.R. Sershen, B. Rivera, R.E. Price, J.D. Hazle, N.J. Halas, and J.L. West, "Nanoshell-mediated near-infrared thermal therapy of tumors under magnetic resonance guidance," *Proc. Natl. Acad. Sci. USA* **100**, 13549–13554 (2003).

[240] C.M. Pitsillides, E.K. Joe, X. Wei, R.R. Anderson, and C.P. Lin, "Selective cell targeting with light-absorbing microparticles and nanoparticles," *Biophys. J.* **84**, 4023–4032 (2003).

[241] H.-P. Berlien and G.J. Mueller, *Applied Laser Medicine*, Springer-Verlag, Berlin (2003).

[242] B. Palpant, Y. Guillet, M. Rashidi-Huyeh, and D. Prot, "Gold nanoparticles assemblies: thermal behaviour under optical excitation," *Gold Bull.* **41**, 105–115 (2008).

[243] G.S. Terentyuk, G.N. Maslyakova, L.V. Suleymanova, N.G. Khlebtsov, B.N. Khlebtsov, G.G. Akchurin, I.L. Maksimova, and V.V. Tuchin, "Laser induced tissue hyperthermia mediated by gold nanoparticles: towards cancer phototherapy," *J. Biomed. Opt.* **14**, 021016-1–9 (2009).

3

Transfection by Optical Injection

David J. Stevenson
SUPA, School of Physics & Astronomy, School of Biology, University of St. Andrews, St. Andrews, Scotland

Frank J. Gunn-Moore
School of Biology, University of St. Andrews, St. Andrews, Scotland

Paul Campbell
Department of Electronic Engineering and Physics, University of Dundee, Dundee, Scotland

Kishan Dholakia
SUPA, School of Physics & Astronomy, University of St. Andrews, St. Andrews, Scotland

This chapter describes the use of light to transiently increase the permeability of mammalian cells to allow membrane impermeable substances to cross the plasma membrane. We refer to this process as optical injection or more specifically optical transfection, when genetic material is introduced into and expressed by a cell. A brief discussion of nonphotonic transfection and injection technologies is presented. A review of the field of optical transfection and the underlying physical mechanisms associated with this process is then described.

Key words: optical transfection, optoporation, optoinjection, photoporation, laserfection, transfection

3.1 Introduction: Why Cell Transfection?

The ability to introduce exogenous DNA into a cell whether that is a prokaryotic cell (transformation) or an eukaryotic cell (transfection) has allowed molecular biologists not only to explore the molecular mechanisms within cells, but also to produce genetically engineered organisms. Földes and Trautner first coined the term transfection in 1964, when they introduced phage SP8 DNA into

B. subtilis bacteria [1]. Since then, there have been a variety of technologies developed to encourage a cell to take up and express foreign nucleic acids. These include optical transfection, lipoplex transfection, polyplex transfection, calcium precipitation, electroporation, the use of the gene gun, hydrodynamic delivery, ultrasound transfection, viral transfection, or even the simple addition of naked DNA. The word "transfection" has historically been used to mean the loading of any membrane impermeable molecule into a cell, but biologists now wish to introduce other molecules into cells such as membrane impermeable drugs, impermeable fluorophores, or even antibodies. Technically the word transfection specifically refers to the internalization and subsequent biological effects of nucleic acids, while the term optical injection is for these other substances. At present the introduction of heterologous nucleic acids into a cell, be it DNA, messenger RNA (mRNA) or small oligonucleotides (as used in iRNA, interference RNA, studies), is by far the most common reason to inject a membrane impermeable molecule.

The field of transfection is wide, and it is not the purpose here to review all of the various technologies but is rather to highlight some of the more commonly used techniques in order to give context to the difficulties associated with transfection. DNA encounters a number of physical and chemical barriers presented by the cell before finally it is able to reach the nuclear machinery required for its expression. If a solution of naked plasmid DNA is mixed with a solution containing mammalian cells, the probability of transfection taking place is very low for almost all cell types. However, exposure of naked plasmid has been demonstrated to transfect some cell types, including the liver [2], skeletal muscle [3, 4], cardiac muscle [5], lung [6], solid tumors, the epidermis, and hair follicles. It is unsurprising that this type of transfection is difficult to achieve as the DNA transcription machinery is located within the nucleus, requiring the plasmid DNA to cross the plasma membrane, a highly viscous cytoplasm, the nuclear membrane, and an even more viscous nucleoplasm. Penetrating these membranes is energetically unfavorable due to the highly hydrophobic nature of the inner lamellae of the lipid bilayers and the hydrophilic nature of the DNA molecule. A typical plasmid DNA molecule is a circular string of double-stranded nucleic acids several kilobase pairs (kbp) long, which depending on the chemical environment can exist in an uncoiled, coiled, or supercoiled state. The volume occupied by plasmid DNA, or its radius of gyration, depends on the number of bases, the extent of its coiling, and whether it has become linearized. For example, the radius of gyration of a 5.2 kbp plasmid ranges between 85 and 120 nm depending on its state of coil [7, 8]. Plasmids of this size may also be compacted into spheres as small as 20–40 nm using polyethylenimine polyplexes [9], the details of which are discussed later.

Crossing the plasma membrane is a necessary but not sufficient requirement for transfection to occur. Once plasmid DNA has crossed the plasma membrane, various physical barriers still exist. Experiments utilizing microinjection within different regions of a cell provide an insight into how these barriers affect naked plasmid DNA. Microinjection was pioneered by Capecchi in 1980, who used an extruded glass needle to inject DNA solution directly into either the nucleus or the cytoplasm of mammalian cells [10]. This simple but powerful technique has changed little since its conception, although some reports have demonstrated micromechanical systems (MEMS) microinjection needles [11]. A small 10–20 bp double-stranded oligomer when microinjected directly into the cytoplasm will rapidly diffuse throughout the cytoplasm and cross the nuclear membrane with ease [12], probably through nuclear pores. The diffusion rate is fast in spite of the viscosity of the cytoplasm. As larger strands are injected, the behavior changes: 500-bp DNA strands will not cross the nuclear membrane, but will diffuse homogenously if directly injected into either the nucleus or the cytoplasm [12]. For comparison, this behavior is also true of 500 kDa dextran spheres [13] with a diameter of 26.6 nm [14]. DNA strands greater than 2,000 bp diffuse very slowly through the highly viscous cytoplasm, at less than 1% of their diffusion rates in water [12]. This size of DNA is still much smaller than that commonly employed in transfection experiments, which tends to range from 5,000 to 10,000 bp. When plasmids of this size range are directly microinjected into the cytoplasm, it has been estimated that as little as 1 in 1,000 are able to make their way to the nucleus for subsequent expression [15]. In contrast only a few (<10) plasmids microinjected into the nucleus are necessary for expression [15]. As well as these physical barriers, there are also chemical barriers such as nucleases present in both the extracellular and intracellular environment that can affect transfection efficiencies by degrading plasmid DNA.

In this chapter, the many advantages of optical injection/transfection will be surveyed. To place this technology within context, a brief review of nonoptical methods of transfection is introduced (section 3.2), followed by a comprehensive review of optical injection and optical transfection since its first discovery in 1984 (3.3). Finally, the physics of transport through a transiently generated photopore (3.4), along with the laser-cell interaction that leads to the generation of a photopore (3.5) are briefly discussed.

3.2 Nonoptical Methods of Transfection

To overcome the various physical and chemical barriers, numerous nano/microparticle mediated transfection techniques have been developed, and a wealth of terminology has been generated to distinguish the various technologies and chemistries from each other.

3.2.1 Lipoplex transfection

Lipoplex transfection was first demonstrated by Felgner et al [16]. Lipoplexes are comprised of cationic lipids containing a polar head and a hydrophobic tail. When mixed in an aqueous solution, these cationic lipids spontaneously aggregate into hollow aqueous nanoshells with a lipid bilayer wall, with the polar heads of one half of the bilayer facing outwards into the surrounding external solution, and the polar heads of the other half of the bilayer facing inwards into the lumen of the lipoplex nanoshell. This behavior is reminiscent of the phospholipid bilayers comprising the plasma membrane of a cell. The polar heads typically consist of amidine, but can also be guanidium or pyridinium. When plasmid DNA, which is negatively charged, is present during this spontaneous lipoplex formation the positively charged groups of the polar heads electrostatically bind to the DNA and sequester it both within the lumen and along the external wall of the lipoplex [17]. The diameters of typical lipoplexes range from 50 nm to over a micron [17, 18]. Exposure of these lipoplexes to cell cultures causes transfection with an efficiency that varies depending on the cell type. Lipoplex mediated transfection is thought to involve a number of steps [17–19]. Briefly, DNA containing lipoplexes bind to cells and become internalized by endocytosis. From there, DNA must escape lysosomal degradation and make its way to the nucleus. The crossing of the nuclear membrane may be in part due to passive entry during division, or by some form of direct nuclear import. The latter is suggested because postmitotic cells are able to be transfected using lipoplexes, so DNA must still be able to get into the nucleus without nuclear membrane degradation.

3.2.2 Polyplex transfection

Polyplex nanoshells are commonly made of cationic polymers such as poly-L-lysine [20] or polyethylenimine [15, 21]. Unlike liposomes, polyplexes are entirely hydrophilic in nature. Delivery of DNA to the nucleus follows a similar route to that of liposomes, with some key differences, including the mechanism of endosomal escape [18], where it has been suggested that there is endosomal swelling and bursting, due to a Cl^- influx brought on by a decrease in pH within the endosome [21]. Polyethylenimine polyplexes also clearly increase the ability of DNA to cross the nuclear membrane [15]. In a variation of polyplex transfection, "magnetofection" uses PEI complexed to superparamagnetic nanoparticles, and high transfection efficiencies have been demonstrated in the presence of high magnetic fields [22].

3.2.3 Gene gun transfection

The gene gun method accelerates micron sized tungsten particles through the cell wall and cell membrane [23]. Although the original incarnations of gene guns used gunpowder, modern versions tend to use high-pressure helium gas [24, 25]. Reasonable injection/transfection efficiencies have also recently been demonstrated using a gene gun without a nano or microparticle carrier: the solution becomes aerosolized and enters a cell culture medium at high velocity [25]. Gene guns come in many architectures, including a vacuum chamber and handheld [26], but the most interesting in terms of direct comparison with optical injection is that of the pneumatic capillary gun. This miniaturized gene gun is capable of delivering highly localized lateral (150 μm) transfection in *Hirudo medicinalis* leech embryos. Notably, the axial resolution is 15 μm, and the penetration depth is variable between 0–50 μm depending on the pressure of the helium.

3.2.4 Ultrasound transfection

Ultrasound is transmitted as a periodic compressive [longitudinal] wave that is characterized by frequencies typically within the range 20 kHz–20 MHz. Interestingly, the biological effects of such waves were recognized almost from the outset of ultrasound technologies during the 1920s, long before imaging applications emerged [27]. Such bioeffects, including uptake, are a function of the ultrasound field parameters of frequency, amplitude and pulse duration, as well as the transmission (and absorption) character of the medium. For sustained application of ultrasound, heating effects typically dominate, and this is the basis of the emerging therapeutic approach of high intensity focused ultrasound. When short pulses of ultrasound are applied to cells (both in suspension and as plated monolayers), it has been found that molecules can be taken up from the locale. Such studies have measured the size dependent uptake from a spectrum of lower molecular weight species [29, 30] as well as intermediate to high molecular weight species, including peptides and protein [28], as well as genetic material such as DNA [29, 30]. A key clue to elucidating the mechanism of uptake arose when insonated cells were inspected by scanning electron microscopy: here, Tachibana and coworkers provided the first evidence that the cell membrane is physically compromised by the process [31]. They illustrated that insonated cells appear to be peppered with pore-like structures and this was later confirmed by high resolution microscopy observations using also confocal [30] and scanning probe microscopy [32]. The process is termed "sonoporation" in recognition of the apparent requirement for membrane disruption to accompany uptake. Notably, while the application of ultrasound could by itself lead to uptake of molecules, when insonation was performed in the presence of microscopic (typically 5 μm in diameter) bubbles (microbubbles), then these effects could be markedly enhanced. Microbubbles thus act to increase the probability of sonoporation [35–37].

3.2.5 Electroporation

Electroporation mediated mammalian cell transfection was first described by Neuman et al [33–35]. The cell membrane can be transiently permeabilized by the application of electrical pulses and because the DNA is charged, the technique has the added advantage that it forces DNA to interact with the cell membrane. In a typical experiment, high concentrations of naked DNA (many 10s of μg/ml) are added to the medium with the cells sitting between two electrodes which are then exposed to an electric field. A typical dose would be 10 pulses of a 0.8 kV/cm field, each 0.1–5.0 ms in duration [36]. Within 3 ms of field application, the sides of the cell facing both the cathode and the anode become permeabilized. Pores in the cell appear volcano-shaped, with the tip of the volcano facing the inside of the cell [37]. These pores range in size from 20–120 nm [37]. Almost immediately, aggregations of negatively charged plasmid DNA can be seen to adhere to the cathode facing side of the cell in "spots" or "islands" on the membrane [36]. The association of the DNA-plasma membrane is strong, and cannot be disrupted by subsequently electroporating the cells with a reversed polarity [36]. Within 10s of seconds, the pores start to disappear [37]. Within 30 min of electroporation, fluorescently labelled DNA can be seen to diffuse into the cytoplasm [36]. It is noteworthy that while electoporation has not been traditionally able to transfect individual cells, single cell electroporation has recently been realized in a microfluidic platform [38].

3.3 Review of Optical Injection and Transfection

Figure 3.1 shows a typical optical injection setup, and Table 3.2 summarizes the optical injection and transfection experiments performed in the literature to date. In a typical optical injection experiment, a solution of membrane impermeable substance such as a fluorophore, macromolecule, or nucleic acid is first exposed to the cells of interest. A laser is then used to transiently permeabilize the cell membrane, during which time the substance of interest either passively diffuses into the cell or is sucked into the cell due to volume exchange resulting from osmotic pressure differences. The mechanism of how the laser permeabilizes the plasma membrane depends on the laser wavelength, whether or not it is pulsed, its pulse duration, the diameter of the focused beam, and how the laser is applied to the cell or cell population.

The use of light to inject or transfect cells enjoys a number of advantages in comparison with the previously described techniques. In comparison with its main alternative for single cell treatment, microinjection, optical injection offers unrivalled simplicity in the ability to target cells in a sterile environment. Microinjection suffers from the requirement of an open solution; optical injection can be performed through a sterile coverslip. Furthermore, the efficiency and viability of optically transfected cells compare favorably with other nonviral technologies. Finally, it is straightforward to integrate an optical injection setup with other microscopic techniques, such as confocal laser scanning microscopy, optical tweezers, and microfluidic systems [39]. Broadly speaking, there are two categories of optical transfection – targeted and untargeted. In targeted transfection, the focal point of the laser is aimed on the plasma membrane, and a tiny hole or pore is generated. Only the targeted cell is transfected; neighboring cells remain completely unaffected. In targeted transfection, an acute reaction to the exposure of the laser is often observed under bright field microscopy. This reaction depends on the dose of exposure, the laser source, and in some cases the presence of chemical absorbers placed in the medium [44, 45]. Below a threshold dose, no reaction to the laser will occur. Above a certain dose, acute morphological changes in the cell occur, including an increase in granularity, "blebbing," and loss of membrane integrity. In an ideal scenario, a therapeutic dose would exist somewhere between the under- and overdose, where the cell membrane permeabilization is transient and the cell recovers within seconds to minutes. In reality, often the therapeutic dose and the overdose slightly overlap, and the postdose viability is $<100\%$ [46, 47]. This overlap is not uncommon in other transfection technologies. A more thorough description of the reaction of the cell to laser irradiation, including a detailed description of the biophysical characteristics of the "photopore" generated, is given in the final section of this document and in the comprehensive reviews of Vogel et al [48, 49].

Typical laser sources for targeted optical injection and transfection include the 800 nm femtosecond (fs) pulsed titanium sapphire laser (Figure 26.2a-d) [13, 50–61], and continuous wave (cw) sources such as 405 nm [40] (Figure 26.2e,f) and 488 nm [41–43]. 1064 nm nanosecond (ns) pulsed Nd:YAG lasers have also been reported in the literature to produce a targeted transfection [44–46].

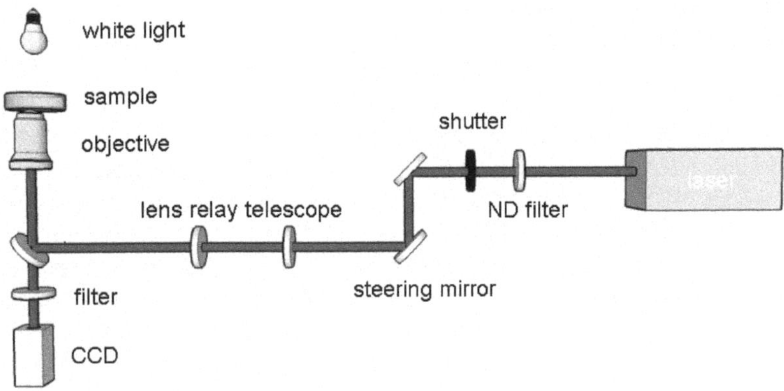

FIGURE 3.1: A typical optical injection/optical transfection setup. A pulsed (femtosecond or nanosecond) or continuous wave laser is passed through a Neutral Density (ND) filter to attenuate the power (typically 10s of mW at focus) and a shutter to provide the correct dose (typically 10s of ms). It is then passed through a steering mirror and a lens relay telescope. The lens relay telescope ensures that the image at the steering mirror is relayed to the back aperture of the microscope objective such that rotation of the steering mirror results in movement of the diffraction limited spot at focus without "clipping" the edge of the back aperture, or beam "walkoff." A filter in front of the CCD camera blocks the beam wavelength but not light in the visible range, allowing the sample but not the laser to be imaged.

However, this laser is more often used for untargeted optical transfection. Figure 26.2a is a fluorescent image showing the optical injection of the membrane impermeable dye, propidium iodide [53]. The image was taken immediately after optical injection. At this stage, the postinjection viability is unknown. 90 minutes later (Figure 26.2b), the same cells were re-stained with propidium iodide. All cells remained viable except one (white arrow), whose membrane was permanently compromised. This demonstrates nicely the dual use of a membrane impermeable dye in optical injection experiments – it can determine both optical injection efficiency and optical injection viability. Figure 26.2c is a fluorescent image showing the optical transfection of a Chinese Hamster Ovary (CHO) cell with plasmid DNA encoding for Green Fluorescent Protein (GFP). The image was taken 48 h after treatment, and cell was co-stained with the blue nuclear dye DAPI. Figure 26.2d is a fluorescent image showing the optical transfection of a primary rat hippocampal neuron with mRNA encoding for *Elk1-GFP* [54]. The image was taken 30 min after treatment. Note typical timings of each event – optical injection occurs immediately, optical transfection of DNA occurs $>$ 24 h after dosing, and optical transfection of mRNA occurs $>$ 30 min after dosing. Finally, Figure 26.2e-f is an example of targeted optical transfection by cw laser. CHO cells were exposed to plasmid DNA encoding for DsRed-Mito and an antibiotic resistance gene and individual cells were dosed with a focused 405 nm laser [40]. After culturing in the antibiotic, only cells that had been optically transfected survived; a stable culture had therefore been established.

In untargeted transfection, groups of 10s to 1000s of cells are membrane permeabilized by direct or indirect laser dosing. This may be achieved directly by simply raster scanning a region of interest with a highly focused laser [4]; directly by dosing a large population with an unfocused nanosecond pulsed beam [47, 48]; or indirectly by creating a laser induced shock wave [49] (Figure 26.3). In shockwave mediated transfection, a laser induced pressure gradient transiently permeabilizes the plasma [50, 51] and/or nuclear membrane [52]. This pressure gradient is formed by hitting a membrane target [50], or the glass-solution interface of the culture dish upon with the cells are sitting [49]. In this respect, it has more in common with the ultrasound genre of transfection.

In the case of targeted optical injection using either cw wavelengths or lasers with a very short pulse-duration, a diverse array of molecules has been loaded into single cells. 800 nm fs pulsed sources have been used to optically inject ethidium bromide [55], sucrose [56], propidium iodide [53, 57–59], lucifer yellow [53, 54, 57], Trypan Blue [39], cascade blue [13], 500 kDa FITC labelled dextran [13], plasmid DNA encoding for GFP [4, 13, 53, 55, 60, 61], luciferase, β-galactosidase, murine erythropoietin [4], and DsRed-Mito [56, 72]. One case in the literature also demonstrated the targeted transfection and translation of a number of mRNA species [54]. 405 nm and 488 nm cw sources have been used to transfect plasmid DNA encoding for GFP [45, 62, 63], beta-galactosidase, and chloramphenicol acetyltransferase [41]. In cases where plasmid DNA was optically injected, both stable and transient transfection have been observed with these targeted laser sources.

Because an optical injection setup is also often capable of optically tweezing, it is also possible to individually manipulate nonadherent cells using a Ti:sapphire laser in cw mode before and after optical injection [39]. Figure 26.4 outlines an experiment demonstrating this. Figure 26.4a-i are screenshots from a movie (available online via reference [39]) where an HL60 cell is optically tweezed into a capillary tube containing a membrane impermeable dye, optically injected, and tweezed back out of the tube once again. Figure 26.4j shows a schematic diagram of the path of the cell. The cell begins its journey by being optically tweezed in the cw regime from the bottom of a microchamber surface (Figure 26.4a), up in z approximately 30 μm (Figure 26.4b), and into the opening of a square Trypan Blue filled capillary tube (Figure 26.4c). It is then positioned next to a blue cell that had previously been fixed, permeabilized, and placed into the capillary tube by tweezing (Figure 26.4d). The cell was then optically injected using the same laser, perturbed to induce a modelocked operation. This resulted in a transient pore (Figure 26.4e, white arrow), and allowed Trypan Blue to stain the cell after six minutes. In the next image (Figure 26.4f), it can be seen that Trypan Blue has entered the cell nucleus. The laser was then unblocked, and in the cw regime the optically injected cell was tweezed out of the capillary tube and placed next to an untreated cell (Figure 26.4g-i).

It is reasonable to assume that almost any membrane impermeable fluorophore, macromolecule, or nucleic acid, below a certain size, is able to be optically injected using a Ti:sapphire 800 nm fs laser source. The utility of targeted optical injection is left to the imagination of the reader. Literally thousands of membrane impermeable fluorophores have been developed or discovered over the years that bind to specific intracellular biochemical targets [62]. These fluorophores are largely useless in investigations involving live cells; the traditional way to load these membrane impermeable substances into cells involves fixation with formaldehyde, followed by permeabilization with Triton-X 100. Another traditional way to overcome this problem is to alter the chemistry of the fluorophore. Smaller organic fluorophores can often be chemically modified to generate an inactive but membrane permeable product, upon which entering the cell undergoes esterase cleavage, activation, and fluorescence [74]. However, this is simply not an option for many larger biological constructs: plasmid DNA, mRNA, siRNA, and antibodies are notable examples. Considering that many commercially available microscopy laboratories, especially those performing multiphoton microscopy, already have an 800 nm titanium sapphire source, we believe that it is only a matter of time before this optical injection is more widely adopted.

In the case of untargeted optical injection with a 1064 ns pulsed Nd:YAG source, a similarly diverse array of molecules have been transfected into small populations of cells. These include Ca^{2+}, Zn^{2+}, Sytox Green, Sytox Blue, tetramethylrhodamine-dextran (MW 3000), Cdc42 binding domain of WASP conjugated to an I-SO dye (55 KDa), quantum dots, siRNA [46], merocyanin 540 [45], Texas Red-glycine, Texas Red-dextran (3kD, 10kD, 40 kD), fluoroscein labelled protein kinase C [49], and plasmid DNA encoding for GFP [45, 48, 63], pSV2-neo [64, 76], pAB6 [64] and Eco-gpt [65]. Where plasmid DNA was used, both stable and transient transfection have been observed with this laser source.

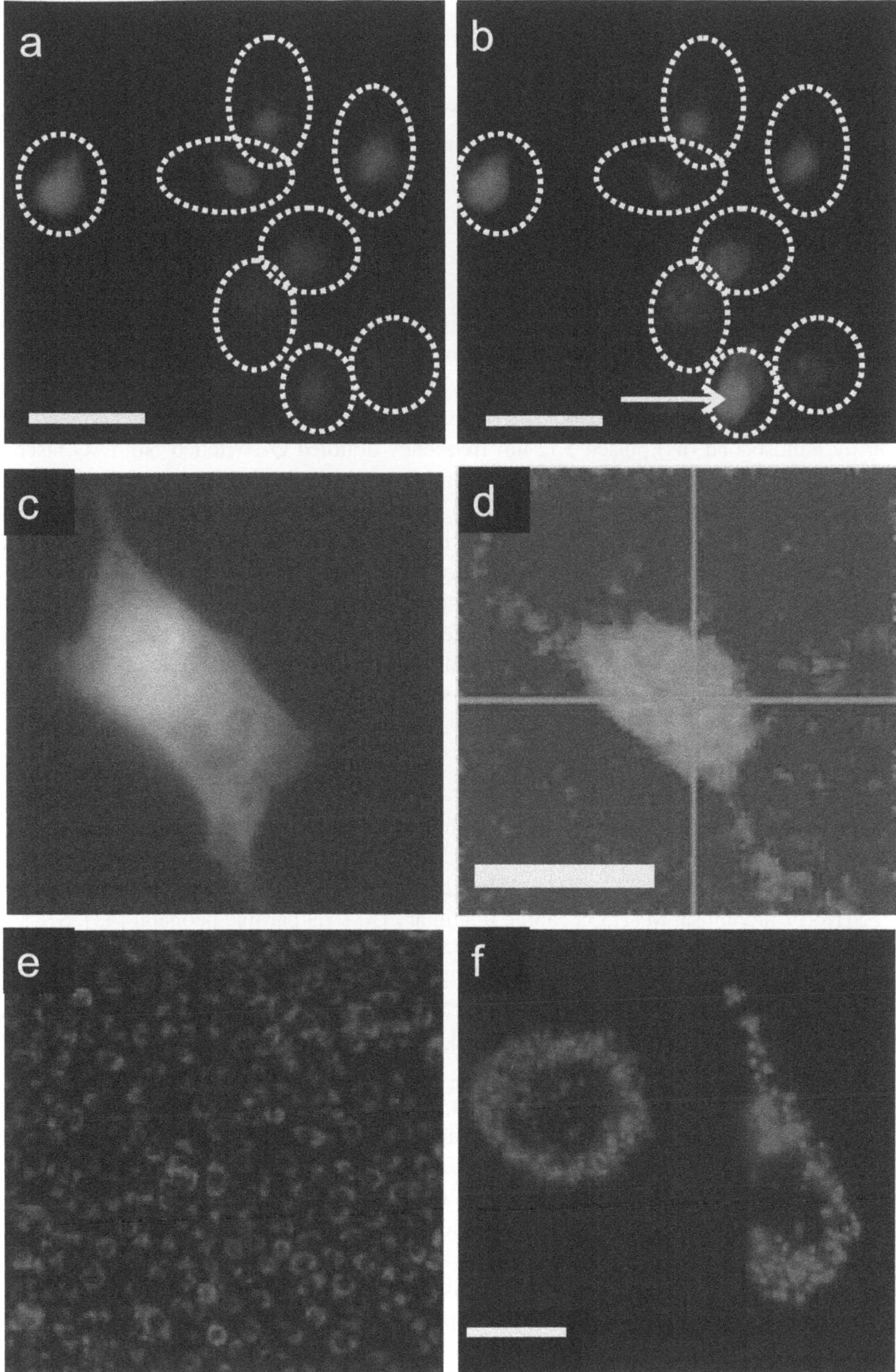

FIGURE 3.2: (See in color after page 572.) Examples of targeted (a,b) optical injection and targeted (b,c) optical transfection using a femtosecond (fs) pulsed Ti:sapphire laser or by (e,f) 405 nm continuous wave (cw) laser. See text for details. Images a and b reprinted with permission (Optical Society of America) [53]. Image d reprinted by permission from Macmillan Publishers Ltd: Nature Methods [54], copyright 2006. Images e and f reprinted with permission (Optical Society of America) [40].

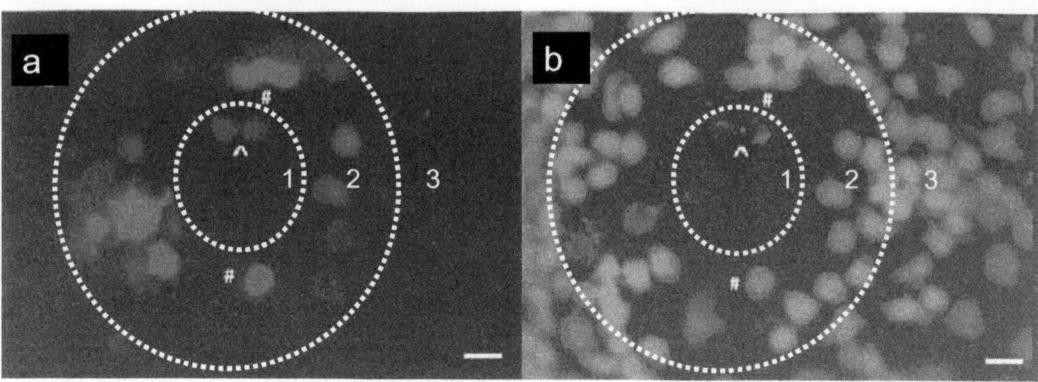

FIGURE 3.3: (See in color after page 572.) Example of untargeted (shockwave mediated) optical injection by nanosecond (ns) pulsed 532-nm frequency doubled Q-switched Nd:YAG laser. a) A single 10 μJ pulse was fired at the coverslip/solution interface in the center of the image in the presence of Texas red-conjugated dextran (3 kD). Red cells therefore represent successful optical injection (but the cells may be also be dead) b) Immediately after irradiation, cells were stained with the viability indicator Oregon green diacetate. Green cells are therefore viable. Three distinct zones were apparent. 1) The central zone (0–30 μm from center) is either devoid of cells due to detachment or contains nonviable optically injected cells ($\wedge$). 2) the pericentral zone (41–50 μm) has >90% viability and also contains some optically injected cells (#). The distal zone (>50 μm) contains 100% viable cells, but no optically injected cells. Reprinted Adapted with permission from [49]. Copyright 2000 American Chemical Society.

A variety of cell-types have been optically transfected, including Chinese Hamster Ovary cells [13, 47, 51, 55, 60, 62, 63], NIH3T3 [66], BAEC [67], HeLa, SU-DLH-4, NTERA-2, MO-2058, PFSK-1, 184-A1, CEM, NIH/3T3, 293T, HepG2, primary rat cardiomyocytes, embryonic C166 [46], GFSHR-17 granulosa, MTH53a canine mammary [53], rat cardiac neonatal [43], SK-Mel 28, NG108-15, T47D clone 11 [13], HT1080-6TG [44], transitional cell carcinoma [48], normal rat kidney [65], tibial mouse muscle (*in vivo*) [4], MCF-7 [45], human salivary gland stem cells, human pancreatic stem cells [55], Madin-Darby Canine Kidney cells [56], PC12, primary rat astrocytes [59], *spisulsa solidissima* oocytes [67], PtK2 [57], rat basophilic leukemia [49], *Triticum aestivum* L. cultivar Giza 164 (wheat) [64], and MatLu rat dorsal prostate adenocarcinoma [47]. However, with specific regard to plasmid DNA transfection, to our knowledge there have been no reports in the literature of a differentiated postmitotic primary cell being optically transfected. The majority of cells transfected by DNA are in fact rapidly dividing established cell-lines, which may highlight the need for the nuclear membrane to dissolve as a requirement for this technique to be successful. It should be re-emphasized however that the optical transfection of mRNA species in postmitotic cells (primary neurons) has been demonstrated [54].

In the case of plasmid DNA transfection, a useful parameter for optimizing the laser dose is that of transfection efficiency. It was mentioned previously that the efficiency and viability of optically transfected cells compare favorably with other nonviral technologies. In this particular field of research, there have been inconsistencies in the methodology used to express transfection efficiency, and as such caution should be used in the interpretation of reported transfection efficiency highlighted in Table 3.2. It is therefore appropriate to highlight some key points about transfection efficiency. There has been a strong trend on this topic to have the following definition of transfection efficiency:

$$N_{un} = (E/D) \cdot 100, \tag{3.1}$$

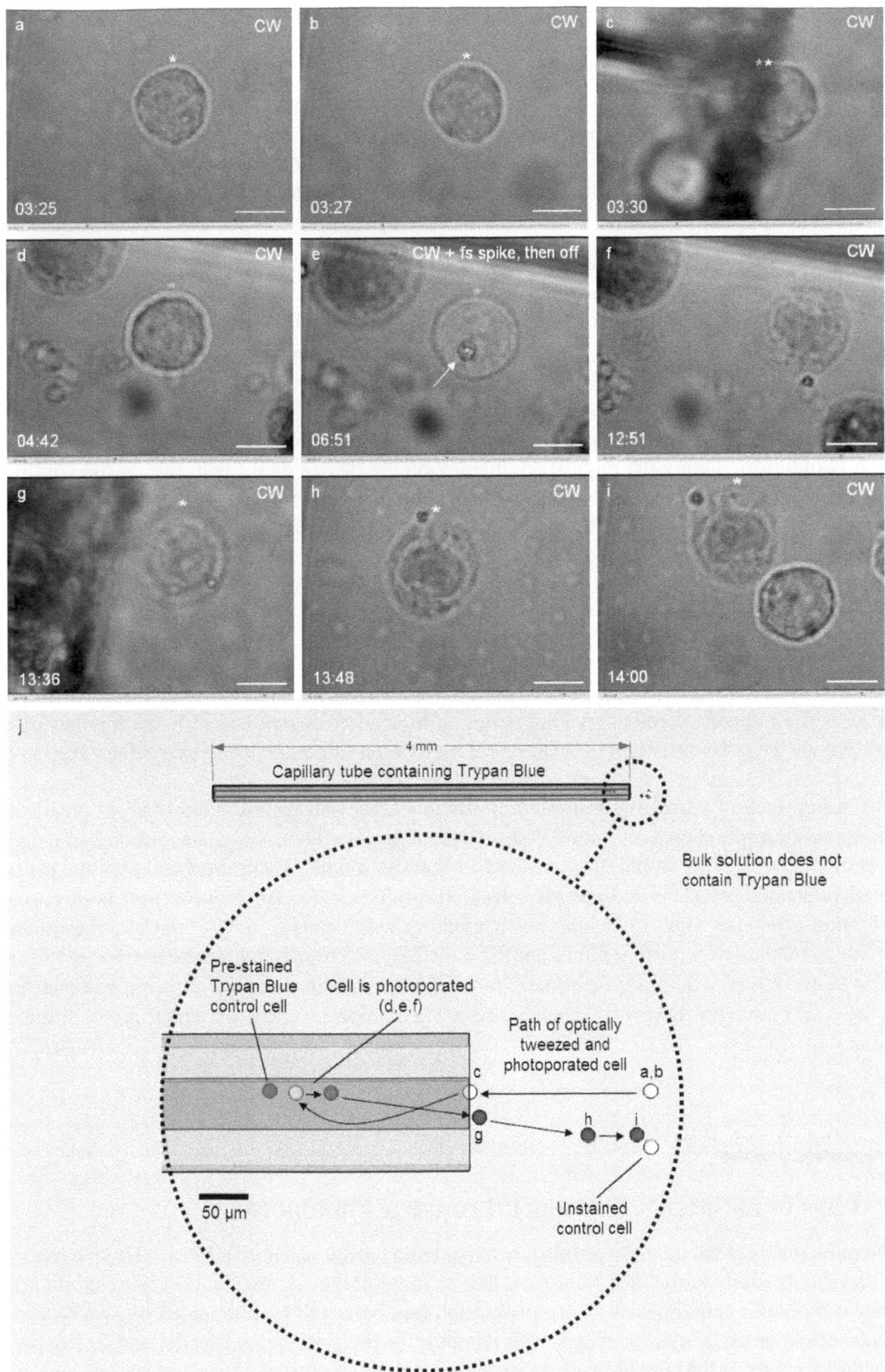

FIGURE 3.4: Example of the targeted optical injection of a nonadherent HL60 cell using a femtosecond (fs) pulsed Ti:Sapphire laser. The same laser was used in cw mode to optically tweeze the injected cell (laser modes are indicated per image). Time is noted in min:sec. Asterisk indicates cell of interest, and color of asterisk indicates solution surrounding the cell (White: RPMI 1640. Blue: 0.4% Trypan Blue). See text for details. Image and caption reprinted with permission from [39]. Original reference also contains a supplementary movie of this experiment online.

where N_{un} is the uncorrected transfection efficiency (%), D is the number of cells dosed on a given experiment in the presence of a plasmid containing a reporter gene, and E is the number of cells transiently expressing the plasmid DNA after a suitable amount of time has passed. The amount of time it takes for transient transfection to occur is on the order of 24–96 hours. Equation 3.1 does not take into account two parameters: 1) some of the dosed cells lose viability; 2) all cells, including those transfected, divide during the time between dosing and analysis. A corrected form of this equation was proposed over 20 years ago by the first group to perform optical transfection, Tsukakoshi et al. [68]:

$$N_{cor} = ((E/D) \cdot 100)/X_D \,, \tag{3.2}$$

where N_{cor} is the population corrected transfection efficiency, D is the number of cells dosed on a given experiment, E is the number of cells transiently expressing the plasmid DNA after a suitable amount of time has passed, and X_D is the ratio of proliferation that has occurred *by the dosed cells* between dosing and the measurement of expression. Although it is difficult to quantify X_D directly without performing long-term image analysis, X_D can be approximated by assuming that the population doubling time of the dosed cells is the same as those of all other cells in the culture dish. It is therefore simply the ratio of confluency on the day of analysis to that of the confluency on the day of dosing. To be clear, confluency is the percentage of the area of a culture dish that cells cover, and can be easily obtained by image analysis or measuring total protein by the Lowry assay [69] or equivalent.

Finally, another extremely important but often overlooked parameter is the postirradiation percentage of viability, V, which for simplicity can be defined as:

$$V = (S/D) \cdot 100, \tag{3.3}$$

where S is the number of cells surviving about an hour after dosing and D is the number of cells dosed. Survival can be measured by a variety of membrane impermeable dyes, such as Trypan Blue [61].

The importance of correcting transfection efficiency for cell proliferation is highlighted in the hypothetical example shown on Table 3.1. Both researchers could report an uncorrected transfection efficiency of 100%. This would disguise the fact that the second researcher had twice the postdose survival rate, and twice the (corrected) transfection efficiency. We propose that both corrected transfection efficiency (Eq. (3.2)) and postirradiation viability (Eq (3.3)) should be the minimum acceptable standard of reporting future optical transfection optimization studies.

It should be stressed that in spite of these minor inconsistencies, optical injection has established itself as a key nonviral transfection technology, the competitiveness of which is not limited by efficiency or viability.

3.4 Physics of Species Transport through a Photopore

The question as to the quantity of substance that enters a cell has been addressed both experimentally and theoretically in the literature. Consider a single, targeted, transiently generated photopore on the surface of a cell. Species transport through this pore will be dominated by two factors: 1) the volumetric changes a cell undergoes in response to the pore being created and 2) the passive diffusion of a species through the pore as it remains open.

TABLE 3.1: Hypothetical example of parameters measured by two typical researchers during a typical optical transfection experiment.

Experimental Details	Researcher 1	Researcher 2
Number of cells dosed (D)	100	100
Number of cells found to be expressing GFP (E)	100	100
Time in hours between dosing and analysis of expression	96	24
Ratio of confluency at time of analysis to that of confluency at time of dosing (X_D)	4	2
Number of cells surviving 1 hour after dosing (by simple dye exclusion)	25	50
Uncorrected transfection efficiency calculated by Eq. (3.1) (N_{un})	100	100
Corrected transfection efficiency calculated by Eq. (3.2) (N_{cor})	25	50
% viability calculated using Eq. (3.3) (V)	25	50

TABLE 3.2: A summary of optical injection and optical transfection experiments performed to date. Abbreviations – cw: continuous wave. fs: femtosecond. ns: nanosecond. Obj: Objective. NA: Numerical Aperture. Spot: diameter of focused laser.

λ (nm)	Ref.	Category of transfection or optical injection	Laser details	Dose	Viability	Obj/NA	Spot (μm)	Cell type	Transfected item (concentration)	Transfection efficiency	Notes
193/308	[72]	Targeted stable ns DNA	Eximer, 6 ns, 200 Hz	2–4 mJ	No mechanical damage or malformation	37×	–	*Triticum aestivum* L. cultivar Giza 164 (plant: wheat)	pAB6 (50 μg/ml)	0.5%	Targeted, high throughput (600,000 cells/hour) wheat transformation.
337	[50]	Targeted ns toxicity assay	N_2 laser, 3 ns	1 pulse	39% 1 h post-irradiation, progressively decreasing to 16% 12 h. > 60% apoptosis (Annexin V Alexa488)	100×/1.3	1.3	NG108	–	–	See 770 nm entry. Irradiation with fs is less toxic than ns pulses.
355	[71]	Targeted ns transient and stable DNA	Frequency tripled Nd:YAG	Single pulse at 0.91–1.10 μJ/pulse	–	100×	–	HuH-7 (highly differentiated hepatocellular carcinoma cells) (transient and stable transfection achieved) NIH/3T3 mouse BALB/C fibroblasts (stable transfection achieved), stimulatory protein 2 (SP2) mouse (BALB/C) myeloma cells (stable transfection achieved)	pEGFP-N1 (20 μg/ml)	HuH-7: 9.4% at 24h, 13% at 48h. NiH/3T3: 10.1% at 24h	Uncorrected transfection efficiency formula used. First example in literature of nonadherent cells being transfected, by employing a separate optical tweezer to hold cell *in situ* during dosing.

Continued on next page

Table 3.2 – continued from previous page

λ (nm)	Ref.	Category of transfection or optical injection	Laser details	Dose	Viability	Obj/NA	Spot (μm)	Cell type	Transfected item (concentration)	Transfection efficiency	Notes
355	[76]	Targeted ns transient and stable DNA	Frequency tripled Nd:YAG 5 ns, 10 Hz	1 mJ	–	32×/0.4	0.5 μm	Normal rat kidney	Eco-gpt gene cloned in pBR- 322 (pSV-2 gpt)	38.8% for uncorrected transient transfection. See Notes	The first case in the literature of laser mediated transfection. Highest transient transfection efficiency was obtained when nucleus, not cytoplasm, was targeted. Transfection efficiency in the original paper was stated as 10.2%, but this was corrected for population doubling. The uncorrected efficiency of 38.8% is stated here for comparison with the current literature, where correcting for population doubling is not commonplace.
355	[44]	Targeted ns DNA	Frequency tripled Nd:YAG	Single 10 ns pulse, 23–67 μJ	–	32×	2	HT1080-6TG (a hyopxanthine phosphoribosyl-transferase deficient human fibrosarcoma	pSV2-neo (12 μg/ml)	$\leq 0.3\%$	Only stable transfection observed; no transient.

Continued on next page

Table 3.2 – continued from previous page

λ (nm)	Ref.	Category of transfection or optical injection	Laser details	Dose	Viability	Obj/NA	Spot (μm)	Cell type	Transfected item (concentration)	Transfection efficiency	Notes
355	[78]	Targeted (raster scanning) transient and stable ns DNA and fluorophore	Frequency tripled Nd:YAG 15 ns, 10 Hz	12000 pulses over 20 min period at 0.2–1.0 mJ	–	40×	1	Embryonic calli of *Oryza sativa* L. cv. Japonica (plant: rice)	pBI221 (encoding for bacterial β-glucuronidase) (50 μg/ml) and other plasmids (including one encoding for hygromycin phosphotransferase), calcein (5 mM)	0.48%	Excellent example of the influence of tonicity on the cell response to photoporation
405	[40]	Targeted cw stable DNA	cw violet diode	0.3 mW for 40 ms	No acute morphological signs of injury	100×	2	Chinese hamster ovary cells K1	pEGFP-N3, DsRed-Mito (3 μg/ml)	–	Only stable transfection observed; no transient.
488	[42]	Targeted cw transient DNA	cw argon ion	1.0–2.0 MW/cm^2 for 1–2.5 s. Optimum dose: 1.0 MJ/cm^2.	69 ± 19% of control colony formation at optimum dose. Similar recovery rates of control and irradiated cells	40×/0.60 or 63×/0.90	1.0 or 0.7	CHO	GFP (8.3 μg/ml)	29 ± 10%	Notably, one of the few cases in the literature to observe *efflux* of a fluorophore (calcein AM) upon irradiation.

Continued on next page

Table 3.2 – continued from previous page

λ (nm)	Ref.	Category of transfection or optical injection	Laser details	Dose	Viability	Obj/NA	Spot (μm)	Cell type	Transfected item (concentration)	Transfection efficiency	Notes
488	[43]	Targeted cw transient DNA	cw argon ion	17 mW for 2 seconds (with medium containing 40 mg/L phenol red) or 10 seconds (with 15 mg/L phenol red)	$94.4 \pm 3.6\%$ (2 second dose), $27.7 \pm 10.5\%$ (10 second dose)	10×/0.45	4	Cardiac neonatal rat cells	GFP (5 μg/ml)	$5.1 \pm 3.5\%$ (2 second dose), $3.25 \pm 2.4\%$ (10 second dose)	Used phenol red and FM 1-43 to increase absorption of laser by plasma membrane. Employed the use of Cyclosporin A, SIN-1, and SNAP to protect against reactive oxygen species generated by laser. Employed the use of 1000x antioxidant supplement, $CaCl_2$, and Pluronic-F68 to promote membrane resealing after dosing.
488	[41]	Targeted cw transient (36h) DNA	cw argon ion	$2{\cdot}10^6$ W/cm^2 for 0.25 s	No acute morphological signs of injury	100×/1.0	5–8	–	*Lac-Z* (beta-galactosidase) (15 μg/ml) or CAT (chloramphenicol acetyltransferase)(8 μg/ml)	–	Employs the use of phenol red solution (15mg/l) to increase laser absorption, which was necessary for the generation of photopores at these doses

Continued on next page

Table 3.2 – continued from previous page

λ (nm)	Ref.	Category of transfection or optical injection	Laser details	Dose	Viability	Obj/NA	Spot (μm)	Cell type	Transfected item (concentration)	Transfection efficiency	Notes
532	[65]	Untargeted ns transient fluorophore, macromolecule, and peptide	Frequency doubled Nd:YAG 5 ns	Single pulse at 10 μJ	Target area was glass/medium interface on monolayer of adherent cells. Central zone (0–30 μm from the center): 0% viability. Pericentral zone (41–50 μm): > 90% viability. Distal zone (> 50 μm): 100% viability	100×/1.3	0.3–0.4	Rat basophilic leukemia	Texas red-glycine (800 D), Texas red-dextran (3 kD, 10kD, 40 kD), fluorescein labelled protein kinase C pseudosubstrate region (RFARKGSLRQ KNV-fluorescein)	30–80%	Up to 40 kD dextran species can be loaded into cells, but loading efficiency at this size is 10x less than for 800 D glycine species. This suggests that passive diffusion through pores may be a mechanism
532	[46]	Targeted or untargeted ns multiple macromolecule	Frequency-doubled Nd:YAG 0.5 ns, 2 kHz	Optimum dose ~60 nJ/μm^2	> 80% for doses ≤60 nJ/μm^2	–	–	HeLa, SU-DLH-4, NTERA-2, MO-2058, PFSK-1, 184- A1, CEM, NIH/3T3, 293T, HepG2, primary rat cardiomyocytes, embryonic C166	Ca^{2+}, Zn^{2+}, Sytox Green, Sytox Blue, tetramethylrhodamine-dextran (MW 3000), Cdc42 binding domain of WASP conjugated to an I-SO dye (55 KDa), quantum dots, SiRNA	–	Pore seals in approximately 30 seconds

Continued on next page

Table 3.2 – continued from previous page

λ (nm)	Ref.	Category of transfection or optical injection	Laser details	Dose	Viability	Obj/NA	Spot (μm)	Cell type	Transfected item (concentration)	Transfection efficiency	Notes
770	[50]	Targeted fs vs targeted ns toxicity assay	Ti:sapphire, 110 fs, 80 MHz	150 mW or 190 mW for 5 ms	79% viability 12 h post-irradiation by MitoTracker Red, ethidium bromide, Annexin V, and morphology	100×/1.3	1.3	NG108	–	–	Compared toxicity of ns pulsed laser at 337 nm with fs pulsed laser at 770 nm. fs laser treatment resulted in higher viability. See 337 nm entry.
780	[51]	Targeted fs single 100 nm particle	Ti:sapphire, 100 fs, 83 MHz	25–70 mW at focus	–	100×/1.4	Diffraction-limited	CHO	Single 100 nm gold nanoparticle	–	The first case in the literature a group has optically tweezed a single nanoscopic object to the surface of a cell prior to optical injection
780	[4]	Untargeted (raster scanning) transient and stable DNA *in vivo*	Ti:sapphire 200 fs, 76 MHz	20 mW raster scan of 95 × 95 μm^2 over the course of 5 s (for optimal dose) at a tissue depth of 2 mm	Morphology: rare fibers observed at 48 h but not 24 h or 70 h post-irradiation. Creatine phosphokinase levels not significantly higher than non-irradiated control 2h post-irradiation. No apoptosis by TUNEL assay observed 7d after irradiation.	50×/0.5	1	Tibial mouse muscle *in vivo*	GFP, luciferase, β-galactosidase, murine erythropoietin (333 μg/ml in 30 μl (i.e. 10 μg total) directly injected into muscle)	About half of the efficiency of electroporation.	Toxicity of electrotransfection is more pronounced than phototransfection. Electroporated cells had significantly increased levels of creatine phosphokinase, higher levels of apoptosis, and demonstrated extensive and irreversible damage by histological observation.

Continued on next page

Table 3.2 – continued from previous page

λ (nm)	Ref.	Category of transfection or optical injection	Laser details	Dose	Viability	Obj/NA	Spot (μm)	Cell type	Transfected item (concentration)	Transfection efficiency	Notes
790	[52]	Targeted fs transient DNA	Ti:sapphire, 120 fs, 100 MHz	3X doses of 40 ms of 70 mW at central core of Bessel beam	–	60×/	1.8	CHO	DsRed-Mito (1.2 μg/ml)	>20% over an axial distance of 100 μm. Peak efficiency was ~47%.	The first incidence in the literature to use a Bessel beam, which has a beam profile akin to a rod of light. This allows targeted fs DNA transfection without the need for focusing, and opens the technique up to automation.
790	[69]	Targeted fs transient DNA	Ti:sapphire, 800 fs, 100 MHz	3X doses of 80 ms of 110 mW at focus	–	see Notes	~5	CHO	DsRed-Mito (3 μg/ml)	25–57%	The first incidence in the literature to use a fiber to deliver targeted optical transfection. Fiber delivery, with custom axicon tip producing a focused beam.
792	[54]	Targeted fs transient DNA	Ti:sapphire, 12 fs, 75 MHz	50–100 ms, 5–7 mW at focus	100%	40×/1.3, 63×/1.25	–	Human salivary gland stem cells (hSGSC), human pancreatic stem cells (hPSC), CHO	Ethidium bromide, pEGFP-N1 (0.4 μg/ml)	hPSC: 75%, hSGCS: 80%, CHO: 90%	<20 fs pulse duration measured *at the focus.*
793	[53]	Targeted fs viability study	Ti:sapphire, 100 fs, 3.8 kHz, regeneratively amplified	5–9 J/cm^2 (low), 14–23 J/cm^2 (medium), and 41–55 J/cm^2, 10–10000 pulses	LD50 (by Trypan Blue exclusion) for 5–9 J/cm^2 is ~100 pulses	40×/0.6	1.6	CHO	–	–	A lower LD50 was observed in cell populations dosed in their cytoplasm compared with those hit in their nuclei.

Continued on next page

Table 3.2 – continued from previous page

λ (nm)	Ref.	Category of transfection or optical injection	Laser details	Dose	Viability	Obj/NA	Spot (μm)	Cell type	Transfected item (concentration)	Transfection efficiency	Notes
800	[55]	Targeted fs DNA	Ti:sapphire, 120 fs, 80 MHz	50–225 mW for 10–250 ms. Optimum dose: 1.2 μJ/cm^2	70 ± 8% by Trypan Blue exclusion 2 h after irradiation.	60×/0.85	1	CHO-K1	pEGFP-N2	50±10% at optimum dose	It is worth reading Brown *et al* (2008) to put this work within context [39].
800	[39]	Targeted fs macromolecule	Ti:sapphire, 100 fs, 82 MHz	Series of < 100 ms doses	No blebbing or acute morphological changes post-irradiation	100×/1.4	0.8	HL60	Trypan Blue (0.4%)	–	Employed the same laser source to optically tweeze (in a cw regime) and optically inject (in a mode-locked regime) an impermeable substance. See Figure 3.4.
800	[56]	Targeted fs transient DNA	Ti:sapphire, 140 fs (∼210 fs at sample), 90 MHz	40 ms, 0.9 nJ per pulse (optimum dose)	90% viability by propidium iodide exclusion.	/0.8	–	GFSHR-17 granulosa/MTH53a canine mammary	pEGFP-C1-HMGB1, pEGFP-C1 (50 μg/ml), propidium iodide (1.5 μM), lucifer yellow (100–1000 μM)	70%	Quantified experimentally the relative volume exchange of about 0.4 during photopore opening.
800	[57]	Targeted fs transient macromolecule	Ti:sapphire, sub-10-fs, 80 MHz	< 10 ms	91.5 ± 8% (by when 0.2 M sucrose employed	/0.95	<1	Madin-Darby canine kidney cells	0.2–0.5 M sucrose (1.6 nm diameter)	For 0.2M sucrose, a theoretical loading efficiency of 72.3% was obtained	The first case in the literature to critically examine the relationship between osmolarity and the amount of species able to be optically injected into a cell. Experiments performed at 4 °C.
800	[58]	Targeted transient fluorophore	Ti:sapphire, 12 fs, 75 MHz	100 ms, 20 mW at focus	91% (PC12), 100% (primary astrocytes)	25×/0.4	2	PC12, primary rat astrocytes	Propidium iodide (5 μg/ml)	–	–

Continued on next page

Table 3.2 – continued from previous page

λ (nm)	Ref.	Category of transfection or optical injection	Laser details	Dose	Viability	Obj/NA	Spot (μm)	Cell type	Transfected item (concentration)	Transfection efficiency	Notes
800	[75]	Targeted fs fluorophore	Ti:sapphire, 130 fs (~200 fs at sample), 82 MHz	0–400 mW for 20 seconds, 4–33 W/cm^2	100% for doses $\leq 12 \cdot 10^{12}$	40×/1.3	0.308	BAEC, *Spisulsa Solidissima Oocytes*	Lucifer yellow, propidium iodide	$\sim$ 100% at doses > $4.0 \cdot 10^{12}$ W/cm^2.	–
800	[13]	Targeted fs transient DNA and fluorophore	Ti:sapphire, 170 fs, 90 MHz	49 or 65 mW for 17 ms	CHO and SK-Mel 28 cells showed good morphology after irradiation; NG108-15 often apoptotic	40×/1.3	–	CHO, SK-Mel 28, NG108-15, T47D clone 11	FITC labelled dextran (MW 500 kDa, which corresponds to a diameter of 26.6 nm [73]), eGFP-N1 (4.7 kB), cascade blue	–	An interesting example of how treating two cells with the same dose can result in the immediate loss of viability of one cell but not the other, highlighting the inherent variability of a cellular response to laser irradiation.
800	[61]	Scanning fs fluorophore	Ti:sapphire, fs, 80 MHz	7 mW	–	40×/1.3	–	PtK2	Propidium iodide (5 μg/ml)	–	–
800	[60]	Targeted fs transient DNA	Ti:sapphire, fs, 80 MHz	50–100 mW for 16 ms	–	"high NA"	–	CHO	pEGFP-N1 (0.4 μg/ml)	100%	Timelapse microscopy indicates no detrimental effects on growth or division.
800	[74]	Targeted fs microbead	Ti:sapphire, 150 fs, 125 Hz, regeneratively ampli fied	20 pulses of 10 nJ/pulse	1	100×/1.25	1	NIH 3T3	200 nm polystyrene	–	–

Continued on next page

Table 3.2 – continued from previous page

λ (nm)	Ref.	Category of transfection or optical injection	Laser details	Dose	Viability	Obj/NA	Spot (μm)	Cell type	Transfected item (concentration)	Transfection efficiency	Notes
840	[62]	Targeted fs mRNA or fluorophore	Ti:sapphire, 100 fs, 80 MHz	8–16 regions on the plasma membrane dosed for 1–5 ms with 24 mW at focus	Transfection with *GFP* mRNA remained viable for at least 24 h	40×/0.8	Diffraction limited	Primary rat neuron	Lucifer yellow, *Elk1*-GFP mRNA, *Elk1* mRNA, *Elk1-ETS* mRNA, *c-fos* mRNA, DS-RED, Venus fluorescent protein, pTRI-Xef (all 10–15 μg/ml)	–	mRNA transfected into the dendrite, but not the cell body, of a primary neuron results in the death of the cell. An excellent example of the power of optical transfection in its ability to transfect specific regions of a cell
840	[63]	Targeted fs mRNA	Ti:sapphire, 100 fs, 80 MHz	16 regions on the plasma membrane dosed for 5 ms with 30–35 mW at the back aperture	Repeated transfection with mRNA over seven day period resulted in high viability over the subsequent four week period	40×/0.8	Diffraction limited	Primary rat neuron	200 μg/ml astrocyte mRNA transcriptome	–	Transfection of astrocyte transcriptome into primary neurons results in phenotype remodelling (i.e. conversion of neurons into astrocytes). First demonstration of multiple rounds of transfection
1064	[45]	Targeted ns transient DNA or fluorophore	Q-switched Nd:YAG, 17 ns, 10 Hz	3–4 J/cm^2	–	100×/	–	MCF-7	merocyanin 540 (7.5 μg/ml), par pEGFPN1 (5 μg/ml)	–	–

Continued on next page

Table 3.2 – continued from previous page

λ (nm)	Ref.	Category of transfection or optical injection	Laser details	Dose	Viability	Obj/NA	Spot (μm)	Cell type	Transfected item (concentration)	Transfection efficiency	Notes
1064 Nd:YAG: 2080 Ho:YAG	[64]	Untargeted transient DNA	Ho:YAG and Nd:YAG, 10 Hz	Nd:YAG: 150 mJ, dosed with 1000, 2000, or 2500 pulses. Ho:YAG: ideal dose was 750 pulses of 2000 mJ	–	See notes	–	Transitional cell carcinoma	pEGFP- N1 (200 μg/ml)	Ho:YAG, ideal dose: 58.26 ± 1.50 Nd:YAG: low efficiency (< 2%) for all doses	The Ho:YAG mechanism may be thermally mediated, as there is strong absorption at this wavelength. Cells increase their temperature to 45–55°C during treatment, which is why the protocol calls for cells to be placed on ice during irradiation. Beam delivered unfocused to cells with 220 μm diameter fiber
1554	[79]	Targeted transient fluorophore and DNA	20 MHz, 170 fs	7 s, 10^{12} W cm^{-2}	Mitochondrial depolarisation (as measured by JC-1) did not occur in cells 1.5h after exposure.	40×/1.0	2	HepG2	Propidium Iodide (1.5 μg/ml). pEGFP-C1 (20 μg/ml)	77.3% (corrected)	The exclusion by cells of Propidium Iodide 10 s after dosing indicates that photopores seal in less than 10 s. DNA transfection efficiency calculated taking proliferation into account.
2080 Ho:YAG	[47]	Untargeted transient DNA	Ho:YAG 10 Hz	2000 mJ, 200 pulses	21.50%	See notes	–	MatLu rat dorsal prostate adenocarcinoma	pEGFP-N1 (300 μg/ml)	41.3%	Beam delivered unfocused to cells

Volumetric changes to a cell occurring during photoporation are mediated largely by the tonicity of its environment. A cell photoporated in a hypertonic environment will undergo solution influx and volumetric increase, in a hypotonic environment solution efflux and volumetric decrease, and in an isotonic environment no net solution flux will occur.

In the case of a hypertonic environment, intracellular water will flow through the plasma membrane out of the cell, resulting in a shrunken cell state prior to dosing. Once a hole in its plasma membrane is generated, the cell will rapidly (< 1 second) expand back to its equilibrium volume (V_{equil}) as solute and solvent flow through the pore. This effect has been demonstrated in Madin-Darby Canine Kidney cells. When suspended in a 1.0 M hypertonic sucrose environment, they shrink to an equilibrium volume of V/V_{equil}=0.578 $\pm$ 0.085. Upon targeted photoporation using a femtosecond laser system, the cells return to a V/V_{equil}=1.000 $\pm$ 0.055 [56]. Theoretical calculations of loading efficiency based on empirical observations of the volume changes of kidney cells exposed to 0.2 M sucrose solution indicate loading efficiencies of 72.3% [56]. The pre-treatment of cells in a hypertonic environment has also been used in the optical transfection of the embryonic calli of *Oryza sativa* L. cv. Japonica (rice) cells [73]. The amount of species entering the cell in this situation can be readily estimated by assuming that the expansion of the cell is due entirely to the volume of the incoming solution.

Conversely, a cell exposed to a hypotonic environment will expand due to the influx of water through the plasma membrane. Once porated, the volume of this cell will shrink to its original volume as solute and solvent rapidly flow *out of* the cell. A hypotonic environment may therefore be disadvantageous in the loading membrane impermeable species into cells by photoporation.

In the case of an isotonic environment, during the time the pore remains open species external to the cell will passively diffuse through the pore until such time as the pore either closes, or the concentration of the species in the intracellular and extracellular environments equal each other. Negligible volumetric changes are observed in cells photoporated in isotonic medium, so the volume equilibration mechanism of species transport does not play a role in this situation. In this case, the flux (number of species travelling through the pore per second) may be calculated according using Fick's first law [75]:

$$flux = -\frac{D \cdot \pi \cdot R^2 \cdot \Delta c \cdot N_A}{L} \cdot k_{corr} \tag{3.4}$$

Where D is the diffusion coefficient ($m^2 s^{-1}$), R is the radius of the photopore (m), Δc is the difference extracellular and intracellular concentration of the solute (mol$\cdot$m^{-3}), N_A is Avogadro's number (6.0221415$\cdot 10^{23}$ molecules$\cdot$mol^{-1}), L is the channel length of the pore (m), and $k_{corr} = (1 - r/R)^2$ is a correction term that extends the validity of Fick's law from point-like objects to objects with the Stokes radius r (m). D may be calculated by the formula:

$$D = \frac{kT}{6\pi\eta r} \tag{3.5}$$

Where k is the Boltzmann constant (1.3806503 $\times$ 10^{-23} m^{2}kg s^{-2} K^{-1}), T is the temperature (K), η is the viscosity of the medium (kg m^{-1}s^{-1}).

The majority of optical injection experiments have been performed in isotonic conditions. Two experiments in particular quantify the amount of species that has been loaded into an optoinjected cell. Stracke et al. (2005) demonstrated that a cell, after being photoporated in a solution of 10 μM FITC labelled dextran (MW: 500 kDa, diameter 26.6 nm) ended up with a cytoplasmic concentration of approximately 0.7 μM [13]. This represents a final cytoplasmic concentration of 7% of that which the surrounding medium contained. Baumgart et al. (2008) elegantly investigated this question using patch pipette techniques [53]. This group manually injected known concentrations of lucifer yellow into canine mammary cells to generate a standard curve of fluorescence. They then used this to calculate the intracellular concentration of lucifer yellow that had been opto-injected using the optimum dose noted in Table 3.2. This, in combination with the changes in membrane potential measured by patch clamping, showed that a final cytoplasmic concentration of about 40% of the surrounding medium was occurring, i.e., if the initial concentration of external fluorophore was 600 μM, one would expect the final intracellular concentration to be 240 μM.

The final section of this review now covers the underlying physical mechanisms associated with the photoporation process.

3.5 Physics of the Laser-Cell Interaction

The mechanisms that govern the technique of optical injection differ based upon the type of laser field employed. This is also an area of much ongoing research and as such we will outline here the main findings. In particular we shall concentrate on the mechanisms governing short pulse laser transfection which presently has proven to be the most effective. In the case of continuous wave light, we have described the use of short wavelength light, notably in the violet-blue range of the spectrum to initiate transfection. Absorption appears to play a key role here. Phenol red, a pH indicator ubiquitously added to cell culture medium, has a molar extinction coefficient of around 10 000 M^{-1} at 488 nm [41]. Concentrations up to 40 mg/L have been added to culture medium during photoporation; using lower concentrations of phenol red necessitates longer laser irradiation times in order to obtain the same optical injection effect [43].

In the case of short pulse sources we distinguish between targeted and untargeted transfection schemes. For untargeted transfection the literature shows that ns pulses have been successfully used to create shock waves in the liquid medium that in turn cause microbubble formation that ultimately leads to cell transfection. While successful, this method is not usually suitable for targeted cell studies.

In the case of targeted cell transfection we concentrate upon the use of femtosecond pulses that have come to the fore. We note however that ns pulses for targeted transfection are now emerging as a viable method that offers a less expensive alternative [44–46]. In the femtosecond domain one has to begin by considering the breakdown processes in water [76].

The numerical studies performed to date elucidate how femtosecond poration may work (Figure 3.5). If we firstly consider the train of ultrashort pulses as having a repetition rate >1 MHz. In this regime we may consider applying average powers and energies such that we are well below the threshold required for bubble formation within the liquid. In this instance – which is the case for the main studies to date in fs transfection – heating effects or thermoelastic stress are not major considerations. However we do see the accumulation of numerous free electrons due to the application of the laser pulses that then react photochemically with the cell membrane causing transient pores to be generated [76–78]. The excitement here is that the fs beams can accumulatively generate enough free electrons without any bubble formation that can "gently" perforate the lipid bilayer and initiate transfection, in essence creating a low density plasma. If we now turn to the instance where a much lower repetition rate is used (<1 MHz) we see the pulse energies are actually much higher and the thermoelastic stress component becomes important – very small cavities (microbubbles) are generated with lifetimes up to 100 ns that mediate the cell poration process. In turn this is somewhat less "gentle" than the high repetition mechanism using the fs beams as too high an energy can cause bubbles that cause cell lysis rather than transfection. Experiments have looked and validated this approach where the size of the bubble generated is monitored *in situ* (online) with the aid of a probe laser beam [79].

It is intriguing to explore if one could replicate the "gentle" transfection at the single cell level that is seen with femtosecond beams with nanosecond pulses – the latter offers far more compact and

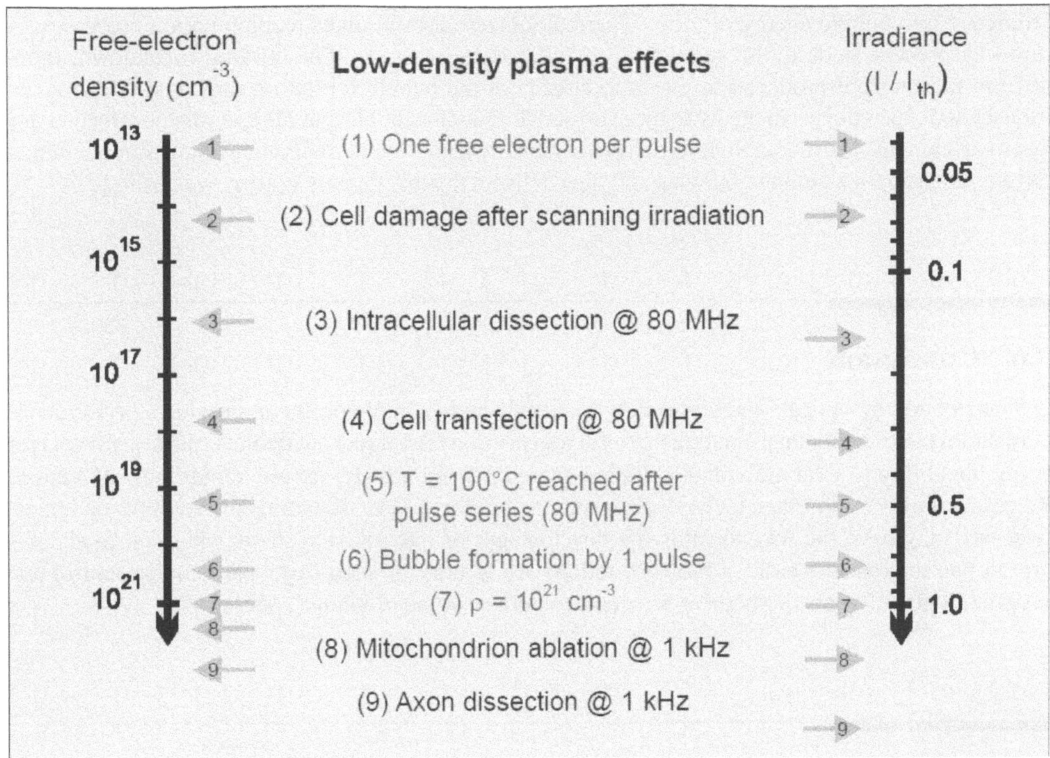

FIGURE 3.5: An overview of the physical breakdown induced by fs pulsed lasers, along with threshold values for the damage, transfection, or dissection of cells. The different effects are depicted together with the corresponding values of free-electron density and irradiance. The irradiance values are normalized to the optical breakdown threshold I_{th} defined by a critical electron density of cr = 1021 cm^{-3}. All data refer to plasma formation in water with femtosecond pulses of about 100-fs duration and 800-nm wavelength [76]. With kind permission from Springer Science+Business Media: *Applied Physics B, Lasers and Optics*, Mechanisms of femtosecond laser nanosurgery of cells and tissues, Vol. 81, 2005, p. 1038, A. Vogel, J. Noak, G. Hüttman, & G. Paltauf, Figure 21, © Springer-Verlag 2005.

inexpensive laser systems making them more practicable. For the case of bulk dielectrics nanosecond optical breakdown is typically associated with significant plasma luminescence and dominant thermomechanical effects — this is not particularly suitable for gentle cell transfection. Very recent studies have explored ultraviolet-visible ns pulses with temporally smooth shapes [77]. Importantly the authors show the existence of a well-defined low-density regime at irradiances below the threshold for luminescent plasma formation. Experimental studies used seeded single-longitudinal-mode pulses with 7–11 ns duration. These are contrasted to more typical (nonseeded) ns pulses that always exhibit picosecond spikes due to longitudinal mode beating (making them less suitable). Using the single longitudinal mode ns laser pulses the study showed that optical breakdown occurred in two distinct steps. First, a nonluminescent low-density plasma is formed the expansion of which creates minute bubbles (typical radii = 500 nm – 10 μm). Electron-hole recombination limits the free-electron density. The conversion efficiency of laser energy into bubble energy is typically very low ($\ll 1\%$). For energies one order of magnitude or more above the bubble formation threshold, the generated plasma suddenly assumes a much larger size. One observes a bright luminescence and large bubbles are produced with radii exceeding 0.1 mm. Accompanying this we have a conversion efficiency into bubble energy of 10%. Thermal ionization overcomes recombination processes. A runaway process leads to the generation of full-density plasmas. For infrared breakdown, luminescent plasmas are produced in one step already at the bubble formation threshold. For regular (nonseeded) laser pulses in the ns regime, no stable low-density plasma regime was observed at any laser wavelength. Most notably microchip lasers in the nanosecond regime may have smooth output pulses: this generates minute bubbles and thus delivers the low density plasma we desire [77].

3.6 Conclusion

Transfection by optical injection is now an established technology. It can provide a performance comparable to or better than existing nonviral transfection techniques, with one of its key advantages being the ability to treat individual cells in a targeted fashion under aseptic conditions. Moreover, its compatibility with other technologies such as confocal laser scanning microscopy or optical tweezers may pave the way towards the development of integrated systems where all modalities are on one microscope base: a bioworkstation. An armory of such exquisite cellular control and visualization will allow biomedical scientists to unravel some of nature's secrets.

Acknowledgments

We thank the UK Engineering and Physical Sciences Research Council for their support and the European Commission network "Photonics4Life." KD is a Royal Society Wolfson Merit Award Holder. Our sincere thanks to Alfred Vogel for very useful discussions.We gratefully acknowledge the observations of Rhodri Morris that led to the work in Figure 3.4.

We thank the UK Engineering and Physical Sciences Research Council for their support and the European Commission network "Photonics 4 Life." KD is a Royal Society Wolfson Merit Award Holder. Our sincere thanks to Alfred Vogel for very useful discussions.

References

[1] J. Földes and T.A. Trautner, "Infectious DNA from a newly isolated B. Subtilis phage," *Z. Vererbungsl.* **95**, 57–65 (1964).

[2] M.A. Hickman, R.W. Malone, T.R. Sih, et al., "Hepatic gene expression after direct DNA injection," *Adv. Drug Deliv. Rev.* **17** (3), 265–271 (1995).

[3] J.A. Wolff, R.W. Malone, P. Williams, et al., "Direct gene transfer into mouse muscle in vivo," *Science.* **247** (4949 Pt 1), 1465–1468 (1990).

[4] E. Zeira, A. Manevitch, A. Khatchatouriants, et al., "Femtosecond infrared laser – an efficient and safe *in vivo* gene delivery system for prolonged expression," *Mol. Ther.* **8** (2), 342–350 (2003).

[5] H. Lin, M.S. Parmacek, G. Morle, et al., "Expression of recombinant genes in myocardium in vivo after direct injection of DNA," *Circulation.* **82** (6), 2217–2221 (1990).

[6] K.B. Meyer, M.M. Thompson, M.Y. Levy, et al., "Intratracheal gene delivery to the mouse airway: characterization of plasmid DNA expression and pharmacokinetics," *Gene Ther.* **2** (7), 450–460 (1995).

[7] A.V. Vologodskii and N.R. Cozzarelli, "Conformational and thermodynamic properties of supercoiled DNA," *Annu. Rev. Biophys. Biomol. Struct.* **23**, 609–643 (1994).

[8] S. Cunha, C.L. Woldringh, and T. Odijk, "Polymer-mediated compaction and internal dynamics of isolated *Escherichia coli* nucleoids," *J. Struct. Biol.* **136** (1), 53–66 (2001).

[9] D.D. Dunlap, A. Maggi, M.R. Soria, et al., "Nanoscopic structure of DNA condensed for gene delivery," *Nucleic Acids Res.* **25** (15), 3095–3101 (1997).

[10] M.R. Capecchi, "High efficiency transformation by direct microinjection of DNA into cultured mammalian cells," *Cell.* **22** (2 Pt 2), 479–488 (1980).

[11] S. Zappe, M. Fish, M.P. Scott, et al., "Automated MEMS-based Drosophila embryo injection system for high-throughput RNAi screens," *Lab. Chip.* **6** (8), 1012–1019 (2006).

[12] G.L. Lukacs, P. Haggie, O. Seksek, et al., "Size-dependent DNA mobility in cytoplasm and nucleus," *J. Biol. Chem.* **275** (3), 1625–1629 (2000).

[13] F. Stracke, I. Rieman, and K. Konig, "Optical nanoinjection of macromolecules into vital cells," *J. Photochem. Photobiol. B, Biol.* **81** (3), 136–142 (2005).

[14] J. Braga, J.M. Desterro, and M. Carmo-Fonseca, "Intracellular macromolecular mobility measured by fluorescence recovery after photobleaching with confocal laser scanning microscopes," *Mol. Biol. Cell.* **15** (10), 4749–4760 (2004).

[15] H. Pollard, J.S. Remy, G. Loussouarn, et al., "Polyethylenimine but not cationic lipids promotes transgene delivery to the nucleus in mammalian cells," *J. Biol. Chem.* **273** (13), 7507–7511 (1998).

[16] P.L. Felgner, T.R. Gadek, M. Holm, et al., "Lipofection: a highly efficient, lipid-mediated DNA-transfection procedure," *Proc. Natl. Acad. Sci. USA.* **84** (21), 7413–7417 (1987).

[17] M.D. Brown, A.G. Schatzlein, and I.F. Uchegbu, "Gene delivery with synthetic (nonviral) carriers," *Int. J. of Pharm.* **229** (1–2), 1–21 (2001).

[18] A. Elouahabi and J.M. Ruysschaert, "Formation and intracellular trafficking of lipoplexes and polyplexes," *Mol. Ther.* **11** (3), 336–347 (2005).

[19] J.M. Bergen, I.K. Park, P.J. Horner, et al., "Nonviral approaches for neuronal delivery of nucleic acids," *Pharm. Res.* **25** (5), 983–998 (2008).

[20] G.Y. Wu and C.H. Wu, "Receptor-mediated in vitro gene transformation by a soluble DNA carrier system," *J. Biol. Chem.* **262** (10), 4429–4432 (1987).

[21] O. Boussif, F. Lezoualc'h, M.A. Zanta, et al., "A versatile vector for gene and oligonucleotide transfer into cells in culture and in vivo: polyethylenimine," *Proc. Natl. Acad. Sci. USA* **92** (16), 7297–7301 (1995).

[22] M. Chorny, B. Polyak, I.S. Alferiev, et al., "Magnetically driven plasmid DNA delivery with biodegradable polymeric nanoparticles," *FASEB. J.* **21(10)**, 2510–2519 (2007).

[23] T.M. Klein, E.D. Wolf, R. Wu, et al., "High-velocity microprojectiles for delivering nucleic acids into living cells," *Nature.* **327** (6117), 70–73 (1987).

[24] A.L. Rakhmilevich, J. Turner, M.J. Ford, et al., "Gene gun-mediated skin transfection with interleukin 12 gene results in regression of established primary and metastatic murine tumors," *Proc. Natl. Acad. Sci. U.S.A.* **93** (13), 6291–6296 (1996).

[25] W.N. Lian, C.H. Chang, Y.J. Chen, et al., "Intracellular delivery can be achieved by bombarding cells or tissues with accelerated molecules or bacteria without the need for carrier particles," *Exp. Cell Res.* **313** (1), 53–64 (2007).

[26] J. O'Brien and S.C. Lummis, "Biolistic and diolistic transfection: using the gene gun to deliver DNA and lipophilic dyes into mammalian cells," *Methods.* **33** (2), 121–125 (2004).

[27] R. Wood and A.L. Loomis, "The physical and biological effects of high frequency sound waves of great intensity," *Phil. Mag. J. Sci.* **4**, 417–436 (1927).

[28] M. Fechheimer, J.F. Boylan, S. Parker, et al., "Transfection of mammalian cells with plasmid DNA by scrape loading and sonication loading," *Proc. Natl. Acad. Sci. USA* **84** (23), 8463–8467 (1987).

[29] D.L. Miller and J. Song, "Tumor growth reduction and DNA transfer by cavitation-enhanced high-intensity focused ultrasound in vivo," *Ultrasound Med. Biol.* **29** (6), 887–893 (2003).

[30] L.J. Weimann and J. Wu, "Transdermal delivery of poly-l-lysine by sonomacroporation," *Ultrasound Med. Biol.* **28** (9), 1173–1180 (2002).

[31] K. Tachibana, T. Uchida, K. Ogawa, et al., "Induction of cell-membrane porosity by ultrasound," *Lancet.* **353** (9162), 1409 (1999).

[32] P.A. Prentice, A. Cuschieri, K. Dholakia, et al., "Membrane disruption by optically controlled microbubble cavitation," *Nat. Phys.* **1**, 107–110 (2005).

[33] E. Neumann, M. Schaefer-Ridder, Y. Wang, et al., "Gene transfer into mouse lyoma cells by electroporation in high electric fields," *EMBO J.* **1** (7), 841–845 (1982).

[34] T.K. Wong and E. Neumann, "Electric field mediated gene transfer," *Biochem. Biophys. Res. Commun.* **107** (2), 584–587 (1982).

[35] R. Heller, "The development of electroporation," *Science.* **295** (5553), 277 (2002).

[36] M. Golzio, J. Teissie, and M.P. Rols, "Direct visualization at the single-cell level of electrically mediated gene delivery," *Proc. Natl. Acad. Sci. USA.* **99** (3), 1292–1297 (2002).

[37] D.C. Chang and T.S. Reese, "Changes in membrane structure induced by electroporation as revealed by rapid-freezing electron microscopy," *Biophys. J.* **58** (1), 1–12 (1990).

[38] M. Khine, C. Ionescu-Zanetti, A. Blatz, et al., "Single-cell electroporation arrays with real-time monitoring and feedback control," *Lab. Chip.* **7** (4), 457–462 (2007).

[39] C.T.A. Brown, D.J. Stevenson, X. Tsampoula, et al., "Enhanced operation of femtosecond lasers and applications in cell transfection," *J. Biophot.* **1** (2), 183–199 (2008).

[40] L. Paterson, B. Agate, M. Comrie, et al., "Photoporation and cell transfection using a violet diode laser," *Opt. Express.* **13** (2), 595–600 (2005).

[41] G. Palumbo, M. Caruso, E. Crescenzi, et al., "Targeted gene transfer in eucaryotic cells by dye-assisted laser optoporation," *J. Photochem. Photobiol. B, Biol.* **36** (1), 41–46 (1996).

[42] H. Schneckenburger, A. Hendinger, R. Sailer, et al., "Laser-assisted optoporation of single cells," *J. Biomed. Opt.* **7** (3), 410–416 (2002).

[43] A.V. Nikolskaya, V.P. Nikolski, and I.R. Efimov, "Gene printer: Laser-scanning targeted transfection of cultured cardiac neonatal rat cells," *Cell Adhes. Commun.* **13** (4), 217–222 (2006).

[44] W. Tao, J. Wilkinson, E.J. Stanbridge, et al., "Direct gene-transfer into human cultured-cells facilitated by laser micropuncture of the cell-membrane," *Proc. Natl. Acad. Sci. USA* **84** (12), 4180–4184 (1987).

[45] S.K. Mohanty, M. Sharma, and P.K. Gupta, "Laser-assisted microinjection into targeted animal cells," *Biotechnol. Lett.* **25** (11), 895–899 (2003).

[46] I.B. Clark, E.G. Hanania, J. Stevens, et al., "Optoinjection for efficient targeted delivery of a broad range of compounds and macromolecules into diverse cell types," *J. Biomed. Opt.* **11** (1) (2006).

[47] S. Sagi, T. Knoll, L. Trojan, et al., "Gene delivery into prostate cancer cells by holmium laser application," *Prostate Cancer Prostatic Dis.* **6**, 127–130 (2003).

[48] A. Vogel and V. Venugopalan, "Mechanisms of pulsed laser ablation of biological tissues," *Chem. Rev.* **103**(5), 2079 (2003). One page!

[49] A. Vogel, J. Noack, G. Huttman, et al., "Mechanisms of femtosecond laser nanosurgery of cells and tissues," *Appl. Phys. B.* **81**(8), 1015–1047 (2005).

[50] M.B. Zeigler and D.T. Chiu, "Laser Selection Significantly Affects Cell Viability Following Single-Cell Nanosurgery," *Photochem. Photobiol.* **85**(5), 1218–1224 (2009).

[51] C. McDougall, D.J. Stevenson, C.T.A. Brown, et al., "Targeted optical injection of gold nanoparticles into single mammalian cells," *J. Biophot.* **2** (2009); http://dx.doi.org/10.1002/jbio.200910030

[52] X. Tsampoula, V. Garces-Chavez, M. Comrie, et al., "Femtosecond cellular transfection using a nondiffracting light beam," *Appl. Phys. Lett.* **91**(5), 053902-1–3 (2007).

[53] M.J. Zohdy, C. Tse, J.Y. Ye, et al., "Optical and acoustic detection of laser-generated microbubbles in single cells," *IEEE Trans. Ultrason. Ferroelectr. Freq. Control.* **53**(1), 117–125 (2006).

[54] A. Uchugonova, K. König, R. Bueckle, et al., "Targeted transfection of stem cells with sub-20 femtosecond laser pulses," *Opt. Express.* **16**(13), 9357–9364 (2008).

[55] D. Stevenson, B. Agate, X. Tsampoula, et al., "Femtosecond optical transfection of cells: viability and efficiency," *Opt. Express.* **14**(16), 7125–7133 (2006).

[56] J. Baumgart, W. Bintig, A. Ngezahayo, et al., "Quantified femtosecond laser based opto-perforation of living GFSHR-17 and MTH53 a cells," *Opt. Express.* **16**(5), 3021–3031 (2008).

[57] V. Kohli, J.P. Acker, and A.Y. Elezzabi, "Reversible permeabilization using high-intensity femtosecond laser pulses: applications to biopreservation," *Biotechnol. Bioeng.* **92**(7), 889–899 (2005).

[58] M. Lei, H. Xu, H. Yang, et al., "Femtosecond laser-assisted microinjection into living neurons," *J. Neurosci. Meth.* **174**(2), 215–218 (2008).

[59] C. Peng, R.E. Palazzo, and I. Wilke, "Laser intensity dependence of femtosecond near-infrared optoinjection," *Phys. Rev. E Stat. Nonlin. Soft Matter Phys.* **75**(4 Pt 1), 041903-1–8 (2007).

[60] U.K. Tirlapur and K. König, "Targeted transfection by femtosecond laser," *Nature.* **418**(6895), 290–291 (2002).

[61] U.K. Tirlapur, K. Konig, C. Peuckert, et al., "Femtosecond near-infrared laser pulses elicit generation of reactive oxygen species in mammalian cells leading to apoptosis-like death," *Exp. Cell Res.* **263**(1), 88–97 (2001).

[62] L.E. Barrett, J.Y. Sul, H. Takano, et al., "Region-directed phototransfection reveals the functional significance of a dendritically synthesized transcription factor," *Nat. Methods.* **3**(6), 455–460 (2006).

[63] J.Y. Sul, C.W. Wu, F. Zeng, et al., "Transcriptome transfer produces a predictable cellular phenotype," *Proc. Natl. Acad. Sci. U.S.A.* **106**(18), 7624–7629 (2009).

[64] T. Knoll, L. Trojan, S. Langbein, et al., "Impact of holmium:YAG and neodymium:YAG lasers on the efficacy of DNA delivery in transitional cell carcinoma," *Lasers Med. Sci.* **19**(1), 33–36 (2004).

[65] J.S. Soughayer, T. Krasieva, S.C. Jacobson, et al., "Characterization of cellular optoporation with distance," *Anal. Chem.* **72**(6), 1342–1347 (2000).

[66] S.E. Mulholland, S. Lee, D.J. McAuliffe, et al., "Cell loading with laser-generated stress waves: the role of the stress gradient," *Pharm. Res.* **16**(4), 514–518 (1999).

[67] M. Terakawa, M. Ogura, S. Sato, et al., "Gene transfer into mammalian cells by use of a nanosecond pulsed laser-induced stress wave," *Opt. Lett.* **29**(11), 1227–1229 (2004).

[68] T.Y. Lin, D.J. McAuliffe, N. Michaud, et al., "Nuclear transport by laser-induced pressure transients," *Pharm. Res.* **20**(6), 879–883 (2003).

[69] X. Tsampoula, K. Taguchi, T. Cizmar, et al., "Fibre based cellular transfection," *Opt. Express.* **16**(21), 17007-1–13 (2008).

[70] R.P. Haughland, *Handbook of Fluorescent Probes and Research Chemicals*, 6th ed., Molecular Probes Inc., Eugene, Oregon (1996).

[71] Y. Shirahata, N. Ohkohchi, H. Itagak, et al., "New technique for gene transfection using laser irradiation," *J. Invest. Med.* **49**(2), 184–190 (2001).

[72] Y.A. Badr, M.A. Kereim, M.A. Yehia, et al., "Production of fertile transgenic wheat plants by laser micropuncture," *Photochem. Photobiol. Sci.* **4**(10), 803–807 (2005).

[73] S. Kurata, M. Tsukakoshi, T. Kasuya, et al., "The laser method for efficient introduction of foreign DNA into cultured-cells," *Exp. Cell Res.* **162**(2), 372–378 (1986).

[74] A. Yamaguchi, Y. Hosokawa, G. Louit, et al., "Nanoparticle injection to single animal cells using femtosecond laser-induced impulsive force," *Appl. Phy. A.* **93**(1), 39–43 (2008).

[75] C. Peng, I. Wilke, and R.E. Palazzo, "Laser-assisted controlled micro-injection into cultured living cells," *CLEO*, **3**(22–27), 2305–2307 (2005).

[76] M. Tsukakoshi, S. Kurata, Y. Nomiya, et al., "A novel method of DNA transfection by laser microbeam cell surgery," *Appl. Phys. B.* **35**(3), 135–140 (1984).

[77] O.H. Lowry, N.J. Rosebrough, A.L. Farr, et al., "Protein measurement with the folin phenol reagent," *J. Biol. Chem.* **193**, 265–275 (1951).

[78] Y. Gao, H. Liang, and M.W. Berns, "Laser-mediated gene transfer in rice," *Physiol. Plant.* **93**(1), 19–24 (1995).

[79] H. He, S.K. Kong, R.K. Lee, et al., "Targeted photoporation and transfection in human HepG2 cells by a fiber femtosecond laser at 1554 nm," *Opt. Lett.* **33**(24), 2961–2963 (2008).

[80] K. Ribbeck and D. Gorlich, "Kinetic analysis of translocation through nuclear pore complexes," *Embo J.* **20**(6), 1320–1330 (2001).

[81] A. Vogel, N. Linz, S. Freidank, et al., "Femtosecond-laser-induced nanocavitation in water: implications for optical breakdown threshold and cell surgery," *Phys. Rev. Lett.* **100**(3), 038102 (2008).

[82] A. Vogel, N. linz, S. Freidank, et al., "Femtosecond and nanosecond laser-induced nanoeffects for cell surgery and modification of glass," Presentation CMHH1, CLEO Workshop Tutorial, Monday May 5 (2008).

[83] P.A. Quinto-Su and V. Venugopalan, "Laser manipulation of cells and tissues," Chapter 4, *Methods Cell Biol.* **82** (2007).

4

Advances in Fluorescence Spectroscopy and Imaging

Herbert Schneckenburger

Hochschule Aalen, Institut für Angewandte Forschung, 73430 Aalen, Germany, Institut für Lasertechnologien in der Medizin und Messtechnik an der Universität Ulm, Helmholtzstr. 12, 89081 Ulm, Germany

Petra Weber, Thomas Bruns, Michael Wagner

Hochschule Aalen, Institut für Angewandte Forschung, 73430 Aalen, Germany

This chapter describes advanced methods of fluorescence imaging and *in vitro* diagnostics with some emphasis on wide-field microscopy with high spatial, spectral, and temporal resolution. Applications are concentrated on tumor diagnosis by autofluorescence, measurements of cholesterol dependent membrane dynamics, and detection of intermolecular interactions in pathogenesis of Alzheimer's disease as well as in sensing of apoptosis.

Key words: Fluorescence microscopy, cells, membranes, diseases

4.1 Introduction

Fluorescence techniques play an increasing role for diagnosis of living cells and tissues. This holds for intrinsic fluorophores, e.g., proteins [1,2] or coenzymes [3,4], as well as for an increasing number of fluorescence markers staining various organelles, e.g., cell nuclei, mitochondria, lysosomes, cytoskeleton, or membranes (for an overview see e.g., [5,6]). In addition, luminescent nanoparticles of semiconductors, so-called quantum dots, were introduced as biological labels, since they revealed to be rather photostable with variable emission spectra depending on the size of these particles [7]. A further important step in developing highly specific probes was the introduction of fluorescent proteins. A green fluorescent protein (GFP) is naturally produced by the jellyfish *Aequorea victoria* [8]. After cloning of the GFP gene, variants emitting light in the blue, yellow, or red spectral region could be generated. By fusion of genes coding for a specific cellular protein and a GFP variant, fluorescent protein chimera were thus created, which permitted a site-specific tracking in living cells or even whole organisms [9,10]. Unfortunately, the application of most fluorescent probes is restricted to *in vitro* systems, and only few dyes are presently used *in vivo*, e.g., fluorescein or indocyanine green for fluorescence angiography [11], or porphyrin derivatives for photodynamic diagnosis and therapy of cancer and other diseases [12]. Therefore, in the future autofluorescence measurements may play an increasing role in the diagnosis of patients.

A further important step for fluorescence diagnosis was the development of highly sensitive detection systems. While a few years ago charge coupled device (CCD) cameras with a sensitivity around 10^{-7} W/cm^2 represented the state of the art, highly sensitive electron multiplying (EM) CCD cameras [13] or image intensifying camera systems with a sensitivity between 10^{-11} and 10^{-14} W/cm^2 have now become available, permitting even single molecule detection [14].

In fluorescence microscopy enormous improvements in resolution have been reported in recent years. This holds on one side for confocal [15] and multiphoton [16,17] laser scanning microscopy (LSM) including 4-Pi and stimulated emission depletion microscopy (STED) with a lateral and axial resolution of less than 60 nm [18]. On the other side, axial resolution in the submicrometer range has been introduced into wide-field microscopy by structured illumination [19] or selective plane illumination (SPIM) [20] techniques. Due to the availability of highly sensitive camera systems, wide-field techniques often require lower doses of exciting light than laser scanning techniques. This may be an important factor for maintaining cell viability, as depicted in Figure 4.1. This figure results from a colony forming experiment upon seeding and irradiation of untreated single glioblastoma cells and evaluation of cell colonies 7 days later [21]. Colony formation was evaluated as a function of wavelength and light, and viability was defined by less than 10 % of reduction of the plating efficiency in comparison with non-irradiated controls. As deduced from Figure 4.1, cells remain viable up to a light dose of about 25 J/cm^2 at an excitation wavelength of 375 nm and up to 100 J/cm^2 at a wavelength of 514 nm. While the first wavelength appears appropriate for excitation of autofluorescence, the second wavelength can hardly excite any intracellular fluorophore, but may be appropriate for Raman microscopy. Cells incubated with fluorescent dyes or transfected with genes encoding for fluorescent proteins are expected to be more sensitive to light than untreated cells, such that the values depicted in Figure 4.1 may represent some upper limit.

Since the maintenance of cell viability under limited light dose is a main prerequisite of the authors' work, most techniques and applications described in this paper are related to wide-field microscopy.

4.2 Techniques and Requirements

4.2.1 Video microscopy and tomography

In wide-field fluorescence microscopy samples are commonly excited by a spectral lamp or a laser whose light is collimated in the entrance plane of a microscope, deflected by a dichroic mirror and focused on the sample by an objective lens. Due to its longer wavelength, fluorescence radiation passes the dichroic mirror and is again focused in the image plane of the microscope, where either an image detector (e.g., CCD or EM-CCD camera) or the entrance slit of a monochromator may be placed. Lateral resolution in the object plane is well defined by the equation $\Delta y = 0.61\ \lambda_{em} / A_N$ (with the emission wavelength λ_{em} and the numeric aperture of the objective lens A_N), whereas axial resolution is given by $\Delta z = n\ \lambda_{em} / A_N^2$ (with the refractive index n of the medium between the object and the objective lens). Although at high numeric aperture both Δy and Δz are in the range of a few hundred nanometers, a clear image from the focal plane of the microscope is generally superposed by out-of-focus parts of the image. To suppress those out-of-focus parts the following approaches have been used:

- Samples are excited by structured illumination, e.g., by imaging an optical grid in the sample plane, as depicted in Figure 4.2. When using different phases for positioning the grid (e.g., $\varphi = 0$, $2\pi/3$ and $4\pi/3$), an image section from the focal plane can be calculated, whereas

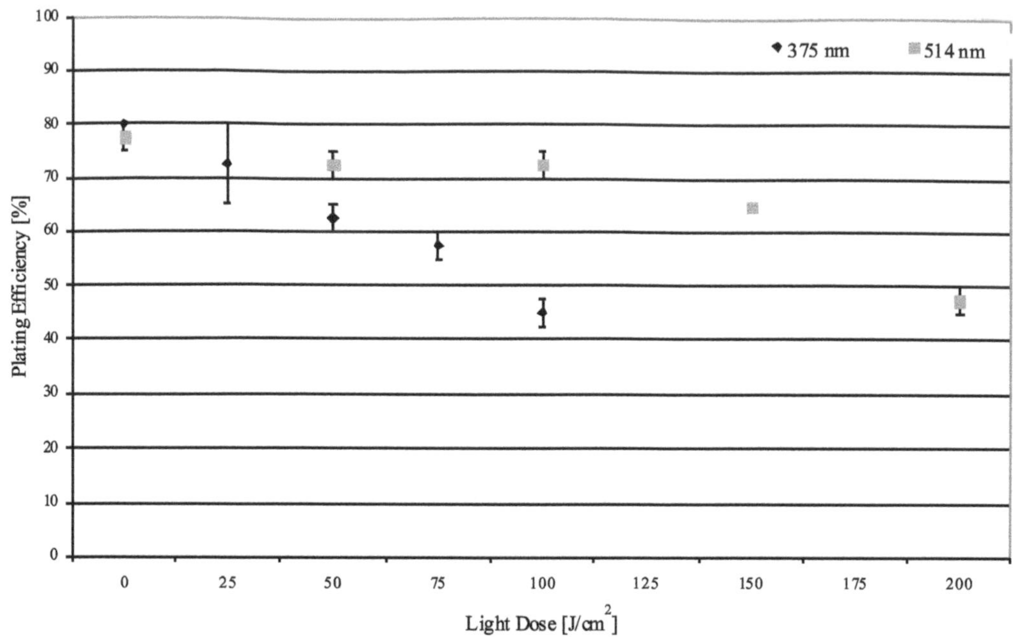

FIGURE 4.1: Percentage of colony formation ("plating efficiency") upon seeding of single U373-MG glioblastoma cells (CLS, Eppelheim, Germany, No. 300366; cells grown in supplemented RPMI 1640 medium) upon exposure to variable excitation wavelengths and light doses. Colonies were evaluated 7 days after seeding. Cells were not incubated with fluorophores nor treated with any other agents.

out-of-focus parts of the image vanish [19]. Axial resolution in this case strongly depends on the frequency of the grid and the numerical aperture of the objective lens, but may reach values around 1 μm or less.

- Only thin layers of a sample are illuminated from the side with a microscope lens of rather high focal depth (Selective Plane Illumination Microscopy, SPIM [20]). By shifting the sample into vertical direction numerous planes can thus be selected, and a 3-dimensional image of high axial resolution in the micrometer or submicrometer range can be reconstructed.

- Sample surfaces are optically excited by an evanescent electromagnetic wave arising upon total internal reflection (TIR) of a laser beam on a cell-substrate interface [22,23]. In this case axial resolution is generally given by the penetration depth of the evanescent wave around 100–150 nm, but can be lowered to only a few nanometers, if the angle of incidence is varied. In this case a tomographic image can be calculated from typically 8–12 individual images recorded at variable angles [24,25].

4.2.2 Spectral imaging

Spectral information has always been valuable in fluorescence imaging, but the importance of spectral signature has increased when quantum dots or fluorescent proteins became available for multicolor staining. Commonly optical gratings, prisms, or interference filters are used to obtain spectral resolution [26], but more recently acousto-optic tunable filters [27] or a combination of

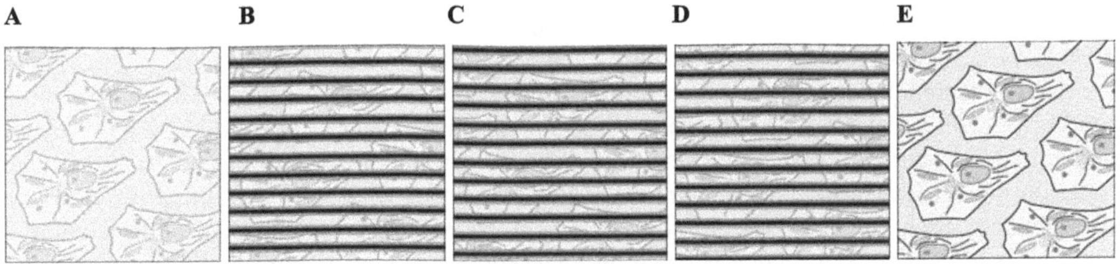

FIGURE 4.2: Principle of structured illumination: (A) conventional image; (B–D) structured images with different phases of a grid; (E) image section from the focal plane, as calculated from the structured images by $[(I_B-I_C)^2 + (I_B-I_D)^2 + (I_C-I_D)]^{1/2}$ [19].

image acquisition and interference techniques [28] have been used increasingly. Of particular interest are those systems where spectral and spatial resolution (Pushbroom imagers or interferometric systems) or spectral and temporal (nanosecond) resolution are combined, e.g., in studies of cell membrane dynamics (see below) or Förster resonance energy transfer (FRET) [29].

4.2.3 Fluorescence anisotropy

Excitation of samples with polarized light is used to select molecules whose optical transition dipole moments are parallel to the electrical field vector. Depolarization of fluorescence is often due to rotational diffusion of these molecules during the lifetime of their excited state, dependent on the size and shape of the molecules as well on the viscosity of their environment. Therefore, fluorescence polarization measurements can be used to examine viscosities of cells or cell membranes [1]. Commonly, fluorescence intensities are measured parallel ($I_\parallel$) and perpendicular ($I_\perp$) to the plane of incidence, and the anisotropy function is determined according to

$$r(t) = \frac{I_\parallel(t) - I_\perp(t)}{I_\parallel(t) + 2I_\perp(t)}.$$

If a sample is excited by a short (picosecond) polarized light pulse, $r(t)$ decreases exponentially (from the initial value r_0) with the rotational diffusion time constant t_r according to $r(t) = r_0 e^{-t/t_r}$, where t_r is proportional to the viscosity of the environment. In the case of free molecular rotation, $r(t)$ decays to 0, whereas in the case that molecular motion is hindered in one direction, $r(t)$ decreases to a final value r_∞, describing some residual anisotropy even at longer times after optical excitation. When measuring fluorescence anisotropies, some specific problems may arise from polarization-dependent sensitivities of detection or from photochemical modification (e.g., photobleaching) of the sample during fluorescence experiments. Whilc the first problem requires calibration of the experimental setup, the second problem can be resolved by simultaneous detection of two polarizations on two different sections of the camera after appropriate splitting of the optical detection path (see Figure 4.3).

4.2.4 Fluorescence lifetime imaging microscopy (FLIM)

In addition to fluorescence intensity, fluorescence lifetime τ revealed to be an important parameter characterizing the interaction of a fluorescent molecule with its molecular and cellular environment. τ is given as the reciprocal of all rates of radiative and nonradiative transitions originating from an excited molecular state. Therefore, τ is affected, e.g., by aggregation, binding to proteins or intermolecular energy transfer (FRET). Samples are commonly excited by short (picosecond) laser pulses, and fluorescence is described by a mono- or multi-exponential decay curve according to $I(t) = \Sigma_i A_i e^{-t/\tau_i}$ (with components i and individual decay times τ_i).

A current technique for fluorescence lifetimes measurement is time-correlated single photon counting (TCSPC) [30, 31]. Single probes or 2-dimensional samples are excited by high repetition picosecond laser pulses, and time delays of all fluorescence photons with respect to the relevant excitation pulse are measured and accumulated. By synchronization with the actual position in a laser scanning microscope (LSM), fluorescence decay curves are thus obtained for all pixels of a 2-dimensional sample [31]. Image acquisition, however, needs considerable light exposure (typically 1 mJ for an area of 100 μm × 100 μm) and prolonged measuring times (often above 20 s). However, light dose may be reduced in wide-field microscopy using ultrasensitive camera systems, as depicted in Figure 3. Here, instead of individual decay curves, fluorescence intensities (I_A, I_B) are commonly measured within two time gates (A,B) shifted by a delay Δt between one another. An effective fluorescence lifetime τ_{eff} can thus be calculated for all pixels of an image according to [32]

$$\tau_{\text{eff}} = \Delta t / \ln(I_A / I_B)$$

.

4.2.5 Fluorescence screening

For more than 10 years fluorescence has played a leading role in high throughput (HTS) and high content screening (HCS) of various diseases as well as for drug discovery [34]. Individual samples are often located in cavities of a so-called microtiter plate, where their fluorescence properties are examined either simultaneously or sequentially. Often fluorescence signals arising from surfaces of a substrate are of particular interest. This holds, e.g., for antigen-antibody reactions, cell-substrate reactions, or cell membrane dynamics. For this case a fluorescence reader with simultaneous total internal reflections (TIR) of a laser beam on 96 individual samples was recently developed [35]. As depicted in Figure 4.4 a laser beam is split into 8 parallel beams, coupled into the glass bottom of a microtiter plate and totally reflected on the surface of 12 cavities each. Therefore, the evanescent electromagnetic field penetrates about 150 μm into all samples and excites membrane associated molecules for fluorescence. Main prerequisites are an appropriate thickness and low absorption of the reflecting glass plate as well as the use of a noncytotoxic adhesive.

In addition to simultaneous fluorescence detection of all cavities, specific properties of individual samples – in particular, fluorescence polarization and decay time – can be assessed by scanning of a miniaturized dual-beam detection system [36]. So screening of a larger number of samples (HTS) is combined with more detailed measurements of smaller sample numbers of particular interest (HCS). Among other applications this setup was used as a reader for apoptosis, where the fluorescence lifetime of membrane associated cyan fluorescent protein (CFP) changed upon cleavage of a CFP-YFP (yellow fluorescent protein) linker by the enzyme caspase-3 [37].

4.3 Applications

4.3.1 Autofluorescence imaging

Intrinsic fluorescence of many organisms is dominated by proteins, extracellular fibers and coenzymes of some oxido-reductases, in particular reduced nicotinamide adenine-dinucleotide (NADH) and flavin molecules [3, 4]. These coenzymes may reflect cell physiology and can be used to probe either hypoxies [38] or deficiencies of the mitochondrial respiratory chain [39]. In addition, autoflu-

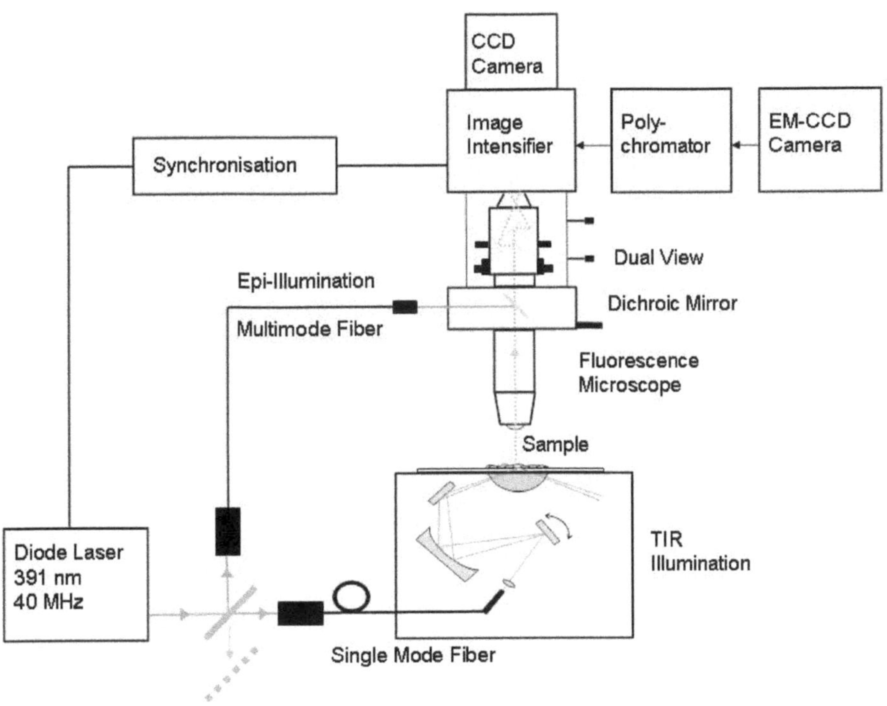

FIGURE 4.3: Setup for wide-field fluorescence microscopy with epi- or total internal reflection (TIR) illumination. For FLIM a time-gated image intensifier is synchronized with the pulses of the exciting laser diode using an adjustable delay line. Fluorescence can be split into 2 beams of different polarization or different spectral signature, detected on 2 sections of the camera system. Fluorescence spectra or cw images are measured optionally (reproduced from [33] with permission by SPIE).

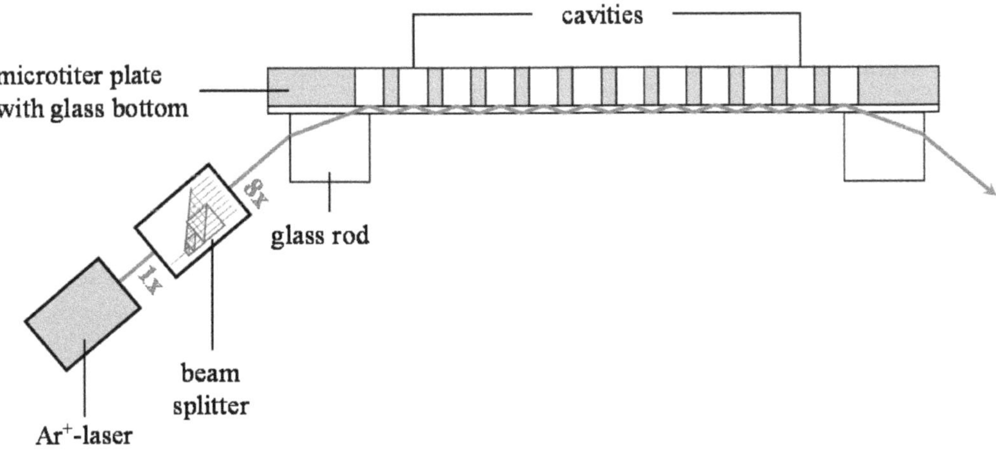

FIGURE 4.4: Setup for TIR fluorescence screening of up to 96 samples in a microtiter plate.

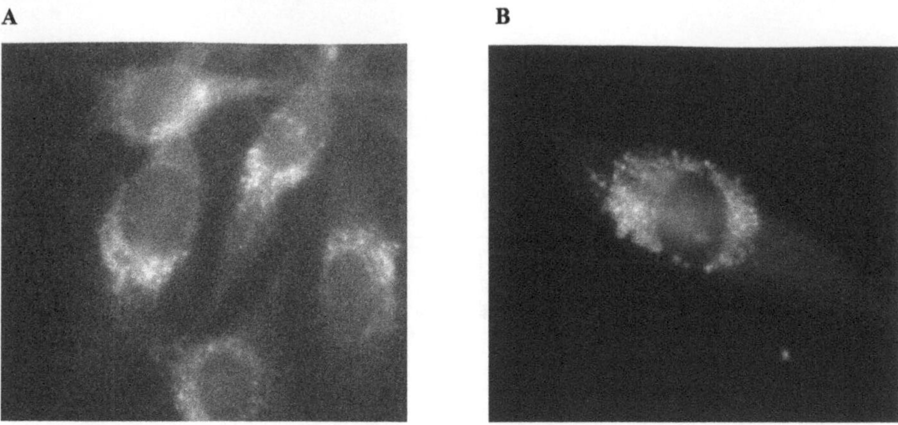

FIGURE 4.5: Autofluorescence of U373-MG glioblastoma cells; (A) reference cells, and (B) cells with activated TP53 suppressor gene. Excitation wavelength: 375 nm, emission measured at $\lambda \geq 415$ nm, image size: $140 \times 120\ \mu\text{m}^2$.

orescence measurements have been used increasingly for early diagnosis of cancer in various organs (e.g., bronchus, colon, bladder, lung, or skin [40–44]). However, a comparison between tumor cells and less malignant cells turned out to be very difficult, since comparable cells lines were hardly available. Possibly this difficulty can now be overcome by use of so-called isogenic cell lines with switchable tumor-specific genes.

In Figure 4.5 autofluorescence of U373-MG human glioblastoma cells is depicted, and images of reference cells and cells with an activated TP53 suppressor gene are compared. Cells were kindly supplied by Prof. Jan Mollenhauer, Dept. of Molecular Oncology, University of South Denmark, Odense. While the tumor (control) cells show a granular as well as a diffuse fluorescence pattern, the granular pattern is dominating in the less malignant cells with the activated suppressor gene. Concomitantly some shift of the emission spectra towards shorter wavelengths and some prolongation of the fluorescence decay times were observed for the less malignant cells (results not shown). These results are presently discussed in connection with differences in mitochondrial metabolism.

4.3.2 Membrane dynamics

Membrane dynamics, including membrane stiffness and fluidity, has a large impact on cell function and may be an important factor for certain diseases, e.g., Morbus Alzheimer [45]. A major component of lipid bilayers is cholesterol, which determines its fluidity and rigidity. Membrane dynamics has so far been deduced from measurements of fluorescence polarization [1] or fluorescence recovery after photobleaching (FRAP) [46] using specific membrane markers. In addition, polarity-sensitive probes, e.g., 6-dodecanoyl-2-dimethylamino naphthalene (laurdan), whose electronic excitation energy is different in polar and nonpolar environments [47,48], proved to be useful for membrane studies. Once incorporated into cell membranes, the fluorescence of laurdan shows a spectral shift towards longer wavelengths when its molecules get into contact with adjacent water molecules of the cytoplasm, e.g., when membrane stiffness decreases or when a phase transition from the tightly packed gel phase to the liquid crystalline phase of membrane lipids occurs.

Fluorescence spectra of single U373-MG cells incubated with laurdan are depicted in Figure 4.6 at $T = 24°\text{C}$. These spectra show two overlapping emission bands with maxima around 440 nm and 490 nm. Without any manipulation of the natural cholesterol content similar fluorescence intensi-

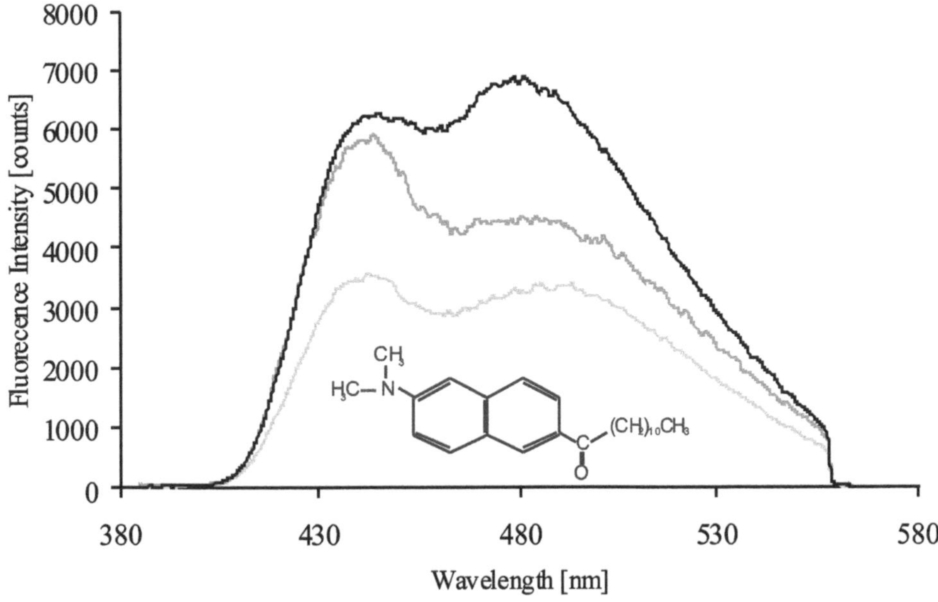

FIGURE 4.6: Molecular structure and fluorescence spectra of laurdan in single U373-MG glioblastoma cells at $T = 24°C$ without further incubation (lower curve), as well as upon cholesterol enrichment (up to 200%, middle curve) or cholesterol depletion (to 50%, upper curve); excitation wavelength: 391 nm. Reproduced from [33] with permission by SPIE.

ties were measured at 440 nm and 490 nm (lower curve). However, after enrichment of cholesterol (by a methyl-β-cyclodextrine:cholesterol complex [50]), the 440 nm band appeared to be more pronounced (middle curve), whereas after depletion of cholesterol (by methyl-β-cyclodextrine) the 490-nm band was predominant (upper curve). This effect was further described by the generalized polarization $GP = (I_{440} - I_{490})/(I_{440} + I_{490})$ [49] with I_{440} corresponding to the fluorescence intensity at 440 nm and I_{490} to that at 490 nm. An increase of GP with the amount of cholesterol, as shown in Figure 4.7, reflects increasing membrane stiffness. GP further decreased with temperature between 16°C and 40°C and was always higher for the plasma membrane (upon TIR illumination) than for intracellular membranes (upon illumination of whole cells, see Figures 4.6 and 4.7) [33].

In addition to spectral imaging, fluorescence anisotropy imaging can give valuable information on membrane dynamics. The faster a molecule rotates between optical excitation and fluorescence, the lower is the expected anisotropy value r (reported in Section 4.2.3). This value can be determined from steady state as well as from time-resolved fluorescence experiments when after short laser pulse excitation $r(t)$ decays exponentially with the rotational relaxation time τ_r, which is a direct measure for the viscosity of the environment. A result of stationary measurements with laurdan is depicted in Figure 4.8 at $T = 24°C$ upon enrichment (left), unaltered concentration (middle), or depletion (right) of cholesterol. This figure demonstrates that fluorescence anisotropy increases with increasing amounts of cholesterol and shows maximum values of r in the plasma membrane as well as minimum values in the nuclear membrane. A decrease of r with increasing temperature is clearly documented and proves a decrease of viscosity in cell membranes (not shown in Figure 4.8).

Fluorescence lifetime of laurdan revealed to be a further parameter of membrane dynamics. Lifetime images, too, were recorded as a function of intracellular cholesterol and proved that τ_{eff} decreased upon depletion of cholesterol in whole cells as well as in plasma membranes (illuminated

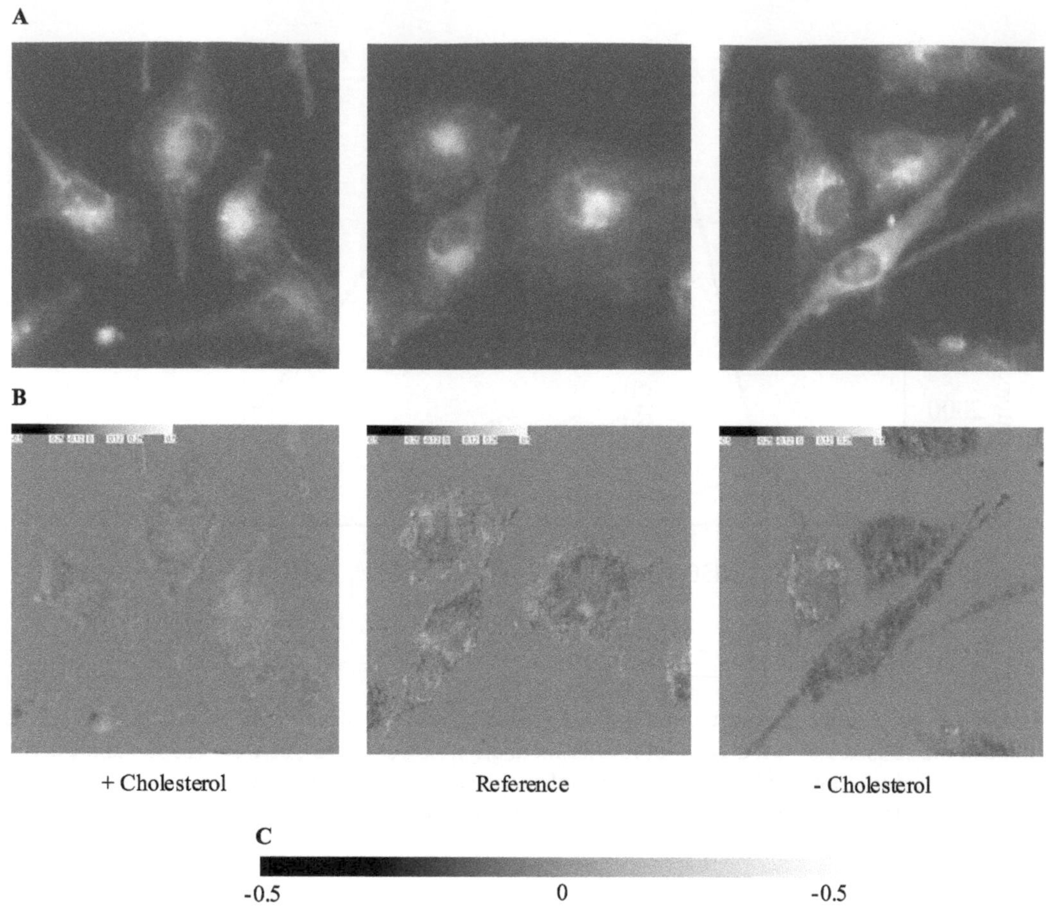

FIGURE 4.7: GP (spectral) imaging of U373-MG glioblastoma cells incubated with laurdan (8 μM; 1 h) upon enrichment of cholesterol (up to 200%, left), without further treatment (middle) or upon depletion of cholesterol (to 50%, right); (A) integral fluorescence intensity at $\lambda \geq 415$ nm; (B) generalized polarization (GP); (C) GP scale. Image size: 140 μm $\times$ 140 μm each.

selectively by TIR; Figure 4.9). In comparison with spectral imaging and anisotropy imaging, FLIM needs only a few mathematical calculations, is comparably fast and permits to record kinetics in the range of few seconds or even less.

In contrast to laser scanning microscopy (LSM) wide-field methods do not possess inherent axial resolution. However, axial resolution can be introduced by structured illumination, SPIM or TIR techniques, as reported in Subsection 4.2.1. A fluorescence image of U373-MG glioblastoma cells incubated with laurdan and exposed to structured illumination is depicted in Figure 4.10(A), whereas a calculated image section (according to [19]) is depicted in Figure 4.10(B) and a conventional wide-field image in Figure 4.10(C). The advantage of structured illumination in comparison with conventional wide-field microscopy is obvious, and a combination with spectral imaging, anisotropy imaging, or FLIM appears possible and advantageous.

As already mentioned, also Total Internal Reflection Fluorescence Microscopy (TIRFM) provides axial resolution, since only a thin layer of about 150 nm is optically excited. In addition, variable-angle TIRFM of fluorophores located either in the plasma membrane or in the cytoplasm

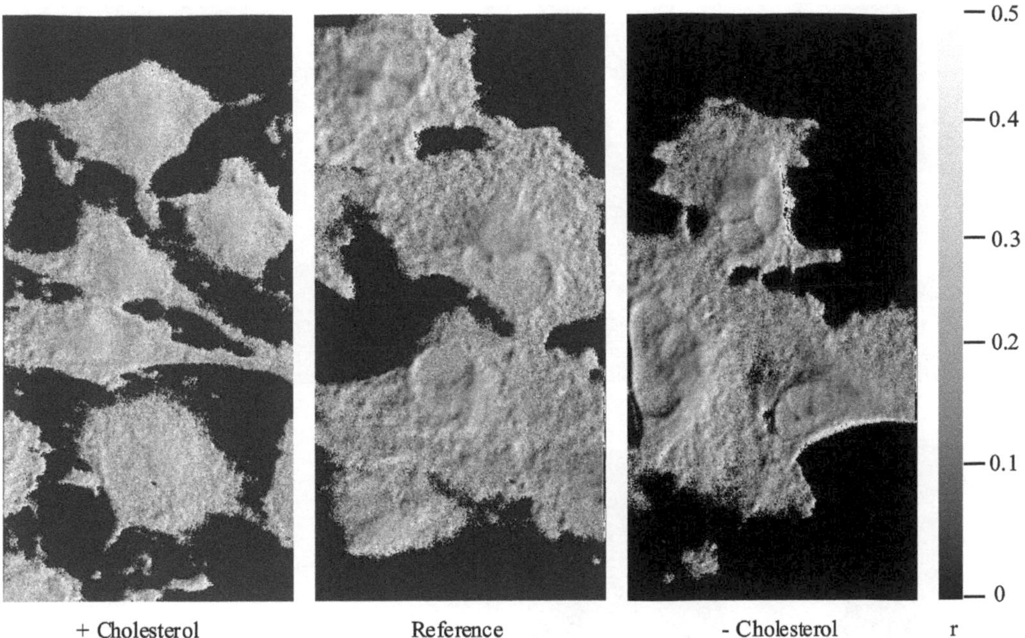

FIGURE 4.8: Steady state fluorescence anisotropy imaging of U373-MG glioblastoma cells incubated with laurdan (8 μM, 1 h) upon enrichment (up to 200%, left), unaltered concentration (middle) or depletion (to 50%, right) of cholesterol (excitation wavelength: 391 nm; emission measured at $\lambda \geq 415$ nm; image size: 70 μm $\times$ 140 μm). Reproduced from [33] with permission by SPIE.

permits fitting of cell-substrate distances with nanometer precision [24, 25]. In Figure 4.11 cell-substrate topologies are depicted for U373-MG glioblastoma cells incubated with laurdan either without further treatment (left) or with 50% depletion of cholesterol. While in the first case cell-substrate distances varied between about 20 nm and 100 nm, those distances increased in the second case to more than 200 nm with only the edges of the cells being closer to the glass substrate. This demonstrates the important role of cholesterol for cell adhesion.

Cell-substrate topologies were further evaluated for photodynamic therapy (PDT) with protoporphyrin IX, and it was proven that cell-substrate distances were maintained or even lowered after application of nonlethal light doses, thus excluding the formation of metastases with rather high probability [25, 51].

4.3.3 FRET-based applications

As already mentioned, Förster resonance energy transfer (FRET) [29] plays an essential role for probing intermolecular interactions, in particular protein-protein interactions. Previously, a sensor for apoptosis, based on FRET from cyan (CFP) to yellow fluorescent protein (YFP) connected by the caspase sensitive amino acid peptide linker DEVD, has been reported in the literature, and a pronounced increase of YFP fluorescence has been observed upon cleavage of the linker [52–54]. In order to increase the sensitivity of this FRET sensor, enhanced cyan fluorescent protein (ECFP) was now anchored in the plasma membrane and optically excited by the evanescent field arising upon total internal reflection of a laser beam [55]. By this way, background fluorescence from

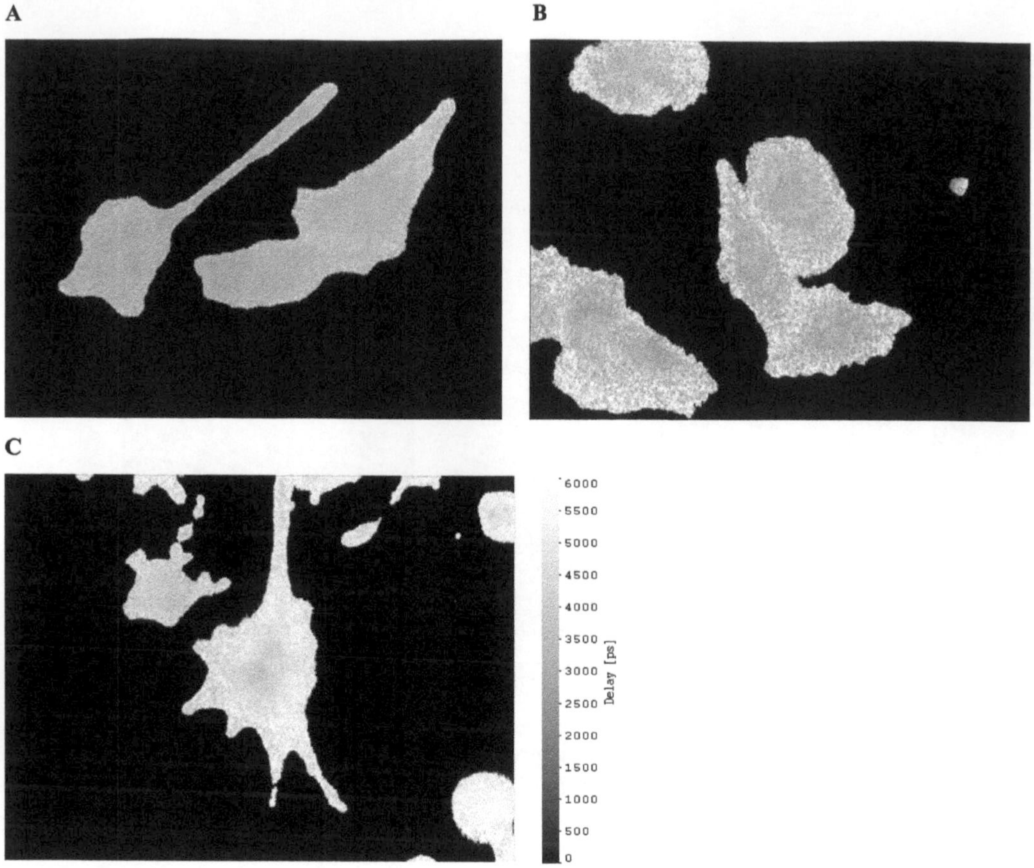

FIGURE 4.9: Fluorescence lifetime images of U373-MG glioblastoma cells incubated with laurdan (8 μM, 1 h) upon depletion (to 50%, A), unaltered concentration (B) or enrichment (up to 200%, C) of cholesterol. Selective illumination of the plasma membrane by TIR (excitation wavelength: 391 nm; emission measured at $\lambda \geq 415$ nm; image size: 130 μm $\times$ 100 μm).

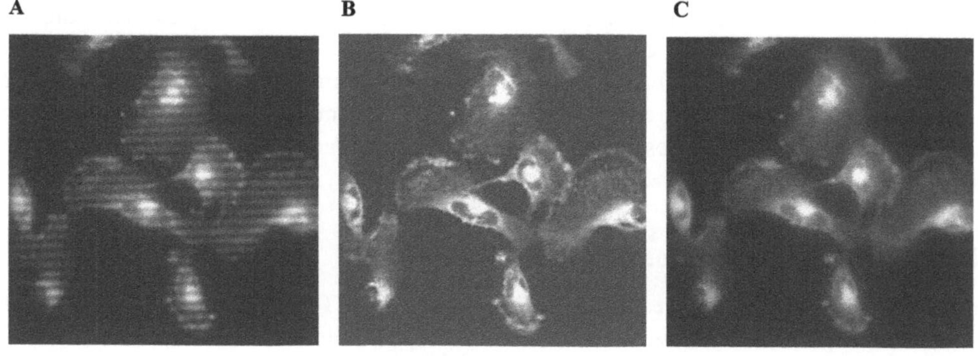

FIGURE 4.10: Fluorescence of U373-MG glioblastoma cells incubated with laurdan (8 μM, 1 h) upon structured illumination (A) with calculated image section (B) and reconstructed conventional image (C) (excitation wavelength: 391 nm; emission measured at $\lambda \geq 415$ nm; image size: 140 μm $\times$ 140 μm).

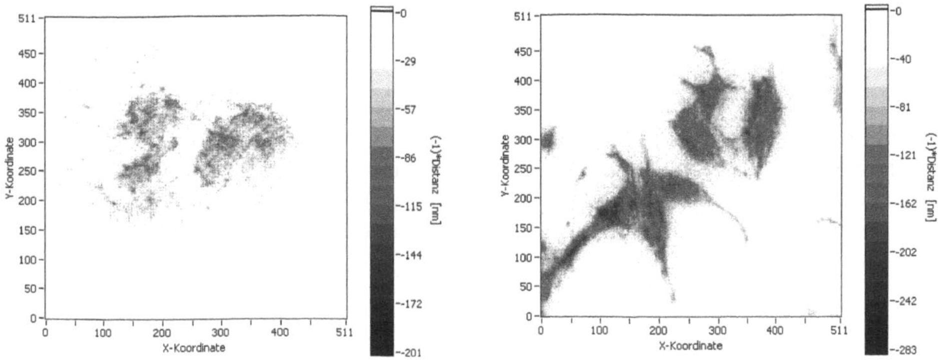

FIGURE 4.11: Cell-substrate topology of U373MG glioblastoma cells upon incubation with laurdan deduced from variable-angle TIRFM either without further treatment of the cells (left) or 50% cholesterol depletion (right) (fluorescence excited at 391 nm and detected at $\lambda \geq 415$ nm; image size: 210 μm $\times$ 210 μm). Spatial coordinates and a scale for cell-substrate distances are indicated.

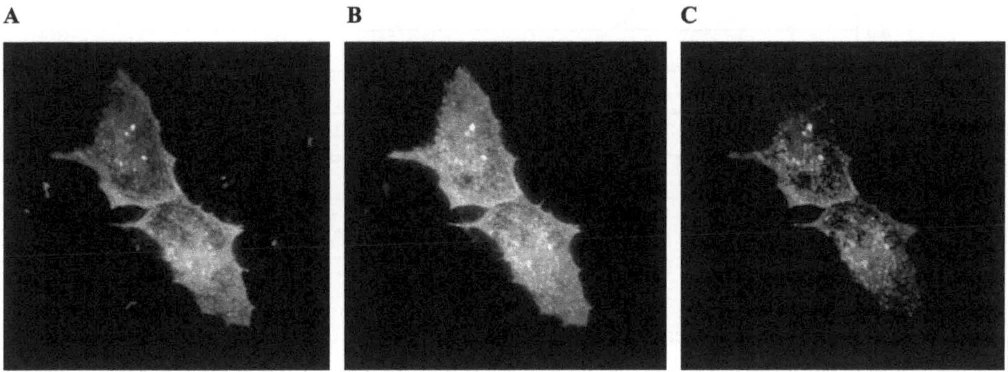

FIGURE 4.12: TIR images of HeLa cells transiently transfected with the MemECFP-DEVD-EYFP plasmid upon excitation of the donor (ECFP) at 391 nm and its detection at 475 $\pm$ 20 nm (A) as well as upon excitation of the acceptor (EYFP) and its detection at $\lambda \geq 515$ nm (B). Figure 4.12(C) mimics the lower EYFP fluorescence upon excitation of the donor and cleavage of the DEVD linker (image size: 140 μm $\times$ 140 μm). Reproduced from [56] with modifications.

the cytoplasm was suppressed efficiently. In addition to fluorescence spectra, fluorescence decay curves were evaluated, since fluorescence lifetime permits to calculate energy transfer rates k_{ET} rather easily according to the relation

$$k_{ET} = 1/\tau - 1/\tau_0$$

with τ corresponding to the lifetime in presence and τ_0 in absence of an acceptor.

In Figure 4.12 fluorescence of HeLa cervix carcinoma cells containing the membrane associated complex MemECFP-DEVD-EYFP (enhanced yellow fluorescent protein) is depicted for the spectral ranges of ECFP (A) and EYFP (B) emission. EYFP emission was similar upon direct optical excitation (at 470 nm) and excitation via energy transfer ECFP→EYFP (with light absorption at 391 nm). After induction of apoptosis (by 2 μM staurosporine) fluorescence intensity of EYFP decreased by a factor of about 2 (C), whereas fluorescence intensity of ECFP increased considerably, thus proving an interruption of intermolecular energy transfer. In the case of FRET (prior to apoptosis) fluorescence lifetime of the donor decreased from 2.9 ns to 2.6 ns for one part of ECFP molecules and from 2.9 ns to 0.5 ns for another part [55]. Therefore, two different energy transfer rates were deduced. In addition to results obtained from single cells in fluorescence microscopy, shortening of fluorescence lifetimes due to energy transfer was also observed for larger cell collectives in a fluorescence reader with a mixed population of apoptotic and nonapoptotic cells as well as a high amount of cells that did not contain the ECPF-DEVD-EYFP complex at all [37]. Therefore, this reader system appears to be quite suitable even for routine applications.

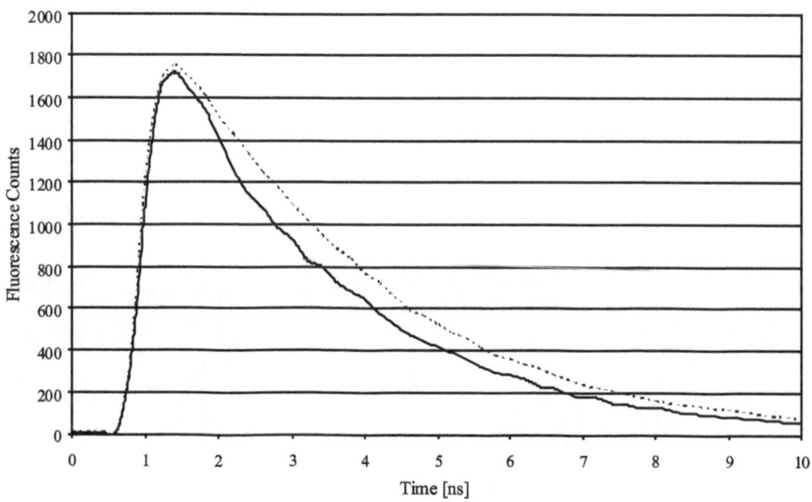

FIGURE 4.13: Fluorescence decay curves of a single U373 MG cell transfected with the BACE-GFP encoding plasmid (dotted) and a cell cotransfected with BACE-GFP and APP-mRFP encoding plasmids (solid) upon whole cell illumination (excitation wavelength: 470 nm; emission measured at $\lambda \geq 515$ nm).

Nonradiative energy transfer was further used to probe the proximity of amyloid precursor protein (APP) and β-secretase (BACE) [56], which play an essential role in the pathogenesis of Alzheimer's disease, since cleavage of APP by BACE is the first proteoloytic step in the production of amyloid peptide. By fusing BACE with green fluorescent protein (GFP) and APP with monomeric red fluorescent protein (mRFP), close proximity of these molecules was proven in the cytoplasm (e.g., within endosomes) as well as in the plasma membrane by FRET experiments. This can be deduced from the fluorescence kinetics in Figure 4.13, according to which fluorescence of BACE-GFP decays faster in presence than in absence of APP-mRFP. Although the difference of fluorescence lifetimes was rather small (2.5 ns v/s 2.7 ns), it could be well evaluated due to the mono-exponential behaviour of GFP fluorescence (in contrast to CFP). It was also proven that intermolecular energy transfer is disrupted upon enrichment as well upon depletion of cholesterol in cell membranes, and therefore seems to depend on membrane fluidity [56].

4.4 Final Remarks

Some advanced methods of fluorescence spectroscopy, microscopy, and imaging have been summarized, and recent results – ranging from tumor diagnosis by autofluorescence, studies of cholesterol dependent membrane dynamics up to intermolecular interactions in the pathogenesis of Alzheimer's disease as well as in sensing of apoptosis – have been reported. The importance of low light doses for irradiation was emphasized, and all methods described in this article were adapted to those low doses. In future work, further shortening of measuring times (for detection of fast kinetics), experimental setups for routine applications and advanced algorithms of evaluation shall be implemented.

Acknowledgments

Current research projects are funded by the Bundesministerium für Bildung und Forschung (BMBF), Ministerium für Wissenschaft, Forschung und Kunst Baden Württemberg as well as Landesstiftung Baden-Württemberg GmbH. Results reported above were achieved in cooperation with B. Angres and H. Steuer (NMI Reutlingen), C. v. Arnim (Dept. of Neurology, Univ. Ulm), W.S.L. Strauss and R. Wittig (ILM Ulm), P. Kioschis (Hochschule Mannheim) and J. Mollenhauer (Dept. of Molecular Oncology, University of South Denmark). Technical assistance by Claudia Hintze is gratefully acknowledged.

References

[1] J.R. Lakowicz, *Principles of Fluorescence Spectroscopy*, Springer Science + Business, 3rd Edition, New York (2006).

[2] J.M. Beechem and L. Brand, "Time-resolved fluorescence of proteins," *Ann. Rev. Biochem.* **54**, 43–71 (1985).

[3] T. Galeotti, G.D.V. van Rossum, D.H. Mayer, and B. Chance, "On the fluorescence of NAD(P)H in whole cell preparations of tumours and normal tissues," *Eur. J. Biochem.* **17,** 485–496 (1970).

[4] J.-M. Salmon, E. Kohen, P. Viallet, J.G. Hirschberg, A.W. Wouters, C. Kohen, and B. Thorell, "Microspectrofluorometric approach to the study of free/bound NAD(P)H ratio as metabolic indicator in various cell types," *Photochem. Photobiol.* **36**, 585–593 (1982).

[5] I. Johnson, "Fluorescent probes for living cells," *Histochem. J.* **30**, 123–140 (1988).

[6] J.M. Mullins, "Overview of fluorochromes," *Methods Mol. Biol.* **115**, 97–105 (1999).

[7] M. Bruchez, M. Moronne, P. Gin, S. Weis, and A.P. Alivisatos, "Semiconductor nanocrystals as fluorescence biological labels," *Science* **281**, 2013–2016 (1998).

[8] C.W. Cody, D.C. Prasher, W.M. Westler, F.G. Prendergast, and W.W. Ward, "Chemical structure of the hexapeptide chromophore of the Aequorea green-fluorescent protein," *Biochemistry* **32**, 1212–1218 (1993).

[9] R. Rizzuto, M. Brini, P. Pizzo, M. Murgia, and T. Pozzan, "Chimeric green fluorescent protein as a tool for visualizing subcellular organelles in living cells," *Curr. Biol.* **5**, 635–642 (1995).

[10] M. Ikawa, S. Yamada, T. Nakanishi, and M. Okabe, "Green fluorescent protein (GFP) as a vital marker in mammals," *Curr. Top. Dev. Biol.* **44**, 1–20 (1999).

[11] R. Brancato and G. Trabucchi, "Fluorescein and indocyanine green angiography in vascular chorioretinal diseases," *Semin. Ophthalmol.* **13**, 189–198 (1998).

[12] T.J. Dougherty, "Photosensitizers: therapy and detection of malignant tumours," *Photochem. Photobiol.* **45**, 879–889 (1987).

[13] C.G. Coates, D.J. Denvir, N.G. McHale, K.D. Thornbury, and M.A. Hollywood, "Optimizing low-light microscopy with back-illuminated electron multiplying charge-coupled device: enhanced sensitivity, speed and resolution," *J. Biomed. Opt.* **9**(6), 1244–1252 (2004).

[14] Y. Sako, S. Minoguchi, and T. Yanagida, "Single-molecule imaging of EGFR signalling on the surface of living cells," *Nature Cell. Biol.* **2**, 168–172 (2000).

[15] J. Pawley, *Handbook of Biological Confocal Microscopy*, Plenum Press, New York (1990).

[16] W. Denk, J.H. Strickler, and W.W. Webb, "Two-photon laser scanning microscope," *Science*, **248**, 73–76 (1990).

[17] K. König, "Multiphoton microscopy in life sciences," *J. Microsc.* **200**, 83–86 (2000).

[18] K.I. Willig, B. Harke, R. Medda, and S.W. Hell, "STED microscopy with continuous wave beams," *Nat. Methods* **4**(11), 915–918 (2007).

[19] M.A.A. Neil, R. Juskaitis, and T. Wilson, "Method of obtaining optical sectioning by structured light in a conventional microscope," *Opt. Lett.* **22**, 1905–1907 (1997).

[20] J. Huisken, J. Swoger, F. Del Bene, J. Wittbrodt, and E.H.K. Stelzer, "Optical sectioning deep inside live embryos by SPIM," *Science* **305**, 1007–1009 (2004).

[21] H. Schneckenburger, P. Weber, M. Wagner, S. Schickinger, T. Bruns, and W.S.L. Strauss, "Dose limited fluorescence microscopy of 5-aminolevulinic acid induced protoporphyrin IX in living cells," *Proc. SPIE* **7164**, 71640F-1–7 (2009).

[22] D. Axelrod, "Cell-substrate contacts illuminated by total internal reflection fluorescence," *J. Cell Biol.* **89**, 141–145 (1981).

[23] H. Schneckenburger, "Total internal reflection fluorescence microscopy: technical innovations and novel applications," *Curr. Opin. Biotechnol.* **16**, 13–18 (2005).

[24] K. Stock, R. Sailer, W.S.L. Strauss, M. Lyttek, R. Steiner, and H. Schneckenburger, "Variable-angle total internal reflection fluorescence microscopy (VA-TIRFM): realization and application of a compact illumination device," *J. Microsc.* **211**, 19–29 (2003).

[25] M. Wagner, P. Weber, W.S.L. Strauss, H.-P. Lassalle, and H. Schneckenburger, "Nanotomography of cell surfaces with evanescent fields," *Adv. Opt. Tech.* **2008**, Article ID 254317 (2008).

[26] W. Schmidt, *Optical Spectroscopy in Chemistry and Life Sciences*, Wiley-VCH, Weinheim (2005).

[27] N. Gupta, "Biosensors technologies: acousto-optic tunable filter-based hyperspectral and polarization imagers for fluorescence and spectroscopic imaging," *Methods Mol. Biol.* **503**, 293–305 (2009).

[28] Y. Garini, I.T. Young, and G. McNamara, "Spectral imaging: principles and applications," *Cytometry* **A69**(8), 735–747 (2006).

[29] T. Förster T, "Zwischenmolekulare Energiewanderung und Fluoreszenz," *Ann. Physik* **2**, 55–75 (1948).

[30] D.V. O'Connor and D. Philipps, *Time-Correlated Single Photon Counting,* Academic Press, London (1984).

[31] W. Becker, *Advanced Time-Correlated Single Photon Counting Techniques*, Springer, Berlin-Heidelberg-New York (2005).

[32] E.P. Buurman, R. Sanders, A. Draijer, H.C. Gerritsen, J.J.F. van Veen, P.M. Houpt, and Y.K. Levine, "Fluorescence lifetime imaging using a confocal laser scanning microscope," *Scanning* **14**, 155–159 (1992).

[33] M. Wagner, P. Weber, T. Bruns, W.S.L. Strauss, and H. Schneckenburger. "Fluorescence and polarization imaging of membrane dynamics in living cells," *Proc. SPIE* **7176**, 717607-1–8 (2009).

[34] G.P. Sabbatini, W.A. Shirley, and D.L. Coffeen, "The integration of high throughput technologies for drug discovery," *J. Biomol. Screen.* **6**, 213–218 (2001).

[35] T. Bruns, W.S.L. Strauss, R. Sailer, M. Wagner, and H. Schneckenburger, "Total internal reflectance fluorescence reader for selective investigations of cell membranes," *J. Biomed. Opt.* **11**(3), 034011-1–7 (2006).

[36] T. Bruns, W.S.L. Strauss, and H. Schneckenburger, "Total internal reflection fluorescence lifetime and anisotropy screening of cell membrane dynamics," *J. Biomed. Opt.* **13**(4), 041317-1–5 (2008).

[37] T. Bruns, B. Angres, H. Steuer, P. Weber, M. Wagner, and H. Schneckenburger, "A FRET-based total internal reflection (TIR) fluorescence reader for apoptosis," *J. Biomed. Opt.* **14**(2), 021003-1–5 (2009).

[38] E.T. Obi-Tabot, L.M. Hanrahan, R. Cachecho, E.R. Berr, S.R. Hopkins, J.C.K. Chan, J.M. Shapiro, and W.W. LaMorte, "Changes in hepatocyte NADH fluorescence during prolonged hypoxia," *J. Surg. Res.* **55**, 575–580 (1993).

[39] M.H. Gschwend, R. Rüdel, W.S.L. Strauss, R. Sailer, H. Brinkmeier, and H. Schneckenburger, "Optical detection of mitochondrial NADH content in human myotubes," *Cell. Mol. Biol.* **47**, OL95–OL104 (2001).

[40] J. Hung, S. Lam, J.C. LeRiche, and B. Palcic, "Autofluorescence of normal and malignant bronchial tissue," *Lasers Surg. Med.* **11**, 99–105 (1991).

[41] B. Banerjee, B. Miedema, and H.R. Chandrasekhar, "Emission spectra of colonic tissue and endogenous fluorophores," *Am. J. Med. Sci.* **315**, 220–226 (1998).

[42] L. Rigacci, R. Albertini, P.A. Bernabei, P.R. Feriini, G. Agati, F. Fusi, and M. Monici, "Multispectral imaging autofluorescence microscopy for the analysis of lymph-node tissues," *Photochem. Photobiol.* **71**, 737–742 (2000).

[43] M.A. D'Hallewin, L. Bezdetnaya, and F. Guillemin; "Fluorescence detection of bladder cancer: a review," *Eur. Urol.* **42**(5), 417–425 (2002).

[44] T. Gabrecht, T. Glanzmann, L. Freitag, B.C. Weber, H. van den Bergh, and G. Wagnières, "Optimized autofluorescence bronchoscopy using additional backscattered red light," *J. Biomed. Opt.* **12**(6), 064016-1-9 (2007).

[45] N.B. Chauhan, "Membrane dynamics, cholesterol homeostasis, and Alzheimer's disease," *J. Lipid Res.* **44**(11), 2019–2029 (2003).

[46] N.L. Thompson, A.W. Drake, L. Chen, and W.V. Broek, "Equilibrium, kinetics, diffusion and self-association of proteins at membrane surfaces: measurement by total internal reflection fluorescence microscopy," *Photochem. Photobiol.* **65**, 39–46 (1997).

[47] T. Parasassi, G. de Stasio, A. d'Ubaldo, and E. Gratton, "Phase fluctuation in phospholipid membranes revealed by laurdan fluorescence," *Biophys. J.* **57**, 1179–1186 (1990).

[48] T. Parasassi, E.K. Krasnowska, L. Bagatolli, and E. Gratton, "Laurdan and prodan as polarity-sensitive fluorescent membrane probes," *J. Fluoresc.* **4**, 365–373 (1998).

[49] T. Parasassi, G. de Stasio, G. Ravagnan, R.M. Rusch, and E. Gratton, "Quantitation of lipid phases in phospholipid vesicles by the generalized polarization of laurdan fluorescence," *Biophys. J.* **60**, 179–189 (1991).

[50] A.E. Christian, M.P. Haynes, M.C. Phillips, and G.H. Rothblat, "Use of cyclodextrins for manipulating cellular cholesterol content," *J. Lipid Res.* **38**, 2264–2272 (1997).

[51] H.-P. Lassalle, H. Baumann, W.S.L. Strauss, and H. Schneckenburger, "Cell-substrate topology upon ALA-PDT using variable-angle total internal reflection fluorescence microscopy (VA-TIRFM)," *J. Environ. Pathol. Toxicol. Oncol.* **26**, 83–88 (2007).

[52] L. Tyas, V.A. Brophy, A.Pope, A.J. Rivett, and J.M. Tavaré, "Rapid caspase-3 activation during apoptosis revealed by fluorescence-resonance energy transfer," *EMBO Reports* **1**(3), 266–270 (2000).

[53] K.Q. Luo, V.C. Yu, Y. Pu, and D.C. Chang, "Application of the fluorescence resonance energy transfer method for studying the dynamics of caspase-3 activation during UV-induced apoptosis in living HeLa cells," *Biochem. Biophys. Res. Commun.* **283**, 1054–1060 (2001).

[54] K. Takemoto, K. Nagai, A. Miyawaki, and M. Miura, "Spatio-temporal activation of caspase revealed by indicator that is insensitive by environmental effects," *J. Cell. Biol.* **160**(2), 235–243 (2003).

[55] B. Angres, H. Steuer, P. Weber, M. Wagner, and H.Schneckenburger, "A membrane-bound FRET-based caspase sensor for detection of apoptosis using fluorescence lifetime and total internal reflection microscopy," *Cytometry* **75A**, 420–427 (2009).

[56] C.A.F. von Arnim, B. von Einem, P. Weber, M. Wagner, D. Schwanzar, R. Spoelgen, W.S.L. Strauss, and H. Schneckenburger, "Impact of cholesterol level upon APP and BACE proximity and APP cleavage" ,*Biochem. Biophys. Res. Commun.* **370**, 207–212 (2008).

5

Applications of Optical Tomography in Biomedical Research

Ana Sarasa-Renedo

Institute for Electronic Structure and Laser, Foundation for Research and Technology-Hellas, Heraklion 71110, Greece

Alex Darrell

Institute for Computer Science, Foundation for Research and Technology-Hellas, Heraklion 71110, Greece

Jorge Ripoll

Institute for Electronic Structure and Laser, Foundation for Research and Technology-Hellas, Heraklion 71110, Greece

This chapter describes different approaches of noncontact optical tomography of biological tissues. Two main techniques will be described, their difference being in that each one operates at different levels of scattering present: Optical Projection Tomography (OPT) for small, nonscattering samples, and Fluorescence Molecular Tomography (FMT) for larger, highly scattering samples. An overview of these techniques and their applications of these techniques are presented.

Key words: optical projection tomography, fluorescence molecular tomography, diffuse optical tomography, scattering, diffusion, radon transform

5.1 Introduction

The search for a noninvasive, inexpensive imaging technique for medical diagnosis has been the main goal of biophysical research in recent years. Ideally, this would be a method capable of detecting a suspicious lesion within biological tissue, localizing and characterizing it without altering the surrounding tissue.

New perspectives in Life Sciences have been recently opened with the parallel development of fluorescent proteins ([1]; for review, see [2]) and novel optical detection tools [3, 4]. In the past years, noninvasive imaging techniques such as MRI and X-ray CT have brought new insights in research and diagnosis [5]. The field of Molecular Imaging goes one step further, coupling the progress in imaging techniques and the fields of Molecular Biology and Genomics, by introducing the use of optical probes in biodetection, thus allowing the monitoring of the dynamics of multiple molecules within living systems [6]. Molecular Imaging is the discipline that studies the *in vivo* visualization of cellular function using noninvasive techniques. There are a number of approaches for Molecular Imaging, such as Magnetic Resonance Imaging (MRI), Ultrasound, Single Photon Emission Computed Tomography (SPECT), Positron Emission Tomography (PET), and Optical Imaging. Optical Imaging can use fluorescence, bioluminescence, absorption, or reflectance as sources of contrast [6].

When working with optical imaging tools, it is important to know that the penetration depth is the main limitation of these techniques, especially when working at wavelengths in the visible. Light absorption and scattering, which generally decrease with increasing wavelength, determine depth of penetration ([3, 7, 8]). For a detailed description of tissue optical properties see http://omlc.ogi.edu/spectra/.

In vivo imaging of small animals provides not only basic knowledge on cell biology and physiology, but also an invaluable tool for drug discovery and gene therapy research.

Specimens of different sizes present different levels of scattering and are thus needed to be imaged optically using different approaches, processing the obtained data stacks in different ways. Depending on their photon time of flight detection, we can distinguish between those techniques related to ballistic (where light retains most of its coherence) and those dealing with diffusive (multiple scattered) photons [9, 10]. Small, nonscattering samples (through which photons will travel with a "ballistic" trajectory) in the range of the millimeter can be imaged using Optical Projection Tomography (OPT), which will be described in Section 5.3. In contrast, large samples of high scattering properties (where light travels in a "diffusive" manner) typically in the range of several centimeters can be imaged using Diffuse Optical Tomography (DOT), or its Fluorescent counterpart Fluorescence Molecular Tomography (FMT) (sometimes also termed fDOT). Moreover, specific theories and tomographic principles are applied differently in each context, namely the diffusion equation for highly scattering samples [11] and the Radon transform for low scattering samples [12].

5.1.1 Fluorescent molecular probes

The use of fluorescent probes in Life Sciences provides dynamic information on location and amount of specific biomolecules. Biological tissues are not only highly scattering for photons, but also strongly attenuate signal differently at different wavelengths mainly due to light absorption by haemoglobin, water, pigments, and fat. To overcome this attenuation, the use of fluorophores emitting at near-infrared (NIR) wavelengths is being favored in biomedical research. More recently, far-red emitting proteins, derived from *Discosoma sp.* red fluorescent protein, have been developed and used for whole-body imaging [1, 13, 14]. Compared to other optical imaging techniques, *in vivo* fluorescence imaging using continuous sources (CW lasers, for example) is relatively inexpensive. With the use of a sensitive camera (CCD) and a laser, whole animal fluorescence emission is detected.

The main fluorophores used in biological imaging are small organic dyes, fluorescent proteins, and quantum dots (for reviews in fluorescent probes and imaging strategies, see [2, 15, 16]).

5.2 Light Propagation in Highly Scattering Media

5.2.1 The diffusion equation

An "ideal" imaging technique should be inexpensive, noninvasive, should provide three-dimensional (3D) structural information *in vivo*, and should also be quantitative. When we consider an optical approach, the light transport through biological tissue (i.e., highly scattering, highly absorbing) containing fluorophores must be theoretically modelled in order to obtain volumetric and quantitative images. The diffuse propagation of light can be modelled using the wave theory (Maxwell's equations) or the transport theory (Radiative Transfer Equation, RTE). Maxwell's equations do not permit a solution extraction in most of the practical systems due to the high complexity of the problems [17]. The RTE or its approximations have been the theoretical basis of light propagation through biological tissues [11, 18, 19]. In transport theory, light is treated as energy propagating through a medium containing particles. The most common approach to the RTE is the diffusion approximation [19, 20]. Light transport within a turbid medium is best described in terms of probability; instead of the electromagnetic vector wavefunction, the magnitudes to determine are in this case the average intensity and the flux intensity.

5.2.2 Fluorescence molecular tomography

Fluorescence Molecular Tomography (FMT) or Fluorescence Diffuse Optical Tomography (fDOT) is an *in vivo*, noninvasive technique that produces quantitative, three-dimensional distributions of fluorescence, and has so far been used in small animal models [3]. FMT combines the use of fluorescent probes as a source of contrast with the principles of optical tomography [11]. The sample bearing fluorescent probes is illuminated with a defined wavelength from different positions, and the emitted light is collected at detectors arranged in a specific order (the CCD chip). The mathematical processing of the data obtained in this way results in the reconstruction of a tomographic image. FMT incorporates the principles of Diffuse Optical Tomography (DOT; [11, 21]), which models light propagation theoretically as being within the diffusion approximation [17], using fluorescence as a source of contrast [22, 23]. Advances in theoretical approaches [24] that enable free space (noncontact) measurements allow the optimization of the full angle FMT imaging, which results in an increase in the accuracy of the tomographic reconstructions [25, 26, 27].

5.2.2.1 The FMT setup

Figure 5.1 shows a schematic view of a setup used for noncontact fluorescence molecular tomography (FMT) data acquisition in transmission mode. Figure 5.2 shows a setup in reflection mode.

A typical FMT system is schematically shown in Figures 5.1 and 5.2. The mouse was placed horizontally on a fixed stage at a known distance from a CCD camera and objective. The lasers are selected such that the fluorophores under study are optimally excited (for example, for the excitation of GFP the 488 nm line of a continuous wave Ar^+-laser is typically used). Light is guided either via fibers or mirrors, producing a $\sim$1 mm spot on the plane in focus of the camera (when in transmission) or a similar spot at the rear of the object under study. Images are typically taken with a 16-bit CCD camera, cooled down to very low temperatures, to which a photography objective is attached. In order to obtain quantitative data for each source position and projection, two measurements were taken: one corresponding to the fluorescence emission and one corresponding to the excitation wavelength. By normalizing fluorescence with excitation measurements the setup is autocalibrated for most constant gain factors and minimizes the error introduced by deviations in the propagation model used for the inverse problem. Details of this approach have been described

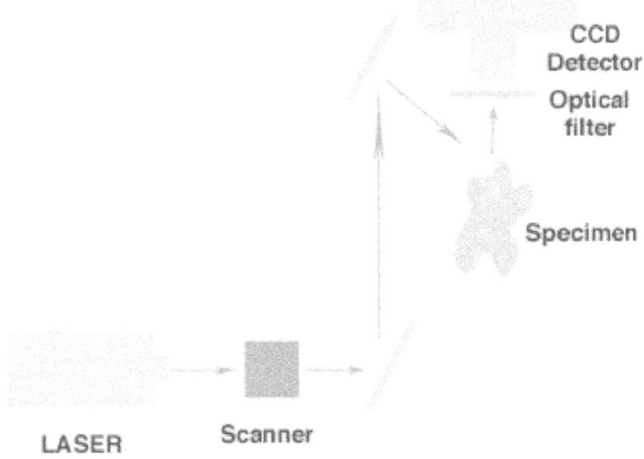

FIGURE 5.1: A Fluorescence Molecular Tomography (FMT) scanning system in transmission mode. In this example, a laser beam is directed to the specimen either via fibers or computer-controlled mirrors order to scan through the sample. Light transmitted through the sample is filtered using a band pass filter and is collected by a CCD detector.

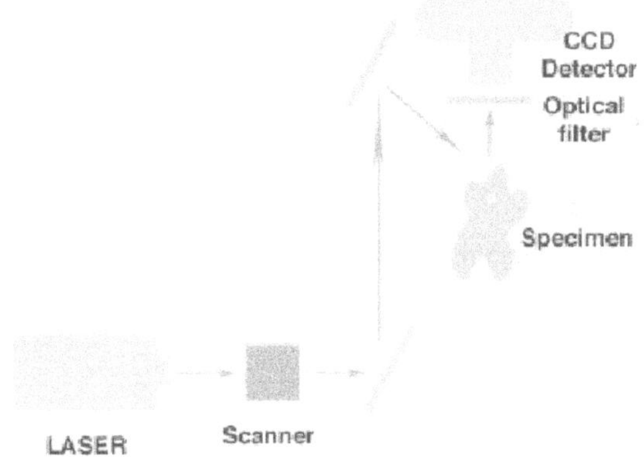

FIGURE 5.2: A Fluorescence Molecular Tomography (FMT) scanning system in reflection mode. In this example, a laser beam is directed via fibers or automated mirrors in order to scan the sample surface. Fluorescence originating from the specimen upon excitation is filtered using a band pass filter and is collected by a CCD detector.

[22, 25, 26, 28]. For recording emission and excitation band pass filters are typically used, centered at the maximum emission wavelength of the fluorophore and laser being used. Data acquisition is performed on a personal computer and the whole experimental procedure is usually controlled by custom-designed software.

5.2.2.2 The inverse problem in diffusive media

Even though it is out of the scope of this chapter to explain in detail how the inverse problem in diffuse imaging is approached, we here show briefly the main steps and equations. In order to obtain 3D images of the fluorophore concentration, we first need to perform what is called the Inverse Problem (see [11]). Let us assume that in the volume V of Figure 1 we have a point source located at $\mathbf{r}_s$ inside the medium with constant intensity (the same can be derived for pulsed or modulated sources). Assuming that the average intensity U detected at $\mathbf{r}$ within V represents a diffuse photon density wave (DPDW) [29] and obeys the Helmholtz equation [17]:

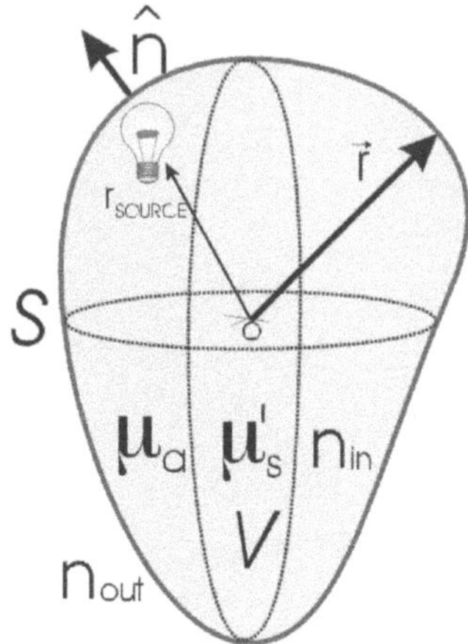

FIGURE 5.3: Scattering geometry for a diffusive volume V of arbitrary shape.

$$\nabla^2 U(\mathbf{r}) + \kappa^2 U(\mathbf{r}) = -\frac{\Lambda(\mathbf{r})}{D} \quad \mathbf{r} \in V, \tag{5.1}$$

with a complex wave-number κ given by:

$$\kappa = \left(-\frac{\mu_a}{D}\right)^{1/2} \tag{5.2}$$

where c is the speed of light in vacuum and $\Lambda(\mathbf{r})$ is the source distribution. In an infinite homogeneous $3D$medium the Green function is given by [17]:

$$g(\kappa|\mathbf{r_s} - \mathbf{r_d}|) = \frac{\exp[i\kappa|\mathbf{r_s} - \mathbf{r_d}|]}{D|\mathbf{r_s} - \mathbf{r_d}|}. \tag{5.3}$$

Taking into account rigorously the boundary S, the average intensity U inside volume V is found through Green's theorem ([30, 31, 32]) as:

$$\begin{aligned} U(\mathbf{r_d}) &= U^{(inc)}(\mathbf{r_d}) - \\ &\frac{1}{4\pi}\int_S \left[U(\mathbf{r}')\frac{\partial g(\kappa|\mathbf{r}'-\mathbf{r_d}|)}{\partial \hat{\mathbf{n}}'} - g(\kappa|\mathbf{r}' - \mathbf{r_d}|)\frac{\partial U(\mathbf{r}')}{\partial \hat{\mathbf{n}}'}\right] dS', \end{aligned} \tag{5.4}$$

where

$$U^{(inc)}(\mathbf{r}) = \int_V \Phi(\mathbf{r}')g\left(\kappa\left|\mathbf{r}' - \mathbf{r}\right|\right) dV \tag{5.5}$$

is the average intensity that is obtained in the absence of the surface. We also need to take into account Fick's Law for the boundaries,

$$J_n(\mathbf{r}) = \mathbf{J}(\mathbf{r}) \cdot \hat{\mathbf{n}} = -D\frac{\partial U(\mathbf{r})}{\partial \hat{\mathbf{n}}}, \tag{5.6}$$

and the boundary condition between the diffusive and nondiffusive medium [33]:

$$U(\mathbf{r})|_S = -C_{nd} D \left.\frac{\partial U(\mathbf{r})}{\partial \hat{\mathbf{n}}}\right|_S, \mathbf{r} \in S \tag{5.7}$$

where the coefficient C_{nd} takes into account the refractive index mismatch between both media [33]. In the case of index matched media, i.e., $n_{out} = n_{in}$, C_{nd}=2, whereas for typical tissue/air index values

(n_{in}=1.333, n_{out}=1) we obtain C_{nd} ~5. Equations (5.4)–(5.7) set the necessary tools for formulating the inverse problem, in this case assuming a homogeneous diffusive volume of arbitrary geometry V. By using the normalized Born expression [34] we can formulate the inverse problem as:

$$P^{nborn}(\mathbf{r}_s,\mathbf{r}_d) = \frac{I^{fluo}(\mathbf{r}_s,\mathbf{r}_d)}{I^{exc}(\mathbf{r}_s,\mathbf{r}_d)} = \frac{\eta\sigma^{fluo}\Delta V}{4\pi C_{nd} D^{fluo}} \sum_{n=1}^{N} \left[\frac{U^{exc}(\mathbf{r}_s,\mathbf{r}_n)G(\mathbf{r}_n,\mathbf{r}_d)}{U^{exc}(\mathbf{r}_s,\mathbf{r}_n)}\right] \tag{5.8}$$

where the top part represents modeling of the fluorescence emission and the bottom part the excitation, and $I^{fluo}$and I^{exc}represent the fluorescence and excitation measurements, respectively. In Eq. (5.8) η is the quantum yield of the fluorochrome and σ^{fluo} is the absorption cross-section of the fluorophore (in cm^{-2}). For a set of M sources and detector pairs, Eq. (5.8) can be more conveniently rewritten in matrix form to yield:

$$\mathbf{W}_{M\times N}\cdot\mathbf{F}_{N\times 1} = \mathbf{P}^{nborn}_{M\times 1} \tag{5.9}$$

where matrix **W** is commonly referred to as the weight or sensitivity matrix, since it represents the probability of a photon to visit each voxel site at $\mathbf{r}_n$ for each source-detector pair. Assuming a homogeneous medium with average scattering and absorbing properties, matrix W may be found numerically, and thus Eq. (5.8) inverted, noting that this problem is ill-posed [11]:

$$\mathbf{F}_{Nx1} = \mathbf{W}^{-1}_{MxN}\mathbf{P}^{nborn}_{Mx1} \tag{5.10}$$

When using CCD detection techniques, **W** tends to be very large (due to a large number of source-detector pairs), and the inversion cannot be performed directly. There are several approaches to invert equation (5.9) and solve for **F** (see, for example, [35, 36]). The approach used in the instances shown in this review is the algebraic reconstruction technique with randomized projection order (R-ART) [37], which was originally developed for x-ray tomography [12]. The selection of this particular inversion method was due to the modest memory requirements for large inversion problems and the calculation speed it attains.

5.2.2.3 Applications of diffuse optical tomography

Advances in the theoretical background soon brought multiple applications for diffuse light in biomedical research. It has been used to quantify oxy- and deoxyhaemoglobin [38, 39] as well as to characterize tissue constituents [40, 41, 42]. In pathological studies, diffuse light has been used to diagnose testicular abnormalities [43] and to detect and characterize breast cancer [44–49], as well as in imaging the brain for cortical studies (see [38]). Diffuse optical tomography (DOT) using near-infrared light was also used in imaging arthritic joint imaging [50, 51], and functional activity in the brain [52–54] and the muscle [55]. In recent years, optical tomography is showing a great potential for drug development [56–58]. The capacity to quantify and resolve molecular function and gene expression by imaging targeted or activable fluorochromes in whole animals has also been proven *in vivo*, opening new insights in cancer research, immunology and pharmacological research.

Fluorescence Molecular Tomography has been applied in very different biomedical research areas [4], being small animal models the main subjects of analyses. Human joints and the breast are also good candidates for combined applications.

-Cancer research and angiogenesis:

Fluorescent molecular probe development, including target-specific and "switchable" probes [59, 60], has broadened the application spectrum of FMT. Overexpression of Cathepsin B (a papain-family cysteine protease) has been associated with tumor progression and arthritis (for review, see [61]). Protease activity has been measured by FMT [22] in 9L glyosarcomas, and activated fluorochrome estimations were achieved [8]. Zacharakis et al. [62] performed tomographic analyses and fluorescent protein quantification of superficially and deep-seated GFP-expressing tumors.

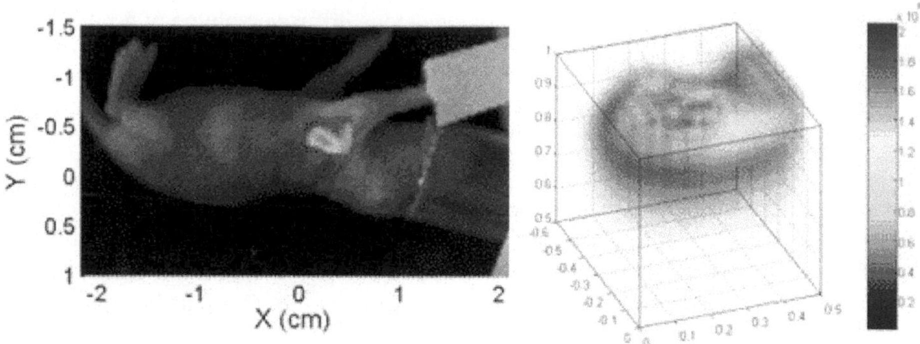

FIGURE 5.4: (See in color after page 572.) Example of an FMT reconstruction of fluorophore concentration, in this case green fluorescent protein (GFP). A mouse puppy expressing GFP in its T lymphocytes was imaged in reflexion mode using a custom-made FMT setup at day 3.5 after birth. At this age, T lymphocytes that migrate from the thymus start seeding the peripheral lymphoid organs, such as the lymph node reconstructed in this figure. Position and relative intensity of fluorescence can be inferred from the right panel.

X-ray computed tomography (CT) images were obtained in parallel, and a good congruence was observed between both techniques.

Montet et al. [63] reported the simultaneous magnetic and fluorescence measurements in tumor-associated angiogenesis and its inhibition by anti-VEGF mAb. Moreover, vascular near infrared probes (AngioSense 680 and/or 750, 2nmol fluorochrome per mouse) were injected into animals bearing CT26 tumors to determine tumor localization [64]. Using Vascular Volume Fraction (VVF), angiogenesis inhibition in response to anti-VEGF mAb was monitored.

-Cardiovascular disease:

Using a magnetofluorescent nanoparticle (CLIO-Cy5.5), myocardial macrophage infiltration upon ligation of the left coronary artery (which resulted in infracted myocardium) was imaged combining FMT and cardiac MRI [65].

-Immunology and inflammatory diseases:

There are different published studies with fluorescent proteins expressed in cells of the Immune System [25, 28]. T lymphocytes expressing Green Fluorescent Protein (GFP) under control of the human CD2 LCR (hCD2-GFP; [66]) were imaged at different tissues using FMT. See Figure 5.4 for such a representative tomographic reconstruction. Optical tomographic approaches have been proposed for detection of synovitis in arthritic finger joints [67].

-Neurodegenerative diseases

AO1987 is a near infrared fluorescent oxazine dye that binds amyloid plaques upon penetration of intact blood-brain barrier [68]. In a murine model of Alzheimer's disease, AO1987 is a very good candidate to monitor disease progression, as well as to evaluate effects of potential drugs on plaque stability.

-Pharmacology:

Whole-body, small animal imaging techniques allow to study a disease and the candidate treatment responses over time in the same animal (for review, see [58]). This accelerates drug research reducing the total number of animals on which experimentation takes place. Moreover, statistical significance is increased by using the same animal prior to administration of the drug as a control.

5.3 Light Propagation in Nonscattering Media

For small samples where scattering can be neglected or large nonscattering samples, new techniques that can provide 3D images of fluorescence and absorption have been developed depending on the applications targeted. Until 2002, the standard technique for mapping three-dimensional gene and protein expression patterns in mouse embryos involved serial sectioning. Reconstructing images from hundreds of serial sections is not only time consuming – physical sectioning also introduces distortions in the final images. In 2002 James Sharpe published a paper in Science Magazine [69] describing an alternative technique based on a 3D imaging system he built. This technique developed into the imaging modality known as Optical Projection Tomography (OPT) (http://genex.hgu.mrc.ac.uk/OPT_Microscopy).

OPT scans can be performed in two modes, generally referred to as the *transmission* and *emission* modes. Transmission mode is generally used for studies of anatomy while emission mode is used to image the fluorescence distribution in fluorescently labelled samples. Transmission and emission data are almost always used together, to visualize fluorescence against a map of anatomy. In *transmission* mode, OPT is similar to x-ray Computed Tomography except that OPT uses light to probe the specimen, rather than x-rays, and the light emerging from the body being focused by a system of optical lenses onto a CCD detector. In *emission* mode, samples are exposed to light of a wavelength known to excite the fluorescent material within the sample, and optical filters and lenses are used to focus the light emitted by the fluorescent material onto a CCD detector.

OPT is suitable for 3D imaging of transparent specimens whose thicknesses in the imaging direction lie in the range 1mm to 15mm. OPT occupies a part of the imaging-gap between MRI and Confocal Microscopy. Other techniques in this imaging gap are: x-ray, Micro-CT, Micro-MRI, Optical Coherance Tomography and Ultrasonic Biomicroscopy. However, all of these other methods suffer from at least one, and usually several, of the following problems: lack of contrast, being inappropriate for fluorescence studies, limited imaging size range, and being prohibitively expensive.

Since its introduction in 2002 [69], applications of transmission and emission OPT have been reported that span various fields of biology. Some of these applications include organ inter-relationship studies in mice [70], three-dimensional imaging and quantification within rodent organs [71, 72], modelling of early human brain development [73], visualizing plant development and gene expression [74], three-dimensional imaging of isolated cell nuclei [75], three-dimensional imaging of xenograft tumors [76], cell tracing [77], three-dimensional imaging of *Drosophila melanogaster* [78], spatiotemporal analysis of the zebrafish [79, 80], developmental embryology and gene expression [69], including a 3D limb patterning atlas [81], *in vitro* 4D quantification of growing mouse limb buds [82], three-dimensional tissue organization and gene expression in *Arabidopsis* [83], gene expression in the adult mouse brain [84], and gene expression in the foregut and lung buds in humans [85].

5.3.1 Optical projection tomography

The physics of light propagation in transmission and emission mode are quite different, and for this reason each mode will be treated separately in the following discussion. Common to both modalities however is the requirement that samples are optically cleared prior to imaging. Without this clearing process, the OPT imaging modality would be similar to Optical Tomography, where scattering, absorption, and a number of other optical effects can make high resolution image acquisition and reconstruction virtually impossible in all but a few cases where samples are naturally transparent.

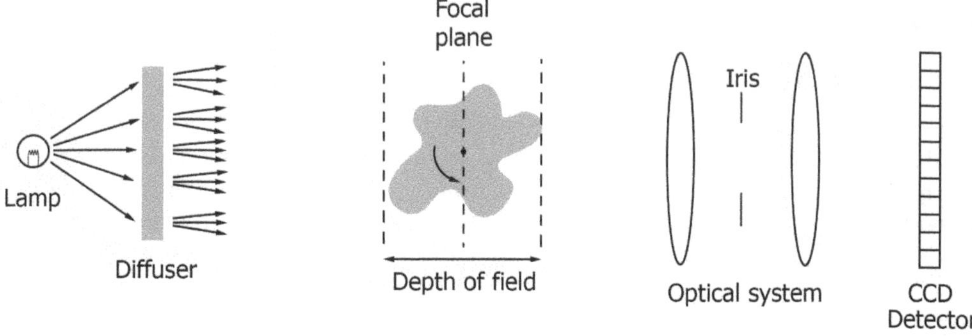

FIGURE 5.5: An OPT scanning system in transmission mode. In this example, light from a lamp is directed through a diffuser and onwards to an optically cleared sample. Light transmitted through the sample is collected by an optical system and focused onto a CCD detector. The depth of field is adjustable by varying the radius of an iris within the optical system. Images (known as projections in this context) of the transmitted light are captured for several hundred angular positions of the sample.

5.3.1.1 Sample clearing and preparation

OPT is capable of producing very high resolution images and reconstructions because of the clearing process used during sample preparation. The clearing process matches the index of refraction of the sample tissue with that of the surrounding medium and renders samples transparent. Samples are fixed with paraformaldehyde/glutaraldehyde, dehydrated with methanol (MeOH), and cleared with benzyl alcohol/benzyl benzoate (BABB).

5.3.1.2 OPT imaging in transmission mode

In transmission mode a wide field homogeneous illumination source is used to illuminate a sample that is rotated stepwise to a number of angular positions. The light transmitted through the sample at each angular position is captured using a CCD microscope. The light source used can be diffuse or nondiffuse, and different OPT scanners employ various light sources ranging from lasers to ultrabright LEDs.

By partially closing the iris of the optical system, the operator can increase the Depth of Field (DOF) of the system and achieve acceptable focus of the entire sample. This also has the effect of reducing the image resolution in projections.

5.3.1.3 OPT imaging in emission mode

In emission mode a wide field illumination source of an appropriate wavelength is used to excite a sample that is rotated to a number of angular positions. The light emitted by the fluorescent at material within the sample at each angular position is captured using a CCD microscope. Different OPT scanners employ various excitation sources ranging from lasers, through mercury arc lamps to ultrabright LEDs. The wide variety of fluorescence filter sets used for fluorescence microscopy allow samples that express or are stained with the various fluorescent proteins and dyes to be imaged using OPT.

As with transmission OPT, by partially closing the iris of the optical system, the operator can increase the DOF and thereby achieve acceptable focus of the entire sample, but at the cost of a reduction in image resolution.

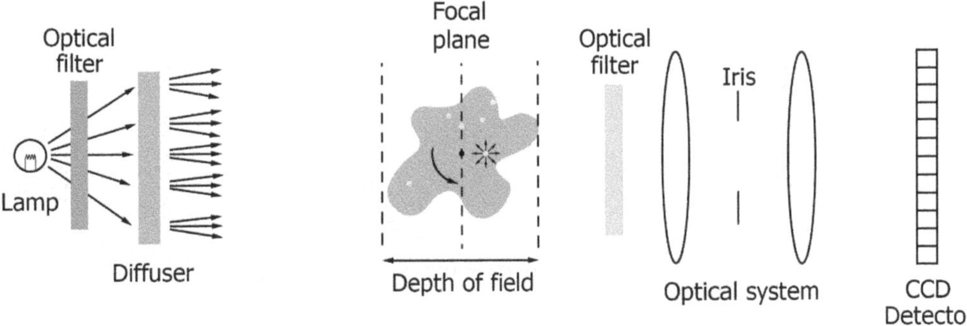

FIGURE 5.6: An OPT scanning system in emission mode. Light from a lamp is directed through an fluorescence excitation filter to excite the fluorescent material within the sample. A diffuser is used to improve the homogeneity of the sample excitation. Fluorophores within the sample absorb the high energy excitation light and isotropically emit lower energy light of a longer wavelength. As in transmission OPT, the depth of field is determined by adjusting the radius of an iris within the optical system. Projections of the light emitted in the direction of the CCD microscope are captured for several angular positions of the sample.

5.3.1.4 Depth of field versus image resolution

The trade-off between resolution and depth of field (DOF) was briefly mentioned above. To put this concept on a firmer footing a clear definition of depth of field and resolution are required. The focal plane of an optical system is a plane perpendicular to the optical axis. Images of objects and points at the focal plane are said to be in perfect focus. The depth of field of an optical system is a region of space distributed equally on either side of the focal plane. Images of objects within the DOF are said to be focused but not in perfect focus, while images outside the DOF are said to be unfocused. According to [86], the DOF is given by:

$$DOF = n_{bath}\left(\frac{n\lambda}{NA^2} + \frac{n}{MNA}e\right), \tag{5.11}$$

where NA is the numerical aperture of the system, n is the index of refraction ($n \approx 1$ for air), λ is the wavelength of the light, M is the lateral magnification of the imaging system, e is the pixel size of the CCD and n_{bath} is the index of refraction of the medium in which the sample is suspended. The Nyquist criterion of sampling frequency requires the detector spacing to be less than half the radius of the Airy disk, which means that the detector spacing must be

$$e \leq M\frac{r_{Airy}}{2} \tag{5.12}$$

Substitution of Eq. (5.12) into Eq. (5.11) leads to a theoretical maximum DOF. Assuming that $n \approx 1$, the theoretical maximum DOF is given by

$$DOF_{\max} = n_{bath}\left(\frac{\lambda}{NA^2} + \frac{r_{Airy}}{2NA}\right). \tag{5.13}$$

The resolution of an optical system is the minimum distance between two physical points such that they can be resolved using the Raleigh criterion [87]. This distance cannot be greater than the radius of the two-dimensional point spread function in the focal plane, i.e., the radius of the Airy disk, given by

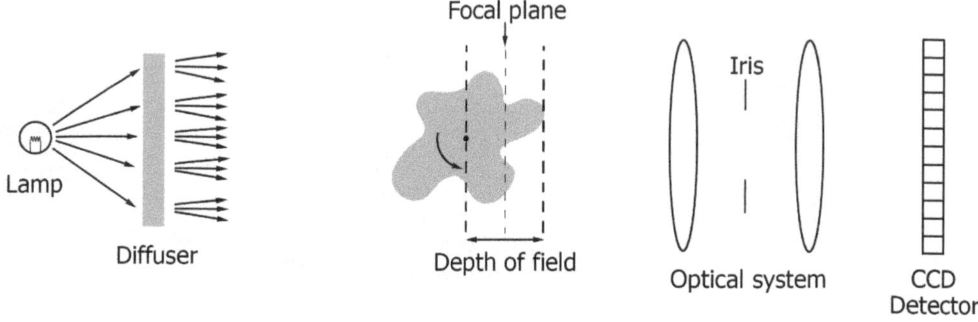

FIGURE 5.7: An OPT scanning system where only half the sample is positioned within the depth of field. For a given sample size, this means that the iris can be further open, which results in a higher resolution in projections.

$$r_{Airy} = \frac{0.61n\lambda}{NA}, \tag{5.14}$$

Substituting Eq. (5.14) into Eq. (5.13) leads to a more succinct expression for $DOF_{\max}$

$$DOF_{\max} = n_{bath}\left(\frac{1.305\lambda}{NA^2}\right). \tag{5.15}$$

From Eq. (5.15), it is clear that by closing the iris of the optical system (thereby decreasing the numerical aperture), the DOF is increased. Equally, Eq. (5.14) shows that by opening the iris of the optical system (thereby increasing the numerical aperture), the resolution limit is increased. Equations (5.14) and (5.15) neatly summarize the compromise between resolution and depth of field.

To reduce the requirement for iris closure, some OPT scanners [88, 89] attempt to keep only one half of the sample in focus, in Figure 5.7.

Such a system requires modifications to the reconstruction algorithm, which for the commonly used Filtered Back Projection algorithm (to be discussed later in the next section) are quite straightforward. However, the contribution from the out of focus portion of the sample in each projection is blurred, and represents a reduction in resolution. Methods of de-emphasizing the signal from the out-of-focus part of the body in each projection are discussed in [90] and are also briefly described in the next section.

5.3.2 Reconstruction methods in OPT

The reconstruction algorithm used in the vast majority of OPT reconstructions is Filtered Back Projection (FBP). The FBP algorithm has the well-known advantage of speed, but the model of image formation implicit in FBP does not closely resemble the process by which OPT images are formed, which is discussed in [91] and includes the effects of diffraction and blurring, even within the depth of field. For cases where quantitatively accurate results are not required, and sufficient resolution can be obtained with an adequate DOF, the FBP algorithm usually performs reasonably well. Other, general disadvantages to FBP are that the filtering process can accentuate noise and can result in streak artifacts, even in areas of the image with otherwise low intensities. The following section describes FBP in terms of transmission tomography, since this is what it was originally intended for.

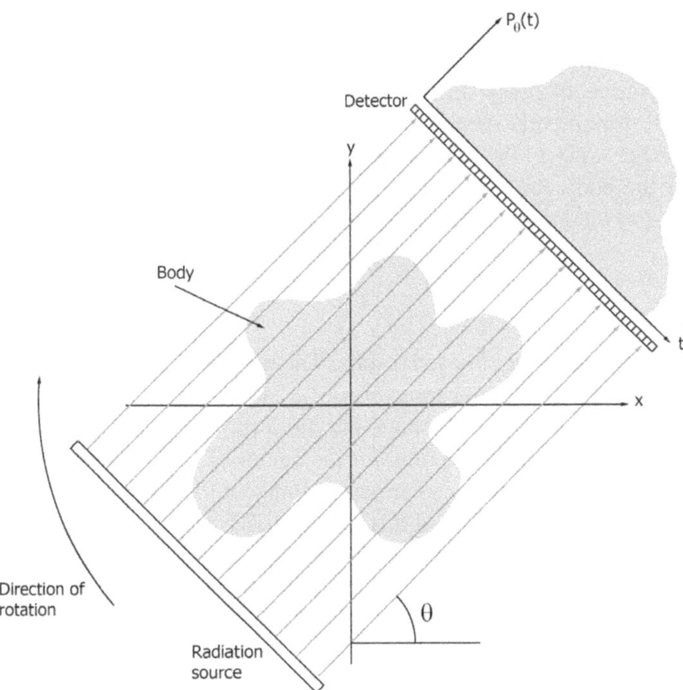

FIGURE 5.8: Formation of projections in transmission tomography. For each projection angle, some property of the body is integrated along a parallel, the value of the integral being registered on a detector.

5.3.2.1 Filtered back projection algorithm

Given several $m \times n$ pixel 2D projections of a three-dimensional body, one can reconstruct a 3D $m \times n \times n$ estimation image of that body using Filtered Back Projection. The process works by reconstructing m two-dimensional $n \times n$ slices of the body, one by one, and stacking them to produce a three-dimensional image. That is, the p-th reconstructed slice is a two-dimensional image obtained by means of applying the FBP algorithm to the p-th row of data in each projection. Therefore, the discussion that follows omits the third dimension and deals with the reconstruction of two-dimensional slices from one-dimensional projection data, the extension to 3D, except for the simple process of stacking, being a repetition of the same process. In addition, for the remainder of this section, the term "body" will be used to describe a two-dimensional slice of the original three-dimensional body.

5.3.2.2 Simple back-projection

Each projection of a body can be considered the combination of several parallel line integrals of a function, $f(x,y)$, describing an aspect of that body, at a particular angle, θ say, through the body. Each projection can be considered a function, $P_\theta(t)$, where t is the distance along a line perpendicular to the direction of the projection. See Figure 5.8.

It can be shown that the one-dimensional Fourier transform of a parallel projection is equal to a slice of the two-dimensional Fourier transform of the body along a radial line parallel to the projection [12]. Because of this, it is in principle possible to estimate the original body by performing two-dimensional inverse Fourier transforms of the one-dimensional Fourier transforms of sufficiently large numbers of projections. In practice however, simple back-projection starts by copying the values of the first projection into an image array, repeatedly, into pixels lying along the line of the first integration/projection. This process is repeated, cumulatively adding contributions to the image array, rotating the image array to the angle θ at which each projection was taken at each repetition. The result is intuitively similar to spreading each projection across the image space, each time rotating the image array to the angle that the projection was taken at. This way, the obvious internal features of the body are repeatedly written into the image space, from various directions, and as more projections are back-projected the internal features of the original body are revealed in detail.

5.3.2.3 Filtering

Simple back-projection produces blurred results. The reason for this is intuitively clear in Fourier space where we recall that the Fourier transform of a projection is equal to a parallel radial line of the Fourier transform of the body. As more and more projections, and so more radial lines, are acquired, the density of information acquired for the center of the image will be far greater than it is towards the outer edges. That is, the center part of image will have been over-sampled in Fourier space. The generally accepted solution to this problem is to use an appropriate function to filter projections before back-projecting. In summary, projections are taken from a body, the Fourier transforms of those projections are found and then filtered, in Fourier space, the inverse Fourier transform of the result is found and back-projected. The overall process is known as Filtered Back Projection.

5.3.2.4 Alternatives to standard FBP

Under certain circumstances, especially when a compromise between resolution and blurring is made, the FBP algorithm can produce reconstructions that are both *qualitatively* poor and *quantitatively* unreliable [90, 91]. Under these circumstances, reconstructions can be improved by pre-processing projections to reduce blurring, and/or incorporating aspects of the process of image formation into the reconstruction algorithm. Published preprocessing steps have included a method of de-emphasising the blurred contribution from unfocused parts of a sample as depicted in Figure 5.7, as well as deconvolving projections with the point spread function in the frequency domain. In addition, noise reduction techniques have been investigated that seek to produce projections with lower noise content by replacing each projection with the average of several projections.

5.3.2.5 Incorporate physics into reconstruction algorithm

The effects of blur and isotropic emission can be separated into quantitative effects and qualitative effects. In [91], the quantitative effects of blur within the DOF were investigated. A common practical way of measuring the DOF of an optical system is to identify an area on either side of the focal plane where images of the same point object have roughly constant Full Width at Half Maximum (FWHM). In practice, in images with roughly constant FWHM, the diameter of the circle of confusion is so small as to make those images indistinguishable from images of a point source in perfect focus. As a point source is moved away from the focal plane, the highest pixel value can be expected to be found at the center of the circle of confusion and is arguably the most indicative of intensity. In [91], an experiment is described where a fluorescent microsphere was excited and imaged at 200 equidistant positions along the optical axis of the system using the full unstopped objective numerical aperture. To a best approximation, these positions were equally distributed in front of and behind the focal plane. The highest pixel value and the Full Width at Half Maximum (FWHM) from each image obtained during the experiment are plotted against defocus in Figure 5.9.

Figure 9 clearly illustrates the dramatic effect that defocus and isotropic emission could have on the apparent measured intensity of a fluorophore, even within the depth of field of the imaging system. Assuming this effect takes place on all axes parallel to the optical axis then the derivation of

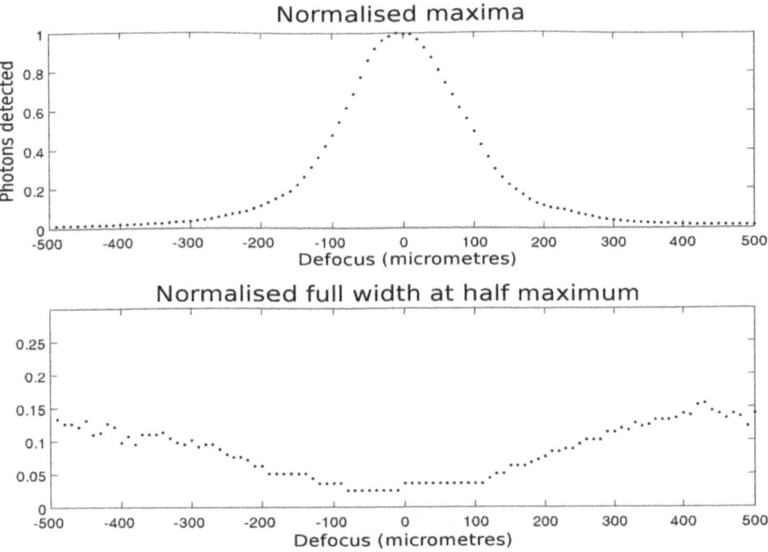

FIGURE 5.9: Top: The intensity (highest pixel value) of a point source in images taken of the point source at various defocus distances is plotted against defocus. Bottom: The FWHM plotted against defocus. Over a focal depth ranging from -200 microns to 200 microns, the apparent intensity changes by about 80% from its maximum value.

the Fourier Slice Theorem [12] requires the introduction of an angle dependent weighting function $w_\theta(t,s)$, which is essentially a stacked repetition of the curve in Figure 5.9 (top). See Figure 5.10.

Consider a slice $f(x,y)$ of a body $f(x,y,z)$. A generic projection of the slice is given by

$$P_\theta(t) = \int_{-\infty}^{\infty} w_\theta(t,s) f(t,s) ds \tag{5.16}$$

where the (t,s) coordinate system is a rotation of the original (x,y) coordinate system through an angle θ and is given by

$$\begin{bmatrix} t \\ s \end{bmatrix} = \begin{bmatrix} \cos\theta & \sin\theta \\ -\sin\theta & \cos\theta \end{bmatrix} \begin{bmatrix} x \\ y \end{bmatrix}. \tag{5.17}$$

The dependence of $w_\theta(t,s)$ on θ can be characterized by the observation that $w_\theta(t,s)$ is always invariant with respect to changes in t.

The Fourier transform of Eq. (5.16) is

$$S_\theta(w) = \int P_\theta(t) e^{i2\pi wt} dt \tag{5.18}$$

which after substituting the definition for $P_\theta(t)$ from Eq. (5.16) is given by

$$S_\theta(w) = \int_{-\infty}^{\infty} \left[\int_{-\infty}^{\infty} w_\theta(t,s) f(t,s) ds \right] e^{i2\pi wt} dt. \tag{5.19}$$

Changing back to the (x,y) coordinate system using the relations gives:

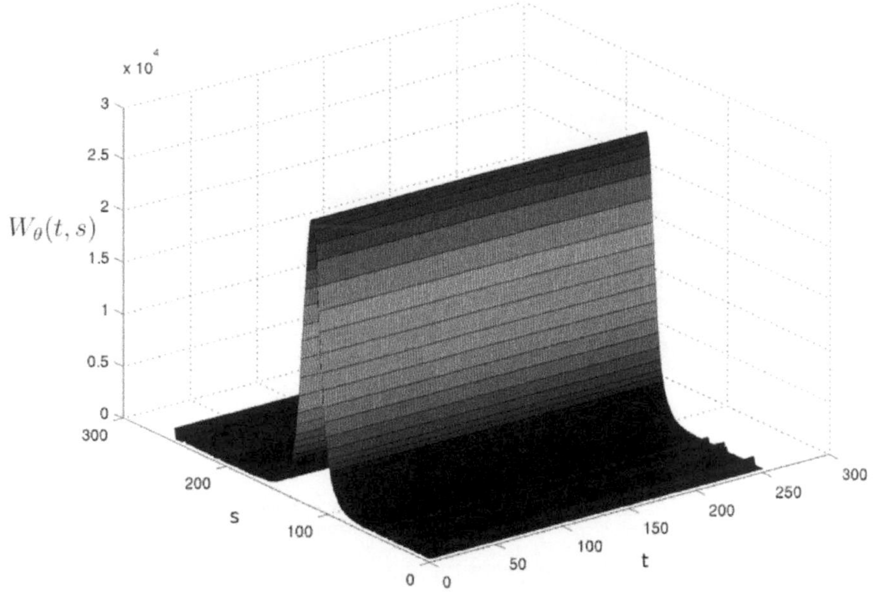

FIGURE 5.10: Surface plot of the angle dependent weighting function $w_\theta(t,s)$, which describes the spatial variation of the contribution of point sources to projections.

$$S_\theta(w) = \int_{-\infty}^{\infty}\int_{-\infty}^{\infty} w_\theta(x,y) f(x,y) e^{i2\pi w(x\cos\theta + y\sin\theta)} dxdy \tag{5.20}$$

which is the windowed Fourier Transform of $f(x,y)$, using $w_\theta(x,y)$ as a window function. By the convolution theorem Eq. (5.20) is given by

$$S_\theta(w) = W_\theta(w\cos\theta, w\sin\theta) \otimes F(w\cos\theta, w\sin\theta) \tag{5.21}$$

where $F(u,v)$ and $W_\theta(u,v)$ are the Fourier transforms of $f(x,y)$ and $w_\theta(x,y)$ respectively.

The fact that the inclusion of the weighting function $w_\theta(x,y)w_\theta(x,y)$ leads to the Fourier transform of a projection not being equal to the Fourier transform of a slice of the body but to the *windowed* Fourier transform of a slice of the body suggests a method of reversing the effect of the weighting function $w_\theta(t,s)$. In the weighted FBP algorithm a weighting is applied to each back projection. Instead of spreading the projection information evenly over the reconstruction image plane, as in standard FBP, the idea is to spread the projection information unevenly, according to the inverse of the profile given by $w_\theta(t,s)$ and shown in Figure 5.10.

The results of this approach are shown in Figure 5.11. In [91] a phantom experiment is described where the quantitative effects of blur and isotropic emission are simulated for a digital phantom similar to that shown in Figure 5.11.

In analysis of a standard FBP reconstruction and a weighted FBP reconstruction are shown side by side in Figure 5.2. The simulation assumed a numerical aperture of 0.064 reduced to 10% of this value by an iris. In the standard FBP reconstruction, the intensities of fluorophores away from the center of the body are reduced by up to 50%. In the weighted FBP reconstruction, all areas are assigned almost equal intensities, without reduced image quality. At this numerical aperture, weighted FBP achieves a high level of *quantitative* accuracy without significant reduction in image quality. At this numerical aperture, the standard FBP reconstruction could be discarded completely,

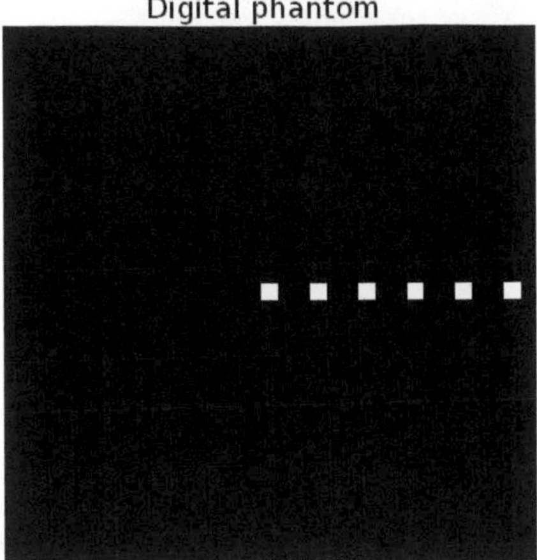

FIGURE 5.11: Digital phantom used to test reconstruction algorithms. The nonrandom distribution of point sources allows precise measurement of intensities assigned in reconstructions. The phantom consists of a 512 by 512 pixel image, primarily composed of zero intensities (black) with six 10×10 regions of pixels where each pixel has value 1000 (white).

in favor of the superior weighted FBP reconstruction. Results for higher numerical apertures exhibit increased noise amplification and in these cases weighted FBP reconstructions are a useful companion to standard FBP reconstructions rather than a replacement.

Acknowledgment

This work was supported by E.U. Integrated Project "Molecular Imaging" LSHG-CT-2003-503-259. A.S.-R. acknowledges support from the Bill and Melinda Gates Foundation (Grant No. 42786). The authors wish to acknowledge invaluable help in building and testing the FMT and OPT setups to J. Aguirre, U. Birk, A. Garofalakis, A. Martin, H. Meyer, R. Favicchio and G. Zacharakis. Mouse transgenic lines were provided by C. Mamalaki and D. Kioussis. A. Darrell acknowledges support from EST Molec-Imag MEST-CT-2004-007643.

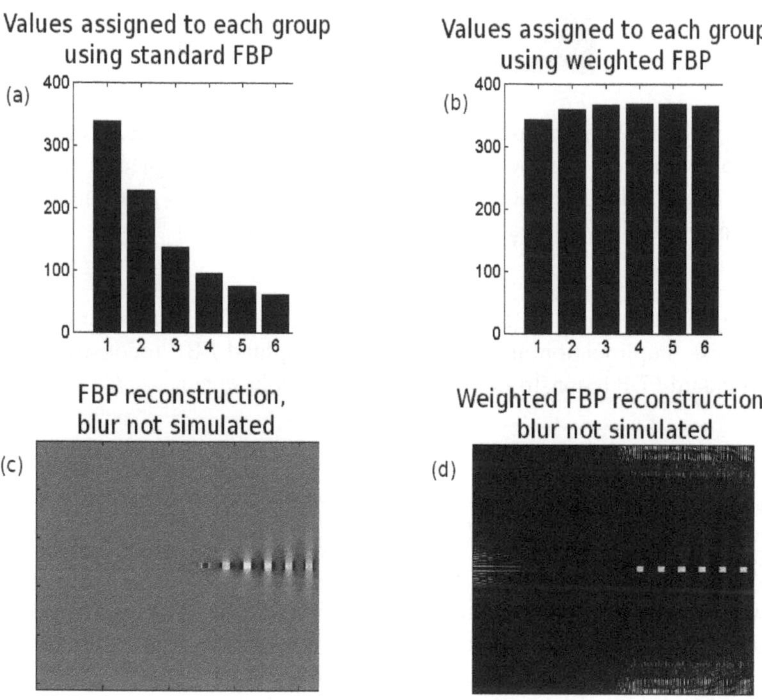

FIGURE 5.12: Results obtained from projections simulated with 50% numerical aperture; (a) the values assigned to the six groups of sources by standard FBP, (b) the values assigned by weighted FBP, (c) the standard FBP reconstruction, and (d) the weighted FBP reconstruction.

References

[1] N.C. Shaner, R.E. Campbell, P.A. Steinbach, et al., "Improved monomeric red, orange and yellow fluorescent proteins derived from *Discosoma sp.* red fluorescent protein," *Nat. Biotechnol.* **22**, 1567–1572 (2004).

[2] B.N. Giepmans, S.R. Adams, M.H. Ellisman, et al., "The fluorescent toolbox for assessing protein location and function," *Science* **312**, 217–224 (2006).

[3] R. Weissleder and V. Ntziachristos, "Shedding light onto live molecular targets," *Nat. Med.* **9**, 123–128 (2003).

[4] V. Ntziachristos, J. Ripoll, L.V. Wang, et al., "Looking and listening to light: the evolution of whole-body photonic imaging," *Nat. Biotechnol.* **23**, 313–320 (2005).

[5] S.R. Cherry, "*In vivo* molecular and genomic imaging: new challenges for imaging physics," *Phys. Med. Biol.* **49**, R13–R48 (2004).

[6] R. Weissleder and U. Mahmood, "Molecular imaging," *Radiology* **219**, 316–333 (2001).

[7] A.E. Cerussi, A.J. Berger, F. Bevilacqua, et al., "Sources of absorption and scattering contrast for near-infrared optical mammography," *Acad. Radiol.* **8**, 211–218 (2001).

[8] V. Ntziachristos, J. Ripoll, and R. Weissleder, "Would near-infrared fluorescence signals propagate through large human organs for clinical studies?" *Opt. Lett.* **27**, 333–335 (2002).

[9] S. Gayen and R. Alfano, "Emerging biomedical imaging techniques," *Opt. Photon. News* **7**, 17–22 (1996).

[10] D.A. Benaron, W.F. Cheong, and D.K. Stevenson, "Tissue optics," *Science* **276**, 2002–2003 (1997).

[11] S. Arridge, "Optical tomography in medical imaging," *Inverse Problems* **15**, R41–R93 (1999).

[12] A. Kak and M. Slaney, *Principles of Computerized Tomographic Imaging*, 1st ed., IEEE Press, New York (1988).

[13] D. Shcherbo, E.M. Merzlyak, T.V. Chepurnykh, et al., "Bright far-red fluorescent protein for whole-body imaging," *Nat. Methods* **4**, 741–746 (2007).

[14] D. Shcherbo, C.S. Murphy, G.V. Ermakova, et al., "Far-red fluorescent tags for protein imaging in living tissues," *Biochem. J.* **418**, 567–574 (2009).

[15] J. Rao, A. Dragulescu-Andrasi, and H. Yao, "Fluorescence imaging *in vivo*: recent advances," *Curr. Opin. Biotechnol.* **18**, 17–25 (2007).

[16] T. Terai and T. Nagano, "Fluorescent probes for bioimaging applications," *Curr. Opin. Chem. Biol.* **12**, 515–521 (2008).

[17] M. Born and E. Wolf, *Principles of Optics*, 7th ed., Cambridge University Press, Cambridge (1999).

[18] S. Chandrasekhar, *Radiative Transfer,* Courier Dover Publications, New York (1960).

[19] A. Ishimaru, *Wave Propagation and Scattering in Random Media*, Vol. 1, Academic Press, New York (1978).

[20] A.G. Yodh and B. Chance, "Spectroscopy and imaging with diffuse light," *Phys. Today* **48**, 38–40 (1995).

[21] V. Ntziachristos, B. Chance, and A.G. Yodh, "Differential diffuse optical tomography," *Optics Express* **5**, 230–242 (1999).

[22] V. Ntziachristos, C.H. Tung, C. Bremer, et al., "Fluorescence molecular tomography resolves protease activity *in vivo*," *Nat. Med.* **8**, 757–760 (2002).

[23] V. Ntziachristos, C. Bremer, E.E. Graves, et al., "*In vivo* tomographic imaging of near-infrared fluorescent probes," *Mol. Imaging* **1**, 82–88 (2002).

[24] J. Ripoll, R.B. Schulz, and V. Ntziachristos, "Free-space propagation of diffuse light: theory and experiments," *Phys. Rev. Lett.* **91**, 103901-1–4 (2003).

[25] A. Garofalakis, G. Zacharakis, H. Meyer, et al., "Three-dimensional *in vivo* imaging of green fluorescent protein-expressing T cells in mice with noncontact fluorescence molecular tomography," *Mol. Imaging* **6**, 96–107 (2007).

[26] H. Meyer, A. Garofalakis, G. Zacharakis, et al., "Noncontact optical imaging in mice with full angular coverage and automatic surface extraction," *Appl. Opt.* **46**, 3617–3627 (2007).

[27] T. Lasser and V. Ntziachristos, "Optimization of 360 degrees projection fluorescence molecular tomography," *Med. Image Anal.* **11**, 389–399 (2007).

[28] A. Martin, J. Aguirre, A. Sarasa-Renedo, et al., "Imaging changes in lymphoid organs *in vivo* after brain ischemia with three-dimensional fluorescence molecular tomography in transgenic mice expressing green fluorescent protein in T lymphocytes," *Mol. Imaging* **7**, 157–167 (2008).

[29] D.A. Boas, M.A. Oleary, B. Chance, et al., "Scattering and wavelength transduction of diffuse photon density waves," *Phys. Rev. E* **47**, R2999–R3002 (1993).

[30] G.B. Arfken and H. J. Weber, *Mathematical Methods for Physicists*, 4th ed., Academic Press, New York (1995).

[31] M. Nieto-Vesperinas, *Scattering and Diffraction in Physical Optics*, 2nd ed., World Scientific, Singapore (2006).

[32] J. Ripoll and M. Nieto-Vesperinas, "Scattering integral equations for diffusive waves: detection of objects buried in diffusive media in the presence of rough interfaces," *J. Opt. Soc. Am. A* **16**, 1453–1465 (1999).

[33] R. Aronson, "Boundary conditions for diffusion of light," *J. Opt. Soc. Am. A* **12**, 2532–2539 (1995).

[34] V. Ntziachristos and R. Weissleder, "Experimental three-dimensional fluorescence reconstruction of diffuse media using a normalized Born approximation," *Opt. Lett.* **26**, 893–895 (2001).

[35] R.J. Gaudette, D.H. Brooks, C.A. DiMarzio, et al., "A comparison study of linear reconstruction techniques for diffuse optical tomographic imaging of absorption coefficient." *Phys. Med. Biol.* **45**, 1051–1070 (2000).

[36] R. Royand and E.M. Sevick-Muraca, "A numerical study of gradient-based nonlinear optimization methods for contrast enhanced optical tomography," *Opt. Exp.* **9**, 49–65 (2001).

[37] X. Intes, V. Ntziachristos, J.P. Culver, et al., "Projection access order in algebraic reconstruction technique for diffuse optical tomography," *Phys. Med. Biol.* **47**, N1–N10 (2002).

[38] D. Boas, D. Brooks, E. Miller, et al., "Imaging the body with diffuse optical tomography," *IEEE Sign. Process. Mag.* **18**, 57–75 (2001).

[39] J.P. Culver, A.M. Siegel, J.J. Stott, et al., "Volumetric diffuse optical tomography of brain activity," *Opt. Lett.* **28**, 2061–2063 (2003).

[40] S. Srinivasan, B.W. Pogue, S. Jiang, et al., "Interpreting hemoglobin and water concentration, oxygen saturation, and scattering measured *in vivo* by near-infrared breast tomography," *Proc. Natl. Acad. Sci. USA* **100**, 12349–12354 (2003).

[41] T. Durduran, R. Choe, J.P. Culver, et al., "Bulk optical properties of healthy female breast tissue," *Phys. Med. Biol.* **47**, 2847–2861 (2002).

[42] A. Torricelli, A. Pifferi, P. Taroni, et al., "*In vivo* optical characterization of human tissues from 610 to 1010 nm by time-resolved reflectance spectroscopy," *Phys. Med. Biol.* **46**, 2227–2237 (2001).

[43] U. Hampel, E. Schleicher, and R. Freyer, "Volume image reconstruction for diffuse optical tomography," *Appl. Opt.* **41**, 3816–3826 (2002).

[44] B.J. Tromberg, N. Shah, R. Lanning, et al., "Non-invasive *in vivo* characterization of breast tumors using photon migration spectroscopy," *Neoplasia* **2**, 26–40 (2000).

[45] V. Ntziachristos and B. Chance, "Probing physiology and molecular function using optical imaging: applications to breast cancer," *Breast Cancer Res.* **3**, 41–46 (2001).

[46] V. Ntziachristos, *Concurrent Diffuse Optical Tomography, Spectroscopy and Magnetic Resonance of Breast Cancer*, Bioengineering, University of Pennsylvania, Philadelphia (2000).

[47] D.J. Hawrysz and E.M. Sevick-Muraca, "Developments toward diagnostic breast cancer imaging using near-infrared optical measurements and fluorescent contrast agents," *Neoplasia* **2**, 388–417 (2000).

[48] V. Ntziachristos, A.G. Yodh, M. Schnall, et al., "Concurrent MRI and diffuse optical tomography of breast after indocyanine green enhancement," *Proc. Natl. Acad. Sci. USA* **97**, 2767–2772 (2000).

[49] B.W. Pogue, S.P. Poplack, T.O. McBride, et al., "Quantitative hemoglobin tomography with diffuse near-infrared spectroscopy: pilot results in the breast," *Radiology* **218**, 261–266 (2001).

[50] A.H. Hielscher, A.D. Klose, A.K. Scheel, et al., "Sagittal laser optical tomography for imaging of rheumatoid finger joints," *Phys. Med. Biol.* **49**, 1147–1163 (2004).

[51] Y. Xu, N. Iftimia, H. Jiang, et al., "Imaging of *in vitro* and *in vivo* bones and joints with continuous-wave diffuse optical tomography," *Opt. Exp.* **8**, 447–451 (2001).

[52] D.A. Benaron, S.R. Hintz, A. Villringer, et al., "Noninvasive functional imaging of human brain using light," *J. Cereb. Blood Flow Metab.* **20**, 469–477 (2000).

[53] J.P. Culver, T. Durduran, C. Cheung, et al., "Diffuse optical measurement of hemoglobin and cerebral blood flow in rat brain during hypercapnia, hypoxia and cardiac arrest," *Adv. Exp. Med. Biol.* **510**, 293–297 (2003).

[54] J.P. Culver, T. Durduran, D. Furuya, et al., "Diffuse optical tomography of cerebral blood flow, oxygenation, and metabolism in rat during focal ischemia," *J. Cereb. Blood Flow Metab.* **23**, 911–924 (2003).

[55] E.M. Hillman, J.C. Hebden, M. Schweiger, et al., "Time resolved optical tomography of the human forearm," *Phys. Med. Biol.* **46**, 1117–1130 (2001).

[56] M. Rudin and R. Weissleder, "Molecular imaging in drug discovery and development," *Nat. Rev. Drug. Discov.* **2**, 123–131 (2003).

[57] V. Ntziachristos, E.A. Schellenberger, J. Ripoll, et al., "Visualization of antitumor treatment by means of fluorescence molecular tomography with an annexin V-Cy5.5 conjugate," *Proc. Natl. Acad. Sci. USA* **101**, 12294–12299 (2004).

[58] J. Ripoll, V. Ntziachristos, C. Cannet, et al., "Investigating pharmacology *in vivo* using magnetic resonance and optical imaging," *Drugs RD* **9**, 277–306 (2008).

[59] R. Weissleder, C.H. Tung, U. Mahmood, et al., "*In vivo* imaging of tumors with protease-activated near-infrared fluorescent probes," *Nat. Biotechnol.* **17**, 375–378 (1999).

[60] S. Achilefu, R.B. Dorshow, J.E. Bugaj, et al., "Novel receptor-targeted fluorescent contrast agents for *in vivo* tumor imaging," *Invest. Radiol.* **35**, 479–485 (2000).

[61] S. Yan and B.F. Sloane, "Molecular regulation of human cathepsin B: implication in pathologies," *Biol. Chem.* **384**, 845–854 (2003).

[62] G. Zacharakis, H. Kambara, H. Shih, et al., "Volumetric tomography of fluorescent proteins through small animals *in vivo*," *Proc. Natl. Acad. Sci. USA* **102**, 18252–18257 (2005).

[63] X. Montet, V. Ntziachristos, J. Grimm, et al., "Tomographic fluorescence mapping of tumor targets," *Cancer Res.* **65**, 6330–6336 (2005).

[64] X. Montet, J.L. Figueiredo, H. Alencar, et al., "Tomographic fluorescence imaging of tumor vascular volume in mice," *Radiology* **242**, 751–758 (2007).

[65] D.E. Sosnovik, M. Nahrendorf, N. Deliolanis, et al., "Fluorescence tomography and magnetic resonance imaging of myocardial macrophage infiltration in infarcted myocardium *in vivo*," *Circulation* **115**, 1384–1391 (2007).

[66] J. de Boer, A. Williams, G. Skavdis, et al., "Transgenic mice with hematopoietic and lymphoid specific expression of Cre," *Eur. J. Immunol.* **33**, 314–325 (2003).

[67] A.K. Scheel, M. Backhaus, A.D. Klose, et al., "First clinical evaluation of sagittal laser optical tomography for detection of synovitis in arthritic finger joints," *Ann. Rheum. Dis.* **64**, 239–245 (2005).

[68] M. Hintersteiner, A. Enz, P. Frey, et al., "*In vivo* detection of amyloid-beta deposits by near-infrared imaging using an oxazine-derivative probe," *Nat. Biotechnol.* **23**, 577–583 (2005).

[69] J. Sharpe, U. Ahlgren, P. Perry, et al., "Optical projection tomography as a tool for 3D microscopy and gene expression studies," *Science* **296**, 541–545 (2002).

[70] A. Asayesh, J. Sharpe, R.P. Watson, et al., "Spleen versus pancreas: strict control of organ interrelationship revealed by analyses of Bapx1-/- mice," *Genes Dev.* **20**, 2208–2213 (2006).

[71] T. Alanentalo, A. Asayesh, H. Morrison, et al., "Tomographic molecular imaging and 3D quantification within adult mouse organs," *Nat. Methods* **4**, 31–33 (2007).

[72] M. Oldham, H. Sakhalkar, Y.M. Wang, et al., "Three-dimensional imaging of whole rodent organs using optical computed and emission tomography," *J. Biomed. Opt.* **12**, 014009-1–10 (2007).

[73] J. Kerwin, M. Scott, J. Sharpe, et al., "3 dimensional modelling of early human brain development using optical projection tomography," *BMC Neurosci.* **5**, 27 (2004).

[74] K. Lee, J. Avondo, H. Morrison, et al., "Visualizing plant development and gene expression in three dimensions using optical projection tomography," *Plant Cell* **18**(9), 2145–2156 (2006).

[75] M. Fauver, E. Seibel, J.R. Rahn, et al., "Three-dimensional imaging of single isolated cell nuclei using optical projection tomography," *Opt. Exp.* **13**, 4210–4223 (2005).

[76] M. Oldham, H. Sakhalkar, T. Oliver, et al., "Three-dimensional imaging of xenograft tumors using optical computed and emission tomography," *Med. Phys.* **33**, 3193–3202 (2006).

[77] C.G. Arques, R. Doohan, J. Sharpe, et al., "Cell tracing reveals a dorsoventral lineage restriction plane in the mouse limb bud mesenchyme," *Development* **134**, 3713–3722 (2007).

[78] L. McGurk, H. Morrison, L.P. Keegan, et al., "Three-dimensional imaging of *Drosophila melanogaster*," *PLoS ONE* **2**, e834 (2007).

[79] R.J. Bryson-Richardson and P.D. Currie, "Optical projection tomography for spatio-temporal analysis in the zebrafish," *Methods Cell. Biol.* **76**, 37–50 (2004).

[80] R.J. Bryson-Richardson, S. Berger, T.F. Schilling, et al., "FishNet: an online database of zebrafish anatomy," *BMC Biol.* **5**, 34 (2007).

[81] A. Delaurier, N. Burton, M. Bennett, et al., "The Mouse Limb Anatomy Atlas: an interactive 3D tool for studying embryonic limb patterning," *BMC Dev. Biol.* **8**, 83 (2008).

[82] M.J. Boot, C.H. Westerberg, J. Sanz-Ezquerro, et al., "*In vitro* whole-organ imaging: 4D quantification of growing mouse limb buds," *Nat. Methods* **5**, 609–612 (2008).

[83] E. Truernit, H. Bauby, B. Dubreucq, et al., "High-resolution whole-mount imaging of three-dimensional tissue organization and gene expression enables the study of Phloem development and structure in *Arabidopsis*," *Plant Cell* **20**, 1494–1503 (2008).

[84] M.K. Hajihosseini, S. De Langhe, E. Lana-Elola, et al., "Localization and fate of Fgf10-expressing cells in the adult mouse brain implicate Fgf10 in control of neurogenesis," *Mol. Cell. Neurosci.* **37**, 857–868 (2008).

[85] H. Sato, P. Murphy, S. Giles, et al., "Visualizing expression patterns of Shh and Foxf1 genes in the foregut and lung buds by optical projection tomography," *Pediatr. Surg. Int.* **24**, 3–11 (2008).

[86] S. Inoue and K.R. Spring, *Video Microscopy: The Fundamentals (The Language of Science)*, 2nd ed., Plenum Publishing Corporation, New York (1997).

[87] J.B. Pawley, "*Handbook of Biological Confocal Microscopy,*" 2nd ed., Springer Science, New York (1995).

[88] J. Sharpe, "Optical projection tomography," *Ann. Rev. Biomed. Eng.* **6**, 209–228 (2004).

[89] J.R. Walls, J.G. Sled, J. Sharpe, et al., "Correction of artefacts in optical projection tomography," *Phys. Med. Biol.* **50**, 4645–4665 (2005).

[90] J.R. Walls, J.G. Sled, J. Sharpe, et al., "Resolution improvement in emission optical projection tomography," *Phys. Med. Biol.* **52**, 2775–2790 (2007).

[91] A. Darrell, H. Meyer, K. Marias, et al., "Weighted filtered backprojection for quantitative fluorescence optical projection tomography," *Phys. Med. Biol.* **53**, 3863–3881 (2008).

6

Fluorescence Lifetime Imaging and Metrology for Biomedicine

Clifford Talbot, James McGinty, Ewan McGhee, Dylan Owen, David Grant, Sunil Kumar, Pieter De Beule, Egidijus Auksorius, Hugh Manning, Neil Galletly, Bebhinn Treanor, Gordon Kennedy, Peter M.P. Lanigan, Ian Munro, Daniel S. Elson, Anthony Magee, Dan Davis, Mark Neil, Gordon Stamp, Christopher Dunsby, and Paul French

Imperial College London, London SW7 2AZ, United Kingdom

This chapter describes recent advances in the development of fluorescence lifetime imaging and related multidimensional fluorescence imaging techniques together with their application to biological tissue and cell cultures. It aims to outline the principles underlying the main fluorescence imaging modalities and describes recent technological advances, with a particular emphasis on fluorescence lifetime imaging. The application of such technology to label-free imaging of biological tissue is then presented with examples from our work at Imperial College London. Applications to imaging molecular cell biology are then reviewed and a brief perspective of future developments is then presented.

Key words: Fluorescence imaging, microscopy, autofluorescence, fluorescence lifetime imaging, hyperspectral imaging, tissues, cells

6.1 Introduction

The chapter is intended to present the potential of fluorescence lifetime imaging (FLIM) and related techniques to obtain useful label-free contrast from intrinsic tissue autofluorescence for clinical and biomedical research applications and to obtain information concerning molecular cell biology by sensing the local environment of fluorescent probes. We review the development of FLIM instrumentation and related imaging technology, with particular reference to our work at Imperial College London developing instrumentation that resolves fluorescence lifetime together with other spectroscopic parameters such as excitation and emission wavelength and polarization. Such technology

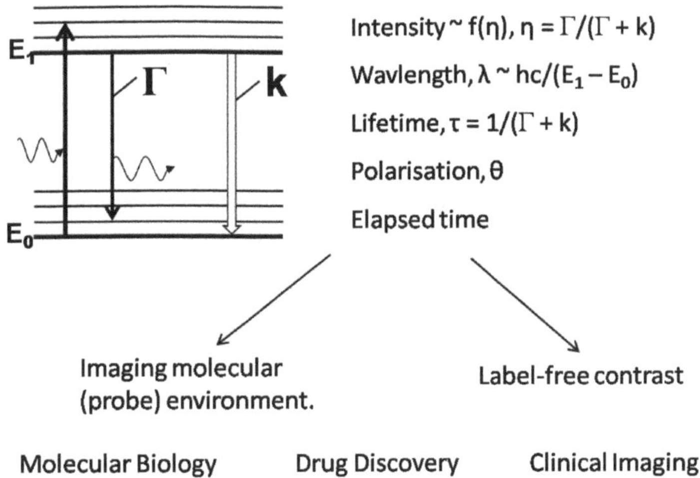

FIGURE 6.1: Overview of multidimensional fluorescence imaging and metrology.

reflects a multidimensional fluorescence imaging (MDFI) and metrology approach that aims to obtain as much useful information as possible from a sample in the shortest possible time. The development of fast multidimensional imaging capabilities is important for real-time clinical diagnostic applications, for higher throughput assays and for the direct investigation of dynamic biological systems. It is hoped that such instrumentation will contribute to the improved understanding and diagnosis of disease and to the development of therapies.

Fluorescence provides a powerful means of achieving optical molecular contrast in single-point (cuvette-based or fiber-optic probe-based) instruments, in cytometers and cell sorters and in microscopes, endoscopes, and multiwell plate readers [1]. Typically for cell biology, fluorescent molecules (fluorophores) are used as "labels" to tag specific molecules of interest. For clinical applications it is possible to exploit the fluorescence properties of target molecules themselves to provide *label-free molecular contrast*, although there is an increasing trend to investigate the use of exogenous labels, including nanoparticles, to both detect diseased tissue, e.g., using photodynamic detection (PDD) and to effect therapeutic intervention, e.g., using photodynamic therapy (PDT) [2] or superheated nanoparticles [3].

Fluorescence imaging is predominantly used to obtain information about *localization* – with imaging providing a mapping of fluorophore distribution and therefore of labelled proteins. By using multiple fluorophores with distinct spectral properties, fluorescence imaging can also elucidate *colocalization* of proteins to within the spatial resolution of the imaging system. The fluorescence process can also provide a *sensing* function since it can be extremely sensitive to the local environment surrounding the fluorophore.

The fluorescence process can be studied with respect to the fluorescence intensity, excitation and emission spectra, quantum efficiency, polarization response, and fluorescence lifetime. These parameters depend on the properties of the fluorophore molecule itself and on its local environment. After a fluorophore has been optically excited to an upper energy level, it may decay back to the ground state either radiatively, with a rate constant, Γ, or nonradiatively, with a rate constant, k, as represented in Figure 6.1. The quantum efficiency, η, of the fluorescence process is defined as the ratio of the number of fluorescent photons emitted compared to the number of excitation photons absorbed and is equal to the radiative decay rate, Γ, divided by the total (radiative + nonradiative) decay rate, $\Gamma + k$, as indicated in Figure 6.1. In general, the quantum efficiency is sensitive to the local fluorophore environment since it can be affected by any factors that change the molecular electronic configuration or de-excitation pathways and therefore the radiative or nonradiative decay rates. Such factors can include the local viscosity, temperature, refractive index, pH, calcium and oxygen concentration, electric field, etc.

Determination of the quantum efficiency, however, requires knowledge of the photon excitation and detection efficiencies, as well as the fluorophore concentration, and quantitative measurements of intensity are hindered by optical scattering, internal re-absorption of fluorescence (inner filter effect) and background fluorescence from other fluorophores present in a sample. Quantitative imaging and metrology based on quantum efficiency is therefore highly challenging, particularly for imaging biological tissue. More robust measurements can be made using *ratiometric* techniques. In the spectral domain, one can assume that unknown quantities such as excitation and detection efficiency, fluorophore concentration, and signal attenuation will be approximately the same in two or more spectral windows and may be effectively "cancelled out" in a ratiometric measurement. This approach is used, for example, with excitation and emission ratiometric calcium sensing dyes [4] and the approach has been demonstrated to provide useful contrast between, e.g., malignant and normal tissue [5]. Fluorescence lifetime measurement is also a ratiometric technique in that it is assumed that the various unknown quantities do not change significantly during the fluorescence decay time (typically ns) and the lifetime determination effectively compares the fluorescence signal at different delays after excitation. The fluorescence lifetime is the average time a fluorophore takes to radiatively decay after having been excited from its ground energy level, and, like the quantum efficiency, it is also a function of the radiative and nonradiative decay rates and so can provide quantitative fluorescence-based molecular contrast. On a slower timescale, fluorescence can also be resolved on "macro" timescales, typically ranging from microseconds to seconds, to observe dynamics, e.g., of calcium levels during signalling processes. Here the absolute analyte concentration is not determined but the temporal variations provide valuable information, particularly for physiological processes. A further dimension of fluorescence is polarization, which is readily measured using polarized excitation sources and can provide information concerning molecular orientation. Furthermore, time-resolved polarization measurements can determine the rotational correlation time – or molecular tumbling time – which can be used to report ligand binding or local solvent properties.

In practice, while fluorescence excitation, emission, lifetime, and polarization-based spectroscopic measurements are widely undertaken in cuvette-based instruments such as spectrophotometers and spectrofluorometers, it is much less common to exploit the rich spectroscopic information available from fluorescence in imaging applications. This has been partly due to instrumentation limitations and partly due to the challenges associated with characterizing fluorescence signals available from typically heterogeneous and often sparsely labelled biological samples. In recent years, however, there have been tremendous changes in imaging tools and technology that are available to biologists and medical scientists. Key drivers have been the dramatic advances in laser and photonics technology, including robust tunable and ultrafast excitation sources, relative low-cost high speed detection electronics and high performance imaging detectors, such as the EM-CCD camera. Technological advances have facilitated the development of powerful new techniques in biophotonics, such as multiphoton microscopy [6], which became widely deployed following its implementation with the conveniently tunable femtosecond Ti:Sapphire lasers [7]. Such instruments confer the ability to excite most of the commonly used fluorophores with a single excitation laser – with computer controlled operation to permit automated excitation spectroscopy [8]. The proliferation of multiphoton microscopes has in turn stimulated the uptake of fluorescence lifetime imaging (FLIM) [9] as a relatively straightforward and inexpensive "add-on" that provides significant new spectroscopic functionality – taking advantage of the ultrafast excitation lasers and requiring only appropriate detectors and external electronic components to implement FLIM (Figure 6.2). In parallel, the development and availability of genetically-expressed fluorescent proteins [10, 11] created unprecedented opportunities to observe many biological processes in live cells and organisms with highly specific labelling.

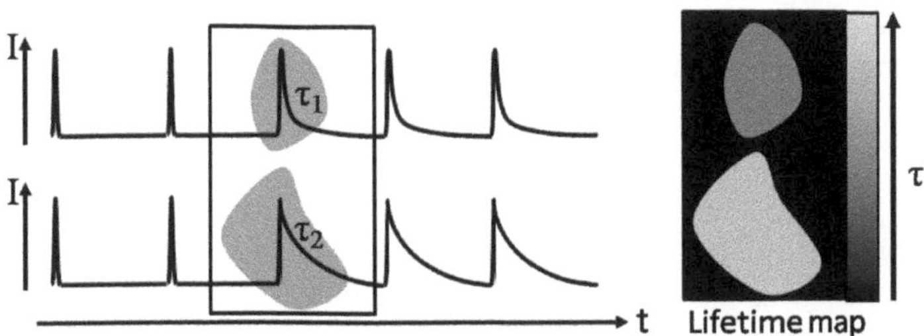

FIGURE 6.2: Schematic of (time domain) fluorescence lifetime imaging. The fluorescence decay profile is measured as a function of time following excitation with a short pulse. Lifetime values can then be encoded in an image by intensity or more commonly, by color.

These opportunities are in turn driving the adoption and development of more sophisticated fluorescence imaging technology. One example in cell biology is the widespread deployment of Förster resonant energy transfer (FRET) [12, 13] techniques that can determine when fluorophores are located within $\sim$ 10 nm of each other, to provide the ultimate in colocalization and the possibility to map protein interactions, such as ligand binding, in cells. In some situations it is possible to read-out the distance between "FRETing" fluorophores – effectively implementing a so-called "*spectroscopic ruler*" [14, 15]. FRET is thus a powerful tool to study molecular cell biology, particularly when combined with genetically-expressed fluorophores, but detecting and quantifying the resonant energy transfer is not straightforward when implemented using intensity-based imaging, for which a number of correction calculations must be performed [16]. Increasingly FLIM and other fluorescence spectroscopic techniques are being applied to improve the reliability of FRET experiments, e.g., [17–19].

For medical applications, the increasing availability of suitable excitation sources, particularly the mode-locked Ti:Sapphire laser, and spectroscopic techniques have prompted an increasing interest in exploiting tissue autofluorescence for label-free clinical applications. Spectrally-resolved imaging of autofluorescence is relatively well established, e.g., [20], and FLIM is now being actively investigated as a means of obtaining or enhancing intrinsic autofluorescence contrast in tissues [21, 22]. There is also a burgeoning interest in applying FLIM to biophotonic devices such as microfluidic systems for lab-on-a-chip applications, e.g., [23, 24], and to utilize FLIM and MDFI to study nanophotonics structures, e.g., [25].

In this chapter we initially review different approaches to FLIM and discuss how it may be extended to multidimensional fluorescence imaging (MDFI) and metrology by combining it with spectrally resolved and polarization-resolved imaging. We then discuss the application of FLIM to imaging biological tissue with an emphasis on label-free contrast based on autofluorescence. Finally there is a discussion of applications of FLIM and MDFI to cell biology, sensing the local fluorophore environment and using FRET to report protein interactions.

6.2 Techniques for Fluorescence Lifetime Imaging and Metrology

6.2.1 Overview

The fluorescence lifetime is the average time the molecule spends in an upper energy level before returning to the ground state. For a large ensemble of identical molecules (or a large number of excitations of the same molecule), the time-resolved fluorescence intensity profile (detected photon histogram) following instantaneous excitation will exhibit a monoexponential decay that may be described as:

$$I(t) = I_0 e^{-\frac{t}{\tau}} + \text{const.} \tag{6.1}$$

where I_0 is intensity of the fluorescence immediately after excitation (at time zero) and the constant term represents any background signal. In practice, the presence of multiple fluorophore species – or multiple states of a fluorophore species arising from interactions with the local environment – often result in more complex fluorescence decay profiles. This is often the case for autofluorescence of biological tissue. Such complex fluorescence decays are commonly modelled by an N-component multi-exponential decay model:

$$I(t) = \sum_{i=1}^{N} C_i e^{-\frac{t}{\tau_i}} + \text{const.} \tag{6.2}$$

where each pre-exponential amplitude is represented by the value C_i. Alternative approaches to model complex fluorescence decay profiles include fitting to a stretched exponential model that corresponds to a continuous lifetime distribution [26], or to a power law decay [27], or to Laguerre polynomials to generate empirical contrast [28]. In general, as more information is used to describe a fluorescence signal, then more detected photons are required to be able to make a sufficiently accurate measurement of it [29, 30]. Thus, fitting fluorescence decay profiles to complex models requires increased data acquisition times, which is often undesirable in terms of temporal resolution of dynamics or considerations of photobleaching or photodamage resulting from extended exposures to excitation radiation. Long acquisition times also require the sample to be stationary, which can be impractical for clinical applications. For many imaging applications, therefore, it is preferable to approximate complex fluorescence decay profiles with a single exponential decay model. The resulting average fluorescence lifetime can still provide a useful contrast since it will usually reflect changes in the decay times or relative contributions of different components, but of course the interpretation of a change in the average lifetime of a complex fluorescence decay profile can be subject to ambiguity.

Where quantitative analysis of complex fluorescence signals is required, it is often sensible to sacrifice image information and utilize all detected photons in a single-point measurement. If multi-component fluorescence lifetime *imaging* of fluorophores exhibiting complex fluorescence is necessary, it is possible to reduce the number of photons required to be detected using *a priori* knowledge or assumptions, e.g., about the magnitude of one or more component lifetimes, i.e., τ_i, or about their relative contributions, i.e., the C_i values of equation (6.2). It can also be useful to assume that some parameters, e.g., τ_i, are global, i.e., they take the same value in each pixel of the image. Such global analysis [31] is often used in the application of FLIM to FRET.

In general, fluorescence lifetime measurement and fluorescence lifetime imaging techniques are categorized as time-domain or frequency-domain techniques, according to whether the instrumentation measures the fluorescence signal as a function of time delay following pulsed excitation or whether the lifetime information is derived from measurements of phase difference between a sinusoidally modulated excitation signal and the resulting sinusoidally modulated fluorescence signal. In principle, frequency and time-domain approaches can provide equivalent information but specific implementations present different trade-offs with respect to cost and complexity, performance and acquisition time and the most appropriate method should be selected according to the target application. Historically, frequency-domain methods were initially developed with simpler electronic instrumentation and excitation source requirements, but time-domain techniques have benefitted from the tremendous advances in microelectronics and ultrafast laser technology and today the different approaches present a similar cost and complexity to most users. A further categorization of fluorescence lifetime measurement techniques can be made according to whether they are sampling

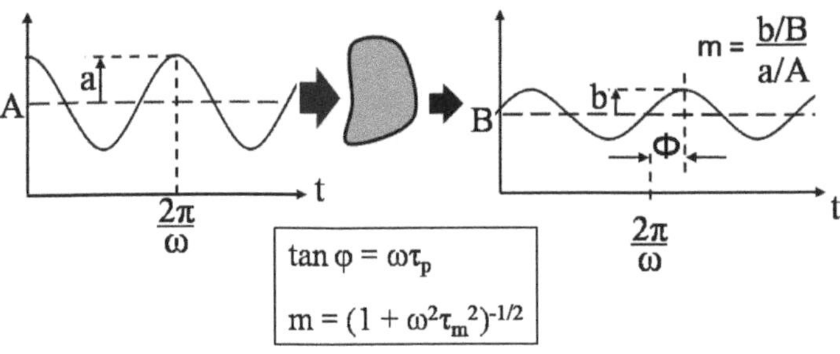

FIGURE 6.3: Schematic of the principle of fluorescence lifetime determination in the frequency-domain.

techniques, which use gated detection to determine the relative timing of the fluorescence compared to the excitation signal, or photon counting techniques, which assign detected photons to different time bins. In general, wide-field FLIM is usually implemented with gated imaging detectors that sample the fluorescence signal while photon counting techniques have been widely applied to single-point lifetime measurements. With respect to the latter, there is an important distinction between single photon counting techniques, such as time-correlated single-photon counting (TCSPC), and techniques based on time-binning (of photoelectrons) or high speed analogue to digital converters that can detect more than one photon per excitation pulse. For laser scanning FLIM, single-point fluorescence lifetime measurement techniques can be readily implemented in laser scanning microscopes as modern electronics makes it straightforward to assign detected photons to their respective images pixels. References [1, 9, 32, 33] provide extensive reviews of the various techniques and in this section we will briefly summarize the main time- and frequency-domain methods are applied to single- and scanning-point fluorescence lifetime measurements and to wide-field fluorescence lifetime imaging.

6.2.2 Single-point and laser-scanning measurements of fluorescence lifetime

6.2.2.1 Frequency-domain measurements

Frequency-domain fluorescence lifetime measurements concern the demodulation of a fluorescence signal excited with modulated light. Initially this approach utilized sinusoidally modulated excitation sources that produced a fluorescence signal that was also sinusoidally modulated but with a different modulation depth and with a phase delay relative to the excitation signal, as illustrated in Figure 6.3.

The fluorescence lifetime can be determined from measurements of the relative modulation, m, and the phase delay, ϕ [1]. These experimental parameters can be conveniently measured using appropriate electronic circuits for synchronous detection, such as a lock-in amplifier. Recent advances in electronics have resulted in relatively low cost synchronous detection [34] and fast digitization circuitry [35]. For a perfect single exponential decay, the lifetime can be calculated from m or ϕ using the following equations:

$$\tau_\varphi = \frac{1}{\omega}\tan\phi, \tag{6.3}$$

$$\tau_m = \frac{1}{\omega}\left(\frac{1}{m^2} - 1\right)^{\frac{1}{2}} \tag{6.4}$$

If the fluorescence excitation is not a pure sinusoid, the frequency-domain approach is still applicable – with the lifetime being calculated from the fundamental Fourier component of a modulated fluorescence signal, which may be excited, for example, by a mode-locked laser [36]. In fact, the implementation of frequency-domain FLIM with periodically pulsed excitation provides an improved signal-to-noise ratio compared to a purely sinusoidal excitation [37]. If the fluorescence does not manifest a single exponential decay, however, the lifetime values calculated from equations (6.3) and (6.4) will not be the same and it is necessary to repeat the measurement at multiple harmonics of the excitation modulation frequency to build up a more complete (multi-exponential) description of the complex fluorescence profile. This is usually calculated by fitting the results to a set of dispersion relationships, e.g., [38, 39]. The increased acquisition and data processing time associated with multiple frequency measurements of complex fluorescence signals can be undesirable. For applications requiring speed, it is possible to measure at only one modulation frequency and obtain an "average" fluorescence lifetime by calculating the mean of the lifetime values from the modulation depth and phase delay measurements.

The ability to use sinusoidally modulated diode lasers or LEDs as excitation sources makes frequency-domain measurements attractive for low-cost applications, for which frequencies of less than 100 MHz are sufficient to provide ns lifetime resolution. To achieve picosecond lifetime resolution, however, does require GHz frequencies and the necessary technology can introduce comparable complexity and expense to that associated with time-domain measurements.

6.2.2.2 Time-domain measurements

Single-point time-domain measurement of fluorescence decay profiles has been implemented using a wide range of instrumentation. Ultrashort pulsed excitation is typically provided by mode-locked solid-state lasers or by gain-switched semiconductor diode lasers. For detection, fast (GHz bandwidth) sampling oscilloscopes and streak cameras have been used for decades but these have been increasingly supplanted by photon counting techniques that build up histograms of the decay profiles. The current most widely used FLIM detection technique is probably time-correlated single-photon counting (TCSPC) [40], which was demonstrated in the first laser scanning FLIM microscope [41]. The development of convenient and relatively low-cost TCSPC electronics, e.g., [42, 43], for confocal and two-photon fluorescence scanning microscopes has greatly increased the uptake of FLIM generally, as well as impacting the development of single-point lifetime fluorometers.

The principle of TCSPC is that, at low fluorescence fluxes, a histogram of photon arrival times can be built up by recording a series of voltage signals that depend on the arrival time of individual detected photons relative to the excitation pulse. By also recording spatial information from the scanning electronics of a confocal/multiphoton microscope, fluorescence lifetime images may be acquired. This is a straightforward way to implement FLIM on a scanning microscope since it only requires adding electronic components after the detector and may be "bolted on" to almost any system. TCSPC is widely accepted as one of the most accurate methods of lifetime determination due to its shot noise-limited detection, high photon economy, low temporal jitter, high temporal precision (offering large number of bins in the photon arrival time histogram), and high dynamic range (typically millions of photons can be recorded without saturation). Its main perceived drawback is a relatively low acquisition rate, owing to the requirement to operate at sufficiently low incident fluorescence intensity levels to ensure single-photon detection at a rate limited by the "dead-time" between measurement events, which is mainly associated with the time-to-amplitude (TAC) circuitry that determine the photon arrival times. However, for modern TCSPC instrumentation this limitation of the electronic circuitry is usually less significant than problems caused by "classical" photon pile up, which limits all single-photon counting techniques. Pulse pile up refers to the issue of more than one photon arriving in a single-photon detection period, which results in apparently shorter lifetimes being "measured." This is avoided by decreasing the excitation power such that the excitation rate is much lower than the pulse repetition rate, which typically limits the maximum detection count rate to approximately 5% of the repetition rate of the laser [44].

An alternative single-point photon-counting time-domain technique is based on temporal photon-binning, for which the photoelectrons arising from the detected photons are accumulated in a number of different time-bins and a histogram is built up accordingly [45]. This does not have the same dead-time and pulse-pile-up limitations as TCSPC, so it may be used with higher photon fluxes to provide higher imaging rates when implemented on a scanning fluorescence microscope. To date, however, it has not been commercially implemented with the same precision as TCSPC and so can provide FLIM at faster rates (approaching real time) but with lower lifetime precision [46]. Of course the temporal resolution of fluorescence lifetime instrumentation is also limited by the temporal impulse response of the detector, as well as the electronic circuitry. Photon counting photomultipliers typically exhibit a response time of ~ 200 ps although faster multichannel plate (MCP) devices can have response times of a few 10's of ps. One significant exception to the above observation is the pump-probe approach where a second (probe) beam, which is delayed with respect to the excitation (pump) beam, interrogates the upper state population. A particularly elegant implementation of this technique for scanning fluorescence microscopy also provides optical sectioning in a manner analogous to two photon microscopy [47].

6.2.2.3 Scanning FLIM microscopy

Laser scanning microscopes are essentially single-channel detection systems and are widely used for biological imaging because of their implementation as confocal or multiphoton microscopes provides improved contrast and optical sectioning compared to wide-field microscopes. Figure 6.4 summarizes some of the more common approaches to realize FLIM in laser scanning microscopes. Having first been implemented using TCSPC [41], scanning FLIM microscopy was subsequently demonstrated using photon-binning [45] and frequency-domain techniques, e.g., [48]. TCSPC is currently the most widely implemented technique for laser scanning FLIM. When imaging typical biological samples, TCSPC typically requires ten's of seconds to acquire sufficient photons for single-photon excited FLIM and often longer for multiphoton excitation, which can limit some of its applications. Although recent technological advances have led to reductions in detector dead time and increased the maximum detection count rates of TCSPC to approach ~ 10 MHz, it is not usually possible in practice to reach the maximum possible count rates before the onset of significant photobleaching and/or phototoxicity. The photon time-binning and frequency-domain approaches are not limited to single-photon detection and so can provide faster imaging of bright samples but photobleaching and/or phototoxicity considerations also limit the maximum practical imaging rates.

For frequency-domain laser scanning FLIM, one can use a sinusoidally modulated excitation laser and apply synchronous detection, e.g., using a "lock-in" amplifier to determine the phase difference and change in modulation depth between the excitation signal and the resulting sinusoidally modulated fluorescence signal. One can also take advantage of pulsed excitation sources, e.g., in two photon microscopes, and exploit the harmonic content of the resulting fluorescence [36]. As discussed in subsection 6.2.2.1, this frequency-domain approach can be implemented using relatively low cost electronic circuitry that is not limited by the dead-time or maximum count rates associated with single-photon counting detection and so can provide high speed FLIM [34, 35].

For all laser scanning microscopes the sequential pixel acquisition means that increasing the imaging speed requires a concomitant increase in excitation intensity, which can be undesirable due to photobleaching and phototoxicity considerations. This is a particular issue for FLIM, for which more photons need to be detected per pixel compared to intensity imaging [46, 49]. For example, it is estimated that detection of a few hundred photons is necessary to determine the lifetime (to $\sim 10\%$ accuracy) using a single exponential fluorescence decay model while a double exponential fit requires $\sim 10^4$ detected photons. For multiphoton FLIM microscopy, for which excitation is relatively inefficient and photobleaching scales nonlinearly with intensity [50], these considerations can result in FLIM acquisition times of many minutes for biological samples. One way to significantly increase the imaging speed of multiphoton microscopy is to use multiple excitation beams in

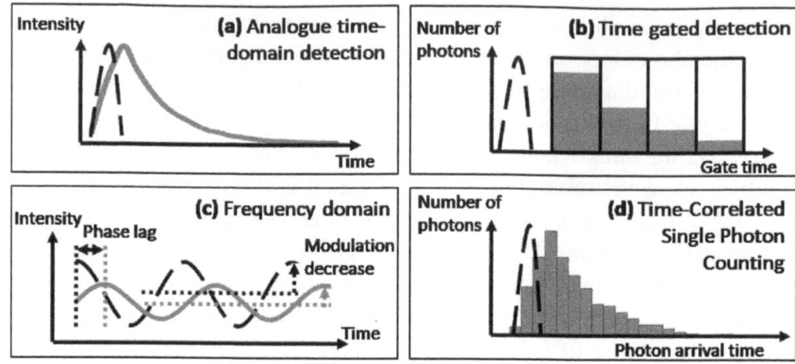

FIGURE 6.4: Techniques for FLIM implemented in scanning microscopes: (a) analogue time-domain measurements (e.g., with a streak camera); (b) time-gated detection (LIMO); (c) frequency-domain measurements (with heterodyne detection) and (d) TCSPC.

parallel [51–53] and this approach can be applied to TCSPC FLIM using 16 parallel excitation and TCSPC detection channels [54, 55]. In general, parallel pixel excitation and detection is a useful approach to increase the practical imaging speed of all laser scanning microscopes. This can be extended to optically-sectioned line-scanning microscopy using a rapidly scanned multiple-beam array to produce a line of fluorescence emission that is relayed to the input slit of a streak camera [55]. FLIM images have been acquired in less than one second using this approach, which is currently limited by the readout rate of the streak camera system. Alternatively, multiple scanning beam excitation can be applied with wide-field time-gated detection, as has been implemented with multibeam multiphoton microscopes [56–58] and with single-photon excitation in spinning Nipkow disc microscopes [59–61]. A direct comparison of TSCPC and wide-field frequency-domain FLIM concluded that TCSPC provided a better signal-to-noise ratio (SNR) for weak fluorescence signals, the frequency-domain approach was faster and more accurate for bright samples [62].

6.2.3 Wide-field FLIM

The parallel nature of wide-field imaging techniques can support FLIM imaging rates of 10's to 100's Hz, e.g., [63, 64], although the maximum acquisition speed is still of course limited by the number of photons/pixel available from the (biological) sample. Wide-field FLIM is most commonly implemented using modulated image intensifiers with frequency- or time-domain approaches, as represented in Figure 6.5.

The frequency-domain approach was established first around 1990, e.g., [65, 66], utilizing frequency-modulated laser excitation and implementing frequency-modulated gain with a microchannel plate (MCP) image intensifier to analyze the resulting fluorescence by acquiring a series of gated "intensified" images acquired at different relative phases between the MCP modulation and the excitation signal. Originally the optical output image from the intensifier was read out using a linear photodiode array but this was rapidly superseded by CCD camera technology [67, 68]. It is possible to implement wide-field FLIM at a single modulation frequency using only three phase measurements [69] to calculate the fluorescence lifetime map although it is common to use 8 or more phase-resolved images to improve the accuracy. This can be necessary if the system exhibits unwanted nonlinear behavior that produces modulated signals at harmonic frequencies.

The development of wide-field frequency-domain FLIM was complemented by the demonstration of a streak camera based approach [70] and by the application of short pulse gated MCP image intensifiers coupled to CCD cameras in an approach described as time-gated imaging for time-

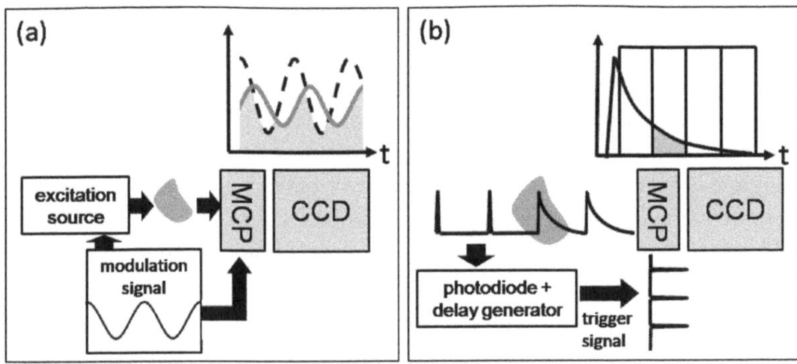

FIGURE 6.5: Wide-field approaches to FLIM: (a) frequency-domain FLIM utilizing a sinusoidal signal to modulate the excitation source and detector at the same frequency (homodyne detection); (b) time-domain FLIM utilizing the excitation source to trigger a wide-field gated detector.

domain wide-field FLIM, e.g., [22, 71, 72]. By gating the MCP image intensifier gain for short (ps–ns) periods after excitation, the fluorescence decay can be sampled such that the fluorescence lifetime image can be calculated. Initially the shortest MCP gate widths that could be applied were over 5 ns but the technology quickly developed to provide subnanosecond gating times [73] and then the use of a wire mesh proximity-coupled to the MCP photocathode led to devices with sub 100 ps resolution [74].

While both frequency and time-domain FLIM offer parallel pixel acquisition compared to scanning microscopy techniques, they both suffer from inherently reduced photon economy owing to the time-varying gain applied to the MCP image intensifier. This can be particularly significant in the time domain if very short ($\sim$ 100 ps) time gates are applied to sample the fluorescence decay profile. In many situations, however, such short time gates are not necessary since the fast rising and falling edges of the time gate provide the required time resolution and so the gate can remain "open" for times comparable to the fluorescence decay time [75]. Thus, the photon economy of time-gated imaging can approach that of wide-field frequency-domain FLIM. For monoexponential fluorescence decay profiles, the minimum resolvable lifetime difference does not depend on the sampling gate width and nanosecond gate widths are sufficient to provide sub-100 ps lifetime discrimination. For both frequency-domain and time-domain measurements, it is necessary to acquire at least three time (phase) gated images in order to obtain a mean fluorescence lifetime in the presence of an offset (background signal). If the background can be determined in a separate measurement and then subtracted from subsequent acquisitions, it is possible to implement high-speed FLIM with just two time (phase)-gated images required for each fluorescence liftime calculation. The required fluorescence lifetimes can be calculated analytically such that real-time fluorescnce lifetime images can be displayed for both frequency- and time-domain FLIM. A frequency-domain FLIM microscope achieved a FLIM rate of 0.7 Hz for a field of view of 300 $\times$ 220 pixels and also provided "lifetime-resolved" images based on the difference between only two phase-resolved images at up to 55 Hz for 164 $\times$ 123 pixels [76]. Frequency-domain FLIM was extended to FLIM endoscopy, acquiring phase-resolved images at 25 Hz to achieve real-time FLIM for a field of view of 32 $\times$ 32 pixels [77].

In the time domain the rapid lifetime determination (RLD) is often implemented using just two time gates (with the background level assumed to be zero or subtracted) with an analytical approach to calculate the single exponential decay lifetime [78]. At Imperial we demonstrated video-rate wide-field RLD FLIM using two sequentially acquired time-gated fluorescence images (subtracting a previously acquired background image) using a gated optical imaging (GOI) intensifier based system, which we also applied to FLIM endoscopy [63]. RLD has been extensively studied and, if both gates are of equal width, the minimum error in lifetime determination is found to be when the gate separation, Δt, is 2.2 times the lifetime being investigated. This optimized measurement is only accurate over a narrow range of sample fluorescence lifetimes and to optimize RLD FLIM over a wider range of sample lifetimes, one can increase the width of the second time gate [79] or use three or more time gates, albeit at the expense of imaging speed. There are also analytic expressions for RLD of a single exponential decay with an unknown background [80] or a biexponential decay using four time-gated images [81]. Further optimisation of RLD FLIM can exploit the potential to vary the CCD integration time in order to collect more photons at the later stages of the decay profile [75].

High-speed FLIM is important for application to dynamic or moving samples. Time-gated imaging (or phase-gated) wide-field imaging, however, is subject to severe artifacts if the sample moves between successive time-gated (or phase-gated) acquisitions [82]. This problem is less severe for TCSPC FLIM where sample motion during an image acquisition will degrade the spatial resolution but not drastically change the apparent fluorescence lifetimes. It is possible to avoid addressing this issue of motion artifacts in RLD FLIM by acquiring the different time-gated (or phase-gated) images simultaneously. This was first implemented in a "single-shot" FLIM system that utilized an optical image splitter to produce two images incident n the same GOI but with the photons from one being delayed with respect to the other by an optical relay [64]. This elegant approach was demonstrated at up to 100 Hz imaging a rat neonatal myocyte stained with the calcium indicator, Oregon green BAPTA-1. Its main drawbacks are that the field of view is reduced with respect to the conventional sequential time-gated imaging acquisition approach and that the complexity of the optical imaging delay line makes it difficult to adjust the delay between the parallel images, in order to accommodate a range of sample fluorescence lifetimes. These issues can be addressed using multiple independently gated GOI detectors to maintain the field of view field [83] at the cost of significantly increased the system size and complexity, or by using a single GOI with a segmented photocathode providing multiple image channels that can be independently gated through the introduction of resistive elements. The latter approach realized wide-field four-channel RLD FLIM at up to 20 Hz and was applied to label-free FLIM of *ex vivo* tissue [82]. This parallel acquisition of multiple time (phase)-gated images has recently been implemented for frequency-domain FLIM using a modulated CMOS detector that can accumulate photoelectrons in two parallel image stores according to a modulated voltage [84]. This permits fast FLIM (with two phase-resolved images) using solid-state camera technology that could potentially be cheaper and faster than current GOI technology, although to date it has only been realized with a 20 MHz modulation frequency and 124×160 pixels in each phase-gated image.

While high-speed FLIM is the compelling application of wide-field time/phase-gated imaging, it is not always applied using RLD. For applications such as FRET it is desirable to image samples exhibiting complex fluorescence decay profiles and to extract information concerning different components of the decay. In the time domain this entails sampling the decay profiles with more time gates and usually fitting the data to a complex decay model using, e.g., a nonlinear least squares Levenberg Marquardt algorithm, although it is possible to use an analytical modified RLD approach to analyse double exponential decay profiles [81]. To characterize complex fluorescence decay profiles using wide-field frequency-domain FLIM, one can phase-gate the resulting fluorescence images for a range of excitation modulation frequencies and then determine multiple exponential decay components, e.g., using Fourier analysis. A particularly elegant approach exploits a nonsinusoidal detector gain (e.g., a fast rectangular time-gated GOI detector function) to simultaneously determine phase delays at multiple harmonic frequencies in the excitation signal [85]. This can reduce the sample exposure time, limiting photobleaching, and photodamage compared to sequential acquisitions at different modulation frequencies.

Sampling and fitting complex decay profiles, however, inevitably increases the FLIM acquisition time since it requires significantly more detected photons for accurate lifetime determination. Where rapid FLIM of samples exhibiting complex fluorescence decays is required, it is usually necessary to use *a priori* information or assumptions such as fixing one or more component lifetimes or assuming that the unknown component lifetimes are the same in each pixel. This is particularly important for FLIM FRET where imaging speed is a consideration.

6.3 FLIM and MDFI of Biological Tissue Autofluorescence

6.3.1 Introduction

In biological tissue, autofluorescence can provide a source of label-free optical molecular contrast and has the potential to discriminate between healthy and diseased tissue. The prospect of detecting molecular changes associated with the early manifestations of diseases such as cancer is particularly exciting. For example, accurate and early detection of cancer allows earlier treatment and significantly improves prognosis [86]. While the exophytic tumors can often be visualized, the cellular and tissue perturbations of the peripheral components of neoplasia may not be apparent by direct inspection under visible light and are often beyond the discrimination of conventional noninvasive diagnostic imaging techniques. In general, label-free imaging are preferable for clinical imaging, particularly for diagnosis, as they avoid the need for administration of an exogenous agent, with associated considerations of toxicity and pharmacokinetics. A number of label-free modalities based on the interaction of light with tissue have been proposed to improve the detection of malignant change. These include fluorescence, elastic scattering, Raman, infrared absorption and diffuse reflectance spectroscopies [20, 87–89]. To date, most of these techniques have been limited to point measurements, whereby only a very small area of tissue is interrogated at a time, e.g., via a fiber-optic contact probe. This enables the acquisition of biochemical information, but provides no spatial or morphological information about the tissue. Autofluorescence can provide label-free "molecular" contrast that can be readily utilized as an imaging technique, allowing the rapid and relatively noninvasive collection of spatially-resolved information from areas of tissue up to tens of centimeters in diameter.

The principal endogenous tissue fluorophores include collagen and elastin cross-links, reduced nicotinamide adenine dinucleotide (NADH), oxidized flavins (FAD and FMN), lipofuscin, keratin, and porphyrins. As shown in Figure 6.6, these fluorophores have excitation maxima in the UV-A or blue (325–450 nm) spectral regions and emit Stokes-shifted fluorescence in the near-UV to the visible (390–520 nm) region of the spectrum. The actual autofluorescence signal excited in biological tissue will depend on the concentration and the distribution of the fluorophores present, on the presence of chromophores (principally hemoglobin) that absorb excitation and fluorescence light, and on the degree of light scattering that occurs within the tissue [20]. Autofluorescence therefore reflects the biochemical and structural composition of the tissue, and consequently is altered when tissue composition is changed by disease states such as atherosclerosis, cancer, and osteoarthritis.

While autofluorescence can be observed using conventional "steady-state" imaging techniques, it is challenging to make sufficiently quantitative measurements for diagnostic applications since the autofluorescence intensity signal may be affected by fluorophore concentration, variations in temporal and spatial properties of the excitation flux, the angle of the excitation light, the detection efficiency, attenuation by light absorption and scattering within the tissue, and spatial variations in the tissue microenvironment altering local quenching of fluorescence. The use of two or more spectral emission windows can be used to improve quantitation through ratiometric imaging but the heterogeneity in the distribution of tissue fluorophores and their broad, heavily overlapping, emission spectra can limit the discrimination achievable, e.g., [90]. More sophisticated spectral unmixing approaches utilizing principal component analysis (PCA) and related techniques applied to large sample number training sets are being developed, e.g., [91] to identify signatures for diagnos-

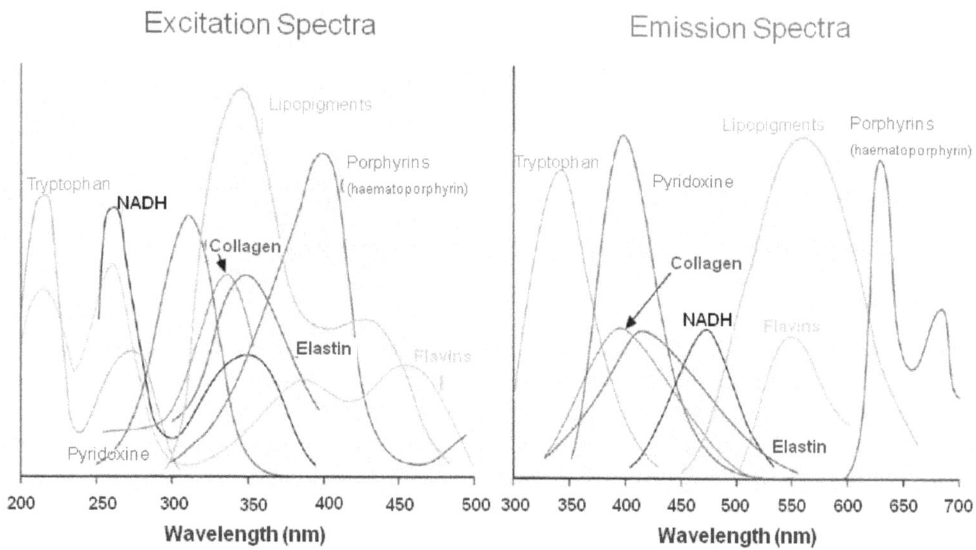

FIGURE 6.6: Excitation and emission spectra of main endogenous tissue fluorophores (adapted from [89]).

tic applications but this it is a highly challenging spectral unmixing problem. These considerations have so far restricted the widespread use of fluorescence imaging for the detection of malignancy, and current fluorescence imaging tools in clinical use are hampered by low specificity and a high rate of false positive findings [92–94].

Increasingly there is interest in exploiting fluorescence lifetime contrast to analyse tissue autofluorescence signals since fluorescence decay profiles depend on relative (rather than absolute) intensity values and fluorescence lifetime imaging (FLIM) is therefore largely unaffected by many factors that limit steady-state measurements [95]. The principal tissue fluorophores exhibit characteristic lifetimes (ranging from hundreds to thousands of picoseconds [20, 96]) that enable spectrally overlapping fluorophores to be distinguished and the sensitivity of fluorescence lifetime to changes in the local tissue microenvironment (e.g., pH, [O_2], [Ca^{2+}]) [1] can provide a readout of biochemical changes indicating the onset or progression of disease.

6.3.2 Application to cancer

Changes in kinetics of time-resolved fluorescence between normal versus malignant tissue were reported as early as 1986 [97] but the complexity and performance of instrumentation associated with picosecond-resolved measurements, and particularly imaging, hindered the development of FLIM as a clinical tool. Single-point measurements of autofluorescence lifetime have revealed differences between normal and neoplastic human tissue in tumours of the oesophagus [98–100], colon [101], brain [90, 95, 102], oral cavity [95], lung [98], breast [103], skin [104], and bladder [98]. Of these results, [95, 98, 100, 101] were obtained *in vivo*.

There have also been a number of FLIM studies of autofluorescence of cancerous tissues. To date, most of these FLIM studies have used laser scanning microscopes and particularly multiphoton microscopes that inherently incorporate an ultrafast excitation laser and for which the extension to FLIM is readily achieved using commercially available time-correlated single-photon counting (TCSPC) electronics. Results obtained using multiphoton FLIM systems to study cancer include *ex*

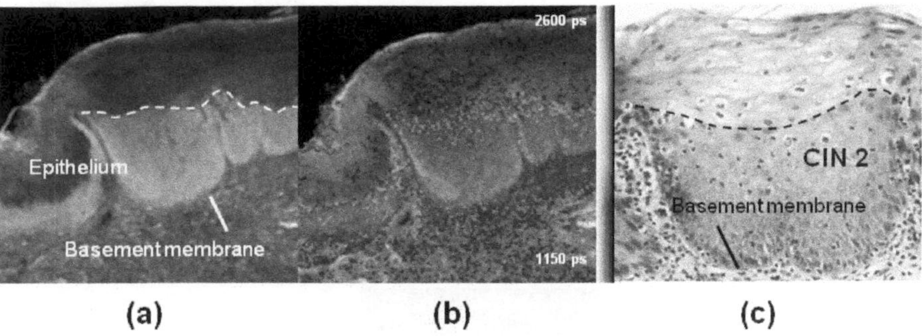

FIGURE 6.7: (See in color after page 572.) Multiphoton fluorescence intensity(a) and lifetime (b) images of fresh section from human cervical biopsy excited at 740 nm, shown with parallel H&E stained section indicating grade CIN2 cancer (adapted from [108]).

vivo studies of brain [105], skin [106, 107], cervix [108], breast [109], and *in vivo* studies in animals [110] and human skin [111]. To illustrate the potential for label-free contrast, Figure 6.7 shows a multiphoton FLIM image of a fresh section of human cervical tissue exhibiting CIN2 cancer, together with the corresponding fluorescence intensity image and a parallel H&E stained section. In general multiphoton excitation provides a convenient approach to image tissue autofluorescence and, to date, the only FLIM microscope licensed for *in vivo* application is a multiphoton microscope (*DermaInspect*, JenLab GmbH).

For many clinical applications such as diagnostic screening for disease, guided biopsy or inter-operative surgery, however, it is desirable to image a larger field of view than is usually possible with multiphoton microscopy. For such applications, wide-field FLIM can provide ~cm fields of view and applied to microscopes, endoscopes, or macroscopes. Ironically, since the first application of FLIM to autofluorescence for imaging cancer tissue [112], which was implemented using a wide-field frequency-domain FLIM system applied to *ex vivo* bladder and oral mucosa, there has been relatively little progress towards clinical wide-field FLIM instrumentation. This situation is changing with recent advances in compact and convenient diode-pumped laser-based u.v. excitation sources, e.g., (*UVP355,* Fianium Ltd.), that are enabling the development of efficient time-gated wide-field FLIM instrumentation [63] for wide-field macroscopic imaging, e.g., of skin [107], liver, colon, and pancreas [113, 114], as well as endoscopy [75, 108]. Figure 6.8 shows wide-field macroscopic FLIM images acquired using compact picosecond excitation laser source at 355 nm and optimized detection using a gated optical intensifier readout using a cooled CCD camera.

6.3.3 Application to atherosclerosis

Autofluorescence techniques can also be applied to study and potentially aid early diagnosis of atherosclerosis – the buildup of lipids on the artery wall to form atherosclerotic plaques, which can impede blood flow and can rupture, causing a blockage leading to a heart attack or stroke. Several studies have shown that autofluorescence of arteries is affected by the presence of plaques. Several studies have utilized spectroscopic data to distinguish normal aortas from those with plaques in varying stages of development, see for example [115–120]. Artificially increasing the levels of collagen types I and III and cholesterol to mimic changes associated with atherosclerosis in normal canine aorta induced significant changes in fluorescence intensity of the tissue [121].

Within arterial tissue the dominant fluorophores are tryptophan, with an emission peak at 325 nm, and collagen and elastin, both with emission spectra peaked near 380 nm. These have been identified as accounting for 95% of the fluorescence when exciting at 310 nm [122]. Subsequent studics

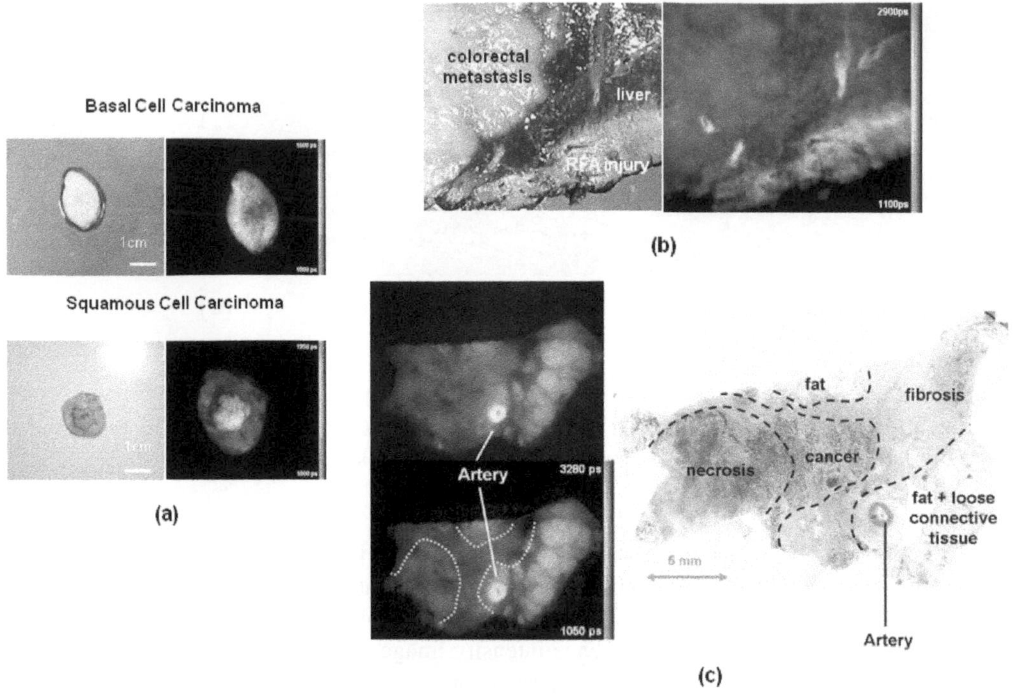

FIGURE 6.8: (See in color after page 572.) Wide-field time-gated FLIM images of freshly resected tissue autofluorescence excited at 355 nm: (a) white light and FLIM images of basal and squamous cell carcinomas (adapted from [107]); (b) white light and FLIM images of liver tissue showing colorectal metastasis and RF ablation damage; and (c) fluorescence intensity and lifetime images of pancreatic tissue showing necrosis, cancer, fat and loose connective tissue with an artery.

have suggested that the best excitation wavelengths for spectrally-resolved studies of atherosclerotic samples are between 314 nm and 344 nm [123, 124] except for calcified plaques, which require excitation at greater than 380 nm [115]. It is important to note, however, that the optimum wavelengths for fluorescence lifetime *contrast* between different tissue components may be different from these excitation maxima. In practice, the excitation wavelength for FLIM is constrained by the availability of suitable excitation sources. Early fluorescence lifetime studies were undertaken using nitrogen lasers operating at 337 nm for single-point measurements and showed an enhanced capability for identification and classification of atherosclerotic plaques [125–127]. We have undertaken a study of the optimum wavelength for FLIM contrast of atherosclerosis in a rabbit model using excitation wavelengths between 355 nm and 520 nm from a tunable ultrafast optical parametric amplifier excitation source and observed that it was possible to differentiate atherosclerotic lesions and artery walls using FLIM when exciting from 355–440 nm [108]. The strongest lifetime contrast was obtained with 355-nm and 400-nm excitation, although contrast was still observed for excitation wavelengths up to 440 nm. It is fortuitous that picosecond pulses at 355 nm can be obtained from compact and convenient frequency tripled Nd-doped solid-state lasers or Yb-doped glass fiber lasers. In our study (unpublished), we compared spectrally-resolved fluorescence emission imaging with FLIM and generally observed that fluorescence lifetime provided superior contrast for excitation between 355 nm and 520 nm.

Using 400 nm excitation, we undertook a study of the variation of autofluorescence lifetime with the different components of atherosclerotic plaques [128]. We applied time-gated imaging in a wide-

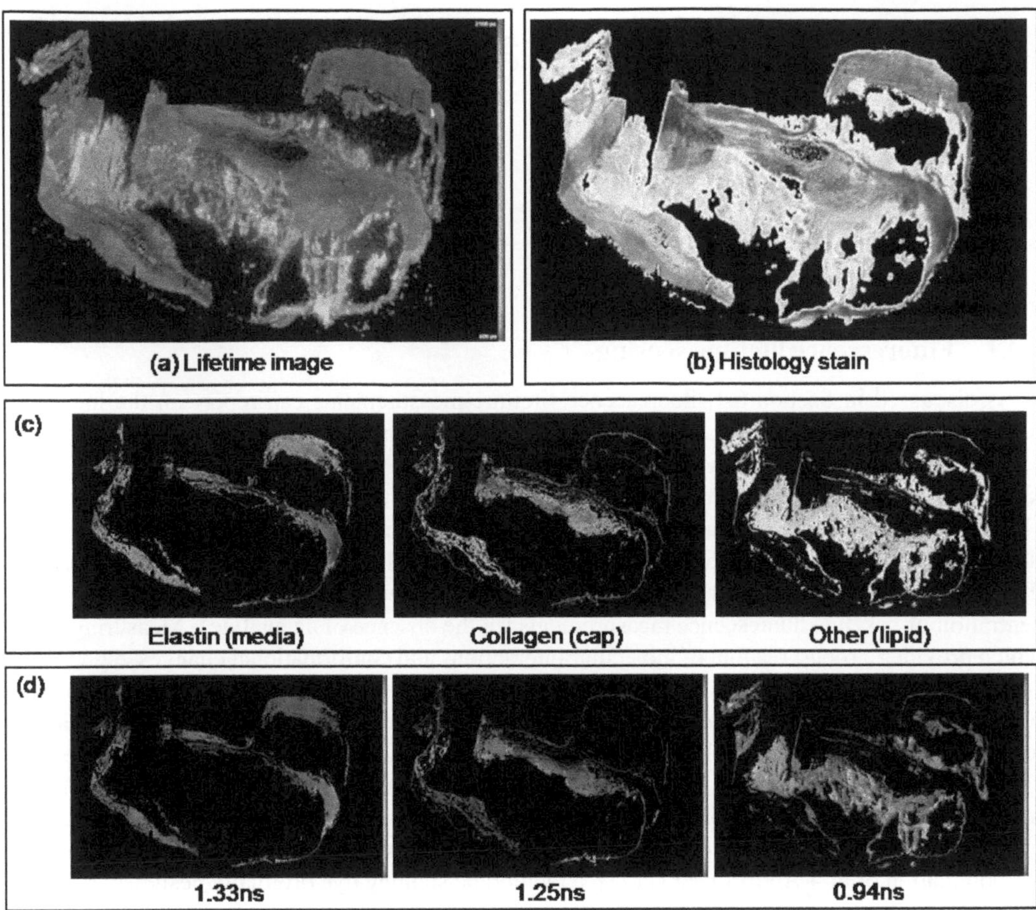

FIGURE 6.9: (See in color after page 572.) Showing images of an artery section obtained from a carotid endarterectomy procedure: (a) wide-field time-gated FLIM montage assembled from 17 images. (b) A micrograph taken with a color camera of the same section stained according to the Elastic van Gieson (EVG) protocol. (c) Masks obtained by color segmentation of the image of the EVG stained section. (d) The masks applied to the FLIM images. The pixels corresponding to each component were binned to provide mean fluorescence lifetimes for each component.

field microscope with a magnification of ×4 to a series of parallel sections of human carotid arteries and correlated the data with histology as illustrated in Figure 6.9. The arteries were obtained from carotid endarterectomy procedures and it is important to note that they were not dipped in formalin or any other fixative following resection, but were instead immediately flash frozen. Lifetime data was acquired from the whole section within 1 hr of defrosting. Because the field of view ($\sim$ 2.5 mm) was smaller than the cross-section of the artery ($\sim$ 10 mm), it was necessary to prepare a montage of the acquired data to produce the FLIM images (Figure 6.9a is a montage of 17 smaller FLIM acquisitions). Following the lifetime data acquisition, the sections were then fixed and stained according to the Elastic van Giesen protocol (Figure 6.9b) to identify the elastin, collagen, and lipid components in the field of view (Figure 6.9c). From these stained images, digital masks corresponding to these components were applied to the autofluorescence lifetime images and the pixels corresponding to each component mask were binned to obtain a mean fluorescence lifetime for each component (Figure 6.9d). Applying this procedure to 5 parallel sections from 4 carotid arteries, we obtained lifetimes in the range ∼ 1.33–1.58 ns, 1.21–1.51 ns and 0.94–1.12 ns for the elastin, collagen and lipid components, respectively. This indicates that FLIM can distinguish collagen rich plaques from lipid rich caps and therefore suggests that lifetime measurements can identify plaques that are vulnerable to rupture.

6.4 Application to Cell Biology

6.4.1 Fluorescence lifetime sensing

As discussed in Section 6.1, fluorescence lifetime measurements can report on the local fluorophore environment and so FLIM can provide molecular functional information concerning fluorescently labelled proteins as well as their localization. Fluorescence lifetime has been applied to read out many perturbations to the local molecular environments, including changes in temperature [129], viscosity [57, 130], refractive index [131, 132], solvent polarity [133] and ionic concentrations, notable calcium [134–136] oxygen [137], and pH [138, 139]. Unlike intensity-based readouts, fluorescence lifetime measurements of such sensors do not require knowledge of fluorophore concentration or baseline fluorescence measurements (in the absence of an analyte). Measuring FRET, which provides a robust readout of molecular interactions and conformational changes, e.g., [17, 18, 140], can also be considered as sensing a change in the local fluorophore environment. The robust nature of lifetime measurements and the increasingly convenient implementation of FLIM have led to numerous applications to cell biology and particularly to the study of cell signalling processes.

Although there is increasing interest in developing new probes for lifetime readouts, FLIM has often been applied to probes originally developed for intensity or spectral readouts, for which it offers the potential for enhanced contrast. An example is illustrated in Figure 6.10, which presents the application of di-4-ANEPPDHQ [141], a membrane staining dye originally designed as a membrane voltage probe and shown to also facilitate the imaging of the distribution of membrane lipid microdomains (MLM) or "lipid rafts," via a spectral shift in fluorescence emission with changes in the order of the lipid bilayer [142]. Lipid rafts are thought to be associated with several important cell processes including signalling. Figure 6.10(a) shows the emission spectra of di-4-ANEPPDHQ labeling two samples of giant unilamellar vesicles (GUV), for which one was incubated with methyl-β-cyclodextrin to extract cholesterol and therefore reduce the lipid order of this artificial membrane. Although this spectral shift can be exploited to map the MLM distribution in cell membranes, our experiments suggest that FLIM can provide superior contrast of membrane lipid order [143]. Figure 6.10(b) shows the fluorescence lifetime histograms for the same GUV samples, illustrating the clear lifetime contrast and Figure 6.10(d) illustrates the application to cell imaging. For comparison, Figure 6.10(c) shows a dual-channel (500–530-nm and 570-nm longpass) intensity image while Figure 6.10(d) shows the corresponding fluorescence lifetime image of the same sample of live HEK cells a in full-growth medium, supplemented with 5 mM of di-4-ANEPPDHQ dye 1 hour before imaging. Both images show the increased lipid order of the plasma membrane compared to the intracellular membranes. FLIM, however, indicates regions in the plasma membrane enriched in a liquid-ordered phase, which seem to be clustered around sites of membrane protrusion. Previous studies have shown that the actin cytoskeleton interacts with the membrane through cholesterol-enriched microdomains [144]. Other dyes have also been reported to show spectral (e.g., Laurdan [145]) or fluorescence lifetime (e.g., PMI-COOH [146]) contrast as a function of membrane lipid order. Such membrane lipid order probes may be multiplexed with other fluorophores including genetically expressed fluorescence proteins, to permit studies correlating MLM accumulation with cell signalling pathways, which may be studied using FRET.

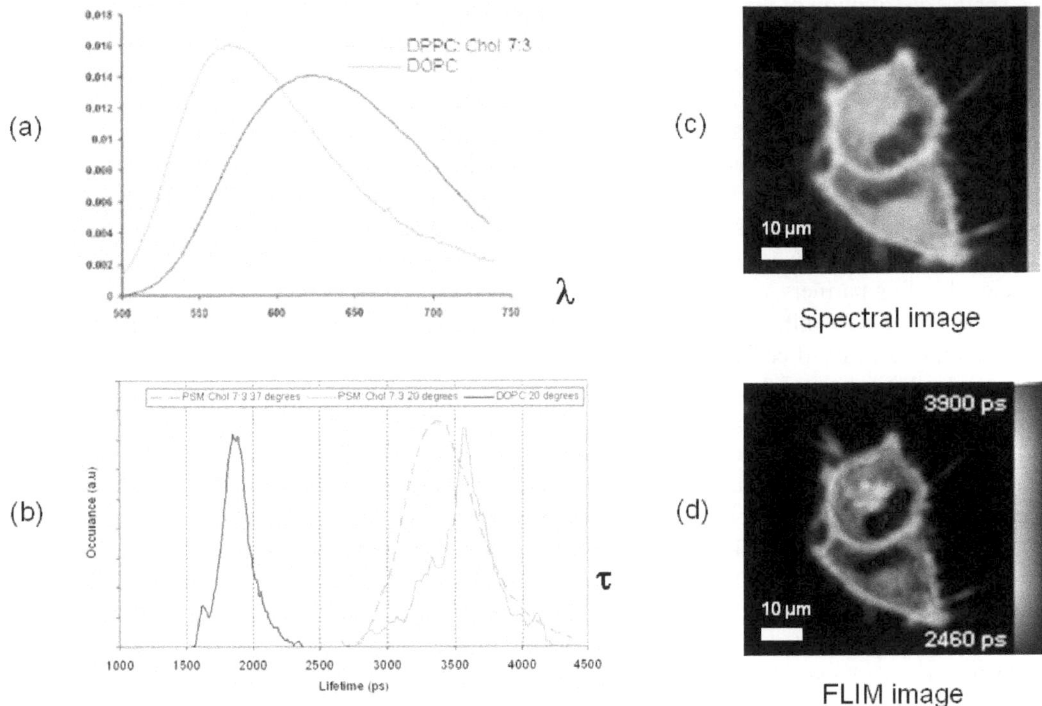

FIGURE 6.10: (See in color after page 572.) Comparison of fluorescence spectral and lifetime readout of the di-4-ANEPPDHQ dye excited at 473 nm: (a) shows change in emission spectra when staining giant unilamellar vesicles with and without extraction of cholesterol by methyl-β-cyclodextrin and (b) shows the corresponding fluorescence lifetime histograms. (c) and (d) show the spectral ratiometric and fluorescence lifetime images respectively of live HEK cells labelled with di-4-ANEPPDHQ. (Figure adapted from [143]).

6.4.2 FLIM applied to FRET

The most widespread application of FLIM in cell biology is probably the mapping of FRET [12] to report the interactions between suitably labelled specific proteins, lipids, enzymes, DNA and RNA, as well as cleavage of a protein, or conformational changes within a protein [13, 147]. A particularly vibrant area of FRET is the application to the study of cell signaling pathways, which is of crucial importance in understanding the means by which cells communicate with one another and respond to different stimuli. New insights into the spatiotemporal organization of protein-protein interactions can open up new possibilities for targeting disease at the molecular level.

FRET is a fluorescence quenching process where the fluorescence of a particular molecular species (the donor) is affected by its proximity to a second chromophore (the acceptor). More precisely, the excitation energy of a donor fluorophore is transferred to a ground state acceptor molecule by a dipole–dipole coupling process [12–14]. Although this process is nonradiative, the emission spectrum of the donor and the absorption spectrum of the acceptor must overlap for FRET to occur [148] and the transition dipole moments of the donor and acceptor must not be perpendicular [149]. The FRET efficiency varies with the inverse sixth power of the distance between donor and acceptor and is usually negligible beyond 10 nm, which is of the order of the distance between proteins bound together in a complex. Measurements of FRET between appropriately labelled species can thus report on binding events, as well as on conformational changes in dual labelled molecules, and the spatial distribution of these subresolution events can be mapped out using fluorescence imaging techniques reading out fluorescence parameters including intensity, excitation and emission spectra, lifetime and polarization anisotropy [1, 18]. In practice, fluorescence lifetime often provides the most robust measurement [17]. The donor fluorescence lifetime is largely independent of factors such as fluorophore concentration, excitation and detection efficiency, inner filter and multiple scattering effects, which can complicate or degrade absolute intensity measurements. In particular, the insensitivity of fluorescence lifetime measurements to fluorophore concentration make FLIM the preferred technique when imaging samples for which the precise stoichiometry of the donor and acceptor species is unknown – as is the case when monitoring interactions between separately labelled species (binding partners, enzyme/substrate pairs etc.).

FLIM FRET was initially implemented using small molecule dyes, e.g., [71, 150, 151], and mainly limited to fixed cell imaging but the advent and widespread deployment of probes based on genetically expressed fluorescent proteins [10, 152, 153] has made it possible to image a host of processes in live cells, e.g., [154–156]. A key example of the application of FLIM FRET to study cell signalling processes was the imaging of protein phosphorylation in live cells by FRET between GFP-tagged PKCa as the donor and a Cy3.5- labelled antibody as the acceptor that was specific to the phosphorylated protein [157]. This technique was extended to be more generally applicable by imaging FRET between a phosphorylated protein, in this case the ErB1 receptor tagged with GFP, and a Cy3- labelled antibody to phosphortyrosine [158, 159]. Other cell signalling processes studied with FLIM FRET include dephosphorylation [160], caspase activity in individual cells during apoptosis [161], NADH [162, 163], and the supramolecular organization of DNA [164].

At Imperial College London we have adapted the approach of [158] to image receptor phosphorylation as a means to study signalling at the immunological synapse (IS). Specifically we have applied FLIM to image FRET between the GFP-tagged KIR2DL1 (KIR) inhibitory receptor and a Cy3-tagged generic antiphosphotyrosine monoclonal antibody. This permitted us to visualize KIR phosphorylation in natural killer (NK) cells interacting with target cells expressing cognate major histocompatibility complex (MHC) class I proteins [165]. Once the live NK cell and target cells formed an IS, they were fixed and then treated with the Cy3-tagged antiphosphotyrosine antibody before imaging. This work revealed that inhibitory KIR signaling is spatially restricted to the IS, that it requires the presence of an Src family kinase and that KIR receptor phosphorylation occurs in microclusters on a spatial scale of the order of the spatial resolution of our confocal microscope, as shown in figure 6.11. The fluorescence intensity and lifetime images were acquired in a confocal microscope with 470 nm excitation and TCSPC detection. The *en face* images of the IS were obtained by acquiring z-stacks of optical sectioned fluorescence lifetime images with 0.5 μm axial separation that were processed using 3D rendering software to obtain the desired projections of the interface between the cells. Each optically sectioned FLIM image was acquired over 300 s and so each z-stack required many tens of minutes. This image acquisition rate was not a problem for imaging the fixed cells in this experiment but would clearly preclude imaging dynamic processes in live cells.

In general, optically sectioned FLIM, such as was required to obtain the FRET data represented in Figure 6.11, has been widely considered as a relatively slow imaging modality because it has mainly been implemented in laser scanning confocal or multiphoton, typically requiring FLIM acquisition times of minutes. Higher FLIM FRET imaging rates have been achieved with wide-field FLIM, e.g., using the frequency-domain approach [159] but this did not provide optical sectioning. This issue can be addressed using optically sectioned Nipkow disc microscopes with wide-field FLIM detection implemented in the time domain and [166] and frequency domain [60]. We have developed this approach, combining a high power supercontinuum excitation source with Nipkow disc microscopy and wide-field time-gated FLIM to acquire depth-resolved fluorescence lifetime images of live cells at rates up to 10 time-gated images per second [61]. This permits us to capture fast time-lapse FLIM FRET sequences of live cell dynamics and to develop higher throughout automated FLIM microscopes [167] for high content analysis and proteomics applications, e.g., using

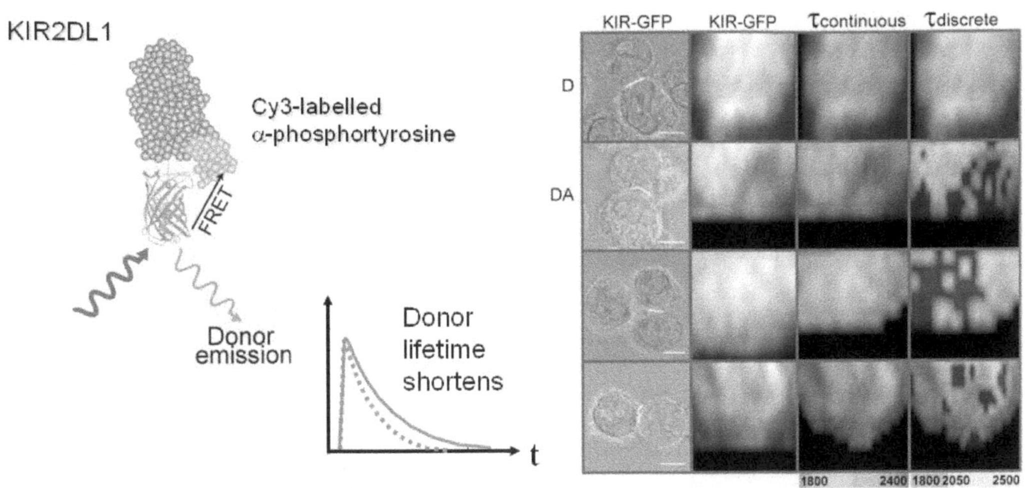

FIGURE 6.11: (See in color after page 572.) Schematic of FRET between EGFP-labelled KIR receptor and Cy3-labelled antiphosphotyosine with inset transmitted light images (where the IS are highlighted in white) and *en face* fluorescence intensity and lifetime images of FRET between NK cells and target cells at the IS with continuous and discrete lifetime scales. The scale bar is 8μm. (Figure adapted from [165]).

FRET to screen protein interactions. Figure 6.12 shows a schematic of this high-speed optically sectioning Nipkow FLIM microscope, together with a FLIM image of live cells exhibiting donor (EGFP) only and FRET constructs (EGFP directly attached to mRFP with a short peptide linker) that was acquired at five frames per second. Also shown is a FLIM FRET image of live cells exhibiting FRET between EGFP-Raf-RBD and H-Ras-mRFP acquired in 5 seconds. This FRET experiment was designed to read out the activation of the Ras protein following stimulation of the EGF receptor signal pathway and illustrates the potential applications to studying cell signalling networks with high-time resolution and/or with high throughout.

6.5 Multidimensional Fluorescence Measurement and Imaging Technology

6.5.1 Overview

In general, when working to maximize the contrast between different fluorophores or states of fluorophores, it is often useful to analyze fluorescence signals with respect to two or more spectral dimensions, as well as three spatial dimensions. For example, utilizing the excitation-emission matrix or applying spectrally-resolved lifetime measurements can improve the capability to unmix signals from different fluorophores, particularly for autofluorescence-based experiments, and to improve quantitation of changes to the local fluorophore environment. Applied to microscopy, endoscopy and assay technology imaging, we believe that a multidimensional fluorescence imaging (MDFI) approach can add significant value for both rapid imaging applications, e.g., for diagnosis and for high-content applications. A vital consideration, however, is that a fluorescence sample will only emit a limited number of fluorescence photons before the onset of photobleaching or damage. There is therefore a limited photon "budget" that must be carefully allocated to appropriate mea-

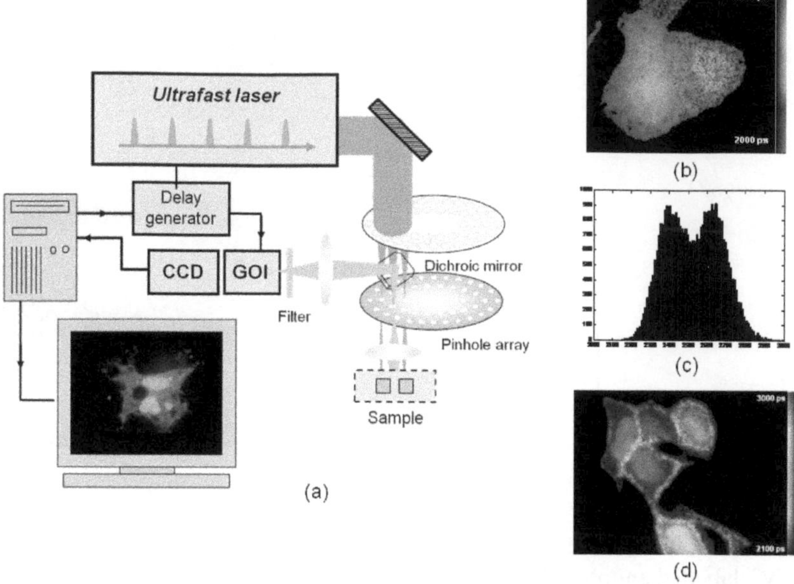

FIGURE 6.12: (See in color after page 572.) (a) Schematic of Nipkow disc FLIM microscope with (b) optically sectioned FLIM image of live cells expressing EGFP (left) and EGFP-mRFP FRET construct with (c) corresponding lifetime histogram and (d) FLIM-FRET image recorded with 5 s acquisition of MDCK cells expressing EGFP-Raf-RBD and H-Ras-mRFP, stimulated with EGF for 10 min. (Figure adapted from [61]).

surement dimensions to maximize the information that can be obtained from a sample. This photon budget may be further constrained for many biological experiments, e.g., those that require imaging at a minimum rate or that can only tolerate limited excitation irradiance. Signal-to-noise ratios will be compromised if the photon budget is allocated over more spectral channels and so it is imperative to maximize the photon efficiency and to judiciously sample the measurement dimensions such that the useful information/photon is also maximized. Optimal performance requires consideration of prior knowledge and of what the investigator wants to learn from the sample. In practice, for any fluorescence imaging experiment including FLIM, spectral discrimination is inherently applied through the choice of filters, dichroic beam splitters, and excitation wavelengths. Different choices can impact the fluorescence intensity or fluorescence lifetime images that are recorded, particularly when multiple fluorophores are present.

6.5.2 Excitation-resolved FLIM

The most important measurement dimension is perhaps the excitation wavelength since this will determine the extent to which different fluorophores are imaged. Ironically, this has traditionally been the most constrained parameter for many fluorescence-based experiments, particularly for confocal laser scanning microscopy that requires spatially coherent sources and so have been typically limited to a few discrete excitation wavelengths, and for FLIM, which requires ultrafast modulated radiation. While convenient tunable cw excitation has long been routinely available for wide-field microscopy using, e.g., filtered lamp sources, the main ultrafast excitation sources available for time-domain FLIM in the visible spectrum were complex c.w. mode-locked dye lasers before the advent of ultrafast Ti:Sapphire lasers. Today there are many all-solid-state ultrafast excitation sources available ranging from modulated LEDs and gain-switched laser diodes through diode-pumped mode-locked fiber lasers and solid-state lasers. The most important of these for FLIM and especially for multiphoton microscopy is the mode-locked femtosecond Ti:Sapphire laser.

Multiphoton excitation permits imaging at significantly increased depths in turbid media such as biological tissue compared to wide-field or confocal microscopy, albeit at the expense of spatial resolution. The increased imaging depth is partly due to the reduced scattering cross-sections at the longer excitation wavelengths and partly to the nonlinear excitation that does not require a confocal detection pinhole for optical sectioning, permitting all of the emitted fluorescence to be collected. Frequently, however, multiphoton excitation is employed with less challenging samples such as cell cultures, for which the convenience and versatility of a (computer-controlled) tunable excitation laser can outweigh the disadvantages of reduced spatial resolution, lower excitation cross-sections, and increased nonlinear photobleaching/photodamage in the focal plane. The ultrafast lasers used for multiphoton excitation are also convenient for FLIM. Nevertheless, there are situations where single-photon excitation is preferable, e.g., to reduce nonlinear photodamage, to achieve the best possible spatial resolution and to realize faster imaging, e.g., of live (dynamic) samples or in a high throughput context. Tunable single-photon excitation may be provided by frequency-doubled Ti:Sapphire lasers over the range $\sim$350–520 nm but this will not excite many important fluorophores. An increasingly popular approach to provide tunable (and ultrafast) excitation over the visible and NIR spectrum is to use ultrafast laser-pumped supercontinuum generation [168] in microstructured optical fibers [169–173]. Supercontinuum sources can be pumped using compact, robust, and relatively inexpensive ultrafast fiber lasers [174, 175] to provide spectral coverage spanning from the ultraviolet to the near infrared [176]. Spectral selection from such a supercontinuum provides a relatively low-cost tunable source of ultrashort optical pulses applicable to single-photon and multiphoton microscopy and FLIM [173].

FLIM can also be conveniently implemented at a range of discrete spectral wavelengths using relatively low-cost semiconductor diode laser and LED technology. Gain-switched picosecond diode lasers have been applied to time-domain FLIM in scanning [177] and wide-field [178] FLIM systems although their relatively low average output power (typically < 1 mW) and sparse spectral coverage limit their range of applications. It is possible to reach higher average powers using sinusoidal modulation for frequency-domain FLIM [34, 179] and, for wide-field FLIM where their low spatial coherence is an advantage, LEDs are increasingly interesting [180], offering higher average powers and extended spectral coverage that extends to the deep ultraviolet (e.g., 280 nm [181]). A LED excitation source has also recently been used to realize frequency-domain FLIM in a scanning microscope [182].

6.5.3 Emission-resolved FLIM

In order to optimize a FLIM experiment, it is desirable to have knowledge of the spectral and temporal properties of the fluorophores to be studied. Indeed, having determined the excitation wavelength, the next consideration for any fluorescence measurement is the spectral discrimination of the detection. It is standard practice to select appropriate filters to optimize detection of a target fluorophore and discriminate against background fluorescence from endogenous fluorophores or other fluorescent labels, e.g., in FRET experiments. To optimize filter selection, however, e.g., for spectral unmixing or ratiometric imaging as well as for FLIM, it is necessary to know the spectral profiles of the fluorophores concerned. Similarly, to optimize FLIM of complex samples, e.g., to image tissue autofluorescence for clinical applications, it is necessary to know how the fluorescence lifetime contrast changes with emission wavelength. A common way to obtain information on the spectral and temporal properties of exogenous fluorophore labels such as organic dyes or fluorescent proteins is to undertake cuvette studies in a fluorometer. At Imperial we have exploited the recently available fiber-laser-based supercontinuum sources to develop a compact (60×90 cm) multidimensional fluorometer able to resolve fluorescence signals with respect to excitation and

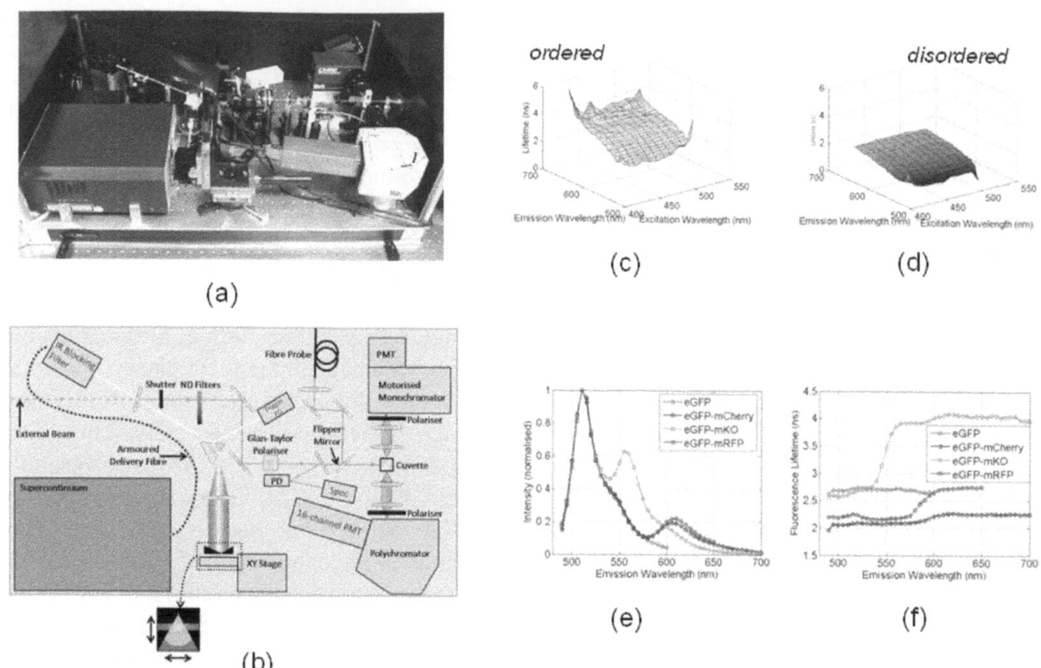

FIGURE 6.13: (a) Photograph and (b) schematic of experimental setup for multidimensional fluorometer; fluorescence excitation-emission-lifetime plots of (c) ordered and (d) disordered lipid order phase of cell membranes labelled with di-4-ANEPPDHQ; (e) emission spectra, and (f) spectral variation of fluorescence lifetime of EGFP alone and linked to notional acceptor molecules (mCherry, mkO, mRFP) in FRET constructs. (Adapted from [183]).

emission wavelength, polarization, and fluorescence lifetime [183]. This instrument is shown in Figure 6.13 together with exemplar data. The upper figures (c, d) show how the variation of fluorescence lifetime of the membrane lipid microdomain marker, di-4-ANEPPDHQ, with excitation and emission wavelength changes between the lipid ordered and disordered phases in artificial unilamellar membranes. The lower figures present the emission spectra (e) and the spectral variation of lifetime (f) of EGFP, alone and linked to candidates for acceptors in potential FRET constructs. It will be seen that, while the EGFP-lifetime decreases in the EGFP-mCherry and EGFP-mRFP constructs due to FRET, there is no similar lifetime reduction for EGFP-mKO, for which FRET is not observed.

It can be useful to obtain similar multidimensional fluorescence information from fluorescence imaging experiments, so that functional spectroscopic information can be correlated with morphology. This can be useful for imaging FRET experiments, where spectral and lifetime measurements can give complementary information [184], for spatially resolved assays of chemical reactions and for studying tissue autofluorescence, particularly when investigating samples with no *a priori* knowledge. FLIM can be combined with multispectral imaging, for which time-resolved images are acquired in a few discrete spectral windows, e.g., [185], or it can be combined with hyperspectral imaging, for which the full time-resolved excitation or emission spectral profile is acquired for each image pixel.

There are several established approaches for implementing hyperspectral imaging. In single beam scanning fluorescence microscopes, the fluorescence radiation can be dispersed in a spectrometer and the spectral profiles acquired sequentially, pixel by pixel. This is readily combined with FLIM in laser scanning microscopes, e.g., using TCSPC systems that incorporate a spectrometer and multi-anode photomultiplier [44, 186]. To increase the imaging rate, it is necessary to move to parallel pixel acquisition, e.g., by combining wide-field FLIM with hyperspectral imaging implemented using filter wheels or acousto-optic [187] or liquid crystal [188] tunable filters to sequentially acquire images in different spectral channels. This approach is not photon efficient because the "out-of-band" light is rejected and the acquisition time will increase with the number of spectral channels. Alternative approaches that can be more efficient include Fourier-transform spectroscopic imaging, e.g., [189] and encoding Hadamard transforms using spatial light modulator technology [190]. In principle, it is possible to combine spectral and 2-D spatial information using a complex diffractive optical element such that it can be detected on a wide-field detector in a single "image" acquisition with the spatial and spectral information being subsequently unmixed in computational postprocessing [191]. In general, however, any approach that requires the final image to be computationally extracted from acquired data can suffer from a reduction in S/N compared to direct image detection. A compromise between photon economy and parallel pixel acquisition that directly detects the spectrally resolved image information is the so-called "push-broom" approach to hyperspectral imaging [192] implemented in a line-scanning microscope. For hyperspectral FLIM, the fluorescence resulting from a line excitation is imaged to the entrance slit of a spectrograph to produce an $(x-\lambda)$ "subimage" that can be recorded on a wide-field FLIM detector. Stage scanning along the y axis then sequentially provides the full $(x-y-\lambda-\tau)$ data set. This approach is photon efficient and the line-scanning microscope configuration provides "semiconfocal" optical sectioning.

We have implemented this "push-broom" approach to develop a line-scanning hyperspectral FLIM microscope utilizing wide-field time-gated imaging [193]. This instrument was developed to study autofluorescence contrast in diseased tissue and Figure 6.14 shows the experimental configuration and presents multidimensional fluorescence data from a hyperspectral FLIM acquisition of unstained fixed section of human cartilage. This MDFI approach can be further extended if we employ a tunable excitation laser to also resolve the excitation wavelength. Using spectral selection of a fiber laser-pumped supercontinuum source to provide tunable excitation from 390 nm to 510 nm with this line-scanning hyperspectral FLIM microscope, we demonstrated the ability to acquire the fluorescence Excitation-Emission-Lifetime (EEL) matrix for each image pixel of a sample. Such a multidimensional data set enables us to subsequently reconstruct conventional "Excitation-Emission Matrices" (EEM) for any image pixel and to obtain the fluorescence decay profile for any point in this EEM space. While this can provide exquisite sensitivity to perturbations in fluorescence emission and enhance the ability to unmix signals from different fluorophores, we note that the size and complexity of such high content hyperspectral FLIM data sets is almost beyond the scope of *ad hoc* analysis by human investigators. Sophisticated bioinformatics software tools are required to automatically analyse and present such data to identify trends and fluorescence "signatures." The combination of MDFI with image segmentation for high content analysis should provide powerful tools, e.g., for histopathology and for screening applications.

6.6 Outlook

The future of FLIM and MDFI looks set to see increasing applications in fluorescence microscopy, particularly for FRET and other approaches to studying protein interactions, as well as in emerging areas such as *in vivo* imaging and high content analysis (HCA). The functionality of MDFI instrumentation will continue to develop and the technological implementations will improve in terms of higher performance, lower cost, and more compact and ergonomic instrumentation.

In terms of functionality, FLIM will be increasingly combined with polarization-resolved imag-

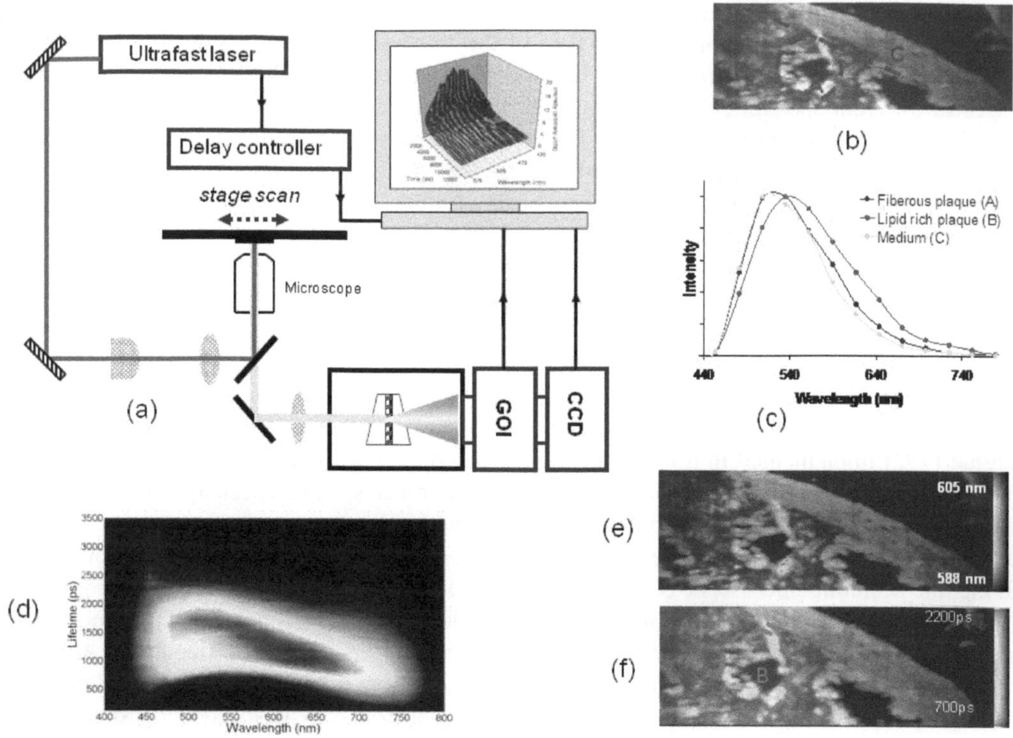

FIGURE 6.14: (See in color after page 572.) (a) Experimental setup for line-scanning hyperspectral FLIM, (b) integrated intensity image of sample of frozen human artery exhibiting atherosclerosis, (c) time-integrated spectra of sample regions corresponding to medium and fiberous and lipid rich plaques, (d) autofluorescence lifetime-emission matrix, (e) map of time-integrated central wavelength and (f) spectrally integrated lifetime map of sample autofluorescence. Adapted from [193].

ing to provide information about rotational mobility, clustering and homoFRET. While steady-state imaging of polarization-resolved fluorescence (i.e., fluorescence anisotropy) is a well-established technique [1] to obtain information concerning fluorophore orientation and to probe resonant energy transfer and rotational decorrelation, time-resolved imaging of fluorescence polarization anisotropy is less widely applied. Ideally the time-resolved, polarization-resolved image data should be acquired in parallel to avoid motion artifacts and this has been realized for wide-field [194] and confocal scanning microscopy [195, 196]. Such instruments can help address the ambiguities associated with steady-state fluorescence anisotropy measurements that are sensitive to static fluorophore orientation and to rotational or energy transfer dynamics. Time-resolved polarization-resolved fluorescence imaging has been applied to study homoFRET [197, 198], where it can provide information concerning protein clustering [196, 197] and fluorophore concentration [199], and to imaging rotational correlation time to map variations in viscosity [57, 194, 200] and ligand binding [201].

A further important trend in fluorescence microscopy is the move to resolution beyond the diffraction limit. Currently superresolved microscopy is mainly concerned with intensity imaging but FLIM has been applied to total internal reflection fluorescence (TIRF) microscopy, e.g., [202], and to single molecule imaging studies, e.g., [203]. A more ambitious goal is to implement FLIM with superresolved imaging techniques like stimulated emission depletion microscopy (STED) [204] and stochastically switched single molecule localization microscopy techniques like PALM [205] or STORM [206]. To this end we have demonstrated a STED FLIM microscope [207] exploiting supercontinuum generation. As yet, there has been no report of FLIM implemented with stochastically switched single molecule localization.

FLIM and MDFI will also be important for drug discovery with the trend to high content analysis and automated imaging for proteomics, offering the potential to screen, e.g., gene knockouts and drug candidates against siRNA libraries. Automated wide-field multiwell plate FLIM was first reported [208] applying frequency-domain FLIM to wide-field FRET and we have reported automated optically sectioned FLIM and FRET using time-gated imaging with a Nipkow confocal multiwell plate reader that is able to acquire optically sectioned FLIM images at a rate of faster than 10 s/well [167]. Although further improvements in hardware are to be anticipated, the next critical issue is to develop appropriate software tools for the acquisition, analysis, and management of high throughput FLIM and MDFI data.

Perhaps the most exciting and challenging prospects for fluorescence imaging and FLIM are associated with the translation to *in vivo* imaging, where the ability to study protein function and interactions may provide tremendously valuable insights into the mechanisms underlying disease. We are approaching this goal by developing FLIM endoscopes and tomographic imaging systems. We have previously reported wide-field FLIM endoscopy [75] and are currently developing a micro-confocal FLIM endoscope, which we have applied to tissue autofluorescence and cells labelled with EGFP [209]. This instrument is intended for clinical diagnostic applications and for *in vivo* studies of disease models. A second direction for *in vivo* clinical FLIM is multiphoton microscopy, which we and others are investigating using the DemaInspect multiphoton microscope that is approved for clinical use. Finally, we are also working on tomographic FLIM imaging systems. We have shown that wide-field time-gated FLIM can be applied to the technique of optical projection tomography (OPT) [210] to provide 3-D fluorescence lifetime maps in optically cleared samples such as mouse embryos [211]. While this provides useful information combining anatomic structure with molecular contrast, it is clearly important to progress to imaging live organisms that cannot tolerate optical clearing. To this end we are working to develop a tomographic FLIM instrument for thick, diffuse samples such as live mice, utilizing time-gated imaging with sample rotation and diffuse fluorescence tomographic reconstruction techniques. In a preliminary experiment we have demonstrated the ability to reconstruct fluorescence lifetime maps of a calcium FRET sensor embedded in a highly scattering phantom [212]. We believe that advances in FLIM technology and software tools can complement the breakthroughs in fluorescent probes to provide unprecedented insights into the fundamental processes of disease and their potential therapies.

Acknowledgments

The authors gratefully acknowledge funding from the Biotechnology and Biological Sciences Research Council (BBSRC), the Department of Trade and Industry (DTI) Beacon award, the Engineering and Physical Sciences Research Council (EPSRC), an EU Framework VI Integrated Project (# LSHG-CT-2003-503259), a Joint Infrastructure Fund Award from the Higher Education Funding Council for England (HEFCE JIF) and a Wellcome Trust Showcase Award.

References

[1] J.R. Lakowicz, *Principles of Fluorescence Spectroscopy*, 2nd ed., Kluwer Academic/Plenum Publishers, New York (1999).

[2] R. Ackroyd, C. Kelty, N. Brown, et al., "The history of photodetection and photodynamic therapy," *Photochem. Photobiol.* **74**(5), 656–669 (2001).

[3] D. Lapotko, E. Lukianova, M. Potapnev, et al., "Method of laser activated nano-thermolysis for elimination of tumor cells," *Cancer Lett.* **239**(1), 36–45 (2006).

[4] G. Grynkiewicz, M. Poenie, and R. Tsien, "A new generation of Ca2+ indicators with greatly improved fluorescence properties," *J. Biol. Chem.* **260**(6), 3440–3450 (1985).

[5] S. Andersson-Engels, J. Johansson, U. Stenram, et al., "Malignant-tumor and atherosclerotic plaque diagnosis using laser-induced fluorescence," *IEEE J. Quantum Electron.* **26**(12), 2207–2217 (1990).

[6] W. Denk, J.H. Strickler, and W.W. Webb, "Two-photon laser scanning fluorescence microscopy," *Science* **248**(4951), 73–76 (1990).

[7] W. Denk and K. Svoboda, "Photon upmanship: Why multiphoton imaging is more than a gimmick," *Neuron* **18**(3), 351–357 (1997).

[8] M.E. Dickinson, E. Simbuerger, B. Zimmermann, et al., "Multiphoton excitation spectra in biological samples," *J. Biomed. Opt.* **8**(3), 329–338 (2003).

[9] R. Cubeddu, D. Comelli, C. D'andrea, et al., "Time-resolved fluorescence imaging in biology and medicine," *J. Phys. D-Appl. Phys.* **35**(9), R61–R76 (2002).

[10] R.Y. Tsien, "The green fluorescent protein," *Ann. Rev. Biochem.* **67**, 509–544 (1998).

[11] M. Zimmer, "Green fluorescent protein (GFP): applications, structure, and related photophysical behavior," *Chem. Rev.* **102**(3), 759–781 (2002).

[12] T. Förster, "Zwischenmolekulare energiewanderung und fluoreszenz," *Ann. Phys.-Berlin* **2**(6), 55–75 (1948).

[13] R.M. Clegg, O. Holub, and C. Gohlke, "Fluorescence lifetime-resolved imaging: measuring lifetimes in an image," in *Biophotonics, Pt A*, Academic Press Inc, San Diego (2003), pp. 509–542.

[14] L. Stryer, "Fluorescence energy transfer as a spectroscopic ruler," *Ann. Rev. Biochem.* **47**, 819–846 (1978).

[15] C.G. Dos Remedios and P.D.J. Moens, "Fluorescence resonance energy transfer spectroscopy is a reliable 'ruler' for measuring structural changes in proteins," *J. Struct. Biol.* **115**(2), 175 (1995).

[16] G.W. Gordon, G. Berry, X.H. Liang, et al., "Quantitative fluorescence resonance energy transfer measurements using fluorescence microscopy," *Biophys. J.* **74**(5), 2702–2713 (1998).

[17] P.I.H. Bastiaens and A. Squire, "Fluorescence lifetime imaging microscopy: spatial resolution of biochemical processes in the cell," *Trend. Cell. Biol.* **9**, 48–52 (1999).

[18] E.A. Jares-Erijman and T.M. Jovin, "FRET Imaging," *Nature Biotech.* **21**(11), 1387–1395 (2003).

[19] K. Suhling, P.M. French, and D. Phillips, "Time-resolved fluorescence microscopy," *Photochem. Photobiol. Sci.* **4**(1), 13–22 (2005).

[20] R. Richards-Kortum and E. Sevick-Muraca, "Quantitative optical spectroscopy for tissue diagnosis," *Annu. Rev. Phys. Chem.* **47**, 555–606 (1996).

[21] B.B. Das, F. Liu, and R.R. Alfano, "Time-resolved fluorescence and photon migration studies in biomedical and model random media," *Rep. Prog. Phys.* **60**(2), 227–292 (1997).

[22] K. Dowling, M.J. Dayel, M.J. Lever, et al., "Fluorescence lifetime imaging with picosecond resolution for biomedical applications," *Opt. Lett.* **23**(10), 810–812 (1998).

[23] R.K.P. Benninger, Y. Koc, O. Hofmann, et al., "Quantitative 3D mapping of fluidic temperatures within microchannel networks using fluorescence lifetime imaging," *Anal. Chem.* **78**(7), 2272–2278 (2006).

[24] Y. Schaerli, R.C. Wootton, T. Robinson, et al., "Continuous-flow polymerase chain reaction of single-copy DNA in microfluidic microdroplets," *Anal. Chem.* **81**(1), 302–306 (2009).

[25] M. Koeberg, D.S. Elson, P.M.W. French, et al., "Spatially resolved electric fields in polymer light-emitting diodes using fluorescence lifetime imaging," *Synth. Met.* **139**(3), 925–928 (2003).

[26] K.C.B. Lee, J. Siegel, S.E.D. Webb, et al., "Application of the stretched exponential function to fluorescence lifetime imaging," *Biophys. J.* **81**(3), 1265–1274 (2001).

[27] J. Wlodarczyk and B. Kierdaszuk, "Interpretation of fluorescence decays using a power-like model," *Biophys. J.* **85**(1), 589–598 (2003).

[28] J.A. Jo, Q.Y. Fang, T. Papaioannou, et al., "Fast model-free deconvolution of fluorescence decay for analysis of biological systems," *J. Biomed. Opt.* **9**(4), 743–752 (2004).

[29] A. Grinwald, "On the analysis of fluorescence decay kinetics by the method of least-squares," *Anal. Biochem.* **59**, 583–593 (1974).

[30] D.R. James and W.R. Ware, "A fallacy in the interpretation of fluorescence decay parameters," *Chem. Phys. Lett.* **120**(4,5), 455–459 (1985).

[31] P.J. Verveer, A. Squire, and P.I.H. Bastiaens, "Global analysis of fluorescence lifetime imaging microscopy data," *Biophys. J.* **78**(4), 2127–2137 (2000).

[32] T.W.J. Gadella (ed.), *FRET and FLIM Imaging Techniques*. Laboratory Techniques in Biochemistry and Molecular Biology. Elsevier: Amsterdam, Netherlands (2008).

[33] A. Esposito, H.C. Gerritsen, and F.S. Wouters, "Optimizing frequency-domain fluorescence lifetime sensing for high-throughput applications: photon economy and acquisition speed," *J. Opt. Soc. Am. A-Opt. Image Sci. Vis.* **24**(10), 3261–3273 (2007).

[34] M.J. Booth and T. Wilson, "Low-cost, frequency-domain, fluorescence lifetime confocal microscopy," *J. Microsc.-Oxf.* **214**, 36–42 (2004).

[35] R.A. Colyer, C. Lee, and E. Gratton, "A novel fluorescence lifetime imaging system that optimizes photon efficiency," *Microsc. Res. Tech.* **71**(3), 201–213 (2008).

[36] P.T.C. So, T. French, W.M. Yu, et al., "Time-resolved fluorescence microscopy using two-photon excitation," *Bioimaging* **3**, 49–63 (1995).

[37] J. Philip and K. Carlsson, "Theoretical investigation of the signal-to-noise ratio in fluorescence lifetime imaging," *J. Opt. Soc. Am. A-Opt. Image Sci. Vis.* **20**(2), 368–379 (2003).

[38] E. Gratton and M. Limkeman, "A continuously variable frequency cross-correlation phase fluorometer with picosecond resolution," *Biophys. J.* **44**(3), 315–324 (1983).

[39] J.R. Lakowicz and B.P. Maliwal, "Construction and performance of a variable-frequency phase-modulation fluorometer," *Biophys. Chem.* **21**(1), 61–78 (1985).

[40] D.V. O'Connor and D. Phillips, *Time-Correlated Single-Photon Counting*, Academic Press, New York (1984).

[41] I. Bugiel, K. Konig, and H. Wabnitz, "Investigation of cells by fluorescence laser scanning microscopy with subnanosecond time resolution," *Lasers Life Sci.* **3**(1), 47–53 (1989).

[42] W. Becker, A. Bergmann, M.A. Hink, et al., "Fluorescence lifetime imaging by time-correlated single-photon counting," *Microsc. Res. Tech.* **63**(1), 58–66 (2004).

[43] Y.L. Zhang, S.A. Soper, L.R. Middendorf, et al., "Simple near-infrared time-correlated single photon counting instrument with a pulsed diode laser and avalanche photodiode for time-resolved measurements in scanning applications," *Appl. Spectrosc.* **53**(5), 497–504 (1999).

[44] W. Becker, *Advanced Time-Correlated Single Photon Counting Techniques*, Springer, Berlin, Heidelberg, N.Y. (2005).

[45] E.P. Buurman, R. Sanders, A. Draaijer, et al., "Fluorescence lifetime imaging using a confocal laser scanning microscope," *Scanning* **14**(3), 155–159 (1992).

[46] H.C. Gerritsen, M.A.H. Asselbergs, A.V. Agronskaia, et al., "Fluorescence lifetime imaging in scanning microscopes: acquisition speed, photon economy and lifetime resolution," *J. Microsc.* **206**(3), 218–224 (2002).

[47] C.Y. Dong, P.T.C. So, and E. Gratton, "Pump-probe fluorescence microscopy: a new method for time- resolved imaging with high spatial resolution," *Biophys. J.* **70**(2), WP280–WP280 (1996).

[48] K. Carlsson and A. Liljeborg, "Simultaneous confocal lifetime imaging of multiple fluorophores using the intensity-modulated multiple-wavelength scanning (IMS) technique," *J. Microsc.* **191**(2), 119–127 (1998).

[49] M. Kollner and J. Wolfrum, "How many photons are necessary for fluorescence-lifetime measurements," *Chem. Phys. Lett.* **200**(1–2), 199–204 (1992).

[50] G.H. Patterson and D.W. Piston, "Photobleaching in two-photon excitation microscopy," *Biophys. J.* **78**(4), 2159–2162 (2000).

[51] J. Bewersdorf, R. Pick, and S.W. Hell, "Multifocal multiphoton microscopy," *Opt. Lett.* **23**(9), 655–657 (1998).

[52] D.N. Fittinghoff, P.W. Wiseman, and J.A. Squier, "Widefield multiphoton and temporally decorrelated multifocal multiphoton microscopy," *Opt. Express* **7**(8), 273–279 (2000).

[53] K.H. Kim, C. Buehler, K. Bahlmann, et al., "Multifocal multiphoton microscopy based on multianode photomultiplier tubes," *Opt. Express* **15**(18), 11658–11678 (2007).

[54] S. Kumar, C. Dunsby, P.A.A. De Beule, et al., "Multifocal multiphoton excitation and time correlated single photon counting detection for 3-D fluorescence lifetime imaging," *Opt. Express* **15**(20), 12548–12561 (2007).

[55] K.V. Krishnan, H. Saitoh, H. Terada, et al., "Development of a multiphoton fluorescence lifetiem imaging microscopy system using a streak camera," *Rev. Sci. Instrum.* **74**(5), 2714–2721 (2003).

[56] M. Straub and S.W. Hell, "Fluorescence lifetime three-dimensional microscopy with picosecond precision using a multifocal multiphoton microscope," *Appl. Phys. Lett.* **73**(13), 1769–1771 (1998).

[57] R.K.P. Benninger, O. Hofmann, J. McGinty, et al., "Time-resolved fluorescence imaging of solvent interactions in microfluidic devices," *Opt. Express* **13**(16), 6275–6285 (2005).

[58] S. Leveque-Fort, M.P. Fontaine-Aupart, G. Roger, et al., "Fluorescence-lifetime imaging with a multifocal two-photon microscope," *Opt. Lett.* **29**(24), 2884–2886 (2004).

[59] D.M. Grant, D.S. Elson, D. Schimpf, et al., "Optically sectioned fluorescence lifetime imaging using a Nipkow disk microscope and a tunable ultrafast continuum excitation source," *Opt. Lett.* **30**(24), 3353–3355 (2005).

[60] E.B. Van Munster, J. Goedhart, G.J. Kremers, et al., "Combination of a spinning disc confocal unit with frequency-domain fluorescence lifetime imaging microscopy," *Cytometry A* **71A**(4), 207–214 (2007).

[61] D.M. Grant, J. McGinty, E.J. McGhee, et al., "High speed optically sectioned fluorescence lifetime imaging permits study of live cell signaling events," *Opt. Express* **15**(24), 15656–15673 (2007).

[62] E. Gratton, S. Breusegem, J. Sutin, et al., "Fluorescence lifetime imaging for the two-photon microscope: time-domain and frequency-domain methods," *J. Biomed. Opt.* **8**(3), 381–390 (2003).

[63] J. Requejo-Isidro, J. McGinty, I. Munro, et al., "High-speed wide-field time-gated endoscopic fluorescence lifetime imaging," *Opt. Lett.* **29**(19), 2249–2251 (2004).

[64] A.V. Agronskaia, L. Tertoolen, and H.C. Gerritsen, "High frame rate fluorescence lifetime imaging," *J. Phys. D-Appl. Phys.* **36**(14), 1655–1662 (2003).

[65] C.G. Morgan, A.C. Mitchell, and J.G. Murray, "Nanosecond time-resolved fluorescene microscopy: principles and practice," *Trans. Roy. Microsc. Soc.* **1**, 463–466 (1990).

[66] E. Gratton, B. Feddersen, and M. Van De Ven, "Parallel acquisition of fluorescence decay using array detectors," *Proc SPIE* **1204**, 21–25 (1990).

[67] J.R. Lakowicz and K.W. Berndt, "Lifetime-selective fluorescence imaging using an RF phase-sensitive camera," *Rev. Sci. Instrum.* **62**(7), 1727–1734 (1991).

[68] T.W.J. Gadella, T.M. Jovin, and R.M. Clegg, "Fluorescence Lifetime Imaging Microscopy (FLIM) – spatial-resolution of microstructures on the nanosecond time-scale," *Biophys. Chem.* **48**(2), 221–239 (1993).

[69] J.R. Lakowicz, H. Szmacinski, K. Nowaczyk, et al., "Fluorescence lifetime imaging," *Anal. Biochem.* **202**, 316–330 (1992).

[70] T. Minami and S. Hirayama, "High-quality fluorescence decay curves and lifetime imaging using an elliptic scan streak camera," *J. Photochem. Photobiol. A-Chem.* **53**(1), 11–21 (1990).

[71] T. Oida, Y. Sako, and A. Kusumi, "Fluorescence lifetime imaging microscopy (flimscopy): methodology development and application to studies of endosome fusion in single cells," *Biophys. J.* **64**(3), 676–685 (1993).

[72] A.D. Scully, A.J. Macrobert, S. Botchway, et al., "Development of a laser-based fluorescence microscope with subnanosecond time resolution," *J. Fluoresc.* **6**(2), 119–125 (1996).

[73] X.F. Wang, T. Uchida, D.M. Coleman, et al., "A two-dimensional fluorescence lifetime imaging system using a gated image intensifier," *Appl. Spectrosc.* **45**(3), 360–366 (1991).

[74] J.D. Hares, "Advances in sub-nanosecond shutter tube technology and applications in plasma physics," *Proc SPIE* **831**, 165–170 (1987).

[75] I. Munro, J. McGinty, N. Galletly, et al., "Towards the clinical application of time-domain fluorescence lifetime imaging," *J. Biomed. Opt.* **10**(5), 051403-1-9 (2005).

[76] O. Holub, M.J. Seufferheld, C. Gohike, et al., "Fluorescence Lifetime Imaging (FLI) in real-time - a new technique in photosynthesis research," *Photosynthetica* **38**(4), 581–599 (2000).

[77] J. Mizeret, T. Stepinac, M. Hansroul, et al., "Instrumentation for real-time fluorescence lifetime imaging in endoscopy," *Rev. Sci. Instrum.* **70**(12), 4689–4701 (1999).

[78] P.D. Devries and A.A. Khan, "An efficient technique for analyzing deep level transient spectroscopy data," *J. Electron. Mater.* **18**(4), 543–547 (1989).

[79] S.P. Chan, Z.J. Fuller, J.N. Demas, et al., "Optimized gating scheme for rapid lifetime determinations of single-exponential luminescence lifetimes," *Anal. Chem.* **73**(18), 4486–4490 (2001).

[80] R.M. Ballew and J.N. Demas, "Error analysis of the rapid lifetime determination method for single exponential decays with a non-zero baseline," *Analytica Chimica Acta* **245**, 121–127 (1991).

[81] K.K. Sharman, A. Periasamy, H. Ashworth, et al., "Error analysis of the rapid lifetime determination method for double-exponential decays and new windowing schemes," *Anal. Chem.* **71**, 947–952 (1999).

[82] D.S. Elson, I. Munro, J. Requejo-Isidro, et al., "Real-time time-domain fluorescence lifetime imaging including single-shot acquisition with a segmented optical image intensifier," *New J. Phys.* **6**(180), 1–13 (2004).

[83] P.E. Young, J.D. Hares, J.D. Kilkenny, et al., "4-frame gated optical imager with 120-ps resolution," *Rev. Sci. Instrum.* **59**(8), 1457–1460 (1988).

[84] A. Esposito, T. Oggier, H.C. Gerritsen, et al., "All-solid-state lock-in imaging for wide-field fluorescence lifetime sensing," *Opt. Express* **13**(24), 9812–9821 (2005).

[85] A. Squire, P.J. Verveer, and P.I.H. Bastiaens, "Multiple frequency fluorescence lifetime imaging microscopy," *J. Microsc.* **197**(2), 136–149 (2000).

[86] *Cancer Facts and Figures*. 2006, American Cancer Society: Atlanta, USA.

[87] K. Sokolov, M. Follen, and R. Richards-Kortum , "Optical spectroscopy for detection of neoplasia," *Curr. Opin. Chem. Biol.* **6**(5), 651–658 (2002).

[88] I.J. Bigio and J.R. Mourant, "Ultraviolet and visible spectroscopies for tissue diagnostics: fluorescence spectroscopy and elastic-scattering spectroscopy," *Phys. Med. Biol.* **42**(5), 803–814 (1997).

[89] G.A. Wagnieres, W.M. Star, and B.C. Wilson, "*In vivo* fluorescence spectroscopy and imaging for oncological applications," *Photochem. Photobiol.* **68**(5), 603–632 (1998).

[90] P.V. Butte, B.K. Pikul, A. Hever, et al., "Diagnosis of meningioma by time-resolved fluorescence spectroscopy," *J. Biomed. Opt.* **10**(6), 064026-1–9 (2005).

[91] D.C. De Veld, M. Skurichina, M.J. Witjes, et al., "Autofluorescence characteristics of healthy oral mucosa at different anatomical sites," *Lasers Surg. Med.* **32**(5), 367–376 (2003).

[92] M.P.L. Bard, A. Amelink, M. Skurichina, et al., "Improving the specificity of fluorescence bronchoscopy for the analysis of neoplastic lesions of the bronchial tree by combination with optical spectroscopy: preliminary communication," *Lung Cancer* **47**(1), 41–47 (2005).

[93] J.F. Beamis, Jr., A. Ernst, M. Simoff, et al., "A multicenter study comparing autofluorescence bronchoscopy to white light bronchoscopy using a non-laser light stimulation system," *Chest* **125**(5 Suppl), 148S–149S (2004).

[94] A. Ohkawa, H. Miwa, A. Namihisa, et al., "Diagnostic performance of light-induced fluorescence endoscopy for gastric neoplasms," *Endoscopy* **36**(6), 515–521 (2004).

[95] C.-P.C. Hsin-Ming Chen, C. You, T.-C. Hsiao, C.-Y. Wang, "Time-resolved autofluorescence spectroscopy for classifying normal and premalignant oral tissues," *Lasers Surg. Med.* **37**(1), 37–45 (2005).

[96] D.S. Elson, J. Requejo-Isidro, I. Munro, et al., "Time-domain fluorescence lifetime imaging applied to biological tissue," *Photochem. Photobiol. Sci* **8**, 795–801 (2004).

[97] D.B. Tata, M. Foresti, J. Cordero, et al., "Fluorescence polarization spectroscopy and time-resolved fluorescence kinetics of native cancerous and normal rat-kidney tissues," *Biophys. J.* **50**(3), 463–469 (1986).

[98] T. Glanzmann, J.P. Ballini, H. Van Den Bergh, et al., "Time-resolved spectrofluorometer for clinical tissue characterization during endoscopy," *Rev. Sci. Instrum.* **70**(10), 4067–4077 (1999).

[99] R.S. Dacosta, B.C. Wilson, and N.E. Marcon, "New optical technologies for earlier endoscopic diagnosis of premalignant gastrointestinal lesions," *J. Gastroenterol. Hepatol.* **17**(Suppl), S85–S104 (2002).

[100] T.J. Pfefer, D.Y. Paithankar, J.M. Poneros, et al., "Temporally and spectrally resolved fluorescence spectroscopy for the detection of high grade dysplasia in Barrett's esophagus," *Lasers Surg. Med.* **32**, 10–16 (2003).

[101] M.A. Mycek, K.T. Schomacker, and N.S. Nishioka, "Colonic polyp differentiation using time-resolved autofluorescence spectroscopy," *Gastrointest. Endosc.* **48**(4), 390–394 (1998).

[102] L. Marcu, J.A. Jo, P.V. Butte, et al., "Fluorescence lifetime spectroscopy of glioblastoma multiforme," *Photochem. Photobiol.* **80**(1), 98–103 (2004).

[103] A. Pradhan, B.B. Das, K.M. Yoo, et al., "Time-resolved UV photoexcited fluorescence kinetics from malignant and non-malignant human breast tissues," *Lasers Life Sci.* **4**(4), 225–234 (1992).

[104] P.A.A. De Beule, C. Dunsby, N.P. Galletly, et al., "A hyperspectral fluorescence lifetime probe for skin cancer diagnosis," *Rev. Sci. Instrum.* **78**(12), 123101 (2007).

[105] S.R. Kantelhardt, J. Leppert, J. Krajewski, et al., "Imaging of brain and brain tumor specimens by time-resolved multiphoton excitation microscopy *ex vivo*," *Neuro-Oncology* **9**(2), 103–112 (2007).

[106] R. Cicchi, D. Massi, S. Sestini, et al., "Multidimensional non-linear laser imaging of basal cell carcinoma," *Opt. Express* **15**(16), 10135–10148 (2007).

[107] N.P. Galletly, J. McGinty, C. Dunsby, et al., "Fluorescence lifetime imaging distinguishes basal cell carcinoma from surrounding uninvolved skin," *British J. Dermat.* **159**(1), 152–161 (2008).

[108] D.S. Elson, N. Galletly, C. Talbot, et al., "Multidimensional fluoresecence imaging applied to biological tissue," in *Reviews in Fluorescence 2006*, C.D. Geddes and J.R. Lakowicz (eds.), Springer Science, New York (2006), pp. 477–524.

[109] P.P. Provenzano, D.R. Inman, K.W. Eliceiri, et al., "Collagen density promotes mammary tumor initiation and progression," *BMC Med.* **6**, 11 (2008).

[110] M.C. Skala, K.M. Riching, A. Gendron-Fitzpatrick, et al., "*In vivo* multiphoton microscopy of NADH and FAD redox states, fluorescence lifetimes, and cellular morphology in precancerous epithelia," *Proc. Natl. Acad. Sci. USA* **104**(49), 19494–19499 (2007).

[111] K. König, "Clinical multiphoton tomography," *J. Biophot.* **1**(1), 13–23 (2008).

[112] J. Mizeret, G. Wagnieres, T. Stepinac, et al., "Endoscopic tissue characterization by frequency-domain fluorescence lifetime imaging (FD-FLIM)," *Lasers Med. Sci.* **12**(3), 209–217 (1997).

[113] N. Galletly, J. McGinty, I. Munro, et al., "Fluorescence lifetime imaging of liver cancer." *Proc. American Gastroenterology Association* **130**, A791–A791 (2006).

[114] N.P. Galletly, J.M. McGinty, P. Cohen, et al., "Detection of gastrointestinal malignancy by fluorescence lifetime imaging of UV laser induced tissue autofluorescence." *Proc. British Soc. Gastroenterol.* **55**, A115–A115 (2006).

[115] G. Filippidis, G. Zacharakis, A. Katsamouris, et al., "Single and double wavelength excitation of laser-induced fluorescence of normal and atherosclerotic peripheral vascular tissue," *J. Photochem. Photobiol. B-Biol.* **56**(2-3), 163–171 (2000).

[116] L.I. Deckelbaum, S.P. Desai, C. Kim, et al., "Evaluation of a fluorescence feedback-system for guidance of laser angioplasty," *Lasers Surg. Med.* **16**(3), 226–234 (1995).

[117] A.J. Morguet, R.E. Gabriel, A.B. Buchwald, et al., "Single-laser approach for fluorescence guidance of excimer laser angioplasty at 308 nm: evaluation *in vitro* and during coronary angioplasty," *Lasers Surg. Med.* **20**(4), 382–393 (1997).

[118] A.J. Morguet, B. Korber, B. Abel, et al., "Autofluorescence spectroscopy using a Xecl excimer-laser system for simultaneous plaque ablation and fluorescence excitation," *Lasers Surg. Med.* **14**(3), 238–248 (1994).

[119] R.T. Strebel, U. Utzinger, M. Peltola, et al., "Excimer laser spectroscopy: influence of tissue ablation on vessel wall fluorescence," *J. Laser Appl.* **10**(1), 34–40 (1998).

[120] S. Warren, K. Pope, Y. Yazdi, et al., "Combined ultrasound and fluorescence spectroscopy for physicochemical imaging of atherosclerosis," *IEEE Trans. Biomed. Eng.* **42**(2), 121–132 (1995).

[121] W.D. Yan, M. Perk, A. Chagpar, et al., "Laser-induced fluorescence 3: quantitative-analysis of atherosclerotic plaque content," *Lasers Surg. Med.* **16**(2), 164–178 (1995).

[122] J.J. Baraga, R.P. Rava, M. Fitzmaurice, et al., "Characterization of the fluorescent morphological structures in human arterial-wall using ultraviolet-excited microspectrofluorimetry," *Atherosclerosis* **88**(1), 1–14 (1991).

[123] A.L. Alexander, C.M.C. Davenport, and A.F. Gmitro, "Comparison of illumination wavelengths for detection of atherosclerosis by optical fluorescence spectroscopy," *Opt. Eng.* **33**(1), 167–174 (1994).

[124] F. Bosshart, U. Utzinger, O.M. Hess, et al., "Fluorescence spectroscopy for identification of atherosclerotic tissue," *Cardiovasc. Res.* **26**(6), 620–625 (1992).

[125] M. Stavridi, V.Z. Marmarelis, and W.S. Grundfest, "Spectro-temporal studies of Xe-Cl excimer laser-induced arterial-wall fluorescence," *Med. Eng. Phys.* **17**(8), 595–601 (1995).

[126] L. Marcu, M.C. Fishbein, J.M.I. Maarek, et al., "Discrimination of human coronary artery atherosclerotic lipid- rich lesions by time-resolved laser-induced fluorescence spectroscopy," *Arterioscler. Thromb. Vasc. Biol.* **21**(7), 1244–1250 (2001).

[127] J.M.I. Maarek, L. Marcu, M.C. Fishbein, et al., "Time-resolved fluorescence of human aortic wall: Use for improved identification of atherosclerotic lesions," *Lasers Surg. Med.* **27**(3), 241–254 (2000).

[128] L. Hegyi, C. Talbot, C. Monaco, et al., "Fluorescence lifetime imaging of unstained human atherosclerotic plaques," *Atheroslcerosis Suppl.* **7**(3), 587 (2006).

[129] N. Kitamura, Y. Hosoda, C. Iwasaki, et al., "Thermal phase transition of an aqueous poly(N-isopropylacrylamide) solution in a polymer microchannel-microheater chip," *Langmuir* **19**(20), 8484–8489 (2003).

[130] J. Siegel, K. Suhling, S. Lévêque-Fort, et al., "Wide-field time-resolved fluorescence anisotropy imaging (TR-FAIM): Imaging the rotational mobility of a fluorophore," *Rev. Sci. Instrum.* **74**(1), 182–192 (2003).

[131] S.J. Strickler and R.A. Berg, "Relationship between absorption intensity and fluorescence lifetime of molecules," *J. Chem. Phys.* **37**(4), 814–820 (1962).

[132] K. Suhling, J. Siegel, D. Phillips, et al., "Imaging the environment of green fluorescent protein," *Biophys. J.* **83**(6), 3589–3595 (2002).

[133] T. Parasassi, E. Krasnowska, L.A. Bagatolli, et al., "Laurdan and Prodan as polarity-sensitive fluorescent membrane probes," *J. Fluoresc.* **8**(4), 365–373 (1998).

[134] J.R. Lakowicz, H. Szmacinski, and M.L. Johnson, "Calcium imaging using fluorescence lifetimes and long-wavelength probes," *J. Fluoresc.* **2**(1), 47–62 (1992).

[135] B. Herman, P. Wodnicki, S. Kwon, et al., "Recent developments in monitoring calcium and protein interactions in cells using fluorescence lifetime microscopy," *J. Fluoresc.* **7**(1), 85–92 (1997).

[136] A.V. Agronskaia, L. Tertoolen, and H.C. Gerritsen, "Fast fluorescence lifetime imaging of calcium in living cells," *J. Biomed. Opt.* **9**(6), 1230–1237 (2004).

[137] H.C. Gerritsen, R. Sanders, A. Draaijer, et al., "Fluorescence lifetime imaging of oxygen in living cells," *J. Fluoresc.* **7**(1), 11–16 (1997).

[138] H.-J. Lin, P. Herman, and J.R. Lakowicz, "Fluorescence lifetime-resolved pH imaging of living cells," *Cytometry A* **52A**, 77–89 (2003).

[139] R. Sanders, A. Draaijer, H.C. Gerritsen, et al., "Quantitative pH imaging in cells using confocal fluorescence lifetime imaging microscopy," *Anal. Biochem.* **227**, 302–308 (1995).

[140] A. Miyawaki, J. Llopis, R. Heim, et al., "Fluorescent indicators for Ca2+ based on green fluorescent proteins and calmodulin," *Nature* **388**(6645), 882–887 (1997).

[141] A.L. Obaid, L.M. Loew, J.P. Wuskell, et al., "Novel naphthylstyryl-pyridinium potentiometric dyes offer advantages for neural network analysis," *J. Neurosci. Methods* **134**(2), 179–190 (2004).

[142] L. Jin, A.C. Millard, J.P. Wuskell, et al., "Cholesterol-enriched lipid domains can be visualized by di-4-ANEPPDHQ with linear and nonlinear optics," *Biophys. J.* **89**(1), L04–L06 (2005).

[143] D.M. Owen, P.M.P. Lanigan, C. Dunsby, et al., "Fluorescence lifetime imaging provides enhanced contrast when imaging the phase-sensitive dye di-4-ANEPPDHQ in model membranes and live cells," *Biophys. J.* **90**(11), L80–L82 (2006).

[144] S. Bodin, C. Soulet, H. Tronchere, et al., "Integrin-dependent interaction of lipid rafts with the actin cytoskeleton in activated human platelets," *J. Cell Sci.* **118**(4), 759–769 (2005).

[145] K. Gaus, E. Gratton, E.P.W. Kable, et al., "Visualizing lipid structure and raft domains in living cells with two-photon microscopy," *Proc. Natl. Acad. Sci. USA* **100**(26), 15554–15559 (2003).

[146] A. Margineanu, J.I. Hotta, M. Van Der Auweraer, et al., "Visualization of membrane rafts using a perylene monoimide derivative and fluorescence lifetime imaging," *Biophys. J.* **93**(8), 2877–2891 (2007).

[147] P.R. Selvin, "The renaissance of fluorescence resonance energy transfer," *Nature Struct. Biol.* **7**(9), 730–734 (2000).

[148] R.P. Haugland, J. Yguerabide, and L. Stryer, "Dependence of kinetics of singlet-singlet energy transfer on spectral overlap," *Proc. Natl. Acad. Sci. USA* **63**, 23–30 (1969).

[149] J. Eisinger and R.E. Dale, "Interpretation of intramolecular energy transfer experiments," *J. Mol. Biol.* **84**, 643–647 (1974).

[150] T.W. Gadella, Jr. and T.M. Jovin, "Oligomerization of epidermal growth factor receptors on A431 cells studied by time-resolved fluorescence imaging microscopy: a stereochemical model for tyrosine kinase receptor activation," *J. Cell. Biol.* **129**(6), 1543–1558 (1995).

[151] P.I.H. Bastiaens and T.M. Jovin, "Microspectroscopic imaging tracks the intracellular processing of a signal transduction protein: fluorescent-labeled protein kinase C beta I," *Proc. Natl. Acad. Sci. USA* **93**(16), 8407–8412 (1996).

[152] J. Lippincott-Schwartz and G.H. Patterson, "Development and use of fluorescent protein markers in living cells," *Science* **300**, 87–91 (2003).

[153] B.N.G. Giepmans, S.R. Adams, M.H. Ellisman, et al., "Review – The fluorescent toolbox for assessing protein location and function," *Science* **312**(5771), 217–224 (2006).

[154] F.S. Wouters, P.J. Verveer, and P.I.H. Bastiaens, "Imaging biochemistry inside cells," *Trend. Cell Biol.* **11**(5), 203–211 (2001).

[155] P. Van Roessel and A.H. Brand, "Imaging into the future: visualizing gene expression and protein interactions with fluorescent proteins," *Nat. Cell. Biol.* **4**(1), E15–E20 (2002).

[156] T. Ng, M. Parsons, W.E. Hughes, et al., "Ezrin is a downstream effector of trafficking PKC-integrin complexes involved in the control of cell motility," *Embo. J.* **20**(11), 2723–2741 (2001).

[157] T. Ng, A. Squire, G. Hansra, et al., "Imaging protein kinase C alpha activation in cells," *Science* **283**(5410), 2085–2089 (1999).

[158] F.S. Wouters and P.I.H. Bastiaens, "Fluorescence lifetime imaging of receptor tyrosine kinase activity in cells," *Curr. Biol.* **9**(19), 1127–1130 (1999).

[159] P.J. Verveer, F.S. Wouters, A.R. Reynolds, et al., "Quantitative imaging of lateral ErbB1 receptor signal propagation in the plasma membrane," *Science* **290**(5496), 1567–1570 (2000).

[160] F.G. Haj, P.J. Verveer, A. Squire, et al., "Imaging sites of receptor dephosphorylation by PTP1B on the surface of the endoplasmic reticulum," *Science* **295**(5560), 1708–1711 (2002).

[161] A.G. Harpur, F.S. Wouters, and P.I. Bastiaens, "Imaging FRET between spectrally similar GFP molecules in single cells," *Nat. Biotechnol.* **19**(2), 167–169 (2001).

[162] J.R. Lakowicz, H. Szmacinski, K. Nowaczyk, et al., "Fluorescence lifetime imaging of free and protein-bound NADH," *Proc. Natl. Acad. Sci. USA* **89**(4), 1271–1275 (1992).

[163] Q.H. Zhang, D.W. Piston, and R.H. Goodman, "Regulation of corepressor function by nuclear NADH," *Science* **295**(5561), 1895–1897 (2002).

[164] S. Murata, P. Herman, and J.R. Lakowicz, "Texture analysis of fluorescence lifetime images of nuclear DNA with effect of fluorescence resonance energy transfer," *Cytometry* **43**(2), 94–100 (2001).

[165] B. Treanor, P.M.P. Lanigan, S. Kumar, et al., "Microclusters of inhibitory killer immunoglobulin like receptor signaling at natural killer cell immunological synapses," *J. Cell Biol.* **174**(1), 153–161 (2006).

[166] D.M. Grant, D.S. Elson, D. Schimpf, et al., "Wide-field optically-sectioned fluorescence lifetime imaging using a Nipkow disk microscope and a tunable ultrafast continuum excitation source," *Opt. Lett.* **30**(24), 3353–3355 (2005).

[167] C.B. Talbot, J. McGinty, D.M. Grant, et al., "High speed unsupervised fluorescence lifetime imaging confocal multiwell plate reader for high content analysis," *J. Biophoton.* **1**(6), 514–521 (2008).

[168] J.K. Ranka, R.S. Windeler, and A.J. Stentz, "Visible continuum generation in air-silica microstructure optical fibers with anomalous dispersion at 800 nm," *Opt. Lett.* **25**(1), 25–27 (2000).

[169] J.E. Jureller, N.F. Scherer, and T.A. Birks, "Widely tunable femtosecond pulses from a tapered fiber for ultrafast microscopy and multiphoton applications," in *Ultrafast Phenomena XIII*, Springer-Verlag, Berlin (2003), pp. 2844–2850.

[170] N. Deguil, E. Mottay, F. Salin, et al., "Novel diode-pumped infrared tunable laser system for multi-photon microscopy," *Microsc. Res. Tech.* **63**(1), 23–26 (2004).

[171] H. Birk and R. Storz, Leica Microsystms Heidelberg GmbH, "Illuminating device and microscope," U.S. Patent No. 6,611,643 (2001)

[172] G. McConnell, "Confocal laser scanning fluorescence microscopy with a visible continuum source," *Opt. Express* **12**(13), 2844–2850 (2004).

[173] C. Dunsby, P.M.P. Lanigan, J. McGinty, et al., "An electronically tunable ultrafast laser source applied to fluorescence imaging and fluorescence lifetime imaging microscopy," *J. Phys. D: Appl. Phys.* **37**, 3296–3303 (2004).

[174] P.A. Champert, S.V. Popov, M.A. Solodyankin, et al., "Multiwatt average power continua generation in holey fibers pumped by kilowatt peak power seeded ytterbium fiber amplifier," *Appl. Phys. Lett.* **81**(12), 2157–2159 (2002).

[175] T. Schreiber, J. Limpert, H. Zellmer, et al., "High average power supercontinuum generation in photonic crystal fibers," *Opt. Commun.* **228**(1–3), 71–78 (2003).

[176] A. Kudlinski, A.K. George, J.C. Knight, et al., "Zero-dispersion wavelength decreasing photonic crystal fibers for ultraviolet-extended supercontinuum generation," *Opt. Express* **14**(12), 5715–5722 (2006).

[177] M. Bohmer, F. Pampaloni, M. Wahl, et al., "Time-resolved confocal scanning device for ultrasensitive fluorescence detection," *Rev. Sci. Instrum.* **72**(11), 4145–4152 (2001).

[178] D.S. Elson, J. Siegel, S.E.D. Webb, et al., "Fluorescence lifetime system for microscopy and multiwell plate imaging with a blue picosecond diode laser," *Opt. Lett.* **27**(16), 1409–1411 (2002).

[179] C.Y. Dong, C. Buehler, P.T.C. So, et al., "Implementation of intensity-modulated laser diodes in time-resolved, pump-probe fluorescence microscopy," *Appl. Opt.* **40**(7), 1109–1115 (2001).

[180] L.K. Van Geest and K.W.J. Stoop, "FLIM on a wide field fluorescence microscope," *Lett. Pept. Sci.* **10**(5–6), 501–510 (2003).

[181] C.D. McGuinness, K. Sagoo, D. McLoskey, and D.J.S. Birch, "A new sub-nanosecond LED at 280 nm: application to protein fluorescence," *Meas. Sci. Technol.* **15**(11), L19–L22 (2004).

[182] P. Herman, B.P. Maliwal, H.J. Lin, et al., "Frequency-domain fluorescence microscopy with the LED as a light source," *J. Microsc.-Oxf.* **203**, 176–181 (2001).

[183] H.B. Manning, G.T. Kennedy, D.M. Owen, et al., "A compact, multidimensional spectrofluorometer exploiting supercontinuum generation," *J. Biophoton.* **1**(6), 494–505 (2008).

[184] W. Becker, A. Bergmann, E. Haustein, et al., "Fluorescence lifetime images and correlation spectra obtained by multidimensional time-correlated single photon counting," *Microsc. Res. Tech.* **69**(3), 186–195 (2006).

[185] J. Siegel, D.S. Elson, S.E.D. Webb, et al., "Whole-field five-dimensional fluorescence microscopy combining lifetime and spectral resolution with optical sectioning," *Opt. Lett.* **26**(17), 1338–1340 (2001).

[186] D.K. Bird, K.W. Eliceiri, C.H. Fan, et al., "Simultaneous two-photon spectral and lifetime fluorescence microscopy," *Appl. Opt.* **43**(27), 5173–5182 (2004).

[187] E.S. Wachman, W.H. Niu, and D.L. Farkas, "Imaging acousto-optic tunable filter with 0.35-micrometer spatial resolution," *Appl. Opt.* **35**(25), 5220–5226 (1996).

[188] D.L. Farkas, C.W. Du, G.W. Fisher, et al., "Non-invasive image acquisition and advanced processing in optical bioimaging," *Comput. Med. Imaging Graph.* **22**(2), 89–102 (1998).

[189] Z. Malik, D. Cabib, R.A. Buckwald, et al., "Fourier transform multipixel spectroscopy for quantitative cytology," *J. Microsc.-Oxf.* **182**, 133–140 (1996).

[190] Q.S. Hanley, P.J. Verveer, and T.M. Jovin, "Spectral imaging in a programmable array microscope by hadamard transform fluorescence spectroscopy," *Appl. Spectrosc.* **53**(1), 1–10 (1999).

[191] C.E. Volin, B.K. Ford, M.R. Descour, et al., "High-speed spectral imager for imaging transient fluorescence phenomena," *Appl. Opt.* **37**(34), 8112–8119 (1998).

[192] R.A. Schultz, T. Nielsen, J.R. Zavaleta, et al., "Hyperspectral imaging: a novel approach for microscopic analysis," *Cytometry* **43**(4), 239–247 (2001).

[193] P. De Beule, D.M. Owen, H.B. Manning, et al., "Rapid hyperspectral fluorescence lifetime imaging,“ *Microsc. Res. Tech.* **70**, 481–484 (2007).

[194] J. Siegel, K. Suhling, S. Lévêque-Fort, et al., "Wide-field time-resolved fluorescence anisotropy imaging (TR-FAIM) - Imaging the mobility of a fluorophore," *Rev. Sci. Instrum.* **74**, 182–192 (2003)

[195] C. Buehler, C.Y. Dong, P.T.C. So, et al., "Time-resolved polarization imaging by pump-probe (stimulated emission) fluorescence microscopy," *Biophys. J* **79**(1), 536–549 (2000).

[196] A.N. Bader, E.G. Hofman, P. Henegouwen, et al., "Imaging of protein cluster sizes by means of confocal time-gated fluorescence anisotropy microscopy," *Opt. Express* **15**(11), 6934–6945 (2007).

[197] I. Gautier, M. Tramier, C. Durieux, et al., "Homo-FRET microscopy in living cells to measure monomer-dimer transition of GFP-tagged proteins," *Biophys. J.* **80**(6), 3000–3008 (2001).

[198] D.S. Lidke, P. Nagy, B.G. Barisas, et al., "Imaging molecular interactions in cells by dynamic and static fluorescence anisotropy (rFLIM and emFRET)," *Biochem. Soc. Trans.* **31**, 1020–1027 (2003).

[199] A.H. Clayton, Q.S. Hanley, D.J. Arndt-Jovin, et al., "Dynamic fluorescence anisotropy imaging microscopy in the frequency domain (rFLIM)," *Biophys. J.* **83**(3), 1631–1649 (2002).

[200] K. Suhling, J. Siegel, P.M. Lanigan, et al., "Time-resolved fluorescence anisotropy imaging applied to live cells," *Opt. Lett.* **29**(6), 584–586 (2004).

[201] R.K.P. Benninger, O. Hofmann, B. Onfelt, et al., "Fluorescence-lifetime imaging of DNA-dye interactions within continuous-flow microfluidic systems," *Angewandte Chemie-Internat. Ed.* **46**(13), 2228–2231 (2007).

[202] H. Schneckenburger, K. Stock, W.S.L. Strauss, et al., "Time-gated total internal reflection fluorescence spectroscopy (TG-TIRFS): application to the membrane marker laurdan," *J. Microsc.-Oxf.* **211**, 30–36 (2003).

[203] J. Widengren, V. Kudryavtsev, M. Antonik, et al., "Single-molecule detection and identification of multiple species by multiparameter fluorescence detection," *Anal. Chem.* **78**(6), 2039–2050 (2006).

[204] S.W. Hell and J. Wichmann, "Breaking the diffraction resolution limit by stimulated emission – Stimulated Emission Depletion Fluorescence Microscopy," *Opt. Lett.* **19**(11), 780–782 (1994).

[205] E. Betzig, G.H. Patterson, R. Sougrat, et al., "Imaging intracellular fluorescent proteins at nanometer resolution," *Science* **313**(5793), 1642–1645 (2006).

[206] M.J. Rust, M. Bates, and X.W. Zhuang, "Sub-diffraction-limit imaging by stochastic optical reconstruction microscopy (STORM)," *Nature Methods* **3**(10), 793–795 (2006).

[207] E. Auksorius, B.R. Boruah, C. Dunsby, et al., "Stimulated emission depletion microscopy with a supercontinuum source and fluorescence lifetime imaging," *Opt. Lett.* **33**(2), 113–115 (2008).

[208] A. Esposito, C.P. Dohm, M. Bahr, et al., "Unsupervised fluorescence lifetime imaging microscopy for high content and high throughput screening," *Mol. Cell. Proteomics* **6**(8), 1446–1454 (2007).

[209] G.T. Kennedy, A.J. Thompson, D.S. Elson, et al., "A fluorescence lifetime imaging scanning confocal endomicroscope," *J. Biophoton.* **2** (2009); DOI 10.1002/jbio.200910065

[210] J. Sharpe, U. Ahlgren, P. Perry, et al., "Optical projection tomography as a tool for 3D microscopy and gene expression studies," *Science* **296**(5567), 541–545 (2002).

[211] J. McGinty, K.R. Tahir, R. Laine, et al., "Fluorescence lifetime optical projection tomography," *J. Biophoton.* **1**(5), 390–394 (2008).

[212] J. McGinty, V.Y. Soloviev, K.B. Tahir, et al., "Three-dimensional imaging of Förster resonance energy transfer in heterogeneous turbid media by tomographic fluorescent lifetime imaging," *Opt. Lett.*, **34** (18), 2772–2774 (2009).

7

Raman and CARS Microscopy of Cells and Tissues

Christoph Krafft

Institute of Photonic Technology, 07745 Jena, Germany

Jürgen Popp

Institute of Photonic Technology, 07745 Jena, Germany,
Institute of Physical Chemistry, University Jena, 07743 Jena, Germany

Raman spectroscopy is based on inelastically scattered light from samples that are illuminated by monochromatic radiation. The Raman spectra contain information on vibrations that provide a sensitive fingerprint of the biochemical composition and molecular structure of samples. The main advantages are that this information can be obtained without additional labels and in a nondestructive way. Coherent anti-Stokes Raman scattering (CARS) offers as a further advantage a signal enhancement of several order of magnitude. Combining with microscopy enables diffraction-limited submicrometer resolution. These properties make Raman microscopic spectroscopy a versatile method to study biological tissues and cells. The contribution summarizes experimental techniques and gives an overview of recent applications in medical science.

Key words: spectroscopy, microscopy, imaging, hard tissues, soft tissues, cells

7.1 Introduction

Raman spectroscopy has been recognized to be a powerful tool for bioanalytical and biomedical applications. The principle is that Raman spectra contain information on vibrations that provide a highly specific fingerprint of the molecular structure and biochemical composition of cells and tissues. The main advantage is that the method provides this information without external markers such as stains or radioactive labels. Compared with another method of vibrational spectroscopy – infrared spectroscopy which excites molecular vibrations due to absorption of mid-infrared radiation (2.5 to 50 μm wavelength) – Raman spectroscopy is performed in the ultraviolet, visible and near infrared spectral range (200–1064 nm). Whereas Raman scattering of water molecules is weak, water strongly absorbs in the mid-infrared range that obscures much of the useful information in the biomolecular range. Infrared spectroscopy is usually restricted to analyze thin, dehydrated samples (tissue sections, dried cell films). As a consequence of the shorter wavelength Raman spectroscopy offers higher spatial resolution than infrared spectroscopy. This property plays an important role in combination with microscopy. Therefore, Raman spectroscopy is believed to have more potential for tissue and cell diagnosis, particular in the area of *in vivo* applications.

To overcome the disadvantage of low Raman signal intensities from most biomolecules, enhancement effects were utilized such as resonance Raman scattering (RRS) and surface enhanced Raman scattering (SERS). Whereas the first technique probes vibrations of chromophores, the second technique depends on short-range interactions with metal surfaces in the nanometer range. Both techniques require careful selection of the excitation wavelength and careful preparation, respectively. Coherent anti-Stokes Raman scattering (CARS) is a nonlinear variant of Raman spectroscopy that combines beside signal enhancement by more than four orders of magnitude further advantages such as directional emission, narrow spectral bandwidth, and no disturbing interference with autofluorescence because the CARS signal is detected on the short wavelength side of the excitation pulses. The energy diagrams in Figure 7.1 compare different scattering schemes after laser excitation. Stokes Raman scattering generates a photon on the long wavelength side similar as fluorescence emission. In case of RRS the excitation wavelength is in resonance with an electronic transition. When using RRS to boost the Raman signals single-photon excited fluorescence has to be coped with. The generally weaker Raman signals can be detected in the presence of intense overlapping fluorescence if the Stokes shift of the fluorescence emission is large or if the faster scattering process is registered before the emission using time-gated detectors. Anti-Stokes Raman scattering does not overlap with fluorescence. However, anti-Stokes Raman scattering is usually weaker than Stokes Raman scattering because an excited vibrational state has to be populated before the scattering process. According to the Botzmann distribution less than 1% of vibrations at 1000 cm^{-1} is in an excited state at room temperature, and the fraction exponentially decreases for higher wavenumbers. The excitation scheme of CARS involving a pump laser, a Stokes laser, and probe laser enables to generate a strongly amplified anti-Stokes signal. The theoretical description of CARS starts with the macroscopic polarization P, which is induced by an electric field E: $P = \chi^{(1)} \cdot \vec{E} + \chi^{(2)} \cdot \vec{E}\vec{E} + \chi^{(3)} \cdot \vec{E}\vec{E}\vec{E}$. Here $\chi^{(n)}$ is the susceptibility tensor of rank n which is a property of matter. For most of bulk samples $\chi^{(2)}$ is close to zero due to symmetry requirement. As a result, $\chi^{(3)}$ becomes the dominant nonlinear contribution. As CARS depends on $\chi^{(3)}$ and $\chi^{(1)} \gg \chi^{(3)}$, the nonlinear effect can only be detected using extraordinary intense electric fields in the MW range that are produced by nanosecond to femtosecond laser pulses. The full theoretical description of CARS is not within the scope of this contribution and can be found elsewhere [1].

To assess the heterogeneity of cells and tissues imaging techniques were developed that combine the spectral information with the spatial information. Raman images are usually recorded in the serial mapping mode. With exposure times of seconds for single spectra, acquisition of images with thousands of spectra takes hours. The key to reduce image acquisition is enhancement of signal intensities by more than three orders of magnitude without losing the Raman advantages including label-free preparation and nondestructivity. Due to its coherent nature, the CARS process gives rise to a stronger and directed signal compared with spontaneous Raman scattering. When coupled to laser scanning microscopes, CARS enables to collect images at video-time rates. Therefore, this variant is to date the most promising candidate for real-time studies of cells and tissues. Stimulated Raman scattering microscopy has recently been reported as another nonlinear vibrational technique for label-free imaging of cells and tissues [2]. This approach may complement Raman and CARS microscopy in the future.

This contribution describes the principles of Raman, RRS, SERS, and CARS microscopy that are expected to gain significance in bioanalytical and biomedical fields, and summarizes major applications. For more comprehensive overviews the reader is referred to reviews on biomedical

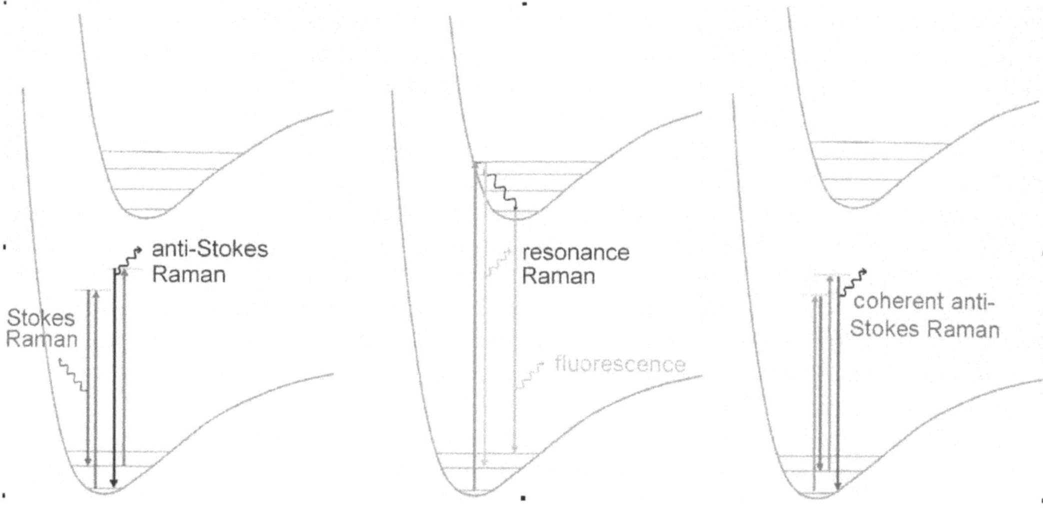

FIGURE 7.1: Scattering schemes of Stokes Raman, anti-Stokes Raman, resonance Raman, fluorescence and coherent anti-Stokes Raman scattering.

applications [3], disease recognition [4], noninvasive analysis of single cells [5], and CARS microscopy in live cell imaging [6] and lipid related diseases [7].

7.2 Experimental Methods

7.2.1 Raman spectroscopy

The basic setup of a dispersive Raman spectrometer consists of a laser as an intense and monochromatic light source, a device that separates the elastically (Rayleigh) scattered light of the sample from the Raman (inelastically) scattered light, a spectrograph, and a detector. As notch and edge filters have higher transmission for the Raman light than premonochromators, these filters are preferably used. Spectrographs for Raman spectrometers are available in Czerny-Turner reflection geometry or in transmission geometry as shown in Figure 7.2a and b, respectively. Charge coupled device (CCD) detectors with typical dimensions of 1024×128 pixels are multichannel detectors that enable to register the whole Raman spectrum simultaneously within a fraction of seconds. Most Raman spectrometers for cell and tissue studies utilize the near infrared (NIR) advantage. Even trace amounts of fluorescent molecules in biological material can cause an intense signal due to several orders of magnitude larger efficiency. Such an autofluorescence background often masks the weaker Raman signals when using visible lasers for excitation. As most tissues and body fluids show minimum absorption in the wavelength interval from 700 to 900 nm, the excited autofluorescence is at minimum and the penetration of the exciting and scattered radiation is at maximum. A typical state-of-the-art Raman spectrometer, which is commercially available from several manufacturers uses a diode laser for excitation at 785 nm, focuses the radiation onto the sample by a microscope objective, collects the scattered light in backscattering geometry, separates the elastically scattered light by a notch or an edge filter, disperses the inelastically scattered radiation in a grating spectrograph, and detects the spectrum by NIR-optimized CCD detectors.

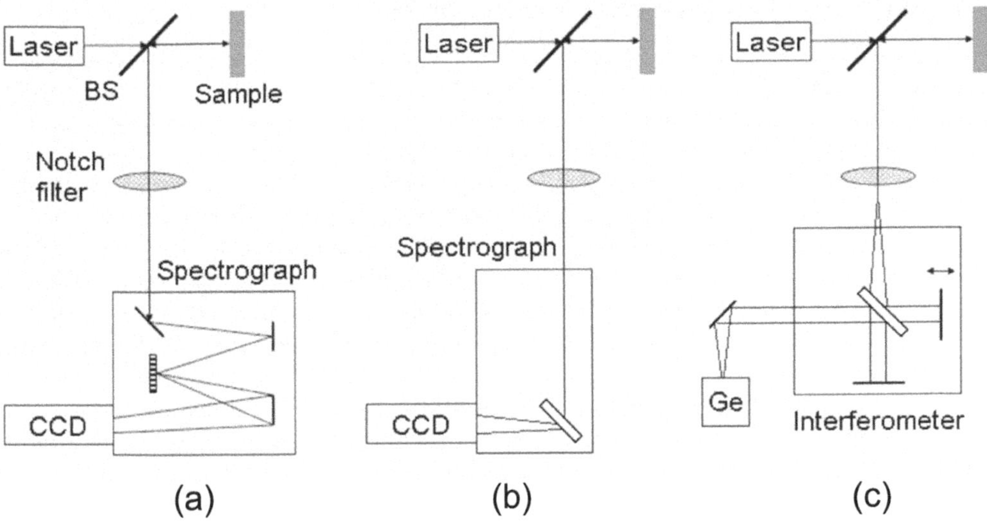

FIGURE 7.2: Schematic setup of Raman spectrometers equipped with a spectrograph with a reflective grating (a), transmissive grating (b), or with an interferometer (c).

Raman spectrometers utilizing the Fourier transform (FT) principle consist of an interferometer instead of a spectrograph (Figure 7.2c). The interferometer converts the Raman scattered radiation into an interferogram that is registered by a fast detector. After Fourier transformation of the interferogram the whole Raman spectrum can be obtained instantaneously. Usually, a solid-state laser at 1064 nm is used for excitation. A further reduced autofluorescence background is advantageous compared with 785-nm excitation. A reduced sensitivity due to the decrease of the scattering intensity with the fourth power of the frequency and due to the lower quantum efficiency of the Germanium-based detector are disadvantages that cause long acquisition times of minutes per spectrum.

7.2.2 Raman microscopy

The coupling of a Raman spectrometer with a microscope offers two main advantages. First, lateral resolutions can be achieved up to the Abbe's limit of diffraction below 1 μm according to the formula $\delta_{lat} = 0.61 \cdot \lambda / NA$. The diffraction limit in the axial dimension can be calculated as $\delta_{ax} = 2 \cdot \lambda n / NA^2$ (wavelength λ, refractive index n, numerical aperture NA). Even lower resolutions have been demonstrated using near field microscopic approaches (see subsection 7.2.4.) High axial resolutions are obtained by confocal microscopes. Second, maximum sensitivity can be achieved because the photon flux (photon density per area) at the focused laser beam onto the sample is at maximum and the collection efficiency of scattered photons from the sample is at maximum. The lateral resolution, the diameter of the focused laser and the collection efficiency depend on the NA of the microscope objective that is the product of the sine of the aperture angle and the diffraction index. High NA objectives have small working distances below 1 mm in air. Water immersion objectives offer a higher NA due to the larger diffraction index of water. Oil immersion objectives in combination with oil should be avoided because the oil can contaminate the sample and the spectral contributions of oil overlap with the Raman spectrum of the sample.

7.2.3 Surface enhanced resonance Raman scattering (SERS)

Since its first observation in 1977 SERS has been gaining popularity in analytical and physical chemistry and also in the biomedical field. The wide applications of SERS in biology, medicine, materials science, and electrochemistry were reported in a special issue of the *Journal of Raman Spectroscopy* [8]. SERS utilizes the optical excitation of surface plasmons, e.g., collective vibrations of the free electrons in metals. The result is an amplification of the electromagnetic field in the close vicinity of the metal surface (few nanometers), which leads to the enhancement of Raman signals from adsorbed molecules up to 15 orders of magnitude and enables to detect even single molecules by SERS [9]. As metal nanostructures SERS often uses spherical nanoparticles made of silver or gold having typical diameters of 20–100 nm. For efficient excitation of SERS the wavelength of the laser excitation has to be tuned to the plasmon absorption of the nanoparticle. Details are written in a recent textbook [10]. The SERS principle can be combined with atomic force microscopy (AFM) if the AFM tip is functionalized with SERS active nanoparticles. The so-called tip enhanced Raman spectroscopy (TERS) enables spatial resolution below the diffraction limit of light while simultaneously chemical information can be acquired. TERS was introduced in 2000. As the theoretical aspects of TERS are a subset of the SERS effect, they are discussed in one section. In a simplified way TERS can be considered as a SERS setup turned upside down and reduced to a single active enhancing site. The enhancing site is localized on an external unit, and a constant enhancement of the Raman signal intensity can be achieved by a single hot spot. TERS experiments with emphasis on life sciences were reviewed [11].

7.2.4 Resonance Raman scattering (RRS)

RRS is another variant of Raman spectroscopy that can be used to enhance its specificity and sensitivity. Here, the excitation wavelength is matched to an electronic transition of the molecule so that vibrational modes associated with the excited state are enhanced (by as much as a factor of 10^6). Electronic transitions of nucleic acids and proteins without prosthetic groups are found below 280 nm. Electronic transitions of proteins with prosthetic groups are also found in the visible spectral region. The enhancement is restricted to vibrational modes of the chromophore that reduces the complexity of Raman spectra from large biomolecules. Chapter 8 describes RRS of human skin to detect carotenoid antioxidant substances. RRS can be combined with SERS in surface enhanced resonance Raman spectroscopy (SERRS) giving even higher enhancement factors. This technique should be used with caution as absorption of the intense laser radiation can bleach the chromophore or cause other photo-induced degradation processes.

7.2.5 Coherent anti-Stokes Raman scattering (CARS) microscopy

Formally, CARS is described as a four-wave mixing. The sample is excited by a pump (ω_p), a Stokes (ω_s), and a probe pulse. These three electromagnetic fields interact to yield anti-Stokes emission at a forth frequency (ω_{as}). The experiments are usually performed in a frequency degenerate manner, such that the pump laser also provides the probe field. Energy conservation dictates the relation $\omega_{as} = 2\,\omega_p - \omega_s$. If the frequency difference $\omega_p - \omega_s$ coincides with a vibrational transition frequency of sample molecules, a strongly enhanced signal is observed, approximately 4 orders of magnitude. Because of its coherent property, the CARS signal increases quadratic with respect to the number of vibrational oscillators in the focal volume. Thus, scanning the sample at a given resonance frequency can be used to determine the spatial distribution of molecules with a vibrational transition at this frequency. As CARS is a coherent process, the phases of the contributing electromagnetic waves have to be matched within the excitation volume.

The demonstration that tight focusing using high-numerical aperture objectives, collinear excitation and the use of NIR excitation leads to very strong CARS signals brought an important

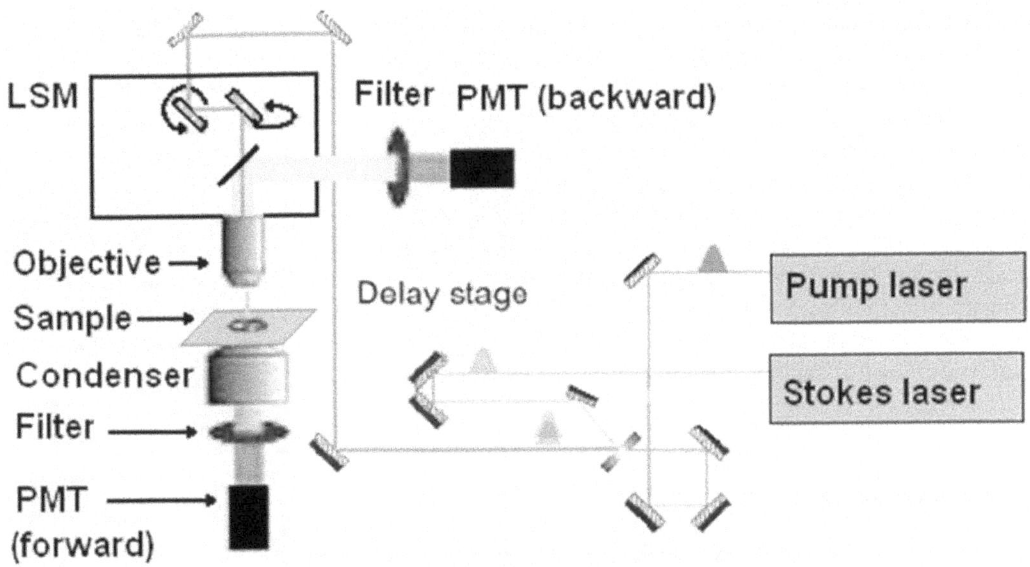

FIGURE 7.3: Schematic CARS setup with pump and Stokes lasers, a laser scanning microscope (LSM), and photomultipliers (PMT) for detection in forward and backward directions.

breakthrough for CARS microscopy 10 years ago [12]. It turns out that under these tight focusing conditions phase matching is provided for the very small excitation volume. The small excitation volume emerging from this tight focusing offers the additional attractive property of a very good spatial resolution along the optical axis of the microscope. For this reason CARS microscopy has an inherent three-dimensional imaging capability without the need to insert a confocal pinhole into the detection path. Unfortunately, CARS microscopy is not a zero-background technique. The suppression of the nonresonant background signal has indeed been the greatest challenge in making CARS microscopy attractive to a broader audience. It has been successfully tackled using time-delayed detection [13], polarization sensitive detection [14, 15], CARS heterodyne spectral interferometry [16], frequency modulation CARS [17], and introducing epi-CARS microscopes that detects signals in the backward direction instead of the usual forward direction (forward-CARS microscopes) [18]. However, as it was detailed in previous work [19] the non-Raman-resonant contribution to the third-order polarizability, which oscillates at the CARS frequency and originates from index-of-refraction inhomogeneity within a sample, can be used to obtain valuable structural, i.e., topological information on the sample. A schematic setup of a CARS setup with a laser scanning microscope and detection in forward and backward directions is displayed in Figure 7.3. Further theoretical and technical consideration can be found elsewhere [15, 20, 21].

7.2.6 Raman imaging

Raman imaging combines the spectral and the spatial information. Most Raman images are collected in the point-mapping mode. Here, the laser is focused onto the sample, the scattered light is registered, and subsequently the focus is moved by a scanning mirror or the sample is moved by a motorized stage to the next position. The total exposure time depends on the number of spectra and the exposure time per spectrum. Accumulating more than a thousand spectra with typical times of a few seconds per spectrum usually takes hours. The signal enhancement by the CARS process allows CARS spectrometers to couple with laser scanning microscopes. The fast scanning mirrors enable image acquisition at video-time rates (24 frames per second). In the case of laser line illumination of the sample, the spatial data can be registered on the detector on a line parallel to the entrance slit of the spectrometer and the spectral information is dispersed perpendicularly. The second spatial dimension of an image is recorded by scanning in the direction perpendicular to that line. The advantage of the so-called line-mapping registration mode is that only one dimension instead of two dimensions in the point-mapping mode has to be scanned. The parallel registration approaches called direct- or wide-field Raman imaging employ intense, global sample illumination. The inelastically scattered light from the sample is projected onto a two-dimensional CCD detector. Most wide-field Raman imaging spectrometers use filters to select the wavelength such as dielectric, acousto-optic tunable, and liquid-crystal tunable filters. All Raman imaging methodologies were compared with respect to acquisition times, image quality, spatial resolution, intensity profiles along spatial coordinates, and spectral signal-to-noise ratios [22].

7.3 Sample Preparation and Reference Spectra

7.3.1 Preparation of tissues

Although no or minimal preparation is required, to date much of the work on Raman spectroscopy of soft tissues had been carried out on *ex vivo* specimens that were either snap frozen at the time of collection or fixed using formalin or transferred to optimal cutting temperature (OCT) medium or embedded in paraffin. Due to the time lapse between tissue excision and spectroscopic examination, such preservations are performed by many groups in order to maintain the biochemical state of the specimens. Fixation by formaldehyde solution (formalin) preserves tissues by preventing autolysis and stabilizing their structure. Formaldehyde reacts with the amino groups of amino acids and promotes coagulation, not precipitation of proteins. Paraffin and OCT serve as support media. Subsequently, thin tissue sections are cut using a microtome in case of paraffin-embedded tissue or a cryotome in case of snap-frozen tissue. Fixation by air drying also works surprisingly well. After the water content has been vaporized in a dry atmosphere, the proteins are precipitated to form an insoluble mass that is resistant to degradation. The effects of *ex vivo* handling procedures (drying, freezing, thawing, and formalin fixation) on mammalian tissue [23], the effects of fixation on human bronchial tissue [24], and the efficacy of dewaxing agents on cervical tissue [25] were described for Raman spectroscopy.

Staining reagents such as hematoxylin for the cell nuclei and eosin for the cytoplasm give additional spectral contributions. Therefore, unstained samples are usually prepared for Raman spectroscopic studies. Due to the nondestructivity of the method, tissues can be stained by hematoxylin and eosin (H&E) stains or by immunohistochemistry after data acquisition, thus allowing visual imaging and pathological examination. For a pre-assessment and selection of sampling areas before vibrational spectroscopic image acquisition, a parallel section can be prepared on standard glass slides, stained and inspected.

The process of paraffin-embedding requires fixation and sample dehydration before embedding and removal of the support medium by organic solvents after sectioning. Tissues are dehydrated by dipping in a graded series of ethanol/water mixtures and then by bathing in xylene or benzene. A danger of this process is that the conformation of biomolecules might be distorted, and the chemical composition of the sample is altered, e.g., hydrophobic constituents, such as tissue lipids, will be partly dissolved by solvent treatment as well. One should be aware that each processing step may change the chemical or molecular properties of the sample. However, these changes induced by the processing procedures may not be a cause for concern as long as one protocol is strictly adopted throughout a study.

7.3.2 Preparation of cells

To prepare cultured cells for Raman studies, it is advantageous to grow cells directly onto the substrate because it circumvents the treatment of cells with trypsin that significantly changes their morphology and introduces spectral artifacts. Substrates made of quartz give lower spectral background than glass. Even lower background, but even more expensive are substrates made of pure CaF_2. However, side effects due to a low solubility of CaF_2 and the presence of Ca^{2+} and F^- ions in cell culture media cannot be excluded.

Squamous epithelial cells can easily be exfoliated from the mucosa, e.g., by gently swiping with a sterile cotton swap. Exfoliated cells are then centrifuged, the supernatant is removed, and the cellular pellet is resuspended. Subsequently, the cell suspension can be deposited onto substrates using the Cytospin system (Thermo, USA). The Cytospin instrument uses centrifugal force to deposit the cells in a sparse monolayer while wicking away the suspension medium, resulting in cellular deposits with cells well separated from their nearest neighbors. This technique is preferable to smearing the swap or brush directly onto the substrate, which typically produces thick clumps of cells that are not suited for either visual or spectroscopic analysis.

7.3.3 Raman spectra of biological molecules

Beside water, which contributes fortunately feebly to Raman spectra, cells and tissues are mainly composed of proteins, nucleic acids, lipids, and carbohydrates. These biological macromolecules consist of defined monomeric subunits such as twenty amino acids and five nucleosides. As a consequence of the low number of different subunits, the spectra of cells and tissues often share some similarities that require data of high signal-to-noise ratios to identify the specific spectral signature. Raman spectra of monomeric subunits can be used to assign bands in spectra of macromolecules. Figure 7.4 compares the Raman spectra of aromatic amino acids and three proteins. Phenylalanine bands at 621, 1003, 1033, 1215, 1586, 1602, and 1610 cm^{-1} (Figure 7.4A) can be identified at similar positions in Raman spectra of concanavalin A (Figure 7.4F), bovine serum albumin (Figure 7.4G), and collagen (Figure 7.4H). The same trend is found for tyrosine bands at 641, 829, 845, 1178, and 1614 cm^{-1} (Figure 7.4B), and tryptophan bands at 755, 874, 1009, 1338, 1358, 1556, 1577, and 1616 cm^{-1}. Some bands of aromatic amid acids are not resolved in the Raman spectra of proteins due to overlap with spectral contributions from other amino acid residues and the protein backbone. In particular, aliphatic side chains contribute to bands at 1126 and 1449 cm^{-1}. Bands at 508 and 550 cm^{-1} in Figure 7.4G indicate disulfide bridges between cysteine residues that stabilize protein structures. The amide I band around 1660 cm^{-1} and the amide III band between 1230 and 1300 cm^{-1} are assigned to vibrations of the peptide group and depend on the protein conformation. Maxima of the amide III band at 1238 cm^{-1} and the amide I band at 1672 cm^{-1} are typical for β-sheet proteins such as concanavalin A. For α helical proteins such as bovine serum albumin the maxima are shifted to 1276 and 1656 cm^{-1}, respectively. The band at 939 cm^{-1} is assigned to C-C_α vibrations of the peptide group with α helical secondary structure. Collagen is characterized by an unusual amino acid composition with high content of prolin, hydroxyprolin and glycin and a triple helix conformation (Figure 7.4H). These properties give rise to bands at 814, 853, 873, 919, 935, 1242, and 1665 cm^{-1}.

Phospholipids are represented by phosphatidylcholin in Figure 7.4D. The twelve most frequent brain lipids were reported elsewhere [26]. Bands at 1063, 1126, 1263, 1297, 1437, and 1656 cm^{-1} are assigned to hydrophobic, unsaturated fatty acid side chains. Further bands at 1734 cm^{-1} to C=O of ester groups, at 1092 cm^{-1} to PO_2 of phosphate groups, and at 714 and 874 cm^{-1} to the hydrophilic choline group $N(CH_3)_3$. The four-ring system of cholesterol is characterized by numerous sharp bands with the most intensive ones at 427, 544, 608, 700, 1439, and 1672 cm^{-1}. Nucleic acids are present in cells and tissues as DNA and RNA. Their Raman spectra depend on the nucleotide composition and the geometry of the phosphate backbone and the ribose or deoxyribose ring. The

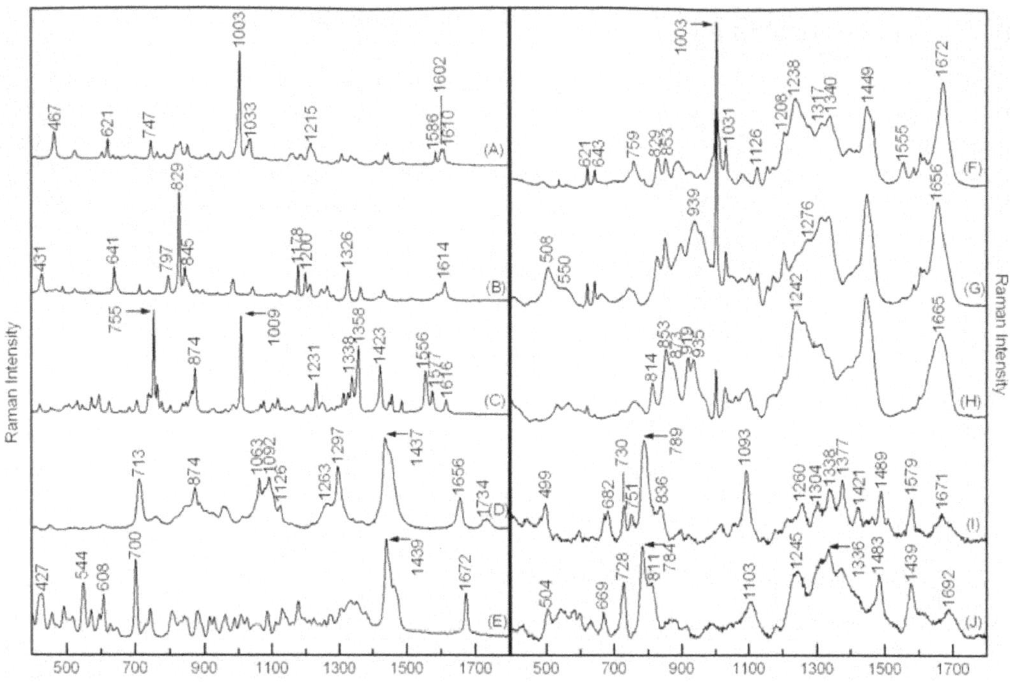

FIGURE 7.4: Raman spectra from 400 to 1800 cm^{-1} of L-phenylalanine(A), L-tyrosine (B), L-tryptophan (C), phosphatidylcholin (D), cholesterol (E), concanavalin A (F), bovine serum albumine (G), collagen (H), DNA (I), and RNA (J).

most prominent marker bands of B-form conformation of DNA (Figure 7.4I) are located at 836 and 1093 cm^{-1}, whereas the bands show different intensities and are shifted to 811 and 1103 cm^{-1} in A-form conformation of RNA (Figure 7.4J). Further band assignments for the nucleobases adenine, guanine, thymine, cytidine, and uridine are given elsewhere [27].

7.4 Applications to Cells

7.4.1 Raman microscopy of microbial cells

Routine clinical microbial identification of pathogenic microorganisms is largely based on nutritional and biochemical tests. In the case of severely ill patients, the unavoidable time delay associated with such identification procedures can be fatal. Simultaneous to the microbial classification by infrared spectroscopy [28], Raman spectroscopy demonstrated its potential in this field [29]. Compared with conventional methods of bacterial identification, Raman and infrared spectroscopy offer a significant time advantage because they can be used without amplification or enhancement steps. In general, sample preparation is easier for Raman spectroscopy than for infrared spectroscopy because bacteria grown in liquid cultures as well as bacteria on agar plates can directly be investigated after only 6 hours of culturing [30]. Bacterial cells are so small that single spectra of an ensemble of cells are usually recorded. Excitation wavelengths in the near infrared were used to suppress autofluorescence background in Raman spectra of bacteria. Recent work of our group includes the online monitoring and identification of bioaerosols in clean room environments [31]. The analysis was divided into several steps: (i) impacting the aerosol on a surface, (ii) presorting the particles with glancing light illumination and fluorescence imaging in order to distinguish between abiotic and biotic particles, and (iii) analyzing the biotic particles using Raman spectroscopy. Support vector machines were applied to classify bacteria on the strain level [32]. Another optical detection approach was presented that combines Raman spectroscopy, fluorescence spectroscopy, digital imaging, and automated recognition algorithms to perform species-level identification [33]. Trace levels of biothreat organisms could be detected in the presence of complex environmental backgrounds. Wide-field illumination and detection enabled imaging of single cells without scanning the laser or moving the sample [34]. A mixture of biomaterial resembling a typical water sample and comprising of Gram-positive, Gram-negative organisms, and proteins could be distinguished at the single cell level.

As the consequence of the higher spatial resolution of Raman spectroscopy compared with infrared spectroscopy, its potential for studies on single bacterial cells has been recognized [35, 36]. A Raman confocal microscope was used to generate a spectral profile from a single microbial cell that enables to differentiate different species and different growth phases [37]. Furthermore, the efficacy as a means to identify cells responsible for the uptake of glucose with varying ratios of $^{13}C_6$ to $^{12}C_6$ was demonstrated. ^{13}C incorporation shifted characteristic peaks to lower wavenumbers. Bacterial identification based on a single cell could reduce the time for analysis down to minutes because only sampling and data acquisition and no cultivation step are required.

7.4.2 Raman spectroscopy of eukaryotic cells

Microscopic techniques based on vibrational spectroscopy have become powerful tools in cell biology because the molecular composition of subcellular compartments can be visualized without the need for labeling. Nonresonant Raman spectroscopic studies on individual human cells were summarized in a chapter of a recent textbook [38]. Figure 7.5 displays typical Raman spectra of the cell nucleus, cell cytoplasm, and lipid vesicle that have been collected from lung fibroblast cells. According to the Raman spectra in Figure 7.4, spectral contributions can be assigned to proteins, DNA, RNA, and lipids in different relative amounts. The nucleus is characterized by intense DNA bands most prominently represented by bands at 786 and 1092 cm^{-1}. Whereas the protein bands are largely unchanged in the cytoplasm the DNA bands are replaced by RNA. This change is represented by less intense bands at 680 and 1375 cm^{-1}, a more intense band at 813 cm^{-1}, and a band shift toward 1100 cm^{-1}. These features are indicated by arrows down. The lipid vesicles show more intense CH deformation bands, phosphate bands, and ester bands relative to protein and nucleic acid bands. These features are indicated by arrows up.

Raman applications include discrimination between normal and transformed lymphocytes [39], identification of cells representing different stages of cervical neoplasia [40], detection of glutamate in optical trapped single synaptosomes [41], monitoring mRNA translation in embryonic stem cells [42], detection of biological changes related to cell life [43] and cell death [44], association of lipid bodies with phagosomes in leukocytes [45], studies of mitotic cells [46], identification of single eukaryotic cells such as yeast species [47], discrimination of live and fixed cultured cervical cells [48], studies of stress-induced changes [49], and analysis of single blood cells for cerospinal fluid diagnostics [50]. The distribution of various cellular components was visualized by Raman imaging such as proteins, DNA, and RNA [51–54], intracellular lipid vesicles [55], mitochondria [56], and chromosomes [57]. Raman images of deuterated and nondeuterated liposomes were not only presented in the lateral dimension, but also in the axial dimension [58] (Figure 7.6). Here, compounds were labeled with deuterium in order to distinguish the species of interest from their cellular environment spectroscopically. The pioneering work by Puppels and coworkers [57] used 514.5-nm laser light. Excitation wavelength in the visible range significantly affected integrity of living cells due to photo-induced degradation processes. It was found that no degradation of cells and chromosomes occurred when laser light of 660 nm was applied [59]. Integrity of cells was also

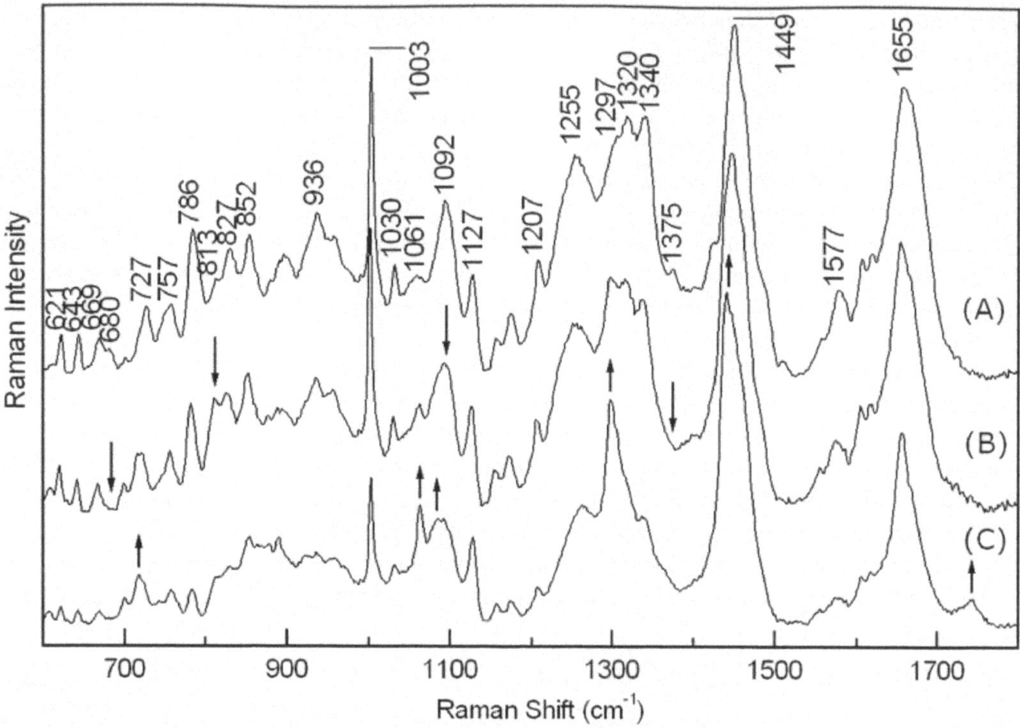

FIGURE 7.5: Raman spectra from 600 to 1800 cm^{-1} of cell nucleus (A), cell cytoplasm (B), and lipid vesicle (C) that have been collected from lung fibroblast cells.

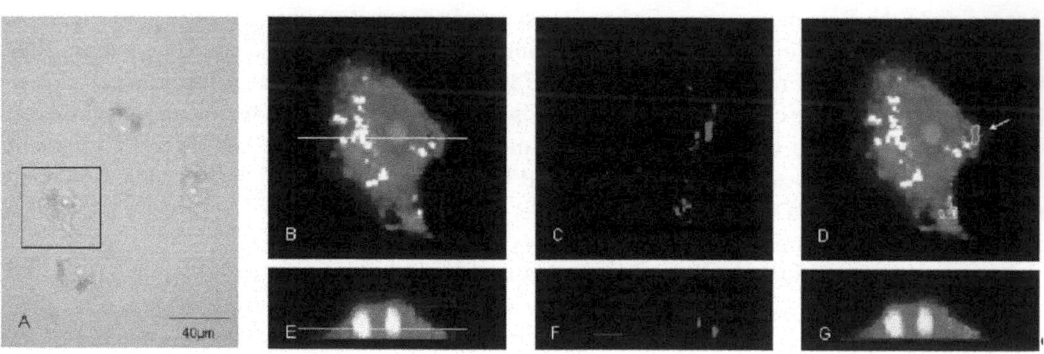

FIGURE 7.6: MCF-7 cells treated with DSPC-d$_{70}$ liposomes at 2 mg/mL (A). Raman images of one cell reconstructed from the C–H (B) and C–D stretching intensities (C). The image in (D) represents an overlay of panels (B) and (C). A depth profile was collected for the same cell (E–G). Scanning positions are indicated by the white bar (from [58] with permission).

confirmed for at least 60 minutes exposure to 115-mW laser power at 785 nm whereas morphological changes occurred in cells with 488- and 514.5-nm laser excitation already at 5-mW power [60].

7.4.3 Resonance Raman spectroscopy of cells

Excitation wavelengths in the ultraviolet and visible range were used to enhance signals from chromophores in microbial and eukaryotic cells utilizing the resonance Raman (RR) effect. Iron-porphyrin complexes, known as hemes, belong to the most versatile redox centers in biology. As widespread prosthetic groups among living systems, hemes are bound covalently or noncovalently to proteins and enzymes. The biological function of heme containing proteins (also called hemoproteins) include (i) oxygen transport by hemoglobin in erythrocytes, (ii) electron transport in cellular respiration by cytochromes in mitochondria, (iii) enzymatic degradation by cytochrome P450, and (iv) innate immunity by flavocytochrome b_{558}. Since the first reported RR spectra of hemoproteins in 1972, a large body of literature has dealt with the structural characterization of metalloporphyrins and heme prosthetic groups in proteins using RR spectroscopy. Overviews in the context of biomedical vibrational spectroscopy have recently been published [61–63]. Raman spectra were reported for oxygenated and deoxygenated hemoglobin within single red blood cells (erythrocytes) *in vivo* using excitation wavelengths of 488, 514, 568, and 632 nm [64]. A setup was introduced to combine RR spectroscopy with optical tweezers to trap erythrocytes [65]. The malaria pigment hemozoin was localized and analyzed by RR imaging [66–68]. The malaria parasites of the genus *Plasmodium* digest a major proportion of red blood cell hemoglobin. As free heme, a product of the hemoglobin digestion, is toxic for the parasite it detoxifies free heme by conversion into an insoluble crystal called hemozoin or malaria pigment. RR imaging was employed to visualize the subcellular distribution of the NADPH oxidase subunit cytochrome b558 in both resting and phagocytosing granulocytes [69]. The Raman signal from cytochrome b558 was resonantly and selectively enhanced in neutrophilic granulocytes by using 413-nm excitation. Using the RR effect of cytochrome C the microbial distribution of nitrifiers and anammox bacteria were determined directly in their natural environment [70]. Significant Raman spectra were recorded in less than 1 second.

Another abundant chromophore in cells absorbing in the visible wavelength region is carotenoid. The intracellular amounts of carotenoids in lymphocytes of lung cancer patients and healthy individuals were compared using Raman spectroscopy [71, 72]. A significant decrease of carotenoids in lung cancer patient was found.

Aromatic amino acids and nucleotides are chromophores in proteins and nucleic acids that occur in almost all cells. These chromophores can be excited by UV light between 220 and 290 nm. The peptide backbone of proteins is mainly responsible for absorption in the deep-UV region below 220 nm. UV RR spectroscopy with 244-nm excitation was used to classify lactic acid bacteria [73], to monitor the bacterial growth and the influence of antibiotics [74], and to characterize bacteria of the strain staphylococcus epidermis [75, 76].

7.4.4 SERS/TERS of cells

Current cell biological research requires better optical labels regarding higher photostability than state-of-the-art fluorescence dyes. The attractive prospects of SERS in cell investigations [77], bacterial identification [78] and in chemical and biochemical diagnostics [79] have recently been summarized. The feasibility to measure SERS spectra from live cells without label molecules was first demonstrated with the cell line HT29 [80]. Cell monolayers were incubated with gold nanoparticles. As SERS only probes molecules in the close vicinity (several nanometers) of metal surfaces and the nanoparticles bind unspecific molecules, Raman spectra were different at almost all places in cells. The capabilities of near infrared SERS using gold nanoparticles were reported to obtain detailed chemical information from single osteosarcoma cells [81]. The authors also characterized

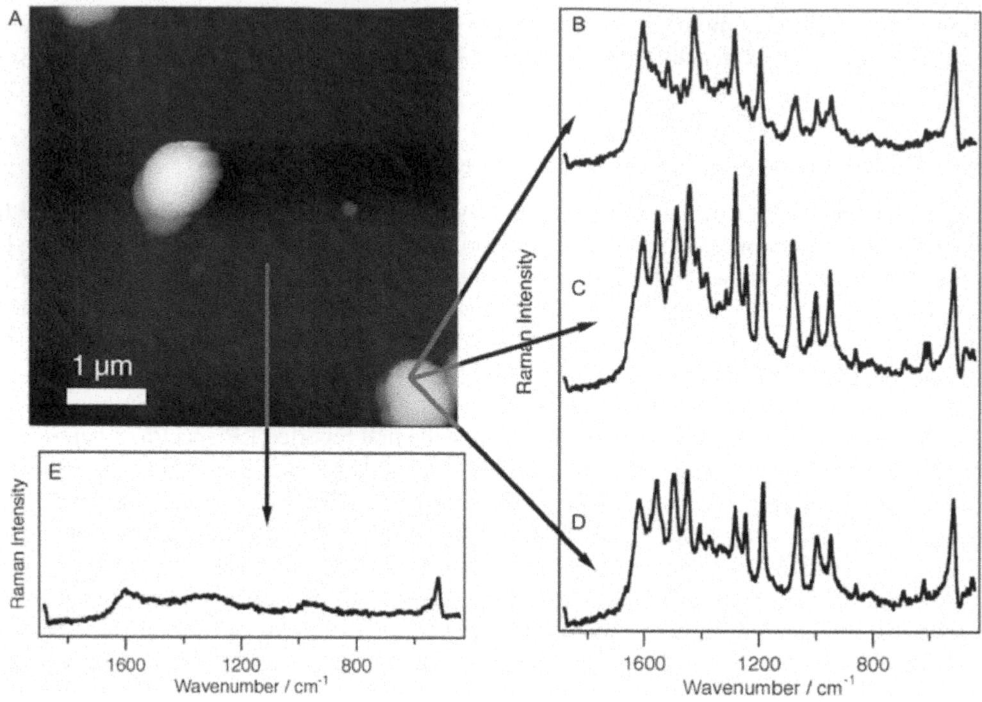

FIGURE 7.7: Topographic image of *S. epidermis* cells with marked locations of the corresponding TERS measurements (A). TERS spectra (B, C, D) measured with a silver coated AFM cantilever on top of a bacterium whereas spectrum (E) corresponds to a reference TERS experiment on the glass surface. The ever present band around 520 cm^{-1} is attributed to Raman scattering of the silicon tip (with permission from [85]).

the distribution of extrinsic molecules that were introduced into the cell, in this case rhodamine 6G. Furthermore, SERS labels were constructed based on indocyanine green on colloidal silver and gold that can deliver information on cellular chemistry [82]. In addition to its own detection by the characteristic indocyanine green SERS signatures, the novel nanoprobe delivers spatially localized chemical information from its biological environment by employing SERS in the local optical fields of the gold nanoparticles.

SERS spectra of bacteria are relatively new and exciting application areas for SERS. Bacteria were treated with sodium borohydride, which served as a nucleating substrate for reduction of silver ions [83]. A rough silver metal coating surrounding the microorganism was formed and intense surface enhanced Raman spectra of the coated bacteria were measured. SERS spectra of bacterial species and strains were obtained on gold nanoparticle covered SiO_2 subtrates [84]. The SERS spectra were spectrally less congested and exhibit greater species differentiation than their corresponding non-SERS Raman spectra at 785 nm excitation.

Gram positive bacteria *Staphylococcus epidermis* ATCC 35984 were studied by TERS to investigate the cell surface with high spatial resolution in order to reveal distinct features [85, 86]. For this purpose a drop of an aqueous bacteria suspension was dried on a glass substrate. In Figure 7.7 TERS spectra of bacteria are presented. The Raman bands were tentatively assigned to polysaccharides and peptides that agree with the reported assembly of staphylococcus surfaces [75]. According to the variation of the TERS signal, the authors assumed that superficial cell dynamics were occurring during spectra acquisition. Whereas phenomena like diffusion usually do not play a role in micro-

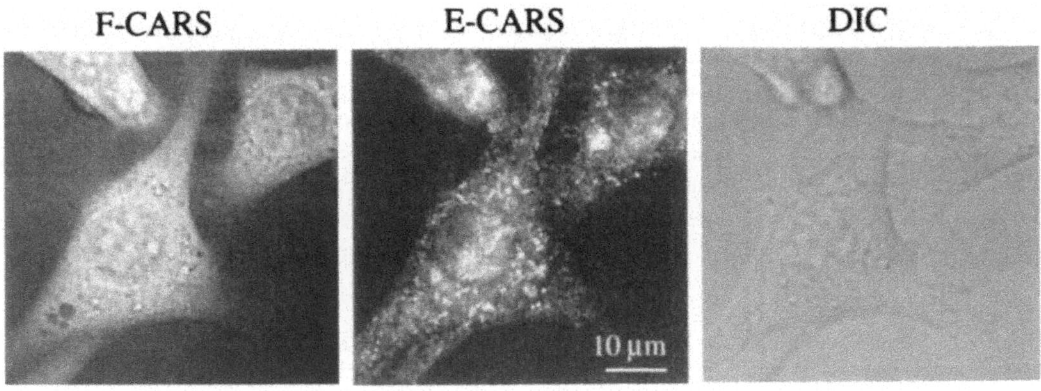

FIGURE 7.8: F-CARS, E-CARS and DIC images of NIH 3T3 cells in interphase. For the CARS images, $\omega_p - \omega_s$ was tuned to the aliphatic C-H stretching vibrational frequency at 2870 cm^{-1}. The pump frequency was 14054 cm^{-1} and the Stokes frequency was 11184 cm^{-1}. The pump and Stokes power were 40 and 20 mW, respectively. The acquisition time was 14 s. The image size is 59×59 μm^2. The DIC image was obtained in 1.7 s by using the laser for the Stokes beam (reprinted from [87] with permission).

scopic Raman experiments, TERS is sensitive to these effects and allows investigating them for the first time.

7.4.5 CARS microscopic imaging of cells

To date most biological applications of CARS have dealt with the investigation of lipids. Lipids play an important role in biological membranes, as energy storage molecules, and as messengers in cellular communications. Lipids are comparatively easy to detect in cells by CARS microscopy due to their high local concentration and their high CH stretching vibration signal (see Figure 7.8).

Lipid vesicles inside HeLa cells [88], the membrane of a lysed erythrocyte cell [89], the growth of lipid droplets in live adipocyte cells [90], the organelle transport in living cells [91] and lipid storage in the nematode caenorhabditis elegans [92, 93] were visualized by CARS microscopy at 2845 cm^{-1}. The nondestructive nature of CARS imaging of cells has been shown over a period of 300 s [91]. In this paper, multiplex CARS was able to probe a spectral window of >300 cm^{-1}. The registration of multiplex CARS images is more time consuming than registration of CARS images at single wavenumbers. The composition and packing of individual cellular lipid droplets were also imaged using multiplex CARS microscopy [94]. The protein distribution in an epithelial cell was visualized through the amide I vibration by CARS microscopy at 1649 cm^{-1} [14]. By use of the OH-stretch vibration of water, cellular hydrodynamics have been studied under the CARS microscope with subsecond time resolution [95]. Strong signals are also observed for the CD stretching vibration. Using deuterated compounds opens up a wide range of applications in lipid research. For this reason, CD labeling has found use in many CARS experiments. Phase segregated lipid domains were observed by making use of deuterated and nondeuterated lipids and tuning to the CD_2 vibrational band at 2090 cm^{-1} [96]. CARS flow cytometry combined a laser-scanning CARS microscope with a polydimethylsiloxane based microfluidic device [97]. This approach measured the population of adipocytes isolated from mouse fat tissues for quantitation of adipocyte size distribution.

Polarization-sensitive detection was used in a multiplex CARS microscope to image lipids in live cells with high sensitivity [98]. Frequency modulated detection was applied to image deuterium-labeled oleic acids in lung cancer cells [17]. The applications of epi- and forward-CARS microscopy to cell biology have previously been demonstrated [87, 99] including high-speed CARS images of live cells in interphase, live cells undergoing mitosis and apoptosis, and the distribution of chromosomes of live cells in metaphase. CARS microscopy was shown to distinguish differentiated versus undifferentiated states of mouse embryonic stem cells [100]. The distinction was based on the CARS intensities over 675 to 950 cm^{-1} Raman shift. This study used a mirrored surface as a substrate that allowed simultaneous collection of both the forward-collected signal and the epi-reflected signal. The significance of CARS in single cell studies is expected to gain in the near future due to the rapid acquisition, high spatial resolution, and three-dimensional imaging capabilities.

7.5 Applications to Tissue

To demonstrate the capability of Raman spectroscopy of tissues, Figure 7.9 shows typical Raman spectra of brain, lung, hydroxyapatite of mineralized samples, and epithelial, connective, and muscular tissue of colon. White matter in brain belongs to the tissue class with the highest lipid content, which is indicated by intense lipid bands. Lung tissue is characterized by high blood perfusion rate that is consistent with significant spectral contributions of hemoglobin. Due to a resonance effect, heme-associated bands are more intense than the bands of the other constituents. The spectrum of mineralized samples is dominated by the inorganic matrix hydroxyapatite. The Raman spectrum of epithelial tissue shares some similarities with spectra of the cell nucleus and cytoplasm because main bands are assigned to proteins and nucleic acids. Comparison with Figure 7.4 reveals that Raman spectra of connective tissue and muscle contain various amounts of collagen. In general, the acquisition of images is advantageous in tissue studies because its inhomogenous nature can be considered better than by acquisition of single spectra. Raman images are able to provide precise morphologic information and enable to determine margins between normal and pathological tissue based on the biochemical fingerprint of the spectra.

7.5.1 Raman imaging of hard tissues

Bone is beside teeth and calculi (concretion of material in organs, usually mineral salts) the main representative of hard tissue. Raman imaging applications to mineralized biological samples to probe spatial variations of the mineral content, mineral crystallinity, and the organic matrix consisting predominantly of collagen type 1 and cells was reviewed [101]. The Raman spectra reveal the extent of mineral formation in tissue using a mineral-to-protein ratio. Other parameters monitor the size and perfection of the crystals and the ratio of carbonate ions and phosphate ions in the hydroxyapatite lattice. In contrast to conventional histological techniques, Raman spectroscopic methods do not require a special sample preparation, e.g., homogenization, decalcification, extraction or dilution. The structural properties of bone were examined by Raman imaging at 785 nm excitation with micrometer range spatial resolution [102, 103]. Raman images from bone were also collected at 532-nm excitation after the fluorescence background was reduced by photochemical bleaching [104]. Advantages of the shorter excitation wavelength are higher quantum efficiency of CCD detectors and more intense Raman signals due the fourth power dependence of scattering intensity with excitation frequency. Both effects result in shorter acquisition times. Typically, Raman images of 300×100 μm^2 areas with 2-μm resolution just require 10 minutes. Raman imaging studies of bone-related diseases include craniosynostosis, which is an abnormal condition of bones making up the skull [105, 106]. Microdamages in bone that were induced using a cyclic fatigue loading regime were investigated by Raman imaging [107].

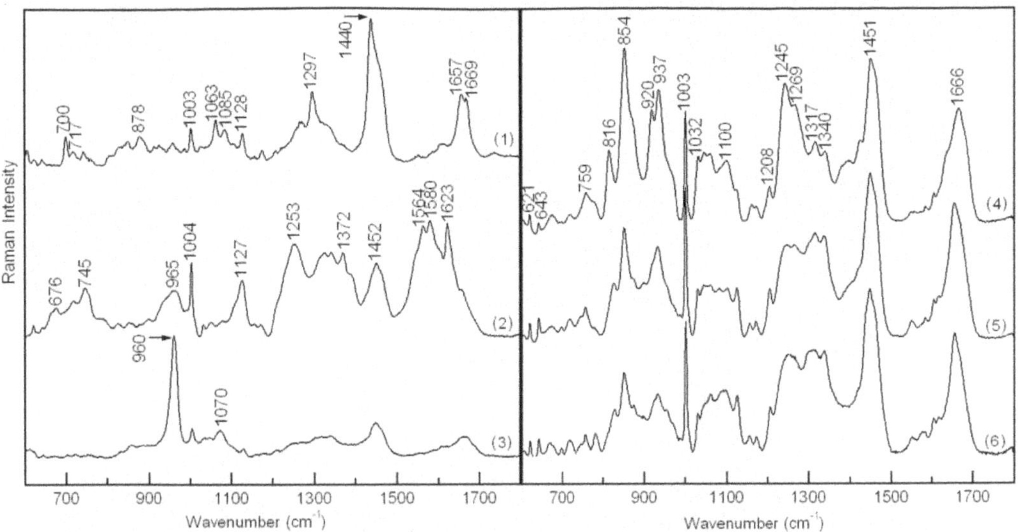

FIGURE 7.9: Raman spectra from 600 to 1800 cm^{-1} of white matter from brain (trace 1), lung tissue (trace 2), mineralized tissue (trace 3), connective tissue (trace 4), muscular tissue (trace 5), and epithelial tissue of colon (trace 6).

7.5.2 Raman imaging of soft tissues

7.5.2.1 Epithelial tissue

The main soft tissue types are epithelium, connective tissue, muscular tissue, and nervous tissue. The epithelium covers the surface of organs, is exposed to the environment, and is in contact to a broad range of potentially aggressive or harmful chemical and physical conditions that can induce a deregulation of cell division in the basal cell layer. Therefore, carcinomas of the epithelium are among the most common forms of cancer. Raman imaging studies were reported for carcinomas of lung (see 7.5.2.3), breast (see 7.5.2.4), skin (see 7.5.2.5), bladder [108], and esophagus [109]. Other normal and abnormal epithelial tissues from the larynx, tonsil, stomach, bladder and prostate were included in *ex vivo* investigations [110, 111, 112]. Raman images of epithelial tissue, connective tissue, and muscular tissue were compared with each other and with corresponding FTIR images [113].

In the context of muscle tissue Raman imaging has been applied less often to skeletal muscle than smooth muscle. Raman imaging of blood vessels that belong to smooth muscles was applied to obtain both chemical and spatial information about atherosclerosis [114]. Using a portable Raman system, Raman spectra were collected from the cervix of 79 patients with exposure time of 5 seconds [115]. Discrimination algorithms were developed to distinguish between normal excto-cervix, squamous metaplasia, and high-grade dysplasia. In a similar approach, 73 gastric tissue samples from 53 patients were diagnosed using Raman spectroscopy with 5 seconds acquisition time and classification and regression tree techniques [116]. The predictive sensitivity and specificity of the independent validation data set to distinguish normal and cancer tissue were 88.9 and 92.9%, respectively.

7.5.2.2 Nervous tissue

Nervous tissues are composed of nerve cells (neurons) and glial cells. These cells form the central nervous system (CNS) and the peripheral nervous system (PNS). Ganglia in colon tissue that were

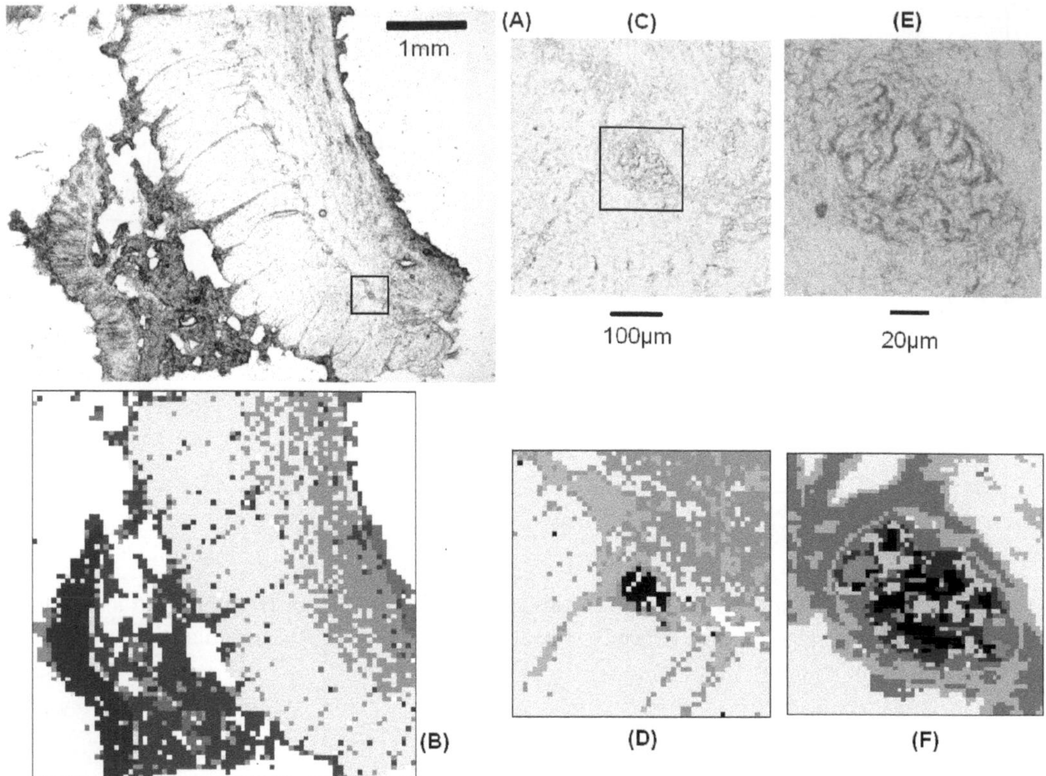

FIGURE 7.10: (See in color after page 572.) Photomicrographs (A, C, E) Raman images (B, D, F) of colon tissue. A step size of 60 μm gives gross overview of the tissue morphology. A step size of 10 μm of the boxed region in (A) resolves a ganglion. A step size of 2.5 μm of the boxed region in (C) displays subcellular features of the ganglion. Raman images were segmented by cluster analysis as previously described [113].

studied by Raman imaging [113] (Figure 7.10) are examples for the PNS. Glial cells in the human brain are smaller and more abundant than neurons. The main functions of glial cells are to surround neurons and hold them in place, to supply nutrients and oxygen to neurons, to insulate one neuron from another, and to destroy pathogens and remove dead neurons. As neurons largely lost the ability to divide, most primary brain tumors are derived from glial cells. The field of neuro-oncological applications of Raman imaging includes primary brain tumors [117], the distinction of vital tumor tissue and necrosis [118], the comparison of Raman and FTIR images from primary brain tumor tissue sections [119], tumors of the meninges called meningeomas [120] and a brain tumor model of rats [121] and mice [122].

7.5.2.3 Lung tissue

Lung cancer is the second most common cancer in humans and the most common cause of cancer deaths in the world. Characterization of the biochemical composition of normal lung tissue is important to understand the biochemical changes that accompany lung cancer development. Frozen sections of normal bronchial tissue were studied by Raman imaging [123]. Subsequent comparison of Raman images with histologic evaluation of stained sections enabled to identify the spectral features of bronchial tissue, mucus, epithelium, fibrocollagenous stroma, smooth muscle, glandular tissue, and cartilage. Raman maps were collected from lung tissue and congenital cystic adenomatoid malformation (CCAM) [124] and bronchopulmonary sequestration (BPS) [125]. The Raman spectra demonstrated that both malformations and normal lung tissue can be distinguished from each other.

7.5.2.4 Breast tissue

Worldwide, breast cancer is the most common cancer among women, with an incidence rate more than twice that of colorectal cancer and cervical cancer and about three times that of lung cancer. Therefore, Raman spectroscopy was applied to complement the existing diagnostic tools. Raman mapping assessed the chemical composition and morphology of breast tissue [126] and lymph nodes in tissue sections of breast cancer patients [127]. A model was presented to fit Raman spectra of breast tissue with a linear combination of basis spectra of the cell cytoplasm, cell nucleus, fat, beta carotene, collagen, calcium hydroxyapatite, calcium oxalate dihydrate, cholesterol-like lipid deposits and water. Another paper showed the ability of Raman spectroscopy to accurately diagnose normal, benign and malignant lesions of the breast *in vivo* in a laboratory setting with high sensitivities and specificities [128]. The differences between normal, invasive ductal carcinoma and ductal carcinoma *in situ* of breast tissue were analyzed in sixty cases by Raman spectroscopy [129]. Kerr-gated Raman spectroscopy enables to probe nondestructively calcifications buried within breast tissue [130]. The method used a picosecond pulsed laser combined with a fast temporal gating of Raman scattered light to collect spectra from a specific depth within scattering media. Spectra characteristic of both hydroxyapatite and calcium oxalate were obtained at depth up to 0.96 mm. As calcium oxalate is associated with benign lesions and hydroxyapatite is found mainly in proliferative lesions including carcinoma, the detection of calcification can indicate the disease state.

7.5.2.5 Skin

According to a review [3], Raman spectroscopic applications to skin had the highest frequency from 1996 to 2006 in the biomedical field, probably because it is the largest human organ and its surface location makes it easily accessible. Applications to skin biology, pharmacology, and disease diagnosis have been reported. Combined approaches of Raman imaging with FTIR imaging studied the permeation of lipids into skin [131, 132]. Confocal Raman microspectroscopy was applied to the stratum corneum, which is the skin's outermost layer, to obtain depth profiles of molecular concentration gradients and transdermal drug delivery up to 200 micrometers [133]. Its combination with the imaging modality confocal scanning laser microscopy provides optical sections of the skin without physically dissecting the tissue [134].

Skin cancers, that include squamous cell carcinoma (SCC), malignant melanoma and basal cell carcinoma (BCC), are the cancers with the highest incidence worldwide. Understanding the molecular, cellular, and tissue changes that occur during skin carcinogenesis is central to cancer research in dermatology. As for many other tissues, vibrational spectroscopy has been used to evaluate these changes [135]. BCC is the most common cancer of the skin. Raman images were acquired from fifteen sections of BCC and compared with histopathology [136]. In this sample set, 100% sensitivity and 93% selectivity were demonstrated. Melanoma, pigmented nevi, BCC, seborrheic keratoses, and normal skin were studied by Raman spectroscopy [137]. The sensitivity and specificity of an artificial neural network classification for diagnosis of melanoma were 85% and 99%, respectively. Nonmelanoma skin cancers including BCC and SCC was diagnosed in nineteen patients *in vivo* using a portable Raman system with a handheld probe [138].

7.5.3 SERS detection of tissue-specific antigens

Immunohistochemistry, immunofluorescence, and chemiluminescence are widely used in pathology and oncology to detect cellular or tissue antigens via specific antigen-antibody interactions. Raman microspectroscopy was combined with immunohistochemistry to detect antigens in tissue specimens [139]. The detection of antigens was achieved by surface-enhanced Raman scattering (SERS) from aromatic Raman labels covalently linked to the corresponding antibody. As an example of this new Raman technique, prostate-specific antigen detection in epithelial tissue from patients with prostate carcinoma was demonstrated. The SERS labels enhance the Raman intensities by several orders of magnitude which significantly increased the sensitivity and decreased the required acquisition time. Compared with the detection of antigen-specific labels by fluorescence spectroscopy, SERS offers advantages for the stability of the labels, resistance to photobleaching and quenching, long-wavelength excitation of multiple labels with a single excitation source. The SERS labels are based on functionalized metal nanoparticles that do not tend to bleach like most fluorescence labels. The intrinsic small linewidths of SERS signals of the labels provide a molecular fingerprint that enables the simultaneous detection of multiple different labels using multivariate algorithms for feature extraction [140]. Extending this approach to immuno-Raman microspectroscopic imaging will allow assessing the distribution of antigens in cells and tissues. Composite organic-inorganic nanoparticles were developed as Raman labels for tissue analysis [141]. The report described the first steps toward application of these nanoparticles to the detection of proteins in human tissue. Two analytes were detected simultaneously using two different nanoparticles in a direct binding assay. In a subsequent report of this group the detection of the prostate-specific antigen in formalin-fixed paraffin embedded prostate tissue sections by SERS nanoparticles and by fluorescence labels were compared [142]. Staining accuracy (e.g., the ability to correctly identify prostate specific antigen expression in epithelial cells) was somewhat less for SERS than fluorescence. However, SERS provided signal intensities comparable to fluorescence, and good intra-, inter- and lot-to-lot consistencies. Overall, both detection reagents possess similar performance with tissue sections, supporting the further development of Raman probes for this application. The use of SERS nanoparticles as a new detection modality for cancer diagnosis was recently reviewed [143]. Very recently, biocompatible and nontoxic nanoparticles for *in vivo* tumor targeting and detection based on SERS have been described [144].

7.5.4 CARS for medical tissue imaging

CARS microscopy is useful for more than imaging unstained cells in culture dishes. Strong epi-CARS signals (see section 7.4.5) offer attractive solutions for medical tissue imaging and enable to probe *in vivo* biomaterials with chemical selectivity. An example is given in Figure 7.11 for brain tissue.

Video rate epi-CARS imaging and spectroscopy of lipid-rich tissue structures in the skin of a live mouse was demonstrated at subcellular resolution, including sebaceous glands, corneocytes, and adipocytes [146]. Axonal myelin in spinal tissue under physiological conditions was probed by both forward and epi-detected CARS [147]. In the same paper, simultaneous CARS imaging of myelin and two-photon excitation fluorescence imaging of intra- and extra-axonal Ca^{2+} was demonstrated. Sciatic nerve tissue and surrounding adipocytes were studied *in vivo* by CARS imaging [148]. Due to the limited penetration depth of CARS (approximately 100 μm), a minimally invasive surgery was performed in this study to open the skin. Recently, penetration of CARS microscopy was extended to 5 mm depth from the surface using a miniature objective lens [149]. This technique was applied to CARS imaging of a rat spinal cord with a minimal surgery on the vertebra. CARS imaging of myelin degradation revealed a calcium dependent pathway [150]. Normal brain structures and primary glioma were identified in fresh unfixed and unstained *ex vivo* brain tissue by CARS microscopy [145]. A combination of three nonlinear optical imaging techniques, CARS, two photon excitation fluorescence and second harmonic generation microscopy, allowed visualization of fibrous astroglial filaments in *ex vivo* spinal tissues [151]. CARS imaged adipocytes, tumor cells, and blood capillaries in mammary tumor tissue [152]. CARS microscopy has recently been applied to visualize molecular composition of arterial walls and atherosclerotic lesions [153, 154].

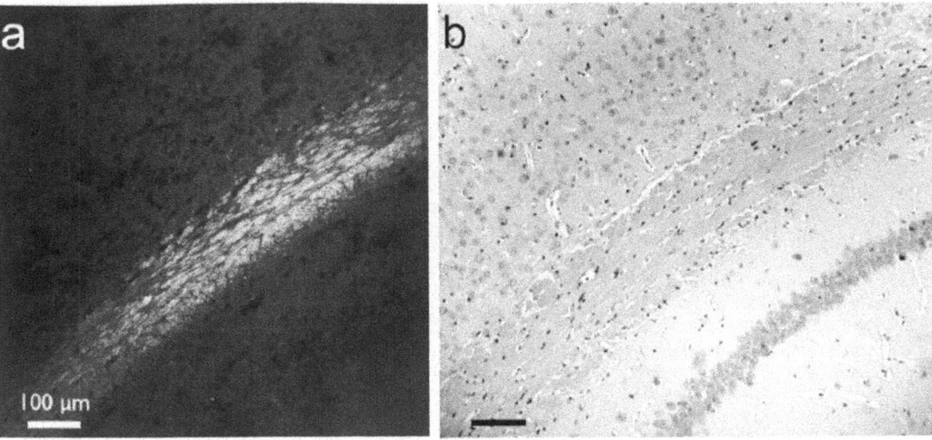

FIGURE 7.11: Single CARS image. The pump and Stokes wavelengths were 816.7 and 1064 nm, respectively. H&E image of the same region of the same mouse brain. Note the structure observable in both images, from upper-left corner to the bottom right are the cortex, corpus callosum, oriens layer, and pyramidal layer [reprinted from [145] with permission].

7.6 Conclusions

Biological applications of CARS imaging are still in their infancy and will continue to grow for years to come. A prohibitive factor in CARS imaging remains the cost and complexity associated with synchronizing two pulsed lasers and effectively coupling them into a scanning microscope. The development of compact, cost-effective laser sources and alternatives to free-space coupling such as delivery by photonic crystal fibers [155] will make such a system more attractive to researchers in the biological sciences. The demonstrated capability of CARS microscopy to image live tissue is expected to stimulate interest in the development of CARS endoscopy for the *in vivo* diagnosis. Fundamental challenges for a wider application of CARS microscopy arise from the nonresonant background and the quadratic dependence on species concentration that hinders the detection of less abundant molecular species. To date, CARS microscopy has provided structural information of large cellular components from highly abundant molecular species such as lipids, proteins, DNA, and water. Advances in sensitivity for CARS are required to realize its potential to detect less abundant molecules. Linear Raman spectroscopy offers the advantage that full spectral information can easily be obtained at high spectral resolution. Such data are often required to detect subtle biochemical differences in cells and tissues. Furthermore, the technique is less expensive, more compact and fiber-optic probes are available for combination with endoscopy. Unless further technological advances improve the applicability of CARS microscopy, linear Raman and its nonlinear variants complement each other as previously shown for colon tissue [156]. Raman and CARS spectroscopy of cells and tissues that are expected to gain significance in the future were recently reviewed [157]. RRS and SERS offer enhancement of Raman signals from cells and tissues that is the key to improve the sensitivity and specificity for detection of cells and tissues. Both techniques require careful selection of the excitation wavelength and careful preparation, respectively. In summary, Raman-based spectroscopic methods have already been successfully applied to numerous bioanalytical problems, and the field is expected to grow in the future with further advances in instrumentation, preparation, and data classification algorithms.

Acknowledgment

C.K. highly acknowledges financial support of the European Union via the Europäischer Fonds für Regionale Entwicklung (EFRE) and the "Thüringer Kultusministerium (TKM)" (project: B714-07037).

References

[1] S. Mukamel, *Principles of Nonlinear Optical Spectroscopy*, 1st ed., Oxford University Press, Oxford, 1995.

[2] C.W. Freudiger, W. Min, B.G. Saar, S. Lu, G.R. Holtom, C. He, J.C. Tsai, J.X. Kang, X.S. Xie, "Label-free biomedical imaging with high sensitivity by stimulated Raman scattering microscopy," *Science* **322**, 1857–1861 (2008).

[3] C. Krafft, V. Sergo, "Biomedical applications of Raman and infrared spectroscopy to diagnose tissues," *Spectroscopy* **20**, 195–218 (2006).

[4] C. Krafft, G. Steiner, C. Beleites, R. Salzer, "Disease recognition by infrared and Raman spectroscopy," *J. Biophoton.* **2**, 13–28 (2008).

[5] R.J. Swain, M.M. Stevens, "Raman microspectroscopy for non-invasive biochemical analysis of single cells," *Biochem. Soc. Trans.* **35**, 544–549 (2007).

[6] L.G. Rodriguez, S.J. Lockett, G.R. Holtom, "Coherent anti-stokes Raman scattering microscopy: a biological review," *Cytometry A* **69**, 779–791 (2006).

[7] J.X. Cheng, H. Wang, T.T. Le, Y. Fu, T.B. Huff, H. Wang, "Chasing lipids in health and diseases by CARS microscopy," *CACS Communications*, 13–18 (2007).

[8] Z.Q. Tian, "Surface-enhanced Raman spectroscopy: advancements and applications," *J. Raman Spectrosc.* **36**, 466–470 (2005).

[9] S. Nie, S.R. Emory, "Probing single molecules and single nanoparticles by surface-enhanced Raman scattering," *Science* **275**, 1102–1106 (1997).

[10] R. Aroca, *Surface-Enhanced Vibrational Spectroscopy*, John Wiley & Sons, Chichester, 2006.

[11] T. Deckert-Gaudig, E. Bailo, V. Deckert, "Perspectives for spatially resolved molecular spectroscopy – Raman on the nanometer scale," *J. Biophoton.* **1**, 377–389 (2008).

[12] A. Zumbusch, G.R. Holtom, X.S. Xie, "Three-dimensional vibrational imaging by coherent anti-Stokes Raman scattering," *Phys. Rev. Lett.* **82**, 4142–4145 (1999).

[13] A. Volkmer, L.D. Book, X.S. Xie, "Time-resolved coherent anti-Stokes Raman scattering microscopy: Imaging based on Raman free induction decay," *Appl. Phys. Lett.* **80**, 1505–1507 (2002).

[14] J.X. Cheng, L.D. Book, X.S. Xie, "Polarization coherent anti-Stokes Raman scattering microscopy," *Opt. Lett.* **26**, 1341–1343 (2001).

[15] J.X. Cheng, X.S. Xie, "Coherent anti-Stokes Raman scattering microscopy: instrumentation, theory, and applications," *J. Phys. Chem. B* **108**, 827–840 (2004).

[16] C.L. Evans, E.O. Potma, X.S. Xie, "Coherent anti-Stokes Raman scattering spectral interferometry: determination of the real and imaginary components of nonlinear susceptibility $\chi(3)$ for vibrational microscopy," *Opt. Lett.* **29**, 2923–2925 (2004).

[17] F. Ganikhanov, C.L. Evans, B.G. Saar, X.S. Xie, "High-sensitivity vibrational imaging with frequency modulation coherent anti-Stokes Raman scattering (FM CARS) microscopy," *Opt. Lett.* **31**, 1872–1874 (2006).

[18] A. Volkmer, J.X. Cheng, X.S. Xie, "Vibrational imaging with high sensitivity via epidetected coherent anti-Stokes Raman scattering microscopy," *Phys. Rev. Lett.* **87**, 023901–023904 (2001).

[19] D. Akimov, S. Chatzipapadopoulos, T. Meyer, N. Tarcea, B. Dietzek, M. Schmitt, J. Popp, "Different contrast information obtained from CARS and nonresonant FWM images," *J. Raman Spectrosc.* **40** (8), 941–947(2009).

[20] A. Volkmer, "Vibrational imaging and microspectroscopies based on coherent anti-Stokes Raman scattering microscopy," *J. Phys. D: Appl. Phys.* **38**, R59–R81 (2005).

[21] J.X. Cheng, "Coherent anti-Stokes Raman scattering microscopy," *Appl. Spectrosc.* **61**, 197–208 (2007).

[22] S. Schlcker, M.D. Schaeberle, S.W. Huffman, I.W. Levin, "Raman microspectroscopy: a comparison of point, line, and wide-field imaging methodologies," *Anal. Chem.* **75**, 4312–4318 (2003).

[23] M.G. Shim, B.C. Wilson, "The effects of ex vivo handling procedures on the near-infrared Raman spectra of normal mammalian tissues," *Photochem. Photobiol.* **63**, 662–671 (1996).

[24] Z. Huang, A. McWilliams, S. Lam, J. English, I. McLean David, H. Lui, H. Zeng, "Effect of formalin fixation on the near-infrared Raman spectroscopy of normal and cancerous human bronchial tissues," *Int. J.Oncol.* **23**, 649–655 (2003).

[25] E.O. Faolain, M.B. Hunter, J.M. Byrne, P. Kelehan, H.A. Lambkin, H.J. Byrne, F.M. Lyng, "Raman spectroscopic evaluation of efficacy of current paraffin wax section dewaxing agents," *J. Histochem. Cytochem.* **53**, 121–129 (2005).

[26] C. Krafft, L. Neudert, T. Simat, R. Salzer, "Near infrared Raman spectra of human brain lipids," *Spectrochim. Acta A* **61**, 1529–1535 (2005).

[27] B. Prescott, W. Steinmetz, G.J. Thomas Jr., "Characterization of DNA structures by laser Raman spectroscopy," *Biopolymers* **23**, 235–256 (1984).

[28] D. Naumann, *Infrared Spectroscopy in Microbiology*, John Wiley & Sons Ltd, Chichester, 2000.

[29] K. Maquelin, C. Kirschner, L.P. Choo-Smith, N. van den Braak, H.P. Endtz, D. Naumann, G.J. Puppels, "Identification of medically relevant microorganisms by vibrational spectroscopy," *J. Microbiol. Methods* **51**, 255–271 (2002).

[30] K. Maquelin, L.-P.i. Choo-Smith, T. Van Vreeswijk, H.P. Endtz, B. Smith, R. Bennett, H.A. Bruining, G.J. Puppels, "Raman spectroscopic method for identification of clinically relevant microorganisms growing on solid culture medium," *Anal. Chem.* **72**, 12–19 (2000).

[31] P. Rösch, M. Harz, K.-D. Peschke, O. Ronneberger, H. Burkhardt, A. Schüle, G. Schmautz, M. Lankers, S. Hofer, H. Thiele, H.-W. Motzkus, J. Popp, "Online monitoring and identification of bio aerosols," *Anal. Chem.* **78**, 2163–2170 (2006).

[32] M. Harz, P. Rösch, K.D. Peschke, O. Ronneberger, H. Burkhardt, J. Popp, "Micro-Raman spectroscopic identification of bacterial cells of the genus *Staphylococcus* and dependence on their cultivation conditions," *Analyst* **130**, 1543–1550 (2005).

[33] K.S. Kalasinsky, T. Hadfield, A.A. Shea, V.F. Kalasinsky, M.P. Nelson, J. Neiss, A.J. Drauch, G.S. Vanni, P.J. Treado, "Raman chemical imaging spectroscopy reagentless detection and identification of pathogens: signature development and evaluation," *Anal. Chem.* **79**, 2658–2673 (2007).

[34] A. Tripathi, R.E. Jabbour, P.J. Treado, J.H. Neiss, M.P. Nelson, J.L. Jensen, A.P. Snyder, "Waterborne pathogen detection using Raman spectroscopy," *Appl. Spectrosc.* **62**, 1–9 (2008).

[35] K.C. Schuster, E. Urlaub, J.R. Gapes, "Single-cell analysis of bacteria by Raman microscopy: spectral information on the chemical composition of cells and on the heterogeneity in a culture," *J. Microbiol. Methods* **42**, 29–38 (2000).

[36] K.C. Schuster, I. Reese, E. Urlaub, J.R. Gapes, B. Lendl, "Multidimensional information on the chemical composition of single bacterial cells by confocal Raman microspectroscopy," *Anal. Chem.* **72**, 5529–5534 (2000).

[37] W.E. Huang, R.I. Griffiths, I.P. Thompson, M.J. Bailey, A.S. Whiteley, "Raman microscopic analysis of single microbial cells," *Anal. Chem.* **76**, 4452–4458 (2004).

[38] M. Romeo, S. Boydston-White, C. Matthus, M. Miljkovic, B. Bird, T. Chernenko, P. Lasch, M. Diem, "Infrared and Raman microspectroscopic studies of individual human cells," in: M. Diem, P.R. Griffiths, J.M. Chalmers (eds.), *Vibrational Spectroscopy for Medical Diagnosis*, John Wiley & Sons, Chichester, 2008, pp. 27–70.

[39] J.W. Chan, D.S. Taylor, T. Zwerdling, S.M. Lane, K. Ihara, T. Huser, "Micro-Raman spectroscopy detects individual neoplastic and normal hematopoietic cells," *Biophys. J.* **90**, 648–656 (2006).

[40] P.R. Jess, M. Mazilu, K. Dholakia, A.C. Riches, C.S. Herrington, "Optical detection and grading of lung neoplasia by Raman microspectroscopy," *Int. J. Cancer* **124**, 376–380 (2008).

[41] K. Ajito, C. Han, K. Torimitsu, "Detection of glutamate in optically trapped single nerve terminals by Raman spectroscopy," *Anal. Chem.* **76**, 2506–2510 (2004).

[42] I. Notingher, I. Bisson, A.E. Bishop, W.L. Randle, J.M. Polak, L.L. Hench, "In situ spectral monitoring of mRNA translation in embryonic stem cells during differentiation in vitro," *Anal. Chem.* **76**, 3185–3193 (2004).

[43] Y.S. Huang, T. Karashima, M. Yamamoto, H.O. Hamaguchi, "Molecular-level investigation of the structure, transformation, and bioactivity of single living fission yeast cells by time- and space-resolved Raman spectroscopy," *Biochemistry* **44**, 10009–10019 (2005).

[44] I. Notingher, S. Verrier, S. Haque, J.M. Polak, L.L. Hench, "Spectroscopic study of human lung epithelial cells (A549) in culture: living cells versus dead cells," *Biopolymers* **72**, 230–240 (2003).

[45] H.J. van Manen, Y.M. Kraan, D. Roos, C. Otto, "Single-cell Raman and fluorescence microscopy reveal the association of lipid bodies with phagosomes in leukocytes," *Proc. Natl. Acad. Sci. USA* **102**, 10159–10164 (2005).

[46] C. Matthus, S. Boydston-White, M. Miljkovic, M. Romeo, M. Diem, "Raman and infrared microspectral imaging of mitotic cells," *Appl. Spectrosc.* **60**, 1–8 (2006).

[47] P. Rsch, M. Harz, K.D. Peschke, O. Ronneberger, H. Burkhardt, J. Popp, "Identification of single eukaryotic cells with micro-Raman spectroscopy," *Biopolymers* **82**, 312–316 (2006).

[48] P.R.T. Jess, D.D.W. Smith, M. Mazilu, K. Dholakia, A.C. Riches, C.S. Herrington, "Early detection of cervical neoplasia by Raman spectroscopy," *Int. J. Cancer* **121**, 2723–2728 (2007).

[49] C. Krafft, T. Knetschke, R.H. Funk, R. Salzer, "Studies on stress-induced changes at the subcellular level by Raman microspectroscopic mapping," *Anal. Chem.* **78**, 4424–4429 (2006).

[50] M. Harz, M. Kiehntopf, S. Stckel, P. Rsch, T. Deufel, J. Popp, "Analysis of single blood cells for CSF diagnostics via a combination of fluorescence staining and micro-Raman spectroscopy," *Analyst* **133**, 1416–1423 (2008).

[51] C. Krafft, T. Knetschke, R.H. Funk, R. Salzer, "Identification of organelles and vesicles in single cells by Raman microspectroscopic mapping," *Vib. Spectrosc.* **38**, 85–93 (2005).

[52] N. Uzunbajakava, A. Lenferink, Y.M. Kraan, B. Willekens, G. Vrensen, J. Greve, C. Otto, "Nonresonant Raman imaging of protein distribution in single human cells," *Biopolymers* **72**, 1–9 (2003).

[53] N. Uzunbajakava, A. Lenferink, Y.M. Kraan, E. Volokhina, G. Vrensen, J. Greve, C. Otto, "Nonresonant confocal Raman imaging of DNA and protein distribution in apoptotic cells," *Biophys. J.* **84**, 3968–3981 (2003).

[54] C. Krafft, T. Knetschke, A. Siegner, R.H. Funk, R. Salzer, "Mapping of single cells by near infrared Raman microspectroscopy," *Vib. Spectrosc.* **32**, 75–83 (2003).

[55] Y. Takai, T. Masuko, H. Takeuchi, "Lipid structure of cytotoxic granules in living human killer T lymphocytes studied by Raman microspectroscopy," *Biochim. Biophys. Acta* **1335**, 199–208 (1997).

[56] C. Matthus, T. Chernenko, J.A. Newmark, C.M. Warner, M. Diem, "Label-free detection of mitochondrial distribution in cells by nonresonant Raman microspectroscopy," *Biophys. J.* **93**, 668–673 (2007).

[57] G.J. Puppels, F.F. de Mul, C. Otto, J. Greve, M. Robert-Nicoud, D.J. Arndt-Jovin, T.M. Jovin, "Studying single living cells and chromosomes by confocal Raman microspectroscopy," *Nature* **347**, 301–303 (1990).

[58] C. Matthus, A. Kale, T. Chernenko, V. Torchilin, M. Diem, "New ways of imaging uptake and intracellular fate of liposomal drug carrier systems inside individual cells, based on Raman microscopy," *Mol. Pharm.* **5**, 287–293 (2008).

[59] G.J. Puppels, J.H. Olminkhof, G.M. Segers-Nolten, C. Otto, F.F. De Mul, and J. Greve, "Laser irradiation and Raman spectroscopy of single living cells and chromosomes: sample degradation occurs with 514.5 nm but not with 660 nm laser light," *Exp. Cell Res.* **195**, 361–367 (1991).

[60] I. Notingher, S. Verrier, H. Romanska, A.E. Bishop, J.M. Polak, and L.L. Hench, "In situ characterisation of living cells by Raman spectroscopy," *Spectroscopy* **15**, 43 (2002).

[61] H.J. Van Manen, C. Morin, and C. Otto, "Resonance Raman microspectroscopy and imaging of hemoproteins in single leukocytes," in: P. Lasch, J. Kneipp (eds.), *Biomedical Vibrational Spectroscopy*, John Wiley & Sons Inc., Hoboken, 2008, pp. 153–179.

[62] B.R. Wood and D. McNaughton, "Resonant Raman scattering of heme molecules in cells and in the solid state," in: P. Lasch, J. Kneipp (eds.), *Biomedical Vibrational Spectroscopy*, John Wiley & Sons, Inc., Hoboken, 2008, pp. 181–208.

[63] B.R. Wood and D. McNaughton, "Resonance Raman spectroscopy of erythrocytes," in: M. Diem, P.R. Griffiths, J.M. Chalmers (eds.), *Vibrational Spectroscopy for Medical Diagnosis*, John Wiley & Sons, Chichester, 2008, pp. 261–309.

[64] B.R. Wood and D. McNaughton, "Micro-Raman characterization of high- and low-spin heme moieties within single living erythrocytes," *Biopolymers* **67**, 259–262 (2002).

[65] K. Ramser, K. Logg, M. Goksör, J. Enger, M. Käll, and D. Hanstorp, "Resonance Raman spectroscopy of optically trapped functional erythrocytes," *J. Biomed. Opt.* **9**, 593–600 (2004).

[66] B.R. Wood, S.J. Langford, B.M. Cooke, F.K. Glenister, J. Lim, and D. McNaughton, "Raman imaging of hemozoin within the food vacuole of *Plasmodium falciparum* trophozoites," *FEBS Letters* **554**, 247–252 (2003).

[67] T. Frosch, S. Koncarevic, L. Zedler, M. Schmitt, K. Schenzel, K. Becker, and J. Popp, "In situ localization and structural analysis of the malaria pigment hemozoin," *J. Phys. Chem. B* **111**, 11047–11056 (2007).

[68] A. Bonifacio, S. Finaurini, C. Krafft, S. Parapini, D. Taramelli, and V. Sergo, "Spatial distribution of heme species in erythrocytes infected with Plasmodium falciparum by use of resonance Raman imaging and multivariate analysis," *Anal. Bioanal. Chem.* **392**, 1277–1282 (2008).

[69] H.J. van Manen, N. Uzunbajakava, R. van Bruggen, D. Roos, and C. Otto, "Resonance Raman imaging of the NADPH oxidase subunit cytochrome b558 in single neutrophilic granulocytes," *J. Am. Chem. Soc.* **125**, 12112–12113 (2003).

[70] R. Ptzold, M. Keuntje, K. Theophile, J. Mueller, E. Mielcarek, A. Ngezahayo, and A. Anders-von Ahlften, "In situ mapping of nitrifiers and anammox bacteria in microbial aggregates by means of confocal resonance Raman microscopy," *J. Microbiol. Meth.* **72**, 241–248 (2008).

[71] T.C. Bakker Schut, G.J. Puppels, Y.M. Kraan, J. Greve, L.L. van der Maas, and C.G. Figdor, "Intracellular carotenoid levels measured by Raman microspectroscopy: comparison of lymphocytes from lung cancer patients and healthy individuals," *Int. J. Cancer* **74**, 20–25 (1997).

[72] G.J. Puppels, H.S. Garritsen, J.A. Kummer, and J. Greve, "Carotenoids located in human lymphocyte subpopulations and natural killer cells by Raman microspectroscopy," *Cytometry* **14**, 251–256 (1993).

[73] K. Gaus, P. Rösch, R. Petry, K.-D. Peschke, O. Ronneberger, H. Burkhardt, K. Baumann, and J. Popp, "Classification of lactic acid bacteria with UV-resonance Raman spectroscopy," *Biopolymers* **82**, 286–290 (2006).

[74] U. Neugebauer, U. Schmid, K. Baumann, U. Holzgrabe, W. Ziebuhr, S. Kozitskaya, W. Kiefer, M. Schmitt, and J. Popp, "Characterization of bacterial growth and the influence of antibiotics by means of UV resonance Raman spectroscopy," *Biopolymers* **82**, 306–311 (2006).

[75] U. Neugebauer, U. Schmid, K. Baumann, W. Ziebuhr, S. Kozitskaya, V. Deckert, M. Schmitt, and J. Popp, "Towards a detailed understanding of bacterial metabolism: spectroscopic characterization of *Staphylococcus epidermidis*," *ChemPhysChem* **8**, 124–137 (2007).

[76] U. Neugebauer, U. Schmid, K. Baumann, W. Ziebuhr, S. Kozitskaya, U. Holzgrabe, M. Schmitt, and J. Popp, "The influence of fluoroquinolone drugs on the bacterial growth of *S. epidermidis* utilizing the unique potential of vibrational spectroscopy," *J. Phys. Chem. A* **111**, 2898–2906 (2007).

[77] J. Kneipp, H. Kneipp, K. Kneipp, M. McLaughlin, and D. Brown, "Surface-enhanced Raman scattering for investigations of eukaryotic cells," in: P. Lasch, J. Kneipp (eds.), *Biomedical Vibrational Spectroscopy*, John Wiley & Sons, Inc., Hoboken, 2008, pp. 243–261.

[78] R.M. Jarvis, R. Goodacre, "Characterisation and identification of bacteria using SERS," *Chem. Soc. Rev.* **37**, 931–936 (2008).

[79] K. Hering, D. Cialla, K. Ackermann, T. Dörfer, R. Möller, H. Schneideweind, R. Mattheis, W. Fritzsche, P. Rösch, and J. Popp, "SERS: a versatile tool in chemical and biochemical diagnostics," *Anal. Bioanal. Chem.* **390**, 113–124 (2008).

[80] K. Kneipp, A.S. Haka, H. Kneipp, K. Badizadegan, N. Yoshizawa, C. Boone, K.E. Shafer-Peltier, J.T. Motz, R.R. Dasari, and M.S. Feld, "Surface-enhanced Raman spectroscopy in single living cells using gold nanoparticles," *Appl. Spectrosc.* **56**, 150–154 (2002).

[81] K.R. Ackermann, T. Henkel, J. Popp, "Quantitative online detection of low-concentrated drugs via a SERS microfluidic system," *Chem. Phys. Chem.* **8**, 2665–2670 (2007).

[82] J. Kneipp, H. Kneipp, W.L. Rice, K. Kneipp, "Optical probes for biological applications based on surface-enhanced Raman scattering from indocyanine green on gold nanoparticles," *Anal. Chem.* **77**, 2381–2385 (2005).

[83] L. Zeiri, B.V. Bronk, Y. Shabtai, J. Czege, S. Efrima, "Silver metal induced surface enhanced Raman of bacteria," *Colloid Surf. A: Physicochem. Eng. Asp.* **208**, 357–362 (2002).

[84] W.R. Premasiri, D.T. Moir, M.S. Klempner, N. Krieger, G. Jones, II, L.D. Ziegler, "Characterization of the surface enhanced Raman scattering (SERS) of Bacteria," *J. Phys. Chem. B* **109**, 312–320 (2005).

[85] U. Neugebauer, P. Rsch, M. Schmitt, J. Popp, C. Julien, A. Rasmussen, C. Budich, V. Deckert, "On the way to nanometer-sized information of the bacterial surface by tip-enhanced Raman spectroscopy," *ChemPhysChem* **7**, 1428–1430 (2006).

[86] C. Budich, U. Neugebauer, J. Popp, V. Deckert, "Cell wall investigations utilizing tip-enhanced Raman scattering," *J. Microsc.* **229**, 533–539 (2008).

[87] J.X. Cheng, Y.K. Jia, G. Zheng, X.S. Xie, "Laser-scanning coherent anti-Stokes Raman scattering microscopy and applications to cell biology," *Biophys. J.* **83**, 502–509 (2002).

[88] T. Hellerer, A. Enejder, A. Zumbusch, "Spectral focusing: high spectral resolution spectroscopy with broad-bandwidth laser pulses," *Appl. Phys. Lett.* **85**, 25–27 (2004).

[89] E.O. Potma, X.S. Xie, "Detection of single lipid bilayers with coherent anti-Stokes Raman scattering (CARS) microscopy," *J. Raman Spectrosc.* **34**, 642–650 (2004).

[90] X. Nan, J.X. Cheng, X.S. Xie, "Vibrational imaging of lipid droplets in live fibroblast cells with coherent anti-Stokes Raman scattering microscopy," *J. Lipid Res.* **44**, 2202–2208 (2003).

[91] X. Nan, E.O. Potma, X.S. Xie, "Nonperturbative chemical imaging of organelle transport in living cells with coherent anti-Stokes Raman scattering microscopy," *Biophys. J.* **91**, 728–735 (2006).

[92] O. Burkacky, A. Zumbusch, C. Brackmann, A. Enejder, "Dual-pump coherent anti-Stokes-Raman scattering microscopy," *Opt. Lett.* **31**, 3656–3658 (2006).

[93] T. Hellerer, C. Axng, C. Brackmann, P. Hillertz, M. Pilon, A. Enejder, "Monitoring of lipid storage in aenorhabditis elegans using coherent anti-Stokes Raman scattering (CARS) microscopy," *PNAS* **104**, 14658–14663 (2007).

[94] H.A. Rinia, K.N. Burger, M. Bonn, M. Mller, "Quantitative label-free imaging of lipid composition and packing of individual cellular lipid droplets using multiplex CARS microscopy," *Biophys. J.* **95**, 4908–4914 (2008).

[95] E.O. Potma, W.P. de Boeij, P.J. van Haastert, D.A. Wiersma, "Real-time visualization of intracellular hydrodynamics in single living cells," *Proc. Natl. Acad. Sci. USA* **98**, 1577–1582 (2001).

[96] E.O. Potma, X.S. Xie, "Detection of single lipid bilayers with coherent anti-Stokes Raman scattering (CARS) microscopy," *J. Raman Spectrosc.* **34**, 642–650 (2003).

[97] H.W. Wang, N. Bao, T.L. Thuc, C. Lu, J.X. Cheng, "Microfluidic CARS cytometry," *Opt. Express* **16**, 5782–5789 (2008).

[98] J.X. Cheng, A. Volkmer, L.D. Book, X.S. Xie, "Multiplex coherent anti-Stokes Raman scattering microspectroscopy and study of lipid vesicles," *J. Phys. Chem. B* **106**, 8493–8498 (2002).

[99] J.X. Cheng, A. Volkmer, L.D. Book, X.S. Xie, "An epi-detected coherent anti-Stokes Raman scattering (E-CARS) microscope with high spectral resolution and high sensitivity," *J. Phys. Chem. B* **105**, 1277–1280 (2001).

[100] S.O. Konorov, C.H. Glover, J.M. Piret, J. Bryan, H.G. Schulze, M.W. Blades, R.F. Turner, "In situ analysis of living embryonic stem cells by coherent anti-stokes Raman microscopy," *Anal. Chem.* **79**, 7221–7225 (2007).

[101] A. Carden, M.D. Morris, "Application of vibrational spectroscopy to the study of mineralized tissues (review)," *J. Biomed. Opt.* **5**, 259–268 (2000).

[102] J.A. Timlin, A. Carden, M.D. Morris, J.F. Bonadio, C.E. Hoffler, K. Kozloff, S.A. Goldstein, "Spatial distribution of phosphate species in mature and newly generated mammalian bone by hyperspectral Raman imaging," *J. Biomed. Opt.* **4**, 28–34 (1999).

[103] A. Carden, R.M. Rajachar, M.D. Morris, D.H. Kohn, "Ultrastructural changes accompanying the mechanical deformation of bone tissue: a Raman imaging study," *Calcif. Tissue Int.* **72**, 166–175 (2003).

[104] K. Golcuk, G.S. Mandair, A.F. Callender, N. Sahar, D.H. Kohn, M.D. Morris, "Is photobleaching necessary for Raman imaging of bone tissue using a green laser?" *Biochim. Biophys. Acta* **1758**, 868–873 (2006).

[105] C.P. Tarnowski, M.A. Ignelzi, W. Wang, J.M. Taboas, S.A. Goldstein, M.D. Morris, "Earliest mineral and matrix changes in force-induced musculoskeletal disease as revealed by Raman microspectroscopic imaging," *J. Bone Miner. Res.* **19**, 64–71 (2004).

[106] N.J. Crane, M.D. Morris, M.A. Ignelzi, G. Yu, "Raman imaging demonstrates FGF2-induced craniosynostosis in mouse calvaria," *J. Biomed. Opt.* **10** (3), 031119-1–8 (2005).

[107] J.A. Timlin, A. Carden, M.D. Morris, R.M. Rajachar, D.H. Kohn, "Raman spectroscopic imaging markers for fatigue-related microdamage in bovine bone," *Anal. Chem.* **72**, 2229–2236 (2000).

[108] B.W. de Jong, T.C. Schut, K. Maquelin, T. van der Kwast, C.H. Bangma, D.J. Kok, G.J. Puppels, "Discrimination between nontumor bladder tissue and tumor by Raman spectroscopy," *Anal. Chem.* **78**, 7761–7769 (2006).

[109] G. Shetty, C. Kendall, N. Shephard, N. Stone, H. Barr, "Raman spectroscopy: elucidation of biochemical changes in carcinogenesis of oesophagus," *Br. J. Cancer* **94**, 1460–1464 (2006).

[110] C. Kendall, N. Stone, N. Shepherd, K. Geboes, B. Warren, R. Bennett, H. Barr, "Raman spectroscopy, a potential tool for the objective identification and classification of neoplasia in Barrett's oesophagus," *J. Pathology* **200**, 602–609 (2003).

[111] N. Stone, C. Kendall, H. Barr, "Raman spectroscopy as a potential tool for early diagnosis of malignancies in esophageal and bladder tissues," in: M. Diem, P.R. Griffiths, J.M. Chalmers (eds.), *Vibrational Spectroscopy for Medical Diagnosis*, John Wiley and Sons Ltd., Chichester, 2008, pp. 203–230.

[112] N. Stone, C. Kendall, N. Shepherd, P. Crow, H. Barr, "Near-infrared Raman spectroscopy for the classification of epithelial pre-cancers and cancers," *J. Raman Spectrosc.* **33**, 564–573 (2002).

[113] C. Krafft, D. Codrich, G. Pelizzo, V. Sergo, "Raman and FTIR microscopic imaging of colon tissue: a comparative study," *J. Biophoton.* **1**, 154–169 (2008).

[114] S.W.E. Van de Poll, T.C. Bakker Schut, A. Van der Laarse, G.J. Puppels, "In situ investigation of the chemical composition of ceroid in human atherosclerosis by Raman spectroscopy," *J. Raman Spectrosc.* **33**, 544–551 (2002).

[115] A. Robichaux-Viehoever, E. Kanter, H. Shappell, D.D. Billheimer, H. Jones, A. Mahadevan-Jansen, "Characterization of Raman spectra measured in vivo for the detection of cervical dysplasia," *Appl. Spectrosc.* **61**, 986–993 (2007).

[116] S.K. Teh, W. Zheng, K.Y. Ho, M. Teh, K.G. Yeoh, Z. Huang, "Diagnosis of gastric cancer using near-infrared Raman spectroscopy and classification and regression tree techniques," *J. Biomed. Opt.* **13**, 034013 (2008).

[117] C. Krafft, S.B. Sobottka, G. Schackert, R. Salzer, "Near infrared Raman spectroscopic mapping of native brain tissue and intracranial tumors," *Analyst* **130** (3), 1070–1077 (2005).

[118] S. Koljenovic, L.P. Choo-Smith, T.C. Bakker Schut, J.M. Kros, H. van den Bergh, G.J. Puppels, "Discriminating vital tumor from necrotic tissue in human glioblastoma tissue samples by Raman spectroscopy," *Lab. Invest.* **82**, 1265–1277 (2002).

[119] C. Krafft, S.B. Sobottka, G. Schackert, R. Salzer, "Raman and infrared spectroscopic mapping of human primary intracranial tumors: a comparative study," *J. Raman Spectrosc.* **37**, 367–375 (2006).

[120] S. Koljenovic, T.C. Bakker Schut, A. Vincent, J.M. Kros, G.J. Puppels, "Detection of meningeoma in dura mater by Raman spectroscopy," *Anal. Chem.* **77**, 7958–7965 (2005).

[121] N. Amharref, A. Beljebbar, S. Dukic, L. Venteo, L. Schneider, M. Pluot, M. Manfait, "Discriminating healthy from tumor and necrosis tissue in rat brain tissue samples by Raman spectral imaging," *Biochim. Biophys. Acta, Biomembranes* **1768**, 2605–2615 (2007).

[122] C. Krafft, M. Kirsch, C. Beleites, G. Schackert, R. Salzer, "Methodology for fiber-optic Raman mapping and FT-IR imaging of metastases in mouse brains," *Anal. Bioanal. Chem.* **389**, 1133–1142 (2007).

[123] S. Koljenovic, T.C. Bakker Schut, J.P. van Meerbeek, A.P. Maat, S.A. Burgers, P.E. Zondervan, J.M. Kros, G.J. Puppels, "Raman microspectroscopic mapping studies of human bronchial tissue," *J. Biomed. Opt.* **9**, 1187–1197 (2004).

[124] C. Krafft, D. Codrich, G. Pelizzo, V. Sergo, "Raman mapping and FTIR imaging of lung tissue: congenital cystic adenomatoid malformation," *Analyst* **133**, 361–371 (2008).

[125] C. Krafft, D. Codrich, G. Pelizzo, V. Sergo, "Raman and FTIR imaging of lung tissue: bronchopulmonary sequestration," *J. Raman Spectrosc.* **46**, 141–149 (2008).

[126] K.E. Shafer-Peltier, A.S. Haka, M. Fitzmaurice, J. Crowe, J. Myles, R.R. Dasari, M.S. Feld, "Raman microspectroscopic model of human breast tissue: implications for breast cancer diagnosis in vivo," *J. Raman Spectrosc.* **33**, 552–563 (2002).

[127] J. Smith, C. Kendall, A. Sammon, J. Christie-Brown, N. Stone, "Raman spectral mapping in the assessment of axillary lymph nodes in breast cancer," *Technol. Cancer Res. Treat.* **2**, 327–332 (2003).

[128] A.S. Haka, K.E. Shafer-Peltier, M. Fitzmaurice, J. Crowe, R.R. Dasari, M.S. Feld, "Diagnosing breast cancer by using Raman spectroscopy," *Proc. Natl. Acad. Sci. USA* **102**, 12371–12376 (2005).

[129] S. Rehman, Z. Movasaghi, A.T. Tucker, S.P. Joel, J.A. Darr, A.V. Ruban, I.U. Rehman, "Raman spectroscopic analysis of breast cancer tissues: identifying differences between normal, invasive ductal carcinoma and ductal carcinoma in situ of the breast tissue," *J. Raman Spectrosc.* **38**, 1345–1351 (2007).

[130] R. Baker, P. Matousek, K.L. Ronayne, A.W. Parker, K. Rogers, N. Stone, "Depth profiling of calcifications in breast tissue using picosecond Kerr-gated Raman spectroscopy," *Analyst* **132**, 48–53 (2007).

[131] C. Xiao, D.J. Moore, C.R. Flach, R. Mendelsohn, "Permeation of dimyristoylphosphatidylcholine into skin: structural and spatial information from IR and Raman microscopic imaging," *Vib. Spectrosc.* **38**, 151–158 (2005).

[132] C. Xiao, D.J. Moore, M.E. Rerek, C.R. Flach, R. Mendelsohn, "Feasibility of tracking phospholipid permeation from IR and Raman microscopic imaging," *J. Invest. Dermatol.* **124**, 622–632 (2005).

[133] P.J. Caspers, G.W. Lucassen, E.A. Carter, H.A. Bruining, G.J. Puppels, "In vivo confocal Raman microspectroscopy of the skin: noninvasive determination of molecular concentration profiles," *J. Invest. Dermatol.* **116**, 434–442 (2001).

[134] P.J. Caspers, G.W. Lucassen, G.J. Puppels, "Combined in vivo confocal Raman spectroscopy and confocal microscopy of human skin," *Biophys. J.* **85**, 572–580 (2003).

[135] N.S. Eikje, K. Aizawa, Y. Ozaki, "Vibrational spectroscopy for molecular characterisation and diagnosis of benign, premalignant and malignant skin tumours," *Biotechnol. An. Rev.* **11**, 191–225 (2005).

[136] A. Nijssen, T.C. Bakker Schut, F. Heule, P.J. Caspers, D.P. Hayes, M.H.A. Neumann, G.J. Puppels, "Discriminating basal cell carcinoma from its surrounding tissue by Raman spectroscopy," *J. Invest. Dermatol.* **119**, 443–449 (2002).

[137] M. Gniadecka, P.A. Philipsen, S. Sigurdsson, S. Wessel, O.F. Nielsen, D.H. Christensen, J. Hercogova, K. Rossen, H.K. Thomsen, R. Gniadecka, L.K. Hansen, H.C. Wulf, "Melanoma diagnosis by Raman spectroscopy and neural networks: structure alteration in proteins and lipids in intact cancer tissue," *J. Invest. Dermatol.* **122**, 443–449 (2004).

[138] C.A. Lieber, S.K. Majumder, D.L. Ellis, D.D. Billheimer, A. Mahadevan-Jansen, "In vivo nonmelanoma skin cancer diagnosis using Raman microspectroscopy," *Lasers Surg. Med.* **40**, 461–467 (2008).

[139] S. Schlücker, B. Küstner, A. Punge, B. R., A. Marx, P. Ströbel, "Immuno-Raman microspectroscopy: *in situ* detection of antigens in tissue specimens by surface-enhanced Raman scattering," *J. Raman Spectrosc.* **37**, 719–721 (2006).

[140] Y.C. Cao, R. Jin, C.A. Mirkin, "Nanoparticles with Raman spectroscopic fingerprints for DNA and RNA detection," *Science* **297**, 1536–1540 (2002).

[141] L. Sun, K.B. Sung, C. Dentinger, B. Lutz, L. Nguyen, J. Zhang, H. Qin, M. Yamakawa, M. Cao, Y. Lu, A.J. Chmura, J. Zhu, X. Su, A.A. Berlin, S. Chan, B. Knudsen, "Composite organic-inorganic nanoparticles as Raman labels for tissue analysis," *Nano Lett.* **7**, 351–356 (2007).

[142] B. Lutz, C. Dentinger, L. Sun, L. Nguyen, J. Zhang, A.J. Chmura, A. Allen, S. Chan, B. Knudsen, "Raman nanoparticle probes for antibody-based protein detection in tissues," *J. Histochem. Cytochem.* **56**, 371–379 (2008).

[143] M.Y. Sha, H. Xu, S.G. Penn, R. Cromer, "SERS nanoparticles: a new optical detection modality for cancer diagnosis," *Nanomed.* **2**, 725–734 (2007).

[144] X. Qian, X.H. Peng, D.O. Ansari, Q. Yin-Goen, G.Z. Chen, D.M. Shin, L. Yang, A.N. Young, M.D. Wang, S. Nie, "In vivo tumor targeting and spectroscopic detection with surface-enhanced Raman nanoparticle tags," *Nat. Biotechnol.* **26**, 83–90 (2008).

[145] C.L. Evans, X. Xu, S. Kesari, X.S. Xie, S.T.C. Wong, G.S. Young, "Chemically-selective imaging of brain structures with CARS microscopy," *Optics Express* **15**, 12076–12087 (2007).

[146] C.L. Evans, E.O. Potma, M. Pouris'haag, D. Cote, C.P. Lin, X.S. Xie, "Chemical imaging of tissue *in vivo* with video-rate coherent anti-Stokes Raman scattering microscopy," *Proc. Natl. Acad. Sci. USA* **102**, 16807–16812 (2005).

[147] H. Wang, Y. Fu, P. Zickmund, R. Shi, J.X. Cheng, "Coherent anti-stokes Raman scattering imaging of axonal myelin in live spinal tissues," *Biophys. J.* **89**, 581–591 (2005).

[148] T.B. Huff, J.X. Cheng, "In vivo coherent anti-Stokes Raman scattering imaging of sciatic nerve tissue," *J. Microsc.* **225**, 175–182 (2007).

[149] H. Wang, T.B. Huff, F. Yan, Y.J. Kevin, J.X. Cheng, "Increasing the imaging depth of coherent anti-Stokes Raman scattering microscopy with a miniature microscope objective," *Opt. Lett.* **32**, 2212–2214 (2007).

[150] Y. Fu, H. Wang, T.B. Huff, R. Shi, J.X. Cheng, "Coherent anti-Stokes Raman scattering imaging of myelin degradation reveals a calcium-dependent pathway in lyso-PtdCho-induced demyelination," *J. Neurosci. Res.* **85**, 2870–2881 (2007).

[151] Y. Fu, H. Wang, R. Shi, J.X. Cheng, "Second harmonic and sum frequency generation imaging of fibrous astroglial filaments in ex vivo spinal tissues," *Biophys. J.* **92**, 3251–3259 (2007).

[152] T.T. Le, C.W. Rehrer, T.B. Huff, M.B. Nichols, I.G. Camarillo, J.X. Cheng, "Nonlinear optical imaging to evaluate the impact of obesity on mammary gland and tumor stroma," *Mol. Imaging* **6**, 205–211 (2007).

[153] T.T. Le, I.M. Langohr, M.J. Locker, M. Sturek, J.X. Cheng, "Label-free molecular imaging of atherosclerotic lesions using multimodal nonlinear optical microscopy," *J. Biomed. Opt.* **12**, 054007 (2007).

[154] H.W. Wang, T.T. Le, J.X. Cheng, "Label-free imaging of arterial cells and extracellular matrix using a multimodal CARS microscope," *Opt. Commun.* **7**, 1813–1822 (2008).

[155] H. Wang, T.B. Huff, J.X. Cheng, "Coherent anti-Stokes Raman scattering imaging with a laser source delivered by a photonic crystal fiber," *Opt. Lett.* **31**, 1417–1419 (2006).

[156] C. Krafft, A. Ramoji, C. Bielecki, N. Vogler, T. Meyer, D. Akimov, P. Rsch, M. Schmitt, B. Dietzek, I. Petersen, A. Stallmach, J. Popp, "A comparative Raman and CARS imaging study of colon tissue," *J. Biophoton.* **2** (5), 303–312 (2009).

[157] C. Krafft, B. Dietzek, J. Popp, "Raman and CARS spectroscopy of cells and tissues," *Analyst* **134** (6), 1046–1057, (2009).

8

Resonance Raman Spectroscopy of Human Skin for the In Vivo Detection of Carotenoid Antioxidant Substances

Maxim E. Darvin, Juergen Lademann

Center of Experimental and Applied Cutaneous Physiology (CCP), Department of Dermatology, Universitätsmedizin Charité Berlin, Germany

This chapter describes the possibility of noninvasive determination of carotenoid antioxidant substances in human skin by utilizing an optical method based on resonance Raman spectroscopy. The possibility of noninvasive measurements of all cutaneous carotenoids as well as the separate determination of the carotenoids beta-carotene and lycopene in the skin is presented.

Furthermore, the importance of carotenoids for the normal functioning of human organism is discussed. The kinetics of carotenoid substances in the human skin, subsequent to a diversity of influencing factors was investigated *in vivo*. It was shown that the level of carotenoids in human skin strongly correlates with specific lifestyle conditions, such as stress factors and the intake of dietary supplementations rich in carotenoids. Stress factors such as irradiations (UV, IR), fatigue, illness, smoking, and alcohol consumption gave rise to a decrease in the carotenoid levels of the skin, while a carotenoid-rich nutrition, based, for instance, on large amounts of fruit and vegetables, increased the carotenoid levels of the skin. Measured decreases occurred rapidly over the course of a few hours (depending on the influenced stress factors and their intensity), while the subsequent increases lasted usually for up to 3 days before levelling. During the summer and autumn months (study was performed in Germany, northern hemisphere), an increase in the average level of cutaneous carotenoids (so-called "seasonal increase") was determined to be 1.26-fold.

Moreover, a strong correlation between the level of cutaneous antioxidant lycopene and skin appearance regarding the quantity and the density of furrows and wrinkles was shown.

The results show that cutaneous carotenoids can serve as marker substances of the antioxidative potential in humans.

It was shown noninvasively with the use of the Raman microscopic method that carotenoids exhibit an inhomogeneous distribution in the skin. The highest concentration was measured in the upper cell layers of the epidermis, i.e., in the stratum corneum. The possible explanation for such an antioxidant distribution is presented. It was additionally shown that topical application of antioxidants clearly increases the cutaneous antioxidant concentration.

Medication of cancer patients with Doxorubicin is frequently accompanied with a side-effect, called palmar-plantar erythrodysesthesia syndrome (PPE). The reason for the development and manifestation of PPE is unclear. The most probable reason based on the free radical interaction with the skin, is discussed. The action of Doxorubicin-induced free radicals, which produced on the skin surface as well as in the epidermis, can be the main purpose for the development of PPE. It was demonstrated that topical application of a cream containing antioxidants, effectively decreased the oxidative action of free radicals on the skin surface, thus possessing a protective function and substantial decreasing of the PPE manifestation.

Key words: resonance Raman spectroscopy, skin, antioxidants, carotenoids, beta-carotene, lycopene, anti-aging

8.1 Introduction

The production of free radicals is a part of the cellular metabolism. On the one hand, free radicals are important for signalling processing in the human organism and for the destruction of viruses and bacteria. On the other hand, they may have a destructive influence. Free radicals are produced in the skin subsequent to internal stress factors, as well as by external sources, such as UV irradiation and environmental hazards. Oxidative stress occurs when the quantities of the produced radicals exceed a critical level. This gives rise to a degradation of the antioxidants and results in the suppression of the antioxidant defense system [1, 2]. Consequently, the defense system of the living cells is reduced, potentially involving the destruction and disorganization of living cells, which may lead to the development of severe diseases, such as cancer, cardiac infarction, arteriosclerosis, Alzheimer's, Parkinson's, etc., [3–5]. Thus, many factors are able to influence the production of free radicals in the organism, such as metabolism, stress, interaction with environmental hazards, and sun irradiation [6].

The antioxidant defense system of the human organism is based on the synergistic action of different antioxidant substances, for instance carotenoids, vitamins, enzymes, and others [7–9]. Carotenoids are powerful fat-soluble antioxidants, which effectively neutralize produced free radicals. Most of these antioxidants, including carotenoids, cannot be synthesized by the organism and therefore require to be supplemented by nourishment, rich in antioxidant substances. Environmental factors, in particular sun irradiation, can produce high amounts of free radicals in the human skin. As a result, high amounts of antioxidants are localized in the skin, which are presumed to neutralize the arising free radicals, in order to prevent a destructive action.

The determination of the concentration of the antioxidants in the human skin is highly important, in order to understand the process of free radical neutralization, the kinetics of the antioxidants under stress conditions, and premature aging. These factors are essential for the effective treatment and prevention of skin diseases, as well as for the development of new protection strategies. Investigation of the kinetics of antioxidants in the human skin and the process of free radical neutralization *in vivo* require the application of noninvasive methods.

In the past, due to the absence of alternative noninvasive measuring methods, high pressure liquid chromatography (HPLC) was used for the determination of carotenoids in biomedia. This method has various disadvantages, e.g., it is highly invasive, time-consuming, and expensive. Recently, a noninvasive spectroscopic method was introduced to measure carotenoid antioxidant substances *in vivo*. This method is based on resonance Raman spectroscopy and enables cutaneous *in vivo* measurements.

8.2 Production of Free Radicals in the Skin

Free radicals are produced by metabolic processes in the human organism and are important for signalling processing and for energy regulation procedures. Additionally, they can be regarded as a defense system against viruses and bacteria. If the concentration of the free radicals reaches a high critical level, they can damage the tissue on a cellular level. This is usually the case, when the skin is exposed to high doses of sun irradiation and environmental hazards. Also, stress factors like illness, fatigue, nicotine, and alcohol consumption are able to produce high amounts of free radicals in the human organism. On the cellular level, mitochondrial oxidative stress, based on the free radical reactions, plays an important role in the process of cell death [10, 11]. Among the variety of free radicals [12], which are produced *inter alia* in the skin, singlet oxygen possesses the highest oxidative activity compared with other free radicals. These highly reactive substances immediately interact with surrounding molecules, thus dissipating the excess of their free energy. The action of an abundant amount of free radicals is able to induce harmful oxidation effects on epidermal proteins, lipids and epidermal DNA, which can be a reason for premature aging and even the formation of skin cancer. Oxidative action of free radicals can damage living cells and their important constituents, as well as elastin and collagen fibers [13, 14]. This results in accelerated skin aging [15].

8.3 Antioxidative Potential of Human Skin

8.3.1 Different types of antioxidants measured in the human skin

The human skin exhibits a well-balanced antioxidant defense system, operating efficiently against the destructive action of free radicals produced in the skin. Protective effects of the cutaneous antioxidants are based on the balanced synergistic action of different antioxidant substances, such as carotenoids (beta-carotene, lycopene, lutein, zeaxanthin), vitamins (A, C, D, and E), enzymes (superoxide dismutase, catalase, glutathione peroxidase) and others (flavonoids, lipoic acid, uric acid, selenium, coenzyme Q10, etc.), which form the antioxidative potential of the human skin.

The human organism is not capable of synthesizing carotenoids and vitamins (A, C and E) as opposed to enzymes. Thus, carotenoids and most vitamins have to be ingested either via nourishment or topically.

The most prevalent cutaneous carotenoids are alpha-carotene, beta-carotene, gamma-carotene, sigma-carotene, lutein, zeaxanthin, and lycopene. Lycopene and the carotenes represent around 70% of the total carotenoid concentration in the human skin [16, 17]. The carotenoids were determined in the stratum corneum and in the epidermis of the skin by Lademann et al. A higher concentration in the upper layer of the stratum corneum in comparison to the epidermis was found [18].

8.3.2 Role of cutaneous carotenoids

Up to now, over 600 carotenoids have been identified and are distributed naturally in different combinations [19]. Carotenoids are abundant, for instance, in fruit and vegetables, in plants, algae, in the yolk of eggs and in milk.

Carotenoids are widely applied as a natural protection against the negative action of free radicals. Among the variety of radicals produced by environmental influences in the skin, the most important is the quenching of highly reactive singlet oxygen. Carotenoids represent powerful antioxidants, which effectively transform the excess of free radical energy into heat, thus essentially minimizing the effect of direct oxidation and, consequently, exert a defense influence [20, 21]. Moreover, among carotenoids, lycopene possesses the highest antioxidative activity, because of its high quenching rate constant [1, 22].

Thus, together with the other types of antioxidant substances, carotenoids form a protective system of living organisms against the negative action of free radicals. The human skin exhibits a substantial role as a boundary between the living organism and the environment. Therefore a high amount of different antioxidants including carotenoids is presented in the human skin.

8.4 Physicochemical Properties of Cutaneous Carotenoids

8.4.1 Antioxidative activity

The carotenoids are frequently used by nature in order to protect plants and living organisms against photo oxidative processes. Carotenoids serve as powerful antioxidants, which efficiently neutralize singlet molecular oxygen and peroxyl radicals [21]. As a result of the interaction between singlet oxygen and the carotenoid molecule, the oxygen yield ground state and the carotenoid molecule is excited into the triplet state. The excited carotenoid molecule quickly returns to its ground state dissipating energy in the form of heat to its surroundings. Among carotenoids, lycopene exhibits the highest antioxidative activity providing the highest physical quenching rate constant with singlet oxygen [22]. Carotenoids are able to repulse several attacks of free radicals before being destroyed [1, 2].

8.4.2 Optical absorption

Carotenoids like alpha-, beta-, gamma-, sigma-carotene, lutein, zeaxanthin, lycopene and their isomers, which are found at different concentrations in the human skin, have approximately the same absorption values in the blue spectral range (for example at 488 nm). In the green region of spectra (514.5 nm) only lycopene has high absorption values compared with other carotenoids [23, 24]. Figure 8.1 shows the absorption spectra of solutions of beta-carotene and lycopene in ethanol. The quantity of conjugated double bonds in the structure of carotenoid molecules constitutes their absorption properties. For instance, beta-carotene has 9 and lycopene 11 conjugated double bonds in their structure.

8.4.3 Solubility

Carotenoids are fat-soluble molecules, which are distributed differently among the lipoprotein classes (very low density lipoproteins, high density lipoproteins and most important of all, low density lipoproteins) in the human organism.

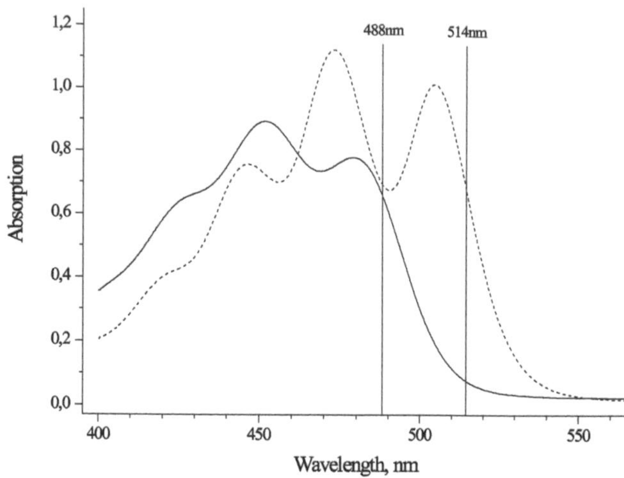

FIGURE 8.1: Absorption spectra of solutions of beta-carotene (solid line) and lycopene (dotted line) in ethanol (taken from [24]).

These molecules can also be solved in organic solvents, such as acetone, alcohol, ethyl ether, chloroform, and ethyl acetate [23].

8.5 Methods for the Detection of Cutaneous Carotenoids

8.5.1 High pressure liquid chromatography (HPLC)

High pressure liquid chromatography (HPLC) is a frequently used chemical method for the detection, separation and quantification of different compounds. Almost all authors, who have performed measurements of carotenoids in biomaterials, used the HPLC method for their detection [16, 25, 26].

Rodriguez-Amaya has described in detail four selected routine HPLC methods, used for the detection of carotenoids in different samples [23].

For industrial applications, mainly in the food industry, the supercritical fluid chromatography (SFC) is frequently applied for separation of carotenoids [27].

Nevertheless, both the HPLC and SFC methods have disadvantages, because they are time-consuming, expensive and highly invasive, which means that the biomaterial has to be removed from the organism for further investigations.

8.5.2 Reflection spectroscopy

Next to invasive chromatography, optical methods have been deployed for the noninvasive determination of carotenoids in the human skin [28].

Reflection measurements in the blue spectral range, which correspond to the maximum of the carotenoid absorption in the skin, enable the determination of the carotenoid concentration by means of reflection spectroscopy. The limitation of this method is a low selectivity between cu-

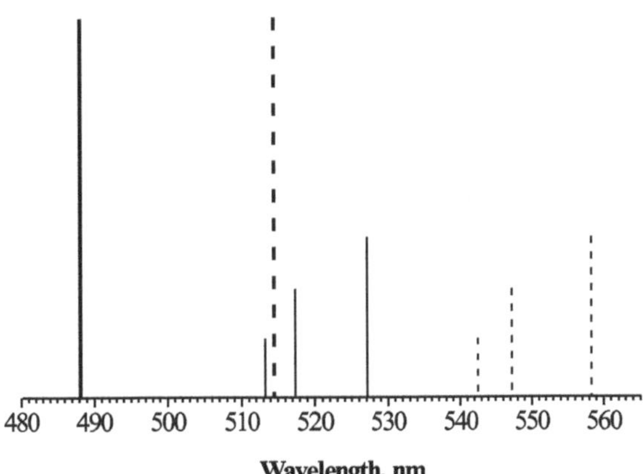

FIGURE 8.2: Scheme of two-wavelength resonance Raman excitations of carotenoid molecules. Solid (dotted) lines showed an excitation wavelength at 488 nm (514.5 nm) and three Raman corresponding lines shifted to the longer wavelengths.

taneous carotenoids and other substances present in the skin, which also absorbs light in the absorption range of carotenoids (~400–520 nm), such as melanin, haemoglobin, oxy-haemoglobin, and bilirubin. The separate determination of carotenoids in the skin is impossible with this method. Additionally, this method has a low sensitivity and, therefore, cannot be used for the measurement of low cutaneous carotenoid concentrations.

8.5.3 Raman spectroscopy

Carotenoids like alpha-, beta-, gamma-, sigma-carotene, lutein, zeaxanthin, lycopene and their isomers, which can be found at different concentrations in the skin, are Raman active molecules. Therefore, they can be easily determined by this noninvasive optical method. The efficiency of Raman scattering strongly depends on the excitation wavelengths and, as a result, on the absorption spectra of the investigated substances [29]. Therefore, to obtain a maximal efficacy of Raman scattering, resonance conditions of excitation were performed in this study.

Figure 8.2 shows the scheme of two-wavelength resonance Raman excitations of carotenoid molecules with wavelengths at 488 nm and 514.5 nm. Three prominent Raman lines at 513.2 nm, 517.3 nm, and 527.2 nm corresponded to the carotenoid molecules under excitation at 488 nm. Raman lines at 542.5 nm, 547.2 nm and 558.3 nm corresponded to the carotenoid molecules under excitation at 514.5 nm. A detailed explanation can be found in paragraph 8.6.3.

The three prominent Raman lines of carotenoid molecules, which correspond to the resonance excitation with wavelengths at 488 nm and 514.5 nm, are shifted to longer wavelengths. They can be easily detected by the utilization of the appropriate receiving filter. A detailed description of this method will be given below in the paragraph 8.6.

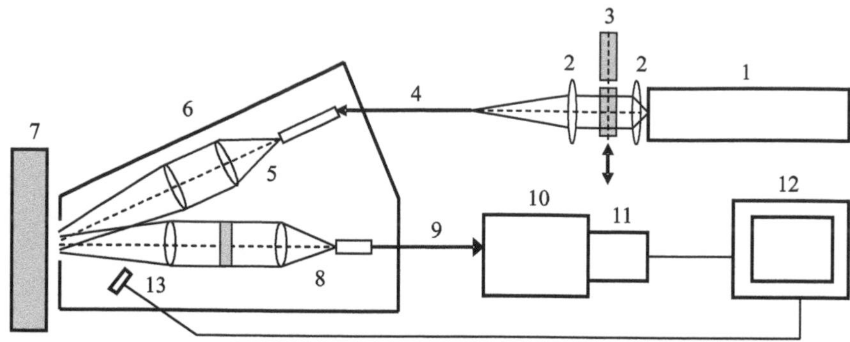

FIGURE 8.3: Scheme of the experimental setup: (1) Ar^+ laser, (2) lens system, (3) excitation filters, (4), (9) optical fibers, (5) excitation channel, (6) optical imaging system, (7) measuring object (skin), (8) receiving channel, (10) spectrograph, (11) CCD camera, (12) personal computer, (13) detector measured back-reflected light from the skin.

8.5.4 Comparison of the methods

The methods for the detection of carotenoids in human skin can be divided into invasive chromatography and noninvasive spectroscopy. In medicine, noninvasive methods are prioritized as being quick and more convenient for the volunteers and patients. Among the noninvasive methods, resonance Raman spectroscopy [29] has a number of advantages compared to reflection spectroscopy [28]. The Raman method allows a rapid and separate measurement of carotenoids in the skin even if their concentration is low. Moreover, this method is quick and, therefore, convenient. The expenditure of time necessary for one measurement varies between 1 to 5 seconds. In terms of sensitivity, precision and reproducibility of the measurements, the Raman method can be considered to be the most efficient procedure.

8.6 Resonance Raman Spectroscopy (RRS)

8.6.1 Setup for *in vivo* resonance Raman spectroscopy of cutaneous carotenoids

The Raman spectroscopic arrangement is presented in Figure 8.3. The Ar^+ laser (1), which operates on a multiline regime, is used as a source for optical excitation of carotenoids. The radiation is collimated by a lens system (2), which is located in front of the laser head and filters by means of specially designed interference filters (3), which transmit just one excitation wavelength at 488 nm or 514.5 nm depending on experimental conditions. The position of the filters is changed manually, as shown by the sign "↕" in the scheme. The filtered excitation light is then focused into an optical fiber (4). This optical fiber (4) is connected to an excitation channel (5) of the optical imaging system (6) where the light is expanded by a lens system and focused onto the skin (7). The power density on the skin surface is less than 60 mW/cm^2, which is well within safety standards.

The special geometry of the receiving channel (8) allows the intensity of back-reflected light from the skin to be minimized. Moreover, utilization of the specially designed filter transmits light at a range between 524 nm and 563 nm and cuts off other wavelengths. Utilization of the present filter enables the transmission of Raman lines at 527.2 nm and 558.3 (corresponding to both excited Raman lines at a shift of 1523 cm^{-1}) without losses (see Figure 8.2). Thus, the Raman signal from the skin is collected by a lens system, filtered and transferred into a fiber bundle (9), which is connected to a spectrograph (10). The obtained spectrum is recorded by a CCD camera (11) and analyzed on a computer (12) as a superposition of the high cutaneous fluorescence intensity and small Raman peaks, which corresponds to the concentration of carotenoids in the skin (see paragraph 8.4.4). The utilization of an additional channel for measurements of the back-reflected signal from the skin (13) allows measurements to be performed on volunteers with all types of skin.

For the purpose of minimizing the size of the measuring device, divided shifted Raman spectroscopy, using light-emitting diodes as a source of excitation, was presented [30]. The authors mentioned a repeatability of over 10% and excellent correlation with laser based resonance Raman spectroscopy.

8.6.2 Optimization of the setup parameters

8.6.2.1 Wavelength selection

In practice, the Raman measurements in cuvette and measurements *in vivo* on the skin are completely different. Regarding the measurements of carotenoids in the skin, resonance conditions of excitation should be performed, because of the very high fluorescence intensity of the skin in the measured wavelength region, which can effectively mask an informative Raman signal and complicate the determination of the carotenoids. Therefore, excitations in the blue-green optical range of the spectra, where the absorption maxima of carotenoids are localized (see Figure 8.1), are always utilized to increase the Raman response. An Ar^+ laser operating at 488 nm and 514.5 nm was used for the measurements of carotenoids in the skin [29]. Under an excitation at 488 nm, all carotenoids are excited resonantly at approximately the same efficacy, whereas an excitation at 514.5 nm, leads to a strong excitation of mainly lycopene (see Figure 8.1). The applied excitation conditions are close to optimal regarding the efficacy of separate excitations of lycopene and the remaining carotenoids. A detailed description can be found in the paragraphs 8.6.4, 8.6.5, and 8.6.6.

The penetration depths of the light at wavelengths of 488 nm and 514.5 nm in to the skin are about 200-250 μm and do not differ strongly [31]. Therefore, Raman measurements are performed in the epidermis and can partially affect the upper part of the dermis (papillary dermis). Additionally, taking into consideration that the distribution of carotenoids in the skin is inhomogeneous and has a strong gradient with a maximal concentration close to the skin surface (depth 4-8 μm) [18], it can be concluded that measurements of carotenoids were performed in the epidermis, while the possible measurement of the carotenoids in the blood is excluded.

8.6.2.2 Stability of the measurements and possible disturbances

The spread of the Raman values obtained from the same skin area of one individual averaged 7–10%. Therefore, the measurement shows a relatively high stability. On the contrary, interindividual differences can be significant after adjustment for age, gender, and diet intake estimates [29].

The reproducibility of the Raman measurements can also be influenced by inhomogeneous pigmentation, roughness and the microstructure of the skin. The size of such cutaneous pigmentation inhomogeneities is in the range of about 1–2 mm, whereas the typical roughness and microstructure inhomogeneities are smaller. Therefore, to average these influences, the excitation laser beam was expanded. It was detected experimentally that the inhomogeneities of irradiated skin area have a minor influence on the Raman measurements when the diameter of the laser spot exceeds 5 mm. The spread of the measured values constitutes 7–10% in this case. Thus, an excitation laser spot of 6.5 mm in diameter on the skin surface (area of $\sim$33 mm^2) was utilized to eliminate these negative influences [24]. The celerity of Raman measurements, performed within about 5 seconds, warrants that the movement of the body has no influence on the signal stability.

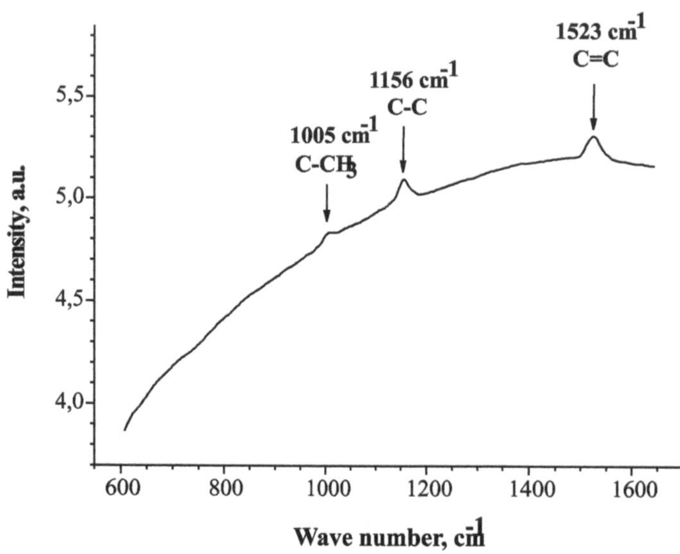

FIGURE 8.4: Typical Raman spectrum obtained from the human skin (taken from [24]) measured *in vivo* under an excitation at 514.5 nm. Excitation with Ar^+ laser, power density on the skin surface ~ 30 mW/cm^2.

8.6.2.3 Bleaching of fluorescence

During the measurements, the Raman signal shows a high stability (around 7–10%), whereas the fluorescence level decreases with time. This effect known as a bleaching of fluorescence [32], is caused by the discoloration of the cutaneous fluorescent centers. Among the endogenous skin fluorophores, the different forms of NAD, keratin located in the epidermis and dermal collagen and elastin, are the most promising [33].

The effect of fluorescence bleaching can be helpful for Raman measurements of volunteers with an extremely low level of carotenoids in the skin. For these volunteers, a slight Raman signal can be masked by a high fluorescence background. During the time of irradiation, fluorescence decreases substantially (up to 50% during 10 minutes), which enables the detection even of a low Raman signal on a large fluorescence background.

8.6.3 Typical RRS spectra of carotenoids obtained from the skin

Figure 8.4 shows a typical Raman spectrum of carotenoids, obtained from the human skin under resonance excitation conditions. As can be seen, three prominent Raman peaks can be well recognized in face of the high fluorescence background, which is produced by collagen, lipids, porphyrins, elastin, etc. The Raman lines at 1005 cm^{-1}, 1156 cm^{-1}, and 1523 cm^{-1} originate from the rocking motions of the methyl groups, from the carbon-carbon single bond and carbon-carbon double-bond stretch vibrations of the conjugated backbone, respectively.

The strong Raman peak, corresponding to the C=C vibration of the conjugated backbone of a carotenoid molecule at 1523 cm^{-1} was used for the measurement of carotenoid concentration. The Gauss approximation line was used to fit the Raman peak and calculate its intensity.

8.6.4 Measurements of the total amount of carotenoids in the skin

As discussed in paragraph 8.2.2, the cutaneous carotenoids alpha-, beta-, gamma-, sigma-carotene, lutein, zeaxanthin, lycopene and their isomers, have approximately the same absorption values in the blue region of the spectra (for example at 488 nm) [23].

Utilizing the excitation wavelength at 488 nm, all cutaneous carotenoids are excited with approximately the same Raman efficacy. This means that the intensity of the Raman peaks will reflect the total carotenoid concentration in the skin [24]. Thus, the total amount of cutaneous carotenoids can be determined by the use of the resonance Raman spectroscopic method under excitation in the blue range of the spectra, corresponding to the absorption maximum of carotenoids, for example, at 488 nm.

8.6.5 Selective detection of cutaneous beta-carotene and lycopene

The noninvasive method based on resonance Raman spectroscopy for the separate determination of beta-carotene and lycopene relative concentrations in the human skin was previously reported by Ermakov et al. [34]. As mentioned in paragraph 8.2.2, absorption spectra of carotenoids presented in the human skin are almost identical, with the exception of lycopene (see Figure 8.1). As a result of the different absorption values in the blue and the green ranges of the spectra, the Raman efficiencies are also different and correlate strongly with the absorption spectra of investigated molecules. These differences were used for the selective detection of the carotenoids beta-carotene and lycopene in human skin [29, 34, 35].

Taking into consideration the main approach that most of the carotenoids in the skin are beta-carotene and lycopene [17], the measured C=C vibration Raman line intensity at 1523 cm^{-1} (see Figures 8.2 and 8.4) under 488 and 514.5 nm excitations (I^{488} and I^{514}), is described by the following equations [29]:

$$I^{488} \sim P^{488}(\delta_{bC}^{488} N_{bC} + \delta_{L}^{488} N_{L}) \tag{8.1}$$

$$I^{514} \sim P^{514}(\delta_{bC}^{514} N_{bC} + \delta_{L}^{514} N_{L}) \tag{8.2}$$

where N_{bC} and N_L are the relative concentrations of beta-carotene and lycopene, P^{488} and P^{514} – are the powers of the 488 nm and 514.5 nm laser excitation lines, σ_{bC}^{488}, σ_{L}^{488}, σ_{bC}^{514}, σ_{L}^{514} are the resonance Raman scattering cross-sections for beta-carotene and lycopene under 488 nm and 514.5 nm excitations, respectively.

After the normalization to the laser power values, equations (8.1) and (8.2) can be transformed to the next equations:

$$I_n^{488} \sim (\delta_{bC}^{488} N_{bC} + \delta_{L}^{488} N_{L})(3) \tag{8.3}$$

$$I_n^{514} \sim (\delta_{bC}^{514} N_{bC} + \delta_{L}^{514} N_{L}) \tag{8.4}$$

where $I_n^{488(514)} = I^{488(514)} / P^{488(514)}$ is a measuring parameter.

The combination of the equations (8.3) and (8.4) gives an analytical expression for the ratio between the beta-carotene and lycopene relative concentrations:

$$\frac{N_{bC}}{N_L} = \frac{\delta_L^{488} - \delta_L^{514} \cdot r}{\delta_{bC}^{514} \cdot r - \delta_{bC}^{488}} \tag{8.5}$$

where $r = I_n^{488} / I_n^{514}$ is a measuring parameter.

The determination of the resonance Raman scattering cross-sections σ for beta-carotene and lycopene under 488 and 514.5 nm excitations were performed experimentally using solutions of beta-carotene and lycopene in ethanol.

The obtained values for σ can be used to simplify expression (8.5) as follows:

$$\frac{N_{bC}}{N_L} = \frac{0.95 - 0.44 \cdot r}{0.06 \cdot r - 1} \tag{8.6}$$

Combining the equations (8.3) and (8.6), expressions for the determination of relative concentrations of beta-carotene and lycopene can be obtained:

$$N_{bC} \sim I_n^{488} \cdot \frac{N_{bC}/N_L}{N_{bC}/N_L + 0.95} \tag{8.7}$$

$$N_L \sim I_n^{488} \cdot \frac{1}{N_{bC}/N_L + 0.95} \tag{8.8}$$

The absolute concentration of carotenoids was determined by calibration measurements using HPLC measurements based on the skin samples obtained from the surgery. The proportional coefficient K_{abs}, between the relative and absolute concentrations can be present as:

$$K_{abs} = \frac{N_{bC(L)}^{abs}}{N_{bC(L)}} = 2000 \frac{nmol \cdot \mathrm{sec}}{counts \cdot mW \cdot g} \tag{8.9}$$

Thus, the equations (8.7) and (8.8) can be presented as:

$$N_{bC}^{abs} = 2000 \cdot I_n^{488} \cdot \frac{N_{bC}/N_L}{N_{bC}/N_L + 0.95} \tag{8.10}$$

$$N_L^{abs} = 2000 \cdot I_n^{488} \cdot \frac{1}{N_{bC}/N_L + 0.95} \tag{8.11}$$

In vivo measurements under real conditions are performed on the skin, which contains not only beta-carotene and lycopene, but also other carotenoids. Taking into consideration the absorption spectra and the applied excitation wavelengths, the measurements allow a separation between the lycopene and other carotenoids present in the skin.

8.6.6 Measurements of cutaneous lycopene

The absorption spectrum of lycopene is different from that of other carotenoids, presented in human skin. Lycopene absorbs in the green range of the spectrum, while other cutaneous carotenoids such as lutein, zeaxanthin, as well as alpha-, gamma-, and sigma-carotenes have their absorption maxima in the blue spectral range (see Figure 8.1). Thus, using the green excitation wavelength at 514.5 nm, the strong Raman excitation of lycopene occurs whereas the contribution of the remaining carotenoids to the Raman signal is negligible. The numerical value of this contribution was obtained experimentally and is equal to 0.06 (see equation (8.6) in paragraph 8.6.5).

Taking into consideration the results obtained with HPLC [1, 16, 36], which show that lycopene present in the skin at high concentrations, it can be concluded that under the excitation at 514.5 nm, mainly lycopene contributes to the intensity of the Raman peak [24].

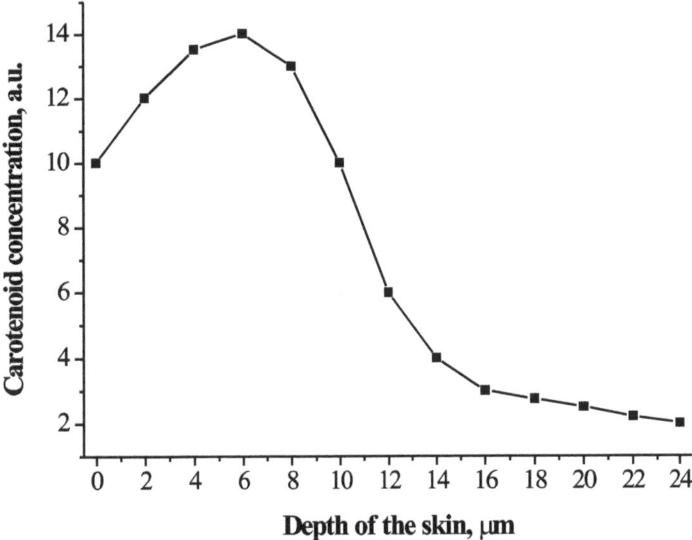

FIGURE 8.5: Distribution of carotenoids in the epidermis of the human skin. Average value, obtained for 3 volunteers.

8.7 Results Obtained by RRS *In Vivo*

8.7.1 Distribution of carotenoids in the human skin

Because carotenoids cannot be synthesized by human organism, they have to be ingested either by nourishment or by topical application. It was shown that the level of carotenoids in the skin of volunteers reflects their lifestyle. A carotenoid-rich nutrition, based on high amounts of fruit and vegetables, substantially increased the level of carotenoids in the human skin [37]. Additionally, the level of cutaneous carotenoids was found to be very high in the vegetarian group, whereas the smokers had an extremely low level of cutaneous carotenoids [29]. The carotenoids exhibit an inhomogeneous distribution in the skin. The highest concentration was measured by Raman microscopy in the upper cell layers of the stratum corneum close to the skin surface. The concentration of carotenoids decreased exponentially in deeper parts of the epidermis [18]. Figurc 8.5 shows the mean distribution of carotenoids in the epidermis of the human skin obtained for 3 volunteers.

The highest concentration of carotenoids was found 4-8 μm distant to the skin surface, whereas the upper 1–3 μm exhibited a slightly lower concentration, probably due to the naturally occurring photochemical degradation of carotenoids by ambient UV light and interaction with environment on the skin surface. Carotenoids can be detected in nearly all human tissues including the skin, but the most important storage is the liver and the adipose tissue. It was shown that the carotenoids beta-carotene and lycopene are variably distributed in the human skin, strongly depending on the body site and the volunteer. The concentration of the carotenoids beta-carotene and lycopene is higher in body sites with a high density of sweat glands [38], such as the palm, planta, and the forehead in comparison to the flexor forearm, back and the neck [35]. This observation is supported by the hypothesis that the carotenoids are probably transported to the skin surface via the sweat, which contains carotenoids [37]. This hypothesis is also confirmed by the detection of fat-soluble vitamin E on the skin surface, which flows out with the sweat [39]. The inhomogeneous distribution of carotenoids in the epidermis with the pronounced maximum close to the skin surface seems to represent further evidence that the carotenoids are delivered from inside the body out to the skin surface via the sweat. The continuously decreasing concentration from the skin surface towards the basal layer shows that carotenoids can penetrate with the sweat into the skin in the same manner as a topically applied formulation.

8.7.2 Stress factors, which decrease the carotenoid level in the skin

Different stress factors are always accompanied by the production of free radicals [40]. Free radicals produced under normal conditions are immediately neutralized by the antioxidative system of the human body. As a result of these interactions, free radicals are neutralized and carotenoid antioxidants are destroyed [1]. The amount of destroyed antioxidants characterizes the intensity of the stress factor and the amount of the neutralized free radicals.

8.7.2.1 Influence of UV irradiation

UV irradiation is a well-known stress factor for the human skin. On the one hand, irradiation of the skin with UV light is useful and essential for the synthesis of vitamin D in the skin and for the activation of the immune system. On the other hand, UV irradiation at high doses can be harmful for the skin due to the production of high amounts of free radicals, which can destroy the antioxidative system of human skin [41]. After the destruction of the antioxidative system, an interaction with living cells can give rise to their destruction and disorganization being the reason for the manifestation of strong diseases and even cancer [4].

Utilizing the noninvasive technique based on resonance Raman spectroscopy, the kinetics of the carotenoid concentration in the human skin, subsequent to irradiation with UV light, was investigated [42]. The skin of the forearm of healthy volunteers was irradiated with a UVB at a power density of 30 mJ/cm^2 during 100 seconds, which was sufficient for the formation of a light erythema. Whereas the lycopene concentration decreased relatively quickly from 0 to 30 minutes after UV irradiation of the skin, beta-carotene decreased after 30 up to 90 minutes. Different quenching rate constants for beta-carotene and lycopene in the reaction of neutralization of free radicals [1, 22] might represent a possible explanation. The degradation time characterizes the time interval between start of decrease and the achieving of the minimum value. The recovery time needed to establish the initial level, which was measured before irradiation, varies from 2 to 4 days depending on the volunteers.

A strong correlation between the individual carotenoid level of volunteers and their magnitude of destruction, which characterizes the quantity of carotenoids destroyed during the degradation time, was determined. Obtained correlation means that volunteers with a high individual carotenoid level results in the destruction of a smaller percentage of carotenoids in comparison to volunteers with a low carotenoid level under the influence of UV irradiation. Thus, volunteers offering higher levels of antioxidants in the skin have an additional defense against the action of free radicals induced by UV radiation.

8.7.2.2 Influence of IR irradiation

In vivo measurements revealed that IR irradiation can substantially reduce the concentration of the carotenoids beta-carotene and lycopene in human skin. IR irradiation was applied at a power density of 190 mW/cm^2 on the skin surface for 30 minutes and gave rise to a degradation of 27% of beta-carotene and 38% of lycopene on average for all volunteers. The recovery time, needed for leveling, was determined at 1–2 days, depending on the volunteers. The carotenoid reduction depended on the dose of applied irradiation and on its spectrum. It has been predicted that carotenoid degradation is related to the neutralization of free radicals in human skin, produced subsequent to IR irradiation [43, 44]. Taking the low energy of the IR photons into consideration, which is insufficient for the formation of free radicals, it can be expected that enzymatic processes in the human skin are involved in the energy transfer of IR irradiation into the molecules of the tissue.

Schroeder et al. showed that IR irradiation-induced side-effects, such as the upregulation of matrix metalloproteinase-1, which can be inhibited in human skin by the topical application of antioxidants [44]. Thus, topical application of antioxidants represents an effective photo-protective strategy and can be adopted, e.g., for the deceleration of premature skin aging, induced by IR irradiation.

IR irradiation of the human skin is a well-established, important, and widely utilized method in medicine and wellness. However, in medicine generally, IR irradiation is used only in cases of lesions and illness. Thus, the positive effects of medical treatment with IR irradiation strongly overcome the possible side-effects. Moreover, antioxidant-rich supplementation can be recommended to increase the initial level in the skin for better protection.

Taking the energy of the visible range photons into consideration, free radical production in human skin can also be predicted.

8.7.2.3 Influence of other stress factors

Despite the fact that it is difficult to measure stress directly, the possibility to measure the influence of a number of stress factors on the antioxidative potential of human skin has been presented. It was found in a one-year *in vivo* study that the level of cutaneous carotenoids of volunteers correlates with the stress factors influencing the organism. Investigations were performed daily, excluding holidays, on 10 volunteers, who were interviewed before each measurement regarding their lifestyle conditions, such as the uptake of carotenoid-rich foods and the influence of possible stress factors, i.e., illness, smoking, alcohol consumption, fatigue, etc. Results showed that illness, fatigue, working stress, alcohol consumption, and smoking reduced the level of carotenoids in the skin of volunteers, while the absence of stress parallel with the consumption of carotenoid-rich food gave rise to an increase in cutaneous carotenoids [37]. The intensity of obtained changes in the level of carotenoids seems to be correlated with the intensity of the influenced stress factor; however, the measurements on volunteers concerning the influence of stress factors were subjective.

8.7.3 Potential methods to increase the carotenoid level in the skin

Principally, there are two possibilities to increase the level of carotenoid antioxidants in the skin: either by systemic or by topical application.

Carotenoids, as well as other antioxidant substances cannot be synthesized by the human organism and have to be taken up by nutrition rich in these substances. Fruit and vegetables contain high amounts of naturally balanced antioxidants such as carotenoids and vitamins.

Raman measurements, performed noninvasively on the skin, showed a strong correlation between the quantity of ingested carotenoids and their level in the skin. One-time supplementation of carotenoid-rich products such as ketchup and tomato paste resulted in an increase in cutaneous carotenoid level. The increase was apparent on the day after supplementation [45, 46]. Supplementation of fruit and vegetables is accompanied by an increase in the carotenoid level, reflecting the lifestyle of volunteers [37]. Moreover, it was shown that vegetarians generally possess a higher level of carotenoids in their skin compared to other volunteers [35].

Within the one-year study, one of the volunteers showed a very low initial concentration of carotenoids in the skin. Over a period of 6 months, this volunteer increased his daily consumption of fruit and vegetables consistently and tried to exclude the influence of possible stress factors. This resulted in an almost twofold increase in the level of carotenoids in the skin at the end of the study, as can be seen from Figure 8.6. A further possibility to increase the level of carotenoids in the human skin represents the topical application of carotenoid containing formulations. Topically

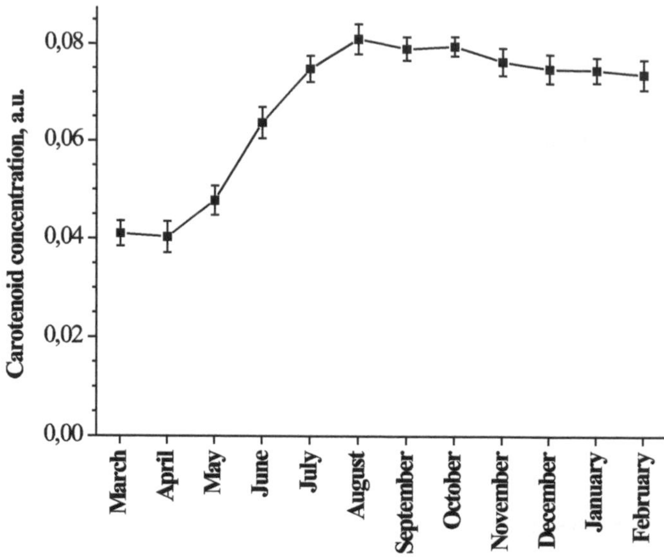

FIGURE 8.6: Influence of a carotenoid-rich diet on the level of carotenoids in the human skin.

applied antioxidants penetrate deeply into the skin down to approx. 24 μm (one hour after application) and increase their level in the layers lying close to the skin surface [18]. Thus, topically applied antioxidants are able to enrich the defense potential of the upper layers of the human skin. This is important for the defense of the skin against the action of environmental hazards and irradiations.

8.7.4 "Seasonal increase" of cutaneous carotenoids

The one-year study on human volunteers additionally revealed that the level of carotenoids in the skin is commonly higher during the summer and autumn months, and lower during the winter and spring months. Figure 8.7 shows average monthly values of the concentration of carotenoids in the skin for one volunteer, obtained during a 12-month period.

The measured, so-called "seasonal increase" was observed for all 10 volunteers participating in the study and was found to be statistically significant. The average "seasonal increase" of the carotenoid content in the skin was determined to be 1.26-fold [37].

The obtained seasonal variation of the carotenoid antioxidant level in the skin ("seasonal increase") was explained by a higher daily consumption of fruit and vegetables, during the summer and autumn months.

The "seasonal increase" can be considered to be an important physiological parameter of the skin, which shows a strengthened cutaneous defense function during these months, when the individuals are increasingly exposed to solar radiation.

8.7.5 Antioxidants and premature aging

The free radical theory of aging [15] is based on the accumulation of irreversible changes on the cell level caused by the oxidative action of free radicals produced in the organism. As a result, cutaneous collagen and elastin fibers can be damaged [13, 14], which gives rise to the appearance of furrows and wrinkles on the skin surface and to premature skin aging.

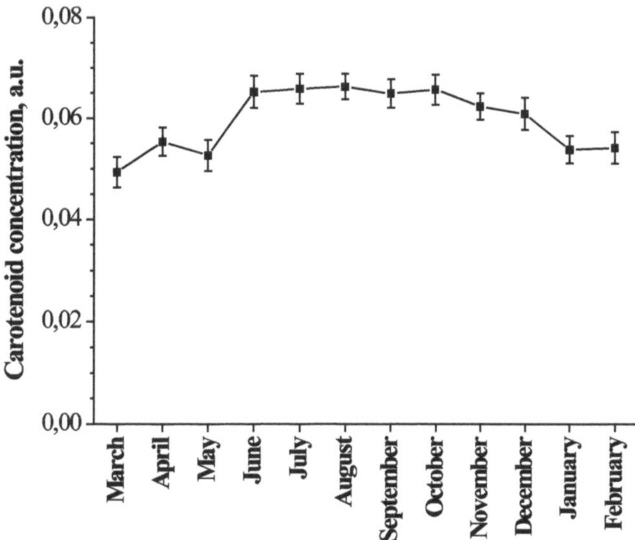

FIGURE 8.7: Average monthly values of the carotenoid concentration in the skin for one volunteer during a 12-month period.

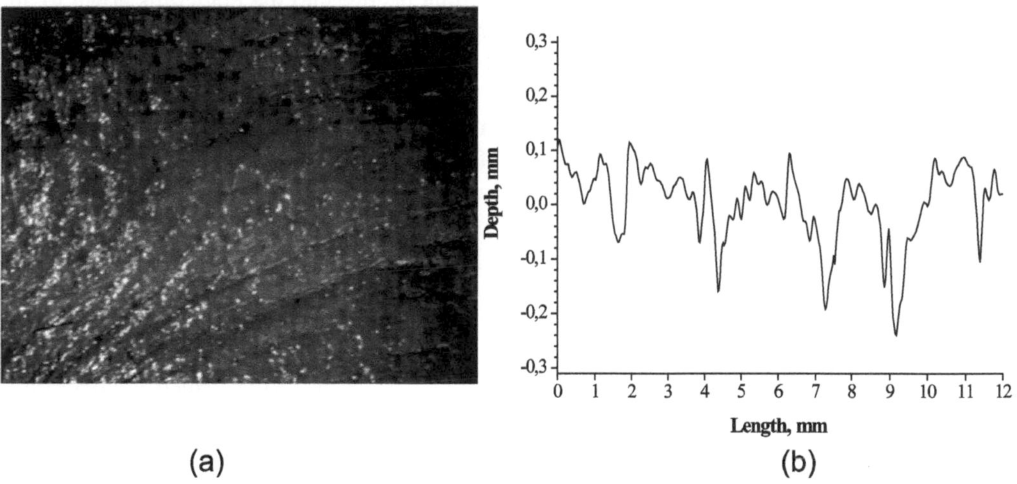

FIGURE 8.8: Typical human skin surface structure image (a) and corresponding skin surface profile obtained by the Primos system (b).

Taking into consideration the effectiveness of the cutaneous antioxidative defense system for the neutralization of free radicals produced in the skin, it can be expected that volunteers, possessing higher levels of antioxidants in their skin, exhibit an additional defense against the negative action of free radicals, and, as a result, against the manifestation of premature aging. Furrows and wrinkles are distributed irregularly on the surface of human skin. Figure 8.8a shows a typical human skin surface image, where the typical distribution of furrows and wrinkles can be seen. The utilization of the noninvasive 3D optical system Primos 4.0 (GFMesstechnik GmbH, Teltow, Germany) as described in detail by Jacobi et al. [47], enabled the analysis of the skin surface regarding the depth and density of furrows and wrinkles. Figure 8b shows the corresponding skin surface profile. The roughness, which is calculated as a combined value of the depth and the density of the furrows and wrinkles of the skin, was determined using the software Primos system.

A strong correlation was observed between the roughness of the skin and the concentration of cutaneous lycopene [48]. Both measurements had no influence on each other and were performed noninvasively on the forehead of 20 volunteers aged between 40 and 50 years. Only those volunteers were included in the study, who had not significantly changed their lifestyle during the last three decades, and who had not applied special food additions and cosmetic formulations with antioxidants. This specific age group was selected as the furrows and wrinkles of the skin are more pronounced at this age, as compared to younger subjects.

The results revealed that those volunteers, exhibiting a high cutaneous antioxidative potential, had significantly lower values of skin roughness. Thus, a continuous high antioxidative potential of the skin protects them against the action of free radicals, which might also result in a better prevention of premature skin aging.

8.7.6 Topical application of antioxidants

During the past years, different antioxidants have been added to cosmetic formulations for the prevention of premature oxidation and to prolong the working life of cosmetic formulation substances. Recently, new data showed that antioxidants can also be useful and essential for protection of the skin.

The cutaneous carotenoid concentration clearly increases after the application of the carotenoid containing cream. Topically applied carotenoids penetrate deeply into the epidermis down to approximately 24 μm, filling up the reservoir of the stratum corneum and, thus, protecting the upper layers of the skin against negative influences of free radicals, which are produced subsequent to stress factors, such as irradiation [18].

Other investigations performed *in vivo* showed that skin, which was pretreated topically with a formulation containing antioxidants has additional defense against the action of free radicals, which are produced on the skin surface subsequent to IR irradiation [44].

8.7.7 Medication with antioxidants

In medical practice, antioxidant substances such as beta-carotene, lycopene, lutein/zeaxanthin, vitamins A, C, D and E are widely applied to improve the therapy of many diseases such as cancer [36], coronary heart disease [49], cardiovascular disease [36, 50], arthritis [51], atherosclerosis [50], Alzheimer's disease [51], skin diseases [52], bone complications [53], asthma [54], age-related cataract and macular degeneration [55], and for photoprotection [56]. The synopsis of the data reveals a distinct correlation between the state of permanent health, a diet rich in fruit and vegetables, and the level of carotenoid antioxidants in the skin of volunteers [37].

Doxorubicin is one of the most potent chemotherapeutic agents against solid and angiomatous tumors. Nevertheless, severe and dose-limiting mucocutaneous side effects, such as palmar-plantar erythrodysesthesia syndrome (PPE) occur. This syndrome is mainly located on the palms and plantae, but it may also affect intertriginous sites. The cause of this syndrome is unclear.

Recently, it was found with the use of laser scanning microscopy that Doxorubicin reaches the skin surface via the sweat glands around one hour after an intravenous injection. Following homogeneous distribution, it penetrates to the skin similar to topically applied.

A high concentration of Doxorubicin was detected on the palms and plantae of patients, where the highest density of sweat glands exists [38]. Furthermore, on the palms and plantae, the stratum corneum is approximately 10 times thicker than on other anatomical sites. The thick stratum

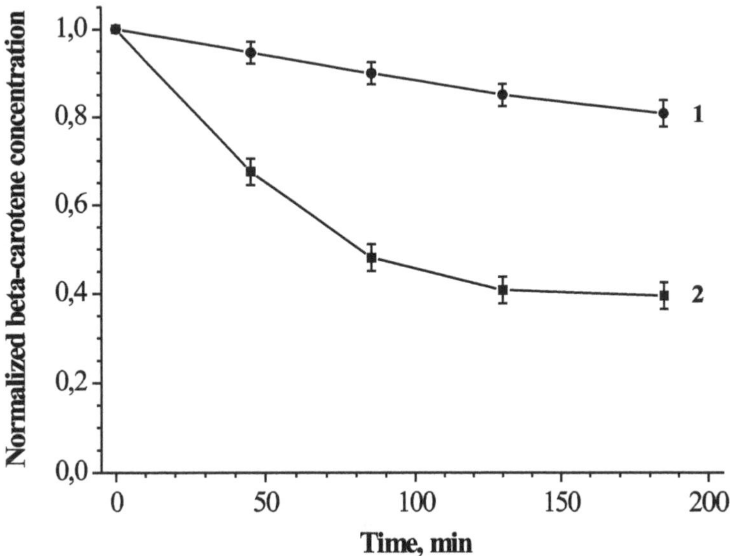

FIGURE 8.9: Degradation dynamic of the beta-carotene, applied topically onto the skin of a healthy volunteer (1) and patient receiving Doxorubicin (2). The time point 0 min corresponds to the end of Doxorubicin injection.

corneum of the palms and plantae represents an effective reservoir for Doxorubicin, where it can be stored for a long period and interact with the surrounding cells.

Cancer treatment with Doxorubicin is partly based on the formation of free radicals, which are supposed to destroy the cancer cells. Therefore, the same effect can occur on the palms and the plantae, where Doxorubicin can interact with the lipid structure of the stratum corneum and the living cells, which might induce their destruction and, as a result, might be a possible reason for the development of the PPE syndrome.

Taking into consideration that topically applied antioxidants are mainly located in the upper skin layers [18], the application of antioxidant containing formulation can be seen as a useful tool for neutralization of Doxorubicin produced radicals in the skin.

Whereas several attempts failed to stop the sweat production after injection of Doxorubicin, topically applied antioxidants, including beta-carotene, were able to neutralize the produced free radicals to a high extent. Using resonance Raman spectroscopy, this effect was analyzed noninvasively on the palms of patients. Figure 8.9 shows that Doxorubicin advances the destruction of beta-carotene, which proves the protective effect of topically applied antioxidants for the skin against the action of Doxorubicin-initiated free radicals [57]. During the period of observations, the PPE syndrome was not manifested in patients utilizing a cream containing a mixture of antioxidants including carotenoids.

Thus, topical application of antioxidant substances seems to be an efficient way to neutralize free radicals and to prevent the PPE syndrome.

A possible explanation for interindividual differences in PPE intensity and incidence might be the individual level of antioxidant substances in the skin of the patients.

8.8 Strategies on the Application of Antioxidant Substances

There are two possibilities to increase the level of carotenoid antioxidants in the skin: either by systemic or by topical application of these substances, which were discussed previously.

Antioxidant substances are widely used in the cosmetic industry for the protection of formulation against oxidation processes, as well as for protection of the skin against the action of free radicals and in medicine for the treatment and prophylaxis of numerous diseases. Many studies have shown a positive impact on health after application of antioxidants, but others also related strong negative results. Different antioxidants at different concentrations were used for these studies. Analyzing the available literature regarding the action of systemically and topically applied artificial and extracted natural antioxidants, allows the conclusion that a combination of topical and systemic applications of a natural mixture of antioxidants, including carotenoids and vitamins applied at low concentrations is more effective for skin protection [9]. Almost no side effects were reported apart from a number of allergic reactions of some volunteers. Taking into consideration that the different antioxidants act as a "protection chain," i.e., all antioxidants possess a synergic action, thus protecting each other against direct destruction during neutralization of free radicals and other reactive species [7, 58], it is understandable that a mixture of antioxidants is more convenient and effective for the organism than a single antioxidant compound.

The utilization of antioxidant substances in cosmetic formulations can be considered to be indispensable for the protection of the skin by means of neutralization of free radicals, produced on the skin surface following medication with Doxorubicin and its analogs (see paragraph 8.6.6), and subsequent to UV and IR irradiations (see paragraphs 8.6.2.1 and 8.6.2.2). In cases of the absence of additional protection, the skin can remain insufficiently protected and accompanying destructive reactions of free radicals can easily occur.

Taking into consideration the high penetration depth of IR irradiation into the skin and the subsequent production of free radicals, an increase in antioxidant concentration in the skin should be recommended not only for skin surface protection, but also for the protection of deeper skin layers, by means of systemic application of well-balanced antioxidant-rich supplements or fruit and vegetables.

By the utilization of such combinations, skin health can be maximally promoted from the outside in and from the inside out.

8.9 Conclusions

Resonance Raman spectroscopy is an optical noninvasive method for the *in vivo* determination of the carotenoid antioxidant substances, such as beta-carotene, lycopene, and others in the human skin. This method has no alternatives in terms of quickness of measurement, high reproducibility of the results and convenience.

The measured kinetics showed that carotenoids can serve as marker substances of the total antioxidative potential of the human skin. Stress factors, like irradiations and illness, which are usually associated with the production of free radicals in the organism, can cause a decrease in the antioxidative potential of the skin, measured by a decrease in the cutaneous carotenoid level. Vice versa, the absence of the influence of stress factors, as well as an antioxidant-rich supplementation, e.g., based on the supplementation of a high amount of fruit and vegetables, gives rise to an increase in the antioxidative potential of the human skin. The kinetic of these two processes is different. The decrease in the carotenoid levels in the skin occurs rapidly over a period of 2 hours, depending on the influenced stress factors and their intensity, while the subsequent recovery to the initial level usually takes up to three days. Thus, the action of stress factors on the skin impairs the antioxidative potential, substantially enfeebling the defense system of the organism.

A one-year study showed that during the summer and autumn months, an increase in the level of cutaneous carotenoids compared with the winter and spring months occur. This so-called "seasonal increase" was determined to be 1.26-fold and explained by the increased amount of carotenoid-rich nourishments in the daily nutrition, e.g., fruit and vegetables, during these months.

It was shown noninvasively that the cutaneous carotenoids beta-carotene and lycopene increased subsequent to UV irradiation of the skin. Beta-carotene and lycopene concentration in the skin does not decrease immediately after UV irradiation. There is a time delay, which varies from 30 up to 90 minutes for beta-carotene and from 0 up to 30 minutes for lycopene, which can be explained by the differences in the quenching rate constants for these substances. A strong correlation between the individual antioxidant level of volunteers and the magnitude of destruction of antioxidants in the skin was found.

IR irradiation of the skin results in the production of free radicals. This effect was measured by the detection of the degradation of the cutaneous carotenoid antioxidant substances beta-carotene and lycopene, subsequent to irradiation of the skin with IR light.

The concentration of carotenoids in the skin reflects the permanent state of the health and lifestyle of volunteers and patients. Nevertheless, the level of the carotenoid antioxidants in the human skin is continuously influenced by a multiplicity of different factors, which often superpose each other.

It was demonstrated that carotenoids are distributed inhomogeneously in human epidermis. The highest concentration was detected in the upper layers of the stratum corneum close to the skin surface. The obtained distribution can be explained by the penetration of sweat containing carotenoids, which is always released to the skin surface, inside the skin. Topically applied antioxidant substances including carotenoids, penetrate deeply into the epidermis down to approximately 24 μm and, thus, increase the concentration of cutaneous carotenoids. Therefore, it is possible to artificially increase the antioxidative potential of human skin by the use of topically applied antioxidant-containing formulations.

It was shown in cancer patients, after an intravenous injection, that Doxorubicin was released to the skin surface with the sweat. The utilization of a carotenoid-containing cream was shown to neutralize free radicals on the skin surface of cancer patients, induced by the action of the chemotherapeutic Doxorubicin. The action of the Doxorubicin-produced radicals on the skin is suggested to be the main cause of the development of PPE syndrome.

The effectiveness of the neutralization of free radicals can be substantially increased, if a balanced mixture of different antioxidants at different concentrations is applied. The question related to the optimization of the mixture and concentration of the applied antioxidants used in cosmetic formulations is still open and should be investigated in the future. For this purpose, utilization of modern noninvasive techniques is expected to be indispensable.

A strong correlation between the level of cutaneous lycopene and the appearance of the skin in regard to density and deepness of furrows and wrinkles was found. The investigations revealed that higher levels of antioxidants in the human skin are related to lower levels of skin roughness, and, as a result, to a better appearance of the skin.

Thus, the important role of carotenoids as powerful antioxidants for the organism is supported by the sufficient uptake of fruit and vegetables, as well as topical applications, which are important protection strategies against the development of different cutaneous diseases and premature skin aging.

References

[1] J.D. Ribaya-Mercado, M. Garmyn, B.A. Gilchrest, et al., "Skin lycopene is destroyed preferentially over beta-carotene during ultraviolet irradiation in humans," *J. Nutr.* **125**(7), 1854–1859 (1995).

[2] H.K. Biesalski, C. Hemmes, W. Hopfenmuller, et al., "Effects of controlled exposure of sunlight on plasma and skin levels of beta-carotene," *Free Radic. Res.* **24**(3), 215–224 (1996).

[3] W.R. Markesbery, "Oxidative stress hypothesis in Alzheimer's disease," *Free Radic. Biol. Med.* **23**(1), 134–147 (1997).

[4] J.S. Moon, C.H. Oh, "Solar damage in skin tumors: quantification of elastotic material," *Dermatology* **202**(4), 289–292 (2001).

[5] E. Koutsilieru, C. Scheller, E. Grünblatt, et al., "Free radicals in Parkinson's disease," *J. Neur.* **249**(2), 1–5 (2002).

[6] I.N. Todorov, G.I. Todorov, *Stress, Aging and Their Biochemical Correction,* Nauka Press, Moscow, (2003).

[7] E.A. Offord, J.C. Gautier, O. Avanti, et al., "Photoprotective potential of lycopene, beta-carotene, vitamin E, vitamin C and carnosic acid in UVA-irradiated human skin fibroblasts,"*Free Radic. Biol. Med.* **32**(12), 1293–1303 (2002).

[8] M. Wrona, W. Korytowski, M. Rozanowska, et al., "Cooperation of antioxidants in protection against photosensitized oxidation," *Free Radic. Biol. Med.* **35**(10), 1319–1329 (2003).

[9] M. Darvin, L. Zastrow, W. Sterry, et al., "Effect of supplemented and topically applied antioxidant substances on human tissue," *Skin Pharmacol. Physiol.* **19**(5), 238–247 (2006).

[10] S. Melov, "Mitochondrial oxidative stress: physiologic consequences and potential for a role in aging," *Ann. N.Y. Acad. Sci.* **908**, 219–225 (2000).

[11] Y.H. Wei, Y.S. Ma, H.C. Lee, et al., "Mitochondrial theory of aging matures – roles of mtDNA mutation and oxidative stress in human aging," *Zhonghua Yi Xue Za Zhi (Taipei)* **64**(5), 259–270 (2001).

[12] B. Halliwell, J.M.C. Gutteridge, *Free Radicals in Biology and Medicine*, 4 th ed., Oxford University Press, (2007).

[13] Y. Kawaguchi, H. Tanaka, T. Okada, et al., "Effect of reactive oxygen species on the elastin mRNA expression in cultured human dermal fibroblasts," *Free Radic. Biol. Med.* **23**(1), 162–165 (1997).

[14] J-C. Monboisse, P. Braquet, and J.P. Borel, "Oxygen free radicals as mediators of collagen breakage," *Agents and Actions* **15**, 49–50 (1984).

[15] D. Harman, "Free radical theory of aging: consequences of mitochondrial aging," *AGE* **6**(3), 86–94 (1983).

[16] F. Khachik, L. Carvalho, P.S. Bernstein, et al.,"Chemistry, distribution, and metabolism of tomato carotenoids and their impact on human health," *Exptl. Biol. Med.*, **227**, 845–851 (2002)

[17] T.R. Hata, T.A. Scholz, I.V. Ermakov, et al., "Non-invasive Raman spectroscopic detection of carotenoids in human skin," *J. Invest. Derm.*, **115**, 441–448 (2000).

[18] J. Lademann, P.J. Caspers, A. van der Pol, et al., "In vivo Raman spectroscopy detects increased epidermal antioxidative potential with topically applied carotenoids," *Laser Phys. Lett.*, **6**(1), 76–79 (2008).

[19] J.A. Olson, I.N. Krinsky, "The colorful fascinating world of carotenoids: important biological modulators," *FASEB J.* **9**, 1547–1550 (1995).

[20] R. Edge, D.J. McGarvey, and T.G. Truscott, "The carotenoids as anti-oxidants – a review," *J. Photochem. Photobiol. B.* **41**(3), 189–200 (1997).

[21] W. Stahl, H. Sies, "Antioxidant activity of carotenoids," *Mol. Asp. Med.* **24,** 345–351 (2003).

[22] P. Di Mascio, S. Kaiser, and H. Sies, "Lycopene as the most efficient biological carotenoid singlet oxygen quencher," *Arch. Biochem. Biophys.* **274**(2), 532–538 (1989).

[23] D.B. Rodriguez-Amaya, *A Guide to Carotenoid Analysis in Food,.* ILSI Press, Washington (2001).

[24] M.E. Darvin, I. Gersonde, M. Meinke, et al., "Non-invasive *in vivo* detection of the carotenoid antioxidant substance lycopene in the human skin using the resonance Raman spectroscopy," *Laser Phys. Lett.* **3**(9), 460–463 (2006).

[25] D. Talwar, K.K. Ha Tom, J. Cooney J, et al., "A routine method for the simultaneous measurement of retinal, α-tocopherol and five carotenoids in human plasma by reverse phase HPLC," *Clinica Chimica Acta* **270,** 85–100 (1998).

[26] F. Khachik, G.R. Beecher, and M.B. Goli, "Separation, identification, and quantification of carotenoids in fruits, vegetables and human plasma by high performance liquid chromatography," *Pure and Appl. Chem.* **63**(1), 71–80 (1991).

[27] E. Vági, B. Simándi, K.P. Vásárhelyiné, et al., "Supercritical carbon dioxide extraction of carotenoids, tocopherols and sitosterols from industrial tomato by-products," *J. Supercrit. Fluids* **40**, 218–226 (2007).

[28] W. Stahl, U. Heinrich, H. Jungmann, et al., "Increased dermal carotenoid level assessed by non-invasive reflection spectrophotometry correlate with serum levels in women ingesting betatene," *J. Nutr.* **128**, 903–907 (1998).

[29] M.E. Darvin, I. Gersonde, M. Meinke, et al., "Non-invasive in vivo determination of the carotenoids beta-carotene and lycopene concentrations in the human skin using the Raman spectroscopic method," *J. Phys. D: Applied Physics* **38,** 1–5 (2005).

[30] S.D. Bergeson, J.B. Peatross, N.J. Eyring, et al., "Resonance Raman measurements of carotenoids using light emitting diodes," *J. Biomed. Opt.* **13**(4), 044026-1–6 (2008).

[31] A. Roggan, O. Minet, C. Schröder, et al., "Determination of optical tissue properties with double integrating sphere technique and Monte Carlo simulations," *Proc. SPIE* **2100**, 42–56 (1994).

[32] A. Splett, Ch. Splett, and W. Pilz, "Dynamics of the Raman background decay," *J. Raman Spectr.* **28**, 481–485, 1997.

[33] V.V. Tuchin, ed. *Optical Biomedical Diagnostics,* **PM107** SPIE Press, Bellingham, WA (2002).

[34] I. Ermakov, M. Ermakova, W. Gellermann W, et al., "Noninvasive selective detection of lycopene and beta-carotene in human skin using Raman spectroscopy," *J. Biomed. Opt.* **9**(2), 332–338 (2004).

[35] M.E. Darvin, I. Gersonde, H. Albrecht, et al., "Determination of beta carotene and lycopene concentrations in human skin using Raman spectroscopy" *Laser Physics* **15**(2), 295–299 (2005).

[36] W. Stahl and H. Sies, "Perspectives in biochemistry and biophysics. Lycopene: a biologically important carotenoid for humans?," *Arch. Biochem. Biophys.* **336**(1), 1–9 (1996).

[37] M.E. Darvin, A. Patzelt, F. Knorr, et al., "One-year study on the variation of carotenoid antioxidant substances in the living human skin: influence of dietary supplementation and stress factors," *J. Biomed. Opt.* **13**(4), 044028-1–9 (2008).

[38] J. Hadgraft, "Penetration routes through human skin," *Proc 10-th EADV Congress Skin and Environment; Perception and Protection* 463–468 (2001).

[39] S. Ekanayake-Mudiyanselage, J. Thiele, "Sebaceous glands as transporters of vitamin E," *Hautarzt* **57**(4), 291–296 (2006).

[40] E. Cadenas, K.J. Davies, "Mitochondrial free radical generation, oxidative stress, and aging," *Free Rad. Biol. Med.* **29**(3/4), 222–230 (2000).

[41] M.J. Jackson, "An overview of methods for assessment of free radical activity in biology," *Proc. Nutr. Soc.* **58**(4), 1001–1006 (1999).

[42] M.E. Darvin, I. Gersonde, H. Albrecht, et al., "In-vivo Raman spectroscopic analysis of the influence of UV radiation on carotenoid antioxidant substance degradation of the human skin," *Laser Physics* **16**(5), 833–837 (2006).

[43] M.E. Darvin, I. Gersonde, H. Albrecht, et al., "In vivo Raman spectroscopic analysis of the influence of IR radiation on the carotenoid antioxidant substances beta-carotene and lycopene in the human skin. Formation of free radicals," *Laser Phys. Lett.* **4**(4), 318–321 (2007).

[44] P. Schroeder, J. Lademann, M.E. Darvin, et al., "Infrared radiation-induced matrix metalloproteinase in human skin: implications for protection," *Journal of Investigative Dermatology*, **128**(10), 2491–2497 (2008).

[45] M.E. Darvin, I. Gersonde, H. Albrecht, et al., "Resonance Raman spectroscopy for the detection of carotenoids in foodstuffs: influence of the nutrition on the antioxidative potential of the skin," *Laser Phys. Lett.* **4**(6), 452–457 (2007).

[46] M.E. Darvin, I. Gersonde, H. Albrecht, et al., "Non-invasive *in-vivo* Raman spectroscopic measurement of the dynamics of the antioxidant substance lycopene in the human skin after a dietary supplementation," *Proc. SPIE*, **6535**, 653502-1–6 (2007).

[47] U. Jacobi, M. Chen, G. Frankowski, et al., "In vivo determination of skin surface topography using an optical 3D device," *Skin Res. Technol.* **10**(4), 207–214 (2004).

[48] M. Darvin, A. Patzelt, S. Gehse, et al., "Cutaneous concentration of lycopene correlates significantly with the roughness of the skin," *Eur. J. Pharm. Biopharm.*, **69**, 943–947 (2008).

[49] S.K. Osganian, M.J. Stampfer, E. Rimm, et al., "Dietary carotenoids and risk of coronary artery disease in women," *Am. J. Clin. Nutr.* **77**, 1390–1399 (2003).

[50] A. Dutta, S.K. Dutta, "Vitamin E and its role in the prevention of atherosclerosis and carcinogenesis: a review," *Am. J. Clin. Nutr.* **22**, 258–268 (2003).

[51] R.B. Parnes, "How antioxidants work," http://home.howstuffworks.com/ antioxidant4.htm, (2002).

[52] T.E. Moon, N. Levine, B. Cartmel, et al., "Effect of retinol in preventing squamous cell skin cancer in moderate-risk subjects: a randomized, double-blind, controlled trial. Southwest Skin Cancer Prevention Study Group," *Cancer Epidemiol. Biomarkers Prev.* **6**, 949–956 (1997).

[53] N.D. Daniele, M.G. Carbonelli, N. Candeloro, et al., "Effect of supplementation of calcium and Vitamin D on bone mineral density and bone mineral content in peri- and post-menopause women: a double-blind, randomized, controlled trial," *Pharmacol. Res.* **50**, 637–641 (2004).

[54] P.J. Pearson, S.A. Lewis, J. Britton, et al., "Vitamin E supplements in asthma: a parallel group randomised placebo controlled trial," *Thorax* **59**, 652–656 (2004).

[55] A. Taylor, "Cataract: relationship between nutrition and oxidation," *J. Am. Coll. Nutr.* **12**, 138–146 (1993).

[56] H. Sies, W. Stahl, "Carotenoids and UV protection," *Photochem. Photobiol. Sci.* **3**, 749–752 (2004).

[57] J. Lademann, A. Martschick, U. Jacobi, et al., "Investigation of doxorubicin on the skin: a spectroscopic study to understand the pathogenesis of PPE," *Suppl. to J. Clin. Oncol.*, **23**(16S), 5093 (2005).

[58] K. Zu, C. Ip, "Synergy between selenium and vitamin E in apoptosis induction is associated with activation of distinctive initiator caspases in human prostate cancer cells," *Cancer Res.* **63**, 6988–6995 (2003).

9

Polarized Light Assessment of Complex Turbid Media Such as Biological Tissues Using Mueller Matrix Decomposition

Nirmalya Ghosh

IISER Kolkata, Mohanpur Campus, PO: BCKV Campus Main Office, Mohanpur 741252, West Bengal, India

Michael Wood, and Alex Vitkin

Ontario Cancer Institute / Department of Medical Biophysics University of Toronto, Toronto, Ontario, Canada

Polarization parameters of light scattered from biological tissue contain rich morphological and functional information of potential biomedical importance. Despite the wealth of interesting parameters that can be probed with polarized light, in optically thick turbid media such as tissues, numerous complexities due to multiple scattering and simultaneous occurrence of many polarization effects present formidable challenges, both in terms of accurate measurement and in terms of extraction/unique interpretation of the polarization parameters. In this chapter, we describe the application of an expanded Mueller matrix decomposition method to tackle these complexities. The ability of this approach to delineate individual intrinsic polarimetry characteristics in tissue-like turbid media (exhibiting multiple scattering, and linear and circular birefringence) was validated theoretically with a polarized-light propagation model and experimentally with a polarization-modulation/synchronous detection technique. The details of the experimental turbid polarimetry system, forward Monte Carlo modeling, inverse polar decomposition analysis, and the results of the validations studies are presented in this chapter. Initial applications of this promising approach in two scenarios of significant clinical interest, that for monitoring regenerative treatments of the heart and for noninvasive glucose measurements, as well as initial *in vivo* demonstration, are discussed.

Key words: polarization, multiple scattering, turbid polarimetry, light transport, Monte Carlo simulations, Stokes vector, Mueller matrix, polar decomposition, biological and medical applications

9.1 Introduction

Polarimetry has played important roles in our understanding of the nature of electromagnetic waves, elucidating the three-dimensional characteristics of chemical bonds, uncovering the asymmetric (chiral) nature of many biological molecules, quantifying protein properties in solutions, supplying a variety of nondestructive evaluation methods, and contributing to remote sensing in meteorology and astronomy [1–3]. The use of polarimetric approaches has also received considerable recent attention in biophotonics [4–6]. This is because polarization parameters of light scattered from biological tissue contain rich morphological and functional information of potential biomedical importance. For example, the anisotropic organized nature of many tissues stemming from their fibrous structure leads to linear birefringence (or linear retardance), manifest as anisotropic refractive indices parallel and perpendicular to the fibers. Muscle fibers and extracellular matrix proteins (such as collagen and elastin) possess this fibrous structure and accordingly exhibit linear birefringence. Changes in this anisotropy resulting from disease progression or treatment response alter the optical birefringence properties, making this a potentially sensitive probe of tissue status [7,8]. Glucose, another important tissue constituent, exhibits circular birefringence due to its asymmetric chiral structure. Its presence in tissue leads to rotation of the plane of linearly polarized light about the axis of propagation (known as optical rotation or optical activity). Measurements of optical rotation may offer an attractive approach for noninvasive monitoring of tissue glucose levels [9–14].

Despite the wealth of interesting properties that can be probed with polarized light, in optically thick turbid media such as tissues, numerous complexities due to multiple scattering present formidable challenges. Multiple scattering causes extensive depolarization that confounds the established techniques. Further, even if some residual polarization signal can be measured, multiple scattering also alters the polarization state, for example by scattering-induced diattenuation and by scattering-induced changes in the orientation of the linear polarization vector that appears as optical rotation [10,14]. Quantitative polarimetry in tissue is further compromised by simultaneous occurrences of many polarization effects.

The Mueller matrix represents the transfer function of an optical system in its interactions with polarized light, the elements reflecting various sample properties of potential interest [15, 16]. However, in complex turbid media such as tissues, many optical polarization effects occur simultaneously (the most common biopolarimetry events are depolarization, linear birefringence, and optical activity), and contribute in a complex interrelated way to the Mueller matrix elements. Hence, these represent several "lumped" effects, masking potentially interesting ones and hindering unique data interpretation. The challenges are thus to minimize or compensate for multiple scattering, and to decouple the individual contributions of simultaneously occurring polarization effects. Each of the individual processes, if separately extracted from the "lumped" system Mueller matrix, can potentially be used as a useful biological metric.

We have recently developed and validated an expanded Mueller matrix decomposition approach for extraction, quantification and unique interpretation of individual intrinsic polarimetry characteristics in complex tissue-like turbid media [17,18]. The ability of this approach to delineate individual intrinsic polarimetry characteristics was validated theoretically with a polarized-light propagation model, and experimentally with a polarization-modulation/synchronous detection technique. In this chapter, we summarize this (and related) research on turbid polarimetry, and discuss initial biomedical applications of this promising approach.

This chapter is organized as follows. In section 9.2, we describe the basics of Mueller matrix algebra and also define the constituent polarization parameters. The mathematical methodology of polar decomposition for extraction of the individual intrinsic polarimetry characteristics from "lumped" Mueller matrix is outlined in section 9.3. Section 9.4 describes the high-sensitivity polarization modulation / synchronous detection experimental system capable of measuring complete Mueller matrix elements from strongly depolarizing scattering media such as tissues. This is followed by the description of the corresponding theoretical model in section 9.5, based on the forward Monte Carlo (MC) modeling, with the flexibility to incorporate all the simultaneous optical (scattering and polarization) effects. Section 9.6 reviews the experimental and theoretical validation results of the polar decomposition approach to delineate individual intrinsic polarimetry characteristics in complex tissue-like turbid media. In section 9.7, we present selected trends of the dependence of decomposition-derived polarization parameters on multiple scattering, propagation path, and detection geometry. In section 9.8, we discuss the initial applications of the Mueller matrix decomposition approach in two scenarios of significant clinical interest, for noninvasive glucose measurements and for monitoring of regenerative treatments of the heart. The proof-of-principle demonstration of *in vivo* use of this method for polarization-based characterization of tissue is also presented in this section. The chapter concludes with a discussion of the prospective biomedical utility of this promising approach.

9.2 Mueller Matrix Preliminaries and the Basic Polarization Parameters

The state of polarization of a beam of light can be represented by four measurable quantities (known as Stokes parameters) that, when grouped in a 4×1 vector, are known as the Stokes vector [15], introduced by G. G. Stokes in 1852. The four Stokes parameters are defined relative to the following six intensity measurements (I) performed with ideal polarizers: I_H, horizontal linear polarizer (0°); I_V, vertical linear polarizer (90°); I_P, 45° linear polarizer; I_M, 135° (−45°) linear polarizer; I_R, right circular polarizer, and I_L, left circular polarizer. The Stokes vector (**S**) is defined as

$$\mathbf{S} = \begin{bmatrix} I \\ Q \\ U \\ V \end{bmatrix} = \begin{bmatrix} IH + IV \\ IH - IV \\ IP - IM \\ IR - IL \end{bmatrix} \tag{9.1}$$

where I, Q, U, and V are Stokes vector elements. I is the total detected light intensity that corresponds to addition of any two orthogonal component intensities, while Q is the portion of the intensity that corresponds to the difference between horizontal and vertical polarization states, U is the portion of the intensity that corresponds to the difference between intensities of linear +45° and −45° polarization states, and V is portion of the intensity that corresponds to the difference between intensities of right circular and left circular polarization states. For a completely polarized beam of light, the Stokes parameters are not all independent [15]

$$I = \sqrt{Q^2 + U^2 + V^2}. \tag{9.2}$$

From Stokes vector elements, the following polarization parameters of partially polarized light can be determined [15]:
degree of polarization

$$DOP = \sqrt{Q^2 + U^2 + V^2}/I, \tag{9.3}$$

degree of linear polarization

$$DOLP = \frac{\sqrt{Q^2 + U^2}}{I}, \tag{9.4}$$

and degree of circular polarization

$$DOCP = \frac{V}{I}. \tag{9.5}$$

While the Stokes vector represents the polarization properties of the *light*, the Mueller matrix (**M**) contains complete information about all the polarization properties of the *medium*. The Mueller matrix **M** (a 4×4 matrix) is a mathematical description of how an optical sample interacts or transforms the polarization state of an incident light beam. In essence, the Muller matrix can be thought of as the "optical fingerprint" or transfer function of a sample. Mathematically, this matrix operates directly on an input or incident Stokes vector, resulting in an output Stokes vector that describes the polarization state of the light leaving the sample. This is described mathematically by the following equation:

$$\mathbf{S}_o = \mathbf{M} \cdot \mathbf{S}_i. \tag{9.6}$$

$$\begin{bmatrix} I_o \\ Q_o \\ U_o \\ V_o \end{bmatrix} = \begin{bmatrix} m_{11} & m_{12} & m_{13} & m_{14} \\ m_{21} & m_{22} & m_{23} & m_{24} \\ m_{31} & m_{32} & m_{33} & m_{34} \\ m_{41} & m_{42} & m_{43} & m_{44} \end{bmatrix} \begin{bmatrix} I_i \\ Q_i \\ U_i \\ V_i \end{bmatrix} = \begin{bmatrix} m_{11}I_i + m_{12}Q_i + m_{13}U_i + m_{14}V_i \\ m_{21}I_i + m_{22}Q_i + m_{23}U_i + m_{24}V \\ m_{31}I_i + m_{32}Q_i + m_{33}U_i + m_{34}V \\ m_{41}I_i + m_{42}Q_i + m_{43}U_i + m_{44}V \end{bmatrix}, \tag{9.7}$$

where $\mathbf{S}_o$ and $\mathbf{S}_i$ are the output and input Stokes vectors, respectively.

The different polarization properties of a medium are coded in the various elements of the Mueller matrix **M**. The three basic polarization properties are diattenuation (differential attenuation of orthogonal polarization), retardance (de-phasing of orthogonal polarization) and depolarization; the functional forms of the corresponding matrices are well known [15].

Diattenuation

Diattenuation (d) by an optical element corresponds to differential attenuation of orthogonal polarizations for both linear and circular polarization states. Accordingly, linear diattenuation is defined as differential attenuation of two orthogonal linear polarization states and circular diattenuation is defined as differential attenuation of right circular polarized light (RCP) and left circular polarized light (LCP). Mathematically, the Mueller matrix for an ideal diattenuator can be defined using two intensity measurements, q and r,for the two incident orthogonal polarization states (either linear or circular). Using this convention, the general Mueller matrix for a linear diattenuator is defined as

$$\begin{pmatrix} q+r & (q-r)\cos 2\theta & (q-r)\sin 2\theta & 0 \\ (q-r)\cos 2\theta & (q+r)\cos^2 2\theta + 2\sqrt{(qr)}\sin^2 2\theta & \left(q+r-2\sqrt{(qr)}\right)\sin 2\theta\cos 2\theta & 0 \\ (q-r)\sin 2\theta & \left(q+r-2\sqrt{(qr)}\right)\sin 2\theta\cos 2\theta & (q+r)\cos^2 2\theta + 2\sqrt{(qr)}\sin^2 2\theta & 0 \\ 0 & 0 & 0 & 2\sqrt{(qr)} \end{pmatrix}. \tag{9.8}$$

where θ is the angle between the diattenuation axis of the sample and the horizontal (laboratory) frame. Briefly, the sample's diattenuation axis is the direction of minimum attenuation (an analogous definition exits for sample's birefringence axis, as detailed further in the next section). Similarly for circular diattenuation, the general form of Mueller matrix is

$$\begin{pmatrix} q+r & 0 & 0 & q-r \\ 0 & 2\sqrt{(qr)} & 0 & 0 \\ 0 & 0 & 2\sqrt{(qr)} & 0 \\ q-r & 0 & 0 & q+r \end{pmatrix}. \tag{9.9}$$

Ideal polarizers that transform incident unpolarized light to completely polarized light are examples of diattenuators (with magnitude of diattenuation $d = 1.0$ for ideal polarizer; d is a dimensionless quantity, ranging from 0 to 1.0). Note that this is analogous to dichroism, which is defined as the differential absorption of two orthogonal linear polarization states (linear dichroism) or of CP states (circular dichroism). The term "diattenuation" is more general in that it is defined in terms of differential attenuation (either by absorption or scattering). Many biological molecules (such as amino acids, proteins, nucleic acids) exhibit dichroism or diattenuation effects.

Retardance

Retardance is the de-phasing of the two orthogonal polarization states. Linear retardance (δ) arises due to difference in phase between orthogonal linear polarization states (between vertical and horizontal or between 45° and −45°). Circular retardance or optical rotation (ψ) arises due to difference in phase between RCP and LCP.

The general form of a Mueller matrix of a linear retarder with retardance δ and orientation angle of retarder axis θ is [15]

$$\begin{pmatrix} 1 & 0 & 0 & 0 \\ 0 & \cos^2 2\theta + \sin^2 2\theta\cos\delta & \sin 2\theta\cos 2\theta(1-\cos\delta) & -\sin 2\theta\sin\delta \\ 0 & \sin 2\theta\cos 2\theta(1-\cos\delta) & \sin^2 2\theta + \cos^2 2\theta\cos\delta & \cos 2\theta\sin\delta \\ 0 & \sin 2\theta\sin\delta & -\cos 2\theta\sin\delta & \cos\delta \end{pmatrix}. \tag{9.10}$$

where θ is the angle between the retardation axis of the sample and the horizontal (laboratory) frame. Similarly, the Mueller matrix for a circular retarder with retardance ψ is

$$\begin{pmatrix} 1 & 0 & 0 & 0 \\ 0 & \cos 2\psi & -\sin 2\psi & 0 \\ 0 & \sin 2\psi & \cos 2\psi & 0 \\ 0 & 0 & 0 & 1 \end{pmatrix}. \tag{9.11}$$

Linear retardance has its origin in anisotropy in refractive indices, which leads to phase retardation between two orthogonal linear polarization states. In tissue, muscle fibers and extracellular matrix proteins (such as collagen and elastin) possess such anisotropy and thus exhibit linear birefringence. Circular retardance or optical rotation arises due to asymmetric chiral structures. In tissue, glucose and other constituents such as proteins and lipids possess such chiral structure and accordingly exhibit circular retardance. Note that while diattenuation is associated with amplitude difference in the two orthogonal field components (either for linearly or circularly polarized field), retardance is associated with phase difference between orthogonal field components.

Depolarization

If an incident state is polarized and the exiting state has a degree of polarization less than one, then the sample exhibits depolarization. Depolarization is intrinsically associated with scattering, resulting in losses of directionality, phase, and coherence of the incident polarized beam. In a turbid medium like biological tissue, multiple scattering is the major source of depolarization. The general form of the depolarization Mueller matrix is

$$M_\Delta = \begin{pmatrix} 1 & 0 & 0 & 0 \\ 0 & a & 0 & 0 \\ 0 & 0 & b & 0 \\ 0 & 0 & 0 & c \end{pmatrix}, \qquad |a|,|b|,|c| \le 1. \tag{9.12}$$

Here $1-|a|$ and $1-|b|$ are depolarization factors for linear polarization, and $1-|c|$ is the depolarization factor for circular polarization. Note that this definition of depolarization factor is different from the Stokes parameter-based definition of degrees of polarization [Eqs. (9.3)–(9.5)]. The latter represents the value of degree of polarization of the emerging beam, and result from several lumped polarization interactions. In contrast, the depolarization factors of Eq. (9.12) represent the pure depolarizing transfer function of the medium.

The operational definition of the Mueller matrices of the polarization properties, described above, enables one to correctly forward model these individual effects of any medium. However, the problem arises when all these polarization effects are exhibited simultaneously in a medium [as is the case for biological tissue that often exhibit depolarization, linear birefringence, optical activity, and diattenuation (the magnitude of diattenuation in tissue is, however, much lower compared to the other effects)]. Simultaneous occurrence of many polarization effects contributes in a complex interrelated way to the resulting Mueller matrix elements. Hence, these represent several "lumped" effects, hindering their extraction / unique interpretation and necessitating additional analysis to decouple the individual sample characteristics. In the following section, we describe a matrix decomposition method to tackle this problem.

9.3 Polar Decomposition of Mueller Matrices for Extraction of the Individual Intrinsic Polarization Parameters

Having described the common polarimetry characteristics of individual elements and their corresponding Mueller matrices for forward modeling, we now turn to the complicated *inverse* problem of separating out the constituent contributions from simultaneous polarization effects.

That is, given a particular Mueller matrix obtained from an unknown complex system, can it be analyzed to extract constituent polarization contributions? Here, we shall discuss an extended Mueller matrix decomposition methodology that enables the extraction of the individual intrinsic polarimetry characteristics from the "lumped" system Mueller matrix [17–19]. In addition to the inverse Mueller matrix decomposition method, we have also developed a polarization sensitive forward Monte Carlo (MC) model capable of simulating all the simultaneous optical (scattering and polarization) effects [20]. This is further supplemented by a high-sensitivity polarization modulation/synchronous detection experimental system capable of measuring the complete Mueller matrix from tissues and tissue-like turbid media [10]. These three methodologies form our comprehensive turbid polarimetry platform. A schematic of this turbid polarimetry platform is shown in Figure 9.1. The (i) experimental polarimetry system and (ii) polarization-sensitive Monte Carlo model are discussed subsequently in Sections 9.4 and 9.5 respectively.

For now, we turn our attention to the inverse Mueller matrix decomposition method [part (iii) of Figure 9.1]. The method consists of decomposing a given Mueller matrix $\mathbf{M}$ into the product of three "basis" matrices [19]

$$\mathbf{M} = \mathbf{M}_\Delta \mathbf{M}_R \mathbf{M}_D, \tag{9.13}$$

representing a depolarizer matrix $\mathbf{M}_\Delta$ to account for the depolarizing effects of the medium, a retarder matrix $\mathbf{M}_R$ to describe the effects of linear birefringence and optical activity, and a diattenua-

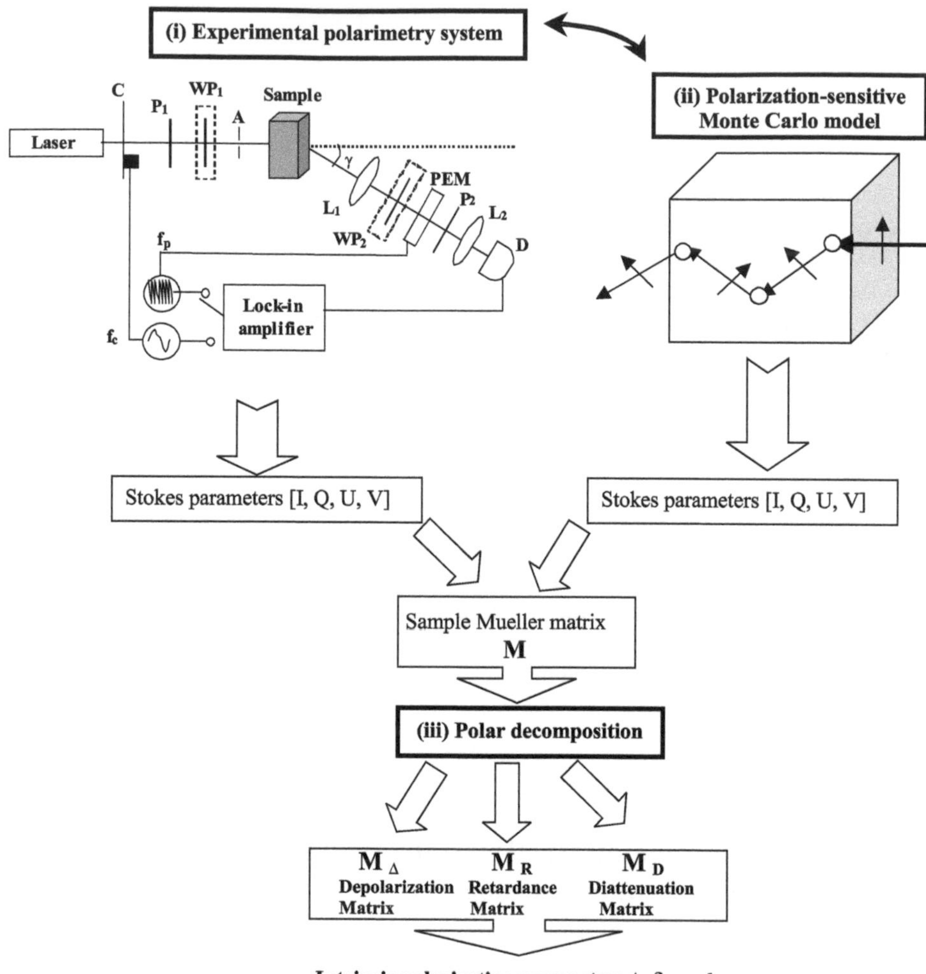

FIGURE 9.1: Schematic of the turbid polarimetry platform: (i) experimental system based on polarization modulation and phase-sensitive synchronous detection, (ii) polarization-sensitive Monte Carlo simulations for forward modeling of simultaneous polarization effects in the presence of turbidity, and (iii) the polar decomposition of the Mueller matrix to inverse calculate the constituent polarization contributions in complex turbid media. In the experimental system (i): C, mechanical chopper; P_1, P_2, polarizers; WP_1, WP_2, removable quarter wave plates; A, aperture; L_1, L_2 lenses; PEM, photoelastic modulator; D, photodetector; f_c, f_p modulation frequencies of mechanical chopper and PEM, respectively. The detection optics can be rotated by an angle γ around the sample.

tor matrix $\mathbf{M}_D$ to include the effects of linear and circular diattenuation. The validity of this decomposition procedure was first demonstrated in optically *clear* media by Lu and Chipman [19]. Here, we extend their analysis to encompass complex tissue-like turbid media. Once calculated, these constituent matrices can further be analyzed to derive quantitative individual polarization medium properties, as summarized below [17, 19, 21]. The diattenuation matrix $\mathbf{M}_D$ is defined as

$$\mathbf{M}_D = \begin{bmatrix} 1 & \bar{d}^T \\ \bar{d} & m_D \end{bmatrix}, \tag{9.14}$$

where m_D is a 3×3 submatrix, the standard form of which is

$$m_D = \sqrt{(1-d^2)}I + (1-\sqrt{(1-d^2)})\, \hat{d}\, \hat{d}^T, \tag{9.15}$$

where I is the 3×3 identity matrix, $\vec{d}$ is diattenuation vector and $\hat{d}$ is its unit vector, defined as

$$\vec{d} = \frac{1}{M(1,1)}[M(1,2)M(1,3)M(1,4)]^T \quad \text{and} \quad \hat{d} = \frac{\vec{d}}{|\vec{d}|}. \tag{9.16}$$

The magnitude of diattenuation $|\vec{d}|$ can be determined as

$$d = \frac{1}{M(1,1)}\sqrt{M(1,2)^2 + M(1,3)^2 + M(1,4)^2}. \tag{9.17}$$

Here $M(i,j)$ are elements of the original sample Mueller matrix $\mathbf{M}$. The coefficients $M(1,2)$ and $M(1,3)$ represent linear diattenuation for horizontal (vertical) and $+45°(-45°)$ linear polarization respectively and the coefficient $M(1,4)$ represents circular diattenuation; this can be seen from the original definition of the Stokes/Mueller formalism [Eq. (9.7)].

Having dealt with diattenuation, the product of the retardance and the depolarizing matrices follows from Eq. (9.13) as

$$\mathbf{M}_\Delta \mathbf{M}_R = \mathbf{M}' = \mathbf{M}\, \mathbf{M}_D^{-1}. \tag{9.18}$$

The matrices $\mathbf{M}_\Delta$, $\mathbf{M}_R$ and $\mathbf{M}'$ have the following form

$$\mathbf{M}_\Delta = \begin{bmatrix} 1 & \vec{0}^T \\ P_\Delta & m_\Delta \end{bmatrix}; \quad \mathbf{M}_R = \begin{bmatrix} 1 & \vec{0}^T \\ \vec{0} & m_R \end{bmatrix}; \quad \text{and} \quad \mathbf{M}' = \begin{bmatrix} 1 & \vec{0}^T \\ P_\Delta & m' \end{bmatrix} \tag{9.19}$$

Here $P_\Delta = (\vec{P} - m\vec{d})/(1-d^2)$, the Polarizance vector $\vec{P} = M(1,1)^{-1}[M(2,1)M(3,1)M(4,1)]^T$ [m_Δ and m_R are 3×3 submatrices of $\mathbf{M}_\Delta$ and $\mathbf{M}_R$]. Similarly, m' is a 3×3 submatrix of $\mathbf{M}'$ and can be written as

$$m' = m_\Delta m_R. \tag{9.20}$$

The submatrix m_Δ can be computed by solving the eigenvalue problem for the matrix $m'\, m'^T$ [19]. This can then be used to construct the depolarization matrix $\mathbf{M}_\Delta$. From the elements of $\mathbf{M}_\Delta$, net depolarization coefficient Δ can be calculated as

$$\Delta = \frac{1}{3}|\mathrm{Tr}(\mathbf{M}_\Delta) - 1|. \tag{9.21}$$

Finally, the expression for the retardance submatrix can be obtained from Eq. (9.20) as

$$m_R = m_\Delta^{-1}\, m'. \tag{9.22}$$

From Eq. (9.19) and (9.22), the total retardance $\mathbf{M}_R$ matrix can be computed.

The value for total retardance (R is a parameter that represents the combined effect of linear and circular birefringence) can be determined from the decomposed retardance matrix $\mathbf{M}_R$ using the relationship

$$R = \cos^{-1}\left\{\frac{\mathrm{Tr}(\mathbf{M}_R)}{2} - 1\right\}. \tag{9.23}$$

$\mathbf{M}_R$ can be further expressed as a combination of a matrix for a linear retarder (having a magnitude of linear retardance δ, its retardance axis at angle θ with respect to the horizontal) and a circular retarder (optical rotation with magnitude of ψ). Using the standard forms of the linear retardance and optical rotation matrices [Eq. (9.10) and (9.11)], the relationship between total retardance (R), optical rotation (ψ) and linear retardance (δ) can be worked out as [19]

$$R = \cos^{-1}\left\{2\cos^2(\psi)\cos^2(\delta/2) - 1\right\}. \tag{9.24}$$

The values for optical rotation (ψ) and linear retardance (δ) can be determined from the elements of the matrix $\mathbf{M}_R$ as [17, 21]

$$\delta = \cos^{-1}\left(\sqrt{(\mathbf{M_R}(2,2) + \mathbf{M_R}(3,3))^2 + (\mathbf{M_R}(3,2) - \mathbf{M_R}(2,3))^2} - 1\right) \tag{9.25}$$

and

$$\psi = \tan^{-1}\left(\frac{\mathbf{M_R}(3,2) - \mathbf{M_R}(2,3)}{\mathbf{M_R}(2,2) + \mathbf{M_R}(3,3)}\right). \tag{9.26}$$

An interesting problem is that the multiplication order in Eq. (9.13) is ambiguous (due to the noncommuting nature of matrix multiplication, $\mathbf{M}_A\mathbf{M}_B \neq \mathbf{M}_B\mathbf{M}_A$), so that six different decompositions (order of multiplication) are possible. It has been shown that the six different decompositions can be grouped in two families, depending upon the location of the depolarizer and the diattenuator matrices [19, 22, 23]. The three decompositions with the depolarizer set after the diattenuator form the first family (of which Eq. (9.13) is a particular sequence). On the other hand, the three decompositions with the depolarizer set before the diattenuator constitute the other family.

$$(M_{\Delta D} \text{ family}) \quad \begin{array}{l} \mathbf{M} = \mathbf{M}_\Delta \mathbf{M}_R \mathbf{M}_D \\ \mathbf{M} = \mathbf{M}_\Delta \mathbf{M}_D \mathbf{M}_R \\ \mathbf{M} = \mathbf{M}_R \mathbf{M}_\Delta \mathbf{M}_D \end{array} \quad (M_{D\Delta} \text{ family}) \quad \begin{array}{l} \mathbf{M} = \mathbf{M}_D \mathbf{M}_R \mathbf{M}_\Delta \\ \mathbf{M} = \mathbf{M}_R \mathbf{M}_D \mathbf{M}_\Delta . \\ \mathbf{M} = \mathbf{M}_D \mathbf{M}_\Delta \mathbf{M}_R \end{array} \tag{9.27}$$

It has been shown that among the six decompositions, product in Eq. (9.13) or its reverse order ($\mathbf{M} = \mathbf{M}_D\ \mathbf{M}_R\ \mathbf{M}_\Delta$) always produce a physically realizable Mueller matrix [23]. The other possible decompositions can be obtained using similarity transformations, for each of the two individual families. It is thus favorable to use these two orders of decomposition when nothing is known *a priori* about an experimental Mueller matrix.

To summarize, we have presented a mathematical methodology of matrix decomposition to separate out the individual intrinsic polarimetry characteristics from "lumped" Mueller matrix obtained from an unknown complex system. Its validation and initial biological applications are described later. In the following section, we describe a high-sensitivity polarization modulation/synchronous detection experimental system for the measurement of Mueller matrix in tissue-like turbid media.

9.4 Sensitive Experimental System for Mueller Matrix Measurements in Turbid Media

In order to measure polarization signals in strongly depolarizing scattering media such as biological tissues, a highly sensitive polarimetry system is required. Multiple scattering leads to depolarization of light, creating a large depolarized source of noise that hinders the detection of the small residual polarization-retaining signal. One possible method to detect these small polarization signals is the use of polarization modulation with synchronous lock-in-amplifier detection. Many types of detection schemes have been proposed with this approach [10, 11, 13]. Some of these perform polarization modulation on the light that is incident on the sample; others modulate the sample-emerging light, by placing the polarization modulator between the sample and the detector. The resultant signal can be analyzed to yield sample-specific polarization properties that can then be linked to the quantities of interest. We describe here a specific experimental embodiment of the polarization modulation/synchronous detection approach. This arrangement carries the advantage of being assumption-independent, in that no functional form of the sample polarization effects is assumed [10]. This is important for polarimetric characterization of complex media such as tissues, since there are typically several polarization-altering effects occurring simultaneously. In such situations, it is preferable to have an approach that does not require assumptions on how tissue alters polarized light, but rather determines it directly.

A schematic of the experimental turbid polarimetry system was shown in Figure 9.1 [part (i)] [10]. Unpolarized light at 632.8 nm from a He-Ne laser is used to seed the system. The light first passes through a mechanical chopper operating at a frequency $f_c \sim 500$ Hz; this is used in conjunction with lock-in amplifier detection to accurately establish the overall signal intensity levels. Recording of the full (4×4) Mueller matrix requires generation of the four input polarization states, 0° (Stokes vector $[1\ 1\ 0\ 0]^T$), 45° (Stokes vector $[1\ 0\ 1\ 0]^T$), and 90° (Stokes vector $[1\ -1\ 0\ 0]^T$) linear polarizations, as well as circular polarization (Stokes vector $[1\ 0\ 0\ 1]^T$), which is enabled by the input optics (a linear polarizer P_1 with/without the quarter wave plate WP_1). The sample-scattered light is detected at a chosen angle (γ) as the detection optics can be rotated around the sample. The detection optics begin with a removable quarter wave plate (WP_2) with its fast axis oriented at $-45°$, when in place allowing for the measurement of Stokes parameters Q and U (linear polarization descriptors), and when removed allowing for the measurement of Stokes parameter V (circular polarization descriptor). The light then passes through a photoelastic modulator (PEM), which is a linearly birefringent resonant device operating at $f_p = 50$ kHz. The fast axis of the PEM is at 0° and its retardation is modulated according to the sinusoidal function $\delta_{\text{PEM}}(t) = \delta_o \sin\omega t$, where $\omega_p = 2\pi f_p$ and δ_o is the user-specified amplitude of maximum retardation of PEM. The light finally passes through a linear analyzer orientated at 45°, converting the PEM-imparted polarization modulation to an intensity modulation suitable for photodetection. The resulting modulated intensity is collected using a pair of lenses (detection area of 1 mm^2 and acceptance angle $\sim 18°$) and is relayed to an avalanche photodiode detector. The detected signal is sent to a lock-in amplifier, with its reference input toggling between the frequencies of the chopper (500 Hz) and the PEM controller (50 kHz and harmonics) for synchronous detection of their respective signals.

For a given polarization state of the incident light, the Stokes vector of light after the analyzing block $[I_f Q_f U_f V_f]^T$, can be related to that of the sample-emerging beam $[IQUV]^T$ as (with detection quarter wave-plate in place) [10]

$$\begin{pmatrix} I_f \\ Q_f \\ U_f \\ V_f \end{pmatrix} = \frac{1}{2} \underbrace{\begin{pmatrix} 1\,0\,1\,0 \\ 0\,0\,0\,0 \\ 1\,0\,1\,0 \\ 0\,0\,0\,0 \end{pmatrix}}_{P_2} \underbrace{\begin{pmatrix} 1\,0 & 0 & 0 \\ 0\,1 & 0 & 0 \\ 0\,0 & \cos\delta & \sin\delta \\ 0\,0 & -\sin\delta & \cos\delta \end{pmatrix}}_{\text{PEM}} \underbrace{\begin{pmatrix} 1 & 0\,0\,0 \\ 0 & 0\,0\,1 \\ 0 & 0\,1\,0 \\ 0 & -1\,0\,0 \end{pmatrix}}_{WP_2} \begin{pmatrix} I \\ Q \\ U \\ V \end{pmatrix}, \tag{9.28}$$

and when the detection quarter wave-plate is removed as

$$\begin{pmatrix} I_{fr} \\ Q_{fr} \\ U_{fr} \\ V_{fr} \end{pmatrix} = \frac{1}{2} \underbrace{\begin{pmatrix} 1\,0\,1\,0 \\ 0\,0\,0\,0 \\ 1\,0\,1\,0 \\ 0\,0\,0\,0 \end{pmatrix}}_{P_2} \underbrace{\begin{pmatrix} 1\,0 & 0 & 0 \\ 0\,1 & 0 & 0 \\ 0\,0 & \cos\delta & \sin\delta \\ 0\,0 & -\sin\delta & \cos\delta \end{pmatrix}}_{\text{PEM}} \begin{pmatrix} I \\ Q \\ U \\ V \end{pmatrix}. \tag{9.29}$$

The detected time-dependent intensities are thus

$$I_f(t) = \frac{I}{2}\left[1 - q\sin\delta + u\cos\delta\right], \tag{9.30}$$

and

$$I_{fr}(t) = \frac{I}{2}\left[1 - v\sin\delta + u\cos\delta\right], \tag{9.31}$$

Here $q = Q/I$, $u = U/I$, and $v = V/I$, and δ is the time-dependent PEM retardation, $\delta = \delta_0 \sin\omega t$. The time-varying circular function in the argument of another circular function of Eq. (9.30) and (9.31) can be expanded in Fourier series of Bessel functions, to yield signals at different harmonics of the fundamental modulation frequency. It can be advantageous in terms of SNR to choose the peak retardance of the PEM such that the zeroth order-Bessel function J_0 is zero [10]; with δ_o = 2.405 radians (resulting in $J_0(\delta_o) = 0$), Fourier-Bessel expansion of Eq. (9.30) and (9.31) gives

$$I_f(t) = \frac{1}{2}\left[1 - 2J_1(\delta_0)q\sin\omega t + 2J_2(\delta_0)u\cos 2\omega t + \ldots\right], \tag{9.32}$$

and

$$I_{fr}(t) = \frac{1}{2}\left[1 - 2J_1(\delta_0)v\sin\omega t + 2J_2(\delta_0)u\cos 2\omega + \ldots\right]. \tag{9.33}$$

The normalized Stokes parameters of the sample-scattered light (q, u, and v) can thus be obtained from synchronously-detected signals at the chopper frequency V_{1fc} (the dc signal level), and at the first and second harmonics of the PEM frequency V_{1fp} and V_{2fp} respectively. The experimentally measurable waveform in terms of the detected voltage signal is,

$$V(t) = V_{1fc} + \sqrt{2}V_{1fp}\sin\omega t + \sqrt{2}V_{2fp}\cos 2\omega t + \ldots, \tag{9.34}$$

with $\sqrt{2}$ factor taking into account the RMS nature of lock-in detection [10]. Applying Eq. (9.34) to the detected signal with the detection wave plate in the analyzer arm [Eq. (9.32)] gives

$$V_{1fc} = \frac{I}{2}k, \qquad \sqrt{2}V_{1f} = -IkJ_1(\delta_0)q, \qquad \sqrt{2}V_{2f} = IkJ_2(\delta_0)u, \tag{9.35}$$

where k is an instrumental constant, same for all equations. The normalized linear polarization Stokes parameters q and u are then obtained from,

$$q = \frac{V_{1fp}}{\sqrt{2}J_1(\delta_o)V_{1fc}}, \tag{9.36}$$

and

$$u = \frac{V_{2fp}}{\sqrt{2}J_2(\delta_o)V_{1fc}}, \tag{9.37}$$

Comparing Eqs. (9.34) and (9.33) when the detection quarter wave plate is removed yields

$$V_{1fc} = \frac{I}{2}k \tag{9.38}$$

$$\sqrt{2}V_{1f} = -IkJ_1(\delta_0)v,$$

and the circular polarization Stokes parameter v is then obtained as,

$$v = \frac{V_{1fp}}{\sqrt{2}J_1(\delta_0)V_{1fc}}. \tag{9.39}$$

The preceding discussion of experimental determination of Stokes vector descriptors deals with quantifying the polarization state of the sample-scattered light; we now turn our attention explicitly to determining the polarization properties of the sample as described by its Mueller matrix. In order to perform measurements of the full (4×4) Mueller matrix, the input polarization is cycled between four states (linear polarization at 0°, 45°, 90°, and right circular polarization) and the output Stokes vector for each respective input state is measured. The elements of the resulting 4 measured Stokes vectors can be analyzed to yield the sample Mueller matrix as [17]

$$\mathbf{M}(i,j) = \begin{bmatrix} \frac{1}{2}(I_H+I_V) & \frac{1}{2}(I_H-I_V) & I_p-\mathbf{M}(1,1) & I_R-\mathbf{M}(1,1) \\ \frac{1}{2}(Q_H+Q_V) & \frac{1}{2}(Q_H-Q_V) & Q_p-\mathbf{M}(2,1) & Q_R-\mathbf{M}(2,1) \\ \frac{1}{2}(U_H+U_V) & \frac{1}{2}(U_H-U_V) & U_p-\mathbf{M}(3,1) & U_R-\mathbf{M}(3,1) \\ \frac{1}{2}(V_H+V_V) & \frac{1}{2}(V_H-V_V) & V_p-\mathbf{M}(4,1) & V_R-\mathbf{M}(4,1) \end{bmatrix}. \tag{9.40}$$

Here, the four input states are denoted with the subscripts H (0°), P (45°), V (90°), and R (right circularly polarized; left circular incidence can be used as well, resulting only in a sign change). The indices $i, j = 1,2,3,4$ denote rows and columns respectively.

The described experimental approach based on polarization modulation and synchronous detection is suitable for sensitive polarimetric detection in turbid media. This experimental system has been used to carry out several fundamental studies on turbid medium polarimetry and for polarization-based characterization of biological tissues. Some of these studies are described in Section 9.6, and 9.8 of this chapter. For now, we turn to the equally challenging problems of accurately forward modeling the polarization signals in turbid media (part (ii) of Figure 9.1, Section 9.5).

9.5 Forward Modeling of Simultaneous Occurrence of Several Polarization Effects in Turbid Media Using the Monte Carlo Approach

To aid in the investigation of polarized light propagation in turbid media such as biological tissue, accurate modeling is enormously useful for gaining physical insight, designing and optimizing experiments, and analyzing measured data. The Maxwell's equations-based electromagnetic theoretical approach is the most rigorous and best-suited method for polarimetric analysis; however, solving Maxwell's equations for polarized light propagation in multiple scattering media is impractical. Alternatively, light propagation through scattering media can in principle be modeled through transport theory; however, transport theory and its simplified variant, the diffusion approximation, are both intensity-based techniques, and hence typically neglect polarization [24]. A more general and robust approach is the Monte Carlo technique [24]. In this statistical approach to radiative transfer, the multiple scattering trajectories of individual photons are determined using a random number generator to predict the probability of each scattering event. The superposition of many photon paths approaches the actual photon distribution in time and space. This approach has the advantage of being applicable to arbitrary geometries and arbitrary optical properties. The first Monte Carlo models were also developed for intensity calculations only and neglected polarization information [25]. More recently, a number of implementations have incorporated polarization into their Monte Carlo models [26]. Currently, the Monte Carlo technique is the most general approach to simulate polarized light propagation in scattering media, although long computation times are often required to generate statistically meaningful results.

A flowchart for polarization-sensitive Monte Carlo model is shown in Figure 9.2. In this modeling, it is assumed that scattering events occur independently and exhibit no coherence effects. The position, propagation direction, and polarization of each photon are initialized and modified as the

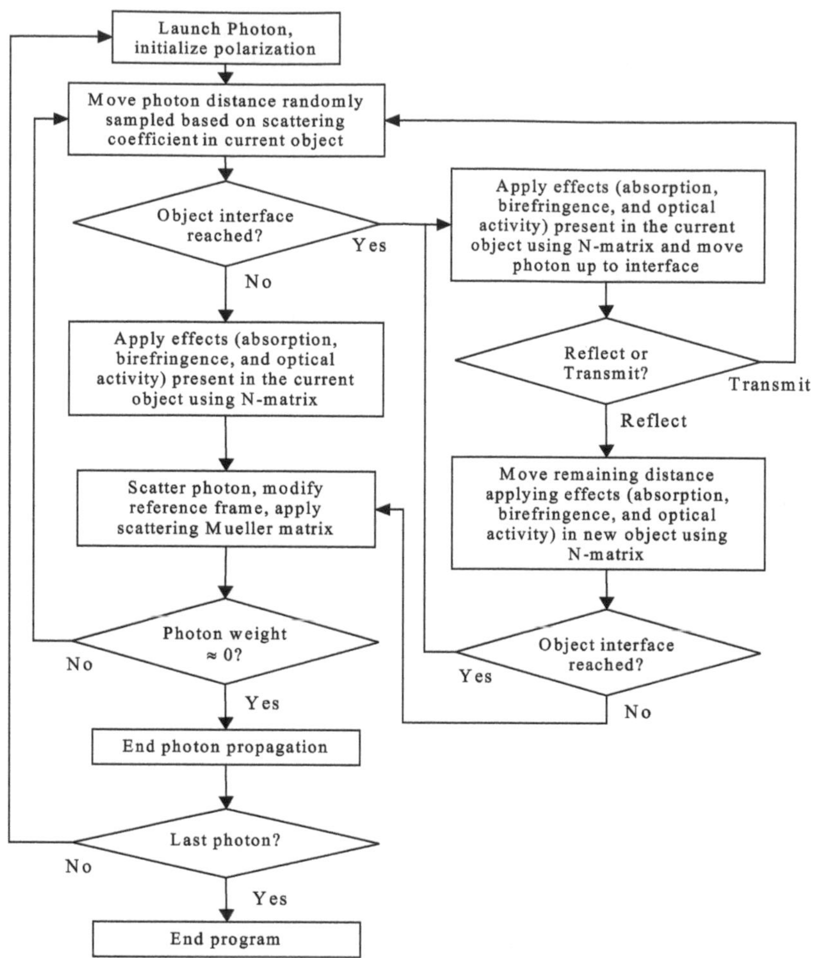

FIGURE 9.2: Flowchart for polarization-sensitive Monte Carlo model. The polarization effects are applied using medium Mueller matrices. The Mueller matrix for scattering is calculated using Mie theory and that for other simultaneously occurring effects (birefringence and optical activity) are calculated using N-matrix approach (see text).

photon propagates through the scattering medium. The photon's polarization, with respect to a set of arbitrary orthonormal axes defining its reference frame, is represented as a Stokes vector **S** and polarization effects are applied using medium Mueller matrices **M** [11, 27]. The photon propagates in the sample between scattering events a distance l sampled from the probability distribution $\exp(-\mu_t l)$. Here, the extinction coefficient μ_t is the sum of the absorption μ_a and scattering μ_s coefficients and l is the distance traveled by the photon between scattering events. When the photon encounters a scattering event, a scattering plane and angle are statistically sampled based on the polarization state of the photon and the Mueller matrix of the scatterer.

The photon's reference frame is first expressed in the scattering plane and then transformed to the laboratory (experimentally observable) frame through multiplication by a Mueller matrix calculated through Mie scattering theory [28]. Upon encountering an interface (either an internal one, representing tissue domains of different optical properties, or an external one, representing external tissue boundary), the probability of either reflection or transmission is calculated using Fresnel coefficients. As no coherence effects are considered, the final Stokes vector for light exiting the sample in a particular direction is computed as the sum of all the appropriate directional photon subpopulations. Various quantities of interest such as detected intensities, polarization properties (Stokes vectors, Mueller matrices), average pathlengths, and so forth, can be quantified once a sufficient number of photon (packets) have been followed and tracked to generate statistically acceptable results (typically 10^7–10^9 photons).

However, most current Monte Carlo models of polarized light propagation do not fully simulate all of the polarization effects of tissue. This is primarily due to the inherent difficulty in formulating *simultaneous* polarization effects, especially in the presence of multiple scattering. As discussed previously, multiplication of the Mueller matrices for individual polarization effects is not commutative, thus, different orders in which these effects are applied will have different effects on the polarization. Ordered multiplication of these matrices does not make physical sense, as in biological tissue these effects (such as optical activity due to chiral molecules and linear birefringence due to anisotropic tissue structures) are exhibited simultaneously, and not one after the other as sequential multiplication implies. Fortunately, a method exists to simulate simultaneous polarization effect in *clear* media through the so-called N-matrix formalism, which combines the effects into a single matrix describing them simultaneously [29]. The N-matrix formalism was thus employed in the polarization-sensitive Monte Carlo simulation code in tissue-like media to model *simultaneous* polarization effects *between* scattering events in the presence of multiple scattering.

The N-matrix approach was first developed by Jones [29], and a more thorough derivation is provided in Kliger et al. [3]. Briefly, in this approach the sample matrix is represented as an exponential function of a *sum* of matrices, where each matrix in the sum corresponds to a single optical polarization effect. The issue of ordering of noncommutative multiplying matrices disappears as matrix addition is always commutative, and applies to differential matrices representing the optical property over an infinitely small optical pathlength. These differential matrices, known as N-matrices, correspond to each optical property exhibited by the sample and are summed to express combined effect. The formalism is expressed in terms of 2×2 Jones matrices applicable to clear nondepolarizing media, rather than the more commonly used 4×4 Mueller matrices. However, a Jones matrix can be converted to a Mueller matrix provided there are no depolarization effects [15]. This is indeed applicable to our Monte Carlo model, as depolarization is caused by the multiple scattering events, and no depolarization effects occur *between* the scattering events. Once converted to the Mueller matrix formalism, this modified N-matrix approach was then applied to the photons as they propagate between scattering events in the MC simulation. This approach enabled the combination of any number of simultaneously occurring polarizing effects [in our case, circular birefringence (optical activity) and linear birefringence were incorporated, since these are the most prominent tissue polarimetry effects] [20].

In the simulations, circular and linear birefringence were modeled through the optical activity χ in degrees per centimeter, and through the anisotropy in refractive indices (Δn), respectively. Here, $\Delta n = (n_e - n_o)$ is the difference in the refractive index along the extraordinary axis (n_e) and the ordinary axis (n_o). For simplicity, it was assumed that the medium is uni-axial and that the direction of the extraordinary axis and the value for Δn is constant throughout the scattering medium. In each simulation, n_e and n_o were taken as input parameters and a specific direction of the extraordinary axis was chosen. As each photon propagates between scattering events, the difference in refractive indices seen by the photon depends on the propagation direction with respect to the extraordinary axis. The effect was modeled using standard formulae describing the angular variation of the refractive index in a uni-axial medium [20].

The ability of the extended polarization-sensitive Monte Carlo model to simulate *simultaneous* polarization effects in the presence of multiple scattering was experimentally validated in solid polyacrylamide phantoms exhibiting simultaneous linear birefringence, optical activity, and depolarization [20]. These were developed using polyacrylamide as a base medium, with sucrose-induced optical activity, polystyrene microspheres-induced scattering (mean diameter $D = 1.40\mu$m, refractive index $n_s = 1.59$), and mechanical stretching to cause linear birefringence (or linear retardance). To apply controllable strain to produce linear birefringence, one end of the polyacrylamide phantoms (dimension of $1 \times 1 \times 4$ cm) was clamped to a mount and the other end to a linear translational stage. The phantoms were stretched along the vertical direction (the long axis of the sample) to introduce varying linear birefringence with its axis along the direction of strain. Measurements of the Stokes parameters (normalized Stokes parameters $q = Q/I,\ u = U/I,\ v = V/I$) of scattered light (in different geometries) from the phantoms were made using the PEM-based polarimeter (described in the previous section), and compared to the results of Monte Carlo simulations run with similar parameters. Measurements were made from phantoms exhibiting turbidity and birefringence (no added sucrose, i.e, no optical activity), as well as for phantoms exhibiting turbidity, birefringence, and optical activity [20]. In both cases, the Monte Carlo-simulated Stokes parameters were in excellent agreement with the controlled experimental results (for details the reader is referred to Ref. 18). These results provided strong evidence for the validity of the Monte Carlo model. The model can therefore accurately simulate complex tissue polarimetry effects, including simultaneous optical activity and birefringence in the presence of scattering. This, in combination with the experimental turbid polarimetry system (and phantoms), was therefore used to test the efficacy of the Mueller matrix decomposition approach to delineate individual intrinsic polarimetry characteristics in complex tissue-like turbid media. Some of these studies are now described.

9.6 Validation of the Mueller Matrix Decomposition Method in Complex Tissue-Like Turbid Media

Previously employed for examination of nonscattering media, the extended Mueller matrix decomposition methodology (described in Section 9.3) has seen only initial use in turbid media, and as such requires validation. The validity of the matrix decomposition approach summarized in Eqs. (9.13)–(9.27) in complex turbid media was therefore tested with both experimental (Section 9.4) and MC-simulated (Section 9.5) Mueller matrices, whose constituent properties are known and user-controlled *a priori* [17,18].

In the experimental studies, the PEM-based polarimeter (Figure 9.1) was used to record Mueller matrices in the forward detection geometry (sample thickness = 1 cm, detection area of 1 mm^2, and an acceptance angle $\sim 18^\circ$ around $\gamma = 0^\circ$ direction were used) from a solid polyacrylamide phantom (discussed in the previous section) that mimics the complexity of biological tissues, in that it exhibits simultaneous linear birefringence, optical activity, and depolarization [10, 17, 20].

Figure 9.3 and Table 9.1 show the experimental Mueller matrix and the corresponding decomposed depolarization ($\mathbf{M}_\Delta$), retardance ($\mathbf{M}_R$), and diattenuation ($\mathbf{M}_D$) matrices from a birefringent (extension = 4 mm), chiral (magnitude of optical activity was χ = 1.96degree cm^{-1}, corresponding to 1 M concentration of sucrose), turbid phantom (scattering coefficient of μ_s = 30 cm^{-1} and anisotropy parameter g = 0.95). The complicated nature of the resultant Mueller matrix $\mathbf{M}$, with essentially all 16 nonzero matrix elements, underscores the problem at hand – how does one extract useful sample metrics from this wealth of intertwined information? In contrast, the three basis matrices derived from the decomposition process exhibit simpler structures with many zero off-diagonal elements, and are directly amenable for further quantification. Equations (9.17), (9.21), (9.25), and (9.26) were applied to the decomposed basis matrices to retrieve the individual polarization parameters (d, Δ, δ, and ψ). The determined values for these are also listed in Table 9.1.

The comparison of the derived and the input control values for the polarization parameters reveals several interesting trends. The expected value for diattenuation d is zero, whereas the decomposition

TABLE 9.1: The values for the polarization parameters extracted from the decomposed matrices (2nd column). The input control values for linear retardance δ and optical rotation ψ (3rd column) were obtained from measurement on a clear ($\mu_s = 0$ cm^{-1}) phantom having the same extension (= 4 mm) and similar concentration of sucrose (1 M) as that of the turbid phantom, and corrected for the increased pathlength due to multiple scattering (determined from Monte Carlo modeling). The expected value for the net depolarization coefficient Δ was determined from the Monte Carlo simulation of the experiment. (Adopted from [17]).

Parameters	Estimated value (from $\mathbf{M}_\Delta$, $\mathbf{M}_R$, $\mathbf{M}_D$)	Expected value
d	0.03	0
δ	1.38 rad	1.34 rad
ψ	2.04°	2.07°
Δ	0.21	0.19

method yields a small but nonzero value of d = 0.03. Scattering-induced diattenuation that arises primarily from singly (or weakly) scattered photons [20] is not expected to contribute here, because multiply scattered photons are the dominant contributors to the detected signal in the forward detection geometry. Hence, the presence of small amount of dichroic absorption (at the wavelength of excitation λ = 632.8 nm) due to anisotropic alignment of the polymer molecules in the polyacrylamide phantom possibly contribute to this slight nonzero value for the parameter d.

The derived decomposition value of $\Delta = 0.21$ seems reasonable, although this is difficult to compare with theory (there is no direct link between the scattering coefficient and resultant depolarization). The value shown in the theoretical comparison column of the Table was determined from the Monte Carlo simulation of the experiment, as described in the previous section. The resultant agreement in the depolarization values is excellent. It is worth noting that decomposition results for an analogous purely depolarizing phantom (same turbidity, no birefringence nor chirality – results not

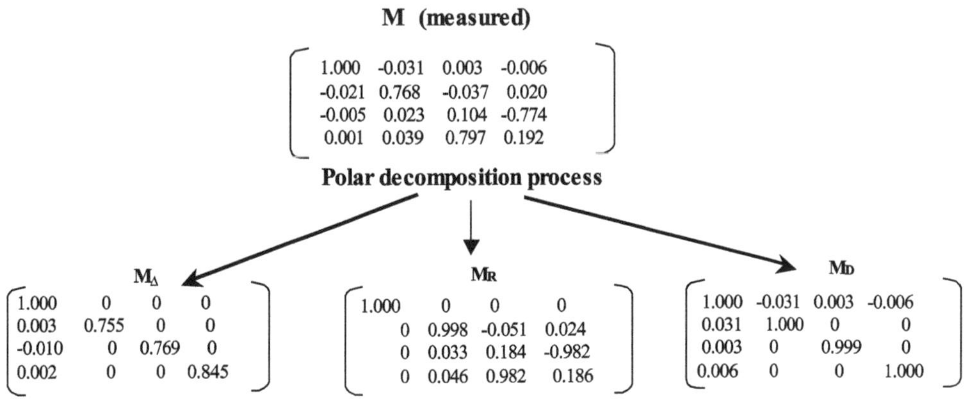

FIGURE 9.3: The experimentally recorded Mueller matrix and the decomposed matrices for a birefringent (extension = 4 mm), chiral (concentration of sucrose = 1 M, $\chi = 1.96°$ / cm), turbid (μ_s = 30 cm^{-1}, g = 0.95, thickness t = 1 cm) phantom (Adopted from [17]).

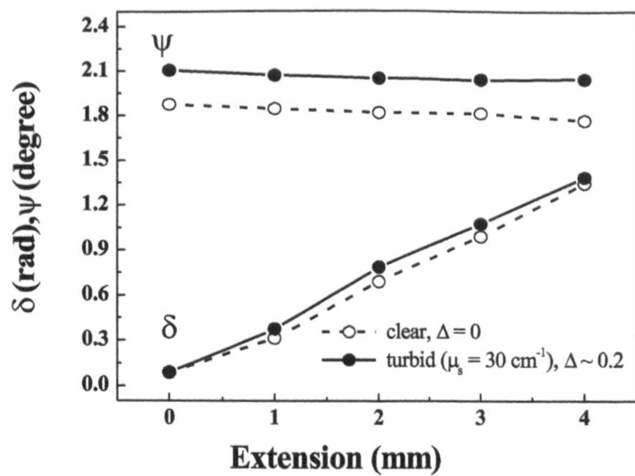

FIGURE 9.4: Linear retardance δ and optical rotation ψ estimated from the decomposition of *experimentally* recorded Mueller matrices from solid chiral (χ = 1.96° cm^{-1}, corresponding to 1 M concentration of sucrose) phantoms having varying degrees of strain-induced birefringence (extension of 0–4 mm, $\delta = 0$–1.345). Results are shown for both clear ($\mu_s = 0$ cm^{-1}) and turbid ($\mu_s = 30$ cm^{-1}, $g = 0.95$) phantoms. The measurements were performed in the forward direction (γ = 0°) through a 1 cm × 1 cm × 4 cm phantom. The points represent decomposition-derived values and the lines are guide for the eye for this and all subsequent figures. (Adopted from [17]).

shown) were within 5% of the above Δ values [17]. This self-consistency implied that decomposition process successfully decouples the depolarization effects due to multiple scattering from linear retardation and optical rotation contributions, thus yielding accurate and quantifiable estimates of the δ and ψ parameters in the presence of turbidity.

The Mueller-matrix derived value of optical rotation $\psi = 2.04°$ of the turbid phantom was, however, slightly larger than the corresponding value measured from a clear phantom having the same concentration of sucrose ($\psi_0 = 1.77°$). This small increase in the ψ value in the presence of turbidity is likely due to an increase in optical path length engendered by multiple scattering. Indeed, the value for ψ, calculated using the optical rotation value for the clear phantom (ψ_0 = 1.77°) and the value for average photon path length ($\langle L \rangle = 1.17$ cm, determined from Monte Carlo simulations [20]) [$\psi = \psi_0 \times \langle L \rangle = 2.07°$] was reasonably close to the decomposition-derived value from the experimental Mueller matrix (ψ = 2.04°).

Although the estimated value for retardance δ of the turbid phantom is slightly larger than that for the clear phantom ($\delta = 1.38$ rad for the turbid phantom, as compared to 1.34 rad for the clear phantom), the value is significantly lower than that one would expect for average photon path length of 1.17 cm ($\delta = 1.34 \times 1.17 = 1.57$ rad). This can be seen from Figure 9.4, where the estimates for δ and ψ of the chiral (χ = 1.96° cm^{-1}) phantoms having varying degree of strain induced birefringence (extension of 0–4 mm) are displayed [17]. The increase in the value for ψ as a result of increased average photon path length in the turbid phantom as compared to the clear phantom is clearly seen. The gradual decrease in the value for ψ with increasing longitudinal stretching is consistent with resulting lateral contraction of the phantom, reducing the effective path length. In contrast to ψ, the expected increase in δ as a result of increased average photon path length in the turbid phantom (compared to clear medium) is not that apparent.

Figure 9.5 shows the derived linear retardance δ and optical rotation ψ parameters, using Monte Carlo-generated Mueller matrices, with chiral molecule concentration as the independent variable [30]. Again, both the clear and turbid values compare well to the input parameter values ($\delta \approx 1.4$ rad

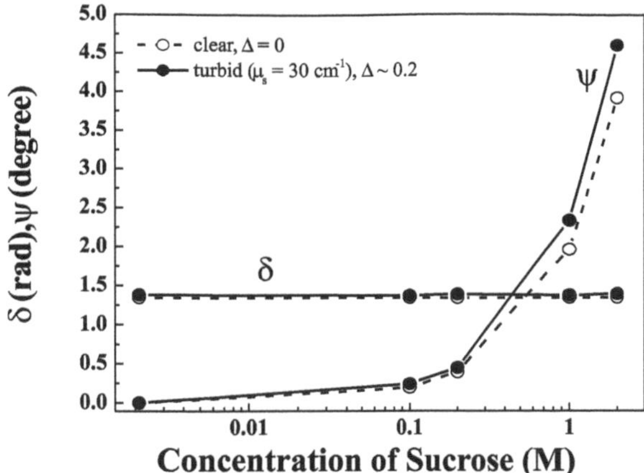

FIGURE 9.5: δ and ψ estimated from the decomposition of *Monte Carlo*-generated Mueller matrices from birefringent media (linear birefringence $\Delta n = 1.36 \times 10^{-5}$, corresponding to $\delta =$ 1.345 rad [4 mm extension of 4 cm long phantom] for a path length of 1 cm, the axis of linear birefringence was kept along the vertical direction, orientation angle $\theta = 90°$) having varying levels of chirality ($\chi = 0, 0.196, 0.392, 1.96$ and $3.92°$ cm^{-1}, corresponding to concentration of sucrose of 0, 0.1, 0.2, 1 and 2 M, respectively). Results are shown for both clear ($\mu_s = 0$ cm^{-1}) and turbid ($\mu_s = 30$ cm^{-1}, $g = 0.95$) media. (Adopted from [30]).

and $\psi \approx 1.96°$ at 1 M sucrose), showing self-consistency in inverse decomposition analysis and successful decoupling. Further, in agreement with the experimental results, while the value for ψ of the turbid media is larger than that for the clear media as a result of the increase in optical path length due to multiple scattering, a similar increase in δ value is not observed.

Note that none of these trends could be gleaned from the lumped Mueller matrix, where at best one would have to resort to semi-empirical comparison of changes in selected matrix elements, which contain contributions from several effects. Derivation and quantification of the absolute linear retardance δ and optical rotation ψ values is enabled exclusively by the polar decomposition analysis. Based on these and other continuing validation studies, this approach appears valid in complex tissue-like turbid media.

9.7 Selected Trends: Path length and Detection Geometry Effects on the Decomposition-Derived Polarization Parameters

An interesting finding of the results presented in the previous section was that while the value for the derived optical rotation ψ of the turbid media was consistently larger than that of the clear media as a result of increased average photon path length, a similar increase in linear retardance δ was not observed. Monte Carlo simulations were carried out further to understand this trend, and to investigate the effect of multiple scattering, propagation path, and detection geometry on the decomposition-derived δ and ψ parameters [17]. Here, we shall discuss some selected results of those studies.

The MC results suggested that the lowering of the value of net retardance δ likely arises because the scattered light does not travel in a straight line but rather along many possible curved zig-zag

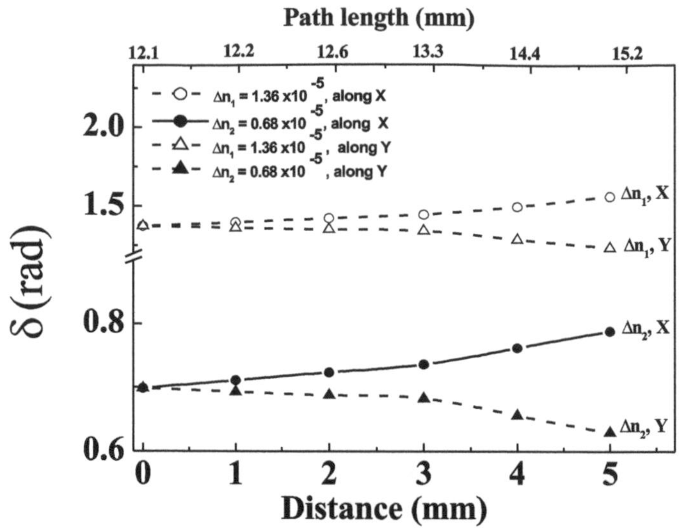

FIGURE 9.6: Variation of δ of transmitted light as a function of distance from ballistic beam position at the exit face of a 1 cm thick birefringent, chiral ($\chi = 1.96^\circ$ cm^{-1}), turbid medium ($\mu_s = 30$ cm^{-1}, $g = 0.95$). The results are shown for two different values of birefringence ($\Delta n =$ 1.36×10^{-5} and 0.68×10^{-5}). The axis of linear birefringence was kept along the vertical (Y-axis) direction and the results are shown for transmitted light collected at different spatial positions along the horizontal (X-axis) and vertical (Y-axis) direction. The Monte Carlo-calculated average photon path length of light exiting the scattering medium is shown on the top axis. (Adopted from [17]).

paths, the curvature being controlled by the values for μ_s and g. While such paths influence ψ by increasing its value through a relationship with the increasing photon path length, the effect of this on the net value for δ is more complex because a component of the curved propagation paths will be along the direction of the linear birefringence axis (along Y in Figs. 9.6, 9.7)

These results are illustrated in Figures 9.6 and 9.7. Figure 9.6 shows the variation of δ values of transmitted light collected at the exit face of a birefringent ($\Delta n = 1.36 \times 10^{-5}$ and 0.68×10^{-5}), chiral ($\chi = 1.96^\circ$ cm^{-1}), turbid medium ($\mu_s = 30$ cm^{-1}, $g = 0.95$, thickness = 1 cm), as a function of distance from the center of the transmitted ballistic beam along the horizontal (X axis, perpendicular to the direction of the axis of birefringence) and vertical (Y axis, parallel to the direction of the axis birefringence) directions. At detector positions along the X axis, δ increases with increasing distance from the center of the ballistic beam (i.e., with increasing average photon path length). In contrast, for detection positions along the Y axis, the value for δ shows gradual decrease with increasing distance from the center of the ballistic beam. This is because a larger component of the photon propagation path is along the axis of birefringence leading to a reduction in net linear retardance δ (because propagation along the direction of the birefringence axis does not yield any retardance [15]). Since such differences in the photon propagation path for the two different detection geometries should have no influence on the value of ψ, the estimates for ψ were found to be identical for similar detection positions either along the X- or the Y-axis at the exit face of the medium (data not shown) [17].

Figure 9.7 demonstrates that for off-axis detection [position coordinate (3,0) or (0,3)], the Mueller matrix-derived δ value vary considerably with a change in the orientation angle (θ) of the birefringence axis. Conversely, for detection around the position of the ballistic beam [coordinate (0, 0)],

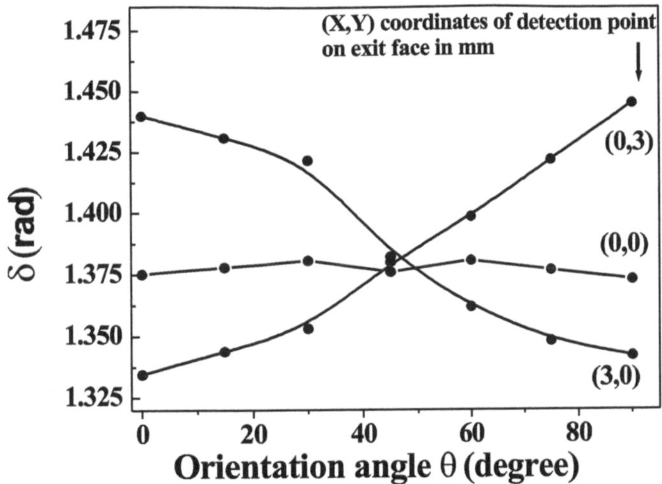

FIGURE 9.7: Variation of δ (estimated from the decomposition of *Monte Carlo-generated* Mueller matrices) as a function of the linear birefringence axis orientation angle θ (with respect to the horizontal) for a birefringent, chiral, turbid medium ($\Delta n = 1.36 \times 10^{-5}$, $\chi = 1.96^\circ$ cm^{-1}, $\mu_s = 30$ cm^{-1}, $g = 0.95$, thickness = 1 cm). Results are shown for three different detection positions at the exit face of the medium, detection around the position of the ballistic beam [position coordinate (0,0)], detection at spatial positions 3 mm away from the ballistic beam position along the horizontal (X) [coordinate (3,0)] and vertical (Y) [position coordinate (0,3)] axis, respectively. (Adopted from [17]).

δ is not influenced significantly by a change in θ. Thus, for simultaneous determination of the intrinsic values for the parameters δ and ψ of a birefringent, chiral, turbid medium in the forward scattering geometry, detection around the direction of propagation of the ballistic beam may be preferable [17].

These and other experimental and MC-simulation results on phantoms having varying optical scattering and polarization properties demonstrated that the Mueller matrix decomposition approach can successfully delineate individual intrinsic polarimetry characteristics in complex tissue-like turbid media in the forward detection geometry. Yet the backward detection geometry may be more convenient for many practical applications (particularly for *in situ* measurements). However, the scattering-induced artifacts are more coupled with the intrinsic polarization parameters in the backward detection geometry, partly because of the increasing contribution of the singly or weakly backscattered photons [10, 14, 21]. For example, backscattering induced changes in the orientation angle of the linear polarization vector can manifest themselves as large apparent optical rotation even in absence of chiral molecules in the medium [10, 14]. Decomposition analysis revealed that this large scattering-induced apparent rotation is due to linear diattenuation (difference in amplitude between the scattered light polarized parallel and perpendicular to the scattering plane) [18, 21]. In addition to diattenuation, backscattered photons yields significant values of linear retardance (differences in phase between the scattered light polarized parallel and perpendicular to the scattering plane) even from isotropic ($\Delta n = 0$) scattering medium [18]. This scattering-induced linear retardance interferes with the actual retardance values of a birefringent turbid medium in a complex interrelated way, thus hindering the determination of the latter in the backward detection geometry.

Our studies have shown that the scattering-induced diattenuation and linear retardance are due

(a)

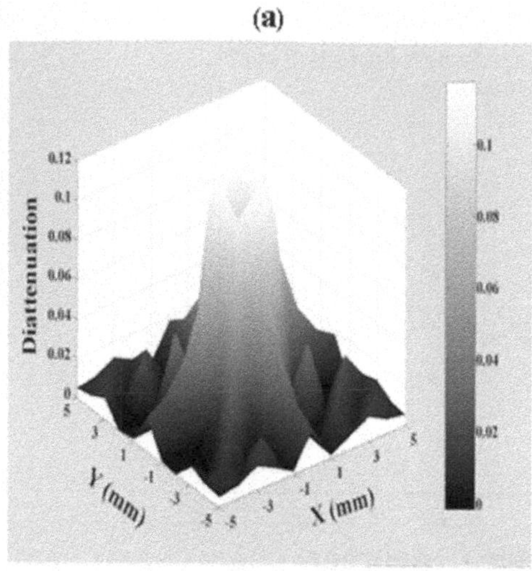

(b)

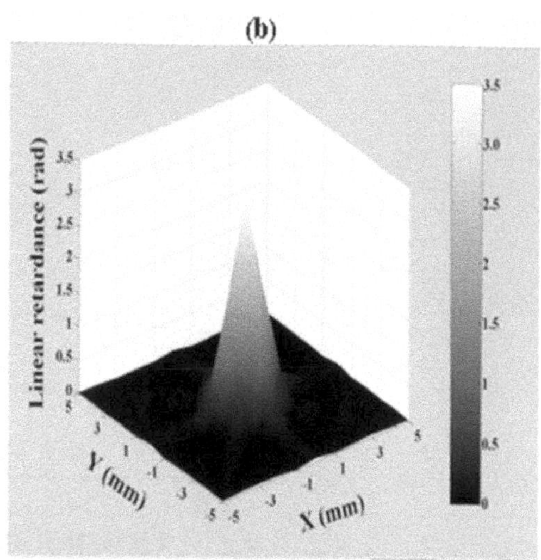

FIGURE 9.8: (a) Diattenuation d and (b) linear retardance δ maps in the backscattering plane ($X-Y$ plane, $Z=0$) derived using decomposition of Monte Carlo-generated Mueller matrices of a nonbirefringent (Δn = 0), achiral ($\chi = 0°$ cm^{-1}), turbid medium (μ_s = 60 cm^{-1}, g = 0.935). The nonzero values for both d and δ results primarily from the confounding effects of singly (weakly) backscattered photons. (Adopted from [18]).

mainly to the singly (or weakly) backscattered photons, and their magnitude is large at detection positions sufficiently close to the exact backscattering direction [18]. As one moves away from the exact backscattering direction, these confounding effects gradually diminish. This is illustrated in Figure 9.8, where the scattering-induced diattenuation (d) and linear retardance (δ) maps in the backscattering plane ($X-Y$ plane, $Z=0$) are shown from a isotropic ($\Delta n = 0$), achiral ($\chi = 0°$ cm^{-1}), turbid medium (μ_s = 60 cm^{-1}, g = 0.935, thickness = 1 cm). These results and further decomposition analyses on Monte Carlo-generated Mueller matrices for turbid media having different scattering coefficients μ_s confirmed that in the backward detection geometry, the effects of scattering-induced linear retardance and diattenuation are weak ($\delta \leq 0.1$ and $d \leq 0.03$) for detection positions located at distances larger than a transport length away from the point of illumination [$r > l_{tr}$, l_{tr} is the transport scattering length $= 1/\mu_s(1-g)$]. Simultaneous determination of the unique intrinsic values of all the polarization parameters from a turbid medium in the backward detection geometry can thus be accomplished by decomposing the Mueller matrix recorded at a distance larger than a transport length away from the point of illumination [18].

Note (as discussed previously in Section 9.3), the multiplication order of the basis matrices in the decomposition analysis [Eq. (9.13)] is ambiguous (due to the noncommuting nature of matrix multiplication). The influence of the order of the matrices in the decomposition analysis on the retrieved polarization parameters was therefore investigated. The experimental and the Monte Carlo-generated Mueller matrices (results presented above) were decomposed following either the order of Eq. (9.13) or its reverse order ($\mathbf{M} = \mathbf{M}_D\, \mathbf{M}_R\, \mathbf{M}_\Delta$). Importantly, the three useful polarization parameters [Δ, ψ, and δ, derived through Eqs. (9.21), (9.25), and (9.26)], were found to be ~independent of the order [difference in their values was in the range 1–5%] [17, 18]. This suggests that the decomposition formalism is self-consistent with respect to the potential ambiguity of ordering. Further work is underway to confirm this initial finding [31].

To summarize Sections 9.6 and 9.7, the applicability of the Mueller matrix decomposition approach in complex tissue-like turbid media (either for the forward or the backward detection geometry) was validated experimentally with optical phantoms having controlled sample-polarizing properties, and theoretically with a polarization sensitive Monte Carlo model capable of simulating complex tissue polarimetry effects. The individual polarization effects can be successfully decoupled and quantified despite their simultaneous occurrence, even in the presence of the numerous complexities due to multiple scattering. The ability to isolate individual polarization properties provides a valuable noninvasive tool for their quantification and may be relevant for biological tissue examinations. In the following, initial applications of this promising approach in two scenarios of significant clinical interest, that for noninvasive glucose measurements and for monitoring regenerative treatments of the heart, as well as initial *in vivo* demonstration, are discussed.

9.8 Initial Biomedical Applications

9.8.1 Noninvasive glucose measurement in tissue-like turbid media

Diabetes Mellitus is a chronic systemic disease, with no known cure, in which the body either fails to produce, or fails to properly respond to the glucose regulator hormone insulin. The most reliable current method for glucose monitoring in diabetic patients necessitates the drawing of blood, usually by a finger prick several times a day – a painful, inconvenient, and poorly compliant procedure. A tremendous need therefore exists for a *noninvasive* glucose monitoring method, as it would increase the determination frequency and enable better insulin and caloric intake, leading to a tighter glucose level control and preventing or delaying long-term complications. A variety of optical methods have been attempted for blood glucose monitoring including, near-infrared (NIR) spectroscopy, Raman spectroscopy, fluorescence, photoacoustics, optical coherence tomography and polarimetry, but none have shown the requisite sensitivity/specificity/accuracy [9]. Polarimetry, based on the chiral (handed) nature of the glucose molecules and their associated optical activity, is particularly promising as it is potentially specific to glucose. In fact, such measurements in clear media have been used for decades in the sugar industry. However, in a complex turbid medium like tissue, polarimetric attempts for glucose quantification have been confounded by several factors. One of the major stumbling blocks is that the optical rotation due to chiral substances in a turbid medium is swamped by the much larger changes in the orientation angle of the polarization vector due to scattering [10, 14, 18, 21].

It is thus essential to isolate the optical rotation caused exclusively by chiral molecules from the (often much larger) apparent rotation caused by the scattering / detection geometry effects. The matrix decomposition methodology is indeed able to perform this task, as shown in Figure 9.9 [30]. The variation of scattering induced rotation α is displayed as a function of distance from the point of illumination of a chiral ($\chi = 0.082^\circ$ cm^{-1}, corresponding to 100 mM concentration of glucose), nonbirefringent ($\Delta n = 0$), turbid medium ($\mu_s = 30$ cm^{-1}, $g = 0.95$, thickness = 1 cm). The incident light was 45° polarized (Stokes vector $[1\ 0\ 1\ 0]^T$) and the rotation of the linear polarization vector (α) was calculated from the recorded Stokes parameters $[I\ Q\ U\ V]$ of light exiting the sample through the backscattering plane (X-Y plane, $Z = 0$) as

$$\alpha = 0.5 \times \tan^{-1}(U/Q) \tag{9.41}$$

As seen, changes in the polarization caused by scattering can manifest themselves as large apparent optical rotation. Decomposition analysis revealed that the large scattering-induced apparent rotation is due to linear diattenuation (also shown in the figure). This confounding effect is due mainly to the singly (weakly) backscattered photons and gradually decreases away from the exact backscattering direction [18] (see Section 9.7). Decomposition of the Mueller matrix can thus decouple this chirality-unrelated rotation from the much smaller ψ rotation values caused by the

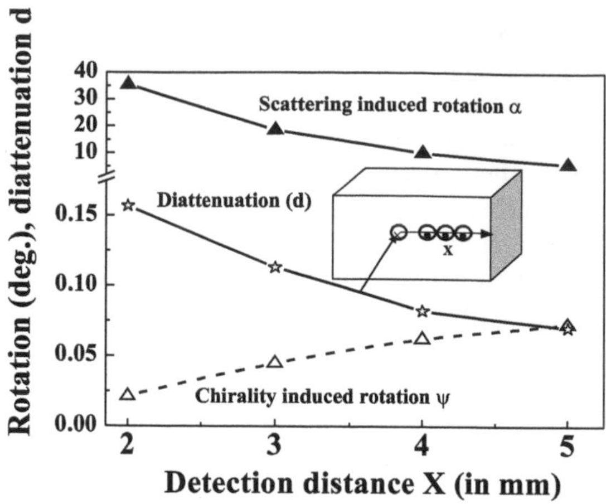

FIGURE 9.9: Calculated optical rotation ψ (derived from the decomposition of Monte Carlo-generated Mueller matrices, open triangles) of scattered light emerging in the backscattering direction as a function of distance from the center of the incident beam from a chiral (χ = 0.082° / cm, corresponding to 100 mM concentration of glucose) isotropic turbid medium (μ_s = 30 cm^{-1}, $g = 0.95$, thickness $t = 1$ cm). The corresponding scattering-induced rotation of the polarization vector derived from the Stokes parameters of scattered light (for incident Stokes vector [1 0 1 0] ^{T}) is shown by solid triangles. The scattering-induced diattenuation d is also shown. The inset shows the backwards detection geometry. The chirality-induced rotation approaches zero in the exact backscattering direction ($X = 0$, data not shown). (Adopted from [30]).

circular birefringence of the medium (which can then be linked to glucose concentration).

The ability to decouple the small optical rotation caused exclusively by chiral molecules even in the presence of numerous complexities due to the scattering/detection geometry and simultaneously occurring polarization effects bodes well for the potential application of this method for noninvasive glucose measurements in tissue. However, this remains to be rigorously investigated. In combination with Monte Carlo-determined path length distributions [32], we are currently exploring methods for extracting chiral molecule concentrations from derived optical rotations. Spectroscopic-based polarimetry combined with chemometric regression analysis is also being investigated to isolate the rotation due to glucose from that caused by other chiral biological constituents [33].

9.8.2 Monitoring regenerative treatments of the heart

The Mueller matrix decomposition method was explored for polarimetric monitoring of myocardial tissue regeneration following stem-cell therapy. The anisotropic organized nature of myocardial tissues stemming from their fibrous structure leads to linear birefringence. After suffering an infarction (heart attack), a portion of the myocardium is deprived of oxygenated blood and subsequently cardiomyocytes die, being replaced by the fibrotic (scar) tissue [34].

Recently, stem-cell-based regenerative treatments for myocardial infarction have been shown to reverse these trends by increasing the muscular and decreasing the scar tissue components [35]. These remodeling processes are expected to affect tissue structural anisotropy, and measurement of linear birefringence may offer a sensitive probe into the state of the myocardium after infarction and report on the success of regenerative treatments. However, these small birefringence alterations must be decoupled from the other confounding polarization effects that are present in the composite

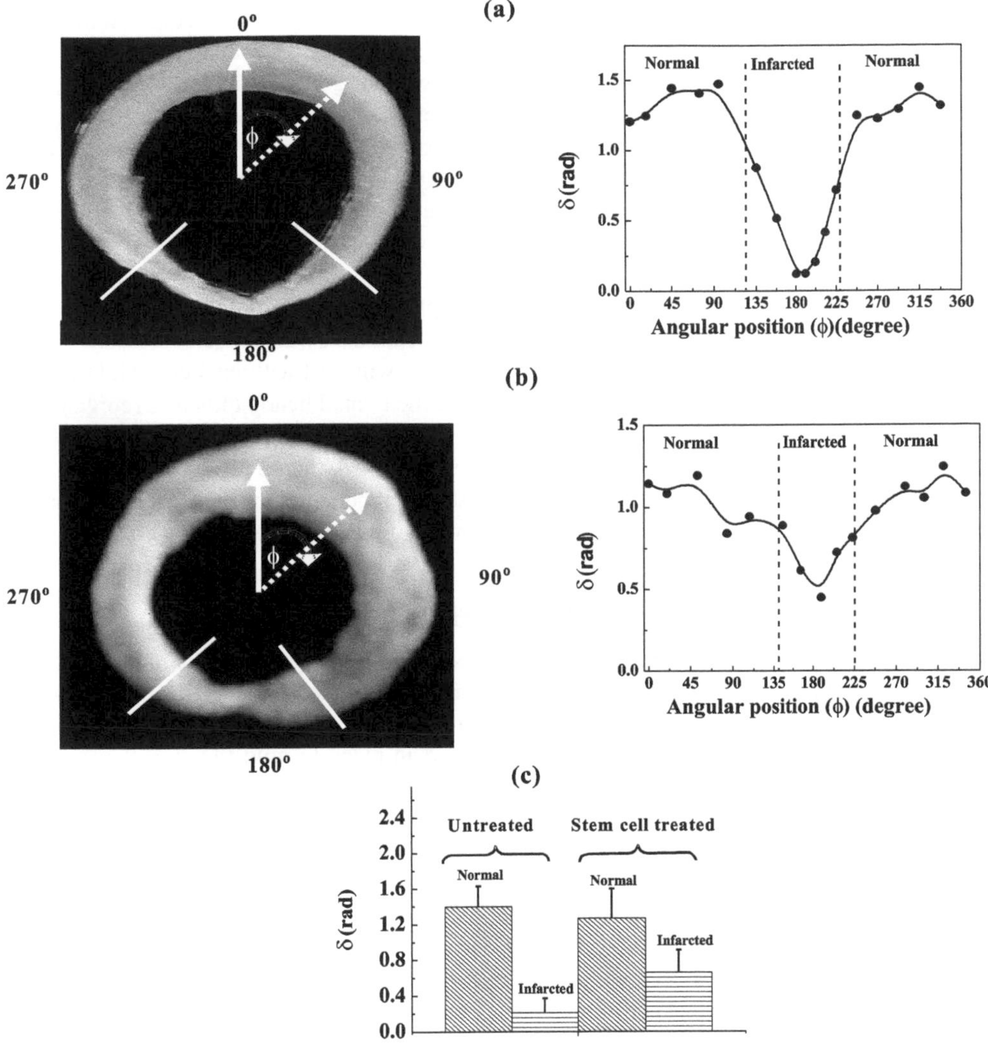

FIGURE 9.10: Linear retardance δ derived from transmission polarization measurements in 1-mm-thick sections from Lewis rat hearts following myocardial infarction. (a) Untreated tissues at four weeks following infarction, (b) tissues following stem-cell treatments, at four weeks following infarction, then two weeks after stem-cell therapy. The marked sector around $\varphi = 180^{\circ}$ indicates the infracted region. Symbols are experimentally derived values, and lines are a guide for the eye. Decreased birefringence levels in the infarcted region compared to normal regions are seen; this difference is reduced following stem-cell therapy, as infracted region retardance values increase towards normal-tissue levels. The results from birefringence measurements from the controls and the stem-cell treated groups of infracted hearts are shown in histogram form in (c). Error bars represent the standard deviation. Both untreated and stem-cell treated groups were comprised of 4 hearts, and 5 measurements were preformed in each region (normal and infarcted). (Adopted from [30]).

signals of the measured Mueller matrix elements from tissue.

Mueller matrices were recorded in the transmission geometry, from 1-mm thick *ex vivo* myocardial samples from Lewis rats after myocardial infarction, both with and without stem-cell treatments [30]. These were analyzed via polar decomposition to obtain linear retardance (δ) values. The δ values for two representative myocardial samples are shown in Figure 9.10.

The normal regions of the myocardium exhibit high levels of anisotropy, with derived δ values in the range of ~ 1.25 rad. A large decrease in linear retardance is seen in the infarcted region ($\delta \sim 0.2$ rad) of the untreated myocardium (Figure 9.10a). This arises because the native well-ordered anisotropic myocardium is replaced with disorganized isotropic scar tissue, likely collagenous in nature. An increase in δ towards the native levels is seen in the infarcted region after stem-cell treatment (Figure 9.10b). The bar graph of Figure 9.10c shows the mean linear retardance values from measurements of treated and untreated heart groups in the infarcted and normal regions. Statistically-significant ($p < 0.05$) differences in derived retardance values were obtained between normal and infarcted regions, and between infarcted regions with and without stem-cell treatments. An increase in retardance seen in the infarcted regions of the treated hearts indicates reorganization and re-growth (this was confirmed by histologic examination and is currently being corroborated by second harmonic generation microscopy for collagen imaging). These results show promise for the use of polarized light monitoring of stem-cell-based treatments of myocardial infarction, and current work is directed towards extending this novel method for *in vivo* biomedical deployment.

9.8.3 Proof-of-principle *in vivo* biomedical deployment of the method

The first *in vivo* use of the Mueller matrix decomposition method for tissue characterization was demonstrated using a dorsal skinfold window chamber mouse model [36]. In this model, the skin layer of an athymic nude mouse (NCRNU-M, Taconic) was removed from a 10 mm diameter region on the dorsal surface, and a titanium saddle was sutured in place to hold the skin flap vertically [37]. A protective glass coverslip (145 ± 15 μm thick) was placed over the exposed tissue plane. This allows for direct optical transmission measurements of polarized light through the ~ 500 μm thick layer. This model enables accurate measurements in an *in vivo* setting, free of many of the challenges inherent in examining fully 3D tissue structures.

Collagenase was injected into a region of dermal tissue to alter the structure of the extracellular matrix. The Mueller matrices were recorded both from the region of collagenase injection and a distant control region. Measurements were made before collagenase treatment and for 5 h postinjection at 30-min intervals, with an additional measurement at 24 h. Photographs of the dorsal skin flap window chamber model and of the experimental system, showing the mouse with its implanted window chamber in the path of the interrogating beam, are displayed in Figure 9.11.

Values for linear retardance (δ), net depolarization coefficient (Δ), optical rotation (ψ) and diattenuation (d) were extracted from the experimentally-derived Mueller matrices at each time point, using the polar decomposition approach [36]. The derived variations of δ and Δ (in both collagenase-treated and control regions) are shown as a function of time following collagenase injection, in Figure 9.12a and 9.12b respectively. Values for optical rotation ψ and diattenuation d did not change appreciably with treatment (data not shown). The value for δ is seen to vary from $\delta \approx 1.2$ rad to $\delta \approx 0.3$ rad in the treated region, in contrast to the control region where the values remain essentially constant at $\delta \approx 1$ rad. The decrease in birefringence is likely due to denaturation of the collagen fibers (collagenase cleaves the collagen fibers by breaking the peptide bonds connecting the monomer peptide units), which reduces the structural anisotropy. This was further confirmed by histology, where a reduction in collagen fibers was observed in the treatment region [36]. Using the approximate light path length, $l \approx 500$ μm (the true path length will be somewhat longer due to scattering [32]), the intrinsic birefringence Δn values were estimated as $\Delta n = \delta\lambda/2\pi l$, where $\lambda = 632.8$ nm. The birefringence values prior to treatment were calculated as $\Delta n \approx 2.2 \times 10^{-4}$, decreasing to $\Delta n \approx 0.6 \times 10^{-4}$, after treatment. These birefringence levels are reasonably close to

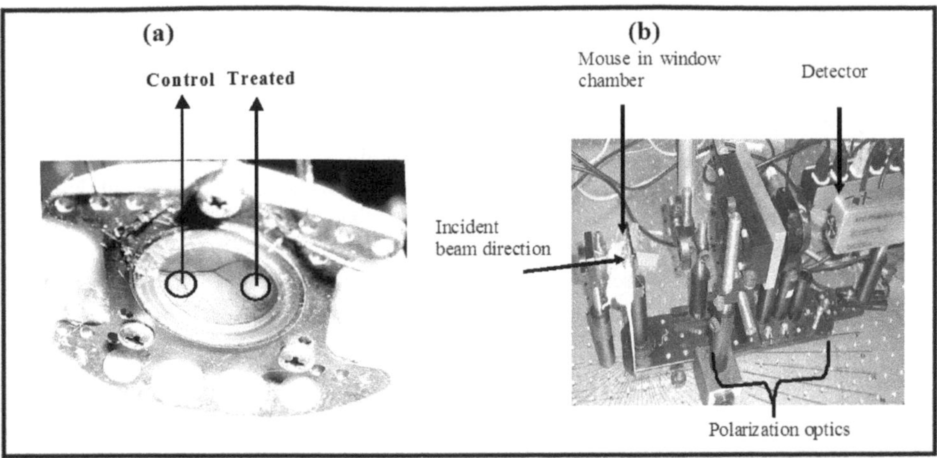

FIGURE 9.11: (a) Photograph of the dorsal skin flap window chamber model in a mouse. Measurements were made in two regions (collagenase treated and control) through the window chamber. (b) Photograph of the experimental system, showing the mouse with its implanted window chamber in the path of the interrogating beam. (Adopted from [36]).

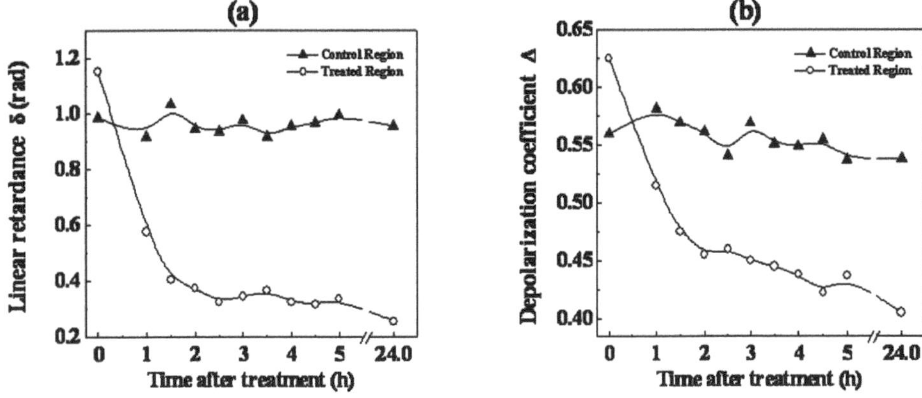

FIGURE 9.12: (a) Linear retardance δ and (b) net depolarization coefficient Δ before and as a function of time after collagenase injection in the treated and control regions. Symbols are experimentally derived values, and lines are a guide for the eye. (Adopted from [36]).

those found in the literature for tissue birefringence, typically $\sim 1 \times 10^{-3}$ [38].

In addition to the observed variation in δ, a decrease in the net depolarization coefficient Δ after treatment, from ~ 0.63 to ~ 0.45, was also noted in the treated region (Figure 9.11b). The values in the control region were again observed to remain essentially constant. This reduction in depolarization is also due to the destruction of the collagen fibers, since these represent one of the primary scattering structures in tissue [24]. Using the changes in the depolarizing properties of the tissue, the scattering coefficient of the tissue was determined using the polarization-sensitive Monte Carlo simulations. The values for μ_s were estimated to be 182 cm^{-1} for the pretreatment tissue and 134 cm^{-1} for the posttreatment tissue, which are in reasonable agreement with literature values [24]. These results demonstrate the ability of the method to quantify changes in tissue structure using polarized light *in vivo*. The interpretation of measured changes in values for birefringence and depolarization show promise for the method's ability to accurately quantify biologically-relevant tissue parameters such as scattering and birefringence.

To conclude this section, the utility of the novel polar Mueller matrix decomposition approach has been initially explored in two important potential biomedical applications, for noninvasive glucose measurement in tissue-like turbid media and for quantification of tissue structural anisotropy. The initial *in vivo* use of the method for polarimetric tissue characterization has also been demonstrated. Results of these studies show promise and warrant further exploration. We are currently expanding our investigations for the use of this promising method *in vivo*, both for noninvasive measurements of glucose and for monitoring the response of infarcted myocardial tissues to stem-cell therapies.

9.9 Concluding Remarks on the Prospect of the Mueller Matrix Decomposition Method in Polarimetric Assessment of Biological Tissues

In this chapter, a novel general method for polarimetry analysis in turbid media based on polar Mueller matrix decomposition has been discussed. The ability of this approach for delineating individual intrinsic polarimetry characteristics in complex tissue-like turbid media was validated theoretically with a polarized-light Monte Carlo model, and experimentally with a polarization-modulation/synchronous detection setup on optical phantoms having controlled sample polarizing properties. The individual polarization effects can be successfully decoupled and quantified despite their simultaneous occurrence, even in the presence of the numerous complexities due to multiple scattering. The ability to isolate individual polarization properties provides a potentially valuable noninvasive tool for biological tissue characterization. Specifically, concentration determination of optically active molecules such as glucose and quantification of tissue structural anisotropy are two important biomedical avenues that have been initially explored. Clearly, there are many other potential applications in biomedicine, both in tissue diagnostics and in treatment response monitoring.

Acknowledgments

Research support from the Natural Sciences and Engineering Research Council of Canada (NSERC) and collaborations with Richard D.Weisel, Ren-Ke Li, Shu-hong Li, Eduardo H. Moriyama, and Brian C. Wilson are gratefully acknowledged. The authors are also thankful to Marika A. Wallenburg for her help during the preparation of the manuscript.

References

[1] V. Ronchi, *The Nature of Light*, Harvard University Press, Cambridge, Massachusetts (1970).

[2] J. Michl and E.W. Thulstrup, *Spectroscopy with Polarized Light*, Wiley-VCH, New York (1986).

[3] D.S. Kliger, J.W. Lewis, and C.E. Randall, *Polarized Light in Optics and Spectroscopy*, Academic Press–Harcourt Brace Jovanovich, New York (1990).

[4] L.V. Wang, G.L Coté, and S. L. Jacques, "Special section guest editorial: tissue polarimetry," *J. Biomed. Opt.* **7**, 278 (2002).

[5] V.V. Tuchin, L. Wang, and D. Zimnyakov, *Optical Polarization in Biomedical Applications*, Springer-Verlag, Berlin, Heidelberg, N.Y., (2006).

[6] V.V. Tuchin (ed.), *Handbook of Optical Sensing of Glucose in Biological Fluids and Tissues*, CRC Press, Taylor & Francis Group, London, (2009).

[7] P.J. Wu and J.T. Walsh Jr., "Stokes polarimetry imaging of rat tail tissue in a turbid medium: degree of linear polarization image maps using incident linearly polarized light," *J. Biomed. Opt.* **11**, 014031 (2006).

[8] J.F. de Boer and T.E. Milner, "Review of polarization sensitive optical coherence tomography and Stokes vector determination," *J. Biomed. Opt.* **7**, 359–371 (2002).

[9] R.J. Mc Nichols and G.L. Coté, "Optical glucose sensing in biological fluids: an overview," *J. Biomed. Opt.* **5**, 5–16 (2000).

[10] X. Guo, M.F.G. Wood, and I.A. Vitkin, "Angular measurement of light scattered by turbid chiral media using linear Stokes polarimeter," *J. Biomed. Opt.* **11**, 041105 (2006).

[11] D. Côte and I.A. Vitkin, "Balanced detection for low-noise precision polarimetric measurements of optically active, multiply scattering tissue phantoms," *J. Biomed. Opt.* **9**, 213–220 (2004).

[12] R.R. Ansari, S. Bockle, and L. Rovati, "New optical scheme for a polarimetric-based glucose sensor," *J. Biomed. Opt.* **9**, 103–115 (2004).

[13] I.A. Vitkin, R.D. Laszlo, and C.L. Whyman, "Effects of molecular asymmetry of optically active molecules on the polarization properties of multiply scattered light," *Opt. Express* **10**, 222–229 (2002).

[14] D. Côté and I.A. Vitkin, "Robust concentration determination of optically active molecules in turbid media with validated three-dimensional polarization sensitive Monte Carlo calculations," *Opt. Express* **13**, 148–163 (2005).

[15] R.A. Chipman, "Polarimetry," Chap. 22 in *Handbook of Optics*, 2nd ed., M. Bass, Ed., Vol. **2**, pp. 22.1–22.37, McGraw-Hill, New York (1994).

[16] C. Brosseau, *Fundamentals of Polarized Light: A Statistical Optics Approach*, Wiley, New York (1998).

[17] N. Ghosh, M.F.G. Wood, and I.A. Vitkin, "Mueller matrix decomposition for extraction of individual polarization parameters from complex turbid media exhibiting multiple scattering, optical activity and linear birefringence," *J. Biomed. Opt.* **13**, 044036 (2008).

[18] N. Ghosh, M.F.G. Wood, and I.A. Vitkin, "Polarimetry in turbid, birefringent, optically active media: a Monte Carlo study of Mueller matrix decomposition in the backscattering geometry," *J. Appl. Phys.* **105**, 102023 (2009).

[19] S. Yau Lu and R.A. Chipman, "Interpretation of Mueller matrices based on polar decomposition," *J. Opt. Soc. Am. A* **13**, 1106–1113 (1996).

[20] M.F.G. Wood, X. Guo, and I.A. Vitkin, "Polarized light propagation in multiply scattering media exhibiting both linear birefringence and optical activity: Monte Carlo model and experimental methodology," *J. Biomed. Opt.* **12**, 014029 (2007).

[21] S. Manhas, M.K. Swami, P. Buddhiwant, N. Ghosh, P.K. Gupta, and K. Singh,"Mueller matrix approach for determination of optical rotation in chiral turbid media in backscattering geometry," *Opt. Express* **14**, 190–202 (2006).

[22] J. Morio and F. Goudail, "Influence of the order of diattenuator, retarder, and polarizer in polar decomposition of Mueller matrices," *Opt. Lett.* **29**, 2234–2236 (2004).

[23] R. Ossikovski, A. De Martino, and S. Guyot, "Forward and reverse product decompositions of depolarizing Mueller matrices," *Opt. Lett.* **32**, 689 (2007).

[24] A.J. Welch and M.J.C. van Gemert, *Optical-Thermal Response of Laser Irradiated Tissue,* Plenum Press, New York (1995).

[25] L. Wang, S.L. Jacques and L. Zheng, "MCML-Monte Carlo modeling of light transport in multi-layered tissues," *Comput. Methods Programs Biomed.* **47**, 131–146 (1995).

[26] M. Moscoso, J.B. Keller, and G. Papanicolaou, "Depolarization and blurring of optical images by biological tissues," *J. Opt. Soc. Am. A* **18**, 949–960 (2001).

[27] D. Côté and I.A. Vitkin, "Pol-MC: a three dimensional polarization sensitive Monte Carlo implementation for light propagation in tissue," available online at http://www.novajo.ca/ont-canc-instbiophotonics/.

[28] C. F. Bohren and D.R. Huffman, *Absorption and Scattering of Light by Small Particles*, Chap. 2, Wiley, New York (1983).

[29] R. Clark Jones, "New calculus for the treatment of optical systems VII. Properties of the N-matrices," *J. Opt. Soc. Am.* **38**, 671–685 (1948).

[30] N. Ghosh, M.F.G. Wood, S.H. Li, R.D. Weisel, B.C. Wilson, Ren-Ke Li, and I.A. Vitkin, "Mueller matrix decomposition for polarized light assessment of biological tissues," *J. Biophotonics* **2** (3), 145–156 (2009).

[31] N. Ghosh, M.F.G. Wood, M.A. Wallenberg and I.A. Vitkin, "Influence of the order of the basis matrices on the Mueller matrix decomposition-derived polarization parameters in complex tissue-like turbid media," *Opt. Commun.* (in press).

[32] X. Guo, M.F.G. Wood, and I.A. Vitkin, "Monte Carlo study of path length distribution of polarized light in turbid media," *Opt. Express* **15**, 1348–1360 (2007).

[33] M.F.G. Wood, D. Côté, and I.A. Vitkin, "Combined optical intensity and polarization methodology for analyte concentration determination in simulated optically clear and turbid biological media," *J. Biomed. Opt.* **13**, 044037 (2008).

[34] Y. Sun, J.Q. Zhang, J. Zhang, and S. Lamparter, "Cardiac remodeling by fibrous tissue after infarction," *J. Lab. Clin. Med.* **135**, 316–323 (2000).

[35] D. Orlic, et al., "Bone marrow cells regenerate infarcted myocardium," *Nature* **410**, 701–705 (2001).

[36] M.F.G. Wood. N. Ghosh, E.H. Moriyama, B.C. Wilson, and I.A. Vitkin, "Proof-of-principle demonstration of a Mueller matrix decomposition method for polarized light tissue characterization *in vivo*," *J. Biomed. Opt.* **14**, 014029 (2009).

[37] M. Khurana, E.H. Moriyama, A. Mariampillai, and B.C. Wilson, "Intravital high-resolution optical imaging of individual vessel response to photodynamic treatment," *J. Biomed. Opt.* **13**, 040502 (2008).

[38] W. Wang and L.V. Wang, "Propagation of polarized light in birefringent media: a Monte Carlo study," *J. Biomed. Opt.* **7**, 350–358 (2002).

10

Statistical, Correlation, and Topological Approaches in Diagnostics of the Structure and Physiological State of Birefringent Biological Tissues

O.V. Angelsky, A.G. Ushenko, Yu.A. Ushenko

Chernivtsi National University, Correlation Optics Department, 2 Kotsyubinskoho Str., Chernivtsi 58012, Ukraine

V.P. Pishak, A.P. Peresunko

Bukovinian State Medical University, 2 Teatralnaya Sq., Chernivtsi, 58000, Ukraine

Optical techniques for investigation of structure of biological tissues (BTs) can be classified into three groups:

- Spectrophotometric techniques [1], based on the analysis of spatial or temporal changes of intensity of a field scattered by BTs;

- Polarization techniques based on the use of the coherency matrix for a complex amplitude [2, 3] and the analysis of the degree of polarization as the factor of correlation of the orthogonal complex components of electromagnetic oscillations at the specified point of a scattered field;

- Correlation techniques based on the analysis of correlations of the parallel polarization components at different points of the object field [4].

Both polarization and correlation characteristics are typically changed for real object fields of BTs, including their images [5, 6].

One of the most promising techniques for obtaining such images is the optical coherent tomography [7] and its new branch — polarization-sensitive optical coherent tomography [8]. This diagnostic approach is based on measuring the coordinate distributions of the Stokes parameters at the BT's images providing important data both on the BT microstructure and on magnitude and coordinate distributions of the parameters of optical anisotropy of the architectonic nets formed by birefringent protein fibrils.

Further progress of the polarization-sensitive optical coherent tomography is connected with the development of new techniques for the analysis and processing of inhomogeneously polarized BT images. That is why, the development of novel approaches to the analysis of laser images of BTs and searching for the new techniques for polarization, interference, and correlation diagnostics of the BT's structure for determining the efficient criteria of estimation of birefringence associated with the BT's physiological state is the topical problem.

Key words: biological tissue, polarization, polarization singularity, correlation, uniaxial crystals, statistical moments, diagnostics

10.1 Introduction

10.1.1 Polarimetric approach

Laser radiation, similar to natural light, can be absorbed and scattered by a BT, and the each of these processes results in filling of the field with the information on micro- and macrostructure of the studied medium and its components. The spectrophotometric techniques for BT diagnostics based on the analysis of the spatial and temporal changes of intensity of the field scattered by such optically inhomogeneous objects are the most widespread and approbated nowadays. At the same time, other diagnostic techniques based on the fundamental concepts, such as "polarization" and "coherence" are also intensively developed.

It has been caused historically that the spatial changes of optical fields are characterized in the terms determining the coherent properties of fields. The concept of "measure of coherence" between two light oscillations is defined as the ability of them to form an interference pattern and is associated with the visibility of interference fringes [9]. This value is the measure of the sum of correlations between parallel components of electrical fields at two points of space. Another type of correlation characteristics of scattered laser fields is the degree of polarization defined as the magnitude of the maximal correlation between the orthogonally polarized components of the electrical field at one point [10]. Polarization properties of light are studied experimentally by measuring the intensity of radiation passing various optically active elements.

The techniques based on the use of the coherency matrix and the degree of polarization are referred to as polarization ones. Let us consider such techniques in more details. Considerable development of the vector approach to the investigations of the morphological structure and physiological state of various BTs [11, 12] calls for the development of the model notions on optically anisotropic and self-similar structure of BTs.

So, the hierarchical self-similar (tropocollagen, microfibril, subfibril, fibril, fascia, etc.) structure of typical connective tissues and tendons have been analyzed in [13]. It is emphasized that the filiform structural elements are discrete being the scale repeated within large interval of "optical sizes", from 1 μm to 10^3 μm. For that, the optical characteristics of such structures of various BTs correspond to "frozen" optically uniaxial liquid crystals [13].

A similar approach to the description of the BT's morphological structure has been used in paper [14], where a BT is considered at the two-component amorphous-crystalline structure. Amorphous component of a BT (fat, lipids, nonstructured proteins) is isotropic in polarization (optically nonactive). The crystalline component of a BT is formed by oriented birefringent protein fibrils (collagen proteins, myosin, elastin etc.). The properties of an isolated fibril are modelled by optically uniaxial crystals whose axis is oriented along the direction of packing at the BT plane, and the birefringence index depends on the fibril matter. Architectonic net formed by differently oriented birefringent bundles is the higher level of the BT organization.

Within the framework of this model one can explain the mechanisms causing polarization inhomogeneity of the fields of various BTs, such as osseous and muscular tissue and tissues of the woman reproductive sphere [14, 15]. The interconnections between the magnitudes of the azimuth of polarization and ellipticity of light oscillations, on the one hand, and the direction (angle) of packing of fibrils and the parameters characterizing birefringence of the corresponding matter, on the other hand, have been also found [16]. It enables improvement of the technique of polarization visualization of the architectonic net of BTs of various morphological types and to use the statistical analysis of the coordinate distributions of the polarization parameters of the scattered fields [17].

The results of the investigation of interconnections between the set of the statistical moments of the first to the fourth orders characterizing microgeometry of a surface and orientation and phase structure of birefringent architectonics of human BT, on the one hand, and the set of the corresponding statistical moments of the coordinate distributions of the magnitudes of the azimuth of polarization and ellipticity (polarization maps) of the images of such objects, on the other hand, have been represented in [18–20]. It has been found that growing magnitudes of kurtosis and skewness of the azimuth of polarization and ellipticity results from the growing dispersion of orientation of the optical axes of birefringent fibrils of BTs. The decreasing magnitudes of kurtosis and skewness corresponds to the increasing dispersion of phase shifts caused by biological crystals of the architectonic nets [18].

Further development of laser polarimetry consists in the development of new techniques for processing of two-dimensional tensors of polarization parameters characterizing the nets of various biological crystals of human tissues [17–22].

So, in part, the statistical analysis of the coordinate distributions of the Stokes parameters provides new information on microstructure (magnitudes and coordinate distributions of the parameters of optical anisotropy of architectonic nets formed by collagen and myosin) of physiologically normal and pathologically changed BTs [18–22].

As a whole, rapid development of the techniques for diagnostic usage of coherent laser radiation has been reflected in optical coherent tomography as the most convenient and elaborated instrument for noninvasive study of the BT structure [23]. The use of a new informative parameter, *viz.* polarization of a laser beam providing the carrying out of contrasting of the BT images stimulates standing out of the new direction of the optical coherent tomography, often referred to as the polarization-sensitive coherent tomography [21–23].

Note, that the peculiarity of "laser polarimetry of distributions of azimuth and ellipticity" consists ina point-by-point analysis of the polarization parameters of the object field and searches for their interconnection with the orientation and anisotropic parameters of the BT architectonics. In this situation, information on the peculiarities (statistical and fractal) of two-dimensional distributions of the polarization parameters of the field, on the one hand, and the orientation and phase characteristics of the object, on the other hand, is left undetermined.

In this connection, the development of laser polarimetry asks for further elaboration of the techniques for nondestructive macrodiagnostics of the BT geometrical-optical structure through improving modern techniques of BT polarization-interference mapping and development of new techniques for reconstruction of BT architectonics. Besides, urgent necessity is present to develop new approaches for both statistical and local topological (singular) analysis of two-dimensional polarization images of BTs.

10.1.2 Correlation approach

It is known [24, 25] that polarization properties of light at some spatial point are comprehensively described by the coherency matrix. This formalism is also adequate for characterization of a beam as a whole, if the characteristics of scattered radiation are unchanged in space, i.e., if the beam is homogeneous. However, it is not sufficient for description of scattered optical fields, spatially inhomogeneous in polarization. Consideration of not only coordinate distributions of the polarization parameters of a field but also searching for interconnections between the state of polarization and the degree of coherence at different points of a field must be carried out in this case.

The initial attempt to describe spatially inhomogeneous in polarization object fields as the direct generalization to the two-point coherency matrix has been performed by Gori [26]. The author shows that some visibility of interference fringes resulting from superposition of radiation from two point sources whose polarization characteristics are formed by the set of polarizers and phase plates corresponds to the each matrix element.

The unified theory of coherence and polarization providing description of changes of the state of polarization of freely propagating light has been developed by Wolf in 2003 [27].

Within the framework of the approach based on the generalized coherency matrix, the new parameter has been proposed [28, 29], namely, the degree of mutual coherence characterizing correlation similarity of the electric vector at two points and experimentally determined by measuring visibility of the corresponding interference distributions.

This approach has been further developed by Ellis and Dogariu [30]. They introduced the complex degree of mutual polarization (CDMP) determining the correlation between the points of a field with different states of polarization and different intensities.

Statistics of intensity fluctuations at the field resulting from scattering of partially coherent beams is analyzed in [31].

The matrix technique proposed by Gori [32] generalizes the description of inhomogeneously polarized coherent fields on the base of the CDMP. It extends the description of light-scattering phenomena to a more general case, when partially polarized light is not obviously completely spatially coherent.

Proceeding from the unified theory of coherence and polarization of random electromagnetic waves, it has been proved theoretically and experimentally that partially polarized light can be described using correlation of unpolarized components [33].

The technique for the determination of the polarization parameters of light from measured intensity fluctuation contrast has been proposed in [34]. This technique presumes parallel measurements of the degree of polarization and the second Stokes parameter directly connected with the azimuth of polarization and ellipticity. Matrix analysis of correlation properties of scattered coherent radiation has been extended on the vector (inhomogeneously polarized) fields. The notion of 3×3 correlation matrix has been introduced [35, 36] for the completely polarized electromagnetic field stationary in a statistical sense.

Another prospective direction is connected with solving the problem of direct measurement of the CDMP for the problems of biomedical optics, namely, processing of coherent inhomogeneously polarized images of BTs obtained by the optical coherent tomography technique.

It has been shown [37] that the CDMP of a coherent image of a BT is the parameter sensitive to orientation and phase changes of the BT's architectonics.

Experimental studies of 2D distributions of the degree of mutual polarization of the BT's laser images for examples of muscular tissue, skin tissue, and osseous shin have proved an interconnection between the coordinate distribution of the CDMP of laser images and geometrical-optical structure of birefringent architectonic nets of physiologically normal and dystrophically changed BTs.

Accounting for the potential diagnostic importance and steadiness of the CDMP, it is necessary to continue searching for the peculiarities of correlation structure of the coordinate distributions of this parameter for various types of BTs. Especial attention must be paid to the development of unified criteria for differentiation of morphological and physiological structure of BTs based on the CDMP. Besides, the combination of the formalism of coordinate distributions of the Mueller matrix elements and the CDMP is also a prospective direction of revealing the new diagnostic criteria of the physiological state of BTs.

10.1.3 Topological or singular optical approach

At inhomogeneous vector electromagnetic fields, lines or surfaces exist at which one of the parameters characterizing the state of polarization is undetermined [38, 39]. Crossing by the observation plane perpendicular to the propagation direction, these singular lines (surfaces) go over to points (lines), respectively. Generally, the object field resulting from the scattering of laser radiation in phase-inhomogeneous media is elliptically polarized. At the transversal cross-section of a beam, however, other (limiting) states of polarization are also taking place [40], *viz.* (i) points where the polarization ellipse degenerates into a circle and polarization is circular (C points); the azimuth of polarization (orientation of the major axis) is undetermined; and (ii) lines with linear polarization (S lines) where the direction of rotation of the electric vector is undetermined.

General principles of the singular analysis of vector fields have been initially formulated by Nye and Hajnal [39, 41–45]. Further consideration of polarization singularities at optical fields has been carried out by Mokhun and Angelsky [46–48]. Detailed theoretical investigation of an optical field with the set of C points has been performed by Freund, Soskin, and Mokhun [49], whose results have been confirmed experimentally [50].

Three morphological forms of distribution of polarization ellipses are possible at speckle fields [51–53], which determine the field behavior in the vicinity of C point: "star" (S), "lemon" (L), and "monostar" (M). These forms differ in the number and positions of lines, at each point of which major and minor axes of the surrounding ellipses are oriented to C point.

Experimental study of the influence of C points on the polarization structure of a field in their closest vicinity has been carried out [54–56]. All three mentioned morphologically allowed forms of the polarization ellipse distributions in the vicinity of isotropic points with circular polarization have been revealed. Their statistical weights were estimated for the example of random vector field resulting from light scattering by inhomogeneous media. The obtained results show skewness and considerable angular dispersion of "forming" straight lines for S and M structures in comparison with their canonical forms. The experimental data directly prove universality of the laws of topology and morphology.

It has been shown within the framework of the topological approach [39, 42] that analysis of the polarization structure of object fields presumes that the areas with right-hand circular and elliptic polarizations are separated from the areas with left-hand circular and elliptical polarizations by the contours (at the observation plane) where polarization is linear; S surfaces correspond to such S lines at 3D space.

The system of singular points of the vector object field determines its structure as a skeleton and reflects in such a manner the polarization properties of a medium. It shows importance for a biomedical optics searching for the conditions for forming polarization singularities by biological objects. Besides, the experimental study of the coordinate distribution of singularities at the polarization images of BTs provides diagnostic data on the polarization properties of media. It shows also the importance of new singular optical criteria for differentiating the physiological state of histological slices of BTs.

The considered approaches [41–43, 54–56] have been taken into account in [57] for the analytical determination of the conditions for forming singly (linear) and doubly (circular) degenerated polarization singularities and experimental investigation of the coordinate distributions of singularities at images of BTs of various morphological structures and states. It has been demonstrated that the third and fourth statistical moments of the distributions of the number of singular points are the most sensitive to changes of the geometrical-optical structure of BTs [58]. It has also been established that the coordinate structure of polarization singularities at images of a physiologically normal BT is self-similar (fractal), while for pathologically changed it is random [57].

Thus, topicality of the materials represented here is caused by the necessity of more general polarization diagnostics of changes of coordinate and local structure (including distributions of optical axes orientations of and magnitude of birefringence) of the nets of organic crystals of BTs with various morphological structures and physiological states.

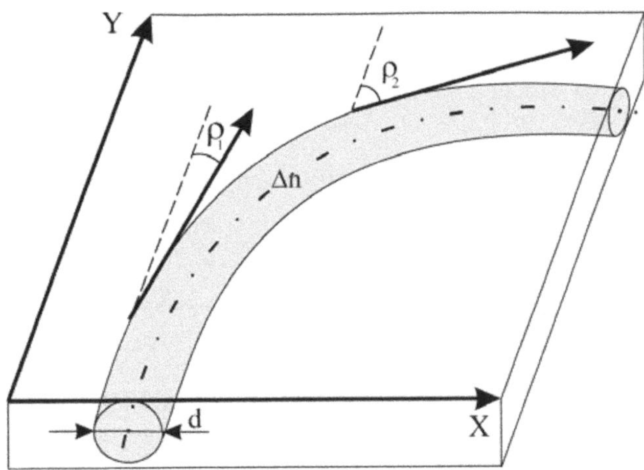

FIGURE 10.1: Birefringent (Δn) fibril with cross-section diameter d; ρ_i – directions of packing of fibrils at the plane of a BT sample.

10.2 Biological Tissue as the Converter of Parameters of Laser Radiation

10.2.1 Crystal optical model of anisotropic component of the main types of biological tissues

It is assumed within the framework of notions on geometrical-optical structure of BT that it consists of two components, *viz.* optically isotropic (amorphous) and anisotropic (architectonic) that is formed by birefringent optically uniaxial fibrils, see Figure 10.1.

Let us characterize the interaction of coherent radiation with the amorphous $\{A\}$ and architectonic $\{C\}$ components of a BT by the following Jones operators:

$$\{A\} = \left\| \begin{matrix} a_{11} & a_{12} \\ a_{21} & a_{22} \end{matrix} \right\| = \left\| \begin{matrix} \exp(-\tau l) & 0 \\ 0 & \exp(-\tau l) \end{matrix} \right\|; \tag{10.1}$$

$$\{C\} = \left\| \begin{matrix} c_{11} & c_{12} \\ c_{21} & c_{22} \end{matrix} \right\| = \left\| \begin{matrix} \cos^2\rho + \sin^2\rho \exp(-i\delta); & \cos\rho \sin\rho \left[1 - \exp(-i\delta)\right]; \\ \cos\rho \sin\rho \left[1 - \exp(-i\delta)\right]; & \sin^2\rho + \cos^2\rho \exp(-i\delta); \end{matrix} \right\|. \tag{10.2}$$

Here τ is the absorption coefficient for laser radiation and amorphous layer of a BT with geometrical thickness l; ρ is the direction of packing of anisotropic (birefringent index Δn) fibrils at the BT plane introducing a phase shift Δn between the orthogonal components (E_x and E_y) of the electrical vector of the probing laser beam with a wavelength λ. It is shown in [1, 17] that the Jones matrix of a geometrically thin layer of a BT where isotropic and anisotropic formations lie at the same plane is represented as the superposition of the operators $\{A\}$and $\{C\}$:

$$\{M\} = \left\| \begin{matrix} a_{11} + c_{11}; & a_{12} + c_{12}; \\ a_{21} + c_{21}; & a_{22} + c_{22}. \end{matrix} \right\|. \tag{10.3}$$

Within the model formalized by Eqs. (10.1) to (10.3) one can find a complex amplitude U of a laser beam transformed by a BT. At the output of the BT layer the magnitudes of the orthogonal components of an amplitude of a laser beam, U_x and U_y, are determined from the following matrix equation:

$$\begin{pmatrix} U_x \\ U_y \end{pmatrix} = \left\| \begin{matrix} a_{11}+c_{11};\ a_{12}+c_{12}; \\ a_{21}+c_{21};\ a_{22}+c_{22}. \end{matrix} \right\| \begin{pmatrix} E_x \\ E_y \end{pmatrix} = \begin{pmatrix} c_{11}E_x + c_{12}E_y \exp(i\delta) \\ c_{21}E_x + c_{22}E_y \exp(i\delta) \end{pmatrix} + \exp(-\tau l) \begin{pmatrix} E_x \\ E_y \end{pmatrix}, \quad (10.4)$$

Solution of Eq. (10.4) with account of Eqs. (10.2) and (10.3) gives the following dependencies for the each with coordinate r at the boundary field:

$$\begin{aligned} U_x(r) &= \left(\cos^2\rho + \sin^2\rho \exp(-i\delta)\right) E_x + \cos\rho \sin\rho \left(1 - \exp(-i\delta)\right) \exp(i\varphi)) E_y; \\ U_y(r) &= \cos\rho \sin\rho \left(1 - \exp(-i\delta)\right) E_x + \left(\sin^2\rho + \cos^2\rho \exp(-i\delta)\right) \exp(i\varphi)) E_y. \end{aligned} \quad (10.5)$$

Let us for simplicity (but without loss of generality of the analysis) consider the mechanisms of formation of polarization structure of the BT's speckle image when a BT is illuminated by a linearly polarized laser beam with the azimuth of polarization $\alpha = 0°$,

$$\begin{pmatrix} E_x \\ E_y \end{pmatrix} \Rightarrow \begin{pmatrix} 1 \\ 0 \end{pmatrix}. \quad (10.6)$$

In this case Eqs. (10.5) are rewritten in the form:

$$\begin{aligned} U_x^A &= \exp(-\tau l); \\ U_y^A &= 0. \end{aligned} \quad (10.7)$$

$$\begin{aligned} U_x^C &= \left[\cos^2\rho + \sin^2\rho \exp(-i\delta)\right]; \\ U_y^C &= \left[\cos\rho \sin\rho \left(1 - \exp(-i\delta)\right)\right]. \end{aligned} \quad (10.8)$$

Here U_x^A, U_y^A and U_x^C, U_y^C are the orthogonal components of amplitude of the object field formed by amorphous (A) and anisotropic (C) components of a BT. To determine the magnitudes of the azimuth of polarization and ellipticity at the points of the boundary field $r(X,Y)$ we write the coherency matrices for the BT's amorphous $\{K^A\}$ and crystalline $\{K^C\}$ components:

$$\{K^A\} = \left\| \begin{matrix} \exp(-\tau l) & 0 \\ 0 & 0 \end{matrix} \right\|; \quad (10.9)$$

$$\{K^C\} = \left\| \begin{matrix} U_x^C U_x^{*C};\ U_x^C U_y^{*C}; \\ U_x^{*C} U_x^C;\ U_y^C U_y^{*C}. \end{matrix} \right\|. \quad (10.10)$$

It follows from Eqs. (10.9) and (10.10) that the coordinate distributions of the azimuth of polarization α and ellipticity β at the BT speckle image consists of two parts, *viz.* homogeneously polarized

$$\left\{ \begin{aligned} \alpha^A &= 0°; \\ \beta^A &= 0° \end{aligned} \right. \quad (10.11)$$

and inhomogeneously polarized

$$\left\{ \begin{aligned} \alpha^C &= 0.5 \arctan \left[\frac{U_x^C U_y^{*C} - U_x^{*C} U_y^C}{U_y^C U_x^{*C} - U_y^C U_y^{*C}} \right]; \\ \beta^C &= 0.5 \arcsin \left[\frac{i\left(U_x^C U_y^{*C} - U_x^{*C} U_y^C\right)}{(q_1+q_2+q_3)^{\frac{1}{2}}} \right], \end{aligned} \right. \quad (10.12)$$

where

$$
\begin{aligned}
q_1 &= \left[U_x^C U_x^{*C} - U_y^C U_y^{*C}\right]^2; \\
q_2 &= \left[U_x^C U_y^{*C} - U_x^{*C} U_y^C\right]^2; \\
q_3 &= i\left[U_x^C U_y^{*C} - U_x^{*C} U_y^C\right]^2.
\end{aligned} \tag{10.13}
$$

Thus, the performed analysis provides the notions on the mechanisms of formation of inhomogeneous polarization structure of the boundary object field of a BT at the each point by (i) decomposition of amplitude of laser wave U into orthogonal linearly polarized coherent components $\begin{pmatrix} U_x^{(r)} \\ U_y^{(r)} \end{pmatrix}$; (ii) formation of a phase shift $\delta(r)$ between these components due to birefringence; and (iii) superposition of these components resulting, generally, into elliptically polarized field with the polarization ellipse described as

$$
\frac{X^2}{U_x^2(r)} + \frac{Y^2}{U_y^2(r)} - \frac{2XY}{U_x(r)U_y(r)}\cos\delta(r) = \sin^2\delta(r). \tag{10.14}
$$

10.2.2 Techniques for analysis of the structure of inhomogeneously polarized object fields

10.2.2.1 Statistical analysis

For estimating the peculiarities of the distributions of the azimuth of polarization, α, and ellipticities, β, of the scattered field we use the statistical moments of the first (S_1), second (S_2), third (S_3), and fourth (S_4) orders calculated following the algorithms [58]:

$$
S_1 = \frac{1}{N}\sum_{i=1}^{N}|z_i|;\ S_2 = \sqrt{\frac{1}{N}\sum_{i=1}^{N}z_i^2};\ S_3 = \frac{1}{\sigma_S^3}\frac{1}{N}\sum_{i=1}^{N}z_i^3;\ S_4 = \frac{1}{\sigma_S^2}\frac{1}{N}\sum_{i=1}^{N}z_i^4, \tag{10.15}
$$

where $N = 800 \times 600$ is the total number of pixels of a CCD-camera registering inhomogeneously polarized field, and z denotes, respectively, random magnitudes of the azimuth of polarization and ellipticity; σ_S is the the dispersion of distribution.

10.2.2.2 Fractal analysis

Fractal analysis of the set of random values $z(\alpha,\beta)$ characterizing the field of scattered light is carried out in the following order: (i) one calculates the autocorrelation functions for the distributions of random values z and finds out the corresponding power spectra $J(\alpha)$ and $J(\beta)$ of the distributions of the azimuth of polarization and ellipticity of the scattered radiation; (ii) one calculates the $\log J(z) - \log(d^{-1})$ dependences of the power spectra, where d^{-1} are the spatial frequencies determined by the size of the structural elements of the BT's architectonics; and (iii) the dependences $\log J(z) - \log(d^{-1})$ are approximated following the least-mean-squares technique for $\Phi(\eta)$, the tangents η_i are determined for linear intervals, and the fractal dimensions of the values z are calculated following the relation

$$
D_i(z) = 3 - tg\eta_i. \tag{10.16}
$$

Classification of the coordinate distributions z is performed following the criteria proposed in [59]: (i) the coordinate distributions z are fractal for the constant slope, $\eta = const$, of the dependence $\Phi(\eta)$for 2 to 3 decades of change of sizes d of the structural elements of architectonics; (ii) the sets z are stochastic for the presence of several slopes of $\Phi(\eta)$; and (iii) the sets z are random values for the absence of constant slopes of $\Phi(\eta)$ within a whole interval of sizes d.

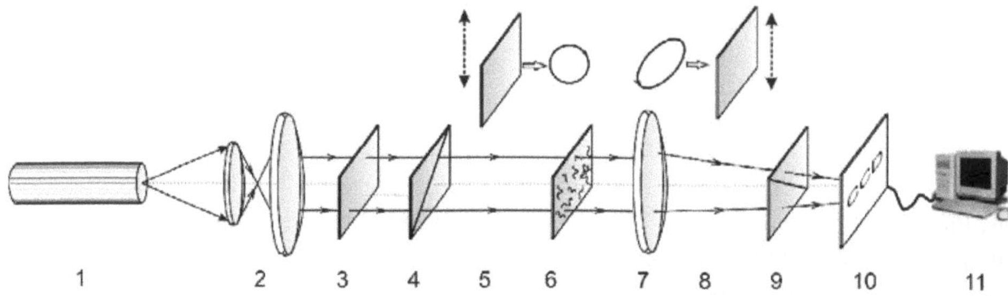

FIGURE 10.2: Optical arrangement of a polarimeter: 1 – He-Ne laser; 2 – collimator; 3 – stationary quarter-wave plate; 5 and 8 – mechanically moving quarter-wave plates; 4 and 9 – polarizer and analyzer, respectively; 6 – studied object; 7 – microobjective; 10 – CCD-camera; 11 – personal computer.

10.3 Laser Polarimetry of Biological Tissues

10.3.1 Polarization mapping of biological tissues: Apparatus and techniques

Optical arrangement for measuring the coordinate distributions of the azimuth of polarization and ellipticity at the images of histological slices of BTs is shown in Figure 10.2. Illumination is performed by a collimated beam ($\otimes = 10^4$ μ m, $\lambda = 0.6328$ μ m, $W = 5.0$ mW). The polarization illuminator consists of quarter-wave plates 3 and 5 and polarizer 4 providing the formation of the laser beam with arbitrary azimuth of polarization $0^\circ \leq \alpha_0 \leq 180^\circ$ and ellipticity $0^\circ \leq \beta_0 \leq 90^\circ$. Polarization images of BTs are projected using the microobjective 7 to the plane of the light-sensitive area of the CCD-camera 10 (800×600 pixels) providing the interval of the measured structural image elements from 2 μm to 2000 μm.

The experimental conditions are chosen to provide minimization of the spatial-angular aperture filtration under BT imaging. It is ensured by matching the angular characteristics of the scattering indicatrix of a BT and the angular aperture of a microobjective.

Continuous distributions of the azimuth of polarization and ellipticity at the BT images have been discreted into a 2D ($m \times n$) tensor of the data with coordinates $\begin{pmatrix} r_{11},\ldots,r_{1m} \\ \ldots\ldots\ldots\ldots \\ r_{n1},\ldots,r_{nm} \end{pmatrix} \equiv r_{m,n}$ within the sensitive area of a CCD-camera. There is the algorithm for determining the coordinate distributions of the azimuth of polarization $\alpha(r_{m,n})$ and ellipticity $\beta(r_{m,n})$ (polarization maps, PMs) of the images of histological slices of a BT: (i) by rotating of the transmission axis of the analyzer, Θ, within the interval from 0° to 180°, one determines the tensors of the maximal and minimal magnitudes of intensity of a laser image, $I_{\min}(r_{m,n})$ and $I_{\max}(r_{m,n})$ for the each pixel (mn) of CCD-camera and the corresponding rotation angles, $\Theta(r_{m,n})$ $(I(r_{m,n}) \equiv I_{\min})$; and (ii) one computes PMs of the BT image using the following equations:

$$\begin{aligned} \alpha(r_{m,n}) &= \Theta\left(I\left(r_{ik}\right) \equiv I_{\min}\right) - \tfrac{\pi}{2}; \\ \beta(r_{m,n}) &= \arctan \tfrac{I(r_{ik})_{\min}}{I(r_{ik})_{\max}}. \end{aligned} \tag{10.17}$$

10.3.2 Statistical and fractal analysis of polarization images of histological slices of biological tissues

Classification of all types of human BTs on the structure of their architectonic component or out-of-cell matrix is given in [13]. So, one distinguishes four main types of BTs whose out-of-cell matrix is the structured nets of optically uniaxial birefringent coaxial protein (collagen, myosin, elastin, etc.) fibrils:

- connective tissue (skin, bones, tendon, etc.);
- muscular tissue;
- epithelial tissue;
- nervous tissue.

On the other hand, there are many BTs forming parenchymatous organs, such as the kidney, liver, spleen, bowel walls, etc. The "island-like" out-of-cell matrix of such tissues is nonstructured, and its polarization properties are much less studied than the properties of the nets of biological crystals of the above-mentioned types of BTs. It makes comparative analysis of PMs of both types of birefringent structures to be topical.

Two types of optically thin (extinction coefficient $\tau \leq 0.1$) histological slices of BTs have been used for such study, namely, structured tissue of osseous shin (Figure 10.3 a, b) and parenchymatous kidney tissue (Figure 10.3 c, d). In all objects chosen for the study the optically anisotropic component with the birefringence index $\Delta n \approx 1.5 \times 10^{-1}$(osseous tissue) and $\Delta n \approx 1.5 \times 10^{-3}$ (kidney tissue) [14, 15] is present and visualized for crossed polarizer and analyzer (Figure 10.3 b, d). Geometric structure of architectonics of such BTs is different. Osseous tissue is formed by quasi-ordered bundles of birefringent collagen fibrils mineralized by crystals of hydroxyapatite [20] (Figure 10.3 b). Parenchymatous kidney tissue contains "island-like" inclusions of anisotropic collagen (Figure 10.2 d).

The results of investigation of the coordinate distributions of $\alpha(r)$ and $\beta(r)$, as well as histograms $W(\alpha)$ and $W(\beta)$ of the magnitudes of the azimuth of polarization and ellipticity at the images of histological slices of physiologically normal osseous tissue (left hand) and kidney tissue (right hand) are shown in Figure 10.4 (fragments a, b and c, d, respectively).

It follows from the obtained data that PMs of an osseous tissue (Figure 10.4 a, b, left hand) are formed by large ensembles (from 100 μm to 200 μm) of homogeneously polarized ($\{\alpha(r_i);\beta(r_i)\} \approx const$) areas (polarization domains). PMs of a kidney tissue (Figure 10.4 a, b, right hand) are formed by homogeneously polarized areas of much less sizes (from 5 μm to 20 μm). The distributions of the azimuth of polarization and ellipticity of the obtained PMs of BTs of both types are characterized by the set of the statistical moments of the first to the fourth orders represented in Table 10.1.

TABLE 10.1: Statistical moments S_i for the coordinate distributions of the state of polarization for images of osseous and kidney BTs.

Osseous tissue (31 samples)				Kidney tissue (27 samples)			
$\alpha(r)$		$\beta(r)$		$\alpha(r)$		$\beta(r)$	
S_1	0.38±0.027	S_1	0.24±0.014	S_1	0.11±0.01	S_1	0.08±0.004
S_2	0.25±0.015	S_2	0.21±0.017	S_2	0.19±0.013	S_2	0.05±0.003
S_3	9.8±0.882	S_3	7.7±0.539	S_3	1.4±0.056	S_3	0.61±0.04
S_4	24.6±2.71	S_4	12.5±1.125	S_4	3.1±0.093	S_4	2.25±0.113

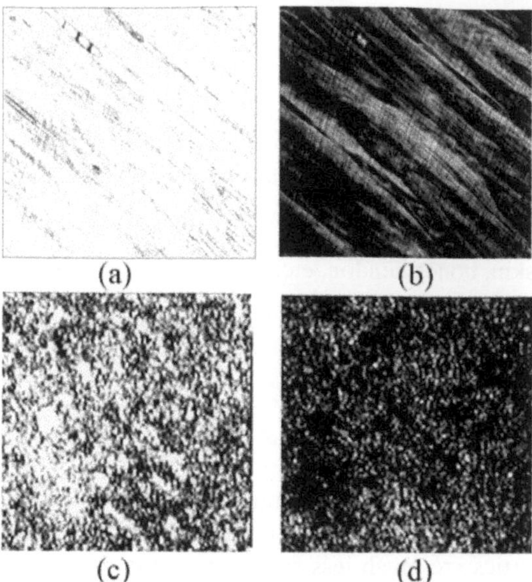

FIGURE 10.3: Polarization images of osseous tissue (a, b) and kidney tissue (c, d) in matched (a, c) and crossed (b, d) polarizer and analyzer, respectively.

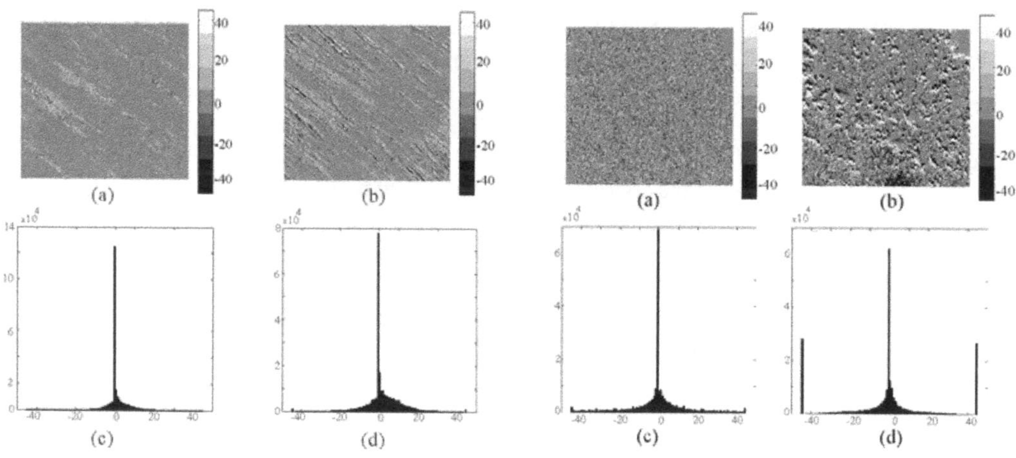

FIGURE 10.4: PMs of osseous tissue (left hand) and kidney tissue (right hand). Fragments (a, b) correspond to 2D distributions (in degrees) of the azimuth of polarization and ellipticity, respectively; fragments (c, d) show histograms of these distributions.

The data represented in Table 10.1 leads to the following conclusions: (i) images of BTs with ordered architectonics are characterized by nonzero magnitudes of skewness and kurtosis of the distributions of the azimuth of polarization and ellipticity of light oscillations; and (ii) statistical structure of the BT's PMs with island architectonics is characterized by the distribution close to the normal law for the azimuth of polarization and ellipticity, so that the statistical moments of the third and fourth orders are less by one order of magnitude in comparison with such parameters of the PMs of osseous tissue.

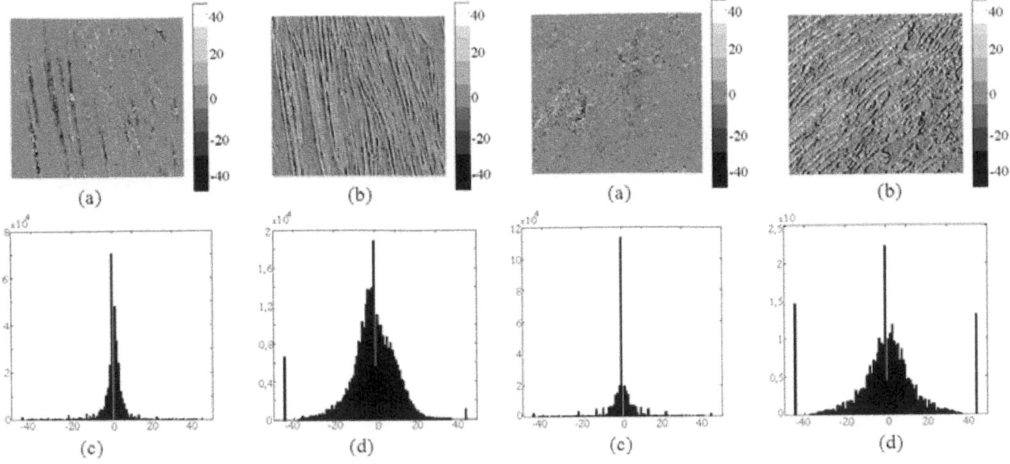

FIGURE 10.5: PMs of physiologically normal (left hand) and dystrophically changed (right hand) muscular tissue. Fragments (a, b) correspond to 2D distributions (in degrees) of the azimuth of polarization and ellipticity, (c, d) – histograms of these polarization parameters.

Such statistical structure of the PMs for BTs of various types can be connected with the peculiarities of morphology of their architectonics (structured or islands).

For osseous tissue, the structures transforming polarization of laser radiation are large scale domains of monooriented birefringent collagen fibrils. In accordance with Eqs. (10.2), (10.8), and (10.12), the coordinate distributions of the azimuth of polarization and ellipticity are homogeneous within such domains, $\rho, \delta\left(r_{m,n}\right) \approx const$. Anisotropic components of the kidney tissue have no pronounced orientation structure. Then, the inhomogeneously polarized component of its image is characterized by the distributions of the azimuth of polarization and ellipticity (Figure 10.4 a, b right hand) close to the normal law.

One can see from the represented data on the statistical moments of the first to the fourth orders for the distributions $W(\alpha)$ and $W(\beta)$ at images of various morphological structures that as birefringent architectonic nets are more structured, the magnitudes of the statistical moments of the third and fourth orders characterizing the distributions of the azimuth of polarization $\alpha(r_{m,n})$ and ellipticity $\beta(r_{m,n})$ are larger.

10.3.3 Diagnostic feasibilities of polarization mapping of histological slices of biological tissues of various physiological states

To search for the feasibilities for differentiation of the geometrical-optical structure of the architectonics of BTs we carried out a comparison study of the statistical and fractal structure of their PMs. The following BTs have been investigated:

- physiologically normal and dystrophically changed muscular tissue (MT) – structured BT with ordered architectonics;
- physiologically normal and pathologically changed skin dermis (SD) – structured BT with disordered architectonics;
- physiologically normal and septically inflamed pulmonary tissue (PT) – BT with island architectonics.

PMs of images of these objects are illustrated in Figs. 10.5–10.7.

Analysis of the obtained results shows that the domain structure of the PMs connected with the discrete structure of the architectonic net of protein crystals is typical for the BTs with ordered (MT)

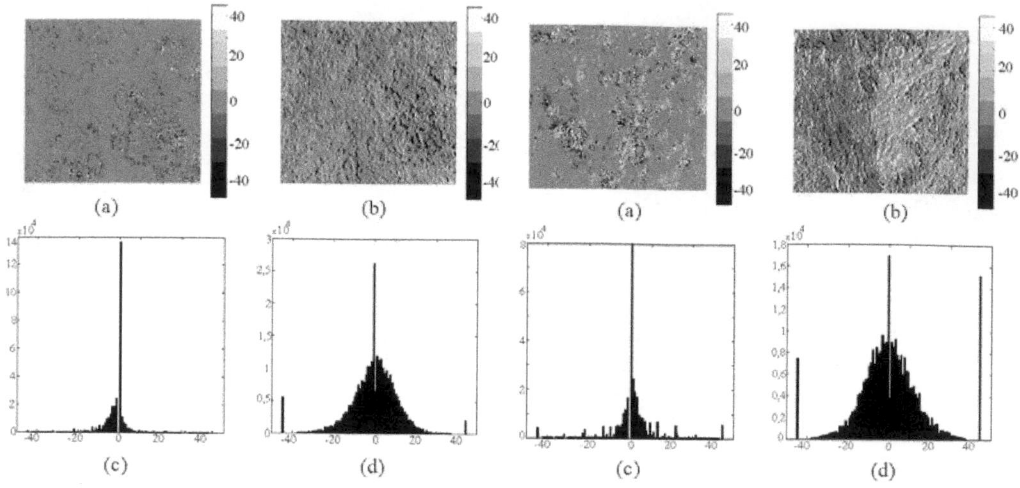

FIGURE 10.6: PMs of physiologically normal (left hand) and pathologically changed (right hand) skin dermis. Fragments (a, b) correspond to 2D distributions (in degrees) of the azimuth of polarization and ellipticity, (c, d) – histograms of these polarization parameters.

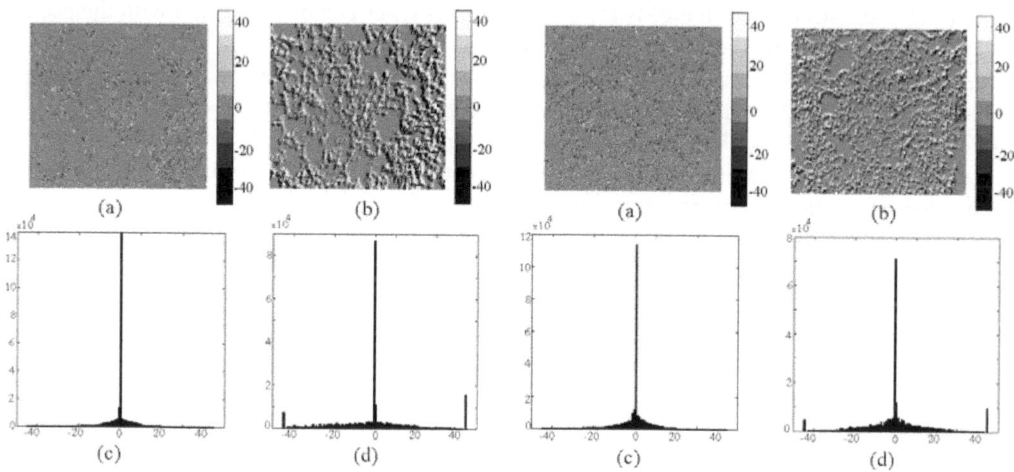

FIGURE 10.7: PMs of physiologically normal (left hand) and septically inflamed (right hand) pulmonary tissue. Fragments (a, b) correspond to 2D distributions (in degrees) of the azimuth of polarization and ellipticity, (c, d) – histograms of these polarization parameters.

and disordered (SD) architectonic nets. Comparison of the polarization azimuth and ellipticity coordinate distributions at images of physiologically normal and pathologically changed samples of the structured BTs shows: (i) decreasing of polarization domains of the PMs for dystrophically changed muscular tissue and narrowing intervals of changes of random magnitudes of the azimuth of polarization and ellipticity at the corresponding image of a histological tome (Figure 10.5); and (ii) scaling of polarization domains of the PMs for SD and increasing intervals of random changes of the azimuth of polarization and ellipticity (Figure 10.6). Such transformation of the PMs coordinate structure can be connected with peculiarities of the geometrical-optical structure changes of the pathologically changed human tissues architectonics. So, increasing interval of orientations (disordering) of the optical axes of birefringent myosin fibrils accompanied with decreasing of birefringence index of their matter [17] is typical for dystrophically changed samples of a BT. Inverse processes of formation of the new directions of growth and increasing anisotropy take place for the architectonic collagen net of pathologically changed samples of SD.

Computer simulation of the geometrical-optical structure changes of the BT architectonics [20] shows that growth of dispersion of birefringent fibrils optical axes orientations (degenerative-dystrophic changes) is typically accompanied by increasing the third and the fourth statistical moments characterizing the distributions of the azimuth of polarization and ellipticity at BT images. Contrary, increasing of the phase shift dispersion (pathologically changed architectonics) is accompanied with decreasing magnitudes of skewness and kurtosis of the distributions of the polarization parameters of BT images.

The feasibilities for statistical differentiation of the PMs are experimentally proved by comparison of the magnitudes of the statistical moments of the first to the fourth orders for the distributions of the azimuth of polarization and ellipticity at images of physiologically normal and pathologically changed BTs (see Tables 10.2–10.4).

TABLE 10.2: Statistical moments of the first to the fourth orders of the coordinate distributions of the azimuth of polarization and ellipticity at images of histological slices of normal and dystrophic muscular tissue.

Norm (21 samples)			Pathology (22 samples)		
S_i	α	β	S_i	α	β
S_1	0.26±0.013	0.12±0.01	S_1	0.21±0.001	0.18±0.011
S_2	0.12±0.01	0.08±0.004	S_2	0.19±0.013	0.12±0.006
S_3	6.7±0.469	4.9±0.294	S_3	9.4±0.846	6.2±0.434
S_4	17.9±1.61	14.5±1.595	S_4	25.4±2.54	21.6±2.592

TABLE 10.3: Statistical moments of the first to the fourth orders of the coordinate distributions of the azimuth of polarization and ellipticity at images of histological slices of normal and pathologically changed skin dermis.

Norm (23 samples)			Pathology (21 samples)		
S_i	α	β	S_i	α	β
S_1	0.16±0.01	0.11±0.005	S_1	0.14±0.06	0.09±0.006
S_2	0.21±0.01	0.15±0.012	S_2	0.17±0.01	0.12±0.012
S_3	4.24±0.334	2.9±0.319	S_3	2.14±0.2	1.76±0.229
S_4	7.14±0.643	6.18±0.803	S_4	4.29±0.515	4.57±0.69

Analysis of the represented data shows that the images of all studied types of BTs are characterized by nonzero magnitudes of the third and fourth statistical moments inherent in distributions of the polarization parameters of histological slices images. So, the magnitudes of skewness and kurtosis of the coordinate distributions of the azimuth of polarization and ellipticity values of the BT

TABLE 10.4: Statistical moments of the first to the fourth orders of the coordinate distributions of the azimuth of polarization and ellipticity at images of histological slices of normal and septically inflamed pulmonary tissue.

Norm (22 samples)			Pathology (22 samples)		
S_i	α	β	S_i	α	β
S_1	0.08±0.004	0.06±0.004	S_1	0.09±0.004	0.08±0.005
S_2	0.19±0.013	0.16±0.02	S_2	0.21±0.01	0.2±0.016
S_3	3.12±0.25	2.64±0.29	S_3	2.04±0.184	1.18±0.142
S_4	6.92±0.76	2.17±0.282	S_4	4.17±0.5	7.28±1.092

images with ordered architectonics (Table 10.2) are three to five times larger than the magnitudes of these parameters characterizing the statistics of the third and fourth orders of inhomogeneously polarized images of BTs with disordered architectonics (Table 10.3). It is proved by the fact that increasing the angular interval of the directions of the optical axes of anisotropic fibrils brings nearer, to a certain extent, their statistics to the normal law. For statistical structure of inhomogeneously polarized images of BTs with island architectonics (Table 10.4), less but nonzero magnitudes of the higher-order statistical moments for the distributions of the azimuth of polarization and ellipticity are typical.

Comparison of the PMs of physiologically normal and pathologically changed BTs shows the following regularities of changes of the statistical moments of the first to the fourth orders for the distributions of the azimuth of polarization and ellipticity.

For the PMs of histological slices of degenerative-dystrophic changed samples of a BT the magnitudes of the statistical moments of the first and second orders of the distributions of the azimuth of polarization and ellipticity differ from such parameters of inhomogeneously polarized images of normal tissue not more than by 5–10%. Much more pronounced (by 30–60%) is the change of the magnitudes of the distributions of the azimuth of polarization and ellipticity (Table 10.2). This result is explained by the fact that increasing dispersion of the orientation angles of the optical axes of myosin fibrils manifests itself in the corresponding increasing of the statistical moments of the third and fourth orders for the distributions of the azimuth of polarization and ellipticity [20].

Contrary tendency takes place for pathological changes (increasing dispersion of phase shifts introduced by the architectonics matter). So, skewness and kurtosis of the coordinate distributions of the azimuth of polarization and ellipticity at the PMs of images of pathologically changed samples of SD decrease by 25–60% (Table 10.3).

Inflammation of pulmonary tissue is accompanied with formation of emphysema, which manifests itself in increasing birefringence of the corresponding anisotropic domains. As a consequence, dispersion of phase shifts increases and, correspondingly, skewness and kurtosis decreases (by 25–50%), first of all, for account of changing distribution of magnitudes of ellipticity at the BT images.

Thus, the statistical analysis of the PMs of BTs with various morphological structures shows the difference in the magnitudes of their statistical moments of the third and fourth orders for the distributions of the azimuth of polarization and ellipticity. The established criteria can be used for differentiation of changes of the geometrical optical properties of BTs connected with their physiological state.

Note, that the use of the statistical approach alone for analysis of the PMs does not provide comprehensive and unambiguous differentiation of physiological states of BTs. As it is seen from Tables 10.2 to 10.4, the magnitudes of skewness and kurtosis for the distributions of the azimuth of polarization and ellipticity within the studied groups of physiologically normal and changed BTs can overlap. On the other hand, structured BTs have hierarchical, self-similar geometry of architectonics

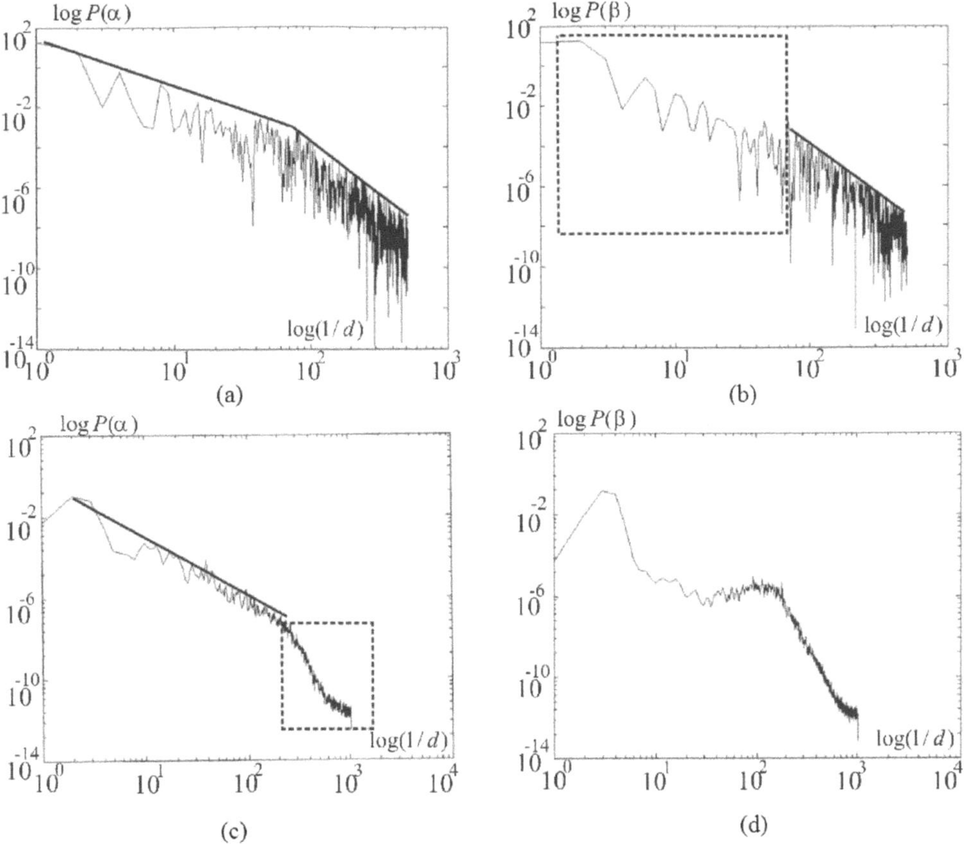

FIGURE 10.8: log-log dependences of the power spectra of the azimuth of polarization and ellipticity at the images of physiologically normal (a, b) and pathologically changed (c, d) skin dermis.

[13]. Then, it is important to determine of optical manifestations of such geometry of the anisotropic component of structured BTs. We have studied for this purpose self-similarity of the BT's PMs for skin dermis. The set of log-log dependences ($\log P(\alpha) - \log(1/d)$ and $\log P(\beta) - \log(1/d)$) for the power spectra of the distributions of the polarization parameters of images of physiologically normal and pathologically changed skin dermis is shown in Figure 10.8.

Analysis of the represented data shows rather complex coordinate distribution of the azimuth of polarization and ellipticity of the PMs for image of normal skin, see Figure 10.8 a, b. $\log P(\alpha) - \log(1/d)$ dependence has two slopes within three decades of changes of the element size of the architectonics. $\log P(\beta) - \log(1/d)$ dependence is the combination of the section with the constant slope and statistical section (Figure 10.8 b). The pathological state of such BT manifests itself in randomization of the distributions of the polarization parameters at images of the corresponding histological slices: constant slope for approximating curves is absent, see Figure 10.8 c, d.

10.3.4 Polarization 2D tomography of biological tissues

This section is devoted to the analysis of the feasibilities for solving the inverse problem of laser polarimetry, *viz.* separate determination of the coordinate distributions of the orientation, $\rho(r)$, and phase, $\delta(r)$, parameters of the BT's architectonics.

10.3.4.1 Orientation tomography of architectonics nets

Based on the proposed model notions on the mechanisms of interaction of laser radiation with the nets of biological crystals (Eqs. (10.1) to (10.12)), one can see that for the crossed polarizer 4 and analyzer 9 (see Figure 10.2) and probing with a linearly polarized laser beam with the azimuth of polarization $\alpha_0 = 0^0$, intensity at the each point of the image of BT's histological tome r can be found by solving the following matrix equation:

$$U(r) = 0.25\{A\}\{\mathrm{M}\}\{P\}D_0 = 0.25\begin{Vmatrix} 1 & -1 \\ -1 & 1 \end{Vmatrix}$$

$$\times\begin{Vmatrix} \cos^2\rho(r) + \sin^2\rho(r)\exp[-i\delta(r)] & \cos\rho(r)\sin\rho(r)\{1-\exp[-i\delta(r)]\} \\ \cos\rho(r)\sin\rho(r)\{1-\exp[-i\delta(r)]\} & \sin^2\rho(r) + \cos^2\rho(r)\exp[-i\delta(r)] \end{Vmatrix}$$

$$\times\begin{Vmatrix} 1 & 1 \\ 1 & 1 \end{Vmatrix}\begin{pmatrix} 1 \\ 0 \end{pmatrix}, \tag{10.18}$$

where D_0, $U(X,Y)$ are the Jones vectors of the probing and object beams, and $\{A\}$, $\{M\}$, and $\{P\}$ are the Jones matrices of the analyzer, BT and polarizer, respectively. Intensity of the object field is determined from (10.18) as

$$I(r) = U(r)U^{\otimes}(r) = I_0\sin^2 2\rho(r)\sin^2\left[\delta(r)/2\right], \tag{10.19}$$

where I_0 is intensity of the probing beam.

Analysis of Eq. (10.19) shows that for rotating azimuth of polarization of the probing beam (ω is the rotation angle of the axes of crossed polarizer and analyzer) the azimuth of polarization may coincide with the optical axis of a biological crystal, $\alpha_0(\omega) = \rho(r)$. In this case, lines of zero intensity, $I^*_\rho(r,\omega) = 0\left|_{\alpha_0(\omega)=\rho(r)}\right.$, or "polarizophotes" occur at the corresponding image of a BT. Thus, for matched rotation of the transmission axes of the crossed polarizer and analyzer one can obtain the set of the distributions of polarizophotes that we call the orientation tomograms of the BT's architectonics. Such a set of the orientation tomograms of the BT's architectonics contains the data on the coordinate distribution of optical axes of protein fibrils. Such information is important for early (preclinical) diagnostics of degenerative-dystrophic (disordering structure) and pathologic (forming the directions of growth) changes of the BT's architectonics [60].

Additional information on the properties of birefringent nets of biological crystals is given by the coordinate distribution of phase shifts introduced by their matter between the orthogonal components of amplitude of laser radiation. Hereinafter, such distributions $\delta(r)$ are referred to as the phase tomograms.

10.3.4.2 Phase tomography of biological tissues

For selection of the phase component (phase tomogram) at the polarization image of a BT, one must place the BT sample into a polarization system consisting of two crossed filters, *viz.* quarter-wave plates 5, 8 and polarizers 4, 9, the angles of transmission axes of which with the axes of the maximal velocity of the plates are +45° and −45°, respectively (see Figure 10.2). In this case, the coordinate intensity distribution at the BT image, $I_\delta(r)$, is determined from the Jones vector of the object field, $E(r)$, through solving the following equations

$$E(r) = 0.25\{A\}\{\Phi_2\}\{\mathrm{M}\}\{\Phi_1\}\{P\}D_0 = 0.25 \begin{Vmatrix} 1 & -1 \\ -1 & 1 \end{Vmatrix} \begin{Vmatrix} i & 0 \\ 0 & 1 \end{Vmatrix} \times$$

$$\times \begin{Vmatrix} \cos^2\rho(r) + \sin^2\rho(r)\exp[-i\delta(r)] & \cos\rho(r)\sin\rho(r)\{1-\exp[-i\delta(r)]\} \\ \cos\rho(r)\sin\rho(r)\{1-\exp[-i\delta(r)]\} & \sin^2\rho(r) + \cos^2\rho(r)\exp[-i\delta(r)] \end{Vmatrix} \times \tag{10.20}$$

$$\times \begin{Vmatrix} 1 & 0 \\ 0 & i \end{Vmatrix} \begin{Vmatrix} 1 & 1 \\ 1 & 1 \end{Vmatrix} \begin{pmatrix} 1 \\ 0 \end{pmatrix},$$

where $\{\Phi_1\}$ and $\{\Phi_2\}$ are the Jones matrices of quarter-wave plates; $\{A\}$ and $\{P\}$ are the Jones matrices of the analyzer and polarizer, respectively. It follows from Eq. (10.20) that the coordinate intensity distribution at filtered in polarization image of a BT is unambiguously connected with the phase shifts as

$$I_\delta(r) = E(r)E^{\otimes}(r) = I_0 \sin^2\left[\delta(r)/2\right]. \tag{10.21}$$

Thus, this approach provides the information on the phase-shifting ability of birefringent matter of a BT irrespectively from the direction of packing of its anisotropic fibrils.

10.3.4.3 Experimental implementation of orientation tomography

The efficiency of the orientation tomography for diagnostics of pathological structures of BTs, e.g., of the architectonics of myometrium (a BT of the woman reproductive sphere), is illustrated by the tomograms shown in Figure 10.9.

This figure shows the orientation tomograms of a collagen net of physiologically normal (fragments a to c) and pathologically changed (fragments d to i) samples of myometrium, namely, with directed (fragments d, e, f) and diffuse (fragments g, h, i) growth of pathologically changed protein fibrils. The left column of the orientation tomograms corresponds to the angle $\omega = 45°$, middle column to $\omega = 25°$, and right column to $\omega = 0°$.

The obtained polarizophotes illustrate in qualitative manner efficiency of the polarization technique for orientation selection of the directions of optical nets of birefringent biological crystals for BTs with various physiological states. These data lead to the following conclusions.

All orientation tomograms for a polarization image of the architectonics of physiologically normal samples of myometrium are the ensembles of small-scale (5 μm to 10 μm) areas of polarizophotes uniformly distributed over a sample area (Figure 10.9 a, b, c). Such structure shows statistical, with practically uniform distribution, orientation structure of the collagen net.

Areas with large-scale (20 μm to 200 μm) optically anisotropic structures, such as the ensembles of similarly oriented collagen fibers correspond to pathological processes of diffuse forming fibromyoma. Similar birefringent domains are optically visualized as the corresponding polarizophotes whose set is shown in Figure 10.9 g, h, i.

Comparison of the set of orientation tomograms shows their topological skewness. It is seen that for a large interval of changes of the angle ω, the ensembles of polarizophotes have individual structures connected with both localization of pathologically changed self-similar (fractal) domains of the architectonics of myometrium and with the directions of their growing.

Asymmetry of polarizophotes is most pronounced, illustrated by comparison of the orientation tomograms for the sample of myometrium with the directed growing fibromyoma (Figure 10.9 d, e, f). It is seen from Figure 10.9 d that for the angle of rotation of the crossed polarized and analyzer $\omega = 45^0$ coinciding with the main direction of growth of collagen fibers, as the corresponding orientation tomogram consists from larger polarizophotes.

The feasibilities of correlation analysis of the orientation structure of the BT's architectonics are illustrated by 2D autocorrelation functions in Figure 10.10 obtained for the orientation tomograms

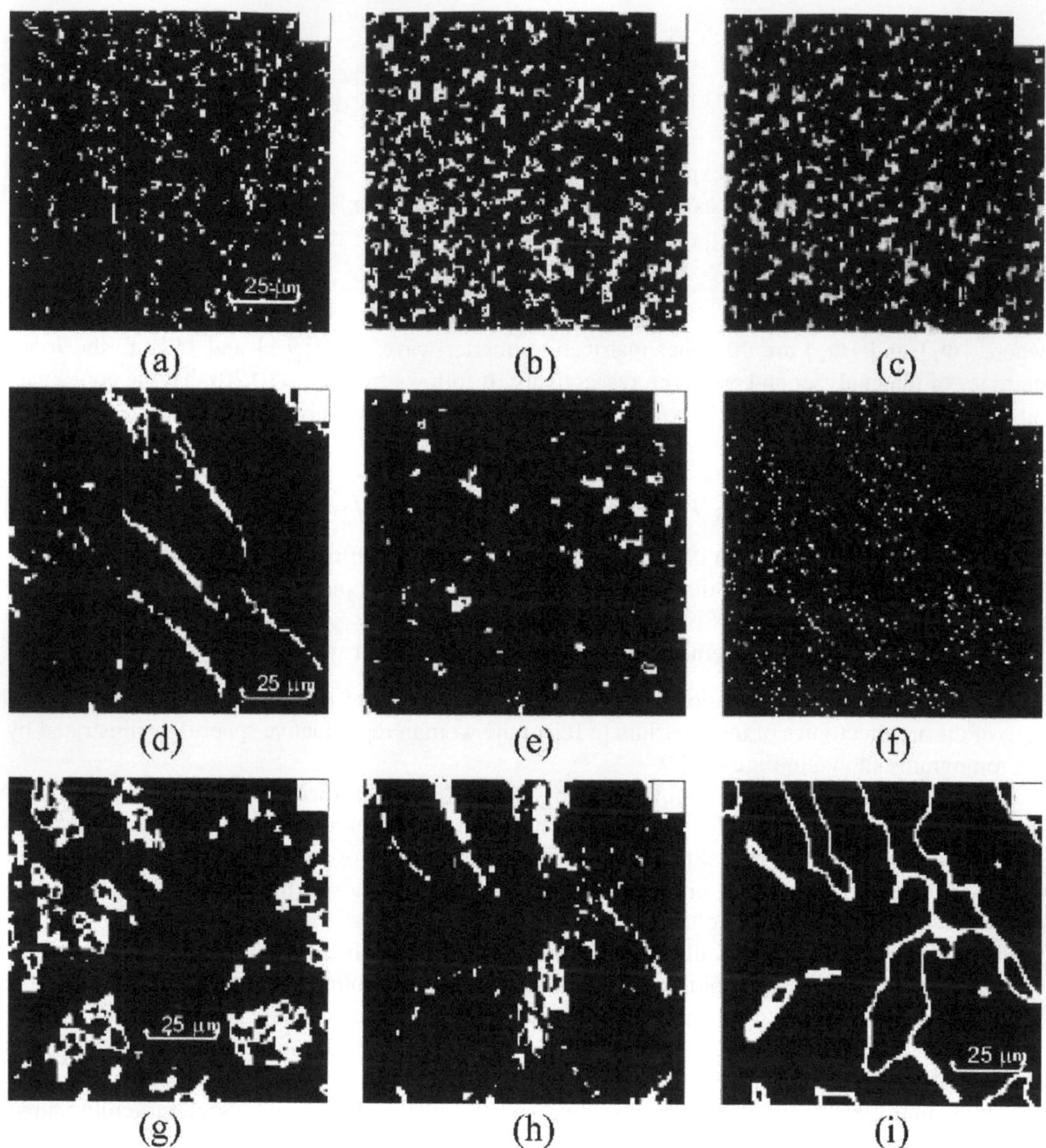

FIGURE 10.9: Orientation tomograms of physiologically normal (a–c) and pathologically changed (d–i) myometrium.

of physiologically normal (Figure 10.9 a, b, c) and pathologically changed (sprouts of fibromyoma, Figure 10.9 d, e, f) myometrium.

Azimuthal structure of 2D autocorrelation functions $\left\{\hat{G}_{xxyy}(\Delta x, \Delta y)\right\}$ (Figure 10.10 b, d) is characterized by the asymmetry coefficient $\zeta(\varphi)$ (Figure 2.9 i, k) of the following form:

$$\zeta(\varphi) = \frac{\Re\left\{\hat{G}_{xxyy}(\varphi)\right\}}{\Re\left\{\hat{G}_{xxyy}(\varphi + 0.5\pi)\right\}}, \tag{10.22}$$

where $\Re\left\{\hat{G}_{xxyy}(\varphi)\right\}$ and $\Re\left\{\hat{G}_{xxyy}(\varphi + 0.5\pi)\right\}$ are half-widths of the function $\left\{\hat{G}_{xxyy}(\Delta x, \Delta y)\right\}$ determined for the orthogonal azimuthal directions of polar angle φ at level $0.5\left[\max\left\{\hat{G}_{xxyy}(\Delta x, \Delta y)\right\}\right]$.

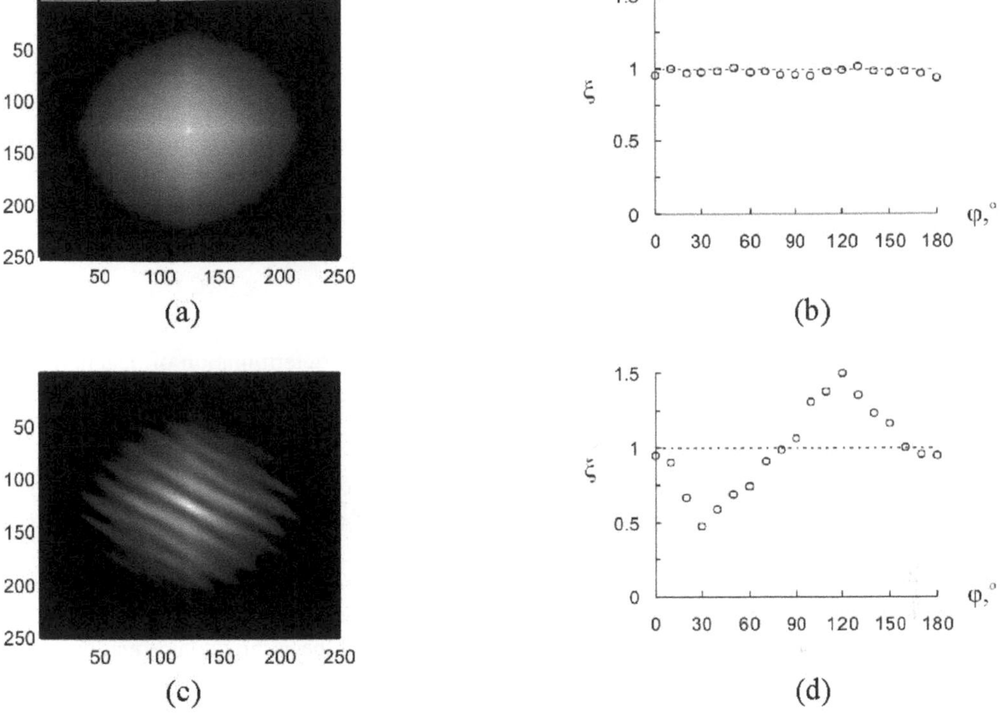

FIGURE 10.10: Autocorrelation functions (a, c) and corresponding asymmetry coefficients (b, d) of the orientation tomograms for samples of myometrium of normal (a) and pathologically changed (sprouts of fibromyoma) (c) tissue.

Architectonics of physiologically normal myometrium is almost uniform on the directions of packing of collagen fibers construction. As so, the function $\left\{\hat{G}_{xxyy}(\Delta x, \Delta y)\right\}$ is very close to the azimuthally symmetric one (Figure 10.10 a), and the coefficient $\zeta(\varphi) \cong 1$ over all angular range (Figure 10.10 b). Tomograms of pathologically changed myometrium are characterized by azimuthally asymmetrical 2D correlation functions (Figure 10.10 c) with the corresponding azimuthal coefficient $\zeta(\varphi)$, see Eq. 10.22 (Figure 10.10 d). Its extremal magnitudes correspond to the angles of the directions of growth of pathologically changed areas of a collagen net of myometrium (Figure 10.9 d and Figure 10.9 h correspondingly).

Thus, orientation tomography of the BT's architectonics enables effective detection of the processes of transformation of its orientation-phase structure, which is promising in the development of medical techniques for pre-clinic diagnostics of physiological state of a BT.

10.3.4.4 Experimental implementation of the phase tomography of the architectonics nets of biological tissues

Pathological and degenerative-dystrophic processes in BTs arising at preclinic stages are accompanied by the following morphological changes of their architectonics [14]: (i) decreasing concentration of mineral matters and disorientation of an architectonic net (rachitis, osteoporosis, etc.); and (ii) growing collagen structures and calcification of soft tissues (psoriasis, collagenoses, myomes, fibromyomes). The tendency of such changes in the first case can be determined as a decrease of the birefringence index $\Delta n \to \Delta n_{\min}$, while in the second case as an increase of its value, as $\Delta n \to \Delta n_{\max}$.

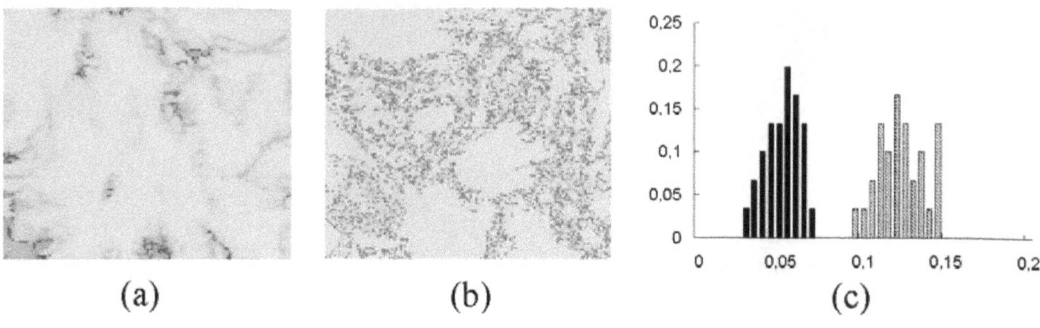

FIGURE 10.11: Diagnostic feasibilities of phase tomography in determination of degenerative-dystrophic changes (osteoporosis) of the structure of osseous tissue.

The results illustrating the feasibilities for selection of the phase component at the images of BTs for pre-clinical diagnostics of degenerative-dystrophic changes of osseous tissue are represented in Figure 10.11.

The data obtained on the coordinate distributions of the magnitudes of the birefringence coefficients leads to the following conclusions. For the sample of a physiologically normal osseous slice (Figure 10.11 a) the magnitude of the birefringence index, $\Delta n(r)$, of the architectonics matter within the areas of its crystalline domains (40 μm to 200 μm) is almost uniform. The phase tomogram for a sample of osseous tissue with forming osteoporosis (decalcified architectonics) has a different structure. The coordinate distribution $\Delta n(r)$ is transformed into the set of small-scaled (5 μm to 30 μm) areas (Figure 10.11 b). The birefringence index of the architectonics of an osseous tissue decreases by one order of magnitude, $\Delta n(r) \approx 0.2 \times 10^{-2} - 1.2 \times 10^{-2}$.

Statistics of such changes is illustrated by the set of diagrams of change of dispersion of fluctuations, $\Delta n(r)$, obtained from the phase tomograms of an osseous tissue (Figure 10.11 c). Experimental data for physiologically normal osseous tissue (solid columns, 30 samples) and degeneratively changed one (dashed columns, 28 samples) are presented. Comparison of these diagrams shows that, within the statistically trustworthy groups of the studied samples, they do not overlap. As so, this statistical parameter of the phase tomograms of the BT architectonics can be used for preclinical diagnostics of degenerative-dystrophic changes of BTs.

10.4 Polarization Correlometry of Biological Tissues

10.4.1 The degree of mutual polarization at laser images of biological tissues

Changes of not only polarization but also correlation characteristics are inherent in the object fields scattered by BTs that can be characterized by the complex degree of mutual polarization [30, 34]:

$$\begin{gathered} W(r_1,r_2) = \\ \left[(U_x(r_1)U_x(r_2) - U_y(r_1)U_y(r_2))^2 + 4U_x(r_1)U_x(r_2)U_y(r_1)U_y(r_2)e^{i(\delta_2(r_2)-\delta_1(r_1))} \right] \times . \\ \left(U_x^2(r_1)+U_y^2(r_1)\right)^{-1}\left(U_x^2(r_2)+U_y^2(r_2)\right)^{-1} \end{gathered} \quad (10.23)$$

Based on Eqs. (10.1) to (10.13), one can reveal the interconnection between the azimuth of polarization, $\alpha(r)$, and ellipticity, $\beta(r)$, at points r of the image of a BT, and the orthogonal components,

$U_x = \sqrt{I_x}$ and $U_y = \sqrt{I_y}$, of the complex amplitudes with the phase shift between them $\delta(r)$:

$$\delta(r_1) = \arctan\left[\frac{\tan 2\beta(r_1)}{\tan\alpha(r_1)}\right]; \delta(r_2) = \arctan\left[\frac{\tan 2\beta(r_2)}{\tan\alpha(r_2)}\right]. \quad (10.24)$$

Using the imaging system shown in Figure 10.2 and measuring the coordinate intensity distributions I_x, I_y, and the coordinate distributions of the states of polarization (α, β) for all $m \times n$ points, one can determine the distribution of the degree of mutual polarization at the boundary field:

$$W(r_2, r_1) = \frac{\left((I_x(r_2)I_x(r_1))^{\frac{1}{2}} - (I_y(r_2)I_y(r_1))^{\frac{1}{2}}\right)^2}{I(r_2)I(r_1)} + \\ + \frac{4\left(I_x(r_2)I_y(r_2)I_x(r_1)I_y(r_1)\right)^{\frac{1}{2}}\cos(\delta_2(r_2) - \delta_1(r_1))}{I(r_2)I(r_1)}. \quad (10.25)$$

Hereinafter, the coordinate distributions of the parameter $W(r_1, r_2)$ are referred to as the polarization-correlation maps (PCMs) of BT images.

10.4.2 Technique for measurement of polarization-correlation maps of histological slices of biological tissues

There is the algorithm for determination of the polarization-correlation structure of a BT image (see Figure 10.2):

- using CCD camera 9 (without the analyzer 8), one measures the coordinate intensity distribution at the BT image, $I(r_{m,n})$, where $r_{m,n}$ are the coordinates of $m \times n$ pixels of the CCD-camera;
- introducing the analyzer 8, whose transmission plane is oriented at the angles $\Theta = 0°$ and $\Theta = 90°$, one measures the intensity tensors $I_x(r_{m,n})$ and $I_y(r_{m,n})$, respectively;
- rotating the analyzer at the angle Θ within the interval $\Theta = 0° \div 180°$, one determines the tensors of the maximal and minimal intensities, $I_{\max}(r_{m,n})$ and $I_{\min}(r_{m,n})$, of the laser image for the each pixel of CCD-camera, as well as the corresponding rotation angles $\Theta(r_{m,n})$ $(I(r_{m,n}) \equiv I_{\min})$;
- one computes the distributions of the phase shifts (Eq. 10.24) $\delta(r_{m,n})$;
- the degree of mutual polarization of the BT image $W(\Delta r)$ is computed following the algorithm (10.25), where $\Delta r = 1$ px;
- the coordinate distribution of the tensor of local magnitudes $W(x, y)$ is determined by scanning of the inhomogeneously polarized image in two perpendicular directions ($x = 800$ px; $y = 600$ px) with the step $\Delta r = 1$ px.

10.4.3 Statistical approach to the analysis of polarization-correlation maps of biological tissues

We choose for our study the samples of histological slices of two groups of BTs, *viz.* (i) normal and degeneratively changed (osteoporosis) osseous shin tissue, and (ii) normal and dysrtrophically changed tissue of skeleton muscle. In Fig 10.12 and Figure 10.13 we show the coordinate distributions of the degree of mutual polarization $W(x, y)$ at the images of histological slices of normal (a) and pathologically changed (b) osseous (Figure 10.12) and muscular (Figure 10.13) tissues.

The difference of the coordinate distributions $W(x, y)$ for the PCMs of BTs of various morphological structures and physiological states is characterized by the set of the statistical moments $S_{i=1;2;3;4}$ whose magnitudes are given in Tables 10.5 and 10.6.

Statistical analysis of the coordinate distributions $W(x, y)$ shows: (i) all statistical moments characterizing the distributions $W(x, y)$ are different for the PCMs of the images of normal and degenerative-dystrophic changed BTs of both types; (ii) differences between averages magnitudes, dispersion, skewness, and kurtosis of the distributions $W(x, y)$ for the PCMs of osseous tissuc arc

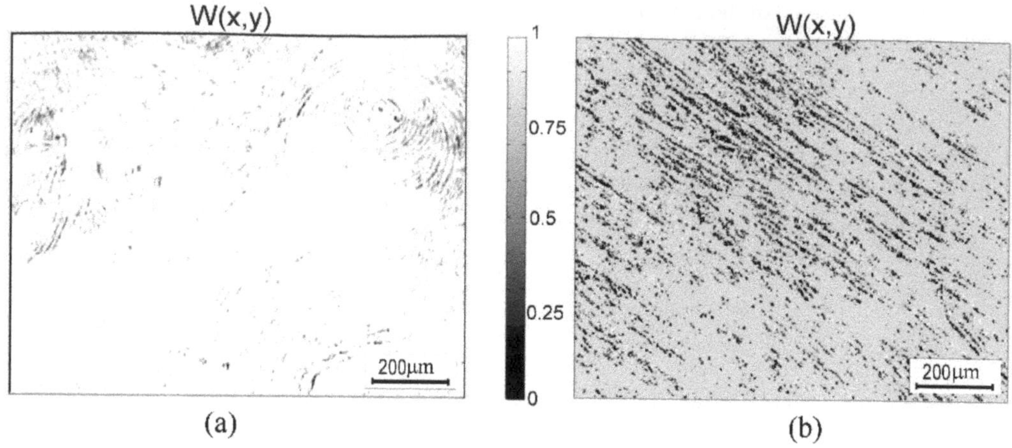

FIGURE 10.12: Coordinate distributions of the degree of mutual polarization $W(x,y)$ at the images of normal (a) and degeneratively changed (b) osseous tissue.

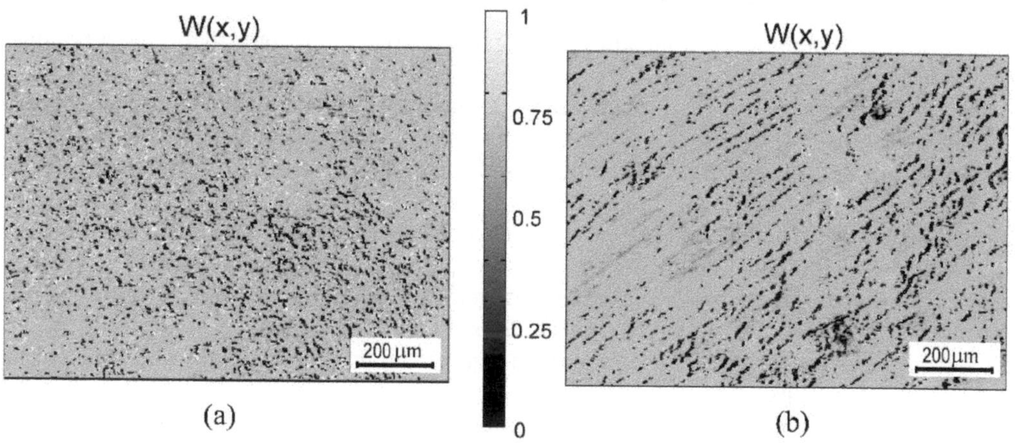

FIGURE 10.13: Coordinate distributions of the degree of mutual polarization $W(x,y)$ at the images of histological slices of normal (a) and dystrophicaly changed (b) tissue of the skeleton muscle.

TABLE 10.5: Statistical moments of the first to the fourth orders for distributions of magnitude $W(x,y)$ for images of histological slices of osseous tissue (normal and degeneratively changed).

S_i	Osseous tissue (norm) 22 samples	Osseous tissue (osteoporosis) 21 samples
S_1	0.9 ± 0.04	0.4 ± 0.02
S_2	0.02 ± 0.001	0.09 ± 0.006
S_3	8 ± 0.64	3.4 ± 0.37
S_4	21 ± 2.5	36 ± 3.2

TABLE 10.6: Statistical moments of the first to the forth orders for distributions of magnitude $W(x,y)$ for images of histological slices of the tissue of skeleton muscle (normal and dystrophic)

S_i	Muscular tissue (norm) 28 samples	Muscular tissue (pathology) 25 samples
S_1	0.62 ± 0.03	0.49 ± 0.02
S_2	0.1 ± 0.06	0.14 ± 0.01
S_3	32 ± 3.2	21 ± 2.31
S_4	81 ± 8.72	43 ± 3.87

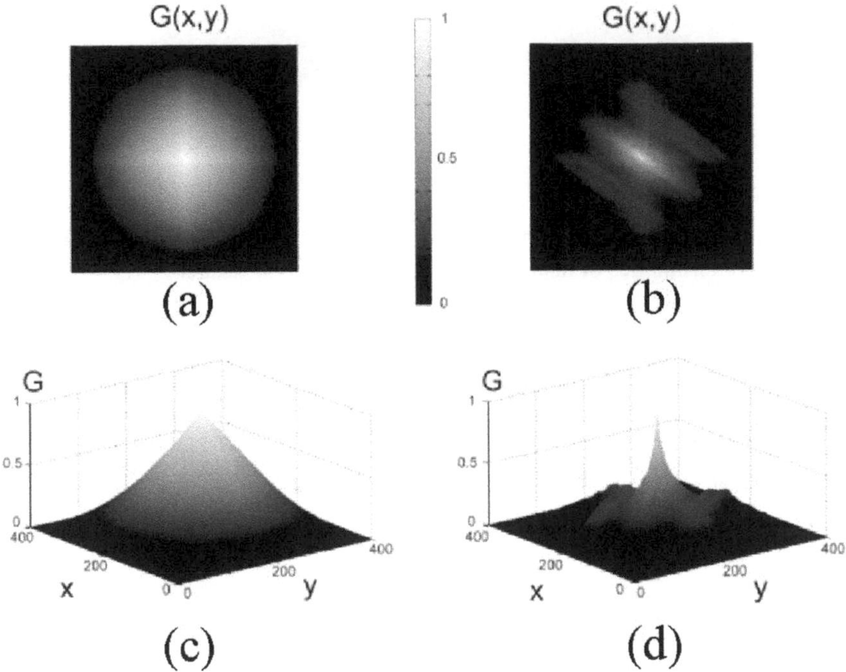

FIGURE 10.14: Autocorrelation functions $G(x,y)$ of 2D distributions of the degree of mutual polarization $W(x,y)$ for images of histological slices of normal (a) and degenerative (b) osseous tissue.

2.5 to 4 times of magnitude; and (iii) differences between the statistical moments of the first to the third orders for the PCMs of skeleton muscle are 30% to 50%, and for kurtosis are 2 to 2.5 times of magnitude.

Autocorrelation functions of the coordinate distributions of the degree of mutual polarization for histological slices of normal and pathologically changed osseous and muscular tissues are shown in Figures 10.14 and 10.15, respectively (fragments a and b for normal and pathologically changed samples, respectively). Structure of the obtained autocorrelation functions has been estimated by two correlation parameters [58], *viz.* a half-width L and dispersion Ω of the distribution of extrema of power spectra of $W(x,y)$, see Table 10.7.

Correlation analysis of the coordinate distributions of magnitudes $W(x,y)$ shows: (i) autocorrela-

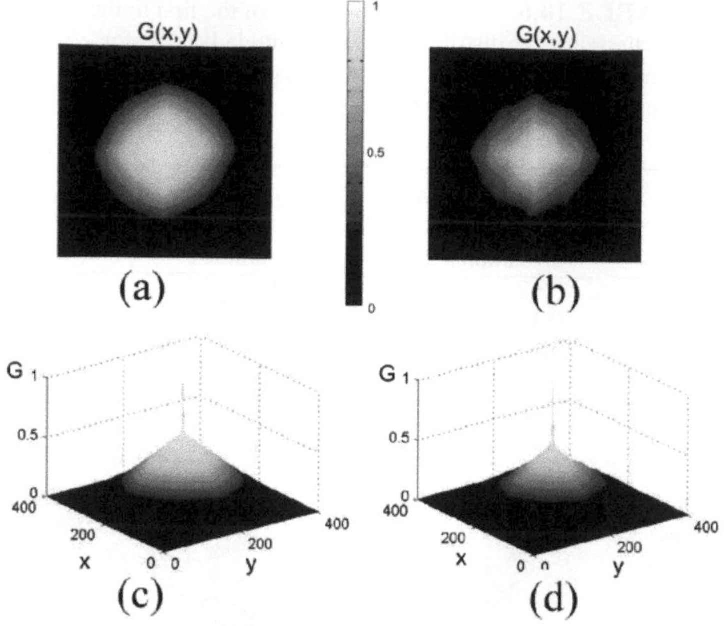

FIGURE 10.15: Autocorrelation functions $G(x,y)$ of 2D distributions of the degree of mutual polarization $W(x,y)$ for images of histological slices of muscular tissue.

TABLE 10.7: Correlation parameters of the distributions of the degree of mutual polarization $W(x,y)$ for images of normal and pathologically changed osseous and muscular tissues.

Par.	Osseous tissue		Par.	Muscular tissue	
	Norm	Pathology		Norm	Pathology
L	0.6 ± 0.04	0.34 ± 0.03	L	$0.25 \pm 0{,}02$	0.13 ± 0.01
Ω	≈ 0	$0.01 \pm 0{,}001$	Ω	$0.07 \pm 0{,}005$	0.005 ± 0.0001

tion functions of the PCMs of images of histological slices for normal and degenerative-dystrophic changed BTs of both types are considerably different in magnitudes L and Ω; (ii) differences in a half-width L for autocorrelation functions of the distributions $W(x,y)$ for the polarization-correlation images of physiologically normal and pathologically changed osseous and muscular tissues are 1.65 to 2 times of magnitude; and (iii) differences in dispersion of the distributions of extrema of power spectra of $W(x,y)$ of physiologically normal and pathologically changed are of one order of magnitude.

10.5 The Structure of Polarized Fields of Biological Tissues

10.5.1 Main mechanisms and scenarios of forming singular nets at laser fields of birefringent structures of biological tissues

Interaction of laser radiation with the set of optically uniaxial anisotropic fibrils (see Figure 10.1) may be considered as the process of generation of polarization singularities (linear and circular states) by a birefringent biological object. In a general case of the elliptically polarized probing beam $E_{0x}+E_{0y}\exp(-i\delta_0)$ (Eqs. (10.1) to (10.13)), the orthogonal amplitudes U_x and U_y of light oscillations at points (r) of an image of the anisotropic components of BT are of the form

$$\begin{aligned} U_x(r) &= \cos\rho\,(E_{0x}\cos\rho+E_{0y}\sin\rho\exp(-i\delta_0))+\\ &+\sin\rho\,(E_{0x}\sin\rho-E_{0y}\cos\rho\exp(-i\delta_0))\exp(-i\delta);\\ U_y(r) &= \sin\rho\,(E_{0x}\sin\rho+E_{0y}\cos\rho\exp(-i\delta_0))+\\ &+\cos\rho\,(E_{0x}\cos\rho-E_{0y}\sin\rho\exp(-i\delta_0))\exp(-i\delta). \end{aligned} \tag{10.26}$$

It follows from Eq. (10.26) that elliptical polarization is the most probable at the image of a BT. Classification of singular states of polarization has been performed in [43–51]. One distinguishes the singly degenerated trajectory when $\beta(r)=0$; $\delta(r)=0$, polarization is linear with an undetermined direction of turning of the electrical vector, and the doubly degenerated trajectory when $\beta(r)=0.25\pi$; $\delta(r)=0.5\pi$, polarization is circular with an undetermined azimuth of polarization. Let us determine the conditions of formation of the singly degenerated polarization singularities:

$$\sin^2\rho\cos^2\rho\,(E_{0x}^2\cos^2\rho-E_{0y}^2\sin^2\rho\exp(-2i\delta_0))\sin^2\delta=0; \tag{10.27}$$

$$\begin{aligned} \rho &= arctg\left[\exp(-2i\delta_0)\tfrac{E_{0y}}{E_{0x}}\right];\\ \delta &= 2q\pi,\quad q=0,1,2... \end{aligned} \tag{10.28}$$

The following relation is justified for the doubly degenerated polarization singularities:

$$\sin^2\rho\cos^2\rho\,(E_{0x}^2\cos^2\rho-E_{0y}^2\sin^2\rho\exp(-2i\delta_0))\sin^2\delta=1, \tag{10.29}$$

from which one determines the conditions of formation of circularly polarized points at BT images:

$$\begin{aligned} \rho &= {}^{\pi}/_{4}+arctg\left[\exp(-2i\delta_0)\tfrac{E_{0y}}{E_{0x}}\right];\\ \delta &= {}^{\pi}/_{2}+2q\pi,\ q=0,\ 1,\ 2,\ 3,... \end{aligned} \tag{10.30}$$

Experimentally the coordinate distributions of such points can be found by selection of linear and circular states of polarization at various points of the BT images from all possible states of the polarization distributions (see the PMs of images of histological slices of BTs in Figure 10.5 to Figure 10.7). The map of the singly degenerated singular points is determined as

$$\begin{aligned} \delta(r) &= \arctan\left[\frac{\tan 2\beta(r)}{\tan\alpha(r)}\right]=2q\pi;\\ \beta(r) &= 0. \end{aligned} \tag{10.31}$$

and the map of the doubly degenerated singular points is determined as

$$\begin{aligned} \delta(r) &= \arctan\left[\frac{\tan 2\beta(r)}{\tan\alpha(r)}\right]=0.5\pi+2q\pi;\\ \beta(r) &= 0.25\pi. \end{aligned} \tag{10.32}$$

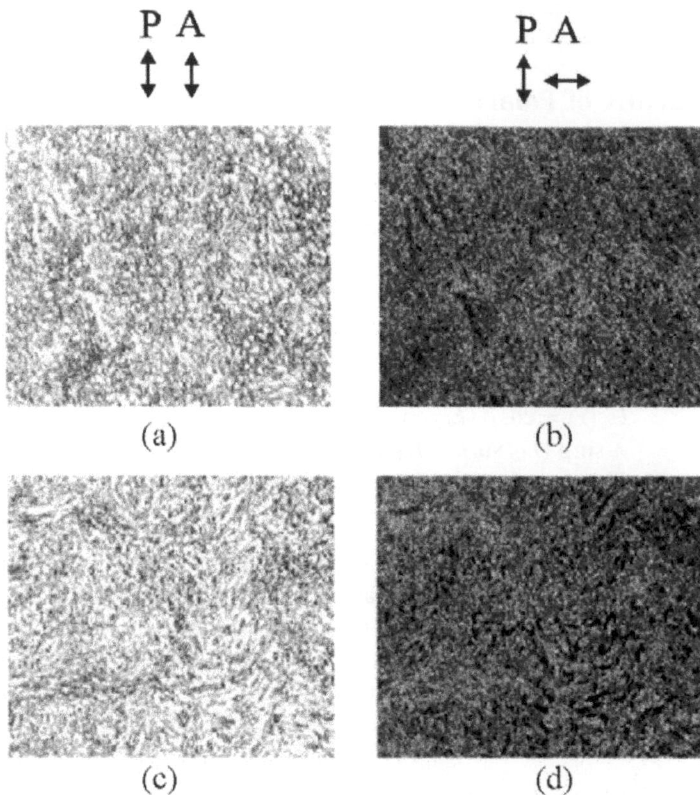

FIGURE 10.16: Polarization images of histological slices of physiologically normal (a, b) and pathologically changed (c, d) kidney tissue. Fragments a, c correspond to the situation with matched polarizer and analyzer; fragments (b, d) – crossed polarizer and analyzer.

10.5.2 Statistical and fractal approaches to analysis of singular nets at laser fields of birefringent structures of biological tissues

In this section we illustrate the feasibilities for polarization singularometry at BT images for preclinic diagnostics of inflammatory processes whose detection by the above considered techniques of laser polarimetry is difficult.

As samples for this study we use optically thin (extinction coefficient $\tau \leq 0.1$) histological slices of normal (Figure 10.16 a, b) and pathologically changed (early stage of sepsis, Figure 10.16 c, d) canine kidney.

The samples of the selected tissues of canine kidney are almost identical from the medical point of view, namely, conventional histochemical techniques do not distinguish them one from another. Pathological changes of kidney were formed experimentally, by inserting special chemical preparations causing septic inflammation. Similarity of the morphological structure of such tissues is proved by the comparative analysis of images (Figure 10.16) obtained with matched and crossed polarizer and analyzer.

For obtaining new and more detailed information on optical anisotropic properties of such objects, one determines the nets of polarization singularities (Eqs. (10.31) and (10.32)) of their laser images based on measurements of the coordinate distributions of the azimuth of polarization, $\alpha(r)$, and ellipticity, $\beta(r)$, see subsection 10.3.1, and following computing (Eq. (10.24)) of 2D tensor of magnitudes of phase shifts $\delta(r)$.

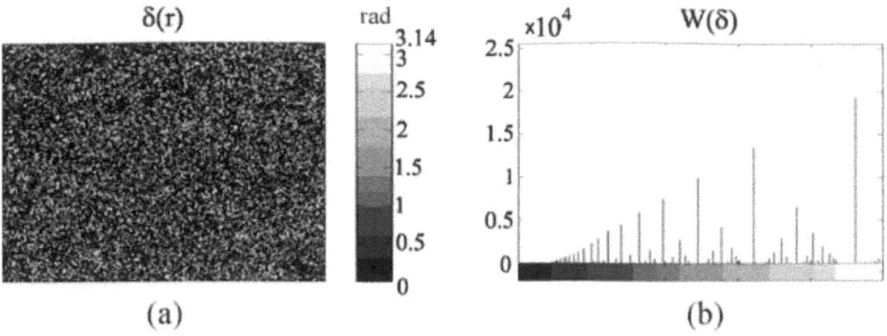

FIGURE 10.17: Coordinate distributions (a) of phase shifts δ and histograms of their magnitudes (b) for the polarization image of a physiologically normal tissue of canine kidney.

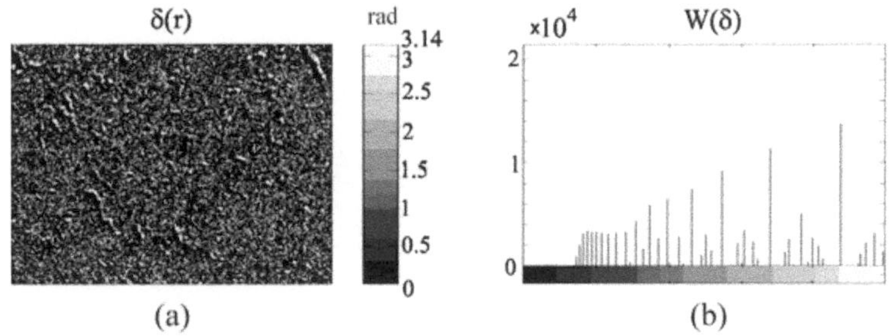

FIGURE 10.18: Coordinate distributions (a) of phase shifts δ and histograms of their magnitudes (b) for the polarization image of a tissue of canine kidney with early septic symptoms.

The coordinate distributions (fragments a) and histograms of the distributions of phase shift magnitudes $\delta(r)$ between the orthogonal components of laser radiation transformed by layers of histological slices (fragments b) are shown in Figure 10.17 and 10.18, respectively, for the normal and septically inflamed canine kidney.

Comparison of the PMs (Figs. 10.17 and 10.18) shows that the intervals of change of phase shifts $\delta(r)$ are almost the same for images of the samples of kidney tissue of both types. Histograms for $W(\delta)$ of the distribution of magnitudes $\delta(r)$ also have a structure.

Much larger differences are observed for the polarization-phase properties of kidney tissues of two types, as it is seen from comparison of the coordinate distributions of singular points at their images, see Figs. 10.19 and 10.20.

There is the algorithm for statistical analysis of the nets of singularly polarized points at the images of kidney tissue:

- determination of the distribution $N(m)$ of the number of singularly polarized points at the BT image, that corresponds to the each column $m_{i=1,2,\ldots 800}$ of pixels of the sensitive area of CCD-camera (Figure 10.21);
- computation of the statistical moments of the first to the fourth orders for $N(m)$ (Table 10.8);
- determination of the log-log dependences $(\log J - \log(1/r))$ of power spectra of J of $N(m)$ of singularly polarized points (Figure 10.22).

The distributions of the number of singularly polarized points $N(m)$ at images of physiologically normal and pathologically changed samples of kidney tissue are shown in Figure 10.21, fragments a and b, respectively.

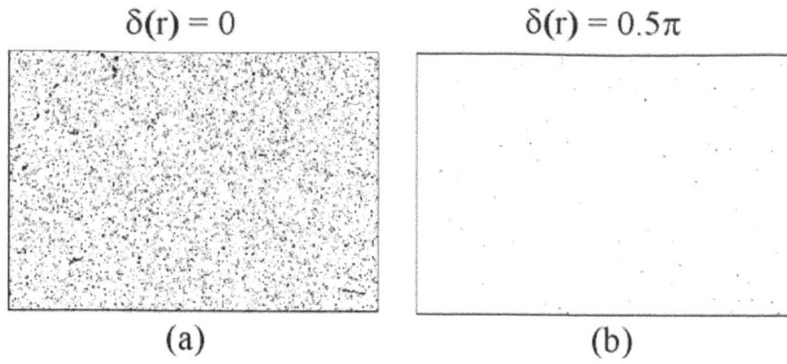

FIGURE 10.19: Coordinate distributions of the singly (a) and doubly (b) degenerated polarization singularities at image of a physiologically normal kidney tissue.

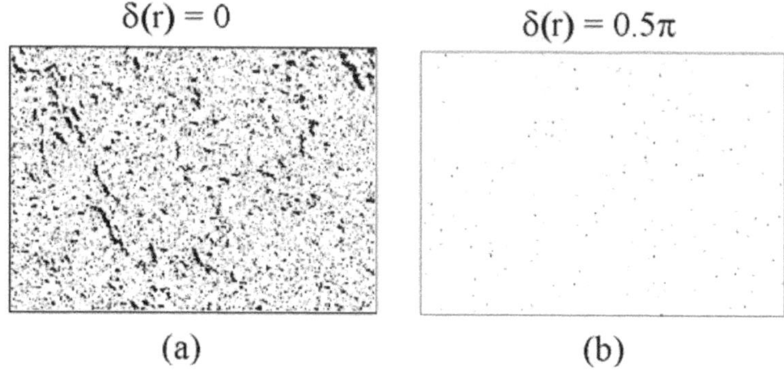

FIGURE 10.20: Coordinate distributions of the singly (a) and doubly (b) degenerated polarization singularities at image of pathologically changed kidney tissue.

One can see that the dependences $N(m)$ determined for images of the samples of kidney tissue of both types have a complex statistical structure. Comparison of the distributions of the number of points with singularly polarized states shows that the number of such points at the image of kidney tissue with early collagenosis (Figure 4.6 b) exceeds their number at the physiologically normal tissue (Figure 4.6 a) by 10%–15%.

More detailed quantitative information on the distribution of polarization singularities $N(m)$ at images of kidney tissue with various physiological states is contained in the statistical moments of the first to the fourth orders shown in Table 10.8.

One can see from these data that the statistical distribution of the number of singularly polarized points at the image of normal kidney tissue is close to the normal one, *viz.* the magnitudes of skewness, S_3, and kurtosis, S_4, of the distributions $N(m)$ are rather small.

Another is observed for the set of the statistical moments characterizing the distribution $N(m)$ of the points with singular states of polarization at the image of pathologically changed kidney tissue.

The statistical moments S_3 and S_4 increase by approximately one order of magnitude in comparison with such parameters for the polarization image of normal kidney tissue. Such a change of the higher-order statistical moments for the distribution of density of singularly polarized points at the image of physiologically changed kidney tissue can be caused by transformation of its optically anisotropic component.

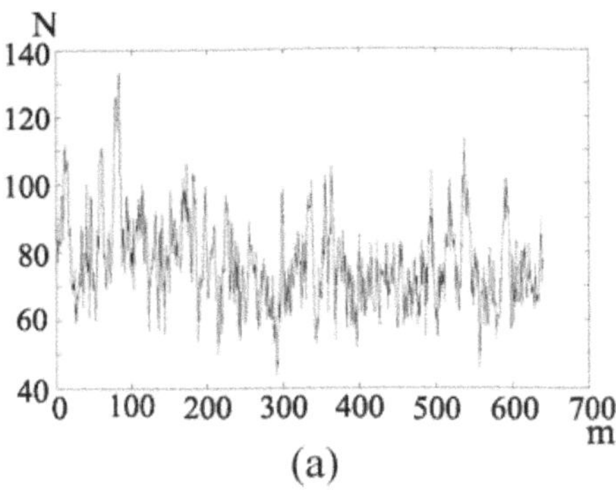

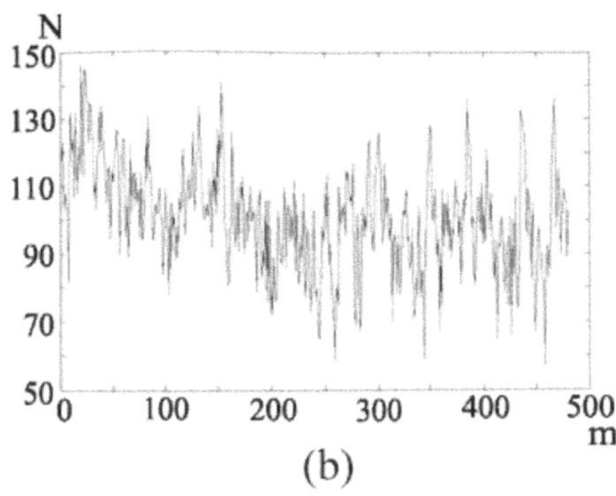

FIGURE 10.21: Distributions on number N of singularly polarized points ($\delta(r) = 0; \delta(r) = 0.5\pi$) at image of physiologically normal (a) and pathologically changed (b) kidney tissue.

TABLE 10.8: Statistical moments of the first to the fourth orders for distribution of the number of polarization singularities at image of the kidney tissue.

S_i	Norm (37 samples)	Pathology (36 samples)
S_1	0.634±0.032	0.706±0.065
S_2	0.198±0.0045	0.149±0.053
S_2	2.689±0.21	21.75±3.11
S_4	3.8±0.34	46.8±5.47

Septic changes of kidney tissue manifest themselves morphologically by growing birefringence due to forming swells, microhematomas, etc. Optically such processes are accompanied by increasing phase shifts $\delta(r)$ between the orthogonal components of amplitude of a laser beam passing through such pathologically changed structures of kidney tissue.

The same phase modulation manifests itself in formation of an additional number of singularly polarized points (Eqs. (10.26) to (10.32)) at image of pathologically changed kidney tissue. As a consequence, skewness S_3 and kurtosis S_4 of the distributions considerably grow.

Additional information on differences in the coordinate distributions of the singularly polarized ($\delta(r) = 0; \delta(r) = 0.5\pi$) points at images of kidney tissue of both types is contained in power spectra of such distributions. $\log(J) - \log(1/r)$ dependences illustrated in Figure 10.22 were obtained for the coordinate distributions of the number $N(m)$ of polarization singularities at images of physiologically normal (a) and pathologically changed (b) kidney tissue.

Analysis of the obtained data shows that interpolation of the $\log(J) - \log(1/r)$ dependencies by the least-square technique for polarization singularities of images of physiologically normal kidney tissue gives one straight line with a constant slope (Figure 10.22 a).

In the polarization sense, the image structure of pathologically changed tissue gives a set of broken lines with several magnitudes of local slopes, *viz.* statistical processing of the $\log(J) - \log(1/r)$ dependencies (Figure 10.22 b).

Thus, it has been shown that the coordinate structure of polarization singularities at the image of physiologically normal BT is self-similar (fractal), while for the pathologically changed one it is random (stochastic).

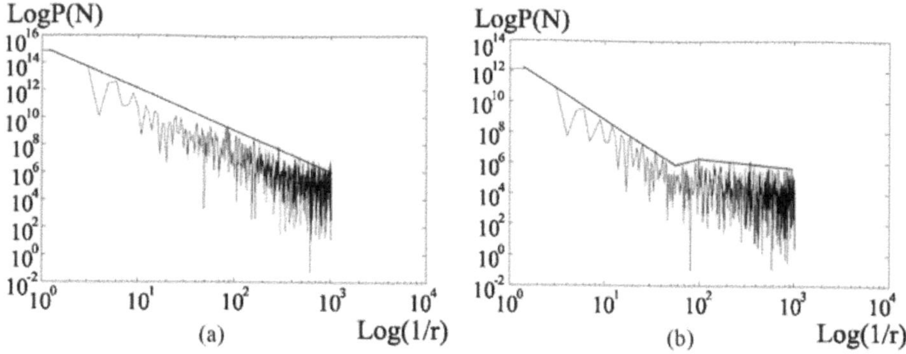

FIGURE 10.22: log-log dependencies of power spectrum J for the distribution of the number $N(m)$ of polarization singularities at image of physiologically normal (a) and pathologically changed (b) kidney tissues.

10.5.3 Scenarios of formation of singular structure of polarization parameters at images of biological tissues

In this subsection we present the results of a computer simulation and experimental study of the processes of formation of S-contours into inhomogeneously polarized images of biological crystals of human tissues.

The model consisting from birefringent curvilinear (with circle-like optical axes) coaxial cylindrical fibrils (Figure 10.23 a) have been used as the virtual net of biological crystal, being the analog of an out-of-cell matrix [13]. Birefringent index was $\Delta n = 1.5 \cdot 10^{-3}$ that corresponds to real birefringence of the most of protein fibrils [13]. The curvature radii of fibrils were changed with the step $\Delta r = r + 0.1r$. Phase shift between the orthogonal amplitude components of a laser beam introduced by single circular-like fibril (Figure 10.23 a) has the form shown in Figure 10.23 b. Such form of the model structure is close to the architectonic nets of most of BTs formed by curvilinear coaxial protein fibrils, fibers, and bundles (connective and muscular tissues of various organs). The polarization structure of the simulated net with 55 curvilinear cylindrical birefringent crystals (Figure 10.23 c) is characterized using the coordinate distributions $S_i(x,y)$ of the Stokes parameters. Such an approach provides the most comprehensive description of the inhomogeneously polarized images [58].

S-lines and $\pm$C-points forming the S-contours [43–57] at the 2D distributions of the Stokes parameter $S_4(x,y)$ at the image of the computer simulated net of biological crystal.

An analysis of an inhomogeneously polarized image of the model net of biological crystals shows the presence of S-contours of various configurations.

10.5.4 Structure of S-contours of polarization images of the architectonic nets of birefringent collagen fibrils

The results of an experimental study of the structure of S-contours as the coordinate distributions of the Stokes parameter S_4 for the images of collagen fibril nets of human skin dermis are shown in Figure 10.25.

It is seen that the experimentally studied configuration of S-contours at the coordinate distributions of the Stokes parameters S_4 of the image of such an object is in agreement with the results of the computer simulation.

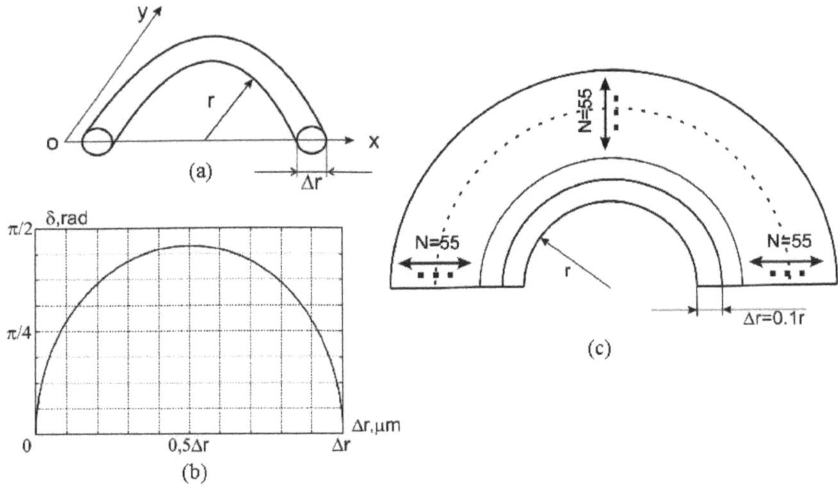

FIGURE 10.23: To the analysis of coordinate distributions of the Stokes parameters $S_i\,(x,y)$ forming at the image of the net of curvilinear cylindrical crystals.

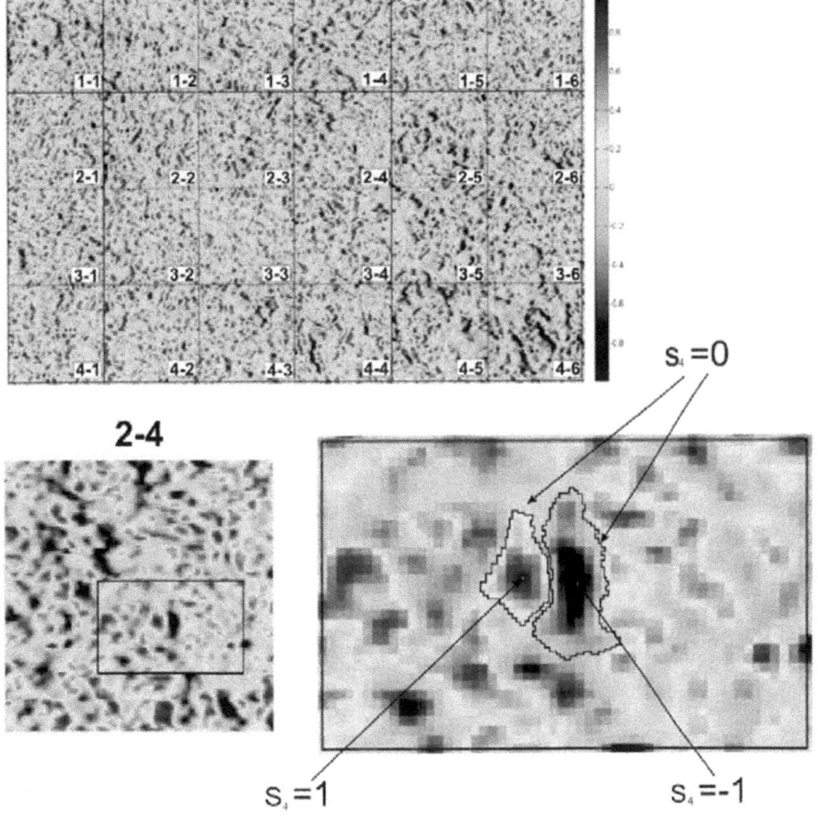

FIGURE 10.24: S-lines and ±C-points forming the S-contours [43-57] at the 2D distributions of the Stokes parameter $S_4\,(x,y)$ at the image of the computer simulated net of biological crystal.

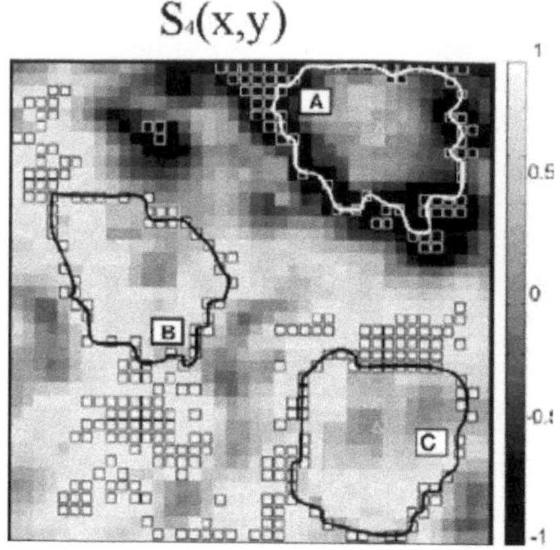

FIGURE 10.25: Structure of S-contours at 2D distributions of the Stokes parameter S_4 of an inhomogeneously polarized image of a histological slice of skin dermis: squares – S-points, $\Delta - +$C ($\delta = 0.5\pi$)-points, $\nabla - -$C ($\delta = -0.5\pi$) - points.

10.5.5 On the interconnection of the singular and statistical parameters of inhomogeneously polarized nets of biological crystals

Analyzing once more the data of Ellis and Dogariu [30, 34], one can state that the complex degree of mutual polarization for two different points, r_1 and r_2, of the object field with linear and circular states of polarization takes the typical magnitude $V(r_1, r_2) = 0,5$ (Eq. (10.25)). On the other hand, the distance $R = r_2 - r_1$ corresponds to the linear size of S-contour at an inhomogeneously polarized field. Thus, a half-width of the autocorrelation function $G(W)$ of the coordinate distribution of the degree of mutual polarization (see subsection 10.4.1), $L = G(W = 0,5)$, must correlate with the average distances $\bar{R}$ between S- and C-points at such an inhomogeneously polarized field.

For experimental proof of this assumption on the interconnection between the correlation and singular approaches we determined the magnitudes of the correlation (L) and singular ($\overline{R}$) parameters of inhomogeneously polarized images of muscular tissue and dermal layer. Polarization images of histological slices of the samples of such tissues with the reconstructed S-contours are shown in Figure 10.26.

It is seen from the figure that polarization images of BTs of both types have a complex structure including the sets of singularities of various types, which form closed S-contours. Comparison of sizes of S-contours shows that for a BT with disordered architectonics (Figure 10.26 b) they are smaller than for the system of ordered birefringent fibrils (Figure 10.26 a).

From the physical point of view, such peculiarities of the structure of inhomogeneously polarized images can be associated with the following mechanisms of interaction of laser radiation with biological crystals of the architectonic nets of human tissues. It has been seen [57] that formation of C-points is determined by two conditions: (i) "phase," $\delta = \frac{2\pi}{\lambda}\Delta n d = \pm^{\pi}/_{2}$, where d is the transversal size of birefringent fibrils, and (ii) "orientation," $\rho = \alpha \pm {}^{\pi}/_{4}$, where α is the azimuth of polarization of the probing laser beam.

The first condition is the same for the BTs of both types, *viz.* the intervals within which geometrical sizes of fibrils of their architectonic nets are changed and are almost the same [13]. The second condition is realized differently. Orientation of the optical axes of biological crystals $\rho = \alpha \pm {}^{\pi}/_{4}$ is more probable for disordered architectonics of the dermal layer than for the ordered net of myosin fibrils of muscular tissue. In other words, more C-points are formed at the plane of polarization image of a histological slice of the dermal layer. Similarly, formation of S-points ($\delta = 0; \alpha = \rho$) by chaotically oriented collagen fibrils of skin dermis is more probable than by myosin fibers of

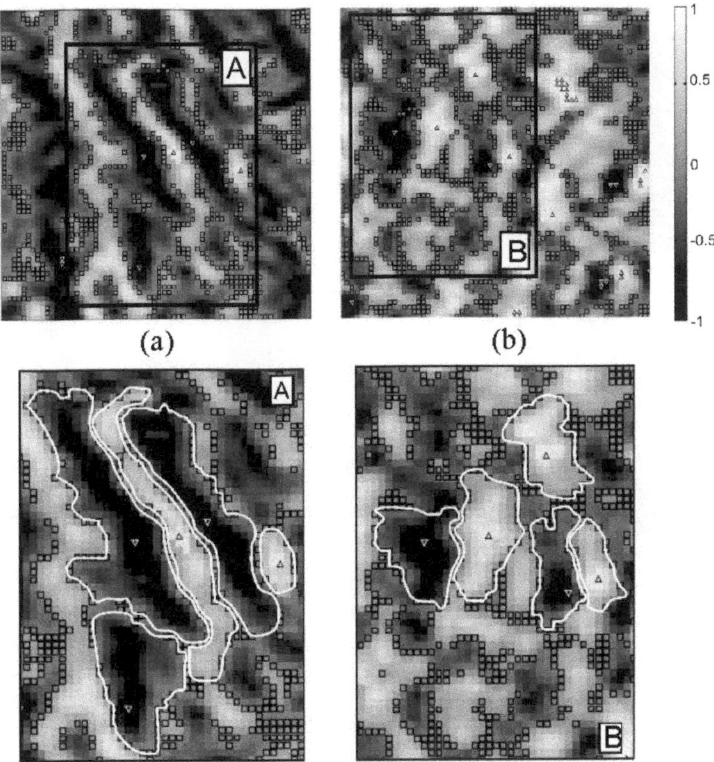

FIGURE 10.26: Coordinate and singular structure of inhomogeneously polarized images of histological slices of muscular tissue (a) and dermal layer (b). Squares – S-points, Δ – +C ($\delta = 0.5\pi$)-points, ∇ – –C ($\delta = -0.5\pi$)-points.

muscular tissue. That is why, distances $\overline{R}$ between S- and C-points determining S-contours at an inhomogeneously polarized image of skin dermis are less than sizes of such topological structures of muscular tissue.

For an estimation of the average size of S-contours at images of muscular tissue and dermal layer we compute the distances $\overline{R}$ at two orthogonal directions for two groups of histological slices of muscular tissue (27 samples – 0.31 ± 0.004 μm) and skin dermis (29 samples – 0.11 ± 0.003 μm). Autocorrelation functions of the coordinate distributions of the degree of mutual polarization at images of histological slices of the dermal layer and muscular tissue are shown in Figure 10.27 a, b, respectively. Comparison of the obtained data shows correlation in tendencies of the changes of average size $\overline{R}$ of S-contours (Fig 10.26 a and Figure 10.26 b) and a half-width of the autocorrelation functions of the distributions of the degree of mutual polarization of images of muscular tissue ($\overline{L} = 0.24 \pm 0.0035$ μm, Figure 10.27) and dermal layer($\overline{L} = 0.07 \pm 0.002$ μm, Figure 10.27 b). There are their ratios: ${}^{\bar{R}_1}/_{\bar{R}_2} \approx 0,29$ and ${}^{\bar{L}_1}/_{\bar{L}_2} \approx 0,32$.

Thus, the interconnection between a half-width of the autocorrelation function of the distributions of the degree of mutual polarization at images of BTs with various morphological structures and average size of topological S-contours is proved experimentally.

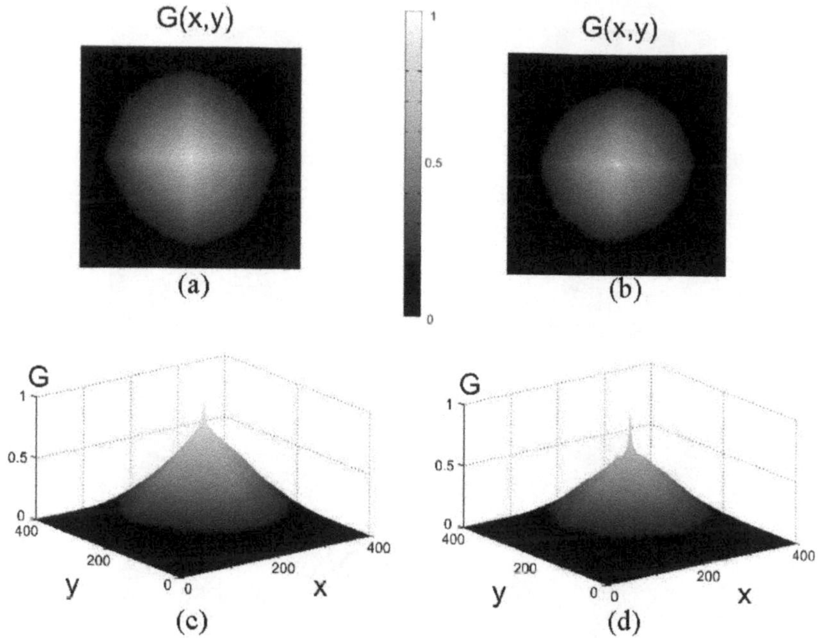

FIGURE 10.27: Autocorrelation functions of the coordinate distributions of the degree of mutual polarization for the images of histological slices of muscular tissue (a, c) and dermal layer (b, d).

10.6 Conclusion

The interconnection of the set of the statistical moments of the first to the fourth orders and fractal parameters (log-log dependencies of power spectra) has been established characterizing the geometrical-optical structure of the architectonic component of a BT, on the one hand, and the set of statistical and fractal characteristics of its polarization map, i.e., 2D distribution of the azimuth of polarization and ellipticity, on the other hand.

The coordinate structure of inhomogeneously polarized images of BTs can be characterized by the degree of mutual polarization for different points of such an image. The connection has been revealed between the magnitude of the degree of mutual polarization and the parameters of anisotropy of various points of a BT, such as the direction of the optical axis and magnitude of a phase shift between the orthogonal components of laser beam amplitude. On this basis, the new technique for direct polarimetric measuring of the coordinate distributions of the degree of mutual polarization at images of BTs has been proposed.

Increasing birefringence of a matter of anisotropic fibrils for unchanged orientation of the optical axes manifests itself in increasing kurtosis of the coordinate distributions of magnitudes of the degree of mutual polarization at images of BTs. Disordering of the optical axes directions of biological crystals causes decreasing skewness of the distribution of the degree of mutual polarization. It provides early (preclinical) differentiation of the optical properties of the anisotropic component of normal and pathologically changed (precancer) connective tissue.

Physical mechanisms and conditions of formation of polarization singularities, such as linearly and circularly polarized states, at BT images have been established, and the technique for polarimetry of singularities at such images has been developed.

The coordinate distributions of the number of singularly polarized points at images of normal BTs have statistical (stochastic) structure. Changing the parameters of optical anisotropy accompanying the formation of pathological states manifests itself in transformation of the log-log dependencies of power spectra of the number of singularly polarized points into self-similar (fractal) dependencies.

Scenarios of formation of topological structure of inhomogeneously polarized images of the nets of biological crystals at human tissues have been analytically substantiated and experimentally investigated. The connection between the peculiarities of orientation and phase structure of the net of biological crystals and polarization singularities of the Stokes parameters of their images has been revealed.

References

[1] V. Tuchin, *Tissue Optics: Light Scattering Methods and Instruments for Medical Diagnosis*, SPIE Press, Bellinghan, WA, 2000 (second edition, SPIE Press, Bellingham, WA, 2007).

[2] V.V. Tuchin, L. Wang, and D.A. Zimnyakov, *Optical Polarization in Biomedical Applications*, Springer-Verlag, Berlin, Heidelberg, N.Y., 2006.

[3] A.G. Ushenko and V.P. Pishak, "Laser polarimetry of biological tissues: principles and applications" in *Coherent-Domain Optical Methods: Biomedical Diagnostics, Environmental and Material Science*, V.V. Tuchin (ed.), Kluwer Academic Publishers, Boston, vol. 1, 93–138 (2004).

[4] E. Wolf, "Unified theory of coherence and polarization of random electromagnetic beams," *Phys. Lett. A* **312**, 263–267 (2003).

[5] M.J. Everett, K. Shoenenberger, B.W. Colston, and L.B. Da Silva, "Birefringence characterization of biological tissue by use of optical coherence tomography," *Opt. Lett.* **23**, 228–230 (1998).

[6] J. Ellis, A. Dogariu, S. Ponomarenko, and E. Wolf, " Interferometric measurement of the degree of polarization and control of the contrast of intensity fluctuations," *Opt. Lett.* **29**, 1536–1538 (2004).

[7] J.F. de Boer, T.E. Milner, M.G. Ducros, S.M. Srinivas, and J.S. Nelson (eds), "Polarization-sensitive optical coherence tomography," in *Handbook of Optical Coherence Tomography*, B.E. Bouma, G.J. Tearney (eds.), Marcel Dekker Inc., New York, 237–274 (2002).

[8] S. Jiao and L.V. Wang, "Two-dimensional depth-resolved Mueller matrix of biological tissue measured with double-beam polarization-sensitive optical coherence tomography," *Opt. Lett.* **27,** 101–103 (2002).

[9] Zernike F., "The concept of degree of coherence and its applications to optical problems," *Physica* **5**, 785–795 (1938).

[10] G. Parrent and P. Roman, "On the matrix formulation of the theory of partial polarization in terms of observables," *Nuovo Cimento* **15**, 370–388 (1960).

[11] A.G. Ushenko, "Polarization structure of scattering laser fields," *Opt. Eng.* **34**, 1088–1093 (1995).

[12] J.F. de Boer, T.E. Milner, M.J.C. van Gemert, and J.S. Nelson, "Two-dimensional birefringence imaging in biological tissue by polarization-sensitive optical coherence tomography," *Opt. Lett.* **22**, 934–936 (1997).

[13] S.C. Cowin, "How is a tissue built?" *J. Biomed. Eng.* **122**, 553–568 (2000).

[14] O.V. Angelsky, A.G. Ushenko, V.P. Pishak, D.N. Burkovets, S.B. Yermolenko, O.V. Pishak, and Y.A. Ushenko, "Polarizing-correlative processing of images of statistic objects in visualization and topology reconstruction of their phase heterogeneity," *Proc. SPIE* **4016**, 419–424 (1999).

[15] O. Angelsky, D. Burkovets, V. Pishak, Yu. Ushenko, and O. Pishak, "Polarization-correlation investigations of biotissue multifractal structures and their pathological changes diagnostics," *Laser Phys.* **10**, 1136–1142 (2000).

[16] A.G. Ushenko, "The vector structure of laser biospeckle fields and polarization diagnostics of collagen skin structures," *Laser Phys.* **10**, 1143–1149 (2000).

[17] O.V. Angelsky, A.G. Ushenko, S.B. Ermolenko, D.N. Burkovets, V.P. Pishak, Yu.A. Ushenko, and O.V. Pishak, "Polarization-based visualization of multifractal structures for the diagnostics of pathological changes in biological tissues," *Opt. Spectrosc.* **89**, 799–804 (2000).

[18] O.V. Angelsky, and P.P. Maksimyak, "Optical diagnostics of random phase objects," *Appl. Opt.* **29**, 2894–2898 (1990).

[19] O.V. Angelsky, P.P. Maksimyak, and S. Hanson, *The Use of Optical-Correlation Techniques for Characterizing Scattering Object and Media* **PM71**, SPIE Press, Bellingham, WA (1999).

[20] Yu.A. Ushenko, "Statistical structure of polarization-inhomogeneous images of biotissues with different morphological structures," *Ukr. J. Phys. Opt.* **6**, 63–70 (2005).

[21] C.E. Saxer, J.F. de Boer, B.H. Park, Y. Zhao, Z. Chen, and J.S. Nelson, "High-speed fiber based polarization-sensitive optical coherence tomography of in vivo human skin," *Opt. Lett.* **25**, 1355–1357 (2000).

[22] J.F. de Boer and T.E. Milner, "Review of polarization sensitive optical coherence tomography and Stokes vector determination," *J. Biomed. Opt.* **7**, 359–371 (2002).

[23] A.F. Fercher, "Optical coherence tomography — principles and applications," *Rep. Prog. Phys.* **66**, 239–303 (2003).

[24] N. Wiener, "Generalized harmonic analysis," *Acta Math.* **55**, 117–258 (1930).

[25] E. Wolf, "Coherence properties of partially polarized electromagnetic radiation," *Nuovo Cimento* **13**, 1165–1181 (1959).

[26] F. Gori, M. Santarsiero, S. Vicalvi, R. Borghi, and G. Guattari, "Beam coherence-polarization matrix," *Pure Appl. Opt.* **7**, 941–951 (1998).

[27] E. Wolf, "Unified theory of coherence and polarization of random electromagnetic beams," *Phys. Lett. A* **312**, 263–267 (2003).

[28] J. Tervo, T. Setala, and A. Friberg, "Degree of coherence for electromagnetic fields," *Opt. Express* **11**, 1137–1143 (2003).

[29] J.M. Movilla, G. Piquero, R. Martínez-Herrero, and P.M. Mejías, "Parametric characterization of non-uniformly polarized beams," *Opt. Commun.* **149**, 230–234 (1998).

[30] J. Ellis and A. Dogariu, "Complex degree of mutual polarization," *Opt. Lett.* **29**, 536–538 (2004).

[31] C. Mujat and A. Dogariu, "Statistics of partially coherent beams: a numerical analysis," *J. Opt. Soc. Am. A* **21**, 1000–1003 (2004).

[32] F. Gori, "Matrix treatment for partially polarized, partially coherent beams," *Opt. Lett.* **23**, 241–243 (1998).

[33] M. Mujat and A. Dogariu, "Polarimetric and spectral changes in random electromagnetic fields," *Opt. Lett.* **28**, 2153–2155 (2003).

[34] J. Ellis, A. Dogariu, S. Ponomarenko, and E. Wolf, "Interferometric measurement of the degree of polarization and control of the contrast of intensity fluctuations," *Opt. Lett.* **29**, 1536–1538 (2004).

[35] M. Mujat, A. Dogariu, and G.S. Agarwal, "Correlation matrix of a completely polarized, statistically stationary electromagnetic field," *Opt. Lett.* **29**, 1539–1541 (2004).

[36] O. Korotkova and E. Wolf, "Spectral degree of coherence of a random three-dimensional electromagnetic field," *J. Opt. Soc. Am. A* **21**, 2382–2385 (2004).

[37] O.V. Angelsky, A.G. Ushenko, and Y.G. Ushenko, "Complex degree of mutual polarization of biological tissue coherent images for the diagnostics of their physiological state," *J. Biomed Opt.* **10**, 060502 (2005).

[38] J. F. Nye and M. Berry, "Dislocations in wave trains," *Proc. R. Soc. Lond.* **A 336**, 165–190 (1974).

[39] J. F. Nye, *Natural Focusing and Fine Structure of Light: Caustics and Wave Dislocations*, Bristol, Institute of Physics (1999)

[40] M. Soskin, V. Denisenko, and R. Egorov, "Topological networks of paraxial ellipse speckle-fields," *J. Opt. A: Pure Appl. Opt.* **6**, S281–S287 (2004).

[41] J.F. Nye, "Polarization effects in the diffraction of electromagnetic waves: the role of disclinations," *Proc. R. Soc. Lond.* **A 387**, 105–132 (1983).

[42] J.F. Nye, "The motion and structure of dislocations in wave fronts," *Proc. R. Soc. Lond.* **A 378**, 219–239 (1981).

[43] J.F. Nye, "Lines of circular polarization in electromagnetic wave fields," *Proc. R. Soc. Lond.* **A 389**, 279–290 (1983).

[44] J.V. Hajnal, "Singularities in the transverse fields of electromagnetic waves. I. Theory," *Proc. R. Soc. Lond.* **A 414**, 433–446 (1987).

[45] J.V. Hajnal, "Singularities in the transverse fields of electromagnetic waves II. Observations on the electric field," *Proc. R. Soc. Lond.* **A 414**, 447–468 (1987).

[46] O.V. Angelsky, I.I. Mokhun, A.I. Mokhun, and M.S. Soskin, "Interferometric methods in diagnostics of polarization singularities," *Phys. Rev.* **E 65**, 036602 (2002).

[47] I. Freund, A.I. Mokhun, M.S. Soskin, O.V. Angelsky, and I.I. Mokhun, "Stokes singularity relations," *Opt. Let.* **27**, 545–547 (2002).

[48] O. Angelsky, A. Mokhun, I. Mokhun, and M. Soskin, "The relationship between topological characteristics of component vortices and polarization singularities," *Opt. Commun.* **207**, 57–65 (2002).

[49] I. Freund, M.S. Soskin, and A. I. Mokhun, "Elliptic critical points in paraxialoptical fields," *Opt. Commun.* **207**, 223–253 (2002).

[50] M.S. Soskin, V. Denisenko, and I. Freund, "Optical polarization singularities and elliptic stationary points," *Opt. Lett.* **28**, 1475–1477 (2003).

[51] M.R. Dennis, "Polarization singularities in paraxial vector fields: morphology and statistics," *Opt. Commun.* **213**, 201–221 (2002).

[52] M.V. Berry and J.H. Hannay, "Umbilic points on Gaussian random surfaces," *J. Phys. A: Math. Gen.* **10**, 1809–1821 (1977).

[53] M.S. Soskin and V.I. Vasil'ev, "Space-time topological dynamics of singularities and optical diabolos in developing generic light fields," *Proc. SPIE* **6729**, 67290B (2007).

[54] M.S. Soskin, V.G. Denisenko, and R.I. Egorov, "Singular elliptic light fields: genesis of topology and morphology," *Proc. SPIE* **6254**, 625404 (2006).

[55] R.W. Schoonover and T.D. Visser, "Polarization singularities of focused, radially polarized fields," *Opt. Express* **14**, 5733–5745 (2006).

[56] M.V. Berry and M.R. Dennis, "Polarization singularities in isotropic random vector waves," *Proc. R. Soc. Lond.* **A 457**, 141–155 (2001).

[57] O.V. Angelsky, A.G. Ushenko, Yu.A. Ushenko, and Ye.G. Ushenko, "Polarization singularities of the object field of skin surface," *J. Phys. D: Appl. Phys.* **39**, 3547–3558 (2006).

[58] O.V. Angelsky, A.G. Ushenko, and Ye.G. Ushenko, "2-D Stokes polarimetry of biospeckle tissues images in pre-clinic diagnostics of their pre-cancer states," *J. Holography Speckle* **2**, 26–33 (2005).

[59] D.J. Whitehouse, "Fractal or fiction?" *Wear* **249**, 345–353 (2001).

[60] A.G. Ushenko, "Laser diagnostics of biofractals," *Quant. Electr.* **29**, 1–7 (1999).

11

Biophotonic Functional Imaging of Skin Microcirculation

Martin J. Leahy

University of Limerick, Ireland

Gert E. Nilsson

Linköping University, Sweden

Visible and near-infrared light, particularly in the wavelength region 600–1000 nm, offer a window into tissues due to the reduced light scattering and absorption. Consequently many methods for functional imaging of the skin microcirculation utilize lasers and other light sources operating in this wavelength range. This chapter reports on modern biophotonic functional imaging methodologies most likely to succeed in general skin testing as well as routine clinical imaging of the microcirculation.

Key words: functional imaging, blood microcirculation, laser Doppler flowmetry, skin, tissue viability imaging (TiVi), photoacoustic tomography (PAT), optical microangiography (OMAG)

11.1 Skin Microvasculature

The most superficial capillaries in the skin reside in the papilla just beneath the epidermis (Figure 11.1). The blood perfusion through the capillaries supplies the tissue with oxygen and nutrients, and removes carbon dioxide and waste metabolites. The red blood cell (RBC) is one of the main chromophores (light absorber) and moves with a velocity of 0.3–1 mm/sec through the capillary. The capillary network is supplied through small arteries and drained by small veins. In some areas of the skin arteriovenous anastomoses (AVA-shunts) play an important role in the thermoregulation of the body. Most of these vessels reside in the upper part of dermis well accessible by visible and near-infrared light impinging on the skin surface.

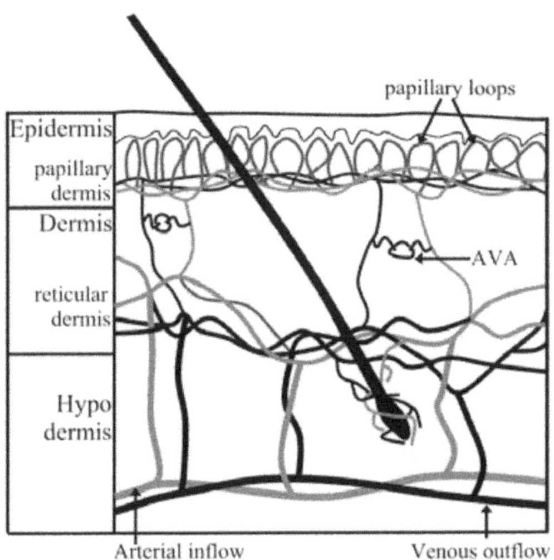

FIGURE 11.1: Microcirculation in the skin.

11.2 Nailfold Capillaroscopy

Capillaries in the nailfold area run in parallel with the skin surface and can be made clearly visible by light microscopy with a magnification of ×200–×600 [1, 2] (Figure 11.2). In spite of the high degree of magnification, movement artifacts can be reduced by fixing the position of the fingernail to the microscope objective, while the transparency of the skin can be increased by use of immersion oil. If a video camera is attached to the microscope the motion of the RBCs can be examined in a frame-by-frame procedure yielding velocity information, which can aid the diagnosis of certain diseases [3]. Cross-correlation techniques have been reported to improve the RBC velocity calculations [4]. Capillary density and diameter are age-dependent and the appearance of abnormal vessels [2] is related to specific diseases. Some conditions can be detected very early as capillary deformations can be observed before other symptoms occur. Ohtsuka et al. [5] demonstrated that capillary loop diameter, width, and length were greater in patients with primary Raynaud's disease that went on to develop undifferentiated connective tissue disease. Similar results were found in patients with systemic sclerosis [2].

Measurements required for analyzing capillary dimensions are generally more time-consuming than the examination itself. Computerized systems have been developed [6] for analyzing capillary network images and the quality has been improved by excluding disturbances caused by hair, liquid, reflections, etc. Combining video-capillaroscopy and various mathematical methods, the statistical properties of the capillaries can be analyzed automatically. The major limitation of capillary microscopy is that it can only be applied to investigation of capillaries in the nailfold, where the vessels run in parallel with the skin surface. Features of the capillaries in this region may not be representa-

FIGURE 11.2: Nailfold capillary pattern of a healthy adult [1]. The fingernail is marked as N. A typical capillary density for a healthy adult is approximately 30–40 mm^{-2}. Reproduced with permission from Wiley-Blackwell Publishing Ltd.

tive for other microvascular networks of the body. Further, when calculating the RBC velocity in a single capillary it must be borne in mind that this velocity is not representative of neighboring capillaries. Ensembles of capillaries must be used to accurately describe the functionality of a specific capillary bed.

11.3 Laser Doppler Perfusion Imaging

Laser Doppler Flowmetry (LDF) refers to the general class of techniques using the optical Doppler effect to measure changes in blood perfusion in the microcirculation noninvasively. Laser Doppler Perfusion Monitoring (LDPM) was developed for point-wise measurement in the 1970s while laser Doppler Perfusion Imaging (LDPI) was developed in the 1980s for imaging of skin blood perfusion. Perfusion is defined as the product of local speed and concentration of blood cells [7]. The principles behind LDF were first developed by Riva et al. [8] who constructed a technique to measure RBC speed in a glass flow tube model while LDPM for assessing tissue microcirculation was first demonstrated in 1975 by Stern [9], and soon developed into clinically useful instruments [10, 11]. The technique operates by using a coherent laser light source to irradiate the tissue surface. A fraction of this light propagates through the tissue in a random fashion and interacts with the different components within the tissue. The light scattered from the moving components such as RBCs will undergo a frequency shift that can be explained by the Doppler effect while the light scattered from the static components will not undergo any shift in frequency. It is virtually impossible to resolve the frequency shifts directly, so the backscattered light from the tissue is allowed to impinge on the surface of a photodetector where beat frequencies are produced. Typical frequencies of these beat notes detected, using a wavelength of 780 nm, range from 0–20 kHz. Contrary to Eq. 11.3, below, the range for a 633 nm source is somewhat smaller, typically 0–12 kHz [7] due to the decreased penetration depth and associated lower RBC velocities. From these beat frequencies it is possible to determine the product of the average speed and concentration (perfusion) of the RBCs within the scattering volume.

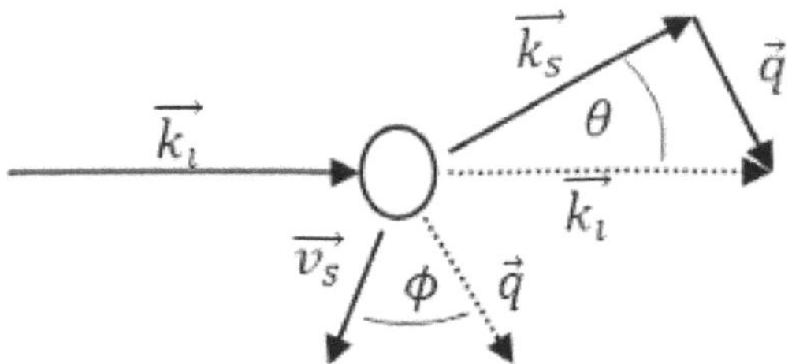

FIGURE 11.3: Process of a single Doppler shifting event.

The frequency shift produced by Doppler scattering is determined by the scattering angle, the wavelength of light in the tissue λ_t and the velocity of the moving scattering object $\vec{v}_s$ (Figure 11.3)

If we assume that a photon injected with propagation vector $\overrightarrow{k_i}$ gets scattered in the direction $\overrightarrow{k_s}$, the resulting scattering vector is given by:

$$\overrightarrow{q} = \vec{k}_i - \vec{k}_s = 2k\sin\left(\frac{\theta}{2}\right) \tag{11.1}$$

The Doppler frequency shift δf between the incident frequency f_i and the scattered frequency f_s for a single scattering event is given by [7]:

$$f_s - f_i = \delta f = \vec{v}_s.\vec{q} = \left(\frac{4\pi}{\lambda_t}\right) v_s \sin\left(\frac{\theta}{2}\right) \cos\phi \tag{11.2}$$

The maximum frequency shift is:

$$\delta f = \left(\frac{4\pi}{\lambda_t}\right) v_s \tag{11.3}$$

which occurs during direct backscattering where the RBC velocity vector is parallel with the photon incident vector. In a system of moving particles such as RBCs in tissue, each photon can undergo more than one Doppler-shift, in which case the detected frequency difference is the sum of the individual scattering events. Individual photons suffer separate and individual resultant Doppler shifts and the resulting Doppler spectrum $P(\omega)$ can be recorded by heterodyne light mixing spectroscopy [12]. This technique makes use of the interference between the shifted and unshifted light to generate a time-varying photocurrent whose spectrum contains only the Doppler shift beat frequencies. The frequency distribution of the fluctuating portion of this photocurrent is related to the RBC speed, while the magnitude is determined primarily by the number of moving RBCs within the scattering volume [13].

The incident laser light beam (632–780 nm) has a penetration depth of below 1 mm depending on the wavelength selected. This implies that it is well suited for examination of the microvascular network of the skin. Since skin microcirculation is a parameter with substantial spatial variation over the skin surface, the inherent limitations of the single-point LDPM technology, employing fiber optics to bring the light to the tissue, was overcome by the introduction of the LDPI technology in the late 1980s [14, 15]. In LDPI the laser beam illuminates the tissue by way of a single scanning point (*Point Raster Scanning*), a scanning line (*Line Scanning*) or by static uniform illumination over the entire skin surface (*Full Field Technology*). The rationale for moving from *Point Raster Scanning* to *Line Scanning* and *Full Field Technology* is to be able to update a complete image at increased speed.

Point Raster Scanning – A raster scan LDPI operates by recording a sequence of point perfusion measurements along a predetermined raster pathway. This was the first type of LDPI system to be developed. The basic schematic is show in Figure 11.4.

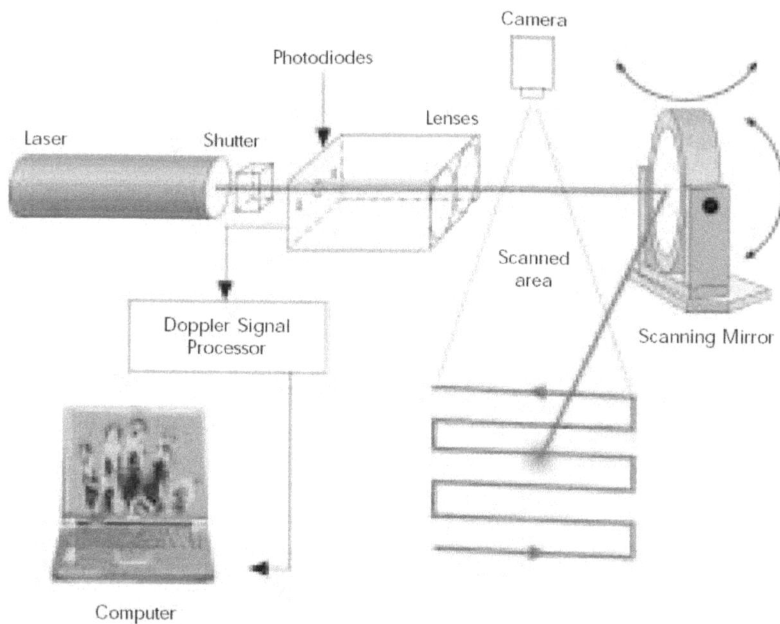

FIGURE 11.4: Schematic of Raster Scan Laser Doppler Imaging System (Moor Instruments Ltd. MoorLDI Brochure).

Imaging is achieved by using a moveable mirror to scan the laser beam over the area in a raster fashion. At each site an individual perfusion measurement is taken. When the scan is complete the system generates a 2D map from the single-point perfusion measurements.

The main drawback with this technique is the time required to produce an image. Each individual measurement can be achieved in 4–50 ms. Lowering the acquisition time further results in increased noise in the signal [16]. For time considerations it is important to minimize the number of measurement points. This is best achieved if the step length between measurements equals the diameter of the laser beam. For smaller step lengths the same physiological information is partly collected by neighboring measurements that increase the overall time without retrieving more information. The moorLDI system reports a capability of capturing a 64 × 64 perfusion image in 20 seconds and has a spatial resolution of 0.2 mm at a distance of 20 cm. The system is capable of scanning a region up to 50 cm ×50 cm and as small as a single-point measurement.

Line Scanning – Laser Doppler Line Scanning Imaging (LDLS) is a new LDPI technology introduced by Moor instruments. This technique does not utilize the standard point raster approach, but works by scanning a laser line over the skin or other tissue surfaces while photodetection is performed in parallel by a photodiode array. This technique allows for numerous perfusion measurements to be performed in parallel. This arrangement greatly reduces acquisition time and is bringing LDPI one step closer to real-time imaging. A schematic of the system is shown in Figure 11.5.

The moorLDLS utilizes a 780 nm laser diode for illumination and 64 diodes to obtain the perfusion measurements in parallel. The system is capable of measuring each line in times of 100 ms – 200 ms. Regions up to 20 cm × 15 cm can be imaged and a 64 × 64 pixel perfusion measurements can be produced in 6 seconds while a maximum resolution image (256 × 64 pixels) is acquired in 26 s. The optimal distance to the object is 15 cm. The device is also capable of single-point multichannel acquisition at 1, 2, 5, 7, and 10 Hz for the entire line of detectors simultaneously.

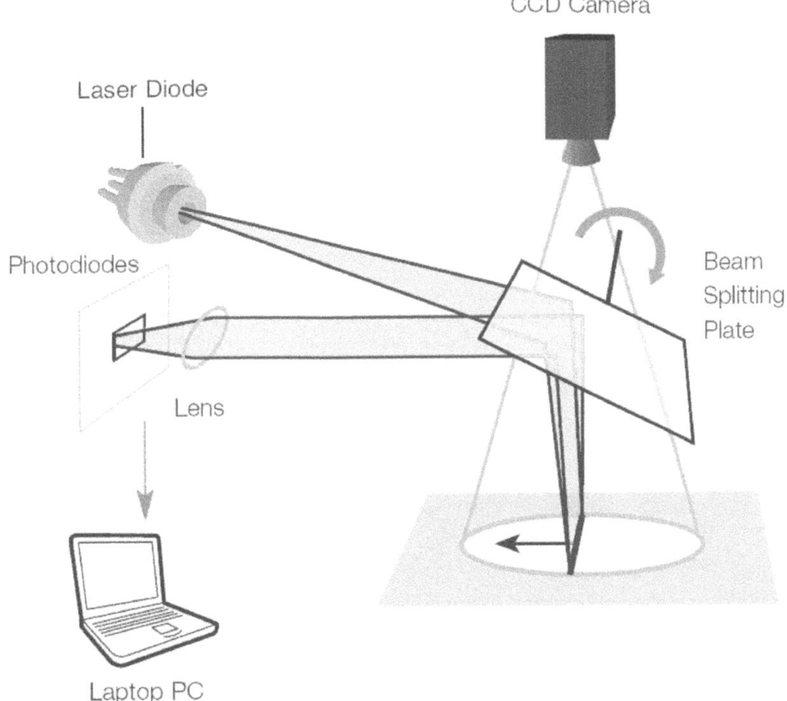

FIGURE 11.5: Schematic of Line Scanning Laser Doppler System (Moor Instruments Ltd. MoorLDI Brochure).

Full Field Technology – A high-speed LDPI system using an integrating CMOS sensor for Full Field illumination has recently been developed by Serov et al. [17]. The basic schematic of the system is shown in Figure 11.6.

A laser source is diverged to illuminate the area of the sample under investigation and the tissue surface is imaged through the objective lens onto a CMOS camera sensor. This new CMOS image sensor has several specific advantages: first, the imaging time is 3–4 times faster than the current commercial raster scan LDPI systems. Second, the refresh rate of the perfusion images is approximately 3.2 s for a 256 × 256 pixel perfusion image. This time includes acquisition, signal processing, and data transfer to the display. For comparison, the specified scan speed of a commercial laser Doppler imaging system, moorLDI^{TM}(Moor Instruments Ltd., UK), is approximately 5 minutes for a 256 × 256 pixel resolution image. However, the scanning imager can measure areas of up to approximately 50 cm × 50 cm in size while the CMOS system at present does not allow imaging of areas larger than approximately 50 mm × 50 mm [18].

The laser Doppler technique has many potential medical applications. However, it still has not been fully integrated into clinical settings and is mainly used in research. Use of LDPI systems have recently been demonstrated in the assessment of burns [19] where it is reported to outperform existing methods of assessment of burn wound depth and provides an objective, real-time method of evaluation. Accuracy of assessment of burn depth is reported to be up to 97% using LDPI, compared with 60–80% for established clinical methods. High perfusion corresponds to superficial dermal burns, which heal with dressings and conservative management. Burns with low perfusion require surgical management i.e., skin grafting. Correct assessment is of particular importance as 1 in 50 grafts fail. If this failure is detected within 24 hours, 50% of these cases can be saved. The use of LDPI has also been investigated in different aspects of cancer research, diagnosis, and

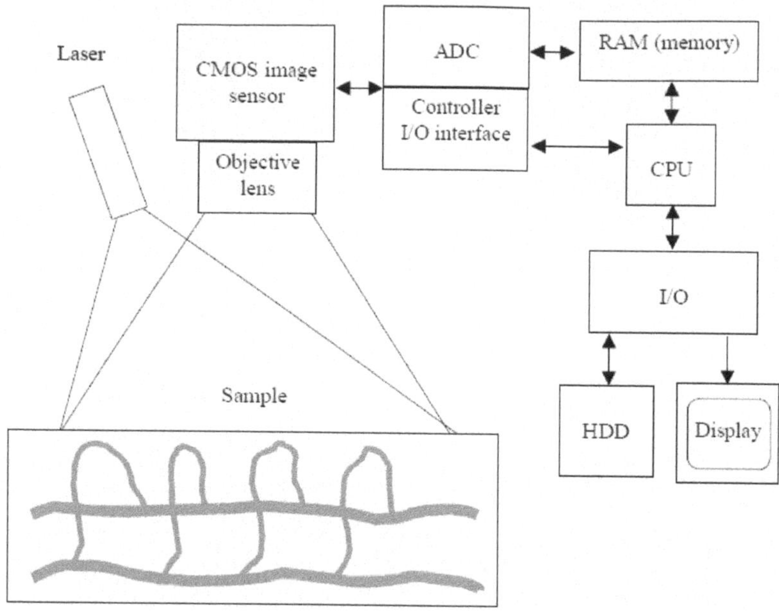

FIGURE 11.6: Schematic of a full-field laser Doppler system [18].)

treatment. It has been found that the higher levels of perfusion in skin tumors can be detected by the LDPI system [20]. The technique has further been applied to examine how a tumor is responding to different treatments such as photodynamic therapy [21]. Wang et al. applied the technique to the assessment of post-operative malignant skin tumors [22]. LDPI has also been applied to many other fields such as examining allergic reactions and inflammatory responses to different irritants [23]. Bjarnason et al. applied the technique in the assessment of patch testing [24, 25]. Ferrell et al. has used the technique to study arthritis and inflammation of joints [26]. LDPI has further been applied in investigating the healing response of ulcers [27] and examining the blood flow pattern in patients at risk from pressure sores [28].

11.4 Laser Speckle Perfusion Imaging

The main drawback with the LDPI technology is the limited update rate that currently does not allow real-time video rate imaging. This can be achieved, however, by use of Laser Speckle Perfusion Imaging (LSPI) that is based on the same fundamental principle as LDPI but with a different and faster way of signal processing. LSPI is generally described as based on the analysis of a time-varying speckle pattern appearing on a remote screen or a detector array being produced by diffusely backscattered laser light from a subject with particles in internal motion (such as RBCs) [29] as demonstrated in Figure 11.7. So far LSPI is identical to Far Field Laser Doppler Perfusion Imaging described in the previous section.

The larger the velocity of the RBCs the quicker the speckle fluctuates. By analyzing the speckle contrast defined as:

$$K = \frac{\sigma}{\langle I \rangle} \leq 1 \tag{11.4}$$

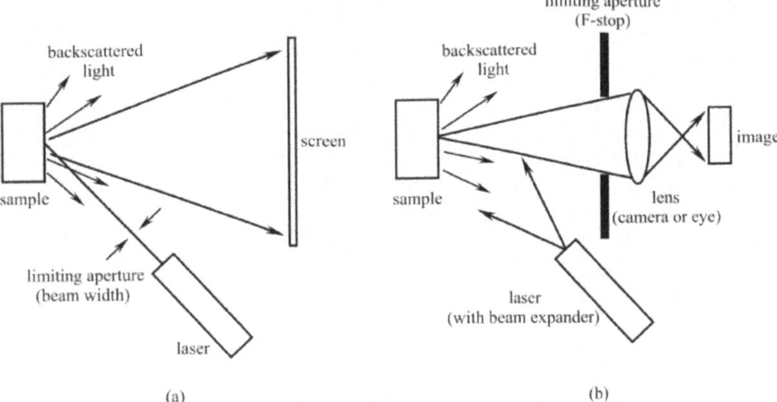

FIGURE 11.7: (a) Far-field speckle. The pattern is formed on a screen some distance away from the object. (b) Image speckle. The pattern at the object itself is observed directly. Reproduced with permission from Institute of Physics Publishing and Professor David Briers [29].

where σ is the standard deviation of the intensity variations and I is intensity. When a photograph is captured with a camera integration time in the order of the fluctuation period, the contrast varies inversely with the speckle fluctuations and thus with average speed. In practice the following equation is used for rapid calculation of the speckle contrast [29]:

$$K = \sqrt{\left\{\frac{\tau_C}{2T}\left[1 - \exp\left({}^{-2T}/_{\tau_C}\right)\right]\right\}} \tag{11.5}$$

where τ_C is the speckle correlation time and T is the camera integration time. This technology is also known under the name LASCA (Laser Speckle Contrast Analysis). Using a 633 nm 30 mW Helium Neon laser as light source expanded over an area of 1 cm^2, Choi et al. [30] recorded images of a rodent skin fold updated every 6 sec with a resolution of 640 $\times$ 480 pixels. The main difficulty with using imaging systems based on eq. 11.5 and the LASCA system is that the relationship between speckle contrast and RBC velocity is nonlinear. A rigorous theory linking the statistics to perfusion does not yet exist. Forrester et al. have developed an improved version of the LASCA system [31], reported to be able to deliver images at a rate of 25 frames per second. This system was used to study changes in the microvasculature in the knee joint capsule of a rabbit before and after femoral artery occlusion, recording a 56.3% decrease after this maneuver. The system was also incorporated into a hand-held endoscope for the simultaneous measurement of the same event. The response was validated by use of an *in vitro* model, showing the linear response over a large flow range (0–800 μl min^{-1}). Equipped with a motion-detection algorithm to limit motion artifacts, the endoscopic setup recorded a similar decrease (58.7%) [31]. Clinical data from human subjects has further been acquired using this endoscopic LSPI system, recording a decrease in blood flow in the medial compartment of the knee after application of a tourniquet. A dose-dependent response to the vasoconstrictor epinephrine was also recorded [32].

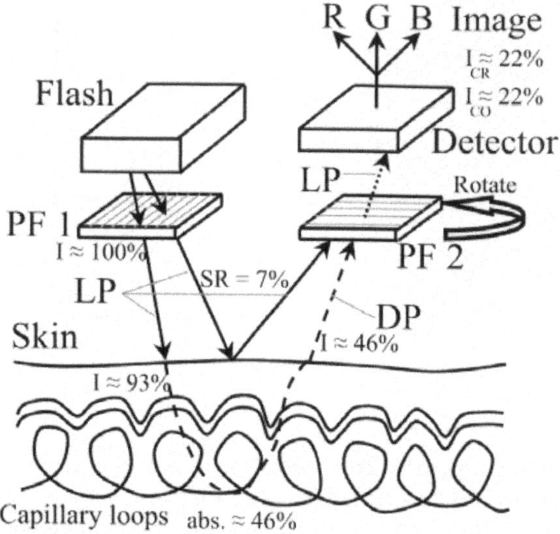

FIGURE 11.8: Fundamental operation of polarization spectroscopy. Remaining percentages of light intensity are illustrated by I, and 100% intensity is observed after the polarization filter. SR represents specular reflection, a combined effect of Fresnel reflection and light returning after few scattering events from the upper layers of the epidermis. LP and DP represent linear and depolarized light respectively, while I_{CO} and I_{CR} represents the remaining intensity that contributes to copolarized (CO) or cross-polarized (CR) data, depending on the 2 possible states of polarization filter PF2. abs. represents the percentage of light absorbed in the tissue. PF2 can be arranged with pass direction parallel (CO) or perpendicular (CR) to PF1.

11.5 Polarization Spectroscopy

Polarization spectroscopy allows the gating of photons returning from different compartments of skin tissue under examination [33]. Figure 11.8 details the operation of basic polarization imaging, showing that by use of simple polarization filters, light from the superficial layers of the skin can be differentiated from light backscattered from the dermal tissue matrix.

When linearly polarized light is incident on the surface of the skin, a small fraction of the light (approximately 5%) is specularly reflected as surface glare (Fresnel reflection) from the skin surface due to refractive index mismatching between the two media. A further 2% of the original light is reflected from the superficial subsurface layers of the *stratum corneum*. These two fractions of light retain the original polarization state, determined by the polarization orientation of the incident light. The remaining portion of light (approximately 93%) penetrates through the epidermis to be absorbed or backscattered by the epidermal or dermal matrix. Approximately 46% of this remaining light is absorbed by the tissue and not re-emitted, while 46% is diffusely backscattered in the dermal tissue. This backscattered portion is exponentially depolarized – more than 10 scattering events are required to sufficiently depolarize light [34] – due to scattering events by chromospheres present in the tissue [35], and also by tissue birefringence due to collagen fibers [36]. The spectral signature of the backscattered light contains information about the main chromophores in the epidermis (melanin) and dermis (hemoglobin), while the surface reflections contain information about the surface topography, such as texture and wrinkles. By placing a second polarization filter in front of the detector with polarization orientation in parallel (coploarized, CO) with or perpendicular (cross-polarized, CR) to that of the filter in front of the light emitter, it is possible to differentiate between the detection of surface reflection and diffusely backscattered light.

To detect events in the microvascular network a coploarization setup is required. There are essentially two different principles employing this setup: Orthogonal Polarization Spectroscopy (OPS) and Tissue Viability Imaging (TiVi).

OPS – This technology employs a single wavelength (548 nm constituting an isobestic point of hemoglobin) in combinations with video microscopy operating at 30 frames per second, thereby increasing contrast and detail by accepting only depolarized backscattered light from the tissue into the probe. Single microscopic blood vessels are imaged *in vivo* at a typical depth of 0.2 mm and magnification of $\times 10$ between target and image, and information about vessel diameter and RBC velocity can be obtained. OPS in human studies is limited only to easily accessible surfaces, but can produce similar values for RBC velocity and vessel diameter as conventional capillary microscopy at the human nailfold plexus [37]. It has also been applied to study superficial and deep burns to the skin [38,39], adult brain microcirculation, and preterm infant microcirculation [40].

TiVi – This technology analyzes the RBC concentration in any skin site of arbitrary size by way of single image or video acquisition by low-cost cameras utilizing CR-technology. For single image acquisition, the flash of a consumer-end RGB digital camera is used as a broadband light source, and the camera CCD acquires an instantaneous image in three 8-bit primary color planes. color filtering is performed on-camera by three 100 nm bandwidth color filters, blue ($\approx$ 400–500 nm), green ($\approx$ 500–600 nm), and red ($\approx$ 600 nm – 700 nm). Using polarization imaging and Kubelka-Munk theory, a spectroscopic algorithm was developed that is not dependent on incident light intensity, taking advantage of the physiological fact that green light is absorbed more by RBCs than red light. For each arriving image, the equation applied is:

$$M_{out} = \frac{M_{red} - M_{green}}{M_{red}} e^{-p\left(\frac{M_{red} - M_{green}}{M_{red}}\right)} \tag{11.6}$$

where M_{red} and M_{green} represent the red and green color planes of the image, and p represents an empirical factor to produce the best linear fit between output variable (called TiVi index) and RBC concentration. M_{out} represents a matrix of maximum 3648 $\times$ 2736 TiVi index values for single-image acquisition. This emerging technology has been commercialized under the name Tissue Viability Imaging by WheelsBridge AB. TiVi represents the first low-cost technique designed to image tissue hematocrit in real time. Figure 11.9 shows an example of a TiVi image color coded in a similar method to laser Doppler images where low and high RBC concentration is represented by blue and red color, respectively.

The technique has high spatial and temporal aspects, with lateral resolution estimated at 50 μm [41]. The average sampling depth in Caucasian skin tissue is estimated to approximately 0.5 mm from Monte Carlo simulations of light transport in tissue [41]. Topical application of vasodilating and vasoconstricting agents demonstrate the capacity to document increases (erythema) and decreases (blanching) in RBC concentration [41]. Due to the fact that frequent calibration is not necessary and no moving parts are included in the design of the TiVi system, excellent stability over time (0.27% instability on the average over three months) as well as uniformity between TiVi-units (less than 4.1 % deviation in sensitivity) have been demonstrated [42]. In contrast to laser Doppler technologies, TiVi is insensitive to movement artifacts, distance to object and ambient light levels. Dedicated software toolboxes have recently been developed for skin damage visualizing, skin color tracking, and spot analysis. A real-time TiVi-technology system is currently under development, whereby RBC concentration can be displayed online. Current technology can capture CR frames at 25 frames per second at a frame size of 400 $\times$ 400 pixels.

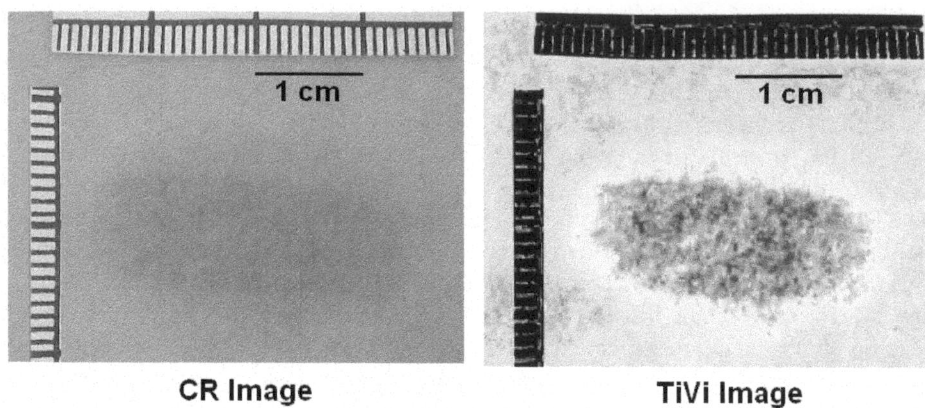

FIGURE 11.9: Example of a CR and TiVi image. The CR image was taken of a UVB burn to the volar forearm of a young healthy volunteer. The image was taken 30 hours after a 32-seconds exposure to an 8-mm-wide UVB source. The TiVi image shows spatial heterogeneity in the blood distribution over the affected area. Notice that hairs have no effect on algorithm performance.

11.6 Comparison of LDPI, LSPI, and TiVi

In order to accurately validate the results of any microcirculation technique, the opinion of a clinical expert is usually required. As of yet there is no modern technology that has supplanted the fundamental visual observation technique in the clinical environment, even though clinicians' interpretations of images can be bypassed entirely by computerized analysis methods such as neural networks. Limitations/precautions of skin assessment visual scoring systems such as the skin blanching assay include the requirement of 1) application of the test substance at equal amounts in each patch in random order, 2) need to make all observations under standardized lighting conditions and 3) the participation of a well-trained and skillful observer. If these requirements are not met, interlaboratory results may vary substantially and be difficult to compare.

In this section the comparison of images produced by three different techniques (LDLS, LSPI, and TiVi) are reported. The following instrumentation was used:

LDLS: MoorLDLS, Moor Instruments, Exeter, UK.

LSPI: MoorFLPI, Moor Instruments, Exeter, UK.

TiVi: TiVi600, WheelsBridge, Linköping, Sweden.

The features of these specific technologies are outlined in Table 11.1.

Study design – Postocclusion reactive hyperaemia tests were performed on 3 nonsmoking healthy subjects (average age of 25 years), with no history of skin diseases, who had not used topical or systemic corticosteroid preparations in the previous 2 months. The subjects were acclimatized in a laboratory at 21°C for 30 minutes and rested in a seated position with the volar forearm placed on a flat table at heart level. The instruments were positioned above the forearm for data acquisition.

Reactive hyperemia caused by the release of brachial arterial occlusion using a sphygmomanometer around the upper arm of the test subjects was used to compare the systems in temporal mode (with the high speed TiVi system used). A baseline level was recorded for 1 minute. Arterial occlusion was then performed by keeping the cuff inflated at 90 and 130 mmHg respectively for 2 minutes. In previous studies with LDPI, upon complete release of the cuff, the resulting increase in blood perfusion is seen, and blood flux oscillations are regularly observed. This phase was observed for a further 3 minutes. In order to further examine the sensitivity of the TiVi system alone

TABLE 11.1: Comparison table outlining mechanical and operational principles of the different devices evaluated.

Property	LDLS	LSPI	TiVi
Principle	Doppler effect	Reduced speckle contrast	Polarization spectroscopy
Variable recorded	Perfusion (speed × concentration)	Perfusion (speed × concentration)	Concentration
Variable units	Perfusion units (P.U.)	Perfusion units (P.U.)	$TiVi_{index}$ (A.U.)
Measurement sites	16,384	431,680	Up to 12 million
Best resolution	≈ 500 μm	≈ 100 μm	≈ 50 μm
Measurement depth	≈ 1 mm	≈ 300 μm	≈ 500 μm
Repetition rate	6 sec (50 × 64)	4 sec (568 × 760)	5 sec (4000 × 3000)
Best capture time	6 sec	4 sec	1/60th sec
Distance dependence	Yes	Yes	No
Zoom function	No	Yes	Yes
Laser driver	Required	Not Required	Not Required
Movement artifact sensitivity	Substantial	Dependent on temporal averaging	None
Video mode	No	Yes	Yes
Video rate	–	25 fps	25 fps
Video size	–	113 × 152	Up to 720 × 576
Pixel resolution	100 per cm^2	40,000 per cm^2	600,000 per cm^2
View angle	Top only	Top only	User selectable
Imager Dimensions	19 cm × 30 cm × 18 cm	22 cm × 23 cm × 8 cm	15 cm × 7 cm × 5 cm
Weight	Scanner 3.2 kg	2 kg	1 kg
Calibration	Motility standard	Motility standard	None
Ambient light sensitivity	Intermediate	Substantial	Negligible

with pressure, a further experiment was conducted where both the high-speed and high-resolution TiVi systems were arranged to image the volar forearm of a healthy individual at increasing and decreasing pressure steps of 10 mmHg.

Physical size prevented all three technologies from being used at the same time on the same test area, meaning that tests with the LSPI had to be carried out separately from TiVi and LDLS studies. The white flash light source on the TiVi did not interfere with the LDLS system, as there are filters present to remove much of the extraneous light. Finally, the TiVi readings were not affected by the scanning beam of the LDLS system as it was outside the sensitive region of the detector. A summary of the results is illustrated in Figure 11.10.

It should be noted that upon application of the sphygmomanometer and subsequent analysis for the TiVi, the curve increased for both 90 mmHg and 130 mmHg. This is because the instrument is sensitive to concentration of RBCs only, and not the speed. This new technology adds a new dimension to the study of post-occlusive reactive hyperemia, with a nontraditional time trace sensitive to the concentration of RBCs only.

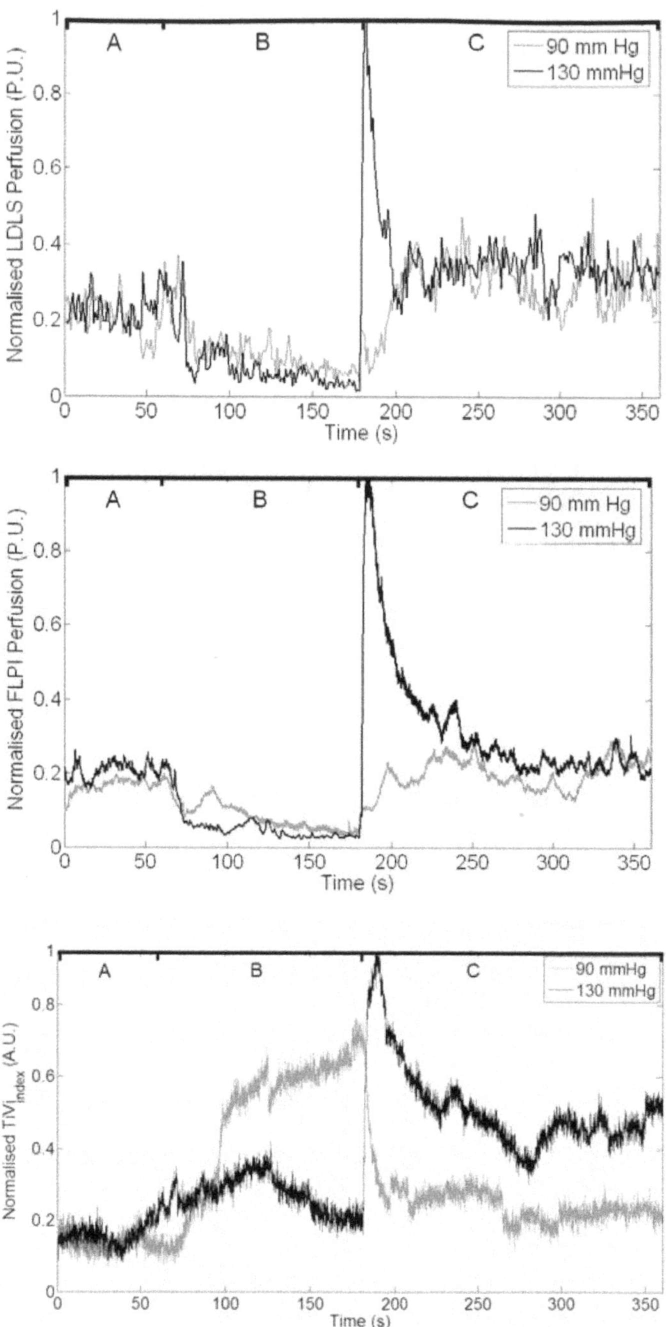

FIGURE 11.10: Time traces of postocclusive reactive hyperemia for LDPI, LSPI, and TiVi. The reaction was sectioned into three; A is the baseline measurement, B is the occlusion for 2 minutes, and C represents the reactive hyperemia upon removal of the occlusion. Pressure was 130 mmHg except for the TiVi curves where 90 mmHg pressure was also used to show the TiVi systems' (high resolution and high speed) sensitivity to pressure.

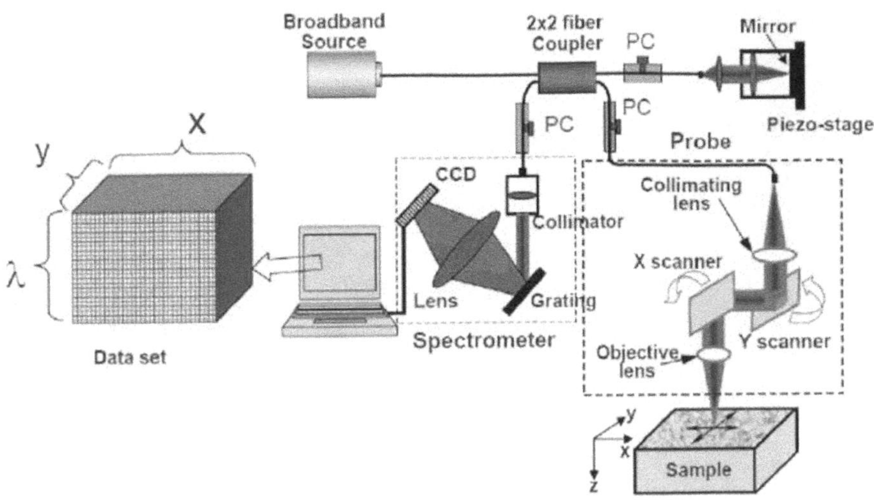

FIGURE 11.11: Schematic of a typical OMAG system [44].

11.7 Optical Microangiography

Optical Microangiography (OMAG) has its origins in Fourier domain OCT and is similar to Doppler OCT. It is a recently developed high-resolution 3D method of imaging blood perfusion and direction at capillary level resolution within microcirculatory beds. A typical setup of an OMAG system is shown in Figure 11.11. The technique itself does not rely on the phase information of the OCT signals to extract the blood flow information. This in turn means that the technique is not sensitive to the environmental phase stability. This allows OMAG to compensate for the inevitable sample movement using the phase data, thus limiting noise through postprocessing of the data [43].

An example of an OMAG image is shown in Figure 11.12. The acquisition time to obtain the 3D OMAG data shown in Figs. 11.12 and 11.13 took about 50 s. However, the processing time required was much larger.

The OMAG technique appears to offer a much greater signal-to-noise ratio compared to standard DOCT allowing the system to visualize more features than standard DOCT as seen in Figure 11.13.

Figure 11.13 compares the two imaging modalities *in vivo* on an adult mouse brain. The images were both obtained on the same mouse with the skull intact. The images were then normalized by the maximum pixel value in each image and log compressed [44]. The improved contrast using this technique is clearly visible. OMAG allows for much smaller features to be visualized deeper into the tissue.

Another advantage of the OMAG technique is that it can determine the direction of the flow by applying a Hilbert transformation for each wavenumber along the positions separately. An example of this is shown in Figure 11.14.

In this experiment the probe beam was shone from top onto the sample. The mirror was then modulated using a triangle waveform. The images were generated for (a) the ascending (b) descending portion of the waveform. When the mirror moves toward the incident beam, blood perfusion that flows away from the beam direction is determined and vice versa [45]. The figure shows that the flow direction is clearly determined using this technique. The minimum flow velocity that can be recorded using OMAG is reported to be $\sim 170\ \mu\text{m s}^{-1}$ while the maximum flow velocity that can be measured is $\sim \pm 2.1$ mm s^{-1}[44].

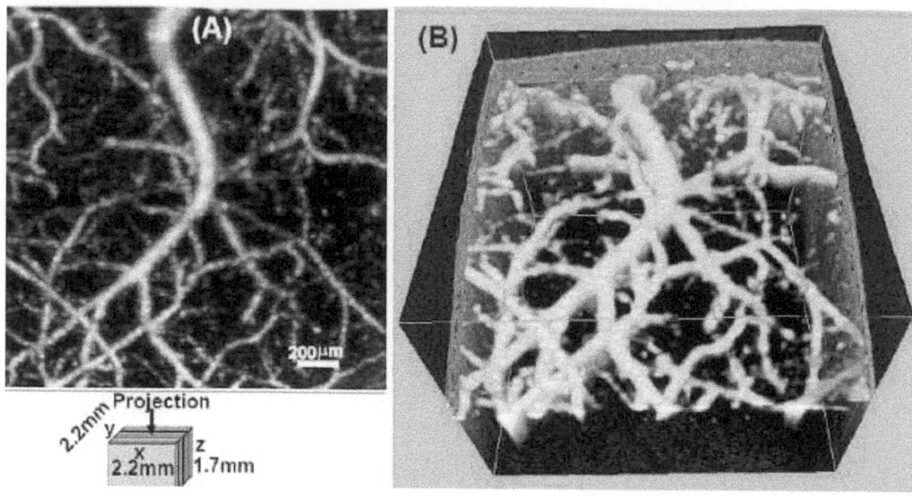

FIGURE 11.12: An adult mouse brain imaged with OMAG in vivo with the skull intact [44].

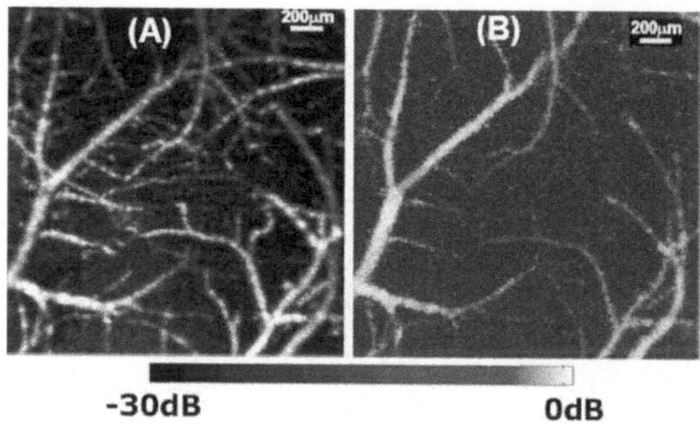

FIGURE 11.13: Comparison (A) OMAG system to (B) standard DOCT. The OMAG system allows visualisation of lower orders of the microvascular network [44].

11.8 Photoacoustic Tomography

In recent years photoacoustic tomography (PAT) has gained considerable interest [46]. This is because the technique seeks to combine the best features of biophotonics with those of ultrasound. It offers high optical contrast as well as high resolution and can image objects located deep in tissue. The technique operates by irradiating a sample with a laser pulse. When the laser light is absorbed it produces pressure waves due to an increase in the temperature and volume. These pressure waves emanate from the source in all directions. A high-frequency ultrasound receiver is used to detect these waves and a 3D picture is built of the structure. This is made possible by the relatively low speed of sound and lack of scattering or diffusion of ultrasound en route to the receiver. The use of all photons is a considerable advantage over techniques such as OCT that are reliant on ballistic photons, since a useful percentage of (diffuse) photons penetrate several cm into the tissue, whereas the number of ballistic photons emanating from a few mm is generally so small

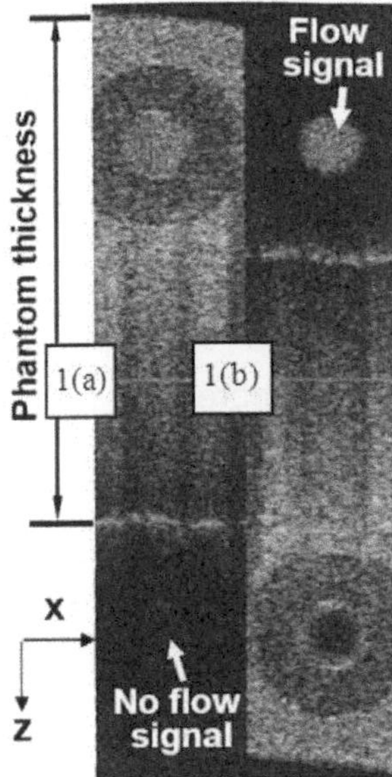

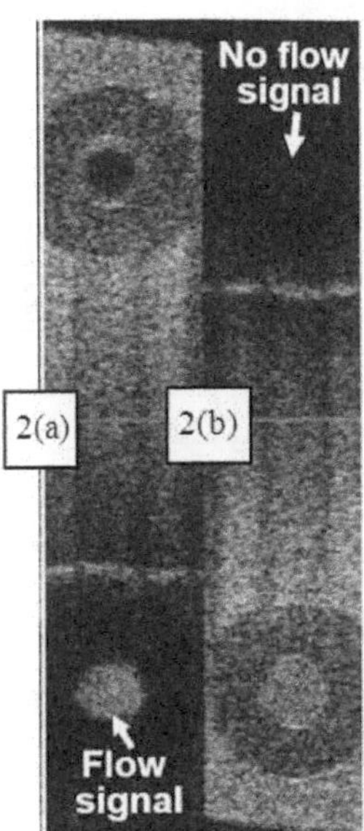

FIGURE 11.14: Determination of directionality of fluid flow using OMAG. (1) flow is orientated towards scanning beam (2) flow is reversed such that is it orientated away from scanning beam. (a) Positive slope of reference mirror, (b) negative slope of reference mirror. The physical dimensions of the image are 1.3×3.1 mm.

as to be unusable. Diffuse, randomized scattering of the light before reaching the microvessels is also useful in providing relatively uniform illumination even to vessels that cannot receive ballistic photons due to other vessels in the direct light path.

In PAT light in the visible to NIR spectrum is used. This wavelength range is selected because hemoglobin in the blood dominates the optical absorption in most soft tissue. If a tissue sample is irradiated by light in this wavelength range, a map of the vessel network within the tissue can be produced. The pulse width is chosen in such a way that it is much shorter than the thermal diffusion time of the system. By this arrangement heat conduction can be neglected [48]. A schematic of the system is shown in Figure 11.15.

The data from PAT is processed by a computer and can be output in an image format. Image reconstruction is achieved by various reconstruction algorithms [49]. The technique is capable of diagnosing different physiological conditions because these can change the light absorption properties of the tissue. Absorption contrast between breast tumors and normal breast tissue as high as 500% in the infrared range has been reported [49]. The technique is also capable of detecting hemoglobin levels in tissue with very high contrast. One group [50] has used the technique to acquire 3D data of the hemoglobin concentration in rat brain. This was achieved using a 532-nm laser source, as the absorption coefficient at this wavelength of whole blood is 100 cm^{-1} and much larger than the averaged absorption coefficient of the gray and white matter of the brain (0.56 cm^{-1}).

The main advantage of combining ultrasound and light is improved resolution (diffraction-limited) and the possibility to image vessels with 60-μm resolution [51]. The Wang lab at University of Washington in St. Louis are building a system that should get down to 5 μm resolution and have already been able to produce exquisite 3D images of microvascular architecture in the rat. One issue with the technique is the time it takes to produce large 3D scans of a region. Times of up to 16 h have been reported to image a region of radius 2.8 cm in the XY plane and a depth 4.1 cm in the Z plane [52]. However, the use of an ultrasonic transducer arrays would greatly reduce imaging time.

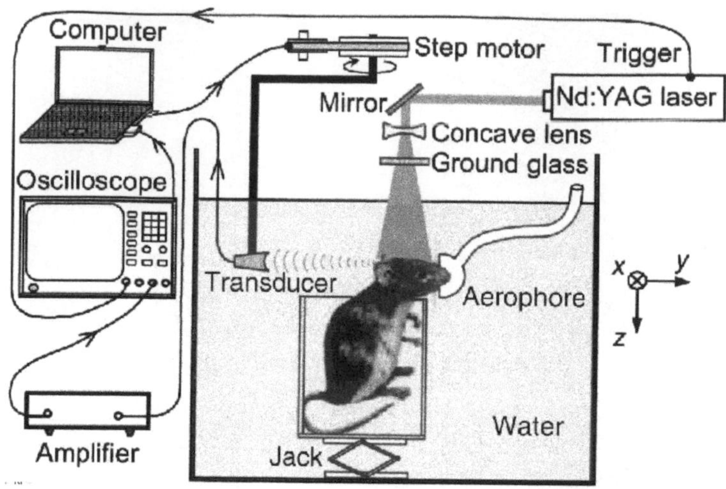

FIGURE 11.15: Photoacoustic tomography system for imaging of the rat brain with the skin and skull intact [47].

11.9 Conclusions

We have discussed the *modus operandi* and applications of capillaroscopy, LDPI, LSPI, TiVi, OMAG, and PAT. Each has had its own advantages and limitations. All retain niche applications. Proper application of each is dependent on user knowledge of light interaction with tissue and the basic workings of the device used. Choice of technology for *in vivo* imaging of the microcirculation is dependent on the sampling depth (nutritional or thermoregulatory vessels), resolution (physical and pixel), field-of-view, and whether structure or function is to be investigated.

TiVi differs from the Doppler techniques in its response to occlusion of the brachial artery. Since this is a global occlusion of the forearm all instruments were sensitive to the effect. LDPI and LSPI showed a large decrease as they are both sensitive to red cell concentration and speed. However TiVi, being sensitive to concentration only, showed stasis when a full arterial occlusion (130 mmHg) was applied quickly and a swelling of red cell concentration when the arterial inflow remained open with the venous outflow occluded (90 mmHg). LSPI and TiVi are both welcome tools in the study of the microcirculation, but care must be taken in the interpretation of the images since blood flow and blood concentration in tissue are essentially different parameters. However, LDPI is the only one of these imaging techniques that up to now has been approved by regulatory bodies including the FDA. Laser Doppler imaging has been shown to accurately assess burn depth and predict wound outcome with a large weight of evidence [19]. Work described here may motivate further studies in comparisons between noninvasive optical techniques for the investigation of tissue microcirculation.

Acknowledgments

The authors are grateful to HEA PRTLI4 (Government of Ireland's national development plan) funding of the National Biophotonics and Imaging Platform Ireland and IRCSET for supporting this research. They also thank Dr. Rodney Gush of Moor Instruments.

References

[1] Z. Nagy and L. Czirják,"Nailfold digital capillaroscopy in 447 patients with connective tissue disease and Raynaud's disease," *J. Eur. Acad. Dermatol. Venereol.* **18**, 62–68 (2004).

[2] T.J. Ryan, "Measurement of blood flow and other properties of the vessels of the skin," in *The Physiology and Pathophysiology of the Skin*, Jarrett A (Ed.), Academic Press, London, vol. **1**, 653–679 (1973).

[3] T. Aguiar, E. Furtado, D. Dorigo, D. Bottino, and E. Bouskela, "Nailfold videocapillaroscopy in primary Sjögren's syndrome," *Angiology* **57**, 593–599 (2006).

[4] D.M. Mawson and A.C. Shore, "Comparison of capiflow and frame by frame analysis for the assessment of capillary red blood cell velocity," *J. Med. Eng. Technol.* **22**, 53–63 (1998).

[5] T. Ohtsuka, T. Tamura A. Yamakage, and S.Yamazaki, "The predictive value of quantitative nailfold capillary microscopy in patients with undifferentiated connective tissue disease," *Brit. J. Dermatol.* **139**, 622–629 (1998).

[6] J. Zhong, C.L. Asker, and G. Salerud, "Imaging, image processing and pattern analysis of skin capillary ensembles," *Skin Res. Technol.* **6**, 45–57 (2000).

[7] M.J. Leahy, F.F. de Mul, G.E. Nilsson, and R. Maniewski. "Principles and practice of the laser-Doppler perfusion technique," *Technol. and Health Care* **7**, 143–162 (1999).

[8] C. Riva, B. Ross, and G.B. Benedek, "Laser Doppler measurements of blood flow in capillary tubes and retinal arteries," *Invest. Ophthalmol.* **11**, 936–944 (1972).

[9] M.D. Stern "In vivo evaluation of microcirculation of coherent light scattering," *Nature* **254**, 56–58 (1975).

[10] G.A. Holloway and D.W. Watkins, "Laser Doppler measurements of cutaneous blood flow," *J. Invest. Dermatol.* **69**, 306–309 (1977).

[11] G.E. Nilsson, T. Tenland, and P.A. Öberg, "A new instrument for continuous measurement of tissue blood flow my light beating spectroscopy," *IEEE Trans. Biomed. Eng.* **27**, 12–19 (1980).

[12] H.Z. Cummins and H.L. Swinney, "Light beating spectroscopy," *Progress in Optics* **8**, 133–200 (1970).

[13] R. Bonner and R. Nossal "Model for laser Doppler measurements of blood flow in tissue," *Appl. Opt.* **20**, 2097–2107 (1981).

[14] K. Wårdell, "Laser doppler perfusion imaging by dynamic light scattering," *IEEE T. Biomed. Eng.* **40**, 309–316 (1993).

[15] T.J. Essex and P. Byrne, "A laser Doppler scanner for imaging blood flow in skin," *J. Biomed. Eng.* **13**, 189–194 (1991).

[16] A. Fullerton, M. Stucker, K.P. Wilhelm, K. Wårdell, C. Anderson, T. Fisher, G.E. Nilsson, and J. Serup, "Guidelines for visualization of cutaneous blood flow by laser Doppler perfusion imaging," *Cont. Derm.* **46**: 129–140 (2002).

[17] A. Serov, W. Steenbergen, and F. de Mul, "Laser Doppler perfusion imaging with an complementary metal oxide semiconductor image sensor," *Opt. Lett.* **27**, 300–302 (2002).

[18] A. Serov, "High-speed laser Doppler perfusion imaging using an integrating CMOS image sensor," *Opt. Express* **13**, 6416–6428 (2005).

[19] S. Monstrey, H. Hoeksema, J. Verbelen, A. Pirayesh, and P. Blondeel, "Assessment of burn depth and burn wound healing potential," *Burns* **34**, 761–769 (2008).

[20] M. Stucker, C. Springer, V. Paech, N. Hermes, M. Hoffmann, and P. Altmeyer "Increased laser Doppler flow in skin tumors corresponds to elevated vessel density and reactive hyperemia," *Skin. Res. Technol.* **12**, 1–6 (2006).

[21] M. Hassan, R. Little, A. Vogel, K. Aleman, K. Wyvill, R. Yarchoan, and A. Gandjbakhche, "Quantitative assessment of tumor vasculature and response to therapy in Kaposi's sarcoma using functional noninvasive imaging," *Technol. Cancer Res. T.* **3**, 451–457 (2004).

[22] I. Wang, S. Andersson-Engels, G.E. Nilsson, G. Wårdell, and K. Svanberg, "Superficial blood flow following photodynamic therapy of malignant non-melanoma skin tumours measured by laser Doppler perfusion imaging," *Brit. J. Dermatol.* **136**, 184–189 (1997).

[23] A. Fullerton, B. Rode, and J. Serup, "Studies of cutaneous blood flow of normal forearm skin and irritated forearm skin based on high-resolution laser Doppler perfusion imaging (HR-LDPI)," *Skin Res. Technol.* **8**, 32–40 (2002).

[24] B. Bjarnason and T. Fischer, "Objective assessment of nickel sulfate patch test reactions with laser Doppler perfusion imaging," *Cont. Derm.* **39**, 112–118 (1998).

[25] B. Bjarnason, E. Flosadottir, and T. Fischer, "Objective non-invasive assessment of patch tests with the laser Doppler perfusion scanning technique," *Cont. Derm.* **40**, 251–260 (1999).

[26] W. Ferrell, P. Balint, and R. Sturrock, "Novel use of laser Doppler imaging for investigating epicondylitis," *Rheumatology* **39**, 1214–1217 (2000).

[27] D. Newton, F. Khan, J. Belch, M. Mitchell, and G. Leese, "Blood flow changes in diabetic foot ulcers treated with dermal replacement therapy," *J. Foot Ankle Surg.* **41**, 233–237 (2002).

[28] J. Nixon, S. Smye, J. Scott, and S. Bond, "The diagnosis of early pressure sores: report of the pilot study," *J. Tissue Viability* **9**, 62–66 (1999).

[29] J.D. Briers, "Laser Doppler, speckle and related techniques for blood perfusion mapping and imaging," *Physiol. Meas.* **22**, R35–R66 (2001).

[30] B. Choi, N.M. Kang, and J.S. Nelson, "Laser speckle imaging for monitoring blood flow dynamics in the in vivo rodent dorsal skin fold model," *Microvasc Res.* **68**, 143–146 (2004).

[31] K.R. Forrester, C. Stewart, C. Leonard, J. Tulip, and R.C. Bray, "Endoscopic laser imaging of tissue perfusion: new instrumentation and technique," *Laser Surg. Med.* **33**, 151–157 (2003).

[32] R.C. Bray, K.R. Forrester, J. Reed, C. Leonard, and J. Tulip, "Endoscopic laser speckle imaging of tissue blood flow: applications in the human knee," *J. Orthop. Res.* **24**, 1650–1659 (2006).

[33] S.G. Demos and R.R. Alfano, "Optical polarization imaging," *Appl. Opt.* **36**, 150–155 (1997).

[34] J.M. Schmitt, A.H. Gandjbakhche, and R.F. Bonnar, "Use of polarized light to discriminate short-pass phonons in a multiply scattering medium," *Appl. Opt.* **31**, 6535–6539 (1992).

[35] A. Ishimaru, "Diffusion of light in turbid media," *Appl. Opt.* **28**, 2210–2215 (1989).

[36] D.F.T. da Silva, B.C. Vidal, D.M. Zezell, T.M.T. Zorn, S.C. Nunez, and M.S. Ribeiro, "Collagen birefringence in skin repair to red-polarized laser therapy," *J. Biomed. Opt.* **11**, 204002-1–6 (2006).

[37] K.R. Mathura, K.C. Vollebregt, K. Boer, J.C. DeGraaff, D.T. Ubbink, and C. Ince, "Comparison of OPS imaging and conventional capillary microscopy to study the human microcirculation," *J. Appl. Physiol.* **91**, 74–78 (2001).

[38] S.M. Milner, S. Bhat, S. Gulati, G. Gherardini, C.E. Smith, and R.J. Bick, "Observations on the microcirculation of the human burn wound using orthogonal polarization spectral imaging," *Burns* **31**, 316–319 (2005).

[39] S. Langer, A.G. Harris, P. Biberthaler, E. von Dobschuetz, and K. Messmer, "Orthogonal polarization spectral imaging as a tool for the assessment of hepatic microcirculation: a validation study," *Transplantation* **71**, 1249–1256 (2001).

[40] O. Genzel-Boroviczeny, J. Strontgen, A.G. Harris, K. Messmer, and F. Christ, "Orthogonal polarization spectral imaging (OPS): a novel method to measure the microcirculation in term and preterm infants transcutaneously," *Pediatr. Res.* **51**, 386–391 (2002).

[41] J. O'Doherty, J. Henricson, C. Anderson, M.J. Leahy, G.E. Nilsson, and F. Sjöberg, "Sub-epidermal imaging using polarized light spectroscopy for assessment of skin microcirculation," *Skin Res. Technol.* **13**, 472–484 (2007).

[42] G.E. Nilsson, H. Zhai, H.P. Chan, S. Farahmand, and H.I. Maibach, "Cutaneous bioengineering instrumentation standardization: the tissue viability imager," *Skin Res. Technol.* **15**, 6–13 (2009).

[43] L. An and R.K. Wang, "Volumetric imaging of microcirculations in human retina and choroids in vivo by optical micro-angiography," *Proc. SPIE* **6855**, 68550A-1–9 (2008).

[44] R.K. Wang, S.L. Jacques, Z. Ma, S. Hurst, S. R. Hanson, and A. Gruber, "Three dimensional optical angiography," *Opt. Express* **15**, 4083–4097 (2007).

[45] R.K. Wang, "Three-dimensional optical micro-angiography maps directional blood perfusion deep within microcirculation tissue beds in vivo," *Phys. Med. Biol.* **52**, N531–N537 (2007).

[46] X.D. Wang, Y.J. Pang, G. Ku G, X.Y. Xie, G .Stoica, and L.H.V. Wang, "Noninvasive laser-induced photoacoustic tomography for structural and functional in vivo imaging of the brain," *Nat. Biotechnol.* **21**, 803–806 (2003).

[47] X. Wang, Y. Pang, G. Ku, G. Stoica, and L.V. Wang, "Three-dimensional laser-induced photoacoustic tomography of mouse brain with the skin and skull intact," *Opt. Lett.* **28**, 1739–1741 (2003).

[48] G.J. Diebold, T. Sun, and M.I. Khan, "Photoacoustic waveforms generated by fluid bodies," in: *Photoacoustic and Photothermal Phenomena III* **69**, D. Bicanic (ed.) – Springer Series in Optical Sciences, Springer-Verlag, Berlin, 263–296 (1992).

[49] M. Xu and L.V. Wang, "Universal back-projection algorithm for photoacoustic computed tomography," *Phys. Rev. E* **71**, 016706-1–7 (2005).

[50] X. Wang, Y. Pang, G. Ku, G. Stoica, and L. Wang, "Three-dimensional laser-induced photoacoustic tomography of mouse brain with the skin and skull intact," *Opt. Lett.* **28**, 1739–1741 (2003).

[51] X. Wang, Y. Xu, M. Xu, S. Yokoo, E.S. Fry, and L.V. Wang, "Photoacoustic tomography of biological tissues with high cross-section resolution: Reconstruction and experiment," *Med. Phys.* **12**, 2799–2805 (2002).

[52] D. Yang, D. Xing, Y. Tan, H. Gu, and S. Yang, "Integrative prototype B-scan photoacoustic tomography system based on a novel hybridized scanning head," *Appl. Phys. Lett.* **88**, 174101-1–3 (2006).

12

Advances in Optoacoustic Imaging

Tatiana Khokhlova and Ivan Pelivanov

Faculty of Physics, Moscow State University, Moscow, 119991, Russia

Alexander Karabutov

International Laser Center, Moscow State University, Moscow, 119991, Russia

This chapter will address the most recent developments in the field of laser optoacoustic (OA) tomography. This imaging modality is based on the thermoelastic effect: absorption of pulsed laser radiation by the biological tissue leads to its heating and subsequent thermal expansion that in turn results in the excitation of the acoustic (or OA) pulse. The waveform of the OA signal contains information on the distribution of laser-induced heat release that in turn depends on the distribution of optical absorption coefficient in tissue. Therefore, detection of the OA signals by an array of transducers allows reconstruction of the distribution of heat release and deduction of information on the distribution of absorbing inclusions in tissue.

OA tomography is applicable in any diagnostic procedure that involves imaging of objects that possess a higher absorption coefficient than the surrounding medium. Such applications include the diagnostics of blood vessels, because blood is the strongest absorber among biological tissues in the near infrared range. Enhanced angiogenesis is typical for the malignant tumors from the beginning of their development, therefore, OA tomography allows for early tumor detection and diagnosis.

The most important problems in OA tomography, as well as in any tomography, are designing the OA signal detection system (the array transducer), and development of image reconstruction algorithms. In recent years a number of array designs, employing both optical and piezoelectric detection of the OA signals, were proposed. This chapter will contain a review of these designs, their potential biomedical applications, advantages, and drawbacks.

As shown in a number of recent works, the inverse problem of optoacoustic tomography has a unique solution that can be found for the case of ideal detection surface. Image reconstruction algorithms based on these solutions have lately been much improved and diversified. Since in the experimental conditions the detection surface is never ideal (finite size and number of detectors, limited viewing angle), image artifacts and distortions are inevitable, whatever an algorithm is used for reconstruction. This chapter will provide the theoretical basis of each algorithm, describe the nature of image artifacts that can be induced, and review the proposed ways to improve image quality.

Key words: optoacoustic tomography, blood, tumor, small animal imaging

12.1 Introduction

Optoacoustic (OA) tomography is a hybrid, laser-ultrasonic method for diagnostics of biological tissues. This method is based on thermoelastic effect: absorption of pulsed laser radiation in the medium under study leads to nonuniform and nonstationary heating of the medium, that in turn enables its thermal expansion and, consequently – generation of the ultrasound (optoacoustic, OA) pulses. The temporal profile of the OA signal contains information on the distribution of laser-induced heat sources in the medium [1], therefore, the reconstruction of this distribution can be performed based on the detected OA signals.

OA tomography, as well as other laser diagnostic methods (optical coherence tomography, OCT, optical diffusion tomography, ODT), is applicable to any problem that requires visualization of an object, that is more optically absorbing than the surrounding medium. The most widespread biomedical applications include imaging of blood vessels, since blood is the strongest optical absorber among other biological tissues in visible and near-infrared range [2], detection of malignant tumors, that are characterized by enhanced vascularization and blood supply [3], and diagnostics of surface or in-depth thermal lesions of biological tissues [4, 5]. It is interesting to determine the place of OA tomography among other optical imaging techniques.

In OCT the information is obtained by detecting coherent radiation backscattered from the medium under study. The requirement of coherence limits the imaging depth to about 2–3 mm, because biological tissues are highly scattering media. Therefore, OCT is mostly applied for diagnostics of skin and mucous tissue malignancies with spatial resolution on the order of tens of microns. On the contrary, in ODT diffuse light that has propagated through the medium is detected. This method allows one to obtain images of absorbing inhomogeneities located at depths of several centimeters within biological tissue. However, spatial resolution of the method significantly decreases with depth due to the random nature of photon trajectories [6]. For example, at depths less than 1 cm the spatial resolution would be on the order of millimeters, at depths of several centimeters the resolution is also on the order of centimeters.

Let us estimate the main characteristics of the OA signals and images. The amplitude of the OA signal, excited in the biological tissue by pulsed laser radiation with fluence that is within a medically safe range, can reach 10^3 Pa, whereas the increase of the tissue temperature will be only $10^{-2}\,^{\circ}$C. Duration of the OA signal corresponds to the characteristic size of absorbing inhomogeneity, a: $\tau \sim a/c_0$, where c_0 is the speed of sound. Spatial resolution of an OA imaging system, δX, is first determined by the frequency bandwidth of the OA signal detectors Δf:

$$\delta X \sim \frac{c_0}{\Delta f}. \tag{12.1}$$

Thus, OA tomography of biological tissues combines the advantages of ODT and conventional ultrasound imaging. On one hand, the contrast of the OA image is determined by the contrast in optical absorption coefficient between the inhomogeneity and the surrounding tissue, which, for example, in case of a malignant tumor can reach up to 200% [7]. On the other hand, it is possible to achieve spatial resolution as high as that in conventional ultrasound (on the order of 0.1–10 mm depending on the frequency bandwidth), whereas imaging depth can reach several centimeters.

Detection of the OA signals in biomedical applications is performed using two major techniques: optical interferometry and piezo-detection. In optical interferometry the Fabry-Perot probe is usually employed: a thin polymer film having reflective coatings at both sides is imposed onto glass substrate [8]. Biological tissue under investigation, in which OA signals are excited by pulsed laser radiation, is brought into acoustic contact with the interferometer, as shown in Figure 12.1(a) [9]. The acoustic wave incident on the interferometer causes local changes in the polymer film thickness and, therefore, changes in the interference pattern. The corresponding changes in the output laser beam intensity are detected by the photodetector. The change of detection site is performed by scanning the laser beam over the interferometer surface.

Frequency bandwidth of an optical interferometry system and its sensitivity mainly depend on the materials the Fabry-Perot probe is made of and on the thickness of the polymer film. For example, a 10 μm film corresponds to the frequency bandwidth of 100 MHz [10]. The sensitivity of the system is determined by the photodetector sensitivity S, the Q-factor of the interferometer F and its frequency bandwidth Δf [11]:

$$\delta L_{\min} \sim \frac{1}{F}\sqrt{\frac{\Delta f}{S}}, \tag{12.2}$$

where $\delta L_{\min}$ is the minimally detectable displacement of the film surface. As seen, in design of an interferometer the compromise has to be found between the interferometer sensitivity and the bandwidth that determines the in-depth spatial resolution. For example, the 25 MHz frequency bandwidth, providing the in-depth spatial resolution of about 40 μm, corresponds to the minimally detectable pressure of 0.3 kPa [9]. The resolution in the lateral direction depends on the width of the scanning laser beam and can be made as small as 10–100 microns.

Optical detection of the OA signals is preferable when the required spatial resolution is on the order of 10–100 microns, and the imaging depth does not exceed 5 mm. In this case the sensitivity of the interferometer exceeds the sensitivity of the piezodetector that can provide the similar spatial resolution [12].

Piezo-detection of the ultrasonic signals has been employed in conventional ultrasound imaging for a long time. The main advantage of piezo-detection over optical interferometry method is higher sensitivity of the detectors with the bandwidth not exceeding 20 MHz. Minimally detectable pressure in this case is on the order of 1–10 Pa. Therefore, the use of piezodetectors is preferable if the required imaging depth is over a centimeter, and the spatial resolution is over hundreds of microns.

Main principles of piezodetector arrays design are known for many years from conventional ultrasound and, to some extent, are valid in OA tomography. However, the major difference of these two tomographic methods is that conventional ultrasound scanners usually operate in pulse-echo mode, and have a possibility of dynamic focusing of the transmitted beam. In OA tomography the laser-induced ultrasound sources are located within the medium under study, and their directivity pattern is not known in advance. The algorithm of OA image reconstruction depends on the design of the receiving transducer array and on the requirements of the particular application. The principles of 2D and 3D OA image reconstruction and the related biomedical applications will be considered in the next sections of this chapter.

12.2 Image Reconstruction in OA Tomography

Before considering image reconstruction algorithms let us formulate the direct and inverse problems of 3D OA tomography. Direct problem of OA tomography is calculation of the pressure field, $p(\vec{r},t)$ excited by laser-induced heat release distribution, $Q(\vec{r})$. This problem can be described by a homogeneous wave equation [1]:

$$\frac{\partial^2 p}{\partial t^2} - c_0^2 \Delta p = \frac{\partial}{\partial t} p_0(\vec{r})\,\delta(t); \quad p_0(\vec{r}) = \Gamma Q(\vec{r}), \tag{12.3}$$

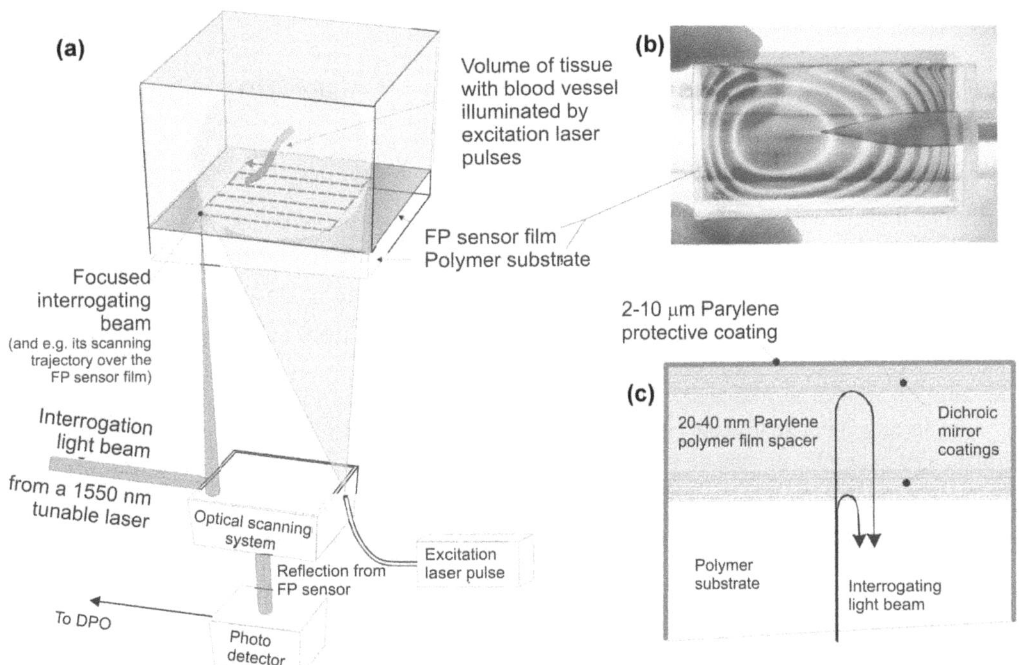

FIGURE 12.1: (*a*) Schematic of the backward mode OA imaging system using optical detection of the OA signals. (*b*) Photograph of the Fabry-Perot sensor. (*c*) Schematic of the Fabry-Perot sensor. © SPIE [9].

where Γ is the efficiency of OA transformation in the medium. The solution of the problem (12.3) can be expressed as [1]:

$$p(\vec{r},t) = \frac{\Gamma}{2\pi c_0^2} \int\limits_{V'} \frac{\frac{\partial}{\partial t}\left[Q(\vec{r}')\ \delta\left(t - \frac{|\vec{r}-\vec{r}'|}{c_0}\right)\right]}{|\vec{r}-\vec{r}'|} dV', \tag{12.4}$$

where OA sources are distributed over the volume V'.

The inverse problem of OA tomography is the reconstruction of the initial pressure distribution, $p_0(\vec{r})$ (or $Q(\vec{r})$ if the value of Γ for the medium is known) from the ultrasonic signals measured, ideally, at each point of a detection surface S. As shown in a number of recent works [13,14], this problem has a unique solution, and it was written in integral form for the three most widespread detection surfaces, namely, a confined sphere, an infinite cylinder and an infinite plane. This problem can be treated in either time- or spatial- frequency domain.

12.2.1 Solution of the inverse problem of OA tomography in spatial-frequency domain

First assume that the detection surface is an infinite plane $z = 0$, and the pressure waveform $p(x,y,z=0,t)$ is detected at each point of the surface. The idea of the method is that the spatial-frequency spectrum of the initial pressure distribution, $p_0(\vec{r})$, and that of the detected signals are related by a certain transform. In case of a plane detection surface it is a cosine transform [14]:

$$P\left(k_x, k_y, \sqrt{(\omega/c_0)^2 - k_x^2 - k_y^2}\right) = \frac{\sqrt{\omega^2 - c_0^2\left(k_x^2 + k_y^2\right)}}{\omega/c_0} \int\limits_0^{\infty} A\left(k_x, k_y, t\right)\cos\left(\omega t\right) dt, \qquad (12.5)$$

where $A\left(k_x, k_y, t\right) = \iint\limits_S p(x, y, z=0, t) e^{-i(k_x x + k_y y)} dxdy$ is the spatial Fourier-spectrum of the detected pressure distribution with the frequencies k_x and k_y, $P\left(k_x, k_y, k_z\right)$ is the three-dimensional Fourier-transform of the initial pressure distribution, $p_0(\vec{r})$, in which the spatial frequencies satisfy the following dispersion relationship: $\omega = c_0\sqrt{k_x^2 + k_y^2 + k_z^2}$, where ω is the frequency. Finally, the three-dimensional Fourier transform of the function P, obtained from (12.5), yields the pressure distribution $p(\vec{r}, t)$ at any time instant, including $t = 0$.

The solution in spatial frequency domain is most effective when the detection surface is plane [15], since in any other detection geometry the spatial spectra are related by a more complicated transform. For example, in case of a cylindrical detection surface it is a Hankel transform.

12.2.2 Solution of the inverse problem of OA tomography in time domain

Let us now consider the main method of the inverse problem solution in time domain. This method is based on the fact that the homogenous wave equation is invariant with respect to time reversal, $t \to -t$ [13] and is often referred to as back propagation method. Assume that the OA source is surrounded by a confined surface S (Figure 12.2(a)), and the emitted OA signal, $p(\vec{r}_S, t)$, is detected at each point of that surface. If the surface S is now represented as a transmitter of the detected signal, reversed in time, $p_{tr}(\vec{r}, t) = p(\vec{r}_S, -t)$, then the resulting acoustic wave will propagate towards the transducer and at the time instant $t = 0$ it will replicate the initial pressure distribution, $p_0(\vec{r})$. Propagation of the time-reversed acoustic wave can be described by the Kirchhoff-Helmholtz integral that relates the acoustic field, radiated by a surface, with the distribution of normal velocity and pressure over that surface. In case of a plane radiating surface the Kirchhoff-Helmholtz integral can be expressed as [16]:

$$p_{tr}(\vec{r}, \omega) = 2\int p(\vec{r}_S, \omega)\frac{\partial G\left(k\left|\vec{r} - \vec{r}_S\right|\right)}{\partial \vec{n}'} dS \qquad (12.6)$$

where $p(\vec{r}_S, \omega)$ is the frequency spectrum of a detected acoustic signal,

$$G\left(k\left|\vec{r} - \vec{r}_S\right|\right) = \frac{e^{ik\left|\vec{r} - \vec{r}_S\right|}}{4\pi\left|\vec{r} - \vec{r}_S\right|}$$

is the Green's function for the free space, $k = \omega/c_0$ is the wave number, $\vec{n}$ is the internal normal to the surface S, dS is an element of the surface S. Strictly speaking, the expression (12.6) is only applicable to flat radiating surfaces, but as an approximation is used for the focused surfaces as well. Physically, this approximation implies that each point of the surface is considered as a source of a spherical wave, however, the reflections of this wave from the surface are not taken into account. In case of continuous ultrasound radiation the superposition of reflected waves may lead to caustic formation, therefore, the discrepancy between the exact solution for the acoustic field and the approximation (12.6) may be significant [16]. However, if the pressure signals are relatively short pulses, like OA signals, caustics cannot be formed and the solution (12.6) coincides with the exact one [17, 18].

Expression (12.6) can be rewritten in time domain:

$$p_{tr}(\vec{r}, t) = -\frac{1}{4\pi^2}\int\left[\frac{1}{cR}\frac{\partial p\left(\vec{r}_S, t - \frac{R}{c}\right)}{\partial t} + \frac{1}{R^2}p\left(\vec{r}_S, t - \frac{R}{c}\right)\right]\left(\vec{e}_R\vec{n}\right) dS, \qquad (12.7)$$

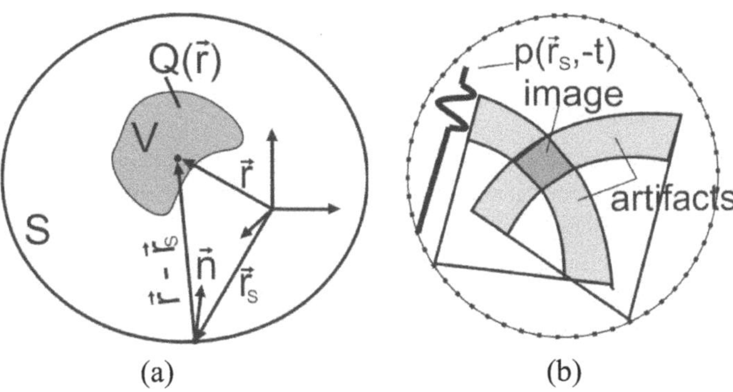

FIGURE 12.2: (*a*) The relative arrangement of the OA source – the distribution $Q(\vec{r})$, and the detection surface S. (*b*) Illustration of the back propagation method of OA image reconstruction.

where $R = |\vec{r} - \vec{r}_S|$, $\vec{e}_R = (\vec{r} - \vec{r}_S)/R$. The solution (12.7) is valid for cylindrical, spherical and, to a constant factor – for the plane surface. For the high frequency components of the OA signal spectrum, i.e., $k|\vec{r} - \vec{r}_S| \gg 1$, the second term of the integrand is small compared to the first one and can be neglected. The resulting expression, $p_{tr}(\vec{r},t) \sim -\int \frac{1}{cR} \frac{\partial p\left(\vec{r}_S, t - \frac{R}{c}\right)}{\partial t} (\vec{e}_R \vec{n})\, dS$, corresponds to the solution, derived empirically in OA tomography a long time ago [19].

12.2.3 Possible image artifacts

All the above-mentioned solutions were derived under assumption of an ideal detection surface, i.e., for the case when each point of the surface is considered as a detector with an infinite bandwidth, and the surface itself is confined or infinite in case of a plane. In reality, however, none of these assumptions holds, which gives rise to the so-called image artifacts or distortions. The nature of these artifacts will be different in cases of temporal and spectral approaches for image reconstruction.

If the spatial frequency domain is used for image reconstruction, in case of a plane detection surface, the distance between the centers of the two neighboring detectors determines the highest spatial frequency, i.e., the highest spatial resolution in the transverse direction [20]. The entire aperture of the array determines the lowest spatial frequency. Due to this limitation, the obtained OA image may contain wave-like artifacts. If the number of the detectors is not large enough, it is difficult to determine between the actual image of the object and its "wave-like" repetitions.

Image reconstruction algorithm in case of the time-domain approach is more obvious. According to Eq. (12.7), each detector is considered as a source of a spherical wave with the temporal profile that corresponds to the temporal derivative of the detected signal reversed in time (see Figure 12.2(b)). Superposition of the spherical wavefronts forms the image that corresponds to the initial laser-induced pressure distribution. However, the image takes up only part of each spherical wavefront, and the remaining part of the wavefront is an arch-like artifact. The ratio of the image brightness to the artifact brightness is proportional to the number of detectors.

If the number of the detectors is large ($\sim$100), the use of the spatial frequency approach results in less-pronounced artifacts than the use of a time-domain algorithm [20]. With the smaller number of the detectors the situation is reversed. As for the computational efficiency, the spatial frequency algorithm is more rapid in case of a plane detection surface due to the use of fast Fourier transform. This advantage is lost in the cases of spherical or cylindrical detection surfaces, because more complex transforms have to be used [15].

12.3 3D OA Tomography

Within the last ten years a number of designs of receiving array transducers for OA tomography were proposed. These designs can be separated into two groups for convenience: plane arrays (the array elements are distributed over a plane surface) [8, 21] and spherical (the array elements are distributed over the spherical surface) [19]. Due to the high complexity of the fabrication of such transducers and a requirement for rapid data acquisition, very few groups use 3D OA tomography in biomedical research. The major applications here are breast cancer detection [22] and imaging of the vasculature of small animals, for instance, rats that are often used as models in biomedical research [9].

In the first OA tomography system for breast cancer diagnostics a 64-element array transducer was used for OA signal detection. The elements were made of piezo-ceramics and were distributed over a spiral curve along a hemisphere with 30-cm diameter, into which the object under study was placed [22]. The size of each element was 13 mm. The OA signals were excited by RF radiation with the frequency 434 MHz, emitted by the antennae built into the bottom of the hemisphere. Due to the low attenuation of the RF waves in tissue it is possible to heat a large tissue volume instantaneously. For obtaining full diagnostic information the array transducer was rotated around its axis that resulted in more than 2000 detection points. Phantom experiments have shown that the spatial resolution of an obtained image was 1–4 mm depending on the position of the OA source and corresponded to the angular aperture of the detectors relative to the source.

The tomography system described above was applied *in vivo* for diagnostics of breast tumors in several dozens of patients [22]. The results of these studies have shown that, in spite of the high promise of this system, the obtained images contain serious errors due to the interference of the RF waves in tissue. Moreover, the contrast of RF absorption in the tumor and healthy tissue turned out quite low. Using the same array with laser excitation of the OA signals would be inefficient, because the attenuation of laser radiation in tissue is much higher, therefore, the heated volume is much smaller, and the effective number of the OA signal detection points is reduced. It is also worth mentioning here that due to mechanical scanning of the transducer array image acquisition time exceeded several hours that is unacceptable in regular clinical practice.

The authors of [21, 23] designed a flat, square array consisting of 590 2×2 mm detectors to address the same medical problem. The array elements were made of piezo-polymer polyvinyldenfluoride (PVDF) film imposed onto backing material. During the OA testing the breast was compressed slightly between the array surface and a glass plate, through which laser irradiation was performed (Figure12.3). Spatial resolution provided by the system, corresponded to the size of a single array element and, depending on the OA source position above the detection plane, was 2–4 mm. The results of the first clinical trials of this system were recently published [23].

An example of the OA mammogram obtained *in vivo* in [23] is shown in Figure 12.4(c). This case refers to a Caucasian woman of 57 years with a palpable mass in the right breast that was pathologically verified after surgery as a ductal carcinoma. The tumor appeared on the X-ray image (Figure 12.4(a)) as a dense oval mass (40 mm on the major axis) with a number of microcalcifications of various size distributed regionally. The breast sonogram [Figure 12.4(b)] reveals a uniformly hypoechoic tumor with well-defined margins (major axis 32 mm) with a few anechoic regions suggestive of cystic features. Although the overall appearance of the tumor in both X-ray and ultrasound images indicated benignity, the presence of heterogeneous calcifications and the age of the patient prompted the radiologist to perform a core biopsy.

In the OA slice images (Figure 12.4(c)) a pronounced ring-shaped structure of higher intensity and, therefore, higher optical absorption is seen. This pattern depicts vascular distribution in the tumor; the periphery of the tumor corresponds to the cancer invasion front with rapid cell proliferation and angiogenesis. On the contrary, the interior of the tumor suffers impaired vascular function due

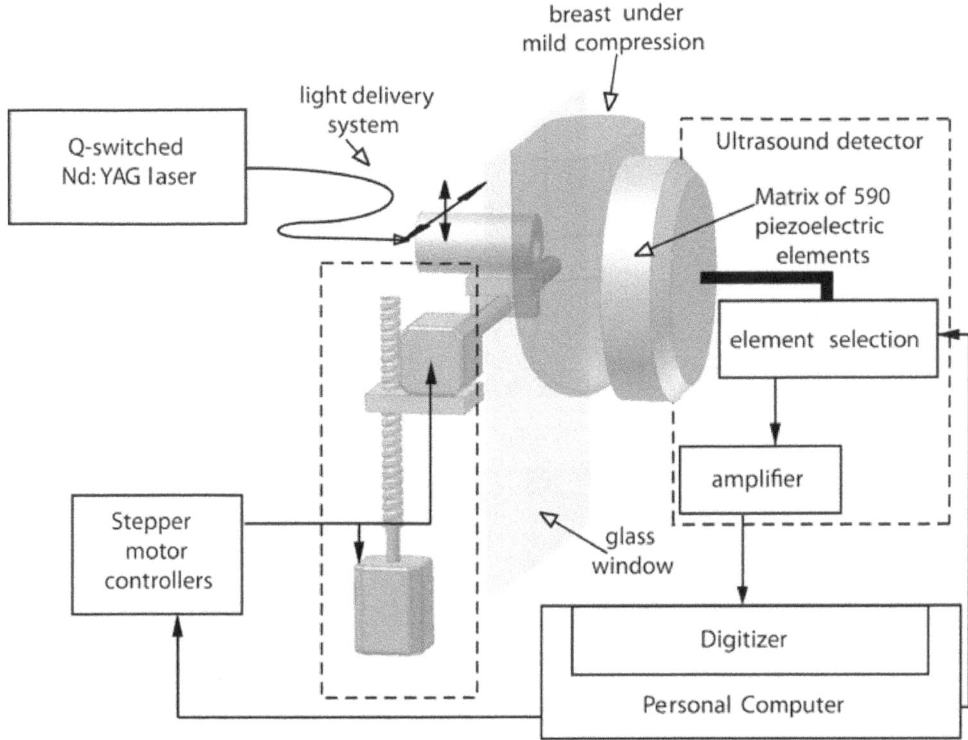

FIGURE 12.3: Schematic of the OA imaging system for breast cancer diagnostics. © IOP [24].

to mechanical forces exerted by proliferating peripheral cells. The outer diameter of the ring structure (30 mm) is close to the sonographically determined lesion size and the pathologically estimated carcinoma size (26 mm). The average image contrast was calculated to be 1.55.

The major drawback of this OA imaging system is a small size and therefore small electric capacity of the array elements and, therefore, high noise level. Minimally detectable pressure in case of these transducers was about 100 Pa, that only allowed to image large (1–2 cm) tumors, that can also be detected by other means.

In the most successful design of the OA tomography system for imaging of small animals vasculature detection of the OA signals was performed by a Fabry-Perot interferometer [9], that, strictly speaking, was not a transducer array: the detection point was changed by fast two-dimensional scanning of the narrow probe laser beam over the interferometer surface (see Fig. 12.1). To evaluate the system, phantom experiments were performed. 3D OA images of the two phantoms shown in Fig. 12.5 were reconstructed from the data acquired over $\sim$10000 spatial points without signal averaging. These studies suggest that the imaging system is capable of obtaining 3D OA images of the superficial blood vessels with spatial resolution of 50 μm. Image acquisition time in this case was 15 minutes.

An important advantage of this tomography system is its ability to work in backward mode that allows for one-sided access to the object under study. For example, vasculature of mouse brain *in vivo* with skin and skull intact were obtained using this system (Figure 12.6). In order to achieve this, the interferometer mirrors were made transparent for the wavelength of pulsed laser radiation that excited the OA signal in tissue, and reflective for the interferometer probe beam.

To summarize this section it is worth mentioning that the largest difficulty in designing transducer arrays for 3D OA tomography is to achieve the required spatial resolution without sacrificing the

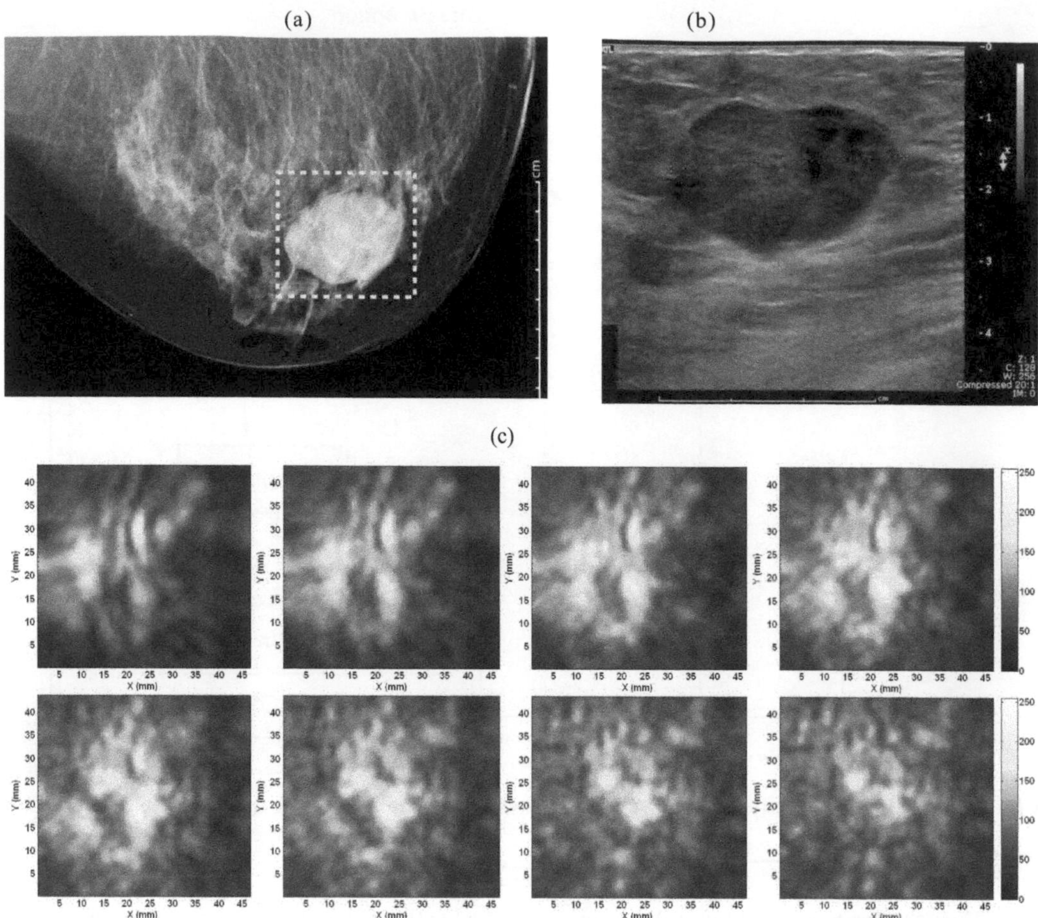

FIGURE 12.4: (See in color after page 572.) X-ray mammography (*a*), ultrasound (*b*), and OA slice images (*c*) of a 57-year-old woman with invasive ductal carcinoma exhibiting neuroendocrine differentiation in right breast (Case 2) [23]. The overall appearance of the tumor in (*a*) and (*b*) indicated benignity. The OA slice images shows high intensity distribution in a ring shape that can be attributed to higher vascular densities at tumor periphery. The interslice spacing is 1 mm with the first slice 9.5 mm below the illuminated breast surface. © OSA [23].

signal-to-noise ratio. Indeed, if the size of a single array element is reduced, the spatial resolution is enhanced, however, due to the decrease in electric capacity of the element the noise level is increased.

12.4 2D OA Tomography

12.4.1 Transducer arrays for 2D OA tomography

2D imaging is often preferable over 3D tomography in the *in vivo* setting, because data acquisition and image reconstruction can be performed in real time. 2D image is a section of the 3D distribution

FIGURE 12.5: (a) Phantom consisting of 300 μm bore silicone rubber tube filled with NIR dye ($\mu_a = 27$ cm^{-1}) tied with human hair immersed to a depth of ~2 mm in a solution of 1.5% Intralipid ($\mu_s' = 10$ cm^{-1}), and its 3D OA image (image volume: 6 mm × 4 mm × 3 mm). Excitation wavelength: 1064 nm; repetition rate: 20 Hz; incident fluence: 15 mJ/cm^2. (b) Phantom submerged in 1.5% Intralipid ($\mu_s' = 1$mm^{-1}) and consisting of polymer tubes filled with human blood (15.2 g/dL), and its 3D OA image (14 mm × 14 mm × 6 mm). Excitation source: 10 Hz OPO pulsed laser at 800 nm. Incident fluence: 6.7 mJ/cm^2. © SPIE [9].

of laser-induced heat sources within the imaging plane. The section thickness corresponds to the resolution in direction, perpendicular to the imaging plane and is determined by the array design.

Figure 12.7(a) shows a diagram of the array transducer that was used for 2D imaging of breast tumors [25]: 32 linear PVDF piezodetectors with 110 mm thickness were imposed onto cylindrically focused surface of backing material. Spatial resolution provided by this system within the imaging plane was determined in phantom experiments: in-depth 0.4 mm, lateral 1–2 mm. Minimally detectable pressure was ~13 Pa. This imaging system was first tested in phantom experiments [26] (Figure 12.7(b),(c)) and then was used in clinical trials for breast tumor imaging. The X-ray mammogram of one of the cases from the trial is presented in Figure 12.8(a): it shows a single large, bright mass that was later proved to be a malignant tumor. Figure 12.8(b),(c) shows mediolateral and craniocaudal OA images of that breast, correspondingly. These were obtained using 755 nm laser illumination. Figure 12.8(b) also shows a single large mass seen in the X-ray image. However, in the craniocaudal section (Figure 12.8(c)) the large mass is reduced in intensity and a second bright object shows up underneath. This bright appearance of the secondary tumor in the craniocaudal section demonstrates that such an OA imaging system can be considered as a useful complement to X-ray mammography setups.

Despite the promising results of the clinical trials, the system developed in [25, 26] suffered an important drawback: a relatively low resolution in a direction perpendicular to the imaging plane due to spatial averaging of the ultrasound wavefront over the surface of each piezodetector. This

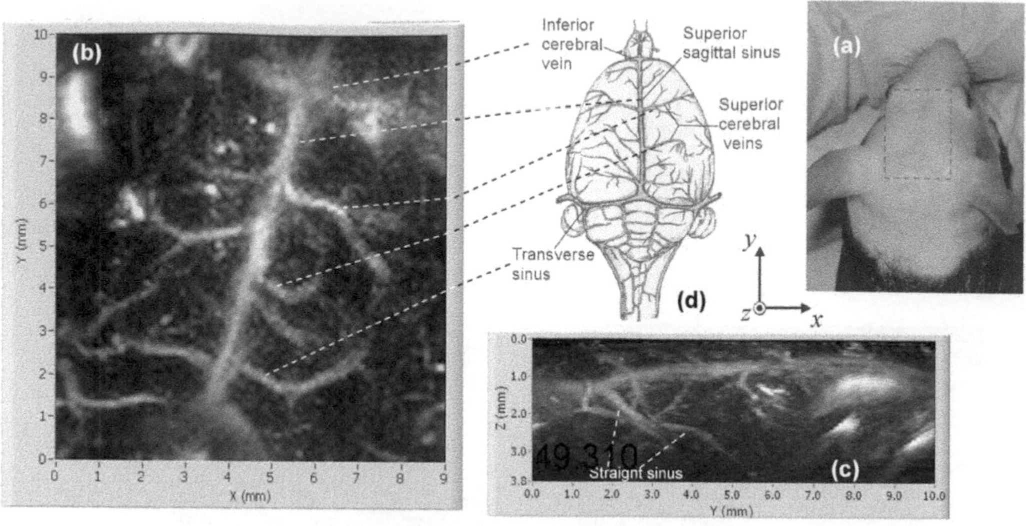

FIGURE 12.6: 3D OA imaging of the vascular anatomy in murine brain. (a) Region over which the OA signals were recorded. The maximum intensity projections (MIP) of the reconstructed 3D OA image of the mouse brain on (b) the lateral plane (x-y plane) and, (c) the vertical y-z plane. (d) A schematic of the anatomy of rodent outer brain circulation system. © SPIE [9].

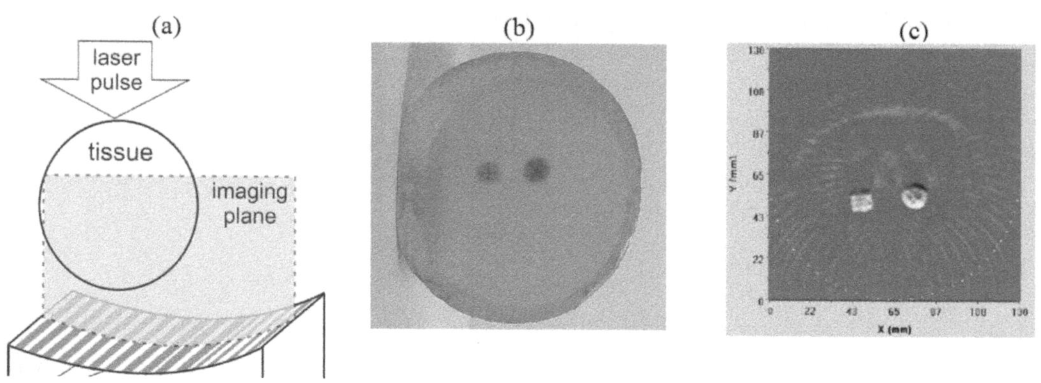

FIGURE 12.7: (a) 2D OA imaging system for imaging of tumors in soft tissues. (b) Plastisol phantom for system testing. The phantom contained a cube and a cylinder (disk) that absorbed preferentially at 757 nm. In addition, each contained two crossed pieces of 0.5-mm pencil lead. (c) OA image of the phantom obtained with laser illumination wavelength of 755 nm. The image shows both the cube and the cylinder with barely discernible lines for the crossed pencil leads. © SPIE [26].

resulted in 15–20 mm resolution in a direction perpendicular to the imaging plane that did not allow obtaining images of tumors less than 10–15 mm in size. However, it is known that mortality rate increases dramatically when the tumor reaches the size of 10 mm [27], therefore, early diagnostics of this disease is crucial.

To solve the problem of poor resolution in direction perpendicular to the imaging plane, the authors of [28] have proposed to use arrays consisting of focused piezodetectors (Figure 12.9). The sensitivity of a single focused piezodetector, the array element, is localized in a narrow elongated

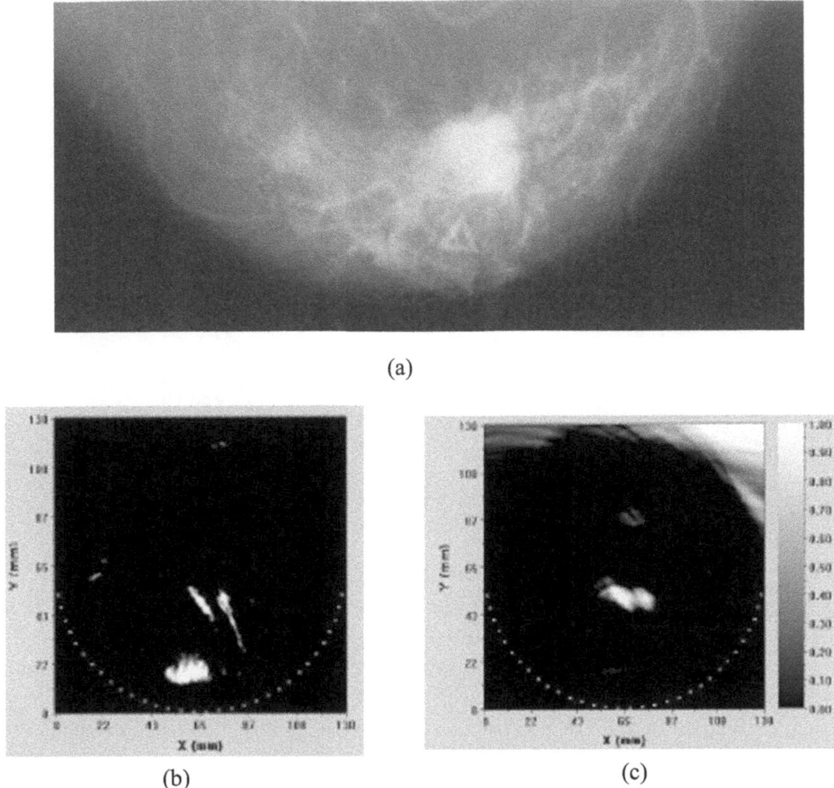

FIGURE 12.8: (a) X-ray mammography image of the cancerous right breast of a patient showing a large mass suspicious for malignancy. Additional tumor was not visible on the mammogram, but was later discovered in the area marked with a triangle. (b), (c) Mediolateral and craniocaudal OA images of the breast obtained with laser illumination at the wavelength of 757 nm, which highlights tissue regions containing hypoxic blood. The additional tumor was revealed beneath the primary lesion. © SPIE [26].

region, therefore, the sensitivity of the entire array is restricted to within the imaging plane. The resolution in direction perpendicular to the imaging plane, δY, corresponds to the width of the focal region and is first determined by the focusing angle of the array element, θ, and its frequency bandwidth, Δf [29]:

$$\delta Y = \frac{0.5\, c_0}{f_0 \sin(\theta/2)}, \tag{12.8}$$

where c_0 is the sound velocity. The increase of the focusing angle of a piezodetector increases its surface and, therefore, its electric capacity; at the same time, as seen from Eq. (12.8), enhances the resolution δY.

An array transducer consisting of 64 such focused piezoelements was used in [30] for phantom experiments aiming at visualization of an absorbing object in scattering medium. The absorber was a 2 mm piece of bovine liver ($\mu_a = 0.42$ cm^{-1}), and the scattering medium was milk diluted with water ($\mu_a = 0.18$ cm^{-1}, $\mu_s' = 1.85$ cm^{-1}). The phantom media were modeling healthy breast tissue containing a small tumor. The absorber was successfully visualized when located at a depth of 4 cm below the laser irradiated surface of scattering medium. The results of these studies suggest that OA tomography is applicable for diagnostics of breast cancer at early stages.

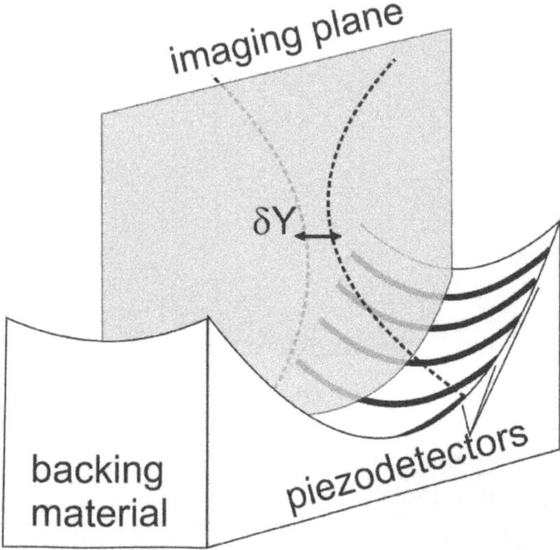

FIGURE 12.9: Array transducer for 2D OA tomography consisting of focused piezodetectors. The sensitivity of each detector is localized within the narrow elongated focal region, therefore, the sensitivity of the whole array is localized within the imaging plane. The lateral size of the focal zone of each detector corresponds to the resolution provided by the array in a direction perpendicular to the imaging plane, δY.

12.4.2 Image reconstruction in 2D OA tomography

The most widespread image reconstruction algorithm in 2D OA tomography is a backprojection algorithm based on the time-reversal method discussed previously. According to Eq. (12.7), the waveform used for backprojection is the first temporal derivative of time-reversed detected signals with an inverted sign. This waveform is projected onto the imaging plane according to the arrival time of the signal to the detector (Figure 12.2(b)), and the so-called arcs of probabilities are plotted that intersect, thus forming an image. However, unlike the case of 3D tomography, the detection of the OA signals is performed by a transducer array, in which the elements are distributed along a line, not a surface. Consequently, the OA wavefront is not fully detected, and the diagnostic information it contains may be partly lost. Therefore, a 2D OA image may not be a correct representation of the heat source distribution in the imaging plane. Examples that illustrate this argument will be considered below.

First consider a uniformly heated cylinder as the OA source and assume that the receiving array transducer consists of point-like elements distributed with a 2-mm period over a circle with a 60-mm radius as in Figure 12.10(a). The output signals of the array elements were calculated numerically using the approach developed in [28], and the OA image was obtained based on these signals. The frequency bandwidth of each element was considered 3 MHz. The length of the OA source, the cylinder, was 60 mm, its diameter – 3 mm. The cylinder axis was perpendicular to the imaging plane, its center coincided with the center of the receiving array, as seen in Fig 12.10(a).

Figure 12.10(b),(c) show the obtained OA image and its cross-section along the white dotted line, correspondingly. As one can see, the OA image contains areas of both positive and negative brightness, whereas the initial heat release distribution is entirely positive. To illustrate the explanation of this phenomenon let us consider the output signal of one of the array elements and its temporal derivative with an inverted sign that was used for image reconstruction (Figure 12.11 (a), (b)). The waveform $-\frac{\partial p(t)}{\partial t}$ has areas of negative values, therefore, areas with negative brightness are also present in the image. However, the superposition of the spherical wavefronts in the backprojection

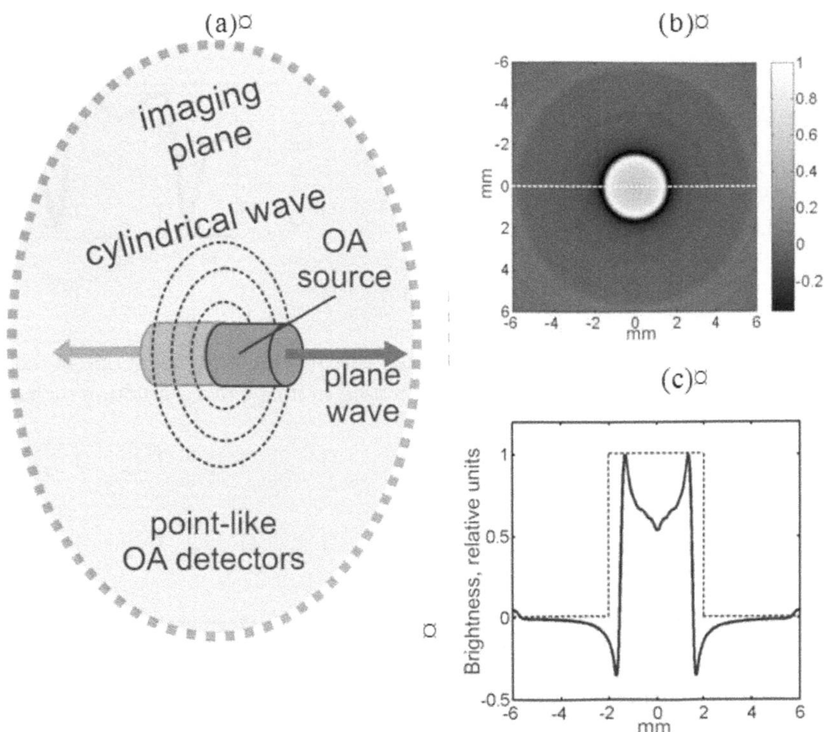

FIGURE 12.10: (a) Relative arrangement of the array of point-like ultrasound detectors and the OA source (the uniformly heated cylinder) used for numerical modeling of the 2D OA image (b). (c) Cross-section of the OA image along the white dotted line (solid line) and of the initial heat release distribution (dashed line). The reconstructed image contains both positive and negative values that result from the insufficient tomographic data.

algorithm results in the interference of the positive and negative areas of the function $-\frac{\partial p(t)}{\partial t}$ from different detectors. Consequently, the relative amplitude of the negative brightness is decreased, as seen from comparison of Figures 12.11(b) and 12.10(c). However, full compensation of the negative brightness can not be achieved in 2D tomography.

The area of positive brightness in Figure 12.10(b) corresponds to the image of the cylinder. The position of its boundary is reproduced correctly within the inaccuracy induced by the limited frequency bandwidth of the detectors. The initial heat release distribution within the cylinder was considered uniform, however, the OA image contains a downward excursion at its center. This phenomenon can be explained by considering the wavefront of the OA signal excited by the cylinder: it is a combination of a cylindrical wave, propagating in a radial direction, i.e., towards the array detectors, and two plane waves propagating along the cylinder axis in two opposite directions (Figure 12.10(a)). The array elements only detect the cylindrical part of the wavefront, therefore, the information contained in the plane waves is lost.

Figure 12.12 shows the cross-section of a 3D OA image of the similar OA source, the uniformly heated cylinder with 2 mm diameter, reconstructed using backprojection algorithm. In this case the detectors were distributed not along a circle, but along the surface of the confined sphere with the same dimensions. As seen from the image, all the negative brightness values are fully compensated, and the distribution within the reconstructed image is uniform, as expected. Thus, correct reconstruction of OA images is possible in 3D case, but not in 2D tomography.

(a)

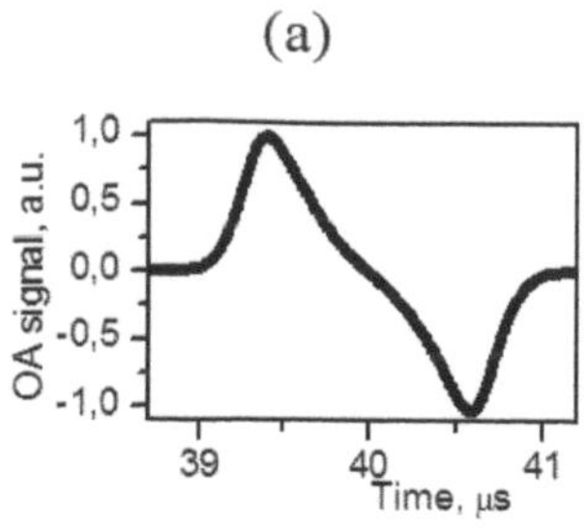

(b)

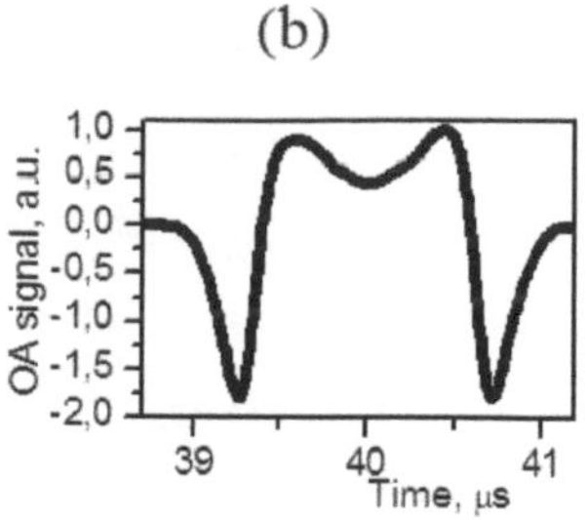

FIGURE 12.11: (a) An example of a numerically calculated OA signal at one of the detectors of the array shown in Figure 12.10. (b) The waveform used in image reconstruction by backprojection algorithm.

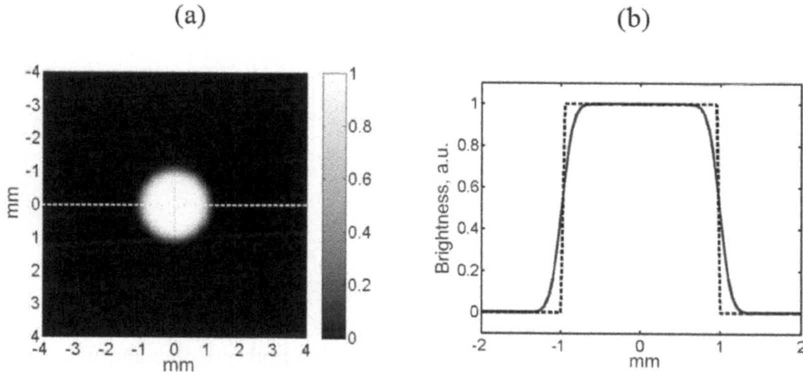

FIGURE 12.12: (a) A cross-section of a numerically calculated 3D OA image of an OA source similar to that shown in Figure 12. 9(a). In this case, unlike the arrangement in Figure 12.9(a), the detectors were considered to be distributed over a spherical surface, not a circular line. The brightness values in the image are in this case entirely positive. (b) The cross-section of the image along the white dotted line (solid line) and of the initial heat release distribution (dashed line).

12.5 Conclusions

The goal of the present paper was to consider the recent achievements and challenges in biomedical applications of OA tomography. Today, the systems for OA signal detection and image reconstruction methods have already achieved the necessary level in order to be employed in specific *in vivo* applications such as diagnostics of soft tissue tumors and lesions in small animals and in humans. However, very few cases of *in vivo* OA diagnostics have been reported so far. Most probably 2D OA imaging will be more feasible in a clinical setting than 3D tomography, because it can be performed in real time although in the 2D case correct image reconstruction is problematic. Sufficient practice and a large literature database will be necessary to interpret such 2D images and extract useful and reliable diagnostic information from them.

Acknowledgment

This work was supported by the Russian Foundation for Basic Research grant 07-02-00940-a, and International Science and Technology Center grant 3691.

References

[1] V.E. Gusev, A.A. Karabutov, *Laser Optoacoustics*, AIP, New York (1993).

[2] F.A. Duck, *Physical Properties of Tissue: A Comprehensive Reference Book*, London, San Diego, N.-Y., Boston: Academic Press (1990).

[3] J. Folkman, "The role of angiogenesis in tumor growth," *Semin. Cancer Biol.* **3**, 65–71 (1992).

[4] M. Yamasaki., S. Sato, H. Ashida, D. Saito, Y. Okada, and M. Obara, "Measurement of burn depths in rats using multiwavelength photoacoustic depth profiling," *J. Biomed. Opt.* **10**, 064011-1–4 (2005).

[5] T.D. Khokhlova, I.M. Pelivanov, O.A. Sapozhnikov, V.S. Solomatin, and A.A. Karabutov, "Opto-acoustic diagnostics of the thermal action of high-intensity focused ultrasound on biological tissues: the possibility of its applications and model experiments," *Quant. Electron.* **36**, 1097–1102 (2006).

[6] A.P. Gibson, J.C. Hebden, and S.R. Arridge, "Recent advances in diffuse optical imaging," *Phys. Med. Biol.* **50**, R1–R43 (2005).

[7] R.L.P. van Veen, H.J.C.M. Sterenborg, A.W.K.S. Marinelli, and M. Menke-Pluymers, "Intraoperatively assessed optical properties of malignant and healthy breast tissue used to determine the optimum wavelength of contrast for optical mammography," *J. Biomed. Opt.* **9**, 129–136 (2004).

[8] E. Zhang and P. Beard, "Broadband ultrasound field mapping system using a wavelength tuned, optically scanned focused laser beam to address a Fabry-Perot polymer film sensor," *IEEE Tran. Ultrason., Ferroelectr. Freq. Control* **53**, 1330–1338 (2006).

[9] E.Z. Zhang, J. Laufer, and P. Beard, "Three dimensional photoacoustic imaging of vascular anatomy in small animals using an optical detection system," *Proc. SPIE* **6437**, 64370S-1–7 (2007).

[10] S. Ashkenazi, Y. Hou, T. Buma, and M. O'Donnell, "Optoacoustic imaging using thin polymer e-talon," *Appl. Phys. Lett.* **86**, 134102–134105 (2005).

[11] J.D. Hamilton, T. Buma, M. Spisar, and M. O'Donnell, "High frequency optoacoustic arrays using etalon detection," *IEEE Tran. Ultrason., Ferroelectr. Freq. Control* **47**, 160–169, (1999).

[12] A. Oraevsky and A. Karabutov, "Ultimate sensitivity of time-resolved opto-acoustic detection," *Proc. SPIE* , 1–12 (2000).

[13] M. Xu, Y. Xu, and L.V. Wang, "Time-domain reconstruction algorithms and numerical simulations for thermoacoustic tomography in various geometries," *IEEE Trans. On Biomed. Eng.* **50**, 1086–1099 (2003).

[14] K.P. Kostli, M. Frenz, H. Bebie, and H.P. Weber, "Temporal backward projection of optoacoustic pressure transients using Fourier transform methods," *Phys. Med. Biol.* **46**, 1863–1872 (2001).

[15] B.T. Cox, S.R. Arridge, and P.C. Beard, "Photoacoustic tomography with a limited-aperture planar sensor and a reverberant cavity," *Inverse Problems* **23**, S95–S112 (2007).

[16] O.A. Sapozhnikov an T.V. Sinilo, "Acoustic field produced by a concave radiating surface with allowance for the diffraction," *Acoust. Phys.* **48**, 720–727 (2002).

[17] O.A. Sapozhnikov, A.E. Ponomarev, and M.A. Smagin, "Transient acoustic holography for reconstructing the particle velocity of the surface of an acoustic transducer," *Acoust. Phys.* **52**, 324–330 (2006).

[18] P. Burgholzer, G.J. Matt, M. Haltmeier, and G. Paltauf, "Exact and approximative imaging methods for photoacoustic tomography using an arbitrary detection surface," *Phys. Rev. E* **75**, 046706-1–10 (2007).

[19] R.A. Kruger, D.R.. Reinecke, and G.A. Kruger, "Thermoacoustic computed tomography – technical considerations," *Med. Phys.* **26**, 1832–1842 (1999).

[20] K.P. Kostli and P.C. Beard, "Two-dimensional photoacoustic imaging by use of Fourier-transform image reconstruction and a detector with an anisotropic response," *Appl. Opt.* **42**, 1899–1908 (2003).

[21] S. Manohar, A. Kharine, J.C.G. van Hespen, W. Steenbergen, and T.G. van Leeuwen, "Photoacoustic mammography laboratory prototype: imaging of breast tissue phantoms," *J. Biomed. Opt.* **9**, 1172–1181 (2004).

[22] R.A. Kruger, W.L. Kiser, A.P. Romilly, and P. Schmidt, "Thermoacoustic CT of the breast: pilot study observations," *Proc. SPIE* **4256**, 1–5 (2001).

[23] S. Manohar, S.E. Vaartjes, J.C.G. van Hespen, J.M. Klaase, F.M. van den Engh, W. Steenbergen, and T.G. van Leeuwen, "Initial results of in vivo non-invasive cancer imaging in the human breast using near-infrared photoacoustics," *Opt. Express* **15**, 12277–12285 (2007).

[24] S. Manohar, A. Kharine, J.C.G. van Hespen, W. Steenbergen, and T.G. van Leeuwen, "The Twente photoacoustic mammoscope: system overview and performance," *Phys. Med. Biol.* **50**, 2543–2557 (2005).

[25] V.G. Andreev, A.A. Karabutov, S.V. Solomatin, E.V. Savateeva, V.L. Aleynikov, Y.V. Zhulin, R.D. Fleming, and A.A. Oraevsky, "Opto-acoustic tomography of breast cancer with arc-array transducer," *Proc. SPIE* **3916**, 36–47 (2003).

[26] T. Khamapirada, P.M. Henrichs, K. Mehta, T.G. Miller, A.T. Yee, and A.A. Oraevsky, "Diagnostic imaging of breast cancer with LOIS: clinical feasibility," *Proc. SPIE* **5697**, 35–44 (2005).

[27] P.M. Webb, M.C. Cummings, C.J. Bain, and C.M. Furnival, "Changes in survival after breast cancer: improvements in diagnosis or treatment?" *The Breast* **13**,7–14 (2004).

[28] V. Kozhushko, T. Khokhlova, A. Zharinov, I. Pelivanov, V. Solomatin, and A. Karabutov, "Focused array transducer for 2D optoacoustic tomography," *J. Acoust. Soc. Am.* **116**, 1498–1506 (2004).

[29] T.D. Khokhlova, I.M. Pelivanov, and A.A. Karabutov, "Optoacoustic tomography utilizing focused transducers: the resolution study," *Appl. Phys. Lett.* **92**, 024105-1–3 (2008).

[30] T.D. Khokhlova, I.M. Pelivanov, V.V. Kozhushko, A.N. Zharinov, V.S. Solomatin, and A.A. Karabutov, "Optoacoustic imaging of absorbing objects in a turbid medium: ultimate sensitivity and application to breast cancer diagnostics," *Appl. Opt.* **46**, 262–272 (2007).

13

Optical-Resolution Photoacoustic Microscopy for In Vivo Volumetric Microvascular Imaging in Intact Tissues

Song Hu, Konstantin Maslov, and Lihong V. Wang

Optical Imaging Laboratory, Department of Biomedical Engineering, Washington University in St. Louis, St. Louis, Missouri, 63130-4899, USA

Microcirculation, the distal functional unit of the cardiovascular system, provides an exchange site for gases, nutrients, metabolic wastes, and thermal energy between the blood and tissues. Pathologic microcirculation reflects the breakdown of homeostasis in organisms, which ultimately leads to tissue inviability. Thus, *in vivo* microvascular imaging is of significant physiological and pathophysiological importance. In this chapter, we report on optical-resolution photoacoustic microscopy (OR-PAM), a reflection-mode micrometer-resolution absorption microscopy recently developed in our laboratory. It permits *in vivo*, noninvasive, label-free, three-dimensional (3D) microvascular imaging down to single capillaries. Using spectroscopic measurements, OR-PAM can differentiate oxy- and deoxy-hemoglobin, allowing the quantification of total hemoglobin concentration (HbT) and hemoglobin oxygenation saturation (sO_2)—the valuable functional parameters involved in neural, hemodynamic, and metabolic activities.

Key words: photoacoustic microscopy, optical-resolution, label-free, microcirculation, capillary, hemoglobin concentration, hemoglobin oxygenation

13.1 Introduction

Cardiovascular disease and its impact on microcirculation are important causes of morbidity and mortality worldwide, accounting for 40% of all deaths [1]. Advances in microvascular imaging facilitate the fundamental understanding of microcirculatory physiology and pathophysiology, revealing the origin and progression of these diseases. Intravital microscopy (IVM) is currently the gold standard for microvascular imaging. With chamber window preparation, conventional optical microscopy can be adopted to quantify the vessel count, diameter, length, density, permeability, and blood flow within the targeted thin tissues. However, IVM generally involves invasiveness and lacks depth information—a crucial morphological parameter. Confocal microscopy and multi-photon microscopy have become powerful tools for noninvasive 3D microvascular imaging, thanks to the development of fluorescent labeling techniques. Nevertheless, for clinical applications that generally do not tolerate labeling, endogenous contrast mechanisms become necessary. Orthogonal polarization spectral (OPS) imaging, taking advantage of blood absorption contrast, enables non-invasive microvascular imaging without the use of fluorescent dyes, paving the way for its clinical use. However, like IVM, OPS provides no depth information.

In contrast, photoacoustic microscopy (PAM) can overcome these limitations, as it relies on endogenous optical absorption contrast and it works in reflection-mode noninvasively with time-resolved depth detection. More importantly, with multi-wavelength measurements, PAM can accurately assess hemoglobin oxygenation within each single vessel, providing a convenient and robust tool for functional imaging. PAM has drawn increasing attention in recent years; however, the spatial resolution in the original dark-field PAM design is limited by ultrasonic parameters [2], making it blind to capillaries as well as small arterioles and venules (<50 μm). In this chapter, we present optical-resolution photoacoustic microscopy (OR-PAM), our newly developed technique that can resolve single capillaries *in vivo* noninvasively.

13.2 Dark-Field PAM and Its Limitation in Spatial Resolution

The photoacoustic effect was first reported in 1880 by Alexander Graham Bell, and later was applied to biomedical imaging thanks to advances in ultrasonic transducers and lasers [3]. Photoacoustic imaging combines optical absorption contrast with ultrasonic detection. Endogenous optical absorption, for example oxy- and deoxy-hemoglobin absorption, is physiologically specific in general; ultrasonic detection enables better resolution in the optical quasi-diffusive or diffusive regime, because ultrasonic scattering per unit path length in biological tissues is 2–3 orders of magnitude weaker than optical scattering. Combining these two attractive features generates a high-resolution functional imaging tool that is able to extend our sight into the diffusive regime. In the dark-field PAM system shown in Figure 13.1(A), a pulsed laser beam delivered by a multimode optical fiber passes through a conical lens to form a ring-shaped illumination, and then is weakly focused into biological tissues, where the optical focus coaxially overlapped with the ultrasonic focus. Figure 13.1(B) shows a detailed schematic of the dark-field illumination and the acoustic-optical confocal configuration. The laser pulse energy deposited into the tissue is partially absorbed and converted into heat, which induces a local pressure rise via transient thermoelastic expansion. The pressure rise travels through the tissue in the form of a wideband ultrasonic wave—referred to as photoacoustic wave, and is detected by an ultrasonic transducer. In dark-field PAM, due to the ring-shaped optical illumination, the photoacoustic signal in the detectable surface area is greatly minimized. As shown in Figure 13.1(B), the tight ultrasonic focus is within the loose optical focus. Thus, the lateral resolution exclusively depends on the ultrasonic parameters, including the center frequency and the numerical aperture of the ultrasonic transducer. Based on the full width at half-maximum (FWHM) of the ultrasonic amplitude receiving (one-way) response function, the lateral resolution is given by

$$\Delta r_R = 0.71\lambda_0/\mathrm{NA}\,, \tag{13.1}$$

where λ_0 and NA denote the center wavelength of the photoacoustic pulse and the numerical aper-

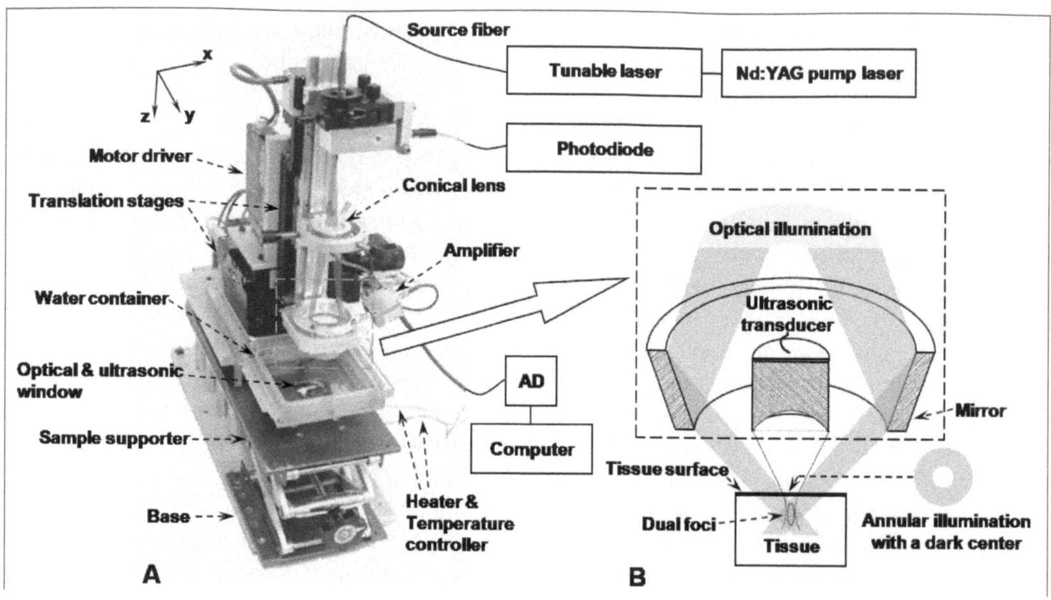

FIGURE 13.1: Schematic of a dark-field PAM system. (A) Photograph of the imaging system. (B) Enlarged sketch demonstrating the formation of the ring-shape optical illumination on the tissue surface and the optical-ultrasonic dual foci inside the tissue [2].

ture of the ultrasonic transducer, respectively. In our recently reported work on dark-field PAM [2], a lateral resolution of $\sim$45 μm was achieved with a center frequency of 50 MHz and an NA of 0.44. This resolution is adequate for many biomedical applications. However, to resolve fine structures such as capillaries, which are 4–9 μm in diameter, higher spatial resolution is desired.

If the lateral resolution is increased acoustically, according to Equation 13.1, an ultrasonic transducer with a center frequency of greater than 400 MHz is required in order to resolve single capillaries. Nevertheless, at such a high frequency, the ultrasonic attenuation ($\sim$35 dB/mm in water and $\sim$100 dB/mm in tissue) limits the penetration depth to less than 100 μm, which is even shallower than the penetration of most optical microscopic techniques. This would limit its applications to the anatomical sites that are visible and close to the tissue surface.

13.3 Resolution Improvement in PAM by Using Diffraction-Limited Optical Focusing

As mentioned above, optical illumination and ultrasonic detection are configured coaxially and confocally in PAM. In this case, the lateral resolution is determined by the product of the two point spread functions (PSF) of the optical illumination and acoustic detection. Since lateral resolution is difficult to push further acoustically, a more feasible alternative is to use fine optical focusing to provide the lateral resolution, while still deriving the axial resolution from time-resolved ultrasonic detection. As shown in Figure 13.2(B), within the acoustic focus, only an optically diffraction-limited spot is illuminated by the focused laser beam, so the detected photoacoustic signal is exclusively generated within the optical focus. Using this technique, referred to as OR-PAM, the lateral resolution of PAM can be easily promoted from 50 μm to 5 μm or even finer.

Although having a depth penetration limit of the same order of magnitude as that of existing high-resolution optical imaging modalities—including optical confocal microscopy, multiphoton microscopy, and optical coherence tomography (OCT)—OR-PAM is primarily sensitive to optical absorption contrast, whereas all the other modalities are dominantly sensitive to optical scattering or fluorescence contrast.

13.4 Bright-Field OR-PAM

13.4.1 System design

In our previous PAM design with ultrasonic resolution, dark-field illumination was employed to minimize the interference due to strong photoacoustic signals from superficial structures [2, 4]. However, a major limitation of dark-field illumination is its suboptimal light-delivery efficiency. Since OR-PAM relies on diffraction-limited optical focusing, significant optical scattering and absorption in biological tissues restrict its penetration to one optical transport mean free path—1 mm. In this sense, OR-PAM works in the ballistic and quasi-ballistic regimes, where surface signal removal is not crucial. Thus, bright-field illumination via a commercial microscope objective is adopted to achieve diffraction-limited optical focusing.

As shown in Figure 13.2(A), a dye laser (CBR-D, Sirah) pumped by an Nd:YLF laser (INNOSLAB, Edgewave) is used as the excitation source. The laser pulse duration is 7 ns, and the pulse repetition rate, controlled by an external trigger, can be as high as 5 kHz. The light beam from the dye laser is attenuated before passing through a spatial filter (25-μm pinhole, P25C, Thorlabs), and then is focused by an objective lens (RMS4X, Thorlabs; NA: 0.1; effective focal length: 45 mm; working distance: 22 mm). The distance between the pinhole and the objective lens is $\sim$400 mm. An optical beam splitter is inserted between the pinhole and the objective lens, enabling focus adjustment and system alignment through an eyepiece. Two right-angle prisms (NT32-545, Edmund Optics) form a cube with a gap of 0.1 mm in between. The gap is filled with silicone oil (1000cSt, Clearco Products), which has an optical refractive index match with glass (silicone oil: 1.4; glass: 1.5) but a large acoustic impedance mismatch (silicone oil: 0.95×10^6 N·s/m^3; glass: 12.1×10^6 N·s/m^3). Thus, this silicone oil layer, optically transparent but acoustically reflective, acts as an optical-acoustic beam splitter. An ultrasonic transducer (V2022-BC, Olympus-NDT) with a center frequency of 75 MHz, a bandwidth of 80%, and an active-element diameter of 6.4 mm is attached to the vertical side of the bottom prism. A plano-concave lens with a 5.2-mm radius of curvature and a 6.4-mm aperture is attached to the bottom of the cube as an acoustic lens (NA: 0.46 in water; focal diameter: 27 μm). Since this lens also functions as a negative optical lens, it is compensated for by a planoconvex optical lens placed on top of the cube. The high NA acoustic lens is immersed into a water tank. A window is opened at the bottom of the water tank and sealed with an ultrasonically and optically transparent polyethylene membrane. The experimental animal is placed under the water tank with the region of interest (ROI) exposed under the transparent window. Ultrasonic gel (Clear Image, SonoTech) is applied to the ROI for acoustic coupling.

The photoacoustic signal detected by the ultrasonic transducer is first amplified by 48 dB (two ZFL 500LN amplifiers, Mini-Circuits) and then digitized by a 14-bit digital acquisition (DAQ) board (CompuScope 14200, Gage Applied Sciences). Two-dimensional (2D) raster scanning, combined with time-resolved ultrasonic detection, offers complete 3D information. The 2D scanner is controlled by a separate PC, which triggers both the data-acquisition PC and the pump laser. The trigger signal is synchronized with the clock-out signal from the DAQ board. For simplicity, the raster scanning is implemented by translating the water tank and the animal together within the horizontal (x-y) plane. A one-dimensional photoacoustic signal (A-line) at each horizontal location is

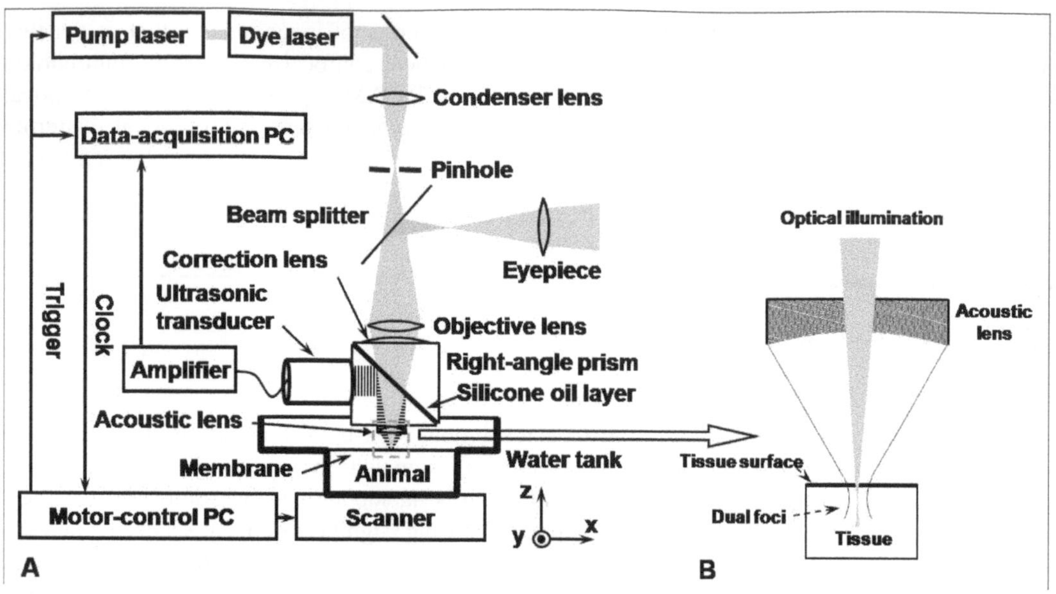

FIGURE 13.2: (A) Schematic of a bright-field OR-PAM system [5]. (B) Enlarged sketch demonstrating the bright-field illumination and acoustic-optical confocal configuration.

recorded for 1 μs at a sampling rate of 200 MS/s. OR-PAM image is formed by combining the time-resolved A-line signals and can be viewed in direct volumetric rendering, cross-sectional (B-scan) images, or maximum amplitude projection (MAP) images.

13.4.2 Spatial resolution quantification

The lateral resolution of the OR-PAM system was experimentally quantified by imaging an Air Force resolution test target (USAF-1951, Edmund Optics) immersed in an optically clear medium. The OR-PAM image shown in Figure 13.3(A) was acquired at the optical wavelength of 590 nm without signal averaging. In the enlarged Figure 13.3(B), the smallest pair that can be well resolved (group 6, element 5) has a gap of 4.9 μm (spatial frequency: 102 mm^{-1}; modulation transfer function value: 0.65). We also quantified the spatial frequency and modulation transfer function values of other two pairs as 64 mm^{-1} with 0.95 and 80 mm^{-1} with 0.8. Nonlinear fitting of the modulation transfer function followed by extrapolation yields a lateral resolution of 5 μm, which is 30% larger than the diffraction limit of 3.7 μm [5].

As a further demonstration of the lateral resolution, a 6-μm-diameter carbon fiber immersed in water was imaged using OR-PAM, as shown in Figure 13.4(A). The photoacoustic signal amplitude profile of a representative cross-section of the carbon fiber in water is illustrated in Figure 13.4(B). The mean FWHM value of the imaged carbon fiber is estimated to be 9.8 μm, which is 3.8 μm wider than the fiber diameter and hence in agreement with the estimated 5-μm resolution. To explore the degradation of the lateral resolution due to the optical scattering of biological tissues, we imaged the same type of carbon fiber covered by the ear of a living nude mouse (thickness: $\sim$200 μm). The photoacoustic signal amplitude profile of a representative cross-section of the carbon fiber under the mouse ear is shown in Figure 13.4(C). The FWHM value is estimated to be 10.2 μm, slightly larger than that in water. Hence, OR-PAM maintains a 5-μm lateral resolution in tissues up to at least 200 μm in depth [6]. The axial resolution can be estimated by the FWHM of the envelope (assumed to be in Gaussian shape) as follows:

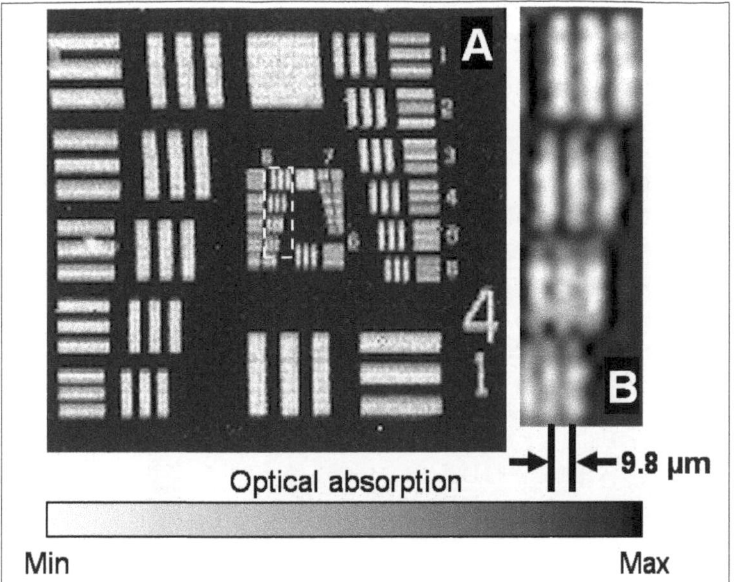

FIGURE 13.3: Photoacoustic MAP images of an USAF-1951 resolution test target. (A) Full view. (B) The high-resolution group highlighted in (A).

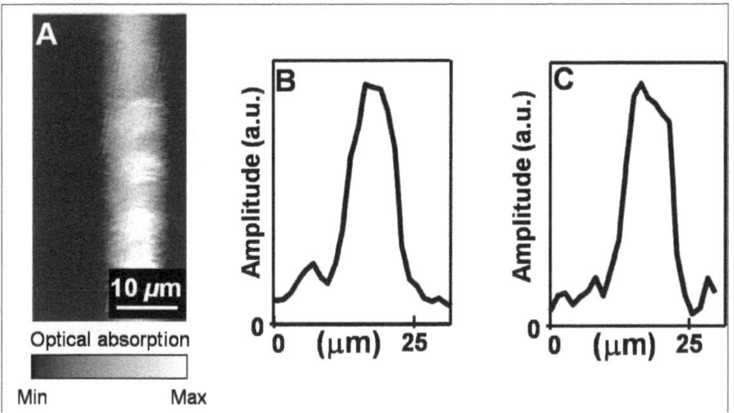

FIGURE 13.4: (A) MAP image of a 6-μm-diameter carbon fiber in water. (B) A representative cross-sectional profile of the carbon fiber in water. (C) A representative cross-sectional profile of a carbon fiber under 200-μm-thick nude mouse ear.

$$\Delta z_R = 0.88 v_s / \Delta f, \tag{13.2}$$

where v_s is the speed of sound, and $\Delta f$is the transducer bandwidth in receiving-only (one-way) mode. In our OR-PAM system, the axial resolution is calculated to be ~15 μm, based on the manufacturer-specified transducer bandwidth (100 MHz) and the speed of sound in tissue (1.5 mm/μs). We also quantified the axial resolution experimentally by imaging a sharp blade (Sterile surgical blade, Butler) immersed in water, which can be considered as a single target surface. The detect photoacoustic signal is bipolar, and the time interval from the positive peak to the negative peak is 10 ns, corresponding to 15 μm in axial distance. The time-domain FWHM of the

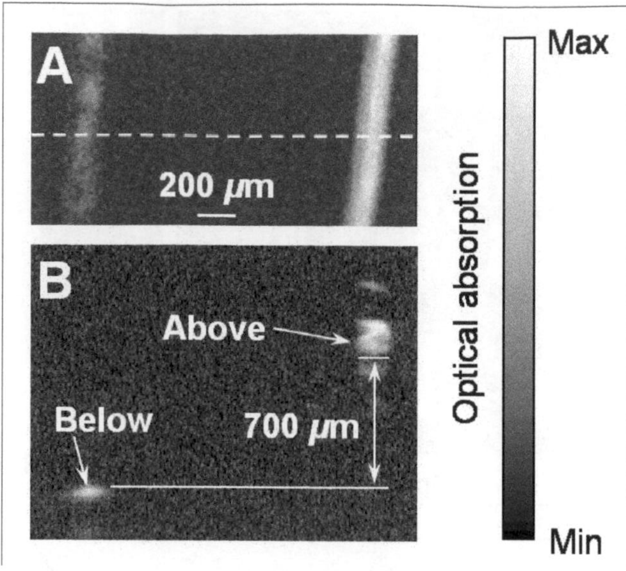

FIGURE 13.5: (A) MAP image of two horse hairs placed above and below a piece of freshly harvested rat scalp. (B) B-scan image at the location marked by the dashed line in panel (A).

detected photoacoustic signal envelope is estimated to be 18 ns, corresponding to 27 μm in axial distance. However, the axial resolutions are expected to deteriorate due to frequency-dependent acoustic attenuation in biological tissues and acoustic coupling medium.

13.4.3 Imaging depth estimation

As mentioned above, the imaging depth of OR-PAM depends on the optical transport mean free path and acoustic attenuation in biological tissues. To quantify this experimentally, two horse hairs (diameter: 200 μm) separately placed above and below a piece of freshly harvested rat scalp were imaged. An OR-PAM image was acquired with 32 times signal averaging at the optical wavelength of 630 nm. As shown in Figure 13.5(A), both hairs are clearly seen. The B-scan image in Figure 13.5(B) shows that the bottom hair is 700 μm deep in the tissue. Therefore, the penetration depth is at least 700 μm although the sensitivity and resolution degrade with depth [6].

13.4.4 Sensitivity estimation

To estimate the sensitivity of OR-PAM to the optical absorption of hemoglobin, we imaged a blood sample *in vitro*. Figure 13.6(A) shows the OR-PAM image of a smear of defibrinated bovine blood (Materials Bio, Inc), where individual RBCs are imaged with high contrast-to-background ratio (~30 dB). A close-up of the boxed region in Figure 13.6(A) clearly demonstrates the capability of resolving single RBCs by OR-PAM, which again validates its 5-μm lateral resolution. For comparison, a transmission optical microscopic image of a bovine blood sample with a similar RBC density is given in Figure 13.6(C). Our results show that OR-PAM has adequate contrast and resolution to image single RBCs. Monitoring the motion and oxygenation of RBCs in capillaries together can be potentially used to estimate the microscopic metabolic rate of oxygen consumption.

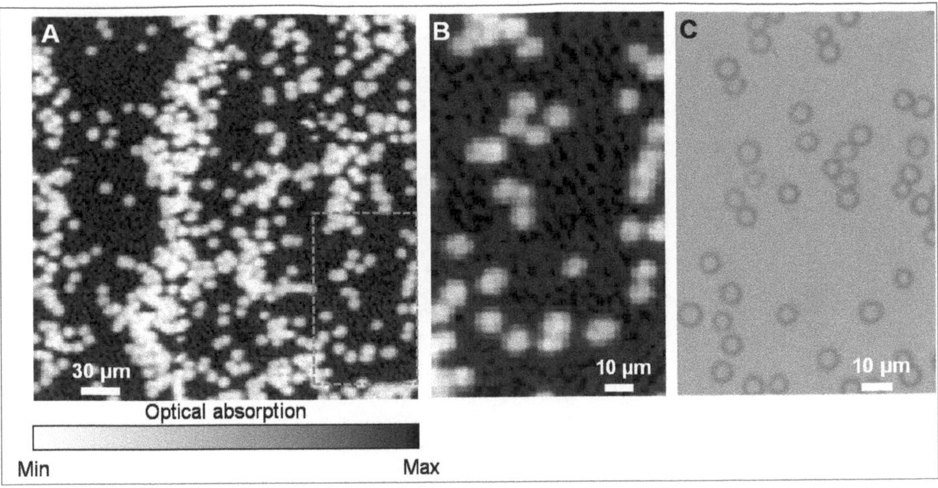

FIGURE 13.6: Images of the bovine RBCs *in vitro*. (A) MAP image acquired by OR-PAM; (B) Further magnified view of the boxed region shown in panel (A); (C) Transmission-mode optical microscopic image.

13.5 *In Vivo* Microvascular Imaging Using OR-PAM

Based on the above characterization, OR-PAM possesses the resolution, penetration depth, and sensitivity required for *in vivo* microvascular imaging. Taking advantage of the physiologically specific hemoglobin absorption contrast and the oxygenation-dependent hemoglobin absorption spectrum, we demonstrated structural and functional sO_2 imaging with micrometer-resolution using OR-PAM as described below.

Mouse ears (thickness: 200-500 μm) were chosen as our first *in vivo* targets because: (i) They have well-developed vasculature and have been widely used for studies of tumor angiogenesis and other microvascular diseases; and (ii) Mouse ears are among the few anatomical sites that are readily imaged by transmission-mode optical microscopy, which can be used to validate the feasibility of OR-PAM.

All experimental animal procedures were carried out in conformance with the laboratory animal protocol approved by the School of Medicine Animal Studies Committee of Washington University in St. Louis.

13.5.1 Structural imaging

A 1 mm-by-1 mm region in a nude mouse ear (Hsd:Athymic Nude-Foxn1^{NU}, Harlan Co.; body weight: ~20 g) was selected and imaged *in vivo* by OR-PAM at the optical wavelength of 578 nm. During image acquisition, the animal was kept warm by a heat lamp and motionless by vaporized isoflurane (1.0–1.5% isoflurane with an airflow rate of 1 liter/min). Unlike in the earlier published study [7], no optical clearing agent was applied to the skin surface. The ROI was scanned with a step size of 1.25 μm without signal averaging. The scanning time for a complete volumetric dataset was ~10 min. After data acquisition, the animal recovered naturally without observable laser damage.

Note that if an isosbestic point—an optical wavelength at which the molar extinction coefficients of the oxy- and deoxy-hemoglobin are equal—is selected for imaging, the photoacoustic signal represents the total hemoglobin concentration, regardless of the hemoglobin oxygenation.

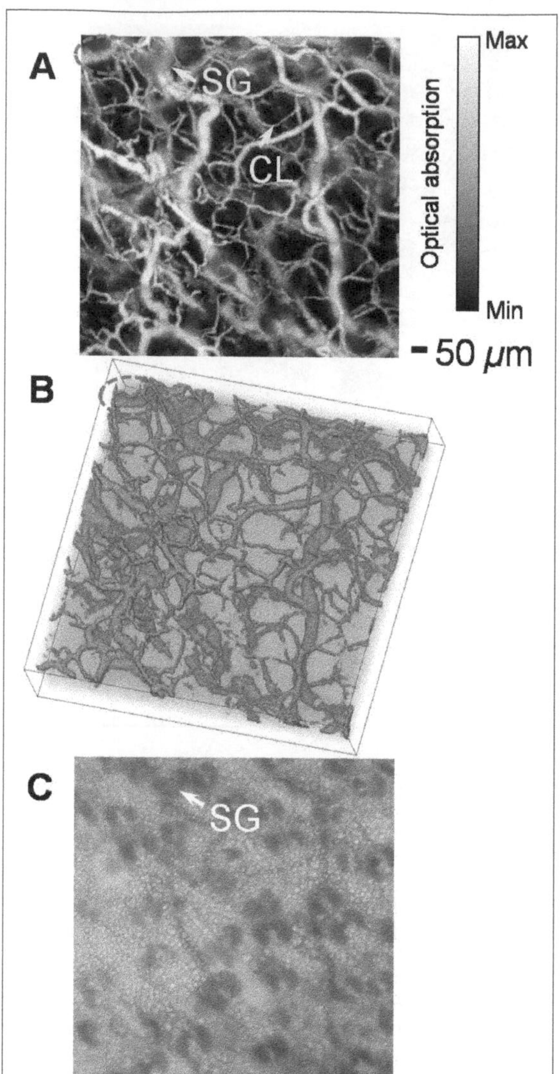

FIGURE 13.7: *In vivo* microvascular imaging of a nude mouse ear. (A) MAP image acquired by OR-PAM; (B) 3D pseudocolor visualization; (C) Photograph taken by a transmission-mode optical microscope. CL, capillary, and SG, sebaceous gland [6].

The OR-PAM image of the microvascular network in Figure 13.7(A) is highly correlated with the photograph in Figure 13.7(C), taken by a transmission-mode optical microscope at a 4× magnification; however, small vessels as well as capillaries are observable only by OR-PAM. The mean ratio of the photoacoustic amplitudes between the blood vessels and the background is 20:1, which demonstrates a high endogenous optical absorption contrast. Some microvessels, such as the tiny vessel labeled with CL in Figure 13.7(A), are single capillaries with diameters of $\sim$5 μm. A volumetric rendering of the photoacoustic data in Figure 13.7(B) shows the 3D positions and connectivity of the blood vessels. For example, at the center of the circled region in Figure 13.7(B), the small-diameter vessel invisible in Figure 13.7(A) is clearly seen $\sim$100 μm above the larger vessel. In addition, vessel bifurcation is clearly observed.

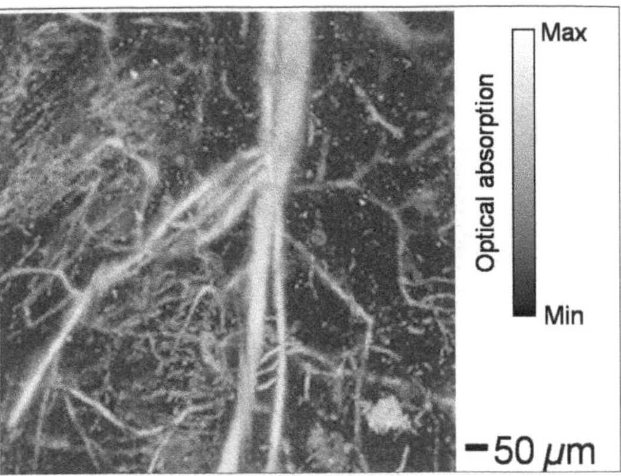

FIGURE 13.8: MAP image of the capillary-beds in a Swiss Webster mouse acquired *in vivo* by OR-PAM.

Figure 13.8 shows an additional example of visualizing capillary beds in the ear of a Swiss Webster mouse (Hsd:ND4, Harlan Co., ~25 g) *in vivo*.

All the results demonstrate that OR-PAM is able to perform *in vivo* 3D microvascular imaging down to single capillaries noninvasively without the aid of exogenous contrast agents.

13.5.2 Microvascular bifurcation

To distribute and collect nutrients, metabolic wastes, and gases in organisms, a branching vascular system is generally required. Although the morphogenesis of a vascular system is likely to be determined by a preprogrammed genetic algorithm, a number of physical, chemical, and biological factors determine the subsequent growth and remodeling during postnatal development [8]. Therefore, the branching pattern and vascular morphology are of great physiological and pathophysiological importance. Possessing capillary-level resolution and high sensitivity, OR-PAM is a valuable tool for microvascular bifurcation studies *in vivo*.

As hypothesized by Murray in 1926 [9], to achieve minimum energy expense, the vascular system should obey

$$Q = kD^{\delta}, \tag{13.3}$$

where Q and $D$are the volumetric flow rate and the diameter of a vessel segment, respectively; k is a proportionality constant; and $\delta = 3$, assuming blood flow is laminar. Later, Uylings modified Murray's law by introducing turbulent flow, under which condition $\delta = 2.33$ [10]. Due to the conservation of the volumetric flow rate at each branching node, the vessel diameters should follow the relationship

$$D^{\delta}_{parent} = \sum D^{\delta}_{daughter}, \tag{13.4}$$

whereD_{parent} and $D_{daughter}$ are the diameters of the parent vessel and the daughter vessels, respectively.

Here, OR-PAM imaged a nude mouse ear *in vivo* at the isosbestic point of 570 nm as shown in Figure 13.9. In this microvascular network, we selected four representative bifurcation nodes consisting of microvessels at different diameters, from arteriole/venule down to the capillary level. Each node connected one parent vessel and two daughter vessels. The vessel diameter was estimated by calculating the FWHM value of the cross-sectional signal of each vessel.

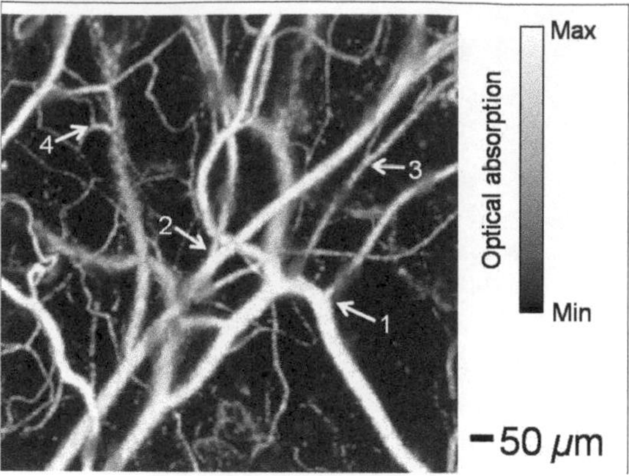

FIGURE 13.9: *In vivo* OR-PAM image of microvascular bifurcation in a nude mouse ear. 1–4 are four representative nodes consisting of microvascular bifurcations.

As shown in Table 13.1, the values of δ ranged from 2.45 to 3.98, in agreement with a previous study [8], which indicates that Murray's law ($\delta = 3$) does not hold perfectly in the microcirculation. Some of the possible reasons are: (i) δ depends on the metabolic-to-viscous dissipation ratio, which is specific to the organ system of interest; and (ii) Murray's law is based on the uniform shear hypothesis, which is not true for most of the vascular trees. In the mesentery, for example, shear stress is amplified in the microcirculation down to capillaries [8].

TABLE 13.1: Microvascular bifurcation quantified *in vivo* by OR-PAM

Node	V1 (μm)	V2 (μm)	V3 (μm)	δ
1	34.8	33.3	13.7	2.45
2	27.3	24.8	15.6	2.66
3	14.9	12.8	11.1	3.22
4	12.3	10.8	9.8	3.98

13.5.3 Functional imaging of hemoglobin oxygen saturation

Thanks to the oxygenation dependence of the hemoglobin absorption spectrum, OR-PAM can use spectroscopic measurements to quantify the relative concentrations of deoxy-hemoglobin (HbR) and oxy-hemoglobin (HbO_2) within single microvessels.

To this end, we first assume HbR and HbO_2 are the two dominant absorbers in blood, which is quite reasonable in the visible spectral range. Thus, the blood absorption coefficient can be calculated by [11],

$$\mu_a(\lambda_i) = \varepsilon_{HbR}(\lambda_i) \cdot [HbR] + \varepsilon_{HbO_2}(\lambda_i) \cdot [HbO_2] \,, \tag{13.5}$$

where $\mu_a(\lambda_i)$ is the blood absorption coefficient at wavelengthλ_i; $\varepsilon_{HbR}(\lambda_i)$ and $\varepsilon_{HbO_2}(\lambda_i)$ are the molar extinction coefficients of HbR and HbO_2, respectively; and $[HbR]$ and $[HbO_2]$are the concentrations of HbR and HbO_2, respectively.

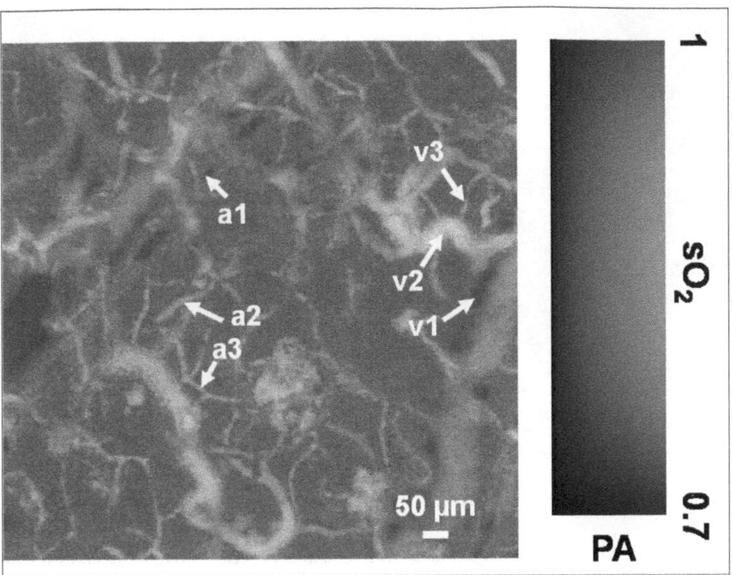

FIGURE 13.10: (See in color after page 572.) Pseudocolor sO_2 mapping of a microvascular network in a nude mouse ear acquired *in vivo* by OR-PAM. a1-a3; arterioles of different diameters; v1–v3: venules of different diameters. PA: photoacoustic signal amplitude.

Alternatively, $\mu_a(\lambda_i)$ can be quantified experimentally by photoacoustic measurements because the detected photoacoustic signal $\Phi(\lambda_i, x, y, z)$ is proportional to the product of $\mu_a(\lambda_i)$ and the local optical fluence $F(\lambda_i, x, y, z)$. Assuming the optical fluence is wavelength-independent, we can replace $\mu_a(\lambda_i)$ by $\Phi(\lambda_i, x, y, z)$ in Equation 13.5 and calculate $[HbR]$ and $[HbO_2]$in relative values. Since $\varepsilon_{HbR}(\lambda_i)$ and $\varepsilon_{HbO_2}(\lambda_i)$ are known parameters [12], only a dual-wavelength measurement is needed to solve $[HbR]$ and $[HbO_2]$. However, using more wavelengths is expected to yield a more accurate result. After obtaining $[HbR]$ and$[HbO_2]$, the sO_2 can be computed as follows:

$$sO_2 = \frac{[HbO_2]}{[HbR] + [HbO_2]}. \tag{13.6}$$

To acquire sO_2 *in vivo*, a nude mouse (Hsd:Athymic Nude-Foxn1^{NU}, Harlan Co.; body weight: ~20 g) was supplied with pure oxygen, and a dual-wavelength measurement at the optical wavelengths of 586 nm (isosbestic point) and 606 nm (deoxy-hemoglobin absorption dominant) was performed. A vessel-by-vessel sO_2 map in the nude mouse ear is shown in Figure 13.10. Different sO_2 levels are visualized with pseudocolors ranging from blue to red, where the photoacoustic signal amplitude measured at the isosbestic point is represented by pixel brightness. Position-dependent patterns of sO_2 change from arteries to veins are revealed by OR-PAM. Usually, sO_2 decreases steadily in pre-capillary arteries with decreasing vessel diameter. For the arterioles indicated by a1-a3 in Figure 13.10, the sO_2 values are 0.97±0.01, 0.92±0.01, and 0.91±0.01 with vessel diameters of >20 μm, 10-20 μm, and <10 μm, respectively. Then, sO_2 continues to decrease in postcapillary venules with increasing vessel diameter. For the venules indicated by v3-v1 in Figure 13.10, the sO_2 values are 0.85±0.01, 0.82±0.01, and 0.73±0.02 with vessel diameters of <10 μm, 30-40 μm, and >50 μm, respectively.

It needs to be pointed out that the measured sO_2 is higher than normal values [13] because the experimental animal was under systemic hyperoxia.

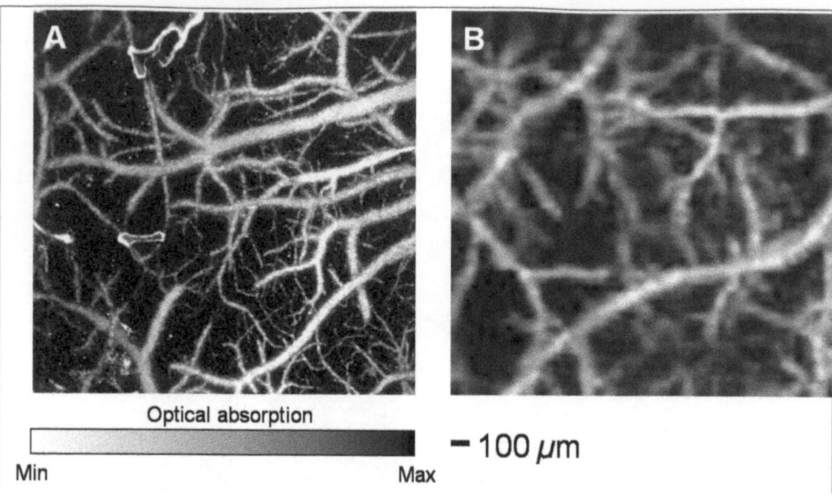

FIGURE 13.11: *In vivo* brain microvascular imaging through intact mouse skulls. (A) OR-PAM image; (B) Doppler OCT image (cropped from Figure 4(A) in reference 16).

13.5.4 *In vivo* brain microvascular imaging

Functional brain mapping is a very important topic in neuroscience. Advances in brain imaging facilitate the fundamental understanding of cognitive reactions and neurological diseases. Most techniques for direct measurements of neuronal activity in the brain, such as electroencephalography, are invasive and poor in spatial resolution. Instead, downstream hemodynamics originating from the brain microcirculation, including changes in hemoglobin concentration and oxygenation, are recorded to map the brain functions [14]. OR-PAM, having been demonstrated as a robust tool for HbT and sO_2 imaging, is potentially a valuable tool for functional brain mapping at cellular levels.

As addressed above, OR-PAM works in reflection-mode with a penetration depth of $\sim$1 mm, which is adequate for functional brain imaging in small animal models. Recently, we have successfully demonstrated *in vivo* brain microvascular imaging down to capillaries in living mice through the intact skull. To the best of our knowledge, OR-PAM is among the very few technologies, if not the only one, that is able to resolve single capillaries through the unthinned intact skull. Two-photon microscopy has been very successful in functional brain imaging, however it relies on fluorescence labeling and requires the skull to be thinned or even removed [15]. Doppler OCT, based on blood flow information, is another high-resolution modality that can image microvasculature through intact mouse skulls without exogenous contrast agents [16]. As a comparison, we imaged the brain microvasculature in a Swiss Webster mouse (Hsd:ND4, Harlan Co., $\sim$30 g) through the intact skull using OR-PAM, and compared our result with the published Doppler OCT data [16]. As shown in Figure 13.11(A), OR-PAM is able to resolve single capillaries with a high signal-to-contrast ratio (SNR), while Doppler OCT presents a resolution of only 20–30 μm as shown in Figure 13.11(B).

13.6 Conclusion and Perspectives

Current optical-resolution microscopic modalities, such as confocal microscopy and multiphoton microscopy, have gained great success in bioscience with the aid of fluorescent labeling. OR-

PAM, relying on optical absorption contrast, is an invaluable complement to existing scattering- and fluorescence-contrast microscopic techniques. With a completely new design, OR-PAM reveals the cellular and molecular features of living tissues, inaccessible via previous photoacoustic imaging modalities, with extraordinary resolution, sensitivity, and accuracy.

On the application side, OR-PAM has potentially very broad applications in microcirculatory physiology, tumor diagnosis and therapy, and laser microsurgery. Assessing HbT and sO_2at microscopic levels using OR-PAM also offers a great opportunity for fundamental biomedical research, such as neurovascular coupling for cognitive reactions and the local metabolic rate of oxygen consumption.

Several technical improvements can be made. First, images can be acquired by scanning the optical-acoustic dual foci instead of the animal and the water tank to reduce possible motion artifacts. Second, it is possible to scan only the optical focus within the acoustic focusing area to reduce the image acquisition time by at least an order of magnitude. Third, the acoustic coupling cube in the current design can be improved to transmit photoacoustic waves much more efficiently (without transformation from *p*-waves into *sv*-waves), so that the SNR can be improved by at least 14 dB. In addition, applying acoustic antireflection coating on the lens should further increase the SNR by another 10 dB.

Acknowledgment

The authors appreciate James Ballard's close reading of the manuscript. This work has been supported in part by National Institutes of Health grants R01 EB000712, R01 NS46214 (Bioengineering Research Partnerships), R01 EB008085, and U54 CA136398 (Network for Translational Research). L.W. has a financial interest in Endra, Inc., which however did not support this work.

References

[1] J.H Barker, G.L. Anderson, and M.D. Menger (eds.), *Clinically Applied Microcirculation Research*. CRC Press, Boca Raton, FL (1995).

[2] H.F. Zhang, K. Maslov, G. Stoica, and L.V. Wang, "Functional photoacoustic microscopy for high-resolution and noninvasive *in vivo* imaging," *Nat. Biotechnol.* **24**, 848–851 (2006).

[3] L.V. Wang, and H.I. Wu, *Biomedical Optics: Principles and Imaging*. John Wiley & Sons, Hoboken, NJ (2007).

[4] K. Maslov, G. Stoica, and L.V. Wang, "*In vivo* dark-field reflection-mode photoacoustic microscopy," *Opt. Lett.* **30**, 625–627 (2005).

[5] K. Maslov, H.F. Zhang, S. Hu, and L.V. Wang, "Optical-resolution confocal photoacoustic microscopy," *Proc. SPIE* **6856**, 68561I (2008).

[6] K. Maslov, H.F. Zhang, S. Hu, and L.V. Wang, "Optical-resolution photoacoustic microscopy for in vivo imaging of single capillaries," *Opt. Lett.* **33**, 929–931 (2008).

[7] V.P. Zharov, E.I. Galanzha, E.V. Shashkov, N.G. Khlebtsov, and V.V. Tuchin, "In vivo photoacoustic flow cytometry for monitoring of circulating single cancer cells and contrast agents," *Opt. Lett.* **31**, 3623–3625 (2006).

[8] G.S. Kassab, "Scaling laws of vascular trees: of form and function," *Am. J. Physiol. Heart. Circ. Physiol.* **290**, H894–H903 (2006).

[9] C.D. Murray, "The physiological principle of minimum work: I. The vascular system and the cost of blood volume," *Proc. Natl. Acad. Sci.* **12**, 207–214 (1926).

[10] H.B.M. Uylings, "Optimization of diameters and bifurcation angles in lung and vascular tree structures," *Bull. Math. Biol.* **39**, 509–520 (1977).

[11] H.F. Zhang, K. Maslov, M. Sivaramakrishnan, G. Stoica, and L.V. Wang, "Imaging of hemoglobin oxygen saturation variations in single vessels in vivo using photoacoustic microscopy," *Appl. Phys. Lett.* **90(5)**, 053901 (2007).

[12] G. Zijlstra, A. Buursma, and O.W. van Assendelft, *Visible and Near Infrared Absorption Spectra of Human and Animal Hemoglobin, Determination and Application.* VSP, Amsterdam, Netherlands (2000).

[13] H. Kobayashi, and N. Takizawa, "Oxygen saturation and pH changes in cremaster microvessels of the rat," *Am. J. Physiol.* **270**, H1453–H1461 (1996).

[14] D. Malonek, and A. Grinvald, "Interactions between electrical activity and cortical microcirculation revealed by imaging spectroscopy: implications for functional brain mapping," *Science* **272**, 551–554 (1996).

[15] N. Nishimura, C.B. Schaffer, B. Friedman, P.S. Tsai, P.D. Lyden, and D. Kleinfeld, "Targeted insult to subsurface cortical blood vessels using ultrashort laser pulses: three models of stroke," *Nat. Methods* **3**, 99–108 (2006).

[16] R.K. Wang, S. Jacques, Z. Ma, S. Hurst, S. Hanson, and A. Gruber, "Three dimensional optical angiography," *Opt. Express* **15**, 4083–4097 (2007).

14

Optical Coherence Tomography Theory and Spectral Time-Frequency Analysis

Costas Pitris, Andreas Kartakoullis, and Evgenia Bousi

Department of Electrical and Computer Engineering, University of Cyprus, Nicosia, 1678, Cyprus

Optical Coherence Tomography (OCT) is an emerging diagnostic imaging technology that has found widespread application in various medical fields. The first part of this chapter is an introduction to OCT theory and applications. It includes an overview of the technology, provides the basic theoretical background, and describes the capabilities of the OCT imaging systems. The second part of the chapter expands on the issue of Spectroscopic OCT (SOCT). After an introduction to the concept, a new spectral time-frequency paradigm is described, based on autoregressive parametric methods. Some initial results are presented which demonstrate the advantages of this technique over the more common FFT-based methods.

Key words: optical coherence tomography, time-frequency analysis

14.1 Introduction

Optical Coherence Tomography (OCT) is analogous to ultrasound imaging, except that it is based on the detection of infrared light waves backscattered (reflected) from different layers and structures within the tissue rather than sound. However, unlike ultrasound, the speed of light is very high, rendering electronic measurement of the echo delay time of the reflected light (time for the signal to return) impossible. Similar measurements can be performed using a technique known as low coherence interferometry. Within the interferometer, the beam leaving the optical light source is split into two parts, a reference and a sample beam. The reference beam is reflected off a mirror at a known distance and returns to the detector. The sample beam reflects off different layers within the tissue and light returning from the sample and reference arms recombines. If the two light beams have travelled the same distances (optical path length), the two beams will interfere. OCT

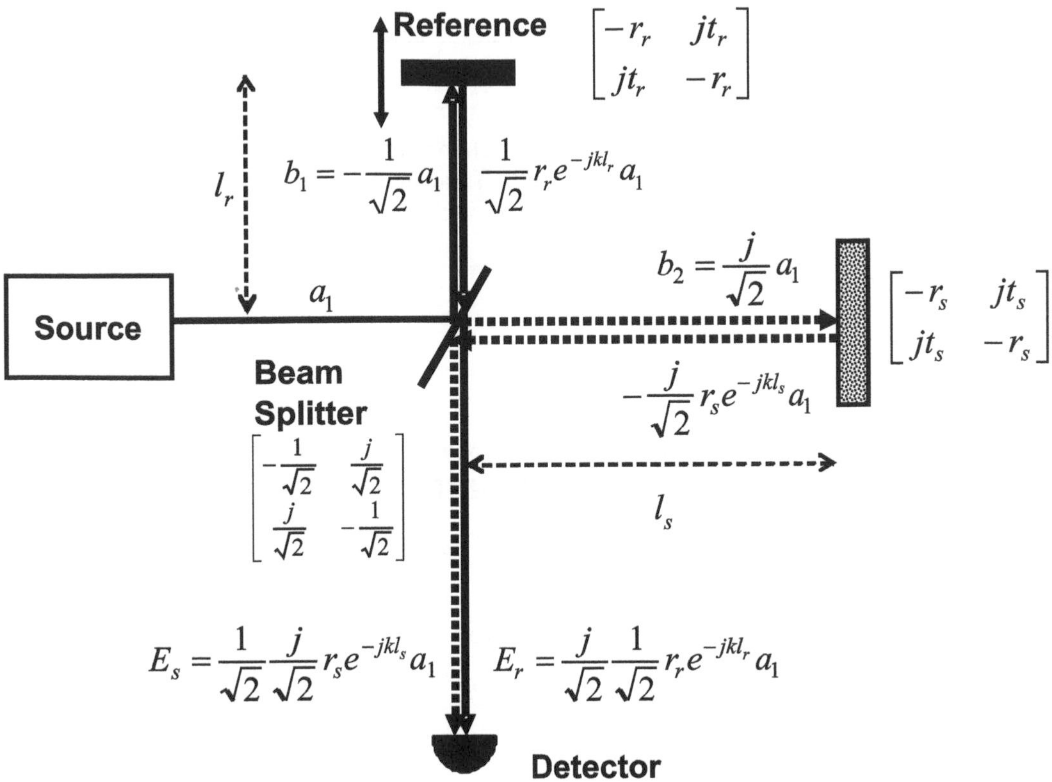

FIGURE 14.1: Michelson interferometer.

measures the intensity of interference obtained from different points within the tissue by moving the mirror in the reference arm that changes the distance light travels in that arm. Low coherence can be used to localize backreflection sites and provide the desired high axial resolution by limiting the interference pattern to a coherence length of mismatch between the two arms. Two- or three-dimensional images are produced by scanning the beam across the sample and recording the optical backscattering versus depth at different transverse positions. The resulting data is a two- or three-dimensional representation of the optical backscattering of the sample on a micron scale. The logarithm of the backscattering signal is represented as a false color or gray scale image.

Although low coherence interferometry for biological measurements has been demonstrated earlier, OCT imaging was first published in 1991 for cross-sectional imaging of the retina and coronary artery [1]. The first *in vivo* OCT imaging studies were performed in the human retina [2, 3] and, since then, ophthalmology remains the most developed application area of OCT. With the advent of new technologies which provide faster acquisition speeds as well as better sensitivity, OCT is developing into a noninvasive diagnostic imaging modality with ubiquitous applications [4–7]. Clinical applications of OCT have emerged in a variety of fields including further advances in ophthalmology, intravascular imaging in cardiology, gastroenterology, dermatology, dentistry, urology, gynecology, and many others.

14.2 Low Coherence Interferometry

The heart of an OCT system is the Michelson interferometer (Figure 14.1) [1–8]. Assuming the simplest case first, we can consider a plane wave (a_1) leaving the source and entering the interferometer, described by

$$a_1 = Ae^{-jkz}, \tag{14.1}$$

where

$$k = \frac{2\pi}{\lambda}. \tag{14.2}$$

This plane wave is split into two parts by a beam splitter ($r = 1/2$, $t = 1/2$) and reflected at the reference and sample arm mirrors. The beam splitter is represented by the scattering matrix:

$$\begin{bmatrix} b1 \\ b2 \end{bmatrix} = \begin{bmatrix} -\frac{1}{\sqrt{2}} & \frac{j}{\sqrt{2}} \\ \frac{j}{\sqrt{2}} & \frac{1}{\sqrt{2}} \end{bmatrix} \begin{bmatrix} a_1 \\ a_2 = 0 \end{bmatrix}. \tag{14.3}$$

The reference and sample fields recombine at the detector resulting in a field of the form:

$$E_d = E_r + E_s = \frac{j}{2}\left[a_1 r_s e^{-2jkl_s} + a_1 r_r e^{-2jkl_r}\right]. \tag{14.4}$$

The detector is sensitive to the intensity of the fields:

$$I_d = |E_d|^2 = \frac{1}{4}|E_r|^2 + \frac{1}{4}|E_s|^2 + \frac{1}{4}E_r^* E_s + \frac{1}{4}E_r E_s^*; \tag{14.5}$$

$$I_d = \frac{1}{4}\left[|Ar_s|^2 + |Ar_r|^2 + real\left\{E_r E_s^*\right\}\right]; \tag{14.6}$$

$$I_d = \frac{1}{4}\left[|Ar_s|^2 + |Ar_r|^2 + 2A^2 r_s r_r \cos(2k\Delta l)\right], \tag{14.7}$$

where

$$\Delta l = l_r - l_s. \tag{14.8}$$

Equation (14.7) implies that for a monochromatic plane wave in free space the interferometric signal consists of a DC term that is proportional to the reflectance from each arm and an AC component that is modulated sinusoidally as a function of path length difference (Δl) between the two arms. The part of this equation that contains backscattering information is the AC component. The OCT signal can therefore by represented by

$$I_{\text{inter}} \propto A_1^2 r_s r_r \cos(2k\Delta l). \tag{14.9}$$

The above treatment assumes a monochromatic source. In OCT, a low coherence source is used so that an interference signal will only be present when the two arms are matched within a coherence length. A low coherence source can be represented as the sum of monochromatic sources:

$$a_1 = \int_{-\infty}^{+\infty} A(k)e^{-jkz}dk. \tag{14.10}$$

The field at the detector for a low coherence source will, therefore, be:

$$E_d = E_r + E_s = \frac{j}{2}\left[\int_{-\infty}^{+\infty} A(k)r_r(k)e^{-2jkl_r}dk + \int_{-\infty}^{+\infty} A(k)r_s(k)e^{-2jkl_s}dk\right]. \quad (14.11)$$

As shown before,

$$I_{\text{inter}} \propto \text{Re}\{E_r E_s^*\}; \quad (14.12)$$

$$I_{\text{inter}} \propto \text{Re}\{\int_{-\infty}^{+\infty} A(k)r_r(k)A^*(k)r_s^*(k)e^{-2jk\Delta l}dk\}; \quad (14.13)$$

$$I_{\text{inter}} \propto \text{Re}\{\int_{-\infty}^{+\infty} |A(k)|^2 r_r(k)r_s^*(k)e^{-2jk\Delta l}dk\}; \quad (14.14)$$

$$I_{\text{inter}} \propto \text{Re}\{\int_{-\infty}^{+\infty} S(k)e^{-2jk\Delta l}dk\}, \quad (14.15)$$

where

$$S(k) = |A(k)|^2 r_r(k)r_s^*(k). \quad (14.16)$$

If the medium is nondispersive, the propagation constants in each arm can be considered equal and can be rewritten using a Taylor series expansion around ω_o, resulting in

$$k(\omega) = k(\omega_o) + k'(\omega_o)(\omega - \omega_o); \quad (14.17)$$

$$I_{\text{inter}} \propto \text{Re}\{e^{-2jk(\omega_o)\Delta l}\int_{-\infty}^{+\infty} S(\omega - \omega_o)e^{-2jk'(\omega-\omega_o)\Delta l}d(\omega - \omega_o)\}. \quad (14.18)$$

If we let the phase delay ($\Delta\tau_p$) and group delay ($\Delta\tau_g$) mismatches be

$$\Delta\tau_p = \frac{2\Delta l}{v_p}, \quad (14.19)$$

where $v_p = \frac{\omega_o}{k(\omega_o)}$ and

$$\Delta\tau_g = \frac{2\Delta l}{v_g}, \quad (14.20)$$

where $v_g = \frac{1}{k'(\omega_o)}$, then I_{inter} can be expressed as

$$I_{\text{inter}} \propto \text{Re}\{e^{-j\omega_o\Delta\tau_p}\int_{-\infty}^{+\infty} S(\omega - \omega_o)e^{-j\Delta\tau_g(\omega-\omega_o)}d(\omega - \omega_o)\}. \quad (14.21)$$

Since the spectrum of the source and the spectral reflectivities of the source and sample are conjugate-symmetric, the Fourier transform is real, resulting in an interferometric signal of the form

$$I_{\text{inter}} \propto \text{Im}\{S(\omega)\}\cos(\omega_o\tau_p). \quad (14.22)$$

14.2.1 Axial resolution

In deriving the axial resolution of an OCT system we can assume, for simplicity, that the OCT source has a Gaussian spectrum. This makes the development of the theory more manageable without compromising the physical understanding of the system. A source with a Gaussian power spectrum, normalized to unit power, such as:

$$S(\omega-\omega_o)=2\sqrt{\frac{\pi}{w_\omega^2}}e^{-\frac{2(\omega-\omega_o)^2}{w_\omega^2}}, \tag{14.23}$$

results in an interference pattern:

$$I_{\text{inter}} \propto e^{-\frac{2\Delta\tau_g^2}{w_t^2}} \cos(\omega_o \Delta\tau_p), \tag{14.24}$$

where

$$w_t = \frac{4}{w_\omega}. \tag{14.25}$$

If we define $\Delta\omega$ and $\Delta\lambda$ to be the FWHM of the frequency and wavelength spectra respectively, with a center wavelength λ, then we can calculate the radius of the a Gaussian spectrum in free space ($v_p = c$):

$$w_\omega = \frac{\Delta\omega}{\sqrt{2\ln 2}} = \frac{2\pi c \Delta\lambda}{\lambda^2\sqrt{2\ln 2}}. \tag{14.26}$$

We can further express the signal intensity in terms of path length mismatch (Δl) results in

$$I_{\text{inter}} \propto e^{-\frac{2\Delta l^2}{w_l^2}} \cos(\omega_o \Delta\tau_p), \tag{14.27}$$

where the radius interferogram in terms of distance, w_l, is

$$w_l = \frac{v_g w_t}{2} = \frac{2v_g}{w_\omega}. \tag{14.28}$$

Therefore, the FWHM of the interferogram, i.e. the OCT axial resolution (dz), in free space ($v_g = c$) is:

$$\text{FWHM} = dz = \frac{2\ln 2}{\pi}\frac{\lambda^2}{\Delta\lambda}. \tag{14.29}$$

In tissue, the presence of a material with an index of refraction, $n > 1$, results in a resolution of dz/n since:

$$\text{FWHM} \propto \frac{\lambda^2}{\Delta\lambda}. \tag{14.30}$$

The inverse relationship of the bandwidth of the source to the point spread function of the system makes development of increasingly better resolution systems more difficult as the sources are pushed to their broadband limit.

The presence of group velocity dispersion (GVD) causes different frequencies to propagate at different velocities. Significant GVD broadens short pulses but, more importantly for OCT, broadens the interferometric autocorrelation causing a degradation in the point spread function of the system. GVD is defined as

$$\frac{dv_g}{d\lambda} = \frac{\omega^2 v_g^2}{2\pi c}k''(\omega_o) \propto k''(\omega_o) \tag{14.31}$$

and can be included in our analysis by further expanding k:

$$k(\omega) = k(\omega_o) + k'(\omega_o)(\omega - \omega_o) + \frac{1}{2}k''(\omega_o)(\omega - \omega_o)^2. \tag{14.32}$$

Assuming the GVD mismatch exists over a distance L in the two arms of the interferometer, then the intensity of the OCT signal has an additional term that includes Δk", the GVD mismatch:

$$I_{inter} \propto \mathrm{Re}\{e^{-j\omega_o\Delta\tau_p} \int_{-\infty}^{+\infty} S(\omega - \omega_o) e^{-j\frac{1}{2}\Delta k''(\omega_o)(\omega-\omega_o)2L} e^{-j\Delta\tau_g(\omega-\omega_o)} d(\omega - \omega_o)\} \tag{14.33}$$

Assuming, again, a Gaussian spectrum the resulting interferometric intensity is

$$I_{inter} \propto \frac{w_t}{W(2L)} e^{-\frac{2\Delta\tau_g^2}{W(2L)^2}} \cos(\omega_o\Delta\tau_p), \tag{14.34}$$

where

$$W(2L)^2 = w_t^2 + 4j\Delta k''(\omega_o)(2L). \tag{14.35}$$

It is obvious from the complex nature of W(2L) that the Gaussian function has now a real and imaginary part defining an envelope change as well as a chirp. Separating the two parts we can get, from the real part, an expression of the new radius of the Gaussian:

$$w_d(2L) = w_t\sqrt{1 + \left(\frac{4\Delta k''(\omega_o)(2L)}{w_t^2}\right)^2}. \tag{14.36}$$

Significant broadening will be present when the second term in the square root becomes greater than 1, which implies that in the case of air vs. some other material even a few mm are enough to significantly broaden the point spread function. From (14–33) we see that a dispersion mismatch also causes a decrease in the peak intensity of the OCT interferogram, which is inversely proportional to the square root of the degree of broadening:

$$\frac{w_t}{|W(2L)|} = \frac{1}{\sqrt[4]{1 + \left(\frac{4\Delta k''(\omega_o)(2L)}{w_t^2}\right)^2}} \propto \frac{1}{\sqrt{\frac{w_d(2L)}{w_t}}}. \tag{14.37}$$

14.2.2 Transverse resolution

The transverse resolution of the OCT systems depends on the delivery optics. Assuming a Gaussian profile for the sample beam the diameter of the beam is given by

$$2w = \sqrt{\frac{2b\lambda}{\pi}}, \tag{14.38}$$

where b is the confocal parameter of the system and λ the center wavelength. We have conventionally called $2w$, i.e., the diameter of the beam at which the intensity falls by e^{-2}, the "resolution of the system." Strictly speaking the resolution should be the FWHM of the Gaussian that is related to the radius by

$$\mathrm{FWHM} = w\sqrt{2\ln(2)}. \tag{14.39}$$

The confocal parameter (b) is defined as twice the Rayleigh range (z_R), which is the distance away from the position of the minimum radius (w_o) to where the radius becomes $\sqrt{2}w_0$.

The system resolution can be measured experimentally by measuring either b or w_o directly. Although measuring b directly is very easy it is not very accurate and it does not provide any measure of the symmetry or the shape of the beam. The radius of the beam can be directly measured using the knife edge method that is harder but a more accurate technique. This involves the recording of the intensity while a sharp edge is scanned through the focal plane. The resulting function, called an "error function" is related to the Gaussian profile by

$$\mathrm{erf}\left(\sqrt{2}\frac{x-x_o}{w_o}\right) = \frac{1}{\sqrt{\pi}} \int_{-\infty}^{\sqrt{2}\frac{x-x_o}{w_o}} e^{-t^2} dt. \tag{14.40}$$

A fit of the "error function" to the data should result in values for w_o, which would only be limited by the resolution of the knife translation. The beam shape can also be recovered by integrating the "error function" while keeping in mind that differentiation is very sensitive to noise. The transverse resolution can also be approximated by: [7]

$$2w_o = \frac{4\lambda l}{\pi D}. \tag{14.41}$$

14.3 Implementations of OCT

14.3.1 Time-domain scanning

All early, i.e., prior to 2003, OCT systems employed mechanical means to vary the path length in the reference arm. Different approaches included linearly scanning galvanometric mirrors, fiber stretchers, optical delay lines, etc. The simplest way to implement a Michelson interferometer is to use a fiber optic coupler and a galvanometer to scan the reference arm in a linear fashion (Figure 14.2.) This scanning will introduce a Doppler shift in the frequency of the light in the reference arm. This frequency will be:

$$f_d = \frac{2v}{\lambda_o} \tag{14.42}$$

and will result in a frequency bandwidth of:

$$\Delta f_d = 2v\frac{\Delta\lambda}{\lambda_o^2} = f_d\frac{\Delta\lambda}{\lambda_o} \tag{14.43}$$

The main limitation of a linearly scanned galvanometer is speed. Due to the large mass of the optics, conversion of rotational motion to linear motion is limited to slow speeds. This makes linear galvanometers unsuited for real-time, *in vivo* imaging. In an attempt to remedy the situation, Tearney et al. have developed an optical phase delay line that was based on rotation of a mirror and can perform individual A-Scans at up to 2 kHz [8]. This device used pulse-shaping techniques developed for ultrafast optics to introduce a phase delay that is proportional to the angle at which the galvanometer is scanned [9, 10].

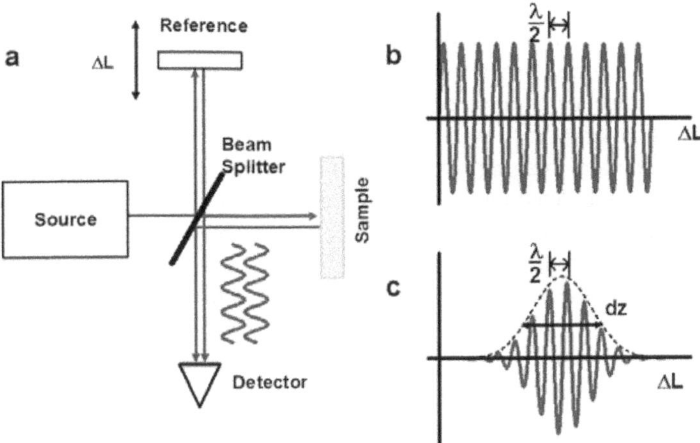

FIGURE 14.2: (a) Basic time domain OCT system with a reference and a sample beam interfering at the detector. (b) Interferometric pattern resulting from moving the reference arm and using a coherent source (e.g. common laser sources.) (c) Interferometric pattern resulting from moving the reference arm and using a low coherence source. The peak of the pattern corresponds to the position of the reference mirror at which the path lengths of the two arms match.

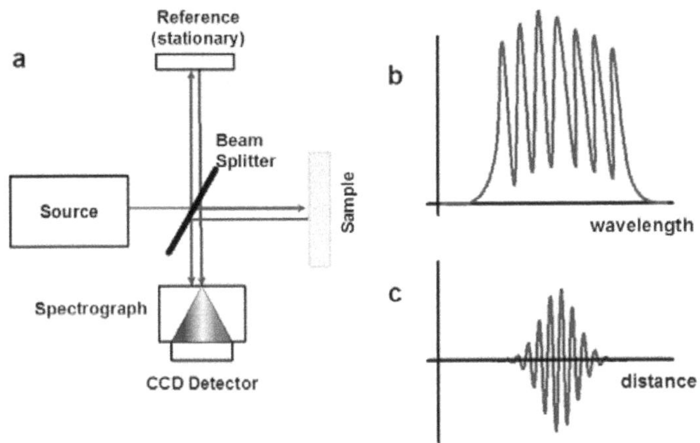

FIGURE 14.3: (a) A basic Fourier Domain OCT system with a reference and a sample beam interfering at the detector. (b) Interferometric pattern resulting from detecting the spectrum from the spectrograph. (c) The interferogram resulting from taking the Fourier transform of (b).

14.3.2 Fourier-domain OCT

Perhaps the most significant technological OCT development in recent years has been the introduction of Fourier domain OCT (FD-OCT) systems. The principles of FD-OCT were known for more than a decade before its value became apparent and practical systems were introduced [11]. FD-OCT can result in 10x to 100x increase in imaging speed and sensitivity over equivalent time-domain systems [12–14]. Real-time 3D and 4D imaging is possible with Fourier-domain systems, resulting in improved visualization and a dramatic increase of the clinical utility of OCT.

In FD-OCT the reference arm of the interferometer is held constant. Instead, the wavelength spectrum of the back-reflected light is detected (Figure 14.3.) The conventional axial scan can be retrieved by taking the Fourier transform of the spectrum. A FD-OCT system can be implemented using two complementary techniques, either by spectral detection (also known as spectral Fourier domain OCT) or by varying the wavelength of the source (also known as swept source OCT or Optical Frequency Domain Imaging, OFDI). In the first case, a broadband source is combined with a spectrograph for the detection of the resulting spectrum in a standard interferometer setup. Alternatively, the source wavelength can be scanned and the resulting spectrum detected by a single detector one wavelength at a time [15–18]. In both cases the resulting signal is similar. FD-OCT requires more complex post-processing but allows for much higher imaging speeds, reduced noise and improved sensitivity.

14.4 Delivery Devices

The success of OCT in clinical applications will depend on a large part on the design and availability of delivery mechanisms that will allow seamless integration with current and new diagnostic modalities. Such integration will lead to the introduction of OCT in mainstream diagnostic applications and allow clinicians to use the enhanced imaging capabilities of this technique to the benefit the patients without the need of extensive retraining or significant increase in procedure time. Such interplay between imaging technologies and clinical diagnostics will both improve outcome and reduce the cost of therapy.

Applications such as imaging during open surgery, laparoscopic imaging and imaging of the skin or oral cavity require forward imaging probes. Such devices can have a variety of shapes and sizes sharing the main characteristic of imaging in the forward direction. Scanning can be achieved with one of a number of different methods. Approaches include piezoelectric bending, torroidal scanning, or galvanometric mirror scanning, and, more recently, micromechanical machine systems (MEMS) [19].

Intraluminal and intravascular imaging requires small fiberoptic catheters [20]. In most cases, the catheter/endoscope consists of a hollow, speedometer-type, cable carrying a single-mode optical fiber. The beam from the distal end of the fiber is focused by a graded index (GRIN) lens and is directed perpendicular to the catheter axis by a microprism or micromirror. The distal optics are encased in a transparent housing. The beam is scanned either circumferentially (by rotation) or linearly (by translation) of the cable inside a transparent, static housing. The catheter/endoscope has a diameter of < 1 mm, which is small enough to allow imaging in a human coronary artery or access through the flush port of a standard endoscope. It can also be made significantly smaller by using smaller components. In fact, OCT imaging can be performed even through a needle-based device, small enough to be introduced anywhere with minimal disruption to the tissue or structure [21]. It can therefore be used to image or deliver light practically anywhere in a living body or nonbiological tissue, even in solid masses.

14.5 Clinical Applications of OCT

Several features of OCT suggest it can be a powerful imaging technology for the diagnosis of a wide range of pathologies.

1. OCT can image with axial resolutions of 1–10 μm, one to two orders of magnitude higher than conventional ultrasound. This resolution approaches that of histopathology, allowing architectural morphology as well as cellular features to be resolved. Unlike ultrasound, imaging can be performed directly through air without requiring direct contact with the tissue or a transducing medium.

2. Imaging can be performed *in situ*, without the need to excise a specimen. This enables imaging of structures in which a biopsy would be hazardous or impossible. It also allows better coverage, reducing the sampling errors associated with excisional biopsy.

3. Imaging can be performed in real time, without the need process a specimen as in conventional biopsy and histopathology. This allows pathology to be monitored on screen and stored in high resolution. Real-time imaging can enable real-time diagnosis, and coupling this information with surgery, can enable surgical guidance.

4. OCT is fiber optically based can be interfaced to a wide range of instruments including catheters, endoscopes, laparoscopes, and surgical probes. This enables imaging of organ systems inside the body.

5. Finally, OCT systems are compact and portable, an important consideration for a clinically viable device.

14.5.1 Ophthalmology

OCT has made its most significant clinical contribution in the field of ophthalmology, where it has become a key diagnostic technology in the areas of retinal diseases and glaucoma [22–24]. Ophthalmic OCT imaging has not only made the transition from the bench to the bedside, but has also contributed to new laboratory investigations by revealing previously hidden clinical features of ocular diseases and the microscopic changes associated with treatment. The technique was first commercialized by Carl Zeiss Meditec, Incorporated and is now considered superior to previous standards of care for the evaluation of a number of retinal conditions [25].

14.5.2 Cardiology

Over the past decade, the application of OCT in the field of cardiology has been extensively investigated. Initially, it was applied to the examination of the structural integrity of the vasculature in the coronary arteries and the evaluation of atherosclerotic plague morphology and stenting complications. Subsequently, cellular, mechanical, and molecular analysis was performed including the estimation of macrophage load [26–29]. The application of OCT to cardiology was greatly enhanced by technological developments such as rotational catheter-based probes, very high imaging speed systems, and functional OCT modalities. Currently, imaging and validation studies are being performed *in vivo*. In comparison to intravascular ultrasound (IVUS), OCT provides higher resolution by approximately an order of magnitude [30, 31].

14.5.3 Oncology

Diagnostic indicators of early neoplastic changes include accelerated rate of growth, mass growth, local invasion, lack of differentiation and anaplasia and metastasis. Evaluation of all of these structural and cellular features is necessary for the correct identification and grading of neoplasias and should be addressed by OCT. Macroscopic changes, such as microstructural changes, architectural morphology, and tumor growth, are relatively easy to identify since they fall within the resolution limits of most of today's OCT systems. Imaging cellular features, however, is more difficult but its diagnostic importance cannot be over emphasized. Fortunately, recent developments in the light sources used for OCT have led to imaging resolution of the order of $\sim 1\ \mu$m, which is approaching that required for cellular imaging.

There are three general application scenarios that are envisioned for OCT in neoplastic diagnosis. First, guiding standard excisional biopsy to reduce sampling errors and false negative results. This can improve the accuracy of biopsy as well as reduce the number of biopsies that are taken, resulting in a cost savings. Second, after more extensive clinical studies have been performed, it may be possible to use OCT to directly diagnose or grade early neoplastic changes. This application will be more challenging since it implies making a diagnosis on the basis of OCT rather than conventional pathology. Applications include situations where OCT might be used to grade early neoplastic changes or determine the depth of neoplastic invasion. Third, there may be scenarios where diagnosis can be made by OCT alone, enabling diagnosis and surgical guidance to be performed in real time. This would enable OCT diagnostic information to be immediately coupled to treatment decisions. The integration of diagnosis and treatment could reduce the number of patient visits, yielding a significant reduction in health care costs and improve patient compliance.

Each of these application scenarios requires a different level of OCT performance not only in its ability to image tissue pathology, but to achieve the required level of sensitivity and specificity in clinical trials for a given clinical situation. Generally speaking OCT can be used to resolve morphological features on several dimension scales ranging from architectural morphology or glandular organization (10–20 μm) to cellular features (1–10 μm). Since cancer is a highly heterogeneous disease, characterized by a spectrum of morphological changes, etiologies, etc., the viability of OCT will be highly dependent upon the details of the specific clinical application.

From very early in the development of OCT, cancer imaging has been an area of intense research interest. Imaging has been performed on a wide range of malignancies including the gastrointestinal, respiratory, and reproductive tracts, skin, breast, bladder, brain, ear, nose, and throat cancers. OCT has been used to evaluate the larynx, and has been found to be effective in quantifying the thickness of the epithelium, evaluating the integrity of the basement membrane, and visualizing the structure of the lamina propria [32]. In addition, preliminary studies have been conducted to evaluate the application of OCT to the oral cavity, oropharynx, vocal folds, and nasal mucosa [33]. An extensive number of studies is devoted to the investigation of OCT imaging in the gastrointestinal tract [34, 35]. The feasibility of using OCT to identify dysplasia includes a recent blinded clinical trial that showed an accuracy of 78% in the detection of dysplasia in patients with Barrett's esophagus [36]. *In vivo* OCT images of the normal human cervix and with moderate dysplasia are shown in Figure 14.4.

14.5.4 Other applications

New applications for OCT are constantly being investigated in addition to the most developed fields of ophthalmology, cardiology, and oncology. New, important applications of OCT include quantitative imaging and 3D mapping of the upper airways, musculoskeletal imaging with OCT and polarization sensitive OCT, and evaluation of dental structures [37–39]. Dental OCT, capable of detecting early decay not visible on standard x-ray, is being developed commercially. Additionally, applications in areas such as neurological, ovarian, and prostate imaging continue to advance, but have not yet matured to the point where they can enter clinical use.

In addition, many OCT-guided procedures are continually being developed, in a variety of areas. The overlying goal is to provide point-of-care screening or diagnosis and to positively affect patient outcome.

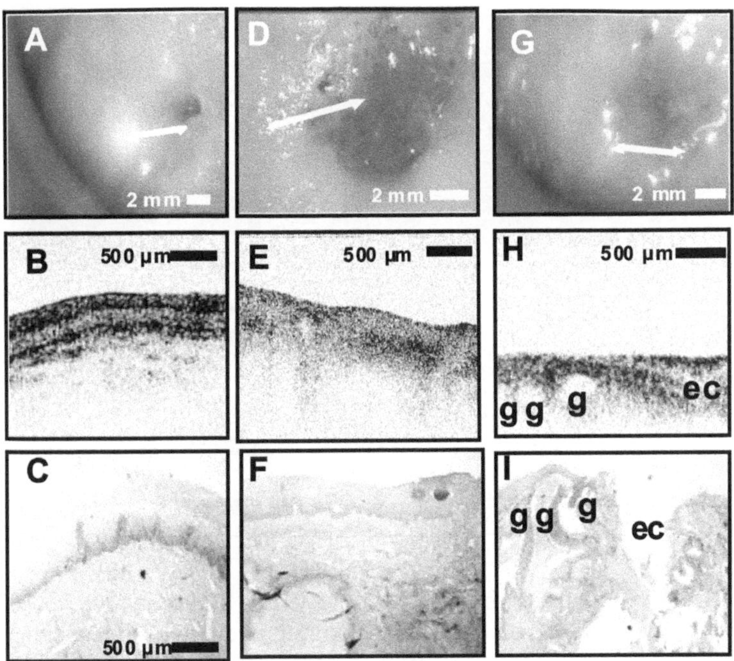

FIGURE 14.4: *In vivo* imaging of the human cervix. The top row shows colposcopic photographs with the line representing the scan range of OCT. The middle row are the OCT images and the bottom row are the corresponding histological sections. (a)–(c) Normal squamous epithelium, (d)–(f) Moderate dysplasia, (g)–(i) Glandular (g) and endocervical (ec) tissue.

14.5.5 OCT in biology

Imaging embryonic morphology is important for investigating the process of development and differentiation as well as elucidating aspects of genetic expression, regulation, and control. The technologies currently used to provide structural information about microscopic specimens include magnetic resonance imaging, computed-tomography, ultrasound, and confocal microscopy. Unfortunately, none of those is optimally suited for imaging laboratory biological samples at high resolution. The first two required an elaborate, complex, and expensive equipment [40]. To effectively image with ultrasound, probes require contact with the tissue. Although confocal microscopy is well suited for optically sectioning specimens, imaging depths are limited to less than 1 mm in scattering tissues [41]. Currently, morphological changes occurring in later stages of development can only be studied with histological preparations at discrete time points. OCT can provide the means to *in vivo* image morphology *in situ* and continuously in the same subject.

One of the most well-explored areas in the application of OCT in developmental biology is embryonic cardiac imaging. Studies have shown that *Xenopus-laevis* hearts can be readily imaged. Estimates of ejection fraction were performed by extrapolating the M-scan image data to a volume (assuming an ellipsoid-shaped ventricle) [42]. Embryonic chick hearts in 2D (*in vivo*) and 3D (fixed) were also imaged [43]. Recently, 4D images of embryonic chick hearts were obtained by pacing excised hearts and gating the image acquisition to the cardiac cycle [44]. Some groups have imaged fixed embryonic murine hearts and imaged paced excised embryonic murine hearts using gated OCT [45]. Recently, genetically altered adult murine skeletal muscle was imaged with OCT but studies demonstrating the use of OCT imaging to compare genetically altered hearts to wild-type hearts is still lacking [46].

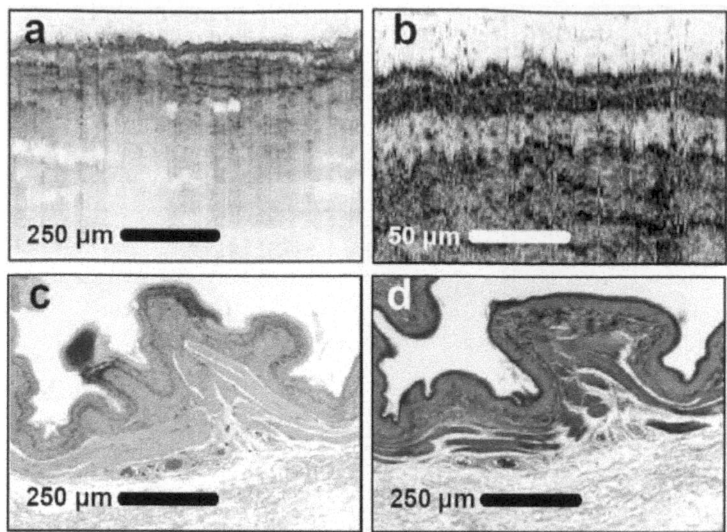

FIGURE 14.5: *In vivo* imaging of the epidermis and dermis of the Hamster cheek pouch. (a) and (b) OCT images at two different magnifications exhibiting imaging of the columnar epithelium of the epidermis and the dermis. (c) and (d) Histological slides of the same area d with H and E (c) and Trichrome (d) satins.

14.6 OCT Image Interpretation

Contrast in OCT images originates from the varying intensity of backreflected light. Signal differences are a direct result of subtle index of refraction variations in the tissue under investigation. As such, an OCT image may not directly correspond to the tissue structure as that would normally appear in a histological section (Figure 14.5). An additional difference between OCT and histopathology is the absence of any exogenous staining dyes. Color contrast can be introduced in OCT images to enhance the visualization of some features, but is only pseudo-color that is still assigned based on intensity variations. However, with the improving sensitivity and resolution of newer OCT systems, more details become apparent and OCT images come ever closer to histopathology.

The correspondence between OCT image features and histological structures is being continuously investigated in order to establish clear diagnostic criteria. In general, tissue microstructure of the order of 20 μm is easily discernible. For example, crypts and villi in the gastrointestinal tract can be directly evaluated from OCT images that resemble those of low magnification histopathology. However, individual cells and nuclei are usually exceedingly difficult to examine with current clinical systems. Even with ultrahigh resolution, some features may still not exhibit enough optical contrast to be discernible in OCT images. Indications of the nuclear status in some tissues can arise from the fact that highly cellular areas appear more intense. In addition, a spectral analysis of such areas can be used to estimate the size of those unresolvable scatterers that give rise to the OCT signal [47].

The visualization and interpretation of OCT images is still an area of intense research and discussion. As with any successful imaging system, presentation and visualization of the OCT images will become increasingly important as the clinical applications of the technique expand. Further, with the advent of clinical studies with acquisitions of large volumes of data, quantitative measurements will have to be extracted from the images. These measurements will require the identification and segmentation of features of interest and will introduce further challenges both in processing the data and presentation of the results. Efficient and effective techniques for visualizing multivariate data will be crucial to the clinical success of the system. For the purposes of OCT, these issues are still mostly unresolved.

When clinicians attempt to interpret OCT images they must be made very aware of some pitfalls that exist. For example, layered and regular structures, such as the layers of collagen, exhibit periodic variations of intensity which are artifacts of polarization variations along the depth of the tissue. This limitation can be alleviated by using polarization sensitive OCT which not only removes the artifact but also estimates the characteristics of the collagen layers. Furthermore, many of the texture characteristics of OCT images are a result of speckle and usually do not correspond to actual microstructure. Speckle is the coherent addition of the backreflected waves that results in maxima and minima in intensity. Speckle can be removed by various techniques [48, 49].

14.7 Spectroscopic OCT

An extension of OCT is Spectroscopic OCT (SOCT) that can perform both cross-sectional tomographic imaging of the structure and imaging of the spectral content. SOCT is able to perform depth-resolved spectroscopy, offering the possibility of reconstructing a 3D spectral map of a sample. SOCT is based on the fact that the envelope of an OCT interferogram is the Fourier transform of the spectrum of the light backreflected from that location. Therefore, an estimate of the spectral content of the light can be estimated from the shape of the OCT interferogram. The first time-domain broadband SOCT technique used the spectral-centroid shift as the indicator of spectral modification, to detected optical absorption by melanin in an African frog tadpole [50]. Later, depth-resolved backscattered spectra and tissue transfer functions were measured with high precision using a Fourier-Domain OCT system, which enabled quantitative estimation of absorber concentration [51]. SOCT has the potential to become a powerful extension of OCT with applications such as the assessment of blood oxygen saturation [52]. One promising application is scatterer size and concentration measurement for the detection of early dysplastic changes using Mie theory [53], or angle-resolved low-coherence interferometry (aLCI), a related technique [54]. It has also been demonstrated that SOCT could be combined with Light Scattering Spectroscopy (LSS) to eliminate the diffuse scattering background [55, 56]. More importantly, however, SOCT can be extend to retrieve information regarding subresolution scatterers based on the spectral variations caused by the scattering process.

14.7.1 Mie theory in SOCT

Depending on the scatterer size, the spectral scattering can be divided into three regimes. For particles much smaller compared to the wavelength, scattering is governed by the Rayleigh process. For particles much larger than the wavelength, scattering can be approximated by geometric optics. For particles comparable to the wavelength, the spectral scattering can be estimated using Mie theory. The most relevant process in OCT is Mie theory because of the relationship between the wavelength of OCT sources, which is in the near-infrared range (800–1300 nm), and many unresolvable tissue variations of diagnostic significance, which in many cases are of the same scale.

When the scatterer size is comparable to the wavelength, there is no simple means to solve for the scattering contribution, other than by the formal solutions of Maxwells equations and applying appropriate boundary equations. These calculations, collectively called Mie theory, dictate that the scattered wave for incident plane waves depends on the distribution of the refractive index of

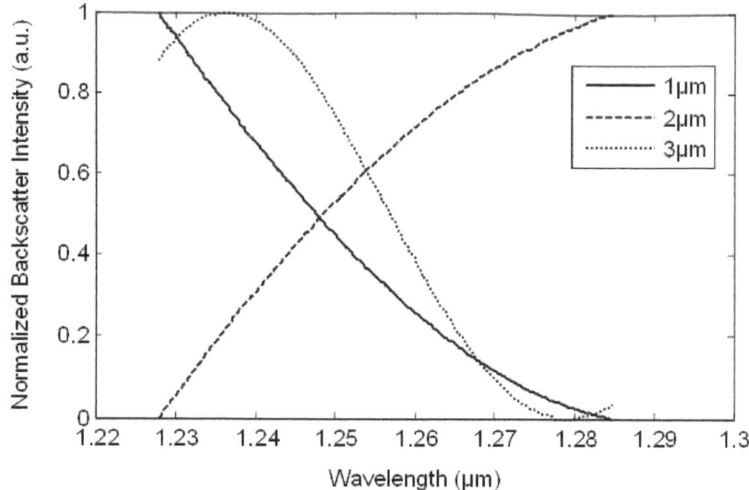

FIGURE 14.6: Example of backscattered intensity predicted by Mie theory between 1220- and 1290-nm wavelength for 1-μm scatterer (solid line), 2-μm scatterer (dash line), and 4-μm scatterer (dot line) diameter.

the scatterers, the incident wave frequency, and the scattering angle [57]. According to Mie theory, scattering is highly spectrally dependent, often forming characteristic modulation patterns as a function of scatter size and wavelength. These variations are apparent even in the relative narrow band of OCT sources. Figure 14.6 shows the characteristic spectral differences of backscattered intensity for scatterers with different size. By quantifying the degree of spectral modulation in the backscattered spectrum, information about the size of the scattering particles can be obtained. This information can be used as contrast modality for providing enhanced differentiation in areas where the scattering particles sizes varies significantly.

14.7.2 Spectral analysis of OCT signals

In SOCT, the information regarding the spectral content of backscattered light is obtained by the time-frequency analysis of the interferometric OCT signal. The backscattered spectrum can be acquired over the entire available optical bandwidth in a single A-scan, allowing the spectroscopic information to be analyzed in a depth-resolved manner [58]. The spectral features can be indicators of scatterer size estimated by spectral analysis methods such as the Fourier transform or the autocorrelation [53,56]. In the literature, the short-time Fourier transform (STFT) and the continuous wavelet transform (Morlet transform) have been used for obtaining the backscattering spectrum. In both of these techniques, performance is complicated by the time-frequency uncertainty principle, which states that there exists an inherent trade-off between the spectral resolution and the time resolution. Improvement in one implies degradation in the other. In OCT, high spatial resolution is required because spectral backscattering is a short-range effect in that large spectral variations can appear within submicrometer scale distance (cell or tissue boundaries). From the work done in LSS there are currently two approaches for sizing the scatterers based on the measured spectra. The first approach is based on pitch detection with the use of Fourier transform or determining the autocorrelation of the transform. The principle behind this approach is that the oscillation frequency in the wavelength-dependent scattering is size dependent, such that larger scatterers tend to produce more oscillatory patterns [59]. The second approach is based on curve fitting using least-square or χ^2

methods [54] using Mie theory to fit the normalized experimental measurement to the theoretical prediction. There are many methods of spectral analysis methods, including the Fast Fourier Transformation (FFT) method, Yule-Walker method, Burg method, Least Squares method, and Maximum likelihood method. Recently, the Burg method was explored for application to OCT data since it offers several advantages over other methods, such as high resolution, faster convergence for shorter segments, and computational efficiency [47].

14.7.3 Spectral analysis based on Burg's method

The traditional, nonparametric methods, such as FFT-based calculations, make no assumption about how the data were generated and the estimation is based entirely on a finite record of data. The finite length of the data, which in the case of OCT time-frequency analysis can be very small, severely limits the frequency resolution and the quality of the estimated spectrum and/or autocorrelation. Parametric or model-based methods, such as Burg's method for AR spectrum parameter estimation, partially alleviate those disadvantages. AR spectrum parameter estimation has been used in data forecasting, speech coding and recognition, model-based spectral analysis, model-based interpolation, signal restoration. The most common biomedical application is ultrasonic tissue characterization [60, 61].

The power spectral density (PSD) of a stochastic process could be expressed as a rational function

$$P_{xx}(\omega) = \sigma^2 \left| \frac{B(\omega)}{A(\omega)} \right|^2 \tag{14.44}$$

where σ^2 is a positive scalar, and the polynomials $(A\omega)$ and $B(\omega)$ are

$$A(\omega) = 1 + \sum_{k=1}^{q} a_k \exp(-jk\omega); \tag{14.45}$$

$$B(\omega) = \sum_{k=0}^{q} b_k \exp(-jk\omega). \tag{14.46}$$

This function can similarly be expressed in the z-domain

$$\phi(z) = \sigma^2 \frac{B(z)B^*(1/z^*)}{A(z)A^*(1/z^*)}, \tag{14.47}$$

where the polynomials $A(z)$ and $B(z)$ have roots that fall inside the unit circle in the z-plane. The rational PSD can be associated with a signal obtained by filtering white noise of power σ^2 with a transfer function [62]

$$H(\omega) = H(z)|_{z=\exp(j\omega)} = \frac{B(z)}{A(z)}. \tag{14.48}$$

The filtering results in a relationship between a signal $x(n)$ and white noise $w(n)$

$$x(n) = \frac{B(p)}{A(p)} w(n). \tag{14.49}$$

The above can be transformed into a difference equation

$$x(n) + \sum_{k=1}^{p} a_k x(n-k) = \sum_{k=0}^{q} b_k w(n-k). \tag{14.50}$$

Three processes can be modeled based on the above equation:

1. Autoregressive (AR) process
Where $b_0 = 1, \quad b_k = 0, \quad k > 0$ resulting in a linear, all-pole filter $H(z) = 1/A(z)$ and the difference equation

$$x(n) + \sum_{k=1}^{p} a_k x(n-k) = w(n). \tag{14.51}$$

The PSD for AR models is

$$P_{xx}^{AR}(\omega) = \frac{|b_0|^2}{|A(\omega)|^2}. \tag{14.52}$$

For the AR model the constant $|b_0|^2$ can be replaced by the error variance σ_p^2 in the PSD giving

$$P_{xx}^{AR}(\omega) = \frac{\sigma_p^2}{|A(\omega)|^2}. \tag{14.53}$$

2. Moving Average (MA) process
Where $a_k = 0, k \leq 1$ resulting in a linear, all zero filter $H(z) = B(z)$ and the relationship

$$x(n) = \sum_{k=0}^{q} b_k w(n-k) \tag{14.54}$$

The PSD for MA models is

$$P_{xx}^{MA}(\omega) = |B(\omega)|^2 \tag{14.55}$$

3. Autoregressive Moving Average (ARMA) process
In this case the linear filter $H(z) = B(z)/A(z)$ has both finite poles and zeros in the z-plane and the corresponding difference equation is given by equation (14.50). The PSD for an ARMA model is

$$P_{xx}^{ARMA}(\omega) = \frac{|B(\omega)|^2}{|A(\omega)|^2} \tag{14.56}$$

It is obvious from the relations above that the estimation of the ARMA parameters, a_k and b_k is essential for the estimation of the PSD. Burg devised a method for estimating the AR parameters as an order-recursive least square lattice method [63]. The AR parameters with Burg's method can be directly estimated from the data without the intermediate step of computing a correlation matrix and solving Yule-Walker equations, like other methods. Burg's method is based on the minimization of the forward and backward in linear errors with the constraint that the AR parameters satisfy the Levinson-Durbin recursion.

Assume we have data measurements $x(n), \quad n = 1, 2, \ldots, N$, we define the forward and backward prediction errors for a pth-order model as

$$\hat{e}_{f,p}(n) = x(n) + \sum_{k=1}^{m} \hat{a}_{p,k} x(n-k) \tag{14.57}$$

$$\hat{e}_{b,p}(n) = x(n-p) + \sum_{k=1}^{m} \hat{a}_{p,k}^{*} x(n+k-m), \tag{14.58}$$

where the AR parameters $\hat{a}_{p,k}$ are related to the reflection coefficients $\hat{K}_p$ with the constraint that they satisfy the Levinson-Durbin recursion given by

$$\hat{a}_{p,k} = \begin{cases} \hat{a}_{p-1,k} + \hat{K}_p \hat{a}^*_{p-1,p-i}, & i = 1, ..., p-1 \\ \hat{k}_p, & i = p \end{cases} \tag{14.59}$$

The next step in Burg's method for estimating the AR parameters is to find the reflection coefficients $\hat{K}_p$ that minimize the least-squares error

$$\varepsilon_m = \min_k \left[\hat{\varepsilon}^f_m(p) + \hat{\varepsilon}^b_m(p)\right], \tag{14.60}$$

where

$$\hat{\varepsilon}^f_m(p) = \sum_{k=p+1}^{N} \left|\hat{e}_{f,p}(k)\right|^2; \tag{14.61}$$

$$\hat{\varepsilon}^b_m(p) = \sum_{k=p+1}^{N} \left|\hat{e}_{b,p}(k)\right|^2. \tag{14.62}$$

Also the prediction errors satisfy the following recursive equations

$$\hat{e}_{f,p}(n) = \hat{e}_{f,p-1}(n) + \hat{K}_p \hat{e}^*_{b,p-1}(n-1); \tag{14.63}$$

$$\hat{e}_{b,p}(n) = \hat{e}_{b,p-1}(n-1) + \hat{K}_p \hat{e}^*_{f,p-1}(n). \tag{14.64}$$

Inserting equations (14.63) and (14.64) in equation (14.59) and performing a minimization of ε_m with respect to the reflection coefficients $\hat{K}_m$, results in

$$\hat{K}_m = \frac{-2 \sum_{k=p+1}^{N} \hat{e}_{f,p-1}(n) \hat{e}^*_{b,p-1}(n-1)}{\sum_{k=p+1}^{N} \left[\left|\hat{e}_{f,p-1}(n)\right|^2 + \left|\hat{e}_{b,p-1}(n-1)\right|^2\right]}. \tag{14.65}$$

The denominator is the least-squares estimate of the forward and backward errors $\hat{\varepsilon}^f_{m-1}(p)$ and $\hat{\varepsilon}^b_{m-1}(p)$, respectively. Thus,

$$\hat{K}_m = \frac{-2 \sum_{k=p+1}^{N} \hat{e}_{f,p-1}(n) \hat{e}^*_{b,p-1}(n-1)}{\left[\hat{\varepsilon}^f_{m-1} + \hat{\varepsilon}^b_{m-1}\right]}, \tag{14.66}$$

where $\hat{\varepsilon}^f_{m-1} + \hat{\varepsilon}^b_{m-1}$ is an estimate of the total error $\hat{\varepsilon}_m$. A recursion could be developed for the estimation of the total least-square error

$$\varepsilon_m = \left(1 - \left|\hat{K}_m\right|^2\right) \varepsilon_{m-1} - \left|\hat{e}_{f,p-1}(m-1)\right|^2 - \left|\hat{e}_{b,p-1}(m-2)\right|^2. \tag{14.67}$$

Burg's method estimates the reflection coefficients from equations (14.66) and (14.67) and Levison-Durbin algorithm is used to obtain the AR model parameters $\hat{a}(k)$. From Eq. (14.53) and the estimates of the AR parameters, the PSD can be formed as

$$\hat{P}^{Bu}_{xx}(\omega) = \frac{\hat{\varepsilon}_p}{\left|1 + \sum_{k=1}^{p} \hat{a}_p(k) \exp(-jk\omega)\right|^2}. \tag{14.68}$$

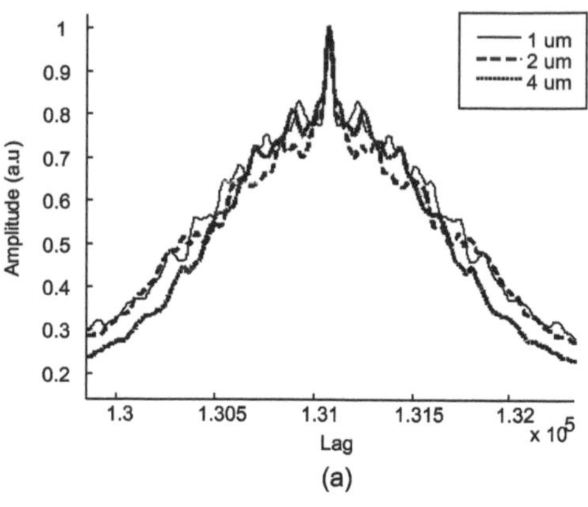

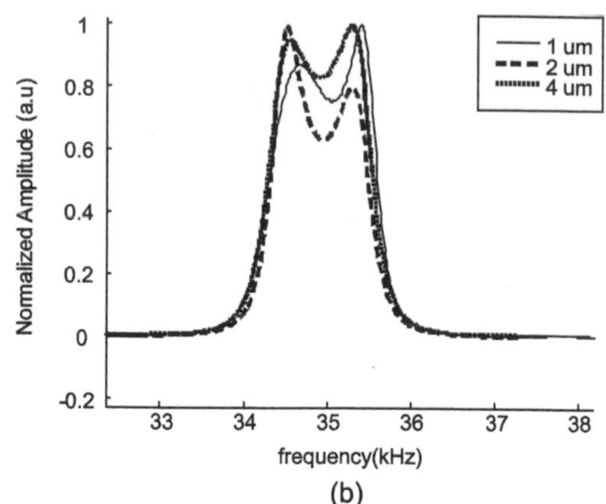

FIGURE 14.7: (a) The autocorrelation of the backscattering spectra obtained with FFT. (b) Backscattered normalized average power spectra obtained with Burg's method.

14.7.4 Experimental demonstration of SOCT for scatterer size estimation

The SOCT technique was applied on samples consisting of polystyrene microspheres embedded in an acrylamide gel. These samples contained 1, 2, and 4 μm diameter spheres at such concentrations to achieve similar backreflected signal strength for the three gels. Images were acquired at $\sim$ 12 times the Doppler frequency to assure adequate sampling. Following the acquisition of the images, the power spectral density, for each small window of the data, was calculated for all three images based on the technique described above (Figure 14.7). Using the PSD either an estimate of the diameter was calculated or the sample was classified in one of three classes ("1μm," "2μm," or "4μm") and a sensitivity and specificity value was calculated as compared to the known class. The results of the analysis are shown in Tables 14.1 and 14.2. For comparison purposes, the traditional FFT-calculated autocorrelation was used for the same analysis (Tables 14.1 and 14.2.) It is obvious that the parametric approach is significantly more accurate than FFT-based techniques.

TABLE 14.1: Classification results for Burg's Method ($p = 120$ AR parameters) and for FFT/Autocorrelation (2^{12}-point FFT)

Diameter (μm)	Burg's Method		FFT/Autocorrelation	
	Sensitivity (%)	Specificity (%)	Sensitivity (%)	Specificity (%)
1	100	100	75	82
2	100	100	69	82
4	85	100	62	82

TABLE 14.2: Diameter estimation results for Burg's Method ($p = 120$ AR parameters) and for FFT/Autocorrelation (2^{12}-point FFT)

	Burg's Method	FFT/Autocorrelation
Mean Error (%)	16.44	65.35
Standard Deviation of the Error (%)	19.10	48.52

14.8 Conclusions

OCT can perform a type of optical biopsy, i.e., imaging of tissue morphology at the micron-scale, *in situ*, and in real time. Image information is available immediately without the need for excision and histologic processing of a specimen. The development of high-resolution and high-speed OCT technology as well as OCT compatible catheter/endoscopes and other delivery devices represent enabling steps for many future OCT imaging clinical applications. More research, including extensive clinical studies, will determine the diagnostic efficacy of OCT as well as the medical fields where it is most applicable. The unique capabilities of OCT imaging suggest that it has the potential to significantly impact the diagnosis and clinical management of many diseases and improve patient prognosis. In addition, OCT can be a powerful research tool for applications in many areas, including developmental biology, where nondestructive, serial imaging, on the same sample, can offer new insights and significant cost and time advantages.

References

[1] D. Huang, E.A. Swanson, C.P. Lin, J.S. Schuman, W.G. Stinson, W. Chang, M.R. Hee, T. Flotte, K. Gregory, C.A. Puliafito, and J.G. Fujimoto, "Optical coherence tomography," *Science* **254**, 1178–1181 (1991).

[2] A.F. Fercher, C.K. Hitzenberger, W. Drexler, G. Kamp, and H. Sattmann, "In vivo optical coherence tomography," *Am. J. Ophthalmol.* **116**, 113–114 (1993).

[3] E.A. Swanson, J.A. Izatt, M.R. Hee, D. Huang, C.P. Lin, J.S. Schuman, C.A. Puliafito, and J.G. Fujimoto, "In vivo retinal imaging by optical coherence tomography," *Opt. Lett.* **18**, 1864–1866 (1993).

[4] J.G. Fujimoto, "Optical coherence tomography for ultrahigh resolution in vivo imaging," *Nat. Biotechnol.* **21**, 1361–1367 (2003).

[5] J.G. Fujimoto, M.E. Brezinski, G.J. Tearney, S.A. Boppart, B. Bouma, M. R. Hee, J.F. Southern, and E.A. Swanson, "Optical biopsy and imaging using optical coherence tomography," *Nat. Med.* **1**, 970–972 (1995).

[6] J.M. Schmitt, "Optical coherence tomography (OCT): a review," *IEEE J. Sel. Top. Quant.* **5**, 1205–1215 (1999).

[7] A.F. Fercher, W. Drexler, C.K. Hitzenberger, and T. Lasser, "Optical coherence tomography-principles and applications," *Rep. Prog. Phys.* **66**, 239–303 (2003).

[8] G.J. Tearney, B.E. Bouma, and J.G. Fujimoto, "High-speed phase- and group-delay scanning with a grating-based phase control delay line," *Opt. Lett.* **22**, 1811–1813 (1997).

[9] J.P. Heritage, A.M. Weiner, and R.N. Thurston, "Picosecond pulse shaping by spectral phase and amplitude manipulation," *Opt. Lett.* **10**, 609–611 (1985).

[10] K.F. Kwong, D. Yankelevich, K.C. Chu, J.P. Heritage, and A. Dienes, "400-Hz mechanical scanning optical delay line," *Opt. Lett.* **18**, 558–560 (1993).

[11] A.F. Fercher, C.K. Hitzenberger, G. Kamp, and S.Y. Elzaiat, "Measurement of intraocular distances by backscattering spectral interferometry," *Opt. Comm.* **117**, 43–48 (1995).

[12] M.A. Choma, M.V. Sarunic, C.H. Yang, and J.A. Izatt, "Sensitivity advantage of swept source and Fourier domain optical coherence tomography," *Opt. Express.* **11**, 2183–2189 (2003).

[13] J.F. de Boer, B. Cense, B.H. Park, M.C. Pierce, G.J. Tearney, and B.E. Bouma, "Improved signal-to-noise ratio in spectral-domain compared with time-domain optical coherence tomography," *Opt. Lett.* **28**, 2067–2069 (2003).

[14] R. Leitgeb, C.K. Hitzenberger, and A.F. Fercher, "Performance of Fourier domain vs. time domain optical coherence tomography," *Opt. Express* **11**, 889–894 (2003).

[15] S.R. Chinn, E.A. Swanson, and J.G. Fujimoto, "Optical coherence tomography using a frequency-tunable optical source," *Opt. Lett.* **22**, 340–342 (1997).

[16] B. Golubovic, B.E. Bouma, G.J. Tearney, and J.G. Fujimoto, "Optical frequency-domain reflectometry using rapid wavelength tuning of a Cr4+:forsterite laser," *Opt. Lett.* **22**, 1704–1706 (1997).

[17] S.H. Yun, G.J. Tearney, J.F. de Boer, N. Iftimia, and B.E. Bouma, "High-speed optical frequency-domain imaging," *Opt. Express* **11**, 2953–2963 (2003).

[18] W.Y. Oh, S.H. Yun, B.J. Vakoc, G.J. Tearney, and B.E. Bouma, "Ultrahigh-speed optical frequency domain imaging and application to laser ablation monitoring," *Appl. Phys. Lett.* **88**, 103902 (2006).

[19] S.A. Boppart, B.E. Bouma, C. Pitris, G.J. Tearney, J.G. Fujimoto, and M.E. Brezinski, "Forward-imaging instruments for optical coherence tomography," *Opt. Lett.* **22**, 1618–1620 (1997).

[20] G.J. Tearney, S.A. Boppart, B.E. Bouma, M.E. Brezinski, N.J. Weissman, J.F. Southern, and J.G. Fujimoto, "Scanning single-mode fiber optic catheter-endoscope for optical coherence tomography," *Opt. Lett.* **21**, 543–545 (1996).

[21] X.D. Li, C. Chudoba, T. Ko, C. Pitris, and J.G. Fujimoto, "Imaging needle for optical coherence tomography," *Opt. Lett.* **25**, 1520–1522 (2000).

[22] R.A. Costa, M. Skaf, L.A.S. Melo, D. Calucci, J.A. Cardillo, J.C. Castro, D. Huang, and M. Wojtkowski, "Retinal assessment using optical coherence tomography," *Prog. Retin. Eye Res.* **25**, 325–353, 2006.

[23] M.R. Dogra, A. Gupta, and V. Gupta, *Atlas of Optical Coherence Tomography of Macular Diseases*, Taylor & Francis, New York (2004).

[24] M.R. Hee, C.A. Puliafito, C. Wong, J.S. Duker, E. Reichel, B. Rutledge, J.S. Schuman, E.A. Swanson, and J.G. Fujimoto, "Quantitative assessment of macular edema with optical coherence tomography," *Arch. Ophthalmol.* **113**, 1019–1029 (1995).

[25] L. Alexander and W. Choate, "Optical coherence tomography: rewriting the standard of care in diagnosis, management and interventional assessment," *Rev. Optometry* **141**, 1CE–8CE (2004).

[26] B.E. Bouma, G.J. Tearney, H. Yabushita, M. Shishkov, C.R. Kauffman, D. DeJoseph Gauthier, B.D. MacNeill, et al., "Evaluation of intracoronary stenting by intravascular optical coherence tomography," *Heart* **89**, 317–320 (2003).

[27] G.J. Tearney, H. Yabushita, S.L. Houser, H.T. Aretz, I.K. Jang, K.H. Schlendorf, C.R. Kauffman, et al, "Quantification of macrophage content in atherosclerotic plaques by optical coherence tomography," *Circulation* **107**, 113–119 (2003).

[28] A.H. Chau, R.C. Chan, M. Shishkov, B. MacNeill, N. Iftimia, G.J. Tearney, R.D. Kamm, et al., "Mechanical analysis of atherosclerotic plaques based on optical coherence tomography," *Ann. Biomed. Eng.* **32**, 1494–1503 (2004).

[29] I.K. Jang, G. Tearney, and B. Bouma, "Visualization of tissue prolapse between coronary stent struts by optical coherence tomography: Comparison with intravascular ultrasound," *Circulation* **104**, 2754 (2001).

[30] J.G. Fujimoto, S.A. Boppart, G.J. Tearney, B.E. Bouma, C. Pitris, and M.E. Brezinski, "High resolution in vivo intra-arterial imaging with optical coherence tomography," *Heart* **82**, 128–133 (1999).

[31] I.K. Jang, B.E. Bouma, D.H. Kang, S.J. Park, S.W. Park, K.B. Seung, K.B. Choi, et al., "Visualization of coronary atherosclerotic plaques in patients using optical coherence tomography: comparison with intravascular ultrasound," *J. Am. Coll. Cardiol.* **39**, 604–609 (2002).

[32] B. Wong, R. Jackson, S. Guo, J. Ridgway, U. Mahmood, J. Shu, T. Shibuya, R. Crumley, M. Gu, W. Armstrong, and Z. Chen, "In vivo optical coherence tomography of the human larynx: normative and benign pathology in 82 patients," *Laryngoscope* **115**,1904–1911 (2005).

[33] W. Armstrong, J. Ridgway, D. Vokes, S. Guo, J. Perez, R. Jackson, M. Gu, J. Su, R. Crumley, T. Shibuya, U. Mahmood, Z. Chen, and B. Wong, "Optical coherence tomography of laryngeal cancer," *Laryngoscope* **116**, 1107–1113 (2006).

[34] C. Pitris, C. Jesser, S.A. Boppart, D. Stamper, M.E. Brezinski, and J.G. Fujimoto, "Feasibility of optical coherence tomography for high-resolution imaging of human gastrointestinal tract malignancies," *J. Gastroenterol.* **35**, 87–92 (2000).

[35] G. Isenberg, M.V. Sivak, A. Chak, R.C.K. Wong, J.E. Willis, B. Wolf, D. Y. Rowland, A. Das, and A. Rollins, "Accuracy of endoscopic optical coherence tomography in the detection of dysplasia in Barrett's esophagus: a prospective, double-blinded study," *Gastrointest. Endosc.* **62**, 825–831 (2005).

[36] J.M. Poneros and N.S. Nishioka, "Diagnosis of Barrett's esophagus using optical coherence tomography," *Gastrointest. Endosc. Clin. N. Am.* **13**, 309–323 (2003).

[37] X. Li, S. Martin, C. Pitris, R. Ghanta, D.L. Stamper, M. Harman, J.G. Fujimoto, and M.E. Brezinski, "High-resolution optical coherence tomographic imaging of osteoarthritic cartilage during open knee surgery," *Arthritis Res. Ther.* **7**, R318–R323 (2005).

[38] J.J. Armstrong, M.S. Leigh, D.D. Sampson, J.H. Walsh, D.R. Hillman, and P.R. Eastwood, "Quantitative upper airway imaging with anatomic optical coherence tomography," *Am. J. Respir. Crit. Care Med.* **173**, 226–233 (2006).

[39] A.Z. Freitas, D.M. Zezell, N.D. Vieira, A.C. Ribeiro, and A.S.L. Gomes, "Imaging carious human dental tissue with optical coherence tomography," J. *Appl. Phys.* **99**, 024906 (2006).

[40] E.J. Morton, S. Webb, J.E. Bateman, L.J. Clarke, and C.G. Shelton, "Three-dimensional X-ray microtomography for medical and biological applications," *Phys. Med. Biol.* **35**, 805–820 (1990).

[41] J.V. Jester, P.M. Andrews, W.M. Petroll, M.A. Lemp, H.D. Cavanagh, "In vivo, real-time confocal imaging," *J. Electron Microsc. Technol.* **18**, 50–60 (1991).

[42] S.A. Boppart, G.J. Tearney, B.E. Bouma, J.F. Southern, M.E. Brezinski, and J.G. Fujimoto, "Noninvasive assessment of the developing Xenopus cardiovascular system using optical coherence tomograpy," *Proc. Natl. Acad. Sci. U.S.A.* **94**, 4256–4261 (1997).

[43] T.M. Yelbuz, M.A. Choma, L. Thrane, M.L. Kirby, and J.A. Izatt, "Optical coherence tomography a new high-resolution imaging technology to study cardiac development in chick embryos," *Circulation* **106**, 2771–2774 (2002).

[44] M.W. Jenkins, F. Rothenberg, D. Roy, Z. Hu, V.P. Nikolski,M. Watanabe, D.L. Wilson, I.R. Efimov, and A.M. Rollins, "4D embryonic cardiography using gated optical coherence tomography," *Opt. Express* **14**, 736–748 (2006).

[45] W. Luo, D.L. Marks, T.S. Ralston, and S.A. Boppart, "Three-dimensional optical coherence tomography of the embryonic murine cardiovascular system," *J. Biomed. Opt.* **11**, 021014 (2006).

[46] J.J. Pasquesi, S.C. Schlachter, M.D. Boppart, E. Chaney, S.J. Kaufman, and S.A. Boppart, "In vivo detection of exerciseinduced ultrastructural changes in genetically-altered murine skeletal muscle using polarization-sensitive optical coherence tomography," *Opt. Express* **14**, 1547–1556 (2006).

[47] A. Kartakoulis and C. Pitris, "Scatterer size-based analysis of Optical Coherence Tomography Signals," Proc. SPIE **6627**, 66270N (2007)

[48] J.M. Schmitt, "Array detection for speckle reduction in optical coherence microscopy," *Phys. Med. Biol.* **42**, 1427–1439 (1997).

[49] A.E. Desjardins, B.J. Vakoc, W.Y. Oh, S.M. Motaghiannezam, G.J. Tearney, and B.E. Bouma, "Angle-resolved optical coherence tomography with sequential angular selectivity for speckle reduction," *Opt. Express* **15**, 6200–6209 (2007).

[50] U. Morgner, W. Drexler, F.X. Kartner, X.D. Li, C. Pitris, E.P. Ippen, and J.G. Fujimoto, "Spectroscopic optical coherence tomography," *Opt. Lett.* **25,** 111–113 (2000).

[51] D.J. Faber, E.G. Mik, M.C.G. Aalders, and T.G. van Leeuwen, "Light absorption of (oxy-) hemoglobin assessed by spectroscopic optical coherence tomography," *Opt. Lett.* **28,** 1436–1438 (2003).

[52] C. Xu, J. Ye, D.L. Marks, and S.A. Boppart, "Near-infrared dyes as contrast-enhancing agents for spectroscopic optical coherence tomography," *Opt. Lett.* **29,** 1647–1649 (2004).

[53] D. Adler, T. Ko, P. Herz, and J.G. Fujimoto, "Optical coherence tomography contrast enhancement using spectroscopic analysis with spectral autocorrelation," *Opt. Express* **12,** 5487–5501 (2004).

[54] A. Wax, C. Yang, V. Backman, K. Badizadegan, C.W. Boone, R.R. Dasari, and M.S. Feld, "Cellular Organization and Substructure Measured Using Angle-Resolved Low-Coherence Interferometry," *Biophys. J.* **82,** 2256–2264 (2002).

[55] A. Wax, C. Yang, and J.A. Izatt, "Fourier-domain low coherence interferometry for light-scattering spectroscopy," *Opt. Lett.* **28**, 1230–1232 (2003).

[56] C. Xu, P.S. Carney, W. Tan, and S.A. Boppart, "Light-scattering spectroscopic optical coherence tomography for differentiating cells in 3D cell culture," Proc. SPIE **6088,** 608804 (2006).

[57] H.C. Van de Hulst, *Light Scattering by Small Particles*, Dover Publications, New York (1981).

[58] R. Leitgeb, M. Wojtkowski, A. Kowalczyk, C.K. Hitzenberger, M. Sticker, and A.F. Fercher, "Spectral measurement of absorption by spectroscopic frequency domain optical coherence tomography," *Opt. Lett.* **25**, 820–822 (2000).

[59] L.T. Perelman, V. Backman, M. Wallace, G. Zonios, R. Manoharan, A. Nusrat, S. Shields, M. Seiler, C. Lima, T. Hamano, I. Itzkan, J. Van Dam, J.M. Crawford, and M.S. Feld, "Observation of periodic fine structure in reflectance from biological tissue: a new technique for measuring nuclear size distribution," *Phys. Rev. Lett.* **80**, 627–630 (1998).

[60] P. Chaturvedi and M. Insana, "Autoregressive Spectral Estimation in Ultrasonic Scatterer Size Imaging," *Ultrasonic Imaging*, **18**, 10–24 (1996).

[61] K.A. Wear, R.F. Wagner, and B.S. Garra, "A comparison of autoregressive spectral estimation algorithms and order determination methods in ultrasonic tissue characterization," *IEEE T. Ultrason. Ferr.* **42**, 709–716 (1995).

[62] P. Stoica and R.L. Moses. *Introduction to Spectral Analysis*. Prentice Hall, Upper Saddle River (1997).

[63] J.P. Burg. *Maximum Entropy Spectral Analysis*, PhD thesis, Stanford University (1975).

15

Label-Free Optical Micro-Angiography for Functional Imaging of Microcirculations within Tissue Beds In Vivo

Lin An, Yali Jia, and Ruikang K. Wang

Division of Biomedical Engineering, School of Medicine, Oregon Health & Science University, 3303 SW Bond Avenue, Portland, Oregon 97239, USA

Optical micro-angiography (OMAG) is a recently developed imaging modality that images the volumetric microcirculations within tissue beds up to 2 mm beneath the surface *in vivo*. Imaging contrast of blood perfusion in OMAG is based on endogenous light scattering from moving blood cells within biological tissue; thus, no exogenous contrast agents are necessary for imaging. The development of OMAG has its origin in Fourier domain optical coherence tomography (OCT). In this chapter, we will first briefly review the perspectives of OCT imaging of blood flow and summarize its advantages and disadvantages in imaging microcirculations within tissue beds *in vivo*. We will then introduce OMAG and its applications in imaging dynamic blood flow, down to capillary level resolution, within the cerebral cortex in small animal models, and within the retina in humans.

Key words: Optical micro-angiography, Doppler optical coherence tomography, blood perfusion, cerebral blood flow, functional imaging

15.1 Introduction

Optical coherence tomography [1,2] is a recently developed imaging technology capable of visualization of tissue microstructures at cellular level resolution up to 2–3 mm deep within highly scattering turbid medium, such as biological tissue, since it was first reported in the 1990s [3]. When combined with laser Doppler velocimetry, a functional extension of OCT, Doppler OCT (DOCT), is then developed that makes it possible to extract the quantitative flow information within functional blood vessels [4,5]. The original development of DOCT was based on the short-time fast Fourier transform (STFFT) analyses on the time-domain interferograms signals [5,6] that is formed between the reference light and the light that is backscattered from tissue. It was quickly realized that the STFFT method has severe limitations on real-time *in vivo* imaging of blood flow, largely due to the coupling issue between the flow imaging resolution and the short-time Fourier window size that can be used for flow analyses [7]. This limitation was then removed by an innovative use of the phase-resolved technique in DOCT [8–10], which was initially developed in the field of ultrasound imaging of blood flow [11]. This use of the phase-resolved technique in DOCT is now commonly referred to as phase-resolved DOCT (PRDOCT). By evaluating the phase difference between adjacent axial OCT scans (A-scan), PRDOCT greatly improves the detection sensitivity for imaging the flow velocity. Recent development of Fourier domain OCT has made an important step further for PRDOCT from laboratory research into visualization and monitoring of the dynamic blood flow within living tissue *in vivo*. This is largely due to the reason that FDOCT has a significant advantage in detection sensitivity over its time-domain counterpart [12–14], leading to increased imaging speed that is essential for any *in vivo* imaging applications. PRDOCT has seen successful developments in many *in vivo* imaging applications [15–18], especially in visualizing micro vasculatures within the human retina [19–23].

Although PRDOCT method could potentially achieve high spatial resolution and high sensitivity for imaging blood flow, its practical *in vivo* imaging performance is however disappointing. The responsible factors that degrade the *in vivo* imaging performance of PRDOCT are that (1) the biological tissue is generally of optical heterogeneous. This sample heterogeneous property imposes a characteristic texture pattern artifact overlaid onto the PRDOCT flow images [24] that masks the slow blood flow signals that would otherwise be detected by the method. (2) *In vivo* sample is always in constant motion, for example due to heartbeat. Thus, the motion artifacts in the PRDOCT blood flow images are inevitable [25]. Several methods have been developed to overcome these problems. One most straightforward method is to increase the axial scan density (A-scan density) over a single B-scan. Although successful, this method increases drastically the experimental time, which is not acceptable for *in vivo* imaging situations. Another method was proposed by Wang et al. [24], which uses a reverse scan pattern for the probe-beam to eliminate the texture pattern artifact. For the similar purpose, Ren et al. [26] used a delay line filter to minimize the artifacts produced by the heterogeneous properties of tissue sample. Though these methods could reduce the texture pattern artifact and increase the velocity sensitivity, they are still based on the phase-resolved method that has strong dependence on system signal-to-noise ratio (SNR) [27–28]. Based on the spatial distribution of the flow velocity, Wojtkowski et al. proposed a method called joint spectral and time-domain OCT [28], which has shown to increase the velocity sensitivity, especially in the low SNR regions within a scanned tissue volume. Still, this method needs a high A-scan density B-scan, which is time-consuming for *in vivo* experiments. To minimize the phase instability caused by the sample movement during the *in vivo* experiments, resonant Doppler imaging [29] was proposed. This method could directly extract the flow velocity information from the intensity distribution without the evaluation of phase distribution. However, this method requires an additional phase modulation in the reference arm, which no doubt increases the complexity of the whole system.

Most recently, based on analyzing the beating frequency caused by the endogenous light backscattered from the moving blood cells, our group has pioneered a novel method, OMAG [30–32] to image the blood microcirculations, down to the capillary level imaging resolution, within a highly scattering tissue sample. It is well known that the light, or precisely the photons, backscattered by a moving particle, for example a moving red blood cell, will carry a Doppler frequency that is generated by its velocity along the light incident direction. This Doppler frequency will generate a Doppler beating frequency, which carries quantitative flow information, within the so-called optical interferograms. To extract this beating frequency, Wang et al. [30–33] used a modulation frequency induced in the interferograms, which was either introduced by a moving reference mirror [30, 34] or by offsetting incident beam in the sample arm [35]. After applying the Hilbert transformation, the statistic signals, originated from the microstructures, and moving signals, generated by the moving particles, can be successfully separated. This method has been proved to be efficient to increase the sensitivity for detecting flow velocity, up to capillary level. The limitation here is that we need to modify the system hardware for the introduction of modulation frequency, either in the reference arm or in the sample arm. For a single B-scan, we can only get one direction flow image. To overcome this problem, Wang [33] proposed a digital modulation algorithm to introduce a digital modulation frequency within a B-scan. Combining the preintroduced modulation and the postdigital modulation method, OMAG is able to achieve directional flow imaging with high spatial resolution and high velocity sensitivity. The importance of OMAG method is that the imaging contrast in flow imaging is based on the flow velocity, intrinsically generated by the moving red blood cells, thus the method is a label-free approach.

Due to the superior performance of OMAG in reconstructing the blood perfusion within the microcirculation tissue bed, it is now widely applied on many *in vivo* applications where accurate visualization of microcirculation is needed to understand how the blood perfusion responds under different physiological conditions, for example, in the studies of stroke in small animal models [30] and the blood perfusion in the human retina and choroid layers [36].

15.2 Brief Principle of Doppler Optical Coherence Tomography

Doppler OCT was developed by essentially combining OCT with the well known laser Doppler velocimetry [4,5]. By relying on the Doppler frequency shift that is imposed on the light when the light interacts with a moving particle, i.e., the Doppler effect, DOCT provides an optical tool to quantitatively evaluate the velocity information of flow within functional blood vessels in a depth-resolved manner. Due to its relevance to OMAG that we will discuss later in this chapter, here we discuss the principle of DOCT briefly. For detailed information on DOCT, please refer to reference [7].

For a typical DOCT as illustrated in Figure 15.1, according to the Doppler principle, the optical wave-vector, $\mathbf{k}_d$, reflected from a moving particle with a velocity of $\boldsymbol{u}$ will be modulated by the velocity projection along the beam direction with a Doppler shift, Δf, compared with the incident optical wave-vector, $\mathbf{k}_0$. Δf can be presented by the following equation:

$$\Delta f = \frac{2}{2\pi}(k_d - k_0) \times u \cos\theta \tag{15.1}$$

If the Doppler shift,g Δf, of the backscattered light from the moving particle is determined and the angle θ is known *a priori*, then the velocity of the moving particles can be directly evaluated by,

$$\mathrm{u} = \lambda_0 \frac{\Delta f}{2\cos\theta} \tag{15.2}$$

where λ_0 is the central wavelength of the light source.

To extract the Doppler shift, a STFFT method was first employed on the interference signals in the time-domain OCT [4, 5]. However, it did not take long to realize that a coupling problem between imaging speed and velocity detection-sensitivity exists using such a method. This problem limits the STFFT approach to be widely used for *in vivo* imaging applications. To solve this problem, the phase-resolved Doppler OCT (PRDOCT) is therefore developed, first in the time-domain OCT [8–10], and later in Fourier-domain OCT [15–20]. Because of the significant advantage in imaging speed and sensitivity of FDOCT over TDOCT, the FDOCT method is now becoming increasingly adopted in many *in vivo* imaging applications. In Fourier-domain PRDOCT, the acquired spectral interferogram, $I(k)$, was Fourier transformed to generate a depth-resolved complex OCT signal,

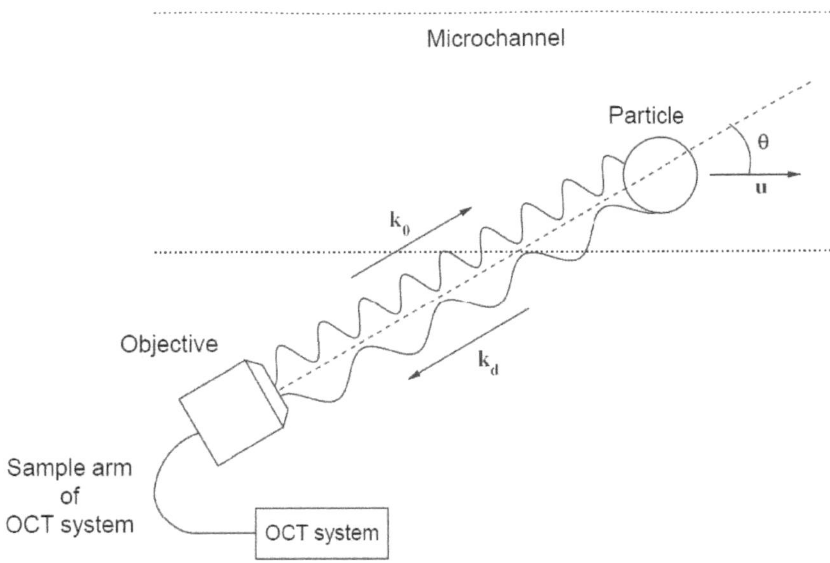

FIGURE 15.1: Schematic for the DOCT operation based on an OCT system of general purpose. The angle, θ, between incident light beam from the sample arm and the flow direction is the Doppler angle. $\mathbf{k}_0$ is the incident optical wave-vector before the light hits the moving particle, while $\mathbf{k}_d$ is the modulated optical wave vector reflected from the moving particle. $\mathbf{u}$ is the velocity of moving particle.

$\tilde{I}(z) = FFT[I(k)] = A(z)\exp[i\Phi(z)]$, where k is the wavenumber and z is the depth coordinate. The magnitude $A(z)$ is used to generate cross-sectional microstructure image of the sample that is the standard OCT image, while the phase difference between adjacent A-lines, $\Delta\Phi(z)$, is used to obtain velocity map [15, 16].

$$\mathrm{u}(z) = \Delta\Phi(z,\tau)\frac{\lambda_0}{4\pi\tau\cos\theta} \tag{15.3}$$

where τ is the time delay between adjacent A-lines.

Using the phase-resolved method, DOCT has shown to achieve high speed, high resolution cross-sectional, microstructural image and the velocity image, simultaneously. The axial resolution is dependent on the coherence length of the light source used, while the lateral resolution is determined by the probe beam spot size on the sample. The velocity sensitivity is dependent on a number of factors, such as the time delay between adjacent A-lines, the system SNR, optical properties of the sample, as well as the Doppler angle θ.

15.3 Optical Micro-Angiography

Optical micro-angiography (OMAG) was developed from the full-range complex Fourier-domain OCT (FDOCT). The OMAG method is capable of effectively separating the optical signals backscattered by the moving particles from those optical signals backscattered by the static background tissue, i.e. the tissue microstructures. In essence, OMAG takes the optical scattering signals originated from the microstructures as the noise signals, and acts to reject these noise signals in order to image

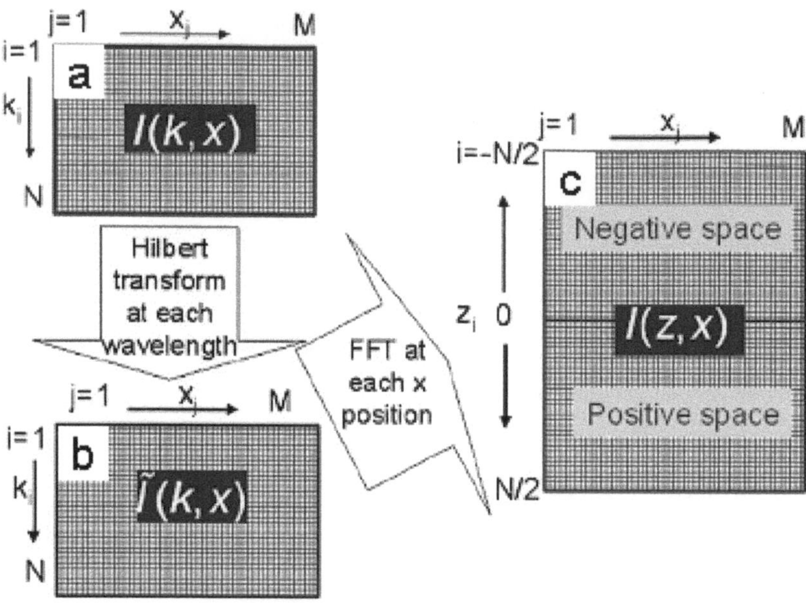

FIGURE 15.2: Flow chart to illustrate the algorithm to achieve full-range complex imaging.

the blood flow. For better illustration of OMAG method, we would like to first briefly introduce the phase-modulation method that makes full-range complex FDOCT imaging possible, and then discuss OMAG imaging of blood perfusion, including its directional flow capability.

15.3.1 *In vivo* full-range complex Fourier-domain OCT

In Fourier-domain OCT, the cross-sectional image of the scanned tissue is reconstructed from the recorded spectral interferograms formed between the reference light and the light that is backscattered from within a tissue sample. There are two methods to realize the interference spectral fringes, i.e., interferograms, formed in a Michelson-like interferometer. The first is to use a broadband light source in an interferometer and to use a high-speed spectrometer to resolve the wavelength dispersed interferograms [spectral domain OCT (SDOCT)] [37, 38]. Usually, the spectrometer employs a high-speed line scan camera to digitize the interferograms for later OCT image reconstruction. Another approach is to employ a swept source laser system to illuminate the interferometer and to use a single detector to digitize the interference signals at each wavelength individually [swept source OCT (SSOCT)] [39, 40]. In either cases, the detector is a square-detector. Thus, the recorded signals are real-valued functions, which produce a complex conjugate artifact in the output plane when Fourier transforming the wavelength information into the distance information. Due to this reason, in traditional FDOCT, only negative- or positive-distance range can be used for imaging purpose, i.e., only half of the retrieved information is useful. To overcome this drawback in FDOCT, Wang [34] proposed a phase-modulation method to reconstruct a complex interferograms in order to achieve full-range complex imaging.

For SDOCT, the interferogram formed between the light backscattered from the sample and reflected from the mirror at each x position is captured by a spectrometer, while in SSOCT it is captured by a point detector at each single wavelength. For simplicity, here we only consider the signals that contribute to the depth localization in the sample. In this case, the interferogram detected at each wavelength λ can be expressed as [34],

$$I(k,x) = 2S(k)\int \sqrt{R(x,z)}\cos[2kz + \phi(k,x,z)]dz \tag{15.4}$$

where $k = 2\pi/\lambda$, z is the relative distance between the reference mirror position and the scatters within the sample, $R(x,z)$ is the normalized intensity backscattered from the scatter at $z, S(k)$ is the spectral density of the light source, and $\phi(k,x,z)$ is the phase that relates to the optical heterogeneity of the sample in the focus coherence volume. Here, we do not consider the possible variances of the reflection in the reference arm during one OCT B-scan. In practice, at each scanning position, the interference spectrum is discretized by N segments, corresponding to the number of pixels in the camera in SDOCT or number of samples captured in one entire sweep in SSOCT. For each one B-scan, the object is sampled by M times in the x direction, which will produce a data matrix with a size of (N, M), see Figure 15.2(a).

As alluded above, for each wavelength the interference recorded by the FDOCT system provides a specific equation for Eq. (15.4), $I_i(k_i,x)$ $(i = 1 : N)$, with the position x as its variable in the lateral direction. Each of these equations is a real-valued function, which determines that the frequency components are Hermitian symmetric according to the zero frequency. And the signal bandwidth of real-valued function is centered at the zero frequency, because of the random phase, $\phi(k,x,z)$, caused by the heterogeneous properties of the scattering sample. For these two reasons, it is difficult to construct the complex signal of the OCT interferogram at each k_i directly from the traditional FDOCT data. However, if a carrier frequency f_c is introduced into $I(k_j, x)$ that can modulate the signal bandwidth sufficiently away from the zero frequency, the complex function can be determined through analytic continuation of the measured interference fringes along the x direction by use of a Hilbert transformation [34] [see Figuress. 15.2(a) and 15.2(b)],

$$\tilde{I}(k_i,x) = 2\frac{i}{\pi}P\int_{-\infty}^{\infty}\frac{I(k_i,\tau)}{\tau - x}d\tau = A(k_i,x)\exp(i(\varphi(x)+\varphi_C)),\ (i=1,...,N), \tag{15.5}$$

where P denotes the Cauchy principle value, $A(k_i,x)$ is the "envelope" or amplitude, φ_c is the phase term caused by the modulation frequency and $\varphi(x)$ is the phase of the analytic signal that is a random fluctuation term caused by the heterogeneous properties of a scattering sample. Next, a Fourier transformation can be performed along k at each position x_j to obtain the B scan OCT structure image, as usually done in the conventional FdOCT [see Figures 15.2(b) and 15.2(c)].

$$I(z,x_j) = FT[\tilde{I}(k,x_j)]\ (j = 1,...,M), \tag{15.6}$$

where the complex conjugate image is completely removed, thus, the entire complex space can be utilized for OCT imaging.

The key element in this full-range complex FDOCT is the introduction of a constant modulation frequency during each B-scan. The most straightforward way to achieve this goal is to mount the reference mirror onto a piezo-stage [34]. An alternative approach is to offset the incident beam in the sample arm, away from the x-scanner pivot point [35] [Figure 15.3(a)]. In the latter method, the introduced modulation frequency can be presented as:

$$f_c = \frac{2k_0\delta}{\pi}\omega \tag{15.7}$$

where k_0 is the central wave number of the light source; δ is the offset distance of the incident light beam from the pivot of the x-scanner; ω is the angular velocity of the scanning mirror. From Eq. (15.7), f_c is linearly proportional to the displacement δ as well as the angular velocity of scanner during the imaging [Figure 15.3(b), where ω was set at ~0.21 rad/s when obtaining this curve]. Compared with the first method that moves the reference mirror in order to introduce the modulation frequency in the interferograms, the latter does not need additional hardware on top of the existing FDOCT system, which is an apparent advantage in practice. The most important

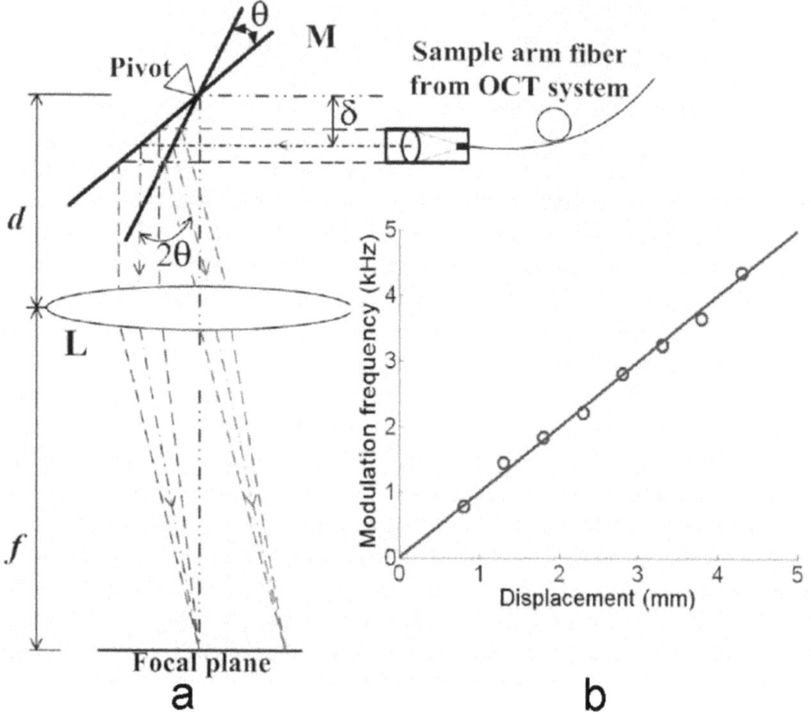

FIGURE 15.3: Schematic (a) showing how the collimated beam in the sample arm deflects from the scanning mirror, and is then focused by the objective lens L. Inset (b), measured modulation frequency (circles) as a function of the displacement δ along with the predictions (solid curve) [35].

advantage for such approach, i.e., constant frequency modulation in the interferograms, is that the FDCT can achieve full-range complex imaging without any limitation on the imaging speed.

15.3.2 OMAG flow imaging

Based on full-range complex FDOCT technology stated in the last section, a constant modulation frequency is introduced into the B-scan of OCT data. The mathematical analysis of the OMAG method essentially maps velocities moving into the tissue away from the surface into one image and velocities moving out of the tissue toward the surface into a second image. Consider a real function that varies with two variables, both of time coordinates, t_1 and t_2

$$B(t_1,t_2) = \cos(2\pi f_0 t_1 + 2\pi(f_M - f_D)t_2 + \phi) \tag{15.8}$$

where f_0, f_M, and f_D are the frequency components, respectively and ϕ is a random phase term. In connection with the discussion below, we assume that f_0 and f_M are two modulation frequency components whereas f_D is a Doppler frequency component. We also assume that there is no correlation between t_1 and t_2, and when t_1 varies t_2 is constant and vice versa. The analytic function of Eq. (15.8) against t_2 can be constructed through the well-known Hilbert transformation if the Bedrosian theorem holds [41, 42], which states that if the modulation frequency $f_M - f_D$ does not overlap the signal bandwidth caused by the random-phase fluctuation term ϕ. Under this condition, the Hilbert transform of Eq. (15.8) is equal to its quadrature representation. Bear in mind that the function $B(t_1,t_2)$ is modulated by the frequency $f_M - f_D$, and $2\pi f_0 t_1$ is a constant phase term. Therefore, if

$f_M - f_D > 0$ the analytic function of Eq. (15.4) can be written as:

$$\tilde{H}(t_1,t_2) = \cos(2\pi f_0 t_1 + 2\pi(f_M - f_D)t_2 + \phi) + j\sin(2\pi f_0 t_1 + 2\pi(f_M - f_D)t_2 + \phi) \tag{15.9}$$

where $j = \sqrt{-1}$; whereas if $f_M - f_D < 0$, it is

$$\tilde{H}(t_1,t_2) = \cos(2\pi f_0 t_1 + 2\pi(f_M - f_D)t_2 + \phi) - j\sin(2\pi f_0 t_1 + 2\pi(f_M - f_D)t_2 + \phi) \tag{15.10}$$

From the mathematical point of view, Eq. (15.10) is clearly the complex conjugate of Eq. (15.9). Now we perform the Fourier transformation against the time variable t_1. (Note that t_2 is now constant.) It is obvious that the frequency component f_0 of Eq. (15.9) is placed in the positive space in the entire Fourier plane, while it sits on the negative space for Eq. (15.10). This is the key element that makes OMAG to image the microvascular flow within the tissue.

Consider the case of OMAG, if we assume that the reference mirror mounted on the piezo-stage moves at a velocity $\vec{v}_{ref}$, with the probe beam proceeding in the B-scan (x-scan) at a velocity of v_x (scalar), and we further assume that a reflecting particle that is detected by OMAG also moves but with its directional velocity projecting onto the probe beam direction being $\vec{v}_s$, then for simplicity we can state the spectral interferogram in the wavelength domain λ as:

$$B(1/\lambda, x) = \cos\left(\frac{4\pi(z_s + (\vec{v}_{ref} + \vec{v}_s)\frac{x}{v_x})}{\lambda} + \varphi(x,z,\lambda)\right) \tag{15.11}$$

Note here that we use vector representations for velocities with the movement towards the incident beam being positive and that opposite being negative. The term z_s is the initial depth position of the reflecting particle (e.g., the red blood cell) at the lateral position x, and $\vec{v}_s$ is the velocity of that reflecting particle, such that the path length difference between sample arm and reference arm is $2(z_s + (\vec{v}_{ref} + \vec{v}_s)t_x)$, where $t_x = x/v_x$ is the scanning time of the probe beam in the B scan, and the factor of 2 accounts for the round trip of the sampling light scattered from the sample back into the interferometer. The term $\phi(x,z,\lambda)$ is a random phase function that relates to the phases of the optical heterogeneous sample. The time $t_x = 0$ would be the start of a B scan. Hence, $B(1/\lambda, x)$ is a sinusoidal oscillation function versus x for each $1/\lambda$ value. It is clear that Eq. (15.8) and Eq. (15.11) are identical if the following substitutions are used:

$$f_0 = z_s,\ t_1 = \tfrac{2}{\lambda},\ f_M = \tfrac{2\vec{v}_{ref}}{\lambda},\ f_D = -\tfrac{2\vec{v}_s}{\lambda},\ t_2 = t_x \tag{15.12}$$

Thus, the values of $\vec{v}_{ref}$ and $\vec{v}_s$ determine whether the analytic function of Eq. (15.11) constructed through the Hilbert transformation is turned into Eq. (15.9) or Eq. (15.10). The analytic function is sequentially constructed through the Hilbert transformation to the B-scan along the x-axis at each $1/\lambda$. During the operation, the factor $4\pi z_s/\lambda$ is simply a constant phase since it does not vary with x.

If $\vec{v}_s = 0$, positive velocities (v_{ref}) would modulate the signal with positive frequencies in x-space, and negative v_{ref} with negative frequencies. The Hilbert transformation converts the information of the modulated signal versus x into a complex number $\tilde{H}(1/\lambda, x)$, but now any subsequent Fast Fourier Transform (FFT) towards $2/\lambda$, $FFT\{\tilde{H}(1/\lambda,x)\}|_{2/\lambda}$ of the Hilbert-encoded information would map positive frequency components into the positive-frequency space and map negative frequency components into the negative-frequency space of the FFT result, leading to that the full-range frequency space can be utilized for imaging [27]. This is in contrast to simply taking $FFT\{B(1/\lambda,x)\}|_{2/\lambda}$, which always maps both positive and negative frequencies into both the positive- and negative-frequency spaces of the transform, resulting in only half of the space being useful for imaging.

Now, consider the effect of a moving particle, $v_s \neq 0$. The particle movement will modify the modulation frequency through the velocity mixing, $\vec{v}_{ref} + \vec{v}_s$, similar to the frequency mixing in the signal processing discipline. An opposite movement of the particle, e.g., blood cells, relative to the movement of the reference mirror results in a decrease of the difference in photon path length between sample and reference arm and decreases the effective frequency of the modulation. If the value of v_s is sufficiently large, the value $\vec{v}_{ref} + \vec{v}_s$ will change its sign. Consequently, after the operations of the Hilbert and Fourier transforms, the corresponding signal due to the particle movement will map into the frequency space opposite to that with $v_s = 0$ However, any small particle movement that is not strong enough to change the sign of the value of $\vec{v}_{ref} + \vec{v}_s$ will still map to the frequency space of when v_s =0. Hence, the signals from the perfused blood cells and the bulk static tissue can be separated in the frequency space of FFT, with the background noise due to small tissue movements rejected in the space that represents the image of blood perfusion.

Figure 15.4 shows an example that was obtained through the OMAG method by imaging a mouse brain cortex with the skull left intact. Figure 15.4(B) shows the image obtained from the raw spectral B-scan interferograms [Figure 15.4(A)]. The entire Fourier space of it is separated into two equal regions. The bottom region is the positive frequency space containing the normal OCT cross-sectional image, static image, within which the histologically important layers such as the cranium and cortex may be clearly demarcated. While on the other hand, the moving signals from all blood vessels, including capillaries, can be precisely localized in the negative space, i.e., the perfusion image [upper region, Figure 15.4(B)] in OMAG. As the positive and negative spaces are exactly mirrored, they can be folded to fuse into a single image to localize with high precision the blood vessels within the tissue [Figure 15.4(C)].

15.3.3 Directional OMAG flow imaging

The previous section introduced the base OMAG method to image blood flow; however, this base method does not give a capability to image the directional flow, which seriously limits the applications for OMAG because the blood flow information regarding its direction is often important in a number of investigations, for example, in the study of complex flow dynamics in the microfluidic mixers and in the investigation of blood flow involvement in cerebrovascular diseases such as ischemia, hemorrhage, vascular dementia, traumatic brain injury, and seizure disorders. To overcome this drawback, Wang [30, 31] proposed a method that forces the reference mirror to move back and forth. But this would significantly increase imaging complexity, for example, the OMAG imaging speed is reduced by half and the computational load on OMAG is doubled to obtain meaningful blood flow images because OMAG needs to acquire two three-dimensional (3D) volumetric spectrogram datasets. It was then quickly realized that a digital modulation frequency (DMF) method can be used to manipulate the modulated interferograms so that the directional flow information can be obtained from one 3D volumetric spectrogram data set [33], without resorting to the hardware approach in [30, 31].

Following the OMAG method described in the last section, assume the frequency modulation in the interferograms is f_M, which is a constant frequency. Here f_M can be provided by a number of approaches, for example, moving the reference mirror at a constant velocity in one direction [30, 32] or offsetting the sample beam at the scanner that gives the B-scan image [35]. For simplicity, the real function of a spectral interferogram can be expressed by Eq. (15.8). If we construct the analytic function of Eq. (15.8) by performing the Hilbert transform in terms of t_1 (note that t_2 is constant), then this analytic function is always Eq. (15.9), because f_0 is always > 0, which is guaranteed by placing the sample surface below the zero delay line. In this case, if we digitally multiplying Eq. (15.9) with a complex function $\exp(-j4\pi f_M t_2)$ and then taking the real part of the results, we can simply transfer Eq. (15.8) into:

$$B'(t_1, t_2) = \cos(2\pi f_0 t_1 - 2\pi(f_M + f_D)t_2 + \phi) \tag{15.13}$$

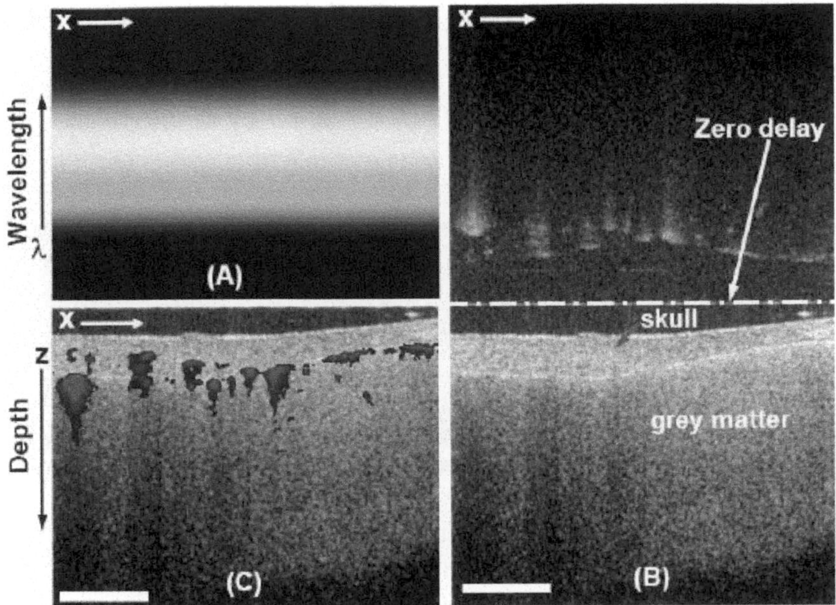

FIGURE 15.4: A cross-section (slice) of an adult mouse brain with the skull intact was imaged with OMAG *in vivo*. (A) The data set representing the slice contains 2-D real-valued spectral interferogram in $x-\lambda$. (B) The imaging result obtained from (A) by OMAG. It is separated by the zero delay line into two equal spaces. The normal OCT image is located in the bottom region (2.2 by 1.7 mm) showing microstructures. The signals from the moving blood cells are located in the upper region (also 2.2 by 1.7 mm). (C) Folded and fused final OAG image (2.2 by 1.7 mm) of the slice showing the location of moving blood (dark spots) within the cortex and mininges where there are more blood vessels than in the skull bone. The scale bar in (B) and (C) represents 500 μm [30].

It is feasible because f_M is known a *priori*. Now, we can construct the analytic function of Eq. (15.13) by considering t_2 a variable, while t_1 is constant. Then, if $f_M + f_D < 0$, the analytic function is

$$\tilde{H}(t_1,t_2) = \cos(2\pi f_0 t_1 - 2\pi(f_M + f_D)t_2 + \phi) + j\sin(2\pi f_0 t_1 - 2\pi(f_M + f_D)t_2 + \phi) \qquad (15.14)$$

but if $f_M + f_D > 0$, the analytic function becomes

$$\tilde{H}(t_1,t_2) = \cos(2\pi f_0 t_1 - 2\pi(f_M + f_D)t_2 + \phi) - j\sin(2\pi f_0 t_1 - 2\pi(f_M + f_D)t_2 + \phi) \qquad (15.15)$$

Consequently, Eqs. (15.13)–(15.15) simulate the exact situation of the mirror moving in an opposite direction to that as described in Section 3.2. This derivation of the analytic function provides a solid mathematical basis for obtaining information on the directions of blood flow from only one 3D spectrogram data set captured by OMAG. As a result, the directional blood flow imaging can be obtained by the following approach: the analysis detailed in section 3.2 to give one image of blood flow direction when the reference mirror moves toward the incident beam, and the analysis described here to give an image of blood flow in the opposite direction, as if the mirror moved away from the incident beam.

Figure 15.5 shows the imaging results obtained through OMAG method and DMF approach. Two microtubes, within which scattering particles are flowing with flow directions against each other,

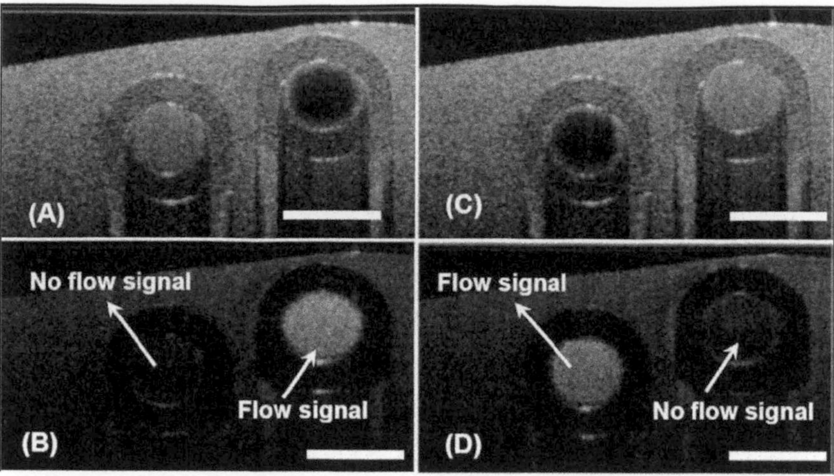

FIGURE 15.5: (A) and (B) are the OMAG structural and flow images calculated from the spectrograms captured when $f_M = 400$ Hz. (C) and (D) are corresponding results computed from the DFM approach applied to the same dataset. Scale bar = 500 μm [33].

were buried under a scattering phantom. Figure 15.5(A) and (B) are the microstructure and flow images respectively, which directly obtained from OMAG algorithm. Applying DMF approach, different direction flow and microstructure images are present in Figure 15.5(C) and (D). It is shown that the directional flow images can be successfully achieved by OMAG and DMF algorithms.

15.4 OMAG System Setup

As analyzed in the above sections, as long as a constant modulation frequency is introduced into the OCT B-scan, OMAG algorithm can be applied to achieve flow image. Both SDOCT and SSOCT configurations can be used to setup an OMAG system.

Figure 15.6 shows a schematic of a typical OMAG system based on the spectral domain Fourier-domain OCT system. In this case, the system typically employs a superluminescent diode (SLD) as the light source, with its bandwidth and central wavelength determining the axial resolution of the system. The light eliminated from the SLD is coupled into a fiber-based Michelson-based optical interferometer where it splits into two arms, the reference arm and the sample arm. In reference arm, the mirror is either stationary or mounted on a piezo-stage. In the sample arm, the light is delivered onto the sample by a probe, in which a pair of X-Y galvanometer scanners is used to achieve 2D scanning. The collimating and objective lens in the probe determine the lateral resolution of the system. The light backscattered from the sample and reflected from the mirror is then collected by the fiber coupler and delivered into a high-speed spectrometer for capturing the spectral interferograms. The spectral resolution in the spectrometer, together with the wavelength of the broadband source, determines the maximum detectable range of the system. Because the whole system is based on the optical fibers, polarization controllers (PC) are often used in the reference arm, sample arm, and/or detection arm in order to maximum the spectral interference fringe contrast at the detector.

For OMAG technology, the key point is the introduction of modulation frequency, which will make it feasible to separate the moving signals from the static background tissue signals. Currently,

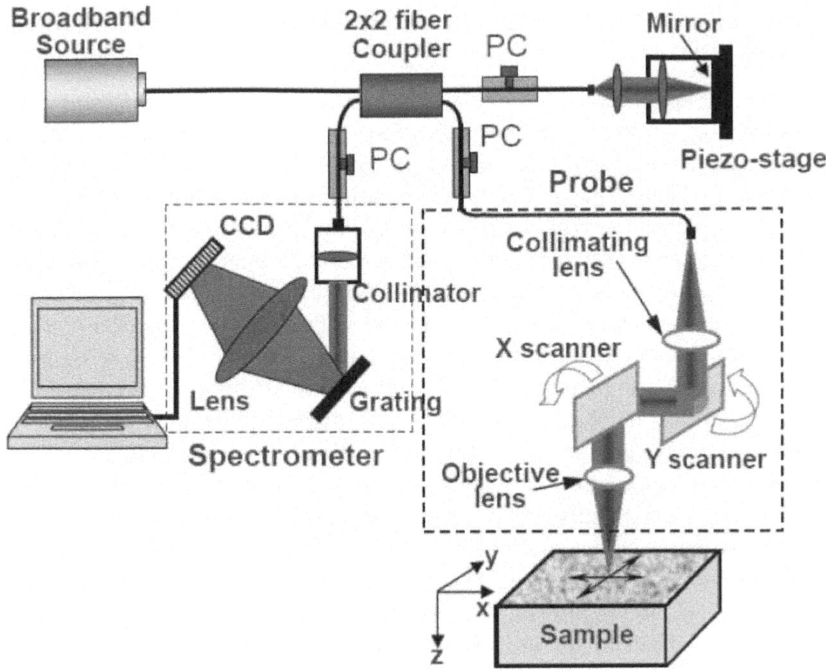

FIGURE 15.6: Schematic of the OAG system used to collect the 3-D spectral interferogram data cube to perform the 3-D angiogram of thick tissue *in vivo*. CCD: the charge coupled device, PC: the polarization controller.

there are two methods from our group to achieve this purpose. The first one is mounting the reference mirror onto a linear pizeo-stage [30, 34], which moves the mirror with a constant velocity during one B-scan. Another approach is to offset the incident beam of the sample arm away from the pivot of the x scanner [35]. In addition, other methods are possible to achieve the same purpose, for example, by placing a phase modulator in either the reference arm or sample arm or both [43, 45], and by using a piezo-electric fiber stretcher like those to achieve full-range complex FDOCT [46].

15.5 OMAG Imaging Applications

Since OMAG has high spatial resolution and velocity sensitivity, it has been used in different applications to visualize the microcirculatory vessel networks within the tissue bed. Below, we give some examples of using OMAG to visualize the vascular perfusion within the human retina and choroids [36], and to image the cerebral blood perfusion within the cortex of the brain in small animal models [30].

15.5.1 *In vivo* volumetric imaging of vascular perfusion within the human retina and choroids

In ophthalmological applications, the blood vessel network map (or angiogram) over the retina and choroids is often useful to aid the diagnosis and treatment for clinicians and physicians. Accurate ocular circulation map, including depth information, could provide quite important clinical information for the ophthalmic diagnosis as well as for the study of a number of eye diseases, such as glaucoma. OMAG presents a great capability for noninvasive, depth-resolved, high-spatial resolution and high-velocity sensitivity for ocular circulation within both the retina and choroid [36].

For OMAG imaging of the ocular blood vessels, a serious problem needs to be considered, that is the bulk motion artifacts caused by the eye movements. This bulk motion could provide an additional modulation frequency to the interferogram captured by the OMAG system. In this case, the background tissue will also be mapped to the perfusion image for its movements. To eliminate these bulk motion artifacts, the phase-resolved optical Doppler tomography technique can be used to estimate the phase change caused by the bulk tissue movements. The estimated phases are then used to compensate the original interferograms that carry the information of motion to obtain the correct ones without motion. Finally, OMAG can be applied to the compensated interferograms to obtain flow images within the scanned tissue volume.

Figure 15.7 shows the *in vivo* imaging results produced by one volume dataset captured at a position near to the optic disk. Figure 15.7(A) gives the OCT fundus projection image obtained by integrating signals along the depth direction from the volumetric OCT structural images [47], where the major blood vessels over the retina can be seen, but not in the choroid. Figure 15.7(B) represents one cross-sectional image within the volumetric OCT image at the position marked by the red line in Figure 15.7(A). Figure 15.7(C) shows the 2D OMAG flow images of the cross-sections after the bulk motion was corrected, where it can be seen that the blood flows within the retina and choroidal layers are clearly delineated. The results also demonstrate that the motion compensation method worked well for the volumetric OMAG imaging. In the retinal region, not only some big vessels, but also the capillary vessels are presented. Because OMAG gives the volumetric structural and flow images simultaneously, the 3D vasculatures can be merged into the 3D structural image. Such a merged volumetric image is illustrated in Figure 15.7(D), where a cut-through view in the center of the volumetric structural image is used to better appreciate how the blood vessels innervate the tissue. To separately view the blood vessels within the retina and choroidal layers, we used the segmentation method [48] to produce two masks, from the three-dimensional OMAG structural image, that represent the retina and choroids. In a cross-sectional view, the mask that represents the retina is shown as the red and yellow lines in Figure 15.7(B), and that represents the choroid as the green and blue lines. To render the volumetric flow image, the blood flow signals within the boundaries of two masks are coded with green color (retina) and red color (choroid). The result is shown in Figure 15.7(E), where the flow image shows us two vessel networks with good connection between vessels.

To view the blood vessels in detail, Figure 15.8(A) gives the *x-y* projection image with the blood vessels in retina coded with green color and those in choroids with red color. Figures 15.8(B) and 15.8(C) illustrate the *x-y* projection images produced from blood vessels in retina and choroids, separately. Compared with Figure 15.7(A), the OMAG image shows not only the retina vessels, but also the choroid vessels. What one should pay attention to is the horizontal lines visible in Figure 15.8(C), the appearance of which are periodical.

15.5.2 Imaging cerebral blood perfusion in small animal models

OMAG has also been employed to visualize the cerebral microvascular networks of a small animal model, such as the mouse. In the treatment and diagnosis of neural vascular diseases, the cerebral blood perfusion is related to the normal and pathophysiologic conditions of brain metabolism

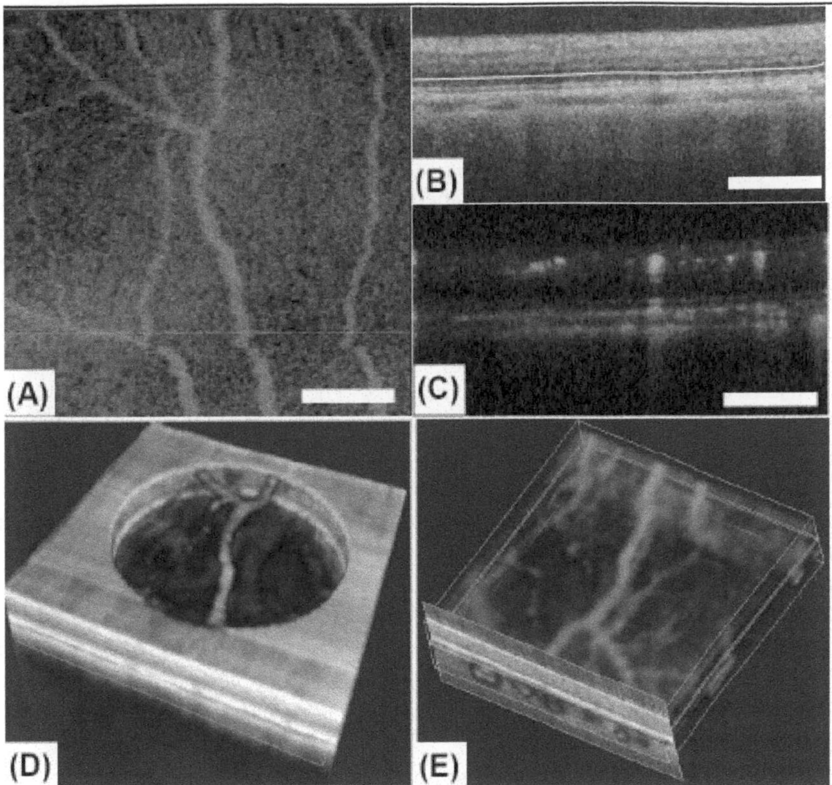

FIGURE 15.7: (See in color after page 572.) *In vivo* volumetric imaging of the posterior chamber of an eye from a volunteer. (A) OCT fundus image of the scanned volume. (B) OCT cross-sectional image at the position marked red in (A), in which four lines separate the blood flows in the retina and choroids. (C) Flying through a movie that represents the 2D OMAG flow images within the scanned volume. (D) Volumetric rendering of the merged structural and flow images with a cut through in the center of the structural image. (E) Volumetric rending of the blood flow image where flows in the retina are coded with green and those in choroids with red. Scale bar = 500 μm [36].

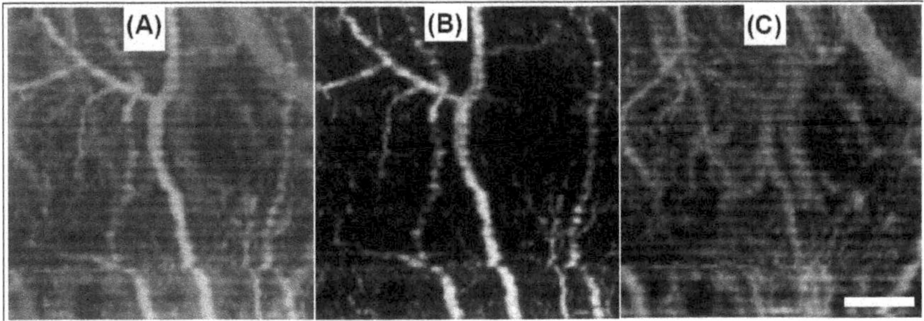

FIGURE 15.8: (See in color after page 572.) *x*-*y* projection images from 3D OMAG blood flow images. (A) Projection image from the whole scanned volume with the blood vessels in the retina are coded with green color, and those in choroids with red color. (B) *x*-*y* projection image from the blood vessels within the retina only. (C) *x*-*y* projection image from the blood vessels within the choroids only. Scale bar = 500 μm [36].

[49, 50]. Numerous techniques have been applied to image the cerebral blood flow (CBF) and volume changes of a small animal model. Autoradiographic methods could not provide dynamic CBF evolution [51], though they contain 3D spatial information. Magnetic resonance imaging [52] and positron emission tomography [53] provide spatial maps of CBF but are limited in their temporal and spatial resolution. Optical intrinsic signal imaging [54], laser speckle imaging [55], and laser-Doppler flowmetry techniques [56] could provide high-resolution images, but are limited to two-dimensional mapping of CBF. Photoacoustic has recently been reported to map vascular structures deep within the brain of small animals [57, 58]. However, the spatial ($\sim$100 μm) resolution of this approach is limited by the viscoelastic filtering of higher acoustic frequencies by the tissue. Moreover, blood flow information is not forthcoming with this method. Confocal microscopy, a widely used technique that is capable of ultrahigh resolution mapping of the cerebrovasculature [59], also requires the removal of intervening bone due to its limited imaging depth (up to 300 μm) and the use of injected fluorescent tissue markers. The OMAG method, which could provide a depth-resolved and noninvasive vascular perfusion image including flow information, makes it a powerful tool for imaging CBF.

Figure 15.9(A) shows the result of the OMAG method of mapping of blood flows within the skin of the mouse, while the cerebrovascular blood perfusion is given in Figure 15.6(B). For comparison, the photograph of the head right after the OMAG imaging and the photograph of the brain cortex after removal of the skin and skull bone are shown in Figure 15.6(C) and Figure 15.6(D), respectively. From Figure 15.6(C), seeing the blood vessels through the skin is almost impossible. Comparing between Figure 15.6(B) and Figure 15.6(D), excellent agreement on the major vascular network over the brain cortex is achieved. Furthermore, the smaller blood vessels not observed in the photograph of Figure 15.6(D) can be seen in the OMAG images, indicating the OMAG capability to delineate the capillary level vessels.

Figure 15.10 shows another OMAG imaging result obtained from an ischemic thrombotic stroke experiment on a mouse model. Figure 15.10(A) shows the blood flow in the cerebral cortex of a mouse with the skull intact. Occlusion of one cerebral artery does not cause cerebral infarction or neurological deficits in the mouse. The image of Figure 15.10(B) was taken from the same animal while the right cerebral artery was blocked for 5 minutes. It is apparent that the blood flow in the cortex, rather than in the right hemisphere only, is reduced as compared to Figure 15.10(A). This is consistent with the well-known collateral circulation between brain hemispheres leading to a total reduction in cortical blood volume due to the occlusion of one common cerebral artery. The capacity of OMAG to achieve such high-resolution imaging of the cerebral cortex – within minutes and without the need for dye injections, contrast agents or surgical craniotomy – makes it an extremely valuable tool for studying the hemodynamics of the brain.

15.6 Conclusions

OMAG is a recently developed imaging technology that can provide high-resolution perfusion imaging of blood vessel networks, down to the capillary level, buried in highly scattering tissue. Based on full-range complex FDOCT, this technology needs an introduction of a constant modulation frequency into one OCT B-scan for separating the moving single from the static tissue signal. OMAG do not need an exogenous agent to provide image contrast, which means it is noninvasive imaging modality. The depth-resolved capability makes it capable of precise localization of the blood perfusion in 3D microstructures. Due to its superior performance on noninvasive mapping blood vessel networks and high imaging speed, OMAG can be widely used in numerous *in vivo* applications. For example, OMAG can be used to *in vivo* monitor the ocular blood perfusion of the

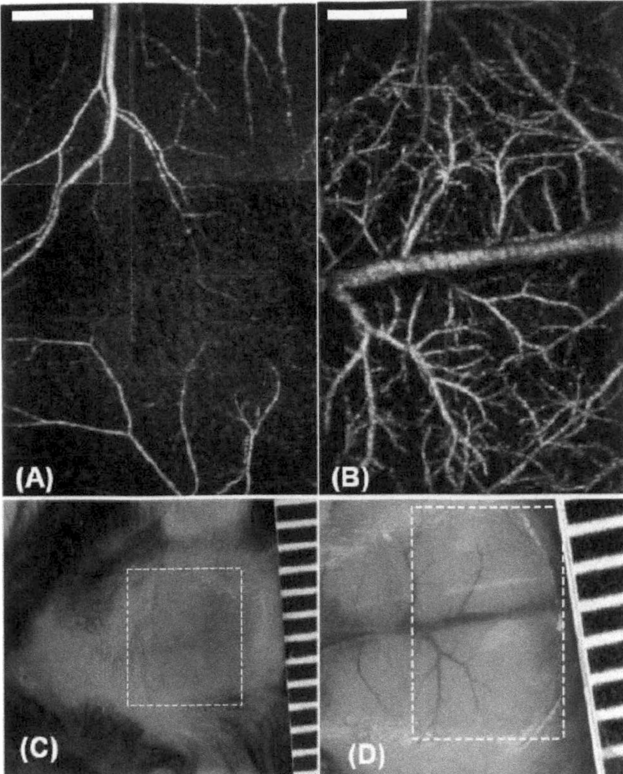

FIGURE 15.9: The head of an adult mouse with the skin and skull intact was imaged with OMAG *in vivo*. (A) and (B) are the projection views of blood perfusion from within the skin and the brain cortex, respectively. (C) Photograph taken right after the experiments with the skin and skull intact. (D) Photograph showing blood vessels over the cortex after the skin and skull of the same mouse were carefully removed. The scale bar indicates 1.0 mm [32].

human eye within both the retina and choroid. And also, it can be used to visualize the cerebral blood flow perfusion in small animal models, such as the mouse, for the studies of stroke, hemorrhage, and neural stimulations *in vivo*. More importantly, the imaging can be performed with the skull left intact.

Acknowledgment

The work presented in this chapter was supported in part by research grants from the National Heart, Lung, and Blood Institute (R01 HL093140, and R01 HL094570), National Institute of Biomedical Imaging and Bioengineering (R01 EB009682), National Institute of Deafness and other Communication Disorders (R01 DC010201), and the American Heart Association (0855733G). The content is solely the responsibility of the authors and does not necessarily represent the official views of the grant-giving bodies.

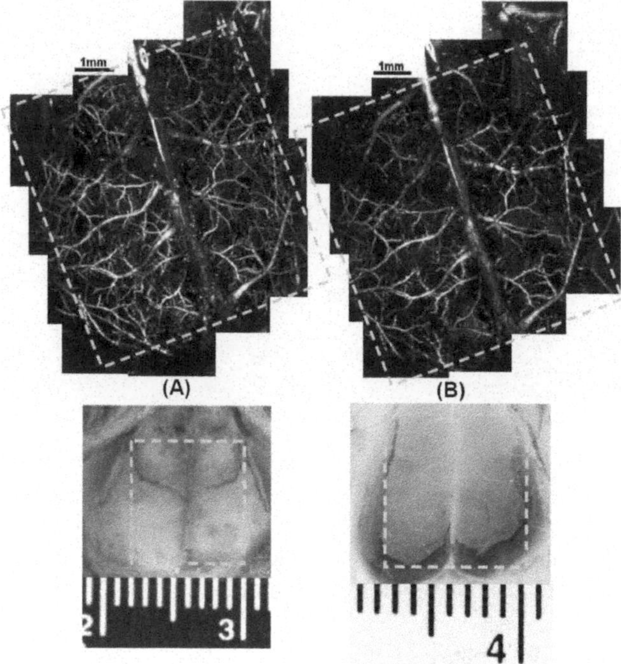

FIGURE 15.10: The entire cerebrovascular flow over the cortex of an adult mouse with the skull intact was imaged with OMAG *in vivo*. (A) and (B) are the projection views of blood perfusion before and after the right cerebral artery was blocked. (C) Photograph of the skull with the skin folded aside, taken right after the experiments. (D) Photograph showing blood vessels over the cortex after the skull and the meninges of the same mouse were carefully removed [30].

References

[1] A.F. Fercher, W. Drexler, C.K. Hitzenberger, and T. Lasser, "Optical coherence tomography – principles and applications," *Rep. Prog. Phys.*, **66**, 239–303 (2003).

[2] P.H. Tomlins and R.K. Wang, "Theory, development and applications of optical coherence tomography," *J. Phys. D: Appl. Phys.* **38**, 2519–2535 (2005).

[3] D. Huang, E. Swanson, C. Lin, J. Schuman, W. Stinson, W. Chang, M. Hee, T. Flotte, K. Gregory, C. Puliafito, and J. Fujimoto, "Optical coherence tomography," *Science*, **254**, 1178–1181 (1991).

[4] Z. Chen, T.E. Milner, S. Srinivas, X.J. Wang, A. Malekafzali, M.J.C. van Gemert, and J.S. Nelson, "Noninvasive imaging of in vivo blood flow velocity using optical Doppler tomography," *Opt. Lett.* **22**, 1119–1121 (1997).

[5] J.A. Izatt, M.D. Kulkarni, S. Yazdanfar, J.K. Barton, and A.J. Welch, "In vivo bidirectional color Doppler flow imaging of picoliter blood volumes using optical coherence tomography," *Opt. Lett.* **22**, 1439–1441 (1997).

[6] Z. Chen, T.E. Milner, D. Dave, and J.S. Nelson, "Optical Doppler tomographic imaging of fluid flow velocity in highly scattering media," *Opt. Lett.* **22**, 64–66 (1997).

[7] Z. Chen, Y. Zhao, S.M. Srinivas, J.S. Nelson, N. Prakash, and R.D. Frostig, "Optical Doppler tomography," *IEEE J. of Selected Topics in Quantum Electronics* **5**, 1134–1141 (1999).

[8] Y. Zhao, Z. Chen, C. Saxer, S. Xiang, J.F. de Boer, and J.S. Nelson, "Phase-resolved optical coherence tomography and optical Doppler tomography for imaging blood flow in human skin with fast scanning speed and high velocity sensitivity," *Opt. Lett.* **25**, 114–116 (2000).

[9] Y.H. Zhao, Z.P. Chen, Z.H. Ding, H. Ren, and J.S. Nelson, "Real-time phase-resolved functional optical coherence tomography by use of optical Hilbert transformation," *Opt. Lett.* **25**, 98–100 (2002).

[10] H.W. Ren, Z.H. Ding, Y.H. Zhao, J. Miao, J.S. Nelson, and Z.P. Chen, "Phase-resolved functional optical coherence tomography: simultaneous imaging of in situ tissue structure, blood flow velocity, standard deviation, birefringence, and Stokes vectors in human skin," *Opt. Lett.* **27**, 1702–1704 (2002).

[11] D.H. Evans and W.N. McDicken, *Doppler Ultrasound: Physics, Instrumentation, and Signal Processing*, John Wiley & Sons, Ltd. Chichester, England (2000).

[12] R. Leitgeb, C.K. Hitzenberger, and A.F. Fercher, "Performance of Fourier domain vs. time domain optical coherence tomography," *Opt. Express* **11**, 889–894 (2003).

[13] J.F. de Boer, B. Cense, B.H. Park, M.C. Pierce, G.J. Tearney, and B.E. Bouma, "Improved signal-to-noise ratio in spectral-domain compared with time-domain optical coherence tomography," *Opt. Lett.* **28**, 2067–2069 (2003).

[14] M.A. Choma, M.V. Sarunic, C. Yang, and J.A. Izatt, "Sensitivity advantage of swept source and Fourier domain optical coherence tomography," *Opt. Express* **11**, 2183–2189 (2003).

[15] R.A. Leitgeb, L. Schmetterer, C.K. Hitzenberger, A.F. Fercher, F. Berisha, M. Wojtkowski, and T. Bajraszewski, "Real-time measurement of in vitro flow by Fourier domain color Doppler optical coherence tomography," *Opt. Lett.* **29**, 171–173 (2004).

[16] J. Zhang and Z. P. Chen, "In vivo blood flow imaging by a swept laser source based Fourier domain optical Doppler tomography," *Opt. Express* **13**, 7449–7459 (2005).

[17] B.J. Vakoc, S.H. Yun, J.F. de Boer, G.J. Tearney, and B.E. Bouma, "Phase-resolved optical frequency domain imaging," *Opt. Express* **13**, 5483–5492 (2005).

[18] Y. Zhao, Z. Chen, C. Saxer, S. Xiang, J.F. de Boer, and J.S. Nelson, "Phase-resolved optical coherence tomography and optical Doppler tomography for imaging blood flow in human skin with fast scanning speed and high velocity sensitivity," *Opt. Lett.* **25**, 114–117 (2005).

[19] R.A. Leitgeb, L. Schmetterer, W. Drexler, A.F. Fercher, R.J. Zawadzki, and T. Bajraszewski, "Real-time assessment of retinal blood flow with ultrafast acquisition by color Doppler Fourier domain optical coherence tomography," *Opt. Express* **11**, 3116–3121 (2003).

[20] B.R. White, M.C. Pierce, N. Nassif, B. Cense, B. Park, G. Tearney, B. Bouma, T. Chen, and J. de Boer, "In vivo dynamic human retinal blood flow imaging using ultra-high-speed spectral domain optical Doppler tomography," *Opt. Express* **11**, 3490–3497 (2003).

[21] R.A. Leitgeb, L. Schmetterer, W. Drexler, A.F. Fercher, R.J. Zawadzki, and T. Bajraszewski, "Real-time assessment of retinal blood flow with ultrafast acquisition by color Doppler Fourier domain optical coherence tomography," *Opt. Express* **11**, 3116–3121 (2003).

[22] B.J. Vakoc, S.H. Yun, J.F. de Boer, G.J. Tearney, and B.E. Bouma, "Phase-resolved optical frequency domain imaging," *Opt. Express* **13**, 5483–5493 (2005).

[23] R.A. Leitgeb, L. Schmetterer, C.K. Hitzenberger, A.F. Fercher, F. Berisha, M. Wojtkowski, and T. Bajraszewski, "Real-time measurement of in vitro flow by Fourier-domain color Doppler optical coherence tomography," *Opt. Lett.* **29**, 171–174 (2004).

[24] R.K. Wang and Z.H. Ma, "Real-time flow imaging by removing texture pattern artifacts in spectral-domain optical Doppler tomography," *Opt. Lett.* **31**, 3001–3003 (2006).

[25] S.H. Yun, G.J. Tearney, J.F. de Boer, and B.E. Bouma, "Motion artifacts in optical coherence tomography with frequency-domain ranging," *Opt. Express* **12**, 2977–2998 (2004).

[26] H.W. Ren, Z.H. Ding, Y.H. Zhao, J.J. Miao, J.S. Nelson, and Z.P. Chen, "Phase-resolved functional optical coherence tomography: simultaneous imaging of in situ tissue structure, blood flow velocity, standard deviation, birefringence, and Stokes vectors in human skin," *Opt. Lett.* **27**, 1702–1704 (2002).

[27] B.H. Park, M.C. Pierce, B. Cense, S.H. Yun, M. Mujat, G. Tearney, B. Bouma, and J. de Boer, "Real-time fiber-based multi-functional spectral-domain optical coherence tomography at 1.3 μm," *Opt. Express* **13**, 3931–3944 (2005).

[28] M. Szkulmowski, A. Szkulmowska, T. Bajraszewski, A. Kowalczyk, and M. Wojtkowski, "Flow velocity estimation using joint spectral and time domain optical coherence tomography," *Opt. Express* **16**, 6008–6025 (2008).

[29] A.H. Bachmann, M.L. Villiger, C. Blatter, T. Lasser, and R.A. Leitgeb, "Resonant Doppler flow imaging and optical vivisection of retinal blood vessels," *Opt. Express* **15**, 408–422 (2007).

[30] R.K. Wang, S.L. Jacques, Z. Ma, S. Hurst, S. Hanson, and A. Gruber, "Three dimensional optical angiography," *Opt. Express* **15**, 4083–4097 (2007).

[31] R.K. Wang, "Three-dimensional optical micro-angiography maps directional blood perfusion deep within microcirculation tissue beds in vivo," *Phys. Med. Biol.* **52**, N531–N537 (2007).

[32] R.K. Wang and S. Hurst, "Mapping of cerebro-vascular blood perfusion in mice with skin and skull intact by optical micro-angiography at 1.3 μm wavelength," *Opt. Express* **15**, 11402–11412 (2007).

[33] R.K. Wang, "Directional blood flow imaging in volumetric optical microangiography achieved by digital frequency modulation," *Opt. Lett.* **33**, 1878–1880 (2008).

[34] R.K. Wang, "In vivo full range complex Fourier domain optical coherence tomography," *Appl. Phys. Lett.* **90**, 054103 (2007).

[35] L. An and R.K. Wang, "Use of scanner to modulate spatial interferogram for in vivo full range Fourier domain optical coherence tomography," *Opt. Lett.* **32**, 3423–3425 (2007).

[36] L. An and R.K. Wang, "In vivo volumetric imaging of vascular perfusion within human retina and choroids with optical micro-angiography," *Opt. Express* **16**, 11438–11452 (2008).

[37] M. Wojtkowski, V.J. Srinivasan, T. Ko, J.G. Fujimoto, A. Kowalczyk, J.S. Duker, "Ultrahigh-resolution, high-speed, Fourier domain optical coherence tomography and methods for dispersion compensation," *Opt. Express* **12**, 2404–2422 (2004).

[38] B. Cense, N. Nassif, T.C. Chen, M.C. Pierce, S.H. Yun, B.H. Park, B.E. Bouma, G.J. Tearney, andJ.F. de Boer, "Ultrahigh-resolution high-speed retinal imaging using spectral-domain optical coherence tomography ," *Opt. Express* **12**, 2435–2447 (2004).

[39] S.H. Yun, G.J. Tearney, J.F. de Boer, N. Iftimia, and B.E. Bouma, "High-speed optical frequency-domain imaging," *Opt. Express* **11**, 2953–2963 (2003).

[40] S.H. Yun, C. Boudoux, G.J. Tearney, and B.E. Bouma, "High-speed optical frequency-domain imaging," *Opt. Lett.* **28**, 1981–1983 (2003).

[41] E.A. Bedrosian, "Product theorem for Hilbert transforms," *Proc. IEEE* **51**, 868–869 (1963).

[42] S.L. Hahn, "Hilbert Transformation" in *The Transforms and Applications Handbook*, A.D. Poularikas, ed., CRC Press, Taylor & Francis Group (2000).

[43] J. Zhang, J.S. Nelson, and Z. Chen, "Removal of a mirror image and enhancement of the signal-to-noise ratio in Fourier-domain optical coherence tomography using an electro-optic phase modulator," *Opt. Lett.* **30**, 147–149 (2005).

[44] E. Gotzinger, M. Pircher, R.A. Leitgeb, and C.K. Hitzenberger, "High speed full range complex spectral domain optical coherence tomography," *Opt. Express* **13,** 583–594 (2005).

[45] A.H. Bachmann, R.A. Leitgeb, and T. Lasser, "Heterodyne Fourier domain optical coherence tomography for full range probing with high axial resolution," *Opt. Express* **14**, 1487–1496 (2006).

[46] S.Vergnole, G. Lamouche, and M.L. Dufour, "Artifact removal in Fourier-domain optical coherence tomography with a piezoelectric fiber stretcher," *Opt. Lett.* **33**, 732–734 (2008).

[47] S. Jiao, R. Knighton, X. Huang, G. Gregori, and C.A. Puliafito, "Simultaneous acquisition of sectional and fundus ophthalmic images with spectral-domain optical coherence tomography," *Opt. Express* **13**, 444–452 (2005).

[48] M. Mujat, R.C. Chan, B. Cense, B.H. Park, C. Joo, T. Akkin, T.C. Chen, and J. F. de Boer, "Retinal nerve fiber layer thickness map determined from optical coherence tomography images," *Opt. Express* **13**, 9480–9491 (2005).

[49] D. Malonek, U. Dirnagl, U. Lindauer, K. Yamada, I. Kanno, and A. Grinvald, "Vascular imprints of neuronal activity: relationships between the dynamics of cortical blood flow,

oxygenation, and volume changes following sensory stimulation," *Proc. Natl. Acad. Sci. USA* **94**, 14826–14831 (1997).

[50] K. Hossmann, "Viability thresholds and the penumbra of focal ischemia," *Ann Neurol* **36**, 557–567 (1994).

[51] O. Sakurada, C. Kennedy, J. Jehle, J.D. Brown, G.L. Carbin, and L. Sokoloff, "Measurement of local cerebral blood flow with iodo [14C] antipyrine," *Am J. Physiol.* **234**, H59–H66 (1978).

[52] F. Calamante, D.L. Thomas, G.S. Pell, J. Wiersma, and R. Turner, "Measuring cerebral blood flow using magnetic resonance imaging techniques," *J. Cereb. Blood Flow Metab.* **19**, 701–735 (1999).

[53] W.D. Heiss, R. Graf, K. Wienhard, J. Lottgen, R. Saito, T. Fujita, G. Rosner, and R. Wagner, "Dynamic penumbra demonstrated by sequential multitracer PET after middle cerebral artery occlusion in cats," *J. Cereb. Blood Flow Metab.* **14**, 892–902 (1994).

[54] A. Grinvald, E. Lieke, R.D. Frostig, C.D. Gilbert, and T.N. Wiesel, "Functional architecture of cortex revealed by optical imaging of intrinsic signals," *Nature* **324**, 361–364 (1986).

[55] A.K. Dunn, H. Bolay, M.A. Moskowitz, and D.A. Boas, "Dynamic imaging of cerebral blood flow using laser speckle," *J. Cereb. Blood Flow Metab.* **21**, 195–201 (2001).

[56] A.N. Nielsen, M. Fabricius, and M. Lauritzen. "Scanning laser-Doppler flowmetry of rat cerebral circulation during cortical spreading depression," *J. Vasc. Res.* **37**, 513–522 (2000).

[57] X.D. Wang, Y. Pang, G. Ku, X. Xie, G. Stoica, and L.-H. Wang, "Non-invasive laser-induced photoacoustic tomography for structural and functional imaging of the brain *in vivo*," *Nat. Biotechnol.* **21**, 803–806 (2003).

[58] H.F. Zhang, K. Maslov, G. Stoica, and L. H. V. Wang, "Functional photoacoustic microscopy for highresolution and noninvasive in vivo imaging," *Nat. Biotechnol.* **24**, 848–851 (2006).

[59] T. Misgeld and M. Kerschensteiner, "*In vivo* imaging of the diseased nervous system," *Nat. Rev. Neurosci.* **7**, 449–463 (2006).

16

Fiber-Based OCT: From Optical Design to Clinical Applications

V. Gelikonov, G. Gelikonov, M. Kirillin, N. Shakhova, A. Sergeev

Institute of Applied Physics RAS, 46 Ulyanov Str., 603950 Nizhny Novgorod, Russia

N. Gladkova, and E. Zagaynova

Nizhny Novgorod State Medical Academy 10/1, Minin and Pozharsky Sq., 603005 Nizhny Novgorod, Russia

This chapter is devoted to the application of a fiber-based OCT device for clinical diagnosis of pathologies. The device design and its customizing by developing optical probes for particular endoscopic applications are discussed. The device and various probes are presented schematically. Examples of using the developed customized OCT modalities in clinical *in vivo* studies are given. The device is shown to be effective for diagnosis of cervical, bladder, and gastrointestinal cancer. In combination with fluorescent cystoscopy OCT is an almost perfect technique for endoscopic diagnosis of early bladder cancer. Because of its high sensitivity to cancer, OCT can also be used for presurgery planning and surgery guidance of the tumor borders.

Key words: optical coherence tomography (OCT), OCT endoscopic probes, polarization-sensitive OCT, cross-polarization OCT, tissue pathology diagnosis, cervical cancer, bladder cancer, gastrointestinal cancer

16.1 Introduction (History, Motivation, Objectives)

The optical coherence tomography (OCT) technique was proposed in 1991 [1] for noninvasive high-resolution (units of microns) imaging of the inner structure of biological objects. The technique is based on the interference selection of the ballistic and snake photons backscattered by a sample under study from a high-intensity multiple scattering fraction. The obtained images characterize spatial distribution of optical inhomogeneities within biotissue specified by its structure and composition. The use of low-coherence CW radiation of a superluminescent diode (SLD) and fiber optics enabled designing a compact Michelson interferometer-based system at the early stage of technique development [1–4]. Even in the pilot work [1] the axial spatial resolution in air was about

17 μm that was provided by the SLD spectrum bandwidth. The transversal spatial resolution of units of micrometers was reached, similarly to confocal microscopy, due to focusing of the probing HE_{11} mode with a given aperture [1].

Since the first publication [1], the fundamental and applied studies of the OCT technique have been carried out resulting in numerous publications. The fundamentals of OCT are described in the book [5]. When developing a device for clinical application one should evidently prefer the fiber optic modality allowing for miniaturization and, hence, convenient access to the areas of interest [5, 6]. The design of a compact device for effective scanning by Michelson interferometer arms mismatch became a key for such miniaturization. One of the most efficient devices is the controlled piezo-fiber delay line [7] with optical path modulation depth up to units of millimeters. A scanning system with such an element can provide high accuracy in maintaining the constant Doppler shift of optical frequencies mismatch in interferometer arms, which is necessary for narrow-band signal detection.

The advantages of fiber-optics OCT modality were also implemented in endoscopic probes design. Catheter-endoscopic probes aimed at transverse vessel scanning [8] were constructed based on the experience gained in designing ultrasound equipment. Advanced potentialities for endoscopic scanning appeared after the development of forward-looking probes [9]. The forward-looking probes may be delivered to tissues of internal organs through the instrumental channels of standard endoscopes. This allows studies of the OCT images of mucous and serous membranes of human organs *in vivo*.

The design of the device for clinical practice must ensure stability of interference patterns of the studied living objects. To avoid the negative effect of induced anisotropy, the fiber arms of the Michelson interferometer were constructed based on polarization maintaining (PM) fibers [4, 9]. Moreover, it was important to provide relatively low sensitivity of the time-domain OCT to object movements, which became one of the advantages of the time-domain modality as compared to the spectral-domain (Fourier-domain) devices. This advantage is gained due to sequential detection of scattered radiation from partial sampling volumes whose size is determined by the coherence length of probing radiation in the time-domain modality, whereas in the spectral-domain modalities each spectral component carries information about the entire scanning depth simultaneously and local movements within the scanning area introduce significant distortions in the reconstructed OCT image.

When applying the OCT technique for clinical studies we obtain wide spectra of OCT image types depicting various pathologies that laid the basis for our OCT image atlas. Analysis gave a set of typical, similarly structureless OCT images of normal and malignant tissue samples [10]. Some of these pathologic cases may be differentiated from normal samples by comparing co- and cross-polarized OCT images. The polarization-sensitive (PS) and cross-polarization (CP) OCT techniques have advanced significantly in recent years [11–13]. The PS OCT [11, 12] allows one to analyze birefringence in biotissues [11, 12], thus enabling diagnosing, for example, the state of tissue burn [12, 14, 15] or early glaucoma [15]. The CP OCT permits one to compare co- and cross-polarization scattering from the sampling volume, which allows differentiating biotissue pathology [13]. This method was implemented almost simultaneously both in open-space [13] and fiber-optics modalities [10, 17].

Note that several PS OCT devices are based on complete analysis of the radiation polarization state in the detection arm of Michelson interferometer giving OCT images of Stokes parameters [18–20]. Direct transfer of such techniques from open-space to PM-fiber optics modality is impossible because of polarization dispersion, as mutual coherence of polarization eigenwaves is lost at the interferometer output. For proper transfer polarization dispersion must be compensated [21] in the low-coherence PM-fiber based interferometer [22]. Note that information about the presence of birefringence in biotissue may be obtained avoiding complete polarization analysis when detecting only a co-polarized component of scattered radiation. In this case, the OCT image exhibits additional stripes, whose frequency is determined by local birefringence distribution.

Co-polarized backscattered radiation may be detected in Michelson interferometers, with arms based on isotropic (single-mode (SM)) or PM fibers. In the case of the SM fiber, the interferometer arms must have polarization isotropy. When the interferometer arms are fixed, this can be achieved by anisotropy compensation with a polarization controller [23]. In the PM-fiber based Michelson interferometer a linearly polarized probing wave is required for birefringence observation [10], and the contrast of stripes obtained in OCT image depends on biotissue orientation [24].

For simultaneous acquisition of both co- and cross-polarized OCT images *in vivo* we have designed a device prototype based on PM fiber with linear polarization of probing wave equipped with a flexible object arm and an endoscopic probe [25]. Two-channel radiation detection in both initial and orthogonal polarizations is based on weak coupling of polarization eigenmodes propagating with different velocities in PM fiber.

The further motivation for practical realizations of the OCT technique was driven by the requirement of compatibility of replaceable flexible probes. Replacement of probes is required for their sterilization and disposable application. In the case of broadband (tens of nanometers) probing radiation, manufacturing of identical flexible optical probes is complicated by fluctuations of PM fiber waveguide dispersion. A novel SM-fiber-based optical scheme was developed to overcome this problem. The optical scheme includes a measuring Fizeau interferometer and an isotropic compensating Michelson interferometer [26]. In this configuration, the probe becomes part of the Fizeau interferometer formed by object layers and surface of the fiber tip. The advantage of the Fizeau interferometer is reproducibility of its optical characteristics.

Optical isotropy is an essential feature of the compensating Michelson interferometer. It is attained by means of 45-degree Faraday cells situated in interferometer arms that provide the "polarization inversion" effect. The method for arbitrary anisotropy compensation in optical SM fibers after double passage in Faraday cells was first proposed and demonstrated in our papers [21, 27]. Further, this method was applied in the works [28–32]. The combination of 45-degree Faraday cells and a reflective mirror is known in the literature as a "Faraday mirror."

The OCT modality based on two interferometers described in the paper [26] ensures replaceability of the probes. However, this excludes performance of polarization (CP and PS) OCT modalities, because of random evolution of the polarization state in the SM fiber under external action affecting probing and detected backscattered radiation. Below we present a novel OCT modality based on the prototype developed in [26], where we overcome the mentioned drawbacks. The optical scheme of the new modality aimed at endoscopic studies allows one to detect changes in polarization characteristics of the backscattered radiation determined by the properties of the biological object.

16.2 Fiber-Based OCT as a Tool for Clinical Application

The novel fiber-based OCT was developed for endoscopic studies in clinical conditions. Similarly to the prototype described in [26], the replaceable endoscopic probes are easily reproducible. The essential feature of the new device is its ability for simultaneous independent detection of backscattered radiation in initial and orthogonal polarizations at random polarization of probing radiation. The physical principles and the design of the device were described in ample detail in [33, 34]; its brief description is given below.

16.2.1 Design of the fiber-based cross-polarization OCT device

The optical scheme of the OCT device is shown in Figure 16.1. Its major part is based on isotropic SM fiber SMF-28 and, similarly to [26], includes a superluminescent diode (SLD) as a low-cohe-

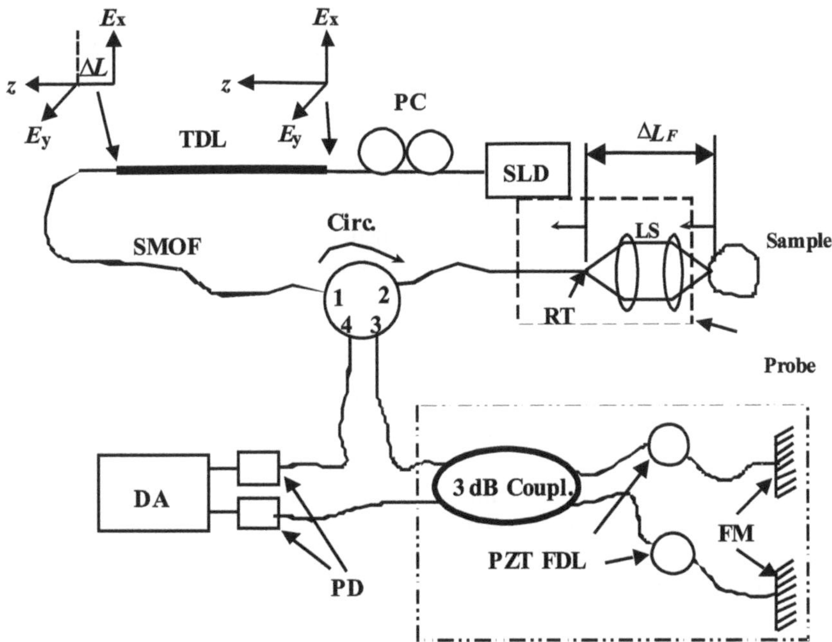

FIGURE 16.1: Optical scheme of CP OCT: SLD – superluminescent diode; PC – polarization controller; TDL – time-delay line consisting of a piece of birefringent fiber; SMOF – single-mode optical fiber; Circ. – 4-port circulator; RT – reflecting tip of the fiber; LS – lens system; Sample – sample under study; 3dB coupler – 0.5 × 0.5 coupler; PZT FDL – piezoelectric fiber delay line; FM – 45° Faraday mirrors; PD – photodiodes; DA – differential amplifier. The dashed box indicates an optical probe that includes Fizeau interferometer, the dashed-dotted box shows a compensating Michelson interferometer that also performs longitudinal A-scanning.

rence source, a measuring Fizeau interferometer, a shared optical path for object and reference waves [34], and a compensating Michelson interferometer with Faraday mirrors.

A time-delay line (TDL) formed by a birefringent PM fiber with length l is a new element. Two orthogonal linearly polarized modes coherently excited at the PM fiber input with amplitudes E_X and E_Y ($E_X = E_Y$) arrive at the TDL output at different times. At the TDL output the fast wave train

$$\vec{e}_X \sim \vec{x}_0 E_X(\omega)\exp\{-i[\beta_X(\omega)\,l - \omega t]\}$$

with group velocity $(v_g)_X$ overtakes the slow wave train

$$\vec{e}_Y \sim \vec{y}_0 E_Y(\omega)\exp\{-i[\beta_Y(\omega)\,l - \omega t]\}$$

with group velocity $(v_g)_Y$ by a given interval with optical length

$$\Delta L = l\left(1 - (v_g)_Y / (v_g)_X\right),$$

which is 10–20% more than the optical depth of A-scan (in this device ΔL = 2 mm). These two waves are transmitted to the SM fiber and excite two orthogonal mutually coherent waves, with the optical path length mismatch remaining unchanged. The polarization state of the two probing radiation waves propagating in the optical scheme generally becomes randomly elliptical.

However, in case of any mechanical perturbation of the fiber leading to phase change, the orthogonality of the waves is preserved if there is no anisotropy of losses. The radiation from the TDL output is impinged through circulator (Circ.) ports 1 and 2 into the probe SM fiber (shown as a dashed line in Figure 16.1) and is further directed as two orthogonally polarized waves $\vec{e}_U \sim \vec{e}_+ E_U(\omega)\exp\{-i[\beta_0 \Delta L]\}$ and $\vec{e}_V \sim \vec{e}_- E_V(\omega)$ to the Fizeau interferometer ($\vec{e}_+$ and $\vec{e}_-$ are the unit vectors of orthogonally polarized elliptic waves with left and right circulations; the factor $\exp(i\omega t)$ is omitted). Like in [26], the Fizeau interferometer of length ΔL_F is formed by the probe fiber surface and object. The shared optical path for the reference and scattered waves from the Fizeau interferometer to the compensating Michelson interferometer, which is also SM fiber-based, equivalently affects the polarization state and optical path lengths of both waves. Thus, the interference signal depends only on the variations of scattered waves parameters relative to the reference ones in the Fizeau interferometer.

In the Fizeau interferometer both, $\vec{e}_U$ and $\vec{e}_V$ waves reflected from the probe surface, produce two orthogonal coherent wave trains with amplitudes $\vec{e}_U^R \sim \vec{e}_U r$ and $\vec{e}_V^R \sim \vec{e}_V r$ (r is the amplitude reflection coefficient). This allows heterodyne detection of scattered radiation having the same polarization states, with stable visibility of the interference pattern. For an anisotropic medium and varying polarization of the scattered waves, the interference will take place if the projections onto the reference waves are nonzero. If the medium anisotropy is irregular, nonzero interference in both channels can be detected due to scattering in initial (co-) and orthogonal (cross-) polarizations.

Let us consider the elliptical coordinate system with unit vectors coinciding with those of the probing waves. In a general case, the backscattered radiation contains four waves. The probing wave $\vec{e}_U$ produces scattered waves (indexed by S) with initial (+) and orthogonal (−) polarizations:

$$\left(\vec{e}_U^S\right)^+ = \vec{e}_+ K_{UU} E_U exp\{-i[k_0(\Delta L + (2\Delta L_F))]\} \tag{16.1}$$

and

$$\left(\vec{e}_U^S\right)^- = \vec{e}_- K_{UV} E_U exp\{-i[k_0(\Delta L + (2\Delta L_F))]\}, \tag{16.2}$$

correspondingly. Analogously, the wave $\vec{e}_V$ also produces waves with initial (+) and orthogonal (−) polarizations:

$$\left(\vec{e}_V^S\right)^+ = \vec{e}_+ K_{VV} E_V exp\{-i[k_0(2\Delta L_F)]\} \tag{16.3}$$

and

$$\left(\vec{e}_V^S\right)^- = \vec{e}_- K_{VU} E_V exp\{-i[k_0(2\Delta L_F)]\}, \tag{16.4}$$

respectively.

Amplitude coefficients $K_{I,J}$ describe polarization transformation in a birefringent medium or scattering to initial ($K_{I,I}$) or orthogonal polarization ($K_{I,J}$) by randomly oriented small-scale anisotropic optical inhomogeneities. The coefficients for a randomly heterogeneous medium are related by the following expressions: $K_{UU} = K_{VV}$ and $K_{UV} = K_{VU}$.

Groups of the backscattered waves at the fiber output have basic path lengths mismatch ΔL the same as the reference waves. Additional mismatch between the group of backscattered and reference waves is equal to double Fizeau interferometer length $2\Delta L_F$. Both wave groups through circulator ports 2 and 3 propagate to the compensating Michelson interferometer. The optical path lengths mismatch in Michelson interferometer arms is modulated by piezo-fiber delay lines (PFDL) [7]. The resulting waves from the first interferometer output are transmitted to the first photodiode, whereas the ones from the second arm are transmitted through circulator ports 3 and 4 to the second photodiode. The signals at the Doppler shift frequency are further subtracted in the differential amplifier, which leads to doubling of antiphased interference signals and to subtraction of the co-phased components of distortions and noises.

The Michelson interferometer is tuned so as to provide compensation of aquired path lengths mismatch. Due to broadband probing radiation with optical spectrum width of $2\sigma_\omega$ the interference signal is proportional to the factor $\exp\left[-\Delta\tau^2/2\sigma_\tau^2\right]$. Here $\Delta\tau$ is the group time-delay mismatch for interfering waves, $2\sigma_\tau = 2/\sigma_\omega$ [5]. As the Michelson interferometer arms mismatch varies at a constant rate, the mutual coherence of wave trains with the corresponding Doppler shift of optical spectra is eventually restored. Constructive interference takes place if the optical path length difference between the object and the reference waves does not exceed $c\tau_g$, which is equal to units of microns if the probing NIR bandwidth is tens of nanometers.

For compensation of the path length mismatch between coherent components in initial and orthogonal polarizations one should select different delays when operating with Michleson interferometer arms mismatch. For the delay $2\Delta L_F - \Delta L$, the reference wave in initial polarization $\vec{e}_V^R$ will interfere with the backscattered wave in orthogonal polarization $\left(\vec{e}_U^S\right)^-$ (the first cross-channel). For the delay $2\Delta L_F$, the reference and backscattered waves will interfere in pairs in initial polarizations: $\vec{e}_U^R$ and $\left(\vec{e}_U^S\right)^+$, as well as $\vec{e}_V^R$ and $\vec{e}_V^R$ and $\left(\vec{e}_V^S\right)^+$, correspondingly (co-channel). The interference between the reference wave $\vec{e}_U^R$ and the backscattered wave $\left(\vec{e}_V^S\right)^-$ will take place for the delay $2\Delta L_F + \Delta L$ (the second cross-channel).

Note that one should account for the fact that with the $2\Delta L_F$ delay two pairs of waves interfere, while with the $2\Delta L_F + \Delta L$ or $2\Delta L_F - \Delta L$ delay, only one pair of waves interferes. It is obvious that if the scanning depth exceeds two or three mentioned delays, there will be two or three interference patterns in a single A-scan, correspondingly.

In our device, both co- and cross-OCT images are displayed on the user interface screen: the cross-OCT image is shown at the top while the co-OCT image is shown at the bottom (examples will be furnished below). The cross-OCT image is detected in the first cross-channel (interference of waves $\vec{e}_V^R$ and $\left(\vec{e}_U^S\right)^-$) and the co-OCT image is detected in the parallel channel (interferences of wave pairs $\vec{e}_U^R$ and $\left(\vec{e}_U^S\right)^+$, and $\vec{e}_V^R$ and $\left(\vec{e}_V^S\right)^+$). Such a limitation of the cross-OCT image representation is specified by the requirements to minimize scanning time.

As a result, the sensitivity of a co-channel is constant and depends neither on the polarization state of probing radiation nor on sample orientation. The cross-channel sensitivity has another manifestation. At circular polarization of probing radiation it does not depend on sample orientation and is two times smaller compared to that in the co-channel. At linear polarization of probing radiation it can be additionally reduced, depending on the sample orientation. At linear polarization of the probing beam, brightness modulation (the fringe pattern) induced by birefringence can be reduced additionally (compared to the co-channel), depending on sample orientation. The visibility of the fringe pattern in this case varies from maximum to zero. This feature can be used for identification of images of a birefringent medium.

In a biological medium which does not contain regular birefringence, appearance of a signal in a cross-channel is induced by the presence of low-scale anisotropic chaotically oriented scatterers. The sensitivities of co- and cross-channels are independent of sample orientation. The sensitivity of the co-channel does not depend on the polarization of the probing radiation either. However, the sensitivity of the cross-channel is two times lower at circular polarization and three times lower at linear polarization than that of the co-channel.

16.2.2 OCT probes: Customizing the device

At the fist stage of setup design a forward-looking endoscopic time-domain OCT probe was elaborated in IAP RAS for the routine endoscopic procedure [9,35]. The diameter of the standard probe is 2.7 mm and the rigid part length is 21 mm (Figure 16.2). The probe is compatible with several endoscopic devices the instrumental (biopsy) channels of which are employed for probe access to the studied tissue. The design of the flexible probe allowed development of a technique for acquisition of OCT images of mucous and serous membranes of human internal organs.

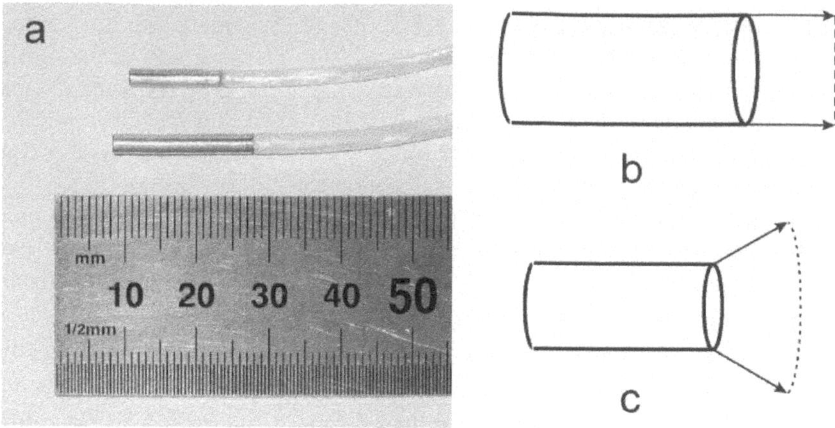

FIGURE 16.2: Standard (bottom) and minimized (top) endoscopic OCT probes (a) and their rectangular (b) and angular (c) scanning configurations.

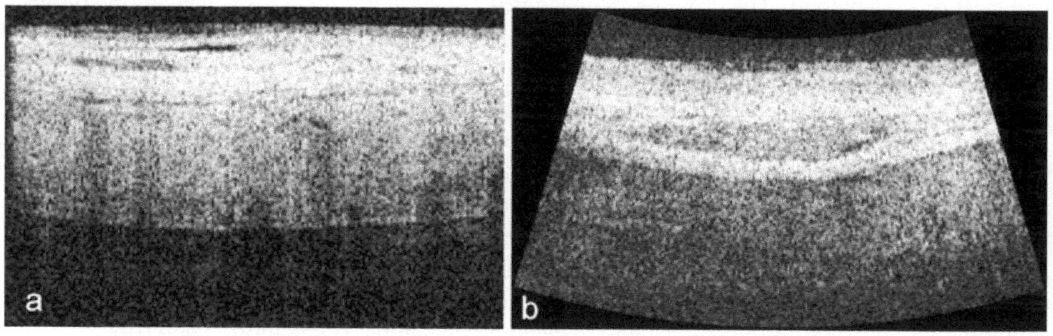

FIGURE 16.3: Typical OCT images obtained with rectangular (a) and angular (b) scanning.

One of the ways to improve endoscopic equipment is its miniaturization. For this the diameter of the OCT probe must be made smaller and at the same time compatible with available endoscopic equipment. Such improvement was performed at the next stage of the setup design.

Test experiments showed that the rigid part of the probe may be diminished appreciably: the diameter may de decreased from 2.7 mm to 2.4 mm, and length from 21 mm to 15 mm (Figure 16.2). Towards this end, the scanning and the focusing systems of the probe were modified. In particular, the two-lens focusing system forming a rectangular image without geometrical distortions (Figure 16.3a) was rejected, as a decrease of the probe size in such a system would lead to diminution of the scanning area. Instead, a single-gradient lens system of the same aperture was adopted. Consequently, rectangular scanning (Figure 16.3a) was replaced by imaging in the angular configuration (Figure 16.3b) with subsequent software correction of geometrical distortions that causes a concave upper boundary.

16.3 Clinical Applications of the Fiber-Based OCT Device

Clinical applications of the endoscopic time-domain OCT equipped with a forward looking probe (EOCT) were focused on the study of mucous membranes of internal organs *in vivo* [36, 37]. We have developed clinical techniques of EOCT application for diagnosis of pathologies of gastrointestinal (GI) tract [38, 39], bladder [40, 41], cervix [42, 43], larynx [44, 45], oral cavity [46, 47], and other organs [48]. The clinical approbation of EOCT was performed in multicenter studies for more than 4000 patients.

Our experience shows that OCT offers solution for a wide spectrum of "classical" problems of medical imaging modalities: detection of pathological changes, including early neoplasia detection; optimization of directed biopsy; differential diagnosis of diseases of different origin but with similar manifestations; revealing localization of pathological changes, including intra-operational real-time planning and organ preserving and reconstructive operations control; and evaluation of dynamics of pathological changes, including treatment control at every stage.

16.3.1 Diagnosis of cancer and target biopsy optimization

The OCT device was tested for diagnosis of neoplasia including its early stages and was shown to provide high (almost 100%) accuracy in differentiating normal tissue from the pathological one. The statistical analysis of the diagnosis results based on data from several hundreds patients demonstrates excellent score in distinguishing various cancer and precancer states (Table 16.1) [49].

TABLE 16.1: Diagnostic accuracy of OCT in neoplasia exposure [49]

Localization	Sensitivity, %	Specificity, %	Diagnostic accuracy, %	Agreement coefficient
Uterine cervix	82	78	81	0.65
Larynx	83	90	87	0.64
Bladder	82	85	85	0.79
Oral cavity	83	98	95	0.76
Conjunctiva	83	92	91	0.62

In spite of its high accuracy, the OCT technique cannot replace the traditional method of morphological verification, but it is a useful complementary tool for optimizing biopsy that allows avoiding unnecessary biopsy procedures in particular cases. For effective real-time evaluation of the obtained OCT images we divide them into three types (Figure 16.4); two of them ("suspicious" and "malignant" states) indicate the need of biopsy procedure. If the OCT image type is identified as "benign," the biopsy procedure can be avoided. The complementary use of OCT in complex examination of cervix allows a 63–65% decrease of the number of unnecessary biopsy procedures in abnormal colposcopic findings.

OCT diagnosis of early stage neoplasia in mucous membranes was also performed for two other clinically important situations: neoplasia of bladder and Barrett esophagus. Conclusions were made based on the results of a large number of multicenter blind recognitions of OCT images obtained during endoscopic diagnosis. We showed that the sensitivity of OCT in diagnosis of early bladder neoplasia (located among the flat "suspicious" areas) is 82%, and the specificity is 85% (Figures 16.5–16.6).

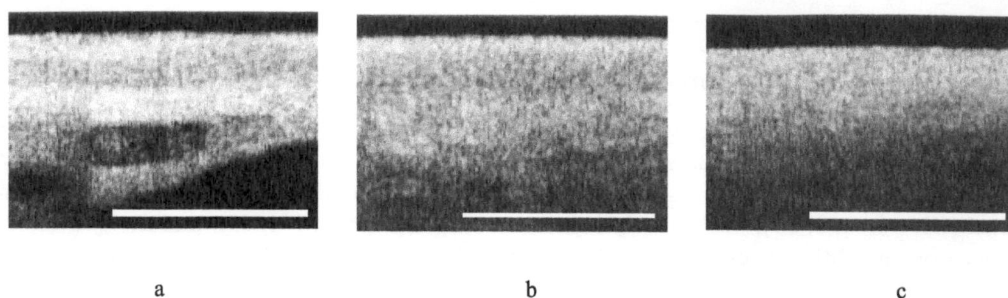

a b c

FIGURE 16.4: Three types of OCT images of the cervical mucosa: a) sharp, high-contrast horizontal layers through more than 75% of lateral range. The upper layer (epithelium) has lower signal intensity than the second (connective tissue) layer – benign images; b) tissue layers are hardly differentiable, increased intensity of the upper layer signal, poorly defined border between the upper and second layers – suspicious images; c) no horizontal layers, or less than 75% – neoplasia images. a) mature metaplastic epithelium (benign condition); b) CIN II; c) invasive cervical carcinoma.

OCT monitoring of Barrett esophagus demonstrated higher diagnostic efficiency compared to endoscopy and traditional imaging techniques. For example, OCT allows a high-accuracy diagnosis of intramucous adenocarcinoma exhibiting sensitivity and specificity of 85%. However, the sensitivity of the technique in diagnosis of high-grade dysplasia in metaplastic epithelium is limited to 68% [35].

The effectiveness of OCT diagnosis can be increased if it is used in combination with other modern diagnostic techniques. OCT combined with fluorescent cystoscopy results in an almost perfect technique for endoscopic diagnosis of bladder cancer at early stages. Its high performance is based on extremely high sensitivity (close to 100%) of fluorescent cystoscopy, which is enforced by high specificity of OCT, while the specificity of fluorescent cystoscopy itself is low. This additional OCT diagnosis of fluorescing areas significantly decreases the false positive (FP) rate. Thus, such a combination increases the positive predictive value from 15% to 43%. Complementary use of OCT imaging and fluorescence cystoscopy can substantially improve diagnostic yield of bladder neoplasia detection [35].

16.3.2 Differential diagnosis of diseases with similar manifestations

The study of mucous membranes at different locations shows that OCT can be effectively used for the differential diagnosis of benign states. For example, it permits differentiating adenomatose polyps of colon from hyperplastic ones with sensitivity of 80% and specificity of 82%, and transmural inflammation of colon from superficial one with sensitivity of 90% and specificity of 83%. The latter allows one to change the treatment strategy avoiding injuring surgical operation [50].

16.3.3 OCT monitoring of treatment

OCT is an attractive technique for clinicians due to its high resolution. The combination of good enough diagnostic accuracy and noninvasiveness makes it an appropriate tool for multiple *in vivo* applications for monitoring the pathological process. So, OCT can be applied for monitoring biotissue structure in conservative and surgical treatments. For example, OCT gives additional information about biotissues in the course of radiation or chemical therapy. It is well known that one of the complications of radiation therapy is mucositis whose signs appear on OCT images earlier

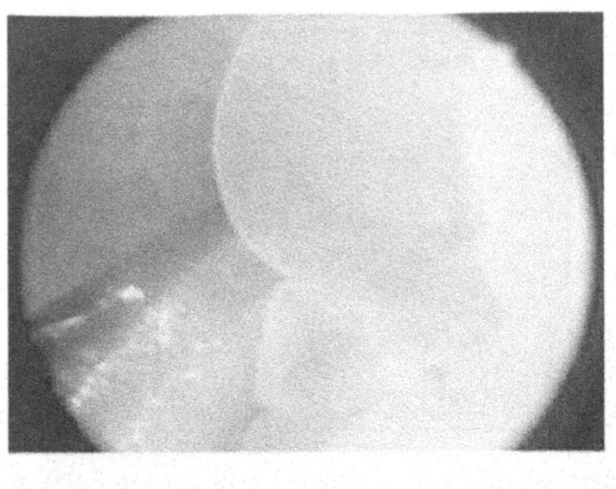

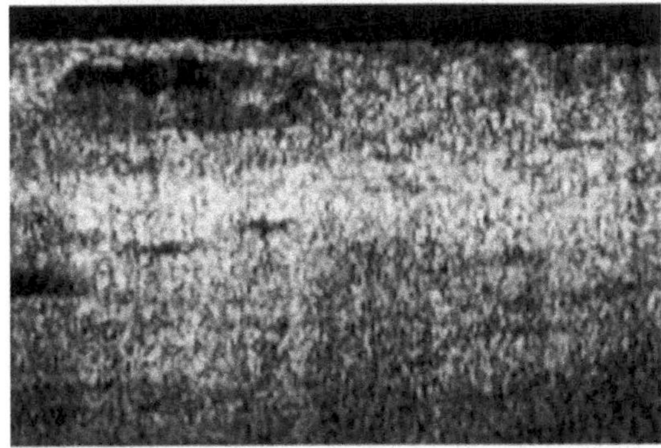

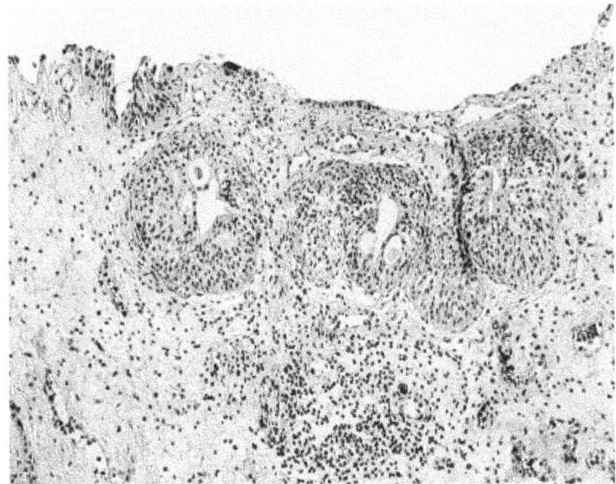

FIGURE 16.5: Cystitis cystica. Cystoscopic image (top); OCT image (benign type) (middle); the corresponding histology, hemotoxylin-eosin (H&E) stained (bottom).

and last longer compared to visual data. The dynamics of the OCT image of the treated area allows one to draw conclusions on the radio-sensitivity of the mucous membrane and predicts mucositis grade [51].

16.3.4 OCT for guided surgery

OCT ability for real-time imaging is extremely attractive for intra-operational application. We assessed OCT capabilities for accurate determination of the proximal border for esophageal carcinoma and the distal border for rectal carcinoma. OCT borders matched the histopathology in 94% cases in the rectum and 83.3% in the esophagus (in the cases of mucosal and submucosal tumor

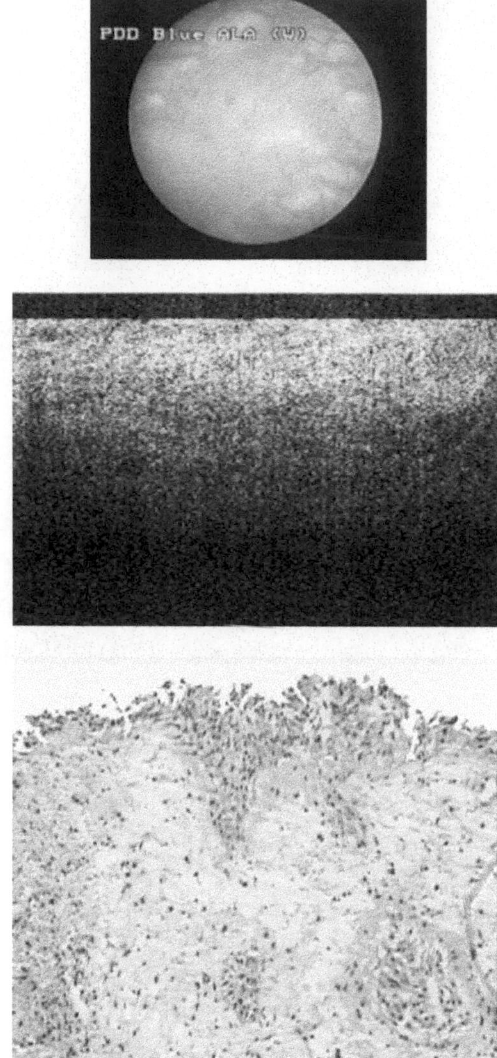

FIGURE 16.6: High-grade dysplasia (cancer *in situ*) in a flat suspicious zone. Cystoscopic image (top); OCT image (neoplasia type) (middle); the corresponding histology (bottom) (H&E).

growth). In the cases of a mismatch between the OCT and histology borders, deep tumor invasion occurred in the muscle layer (esophagus, rectum). OCT can be used for presurgery planning and surgery guidance of the proximal border for esophageal carcinoma and the distal border for rectal carcinoma with superficial stages. However, deep invasion in the rectum or esophageal wall must be controlled by alternative diagnostic modalities (Figure 16.7) [35].

The application of OCT in clinical practice described above proves that it is a useful tool for diagnosis and treatment control in endoscopy and determines a further way of its development. Enhancement of OCT efficiency is bound up with technological upgrade (design of new OCT modalities and engineering customizing for particular applications) and development of clinical techniques accounting for particular problems. Further, we describe the application of the OCT device enhanced by cross-polarization modality and miniprobe operation.

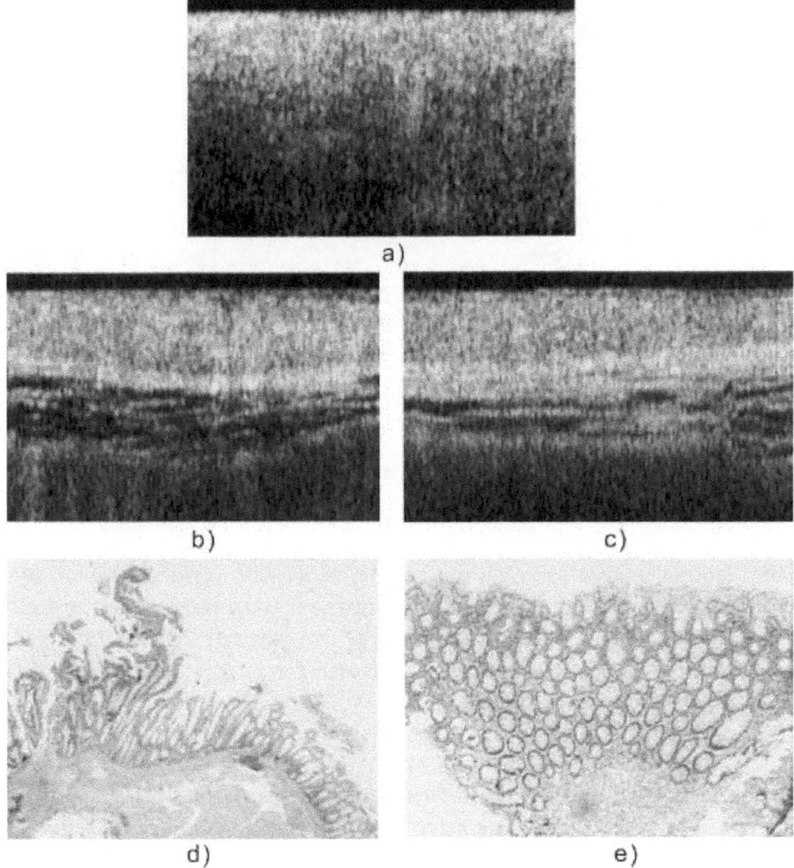

FIGURE 16.7: Determination of the border of rectal adenocarcinoma growing within the bound of mucous membrane. The histologic specimen (d) demonstrates the border between tumor and normal mucosa (on the left – tumor; on the right – healthy mucosa). OCT image without structure (a) from the region of histological tumor border and healthy tissue (b). Reference OCT image (c) and histologic specimen (e) 1 cm from the border without pathology.

16.3.5 Cross-polarization OCT modality for neoplasia OCTdiagnosis

In some cases, benign and malignant process may be hard to distinguish by the traditional OCT modality, which decreases its specificity.

The polarization properties of biotissue are known to be connected with collagen state and depolarization capabilities decrease as a result of malignant transformation. Consequently, detection of tissue birefringent and depolarization properties by cross-polarization OCT (CP-OCT) may increase specificity.

The combination of histologic biotissue staining with picrosirius red (PSR) and its inspection by means of polarization microscopy is known to be a specific technique for collagen state evaluation [52] and thus for estimation of polarization properties. In our experimental study of *ex vivo* biotissue samples we show that the areas of maximal brightness in CP-OCT images obtained in orthogonal polarization coincide with areas of structured birefringent collagen in histologic specimens stained with PSR (yellow and red in histological images) (Figures 16.8–16.9, color not shown).

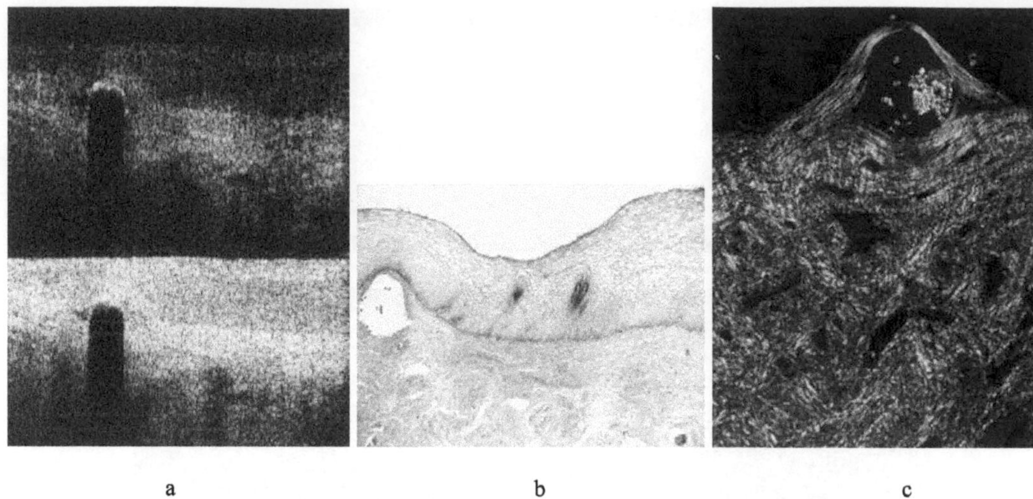

FIGURE 16.8: Normal uterine cervical mucosa (*ex vivo*) with opaque surgical suture as a reference. OCT image (a) in parallel (bottom) and orthogonal (top) polarization, histological image (1.7×1 mm) stained with H&S (b); histological image in polarized light (0.45 × 0.3 mm) stained with PSR (c). A surgical suture was used as a reference for proper comparison of CP-OCT and histology images.

CP-OCT images of normal mucous membrane of uterine cervix and bladder demonstrate a layered tissue structure. The brightness of the image obtained in orthogonal polarization indicates high object depolarization capability.

The images of benign states mimicking neoplasia areas in the colposcopic study (uterine cervix mucosa with metaplasia) are presented in Figure 16.10.

Connective tissue in the metaplasia area gives signal in orthogonal polarization due to nondeformed type I and type III collagen (yellow, red, or green in histological images in polarized light), which is indicative of benign state.

CP-OCT allows detecting an invasive cancer process among nonstructured (nonstratified) images because stroma in developed neoplasia does not contain structured collagen and does not provide any signal in orthogonal polarization (Figure 16.11).

16.3.6 OCT miniprobe application

The former modification of the endoscopic OCT probe (2.7 mm in diameter, 21 mm in distal operating part length) limited the choice of the endoscopes to be used as a basic instrument. This limitation is induced by diameter (the majority of the available endoscopes have a 2.8 mm channel) and distal part length limiting probe penetration through the endoscope curvature parts.

In order to overcome these limitations we have developed an OCT miniprobe with a diameter of 2.4 mm and distal part of 13 mm, which is compatible with all standard working channels (not less than 2.6 mm in diameter) of rigid and flexible endoscopes used in gastroenterology, urology, pulmonology, laryngology, and gynecology. Figure 16.12 demonstrates an OCT miniprobe combined with flexible gastroscope "Olympus GIF-Q 40" (working channel 2.8 mm in diameter) as well as with rigid hysteroscope "Storz" (same size of working channel).

Clinical application of the OCT system equipped with the miniprobe demonstrated that the obtained transversal scanning range is sufficient for evaluation of pathological changes in tissue. This

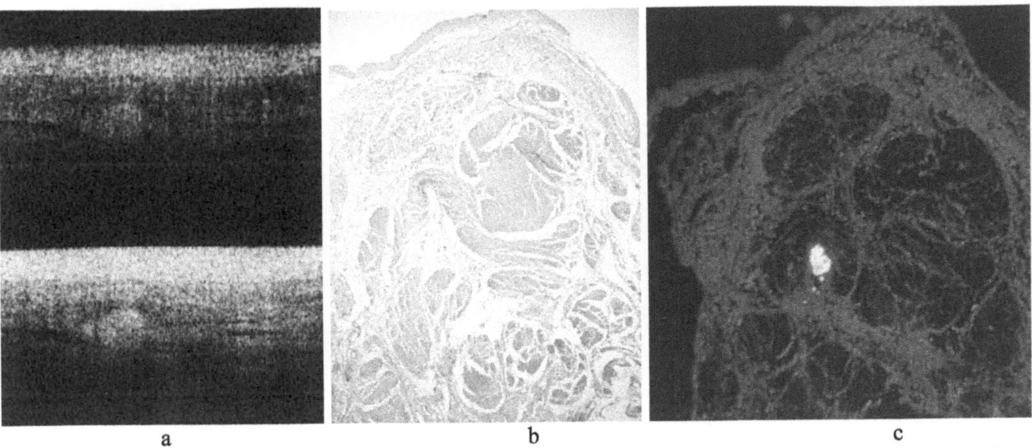

FIGURE 16.9: Bladder left wall with a reference surgical suture: OCT image (a) in parallel (bottom) and orthogonal (top) polarization, histological image (1.8 × 2.3 mm) stained with H&E (b); histological image (1.8 × 2.3 mm) in polarized light stained with PSR (c). The bladder tissue is compressed for OCT study and the thicknesses of the layers do not coincide in the OCT and histology images.

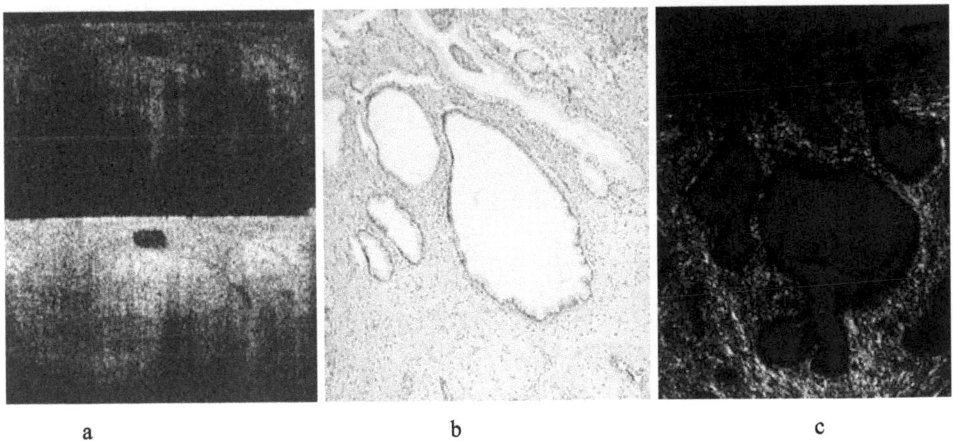

FIGURE 16.10: Uterine cervix mucosa (metaplasia): OCT image (a) in initial (bottom) and orthogonal (top) polarization; histological image (b) stained with H&E, histological image (1.8 × 2.3 mm) in polarized light stained with PSR (c).

can be illustrated by a set of images of benign and neoplastic states obtained by means of cystoscopy. OCT images of chronic inflammation exhibit structured, layered media (Figure 16.13). This state is characterized by the increase of urothelium thickness. The structured (layered) image indicates a benign state in the sample under examination, thus avoiding a biopsy procedure.

The lack of structured (layered) image indicates a malignant state in the sample under examination (Figure 16.14).

Similar results were obtained in examination of GI tissues (Figure 16.15). The OCT images of the areas suspicious for cancer exhibit no pronounced structure (Figure 16.16).

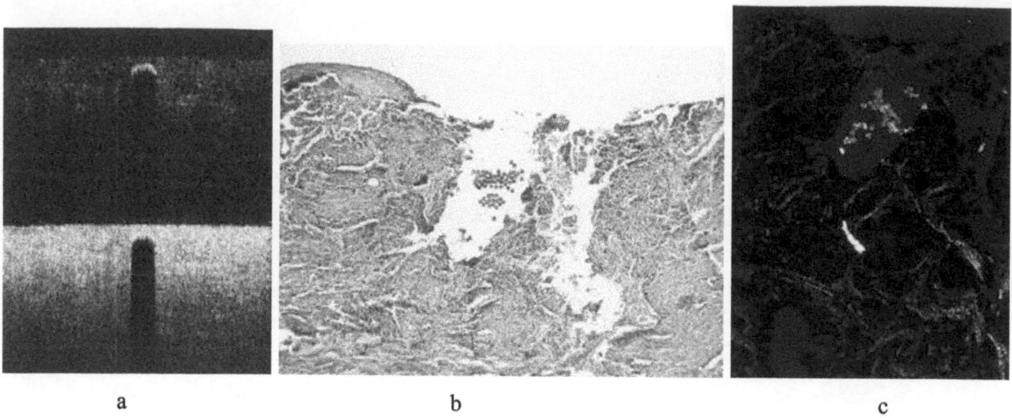

FIGURE 16.11: Uterine cervix mucosa (area of invasive carcinoma): OCT image (a) in initial (bottom) and orthogonal (top); histological image (1.7 × 1 mm) stained with H&E (b); histological image (0.45 × 0.3 mm) in polarized light stained with PSR (c).

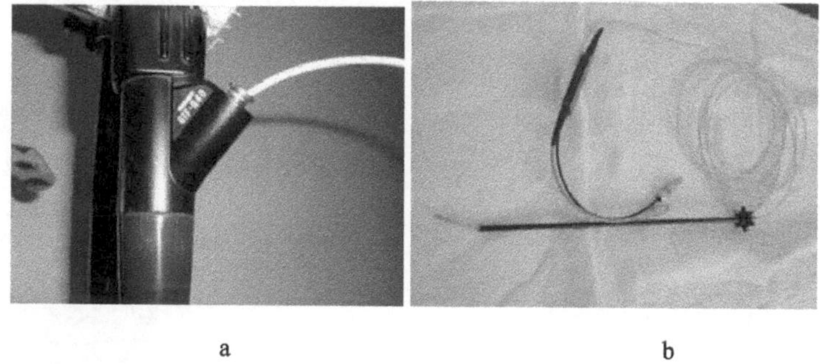

FIGURE 16.12: OCT probe inserted into the fibroendoscope working channel (a), OCT probe in a standard hysteroscopic tube (rigid) (b).

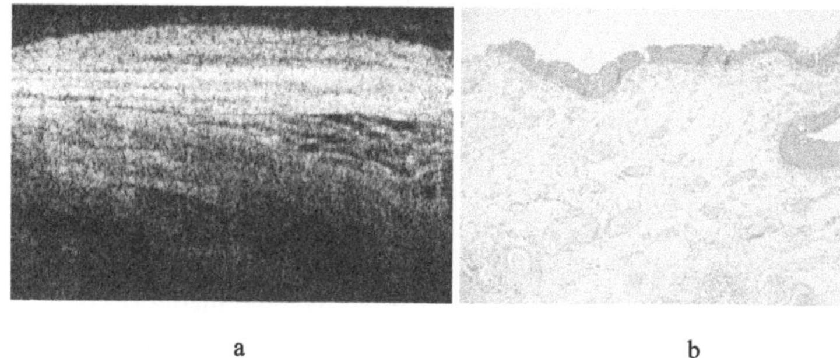

FIGURE 16.13: Area of pronounced inflammation of bladder wall. OCT image (a) and histological image (1.7 × 1 mm) stained with H&E (b).

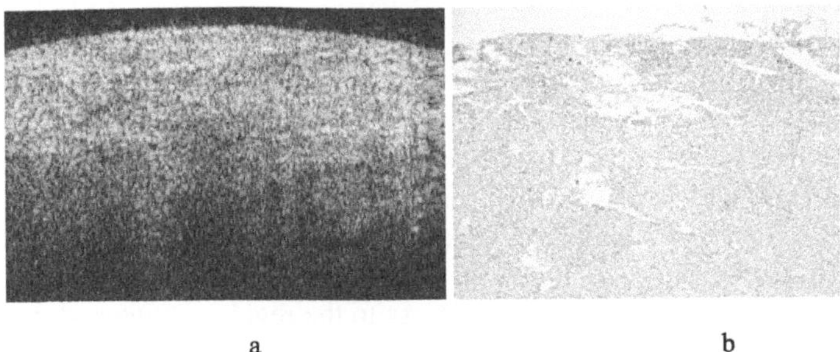

a b

FIGURE 16.14: Bladder cancer: OCT image (a) in initial (bottom) and orthogonal (top) polarizations; histology image (1.7 × 1 mm) stained with H&E (b).

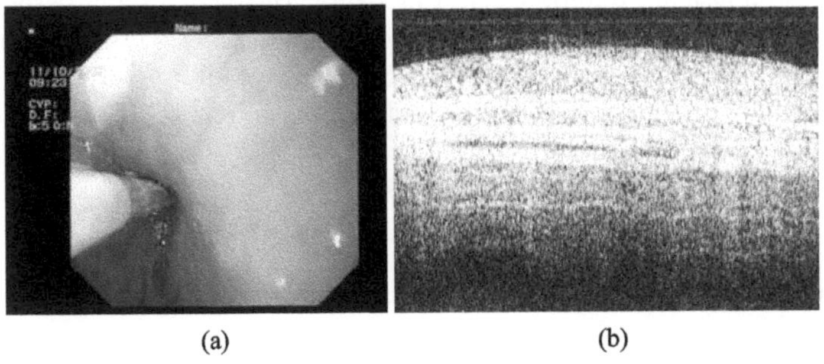

(a) (b)

FIGURE 16.15: Normal esophagus mucosa. OCT miniprobe in esophagus (a) and OCT image (b).

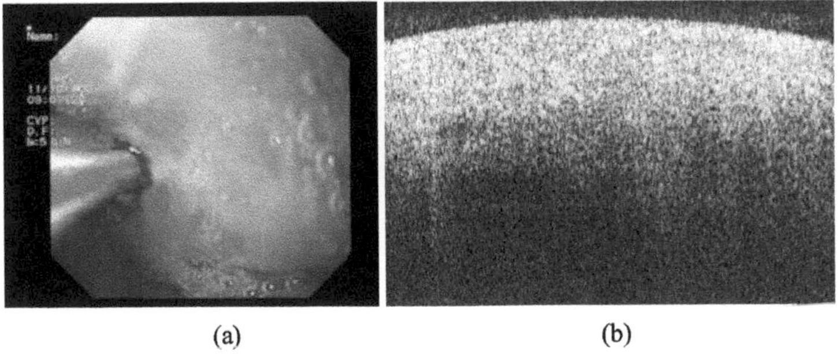

(a) (b)

FIGURE 16.16: Esophagus tumor surface (adenocarcinoma). OCT miniprobe in esophagus (a) and OCT image (b).

Thus, the developed OCT setup clinically approbated on more than 4000 patients revealed its capabilities for diagnosis of pathological changes (including early neoplasia detection), optimization of target biopsy, differential diagnosis of diseases of different origin and similar manifestations, localization of pathological changes (including intra-operational real-time planning and organ preserving and reconstructive operations control), and treatment monitoring.

16.4 Conclusion

This chapter describes how the fiber-based OCT setup applicable for endoscopic studies can be constructed. The principles of OCT signal formation and scheme of the constructed setup are given. The examples of its various clinical applications are discussed. The chapter also shows how the developed fiber-based OCT systems were successfully customized for various applications in clinics. This customization allowed easy probe access to the regions of interest providing fast and precise diagnosis of pathological cases.

Clinical application of the novel endoscopic CP-OCT system equipped with a miniprobe proved this system to be more compatible with standard endoscopic equipment compared to the traditional one. As a standard working channel is 2.8 mm in diameter, the miniprobe with a diameter of 2.4 mm is better adapted to the certified devices and does not require additional conditions for operation.

Implementation of cross-polarization OCT modality allowed detecting different pathologies that are difficult to distinguish in a traditional OCT modality. The benign states tissues exhibit a stratified image structure, whereas malignant states give nonstratified images in initial polarization and low-level signals in an orthogonal channel. Some benign states exhibiting nonstratified image structure in initial polarization provide a high-level signal in an orthogonal channel increasing OCT specificity in neoplasia recognition.

Acknowledgments

This work was supported in part by the State Contract of the Russian Federation No. 02.522.11. 2002, the Presidium of the RAS Program "Fundamental Sciences for Medicine," and the RFBR grants No. 08-02-01152 and 08.04.97098. The authors are grateful to the medical staff for clinical investigations.

References

[1] D. Huang, E.A. Swanson, C.P. Lin, J.S. Schuman, W.G. Stinson, W.Chang, M.R. Hee, T. Flotte, K.Gregory, C.A. Puliafito, and J.G. Fujimoto, "Optical coherence tomography," *Science* **254**, 1178–1181 (1991).

[2] E.A. Swanson, J.A. Izatt, M.R. Hee, D. Huang, C.P. Lin, J.S. Schuman, C.A. Puliafito, and J.G. Fujimoto, "*In vivo* retinal imaging by optical coherence tomography," *Opt. Lett.* **18,** 1864–1866 (1993).

[3] A.M. Sergeev, V.M. Gelikonov, G.V. Gelikonov, F.I. Feldchtein, K.I. Pravdenko, D.V. Shabanov, N.D. Gladkova, V.V. Pochinko, V.A. Zhegalov, G.I. Dmitriev, I.R. Vazina, G.A. Petrova, and N.K. Nikulin, "*In vivo* optical coherence tomography of human skin microstructure," Proc. SPIE **2328**, 144–150 (1994).

[4] V.M. Gelikonov, G.V. Gelikonov, R.V. Kuranov, N.K. Nikulin, G.A. Petrova, V.V. Pochinko, K.I. Pravdenko, A.M. Sergeev, F.I. Feldchtein, Y.I. Khanin, and D.V. Shabanov, "Coherent optical tomography of microscopic inhomogeneities in biological tissues," *Lett. J. Exp. Theor. Phys.* **61**, 149–153 (1995).

[5] B.E. Bouma and G.J. Tearney (Eds.), *Handbook of Optical Coherence Tomography,* Marcel Dekker, New York (2002).

[6] V.M. Gelikonov and N.D. Gladkova, "A decade of optical coherence tomography in Russia: from experiment to clinical practice," *Radiophys. Quant. Electron.* **47**, 835–847 (2004).

[7] F.I. Feldchtein, V.M. Gelikonov, G.V. Gelikonov, N.D. Gladkova, V.I. Leonov, A.M. Sergeev, and Ya.I. Khanin, "Optical fiber interferometer and piezoelectric modulator," Patent US 5835642 (1998).

[8] G.J. Tearney, S.A. Boppart, B.E. Bouma, M.E. Brezinski, N.J. Weissman, J.F. Southern, and J.G. Fujimoto, "Scanning single-mode fiber optic catheter-endoscope for optical coherence tomography," *Opt. Lett.*, **21**, 543–545 (1996).

[9] A.M. Sergeev, V.M. Gelikonov, G.V. Gelikonov, F.I. Feldchtein, R.V. Kuranov, N.D. Gladkova, N.M. Shakhova, L.B. Snopova, A.V. Shakhov, I.A. Kuznetzova, A N. Denisenko, V.V. Pochinko, Y.P. Chumakov, and O.S. Streltzova, "*In vivo* endoscopic OCT imaging of precancer and cancer states of human mucosa," *Opt. Exp.* **1**, 432–440 (1997).

[10] R.V. Kuranov, V.V. Sapozhnikova, I.V. Turchin, E.V. Zagainova, V.M. Gelikonov, V.A. Kamensky, L.B. Snopova, and N.N. Prodanetz, "Complementary use of cross-polarization and standard OCT for differential diagnosis of pathological tissues," *Opt. Exp.* **10**, 707–713 (2002).

[11] M.R. Hee, D. Huang, E.A. Swanson, and J.G. Fujimoto, "Polarization-sensitive low-coherence reflectometer for birefringence characterization and ranging," *J. Opt. Soc. Am.* **9**, 903–908 (1992).

[12] J.F. de Boer, T.E. Milner, M.J.C. van Gemert, and J.S. Nelson, "Two-dimensional birefringence imaging in biological tissue by polarization-sensitive optical coherence tomography," *Opt. Lett.* **22**, 934–936 (1997).

[13] J.M. Schmitt and S.H. Xiang, "Cross-polarized backscatter in optical coherence tomography of biological tissue," *Opt. Lett.* **23**, 1060–1062 (1998).

[14] J.F. de Boer, S.M. Srinivas, A. Malekafzali, Z.P. Chen, and J S. Nelson, "Imaging thermally damaged tissue by polarization sensitive optical coherence tomography," *Opt. Exp.* **3**, 212–218 (1998).

[15] M.J. Everett, K. Schoenenberger, B.W. Colston, Jr., and L.B. Da Silva, "Birefringence characterization of biological tissue by use of optical coherence tomography," *Opt. Lett.* **23**, 228–230 (1998).

[16] B. Cense, H.C. Chen, B.H. Park, M.C. Pierce, and J.F. de Boer, "*In vivo* birefringence and thickness measurements of the human retinal nerve fiber layer using polarization-sensitive optical coherence tomography," *J. Biomed. Opt.* **9**, 121–125 (2004).

[17] F.I. Feldchtein, G.V. Gelikonov, V.M. Gelikonov, R.R. Iksanov, R.V. Kuranov, A.M. Sergeev, N.D. Gladkova, M.N. Ourutina, J.A. Warren, Jr., and D.H. Reitze, "*In vivo* OCT imaging of hard and soft tissue of the oral cavity," *Opt. Exp.* **3**, 239–250 (1998).

[18] J.F. De Boer, T.E. Milner, and J.S. Nelson, "Determination of the depth-resolved Stokes parameters of light backscattered from turbid media by use of polarization- sensitive optical coherence tomography," *Opt. Lett.* **24**, 300–302 (1999).

[19] C.E. Saxer, J.F. de Boer, B.H. Park, Y.H. Zhao, Z.P. Chen, and J.S. Nelson, "High-speed fiber-based polarization-sensitive optical coherence tomography of in vivo human skin," *Opt. Lett.* **25**, 1355–1357 (2000).

[20] J.F. de Boer and T.E. Milner, "Review of polarization sensitive optical coherence tomography and Stokes vector determination," *J. Biomed. Opt.* **7**, 359–371 (2002).

[21] V.M. Gelikonov, D.D. Gusovskii, V.I. Leonov, and M.A. Novikov, "Birefringence compensation in single-mode optical fibers," *Sov. Tech. Phys. Lett.* (*USA*) **13**, 322–323 (1987).

[22] D.P. Dave, T. Akkin, and T.E. Milner, "Polarization-maintaining fiber-based optical low-coherence reflectometer for characterization and ranging of birefringence," *Opt. Lett.* **28**, 1775–1777 (2003).

[23] Y.C. Lefevre, "Single-mode fiber fractional wave devices and polarization controllers," *Electron. Lett.* **15**, 778–780 (1980).

[24] T. Xie, S. Guo, J. Zhang, Z. Chen, and G.M. Peavy, "Use of polarization-sensitive optical coherence tomography to determine the directional polarization sensitivity of articular cartilage and meniscus," *J. Biomed Opt.* **11**, 064001 (2006).

[25] V.M. Gelikonov and G.V. Gelikonov, "Fibreoptic methods of cross-polarisation optical coherence tomography for endoscopic studies," *Quant. Electr.* **38**, 634–640 (2008).

[26] F. Feldchtein, J. Bush, G. Gelikonov, V. Gelikonov, and S. Piyevsky, "Cost-effective, all-fiber autocorrelator based 1300 nm OCT system," *Proc SPIE* **5690**, 349–354 (2005).

[27] V.M. Gelikonov, V.I. Leonov, and M.A. Novikov, "Fiber-Optical Sensor," USSR Patent 1315797 A1, USSR, Bulletin **21**, 3 (1987)

[28] N.A. Olsson, "Polarisation-independent configuration optical amplifier," *Electron. Lett.*, **24**, 1075–1076 (1988).

[29] M. Martinelli, "A universal compensator for polarization changes induced by birefringence on a retracing beam," *Opt. Commun.* **72**, 341–344 (1989).

[30] N.C. Pistoni and M. Martinelli, "Birifringence effects suppression in optical fiber sensor circuit," *Proc. 7th Optical Fiber Sensors Conf., Sydney, Australia: IEEE.*, 125–128 (1990).

[31] N.C. Pistoni and M. Martinelli, "Polarization noise suppression in retracing optical fiber circuits," *Opt. Lett.* **15**, 711–713 (1991).

[32] M.D. Van Deventor, "Preservation of polarization orthogonality of counterpropagation waves through dichroic birefringent optical media: proof and application," *Electron. Lett.*, **27**, 1538–1540 (1991).

[33] V.M. Gelikonov and G.V. Gelikonov, "New approach to cross-polarized optical coherence tomography based on orthogonal arbitrarily polarized modes," *Laser Physics Lett.* **3**, 445–451 (2006).

[34] A.D. Drake and D.C. Leiner, "Fiber-optic interferometer for remote subangstrom vibration measurement," *Rev. Sci. Instrum.*, **55**, 152–155 (1984).

[35] E. Zagaynova, N. Gladkova, N. Shakhova, G. Gelikonov, and V. Gelikonov, "Endoscopic OCT with forward-looking probe: clinical studies in urology and gastroenterology," *J. Biophotonics* **1**, 114–128 (2008).

[36] A.M. Sergeev, V.M. Gelikonov, G.V. Gelikonov, F.I. Feldchtein, R.V. Kuranov, N.D. Gladkova, N.M. Shakhova, L.B. Snopova, A.V. Shakhov, and I.A. Kuznetzova, "*In vivo* endoscopic OCT imaging of precancer and cancer states of human mucosa," *Opt. Exp.* **1**(13) 432–440 (1997).

[37] F.I. Feldchtein, G.V. Gelikonov, V.M. Gelikonov, R.V. Kuranov, A.M. Sergeev, N.D. Gladkova, A.V. Shakhov, N.M. Shakhova, L.B. Snopova, and A.B. Terentieva, "Endoscopic applications of optical coherence tomography," *Opt. Exp.* **3**, 257–269 (1998).

[38] S. Jackle, N.D. Gladkova, F.I. Feldchtein, A.B. Terentieva, B. Brand, G.V. Gelikonov, V.M. Gelikonov, A.M. Sergeev, A. Fritscher-Ravens, and J. Freund, "*In vivo* endoscopic optical coherence tomography of the human gastrointestinal tract – toward optical biopsy," *Endoscopy* **32,** 743–749 (2000).

[39] G. Zuccaro, N.D. Gladkova, J. Vargo, F.I Feldchtein, E.V. Zagaynova, D. Conwell, G.W. Falk, J.R. Goldblum, J. Dumot, and J. Ponsky, "Optical coherence tomography of the esophagus and proximal stomach in health and disease," *Am. J. Gastroenterol.* **96**, 2633–2639 (2001).

[40] E.V. Zagaynova, O.S. Strelzova, N.D. Gladkova, L.B. Snopova, G.V. Gelikonov, F.I. Feldchtein, and A.N. Morozov, "*In vivo* optical coherence tomography feasibility for bladder disease," *J. Urology*. **157**, 1492–1497 (2002).

[41] M.J. Manyak, N. Gladkova, J.H., Makari, A. Schwartz, E. Zagaynova, L. Zolfaghari, J. Zara, R. Iksanov, and F. Feldchtein, "Evaluation of superficial bladder transitional cell carcinoma by optical coherence tomography," *J. Endourology* **19,** 570–574 (2005).

[42] N.M. Shakhova, F.I. Feldchtein, and A.M. Sergeev, "Applications of optical coherence tomography in gynecology" in *Handbook of Optical Coherence Tomography,* B.E. Bouma and G.J. Tearney (Eds.), Marcel Dekker, New York, Basel, 649–672 (2002).

[43] P.F. Escobar, J.L. Belinson, A. White, N.M. Shakhova, F.I. Feldchtein, M.V. Kareta, and N.D. Gladkova, "Diagnostic efficacy of optical coherence tomography in the management of preinvasive and invasive cancer of uterine cervix and vulva," *Int. J. Gynecolog. Cancer* **14**, 470–474 (2004).

[44] A.V. Shakhov, A.B. Terentjeva, V.A. Kamensky, L.B. Snopova, V.M. Gelikonov, F.I. Feldchtein, and A.M. Sergeev, "Optical coherence tomography monitoring for laser surgery of laryngeal carcinoma," *J. Surg. Oncol.* **77**, 253–259 (2001).

[45] N.D. Gladkova, A.V. Shakhov, F.I. Feldchtein, "Capabilities of optical coherence tomography in laryngology" in *Handbook of Optical Coherence Tomography,* B.E. Bouma and G.J. Tearney (Eds.), Marcel Dekker, Inc., New York, Basel, 705–724 (2002).

[46] J.V. Fomina, N.D. Gladkova, L.B. Snopova, N.M. Shakhova, F.I. Feldchtein, and A.V. Myakov, "*In vivo* OCT study of neoplastic alterations of the oral cavity mucosa" in *Coherence Domain Optical Methods and Optical Coherence Tomography in Biomedicine VIII, Proc. SPIE* **5315**, 41–47 (2004).

[47] N.D. Gladkova, J.V. Fomina, A.V. Shakhov, A.B. Terentieva, M.N. Ourutina, V.K. Leontiev, F.I. Feldchtein, and A.V. Myakov, "Optical coherence tomography as a visualization method for oral and laryngeal cancer," *Oral Oncology,* A.K. Varma and P. Reade (Eds.), **9**, 272–281 (2003).

[48] L.S. Dolin, F.I. Feldchtein, G.V. Gelikonov, V.M. Gelikonov, N.D. Gladkova, R.R. Iksanov, V.A. Kamensky, R.V. Kuranov, A.M. Sergeev, and N.M. Shakhova, "Fundamentals of OCT and clinical applications of endoscopic OCT," in *Handbook of Coherent Domain Optical Methods*, V.V. Tuchin (Ed.), Kluwer Academic Publishers, Boston, **2**, 211–270 (2004).

[49] F.I. Feldchtein, N.D. Gladkova, L.B. Snopova, E.V. Zagaynova, O.S. Streltzova, A.V. Shakhov, A.B. Terentjeva, N.V. Shakhova, I.A. Kuznetsova, and G.V. Gelikonov, "Blinded recognition of optical coherence tomography images of human mucosa precancer" in *Coherence Domain Optical Methods in Biomedical Science and Clinical Applications VII, Proc. SPIE* **4956**, 89–94 (2003).

[50] B. Shen, G. Zuccaro, T.L. Gramlich, N. Gladkova, P. Trolli, M. Kareta, C.P. Delaney, J.T. Connor, B.A. Lashner, and C.L. Bevins, "*In vivo* colonoscopic optical coherence tomography for transmural inflammation in inflammatory bowel disease," *Clin. Gastroenterol. Hepatol.* **2,** 1080–1087 (2004).

[51] N. Gladkova, A. Maslennikova, I. Balalaeva, F. Feldchtein, E. Kiseleva, M. Karabut, and R. Iksanov, "Application of optical coherence tomography in the diagnosis of mucositis in patients with head and neck cancer during a course of radio(chemo) therapy," *Med. Laser Appl.* **23**, 186–195 (2008).

[52] L..U. Junqueira, G. Bignolas, and R.R. Brentani, "Picrosirius staining plus polarization microscopy, a specific method for collagen detection in tissue sections," *Histochemical J.* **11**, 447–455 (1979).

17

Noninvasive Assessment of Molecular Permeability with OCT

Kirill V. Larin

University of Houston, Houston, TX 77204, USA, Institute of Optics and Biophotonics, Saratov State University, Saratov, 410012, Russia

Mohamad G. Ghosn

University of Houston, Houston, TX 77204, USA

Valery V. Tuchin

Institute of Optics and Biophotonics, Saratov State University, Saratov, 410012, Russia, Institute of Precise Mechanics and Control of RAS, Saratov 410028, Russia

Noninvasive imaging, monitoring, and assessment of molecular transport in epithelial tissues are extremely important for many biomedical applications. For example, successful management of many devastating diseases (e.g., cancer, diabetic retinopathy, glaucoma, and different cardiovascular disorders) requires long-term treatment with drugs. In contrast to the traditional oral route, topical or local drug delivery through epithelial tissues (e.g., skin or cornea and sclera of the eye) is currently accepted as a preferred route for drug administration because the hepatic first pass effect of orally administered drugs and problems of the acidic environment and pulsed absorption in the stomach can be avoided and complications and side effects can be reduced. However, topical delivery of therapeutic agents to target tissues in effective concentrations remains a significant challenge due to low permeability of epithelial tissues and drug washout. Significant research effort is now devoted to the development of therapeutically effective topical formulations such as drug delivery systems (patches), gels, creams, ointments, lotions, as well as application of various diffusion enhancers. Successful developments of these formulations require advanced understanding of the kinetics of drug distribution in epithelial tissues for better manipulation and optimization of therapeutic processes and outcomes. This chapter describes recent progress made on the development of a noninvasive biosensor, based on optical coherence tomography (OCT) technique, for assessment of molecular and drug diffusion in epithelial tissues. High in-depth and transverse resolution (up to a few m) and high dynamic range (up to 120 dB) of the OCT allows sensitive and accurate real-time monitoring of the diffusion processes due to precise depth-resolved measurements of tissue optical properties at different layers.

Key words: molecular and drug diffusion, epithelial tissues, noninvasive, Optical Coherence Tomography

17.1 Introduction

The development of a functional imaging method for noninvasive assessment of molecular diffusion and drug biodistribution in epithelial tissues would significantly benefit several basic research and clinical areas such as clinical pharmacology, ophthalmology, and structural and functional optical imaging of tissues and cells. For example, there is a great need for a noninvasive method to assess pharmacokinetics of topically applied drugs in ocular tissues in order to achieve desired and controlled therapeutic outcomes [1–7]. Additionally, application of permeation enhancers in topical formulations may change the time course of clinical effects and requires understanding of kinetics of drug distribution in tissues for better manipulation and optimization of the therapeutic processes and outcomes [7–10]. Specific recommendations for the pharmacokinetics or biopharmaceutical data relevant to a new drug or new dosage form have been defined by the FDA in its guidelines and greatly depend on the availability of a method for *in vivo* assessment of the diffusion of topically applied drugs and molecules. Therefore, the availability of a noninvasive instrumentation and method will allow researchers and clinical investigators a means of measuring tissue drug diffusions proficiently and completely noninvasively. Furthermore, such a technique will present the possibility of more efficiently determining the effectiveness of a topical drug administration during clinical trials and animal research. For example, recently a research group performed experiments to monitor the diffusion of an antibody-based drug in live rabbits to determine its effectiveness using the ELISA assay [11]. However, to quantify the amount of the drug that permeated into the eye, the rabbit had to be sacrificed and the eye enucleated. With a real-time, high-resolution, depth-resolved technique, such sacrifice would be superfluous, as the drug diffusion rate would be assessable directly within the live subject rather than postmortem.

Additionally, mathematical pharmacokinetic models are commonly used to predict the effectiveness of newly developed topical drugs [12, 13]. A noninvasive biosensor can aid in the understanding of pharmacokinetics for drug development, making it possible to quickly and efficiently determine the diffusion of different drugs to verify a proposed model. A deeper, quantitative understanding of topical drug diffusion can aid in the development of new treatments by providing insight into tissue permeability barriers and mechanisms behind drug transport across tissue layers.

In the past decades, several scientific groups have developed and applied different experimental techniques to study drug diffusion in tissues *in vitro*, including spectrofluorometers, Ussing apparatus [14–16], fluorescence microscopy [17, 18], and microdialysis [19]. Although these respective techniques have enriched our knowledge about drugs' diffusion properties in tissues, they are basically limited for *in vitro* studies only.

Recently, several additional methods and techniques have been employed to study diffusion processes in tissues *in vivo*. For example, the diffusion of glucose molecules in rabbit sclera was estimated precisely using reflection spectroscopy [20]. The diffusion coefficients for water and glucose were elucidated from the Monte-Carlo (MC) numerical simulation method of experimentally measured optical clearing of the tissue. While the power of this method has been demonstrated in relatively homogeneous tissues, it poses a difficulty in estimating diffusion coefficients in multi-layered structures without prior known information of tissues layers. Magnetic resonance imaging (MRI) has been also proposed for *in vivo* imaging of drug diffusion in eyeballs [21, 22]. However, several drawbacks of this technology, such as the scarcity of actual drugs possessing paramagnetic properties, the extensive time required for image acquisition and signal processing, and the low imaging resolution limit its application for drug diffusion studies in epithelial tissues, such as cornea and sclera of the eye. Ultrasound has also been utilized in transdermal diffusion studies [23]. But factors such as low resolution and the need for contact procedure pose as significant drawbacks, limiting its application as well. On the other hand, optical coherence tomography (OCT) presents a technique that can potentially overcome limitations posed by the above-mentioned methods due to its capability of real-time imaging with high in-depth resolution (up to a few micrometers) and may be used completely noninvasively.

OCT is a relatively new noninvasive optical diagnostic technique that provides depth-resolved images of tissues with resolution of a few μm at depths of up to several mm. This technique was introduced in 1991 to perform tomographic imaging of the human eye [24]. Since then, OCT is being actively developed by several research groups for many clinical diagnostic applications (reviewed in, e.g., [25–27]). The basic principle of the OCT is to detect backscattered photons from a tissue of interest within a coherence length of the source by using a two-beam interferometer. OCT imaging is somewhat analogous to ultrasound B-mode imaging except that it uses light, not sound. Briefly, light from a broadband source (a laser with low coherence) is aimed at objects to be imaged using a beam splitter. Light scattered from the tissue is combined with light returned from the reference arm, and a photodiode detects the resulting interferometric signal. Interferometric signals can be formed only when the optical path length in the sample arm matches the reference arm length within coherence length of the source. By gathering interference data at points across the surface, cross-sectional 2D images can be formed in real time with resolution of about a few μm at depths up to several millimeters, depending on the tissue optical properties. By averaging the 2D OCT images into a single 1D distribution of light in depth in logarithmic scale, one can measure the optical properties of the object by analyzing the profile of light attenuation [28–31].

Recently, the power of OCT technology for noninvasive assessment of molecular diffusion has been demonstrated in epithelial tissues including sclera, cornea, skin, and vasculature [32–41]. Because of OCT's ultrahigh resolution and ability to reach deep tissue layers (up to a few mm), it is well suited for measuring the molecular permeation in epithelial biological tissues. In this chapter, we summarize recent studies on the application of OCT for monitoring and quantifying of molecular diffusion in sclera and cornea of the whole rabbit eye as well as the permeation of protein molecules in human carotid tissue measured *ex vivo*.

17.2 Principles of OCT Functional Imaging

The ability of a chemical compound to change tissue scattering properties is based on a number of biophysical processes [28–30, 42–46]. The concept of two fluxes – diffusion of molecules into tissue and movement of bulk tissue water out, which may be relatively independent or interacting fluxes, defines the kinetics and dynamics of tissue optical properties [46]. Diffusion of hyperosmotic molecules into interstitial space of tissues leads to an increase of the refractive index of the interstitial fluid (ISF), thus to refractive index matching of collagen fiber (and other tissue components – scatters) and ISF. Replacement of tissue fluids with the molecules (and drugs) may induce an intensive water flow from a tissue, which may cause dehydration and corresponding alteration of tissue morphological and optical properties by (1) refractive index matching or mismatching, (2) increase of tissue collagen fiber packing density (ordering of scatterers), (3) decrease of tissue thickness, and (4) collagen reorganization.

Similarly to all optical methods, as the OCT beam penetrates deeper into a tissue, the signal strength diminishes due to optical attenuation. Therefore, there is a relationship between the depth of penetration and the signal strength at a particular depth. This can be described in terms of the

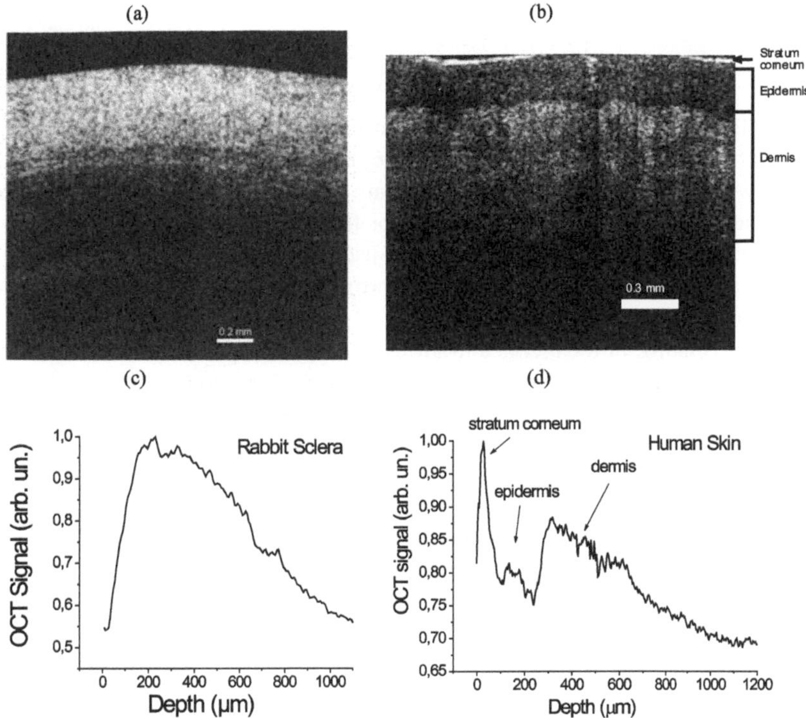

FIGURE 17.1: Typical OCT image and corresponding 1D signal recorded from rabbit sclera (a and c) and human skin (b and d).

OCT signal slope (OCTSS). In the first approximation, the OCTSS plotted on a logarithmic scale (Figure 17.1b, d), is proportional to the total attenuation coefficient of the tissue, μ_t:

$$ln\left(\frac{I(z)}{I_0}\right) \equiv \text{OCTSS} = -\mu_t z = -(\mu_a + \mu_s)z \tag{17.1}$$

where μ_s and μ_a are the scattering and absorption coefficients, respectively. Since $\mu_s \gg \mu_a$ in the near infrared (NIR) spectral range:

$$\ln\left(\frac{I(z)}{I_0}\right) \equiv \text{OCTSS} = -\mu_s z \tag{17.2}$$

The scattering coefficient of a tissue depends on the refractive index mismatch between the ISF and the tissue components (fibers, cell components). In a simple model of scattering dielectric spheres, μ_s can be approximated as [47]:

$$\mu_s = \frac{3.28\pi r^2 \rho_s}{1-g}\left(\frac{2\pi r}{\lambda}\right)^{0.37}\left(\frac{n_s}{n_{\text{ISF}}} - 1\right)^{2.09}, \tag{17.3}$$

where g is the tissue anisotropy factor, r is the radius of scattering centers, ρ_s the volume density of the scatterers, λ the wavelength of the incident light, and n_s and n_{ISF} the refractive indices of the scatterers and ISF, respectively. If the refractive index of the scatterers remains constant and is higher than the refractive index of ISF, the diffusion of molecules inside the medium reduces the refractive index mismatch, $\Delta n = n_s - n_{\text{ISF}}$

$$\mu_s = \frac{3.28\pi r^2 \rho_s}{1-g}\left(\frac{2\pi r}{\lambda}\right)^{0.37}\left(\frac{n_s}{n_{\mathrm{ISF}}+\delta n_{\mathrm{drug}}}-1\right)^{2.09}, \quad (17.4)$$

where δn_{drug} is the drug-induced increase of the refractive index of ISF. Therefore, an increase of tissue drug concentration will raise the refractive index of ISF that will decrease the scattering coefficient as a whole. Note, that these equations are derived from Mie theory of light scattering on dielectric spheres and, strictly speaking, do not fully describe the complexity of light-tissue interactions. However, the equations are frequently used as the first approximation and are utilized here to demonstrate the effect; more rigorous approximation of light-tissue interaction requires Monte Carlo calculations.

Alternatively, the change in local concentrations of scattering particles such as cell components or collagen fibers could alter scattering as well. A micro-optical model developed by Schmitt and Kumar [48] can be used to describe changes in tissues scattering caused by tissue dehydration (tissue shrinkage) because it treats the tissue as a collection of differently sized scattering particles with distribution function modified to account for correlated scattering among densely packed particles. Assuming that the waves scattered by the individual particles in a thin slice of the tissue volume add randomly, the scattering coefficient and scattering anisotropy factor of the volume can then be approximated as the following expressions [48]

$$\mu_s = \sum_{i=1}^{N_p} \frac{\eta(2a_i)}{v_i}\sigma_s(2a_i), \qquad g = \frac{\sum_{i=1}^{N_p} \mu_s(2a_i) g_i(2a_i)}{\sum_{i=1}^{N_p} \mu_s(2a_i)}, \quad (17.5)$$

where N_p is the number of particle diameters; $\eta(2a_i)$ is the volume fraction of particles of diameter $2a_i$; $\sigma_s\,(2a_i)$ is the optical cross-section of an individual particle with diameter $2a_i$ and volume v_i. At the limit of an infinitely broad distribution of particle sizes,

$$\eta(2a) \approx (2a)^{3-D_f} \quad (17.6)$$

where D_f is the (volumetric) fractal dimension; for $3 < D_f < 4$, this power-law relationship describes the dependence of the volume fractions of the subunits of an ideal mass fractal on their diameter $2a$.

To account for the interparticle correlation effects which is important for systems with volume fractions of scatters higher than 1–10% (dependent on particle size), the following correction should be done [48]

$$\eta(2a) \rightarrow \frac{[1-\eta(2a)]^{p+1}}{[1+\eta(2a)(p-1)]^{p-1}}\eta(2a), \quad (17.7)$$

where p is the packing dimension that describes the rate at which the empty space between scatters diminishes as the total density increases. For spherical particles $p = 3$, for rod-shaped and sheet-like particles $p \rightarrow 2$ and 1, respectively. Since the elements of tissue have all of these different shapes and may exhibit cylindrical and spherical symmetry simultaneously, the packing dimension may lie anywhere between 1 and 5.

Tissues mean refractive index can be calculated by the law of Gladstone and Dale as a weighted average of refractive indices of interstitial fluid (n_{ISF}) and collagen fibers or cell components (n_s):

$$\bar{n} = \phi_s n_s + (1-\phi_s) n_{\mathrm{ISF}}, \quad (17.8)$$

where ϕ_s is the volume fraction of collagen fibers and/or cell components in tissues. Therefore, changes in the volume fraction of tissue components ϕ_s (e.g., by shrinkage or swelling of the tissues) will change the overall refractive index of the tissues.

Therefore, since OCT measures the in-depth light distribution with high resolution, changes in the in-depth distribution of the tissue scattering coefficient and/or refractive index are reflected in changes in the OCT signal [28–30, 38–41, 49]. Thus, since the diffusion of macromolecules in a tissue introduces local changes in the tissue optical properties (scattering coefficient and refractive index), one can monitor and quantify the diffusion process by depth-resolved analyses of the changes in the OCT signal recorded from a sample.

17.3 Materials and Methods

17.3.1 Experimental setup

The experiments were performed using a time-domain OCT system (Imalux Corp, Cleveland, OH), schematically shown in Figure 17.2. The optical source used in this system is a low-coherent broadband, near-infrared (NIR) light source with wavelength of 1310$\pm$15 nm, output power of 3 mW, and resolution of 25 μm (in air). The sample arm of the interferometer directs light into tissue through a single-mode optical fiber within a specially designed miniature endoscopic probe. The endoscopic probe allowed for scanning of the sample surface in the lateral direction (X axis). Light scattered from the sample and light reflected from the reference arm mirror formed an interferogram, which was detected by a photodiode. In-depth scanning was produced electronically by piezo-electric modulation of the fiber length. Two-dimensional images were obtained by scanning the incident beam over the sample surface in the lateral direction and in-depth (Z-axial) scanning by the interferometer (Figure 17.1a, b). The acquired images were 2.2 by 2.4 mm and required an acquisition rate of approximately 4.2 seconds per image. The obtained 2D images were averaged in the lateral direction (over $\approx$ 1 mm, that was sufficient for speckle-noise suppression) into a single curve to obtain an OCT signal that represents 1-D distribution of light in-depth in logarithmic scale (Figure 17.1c, d). The depth scales shown in the all figures and graphs are calculated by dividing the optical path length given by the OCT system to the mean refractive index of the tissue that was assumed to be as 1.4.

17.3.2 Ocular tissues

Fresh rabbit and porcine eyes were obtained from a local supplier and kept cooled while stored in 0.9% sodium chloride solution to prevent dehydration. The eyes were freshly enucleated and experiments were performed in the first three days upon arrival to guarantee minimal changes in the physiological status of the tissues. Before experiments, the eyes were placed in a specially designed dish submersed in saline and allowed to acclimate to the room temperature. Experiments were conducted at 20°C. Constant temperature was maintained for the duration of the experiments. OCT images were continuously acquired for 60–180 minutes. The light beam was delivered perpendicular to the epithelial surface of the sclera or the cornea. Before adding any agent, images of the sclera or the cornea were taken for about 8–10 minutes to record a baseline. Diffusion of different molecules was studied during these experiments. Diffusion of purified and distilled water and 20%-glucose solution was monitored and permeability rate was calculated. Glucose solutions were prepared by dissolving different amounts (20 g and 40 g) of glucose (Mallinckrodt Baker, Inc., Phillisbugh, NJ) in 100 ml of pure water to obtain 20% and 40% glucose solutions, respectively. The diffusion of several ophthalmic drugs including metronidazole, ciprofloxacin, dexamethasone, and mannitol were studied as well. The concentrations of metronidazole (Abbott laboratories, North Chicago, IL), ciprofloxacin (Alcon laboratories, Fort Worth, TX), and dexamethasone (Phoenix Scientific, Inc., St. Joseph, MO) were 0.5%, 0.3%, and 0.2%, respectively, and mannitol – 20%.

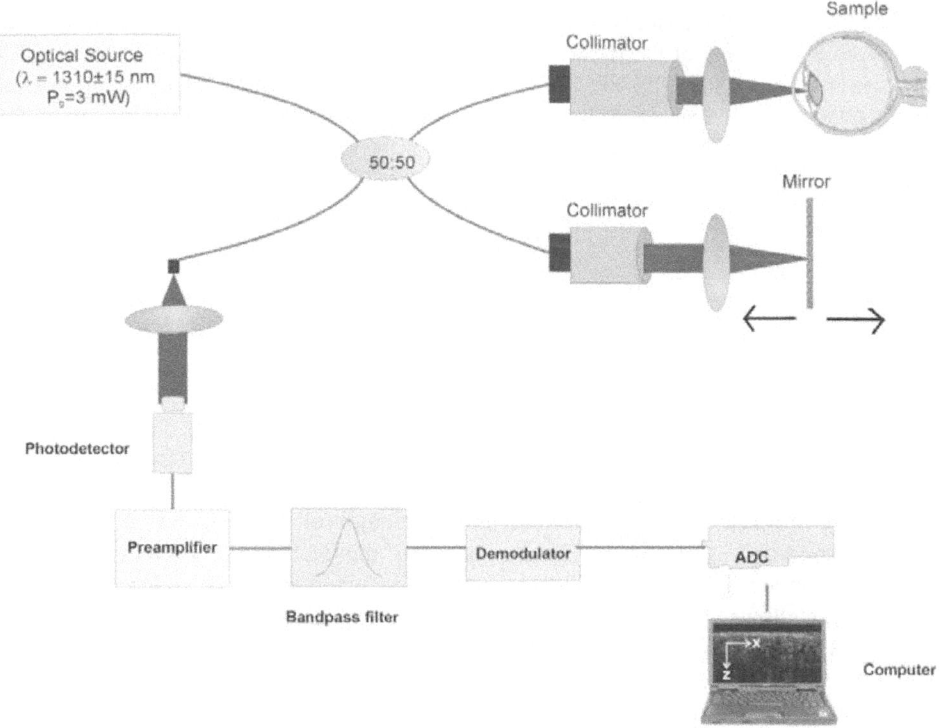

FIGURE 17.2: Schematic diagram of experimental setup.

17.3.3 Vascular tissues

Healthy human carotid endarterectomy (CEA) tissues were resected during a surgery and transferred to phosphate-buffered saline (PBS) at 5°C. For experimentation, the tissues were cut with a hole borer into 5 mm diameter discs and transferred to a 96 well (6 mm diameter well) microtiter plate with the intimal surface facing up (Figure 17.3). One tissue disk was used for each OCT measurement that was started within 6 hours after tissue acquirement. Diffusion of glucose and low density lipoproteins (LDL) were studies in these experiments. LDL was isolated by sequential ultracentrifugation as described previously [50], and dialyzed vs. PBS then concentrated by ultrafiltration. The stock solution of LDL had a protein concentration of 37.8 mg/ml. An aliquot (100 μl) of the stock solution was added to each well containing a tissue disk and 100 μl PBS, giving a final LDL concentration of 18.9 mg/ml. The 100 μl of PBS was sufficient to totally immerse the tissue; thereby maintaining its hydration condition. The samples were imaged in 100 μl of PBS for 5 minutes to acquire baseline data. The 100 μl of the agents (40%-glucose or 39 mg/ml LDL) was then added to the well, giving a final concentration of 20%-glucose or 19.5 mg/ml LDL. The sample was then imaged for another 40 minutes. Multiple independent experiments at both 20°C and 37°C were performed.

17.3.4 Data processing

The permeability rate of the studied molecules in cornea, sclera, and CEA was calculated using two methods – OCT signal slope (OCTSS) and amplitude (OCTA) measurements with a specially developed C++ program. The two-dimensional OCT images were averaged in the lateral (X axial) direction into a single curve to obtain an OCT signal that represents one-dimensional distribution of

FIGURE 17.3: Magnified photograph of a 5 mm diameter carotid tissue sample immersed in PBS buffer in a well of a 96-well microtiter plate.

intensity in-depth (in logarithmic scale) (Figure 17.4). In the OCTSS method, a region in the tissue, where minimal alterations to OCT signals had occurred, was selected and its thickness (z_{region}) was measured. The diffusion of the agents in the chosen region was monitored and time of diffusion was recorded (t_{region}). The time of diffusion was calculated as time when the saturation stage was reached minus the time when the OCTSS started to change (see, e.g., Figure 17.5). The permeability rate ($\overline{P}$) was calculated by dividing the measured thickness of the selected region by the time it took for the molecules to diffuse through (as $\overline{P} = \frac{z_{\text{region}}}{t_{\text{region}}}$).

For depth-resolved measurements, OCT amplitude (OCTA) method was used to calculate the permeability rate at the specific depths in the tissues as $P(z) = z_i/t_{z_i}$, where z_i is the depth at which measurements were performed (calculated from the front surface of the tissue) and t_{z_i} is the time of molecular diffusion to this depth. The time, t_{z_i}, was calculated from the moment the agent was added to the tissue until agent-induced change in the OCT amplitude was commenced. Note that unlike in OCTSS method, the depth z_i was calculated from the epithelial surface of the tissues to the specific depth in tissues' stroma in the OCTA method.

17.4 Results

17.4.1 Diffusion in the cornea

Figure 17.5 shows a typical OCTSS graph obtained from a rabbit cornea during a dexamethasone diffusion experiment. The permeability rate was calculated from 90–370 μm regions from the surface of the cornea using the OCTSS method. After topical application, the diffusion of dexamethasone inside the cornea altered the local scattering coefficient and was detected by the OCT system. The increase in the local in-depth concentration of dexamethasone resulted in the decrease of scattering coefficient and, consequently, in the decrease of the measured OCT signal slope. It can be reasonably deduced that after approximately 15 minutes from application, a droplet of dexamethasone had entirely penetrated through the tissue. In theory, when the drug reached the aqueous humor of the eye, a reverse process took place in order to balance the osmotic gradient. Based on the physiology of the tissues, the concentration gradient between two sides of a tissue creates a driving force for a fluid to travel from a medium of high concentration to that of a lower one. This force is

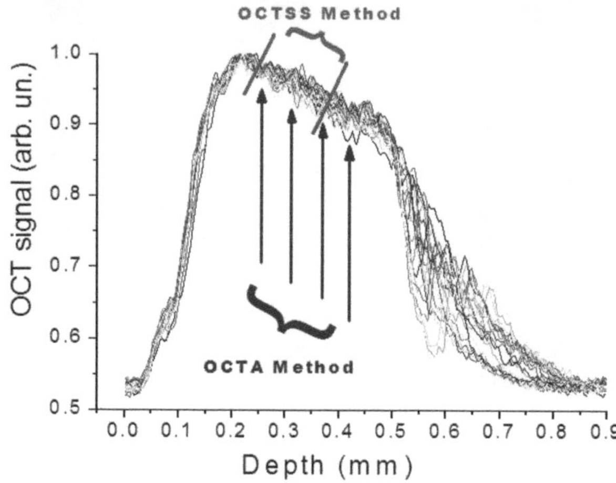

FIGURE 17.4: OCT signals recorded from a rabbit sclera at different times indicating region and depths for OCTSS and OCTA measurements.

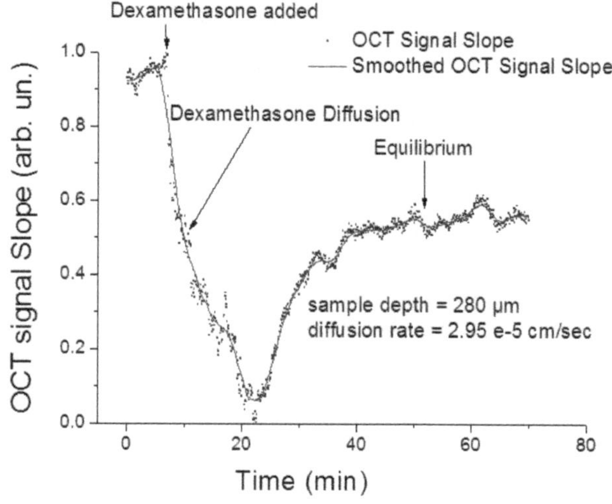

FIGURE 17.5: OCT signal slope as a function of time recorded from a rabbit cornea during dexamethasone diffusion experiments.

likely to be the source of the reverse process seen in Figure 17.5 and all other similar graphs. The driving force remains an active component until equilibrium has reached.

The permeability rates of different molecules in the cornea were measured in twenty eight experiments. The permeability rate for water was found to be $(1.68 \pm 0.54) \times 10^{-5}$ cm/sec. Ciprofloxacin (0.3%), dexamethasone (0.2%), and metronidazole (0.5%) had permeability rates of $(1.85 \pm 0.27) \times 10^{-5}$, $(2.42 \pm 1.03) \times 10^{-5}$, and $(1.59 \pm 0.43) \times 10^{-5}$cm/sec, respectively. A summary of the permeability rates for different drugs in the rabbit cornea is presented in Table 17.1.

Previously, the permeability rate of water in rabbit cornea of 1.5×10^{-4} cm/s was reported by Grass based on *in vitro* measurements on isolated tissues [51]. Also, Ciprofloxacin was measured to have a permeability rate of 0.29×10^{-5} cm/sec [52]. The discrepancy between the literature

TABLE 17.1: Permeability rate of different agents in the rabbit cornea.

Agent	Permeability rate ± standard deviation (cm/sec)	Number of independent experiments
Water	$(1.68 \pm 0.54)\times10^{-5}$	8
Ciprofloxacin	$(1.85 \pm 0.27)\times10^{-5}$	4
Mannitol	$(1.46 \pm 0.08)\times10^{-5}$	4
Dexamethasone	$(2.42 \pm 1.03)\times10^{-5}$	7
Metronidazole	$(1.59 \pm 0.43)\times10^{-5}$	5

and these experimental values is most likely due to distinctions in experimental and calculation methods used for estimation of the permeability rates as well as tissue preparation (e.g., isolated versus intact tissues, hydration factors, etc.). Furthermore, it was recently verified that permeability rates of several agents might be quite different if measured in an isolated cornea or a cornea in the whole eyeball (for instance, calculated permeability rate of pure water in an isolated cornea was $1.39\pm0.24\times10^{-4}$cm/sec that is very close to the previously reported values) [40].

17.4.2 Diffusion in the sclera

Figure 17.6 illustrates a typical OCTSS graph obtained from a rabbit sclera during a glucose diffusion experiment. The permeability rate was calculated from a 105 μm thick region (70–175 μm depth from the surface of the sclera). Similar to the experiment with the cornea, the increase in the local in-depth concentration of glucose in the sclera initiated a decrease in the scattering coefficient and, correspondingly, in OCTSS. The same reverse process occurred in these sets of experiments as well. The amplitude and speed of the reversal process could be governed by the volume-to-volume ratio between the eye and the solution it is residing in. The volume of the eye was estimated to be 8 cm^3. For that, we maintained 8 cm^{3}volume of drug solutions to keep a 1:1 ratio. As a result, the equilibrium stage was reached at about 50% of initial value.

Thirty-one experiments were performed with the sclera of different rabbit eyes. The permeability rate of water was found to be $(1.33 \pm 0.28)\times10^{-5}$cm/sec. Ciprofloxacin, mannitol, glucose 20%, and metronidazole had a permeability rate of $(1.41 \pm 0.38)\times10^{-5}$, $(6.18 \pm 1.08)\times10^{-6}$, $(8.64 \pm 1.12)\times10^{-6}$, and $(1.31 \pm 0.31)\times10^{-5}$cm/sec, respectively. Results of the permeability rates of these agents in sclera are summarized in Table 17.2 and close to those rates published in earlier studies. For example, the permeability rate of ciprofloxacin in a rabbit sclera was measured to be $(1.88 \pm 0.617)\times10^{-5}$cm/sec [52]. Additionally, the results for glucose diffusion are in a good correlation with results for a human sclera [53].

TABLE 17.2: Permeability rate of different agents in the rabbit sclera.

Agent	Permeability rate ± standard deviation (cm/sec)	Number of independent experiments
Water	$(1.33 \pm 0.28)\times10^{-5}$	5
Ciprofloxacin	$(1.41 \pm 0.38)\times10^{-5}$	3
Mannitol	$(6.18 \pm 1.08)\times10^{-6}$	5
Glucose	$(8.64 \pm 1.12)\times10^{-6}$	14
Metronidazole	$(1.31 \pm 0.29)\times10^{-5}$	4

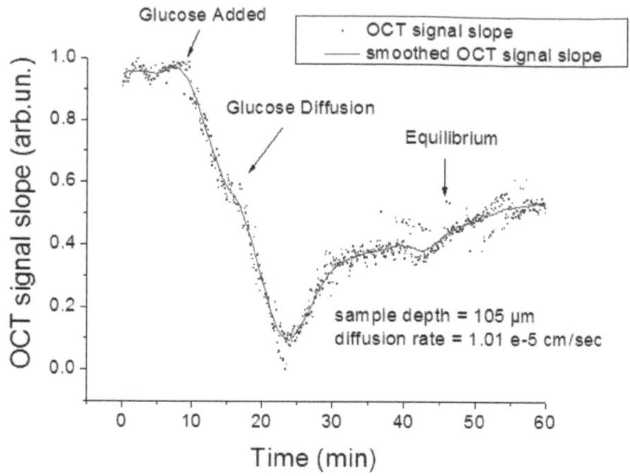

FIGURE 17.6: OCT signal slope as a function of time recorded from a sclera during glucose 20% diffusion experiments.

Therefore, the obtained results demonstrate the capability of the OCT technique for accurate monitoring and quantification of molecular diffusion in epithelial ocular tissues. The local optical properties of tissues were dynamically changed as a function of molecular diffusion and were depicted by depth-resolved analysis of the OCT signals. Additionally, one can see from these tables that the permeability rates of small-concentrated drugs dissolved in water (ciprofloxacin-0.3% and metronidazole-0.5%) lie within the range of that of pure water. Hence, the changes in the optical properties of the tissues were induced, most likely, by the carrier of the drug solution rather than the drug itself in this case. Due to a distinct carrier – polyethylene glycol 400 of 50% concentration, the permeability rates for the small-concentrated dexamethasone (0.2%) were significantly different from that of water. Nevertheless, monitoring and quantification of diffusion processes of small-concentrated agents in tissues might be possible by application of different methods of OCT signal acquisition and processing. For example, as mentioned above, the diffusion of drugs in tissues might affect its collagen organization [9, 10, 16, 43, 54, 55]. One of the most effective noninvasive methods for monitoring the structural rearrangements is polarization-sensitive OCT [56–59]. OCT images in orthogonal polarization can potentially provide higher sensitivity and specificity of drug diffusion monitoring. Another possible approach is by application of drugs in different formulations (e.g., ointments, gels, and various sustained and controlled-release substrates).

The results summarized in Tables 17.1 and 17.2 also demonstrate that permeability rates for the same molecules in the cornea and sclera are different. The structural and physiological difference of the two tissue types plays a significant role in explaining the variance elicited by the study. A key factor in the study of molecular permeability rates across biological tissues is the degree of hydration. As a result of hydration the tissue fibrils can move further apart, thus creating extra space between them [60]. Swelling could come from the glycosaminoglycan (GAGs) or from anion bindings occurring in the tissue [61, 62]. The sclera has 10 times less GAGs than the cornea and the scleral stroma has a greater degree of fibrillar interweave than corneal stroma [63]. Thus, the hydration of sclera increases only 20–25% even though the cornea swells to many times its original weight in aqueous solution [61]. Also, recent investigations show that the permeability rate is proportional to the tissues' hydration: as the hydration in a tissue increases, the permeability rate increases [64]. This could serve as a reason as to why the cornea had a higher permeability rate to that of the sclera for the same molecules in these experiments.

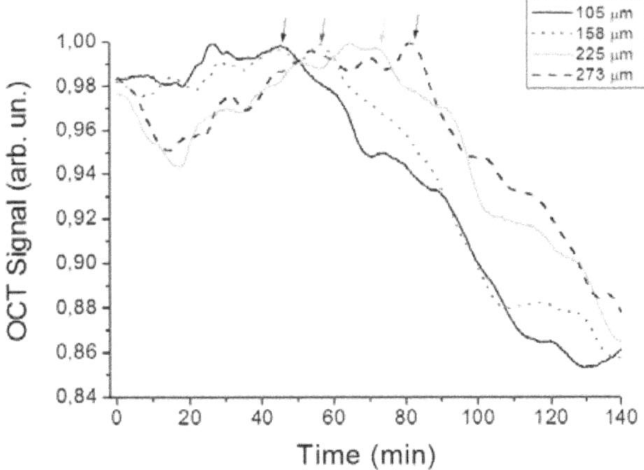

FIGURE 17.7: OCT signal as a function of time recorded at different depths during a mannitol diffusion experiment in the sclera. Arrows indicate the mannitol's front reaching a different depth inside a sclera.

17.4.3 In-depth diffusion monitoring

The high resolution and ability for noninvasive depth-resolved imaging of OCT provides unique opportunities for investigating the molecular diffusion not only as a function of time but also as a function of depth. The diffusion process of different molecules at varying depths from the surface of the tissue was calculated using the previously described OCTA method. Figure 17.7 shows typical OCT signal measured at depths 105, 158, 225, and 273 μm away from the surface of a rabbit sclera during mannitol diffusion experiment. The arrows on each of the OCT signals depict the time the drug action reached that particular depth, as illustrated by a sharp decrease of the OCT signal. Figure 17.8 shows the calculated permeability rates of 20%-mannitol measured at different depth in the sclera. This graph demonstrates that permeability rate inside the sclera is not homogenous and is increasing with increase of the depth as it expected for multilayered objects with different tissue density.

Therefore, the results for depth-resolved quantification of molecular diffusion in tissues, shown in Figures 17.7 and 17.8, demonstrate that the permeability rate inside epithelial tissues is increasing with the increase of depth. Several factors contribute for such an increase of the permeability rates with the tissue depth. Layered structure of the tissues, the difference in the diameter sizes of the collagen fibers in each layer, and the diverse organizational patterns of the collagen bundles at different depths are likely to contribute to the observed trend. For instance, the sclera possesses three distinct layers that include the episclera, the stroma, and lamina fusca. The outer layer, the episclera, consists primarily of collagen bundles that intersect at different angles along the surface of the sclera. This inconsistency in organization of the collagen bundles causes a lower permeability rate of the molecules, imposing an effective resistance force that potentially reduces the speed of molecular penetration into the tissue. In the stroma and the lamina fusca of the sclera, the collagen fibrils are more organized and oriented in two patterns; meridionally or circularly, reflecting the observed increase in molecular permeability in the stroma. Additionally, the collagen bundles differ in size throughout the different layers of the sclera: the collagen bundles in the external region of the sclera are narrower and thinner than those in the inner region [65]. The different diameters of the collagen fibers in different tissue depths may also influence the molecular diffusion in tissues.

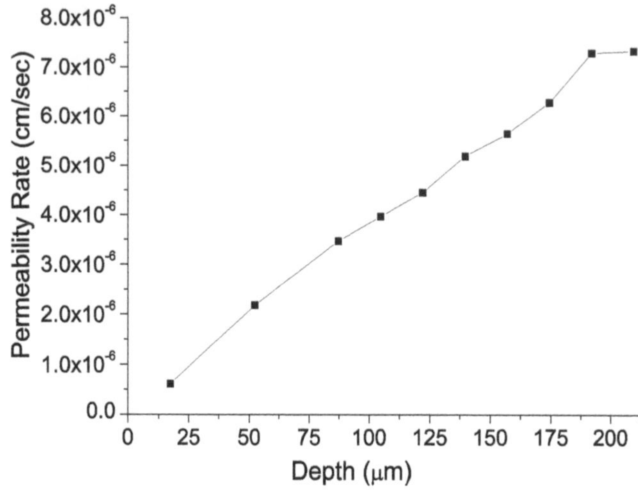

FIGURE 17.8: Permeability rates recorded at different depths in a sclera during the diffusion of 20%-mannitol.

17.4.4 Diffusion in the carotid

For the carotid permeation study, a baseline was established for the first five minutes of each experiment using only PBS after which the drugs were added and their permeability coefficients were quantified similarly as described above at different temperatures. The permeability rate of 20%-glucose was determined to be $(3.51 \pm 0.27)\times10^{-5}$cm/sec ($n = 13$) at 20°C and $(3.70 \pm 0.44)\times10^{-5}$ cm/sec ($n = 5$) at 37°C. However, the permeability rate for LDL was found to be $(2.42 \pm 0.33)\times10^{-5}$ cm/sec ($n = 5$) at 20°C and $(4.77 \pm 0.48)\times10^{-5}$ cm/sec ($n = 7$) at 37°C. Figure 17.9 illustrates the typical OCTSS graphs of glucose permeation at 20°C and 37°C (Figure 17.9a, b) and LDL permeation at 20°C and 37°C (Figure 17.9c, d) in human carotid tissue, respectively. As shown by the data above, there is a significant difference between glucose and LDL permeability rates with temperature increase. Accordingly, the permeability rate nearly doubled for LDL while glucose did not show any significant change in the permeability rate at the higher temperature (Figure 17.10). This phenomenon may be explained by the fact that glucose (MW = 180) is significantly smaller in size than LDL ($MW = 2\times10^6$). The increase in the permeability rate of LDL beyond that of glucose could be explained by a possible separate transport mechanism exclusive to LDL as opposed to small molecules [66]. Surprisingly, at the physiological temperature of 37°C, the larger LDL molecules permeated faster (1.3-fold) through the arterial tissue than glucose, again hinting at the functionality of a possible active LDL transport mechanism present in healthy tissue.

Previous studies demonstrated that the diffusion of molecules into the arterial wall is dependent on the wall permeability and the solute concentration [36, 37]. Factors such as cell death, wounding of endothelial cell membranes, variations in shear stress at branch sites of arteries, dysfunction of the intact endothelial layers, and blood pressure serve as components that influence permeation in the arterial wall. As confirmed by the results of this study, temperature is also a notable factor in molecular diffusion in biological tissues. Several studies have proven an enhanced diffusion rate in biological tissues with an increase in temperature. One such study using human cadaver skin demonstrated an increased permeability rate of hydrophilic Calcein from 25°C to 31.5°C. The disordering of the stratum corneum lipid structure, the disruption of the keratin network structure in the stratum corneum, and the decomposition and vaporization of keratin at high temperatures all attributed to the significant increase [67].

In a similar manner, heat may cause the permeability of arteries to increase by affecting the ar-

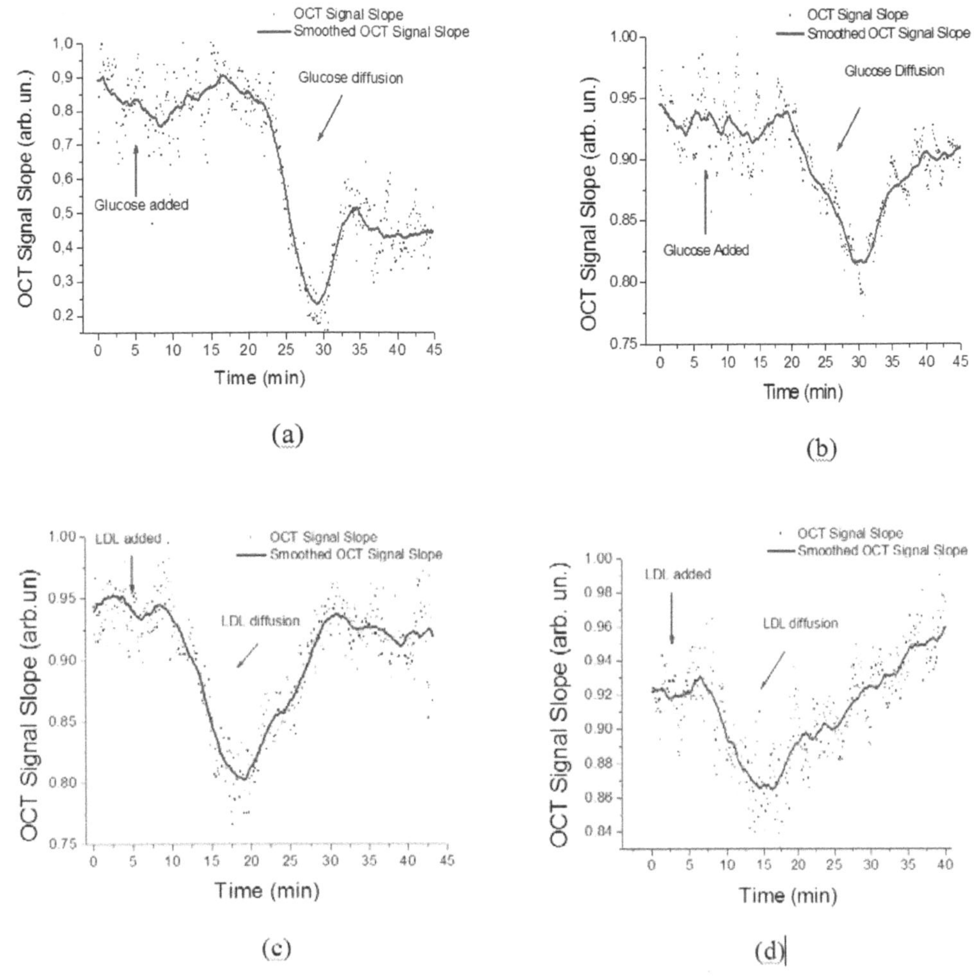

FIGURE 17.9: OCTSS graphs of molecular diffusion in human carotid tissues. (a) Glucose at 20°C , (b) Glucose at 37°C; (c) LDL at 20°C; (d) LDL at 37°C.

terial collagen organization. LDL is able to enter the arterial wall as intact particles by vesicular ferrying through endothelial cells and by passive sieving through the pores that are in or between the endothelial cells [68]. In investigation of the transport of LDL through the arterial endothelium in a rat aorta and coronary artery, the receptor-mediated process that involved vesicles decreased at a lower temperature, which thus caused a decrease in the permeability of the artery [69]. Another study examining the effect of temperature and hydrostatic pressure on LDL transport across micro-vessels using quantitative fluorescence microscopy reported a large reduction in LDL permeability rates when microvessels were cooled down: the permeability rate of LDL decreased about 80.9% when the microvessels' temperature decreased from approximately 20°C to 5°C. The temperature decrease causes closure of the hydraulic pathway and transcellular vesicular pathways of LDL diffusion, directly leading to a decrease in its permeability rate [70].

17.5 Conclusions

The experimental results obtained in these studies demonstrate that OCT can be an effective tool in the study of molecular diffusion in tissues. Different analytes and chemical compounds, such as water, glucose, ciprofloxacin, dexamethasone, metronidazole, and mannitol were tested and corresponding permeability rates were calculated for both the sclera and the cornea of the eye. Diffusion across the sclera has been studied extensively as a function of molecular weight and other parameters. In-depth permeability rates were also estimated in both the sclera and the cornea. An increase in the permeability rate as a function of depth was observed resulting from the layered organization of the tissue. The capability of in-depth measurements pose a new trend in the study of drug diffusion and might be beneficial and of great importance in a variety of different fields. The permeation of a small molecule (glucose) and a large particle (LDL) through normal CEA tissue has been measured at room (20°C) and physiological (37°C) temperatures. The permeation of glucose did not change significantly when the temperature was elevated from 20°C to 37°C, suggesting that the tissue is freely permeable to small molecules over this temperature range. In contrast, the permeation of LDL increased almost twice with this temperature elevation, suggesting that this increase alters the structure of the arterial intima in a way that allows it to become much more permeable to large particles on the order of 20 nm in diameter. Unexpectedly, at 20°C the permeation of LDL was 1.3-fold greater than that of glucose, suggesting that at this physiological temperature the movement of LDL across the intimal boundary is enhanced by an active mechanism in addition to passive diffusion. Therefore, this study describes a new approach for quantitative measurement of lipoprotein movement into the subintimal space where multiple processes causing progression and regression of atherosclerosis might occur.

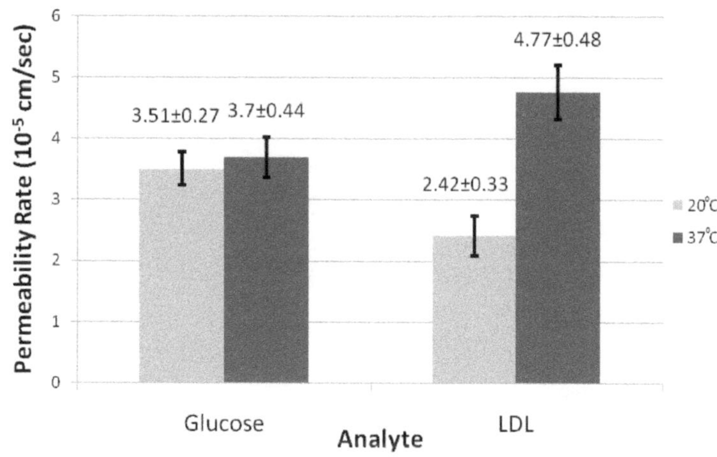

FIGURE 17.10: The permeability rate of glucose and LDL at 20°C and 37°C in a human carotid endarterectomy tissue.

It is important to note that in these studies the refractive index was assumed to be 1.4 for all calculations of the permeability rates. This assumption is not entirely accurate since molecular diffusion into the tissue will change its average refractive index, and, therefore, will affect the size of the tissue region used for the calculations. This change in the tissues' optical size was small enough to be neglected in the calculations; however, the development of correction models for the change of the local refractive index and the estimation of its influence on accuracy of permeability rates calculations is needed to increase accuracy of future studies.

Acknowledgments

This study was funded in part by grants from ONR, The Institute of Biomedical Imaging Sciences, CRDF (RUB1-2932-SR-08), and NIH. VVT was supported by grants 224014 PHOTONICS4LIFE-FP7-ICT-2007-2, RF President's 208.2008.2, 2.1.1/4989, 2.2.1.1/2950 of RF Program on the Development of High School Potential, and the Special Program of RF "Scientific and Pedagogical Personnel of Innovative Russia," Governmental contract 02.740.11.0484.

The authors thank Dr. Joel Morrisett (Baylor College of Medicine, Houston, Texas) for his invaluable contribution to these studies. Also students of Biomedical Optics Laboratory at the University of Houston Saba Syed, Michael Leba, Astha Vijayananda, and Panteha Rezaee contributed to these studies.

References

[1] E. Anglade, R. Yatscoff, R. Foster, et al., "Next-generation calcineurin inhibitors for ophthalmic indications," *Expert Opin. Invest. Drugs* **16**, 1525–1540 (2007).

[2] D.D.S. Tang-Liu and A. Acheampong, "Ocular pharmacokinetics and safety of ciclosporin, a novel topical treatment for dry eye," *Clin. Pharmacokin.* **44**, 247–261 (2005).

[3] D.H. McGee, W.F. Holt, P.R. Kastner, et al., "Safety of moxifloxacin as shown in animal and in vitro studies," *Survey Ophthal.* **50**, S46–S54 (2005).

[4] C.G. Wilson, "Topical drug delivery in the eye," *Exp. Eye Res.* **78**, 737–743 (2004).

[5] S.B. Koevary, "Pharmacokinetics of topical ocular drug delivery: potential uses for the treatment of diseases of the posterior segment and beyond," *Current Drug Metabol.* **4**, 213–222 (2003).

[6] T. Sideroudi, N. Pharmakakis, A. Tyrovolas, et al., "Non-contact detection of ciprofloxacin in a model anterior chamber using Raman spectroscopy," *J. Biomed. Opt.* **12**, 034005 (2007).

[7] B.K. Nanjawade, F.V. Manvi, and A.S. Manjappa, "In situ-forming hydrogels for sustained ophthalmic drug delivery," *J. Control. Release* **122**, 119–134 (2007).

[8] Y. Shirasaki, "Molecular design for enhancement of ocular penetration," *J. Pharm. Sci.* **97**, 2462–2496 (2008).

[9] P.K. Pawar and D.K. Majumdar, "Effect of formulation factors on in vitro permeation of moxifloxacin from aqueous drops through excised goat, sheep, and buffalo corneas," *AAPS Pharm. Sci. Tech.* **7**, E89–E94 (2006).

[10] B.H. Rohde and G.C. Chiou, "Effect of permeation enhancers on beta-endorphin systemic uptake after topical application to the eye," *Ophthal. Res.* **23**, 265–271 (1991).

[11] M. Ottiger, M.A. Thiel, U. Feige, et al., "Efficient intraocular penetration of topical anti-TNF-{alpha} single-chain antibody (ESBA105) to anterior and posterior segment without penetration enhancer," *Invest. Ophthal. Vis. Sci.* **50**, 779–786 (2008).

[12] R. Avtar and D. Tandon, "Modeling the drug transport in the anterior segment of the eye," *Eur. J. Pharm. Sci.* **35**, 175–182 (2008).

[13] W. Zhang, M.R. Prausnitz, and A. Edwards, "Model of transient drug diffusion across cornea," *J. Control. Release.* **99**, 241–258 (2004).

[14] J.-W. Kim, J.D. Lindsey, N. Wang, et al., "Increased human scleral permeability with prostaglandin exposure," *Invest. Ophthal. Vis. Sci.* **42**, 1514–1521 (2001).

[15] J. Ambati, C.S. Canakis, J.W. Miller, et al., "Diffusion of high molecular weight compounds through sclera," *Invest. Ophthal. Vis. Sci.* **41**, 1181–1185 (2000).

[16] K. Okabe, H. Kimura, J. Okabe, et al., "Effect of benzalkonium chloride on transscleral drug delivery," *Invest. Ophthal. Vis. Sci.* **46**, 703–708 (2005).

[17] T.W. Kim, J.D. Lindsey, M. Aihara, et al., "Intraocular Distribution of 70-kDa Dextran after subconjunctival injection in mice," *Invest. Ophthal. Vis. Sci.* **43**, 1809–1816 (2002).

[18] J.D. Lindsey and R.N. Weinreb, "Identification of the mouse uveoscleral outflow pathway using fluorescent Dextran," *Invest. Ophthal. Vis. Sci.* **43**, 2201–2205 (2002).

[19] A.K. Mitra and S. Macha, "Ocular pharmacokinetics in rabbits using a novel dual probe microdialysis technique," *Invest. Ophthal. Vis. Sci.* **42**, S174–S174 (2001).

[20] E.A. Genina, A.N. Bashkatov, Yu.P. Sinichkin, and V.V. Tuchin, "Optical clearing of the eye sclera in vivo caused by glucose," *Quant. Electr.* **36**, 1119–1124 (2006).

[21] H. Kim, M.J. Lizak, G. Tansey, et al., "Study of ocular transport of drugs released from an intravitreal implant using magnetic resonance imaging," *Ann. Biomed. Eng.* **33**, 150–164 (2005).

[22] S.K. Li, E.K. Jeong, and M.S. Hastings, "Magnetic resonance imaging study of current and ion delivery into the eye during transscleral and transcorneal iontophoresis," *Invest. Ophthal. Vis. Sci.* **45**, 1224–1231 (2004).

[23] D. Bommannan, H. Okuyama, P. Stauffer, et al., "Sonophoresis. I. The use of high-frequency ultrasound to enhance transdermal drug delivery," *Pharm. Res.* **9**, 559–564 (1992).

[24] D. Huang, E.A. Swanson, C.P. Lin, et al., "Optical coherence tomography," *Science* **254**, 1178–1181 (1991).

[25] A.F. Fercher, W. Drexler, C.K. Hitzenberger, et al., "Optical coherence tomography – principles and applications," *Rep. Prog. Phys.* **66**, 239–303 (2003).

[26] J.M. Schmitt, "Optical coherence tomography (OCT).: a review," *IEEE J. Select. Tops. Quant. Electr.* **5**, 1205–1215 (1999).

[27] P.H. Tomlins and R.K. Wang, "Theory, developments and applications of optical coherence tomography," *J. Phys. D: Appl. Phys.* **38**, 2519–2535 (2005).

[28] R.O. Esenaliev, K.V. Larin, I.V. Larina, et al., "Noninvasive monitoring of glucose concentration with optical coherence tomography," *Opt. Lett.* **26**, 992–994 (2001).

[29] K.V. Larin, M.S. Eledrisi, M. Motamedi, et al., "Noninvasive blood glucose monitoring with optical coherence tomography – a pilot study in human subjects," *Diab. Care* **25**, 2263–2267 (2002).

[30] K.V. Larin, M. Motamedi, T.V. Ashitkov, et al., "Specificity of noninvasive blood glucose sensing using optical coherence tomography technique: a pilot study," *Phys. Med. Biol.* **48**, 1371–1390 (2003).

[31] M. Kinnunen, R. Myllylä, T. Jokela, et al., "In vitro studies toward noninvasive glucose monitoring with optical coherence tomography," *Appl. Opt.* **45**, 2251–2260 (2006).

[32] M.G. Ghosn, S.H. Syed, N.A. Befrui, et al., "Quantification of molecular diffusion in arterial tissues with optical coherence tomography and fluorescence microscopy," *Laser Phys.* **19**, 1272–1275 (2009).

[33] M.G. Ghosn, M. Leba, A. Vijayananda, et al., "Effect of temperature on permeation of low density lipoprotein particles through human carotid artery tissues," *J. Biophot.* DOI: 10.1002/jbio.200810071 (2009).

[34] K.V. Larin and V.V. Tuchin, "Functional imaging and assessment of the glucose diffusion rate in epithelial tissues with optical coherence tomography," *Quant. Electr.* **38**, 551–556 (2008).

[35] M.G. Ghosn, E.F. Carbajal, N. Befrui, et al., "Differential permeability rate and percent clearing of glucose in different regions in rabbit sclera," *J. Biomed. Opt.* **13**, 021110 (2008).

[36] M.G. Ghosn, E.F. Carbajal, N.Befrui, et al., "Permeability of hyperosmotic agent in normal and atherosclerotic vascular tissues," *J. Biomed. Opt.* **13**, 010505 (2008).

[37] M.G. Ghosn, E.F. Carbajal, N. Befrui, et al., "Concentration effect on the diffusion of glucose in ocular tissues," *Opt. Lasers Eng.* **46**, 911–914 (2008).

[38] K.V. Larin, M.G. Ghosn, S.N. Ivers, et al., "Quantification of glucose diffusion in arterial tissues by using optical coherence tomography," *Laser Phys. Lett.* **4**, 312–317 (2007).

[39] M.G. Ghosn, V.V. Tuchin, and K.V. Larin, "Non-destructive quantification of analytes diffusion in cornea and sclera by using optical coherence tomography," *Invest. Ophthal. Vis. Sci.* **48**, 2726–2733 (2007).

[40] K.V. Larin and M.G. Ghosn, "Influence of experimental conditions on drug diffusion in cornea," *Quant. Electr.* **36**, 1083–1088 (2006).

[41] M.G. Ghosn, V.V. Tuchin, and K.V. Larin, "Depth-resolved monitoring of glucose diffusion in tissues by using optical coherence tomography," *Opt. Lett.* **31**, 2314–2316 (2006).

[42] V.V. Tuchin, I.L. Maksimova, D.A. Zimnyakov, et al., "Light propagation in tissues with controlled optical properties " *J. Biomed. Opt.* **2**, 401–417 (1997).

[43] A.T. Yeh and J. Hirshburg, "Molecular interactions of exogenous chemical agents with collagen-implications for tissue optical clearing," *J. Biomed. Opt.* **11**, 14003 (2006).

[44] G. Vargas, K.F. Chan, S.L. Thomsen, et al., "Use of osmotically active agents to alter optical properties of tissue: effects on the detected fluorescence signal measured through skin," *Lasers Surg. Med.* **29**, 213–220 (2001).

[45] V.V. Tuchin, *Optical Clearing of Tissues and Blood*, SPIE Press, Bellingham, WA (2006).

[46] V.V. Tuchin, "Optical clearing of tissues and blood using the immersion method," *J. Phys. D-Appl. Phys.* **38**, 2497–2518 (2005).

[47] R. Graaff, J.G. Aarnoudse, J.R. Zijp, et al., "Reduced light-scattering properties for mixtures of spherical-particles - a simple approximation derived from Mie calculations," *Appl. Opt.* **31**, 1370–1376 (1992).

[48] J.M. Schmitt and G. Kumar, "Optical scattering properties of soft tissue: a discrete particle model," *Appl. Opt.* **37**, 2788–2797 (1998).

[49] A.I. Kholodnykh, I.Y. Petrova, K.V. Larin, et al., "Precision of measurement of tissue optical properties with optical coherence tomography," *Appl. Opt.* **42**, 3027–3037 (2003).

[50] G.J. Nelson, *Blood Lipids and Lipoproteins: Quantitation, Composition, and Metabolism,* Wiley-Interscience, New York (1972).

[51] G.M. Grass and J.R. Robinson, "Mechanisms of corneal drug penetration. I: in vivo and in vitro kinetics," *J. Pharm. Sci* .**77**, 3–14 (1988).

[52] T.L. Ke, G. Cagle, B. Schlech, et al., "Ocular bioavailability of ciprofloxacin in sustained release formulations," *J. Ocul. Pharm. Ther.* **17**, 555–563 (2001).

[53] A.N. Bashkatov, E.A. Genina, Y.P. Sinichkin, et al., "Estimation of the glucosa diffusion coefficient in human eye selera," *Biophysics* **48**, 292–296 (2003).

[54] R.R. Pfister and N. Burstein, "The effects of ophthalmic drugs, vehicles, and preservatives on corneal epithelium: a scanning electron microscope study," *Invest. Ophthal. Vis. Sci.***15**, 246–259 (1976).

[55] K. Vizarova, D. Bakos, M. Rehakova, et al., "Modification of layered atelocollagen by ultraviolet irradiation and chemical cross-linking: structure stability and mechanical properties," *Biomaterials* **15**, 1082–1086 (1994).

[56] J.F. deBoer, T.E. Milner, M.J.C. vanGemert, et al., "Two-dimensional birefringence imaging in biological tissue by polarization-sensitive optical coherence tomography," *Opt. Lett.* **22**, 934–936 (1997).

[57] W. Drexler, D. Stamper, C. Jesser, et al., "Correlation of collagen organization with polarization sensitive imaging of in vitro cartilage: Implications for osteoarthritis," *J. Rheumat.* **28**, 1311–1318 (2001).

[58] S.L. Jiao, W.R. Yu, G. Stoica, et al., "Optical-fiber-based Mueller optical coherence tomography," *Opt. Lett.* **28**, 1206–1208 (2003).

[59] B.E. Applegate, C.H. Yang, A.M. Rollins, et al., "Polarization-resolved second-harmonic-generation optical coherence tomography in collagen," *Opt. Lett.* **29**, 2252–2254 (2004).

[60] R.A. Farrell and R.L. McCally, *Corneal Transparency. In Principles and Practice of Ophthalmology*, WB Saunders, Philadelphia, PA (2000).

[61] G.F. Elliott, J.M. Goodfellow, and A.E. Woolgar, "Swelling studies of bovine corneal stroma without bounding membranes," *J. Physiol.* **298**, 453–470 (1980).

[62] S. Hodson, D. Kaila, S. Hammond, et al., "Transient chloride binding as a contributory factor to corneal stromal swelling in the ox," *J. Physiol.* **450**, 89–103 (1992).

[63] D.M. Maurice, "The cornea and the sclera," in *The Eye,* Academic Press, New York and London (1969).

[64] O.A. Boubriak, J.P. Urban, S. Akhtar, et al., "The effect of hydration and matrix composition on solute diffusion in rabbit sclera," *Exp. Eye. Res.* **71**, 503–514 (2000).

[65] Y. Komai and T. Ushiki, "The three-dimensional organization of collagen fibrils in the human cornea and sclera," *Invest. Ophthal. Vis. Sci.* **32**, 2244–2258 (1991).

[66] E.M. Renkin, "Multiple pathways of capillary permeability," *Circul. Res.* **41**, 735–743 (1977).

[67] J.H. Park, J.W. Lee, Y.C. Kim, et al., "The effect of heat on skin permeability," *Int. J. Pharm.* **359**, 94–103 (2008).

[68] L.B. Nielsen, "Transfer of low density lipoprotein into the arterial wall and risk of atherosclerosis," *Atherosclerosis* **123**, 1–15 (1996).

[69] E. Vasile, M. Simionescu, and N. Simionescu, "Visualization of the binding, endocytosis, and transcytosis of low-density lipoprotein in the arterial endothelium in situ," *J. Cell Biol.* **96**, 1677–1689 (1983).

[70] J. C. Rutledge, "Temperature and hydrostatic pressure-dependent pathways of low-density lipoprotein transport across microvascular barrier," *Am. J. Physiol.* **262**, H234–H245 (1992).

18

Confocal Light Absorption and Scattering Spectroscopic Microscopy

Le Qiu and Lev T. Perelman

Biomedical Imaging and Spectroscopy Laboratory, Beth Israel Deaconess Medical Center, Harvard University, Boston, Massachusetts 02215 USA

This chapter reviews biomedical applications of confocal light absorption and scattering spectroscopic microscopy, an optical imaging technique that uses light scattering spectra as a source of the highly specific native contrast of internal cell structures. Confocal light absorption and scattering spectroscopic microscopy (CLASS) combine the principles of microscopy with light scattering spectroscopy (LSS), an optical technique that relates the spectroscopic properties of light elastically scattered by small particles to their size, refractive index, and shape. The multispectral nature of LSS enables it to measure internal cell structures much smaller than the diffraction limit without damaging the cell or requiring exogenous markers, which could affect cell function. CLASS microscopy approaches the accuracy of electron microscopy but is nondestructive and does not require the contrast agents common to optical microscopy. In this chapter we discuss basic physical principles of LSS and CLASS microscopy. We also devote a significant amount of space to the discussion of applications of CLASS microscopy in such diverse areas as obstetrics, neuroscience, ophthalmology, cellular and tissue imaging with nanoparticulate markers, and drug discovery.

Key words: microscopy, spectroscopy, light scattering, confocal, cell

18.1 Introduction

The electron microscope can resolve the subcellular structure with very high resolution, but it can only work with nonviable cells. Standard optical microscopy, though nondestructive, lacks contrast in cells and thus requires the introduction of fluorophores or other exogenous compounds. There is a clear need for a tool that can monitor cells and subcellular organelles on a submicrometer scale without damaging them or using exogenous markers that could affect cellular metabolism. To address this problem we developed CLASS microscopy by combining the principles of LSS and confocal microscopy that requires no exogenous labels, thus avoiding their potential interference with cell processes [1, 2]. At the same time the multispectral nature of LSS [3–7] enables the CLASS microscope to measure internal cell structures much smaller than the diffraction limit.

When imaging the cell, the origin of the CLASS signal is light scattering from the subcellular compartments. Though there are hundreds of cell types, the subcellular compartments in different cells are rather similar and are limited in number [6]. Any cell is bounded by a membrane, a phospholipid bilayer approximately 10 nm in thickness. Two major cell compartments are the nucleus, which has a size of 7 to 10 μm, and the surrounding cytoplasm. The cytoplasm contains various other organelles and inclusions. The large dimension of a mitochondrion may range from 1 μm to 5 μm and the diameter typically varies between 0.2 μm to 0.8 μm. Other smaller organelles include lysosomes, which are 250 to 800 nm in size and of various shapes and peroxisomes that are 200-nm to 1.0-μm spheroidal bodies of lower densities than the lysosomes. Peroxisomes are more abundant in metabolically active cells such as hepatocytes where they are counted in the hundreds.

Most cell organelles and inclusions are themselves complex objects with spatially varying refractive indices [8]. Many organelles such as mitochondria, lysosomes, and nuclei possess an average refractive index substantially different from that of their surrounding matter. Therefore, an accurate model acknowledges subcellular compartments of various sizes with a refractive index different from that of the surrounding.

Sizes, shapes, and refractive indices of major cellular and subcellular structures are presented in Figure 18.1. On this figure we also provide the information on the relevant approximations that can be used to describe light scattering from these objects.

Studies of light scattering by cells have a long history. Brunsting et al. initiated a series of experiments relating the internal structure of living cells with the scattering pattern by measuring forward and near forward scattering in cell suspensions using a rigorous quantitative approach [9]. The researchers used cells of several types such as Chinese hamster's oocytes (CHO), HeLa cells, and nucleated blood cells. They compared the resulting angular distribution of the scattered light with one predicted by the Mie theory and found very good agreement between theory and experiment was achieved by approximating a cell as a denser sphere imbedded into a larger less dense sphere. The sizes of these spheres corresponded to the average sizes of the cell nuclei and cells, respectively. The results agree well with scattering theory. Particles large compared to a wavelength produce a scattered field that peaks in the forward and near backward directions in contrast to smaller particles, which scatter light more uniformly. Despite nonhomogeneity and the lack of a perfectly spherical shape of cells and their nuclei, experimental results were well explained using the Mie theory, which deals with uniform spheres. This result was supported by Sloot et al. [10] in experiments with white blood cells (leukocytes) who found that light scattering by the leukocytes in the near forward direction can be explained if each cell was approximated as being composed of two concentric spheres, one being the cell itself and the other being the nucleus, and Hammer et al. [11], who showed that near-forward scattering of light by red blood cells can be accurately described using the van de Hulst approximation, which is derived for large particles of spherical shapes rather than the actual concave-convex disks that are red blood cells.

Studies of the angular dependence of light scattering by cells were carried out by Mourant et al. [12]. These studies showed that the cell structures responsible for light scattering can be correlated with the angle of scattering. When a cell is suspended in a buffer solution of lower refractive index, the cell itself is responsible for small angle scattering. This result has been used in flow cytometry to estimate cell sizes [13]. At slightly larger angles the nucleus is primarily responsible for scattering. It is the major scatterer in forward directions when the cell is a part of a contiguous layer.

Smaller organelles, cell inclusions, suborganelles, and subnuclear inhomogeneities are responsible for scattering at larger angles. Scattering may originate from organelles themselves or their internal components. Angular dependence may elucidate whether the scattering originates from the

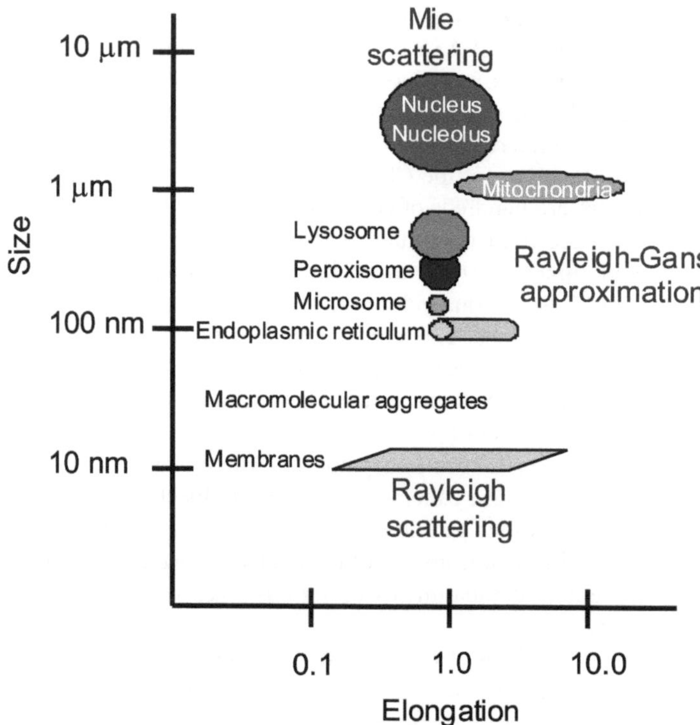

FIGURE 18.1: Hierarchy of scales inside of a cell and optical properties of subcellular structures along with the relevant approximations that can be used to describe light scattering from those objects. Refractive index of cytoplasm is approximately in the 1.36–1.37 and organelles in 1.41–1.44 range [7, 8, 10].

objects of regular or irregular shape, spherical or elongated, inhomogeneous or uniform. In some cases large angle scattering can be attributed to a specific predominant organelle. Research conducted by Beavoit et al. [14] provided strong evidence that mitochondria are primarily responsible for light scattering from hepatocytes.

Components of organelles can also scatter light. Finite-difference time-domain (FDTD) simulations provide means to study spectral and angular features of light scattering by arbitrary particles of complex shape and density. Using FDTD simulations Drezek et al. [15] investigated the influence of cell morphology on the scattering pattern and demonstrated that as the spatial frequency of refractive index variation increases, the scattering intensity increases at large angles.

18.2 Light Scattering Spectroscopy

Not only does light scattered by cell nuclei have a characteristic angular distribution peaked in the near-backward directions but it also exhibits spectral variations typical for large particles. This information has been used to study the size and shape of small particles such as colloids, water droplets, and cells [16]. The scattering matrix, a fundamental property describing the scattering event, depends not only on the scatterer's size, shape, and relative refractive index, but also on the wavelength of the incident light. This method is called light scattering spectroscopy or LSS, and can be very useful in biology and medicine.

Bigio et al. [17] and Mourant et al. [18] demonstrated that spectroscopic features of elastically scattered light can detect transitional carcinoma of the urinary bladder, adenoma and adenocarcinoma of the colon and rectum with good accuracy. In 1997 Perelman et al. observed characteristic LSS spectral behavior in the light backscattered from the nuclei of human intestinal cells [3]. Comparison of the experimentally measured wavelength-varying component of light backscattered by the cells with the values calculated using Mie theory, and the size distribution of cell nuclei determined by microscopy, demonstrated that both spectra exhibit similar oscillatory behavior. This method was successfully applied to diagnose precancerous epithelia in several human organs *in vivo* [4, 5].

One very important aspect of LSS is its ability to detect and characterize particles smaller than the diffraction limit. Particles much larger than the wavelength of light show a prominent backscattering peak and the larger the particle, the sharper the peak [6]. Measurement of 260 nm particles was demonstrated by Backman et al. [19, 20] and 100-nm particles by Fang et al. [21]. Scattering from particles with sizes smaller than a wavelength dominates at large angles and does not require an assumption that the particles are spherical or homogenous. Not only is submicron resolution achievable, but can also be done with larger NA confocal optics. By combining LSS with confocal scanning microscopy we recently identified submicron structures within the cell [1, 2].

These conclusions imply that not only is submicron resolution achievable, but also that it can be done with the larger NA optics characteristic of confocal arrangements. Thus, by combining LSS with confocal microscopy one can develop a microscopy technique, which is capable of identifying submicron structures within the cell.

18.3 Confocal Microscopy

The key principle of confocal laser scanning microscopy (CLSM) is that the sample is illuminated with a focused spot of laser light and the image is built up by scanning the spot over the field of view. This optical arrangement offers great flexibility in image acquisition strategies. In particular it enables obtaining optical section images – that is, images in which light from out-of-focus regions does not contribute to the image. Optical section imaging has opened up a wide range of applications in microscopy and allows the production of animated 3D projections.

Several research groups have demonstrated the use of confocal scanning microscopes for imaging human and animal tissues both *in vivo* and *ex vivo* [22, 23]. Commercial units are now available. Skin and oral mucosa are easily accessible for such an imaging technique. The technique has also been applied to the bladder, embryo, and foliculitis [24–26]. Rajadhyaksha et. al. [27] have imaged human skin to a depth of 350 microns and 450 microns in oral mucosa with a lateral resolution of 0.5–1 microns and axial resolution of 3–5 microns.

To obtain high spatial resolution, Webb et al. [28] utilized two-photon induced fluorescence. In experiments with *in vitro* rabbit ear mucosa So et al. [29] showed that micron scale resolution and a few hundred micron penetration depths could be achieved. Similarly two-photon excitation was used to image living cells by Piston [30].

Masters et al. [31] combined confocal microscopy with fluorescence spectroscopy to create two dimensional cross-sectional images of *in vitro* cornea. In these experiments a confocal microscope served for observing a single focal plane of thick objects with a high resolution and contrast compared with that of standard microscopes. In contrast to conventional confocal microscopy, the confocal image was formed at a longer wavelength than that of the laser illumination. Recently, confocal

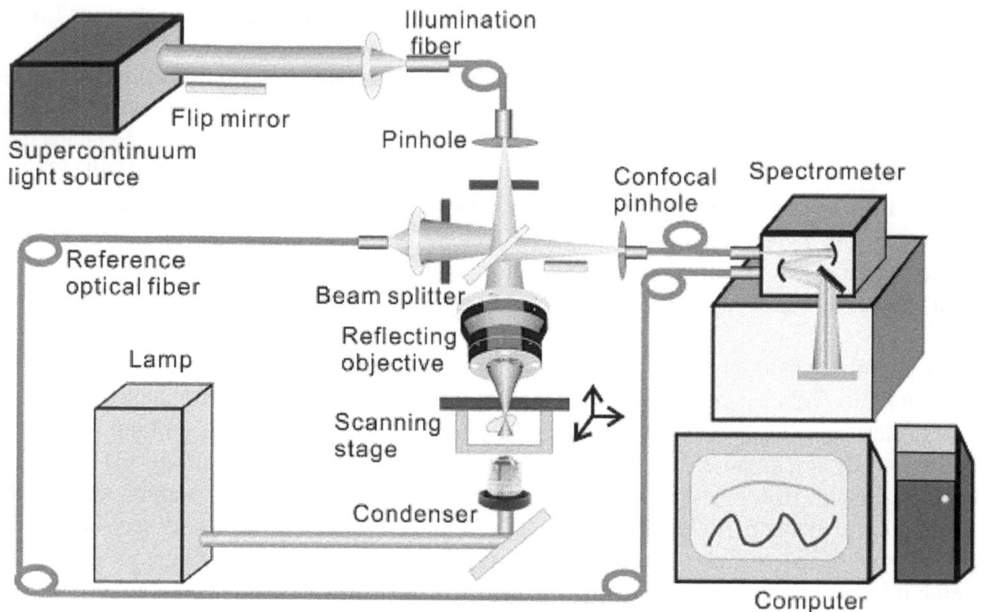

FIGURE 18.2: Schematic of the prototype CLASS microscope.

fluorescence microscopy was used to image renal interstitial cells *in situ* [32]. Three-dimensional reconstructions of serial sections revealed spiral arrangements in some bladders of renal medullary interstitial cells.

In general, a confocal microscope that is used correctly will always give a better image than can be obtained with a standard microscope. This improvement essentially comes from the rejection of out-of-focus interference. The improvement can vary between marginal, in the case of very flat specimens like chromosome squashes, to spectacular, in the case of large, whole-mount specimens, such as embryos. This feature, rejecting out-of-focus light, is very important in our case of combining LSS and confocal microscopy, since it allows observing the light scattering spectrum coming mainly from the confocal volume.

18.4 CLASS Microscopy

The schematic of the prototype CLASS microscope is presented in Figure 18.2. System design of the CLASS microscope provided for broadband illumination with either a Xe arc lamp for the measurements performed on extracted organelles in suspension, or a supercontinuum laser (Fianium SC-450-2) for the measurements performed on organelles in living cells. Both sources used an optical fiber to deliver light to the sample.

Depth-sectioning characteristics of a CLASS microscope can be determined by translating a mirror located near the focal point and aligned normal to the optical axis of the objective using five wavelengths spanning the principal spectral range of the instrument (Figure 18.3). The half-width of the detected signal is approximately 2 μm, which is close to the theoretical value for the 30 μm pinhole and 36× objective used [33–35]. In addition, the shapes of all five spectra shown in Figure 18.3 are almost identical (500 nm, 550 nm, 600 nm, 650 nm, and 700 nm), which demonstrates the

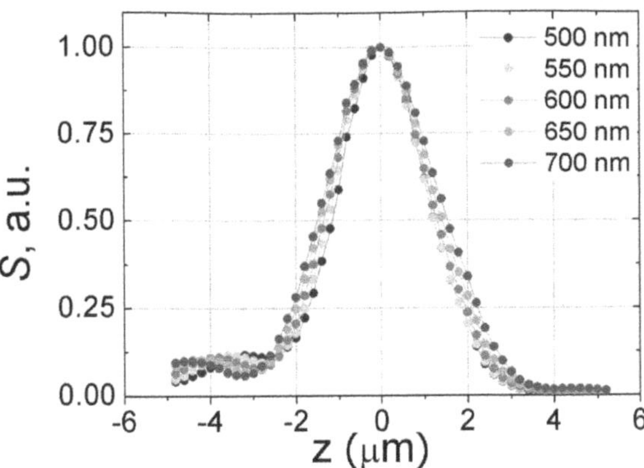

FIGURE 18.3: Depth sectioning of the CLASS microscope along the vertical axis at five different wavelengths (500 nm, 550 nm, 600 nm, 650 nm, and 700 nm). The almost identical nature of the spectra demonstrates the very good chromatic characteristics of the instrument [1].

excellent chromatic characteristics of the instrument. Small maxima and minima on either side of the main peak are due to diffraction from the pinhole. The asymmetry is due to spherical aberration in the reflective objective [36].

To check the characteristics of the microscope, calibrate it, and test the size-extraction algorithms, several experiments employing polystyrene beads in suspensions were performed. The samples were created by mixing beads of two different sizes in water and in glycerol in order to establish that the technique can separate particles of multiple sizes. Glycerol was also used in addition to water because the relative refractive index of the polystyrene beads in water (1.194 at 600 nm) is substantially higher than that of subcellular organelles in cytoplasm, which is in the range of 1.03 to 1.1 at visible wavelengths. By suspending the beads in glycerol one can decrease the relative refractive index to 1.07 to 1.1 in the visible range, a closer approximation to the biological range. (The refractive index of polystyrene in the working range can be accurately described by the expression $n = 1.5607 + 10002/\lambda^2$ where λ is in nanometers [37].)

Beads with a nominal mean size (diameter) of $\delta = 175$ nm and a standard deviation of size distribution of 10 nm were mixed with the beads with a mean size of $\delta = 356$ nm and a standard deviation of 14 nm. Figure 18.4(a) shows the CLASS spectra of polystyrene bead mixtures in water and glycerol and a comparison with the theoretical fit. In these experiments Brownian motion moved the beads in and out of the microscope focus. Therefore, the data were taken by averaging over a large number of beads. This was necessary to improve the statistics of the measurements. The difference between the experimental measurements and the manufacturer's labeling is less than 1% for both cases.

Figure 18.4(b) shows the extracted-size distributions. The parameters of the extracted-size distributions are very close to the parameters provided by the manufacturer (see Table 18.1). For example, the extracted mean sizes of the 175 nm beads are within 15 nm of the manufacturer's sizes, and the mean sizes of the 356 nm beads are even better, within 4 nm of the manufacturer's sizes.

Another test involving beads (or microspheres) was performed to establish imaging capabilities of the CLASS microscope. To insure that CLASS microscopy detects and correctly identifies objects in the field of view it was modified by adding a wide-field fluorescence microscopy arm, which shares a major part of the CLASS optical train. The instrument was tested on suspensions of carboxylate-modified Invitrogen microspheres, which exhibit red fluorescence emission at a wave-

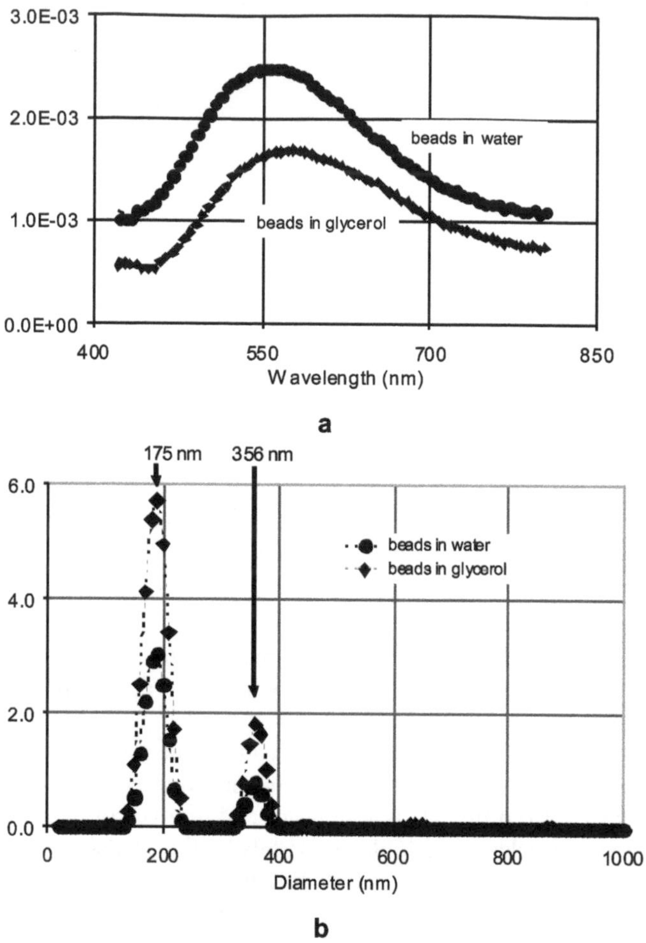

FIGURE 18.4: CLASS spectra (a) and extracted-size distributions (b) for polystyrene beads in water and glycerol. For figure (a), dots are for experimental data points and solid curves are for the spectra reconstructed from the theoretical model; for figure (b), the points are calculated values with solid curves as a guide for the eye [2].

TABLE 18.1: Size-distribution parameters for polystyrene beads.

	Mean size, nm	Standard deviation, nm	Mean size, nm	Standard deviation, nm
Manufacturer's data	175	10	356	14
CLASS Microscopy in water	185	40	360	30
CLASS Microscopy in glycerol	190	40	360	30

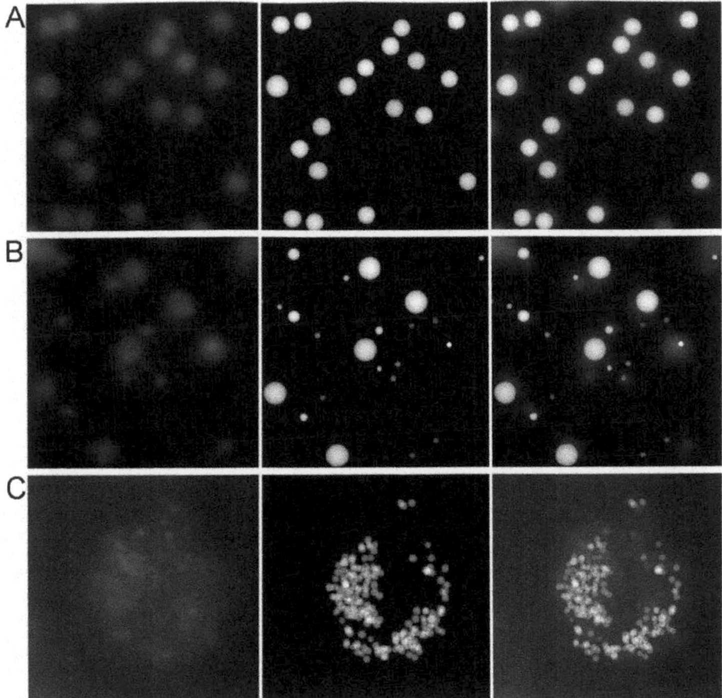

FIGURE 18.5: (See in color after page 572.) Fluorescence image of the suspensions of carboxylate-modified 1.9 μm diameter microspheres exhibiting red fluorescence (left side), the image reconstructed from the CLASS data (middle) and the overlay of the images (right side) (a). Same set of the images of the mixture of three sizes of fluorescent beads with sizes 0.5 μm, 1.1 μm, and 1.9 μm mixed in a ratio of 4:2:1 (a). Same set of the images of live 16HBE14o- human bronchial epithelial cells with lysosomes stained with lysosome-specific fluorescence dye (a) [1].

length of 605 nm with excitation at 580 nm. The microspheres were effectively constrained to a single-layer geometry by two thin microscope slides coated with a refractive index matching optical gel. Figure 18.5(a) shows (from left to right) the fluorescence image of the layer of 1.9 μm diameter microspheres, the image reconstructed from the CLASS data and the overlay of the images. Figure 18.5(b) shows a mixture of three sizes of fluorescent beads with sizes 0.5 μm, 1.1 μm, and 1.9 μm mixed in a ratio of 4:2:1. Note the misleading size information evident in the conventional fluorescence images. A 0.5 μm microsphere that is either close to the focal plane of the fluorescence microscope, or carries a high load of fluorescent label, produces a spot which is significantly larger than the microsphere's actual size. The CLASS image (middle of Figure 18.5(c)), on the other hand, does not make this error and correctly reconstructs the real size of the microsphere. We also can see that prior fluorescence labeling does not affect the determination of these objects with CLASS measurements.

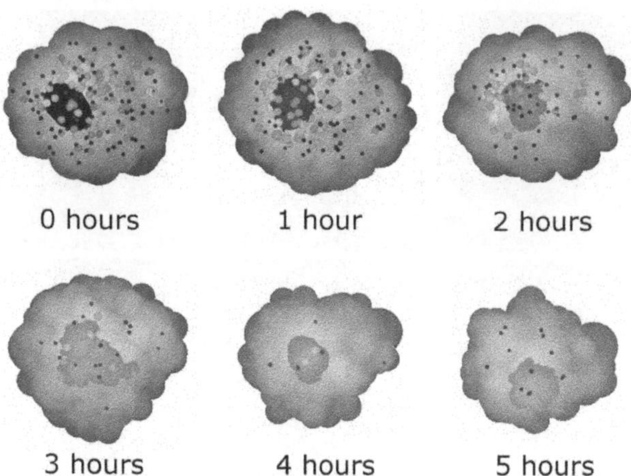

FIGURE 18.6: The time sequence of the CLASS microscope reconstructed images of a single cell. The cell was treated with DHA and incubated for 21 hours [1].

18.5 Imaging of Live Cells with CLASS Microscopy

To confirm the ability of CLASS to detect and identify specific organelles in a live cell simultaneous CLASS and fluorescence imaging of live 16HBE14o- human bronchial epithelial cells, with the lysosomes stained with a lysosome-specific fluorescent dye was performed using combined CLASS/Fluorescence instrument. The fluorescence image of the bronchial epithelial cell, the CLASS reconstructed image of the lysosomes and the overlay of two images are provided in Figure 18.5(c).

The overall agreement is very good, however, as expected, there is not always a precise, one-to-one correspondence between organelles appearing in the CLASS image and the fluorescence image. This is because the CLASS image comes from a single, well-defined confocal image plane within the cell, while the fluorescence image comes from several focal "planes" within the cell, throughout the thicker depth of field produced by the conventional fluorescence microscope. Thus, in the fluorescence image, one observes the superposition of several focal "planes" and, therefore, additional organelles above and below those in the single, well-defined confocal image plane of the CLASS microscope.

Another interesting experiment is to check the ability of CLASS microscopy to do time sequencing on a single cell. The cell was incubated with DHA a substance that induces apoptosis, for 21 hours. The time indicated in each image is the time elapsed after the cell was removed from the incubator. In Figure 18.6, the nucleus, which appears as the large blue organelle, has its actual shape and density reconstructed from the CLASS spectra obtained using point-by-point scanning. The remaining individual organelles reconstructed from the CLASS spectra are represented simply as spheroids whose size, elongation, and color indicate different organelles. The small red spheres are peroxisomes and the intermediate-size green spheres are lysosomes.

Organelles with sizes in the 1000-nm to 1300-nm range are mitochondria, and are shown as large yellow spheroids. The shape of the nucleus has changed dramatically by the third hour and the nuclear density, indicated by color depth, has decreased with time. The organelles have almost completely vanished by 4 hours.

18.6 Characterization of Single Gold Nanorods with CLASS Microscopy

Recently, significant attention has been directed toward the applications of metal nanoparticles, such as gold nanorods, to medical problems, primarily as extremely bright molecular marker labels for fluorescence, absorption, or scattering imaging of living tissue [38]. Nanoparticles with sizes small compared to the wavelength of light made from metals with a specific complex index of refraction, such as gold and silver, have absorption and scattering resonance lines in the visible part of the spectrum. These lines are due to in-phase oscillation of free electrons and are called surface plasmon resonances.

However, samples containing a large number of gold nanorods usually exhibit relatively broad spectral lines. This observed linewidth did not agree with theoretical calculations, which predict significantly narrower absorption and scattering lines. As we have shown in [39], the spectral peak of nanorods is dependent on their aspect ratio, and this discrepancy is explained by the inhomogeneous line broadening caused by the contribution of nanorods with various aspect ratios.

This broadened linewidth limits the use of nanorods with uncontrolled aspect ratios as effective molecular labels, since it would be rather difficult to image several types of nanorod markers simultaneously. However, this suggests that nanorod-based molecular markers selected for a narrow aspect ratio and, to a lesser degree, size distribution, should provide spectral lines sufficiently narrow for effective biomedical imaging.

We performed [39] optical transmission measurements of gold nanorod spectra in an aqueous solutions using a standard transmission arrangement for extinction measurements described in [40]. Concentrations of the solutions were chosen to be close to 10^{10} nanoparticles per milliliter of the solvent to eliminate optical interference. The measured longitudinal plasmon mode of the nanorods is presented as a dotted curve on Figure 18.7. It shows that multiple nanorods in aqueous solution have a width at half maximum of approximately 90 nm. This line is significantly wider than the line one would get from either T-matrix calculations or the dipole approximation. The solid line in Figure 18.7 shows the plasmon spectral line calculated using the T-matrix, for nanorods with a length and widths of 48.9 nm and 16.4 nm, respectively. These are the mean values of the sizes of the multiple nanorods in the aqueous solution. The theoretical line is also centered at 700 nm but has a width of approximately 30 nm. The ensemble spectrum is three times broader than the single particle spectrum.

The CLASS microscope described above is at present uniquely capable of performing single nanoparticle measurements. To determine experimentally that individual gold nanorods indeed exhibit narrow spectral lines single gold nanorods were selected and their scattering spectra measured using the CLASS microscope.

The nanorods were synthesized in a two-step procedure adapted from Jana et al. [41]. A portion of one of the transmission electron microscope (TEM) images of a sample of gold nanorods synthesized using the above procedure is shown in Figure 18.8. We measured the sizes of 404 nanorods from six different TEM images. We evaluated the average length and width of the nanorods and standard deviations and obtained 48.9$\pm$5.0 nm and 16.4$\pm$2.1 nm respectively, with an average aspect ratio of 3.0$\pm$0.4.

The concentration of gold nanorods used was approximately one nanorod per 100 μm^3 of solvent. The dimensions of the confocal volume have a weak wavelength dependence and were measured to be 0.5 μm in the lateral direction and 2 μm in the longitudinal direction at 700 nm, which gives a probability of about 0.5% to find a particle in the confocal volume. Thus, it is unlikely that more than one nanorod is present in the confocal volume. To locate individual nanorods, a 64 by 64 raster scan was performed with 0.5 μm steps. The integrated spectral signal was monitored from 650 nm to 750 nm and when a sudden jump in the magnitude of a signal was observed it was clear a nanorod is present in the confocal volume. Then a complete spectrum for this particle was collected.

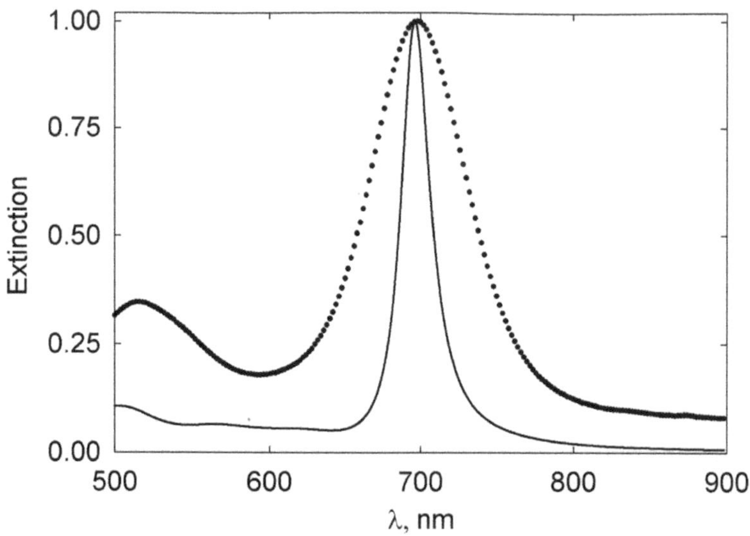

FIGURE 18.7: Optical properties of an ensemble of gold nanorods. Normalized extinction of the same sample of gold nanorods in an aqueous solution as in the TEM image. Dots – experiment, dashed line – T-matrix calculation for a single-size nanorod with length and width of 48.9 nm and 16.4 nm, respectively [39].

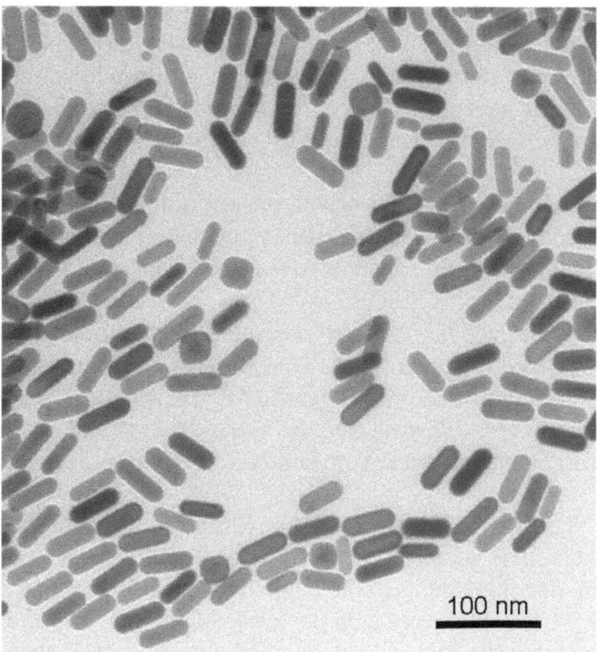

FIGURE 18.8: TEM image of a sample of gold nanorods with an average length and standard deviation of 48.9±5.0 nm and an average diameter and standard deviation of 16.4±2.1 nm [39].

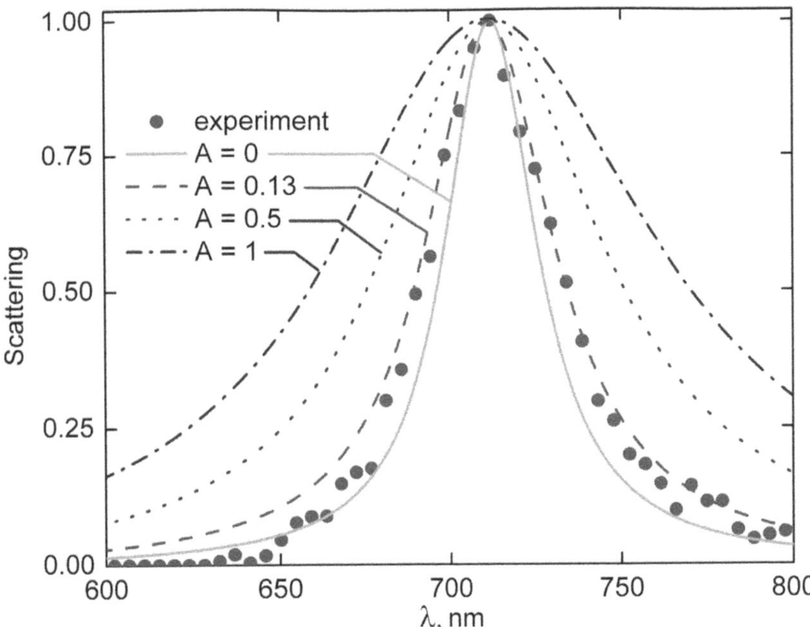

FIGURE 18.9: Normalized scattering spectrum for a single gold nanorod. Dots – CLASS measurements. Other lines are T-matrix calculations for a nanorod with an aspect ratio of 3.25 and a diameter of 16.2 nm and various A values. A solid line is for the natural linewidth, $A = 0$. Also included are lines for $A = 0.5$ and $A = 1$. The curve for $A = 0.13$ is the best fit for measurements made on eight different nanorods [42].

Scattering spectra from nine individual gold nanorods, all of which had a linewidth of approximately 30 nm, was measured [42]. The experimental data from one of these spectra is shown in Figure 18.9. These measurements were compared with numerical calculations that use the complex refractive index of gold [43] and various values of the phenomenological A-parameter correction [44] used to account for finite size and interface effects. The curve for $A = 0.13$ is the best fit for measurements made on eight different nanorods. This agrees very well with an A-parameter calculated using a quantum mechanical jellium model. Thus, using the CLASS microscope we have detected the plasmon scattering spectra of single gold nanorods [39, 42]. From these measurements one can draw the conclusion that single gold nanorods exhibit a scattering line significantly narrower than the lines routinely observed in experiments that involve multiple nanorods. Narrow, easily tunable spectra would allow several biochemical species to be imaged simultaneously with molecular markers which employ gold nanorods of several different controlled aspect ratios as labels. These markers could be used for cellular microscopic imaging where even a single nanorod could be detected. Minimizing the number of nanoparticles should reduce possible damage to a living cell. For optical imaging of tumors multiple gold nanorods with a narrow aspect ratio distribution might be used. A possible technique for obtaining a narrow aspect ratio distribution might employ devices already developed for cell sorting. These would use the position of the narrow plasmon spectral line for particle discrimination.

18.7 Conclusion

This chapter shows that CLASS microscopy is capable of reconstructing images of living cells with submicrometer resolution without the use of exogenous markers. Fluorescence microscopy of living cells require the application of molecular markers that can affect normal cell functioning. In some situations, such as studying embryo development, phototoxicity caused by fluorescent tagged molecular markers is not only undesirable but unacceptable. Another potential problem with fluorescence labeling is related to the fact that multiple fluorescent labels might have overlapping lineshapes and this limits the number of species that can be imaged simultaneously in a single cell.

CLASS microscopy is not affected by these problems. It requires no exogenous labels and is capable of imaging and continuously monitoring individual viable cells, enabling the observation of cell and organelle functioning at scales on the order of 100 nm. The CLASS microscope can provide not only size information but also information about the biochemical and physical properties of the cell since light scattering spectra are very sensitive to absorption coefficients and the refractive indices, which in turn are directly related to the organelle's biochemical and physical composition (such as the chromatin concentration in nuclei or the hemoglobin concentration and oxygen saturation in red blood cells).

CLASS microscopy can also characterize individual nanoparticles that have been used recently for high-resolution specific imaging of cancer and other diseases. Studies with the CLASS microscope demonstrated that individual gold nanoparticles indeed exhibit narrow spectral lines usually not observed in experiments involving ensembles of nanorods. Those studies also reveal that gold nanorod scattering-based biomedical labels with a single, well-defined aspect ratio might provide important advantages over the standard available absorption nanorod labels.

Acknowledgment

This study was supported by the National Institutes of Health grants EB003472 and RR017361, the National Science Foundation grant BES0116833.

References

[1] I. Itzkan, L. Qiu, H. Fang, et al., "Confocal light absorption and scattering spectroscopic microscopy monitors organelles in live cells with no exogenous labels," *Proc. Natl. Acad. Sci. U. S. A.* **104**, 17255–17260 (2007).

[2] H. Fang, L. Qiu, E. Vitkin, et al., "Confocal light absorption and scattering spectroscopic microscopy," *Appl. Opt.* **46**, 1760–1769 (2007).

[3] L.T. Perelman, V. Backman, M. Wallace, et al., "Observation of periodic fine structure in reflectance from biological tissue: a new technique for measuring nuclear size distribution," *Phys. Rev. Lett.* **80**, 627–630 (1998).

[4] V. Backman, M.B. Wallace, L.T. Perelman, et al., "Detection of preinvasive cancer cells," *Nature* **406**, 35–36 (2000).

[5] R.S. Gurjar, V. Backman, L.T. Perelman, et al., "Imaging human epithelial properties with polarized light-scattering spectroscopy," *Nat. Med.* **7**, 1245–1248 (2001).

[6] L.T. Perelman and V. Backman, "Light scattering spectroscopy of epithelial tissues: principles and applications," in *Handbook on Optical Biomedical Diagnostics*, V.V. Tuchin (Ed.), SPIE Press, Bellingham, 675–724 (2002).

[7] V. Backman, R. Gurjar, K. Badizadegan, et al., "Polarized light scattering spectroscopy for quantitative measurement of epithelial cellular structures in situ," *IEEE J. Sel. Top. Quant. Elect.* **5**, 1019–1026 (1999).

[8] J. Beuthan, O. Minet, J. Helfmann, et al., "The spatial variation of the refractive index in biological cells," *Phys. Med. Biol.* **41**, 369–382 (1996).

[9] A. Brunstin and P.F. Mullaney, "Differential light-scattering from spherical mammalian-cells," *Biophys. J.* **14**, 439–453 (1974).

[10] P.M.A. Sloot, A.G. Hoekstra, and C.G. Figdor, "Osmotic response of lymphocytes measured by means of forward light-scattering – theoretical considerations," *Cytometry* **9**, 636–641 (1988).

[11] M. Hammer, D. Schweitzer, B. Michel, et al., "Single scattering by red blood cells," *Appl. Opt.* **37**, 7410–7418 (1998).

[12] J.R. Mourant, J.P. Freyer, A.H. Hielscher, et al., "Mechanisms of light scattering from biological cells relevant to noninvasive optical-tissue diagnostics," *Appl. Opt.* **37**, 3586–3593 (1998).

[13] J.V. Watson, *Introduction to Flow Cytometry*, Cambridge University Press, Cambridge (1991).

[14] B. Beauvoit, T. Kitai, and B. Chance, "Contribution of the mitochondrial compartment to the optical-properties of the rat-liver – a theoretical and practical approach," *Biophys. J.* **67**, 2501–2510 (1994).

[15] R. Drezek, A. Dunn, and R. Richards-Kortum, "Light scattering from cells: finite-difference time-domain simulations and goniometric measurements," *Appl. Opt.* **38**, 3651–3661 (1999).

[16] R.G. Newton, *Scattering Theory of Waves and Particles*, McGraw-Hill, New York (1966).

[17] I.J. Bigio and J.R. Mourant, "Ultraviolet and visible spectroscopies for tissue diagnostics: fluorescence spectroscopy and elastic-scattering spectroscopy," *Phys. Med. Biol.* **42**, 803–814 (1997).

[18] J.R. Mourant, I.J. Bigio, J. Boyer, et al., "Spectroscopic diagnosis of bladder cancer with elastic light scattering," *Lasers Surg. Med.* **17**, 350–357 (1995).

[19] V. Backman, V. Gopal, M. Kalashnikov, et al., "Measuring cellular structure at submicrometer scale with light scattering spectroscopy," *IEEE J. Sel. Top. Quant. Elect.* **7**, 887–893 (2001).

[20] V. Backman, R. Gurjar, L.T. Perelman, et al., "Imaging and measurement of cell structure and organization with submicron accuracy using light scattering spectroscopy," *Proc. SPIE* **4613**, 101–110 (2002).

[21] H. Fang, M. Ollero, E. Vitkin, et al., "Noninvasive sizing of subcellular organelles with light scattering spectroscopy," *IEEE J. Sel. Top. Quant. Elect.* **9**, 267–276 (2003).

[22] S. Nioka, Y. Yung, M. Shnall, et al., "Optical imaging of breast tumor by means of continuous waves," *Adv. Exp. Med. Biol.* **411**, 227–232 (1997).

[23] W.M. Petroll, J.V. Jester, and H.D. Cavanagh, "In vivo confocal imaging," *Int. Rev. Exp. Pathol.* **36**, 93–129 (1996).

[24] P.M. Kulesa and S.E. Fraser, "Confocal imaging of living cells in intact embryos," *Methods Mol. Biol.* **122**, 205–222 (1999).

[25] F. Koenig, S. Gonzalez, W.M. White, et al., "Near-infrared confocal laser scanning microscopy of bladder tissue in vivo," *Urology* **53**, 853–857 (1999).

[26] S. Gonzalez, M. Rajadhyaksha, A. Gonzalez-Serva, et al., "Confocal reflectance imaging of folliculitis in vivo: correlation with routine histology," *J. Cutan. Pathol.* **26**, 201–205 (1999).

[27] M. Rajadhyaksha, R.R. Anderson, and R.H. Webb, "Video-rate confocal scanning laser microscope for imaging human tissues in vivo," *Appl. Opt.* **38**, 2105–2115 (1999).

[28] R.M. Williams, D.W. Piston, and W.W. Webb, "Two-photon molecular excitation provides intrinsic 3-dimensional resolution for laser-based microscopy and microphotochemistry," *FASEB J.* **8**, 804–813 (1994).

[29] K. König, P.T. So, W.W. Mantulin, et al., "Two-photon excited lifetime imaging of autofluorescence in cells during UVA and NIR photostress," *J. Microsc.* **183**, 197–204 (1996).

[30] D.W. Piston, "Imaging living cells and tissues by two-photon excitation microscopy," *Trends Cell Biol.* **9**, 66–69 (1999).

[31] B.R. Masters, "Confocal microscopy of the in-situ crystalline lens," *J. Microsc.* **165**, 159–167 (1992).

[32] M.M. Kneen, D.G. Harkin, L.L. Walker, et al., "Imaging of renal medullary interstitial cells in situ by confocal fluorescence microscopy," *Anat. Embryol. (Berl.)* **200**, 117–121 (1999).

[33] R.H. Webb, "Confocal optical microscopy," *Rep. Prog. Phys.* **59**, 427-471 (1996).

[34] T. Wilson and A.R. Carlini, "Size of the detector in confocal imaging-systems," *Opt. Lett.* **12**, 227–229 (1987).

[35] V. Drazic, "Dependence of 2-dimensional and 3-dimensional optical transfer-functions on pinhole radius in a coherent confocal microscope," *J. Opt. Soc. Am. A* **9**, 725–731 (1992).

[36] B.A. Scalettar, J.R. Swedlow, J.W. Sedat, et al., "Dispersion, aberration and deconvolution in multi-wavelength fluorescence images," *J. Microsc.-Oxford* **182**, 50–60 (1996).

[37] E. Marx and G.W. Mulholland, "Size and refractive-index determination of single polystyrene spheres," *J. Res. Nat. Bur. Stand.* **88**, 321–338 (1983).

[38] N.J. Durr, T. Larson, D.K. Smith, et al., "Two-photon luminescence imaging of cancer cells using molecularly targeted gold nanorods," *Nano. Lett.* **7**, 941–945 (2007).

[39] L. Qiu, T.A. Larson, D.K. Smith, et al., "Single gold nanorod detection using confocal light absorption and scattering spectroscopy," *IEEE J. Sel. Top. Quant. Elect.* **13**, 1730–1738 (2007).

[40] C.F. Bohren and D.R. Huffman, *Absorption and Scattering of Light by Small Particles*, Wiley, New York (1983).

[41] N.R. Jana, L. Gearheart, and C.J. Murphy, "Wet chemical synthesis of high aspect ratio cylindrical gold nanorods," *J. Phys. Chem. B* **105**, 4065–4067 (2001).

[42] L. Qiu, T.A. Larson, D. Smith, et al., "Observation of plasmon line broadening in single gold nanorods," *Appl. Phys. Lett.* **93**, 153106 (2008).

[43] P.B. Johnson and R.W. Christy, "Optical-constants of noble-metals," *Phys. Rev. B* **6**, 4370–4379 (1972).

[44] U. Kreibig and M. Vollmer, *Optical Properties of Metal Clusters*, Springer-Verlag, Berlin, New York (1995).

19

Dual Axes Confocal Microscopy

Michael J. Mandella
Stanford University, Stanford, CA

Thomas D. Wang
University of Michigan, Ann Arbor, MI

This chapter describes the theory, design, and implementation of a novel confocal architecture that uses separate optical axes for illumination and collection of light. This configuration can achieve subcellular resolution (< 5 μm) with a deep tissue penetration (> 500 μm) and a high dynamic range. Moreover, a large field-of-view can be generated with a diffraction-limited focal volume using post-objective scanning that also allows for this configuration to be scaled down in size to millimeter dimensions without loss of performance. Endoscope compatibility can be achieved with use of a MEMS (micro-electro-mechanical systems) scanner and fiber coupling. An instrument of this size can be used to perform real-time histopathology from the epithelium of hollow organs and longitudinal studies in small animals.

Key words: confocal, microscopy, scattering, tissue, imaging

19.1 Introduction

The medical interpretation of human disease is currently being performed by taking tiny tissue specimens (biopsies) to the microscope for histological evaluation [1]. This process can be time-consuming, and has limitations due to processing artifact, sampling error, and interpretive variability. Alternatively, there has recently been a substantial effort to develop miniature confocal microscopes that can be taken to the tissue via medical endoscopes to perform real-time *in vivo* imaging.

19.1.1 Principles of Confocal Microscopy

Confocal microscopes provide clear images in "optically thick" biological tissues using the technique of optical sectioning that has traditionally been used as a tabletop laboratory instrument to evaluate cells and tissues [2,3]. A pinhole or single mode optical fiber is placed in between the objective lens and the detector to allow only the light that originates from within a tiny focal volume below the tissue surface to be collected. Sources of light from all other depths become spatially filtered by the pinhole. Recently, significant progress has been made with endoscope compatible confocal instruments for visualizing inside the human body. This direction has been accelerated by the increasing availability, variety, and low cost of optical fibers and light sources. These methods are being developed for clinical use as an adjunct to endoscopy to guide tissue biopsy for surveillance of cancer, improve diagnostic yield, and reduce pathology costs. These efforts are technically challenged by demanding performance requirements, including small instrument size, high axial resolution, large penetration depth, and fast frame rate.

The required performance parameters for miniature confocal instruments to image *in vivo* are defined by the specific application. In particular, improved capabilities are needed for the early detection of cancer in the digestive tract. Medical endoscopes can easily access the lumen of hollow organs to image the epithelium in the digestive tract, including that of the esophagus and colon, to detect dysplasia, a premalignant condition. As shown in Figure 19.1, dysplasia occurs exclusively in the epithelium and represents an intermediate step in the transformation of normal mucosa to carcinoma. Dysplasia has a latency period of 7 to 14 years before progressing to cancer and offers a window of opportunity for evaluating patients by endoscopy who are at increased risk for developing cancer. The early detection and localization of dysplastic lesions can guide tissue resection and prevent future cancer progression. Dysplastic glands are present on the mucosal surface and extend down toward the muscularis. Thus, an imaging depth of $\sim$500 μm is sufficient to evaluate most early disease processes in the epithelium. Also, premalignant cells are frequently defined by abnormal morphology of the cell nucleus, requiring subcellular axial and transverse resolution on the order of <5 μm. Furthermore, motion artifacts can be introduced by organ peristalsis, heartbeats, and respiratory displacement, requiring a frame rate of > 4 per second. As will be discussed in more detail in this chapter, the novel dual axes confocal architecture performs off-axis illumination and collection of light to overcome tissue scattering. As a result, greater dynamic range and tissue penetration can be achieved, allowing for the collection of vertical cross-sectional images in the plane perpendicular to the mucosal surface, as shown in Figure 19.1. This view allows for greater sensitivity to subtle changes that alter the normal tissue growth patterns and reveals early signs of the disease processes compared to that of horizontal cross-sectional images.

19.1.2 Role for dual axes confocal microscopy

In this chapter, we discuss a novel confocal microscope design that uses a dual axes rather than single axis architecture. This configuration represents a new class of imaging instrument that is designed to overcome tissue scattering by collecting light off-axis [4–8]. In the conventional single axis confocal microscope, a high NA objective is needed to achieve subcellular axial resolution. This large collection angle collects much of the light backscattered by the intense illumination beam, reducing the dynamic range of detection. Furthermore, the objective cannot be scaled down in size for *in vivo* imaging without also reducing the working distance (WD), and since scanning is conventionally performed in the preobjective position, the field-of-view (FOV) will also be reduced proportionally. As a consequence, the scanning beam passes through the high NA objective at off-axis angles, limiting the field-of-view and increasing sensitivity to off-axis aberrations, which include astigmatism and coma. The limited dynamic range and working distance of the single axis architecture results in optical sections collected within the tissue along a horizontal cross-section, as shown by the horizontal white dashed line in Figure 19.1. The epithelial biology that can be studied

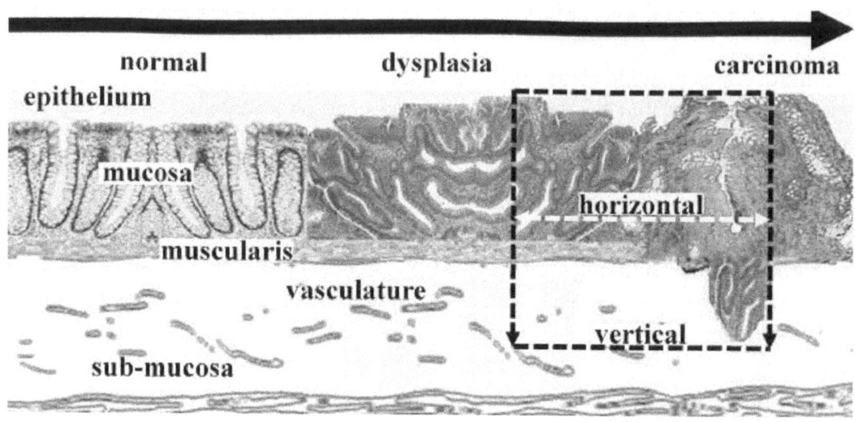

FIGURE 19.1: Dysplasia represents a premalignant condition in the epithelium of hollow organs, such as the colon and esophagus. The dual axes confocal architecture has the dynamic range to image in the vertical cross-section (dashed black lines) to visualize disease processes with greater depth.

in this orientation is usually limited to tissue that has the same degree of differentiation.

However, in the dual axes configuration, two fibers are centered along the optical axes of low NA objectives for spatially separated light paths for illumination and collection. The region of overlap between the two beams defines defines the approximately spherical focal volume, hence the resolution, and can achieve subcellular levels (<5 μm). Very little of the light that is scattered by tissue along the illumination path enters the low NA collection objective, thus significantly improving the dynamic range of detection. Furthermore, the low NA objectives create an increased working distance so that the scanning mirror can be placed on the tissue side of the lens in the postobjective position [9]. In this configuration, the beams always pass through the simple objectives on-axis, resulting in a diffraction limited focal volume that can then be scanned over an arbitrarily large field-of-view. This design feature allows the instrument to be scaled down in size to millimeter dimensions without losing performance. The reduction in instrument size can be implemented using devices fabricated with MEMS (micro-electro-mechanical systems) technology for compatibility with medical endoscopes [10,11]. Moreover, the use of low NA objectives results in less sensitivity to off-axis aberrations.

19.2 Limitations of Single Axis Confocal Microscopy

Recent advances in the development of microlenses and miniature scanners have resulted in the development of fiber optic coupled instruments that are endoscope compatible [12,13]. This technique has been demonstrated *in vivo* in the clinic for a number of medical applications. Most confocal microscopes use a single axis design, where the pinhole (fiber) and objective are located along one main optical axis. A high NA objective is used to achieve both axial and transverse subcellular resolution and maximum light collection, and the same objective is used for both the illumination and collection of light. In order to scale down the dimension of these instruments for endoscope compatibility, the diameter of the objective must be reduced to ~5 mm or less. As a consequence, the working distance as well as the field-of-view is also decreased, as shown by the progression of the 3 different objective diameters in Figure 19.2. The tissue penetration depth is typically inade-

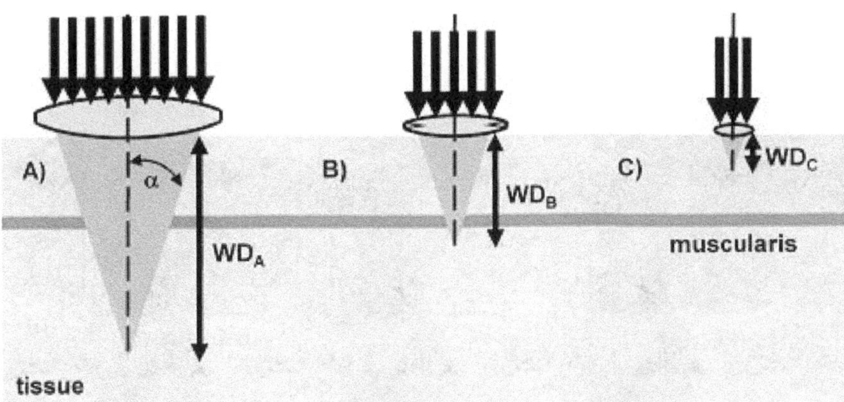

FIGURE 19.2: For endoscope compatibility, the diameter of a single axis confocal microscope must be scaled down in size (A→B→C), resulting in a reduced working distance (WD) and limited imaging depth.

quate to assess the tissue down to the muscularis, which is located at a depth of ~500 μm and is an important landmark for defining the early prescence of epithelial cancers.

19.2.1 Single axis confocal design

For the conventional single axis architecture, the transverse, Δr_s, and axial, Δz_s, resolutions between full-width-half-power (FWHP) points for uniform illumination of the lenses are defined by the following equations [3]

$$\Delta r_s = \frac{0.37\lambda}{n\sin\alpha} \approx \frac{0.37\lambda}{n\alpha}; \quad \Delta z_s = \frac{0.89\lambda}{n(1-\cos\alpha)} \approx \frac{1.78\lambda}{n\alpha^2}, \tag{19.1}$$

where λ is the wavelength, n is the refractive index of the medium, $\alpha$is the maximum convergence half-angle of the beam, NA = $n\sin\alpha$ is the numerical aperture, and $\sin\alpha \approx \alpha$ for low NA lenses. Eq. (19.1) implies that the transverse and axial resolution varies as 1/NA and 1/NA^2, respectively. A resolution of less than 5 μm is adequate to identify subcellular structures that are important for medical and biological applications. To achieve this resolution in the axial dimension, the objective lens used requires a relatively large NA (>0.4). The optics can be reduced to the millimeter scale for *in vivo* imaging, but requires sacrifice of resolution, field-of-view, or working distance. Also, a high NA objective limits the available working distance, and requires that the scanning mechanism be located in the preobjective position, restricting the FOV and further increasing sensitivity to aberrations.

19.2.2 Single axis confocal systems

Two endoscope compatible confocal imaging systems are commercially available for clinical use. The EC-3870K (Pentax Precision Instruments, Tokyo, Japan) has an integrated design where a confocal module (Optiscan Pty Ltd, Victoria, Australia) is built into the insertion tube of the endoscope, and results in an overall diameter of 12.8 mm, as shown in Figure 19.3A [14]. This module uses the single axis optical configuration where a single mode optical fiber is aligned on-axis with an objective that has an NA $\approx$ 0.6. Scanning of the distal tip of the optical fiber is performed mechanically by coupling the fiber to a tuning fork that vibrates at resonance. Axial scanning is performed with a shape memory alloy (nitinol) actuator that can translate the focal volume over a

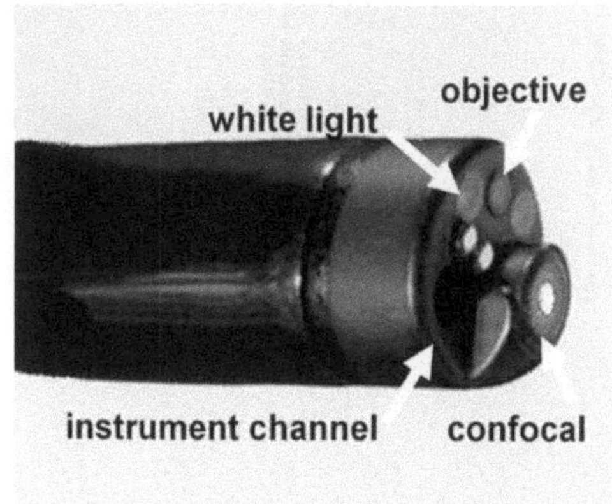

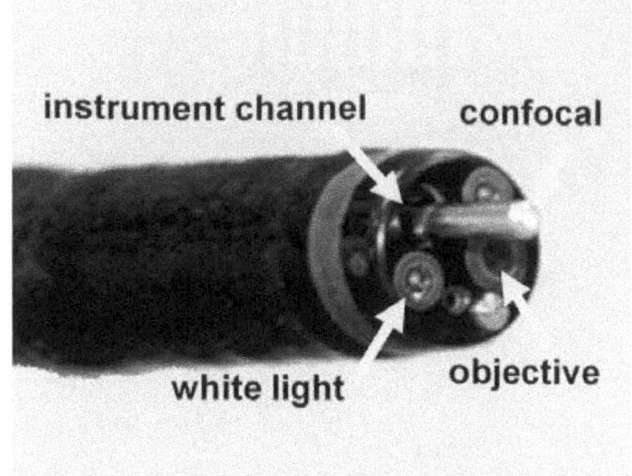

A) Pentax/Optiscan **B) Cellvizio® GI**

FIGURE 19.3: Endoscope compatible single axis confocal instruments. A) The EC-3870K (Pentax) has a confocal module (Optiscan) integrated into the endoscope insertion tube. B) The Cellvizio® GI is a confocal miniprobe that passes through the instrument channel of the endoscope. Reused with permission.

distance of 0 to 250 μm below the tissue surface. Excitation is provided at 488 nm (peak absorption of fluorescein) by a semi-conductor laser, and a transverse and axial resolution of 0.7 and 7 μm, respectively, has been achieved. The images are collected at a frame rate of either 0.8 or 1.6 Hz to achieve a field-of-view either 500 x 500 or 500 x 250 microns, which are then displayed by 1024×1024 or 1024×512 pixels, respectively. The dimension of the confocal module by itself is $\sim$5 mm. When a suspicious lesion is identified, the confocal window located on the distial tip is placed into contact with the tissue to collect images. A separate instrument channel can be used to obtain pinch biopsies of tissue.

The Cellvizio® GI (Mauna Kea Technologies, Paris, France) is a miniprobe with diameters that range from 1.5 to 2.5 mm, and passes through the standard instrument channel of medical endoscopes, as shown in Figure 19.3B. This instrument moves independently of the endoscope, and its placement onto the tissue surface can be guided by the conventional white light image [13, 15]. This miniprobe consists of a fiber bundle with $\sim$30,000 individual fibers that is aligned on-axis with an objective that has an NA $\approx$ 0.6. The core of each individual fiber acts as a collection pinhole to reject out of focus light. Scanning is performed at the proximal end of the bundle in the instrument control unit with a 4 kHz oscillating mirror for horizontal lines and a 12 Hz galvo mirror for frames. In this design, axial scanning cannot be performed. Instead, separate miniprobes that have different working distances are needed to optically section at different depths. Excitation is provided at 488 nm, and the transverse and axial resolution of these instruments ranges from 2.5 to 5 μm and 15 to 20 μm, respectively. Images are collected at a frame rate of 12 Hz with a field-of-view of either 600×500 μm or 240×200 μm.

19.3 Dual Axes Confocal Architecture

The dual axes confocal architecture overcomes many of the limitations of the single axis configuration for purposes of miniaturization and *in vivo* imaging. Spatial separation of the illumination

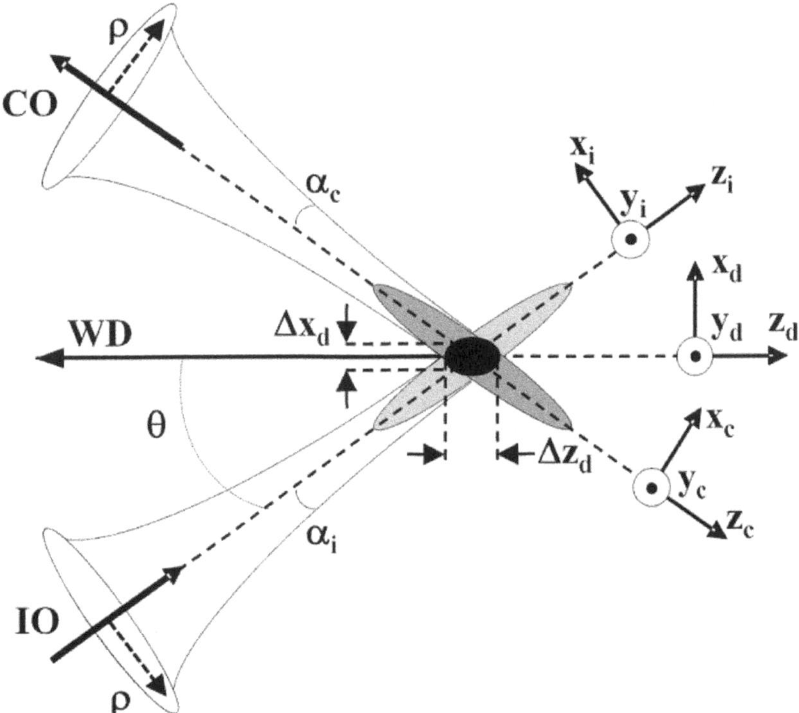

FIGURE 19.4: Dual axes confocal architecture. Separate objectives are used for off-axis illumination (IO) and collection (CO) of light having maximum convergence angles α_i and α_c, respectively. The overlap of the two beams (black oval) provides the system response, resulting in a significant improvement in the transverse and axial dimensions, Δx_d, and Δz_d, respectively. Reused with permission.

and collection axes is needed to achieve subcellular resolution with a sufficient working distance to provide submucosal tissue penetration. Furthermore, the introduction of postobjective scanning allows for this design to be scaled down to millimeter dimensions for endoscope compatibility without loss of performance.

19.3.1 Dual axes design

In the dual axes configuration, separate low NA objectives are used for illumination (IO) and collection (CO), as shown in Figure 19.4. The maximum convergence half-angles of the illumination and collection beams are represented by α_i and α_c, respectively. The optical axes for illumination and collection are defined to cross the z-axis (z_d) at an angle θ. The main lobe of the PSF of the illumination objective is represented by the light gray oval. This lobe has a narrow transverse but a wide axial dimension. Similarly, the main lobe of the PSF of the collection objective is similar in shape but symmetrically reflected about z_d, as represented by the dark gray oval. For dual axes, the combined PSF is represented by the overlap of the two individual PSF's, represented by the black oval. This region is characterized by narrow transverse dimensions, Δx_d and Δy_d (out of the page), and by a significantly reduced axial dimension, Δz_d, which depends on the transverse rather than the axial dimension of the individual beams where they intersect.

19.3.2 Dual axes point spread function

The PSF for the dual axes confocal architecture can be derived using diffraction theory with paraxial approximations [16]. As shown in Figure 19.4, the coordinates for the illumination (x_i, y_i, z_i) and collection (x_c, y_c, z_c) beams are defined in terms of the coordinates of the main optical axis (x_d, y_d, z_d), and may be expressed as follows:

$$\begin{array}{ll} x_i = x_d \cos\theta - z_d \sin\theta & x_c = x_d \cos\theta + z_d \sin\theta \\ y_i = y_d & y_c = y_d \\ z_i = x_d \sin\theta + z_d \cos\theta & z_c = -x_d \sin\theta + z_d \cos\theta \end{array}. \tag{19.2}$$

The maximum convergence half-angles of the focused illumination and collection beams in the sample media are represented as α_i and α_c, respectively. The angle at which the two beams intersect the main optical axis is denoted as θ. A set of general dimensionless coordinates may be defined along the illumination and collection axes, as follows:

$$\begin{array}{ll} u_i = k_i n z_i \sin^2\alpha_i & u_c = k_c n z_c \sin^2\alpha_c \\ v_i = k_i n\sqrt{x_i^2 + y_i^2}\sin\alpha_i & v_c = k_c n\sqrt{x_c^2 + y_c^2}\sin\alpha_c \end{array}. \tag{19.3}$$

The wavenumbers for illumination and collection are defined as $k_i = 2\pi/\lambda_i$ and $k_c = 2\pi/\lambda_c$, respectively, where λ_i and λ_c are the wavelengths, and n is the index of refraction of the media.

The amplitude PSF describes the spatial distribution of the electric field of the focused beams. Diffraction theory may be used to show that the PSF of the illumination and collection beams is proportional to the Huygens-Fresnel integrals below [16]:

$$U_i(v_i, u_i) \propto \int_0^1 W_i(\rho) J_0(\rho v_i) e^{-ju_i\rho^2/2}\rho d\rho; \tag{19.4}$$

$$U_c(v_c, u_c) \propto \int_0^1 W_c(\rho) J_0(\rho v_c) e^{-ju_c\rho^2/2}\rho d\rho, \tag{19.5}$$

where J_0 is the Bessel function of order zero, and ρ is a normalized radial distance variable at the objective aperture. The weighting function, $W(\rho)$, describes the truncation (apodization) of the beams. For uniform illumination, $W(\rho) = 1$. For Gaussian illumination, the objectives truncate the beams at the $1/e^2$ intensity, resulting in a weighting function of $W(\rho) = e^{-\rho^2}$. In practice, the beams are typically truncated so that 99% of the power is transmitted. For a Gaussian beam with a radius ($1/e^2$ intensity) given by w, an aperture with diameter πw passes ~99% of the power. In this case, the weighting function is given as follows:

$$W(\rho) = e^{-(\pi\rho/2)^2}. \tag{19.6}$$

For the single axis configuration, the illumination and collection PSF's at the focal plane ($u_i = u_c = 0$) are identical functions of the radial distance ρ, and can both be given by U_s using the substitution $v = knr\sin\alpha$, as follows:

$$U_s(v) \propto \int_0^1 W_i(\rho) J_0(\rho v)\rho d\rho. \tag{19.7}$$

The resulting signal at the detector V from a point source reflector in the media is proportional to the power received, and is given by the square of the product of the overlapping PSF's as follows:

$$V = A\left|U_i U_c\right|^2, \tag{19.8}$$

where A is a constant.

For uniform illumination ($W = 1$), the detector output V_s for the single axis design is:

$$V_s \propto \left(\frac{2J_1(v)}{v}\right)^4. \tag{19.9}$$

Similarly, since the depth of focus for each individual beam, described within the exponential term in the integral product of Eqs. (19.4) and (19.5), is much larger than that of the transverse width, the exponential term may be neglected. As a result, the detector output V_d for the dual axes configuration for uniform illumination is given as follows:

$$V_d \propto \left(\frac{2J_1(v_i)}{v_i}\right)^2 \left(\frac{2J_1(v_c)}{v_c}\right)^2. \tag{19.10}$$

This expression can be combined with Eqs. (19.2) and (19.3) to derive the result for transverse and axial resolution with uniform illumination as follows [8]:

$$\Delta x_d = \frac{0.37\lambda}{n\alpha\cos\theta}; \quad \Delta y_d = \frac{0.37\lambda}{n\alpha}; \quad \Delta z_d = \frac{0.37\lambda}{n\alpha\sin\theta}. \tag{19.11}$$

Note that for the dual axes configuration, the axial resolution is proportional to $1/\text{NA}$, where $\text{NA}= n\sin\alpha \approx n\alpha$, rather than $1/\text{NA}^2$, as is the case for the single axis design, shown in Eq. (19.1). This relationship results from the overlap of the illumination and collection beams, and explains the high axial resolution that can be achieved using low NA objectives. For example, with uniform illumination and the following parameters: $\alpha = 0.21$ radians, $\theta = 30$ degrees, $\lambda = 0.785$ μm and $n = 1.4$ for tissue, Eq. (19.1) reveals a result for single axis of Δx_s = 1.0 μm, Δy_s = 1.0 μm, and Δz_s = 22.6 μm for the transverse and axial resolutions, respectively. On the other hand, Eq. (19.11) reveals a result for dual axes of Δx_d = 1.1 μm, Δy_d = 1.0 μm, and Δz_d = 2 μm for the transverse and axial resolutions, respectively. Thus, subcellular resolution can be achieved in both the transverse and axial dimensions with dual axes using low NA optics but not in the single axis configuration.

For an endoscope compatible instrument, delivery of the illumination and collection light is performed with use of optical fibers and is more appropriately modeled by a Gaussian rather than a uniform beam. With this apodization, the detector response for the dual axes configuration from a point source reflector in the media, given by Eq. (19.8), may be solved numerically as a function of transverse (x_d and y_d) and axial (z_d) dimensions. The integrals are calculated in Matlab, and use the weighting function with 99% transmission, given in Eq. (19.6). In comparison, this model reveals a result of Δx_d = 2.4 μm, Δy_d = 2.1 μm, and Δz_d = 4.2 μm for the transverse and axial resolutions, respectively. Thus, the use of optical fibers, modeled by a Gaussian beam, produces results that are slightly worse but still comparable to that of uniform illumination [17].

Differences in the dynamic range between the single axis and dual axes configurations can also be illustrated with this model. The calculated axial response for the single axis design with Gaussian illumination is shown by the dashed line in Figure 19.5A, where optical parameters are used that achieve the same axial resolution (FWHM) of 4.2 μm. The result reveals that the main lobe falls off in the axial (z-axis) direction as $1/z^2$, and reaches a value of approximately -25 dB at a distance of 10 μm from the focal plane ($z = 0$). In addition, a number of sidelobes can be appreciated. In comparsion, the response for the dual axes configuration, shown by the solid line in Figure 19.5A, reveals that the main lobe rolls off in the axial (z-axis) direction as $\exp(-kz^2)$, and reaches a value of -60 dB at a distance of 10 μm from the focal plane ($z = 0$). Thus, off-axis illumination and collection of light in the dual axes confocal architecture results in a significant improvement in dynamic range and in an exponential rejection of out-of-focus scattered light in comparison to that for single axis. This advantage allows for the dual axes configuration to collect images with deeper tissue penetration and with a vertical cross-section orientation. The results for the numerically

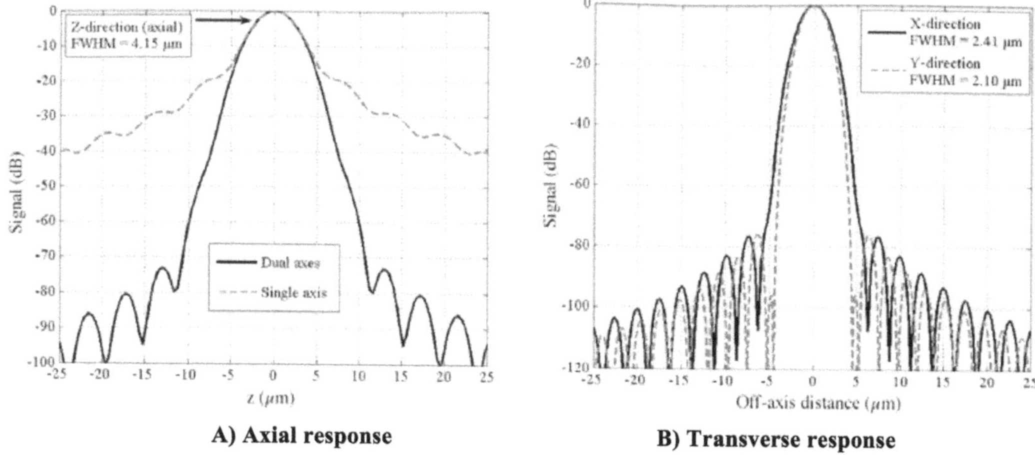

FIGURE 19.5: Gaussian beam model. A) The axial response of the single axis (dashed line) configuration falls off as $1/z^2$ and that for the dual axes (solid line) design falls off as $\exp(-kz^2)$, resulting in a significant improvement in dynamic range. B) Transverse (X-Y directions) response. Reused with permission.

simulated transverse response of the dual axes configuration with Gaussian illumination are shown in Figure 19.5B.

19.3.3 Postobjective scanning

In confocal microscopes, scanning of the focal volume is necessary to create an image. In the single axis architecture, the high NA objectives used limit the working distance, thus the scan mirror is by convention placed on the pinhole (fiber) side of the objective, or in the preobjective position, as shown in Figure 19.6A. Scanning orients the beam at various angles to the optical axis and introduces off-axis aberrations that expand the focal volume. In addition, the FOV of preobjective scanning systems is proportional to the scan angle and the focal length of the objective. The diameter of the objective limits the maximum scan angle, and as this dimension is reduced for endoscope compatibility, the focal length and FOV are also diminished.

In the dual axes configuration, the low NA objectives used creates a long working distance that allows for the scanner to be placed on the tissue side of the objective, or in the postobjective position [9]. This design feature is critical for scaling the size of the instrument down to millimeter dimensions for *in vivo* imaging applications without losing performance. As shown in Figure 19.6B, the illumination light is always incident on-axis to the objective. In the postobjective location, the scan mirror can sweep a diffraction-limited focal volume over an arbitrarily large FOV, limited only by the maximum deflection angle of the mirror. Moreover, the scanner steers the illumination and collection beams together with the intersection of the two beams oriented at a constant angle θ and with the overlapping focal volume moving without changing shape along an arc-line. This property can be conceptualized by regarding the dual axes geometry as being equivalent to two separate beams produced from two circles in the outer annulus of a high NA lens containing a central obstruction (or a large central hole). A flat scan mirror deflects both beams equally, and thereby preserves the overlapping region without introducing aberrations to the beams.

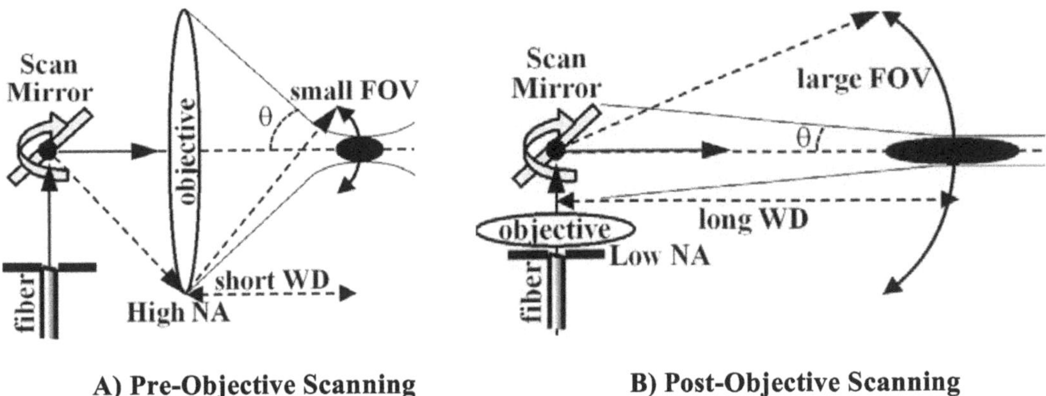

FIGURE 19.6: Scanning geometries. A) For preobjective scanning, illumination light is incident on the objective off-axis, resulting in more sensitivity to aberrations and limited FOV. B) With postobjective scanning, the incident light is on-axis, less sensitive to aberrations, and achieves large FOV. postobjective scanning is made possible by the long working distance (WD) produced by the low NA objectives used in the dual axes architecture.

19.3.4 Improved rejection of scattering

In the dual axes confocal architecture, the off-axis collection of light significantly reduces the deleterious effects of tissue scattering on the dynamic range of detection and allows for deeper ballistic photons to be resolved [19]. These features provide the unique capacity to collect vertical cross-sectional images in the plane perpendicular to the tissue surface. This is the preferred view of pathologists because differences from the normal patterns of tissue differentiation are revealed in the direction from the lumen to the submucosa.

19.3.4.1 Optical configurations

The improvement in rejection of light scattered by tissue can be illustrated by comparing the dynamic range of detection between the single and dual axes optical configurations with equivalent axial resolution, as shown in Figure 19.7A and B. The incident beams are modeled with a Gaussian profile because this is representative of light delivered through an optical fiber. For the single axis configuration, this beam is focused into the tissue by an ideal lens (L_1). A mirror (M) is embedded in the tissue at the focal plane (parallel to the x-y plane) of the objective lens. In this scheme, the rays that reflect from the mirror pass back through the lens L_1, deflect at an angle off the beam splitter, and are focused by an ideal lens (L_2) on to a pinhole detector. For the dual axes set-up, the incident Gaussian beam is focused into the tissue by an ideal lens (L_3) with its axis oriented at an angle $\theta = 30^\circ$ to the $z-$axis, and an ideal lens (L_4) focuses the backscattered beam, with its axis z'at an angle -30° to the z-axis, onto the pinhole detector. As before, a mirror (M) with its plane perpendicular to the z-axis and passing through the coincident focuses of the lenses is embedded in the tissue to reflect the incident light to the detector. In both configurations, the lens system has a magnification of 1 from the focal plane to the pinhole detector.

In order to achieve an equivalent -3 dB axial resolution (FWHM), the NA's for the single and dual axes configurations are defined to be 0.58 and 0.21, respectively. From diffraction theory, discussed above, the theoretical transverse and axial resolutions for the PSF for dual axes at a wavelength $\lambda = 633$ nm with an average tissue refractive index of 1.4 and NA = 0.21 are found to be Δx= 1.16 μm, Δy= 1.00 μm, and Δz= 2.00 μm. The mirror is placed at a distance of 200 μm

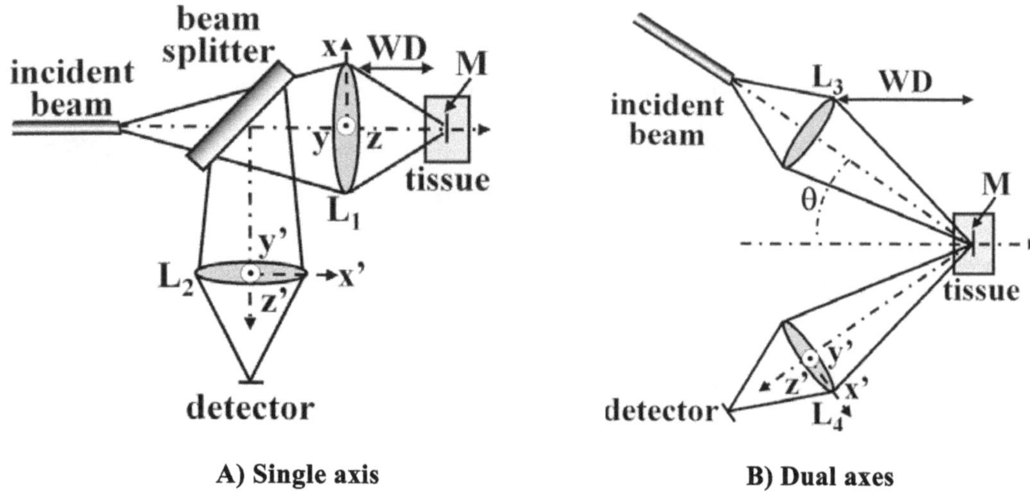

FIGURE 19.7: Schematic diagrams for tissue scattering model. The A) single axis and B) dual axes optical configurations are used to evaluate the axial response at the detector. Reused with permission.

below the tissue surface in the focal plane of the objective lenses for both the single and dual axes configurations. This depth is representative of the imaging distance of interest in the epithelium of hollow organs. The calculations performed to analyze the effects of tissue scattering on light are based on Monte Carlo simulations using a nonsequential ray tracing program (ASAP® 2006 Breault Research Organization, Tucson, AZ). Three assumptions are made in this simulation study: 1) multiple scattering of an incoherent beam dominates over diffraction effects, 2) the nonscattering optical medium surrounding the lenses and the tissue (the scattering medium) is index matched to eliminate aberrations, and 3) absorption is not included to simplify this model and because there is much larger attenuation due to the scattering of ballistic photons.

19.3.4.2 Mie scattering analysis

We use Mie theory with the Henyey-Greenstein phase function $p(\theta)$ to model the angular dependence of tissue scattering, as follows [20,21]:

$$p(\theta) = \frac{1}{4\pi} \frac{1-g^2}{(1+g^2-2g\cos\theta)^{3/2}}, \tag{19.12}$$

where g, the anisotropy factor, is defined as

$$g = \langle \cos\theta \rangle = \int_0^{2\pi} \int_0^{\pi} \cos\theta \cdot p(\theta) \sin\theta d\theta d\varphi. \tag{19.13}$$

Given the average scatterer size, refractive index, and concentration, the attenuation coefficient μ_s and anisotropy g are determined and provided to the ASAP program as simulation parameters. For a tissue phantom composed of polystyrene spheres with a diameter of 0.48 μm, refractive index 1.59, and a concentration of 0.0394 spheres/μm^3 in water, the values g = 0.81 and μ_s = 5.0 mm^{-1} at λ = 633 nm are calculated from Mie theory [22].

For single axis, $P(y')$ is defined as the photon flux distribution along the y'-axis at the detector. The photon flux can be normalized by defining $P^*(y') = P(y')/P_{max}$, where P_{max} is the maximum flux. The normalized flux $P^*(y')$ consists of ballistic (signal) and multiple scattered (noise) photons, as shown in Figure 19.8A [23]. The maximum flux for both the signal and noise components arrive

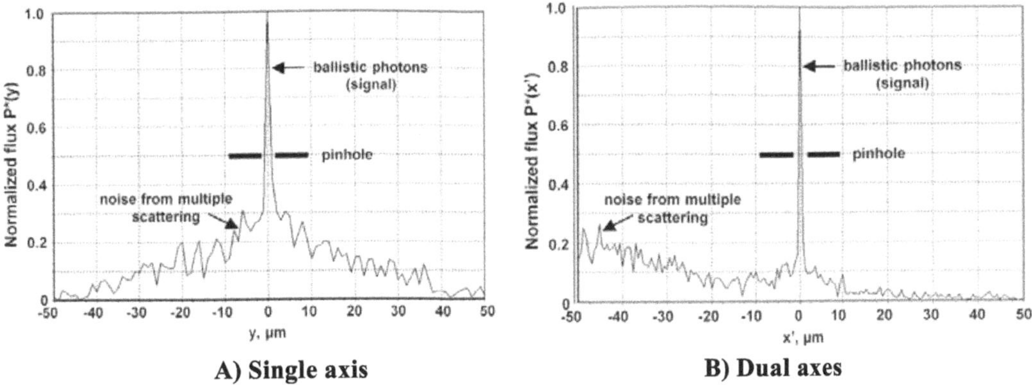

FIGURE 19.8: Distributions of photon flux in tissue scattering model. The peak value of multiple scattered photons for A) single axis is co-located with the confocal pinhole while that for B) dual axes is separated by ~50 μm. As a consequence, the ballistic photons for dual axes result in greater signal-to-noise and dynamic range. Reused with permission.

at center of the detector. A confocal pinhole placed in front of the detector can filter out some but not all of this "noise," resulting in a reduced signal-to-noise ratio (SNR). For dual axes, the detector is angled off the optical axis by 30 deg. $P(x')$ is defined as the photon flux distribution along the x'-axis at the detector. The photon flux can be normalized by defining $P^*(x') = P(x')/P_{max}$, where P_{max} is the maximum flux at the detector. Figure 19.8B shows that normalized photon flux distribution for dual axes also exhibits a ballistic and multiple scattered components. However, for dual axes, the peak flux of multiple scattered photons arrives ~50 μm lateral to the center of the detector where the ballistic photons arrive, a consequence of off-axis collection. Thus, there is much less "noise" for the confocal pinhole (diameter ~1 μm) to filter out, resulting in a higher SNR and dynamic range.

19.3.4.3 Improvement in dynamic range

An implication of this result is that the dual axes configuration has improved dynamic range compared to that of single axis. This difference can be quantified by determining the axial response at the detector. This can be done by calculating the photon flux $f(\Delta z)$ as the mirror is displaced along the z-axis in the tissue. The flux is calculated using Monte-Carlo simulations in ASAP® with the mirror at distances in the range $-10\ \mu\text{m} < \Delta z < 10\ \mu\text{m}$ with respect to the focal plane at $z = 0$, which is located at 200 μm below the tissue surface. The flux is then normalized according to $F(\Delta z) = f(\Delta z)/f(0)$. The axial response is shown in Figure 19.9A for various pinhole diameters D, including 1, 2 and 3 μm, which correspond to typical fiber core dimensions. Note that for each pinhole diameter, the dual axes (DA) configuration has significantly better dynamic range than that of single axis (SA). Note that the introduction of tissue scattering results in a reduction of the dynamic range compared to that found in free space, as shown by Figure 19.5A.

We can also determine the axial response of the detector for various optical lengths in tissue. This analysis reveals differences in the dynamic range between the single and dual axes configuration for tissues with various scattering properties. The total optical length L is defined as twice the product of the scattering coefficient μ_s and the tissue depth t, or $L = 2\mu_s t$. The factor of two originates from the fact that the total path length is twice the tissue depth. The axial response is shown in Figure 19.9B for various optical lengths L, including 4.8, 6.4, and 8.0. Note that for each optical length L, the dual axes (DA) configuration has significantly better dynamic range than that of single axis (SA). These values of L are typical parameters of gastrointestinal epithelium. At $\lambda = 633$ nm, μ_s is

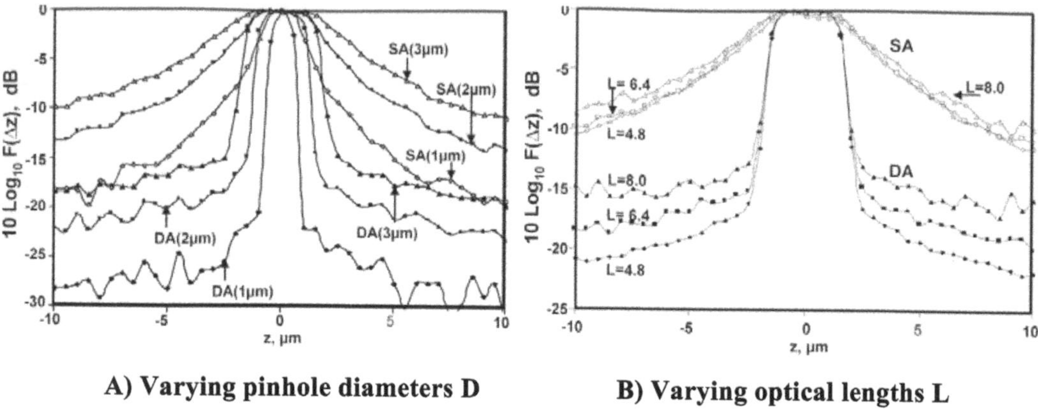

FIGURE 19.9: Axial response for single and dual axes geometries. The dual axes (DA) configuration has a much greater dynamic range than that for single axis (SA) given different A) pinhole diameters (1, 2, and 3 μm) and B) optical lengths L (4.8, 6.4 and 8.0). Reused with permission.

about 7 mm^{-1} for esophagus tissue [23] and about 20 mm^{-1} for normal colonic mucosa [24]. The range of tissue depths spanned by $L = 4.8$ to 8 for esophagus and colon is 340 μm to 570 μm and 120 μm to 200 μm, respectively. In addition, these results shows that for single axis only minimal changes occur in the dynamic range with approximately a factor of 2 difference in optical thickness L, while for dual axes significant changes occur over this thickness range. Furthermore, scattering does not appear to alter the FWHM of the axial response for either single or dual axes over this range of lengths.

19.3.4.4 Geometric differences produced by off-axis detection

The superior axial response of the dual axes confocal architecture has a simple geometric explanation. When the mirror moves away from the focal plane by $\pm\Delta$, the centroid of the beam is steered away from the optical axis by $\pm 2\Delta \sin\theta$ from where the center of the pinhole is located [19]. Even taking into consideration diffraction and the broadening of the out-of-focus beam, the beam intensity decreases exponentially when $\Delta > D/2$ (for $\theta = 30^\circ$). But in the single axis case, many of the photons scattered near the vicinity of the focal plane ($\pm\Delta$) are collected by the detector through the pinhole. Thus, the spatial filtering effect by a pinhole for the single axis configuration is not as effective as that for dual axes. The implication of this effect for imaging deep in tissue is evident. In the single axis case, many of the multiple scattered photons that arrive from the same direction as that of the ballistic photons, starting from the surface to deep within the tissue, are collected by the detector despite the presence of a pinhole to filter the out-of-focus light. This explains why in Figure 19.8A the single axis configuration has a large noise component alongside the ballistic component. Thus, the dual axes confocal architecture provides optical sectioning capability that is superior to that of the conventional single axis design in terms of SNR and dynamic range, and this result can be generalized to a range of relevant pinhole sizes. As a result, the dual axes architecture allows for imaging with greater tissue penetration depth, thus is capable of providing images in the vertical cross-section with high contrast. The implementation of the dual axes confocal configuration into an endoscope-compatible instrument for collection of both reflectance and fluorescence has significant implications for *in vivo* imaging by providing both functional and structural information deep below the tissue surface.

19.4 Dual Axes Confocal Imaging

The dual axes confocal architecture was first implemented as a tabletop instrument using readily available optical components to demonstrate the proof of concept of off-axis light illumination and collection with postobjective scanning. In particular, the primary advantages of the dual axes configuration including high dynamic range and deep tissue penetration are revealed by vertical cross-sectional images with either reflectance or fluorescence. The combination of these two imaging modes forms a powerful strategy for integrating structural with functional information.

19.4.1 Solid immersion lens

The dual axes optical design incorporates a solid immersion lens (SIL) made from a fused-silica hemisphere at the interface where the two off-axis beams meet the tissue. This refractive element minimizes spherical aberrations that occur when light undergoes a step change in refractive index between two media. The curved surface of the SIL provides a normal interface for the two beams to cross the air-glass boundary. Fused silica is used because its index of refraction of $n = 1.45$ is closely matched to that of tissue. Note that as the beams are scanned away from their neutral positions, they will no longer be incident to the surface of the SIL and small aberrations will occur. Another feature of the SIL is that its curved surface increases the effective NA of the beams in the tissue by a factor of n, the index of refraction, and produces higher resolution and light collection efficiency. On the other hand, the SIL acts to reduce the scanning displacement of the beams in the tissue by a factor of $1/n$ so that larger deflections are needed to achieve the desired scan range.

19.4.2 Horizontal cross-sectional images

Reflectance imaging takes advantage of subtle differences in the refractive indices of tissue microstructures to achieve contrast. The backscattered photons can provide plenty of signal to overcome the low NA objectives used for light collection in the dual axes configuration. The first reflectance images were collected with a tabletop system that used a 488 nm semiconductor laser for producing illumination light that was coupled into a single mode optical fiber and focused by a set of collimating lenses with NA$= 0.16$ to a spot size with $\sim$400 μW of power [8]. These parameters produced a transverse and axial resolution of 1.1 and 2.1 μm, respectively. The reflected light was collected by an analogous set of optics. The off-axis illumination and collection was performed at $\theta = 30$ deg. to the main optical axis. Reflectance images were collected in horizontal cross-sections of freshly excised specimens of esophagus *ex vivo*. As shown in Figure 19.10A, the cell membrane and individual nuclei of squamous (normal) esophageal mucosa can be appreciated in the image collected at $z = 0$ μm, scale bar 20 μm.

Much greater image contrast can be achieved with fluorescence imaging where the use of optical reporters, such as GFP, and exogenous probes can reveal the over expression of molecular targets. The same tabletop dual axes microscope can also be used to collect fluorescence images by using a long pass filter to block the excitation light and photomultiplier tube (PMT) for detection [25]. In Figure 19.10B, a fluorescence image of the cerebellum of a transgenic mouse that constituitively expresses GFP under the control of a β-actin-CMV promoter-enhancer at a depth of $z = 30$ μm is shown, scale bar 50 μm. Purkinje cell bodies (arrows) can be seen as large round structures aligned side by side in a row, separating the granule layer and the molecular layer.

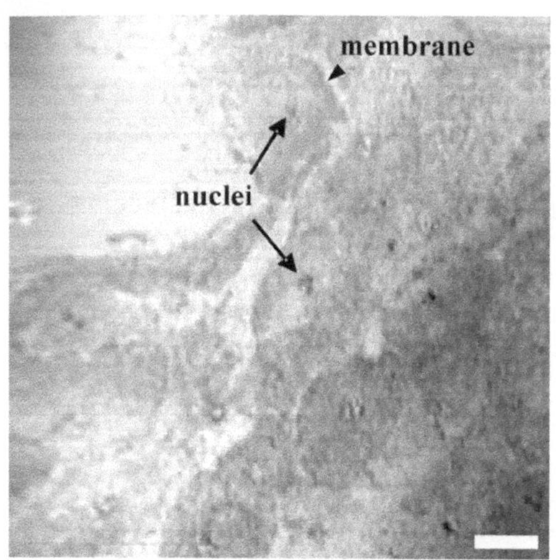

A) Reflectance (esophagus)

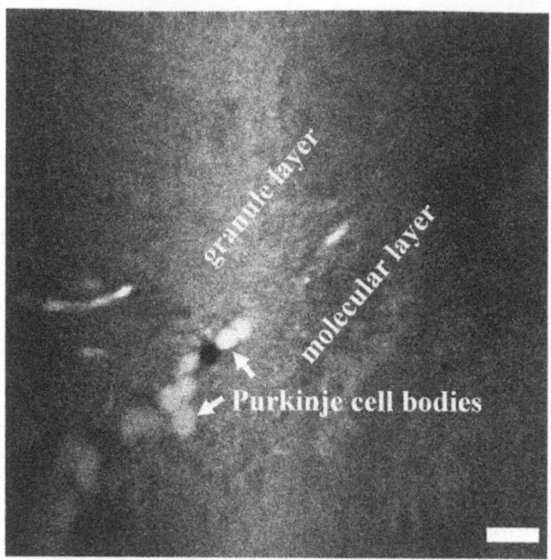

B) Fluorescence (cerebellum)

FIGURE 19.10: Horizontal cross-sectional dual axes images *ex vivo*. A) Squamous esophageal mucosa collected at $z = 0\ \mu$m with $\lambda = 488$ nm reveals subcellular features, including cell nuclei (arrows) and membrane (arrowhead), scale bar 20 μm. B) Fluorescence image of excised cerebellum from genetically engineered mouse that constituitively expresses GFP in horizontal cross-section at a depth of $z = 30\ \mu$m, scale bar 50 μm. Purkinje cell bodies (arrows) can be seen aligned in a row, separating the granule layer and the molecular layer.

19.4.3 Vertical cross-sectional images

In order to collect vertical cross-sectional images, heterodyning can be performed to provide a coherence gate to filter out illumination photons that are multiply-scattered and travel over longer optical paths within the tissue [9]. This approach is demonstrated with a fiber optic Mach-Zehnder interferometer, shown in Figure 19.11A. A broadband near-infrared source produces light centered at $\lambda = 1345$ nm with a 3 dB bandwidth of 35 nm and a coherence length in tissue of $\sim$50 μm. A fiber coupler directs $\sim$99% of the power to the illumination path, which consists of a single mode optical fiber (SMF_1) with a collimating (CL_1) and focusing lens (FL_1) with NA = 0.186. The axes of illumination and collection are oriented at $\theta = 30^o$ to the midline. Light reflected from the tissue is collected by the second set of focusing (FL_2) and collimating (CL_2) lenses into another single mode fiber (SMF_2). The lens and fiber parameters are the same for both the illumation and collection beams. The fiber optic coupler directs $\sim$1% of the source into a reference beam which is frequency shifted by an acousto-optic modulator at 55 MHz for heterodyne detection. An adjustable optical delay is used to increase the signal by matching the optical path length of the reference beam to that of the ballistic photons. An adjustable optical delay is used to increase the signal. In addition, a polarization controller consisting of two half-wave plates and a single quarter-wave plate is used to maximize the signal. The reference and collection beams are combined by a 50/50 coupler and the resulting heterodyne signal is detected by a balanced InGaAs detector (D_1,D_2) with a bandwidth of 80 MHz. The resulting electronic signal is then processed with a band pass filter (BPF) with a 3 MHz bandwidth centered at 55 MHz, then demodulated (DM), digitized by a frame grabber (FG), and displayed (D).

In this heterodyne detection scheme, the reference beam essentially provides amplification of the weak collection beam via coherent optical mixing, and enables the measurement of reflected light

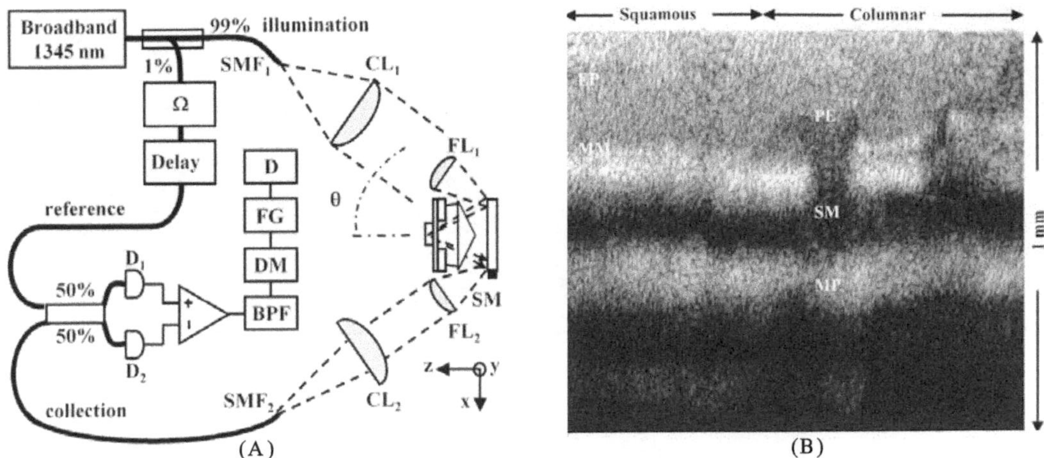

FIGURE 19.11: Vertical cross-sectional dual axes confocal reflectance images *ex vivo*. A) Schematic of optical circuit for heterodyne detection, details discussed in the text. B) Reflectance image of squamo-columnar junction in esophagus with vertical depth of 1 mm. Squamous (normal) mucosa reveals epithelium (EP) and muscularis mucosa (MM) over the left half. Columnar (intestinal metaplasia) mucosa shows pit epithelium (PE) over the right half. Submucosa (SM) and musclaris propria (MP) are seen on both sides. Reused with permission.

with a dynamic range larger than 70 dB. postobjective scanning is performed with the scan mirror (SM) placed distal to the objective lenses. Reflectance images were collected from fresh biopsy specimens taken from the squamo-columnar junction of human subjects with Barrett's esophagus. Specimens with dimensions of ~3 mm were resected with jumbo biopsy forceps, and the mucosal surface was oriented normal to the z-axis. Vertical cross-sectional images were collected with depth of 1 mm. From Figure 19.11B, squamous (normal) mucosa is present over the left half of the image with an intact epithelium (EP). The other structures of normal esophageal mucosa, including the muscularis mucosa (MM), submucosa (SM), and muscularis propria (MP), can also be identified. Columnar mucosa consistent with intestinal metaplasia is seen over the right half of the image, and reveals the presence of pit epithelium (PE) [9]. These findings correlate with the tissue microstructures seen on histology.

19.4.4 Dual axes confocal fluorescence imaging

Fluorescence detection adds an entirely new dimension to the imaging capabilities of the dual axes architecture. Detection in this mode offers an opportunity to achieve much higher image contrast compared to that of reflectance and is sensitive to molecular probes that identify unique tissue and cellular targets. These features provide a method to perform functional as well as structural imaging, and open the door for the study of a wide variety of molecular mechanisms. Although the use of low NA objectives in the dual axes configuration reduces the collection efficiency of the optics by a factor of $\sim NA^2$, this deficit can be overcome by the use of bright fluorophores as labels. In order to achieve the deep tissue penetration depths possible with the off-axis collection of light, a near-infrared laser at 785 nm is used as the source and a photomultiplier (PMT) tube with a long pass filter to block the excitation light is used for detection [26]. The large dynamic range (>40 dB) of the dual axes confocal architecture encountered with collection of vertical cross-sectional images requires modulation of the PMT gain to compensate for the rapid decrease in fluorescence signal with axial depth due to tissue absorption and scattering.

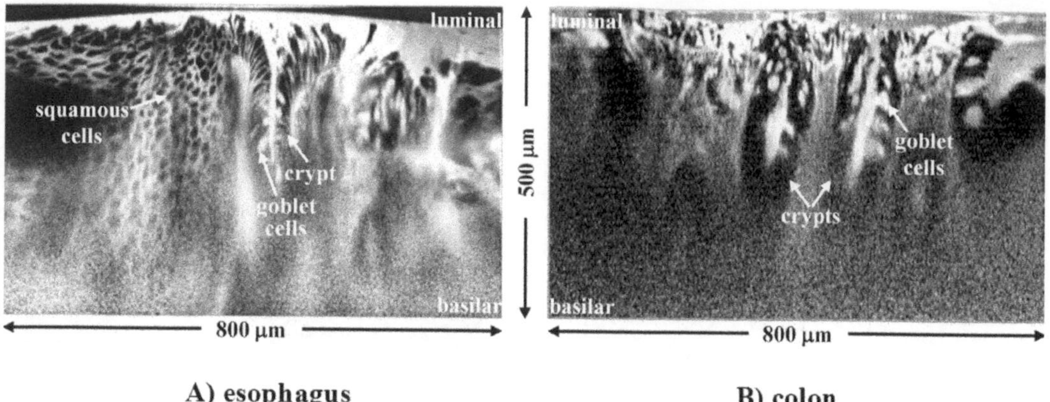

FIGURE 19.12: Vertical cross-sectional dual axes confocal fluorescence images *ex vivo*. A) Squamo-columnar junction in an esophagus with a vertical depth of 500 μm. Squamous mucosa present over the left half. Columnar (intestinal metaplasia) mucosa over the right half shows crypts with goblet cells. B) Colon. Many goblet cells can be seen in dysplastic crypts from a flat colonic adenoma. Reused with permission.

In Figure 19.12, vertical cross-sectional fluorescence images of A) esophagus and B) colon collected with a tabletop dual axes confocal microscope are shown [26]. These specimens were incubated with a near-infrared dye (IRDye®800CW, LI-COR Biosciences) prior to imaging after being freshly excised during endoscopy. These images were collected at 2 frames per second with a transverse and axial and resolution of 2 and 3 μm, respectively. With use of postobjective scanning, a very large field-of-view of 800-μm wide by 500-μm deep was achieved. In Figure 19.12A, the specimen was collected from the squamo-columnar junction of a patient with Barrett's esophagus. Over the left half of the image, the individual squamous cells from normal esophageal mucosa can be seen in the luminal to the basilar direction down a depth of 500 μm. Over the right half of the image, vertically oriented crypts with individual mucin-secreting goblet cells associated with intestinal metaplasia can be appreciated as brightly stained vacuoles. This diseased condition is associated with greater than 100-fold relative risk of developing cancer. In Figure 19.12B, the specimen was collected from a flat colonic adenoma, and the image reveals vertically oriented dysplastic crypts with individual goblet cells.

Volume rendering can also be performed with the dual axes confocal microscope to illustrate 3D imaging capabilities. These views are important for tracking cell movements, observing protein-protein interactions, and monitoring angiogenic development. A xenograft mouse model of glioblastoma multiforme has been developed by subcutaneously implanting $\sim 10^7$ human U87MG glioblastoma cells in the flank of a nude mouse. Horizontal cross-sectional fluorescence images were collected with a tabletop instrument when the tumors reached $\sim$1 cm in size. The mice were anesthetized for the *in vivo* imaging session, and indocyanine green (ICG) at a concentration of 0.5 mg/ml was injected intravenously to produce contrast. A skin flap overlying the tumor was exposed, and horizontal cross-sectional images were collected with a field-of-view of 400×500 μm^2. A fluorescence image collected at 50 μm below the tissue surface, shown in Figure 19.13A, reveals that the glioblastoma has developed a dense, complex network of tortuous vasculature. A total of 400 horizontal cross-sectional images acquired at 1 μm increments were used to generate the 3D volumetric image, shown in Figure 19.13B. Volume rendering was performed using Amira® modeling software.

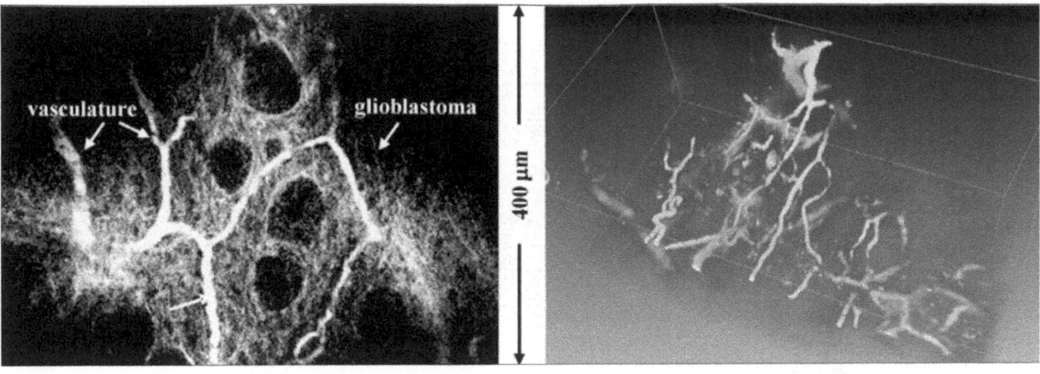

FIGURE 19.13: Dual axes confocal fluorescence images in small animal models. A) A horizontal cross-sectional fluorescence image of a human U87MG xenograft glioblastoma tumor implanted subcutaneously in the flank of a nude mouse was collected *in vivo* at 50-μm depth using i.v. indocyanine green (ICG). B) A 3D volumetric image is generated from a z-stack of 400 sections collected at 1-μm intervals.

19.5 MEMS Scanning Mechanisms

The long working distance created by the low NA objectives in the dual axes architecture provides space for a scanning mechanism to be placed in the postobjective position. This location is a key feature of the design that allows for scaling down of the optics to millimeter dimensions. As a result, endoscope compatibility can be achieved without loss of performance. For *in vivo* imaging, a fairly rapid scan rate is needed to overcome motion artifacts introduced by organ peristalsis, heart beating, and respiratory activity, requiring a frame rate of > 4 per second. Scanners fabricated with MEMS (micro-electro-mechanical systems) technology are well suited to meet the size and speed requirements for *in vivo* applications [10, 11, 27].

19.5.1 Scanner structure and function

In the dual axes configuration, the MEMS mirror performs high speed scanning while maintaining a fixed intersection of the two beams below the tissue surface. The design of the mirror uses a gimbal geometry to perform scanning in the horizontal (X-Y) plane, and rotation around an inner and outer axes defined by the location of the respective springs, as shown by the scanning electron micrograph (SEM) in Figure 19.14A, scale bar 500 μm [28, 29]. The overall structure has a barbell shape with two individual mirrors that have active surface dimensions of 600×650 μm^2. A 1.51 mm long strut connects these two mirrors so that the illumination and collection beams preserve the overlapping focal volume in the tissue. Electrostatic actuation in each direction is provided by two sets of vertical combdrives that generate a large force to produce large deflection angles. There are 4 actuation voltages (V_1, V_2, V_3, and V_4) that provide power to the device and are connected to the bondpads via an ultrasonic wire bonding technique. For two dimensional (*en face*) imaging, rotation around the outer axis (V1 and V2) is resonantly driven with a sine wave, while rotation around the inner axis (V3 and V4) is driven at DC with a sawtooth waveform.

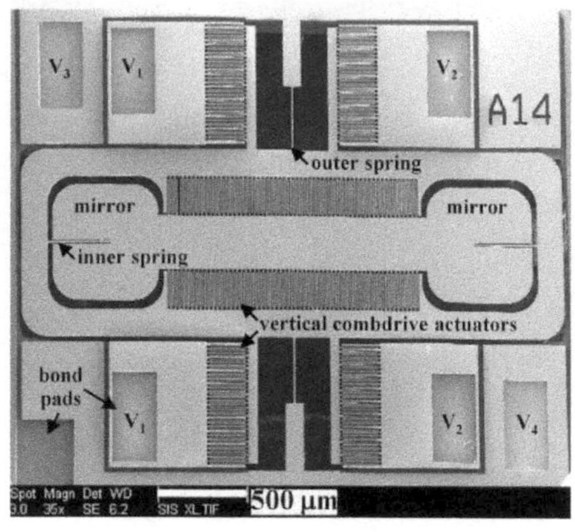

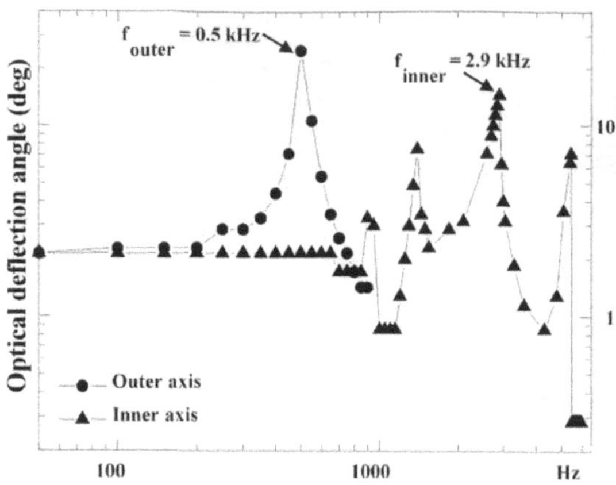

A) MEMS 2D scanner **B) Frequency response**

FIGURE 19.14: Dual axes scanner. A) Gimbaled micromirror rotates around about inner and outer springs and is driven by vertical combdrive actuators, scale bar 500 μm. B) Frequency response of structure show resonant peaks. Reused with permission.

Fabrication of this device involves 4 deep-reactive-ion etching (DRIE) steps to achieve self-alignment of the combdrive fingers in the device layers by transferring mask features sequentially from the upper to lower layers and to remove the backside of the substrate behind the mirror and release the gimbal for rotation. Images can be acquired at rates up to 30 frames/sec with a maximum field-of-view of $800 \times 400\ \mu\text{m}^2$. The mirror surface is coated with a 12-nm-thick layer of aluminum to increase reflectivity. This is particularly important for fluorescence imaging because of the low collection efficiency of the dual axes architecture. The reflectivity increases a factor of two from 37% to 74% at visible wavelengths compared to bare single crystalline silicon. Since the imaging beam reflects off the MEMS scanner twice before collection, the effective increase in photon collection is a factor of four.

19.5.2 Scanner characterization

The parameters of the scanner are characterized for quality control purposes prior to use in the microscope. First, the flatness of the mirror is measured by an interferometric surface profiler to identify micromirrors that have a peak-to-valley surface deformation $< 0.1\ \mu$m. The root mean square (RMS) roughness of the mirror surface itself should be $\sim$25 nm. Identifying devices with this surface profile ensures a high optical quality of the reflected beams. Then, the static deflection curve is measured, and optical deflections of $\pm 4.8^\circ$ are achieved at 160 V for the outer axis, and $\pm 5.5^\circ$ at 170 V for the inner axis. The difference between the two is expected to be mainly from the nonuniform etching profile of DRIE. The frequency response of the micromirror is obtained using a driving voltage of $77 + 58 \sin 2\pi\omega f t$ V. The torsional resonance frequencies (f_0) are at 500 Hz with ± 12.4 deg optical deflection, and 2.9 kHz with ± 7.2 deg optical deflection for the outer and inner axes, respectively, as shown in Figure 19.14B. The parametric resonances [18] can sometimes be observed in the inner axis near frequencies of $2f_0/n$, where n is an integer ≥ 1. This phenomenon is caused by the nonlinear response of the torsional combdrives, which leads to subharmonic oscillations.

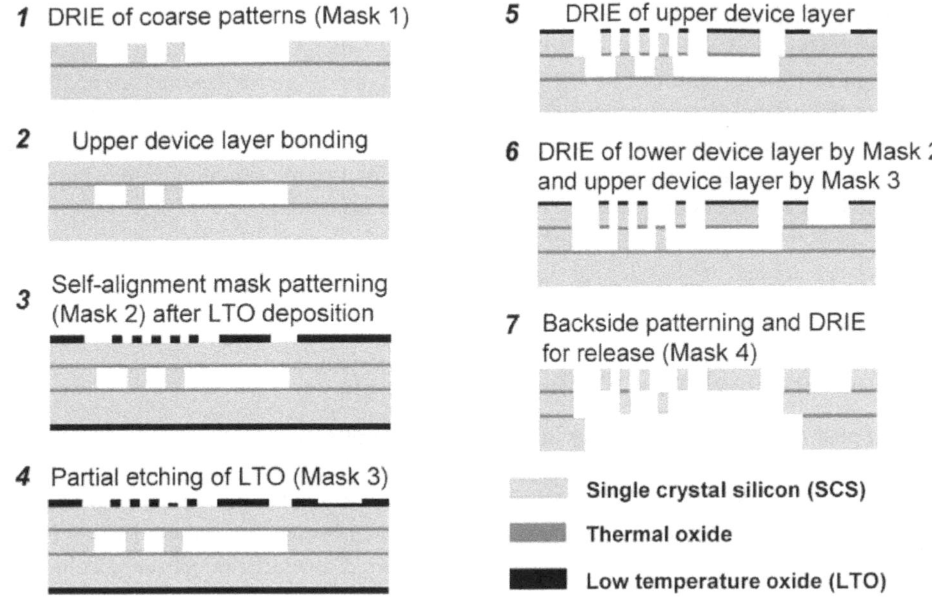

FIGURE 19.15: Fabrication process flow for the MEMS scanner. Details provided in the text. Reused with permission.

19.5.3 Scanner fabrication process

The fabrication process, shown in Figure 19.15, starts with a silicon-on-insulator (SOI) wafer composed of a silicon substrate, buried oxide, and silicon lower device layers that are 530, 1, and 30 μm thick, respectively [28]. A deep-reactive-ion-etch (DRIE) of coarse patterns, including the combdrives and trenches, is performed on the SOI wafer with Mask 1 (step 1). Next, an oxide layer is grown on a plain silicon wafer using a wet oxidation process. This wafer is then fusion bonded onto the etched surface of the SOI wafer. (step 2). The yield is increased by bonding in vacuum, and the bonded plain wafer is ground and polished down to 30 μm thickness, becoming the upper device layer. The two oxide layers between the silicon layers provide electrical isolation, and act as etch stops for DRIE allowing precise thickness control. The frontside of the double-stacked SOI wafer is patterned and DRIE etched to expose the underlying alignment marks in the lower device layer. Then, a low temperature oxide (LTO) layer is deposited on both sides of the wafer. The frontside layer is patterned by two masks. The first mask (Mask 2) is the self-alignment mask (step 3), and is etched into the full thickness of the upper LTO layer. The second mask (Mask 3) is mainly for patterning the electrodes for voltage supplied to the lower device layer (step 4). It goes through a partial etch leaving a thin layer of LTO. The alignment accuracy of each step needs to be better than $g/2$, where g is the comb gap. Since most devices have 6 μm comb gaps, this leads to a required alignment accuracy of better than 3 μm.

Good alignment accuracy is important in minimizing failures due to electrostatic instability during actuation. These three masks eventually define the structures in the upper, lower, and double-stacked layers. After the frontside patterning is done, the LTO layer at the wafer backside is stripped (step 5). The wafer is cleaned and photoresist is deposited on the backside. Then, frontside alignment marks are patterned. Next, the upper silicon layer is etched with the features of Mask 2 in DRIE. Then, a thin LTO and buried oxide layer is anisotropically dry-etched. Finally, the lower and upper silicon layers are etched (DRIE) simultaneously with features patterned by Mask 2 and 3, respectively (step 6). For backside processing, the wafer is bonded to an oxidized handle wafer with photoresist. The backside trenches are patterned with Mask 4 on photoresist (step 7). The backside trench should etch through the substrate to release the gimbal structure, so handle wafer bonding and thick resist is required for DRIE. Alignment to the frontside features are accomplished by aligning to the previously etched patterns. After the substrate (530 μm) is etched by DRIE, the process wafer is separated from the handle wafer with acetone. After cleaning, the exposed oxide layer is directionally dry-etched from the backside. Finally, the remaining masking LTO and exposed buried oxide layer is directionally etched from the frontside.

19.6 Miniature Dual Axes Confocal Microscope

The use of postobjective scanning in the dual axes configuration allows for this design to be scaled down in size to millimeter dimensions, and the long working distance created by low NA objectives create space for a MEMS micromirror to perform beam scanning. The first miniature prototype was developed with a 10 mm diameter package to facilitate assembly and alignment [30]. The second generation instrument is further scaled down in size to 5 mm diameter for endoscope compatibility.

19.6.1 Imaging scanhead

Systems integration is a very challenging part of the development of the miniature dual axes confocal microscope because of the small size required for compatibility with medical endoscopes. This process requires a package design that allows for precise mounting of the following optical elements: two fiber-coupled collimator lenses, a 2D MEMS scanning mirror, a parabolic focusing mirror, and a hemispherical index-matching SIL element [30]. The basic optical design of the miniature scanhead is shown in Figure 19.16. Two collimated beams are focused at an inclination angle θ to the z-axis by a parabolic mirror with a maximum cone half-angle α to a common point in the tissue after being deflected by a 2D MEMS scan mirror. The flat side of the SIL is placed against the tissue surface to couple the incident beams with minimal aberrations. The parabolic mirror is fabricated using a molding process that provides a surface profile and smoothness needed for diffraction-limited focusing of the collimated beams. Once the collimated beams are aligned parallel to each other, the parabolic mirror then provides a "self-aligning" property in this optical system, which forces the focused beams to intersect at a common focal point within the tissue. Focusing is performed primarily by the parabolic mirror which is a nonrefractive optical element and produces beams with an NA of 0.12. This feature allows for light over a broad spectral regime to become focused to the same point below the tissue surface simultaneously, allowing for multispectral confocal imaging to be performed. The folded path in this design effectively doubles the axial (z-axis) movement of the focal volume below the tissue relative to that of the scan mirror. As a result, the MEMS chip requires a smaller range of travel, and faster axial scanning can be achieved.

19.6.2 Assembly and alignment

Alignment of the two beams in the dual axes configuration is a key step to maximizing the imaging performance of this instrument. This step can be accomplished by locating the two fiber-pigtailed collimators in a pair of v-grooves that are precision machined into the housing, as shown in Figure 19.17A [30]. An accuracy of 0.05 deg can be achieved in aligning the two beams parallel to one another using the v-grooves with preassembled fiber collimators. Additional precision in alignment can be attained with use of Risley prisms to provide fine steering of the collimated beams to bring the system into final alignment. These wedges are angled at 0.1 deg, and can be

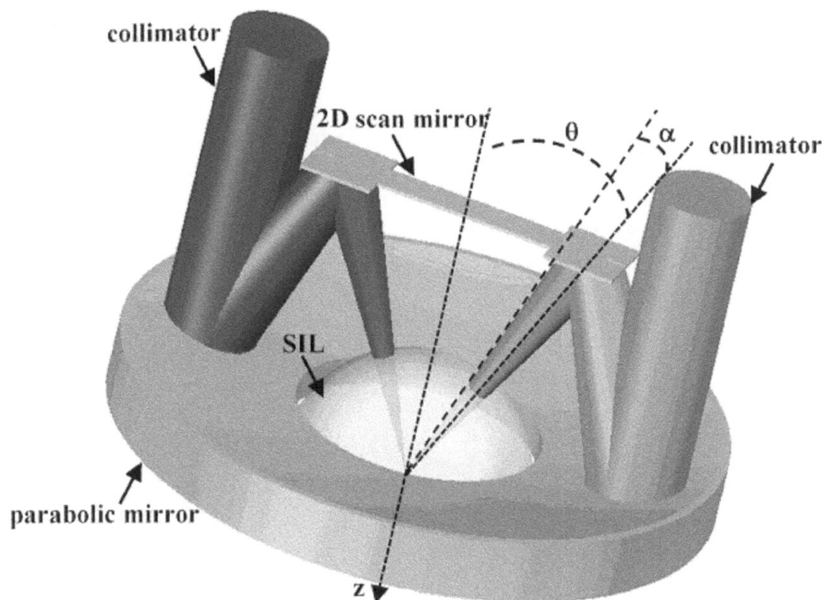

FIGURE 19.16: Miniature dual axes scanhead. Optical design consist of two collimated beams that are focused at convergence angle α by a parabolic mirror at angle θ to the z-axis. Beams overlap within the tissue after being deflected by a MEMS micromirror. The SIL minimizes spherical aberrations and provides index-matching to the tissue. Reused with permission.

rotated to steer the collimated beam in an arbitrary direction over a maximum range of ~0.05 deg. This feature maximizes the overlap of the two beams after they are focused by the parabolic mirror. Two wedges are used in each beam so that complete cancellation of the deflection by each can be achieved, if needed, to provide maximum flexibility.

Axial (z-axis) displacement perpendicular to the parabolic mirror is performed with a computer-controlled piezoelectric actuator (Physik Instrumente GmbH) that moves a slider mechanism, which consists of 3 mechanical supports, as shown in Figure 19.17B. This feature adjusts the imaging depths for collection of 3D volumetric images. The distal end of the slider has a mounting surface for a printed circuit board (PCB), which supports the MEMS chip, wire bonding surfaces (bondpads), and soldering terminals. Power is delivered to the mirror via wires that run through the middle of the housing, and are soldered to the PCB terminals. The z-translation stage is actuated by the closed-loop piezoelectric linear actuator. Finally, the scanhead assembly is covered and sealed from the environment using UV-curing glue to prevent leakage of bodily fluids.

19.6.3 Instrument control and image acquisition

Instrument control and data acquisition are performed using a LabVIEW platform and two control boards (National Instruments PXI-6711 and PXI-6115). The image field-of-view and frame rate are determined by the frequency and amplitude of the 4 actuation signals delivered to the MEMS scanner. For each 2D horizontal cross-sectional image, the micromirror is driven in resonance around the outer (fast) axis using a unipolar sine wave. Rotation of the inner (slow) axis is driven at the DC mode with a unipolar sawtooth waveform. This waveform is smoothed at the transition edges to avoid the introduction of high frequency ringing by the inner axis. The inner axis has its opposing combdrive actuator banks driven 180 deg out of phase to maximize the linear region of the angular deflection while the outer axis is driven with only one side of the combdrive banks in resonance [28].

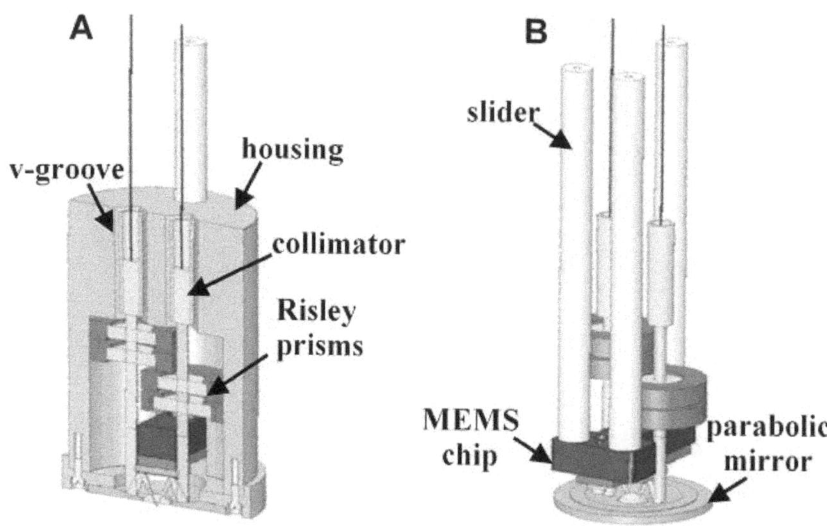

FIGURE 19.17: Alignment and assembly of dual axes scanhead. A) Precision machined v-grooves and Risley prisms provide coarse and fine alignment, respectively, of two beams. B) Axial (z-axis) displacement of the MEMS chip perpendicular to the parabolic mirror is made along a slider mechanism. Reused with permission.

The step size and range of z-axis displacement of the piezoelectric actuator is computer-controlled to generate 3D volumetric images that are created with Amira software via postprocessing. The PMT gain is also modulated electronically to adjust for the decreasing fluorescence intensity at greater tissue depths. Frame averaging is performed to reduce noise at low signal levels. All images are saved in 16 bit format.

19.6.4 *In vivo* confocal fluorescence imaging

Confocal fluorescence images have been collected *in vivo* with the miniature dual axes instrument in an anesthetized mouse following intravenous administration of fluorescein isothiocyanate–dextran (FITC-dextran, 100μL) to provide contrast. The mouse was placed on its side onto a translational stage so that its intact ear could be flattened against the hemispherical window of the miniature dual axes confocal microscope, which was arranged in an inverted orientation. A horizontal cross-sectional image was collected with 488 nm excitation, and the vasculature of the mouse's ear can be clearly delineated in Figure 19.18A. A 3D volume rendered image of the vessels in the mouse's ear was generated from a z-stack of images collected from the surface of the ear to 90 μm below in 2 μm intervals, shown in Figure 19.18B. The piezoelectric actuator was used to axially scan the focal volume over this range. Images were collected at 4 frames per second with 4 frame averaging, and the full 3D volume was generated in 40 seconds. The intensity of the fluorescence signal decayed with time as the constrast solution slowly seeps out of the vessel walls.

19.6.5 Endoscope compatible prototype

The next goal for miniaturization of the dual axes microscope for purposes of *in vivo* imaging is to further reduce the form factor to a size that is endoscope compatible. The endoscope compatible prototype uses the same replicated parabolic focusing mirror as that shown in Figure 19.16 but with

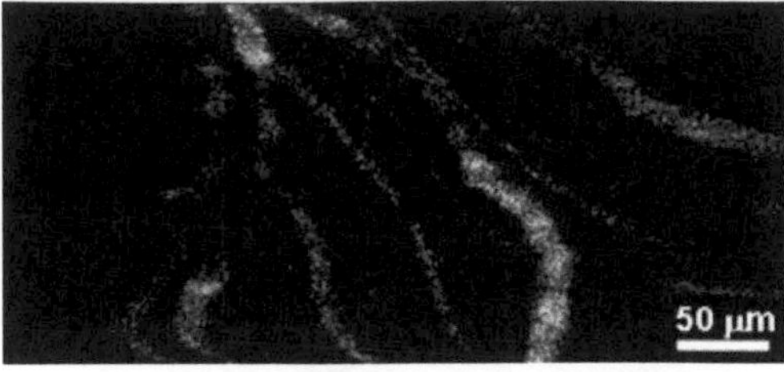

A) *in vivo* mouse ear vasculature **B) 3D volumetric image**

FIGURE 19.18: Dual axes confocal fluorescence image *in vivo*. A) Mouse ear imaged with miniature microscope. B) A 3D volumetric image is generated from a z-stack of 45 sections collected at 2-μm intervals.

an outer diameter of 5 mm to allow for the package to be inserted into the instrument channel of the endoscope. The same MEMS scanner that was developed for the first microscope and described in Sec. 19.4 was used in the endoscope compatible version as well. This next miniaturization step of the dual-axes confocal architecture was accomplished using similar optical elements as that used in the previous 10 mm design. The main components consist of the following: two 1 mm diameter fiber-coupled GRIN collimator lenses, a 3.2 mm $\times$ 2.9 mm MEMS scanner, a 5 mm diameter parabolic focusing mirror, and a 2 mm diameter hemispherical index-matching SIL. Alignment is provided by a pair of 1 mm diameter rotating wedges (Risley prisms), which are inserted in one of the collimated beams. The collimators and Risley prisms are both located by precision wire-EDM machined v-grooves and glued into place with UV curing glue. As in the larger dual axes microscopes, the combined precision of the v-grooves and the pointing accuracy of the preassembled fiber collimators, allow for the collimated beams to become parallel to each other to within 0.05 deg accuracy. The alignment wedges have a small (0.1 deg) wedge angle, which allows for steering of a collimated beam over a range of about 0.05 deg in any direction as each wedge is rotated. Once the collimated beams are aligned parallel to each other, then the parabolic mirror provides a "self-aligning" property for the rest of the system, which forces the focused beams to intersect at a common focal point below the tissue surface. This property allows for the parabolic and MEMS mirrors to be repeatedly removed and replaced if necessary without requiring repeated realignment of the two beams.

This smaller package design also accommodates a slider mechanism, which is used for axial (z-axis) scanning of the MEMS chip to provide variable imaging depths within the tissue and for generating 3D volumetric images. This smaller slider mechanism comprises a single rod, which slides within a precision hole drilled through the housing. The MEMS chip is mounted by an adhesive to a PCB, which is in-turn mounted onto the slider. The PCB provides bondpads to accommodate wire bonding to the MEMS chip and also to provide soldering terminals for the external control wires that power the scan mirror. The endoscope-compatible dual axes confocal microscope can be inserted through the 6 mm diameter instrument channel of a special therapeutic upper GI endoscope (Olympus GIF XT-30) that is available for clinical use, as shown in Figure 19.19A. An magnified view of the distal tip is shown in Figure 19.19B. The axial (z-axis) scanner is activated by a 3 mm diameter linear actuator driven by a micromotor, which can be seen at the proximal end (left) of the unsealed assembly in Figure 19.19C.

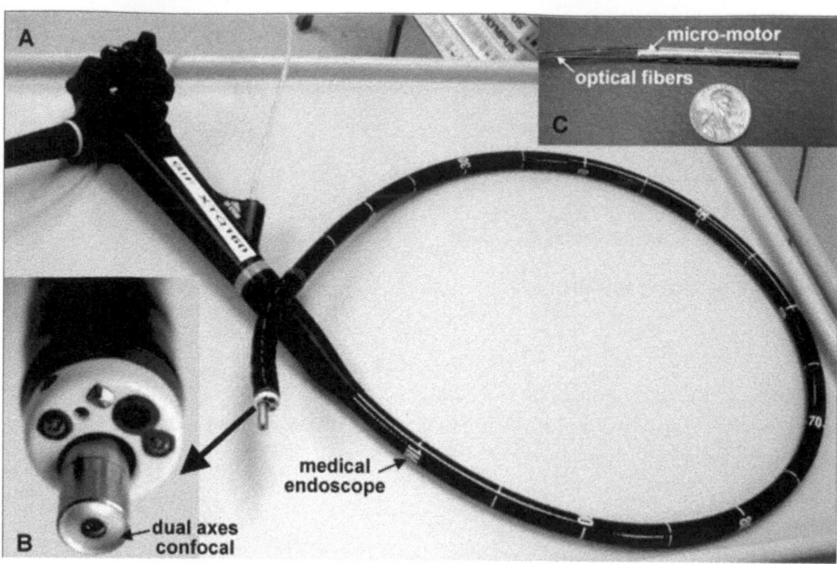

FIGURE 19.19: Endoscope compatible dual axes confocal microscope. A) Microscope passes through the instrument channel of Olympus GIF XT-30 therapeutic upper endoscope. B) Distal tip of endoscope. C) Axial scanning is performed by a linear actuator driven by a stepper micromotor on the proximal end (left) of the unsealed assembly.

19.7 Conclusions and Future Directions

In this chapter, we present the theory, design, and implementation of a novel confocal architecture that uses spatially separated optical axes for the illumination and collection of light. This innovative approach overcomes many of the limitations of the conventional single axis configuration that are encountered with miniaturization and *in vivo* imaging. The use of off-axis light illumination and collection with low NA objectives allows for better filtering of scattered light, postobjective scanning, and system scalability while preserving high axial resolution. Moreover, the superior dynamic range of this geometry allows for the collection of vertical cross-sectional images, a preferred orientation for evaluating tissue differentiation patterns to detect the presence of disease. Scaling down the size of the instrument for endoscope compatibility has been demonstrated using a replicated parabolic focusing mirror and MEMS micromirror. Future development of this technique will focus on achieving the expected levels of performance in the miniature instrument package, including addressing issues of repeatability and reliability. In addition, extension of this approach to multispectral detection, real time z-axis scanning, and molecular imaging are planned. As this novel approach matures, we will soon be able to perform clinical investigation in human subjects and longitudinal studies in small animal models with the ability to visualize tissue microstructures with greater axial depths. This capability can be combined with fluorescence-labeled probes to unravel the previously unknown molecular mechanisms of disease.

Acknowledgments

This work was funded in part by grants from the National Institutes of Health, including K08 DK67618, P50 CA93990, U54 CA136429, and U54 CA105296. We thank Christopher H. Contag, Shai Friedland, Gordon S. Kino, Jonathon T.C. Liu, Wibool Piyawattanametha, Hyejun Ra, Roy M. Soetikno, Olav Solgaard, and Larry K. Wong for their technical support.

References

[1] R.S. Cotran, V. Kumar, and T. Collins (eds.), *Robbins and Cotran Pathologic Basis of Disease*, 7th ed., Saunders, Philadelphia, PA (2005).

[2] J. Pawley, *Handbook of Biological Confocal Microscopy*, 3rd ed., Plenum, New York, NY (1996).

[3] T.R. Corle and G.S. Kino, *Confocal Scanning Optical Microscopy and Related Imaging Systems*, Academic Press, Boston (1996).

[4] S. Lindek and E.H.K. Stelzer, "Optical transfer functions for confocal theta fluorescence microscopy," *J. Opt. Soc. Am.* **13**, 479–482 (1996).

[5] E.H.K. Stelzer and S. Lindek, "Fundamental reduction of the observation volume in far-field light microscopy by detection orthogonal to the illumination axis: confocal theta microscopy," *Opt. Commun.* **111**, 536–547 (1994).

[6] E.H.K. Stelzer and S. Lindek, "A new tool for the observation of embryos and other large specimens: confocal theta fluorescence microscopy," *J. Microsc.* **179**, 1–10 (1995).

[7] R.H. Webb and F. Rogomentich, "Confocal microscope with large field and working distance," *Appl. Opt.* **38**, 4870–4875 (1999).

[8] T.D. Wang, M.J. Mandella, C.H. Contag, and G.S. Kino, "Dual axes confocal microscope for high resolution *In Vivo* Imaging," *Opt. Lett.* **28**, 414–416 (2003).

[9] T.D. Wang TD, M.J. Mandella, C.H. Contag, and G.S. Kino, "Dual axes confocal microscope with post-objective scanning and low coherence heterodyne detection," *Opt. Lett.* **28**, 1915–1917 (2003).

[10] D.L. Dickensheets and G.S. Kino, "Micromachined scanning confocal optical microscope," *Opt. Lett.* **21**, 764–766 (1996).

[11] Y. Pan, H. Xie, and G.K. Fedder, "Endoscopic optical coherence tomography based on a microelectromechanical mirror," *Opt. Lett.* **26**, 1966–1968 (2001).

[12] P.M. Delaney, M.R. Harris, and R.G. King, "Fiber-optic laser scanning confocal microscope suitable for fluorescence imaging," *Opt. Lett.* **33**, 573–577 (1994).

[13] T.D. Wang, S. Friedland, P. Sahbaie, et al, "Functional imaging of colonic mucosa with a fibered confocal microscope for real time *in vivo* pathology," *Clin. Gastro. & Hep.* **5**, 1300–1305 (2007).

[14] R. Kiesslich, J. Burg, M. Vieth, et al, "Confocal laser endoscopy for diagnosing intraepithelial neoplasias and colorectal cancer in vivo," *Gastro.* **127**, 706–713 (2004).

[15] L. Thiberville, S. Moreno-Swirc, T. Vercauteren, et al, "*In vivo* imaging of the bronchial wall microstructure using fibered confocal fluorescence microscopy," *Am. J. Respir. Crit. Care Med.* **175**, 22–31 (2007).

[16] M. Born and E. Wolf, *Principles of Optics*, 7th ed., Cambridge Press, Cambridge, U.K. (1999).

[17] J.T.C. Liu, M.J. Mandella, J.M. Crawford, et al, "A dual-axes confocal reflectance microscope for distinguishing colonic neoplasia," *J. Biomed. Opt.* **11**, 054019-1–10 (2006).

[18] I.S. Gradshteyn and I.M. Ryzhik, *Table of Integrals, Series, and Products*, 6th ed., A. Jeffrey and D. Zwillinger, Eds., Academic Press, San Diego (2000).

[19] L.K. Wong, M.J. Mandella, G.S. Kino, and T.D. Wang, "Improved rejection of multiply-scattered photons in confocal microscopy using dual-axes architecture," *Opt. Lett.* **32**, 1674–1676 (2007).

[20] L. Henyey and J. Greenstein, "Diffuse radiation in the galaxy," *Astrophys. Journal* **93**, 70–83 (1941).

[21] K.T. Mehta and H.S. Shah, "Correlating parameters of the Henyey-Greenstein phase function equation with size and refractive index of colorants," *Appl. Opt.* **24**, 892–896 (1985).

[22] C. Bohren and D. Huffman, *Absorption and Scattering of Light by Small Particles*, John Wiley & Sons, Hoboken, NJ (1983).

[23] S.A. Prahl, M. Keijzer, S.L. Jacques, and A.J. Welch, "Dosimetry of laser radiation in medicine and biology," *Proc. SPIE* **5**, 102–111 (1989).

[24] R. Bays, G. Wagnières, D. Robert, et al., "Clinical determination of tissue optical properties by endoscopic spatially resolved reflectometry," *Appl. Opt.* **35**, 1756–1766 (1996).

[25] T.D. Wang, C.H. Contag, M.J. Mandella, et al., "Confocal fluorescence microscope with dual-axis architecture and biaxial postobjective scanning." *J. Biomed. Opt.* **9**, 735–742 (2004).

[26] J.T.C. Liu, M.J. Mandella, J.M. Crawford, et al, "Efficient rejection of scattered light enables deep optical sectioning in turbid media with low-NA optics in a dual-axes confocal architecture," *J Biomed. Opt.* **13**, 034020 (2008).

[27] W. Piyawattanametha, R.P. Barretto, T.H. Ko, et al., "Fast-scanning two-photon fluorescence imaging based on a microelectromechanical systems two-dimensional scanning mirror," *Opt. Lett.* **31**, 2018–2020 (2006).

[28] H. Ra, W. Piyawattanametha W, M.J. Mandella et al., "Three-dimensional in vivo imaging by a handheld dual-axes confocal microscope," *Opt. Expr.* **16**, 7224–7232 (2008).

[29] U. Krishnamoorthy, D. Lee, and O. Solgaard, "Self-aligned vertical electrostatic combdrives for micromirror actuation," *J. Microelectromech. Syst.* **12**, 458–464, (2005).

[30] J.T.C Liu, M.J. Mandella, H. Ra, et al., "Miniature near-infrared dual-axes confocal microscope utilizing a two-dimensional microelectromechanical systems scanner," *Opt. Lett.* **32**, 256–258 (2007).

20

Nonlinear Imaging of Tissues

Riccardo Cicchi, Leonardo Sacconi, and Francesco Pavone

European Laboratory for Nonlinear Spectroscopy (LENS) and Department of Physics, University of Florence, Sesto Fiorentino, 50019, Italy

During the last two decades, nonlinear imaging techniques experienced an impressive growth in biological and biomedical imaging applications. New techniques have been developed and applied to several topics in modern biology and biomedical optics. The nonlinear nature of the excitation provides an absorption volume spatially confined to the focal point. The localization of the excitation is maintained even in strongly scattering tissues, allowing deep-tissue high-resolution microscopy. This chapter gives a general overview of two-photon fluorescence, second-harmonic generation, and fluorescence lifetime imaging techniques, including a theoretical approach to the physical background, a brief description of the methodologies used, and a list of biological and biomedical applications on tissue imaging.

Key words: nonlinear optics, tissue imaging, two-photon fluorescence, second harmonic generation, fluorescence lifetime

20.1 Introduction

Two-photon fluorescence (TPF) microscopy is a laser scanning imaging technique based on a nonlinear optical process in which a molecule can be excited by simultaneous absorption of two photons in the same quantum event. The first theoretical prediction of a two-photon excited electronic transition was introduced for the first time by Maria Göppert-Mayer more than 70 years ago [1]. Nevertheless, the application of this theory to fluorescence microscopy had to wait the development of high-power ultrafast laser sources. The first TPF laser scanning microscope was realized in 1990 by Denk, Strickler, and Webb [2].

TPF microscopy offers two big advantages with respect to conventional fluorescence techniques. First, the fluorescence signal depends nonlinearly on the density of photons, providing an absorption volume spatially confined to the focal point. Second, for both tissue autofluorescence and commonly used fluorescence probes, two-photon absorption occurs in the near IR wavelength range, allowing deeper penetration into highly scattering media with respect to the equivalent single-photon techniques. The localization of the excitation is maintained even in strongly scattering tissues because the scattered photon density is too low to generate significant signal [3]. For these reasons nonlinear microscopy allows micron-scale resolution in deep tissue.

During the last two decades, the application of TPF microscopy in biological and life sciences has endured an impressive growth, above all in biomedical optics and imaging fields. TPF microscopy has already been successfully used as a powerful technique for many imaging applications in life science [4–6], including fluorescent proteins investigation and spectroscopy [7–10] and for studying biological mechanisms inside cells [11] and tissues [12]. Recently, it has also been applied to *in vivo* imaging, using common microscopes [13, 14] or combined with optical fibres [15, 16] to realize movable or endoscopic [17] microscopes.

TPF microscopy is particularly useful in human tissue imaging and optical biopsy [18–21]. Human tissues intrinsically contains many different fluorescent molecules that allow imaging without any exogenously added probe. Tissue intrinsic fluorophores, as NADH, keratins, flavins, melanin, elastin, cholecalciferol (vitamin D_3), can be excited by two- or three-photon absorption [22] using the Ti:Sapphire laser wavelength (typically between 700 and 1000 nm), which is comprised in the so-called *optical therapeutic window of tissues.*

An immediate implementation for TPF microscopy is represented by second-harmonic generation (SHG) microscopy. SHG is a nonlinear second-order optical process occurring in materials without a center of symmetry, having a large hyperpolarizability. SHG has already been largely used for imaging noncentrosymmetric molecules inside cells [23–25], cellular membranes [26–31], brain [32], and biological tissues [22, 33, 34]. In particular, because of its fibrillar structure [35], collagen intrinsically has a high hyperpolarizability, providing a strong second-harmonic signal [36] and it can be imaged inside skin dermis tissue using SHG microscopy [22, 37]. Recently, SHG microscopy was also used for investigating collagen orientation and its structural changes in healthy tissues [38] as human dermis [39, 40] or cornea [41], and also for studying its dynamical modulation in tumors [42–45].

A further development in fluorescence investigation is represented by the analysis of fluorescence dynamics. Once excited, a population of fluorescent emitting molecules exhibits a fluorescence intensity that exponentially decays in time. Lifetime measurement can yield information on the molecular microenvironment of a fluorescent molecule or on its energy exchanges. Fluorescence lifetime imaging microscopy (FLIM) is a relatively new laser scanning imaging technique. It consists in representing "pixel-by-pixel" in the image the measured fluorescence lifetimes instead of the fluorescence intensities as in traditional laser scanning imaging techniques [46–48]. FLIM is a useful technique in studying protein localization [49], protein-protein interaction [50, 51] and fluorescent molecular environment [52]. It has already been used on tissue imaging, coupled to TPF, in application involving skin tissue characterization [53] and horny layer pH gradient monitoring [54], and for Ca^{2+} detection and imaging [55].

20.2 Theoretical Background

20.2.1 Two-photon excitation fluorescence microscopy

In 1931 Maria Göppert-Mayer [1] predicted for the first time the possibility that an electronic transition occurs not only by absorption of a single resonant photon, but also by simultaneous absorption of two photons, each having an energy equal to a half of the transition energy, in the same quantum event. The first experimental observation of this phenomenon was performed 30 years later, when Kaiser and Garret measured a two-photon transition in a crystal of CaF_2:Eu^{2+} [56]. A two-photon microscope was not immediately built. Because of the nonlinear properties of the process, two-photon absorption requires high spatial and temporal density of photons, not achievable with old light sources without biological damage of the samples. The development of femtosecond pulsed laser sources overlapped the problem. However, since the experiment of Kaiser and Garret, other 29 years passed prior to the realization of the first two-photon laser scanning microscope by Denk, Strickler, and Webb [2] in 1990.

The main advantages demonstrated by this tool were the possibility of performing an intrinsic optical sectioning, a reduced photodamage induced on the sample, an increased penetration depth and a higher spatial resolution with respect to the confocal microscope. In the last 18 years, the TPF microscope underwent an impressive growth, making it one of the most used tools in biological imaging of cells and tissues.

Let us consider a fluorescent molecule as an unperturbed system and the oscillating electromagnetic field as perturbation. The overall system is described by the Schrödinger equation:

$$i\hbar\frac{\partial\psi}{\partial t} = \hat{H}\psi, \tag{20.1}$$

where ψ is the wave function of the system and H is the hamiltonian of the system, which is the sum of two terms:

$$\hat{H} = \hat{H}_0 + \lambda V. \tag{20.2}$$

The first one, H_0, is the hamiltonian of the unperturbed system and the second one, λV, is made by two terms: λ is a constant related to the perturbation order and V is the perturbing electric potential. The generic state of the perturbed system can be expressed, using ψ, as a linear combination of eigenstates of the unperturbed system:

$$\psi = \sum c_n(t) u_n \exp(-i\omega_n t), \tag{20.3}$$

where u_n and $\varepsilon_n = \hbar\omega_n$ are the eigenstates and the eigenvalues of the unperturbed system, respectively, and the coefficients $c_n(t)$ can be expanded in power series of λ, as follows:

$$c_k(t) = c_k^{(0)}(t) + \lambda c_k^{(1)}(t) + \lambda^2 c_k^{(2)}(t) + \ldots \tag{20.4}$$

Using the last two equations for wave function and coefficient, respectively, and substituting them into the Schrödinger equation, at the first order of the perturbation theory, and assuming the electric dipole approximation,[1] the probability to find the system in a state u_k (different from the initial state u_m) after a time t assumes the following form:

$$\left|c_k^{(1)}(t)\right|^2 \cong \frac{E_0^2}{\hbar^2}\left|\mu_{km}^*\right|^2 \frac{\sin^2\left(\frac{\omega_{km}-\omega}{2}t\right)}{(\omega_{km}-\omega)^2}. \tag{20.5}$$

This is the probability to find the system in the state u_k after a time t. The last expression can be divided by the time t:

$$\frac{\left|c_k^{(1)}(t)\right|^2}{t} \cong \frac{E_0^2}{\hbar^2}\left|\mu_{km}^*\right|^2 \frac{\sin^2\left(\frac{\omega_{km}-\omega}{2}t\right)}{(\omega_{km}-\omega)^2 t} \tag{20.6}$$

[1] The electric dipole approximation can be used when the electric field wavelength is much longer with respect to the atomic or molecular radius. It consists in approximating $\exp(i\mathbf{k}\cdot\mathbf{r}) \approx 1$, where $\mathbf{k}$ is the wave vector of the perturbing electric field and $\mathbf{r}$ is the radius vector of a point in an atom or molecule. This assumption is valid when $\lambda \gg r$.

to obtain the probability per unit of time, that corresponds to the transition probability for one-photon absorption. In particular, the transition probability depends on the square amplitude of the electric field, corresponding to the light intensity.

The transition from the state u_m to the state u_k is also possible through simultaneous absorption of two photons. In order to calculate this probability, the perturbation theory at the second order has to be used. By considering the existence of an intermediate virtual state u_i, the transition probability is represented by the product of the first order transition probability for $m-i$ transition times the first order transition probability for $i-k$ transition. The square module of the second order coefficient assumes the following form:

$$\left|c_k^{(2)}(t)\right|^2 \propto \frac{E_0^4}{16\hbar^4}. \tag{20.7}$$

In particular, it is worth noting that the transition probability for a two-photon transition through an intermediate virtual state is proportional to the fourth power of the electric field amplitude. At the second order of the perturbation theory, the two-photon transition probability scales with the square of the light intensity, instead of with intensity, as demonstrated for one-photon transition.

$$\left|c_k^2(t)\right|^2 \propto I_0^2. \tag{20.8}$$

20.2.2 Second-harmonic generation microscopy

Two-photon absorption is not the only nonlinear optical process useful for imaging biological samples. Another possibility is to take advantage of the harmonic up-conversion. Second-harmonic generation (SHG) were used for several years as spectroscopic tool and, during the years 70s of the last century, was also applied to microscopic imaging [57–59].

SHG is a nonlinear second-order optical process occurring in materials without a center of symmetry, having a large hyperpolarizability. In nonlinear optics a molecule interacts with a driving optical electric field, not only through its intrinsic electric dipole moment, but also through the induced electric dipole moment. In general, the electrical dipole moment depends on a power series of the electric field:

$$\mu = \mu_0 + \alpha E + \frac{1}{2}\beta E^2 + \frac{1}{6}\gamma E^3 + \ldots, \tag{20.9}$$

where α is the polarizability, related to the linear process as single-photon absorption and reflection; β is the first hyperpolarizability, related to the second-order phenomena as second-harmonic generation, sum and difference frequency generation; γ is the second hyperpolarizability, related to the third-order phenomena as light scattering, third-harmonic generation and both two- and three-photon absorption.

Harmonic generation is an optical phenomenon involving coherent radiative scattering, whereas fluorescence generation involves incoherent radiative absorption and re-emission. As such, fluorescence and harmonic images are derived from fundamentally different mechanisms. Because of its coherent nature, harmonic radiation is usually highly directional and depends critically on the spatial extent of the emission source, making a full description of harmonic generation more complicated than fluorescence generation. The description will be limited to the molecular SHG generated by a tightly focused beam, as in a high resolution microscope, and to a particular distribution of SHG scatterers.

The SHG cross-section of a single molecular scatterer can be calculated by assuming that the excitation field is linearly polarized in the z direction and that the hyperpolarizability tensor has only the β_{zzz} component. If the excitation light has a frequency ω, the induced dipole moment at frequency 2ω is given by:

$$\vec{\mu}_{2\omega} = \frac{1}{2}\beta_{zzz}E_{\omega}^{2}\hat{z}, \tag{20.10}$$

where E_ω is the excitation field amplitude. The resulted power per differential solid angle at an inclination ψ from the z axis, in units of photons/second, may be expressed as:

$$P_{2\omega}(\psi) = \frac{3}{16\pi}\frac{4n_{2\omega}\hbar\omega^5}{3\pi n_{\omega}^{2}\varepsilon_0^3 c^5}\left|\beta_{zzz}\right|^2 \sin^2(\psi)\, I_{\omega}^{2}, \tag{20.11}$$

where ε_0 is the free-space permittivity, c is the speed of light, n_ω and $n_{2\omega}$ are the medium refractive indices at frequency ω and 2ω, respectively. As in two-photon fluorescence, the emitted power is proportional to the square of the excitation light intensity.

If we consider the SHG emitted by a generic spatial distribution of radiating dipoles, driven by a focused gaussian beam, because SHG is a coherent process, the total emitted power and the emitted radiation pattern depend on how the scatterers are spatially distributed inside the focal volume. Here, the case where the excitation laser is polarized along the y axis and the scatterers hyperpolarizabilities are uni-axial, also directed along the y axis, is considered. By assuming that the scatterers are uniformly distributed along the z axis, with constant density N_1 per unit length, the resultant SHG power emitted per unit of solid angle is given by:

$$P_{2\omega}(\Omega) = \frac{\pi n\varepsilon_0 c r^2 w_z^2}{4} N^2 \left|\vec{E}_{2\omega}^{(0)}(\Theta)\right|^2 \exp\left(\frac{-k_{2\omega}^2 w_z^2(\cos\theta - \xi)^2}{4}\right), \tag{20.12}$$

where n is the refractive index of the medium in which the molecule is immersed, ε_0 is the free-space electrical permittivity, c is the speed of light in vacuum, w_z is the waist length along the optical axis, N is the number of harmonophores, $k_{2\omega}$ is the wavenumber at frequency 2 ω, and ξ is the De-Gouy shift or phase anomaly.

The resultant SHG radiation power is confined to a forward directed off-axis cone. The angular deviation from the optical axis is given by:

$$\theta_{peak} \approx \cos^{-1}\xi, \tag{20.13}$$

that also corresponds to the angle with maximum emitted power. In this case it is worth noting that the SHG radiated power scale with the square of the radiating dipole number (N^2), instead of with N as in two-photon fluorescence.

In general, SHG light can be emitted in both forward and backward direction with respect to the excitation light direction. The angular distribution of the emitted SHG depends on the angle between the direction of the excitation field and the direction of the hyperpolarizability. The ratio between the forward-directed and the backward-directed SHG varies with the same angle. The main SHG contribution is directed in the same direction as the excitation light and usually SHG is detected in a forward direction. In the case of collagen, SHG light can be detected in both forward and backward directions. Usually, forward detection is used in thin tissue samples imaging because of the low backward-directed contribution. In thick tissue samples, only backward detection is possible because the sample thickness prevents a forward-directed detection of the SHG signal. Anyway, in this case the backward emitted SHG signal can be strongly enhanced by the backscattered and reflected forward-emitted SHG light. This particular effect requires highly scattering media to occur, as biological tissues, which fit all the requirements to be imaged with backward-detected SHG.

20.2.3 Fluorescence lifetime imaging microscopy

Since the end of the 80s, time-resolved fluorescence spectroscopy is a well-established technique for studying the emission dynamics of fluorescent molecules, as the distribution of times between the electronic excitation of a molecule and the radiative decay of the electron from the excited state producing an emitted photon [60, 61]. The temporal extent of this distribution is referred to as the fluorescence lifetime of the molecule. Lifetime measurement can yield information on the molecular microenvironment of a fluorescent molecule. Many factors such as ionic strength, oxygen concentration, pH, binding to macromolecules, and the proximity of other molecules can all modify the lifetime of a fluorescent molecule.

At the beginning of the years 90s, an exciting new development of the field has been the development of the technique of fluorescence lifetime imaging microscopy (FLIM) [62, 63]. In this technique lifetimes are measured at each image pixel and displayed as contrast. Lifetime imaging systems have been demonstrated using wide-field [62], confocal [64], and also multiphoton [65] microscopy. FLIM couples the advantages of fluorescence microscopy and time-resolved spectroscopy by revealing the spatial distribution of a fluorescent molecule together with information about its microenvironment.

The fluorescence lifetime of a molecule usually represents the average amount of time the molecule remains in the excited state prior to its return to the ground state. If we consider an ensemble of molecules, excited with an infinitely short pulse of light, resulting in an initial population N_0 in the excited state, then the decay rate of the initially excited population is proportional to the number of molecules in the excited state:

$$\frac{dN(t)}{dt} = -(\Gamma + k)N(t), \tag{20.14}$$

where $N(t)$ is the number of molecules in the excited state at time t, Γ and k are the radiative and nonradiative decay rate constants, respectively. Integrating the equation, with the initial conditions $N(t{=}0) = N_0$, we obtain:

$$N(t) = N_0 \exp\left(-\frac{t}{\tau}\right), \tag{20.15}$$

where:

$$\tau = \frac{1}{\Gamma + k} \tag{20.16}$$

is the lifetime of the excited state. Hence, we expect the fluorescence intensity $I(t)$, which is proportional to the population in the excited state, to decay exponentially. In general the lifetime of the excited state of a population of molecules can be calculated as the ensemble-averaged time the population remains in the excited state:

$$\langle t \rangle = \frac{\int_0^\infty tN(t)dt}{\int_0^\infty N(t)dt}. \tag{20.17}$$

If $N(t)$ is an exponential decay, then:

$$\langle t \rangle = \frac{\int_0^\infty tN_0 \exp\left(-\frac{t}{\tau}\right) dt}{\int_0^\infty N_0 \exp\left(-\frac{t}{\tau}\right) dt} \tag{20.18}$$

and

$$\langle t \rangle = \tau. \tag{20.19}$$

Hence, for an exponential decay, the average amount of time a molecule remains in its excited state is equal to the fluorescence lifetime.

There are essentially two methods to measure the fluorescence lifetime. These are related to measurements performed in the time domain and in the frequency domain, respectively. The former is based on the measurement of the time-dependence of the fluorescence decay when the sample is excited by a pulse; the latter is based on the measurement of the fluorescence phase shift with respect to a modulated excitation. In the time domain the most common measuring technique consists of time-correlated single-photon counting (TCSPC). The fluorescence emission of molecules excited by a pulse has an exponential distribution in time, as shown in the previous paragraph. The TCSPC method can record this distribution by measuring the time delays of individual fluorescence photons with respect to the excitation pulse.

Let's suppose to divide the fluorescence emission distribution in a certain number N of time-channels of amplitude Δt. If we consider a time-channel i, with $i < N$, contained within the fluorescence decay, then this channel corresponds to a mean time delay t_i and interval Δt. If, following one excitation pulse, the average number of photons detected in the i-th time-channel is equal to n_i, corresponding to an average number of detected photoelectrons x_i, then

$$x_i = n_i q, \tag{20.20}$$

where q is the detector quantum efficiency. If there are a large number of excitation pulses for every count registered in the time-channel i, then the probability $P_x(i)$ of generating other than x_i photoelectrons in the i time-channel is given by the Poisson distribution

$$P_x(i) = \frac{x_i^x \exp(-x_i)}{x!} \tag{20.21}$$

with

$$\sum_{x=0}^{\infty} P_x(i) = 1. \tag{20.22}$$

Therefore, the probability of detecting at least one photoelectron per excitation pulse is given by

$$P_{x\geq 1}(i) = 1 - P_0(i) = 1 - \exp(-x_i) = 1 - \left(1 - x_i + \frac{x_i^2}{2} + \ldots\right) \tag{20.23}$$

At very low light levels $x_i \gg x_i^2/2$, and the previous expression gives

$$P_{x\geq 1}(i) \simeq x_i = n_i q. \tag{20.24}$$

Hence, under these conditions, the probability of detecting one or more photons per excitation pulse in the time-channel i is proportional to the fluorescence intensity at a delay time t_i. This is the general required condition for accurate TCSPC measurement. Assuming that the repetition rate of the source is R_s and the total rate of the detected pulses due to fluorescence photons over all delay times is R_d, if the condition

$$\frac{R_d}{R_s} = \alpha \ll 1 \tag{20.25}$$

is respected, then the probability of detecting two fluorescence photons per excitation pulse is negligible. A value $\alpha < 0.01$ is typically required for TCSPC. For δ-function excitation (this condition is true for excitation provided by femtoseconds lasers), the number of counts (equivalent to the number of detected fluorescence photons) Y_i acquired in the time-channel i, during a measurement time T for a single exponential decay of the fluorescence intensity with lifetime τ is given by

$$Y_i = \alpha R_s T \frac{\Delta t}{\tau} \exp\left(-\frac{i\Delta t}{\tau}\right) \tag{20.26}$$

This expression demonstrates another advantage of the TCSPC technique; the measurement precision can be enhanced by simply increasing the acquisition time T.

20.3 Morphological Imaging

20.3.1 Combined two-photon fluorescence-second-harmonic generation microscopy on skin tissue

TPF microscopy has already been successfully used as a powerful technique in skin imaging and optical biopsy by means of a morphological characterization [18, 20, 21, 66]. Additional morphological information can be provided by second-harmonic generation (SHG) microscopy, which can be combined with TPF microscopy using the same laser source. Combined TPF-SHG microscopy has been applied to skin physiology and pathology, and specifically to the study of normal skin [18, 20, 21, 39, 53], cutaneous photoaging [67, 68], psoriasis [53], and selected skin tumors, including basal cell carcinoma (BCC) [42, 44, 69], and malignant melanoma (MM) [53, 70].

The layered structure of human skin can be resolved using combined TPF and SHG microscopy. The epidermis provides high two-photon autofluorescence, whereas dermis provides both TPF (elastin) and SHG signal (collagen). SHG contribution is vanishing in the epidermis. The outer epidermal layer (horny layer) is particularly fluorescent with respect to other layers because of its high concentration of keratins. Horny cells are dead cells with high keratinization. Geometry and morphology of cells can be resolved by TPF microscopy and compared to a thin, *en face* cut, slice of a skin sample (Figure 20.1b) stained with hematoxylin-and-eosin in a nonconventional histological routine. Polygonal structure of cells is clearly distinguishable in both images. In this layer the autofluorescence contribution is mainly due to the keratinized cellular membranes. The inner layers of skin (granular layer, spinal layer) are constituted by living cells. In these cellular layers tissue autofluorescence is lower than in the horny layer, but the cells are better resolved, as shown in Figure 20.1c, because of their higher metabolic activity. Cytoplasm provides a high fluorescence signal, mainly coming from NADH and cholecalciferol for both layers, from loricrin, profilaggrin, and from cytokeratins 1 and 5 for the granular and spinal layers, respectively. The comparison with nonconventional, *en face* cut histology (Figure 20.1d) shows a good correlation: cells appear with highly fluorescent cytoplasms and dark nuclei and they show high contrast with respect to the surrounding interstitial medium. The innermost epidermal layer (basal layer) contains the youngest cells having the largest metabolic activity. Basal cells are smaller and more fluorescent than other cells in the epidermis because of a higher concentration of fluorescent proteins. Cytokeratins 5 and 14 characterize this cellular layer. A combined TPF-SHG image of the basal layer (Figure 20.1e) shows both cellular TPF (in green) and a SHG signal coming from the collagen fibers inside the dermis (in red). Basal cells are well resolved as in the corresponding nonconventional histological image (Figure 20.1f), whereas collagen fibers are better resolved and detailed.

The dermis is a connective tissue under the epidermis, mainly composed of elastin and collagen fibers. Elastin and collagen produce high TPF and SHG signals, respectively. Combined TPF-SHG imaging is a powerful technique for dermis imaging, as represented in Figure 20.1g in which a TPF image of elastic fibers (in green) and a SHG image of collagen fibers (in red) of the same field are superimposed. The selectivity in the excitation is demonstrated by the high signal-to-noise ratio obtained. Such an image shows a good correlation with that obtained with a Verhoeff-Van Gieson stain on a formalin-fixed, paraffin-embedded normal skin sample (Figure 20.1h).

20.3.2 Combined two-photon fluorescence-second-harmonic generation microscopy on diseased dermis tissue

Cutaneous keloids (K) and hypertrophic scars (HS) are examples of abnormal wound healing and are characterized by excessive dermal fibrosis. K is a pathological scar arising in the skin following a previous surgical procedure. K develop in genetically susceptible individuals and, unlike normal

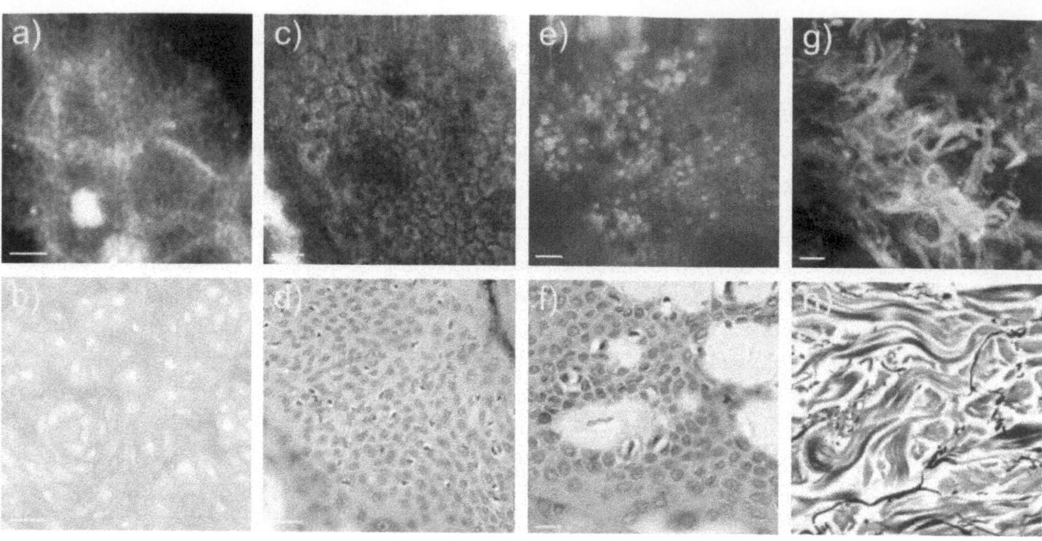

FIGURE 20.1: (See in color after page 572.) Comparison between *en face* optically sectioned human skin sample imaged with combined TPE intrinsic fluorescence (green) and SHG (red) microscopy and corresponding microphotographs taken with optical bright field microscopy on 5 μm thick *en face* sliced human skin samples stained with hematoxylin-and-eosin. Horny layer in human skin at 10 μm depth acquired by TPF (a) and corresponding histological image (b). Scale bars, 20 μm. Spinal layer in human skin at 45-μm depth acquired by TPF (c) and corresponding histological image (d). Scale bars, 20 μm. Dermo-epidermal junction in human skin at 70-μm depth acquired by TPF and SHG (e) and corresponding histological image (f). Scale bars, 25 μm. Elastic (green) and collagen (red) fibers in human skin at 120 μm depth acquired by TPF and SHG (g) and corresponding histological image (h) stained with Verhoeff-Van Gieson stain. Scale bars, 25 μm.

scars (NS), do not regress. By definition, in K the superfluous fibrous tissue extends beyond the margins of the original wound while in HS remains within the boundaries of the initial wound. K and HS scars are difficult to manage and the treatment must be carefully tailored for each patient. In Figure 20.2 four histologically different regions are represented. The healthy epidermis (E) and the healthy dermis (D) are located in the first layers of the sample (Figure 20.2c). The fibroblastic/myofibroblastic cellular proliferation (F) and the so-called "amiantoid collagen fibers," typical of K, can be found in the deepest layers of the skin lesion. The morphology of these four different tissue regions can be scored by measuring the so-called "Second Harmonic to Autofluorescence Aging Index of Dermis" (SAAID) score for each region. The SAAID value is a measure of the ratio between collagen and elastic tissue. SAAID is defined by:

$$SAAID = \frac{I_{SHG} - I_{TPE}}{I_{SHG} + I_{TPE}}$$

It can be used to evaluate intrinsic and extrinsic skin aging [67, 68, 71], as well as to give a measure of the fibrotic status of the dermis. A similar analysis, based on the SAAID score evaluation, can be performed in skin samples representative of normal dermis and BCC. In Figure 20.3 BCC (B), tumor-stroma interface (S), and surrounding connective tissue (D) are resolved by combined TPF-SHG imaging. The SAAID score differs between BCC, tumor-stroma interface, and normal dermis in all the investigated samples. A morphological modification has been observed inside the tumor-stroma interface. In fact, as depicted in Figure 20.3c, a thin layer with a high SHG signal, corresponding to a region (S) containing only collagen can be observed. The corresponding SAAID

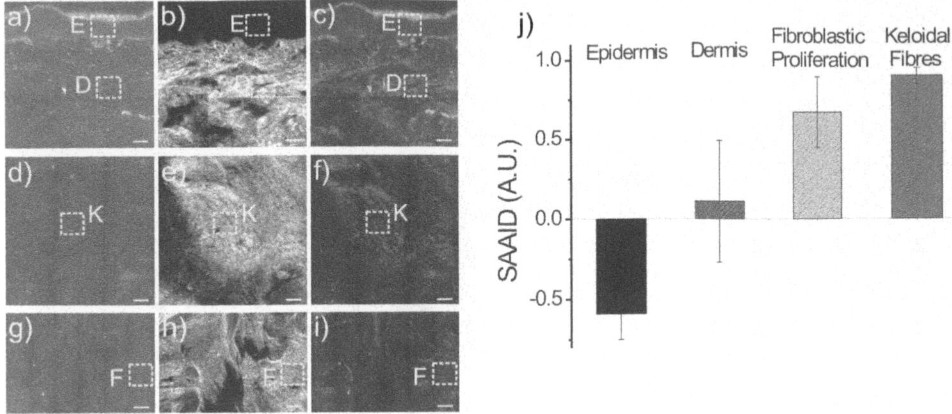

FIGURE 20.2: (See in color after page 572.) Transversal optically sectioned *ex vivo* skin sample, containing both epidermis (E) and healthy dermis (D) tissue, imaged using TPF microscopy (a), SHG microscopy (b), and the merge between the SHG and TPF image (c). Scale bars, 60 μm. Transversal optically sectioned *ex vivo* scar sample, containing keloidal fibers (K), imaged using TPF microscopy (d), SHG microscopy (e), and the merge between SHG and TPF image (f). Scale bars, 60 μm. Transversal optically sectioned *ex vivo* skin sample, containing fibrotic dermis tissue (F), imaged using TPF microscopy (g), SHG microscopy (h), and the merge between SHG and TPF image (i). Scale bars, 120 μm. Epidermis (E), healthy dermis (D), fibroblastic proliferation (F) and keloidal fibers (K) SAAID scores, calculated in five different zones for each of the four regions, are plotted in a bar graph (j).

value is positive, because of the high collagen SHG contribution and the reduced TPF signal arising from this region.

The results obtained by Cicchi et al. [71], together with those ones obtained on skin photo-ageing [67, 68] (for which the SAAID score has been introduced for the first time), has shown SAAID index as a good parameter to discriminate between altered connective tissue regions (normal dermis, vs. pathological wound healing, photo-aged skin, tumor stroma). This simple scoring method for connective tissue could be also used to assess the effectiveness of a laser treatment acting on collagen or elastic fibers.

20.3.3 Combined two-photon fluorescence-second-harmonic generation microscopy on bladder tissue

Morphological features of bladder urothelium can be highlighted on fresh biopsies by taking advantage of the autofluorescence of NADH, which is still present inside a tissue up to 4–5 hours from excision. Both healthy bladder cells and carcinoma *in situ* (CIS) cells give a high fluorescent signal if imaged within 2–3 hours from excision (see Figure 20.4). In these measurements an excitation wavelength of 740 nm, which is adequate to excite NADH fluorescence by two-photon absorption, has been used. TPF allows subcellular spatial resolution, enabling bladder urothelium imaging at the subcellular level.

A morphological analysis can be accomplished from the acquired images. In particular, cells of healthy bladder mucosa (Figure 20.4a) appear more regular in distribution and shape with respect to the corresponding CIS cells (Figure 20.4b). Moreover, the nucleus (dark in the images) to cytoplasm (bright in the images) ratio of the represented cells is quite different between healthy bladder mucosa and CIS. The fluorescence intensity is higher in healthy samples with respect to the corresponding

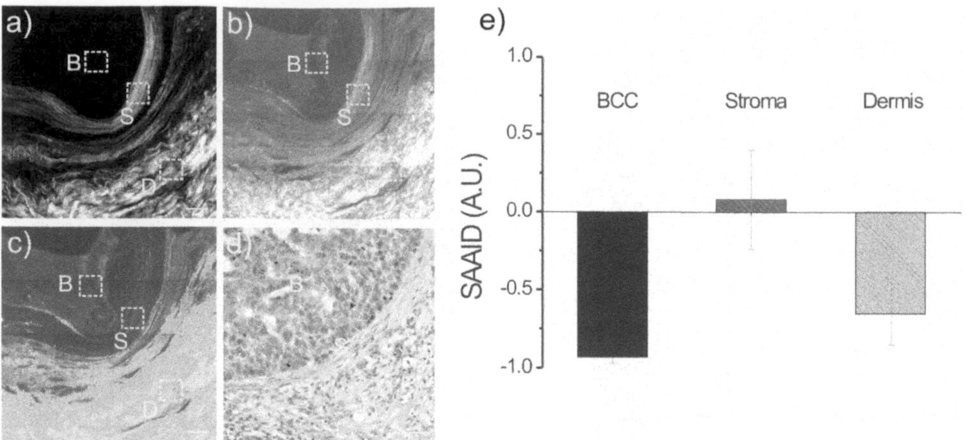

FIGURE 20.3: (See in color after page 572.) Transversal optically sectioned *ex vivo* BCC sample imaged using SHG microscopy (a), TPF microscopy (b), merging the SHG and the TPF image (c), and the corresponding histological image (d). Scale bars, 80 μm. BCC (B), tumor-stroma interface (S) and surrounding connective tissue (D) are highlighted by the corresponding letters. The SAAID score, calculated for the three highlighted regions (dashed line), is plotted in the bar graph in (e).

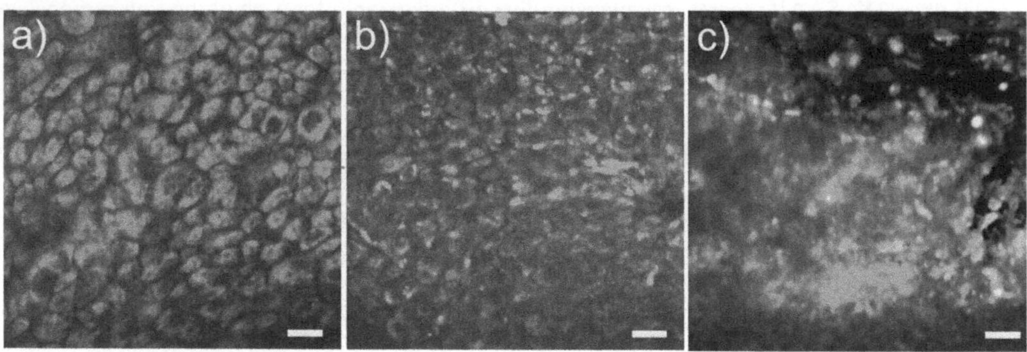

FIGURE 20.4: (See in color after page 572.) Two representative images of healthy bladder mucosa (a) and carcinoma *in situ* (b) acquired at 30-μm depth from tissue surface on *ex vivo* fresh biopsies. Images have been acquired using a two-photon autofluorescence of NADH with an 740 nm excitation wavelength. Scale bars, 10 μm. Tumor-connective tissue margin (c) for a carcinoma *in situ*, imaged by combined TPF microscopy (green) and SHG microscopy (blue). Scale bars, 15 μm.

CIS sample, even if this datum is strongly affected by the biological variability and the time between excision and imaging.

The tumor margins of CIS can be highlighted by combining TPF of NADH with SHG of collagen. As depicted in Figure 20.4c, collagen stroma surrounding CIS cells can be imaged with SHG microscopy, whereas CIS cells can be imaged by using a two-photon autofluorescence of NADH, giving a high-resolution, high-contrast image of tumor margins.

20.3.4 Second-harmonic generation imaging on cornea

As described in detail in the introduction, SHG imaging is an emerging microscopy technique, which is particularly well suited to analyze connective tissues due to the significant second order nonlinear susceptibility of collagen, and may become a noninvasive complement to traditional structural methods such as electron microscopy, X-ray diffraction, and histological analysis.

The corneal stroma displays a very regular assembly of collagen fibrils (<30 nm in size) that are arranged parallel to each other into lamellar domains (i.e., 0.5–2.5 μm thick planar structures running parallel to the corneal surface). The highly ordered nature of this tissue makes it a convenient model to study disorganization events of the normal connective matrix. Matteini et al. have used SHG imaging to quantify the photothermally-induced modifications in the fibrillar collagen assembly of laser-treated porcine corneas. When the corneal stroma undergoes low-power continuous wave diode-laser treatment, a controlled thermal effect can be induced within the irradiated volume. The result is a mild perturbation of the regular fibrillar arrangement, while uncontrolled denaturation of collagen is avoided. Moreover, the extent of lattice disorder decreases smoothly and progressively with the distance from the center of the irradiated area.

This model can be successfully imaged and analyzed by using SHG microscopy, as demonstrated in Figure 20.5. Both control and a laser-irradiated areas have been analyzed by using polarization-modulated SHG signal. The polarization of excitation light has been varied between 0 and 360 degrees in 36 steps. A SHG image has been acquired for each polarization step. The sums of the all acquired images are represented in Figures 20.5a and 20.5b for control and laser-irradiated regions, respectively. The corresponding polarization graphs, calculated by averaging on a 0.5 μm area (corresponding to the Point Spread Function) and on a 3 μm area (corresponding to a single lamellar domain), are represented in Figures 20.5c and 20.5d for control and laser-irradiated regions, respectively. The modulation depth is independent on the averaging area in the control sample, whereas it is strongly dependent on it for the laser-irradiated area. This result suggests that laser-irradiation is causing a loss in the organization of lamellae. Anyway, the laser-induced damage is not enough to cause collagen denaturation, which would be corresponding to the vanishing of the SHG signal.

20.3.5 Improving the penetration depth with two-photon imaging: Application of optical clearing agents

In microscopic imaging techniques that perform optical sectioning of the sample, the penetration depth in biological tissue is commonly limited by high turbidity of the medium, which is the case of human skin tissue. [72]. In the *optical therapeutic window of tissues* (600–1600 nm spectral range), scattering is the main phenomenon limiting the penetration depth. Greater penetration or, alternatively, enhancement of image contrast are desirable objectives in all such imaging techniques.

Clearing agents have long been used to reduce scattering in fixed animal and plant tissue sections [73], but their *in vivo* use is at a much earlier stage of development. One of the main reasons is that the commonly used agents for fixed sections (dimethyl-sulfoxyde for example) are toxic; therefore, a need has arisen for biocompatible alternative optical clearing agents (OCAs). The study of such OCAs and other exogenous agents capable of reducing scattering, enhancing contrast, and increasing penetration depth has been relatively recent [74–77]. An understanding of the clearing process at the microscopic level is still far from being complete. A common hypothesis [74, 76, 78, 79] is based on the assumption of diffusion of the OCA into the tissue and the subsequent outflow of water under osmotic pressure. The combined effects of OCA ingress and water egress are generally thought to provide better refractive index matching of the remaining ground matter, thus reducing scattering. The use of hyperosmotic agents, such as glycerol, propylene glycol, and acetic acid, has been reported in optical coherence tomography [77, 78], second-harmonic generation microscopy

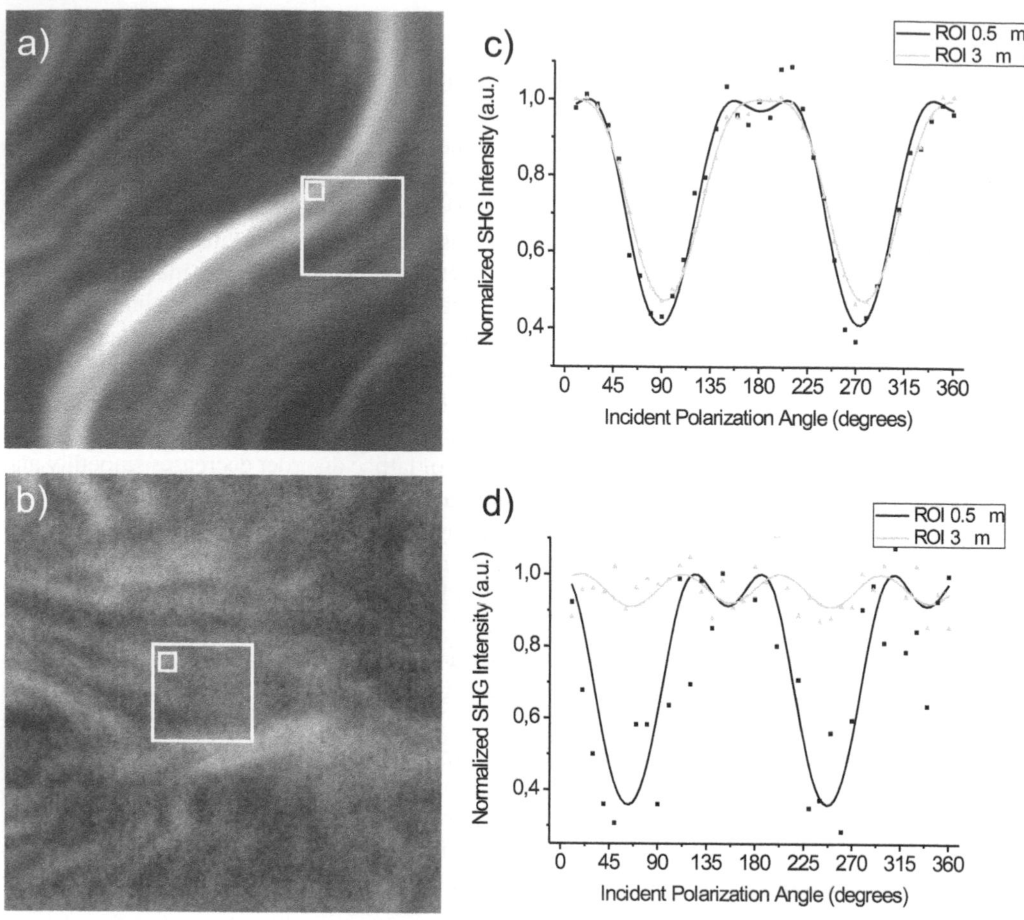

FIGURE 20.5: SHG images of laser-irradiated corneal sample. Images are taken in a control non-irradiated region (a) and in a laser-irradiated region (b). Field of view: 20 × 20 μm. Polarization-modulated SHG intensity versus polarization angle, averaged on a region of 0.5 μm and on a region of 3 μm, corresponding to the Point Spread Function and to a single lamellar domain, for the control region (c) and for the laser-irradiated region (d).

[80, 81], confocal reflectance microscopy [79, 82], and recently Cicchi et al. have exploited their effectiveness in TPF microscopy [83].

TPF microscopy has generated impressive *in vivo* images of human skin [19, 20, 53], and is applicable to many other tissue types. The application of OCAs may lead to the enhancement of TPF microscopy, since it has been demonstrated that the effect of scattering causes drastic reduction in the penetration depth to less than that of the equivalent single-photon fluorescence whilst largely leaving resolution unchanged [84, 85]. Improvement in the penetration depth of TPF microscopy has been obtained by optimizing the pulse shape and repetition rate for the sample under investigation [86], or limiting aberrations introduced by biological tissues using adaptive optics [87, 88]. The first use of OCAs in TPF microscopy has been reported by Cicchi et al. [83] in 2005. In this work, images of connective tissue inside human dermis have been acquired at different depths measured from the sectioned surface of the sample. Images have been collected in stacks, each comprising four images (each of 500 × 500 pixels resolution, 100 μm ×100 μm field of view) taken at depths of 20, 40, 60, and 80 μm from the sectioned surface. An excitation wavelength of 750 nm, a high-magnification (60×) objective lens, and a laser power at the sample of approximately 4 mW have been used. Before acquisition, the sample has been immersed in 0.1 ml of phosphate buffered saline (PBS) in order to prevent drying and shrinkage. Then, the tissue has been immersed in 0.5 ml of an OCA and one image stack has been acquired every 30 seconds for 6–7 minutes. Finally, the OCA has been removed and the sample has been immersed again in 0.5 ml of PBS, in order to observe the reversibility of the clearing process. Cicchi et al. have investigated three different biocompatible OCAs: glycerol, propylene glycol, both in anhydrous form, and glucose (aqueous solution, 5M). Aqueous dilutions of these agents have also been investigated. After application of the OCA, a check for axial shrinkage of the samples has been performed every 30 seconds by checking the measured distance from the sample surface and comparing the acquired image with previously acquired images. An upper limit of 2% shrinkage in the 6–7 minutes duration of the experiments has been estimated. The detected signal has been confirmed to be predominantly two-photon autofluorescence by observing the complete loss of signal after insertion of a band pass filter centered at half the excitation wavelength in the fluorescence path. In this case backscattered second-harmonic light is negligible because the sample thickness is much less than a mean free path for a visible photon in the tissue. To estimate the average contrast in each image, a contrast function C has been defined as follows:

$$C = \text{Contrast} = \sum_{i,j=1}^{N_{lines}} \left| I_{ij} - \langle I_{ij} \rangle \right|, \tag{20.27}$$

where $\langle I_{ij} \rangle$ is the mean intensity of the eight pixels surrounding the ij-th pixel and $N_{lines} = N - 2$, with N equal to the number of lines of the image. This contrast function is linearly dependent on the detected signal intensity and varies according to the structures in the image. Hence, its usefulness is primarily to enable the comparison between images of the same sample at the same depth maintaining the same field of view. Normalization to the total intensity would be required in order to compare different images.

The images in Figure 20.6 show connective tissue in the human dermis, which is primarily composed of elastin and collagen fibers. The contrast enhancement and the increased penetration depth (from 40 μm to 80 μm), as well as a higher intensity signal, are clearly distinguishable from the images. In order to compare different OCAs it is helpful to define the relative contrast (RC), as:

$$RC = 100\frac{\Delta C}{C} = 100\frac{C[OCA] - C[PBS]}{C}[PBS], \tag{20.28}$$

where $C[OCA]$ and $C[PBS]$ are calculated using the Equation 20.27 for OCA and PBS immersion, respectively. With glycerol a RC value of 300 at 40 μm has been measured, together with a dramatic increase with increasing depth. In deeper sections the effect is greater because of the cumulative effect of the reduction in scattering in the sections closer to the tissue surface. In addition, this effect is further enhanced by the dependence of the contrast on the intensity of the fluorescence signal, which is in turn dependent on the square of the excitation intensity, by means of the quadratic dependence of the two-photon excitation cross-section. A comparison of the effectiveness of the three different OCAs used (glycerol, propylene glycol, glucose 5M in water) is reported in graph (Figure 20.6).

In conclusion, the use of a biocompatible optical clearing agent in *in vivo* applications of TPF microscopy or other nonlinear optical biopsies could provide an option for deep tissue imaging. However, much work remains to be done to establish a detailed understanding of the action of the agent as well as which are suitable biocompatible concentrations [89] and the means of delivery. Cicchi et al. [83] have shown that at high concentration, the hyperosmotic agents glycerol, propylene glycol, and glucose, are effective in improving the image contrast and penetration depth (by up to a factor of two) in TPF microscopy of *ex vivo* human dermis. Such improvements have been obtained within a few minutes of application.

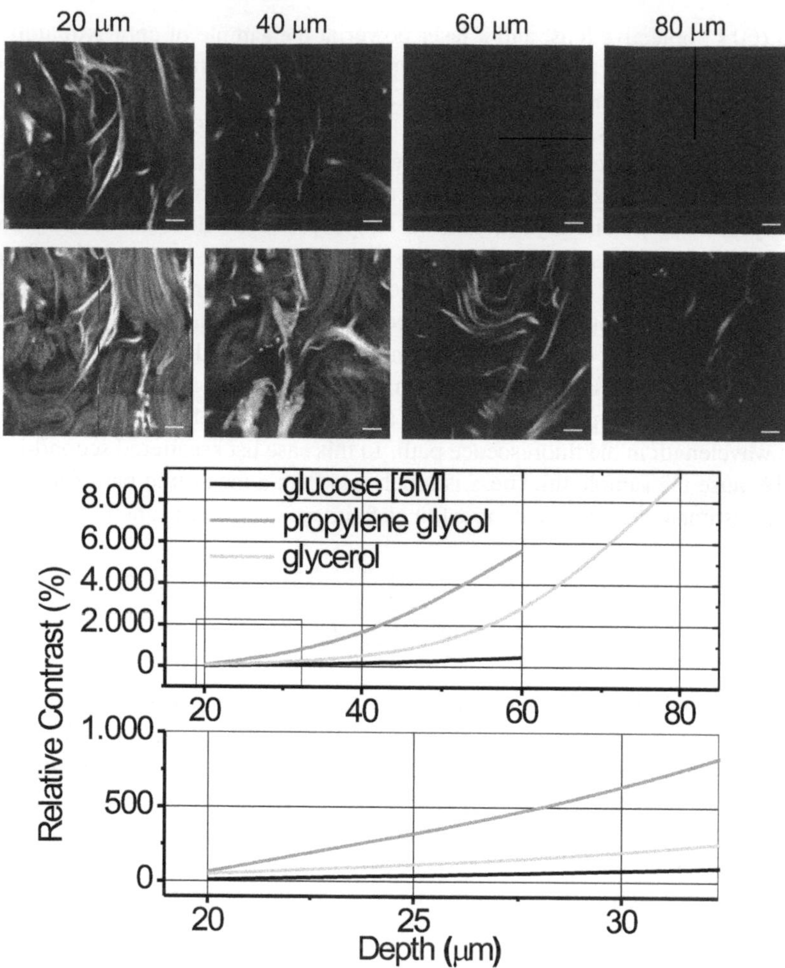

FIGURE 20.6: Image stacks for a human skin tissue sample immersed in 0.1 ml PBS (upper row) and after 7 minutes immersion in glycerol (lower row). The depths from tissue surface are indicated. Scale bars, 10 μm. Relative contrast (RC), calculated on skin dermis tissue images, versus depth for three different OCAs: glucose [5M] (black line), propylene glycol (gray line), and glycerol (light gray line). The cyan area in the upper graph is represented on a magnified scale in the lower graph.

20.4 Chemical Imaging

20.4.1 Lifetime imaging of basal cell carcinoma

The cutaneous basal cell carcinoma (BCC) is the most frequent skin cancer in caucasian populations. The diagnosis of BCC is commonly based on clinical and histopathological examination on tissue biopsy. Recently however, noninvasive diagnostic tools also enabling the *in vivo* study of skin tumors at a nearly histological resolution are becoming increasingly reliable. Among them, fluorescence lifetime imaging microscopy (FLIM) and multispectral two-photon fluorescence (MTPE), based on two-photon excitation, enable imaging of optically thick tissues with deep penetration and high spatial resolution. The dependence of fluorescent emission on the biochemical and physio-

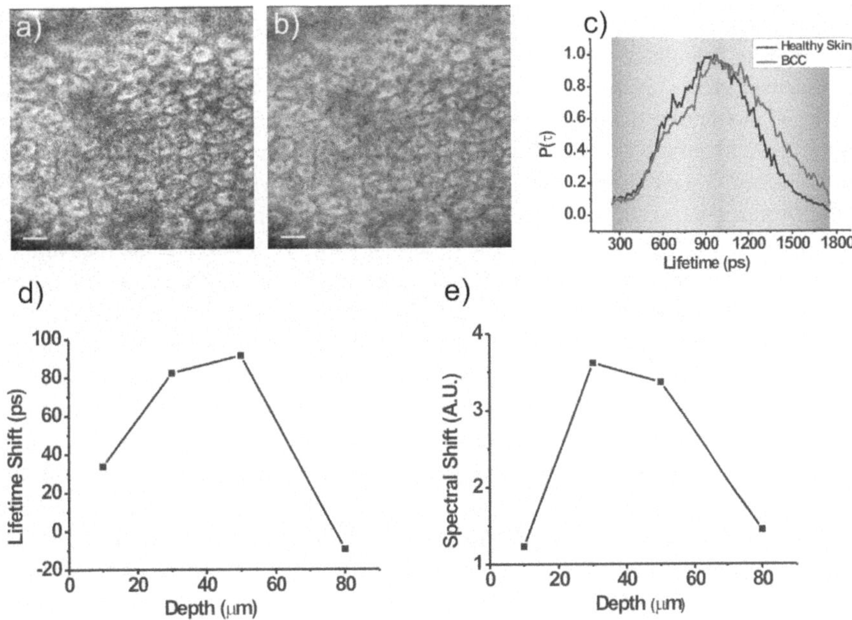

FIGURE 20.7: (See in color after page 572.) TPF image of human healthy epidermis spinal layer (a), acquired by photon counting method at 50-μm depth from the skin surface with an excitation wavelength of 740 nm and the corresponding FLIM image (b), obtained after system response de-convolution and single-exponential fit. Field of view, 200 μm. Scale bars, 20 μm. The color-coded scale for FLIM image is represented in (c) where normalized lifetime distributions for images acquired at the same depth for healthy skin (black line) and BCC (red line) are plotted. The lifetime shift and the spectral shift (measure of the area enclosed between healthy skin and BCC emission spectra) are plotted versus the depth of recording in (d) and (e), respectively.

chemical properties of the substrate enables these techniques to discriminate tissue types of different morpho-functional characteristics, such as normal from neoplastic tissue. In fact, cancerous tissue differs from normal tissue in the distribution of autofluorescence lifetime, the relative concentration of intrinsic fluorophores and the modifications of the histological components. Due to these properties, these techniques are well suited for deep, high resolution imaging of thick tissue sections, with enhanced contrast, and may represent two powerful imaging tools for the diagnosis of basal cell carcinoma and other skin cancers and lesions.

From the morphological point of view, epidermis tissue is characterized by a series of cellular layers each having its histological features. When performing an optical sectioning of the tissue, it's hard to observe significant differences from the morphological point of view in each layer between healthy skin and BCC. In general, BCC cells are slightly different in shape with respect to healthy cells and have a different nucleus to cytoplasm ratio [90]. Anyway, the biological variability does not allow discrimination between healthy skin and BCC, based only on a morphological analysis. Some larger differences are revealed when observing the fluorescence lifetime distributions and the spectral emissions of different tissue layers, as demonstrated in [69].

Figure 20.7a shows a TPF image acquired inside the spinal layer of healthy skin at a depth of 50 μm from the skin surface. In this layer cells have a fluorescent cytoplasm and a dark nucleus. The corresponding lifetime image (in color-coded scale) is shown in Figure 20.7b. Looking at the normalized mean lifetime distribution of the image pixels (black line in Figure 20.7c) a broad behavior is noted, centered at 950 ps. Compared to the normalized mean lifetime distribution of the image acquired at the same depth inside BCC tissue (red line in Figure 20.7c), a shift in the mean lifetime distribution, calculated as the difference between the barycenters of the two distributions, of 91 ps is observed in this layer. In other skin cellular layers a shift towards shorter mean lifetimes for the healthy skin lifetime distribution with respect to BCC is also observed. The lifetime shift for each layer, plotted versus depth in Figure 20.7d, has a higher value in the intermediate layers, whereas it is reduced in the extreme ones. In [69], the authors have found a similar behavior for the spectral shift. In particular, BCC exhibits a fluorescence emission spectrum shifted towards shorter wavelengths with respect to healthy skin. The "spectral shift" is defined as the measure of the area enclosed between the healthy skin and the BCC fluorescence emission spectra. It is plotted versus depth in Figure 20.7e, showing that this quantity has a larger value in the intermediate layers of the epidermis than in the extreme ones. The difference mainly involves the layers containing living cells. The large shift and the small shift observed in the intermediate and extreme epidermal layers, respectively, together with the depth dependent behavior of the "spectral shift," confirm the results obtained using lifetime imaging: the differentiation between healthy skin and BCC occurs in the intermediate epidermal layers, whereas similar characteristics have been observed in the extreme layers.

In conclusion, combined time- and spectral-resolved TPF microscopy has been used as a novel powerful method towards *ex vivo* imaging, characterization and evaluation of BCC. Cicchi et al. [69] have shown that this multidimensional imaging technique is able to discriminate between normal and neoplastic tissues, in a good correlation with histopathology. The multiple nonlinear approach has characterized different tissue regions having different morphological and functional features. All the results obtained with the described techniques were confirmed by corresponding histopathological examination, supporting the diagnostic accuracy of the method. This evaluation method, based on spectral and lifetime analysis could be successfully used also in *in vivo* skin cancer tissue evaluation and extended to other inflammatory and neoplastic skin conditions.

20.4.2 Enhancing tumor margins with two-photon fluorescence by using aminolevulinic acid

The selective tumor localization of endogenous protoporphyrines induced by δ-aminolevulinc acid (ALA) was first discovered in 1990 [91]. In the immediate following years, scientists started to treat malignant tumors in the skin of humans by administering the ALA topically [92, 93]. ALA is a naturally occurring, 5-carbon, straight-chain amino acid. ALA is a precursor in the synthesis of haem, which is the protoporphyrin IX (PpIX) molecule with an iron atom in its ferrous (Fe^{2+}) state incorporated into its core. The haem biosynthetic pathway takes place partly in the mitochondria and partly in the cytosol of the cell. In this process, a total of eight straight ALA molecules are transformed to the tetrapyrrolic macrocycle of PpIX [94]. The incorporation process of Fe^{2+} into PpIX by the enzyme ferrochelatase has the major rate-limiting function. In situations with iron depletion, ferrochelatase can instead incorporate zinc into the PpIX molecule, yielding the fluorescent Zn-protoporphyrin. All enzymatic steps in the process are irreversible. There is a negative feedback in the system, accomplished by the final product, namely haem, which exhibits a negative feedback on the enzyme ALA-S. When supplying ALA in excess, the negative feedback in the system is bypassed. In addition, in some premalignant and malignant lesions, the enzyme catalysing the third step, porphobilinogen deaminase (PBGD), has been shown to exhibit an increased activity [95–97], and ferrochelatase has been shown to have a reduced activity [97–100]. The result is a relative selective accumulation of porphyrines, mainly PpIX, in malignant tissue [101, 102]. PpIX is an intrinsically fluorescent molecule causing the effect of enhancing the fluorescence response of malignant tissue with respect to the normal healthy tissue. For skin cancers, the lifetime of the described biosynthetic pathway is typically 3–4 hours for topically applied creams containing 20% ALA. ALA is commonly used in photodynamic therapy as a photosensitizer, and its effectiveness in enhancing tumor fluorescence and in increasing sensitivity in the detection of tumor margins has been already demonstrated [69].

In Figure 20.8(a,b) a comparison between a transverse optically sectioned TPF image (a) acquired from an ALA-treated sample and a microphotograph (b) obtained by optical microscope after histology on the same sample is shown. In both images, the borders between normal and cancerous tissue are well defined and the cells are resolved. The main parameter in distinguishing between normal epidermis tissue (E) and BCC tissue (B) is the fluorescence light intensity that is higher in BCC. This datum is correlated to the higher fluorescence of tumor tissue caused by the accumulation of fluorescent PpIX caused by topical application of ALA. The higher fluorescence signal observed in ALA-treated samples allows not only an easier BCC discrimination but also an increased penetration depth, and hence the capability of a deeper detection of BCC with respect to imaging technique using only tissue autofluorescence.

ALA-treated samples do not only show enhanced fluorescence from BCC tissue because of PpIX accumulation, but also a large difference in lifetime distribution. ALA application causes the accumulation of PpIX inside tumor tissue. Protoporphyrines are more complex molecules with respect to skin intrinsic fluorescent molecules, having a longer fluorescence lifetime. For this reason in ALA-treated samples imaged using FLIM, a multiexponential fitting function has been used in order to take into account both the fast (corresponding to intrinsic fluorescent proteins) and the slow (corresponding to PpIX) lifetime components. A TPE image (Figure 20.8c) confirms the results shown in Figure 20.8a: BCC tissue (B) is more fluorescent than healthy skin (H) if ALA has been applied. The corresponding FLIM image (Figure 19.8d) reveals a higher contrast between healthy skin and BCC because of the accumulation inside cancerous tissue of PpIX, having a longer lifetime (more than 8 ns) with respect to skin tissue intrinsic fluorescent molecules (typically within 2 ns). This PpIX accumulation inside BCC tissue is confirmed when looking at the fluorescence lifetime distribution graph (Figure 20.8e), which exhibits three peaks. The peak centered on the shortest lifetimes corresponds to the skin tissue intrinsic fluorescent molecules, whereas the peak centered on the longest lifetimes is caused by the accumulated PpIX contribution. The peak in the middle is probably due to the presence of PpIX photoproducts, generated by skin tissue illumination, and of PpIX aggregates [103].

20.5 Morpho-Functional Imaging

20.5.1 Single spine imaging and ablation inside brain of small living animals

In combination with fluorescent protein expression techniques, TPF microscopy has become an indispensable tool to image cortical plasticity in living mice [104, 105]. In parallel to its application in imaging, multiphoton absorption has also been used as a tool for the selective disruption of cellular and intracellular structures [106, 107]. For example, Tolic-Norelykke et al. have used multiphoton absorption to perform selective laser dissection of the cytoskeleton in yeast cells labelled with GFP to investigate the force balance in the mitotic spindle [107]. A similar approach has also been applied *in vivo* where two-photon imaging and laser-induced lesions have been combined to study the microvasculature of the brain. Laser-induced lesions have been used to breach the blood-brain barrier and demonstrate the surveillant role of microglia [108] and produce targeted photodisruptions as a model of stroke [109].

Other groups have taken advantage of multiphoton absorption to ablate or dissect individual neurons. Multiphoton nanosurgery has been performed in worms to study axon regeneration [110] and dissect the role of specific neurons in behavior [111]. Recently, Sacconi et al. have demonstrated a method for performing multiphoton nanosurgery in the central nervous system of mice and discuss

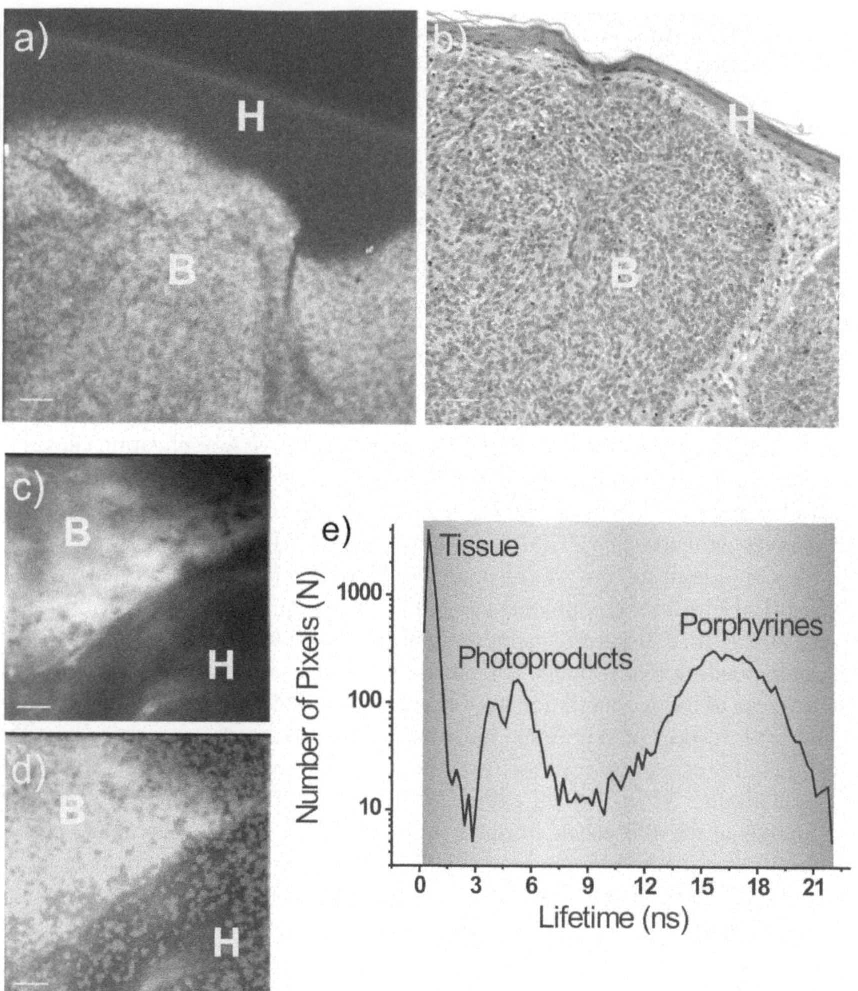

FIGURE 20.8: (See in color after page 572.) Transversal optically sectioned human skin tissue image acquired using TPF on a fresh ALA-treated sample (a) and the microphotograph taken after conventional histopathological examination and visualization by optical microscope on the same sample (b). Scale bars, 30 μm. *En face* optically sectioned human skin tissue image acquired using TPF on a fresh sample, excised 4 hours after ALA application (c) with the corresponding high contrast FLIM image (d), obtained after system response de-convolution and three-exponential fit. Scale bars, 20 μm. B: Basal Cell Carcinoma. H: healthy tissue. Mean lifetime distribution (e) of the image pixels of the image in (d).

the widespread applications in neurobiology research [112]. The authors exploited the spatial localization and deep penetration of multiphoton excitation to perform selective lesions on the neuronal processes of cortical neurons in mice expressing fluorescent proteins (YFP H-line and GFP M-line). Neurons have been irradiated with a focused, controlled dose of femtosecond-pulsed laser energy delivered through a cranial optical window. The morphological changes have been then analyzed with time lapse 3D two-photon imaging over a period of minutes to days after the procedure. The TPF imaging and nanosurgery has been performed using a custom-made, upright, scanning microscope [70]. The system has been designed for maximum stability during laser neurosurgery by incorporating a closed-loop feedback system to precisely control beam positioning in the x, y, and z axes. First, a z-stack has been acquired with two-photon microscopy to obtain a 3D-reconstruction of the dendritic structure and select an x-, y-, and z-coordinate for our desired lesion site. Second, laser-dissection has been carried out by parking the laser beam at the chosen point and increasing the power from 30 mW to 170 mW (measured after the objective lens). The dose of laser energy has been set by opening the shutter for a prescribed period with effective doses for lesions ranging from 150 to 300 ms depending on the depth of the dendritic arbor. Finally, the laser power has been decreased back to 30 mW and a 3D-image of the neuron has been acquired to visualize the effect of laser irradiation. The excitation wavelength has been 935 nm for both imaging and neurosurgery operation on transgenic mice expressing green fluorescent protein (GFP, $n = 15$) and yellow fluorescent protein (YFP, $n = 18$). These experimental parameters have been selected based on previous investigations in cultured cells [113].

In a first series of experiments, this methodology has been applied to irradiate a single dendrite in the parietal cortex. Briefly, the authors have found that the morphological consequences of irradiating a single dendrite can be grouped into two categories of response: 1) transient swelling with recovery and 2) complete dendritic dissection. In the former case, they have observed a swelling of the dendrite extending 5 to 25 μm from the point of irradiation along the dendrite in both directions. The spines present on the irradiated dendrite temporarily disappeared with the swelling (data not shown), but returned after a period of minutes to hours with the original shape of the dendrite. In the latter class of responses, the dendrite has been completely severed. As shown in Figure 20.9a, after laser irradiation the terminal end of the dendrite distal to the dissection point follows a sequence of swelling, degeneration, and disappearance. We can clearly observe (Figure 20.9b) that the portion of dendrite no longer attached to the soma disappears. As shown in Figure 20.9a, the laser dissection can be performed with submicrometric precision and without any visible collateral damage to the surrounding neuronal structures. In addition, no damage or morphological changes have been observed in astrocytes around the lesion site (see Figure 20.9c).

The dendrite proximal to the dissection point and still connected to the soma has been observed to regain its original shape and spine density (Figure 20.10a). Several days after dendritic dissection, the authors did not observe any changes in the remaining dendrite and surrounding area. On the other hand, they have observed morphological changes in the dendritic spines on the remaining part of the severed dendrite (Figure 20.10b). As shown in Figure 20.10c, the number of spines (Ns in Figure 20.10c) changes in the days following the nanosurgery. The continued expression of fluorescent proteins shows that the neuron is alive and, further, the remodelling of spines following the lesion demonstrates that the neuron is morpho-functionally active. These observations are in agreement with previous studies of cultured neurons following dendritic dissection [114]. The precise physical mechanism of the laser-induced dissection of fluorescent structures is not easily understood. However, the most likely process driving nanosurgery is nonlinear photo-ionization [115]. A full characterization of this nonlinear process is not feasible *in vivo*. Therefore, we have focused our attention on providing practical guidelines for making use of this phenomenon. Sacconi et al. have found that the efficiency of laser dissection strongly depends on the level of fluorescent protein expression in the neuronal structure. Considering the maximum laser power available at the objective lens of their system ($\approx$ 170 mW), they have been able to perform laser-induced dendrite dissection at a maximum depth of 200–300 μm in neurons with highly fluorescent dendrites. In experiments performed at depths greater than 100 μm, more than one pulse of laser energy was occasionally required to completely dissect the dendrite. The spatial precision and the specificity of this method have been demonstrated by irradiating an area very near the dendrite (1–0.5 μm) but with no fluorescent features. As shown in Figure 20.11a, in this case they did not induce any visible alteration of the dendrite. This result demonstrates the high specificity and submicrometric precision of the laser dissection method. Commensurate with previous findings [113], they have been only able to ablate or dissect structures expressing fluorescent proteins. The spatial localization of multiphoton nanosurgery is maximally demonstrated by the ablation of individual dendritic

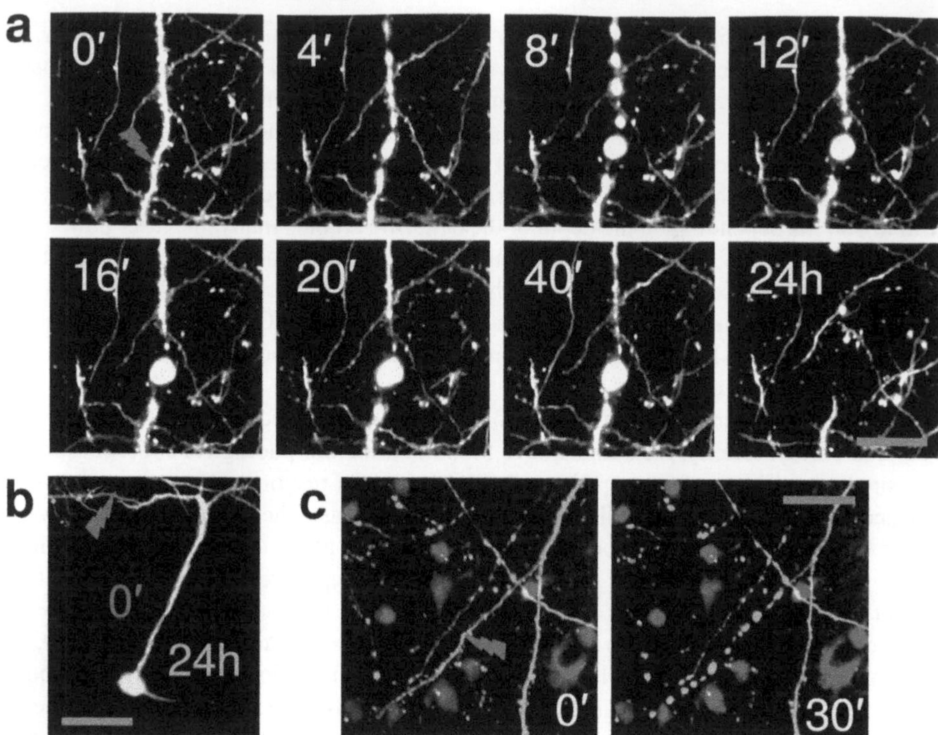

FIGURE 20.9: (See in color after page 572.) Laser-induced lesion of a single dendrite. (a) Time-lapse images of an irradiated dendrite (at $\approx$ 100 μm depth) in a GFP-M transgenic mouse. The dendrite has been irradiated (as indicated by the tip of the red lightning) just after the acquisition of the first image. Each image is a maximum-intensity projection of a set of optical sections acquired at a 1 μm z-step. Yellow numbers refer to the time from irradiation. Scale bar, 25 μm. (b) Overlay of maximum-intensity z-projections (from 500 to 100-μm depth) of a layer-5 pyramidal neuron before (red) and 24 h after (green) dendritic dissection (GFP-M line). This merge shows the integrity of the remaining structure after dissection. Scale bar, 60 μm. (c) The specificity of the laser dissection is demonstrated in two-color imaging experiments with labelled astrocytes. Laser nanosurgery is shown to disrupt only the GFP expressing dendrite (green), while the surrounding astrocytes (red) loaded with an intravital morphological dye (sulforhodamine 101) remain unaffected. Scale bar, 25 μm.

spines (Figure 20.11b). As clearly shown in Figure 20.11b, they have been able to remove a single dendritic spine without causing any visible collateral damage to the adjacent spines or to the parent dendrite.

The use of nonlinear optical methods in biology is continually undergoing developments and refinements [6]. TPF microscopy is particularly useful in neuroscience where *in vivo* imaging has shown great potential in studying the structural correlates of learning and memory [105].

Here, we have described a method to perform selective lesions on cortical neurons in mice expressing fluorescent proteins. This methodology has been applied to dissect single dendrites and to ablate individual dendritic spines without causing any visible collateral damage to the surrounding neuronal structures. The combination of multiphoton nanosurgery and *in vivo* imaging represents a promising tool for probing and disrupting neuronal circuits. The potential of using this precise optical method to perturb individual synapses cannot be overstated. Using multiphoton nanosurgery, the synaptic organization of the brain can now be teased apart *in vivo* to understand microcircuitry,

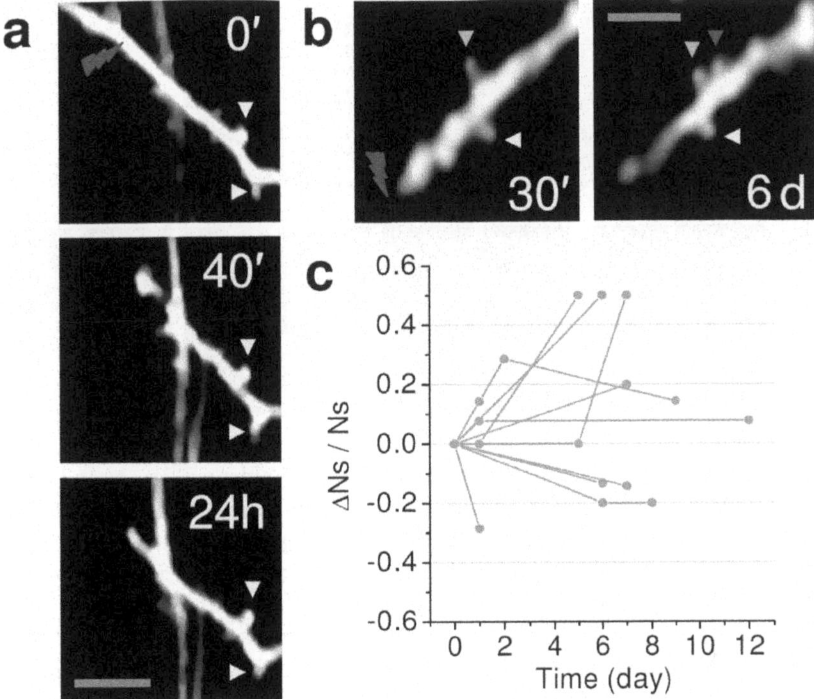

FIGURE 20.10: Spine density following dendritic dissection (a) Time-lapse images of an irradiated dendrite (at ≈ 200 μm depth; YFP-H mouse). The arrowheads highlight the stability of the dendritic spines in the remaining part of the severed dendrite. Scale bar, 8 μm. (b) Long-term morphological changes of dendritic spines in an irradiated dendrite (at ≈ 150-μm depth; YFP-H P30 mouse). Arrowheads: light gray, no change; gray, morphological changes; dark gray, new spine. Scale bar, 7 μm. (c) Percent variation of the number of spines (Ns) several days after laser dissection in 10 young mice (P30; YFP-H). The analysis has been performed in the remaining portions (10–50 μm) of the irradiated dendrites.

in the same way that chemical and electrical ablation methods have linked brain and behavior in the previous century. In particular, we envisage that this technique will provide insights in the mechanisms and dynamics of morphological remodelling in the brain following focal ischemic lesions. In addition, the response of the surrounding cells to this specific perturbation can be studied as a model of the reactive and degenerative changes in microstructure observed in certain neurodegenerative diseases [116]. The disruption of individual spines may also provide insight into neuronal-glial interactions at the synapse [117]. Whilst we do not expect to measure the behavioral consequences of deleting a single dendritic spine, it would be interesting to study the effect of disrupting a group of spines acquired during a learning task. The electrical consequences of these focal disruptions can also be explored with conventional and optical electrophysiological techniques [118–120]. Finally, the combination of nanosurgery and *in vivo* imaging is an exciting tool for neuropharmacology research because it can be used to evaluate the efficacy of systemic drugs on focal damage to the nervous system *in vivo*.

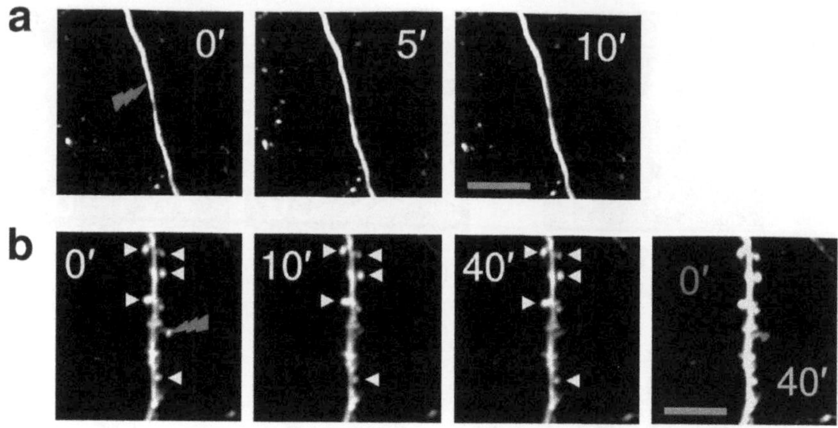

FIGURE 20.11: Spatial resolution: single dendritic spine ablation. (a) Time-lapse images of a dendrite (in GFP-M line) where, as indicated by the tip of the red lightning, the laser beam was parked at $\approx 0.5\ \mu$m from the edge of the dendrite. This experiment was performed at $\approx 100\ \mu$m depth. Scale bar, 25 μm. (b) Time-lapse images of a section of the dendrite where several spines are present (GFP-M). The single spine indicated by the red lightning symbol was irradiated just after the acquisition of the first image. The yellow arrowheads highlight the stability of the surrounding dendritic spines. The last panel shows an overlay of a pair of images from the sequence. The first image (t = 0′) is shown in red and the last one (t = 40′) in green. This panel emphasizes the spine stability and the absence of any swelling in the dendrite. Scale bar, 15 μm.

20.5.2 Optical recording of electrical activity in intact neuronal network by random access second-harmonic (RASH) microscopy

The central nervous system can process a tremendous amount of information, which is encoded in terms of spikes and transmitted between neurons at synapses. A central question in neuroscience is how simple processes in neurons can generate cognitive functions and form complex memories like those experienced by humans and animals. In principle, if one would be able to record from all the neurons in a network involved in a given behavior, it would be possible to reconstruct the related computations. Unfortunately, this is not possible with current techniques for several reasons. Generally, the more precise the method of neuronal recording is (e.g., with the patch-clamp technique), the more limited the number of simultaneously recorded neurons becomes. Conversely, global recordings (e.g., field recordings) collect activity from many neurons but lose information about the computational role of single neurons.

Current optical techniques of recording membrane potential (Vm) can potentially overcome these problems [121, 122]. Most approaches to the optical recording of fast Vm events in neural systems rely on one-photon methods [123–127]. These methods can be used to generate high signal-to-noise ratio (S/N) measurements of action potentials (APs) from subcellular regions in a single trial and subthreshold events with averaging. However, in intact tissue slices their effectiveness in detecting AP in multiple deep neurons is markedly limited because strong multiple light scattering blurs the images. In order to record deep Vm activity in intact systems maintaining a high spatial resolution, nonlinear optical methods are required [3, 6, 26]. Fast SHG recordings of Vm have been achieved in model membranes [24], in Aplysia neurons in culture [120, 128] and in intact mammalian neural systems [118, 119]. The next challenge is to record multiple simple APs simultaneously from assemblies of neurons in intact systems.

The major limiting factor of nonlinear microscopes is their scanning time. For this reason, the optical recording of neuronal APs is possible only in a single position of the neuron by using a line

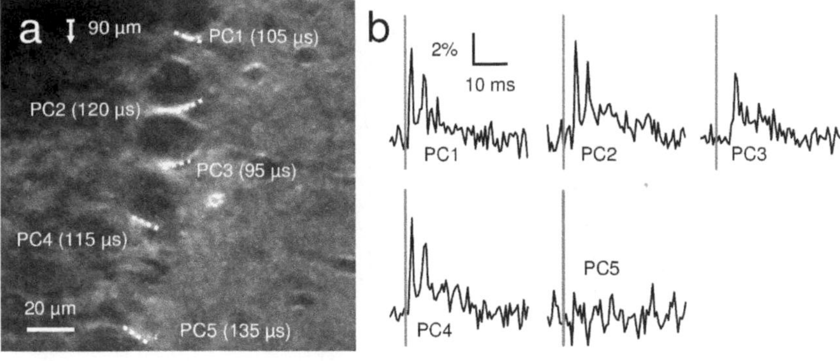

FIGURE 20.12: Optical multiunit recording of stimulated electrical activity. a) SHG image of a cerebellar slice taken at a depth of 90 μm. The multiunit SHG recording has been carried out from the lines drawn (dotted) on the 5 PCs, with the integration time per membrane pass indicated. b) Multiplexed recording of APs from the 5 PCs following stimulation of climbing fibers (gray lines). Each trace represents the average of 20 trials for each PCs. The time resolution is 0.47 ms.

scanning procedure [128]. In principle, the optical measurement of time-dependent processes does not require the production of images at all. Instead, more time should be spent collecting as many photons as possible from selective positions, where the image plane intersects the biological objects of interest. Using this approach, fast physiological processes, like APs in the soma of multiple neurons, can be recorded at a sampling frequency of more than 1 kHz. This cannot be achieved with a standard galvanometer mirror scanner since about 1 ms is the time required to reach and stabilize a new position. Scanning a set of points within a plane at high speed is possible with two orthogonal acousto-optic deflectors (AODs). In an AOD, a propagating ultrasonic wave establishes a grating that diffracts a laser beam at a precise angle that can be changed within a few microseconds. Two-photon microscopy using AODs was used to perform fast calcium imaging in cultured neurons [129] and in brain slices [130].

Recently, Sacconi et al. have combined the advantages of SHG with an AOD-based random access (RA) laser excitation scheme to produce a new microscope (RASH) capable to optically record fast Vm events ($\sim$ 1 ms). This system is capable of resolving APs occurring simultaneously in several neurons in a wide-field ($150 \times 150\ \mu m^2$) configuration, and with deep tissue penetration in living brain slices. The multiple single-neuron electrical recordings have been performed in the Purkinje cell layer of the cerebellum [131, 132]. Acute cerebellar slices have been used and labelled by bulk loading of the styryl dye FM4-64. This staining procedure has allowed large-scale SHG imaging without causing any perturbations to the electrophysiological properties of the neurons. The RASH microscope has been used, in combination with patch-clamping, to detect complex spikes and simple spikes demonstrating the capability of this system to optically record single APs in a cluster of PCs. These results show that RASH microscopy provides a new powerful tool for investigating neural circuit activity by simultaneously monitoring APs in arbitrarily selected multiple single neurons [133].

RASH microscopy has been used to record membrane potential from multiple PCs with near simultaneous sampling. APs have been triggered by stimulation of afferent fibers. The RASH system AODs has been rapidly scanned between lines drawn in the membranes of neurons to perform multiplex measurements of the SHG signal. Including the commutation time of the AODs ($\approx 4\ \mu$s), and a signal integration time per membrane pass in accordance with the previous section ($\approx 100\mu$s), the authors have been able to perform multiple line-scan recordings of up to 5 neurons and to achieve a sampling frequency sufficient to detect APs ($\approx$ 2 kHz). Figure 20.12a shows a field of view in-

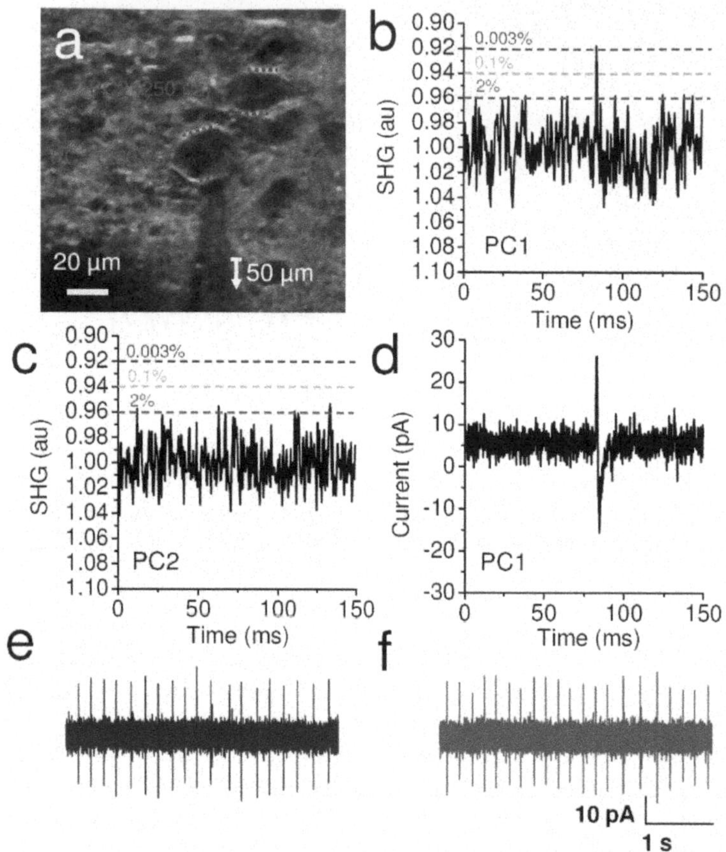

FIGURE 20.13: Optical multiunit recording of spontaneous electrical activity. a) SHG image from a cerebellar slice taken at a depth of 50 μm. The multiunit SHG recording has been carried out from the lines drawn (dotted) on PC1 and PC2, with the integration time per membrane pass indicated. PC1 has been also measured simultaneously by electrophysiology (shadow of electrode can be seen below PC1). b) SHG signal from PC1 showing a spontaneous AP recorded in a single trial. Each point represents 0.535 ms. Confidence intervals are drawn indicating the probability of the noise crossing thresholds. The probability of the noise crossing a threshold of 0.92 is 0.003%, indicating the event shown is an AP (see panel e). c) SHG signal from PC2, suggesting that no spontaneous activity has been detected in this 150 ms sampling time. d) Simultaneous electrical recording of PC1, corresponding to the SHG trace shown in panel b. The electrical recording of spontaneous activity in PC1 before e) and after f) the SHG signal collection.

cluding 5 PCs at a depth of 90 μm. The lines drawn through the membrane of each neuron and the signal integration time per membrane pass are indicated beside each cell in Figure 20.12a. The individual raw SHG traces are shown in Figure 20.12b and the real-time traces have been corrected for multiplexing delay, as summarized in Figure 20.12c. Complex spikes have been observed in PCs and evidence of synchrony and asynchrony have been observed across the population of neurons. PC5 in Figure 20.12b has been found not active across trials, where the membrane SHG signal would have been sufficient to detect an AP. Assuming that all cells are viable, this quiescence could be due to multiple reasons, including low release probability leading to the generation of few and jittered spikes, or that the stimulation of a set of fibers not directly connected with this PC, since it has been shown that each climbing fiber, branches into a final subset of fibers contacting on average 5 adjacent cells [134, 135]. The results in Figure 20.12 shows that the averaging procedure is useful for studying triggered responses, however, the autorhythmic simple spike firing of PCs would be effectively cancelled-out by the process of averaging. In order to record spontaneous activity, APs must be detected in a single trial without averaging.

All the efforts to perform 5 simultaneous acquisitions of APs from 5 different Purkinje cells with a single trial and an adequate S/N ratio have been unsuccessful because the laser intensity required was near the photodamage threshold. Another approach to improve the S/N ratio consists in increasing the membrane integration time, rather than increasing laser power and risking photodamage [120]. To implement this strategy and maintain a sampling frequency of 2 kHz, the number of scanned neurons has been decreased. As shown in Figure 20.13, two PCs have been sampled at this frequency and spontaneous APs have been resolved without averaging. The membrane integration time per neuron has been $\sim 250\ \mu s$. The electrical activity of one neuron has been monitored by a patch-clamp electrode in loose cell attached configuration (PC1 in Figure 20.13a). Figures 20.13b and 20.13c show the SHG signals from neurons PC1 and PC2, respectively. The simultaneous electrical recording of PC1 is shown in Figure 20.13d, where a single, spontaneous AP has been observed. As reported in Figure 19.13b, this AP has been also clearly observed in the SHG trace as a point significantly above the background noise. Given that the noise in the SHG signal is Poisson distributed, we can establish the probability of one point overcoming an arbitrary threshold, as shown in Figures 20.13b and 20.13c. Considering a $\Delta S/S \approx 10\ \%$, a spontaneous AP should correspond to a $\Delta SHG/SHG \approx 6$ - 8%. The probability of one point exceeding a threshold of 0.94 ($\Delta SHG/SHG = 6\%$) is 0.1% and, with a threshold set at 0.92 ($\Delta SHG/SHG = 8\%$), this estimate decreases to 0.003%. This means that the signal exceeding this threshold is an AP with a level of confidence of 99.997%, as seen in Figure 20.4b. Setting a threshold for spike detection at a $\Delta SHG/SHG \approx 6\%$ therefore permits the detection of spontaneous APs with a good level of confidence. Consequently, the SHG recording from PC2 shows evidence of quiescence during this sampling period, since the S/N would not be sufficient to detect any APs. Furthermore, the experimental parameters used in this approach do not lead to photodamage, as shown by the similarity of spontaneous activity before (Figure 20.13e) and after (Figure 20.7f) the optical recording period.

Here, we have presented a novel optical method for recording multiunit electrical activity in intact neuronal networks in brain slices. The bulk loading of FM4-64, combined with RASH microscopy, provides a fast and noninvasive approach to measure APs in neuronal assemblies. Local network activity in PCs, in response to triggered stimulation, has been successfully recorded by averaging across a series of trials. Although the authors have presented recordings of APs induced by afferent fiber stimulation in clusters of up to 5 neurons, this is by no means a limitation of the apparatus. For example, the number of recordable neuron can be enhanced by decreasing the integration time of each membrane pass. To reach a comparable S/N, one must simply increase the number of trials to average. A sampling frequency comparable to conventional electrophysiology ($\sim$10 kHz) can easily be achieved by reducing the number of neurons simultaneously acquired and/or the integration time per membrane pass. The recording of APs in clusters of deep neurons achieved in this work has not been possible with previous techniques.

We have reported that RASH microscopy could be used to detect the spontaneous activity of two PCs in single trial recordings. Although at the moment the detection of spontaneous firing activity is limited to pairs of neurons, a task that can also be performed with double patch-clamp recordings, RASH microscopy allows more freedom in choosing neurons and rapidly switching between cells without changing and repositioning electrodes. Presently, the main limitation of RASH microscopy is photodamage. Because the SHG signal voltage response is small, high illumination intensities and/or laser integration times must be used to attain useful S/N. The development of more efficient SHG probes [136] or laser excitation strategies to minimize photodamage [137] could enable this technique to be used to optically record spontaneous electrical activity in larger assemblies of neurons.

RASH microscopy is a very versatile and promising method complementary to existing electrophysiological techniques. For example, multielectrode arrays (MEA) can measure the single unit activity from large populations of neurons (e.g., PC in cerebellar cortex [138, 139]), but the electrodes are placed in fixed positions and spaced by distances usually larger than $\approx 50\ \mu$m. This constraint can be overcome by RASH microscopy, which can detect the activity of adjacent neurons located at any position in a chosen field of view. While this field of view may be sufficient to investigate characteristics of local networks, a combination of RASH microscopy and MEA techniques would provide a more complete description of global network dynamics.

Although the 2-D RA excitation scheme used here was appropriate for the well-defined planar geometry of the cerebellum, RASH microscopy could easily be implemented with a 3-D RA scanning system [140] to record the electrical activity of more complex, three-dimensional networks in intact brain slices.

There is also a significant potential to combine RASH microscopy with laser stimulation methods to determine the connectivity of neuronal networks. For example, the approach of using patterned laser uncaging of neurotransmitters [141] or two-photon stimulation of neurons [142] would be ideal complements to multi-unit recording with RASH microscopy. Similarly, the optical control of neural activity with opsin-based genetic tools [143], would be an excellent application with RASH to rapidly interrogate the connectivity of neuronal circuitry.

20.6 Conclusion

This chapter has provided a detailed description of nonlinear laser imaging techniques used to image biological tissues. Starting from a theoretical approach, giving a strong theoretical background of the described techniques, we have shown several modern applications involving nonlinear laser scanning imaging techniques.

In particular, we have shown that a combination of two-photon fluorescence (TPF) and second-harmonic generation (SHG) microscopy is a powerful tool to image biological tissues without the needing of any exogenous probe. The combination of this two techniques is able to deeply highlight tissue morphology with subcellular spatial resolution. Combined TPF-SHG have been used in skin tissue imaging, bladder mucosa imaging, and in laser-irradiated cornea imaging. These are only three particular examples of tissues, chosen to show the capabilities of the nonlinear imaging approach on tissues. Imaging can be easily extended to other kinds of tissue, taking advantage of the NADH autofluorescence and of second-harmonic generated by collagen to image cellular and connective tissues, respectively.

Morphology is not the only feature we have been able to highlight by using nonlinear laser imaging. A chemical information can be revealed by means of time-resolved detection of tissue fluorescence. We have seen FLIM as a good characterization tool for discerning different epidermal layers, as well as for discriminating between healthy skin tissue and Basal Cell Carcinoma. We have also shown that FLIM is particularly powerful in highlighting skin cancer margins, when coupled with the topical administration of protoporphyrines precursors as such Aminolevulinic Acid. Due to its environment-sensitive capability, FLIM imaging and characterization can be used to spectroscopically characterize different types of biological tissues, including tumors and noncancer lesions and diseases.

Moreover, nonlinear imaging can be used not only to acquire images and detect a morphological information, but also to determine a particular functionality of a tissue. In the last part of this chapter we have demonstrated the capabilities of both TPE and SHG in providing morpho-functional brain imaging in living mice and in intact slices.

We have shown that TPF is useful to detect single spines inside brain in living mice, as well as to induce microlesions and detect the effects on the brain plasticity. The combination of multiphoton nanosurgery and *in vivo* imaging represents a promising tool for probing and disrupting neuronal circuits. The potential of using this precise optical method to perturb individual synapses cannot be overstated. Using multiphoton nanosurgery, the synaptic organization of the brain can now be teased apart *in vivo* to understand the microcircuitry of neuronal networks. We also envisage that this technique will provide insights in the mechanisms and dynamics of morphological remodelling in the brain following focal ischemic lesions. In addition, the response of the surrounding cells to this specific perturbation can be studied as a model of the reactive and degenerative changes in microstructure observed in certain neurodegenerative diseases. Finally, the combination of nanosurgery and *in vivo* imaging is an exciting tool for neuropharmacology research, because it can be used to evaluate the efficacy of systemic drugs on focal damage to the nervous system *in vivo*.

RASH microscopy provides a fast and noninvasive approach to measure APs in neuronal assemblies. The reported results have shown the strength of this technique in describing the temporal dynamics of neuronal assemblies, opening promising perspectives in understanding the computations of neuronal networks. There is also a significant potential to combine RASH microscopy with laser stimulation methods to determine the connectivity of neuronal networks. For example, the approach of using patterned laser uncaging of neurotransmitters or two-photon stimulation of neurons would be ideal complements to multiunit recording with RASH microscopy. Similarly, the optical control of neural activity with opsin-based genetic tools, would be an excellent application with RASH to rapidly interrogate the connectivity of neuronal circuitry.

Acknowledgment

The research projects presented in this chapter have been supported by the European Community's Sixth Framework Program (Marie Curie Transfer of Knowledge Fellowship MTKD-CT-2004-BICAL-509761), by "Consorzio Nazionale Interuniversitario per le Scienze Fisiche della Materia" (CNISM), by "Ministero dell'Università e della Ricerca" (MIUR), by "Ente Cassa di Risparmio di Firenze" (private foundation), and by "Agenzia Spaziale Italiana" (ASI).

The authors thank all the collaborators who have been involved in the research projects. In particular, we thank Dr. Despoina Stampouli for her collaboration in the skin imaging project, Prof. Daniela Massi and Dr. Enza Maio (Department of Human Pathology and Oncology, University of Florence) for having provided histological samples and for having performed histology; Prof. Torello Lotti, Prof. Vincenzo De Giorgi and Dr. Serena Sestini (Department of Dermatology, University of Florence) for having provided skin samples; Prof. Marco Carini, Dr. Alfonso Crisci, Dr. Gabriella Nesi, and Dr. Saverio Giancane (Division of Urology, University of Florence) for having provided bladder samples; Prof. Roberto Pini and Dr. Paolo Matteini (Consiglio Nazionale delle Ricerche CNR, Florence) for having provided samples of cornea.

References

[1] M. Göppert-Mayer, "Uber elementarekte mit zwei Quantensprunger," *Ann. Phys.* **9**, 273–295 (1931).

[2] W. Denk, H.J. Strickler, and W.W. Webb, "Two-photon laser scanning fluorescence microscope," *Science* **248**, 73–76 (1990).

[3] F. Helmchen and W. Denk, "Deep tissue two-photon microscopy," *Nat. Methods* **2**, 932–940 (2005).

[4] K. König, "Multiphoton microscopy in life sciences," *J. Microsc.* **200**, 83–104 (2000).

[5] R.M. Williams, W.R. Zipfel, and W.W. Webb, "Multiphoton microscopy in biological research," *Curr. Opin. Chem. Biol.* **5**, 603–608 (2001).

[6] W.R. Zipfel, R.M. Williams, and W.W. Webb, "Nonlinear magic: multiphoton microscopy in the biosciences," *Nat. Biotechnol.* **21**, 1369–1377 (2003).

[7] S. Huang, A.A. Heikal, and W.W. Webb, "Two-photon fluorescence spectroscopy and microscopy of NAD(P)H and flavoprotein," *Biophys. J.* **82**, 2811–2825 (2002).

[8] Q. Ruan, Y. Chen, E. Gratton, M. Glaser, and W.W. Mantulin, "Cellular characterization of adenylate kinase and its isoform: two-photon excitation fluorescence imaging and fluorescence correlation spectroscopy," *Biophys. J.* **83**, 3177–3187 (2002).

[9] A. Volkmer, V. Subramaniam, D.J.S. Birch, and T.M. Jovin, "One and two-photon excited fluorescence lifetimes and anisotropy decays of green fluorescent proteins," *Biophys. J.* **78**, 1589–1598 (2000).

[10] C. Xu, W. Zipfel, J.B. Shear, R.M. Williams, and W.W. Webb, "Multiphoton fluorescence excitation: new spectral windows for biological nonlinear microscopy," *Proc. Natl. Acad. Sci. USA* **93**, 10763–10768 (1996).

[11] M. Rubart, E. Wang, K.W. Dunn, and L.J. Field, "Two-photon molecular excitation imaging of Ca^{2+} transients in Langendorff-perfused mouse hearts," *Am. J. Physiol. Cell Physiol.* **284**, C1654–C1668 (2003).

[12] M. Oheim, E. Beaurepaire, E. Chaigneau, J. Mertz, and S. Charpak, "Two-photon microscopy in brain tissue: parameters influencing the imaging depth," *J. Neurosci. Methods* **111**, 29–37 (2001).

[13] T. Ota, H. Fukuyama, Y. Ishihara, H. Tanaka, and T. Takamatsu, "In situ fluorescence imaging of organs through compact scanning head for confocal laser microscopy," *J. Biomed. Opt.* **10**, 024010 (2005).

[14] G.A. Tanner, R.M. Sandoval, and K.W. Dunn, "Two-photon in vivo microscopy of sulfonefluorescein secretion in normal and cystic rat kidneys," *Am. J. Physiol. Renal Physiol.* **286**, F152–F160 (2004).

[15] F. Helmchen, D.W. Tank, and W. Denk, "Enhanced two-photon excitation through optical fiber by single-mode propagation in a large core," *Appl. Opt.* **41**, 2930–2934 (2002).

[16] S.P. Tai, M.C. Chan, T.H. Tsai, S.H. Guol, L.J. Chen, and C.K. Sun, "Two-photon fluorescence microscope with a hollow-core photonic crystal fiber," *Opt. Express* **12**, 6122–6128 (2004).

[17] M.T. Myaing, D.J. MacDonald, and X. Li, "Fiber-optic scanning two-photon fluorescence endoscope," *Opt. Lett.* **31**, 1076–1078 (2006).

[18] J.C. Malone, A.F. Hood, T. Conley, J. Nürnberger, L.A. Baldridge, J.L. Clendenon, K.W. Dunn, and C.L. Philips, "Three-dimensional imaging of human skin and mucosa by two-photon laser scanning microscopy," *J. Cutan. Pathol.* **29**, 453–458 (2002).

[19] B.R. Masters, P.T.C. So, and E. Gratton, "Multi photon excitation fluorescence microscopy and spectroscopy of in vivo human skin," *Biophys. J.* **72**, 2405–2412 (1997).

[20] B.R. Masters, P.T.C. So, and E. Gratton, "Optical biopsy of in vivo human skin: multi-photon excitation microscopy," *Lasers Med. Sci.* **13**, 196–203 (1998).

[21] P.T.C. So, H. Kim, and I.E. Kochevar, "Two-Photon deep tissue ex vivo imaging of mouse dermal and subcutaneous structures," *Opt. Express* **3**, 339–351 (1998).

[22] W.R. Zipfel, R.M. Williams, R. Christie, A.Y. Nikitin, B.T. Hyman, and W.W. Webb, "Live tissue intrinsic emission microscopy using multiphoton-excited native fluorescence and second harmonic generation," *Proc. Natl. Acad. Sci. USA* **100**, 7075–7080 (2003).

[23] P.J. Campagnola, M. Wei, A. Lewis, and L.M. Loew, "High-resolution nonlinear optical imaging of live cells by second harmonic generation," *Biophys. J.* **77**, 3341–3349 (1999).

[24] T. Pons, L. Moreaux, O. Mongin, M. Blanchard-Desce, and J. Mertz, "Mechanisms of membrane potential sensing with second-harmonic generation microscopy," *J. Biomed. Opt.* **8**, 428–431 (2003).

[25] A. Zoumi, A.T. Yeh, and B.J. Tromberg, "Imaging cells and extracellular matrix in vivo by using second-harmonic generation and two-photon excited fluorescence," *Proc. Natl. Acad. Sci. USA* **99**, 11014–11019 (2002).

[26] P.J. Campagnola and L.M. Loew, "Second-harmonic imaging microscopy for visualizing biomolecular arrays in cells, tissues and organisms," *Nat. Biotechnol.* **21**, 1356–1360 (2003).

[27] J. Mertz and L. Moreaux, "Second-harmonic generation by focused excitation of inhomogeneously distributed scatterers," *Opt. Comm.* **196**, 325–330 (2001).

[28] L. Moreaux, O. Sandre, S. Charpak, M. Blanchard-Desce, and J. Mertz, "Coherent scattering in multi-harmonic light microscopy," *Biophys. J.* **80**, 1568–1574 (2001).

[29] L. Moreaux, O. Sandre, and J. Mertz, "Membrane imaging by second-harmonic generation microscopy," *J. Opt. Soc. Am. B* **17**, 1685–1694 (2000).

[30] L. Moreaux, O. Sandre, and J. Mertz, "Membrane imaging by second-harmonic generation and two photon microscopy," *Opt. Lett.* **25**, 320–322 (2000).

[31] L. Sacconi, M. D'Amico, F. Vanzi, T. Biagiotti, R. Antolini, M. Olivotto, and F.S. Pavone, "Second harmonic generation sensitivity to transmembrane potential in normal and tumor cells," *J. Biomed. Opt.* **10**, 024014 (2005).

[32] D.A. Dombeck, K.A. Kasischke, H.D. Vishwasrao, M. Ingelsson, B.T. Hyman, and W.W. Webb, "Uniform polarity microtubule assemblies imaged in native brain tissue by second-harmonic generation microscopy," *Proc. Natl. Acad. Sci. USA* **100**, 7081–7086 (2003).

[33] P.J. Campagnola, A.C. Millard, M. Terasaki, P.E. Hoppe, C.J. Malone, and W.A. Mohler, "Three-dimensional high-resolution second-harmonic generation imaging of endogenous structural proteins in biological tissues," *Biophys. J.* **81**, 493–508 (2002).

[34] Y. Guo, P.P. Ho, H. Savage, D. Harris, P. Sacks, S. Schantz, F. Liu, N. Zhadin, and R.R. Alfano, "Second-harmonic tomography of tissues," *Opt. Lett.* **22**, 1323–1325 (1997).

[35] M.J. Buehler, "Nature designs tough collagen: explaining the nanostructure of collagen fibrils," *Proc. Natl. Acad. Sci. USA* **103**, 12285–12290 (2006).

[36] S. Roth and I. Freund, "Second harmonic generation in collagen," *J. Chem. Phys.* **70**, 1637–1643 (1979).

[37] R.M. Williams, W.R. Zipfel, and W.W. Webb, "Interpreting second-harmonic generation images of collagen I fibrils," *Biophys. J.* **88**, 1377–1386 (2005).

[38] P. Stoller, K.M. Reiser, P.M. Celliers, and A.M. Rubenchik, "Polarization-modulated second harmonic generation in collagen," *Biophys. J.* **82**, 3330–3342 (2002).

[39] Y. Sun, W.L. Chen, S.J. Lin, S.H. Jee, Y.F. Chen, L.C. Lin, P.T.C. So, and C.Y. Dong, "Investigating mechanisms of collagen thermal denaturation by high resolution second-harmonic generation imaging," *Biophys. J.* **91**, 2620–2625 (2006).

[40] T. Yasui, Y. Tohno, and T. Araki, "Characterization of collagen orientation in human dermis by two-dimensional second-harmonic-generation polarimetry," *J. Biomed. Opt.* **9**, 259–264 (2004).

[41] M. Han, G. Giese, and J.F. Bille, "Second harmonic generation imaging of collagen fibrils in cornea and sclera," *Opt. Express* **13**, 5791–5797 (2005).

[42] E. Brown, T. McKee, E. DiTomaso, A. Pluen, B. Seed, Y. Boucher, and R.K. Jain, "Dynamic imaging of collagen and its modulation in tumors in vivo using second-harmonic generation," *Nat. Med.* **9**, 796–800 (2003).

[43] Y. Guo, H.E. Savage, F. Liu, S.P. Schantz, P.P. Ho, and R.R. Alfano, "Subsurface tumor progression investigated by noninvasive optical second harmonic tomography," *Proc. Natl. Acad. Sci. USA* **96**, 10854–10856 (1999).

[44] S.J. Lin, S.H. Jee, C.J. Kuo, R.J. Wu, W.C. Lin, J.S. Chen, Y.H. Liao, C.J. Hsu, T.F. Tsai, Y.F. Chen, and C.Y. Dong, "Discrimination of basal cell carcinoma from normal dermal stroma by quantitative multiphoton imaging," *Opt. Lett.* **31**, 2756–2758 (2006).

[45] P.P. Provenzano, K.W. Eliceiri, J.M. Campbell, D.R. Inman, J.G. White, and P.J. Keely, "Collagen reorganization at the tumor-stromal interface facilitates local invasion," *BMC Med.* **4**(1), 38–53 (2006).

[46] P.I.H. Bastiaens and A. Squire, "Fluorescence lifetime imaging microscopy: spatial resolution of biochemical processes in the cell," *Trends in Cell. Biol.* **9**, 48–52 (1999).

[47] W. Becker, A. Bergmann, C. Biskup, L. Kelbauskas, T. Zimmer, N. Klöcker, and K. Benndorf, "High resolution TCSPC lifetime imaging," *Proc. SPIE* **4963**, 175–184 (2003).

[48] W. Becker, A. Bergmann, C. Biskup, T. Zimmer, N. Klöcker, and K. Benndorf, "Multi-wavelength TCSPC lifetime imaging," *Proc. SPIE* **4620**, 79–84 (2002).

[49] Y. Chen and A. Periasamy, "Characterization of two-photon excitation fluorescence lifetime imaging microscopy for protein localization," *Microsc. Res. Tech.* **63**, 72–80 (2004).

[50] F.G. Cremazy, E.M. Manders, P.I. Bastiaens, G. Kramer, G.L. Hager, E.B.V. Munster, P.J. Verschure, T.J.G. Jr, and R.V. Driel, "Imaging in situ protein-DNA interactions in the cell nucleus using FRET-FLIM," *Exp. Cell Res.* **309**, 390–396 (2005).

[51] R.R. Duncan, A. Bergmann, M.A. Cousins, D.K. Apps, and M.J. Shipston, "Multi-dimensional time-correlated single photon counting (TCSPC) fluorescence lifetime imaging microscopy (FLIM) to detect FRET in cells," *J. Microsc.* **215**, 1–12 (2004).

[52] S.Y. Breusegem, M. Levi, and N.P. Barry, "Fluorescence correlation spectroscopy and fluorescence lifetime imaging microscopy," *Nephron Exp. Nephrol.* **103**, e41–e49 (2006).

[53] K. Knig and I. Riemann, "High-resolution multiphoton tomography of human skin with subcellular spatial resolution and picosecond time resolution," *J. Biomed. Opt.* **8**, 432–439 (2003).

[54] K.M. Hanson, M.J. Behne, N.P. Barry, T.H. Mauro, E. Gratton, and R.M. Clegg, "Two-photon fluorescence lifetime imaging of the skin stratum corneum pH gradient," *Biophys. J.* **83**, 1682–1690 (2002).

[55] C.D. Wilms, H. Schmidt, and J. Eilers, "Quantitative two-photon Ca^{2+} imaging via fluorescence lifetime analysis," *Cell Calcium* **40**, 73–79 (2006).

[56] W. Kaiser and C.G.B. Garret, "Two-photon excitation in CaF_2:Eu^{2+}," *Phys. Rev. Lett.* **7**, 229–231 (1961).

[57] J.N. Gannaway and C.J.R. Sheppard, "Second-harmonic imaging in the scanning optical microscope," *Opt. Quant. Elect.* **10**, 435–439 (1978).

[58] R. Hellwarth and P. Christensen, "Nonlinear optical microscopic examination of structure in polycristalline ZnSe," *Optics Comm.* **12**, 318–322 (1974).

[59] C.J.R. Sheppard, R. Kompfner, J. Gennaway, and D. Walsh, "Scanning harmonic optical microscope," *IEEE J. Quantum Electron.* **13**, 912–912 (1977).

[60] E.P. Diamandis, "Immunoassays with time-resolved fluorescence spectroscopy: principles and applications," *Clin. Biochem.* **21**, 139–150 (1988).

[61] G.A. Elve, A.F. Chaffotte, H. Roder, and M.E. Goldberg, "Early steps in cytochrome *c* folding probed by time-resolved circular dichroism and fluorescence spectroscopy," *Biochemistry* **31**, 6876–6883 (1992).

[62] J.R. Lakowicz and K.W. Berndt, "Lifetime-selective fluorescence imaging using an rf phase sensitive camera," *Rev. Sci. Instrum.* **62**, 1727–1734 (1991).

[63] J.R. Lakowicz, H. Szmacinski, K. Novaczyk, K.W. Berndt, and M. Johnson, "Fluorescence lifetime imaging," *Anal. Biochem.* **202**, 313–330 (1992).

[64] R. Sanders, A. Draaijer, H.C. Gerritsen, P.M. Houpt, and Y.K. Levine, "Quantitative pH imaging in cells using confocal fluorescence lifetime imaging microscopy," *Anal. Biochem.* **227**, 302–308 (1995).

[65] T. French, P.T.C. So, D.J. Weaver Jr., T. Coelho-Sampaio, E. Gratton, E.W. Voss Jr., and J. Carrero, "Two-photon fluorescence lifetime imaging microscopy of macrophage-mediated antigen processing," *J. Microsc.* **185**, 339–353 (1997).

[66] B.R. Masters and P.T.C. So, "Confocal microscopy and multi-photon excitation microscopy of human skin in vivo," *Opt. Express* **8**, 2 (2001).

[67] M.J. Koehler, K. Knig, P. Elsner, R. Buckle, and M. Kaatz, "In vivo assessment of human skin aging by multiphoton laser scanning tomography," *Opt. Lett.* **31**, 2879–2881 (2006).

[68] S.J. Lin, R.J. Wu, H.Y. Tan, W. Lo, W.C. Lin, T.H. Young, C.J. Hsu, J.S. Chen, S.H. Jee, and C.Y. Dong, "Evaluating cutaneous photoaging by use of multiphoton fluorescence and second-harmonic generation microscopy," *Opt. Lett.* **30**, 2275–2277 (2005).

[69] R. Cicchi, D. Massi, S. Sestini, P. Carli, V. De Giorgi, T. Lotti, and F.S. Pavone, "Multidimensional non-linear laser imaging of Basal Cell Carcinoma," *Opt. Express* **15**(16), 10135–10148 (2007).

[70] R. Cicchi, L. Sacconi, A. Jasaitis, R.P. O'Connor, D. Massi, S. Sestini, V. De Giorgi, T. Lotti, and F.S. Pavone, "Multidimensional custom-made non-linear microscope: from ex-vivo to in-vivo imaging," *Applied Physics B* **92**, 359–365 (2008).

[71] R. Cicchi, S. Sestini, V. De Giorgi, D. Massi, T. Lotti, and F.S. Pavone, "Nonlinear laser imaging of skin lesions," *Journal of Biophotonics* **1**(1), 62–73 (2008).

[72] W.F. Cheong, S.A. Prahl, and A.J. Welch, "A review of the optical properties of biological tissues," *IEEE J. Quantum Electron.* **26**, 2166–2185 (1990).

[73] J.A. Kiernan, *Histological and Histochemical Methods*, 3rd ed. New York: Oxford University Press (1999).

[74] H. Liu, B. Beauvoit, M. Kimura, and B. Chance, "Dependence of tissue optical properties on solute induced changes in refractive index and osmolarity," *J. Biomed. Opt.* **1**, 200–211 (1996).

[75] V.V. Tuchin, *Optical Clearing of Tissues and Blood.* Bellingham: SPIE Press (2006).

[76] V.V. Tuchin, I.L. Maksimova, D.A. Zimnyakov, I.L. Kon, A.H. Mavlutov, and A.A. Mishin, "Light propagation in tissue with controlled optical properties," *J. Biomed. Opt.* **2**, 401–417 (1997).

[77] G. Vargas, E.K. Chan, J.K. Barton, H.G. Rylander, and A.J. Welch, "Use of an agent to reduce scattering in skin," *Lasers Surg. and Med.* **24**, 133–141 (1999).

[78] Y. He and R.K. Wang, "Dynamic optical clearing effect of tissue impregnated with hyperosmotic agents and studied by optical coherence tomography," *J. Biomed. Opt.* **9**, 200–206 (2004).

[79] I.V. Meglinski, A.N. Bashkatov, E.A. Genina, D.Y. Churmakov, and V.V. Tuchin, "The enhancement of confocal images of tissue at bulk optical immersion," *Laser Physics* **13**, 65–69 (2003).

[80] S. Plotnikov, V. Juneja, A.B. Isaacson, W.A. Mohler, and P.J. Campagnola, "Optical clearing for improved contrast in second harmonic generation imaging of skeletal muscle," *Biophys. J.* **90**, 328–339 (2006).

[81] A.T. Yeh and J. Hirshburg, "Molecular interactions of exogenous chemical agents with collagen: implications for tissue optical clearing," *J. Biomed. Opt.* **11**, 014003 (2006).

[82] A.F. Zuluaga, R. Drezek, T. Collier, M.F. R. Lotan, and R. Richards-Kortum, "Contrast agent for confocal microscopy: how simple chemicals affects confocal images of normal and cancer cells in suspension," *J. Biomed. Opt.* **7**, 398–403 (2002).

[83] R. Cicchi, D. Massi, D.D. Sampson, and F.S. Pavone, "Contrast and depth enhancement in two-photon microscopy of human skin ex vivo by use of optical clearing agents," *Opt. Express* **13**, 2337–2344 (2005).

[84] A.K. Dunn, V.P. Wallace, M. Coleno, M.W. Berns, and B.J. Tromberg, "Influence of optical properties on two-photon fluorescence imaging in turbid samples," *Appl. Opt.* **39**, 1194–1201 (2000).

[85] M. Gu, X. Gan, A. Kisteman, and M.G. Xu, "Comparison of penetration depth between two-photon excitation and single-photon excitation in imaging through turbid tissue media," *Appl. Phys. Lett.* **77**, 1551–1553 (2000).

[86] E. Beaurepaire, M. Oheim, and J. Mertz, "Ultra-deep two-photon fluorescence excitation in turbid media," *Opt. Commun.* **188**, 25–29 (2001).

[87] A. Leray and J. Mertz, "Rejection of two-photon fluorescence background in thick tissue by differential aberration imaging," *Opt. Express* **14**, 10565–10573 (2006).

[88] P.N. Marsh, D. Burns, and J.M. Girkin, "Practical implementation of adaptive optics in multiphoton microscopy," *Opt. Express* **11**, 1123–1130 (2003).

[89] G. Vargas, K.F. Chan, S.L. Thomsen, and A.J. Welch, "Use of osmotically active agents to alter optical properties of tissue: effects on the detected fluorescence signal measured through skin," *Lasers Surg. and Med.* **29**, 213–220 (2001).

[90] J. Paoli, M. Smedh, A.-M. Wennberg, and M.B. Ericson, "Multiphoton laser scanning microscopy on non-melanoma skin cancer: morphologic features for future non-invasive diagnostics," *J. Invest. Derm.* **128**, 1248-1255 (2008).

[91] M. El-Far, M. Ghoneim, and E. Ibraheim, "Biodistribution and selective in vivo tumor localization of endogenous porphyrins induced and stimulated by 5-amino-levulinic acid: a newly developed technique," *J. Tumor Mark. Onc.* **5**, 27–34 (1990).

[92] J.C. Kennedy and R.H. Pottier, "Endogenous protoporphyrin IX, a clinically useful photosensitizer for photodynamic therapy," *J. Photochem. Photobiol. B* **14**, 275–292 (1992).

[93] J.C. Kennedy, R.H. Pottier, and D.C. Pross, "Photodynamic therapy with endogenous protoporphyrin IX: basic principles and present clinical experience," *J. Photochem. Photobiol. B* **6**, 143–148 (1990).

[94] S.S. Bottomley and U. Muller-Eberhard, "Pathophysiology of heme synthesis," *Semin. Hematol.* **25**, 282–302 (1988).

[95] P. Hinnen, F.W.M. De Rooij, M.L.F. Van Velthuysen, A. Edixhoven, R. Van Hillegersberg, H.W. Tilanus, J.H.P. Wilson, and P.D. Siersema, "Biochemical basis of 5-aminolaevulinic acid-induced protoporphyrin IX accumulation: a study in patients with (pre)malignant lesions of the oesophagus," *Br. J. Cancer* **78**, 679–682 (1998).

[96] M. Kondo, N. Hirota, T. Takaoka, and M. Kajiwara, "Heme-biosynthetic enzyme activities and porphyrin accumulation in normal liver and hepatoma cell lines of rat," *Cell Biol. Toxicol.* **9**, 95–105 (1993).

[97] L. Leibovici, N.I.L.I. Schoenfeld, H.A. Yehoshua, R. Mamet, R. Rakowski, A. Shindel, and A. Atsmon, "Activity of porphobilinogen deaminase in peripheral blood mononuclear cells of patients with metastatic cancer," *Cancer* **62**, 2297–2300 (1988).

[98] H.A. Dailey and A. Smith, "Differential interaction of porphyrins used in photoradiation therapy with ferrochelatase," *Biochem. J.* **223**, 441–445 (1984).

[99] M.M.H. El-Sharabasy, A.M. El-Waseef, M.M. Hafez, and S.A. Salim, "Porphyrin metabolism in some malignant diseases," *Br. J. Cancer* **65**, 409–412 (1992).

[100] R. Van Hillegersberg, J.W. Van den Berg, W.J. Kort, O.T. Terpstra, and J.H. Wilson, "Selective accumulation of endogenously produced porphyrins in a liver metastasis model in rats," *Gastroenterology* **103**, 647–651 (1992).

[101] H. Heyerdahl, I. Wang, D.L. Liu, R. Berg, S. Andersson-Engels, Q. Peng, J. Moan, S. Svanberg, and K. Svanberg, "Pharmacokinetic studies on 5-aminolevulinic acid-induced protoporphyrin IX accumulation in tumours and normal tissues," *Cancer Lett.* **112**, 225–231 (1997).

[102] J. Regula, A.J. MacRobert, A. Gorchein, G.A. Buonaccorsi, S.M. Thorpe, G.M. Spencer, A.R. Hatfield, and S.G. Bown, "Photosensitisation and photodynamic therapy of oesophageal, duodenal and colorectal tumours using 5 aminolaevulinic acid induced protoporphyrin IX – a pilot study," *Gut* **36**, 67–75 (1995).

[103] M. Kress, T. Meier, R. Steiner, F. Dolp, R. Erdmann, U. Ortmann, and A. Rck, "Time-resolved microspectrofluorometry and fluorescence lifetime imaging of photosensitizers using picosecond pulsed diode lasers in laser scanning microscopes," *J. Biomed. Opt.* **8**, 26–32 (2003).

[104] W.C. Lee, H. Huang, G. Feng, J.R. Sanes, E.N. Brown, P.T. So, and E. Nedivi, "Dynamic remodeling of dendritic arbors in GABAergic interneurons of adult visual cortex," *PLoS Biol.* **4**, e29 (2006).

[105] J.T. Trachtenberg, B.E. Chen, G.W. Knott, G. Feng, J.R. Sanes, E. Welker, and K. Svoboda, "Long-term in vivo imaging of experience-dependent synaptic plasticity in adult cortex," *Nature* **420**, 788–794 (2002).

[106] J.A. Galbraith and M. Terasaki, "Controlled damage in thick specimens by multiphoton excitation," *Mol. Biol. Cell.* **14**, 1808–1817 (2003).

[107] I.M. Tolic-Norrelykke, L. Sacconi, G. Thon, and F.S. Pavone, "Positioning and elongation of the fission yeast spindle by microtubule-based pushing," *Curr. Biol.* **14**, 1181–1186 (2004).

[108] A. Nimmerjahn, F. Kirchhoff, and F. Helmchen, "Resting microglial cells are highly dynamic surveillants of brain parenchyma in vivo," *Science* **308**, 1314–1318 (2005).

[109] N. Nishimura, C.B. Schaffer, B. Friedman, P.S. Tsai, P.D. Lyden, and D. Kleinfeld, "Targeted insult to subsurface cortical blood vessels using ultrashort laser pulses: three models of stroke," *Nat. Methods* **3**, 99–108 (2006).

[110] M.F. Yanik, H. Cinar, H.N. Cinar, A.D. Chisholm, Y. Jin, and A. Ben-Yakar, "Neurosurgery: functional regeneration after laser axotomy," *Nature* **432**, 822–822 (2004).

[111] S.H. Chung, D.A. Clark, C.V. Gabel, E. Mazur, and A.D. Samuel, "The role of the AFD neuron in *C. Elegans* thermotaxis analyzed using femrosecond laser ablation," *BMC Neurosci.* **7**, 30 (2006).

[112] L. Sacconi, R.P. O'Connor, A. Jasaitis, A. Masi, M. Buffelli, and F.S. Pavone, "In vivo multiphoton nanosurgery on cortical neurons," *J. Biomed. Opt.* **12**, 050502 (2007).

[113] L. Sacconi, I.M. Tolic-Norrelykke, R. Antolini, and F.S. Pavone, "Combined intracellular three-dimensional imaging and selective nanosurgery by a nonlinear microscope," *J. Biomed. Opt.* **10**, 014002 (2005).

[114] G.W. Gross and M.L. Higgins, "Cytoplasmic damage gradients in dendrites after transection lesions," *Exp. Brain Res.* **67**, 52–60 (1987).

[115] A. Vogel and V. Venugopalan, "Mechanisms of pulsed laser ablation of biological tissues," *Chem. Rev.* **103**, 577–644 (2003).

[116] J.C. Fiala, J. Spacek, and K.M. Harris, "Dendritic spine pathology: cause or consequence of neurological disorders?," *Brain Research Reviews* **39**, 29–54 (2002).

[117] A. Araque, G. Carmignoto, and P.G. Haydon, "Dynamic signaling between astrocytes and neurons," *Annu. Rev. Physiol.* **63**, 795–813 (2001).

[118] D.A. Dombeck, L. Sacconi, M. Blanchard-Desce, and W.W. Webb, "Optical recording of fast neuronal membrane potential transients in acute mammalian brain slices by second-harmonic generation microscopy," *J. Neurophysiol.* **94**, 3628–3636 (2005).

[119] M. Nuriya, J. Jiang, B. Nemet, K.B. Eisenthal, and R. Yuste, "Imaging membrane potential in dendritic spines," *Proc. Natl. Acad. Sci. USA* **103**, 786–790 (2006).

[120] L. Sacconi, D.A. Dombeck, and W.W. Webb, "Overcoming photodamage in second-harmonic generation microscopy: real-time optical recording of neuronal action potentials," *Proc. Natl. Acad. Sci. USA* **103**, 3124–3129 (2006).

[121] A. Grinvald and R. Hildesheim, "VSDI: a new era in functional imaging of cortical dynamics," *Nat. Rev. Neurosci.* **5**, 874–885 (2004).

[122] M. Zochowski, M. Wachowiak, C.X. Falk, L.B. Cohen, Y.W. Lam, S. Antic, and D. Zecevic, "Imaging membrane potential with voltage-sensitive dyes," *Biol. Bull.* **198**, 1–21 (2000).

[123] S.D. Antic, "Action potentials in basal and oblique dendrites of rat neocortical pyramidal neurons," *J. Physiol.* **550**, 35–50 (2003).

[124] J.E. Gonzalez and R.Y. Tsien, "Improved indicators of cell membrane potential that use fluorescence resonance energy transfer," *Chem. Biol.* **4**, 269–277 (1997).

[125] T. Knopfel, K. Tomita, R. Shimazaki, and R. Sakai, "Optical recording of membrane potential using genetically targeted voltage-sensitive fluorescent proteins," *Methods* **30**, 42–48 (2003).

[126] M.S. Siegel and E.Y. Isacoff, "A genetically encoded optical probe of membrane voltage," *Neuron* **19**, 735-741 (1997).

[127] R.A. Stepnoski, A. LaPorta, F. Raccuia-Behling, G.E. Blonder, R.E. Slusher, and D. Kleinfeld, "Noninvasive detection of changes in membrane potential in cultured neurons by light scattering," *Proc. Natl. Acad. Sci. USA* **88**, 9382–9386 (1991).

[128] D.A. Dombeck, M. Blanchard-Desce, and W.W. Webb, "Optical recording of action potentials with second-harmonic generation microscopy," *J. Neurosci.* **24**, 999–1003 (2004).

[129] R. Salome, Y. Kremer, S. Dieudonne, J.F. Leger, O. Krichevsky, C. Wyart, D. Chatenay, and L. Bourdieu, "Ultrafast random-access scanning in two-photon microscopy using acousto-optic deflectors," *J. Neurosci. Methods* **154**, 161–174 (2006).

[130] V. Iyer, T.M. Hoogland, and P. Saggau, "Fast functional imaging of single neurons using random-access multiphoton (RAMP) microscopy," *J. Neurophysiol.* **95**, 535–545 (2006).

[131] J.C. Eccles, "The cerebellum as a computer: patterns in space and time," *J. Physiol.* **229**, 1–32 (1969).

[132] D. Marr, "A theory of cerebellar cortex," *J. Physiol.* **202**, 437–470 (1969).

[133] L. Sacconi, J. Mapelli, D. Gandolfi, J. Lotti, R.P. O'Connor, E. D'Angelo, and F.S. Pavone, "Optical recording of electrical activity in intact neuronal networks with random access second-harmonic generation microscopy," *Opt. Express* **16**, 14910–14921 (2008).

[134] J.C. Eccles, M. Ito, and J. Szentagothai, *The Cerebellum as a Neuronal Machine*. Springer (1967).

[135] H. Nishiyama and D.J. Linden, "Differential maturation of climbing fiber innervation in cerebellar vermis," *J. Neurosci.* **24**, 3926–3932 (2004).

[136] T.Z. Teisseyre, A.C. Millard, P. Yan, J.P. Wuskell, M.D. Wei, A. Lewis, and L.M. Loew, "Nonlinear optical potentiometric dyes optimized for imaging with 1064-nm light," *J. Biomed. Opt.* **12**, 044001 (2007).

[137] N. Ji, J.C. Magee, and E. Betzig, "High-speed, low-photodamage nonlinear imaging using passive pulse splitter," *Nat. Methods* **5**, 197–202 (2008).

[138] U. Egert, D. Heck, and A. Aertsen, "Two-dimensional monitoring of spiking networks in acute brain slices," *Exp. Brain Res.* **142**, 268–274 (2002).

[139] J. Mapelli and E. D'Angelo, "The spatial organization of long-term synaptic plasticity at the input stage of cerebellum," *J. Neurosci.* **27**, 1285–1296 (2007).

[140] G.D. Reddy and P. Saggau, "Fast three-dimensional laser scanning scheme using acousto-optic deflectors," *J. Biomed. Opt.* **10**, 064038 (2005).

[141] S. Shoham, D.H. O'Connor, and R. Segey, "How silent is the brain: is there a 'dark matter' problem in neuroscience?" *J. Comp. Physiol. A: Neuroethol. Sens. Neural Behav. Physiol.* **192**, 777–784 (2006).

[142] V. Nikolenko, K.E. Poskanzer, and R. Yuste, "Two-photon photostimulation and imaging of neural circuits," *Nat. Methods* **4**, 943–950 (2007).

[143] E.S. Boyden, F. Zhang, E. Bamberg, G. Nagel, and K. Deisseroth, "Millisecond-timescale, genetically targeted optical control of neural activity," *Nat. Neurosci.* **8**, 1263–1268 (2005).

21

Endomicroscopy Technologies for High-Resolution Nonlinear Optical Imaging and Optical Coherence Tomography

Yicong Wu and Xingde Li

Department of Biomedical Engineering, Johns Hopkins University, Baltimore, MD, 21205 U.S.A.

Recent advances in fiber-optics, micro-optics and miniaturized optical and/or mechanical scanners have promoted rapid development and clinical translation of nonlinear optical (e.g., two-photon fluorescence and second harmonic generation) microscopy and optical coherence tomography (OCT), enabling depth-resolved endomicroscopic imaging of internal organs with unprecedented resolution. This paper presents a review of nonlinear optical imaging and OCT endomicroscopy technologies with emphasis on their major building blocks, including the commonly used 2D/3D beam scanning mechanisms and miniature objective lenses. The advantages and challenges associated with various endomicroscopes are also discussed. Special designs and engineering considerations are presented for a few representative endomicroscopes, including an all-fiber-optic rapid scanning nonlinear optical imaging endomicroscope, a high-resolution OCT balloon imaging catheter, and an ultrathin OCT imaging needle. Some exemplary imaging results achieved with these endomicroscopy technologies are presented, demonstrating the promising role of these emerging endomicroscopy technologies for basic laboratory research and for translational early disease detection and image-guided interventions.

Key words: endomicroscopy; nonlinear optical imaging; two-photon fluorescence; second harmonic generation; optical coherence tomography; micro-optics; fiber-optics; MEMS scanning mirrors; micro-lens; micro-motor; focus tracking; resonant fiber scanners; double-clad fiber; photonic crystal fiber; OCT balloon catheter; OCT imaging needle; internal organ imaging

21.1 Introduction

Endomicroscopy is a vital medical procedure for diagnosis, guidance of biopsy, and surgery and has experienced more than 200 years of innovation [1–3]. Endoscopic magnetic resonance imaging (MRI) and endoscopic ultrasonography (EUS) can sample a large tissue volume with a resolution in the order of several hundreds of microns [4–7]. The diameter of these endoscopes is typical around 1–10 mm. Reduction in the size of the MRI coil or ultrasonic probe can deteriorate the resolution and sensitivity for a given magnetic field strength or the ultrasonic operating frequency. In comparison, optical endoscopy, which was first introduced by Philip Bozzini in 1806 [1], can offer a higher sensitivity as well as higher spatial resolution. Based on tissue optical scattering and absorption properties, conventional white-light or narrowband optical endomicroscopes produce magnified images of tissue surface [8–10]. New advances in optical endomicroscopy are going beyond white light, and take full advantage of light's properties. For example, the addition of intrinsic fluorescence or Raman detection to conventional endomicroscopes provides functional assessment of biological tissues [11, 12]. Furthermore, the use of nonradioactive exogenous optical contrast agents (fluorescent, absorbing and/or scattering) can enhance the sensitivity for targeted imaging of regions of interest [13, 14]. However, these optical endomicroscopes do not provide depth-resolved information about the tissue structure and function. It has long been recognized that endomicroscopy with depth-resolving capability can possibly perform "optical biopsy" *in situ* but without the need for tissue removal, and provide better diagnostic information.

Confocal microscopy, nonlinear optical microscopy (such as multiphoton fluorescence and higher harmonic generation microscopy), and optical coherence tomography (OCT) are recently developed high-resolution imaging technologies capable of optical sectioning and three-dimensional assessment of biological samples [15–24]. The confocal technique utilizes a pinhole in front of the photo detector to reject multiply scattered photons, which otherwise would obscure the images [15]. The confocal imaging depth is generally about a few hundreds of microns. Nonlinear optical microscopy relies on the high-order light-matter interaction processes including multiphoton absorption, higher harmonic generation, and coherent anti-Stokes Raman scattering (CARS) [17, 21, 25], where the high-order dependence of the interaction cross-section on the incident light intensity results in the inherent ability of optical sectioning without the use of a spatial filter or a pinhole. Different from confocal microscopy, nonlinear microscopy permits the use of near-infrared light and wide-field detection. Again, the multiple scattering in most biological tissues quickly degrades the tight beam focus, reducing the nonlinear interaction cross-section and resulting in a practical imaging depth to about a few hundred microns. Among the nonlinear optical imaging modalities, two-photon fluorescence (TPF) and second harmonic generation (SHG) are commonly used in tandem to provide complementary information on tissue morphology and physiology [26, 27]. Different from confocal and nonlinear optical techniques, OCT employs low coherence light and interferometry techniques to provide optical sectioning or optical ranging [23, 24]. Similar to confocal microscopy, the image contrast in OCT arises from the optical properties (scattering and absorption) of biological tissues (and generally dominated by scattering). The use of low coherence gating permits more effective rejection of multiply scattered photons compared to confocal, and thus enabling deeper imaging depth up to 1–3 mm when near infrared light (800–1300 nm) is used.

Confocal, nonlinear microscopy and OCT have become powerful tools for 3D high-resolution imaging of biological tissues. The superb spatial resolution and sensitivity to structurally, physiologically or biochemically relevant information (such as blood flow, NADH, FAD, collagen, etc.) suggest their great potential for early disease diagnosis and image-guided interventions. However, the 0.2–3 mm imaging depth limits their noninvasive clinical applications only to skin with readily available bulk imaging devices (such as a microscope). For imaging of internal luminal organs (such as the epithelial linings where many cancers originate).[1], new types of miniature probes (termed as endomicroscope) are required to perform confocal, nonlinear optical, and OCT imaging. It is envisioned that such a probe can be integrated with standard clinical endoscopes and endomicroscopic confocal, nonlinear, or OCT imaging can be performed during routine clinical endoscopy procedures. Development of miniature probes for confocal, nonlinear or OCT imaging faces many technical challenges. Such a probe needs to integrate all the essential functions of a scanning microscope into a compact size, including light delivery, signal collection, fast 2D or 3D beam scanning and focusing. With the advances in micro-optics and micro-electromechanical technologies, the past decade, in particular the past several years, have witnessed explosive development in the field of endomicroscopy [30–35].

Various prototype confocal endomicroscopes have been developed for three-dimensional imaging of biological tissues that have been detailed in many publications [32, 36–41]. In this paper, we will only focus on the development of nonlinear optical endomicroscopy and endoscopic OCT technologies and survey their applications toward *in vivo* imaging. There is a nearly universal feature in these two imaging endomicroscopy systems, i.e., a scanning mechanism moving the focal spot across the field of view to form a two-dimensional (2D) or three-dimensional (3D) image, and a beam focusing mechanism allowing the adjustment of the imaging plane for depth-resolved imaging in nonlinear optical endomicroscopy or focus tracking in OCT endomicroscopy. Therefore, we will first discuss the common technical issues in achieving 2D and 3D imaging in both nonlinear and OCT endomicroscopy. Other considerations specific to each imaging technology will then be presented such as special optic fibers, objective lenses, and system assembling in a nonlinear optical endomicroscope and OCT endomicroscope. Finally, some representative technological embodiments and their applications will be illustrated.

21.2 Beam Scanning and Focusing Mechanisms in Endomicroscopes

According to the imaging beam direction with respect to the longitudinal axis of endomicroscopes, endomicroscopic probes can be divided into two groups, forward-viewing probes and side-viewing probes (see Figure 21.1). Forward-viewing endomicroscopes emit and collect light in front of the probes, whereas side-viewing endomicroscopes emit and collect light from the side of the probes by using small reflectors. The additional optical path in side-viewing probes generally requires different beam scanning and focusing configurations compared to those in forward-viewing probes. In the following sections, we will respectively discuss the scanning and focusing mechanisms in these two kinds of probes.

21.2.1 Mechanical scanning in side-viewing endomicroscopes

In side-viewing endomicroscopes such as conventional OCT catheters, one easy way to achieve beam scanning is by either circumferential rotation or linear translation of the whole probe with an actuation device situated at the proximal end of the probe. As shown in Figure 21.2(a), a rotational fiber-optic junction (or rotary joint) is needed to couple the stationary fiber from the light source and the rotating endoscope by using a pair of lenses or a capillary coupler [42]. By scanning the rotary joint with a DC motor at the proximal end, the rotary torque can be transferred to the dis-

[1]It should be noted that some of the optical endomicroscopy technologies can also be applied for intravascular imaging of vulnerable plaques [28, 29] But this review paper will only focus on endomicroscopy applications in other internal luminal organs such as the gastrointestinal tract, respiratory tract, oral cavity and bladder etc.

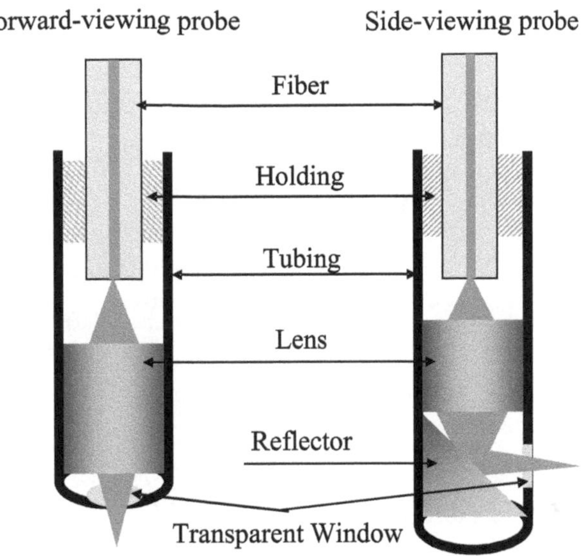

FIGURE 21.1: The schematic of forward-viewing and side-viewing endomicroscopes.

tal end through a hollow metal coil (e.g.,, speedometer cable) [42, 43]. The advantages of this type side-viewing scanning endoscope include extremely compact size and simple probe assembling procedures. However, beam intensity coupling through the rotary joint and polarization may fluctuate during rotation that will result in image artifacts. To avoid rotating the entire endoscope and the above-mentioned potential artifacts, another type of side-viewing probe has been developed by employing micro-motors at the distal end to scan the beam (as opposed to the endoscope) [44, 45]. The associated disadvantage of this type of distal scanning endoscope is the relatively large probe size that may complicate the delivery through the Y-shape entrance of a standard gastrointestinal (GI) endoscope, and the beam blockage caused by the motor drive wires.

21.2.2 Scanning mechanisms in forward-viewing endomicroscopes

In comparison with side-viewing endomicroscopes, there are a variety of scanning methods in forward-viewing endomicroscopes, including (but limited to) proximal-end beam scanning when an imaging fiber bundle is used in the probe, distal end beam scanning by MEMS micromirrors, microlens stack and resonance fiber-optic scanners. It should be noted that some of these scanning mechanisms can also be implemented in a side-viewing configuration. In the following section, we will discuss these scanning methods in details.

21.2.2.1 Proximal end beam scanning with an imaging fiber bundle

In forward-viewing imaging endomicroscopes, scanning can be realized at the proximal end of the probes where an imaging (or coherent) fiber bundle is used for illumination, as shown in Figure 21.2(b). A fiber bundle generally consists of up to $\sim$100,000 individual single fibers closely packed within a diameter of several hundred microns to a few millimeters. The coherent fiber bundle functions as an image conduit and relays an intensity image of the sample from its distal to proximal end, where conventional illumination and detection schemes as in confocal microscopy can be applied [37, 46–50]. Typically, bulk galvanometer-based scanning mirrors are used to achieve high-speed imaging at the proximal end of the imaging fiber bundle up to 15 frames/s [51]. The use of a co-

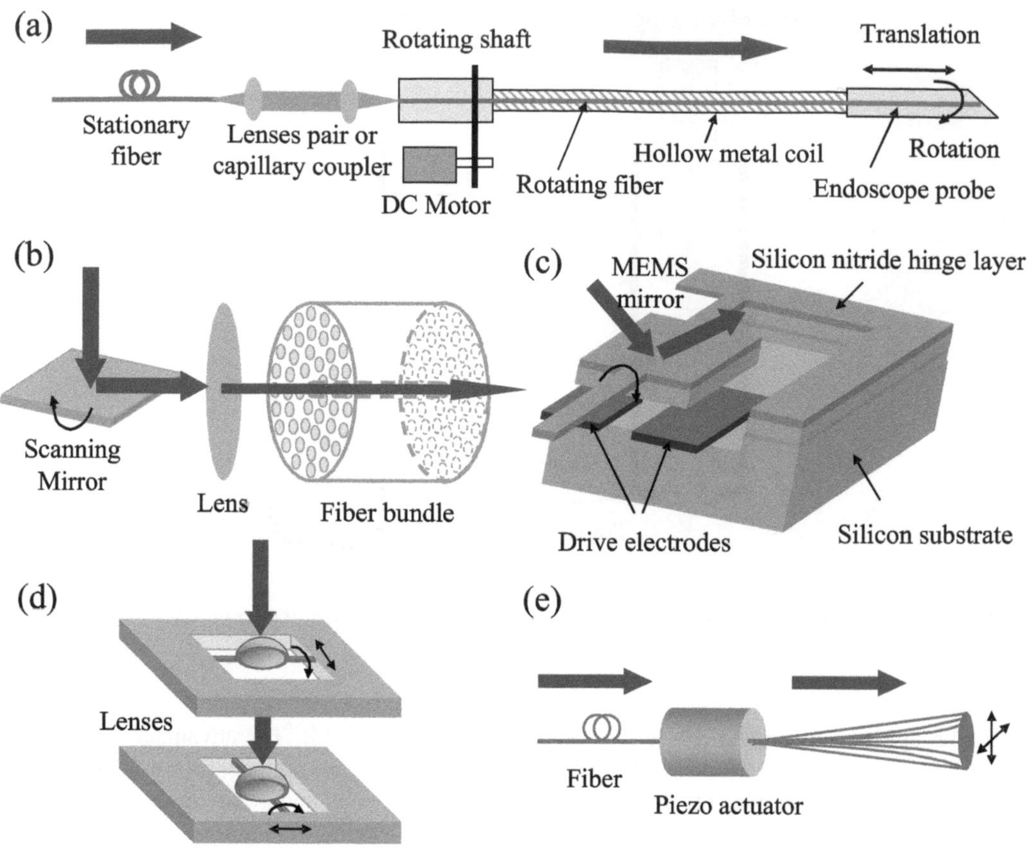

FIGURE 21.2: The schematic of several commonly used scanning mechanisms in miniature endomicroscopes: (a) Mechanical scanning; (b) Imaging fiber bundle; (c) MEMS scanning mirror; (d) Stack of micro-lenses or GRIN lenses; (e) Resonant fiber-optic scanner.

herent imaging fiber bundle eliminates the need for integrating the scanners to the distal end of the probe, facilitating dramatic imaging probe miniaturization. The fiber bundle naturally permits wide-field collection as an extra bonus for nonlinear optical imaging. However, there are several major challenges when using coherent fiber bundles, including (1) image pixilation and limited lateral resolution caused by the nonnegligible dead packing space between adjacent fibers and the finite size of individual fiber; (2) optical cross-talk between adjacent fibers (excitation cross-talk in nonlinear optical imaging and both incident and back-reflection cross-talk in OCT imaging) causing image blurring; and (3) the limited light coupling and transmission in individual fiber due to the small fiber diameter. Incoherent fiber bundles have been investigated to reduce the cross-talk between neighboring pixels as long as the relative positions of individual fibers are traceable and reconstructible [37]. The leakage can also be minimized by using a spatial light modulator to provide sequential illumination for each fiber [47].

Nevertheless, to eliminate the pixilation artifact and optical cross-talk, a single fiber for both light delivery and signal collection has become a more attractive alternative. The use of a single fiber requires a distal scanning mechanism in a compact configuration for a forward-viewing endomicroscope. Several methods have been developed for the distal scanning, e.g., MEMS mciromirrors, microlens stack and resonance fiber-optic scanners, as shown in Figures 21.2(c–e), respectively. Table 21.1 summarizes performance characteristics of these different scanning methods that are

TABLE 21.1: Micro-scanning mechanisms.

Scanning mechanisms	Mirror-based scanners	Lens-based scanners	Fiber-based scanners
Technology	MEMS micro-mirror	MEMS micro-lens, Angle-cut GRIN lens	Resonant fiber
Actuation	Electrostatic, Piezoelectric, Magnetic, Thermal	Electrostatic, Magnetic, Proximal rotation	Piezoelectric
2D scan	Two cascaded 1-D mirrors or a 2D mirror	Stacked two lenses	Orthogonal driving, Lissajous or spiral scan
Key factors	Driving force, Mirror dimensions	Driving force, Lens dimensions	Driving force, Free-stand fiber length
Size	0.5–2 mm plate	0.5–3 mm plate	0.5–1.5 cm fiber, 1–2 mm PZT
Frequency	100–10 kHz	200–4.5 kHz	300–1.5 kHz
Range	Up to 30° rotation angle	Up to 38° rotation angle	Up to 40° vibration angle
Advantages	High integration, Low power consumption	Easy alignment, No extra focus lens	Easy fabrication, Easy assembly
Limitations	Complicated fabrication, Reflection loss	Chromatic aberration, Complicated design	Image distortion, Calibration required

feasible for use in a miniature imaging probe setting. In the following sub-sections we will discuss these methods in details.

21.2.2.2 MEMS scanning micromirrors

Similar to a pair of galvanometer scanning mirrors used in a standard laser scanning microscope, micro-electromechanical system (MEMS) scanning mirrors have been used to scan the imaging beam at the distal end of both nonlinear optical endomicroscopes and OCT endomicroscopes [52–60]. Various micro-fabricated torsional mirrors for optical beam deflection have been reported [61–64]. Silicon-based microfabrication technologies enable high degree of integration of scanning mirrors, torsional hinges, a supporting substrate and control circuits on the same chip (Figure 21.2(c)). The use of electrostatic actuation, in particular those with a comb drive structure, permits low power consumption and strong actuation force [65]. A wide range of frequency response from 100 Hz to 10 kHz can be achieved with MEMS. Typical micro-mirrors with a 0.5–2 mm diameter can have a mechanical scanning angle up to $\sim 30^\circ$ with reasonably low driving voltages ($\sim$10–120 V) [66, 67]. Many efforts have been concentrated on increasing the scan frequency, scan angle and robustness of MEMS scanners, while others on reducing and calibrating the highly nonlinear sweep of the mirror due to the fact that the electrostatic force is proportional to the square of the drive voltage and dependent on the deflection angle [68].

Two cascaded 1D micro-scanners can be arranged to form a 2D raster scanner. 2D scanning mirrors can also be micro-fabricated on a single silicon plate [69]. The advantage of two cascaded 1D micro-scanners is that the orthogonal scanning can be independently controlled, but the light spot might walk off the second mirror when the scan angle of the first mirror is large. Thus, a single 2D scanner is more attractive. Various two-axis mirrors have been developed based on the electrothermal and electrostatic vertical comb actuators [54, 70, 71]. Feasibilities of 1D and 2D MEMS scanners have been demonstrated in nonlinear and OCT endomicroscopes [52, 58–60, 64, 72]. Overall, MEMS scanners have a great potential to be integrated in a compact endomicroscope yet the relatively large substrates with the drive circuits still present some engineering challenges in their endomicroscopic applications.

21.2.2.3 Stack of scanning micro-lenses or GRIN lenses

In contrast to reflective scanning micro-mirrors that use reflection to achieve spatial beam scanning, imaging beam scanning can also be achieved by translating the refractive lenses off the optical axis (Figure 21.2(d)). These transmissive micro-lens scanners have a compact package size and alignment advantages, and naturally integrate beam scanning with beam focusing since one of the lenses can be used as the objective lens. The actuation mechanism of the micro-lens scanners are similar to MEMS scanning mirrors, e.g., by using either electrostatic or magnetic force, with a resonant frequency from 550 Hz to 4.5 kHz with a scanning range of 100 μm [73–75]. Recently, a micro-lens based on ferromagnetic nickel platform and thermosetting elastomer polydimethylsiloxane (PDMS) has been demonstrated to achieve an $\sim$125 μm lateral scan range when actuated by an external magnetic field of 22.2×10^{-3} Tesla [76], eliminating the need for a high voltage drive. Considering the imaging beam has to go off axes of the micro-lenses, it remains technically challenging to design high quality micro-lenses which have to be conveniently integrated to the MEMS substrate.

In addition to scanning micro-lens stacks, a pair of angle-cut rotating gradient-index (GRIN) lenses can also be used to deflect and scan the optical beam across the specimen [77], with the same principle as a pair of rotating wedges [78, 79]. Light from a single-mode fiber enters the flat side of the first GRIN lens and exits from the other side with an angle cut. The light then enters the second GRIN lens through an identical angle-cut surface, and finally exits from the flat surface of the second GRIN lens and focuses on the sample. A fan sweep beam scanning pattern is generated when the two GRIN lenses are rotated in opposite directions at the same angular speed. A prototype miniature probe with a 1.65 mm diameter based on a pair of scanning wedge lenses has been successfully demonstrated for OCT imaging [77]. Accurate beam scanning requires excellent mechanical stability and special engineering attention for the two optical axes of the two rotating lenses.

21.2.2.4 Resonant fiber scanners

Although the two micro-scanners mentioned earlier can achieve a large lateral beam scan at a high speed, they require complicated fabrication processes and control mechanisms. A simpler method has been recently developed by scanning the distal tip of an optical fiber cantilever at its mechanical resonant frequency with a piezoelectric actuator (Figure 21.2(e)) [80–84]. Compared to a magnetically actuated forward-imaging probe [85, 86], the resonant fiber-optic scanner is more compact with a much higher scanning speed. Small piezo vibration at the cantilever base (on the order of a few microns) can be dramatically amplified at the sweeping fiber tip when the piezo is driven at the mechanical resonance frequency of the fiber cantilever, and a mechanical scanning angle larger than $\pm$ 20 degrees can be conveniently achieved.

One type of resonant fiber-optic piezoelectric scanners is based on tubular piezoelectric actuators [82, 84]. The cylindrical symmetry makes this type of actuators very suitable for endomicroscopic settings. The outer surface of the PZT tube is divided into four quadrants, forming two pairs of the drive electrodes. The optical fiber cantilever is glued to the actuator tip. One-dimensional resonant line scan is achieved when only one pair of electrodes are driven with a sinusoidal waveform at the first mechanical resonant frequency of the cantilever, and the frequency is given by [87]

$$f_r = \frac{\beta}{4\pi}\left(\frac{E}{\rho}\right)^{1/2}\left(\frac{R}{L^2}\right) \tag{21.1}$$

Here f_r is the resonant frequency of the cantilever; β is a constant of 3.52 at zeroth-order vibration mode; E and ρ are the Young's modulus and mass density of the cantilever, respectively; R and L are the radius and length of the cantilever, respectively. Eq. (21.1) tells that the beam (i.e., the cantilever tip) scanning frequency is inversely proportional to the square of the cantilever length. For a single-mode fiber cantilever with $E \approx 75$ GPa, $\rho \approx 2.3$ g/cm^3 and $R \approx 62.5$ μm [88], the rule of thumb is that a 10-mm long cantilever corresponds to a resonant scanning frequency of about 1 kHz. The deflection angle of the fiber cantilever (α) is determined by the PZT driving voltage and the driving frequency. The deflection range of the fiber tip is dependent on the length of the fiber, i.e., $2L\sin\alpha$. The sweeping fiber tip is imaged onto the sample by a micro-lens, producing lateral beam scanning. Finally, the lateral beam scanning range, D, is given by

$$D = 2LM\sin\alpha \tag{21.2}$$

where M is the magnification of the lens from the fiber tip to the sample. When using a 8.5-mm long fiber cantilever and a lens with a magnification of $\sim$1.8, an approximately 2.5 mm scanning range on the sample can be achieved with a peak-to-peak drive voltage of $\sim$ 60 V near the resonance frequency of 1.4 kHz [82].

To generate a circular scan, one pair of the electrodes (e.g., the X-pair) will be driven with a sine wave while the other pair (e.g., the Y-pair) driven by a cosine wave with the same amplitude and frequency as the X-drive, i.e., $V_x = V\sin 2\pi f_r t$, $V_y = V\cos 2\pi f_r t$, where V is the maximum drive voltage, and f_r is the resonant frequency of the fiber-optic cantilever. By triangularly modulating the drive voltage, an open-close spiral scanning pattern can be formed (see Figure 21.3(a)). The method has been successfully adopted in a nonlinear optical endomicroscope and OCT endomicroscope for real-time imaging in the forward direction [82, 84].

Another resonant PZT scanner utilizes a bi-morph piezoelectric plate to vibrate a freely-standing fiber-optic cantilever that is fixed on the plate. A short-piece stiffening rod is used to provide extra support for the cantilever along one direction (e.g., X-direction), which produces a mechanical resonance frequency slightly different from the resonance frequency along the Y-direction. A Lissajous scanning pattern by the cantilever tip will be achieved (see Figure 21.3(b)) when the cantilever is excited with the superposition of the two resonance frequencies, f_x and f_y [80]

$$L = V(\sin 2\pi f_x t + \cos 2\pi f_y t) \tag{21.3}$$

The Lissajous pattern is determined by the ratio of the two frequencies. A repetitive pattern can be obtained only if this ratio is a rational number, i.e., $\frac{f_x}{f_y} = \frac{n_x}{n_y}$, where n_x and n_y are the smallest possible integers. The numbers n_x and n_y roughly determines the spatial resolution of the pattern (the density of the intersections of the trajectory), and the pattern repeat frequency, f_R, is given by $f_R = \frac{f_x}{n_x} = \frac{f_y}{n_y}$. Similar to the tubular PZT scanner discussed above, the drive voltage V determines the scanning range of the Lissajous pattern. This resonant fiber scanner can also achieve large lateral deflections of the tip by about 1 mm at a high scanning speed (e.g., for f_x and f_y in the range of 300–800 Hz, n_x, and n_y in the range of $\sim$150–400 and therefore a frame rate of 2 fps) with a reasonably compact form (e.g., 8 mm long $\times$ 2 mm wide $\times$ 0.5 mm thick). It is understood that the nonlinear dynamics of the response of a fiber-optic scanner to its mechanical resonance can cause severe imaging distortion; but such distortion can be remedied by image processing or a nonlinear controller to force the scan to follow a defined scanning pattern [89].

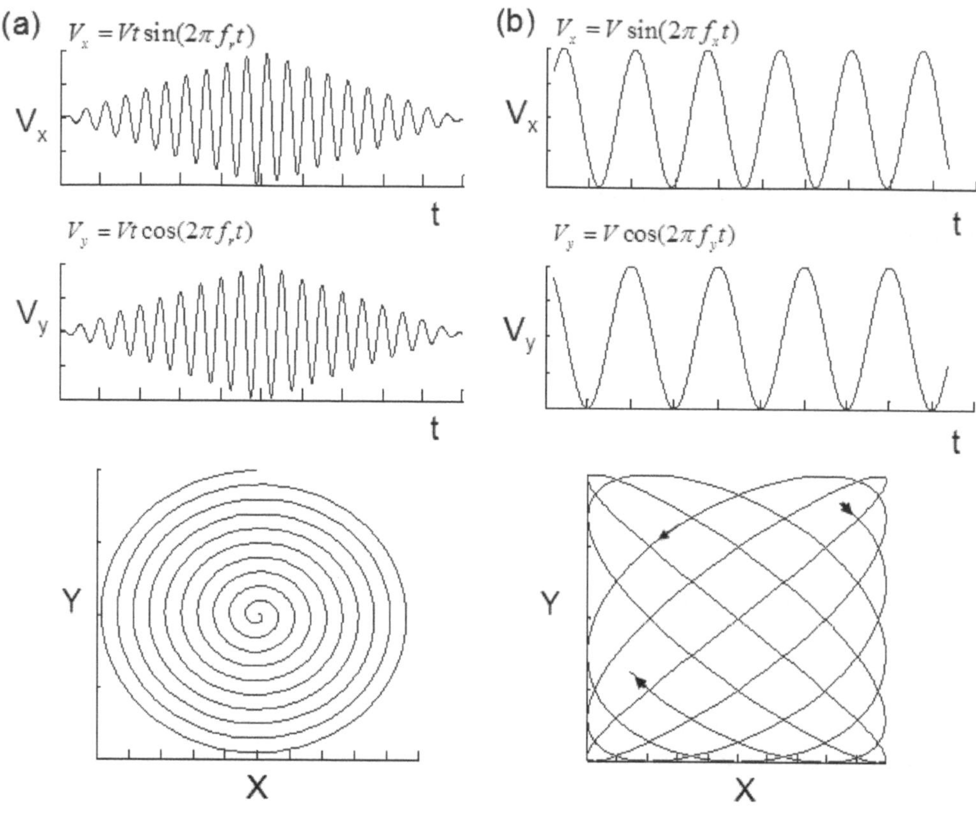

FIGURE 21.3: (a) Drive waveforms for a tubular piezoelectric resonant actuator and the resultant spiral scanning pattern; (b) Drive waveforms for a bi-morph piezoelectric plate and the resultant Lissajous scanning pattern.

21.2.3 Compact objective lens and focusing mechanism

An objective lens is another key component in an endomicroscope. The objective lens determines the lateral and axial resolution (except for OCT where the axial resolution is governed by the spectrum bandwidth of the light source). In side-viewing endomicroscopes, it is challenging to use a high NA and short working-distance objective lens because the additional optical path bent by a micro-reflector generally reduces the effective working distance between the probe and the tissue surface. In comparison, with the forward-viewing configuration, an objective lens with a high NA and short working-distance can be used to obtain ultrahigh resolution without compromising the overall diameter of the probe. In addition to a compact size, the objective lens should have good optical performance over a broad spectral range.

GRIN lenses have been used in many miniaturized endomicroscope prototypes owing to their small diameter and cylindrical geometry [42, 58, 83, 90–92]. A more compact version of GRIN lenses have been demonstrated as GRIN fiber lenses, which can be robustly connected with other optical components such as a single-mode fiber and glass rod by thermal fusion [93]. Specially designed lens systems such as a ball-lens system without the use of micro-reflectors have been used in side-viewing OCT endomicroscopes [94]. The formation of a lens on the tip of a photonic crystal fiber by means of arc discharge has been reported [95]. All these fiber-based lenses greatly reduce the lens size and facilitate optics alignment and probe engineering process. However, conventional

GRIN or fiber lenses suffer severe chromatic aberration, resulting in a suboptimal imaging quality. Thus, a high performance achromatic lens is critical for performing ultrahigh-resolution imaging. Recently, customized super-achromatic compound micro-lens with 1.5 mm diameter has been reported for scanning OCT endomicroscope [96, 97]. For nonlinear optical endomicroscopy, the wide separation between the excitation and emission wavelengths (e.g., near infrared excitation at ∼800 nm to visible fluorescence or SHG signals at ∼400–500 nm) also require the miniature objective lens to be achromatic over a broad spectrum range. Chromatic aberration will cause a considerable focal shift between the excitation and the TPF/SHG signals and thus result in dramatic reduction in the TPF/SHG collection efficiency. Recently, we have demonstrated a flexible scanning fiber-optic nonlinear optical endomicroscope using a miniature compound aspherical lens, which exhibited reduced chromatic aberration and increased the collection efficiency of the nonlinear optical signals by a factor of ∼5 over a GRIN lens configuration [98].

To achieve depth-resolved imaging in nonlinear optical endomicroscopy or focus tracking in OCT endomicroscopy, a built-in mechanism for changing the focal plane is necessary. When using a rigid rod lens relay system, the depth scanning mechanisms can be implemented at the proximal end by simply translating an external objective lens along the axial direction [90, 99–101]. For fiber-optic endomicroscopes, focus tracking has to be implemented at the distal end. A hydraulic or pneumatic system has been demonstrated to move the end of a fiber relative to the objective lens [36, 102]. An alternative approach is to utilize a miniature motor to alter the distance between the fiber and the objective lens [83]. In addition, deformable MEMS lenses and reflecting membranes have also demonstrated a great potential for varying the beam focus at the distal end of the probe by using electrowetting or electrostatic force [103–105]. Overall, many challenges remain with distal end beam focus varying/tracking in endomicroscopy including system miniaturization, reduction in the drive voltage and improvement on focus varying/tracking accuracy and repeatability over a sufficient range.

21.3 Nonlinear Optical Endomicroscopy

21.3.1 Special considerations in nonlinear optical endomicroscopy

In addition to beam scanning, another major issue in endomicroscopic implementation of nonlinear optical imaging is how to efficiently deliver the single-mode excitation light and collect the (multimode) nonlinear optical signals. Although a single-mode fiber (SMF) can be focused to a near diffraction limited spot to produce efficient nonlinear excitation, SMF suffers a poor collection efficiency for the nonlinear optical signals due to the small collection area (of a typical diameter ∼3–5 μm) and limited numerical aperture (NA $\sim$ 0.12). The small core diameter and low NA make the fiber particularly sensitive to spherical and chromatic aberration of the objective lens, causing significant reduction in collection of the nonlinear signals from the focal volume. In addition, the small core diameter and low NA also preclude the advantage of full-field collection for the nonlinear optical signals. Multimode fiber is superior for large-area signal collection, but with poor focusing beam quality to produce efficient nonlinear excitation. This dilemma can be resolved by the use of a double-clad fiber, which allows single-mode laser delivery with the single-mode core and large-area collection of the nonlinear optical signals with the multimode inner clad (in addition to the core).

A standard double-clad fiber (DCF), as shown in Figure 21.4(a), has been made as a step-index fiber, which is composed of a single-mode core typically with germanium-doped silica, an inner cladding layer typically with fluorine-doped silica, an outer cladding layer typical with pure silica, and a coating layer typical with low-index polymer. This kind of DCF allows the single-mode core to deliver single-mode fs excitation light whereas the inner cladding to transmit multimode fluorescence or SHG signals. DCF is often used in fiber lasers and optical amplifiers and readily available. For example, a double-clad fiber (SMM900) with a core/inner clad diameter of 3.5/103 μm and NA of 0.19/0.24 is commercially available (Fibercore Ltd.). Such a DCF has been successfully implemented in a scanning fiber-optic nonlinear optical endomicroscope with excellent imaging ability [84]. Compared to a single-mode fiber, the DCF greatly improves the collection efficiency TPF/SHG signals by 2–3 orders.

Another type of double-clad fiber is photonic crystal double-clad fiber (PC-DCF) as shown in Figure 21.4(b). The PC-DCF comprises a pure silica core surrounded by a hybrid air-silica structure severing as the inner cladding layer and following by an air hole ring as the outer cladding layer [106]. The hole-to-hole pitch ratio can be tuned to ensure the endlessly single-mode guidance with large effective mode area (e.g., up to 35 μm in diameter). The inner clad area made of air-silica structure can provide large NA (e.g., up to 0.8). A customized PC-DCF (6-μm core and 55-μm inner clad) has enabling two orders of magnitude of the enhancement of the collection efficiency of TPF signals compared to a single-mode fiber [107]. Recently PC-DCF has become commercially available. For example, a PC-DCF (Crystal Fiber, DC-165-16-Passive) has a core/inner clad diameter of 16/165 μm and NA of 0.04/0.6 and it has been used in developing nonlinear optical endomicroscopy technologies [57, 108]. The large single-mode core of the PC-DCF reduces the nonlinear effects (such as self-phase modulation) for a given delivery power. But the large core diameter and low NA make it challenging to focus the excitation beam to a small spot size with a given miniature objective lens, thus reducing the nonlinear excitation efficiency. Ray tracing analyses on a GRIN lens configuration (of a magnification 0.5) show that the ratio of the TPF signals captured by the commercial PC-DCF versus the standard DCF is $\sim$0.03, which is very low. The use of a PC-DCF in an endomicroscope would also increase the rigid portion length of the endomicroscope due to the requirement of beam expansion and refocusing mechanisms, etc. Generally speaking, in engineering a compact TPF/SHG endomicroscope, the core size of the DCF has to be carefully chosen with a tradeoff among the excitation/collection efficiency, the nonlinear effects and the rigid length of the endoscope.

Both the DCF and PC-DCF have normal dispersion, resulting in temporal pulse width broadening due to group-velocity dispersion (GVD) and nonlinear effects such as self-phase modulation (SPM) [109]. In TPF/SHG imaging, the excitation efficiency is inversely proportional to the temporal pulse width, and thus the temporal pulse broadening will lead to reduction in the efficiency of nonlinear excitation. Therefore, pre-chirping is necessary for fiber-optic nonlinear optical imaging endomicroscopes. A conventional grating/lens-based pulse stretcher can be utilized to provide anomalous GVD to prechirp the optical pulses [110]. Recently, photonic crystal fibers based on photonics bandgap effects to guide light propagation have been developed. These fibers exhibit anomalous dispersion over certain wavelength range and can be used for prechirping [111]. For example, a hollow-core photonic band gap (PBF) from Crystal Fibre (HC-800-02, see Figure 21.4(c)) uses a micro-structured cladding region with air holes to guide light in a hollow core, and provides negative dispersion for wavelengths longer than 800 nm. It has been employed for dispersion compensation in our endomicroscope [98, 112, 113]. Our studies have shown that, for excitation pulses at 810$\pm$18 nm with an initial pulse width 60-fs, the PBF fiber can well compensate the dispersion of DCF in the endomicroscope when the length ratio of PBF to DCF is about 1.1. The achievable pulse width is about 130 fs with 20 mW delivered through the DCF core. With higher power laser (50 mW) transmitting in the DCF core, the pulses suffer self-phase modulation and other nonlinear effects, and the temporal pulse duration broadens to about 200 fs.

21.3.2 Nonlinear optical endomicroscopy embodiments and applications

By integrating the technologies summarized above, several flexible prototype endomicroscopes have been developed for TPF and SHG imaging of biological samples, as shown in Figures 21.5(a–d). The first one (Figure 21.5(a)) is based on proximal scanning using an imaging fiber bundle and

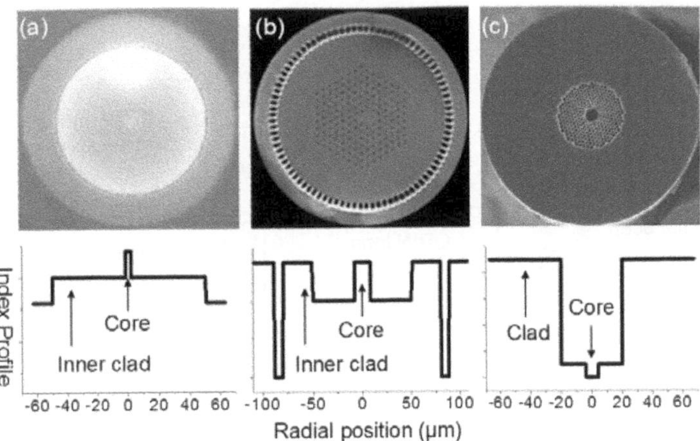

FIGURE 21.4: The schematic of special fibers employed in nonlinear optical endomicroscopy systems: (a) Double-clad fiber; (b) Photonic crystal double-clad fiber; (c) Hollow-core photonic bandgap fiber. The bottom cures shows the schematic of the refractive index profile of each kind of fiber, respectively.

has been used for monitoring brain microvasculature in intact neocortex of living rats [48]. It is clear that the images suffer from the pixilation artifact. The second one (Figure 21.5(b)) utilizes a single-mode fiber or hollow-core PBF resonant scanner based on bi-morph piezoelectric plate or tubular piezoelectric actuators for both beam scanning and excitation delivery, whereas an additional high NA multimode fiber is employed for nonlinear optical signal collection. This kind of TPF endomicroscopes has been demonstrated as a head-mounted probe for studying dendritic morphology, cerebral blood vessels, and calcium transients in *in vivo* animal models [80, 83]. The constraint of this type of endomicroscopes is the requirement of two separate fibers for excitation light delivery and nonlinear signal collection plus a dichroic mirror/prism, making it difficult to achieve an endomicroscope of a compact size.

To create a more compact and flexible endomicroscope, it is ideal to use a single fiber for both single-mode laser excitation delivery and effective nonlinear signal collection. A compact endomicroscope based on a standard double-clad fiber has been developed, as shown in Figure 21.5(c) [84]. The probe basically consisted only of a standard DCF, a small tubular piezoelectric actuator for resonantly sweeping the DCF tip for 2D beam scanning, and a GRIN lens for beam focusing. This design requires simple alignment and makes the endomicroscope flexible and compact with an overall diameter of 2.4 mm. The nonlinear optical endomicroscope can perform real-time TPF/SHG imaging of biological samples at $\sim$3 frames/second. The imaging ability of this system was demonstrated on *in vitro* TPF imaging of cancer cells, *ex vivo* SHG imaging of rat tail tendon, and depth-resolved *ex vivo* TPF imaging of rat interstitial tissue, as shown in Figures 21.6(a), (b) and (c–f), respectively. In Figure 21.6(a), breast cancer cells SK-BR-3 were targeted by FITC-tagged anti Here2/Nu monoclonal antibodies and strong fluorescence signals from cell membranes were easily observed. In Figure 21.6(b), collagen fiber bundles in rat tail tendon were clearly identifiable from the SHG images, in which the intrinsic signals are solely attributed to type I collagen. In Figures 21.6(c–f), the rat oral tissue was stained with acridine orange for enhancing the nuclei contrast and the depth-resolved TPF images reveal different morphologies at different tissue depths. Overall, these preliminary results strongly suggest the potential of the TPF/SHG endomicroscope for real-time assessment of epithelial tissues and collagen fiber network under various clinically relevant conditions.

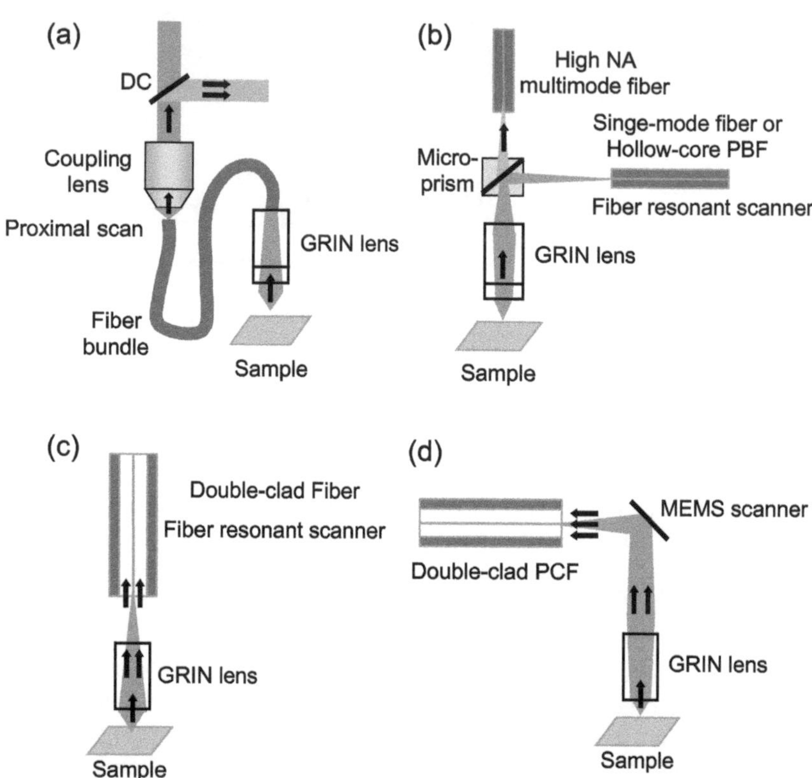

FIGURE 21.5: The schematic of TPF/SHG endomicroscopes: (a) Proximal scanning via an imaging fiber bundle; (b) Distal scanning using two separate fibers for pulse delivery and signal collection; (c) Resonant fiber scanner using a single double-clad fiber; (d) MEMS-mirror-based distal scanning with a single photonic crystal double-clad fiber.

Another nonlinear optical endomicroscope based on a photonic crystal double-clad fiber has been reported, as shown in Figure 21.5(d) [72, 114]. The probe consisted of a PC-DCF, a 2D MEMS mirror, and a GRIN lens, and the size was about 5 mm in diameter [72]. The 2D MEMS mirror performed raster beam scanning at a speed of 7 lines/s over an area of $80 \times 130\ \mu m^2$. The imaging ability of this system was demonstrated on *ex vivo* imaging of rat large intestine tissues and breast cancer tissues both stained with acridine orange. The results showed that the endomicroscopy can visualize 3D tissue morphologies with subcellular resolution.

In most of the nonlinear optical endomicroscopes, a GRIN lens has been used for beam focusing, which gives lateral and axial resolutions (FWHM) at the order of 1–2 μm and 5–20 μm, respectively. Ray tracing analysis has shown that GRIN lens suffers severe chromatic aberration, causing a considerable focal shift between the excitation (near infrared) and the TPF/SHG signals and thus resulting in dramatic reduction in the TPF/SHG collection efficiency. Recently, we have reported the use of a miniature compound lens consisting of a pair of aspherical lenses to minimize the chromatic aberration and increase the TPF/SHG collection efficiency. In addition, a short piece of multimode-fiber (MMF) was introduced at the distal end of the DCF to mitigate the influence of chromatic aberration and further improve TPF/SHG collection. Both ray tracing simulations and experiments on TPF imaging of fluorescent beads and SHG imaging of rat tail tendon demonstrated $\sim$9 ($\sim$4) times improved collection efficiency for TPF (SHG) with the new endomicroscope design, which utilized a compound lens and a multimode-fiber collector.

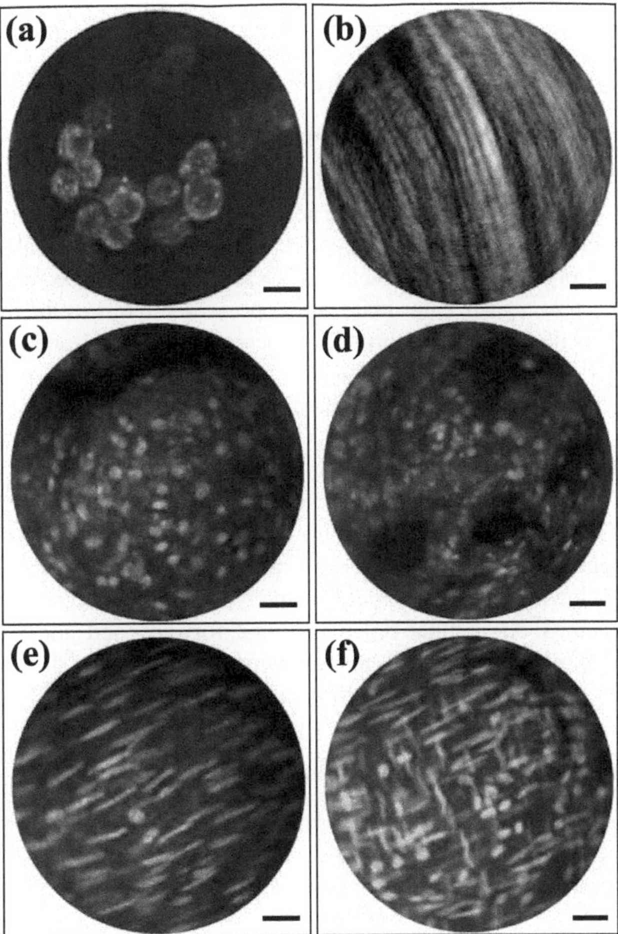

FIGURE 21.6: Typical nonlinear optical images obtained with a scanning fiber-optic endomicroscope system: (a) TPF image of breast cancer cells SK-BR-3 targeted by FITC-tagged monoclonal antibodies (anti Here2/Nu); (b) SHG image of tail tendon of rats; (c–f) Typical depth-resolved TPF images of rat small intestine tissue stained with acridine orange: (c) At the surface of the intestinal epithelium; (d) At the depth of 70 μm below the surface of intestinal epithelium; (e) At the surface of the muscularis externa; (f) At the depth of 30 μm below the surface of muscularis externa. Scale bar: 20 μm.

Besides TPF/SHG endomicroscopy, the development of Coherent Anti-Stokes Raman Scattering (CARS) endomicroscopy has started, making it potentially possible to perform simultaneous endomicroscopic CARS and TPF imaging [115]. The development of the nonlinear endomicroscopy technology is still at its early stage. The initial results have shown tremendous promise of this technology for *in vivo* animal model imaging studies and clinical applications.

21.4 Optical Coherence Tomography Endomicroscopy

21.4.1 Special considerations in OCT fiber-optic endomicroscopy

OCT is capable of assessing depth-resolved high-resolution tissue micro-anatomy with an imaging depth up to 1–3 mm, and it has emerged as a new imaging modality potentially with a variety of clinical applications. Fiber-optic endomicroscopes/catheters are instrumental to enable OCT imaging of internal luminal organs. It is desirable that OCT imaging can be integrated with standard clinical instruments such as a gastrointestinal endoscope, which requires the fiber-optic OCT probe be sufficiently small and flexible to pass through the accessory port of standard endoscopes. Different from nonlinear optical imaging endomicroscopy, OCT requires single-mode for both imaging beam delivery and backscattered light collection, and thus a single-mode fiber is used in the fiber-optic OCT probe. To minimize the probe diameter and facilitate the probe assembling process, a GRIN lens is typically employed for beam focusing. To achieve ultrahigh-resolution where a source of a broad spectrum is required, a high-performance achromatic lens suitable for the entire source spectrum is crucial. Other key parameters to consider when designing OCT probes include numerical aperture, depth of focus, working distance, beam scanning mechanism, scanning range, and imaging speed. The following section will detail some typical endomicroscopic OCT embodiments and some OCT applications. As summarized in Section 2, forward- and side-viewing probes involve different beam scanning mechanisms; thus, the following discussions on the endomicroscopic OCT embodiments will start with the scanning mechanisms among all other key elements in OCT probes.

21.4.2 Endomicroscopic OCT embodiments and the applications

Figure 21.7(a) shows the schematic of a forward-viewing OCT probe based on a MEMS scanning mirror. Typically, light from a single-mode fiber is collimated and strikes the MEMS mirror. The optical beam is then scanned by the MEMS mirror and focused by a scan lens to the sample. A probe by using a 1 mm $\times$ 1 mm MEMS mirror with electro-thermal actuation has been reported, which had a lateral scan range of 3–4 mm with a 20 μm focused spot size [52], and the probe has been demonstrated for imaging normal and cancerous bladder on a rat model. The overall diameter of the probe was still very large (5 mm) mainly due to the large substrate of the MEMS mirror [53]. Recently, a two-axis MEMS mirror based on vertical comb-driven structure has been utilized to perform 3-D OCT imaging [56, 58]. In addition to forward-viewing endomicroscopes, side-viewing OCT endoscopes have also been demonstrated [59]. Overall, MEMS-based OCT probes can provide fast imaging rate. Further engineering development is still required to reduce the dimension of MEMS scanners and to improve their mechanical robustness, making them suitable for *in vivo* endomicroscopic OCT applications.

Compared to MEMS scanners, fiber-optic scanners can make OCT probe more flexible and compact. A forward-viewing OCT probe was reported with electromechanical principle to move a fiber tip in front of a stationary lens [85]. The fiber was mounted on a coil that can be actuated in a small stationary magnet when a varying current running through the coil. About 2 mm swinging along the tissue surface with a 20-μm spot size was achieved. This probe had a 2.2 mm diameter that could be fitted into the biopsy channel of endoscopes. The probe has been demonstrated *in vivo* for imaging cancerous human mucosa in larynx, esophagus, uterine cervix and urine bladder. However, the lateral scanning speed was slow (about 2 mm/s), which was limited by the mass of the coil attached to the fiber. Recently, a fast fiber-optic resonant scanning OCT probe has been reported for forward-view imaging [82]. As shown in Figure 21.7(b), a miniature tubular PZT actuator of a 1.5 mm diameter was used to resonantly scan the fiber-optic cantilever. The PZT, driving electrodes and focusing lens were assembled in a hypodermic tube with 2.4-mm outer diameter. High-speed

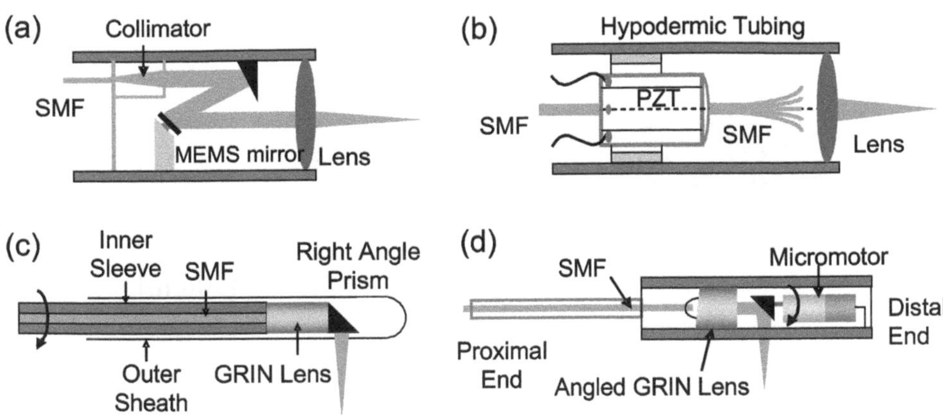

FIGURE 21.7: The schematic of forward-viewing OCT endoscope probe (a and b) and side-viewing OCT endoscope probe (c and d) based on (a) MMES mirror scanning; (b) Fiber resonant scanning; (c) Proximal rotation; (d) Distal rotation using a micro-motor.

beam scanning (e.g., in the range of kHz) and over a large lateral scan range (e.g., 2–3 mm) can be achieved with the resonance fiber-optic scanner. It should be noted that a probe of a smaller diameter (1.5–2.0 mm) is possible (and has been demonstrated) but with a tradeoff of a reduced lateral scanning range. To accommodate a broad source spectrum bandwidth for achieving an ultrahigh axial resolution, a customized super-achromatic compound micro-lens has also been developed to improve the imaging quality [96]. Figures 21.8(a) and (b) compare the representative images of rabbit trachea acquired by the rapid scanning OCT endomicroscopes equipped with a micro compound lens and with a GRIN lens, respectively. Fine tissue structures such as the epithelium (E), lamina propria (LP), submucosa (SM) and glands (G) are evident in both images. However, the image quality achieved with the micro compound lens endoscope was better than the GRIN lens endoscope. It is also noted that the image quality in Figure 21.8(a) deteriorates much slower than the one acquired by the GRIN lens endoscope (Figure 21.8(b)). Further studies also showed that the images acquired by the compound-lens endoscope were comparable to the those acquired by a state-of-the-art bench-top OCT system, which afforded to use a large achromatic objective lens and lower data acquisition speed. It should be mentioned that the image acquisition sequence with the scanning endoscope is to have fast lateral scanning first, followed by a slow depth scanning [82, 116]. This is opposite to the image acquisition sequence used in a conventional OCT system that performs fast depth scanning first, followed by slow lateral scanning. The new image acquisition sequence permits real-time focus tracking, which is essential for maintaining a high lateral resolution at various depths [116–118]. Real-time focus tracking has been demonstrated on an OCT endoscope setting enabled by the rapid scanning fiber-optic resonant scanner [116].

Side-viewing OCT endomicroscope probes can be developed by modifying the design of forward-viewing probes. Generally speaking, beam scanning mechanisms are simpler in side-viewing probes as the actuation mechanism can be implemented at the proximal end of the probe to achieve linear translation or circumferential rotation. The first side-viewing OCT catheter endomicroscope with circumferential scanning was reported in 1996 [43]. Figure 21.7(c) shows the schematic of the probe. It consisted of an optical rotary joint at its proximal end, a single-mode fiber running through the length of the catheter, and an optical focusing (GRIN lens) and a beam directing element (microprism) at the distal end. The single-mode fiber, GRIN lens and microprism were all encased in a flexible speedometer cable (i.e., a hollow metal coil). At the proximal end, a DC motor was used to drive the catheter through a gear mechanism and the body of the catheter including the hous-

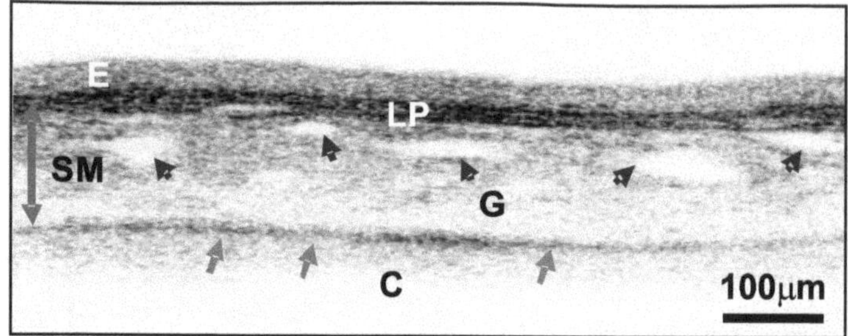

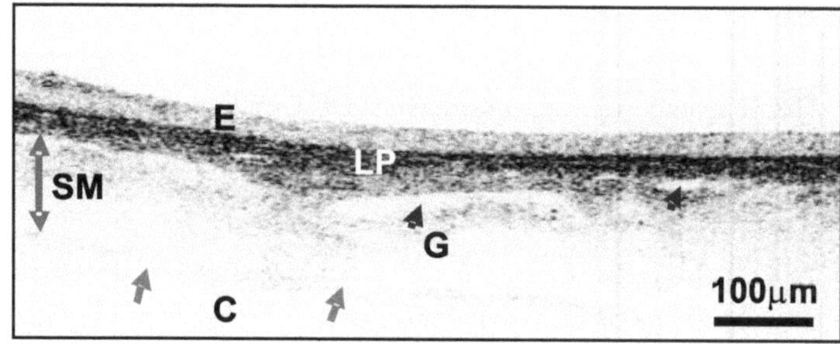

FIGURE 21.8: Typical endomicroscopic OCT images of rabbit trachea obtained with a rapid fiber scanning probe composed of: (a) a compound microlens; (b) a GRIN lens. E: epithelium; LP: lamina propria; SM: submucosa; G: gland (indicated by the short arrows); C: cartilage ring (indicated by the long arrow). (Images are adapted from Ref. [96] with permission.)

ing metal coil can freely rotate inside a stationary transparent plastic sheath with an outer diameter of 1.1 mm. In addition to circumferential imaging, a side-viewing OCT endomicroscope can also perform linear imaging by linearly translating the entire probe along the longitudinal direction of the endoscope in a push-and-pull mode by an actuator placed at the proximal end [119, 120]. The conventional side-viewing OCT endomicroscope is easy to engineer and has played a critical role in translating OCT for *in vivo* applications in animal models, and on human patients for intraluminal imaging of the GI tract, cardiovascular system, and the respiratory tract [28, 120–124].

The associated challenges of proximal end actuation is that the fiber may experience varying stress, resulting in the change of polarization to the light propagating through the fiber-optic probe and consequently producing polarization-sensitive artifacts. The introduction of Faraday rotator between the GRIN lens and microprism in the distal end could compensate bending-induced fiber birefringence [125]. To completely avoid the artifacts, a probe with a rotating reflector driven by a small motor implemented at the distal end has been demonstrated (Figure 21.7(d)), in which the entire probe can remain stationary during imaging. This kind of OCT probe was firstly reported using a commercial micromotor coupled with a rod mirror [44]. The endomicroscope had a 5 mm outer diameter and permitted the rotation speeds varying from 1 to 100 Hz. A scanning probe with a similar design but a smaller footprint (i.e., 2.4 mm diameter) and higher scanning speed (up to 1 kHz) was also developed [45]. In general, distal actuation/scanning provides better beam scanning uniformity with reduced image distortion. The major drawbacks of this type of micromotor-based scanning endoscopes include (1) the increased rigid length of the probe at the distal end that may complicate the delivery through the Y-shape entrance of a standard GI endoscope, and (2) the blockage to the imaging beam by the motor drive wires.

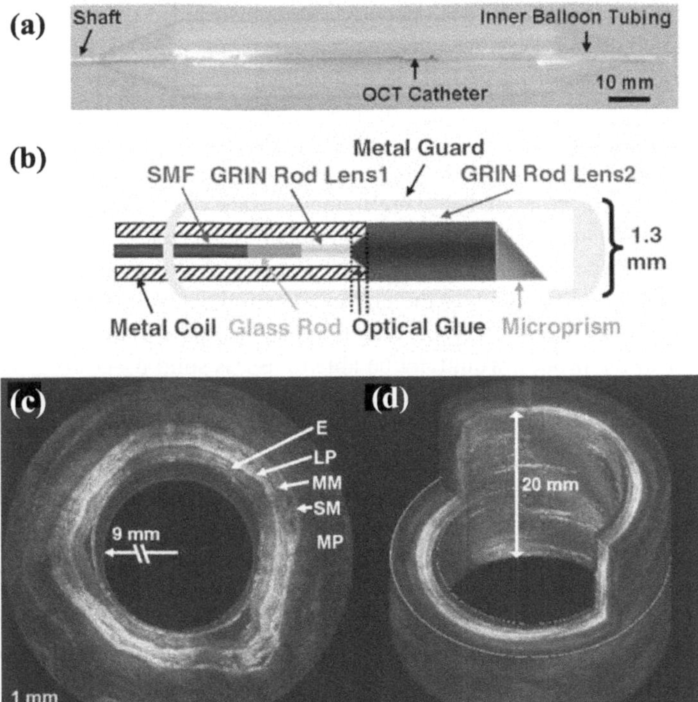

FIGURE 21.9: (a) Photograph of an OCT imaging catheter inside a double-lumen balloon; (b) Schematic of the distal optics design of an OCT imaging catheter using a micro-compound lens; (c) Representative full circumferential OCT images of pig esophagus acquired *in vivo* with the balloon imaging catheter. Different stratified layers of the esophagus can be clearly identified, including epithelium (E), lamina propria (LP), muscularis mucosae (SM), submucosae (SM), and muscularis propria (MP). (d) Comprehensive 3D image constructed from a series of spiral scans of the esophagus. The longitudinal scan distance is 20 mm. (Images are adapted from Ref. [126] with permission.)

Most of the OCT endoscopes discussed above have a short working distance, (e.g., 1 to 4 mm). These endoscopes are well suited for imaging small lumens or a small sector of large lumens when the probes are close to or in direct contact with the target tissue surface. The short working distance, however, precludes full circumferential imaging of large luminal organs such as human esophagus. There is an increasing clinical need for systematic assessment of the entire esophagus for Barrett's surveillance and early cancer detection, inspiring the development of a new type of OCT endoscopes – a balloon imaging catheter [120, 126, 127]. Such a catheter basically integrates a miniature OCT imaging probe within the inner lumen of a double-lumen balloon catheter with the balloon (when inflated) to flatten the natural folds of the esophagus (see Figure 21.9(a)). The major challenge in developing such an OCT imaging probe is to achieve high transverse resolution at a large working distance (e.g., ∼10 mm) while keeping the optical components small (e.g., ∼1–2 mm diameter), so that the entire balloon imaging catheter (with the OCT probe situated within the inner lumen of the balloon) can be delivered into the esophagus through the accessory port of a standard endoscope. Recently, we have presented a new optics design for an OCT balloon imaging catheter [126], which utilized a micro-compound lens as shown in Figure 21.9(b). The beam from the SMF was first tightly focused by a GRIN fiber lens and then relayed to the sample by a GRIN rod lens to achieve a 9.6-mm working distance and a 39-μm spot size. The performance of the OCT balloon catheter has been demonstrated in *in vivo* pig esophagus imaging along with a swept source OCT system. Figure 21.9(c) shows a representative full-circumferential cross-sectional image, whereas Figure 21.9(d) displays the reconstructed 3D image from a series of circumferential images with a 20-μm pitch. As shown in Figure 21.9(c), the OCT balloon catheter can clearly differentiate the normal layered structures of the esophagus, including the epithelium, lamina propria, muscularis mucosae, submucosae, and muscularis propria. These preliminary results demonstrated the capability of OCT balloon catheters for systematic *in vivo* imaging of human esophagus.

As mentioned before, the imaging depth of OCT is about 1–3 mm below tissue surface. This imaging depth is sufficient for assessing the epithelial/endothelial linings of internal luminal and cardiovascular systems. However, the OCT endomicroscopes we discussed so far, including both forward- and side-viewing probes, would not be able to assess solid tissues/organs beyond the 1–3 mm imaging depth. New technologies and devices are very desirable for enabling interstitial imaging of solid tissues/organs. A 27 gauge ($\sim$410 μm) OCT imaging needle capable of high-resolution interstitial imaging of solid tissues in a minimally invasive fashion has thus been developed to meet the needs [42]. One potential clinical application of the OCT needle is to image tissue pathologic changes and guide biopsy in solid tissues by integrating an OCT needle within the core of a biopsy needle. This procedure could potentially increase the accuracy of core biopsy by detecting tissue architectural abnormalities prior to actual tissue excision. In addition to providing image guidance, OCT imaging needle itself can also be potentially used as a diagnostic tool to evaluate tissue status without the need for biopsy. The schematic and scanning mechanism of the first OCT imaging needle are similar to a side-viewing probe (see Figure 21.7(c)). At the distal end, the single-mode fiber was angle cleaved at 8° and attached to a 0.486-pitch GRIN lens with a 250-μm diameter, producing a focused beam with a confocal parameter of $\sim$380 μm corresponding to a $\sim$17-μm spot size. A microprism was customized to fit within the 300 μm inner diameter of a 27-guage hypodermic tube and deflect the focused beam perpendicularly to the needle axis. The focal position of the needle was further adjusted by varying the distance between the GRIN lens and the prism. The final focus was controlled at $\sim$80 μm outside of the needle. The single-mode fiber, the GRIN lens and the prism were attached with each other by UV-curing optical cement to form a single unit and housed concentrically within a 27-gauge hypodermic tube. During the imaging, the needle was in direct contact with the tissue and circumferentially rotated by a DC motor at the proximal end. High-resolution imaging of hamster leg muscle with the fine OCT needle was demonstrated *in vivo*. Initial results with the OCT imaging needle were very encouraging. With further development on the needle optical and mechanical designs, a robust imaging needle with better image quality is possible to address clinical needs such as detection of neoplasia in the pancreas [128].

21.5 Summary

In summary, endomicroscopy technologies for nonlinear optical imaging and optical coherence tomography have been under rapid development, aiming for noninvasive or minimally invasive, high-resolution and high-speed imaging of tissue microstructures and assessment of tissue pathology *in vivo*. In this review, we discussed some general technological and engineering issues in developing an endomicroscope, including beam scanning and focusing mechanisms, and the pros and cons of various endomicroscope designs. We can by no means exhaust all innovative designs that have been reported in literature. We hope this book chapter will serve as a general introduction and guidance to readers who would like to know more about endomicroscopy for both OCT and nonlinear optical imaging.

Overall, nonlinear optical endomicroscopes and OCT endoscopes have shown great flexibility and reliability for high-resolution imaging of internal luminal organs. Optical fibers including imaging fiber bundle, GRIN fiber lens, double-clad fiber and photonic crystal fiber provide convenient and efficient means to bring the light to tissue and collect the optical signals back from the tissue. The development of MEMS technologies, fiber-optic resonant scanners and microlenses greatly facilitates probe miniaturization and functional integration for depth-resolved high-resolution imaging. With further technology innovation, endomicroscopy will soon be able to generate high-quality images approaching to those achieved by standard bench-top laser scanning microscopy. Miniature, high-resolution and high-speed nonlinear optical endomicroscope and OCT endoscope will play an important role in medical diagnosis, guidance of biopsy, and minimally invasive surgery.

Acknowledgment

Our research and development in endomicroscopy technologies for high-resolution nonlinear optical imaging and optical coherence tomography would not be possible without the invaluable contributions of other postdoctoral fellows, research scientists, Ph.D students and collaborators, including Dr. Yongping Chen, Dr. Yuchuan Chen, Dr. Michael J. Cobb, Dr. Kevin Hsu, Henry L. Hu, Dr. Li Huo, Dr. Joo Ha Hwang, Dr. Michael B. Kimmey, Dr. Yuxin Leng, Dr. Ming-Jun Li, Dr. Xiumei Liu, Daniel J. MacDonald, Jeff Magula, Dr. Mon T. Myaing, Dr. Hongwu Ren, Dr. Danling Wang, Addie Warsen, Jiefeng Xi, and Dr. Desheng Zheng. This work has been supported in part by the Whitaker Foundation, Coulter Foundation Translational Research Award, National Institutes of Health, and National Science Foundation (Career Award XDL).

References

[1] J. Shah, "Endoscopy through the ages," *BJU Int.* **89**, 645–652 (2002).

[2] J.E. Moore and G. Zouridakis, (eds.), *Biomedical Technology and Devices Handbook,* CRC Press, Boca Raton (2004).

[3] R. Kiesslich, P.R. Galle, and M.F. Neuratch, *Atlas of Endomicroscopy,* Springer, New York (2007).

[4] K. Inui, S. Nakazawa, J. Yoshino, K. Yamao, H. Yamachika, T. Wakabayashi, N. Kanemaki, and H. Hidano, "Endoscopic MRI – preliminary results of a new technique for visualization and staging of gastrointestinal tumors," *Endoscopy* **27**, 480–485 (1995).

[5] S.G. Worthley, G. Helft, V. Fuster, Z.A. Fayad, M. Shinnar, L.A. Minkoff, C. Schechter, J.T. Fallon, and J.J. Badimon, "A novel nonobstructive intravascular MRI coil – In vivo imaging of experimental atherosclerosis," *Arterioscler. Thromb. Vasc. Biol.* **23**, 346–350 (2003).

[6] T.E. Yusuf and M.S. Bhutani, "Role of endoscopic ultrasonography in diseases of the extrahepatic biliary system," *J. Gastroenterol. Hepatol.* **19**, 243–250 (2004).

[7] T.E. Yusuf, S. Tsutaki, M.S. Wagh, I. Waxman, and W.R. Brugge, "The EUS hardware store: state of the art technical review of instruments and equipment (with videos)," *Gastrointest. Endosc.* **66**, 131–143 (2007).

[8] M. Andrea, O. Dias, and A. Santos, "Contact endoscopy during microlaryngeal surgery – a new technique for endoscopic examination of the larynx," *Ann. Otol. Rhinol. Laryngol.* **104**, 333–339 (1995).

[9] H. Inoue, S. Kudo, and A. Shiokawa, "Novel endoscopic imaging techniques toward in vivo observation of living cancer cells in the gastrointestinal tract," *Dig. Dis.* **22**, 334–337 (2004).

[10] H. Machida, Y. Sano, Y. Hamamoto, M. Muto, T. Kozu, H. Tajiri, and S. Yoshida, "Narrow-band imaging in the diagnosis of colorectal mucosal lesions: a pilot study," *Endoscopy* **36**, 1094–1098 (2004).

[11] S. Andersson-Engels, C. Klinteberg, K. Svanberg, and S. Svanberg, "In vivo fluorescence imaging for tissue diagnostics," *Phys. Med. Biol.* **42**, 815–824 (1997).

[12] M.G. Shim, L.M.W.M. Song, N.E. Marcon, and B. C. Wilson, "In vivo near-infrared Raman spectroscopy: demonstration of feasibility during clinical gastrointestinal endoscopy," *Photochem. Photobiol.* **72**, 146–150 (2000).

[13] C. Ell, "Improving endoscopic resolution and sampling: fluorescence techniques," *Gut* **52**, 30–33 (2003).

[14] J.F. Rey, H. Inoue, and M. Guelrud, "Magnification endoscopy with acetic acid for Barrett's esophagus," *Endoscopy* **37**, 583–586 (2005).

[15] W.B. Amos, J.G. White, and M. Fordham, "Use of confocal imaging in the study of bological structures," *Appl. Opt.* **26**, 3239–3243 (1987).

[16] J.B. Pawley, *Handbook of Biological Confocal Microscopy,* Plenum Press, New York (1996).

[17] W. Denk, J.H. Strickler, and W.W. Webb, "2-Photon laser scanning fluorescence microscopy," *Science* **248**, 73–76 (1990).

[18] W.R. Zipfel, R.M. Williams, and W.W. Webb, "Nonlinear magic: multiphoton microscopy in the biosciences," *Nat. Biotechnol.* **21**, 1368–1376 (2003).

[19] P.A. Franken, G. Weinreich, C.W. Peters, and A.E. Hill, "Generation of optical harmonics," *Phys. Rev. Lett.* **7**, 118–119 (1961).

[20] Y.C. Guo, P.P. Ho, H. Savage, D. Harris, P. Sacks, S. Schantz, F. Liu, N. Zhadin, and R.R. Alfano, "Second-harmonic tomography of tissues," *Opt. Lett.* **22**, 1323–1325 (1997).

[21] M.D. Duncan, J. Reintjes, and T.J. Manuccia, "Scanning coherent anti-Stokes Raman microscope," *Opt. Lett.* **7**, 350–352 (1982).

[22] J.X. Cheng and X.S. Xie, "Coherent anti-Stokes Raman scattering microscopy: instrumentation, theory, and applications," *J. Phys. Chem. B* **108**, 827–840 (2004).

[23] D. Huang, E.A. Swanson, C.P. Lin, J.S. Schuman, W.G. Stinson, W. Chang, M.R. Hee, T. Flotte, K. Gregory, C.A. Puliafito, and J.G. Fujimoto, "Optical coherence tomography," *Science* **254**, 1178–1181 (1991).

[24] J.G. Fujimoto, "Optical coherence tomography for ultrahigh resolution in vivo imaging," *Nat. Biotechnol.* **21**, 1361–1367 (2003).

[25] Y.R. Shen, *The Principles of Nonlinear Optics,* John Wiley, New York (1984).

[26] A. Zoumi, A. Yeh, and B.J. Tromberg, "Imaging cells and extracellular matrix in vivo by using second-harmonic generation and two-photon excited fluorescence," *Proc. Natl. Acad. Sci. U. S. A.* **99**, 11014–11019 (2002).

[27] W.R. Zipfel, R.M. Williams, R. Christie, A.Y. Nikitin, B.T. Hyman, and W.W. Webb, "Live tissue intrinsic emission microscopy using multiphoton-excited native fluorescence and second harmonic generation," *Proc. Natl. Acad. Sci. U. S. A.* **100**, 7075–7080 (2003).

[28] I.K. Jang, B.E. Bouma, D.H. Kang, S.J. Park, S.W. Park, K.B. Seung, K.B. Choi, M. Shishkov, K. Schlendorf, E. Pomerantsev, S.L. Houser, H.T. Aretz, and G.J. Tearney, "Visualization of coronary atherosclerotic plaques in patients using optical coherence tomography: comparison with intravascular ultrasound," *J. Am. Coll. Cardiol.* **39**, 604–609 (2002).

[29] S.H. Yun, G.J. Tearney, B.J. Vakoc, M. Shishkov, W.Y. Oh, A.E. Desjardins, M.J. Suter, R.C. Chan, J.A. Evans, I.K. Jang, N.S. Nishioka, J.F. de Boer, and B.E. Bouma, "Comprehensive volumetric optical microscopy in vivo," *Nat. Med.* **12**, 1429–1433 (2006).

[30] R.S. Dacosta, B.C. Wilson, and N.E. Marcon, "New optical technologies for earlier endoscopic diagnosis of premalignant gastrointestinal lesions," *J. Gastroenterol. Hepatol.* **17**, S85–S104 (2002).

[31] A.D. Mehta, J.C. Jung, B.A. Flusberg, and M.J. Schnitzer, "Fiber optic in vivo imaging in the mammalian nervous system," *Curr. Opin. Neurobiol.* **14**, 617–628 (2004).

[32] J. A. Evans and N. S. Nishioka, "Endoscopic confocal microscopy," *Curr. Opin. Gastroenterol.* **21**, 578–584 (2005).

[33] B.A. Flusberg, E.D. Cocker, W. Piyawattanametha, J.C. Jung, E.L.M. Cheung, and M.J. Schnitzer, "Fiber-optic fluorescence imaging," *Nat. Methods* **2**, 941–950 (2005).

[34] Z. Yaqoob, J.G. Wu, E.J. McDowell, X. Heng, and C.H. Yang, "Methods and application areas of endoscopic optical coherence tomography," *J. Biomed. Opt.* **11**, 063001 (2006).

[35] L. Fu and M. Gu, "Fibre-optic nonlinear optical microscopy and endoscopy," *J. Microsc. (Oxf)* **226**, 195–206 (2007).

[36] A.R. Rouse and A.F. Gmitro, "Multispectral imaging with a confocal microendoscope," *Opt. Lett.* **25**, 1708–1710 (2000).

[37] C.P. Lin and R.H. Webb, "Fiber-coupled multiplexed confocal microscope," *Opt. Lett.* **25**, 954–956 (2000).

[38] K.B. Sung, C.N. Liang, M. Descour, T. Collier, M. Follen, and R. Richards-Kortum, "Fiber-optic confocal reflectance microscope with miniature objective for in vivo imaging of human tissues," *IEEE Trans. Biomed. Eng.* **49**, 1168–1172 (2002).

[39] G.J. Tearney, M. Shishkov, and B.E. Bouma, "Spectrally encoded miniature endoscopy," *Opt. Lett.* **27**, 412–414 (2002).

[40] T.F. Watson, M.A.A. Neil, R. Juskaitis, R.J. Cook, and T. Wilson, "Video-rate confocal endoscopy," *J. Microsc. (Oxf)* **207**, 37–42 (2002).

[41] T.D. Wang, M.J. Mandella, C.H. Contag, and G.S. Kino, "Dual-axis confocal microscope for high-resolution in vivo imaging," *Opt. Lett.* **28**, 414–416 (2003).

[42] X.D. Li, C. Chudoba, T. Ko, C. Pitris, and J.G. Fujimoto, "Imaging needle for optical coherence tomography," *Opt. Lett.* **25**, 1520–1522 (2000).

[43] G.J. Tearney, S.A. Boppart, B.E. Bouma, M.E. Brezinski, N.J. Weissman, J.F. Southern, and J.G. Fujimoto, "Scanning single-mode fiber optic catheter-endoscope for optical coherence tomography," *Opt. Lett.* **21**, 543–545 (1996).

[44] P.R. Herz, Y. Chen, A.D. Aguirre, K. Schneider, P. Hsiung, J.G. Fujimoto, K. Madden, J. Schmitt, J. Goodnow, and C. Petersen, "Micromotor endoscope catheter for in vivo, ultrahigh-resolution optical coherence tomography," *Opt. Lett.* **29**, 2261–2263 (2004).

[45] P.H. Tran, D.S. Mukai, M. Brenner, and Z.P. Chen, "In vivo endoscopic optical coherence tomography by use of a rotational microelectromechanical system probe," *Opt. Lett.* **29**, 1236–1238 (2004).

[46] T. Dabbs and M. Glass, "Fiberoptic confocal microscope – focon," *Appl. Opt.* **31**, 3030–3035 (1992).

[47] P.M. Lane, A.L.P. Dlugan, R. Richards-Kortum, and C.E. MacAulay, "Fiber-optic confocal microscopy using a spatial light modulator," *Opt. Lett.* **25**, 1780–1782 (2000).

[48] W. Gobel, J.N.D. Kerr, A. Nimmerjahn, and F. Helmchen, "Miniaturized two-photon microscope based on a flexible coherent fiber bundle and a gradient-index lens objective," *Opt. Lett.* **29**, 2521–2523 (2004).

[49] T.Q. Xie, D. Mukai, S.G. Guo, M. Brenner, and Z.P. Chen, "Fiber-optic-bundle-based optical coherence tomography," *Opt. Lett.* **30**, 1803–1805 (2005).

[50] M. Lelek, E. Suran, F. Louradour, A. Barthelemy, B. Viellerobe, and F. Lacombe, "Coherent femtosecond pulse shaping for the optimization of a non-linear micro-endoscope," *Opt. Express* **15**, 10154–10162 (2007).

[51] K.B. Sung, C. Liang, M. Descour, T. Collier, M. Follen, A. Malpica, and R. Richards-Kortum, "Near real time in vivo fibre optic confocal microscopy: sub-cellular structure resolved," *J. Microsc. (Oxf)* **207**, 137–145 (2002).

[52] Y.T. Pan, H.K. Xie, and G.K. Fedder, "Endoscopic optical coherence tomography based on a microelectromechanical mirror," *Opt. Lett.* **26**, 1966–1968 (2001).

[53] T.Q. Xie, H.K. Xie, G.K. Fedder, and Y.T. Pan, "Endoscopic optical coherence tomography with a modified microelectromechanical systems mirror for detection of bladder cancers," *Appl. Opt.* **42**, 6422–6426 (2003).

[54] A. Jain, A. Kopa, Y.T. Pan, G.K. Fedder, and H.K. Xie, "A two-axis electrothermal micromirror for endoscopic optical coherence tomography," *IEEE J. Sel. Top. Quantum Electron* **10**, 636–642 (2004).

[55] J.T.W. Yeow, V.X. D. Yang, A. Chahwan, M.L. Gordon, B. Qi, I.A. Vitkin, B.C. Wilson, and A.A. Goldenberg, "Micromachined 2-D scanner for 3-D optical coherence tomography," *Sens. Actuators, A* **117**, 331–340 (2005).

[56] W. Jung, D.T. McCormick, J. Zhang, L. Wang, N.C. Tien, and Z.P. Chen, "Three-dimensional endoscopic optical coherence tomography by use of a two-axis microelectromechanical scanning mirror," *Appl. Phys. Lett.* **88**, 163901 (2006).

[57] L. Fu, A. Jain, H.K. Xie, C. Cranfield, and M. Gu, "Nonlinear optical endoscopy based on a double-clad photonic crystal fiber and a MEMS mirror," *Opt. Express* **14**, 1027–1032 (2006).

[58] W.G. Jung, D.T. McCormick, Y.C. Ahn, A. Sepehr, M. Brenner, B. Wong, N.C. Tien, and Z.P. Chen, "In vivo three-dimensional spectral domain endoscopic optical coherence tomography using a microelectromechanical system mirror," *Opt. Lett.* **32**, 3239–3241 (2007).

[59] K.H. Kim, B.H. Park, G.N. Maguluri, T.W. Lee, F.J. Rogomentich, M. G. Bancu, B.E. Bouma, J.F. de Boer, and J.J. Bernstein, "Two-axis magnetically-driven MEMS scanning catheter for endoscopic high-speed optical coherence tomography," *Opt. Express* **15**, 18130–18140 (2007).

[60] C.L. Hoy, N.J. Durr, P.Y. Chen, W. Piyawattanametha, H. Ra, O. Solgaard, and A. Ben-Yakar, "Miniaturized probe for femtosecond laser microsurgery and two-photon imaging," *Opt. Express* **16**, 9996–10005 (2008).

[61] Z.J. Yao and N.C. MacDonald, "Single crystal silicon supported thin film micromirrors for optical applications," *Opt. Eng.* **36**, 1408–1413 (1997).

[62] D.L. Dickensheets and G.S. Kino, "Micromachined scanning confocal optical microscope," *Opt. Lett.* **21**, 764–766 (1996).

[63] H.Y. Lin and W.L. Fang, "A rib-reinforced micro torsional mirror driven by electrostatic torque generators," *Sens. Actuators, A* **105**, 1–9 (2003).

[64] J.M. Zara, S. Yazdanfar, K. D. Rao, J.A. Izatt, and S.W. Smith, "Electrostatic micromachine scanning mirror for optical coherence tomography," *Opt. Lett.* **28**, 628–630 (2003).

[65] D. Hah, S.T.Y. Huang, J.C. Tsai, H. Toshiyoshi, and M.C. Wu, "Low-voltage, large-scan angle MEMS analog micromirror arrays with hidden vertical comb-drive actuators," *J. Microelectromech. Syst.* **13**, 279–289 (2004).

[66] W. Lang, H. Pavlicek, T. Marx, H. Scheithauer, and B. Schmidt, "Electrostatically actuated micromirror devices in silicon technology," *Sens. Actuators, A* **74**, 216–218 (1999).

[67] H. Schenk, P. Durr, T. Haase, D. Kunze, U. Sobe, H. Lakner, and H. Kuck, "Large deflection micromechanical scanning mirrors for linear scans and pattern generation," *IEEE J. Sel. Top. Quantum Electron* **6**, 715–722 (2000).

[68] H. Toshiyoshi, W. Piyawattanametha, C.T. Chan, and M.C. Wu, "Linearization of electrostatically actuated surface micromachined 2D optical scanner," *J. Microelectromech. Syst.* **10**, 205–214 (2001).

[69] P.M. Hagelin and O. Solgaard, "Optical raster-scanning displays based on surface-micromachined polysilicon mirrors," *IEEE J. Sel. Top. Quantum Electron* **5**, 67–74 (1999).

[70] M.H. Kiang, O. Solgaard, K.Y. Lau, and R.S. Muller, "Electrostatic combdrive-actuated micromirrors for laser-beam scanning and positioning," *J. Microelectromech. Syst.* **7**, 27–37 (1998).

[71] D. Hah, C.A. Choi, C.K. Kim, and C.H. Jun, "A self-aligned vertical comb-drive actuator on an SOI wafer for a 2D scanning micromirror," *J. Micromech. Microeng.* **14**, 1148–1156 (2004).

[72] L. Fu, A. Jain, C. Cranfield, H.K. Xie, and M. Gu, "Three-dimensional nonlinear optical endoscopy," *J. Biomed. Opt.* **12**, 040501 (2007).

[73] S. Kwon and L. Lee, "Stacked two dimensional microlens scanner for micro confocal imaging array," in *15th Annual IEEE International MEMS 2002 Conference* (Las Vegas, 2002).

[74] S. Kwon and L.P. Lee, "Micromachined transmissive scanning confocal microscope," *Opt. Lett.* **29**, 706–708 (2004).

[75] K. Takahashi, H.N. Kwon, K. Saruta, M. Mita, H. Fujita, and H. Toshiyoshi, "A two-dimensional f-theta micro optical lens scanner with electrostatic comb-drive XY-stage," *IEICE Electron Expr.* **2**, 542–547 (2005).

[76] C.P.B. Siu, H. Zeng, and M. Chiao, "Magnetically actuated MEMS microlens scanner for in vivo medical imaging," *Opt. Express* **15**, 11154–11166 (2007).

[77] J.G. Wu, M. Conry, C.H. Gu, F. Wang, Z. Yaqoob, and C. H. Yang, "Paired-angle-rotation scanning optical coherence tomography forward-imaging probe," *Opt. Lett.* **31**, 1265–1267 (2006).

[78] C.T. Amirault and C.A. Dimarzio, "Precision pointing using a dual-wedge scanner," *Appl. Opt.* **24**, 1302–1308 (1985).

[79] W.C. Warger and C.A. DiMarzio, "Dual-wedge scanning confocal reflectance microscope," *Opt. Lett.* **32**, 2140–2142 (2007).

[80] F. Helmchen, M.S. Fee, D.W. Tank, and W. Denk, "A miniature head-mounted two-photon microscope: high-resolution brain imaging in freely moving animals," *Neuron* **31**, 903–912 (2001).

[81] E.J. Seibel and Q.Y.J. Smithwick, "Unique features of optical scanning, single fiber endoscopy," *Lasers Surg. Med.* **30**, 177–183 (2002).

[82] X.M. Liu, M.J. Cobb, Y.C. Chen, M.B. Kimmey, and X.D. Li, "Rapid-scanning forward-imaging miniature endoscope for real-time optical coherence tomography," *Opt. Lett.* **29**, 1763–1765 (2004).

[83] B.A. Flusberg, J.C. Lung, E.D. Cocker, E.P. Anderson, and M.J. Schnitzer, "In vivo brain imaging using a portable 3.9 gram two-photon fluorescence microendoscope," *Opt. Lett.* **30**, 2272–2274 (2005).

[84] M.T. Myaing, D.J. MacDonald, and X.D. Li, "Fiber-optic scanning two-photon fluorescence endoscope," *Opt. Lett.* **31**, 1076–1078 (2006).

[85] A.M. Sergeev, V.M. Gelikonov, G.V. Gelikonov, F.I. Feldchtein, R.V. Kuranov, N.D. Gladkova, N.M. Shakhova, L.B. Snopova, and A.V. Shakhov, "In vivo endoscopic OCT imaging of precancer and cancer states of human mucosa," *Opt. Express* **1**, 432–440 (1997).

[86] F.I. Feldchtein, V.M. Gelikonov, and G.V. Gelikonov, "Design of OCT scanners" in *Handbook of Optical Coherence Tomography,* B. E. Bouma and G. J. Tearney (eds.), Marcel Dekker, New York (2001), pp. 125–142.

[87] L.E. Kinsler, A.R. Frey, A.B. Coppens, and J.V. Sanders, *Fundamentals of Acoustics,* John Wiley, New York (1982).

[88] R. Isago, S. Domae, D. Koyama, K. Nakamura, and S. Ueha, "High-frequency optical scanner based on bending vibration of optical fiber," *Jpn. J. Appl. Phys.* **45**, 4773–4779 (2006).

[89] Q.Y.J. Smithwick, P.G. Reinhall, J. Vagners, and E.J. Seibel, "A nonlinear state-space model of a resonating single fiber," *J. Dyn. Syst. Meas. Contr.* **126**, 88–101 (2004).

[90] J.C. Jung and M.J. Schnitzer, "Multiphoton endoscopy," *Opt. Lett.* **28**, 902–904 (2003).

[91] J.C. Jung, A.D. Mehta, E. Aksay, R. Stepnoski, and M.J. Schnitzer, "In vivo mammalian brain imaging using one- and two-photon fluorescence microendoscopy," *J. Neurophysiol.* **92**, 3121–3133 (2004).

[92] H. Li, B.A. Standish, A. Mariampillai, N.R. Munce, Y.X. Mao, S. Chiu, N.E. Alarcon, B.C. Wilson, A. Vitkin, and V.X.D. Yang, "Feasibility of interstitial Doppler optical coherence tomography for in vivo detection of microvascular changes during photodynamic therapy," *Lasers Surg. Med.* **38**, 754–761 (2006).

[93] W.A. Reed, M.F. Yan, and M.J. Schnitzer, "Gradient-index fiber-optic microprobes for minimally invasive in vivo low-coherence interferometry," *Opt. Lett.* **27**, 1794–1796 (2002).

[94] V.X.D. Yang, Y.X. Mao, N. Munce, B. Standish, W. Kucharczyk, N.E. Marcon, B.C. Wilson, and I.A. Vitkin, "Interstitial Doppler optical coherence tomography," *Opt. Lett.* **30**, 1791–1793 (2005).

[95] G.J. Kong, L. Kim, H.Y. Choi, L.E. Im, B.H. Park, U.C. Paek, and B.H. Lee, "Lensed photonic crystal fiber obtained by use of an arc discharge," *Opt. Lett.* **31**, 894–896 (2006).

[96] D.L. Wang, B.V. Hunter, M.J. Cobb, and X.D. Li, "Super-achromatic rapid scanning microendoscope for ultrahigh-resolution OCT imaging," *IEEE J. Sel. Top. Quantum Electron* **13**, 1596–1601 (2007).

[97] A.R. Tumlinson, B. Povazay, L.P. Hariri, J. McNally, A. Unterhuber, B. Hermann, H. Sattmann, W. Drexler, and J.K. Barton, "In vivo ultrahigh-resolution optical coherence tomography of mouse colon with an achromatized endoscope," *J. Biomed. Opt.* **11**, 064003 (2006).

[98] Y. Wu, J. Xi, M.J. Cobb, and X.D. Li, "Scanning fiber-optic nonlinear endomicroscopy with miniature aspherical compound lens and multimode fiber collector," *Opt. Lett.* **34**, 953–955 (2009).

[99] D.M. Rector, R.F. Rogers, and J.S. George, "A focusing image probe for assessing neural activity in vivo," *J. Neurosci. Methods* **91**, 135–145 (1999).

[100] M.J. Levene, D.A. Dombeck, K.A. Kasischke, R.P. Molloy, and W.W. Webb, "In vivo multiphoton microscopy of deep brain tissue," *J. Neurophysiol.* **91**, 1908–1912 (2004).

[101] T.Q. Xie, S.G. Guo, Z.P. Chen, D. Mukai, and M. Brenner, "GRIN lens rod based probe for endoscopic spectral domain optical coherence tomography with fast dynamic focus tracking," *Opt. Express* **14**, 3238–3246 (2006).

[102] A.R. Rouse, A. Kano, J.A. Udovich, S.M. Kroto, and A.F. Gmitro, "Design and demonstration of a miniature catheter for a confocal microendoscope," *Appl. Opt.* **43**, 5763–5771 (2004).

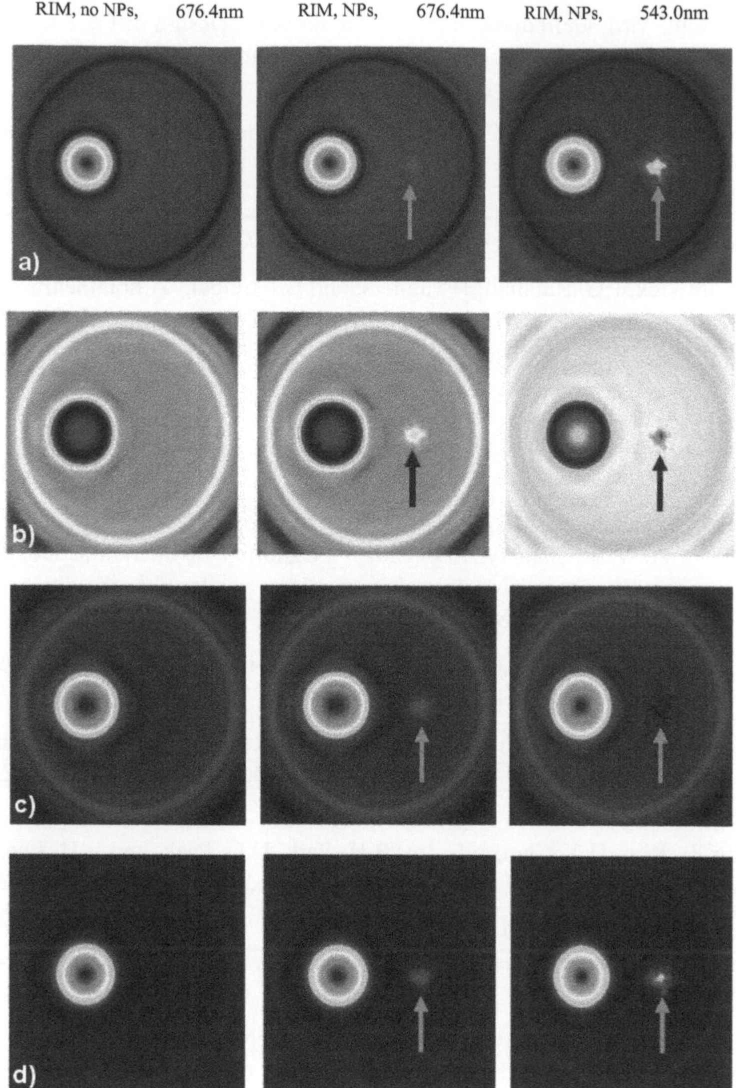

FIGURE 1.13: OPCM images of a single cell for different values (a: -150°, b: -90°, c: $+90^{\circ}$, d: $+180^{\circ}$) of the phase offset Ψ between the reference and diffracted beam of the OPCM at RIM (optical immersion) conditions including a cluster of 42 gold NPs located in a position symmetrically opposite to the nucleus. The arrows indicate the position of the cluster. The left column corresponds to a cell without NPs. The other columns correspond to a cell with NPs at resonant (right) and nonresonant (middle) conditions.

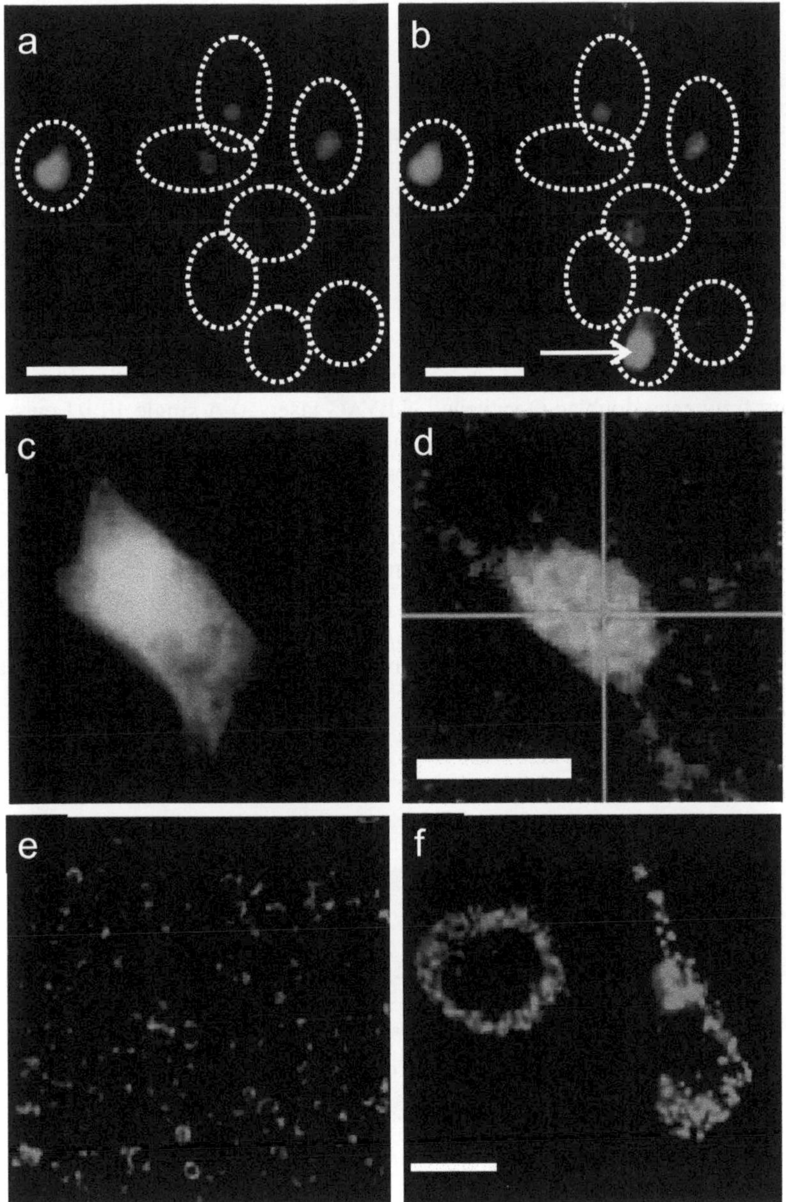

FIGURE 3.2: Examples of targeted (a,b) optical injection and targeted (b,c) optical transfection using a femtosecond (fs) pulsed Ti:sapphire laser or by (e,f) 405 nm continuous wave (cw) laser. See text for details. Images a and b reprinted with permission (Optical Society of America) [53]. Image d reprinted by permission from Macmillan Publishers Ltd: Nature Methods [54], copyright 2006. Images e and f reprinted with permission (Optical Society of America) [40].

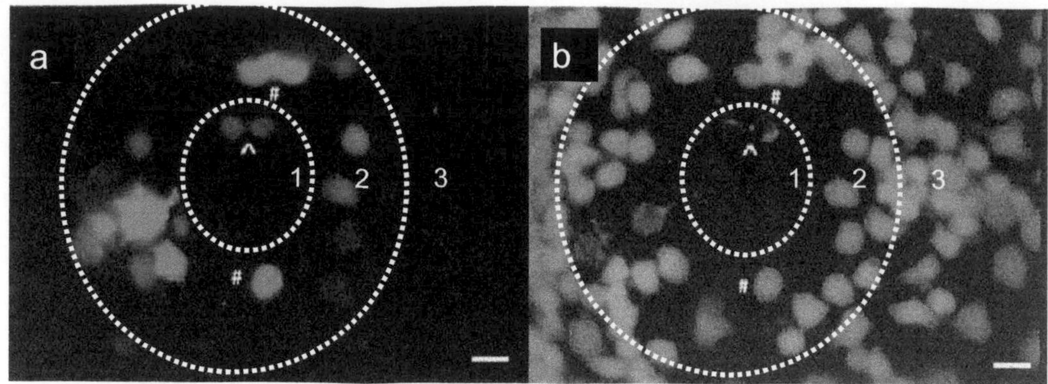

FIGURE 3.3: Example of untargeted (shockwave mediated) optical injection by nanosecond (ns) pulsed 532-nm frequency doubled Q-switched Nd:YAG laser. a) A single 10 μJ pulse was fired at the coverslip/solution interface in the center of the image in the presence of Texas red-conjugated dextran (3 kD). Red cells therefore represent successful optical injection (but the cells may be also be dead) b) Immcdiatcly after irradiation, cells were stained with the viability indicator Oregon green diacetate. Green cells are therefore viable. Three distinct zones were apparent. 1) The central zone (0–30 μm from center) is either devoid of cells due to detachment or contains nonviable optically injected cells ($\wedge$). 2) the pericentral zone (41–50 μm) has >90% viability and also contains some optically injected cells (#). The distal zone (>50 μm) contains 100% viable cells, but no optically injected cells. Reprinted Adapted with permission from [49]. Copyright 2000 American Chemical Society.

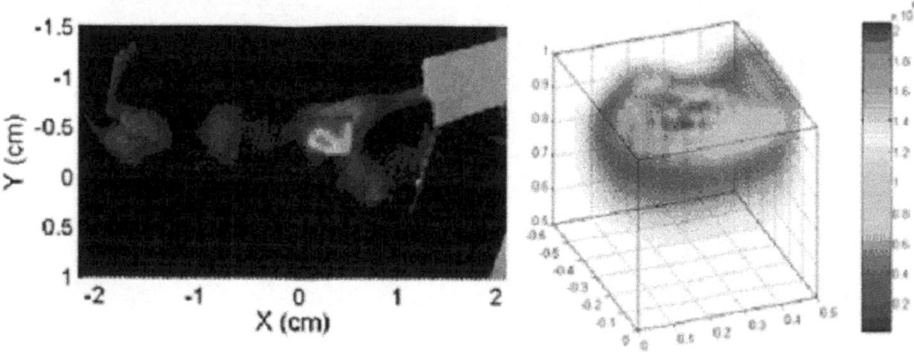

FIGURE 5.4: Example of an FMT reconstruction of fluorophore concentration, in this case green fluorescent protein (GFP). A mouse puppy expressing GFP in its T lymphocytes was imaged in reflexion mode using a custom-made FMT setup at day 3.5 after birth. At this age, T lymphocytes that migrate from the thymus start seeding the peripheral lymphoid organs, such as the lymph node reconstructed in this figure. Position and relative intensity of fluorescence can be inferred from the right panel.

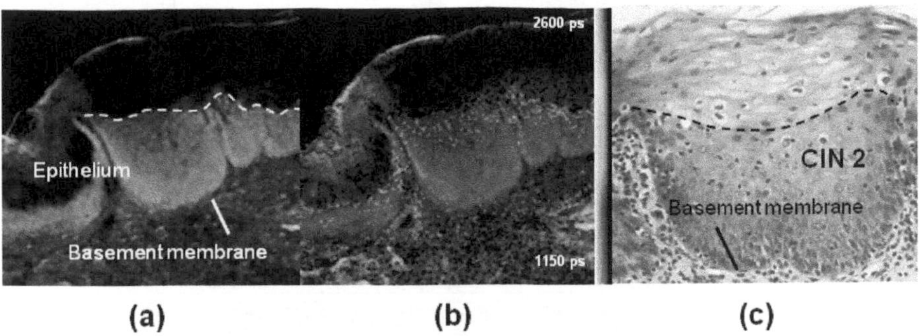

FIGURE 6.7: Multiphoton fluorescence intensity(a) and lifetime (b) images of fresh section from human cervical biopsy excited at 740 nm, shown with parallel H&E stained section indicating grade CIN2 cancer (adapted from [108]).

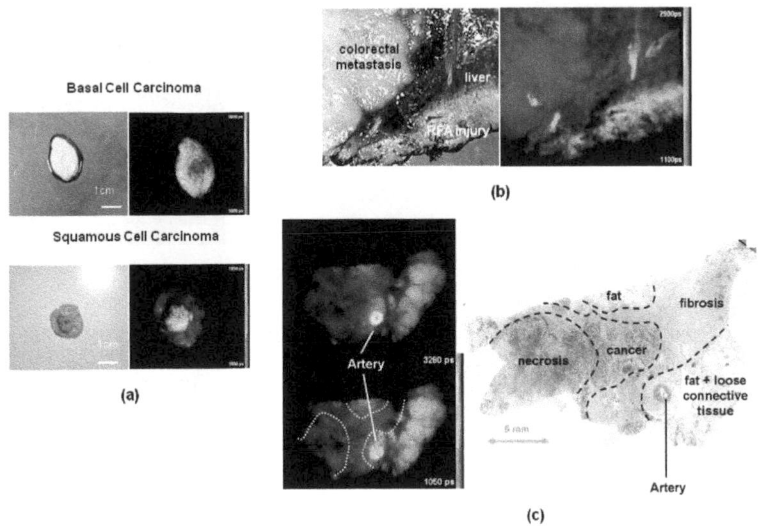

FIGURE 6.8: Wide-field time-gated FLIM images of freshly resected tissue autofluorescence excited at 355 nm: (a) white light and FLIM images of basal and squamous cell carcinomas (adapted from [107]); (b) white light and FLIM images of liver tissue showing colorectal metastasis and RF ablation damage; and (c) fluorescence intensity and lifetime images of pancreatic tissue showing necrosis, cancer, fat and loose connective tissue with an artery.

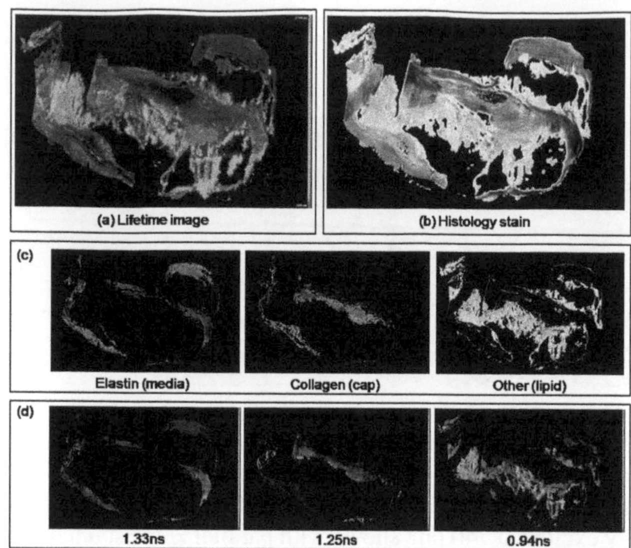

FIGURE 6.9: Showing images of an artery section obtained from a carotid endarterectomy procedure: (a) wide-field time-gated FLIM montage assembled from 17 images. (b) A micrograph taken with a color camera of the same section stained according to the Elastic van Gieson (EVG) protocol. (c) Masks obtained by color segmentation of the image of the EVG stained section. (d) The masks applied to the FLIM images. The pixels corresponding to each component were binned to provide mean fluorescence lifetimes for each component.

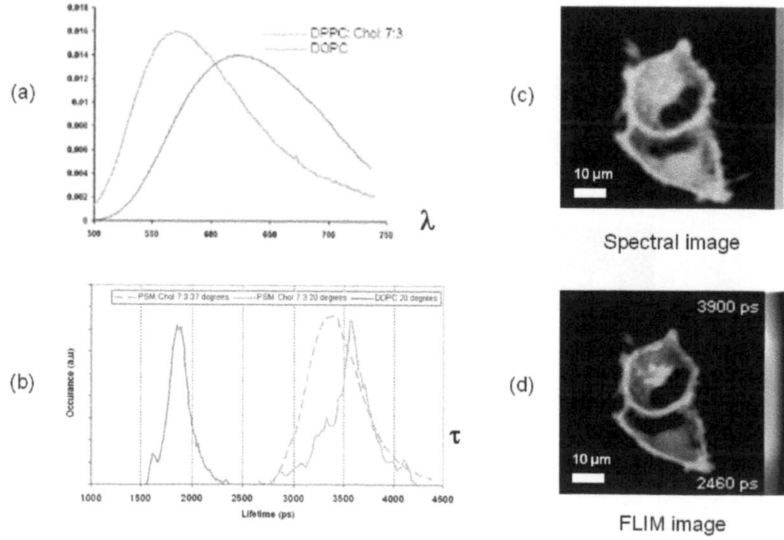

FIGURE 6.10: Comparison of fluorescence spectral and lifetime readout of the di-4-ANEPPDHQ dye excited at 473 nm: (a) shows change in emission spectra when staining giant unilamellar vesicles with and without extraction of cholesterol by methyl-β-cyclodextrin and (b) shows the corresponding fluorescence lifetime histograms. (c) and (d) show the spectral ratiometric and fluorescence lifetime images respectively of live HEK cells labelled with di-4-ANEPPDHQ. (Figure adapted from [143]).

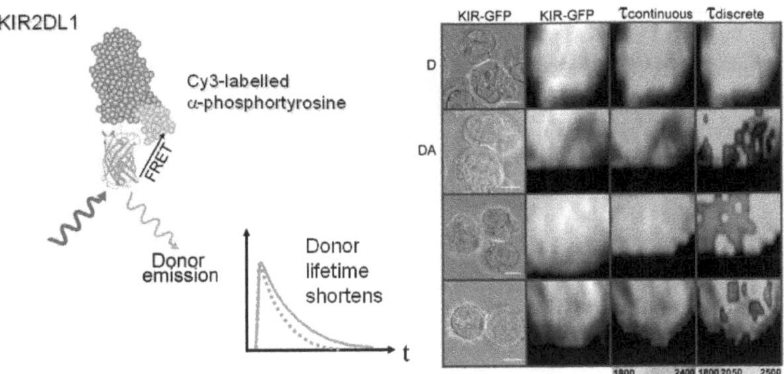

FIGURE 6.11: Schematic of FRET between EGFP-labelled KIR receptor and Cy3-labelled antiphosphotyosine with inset transmitted light images (where the IS are highlighted in white) and *en face* fluorescence intensity and lifetime images of FRET between NK cells and target cells at the IS with continuous and discrete lifetime scales. The scale bar is 8μm. (Figure adapted from [165]).

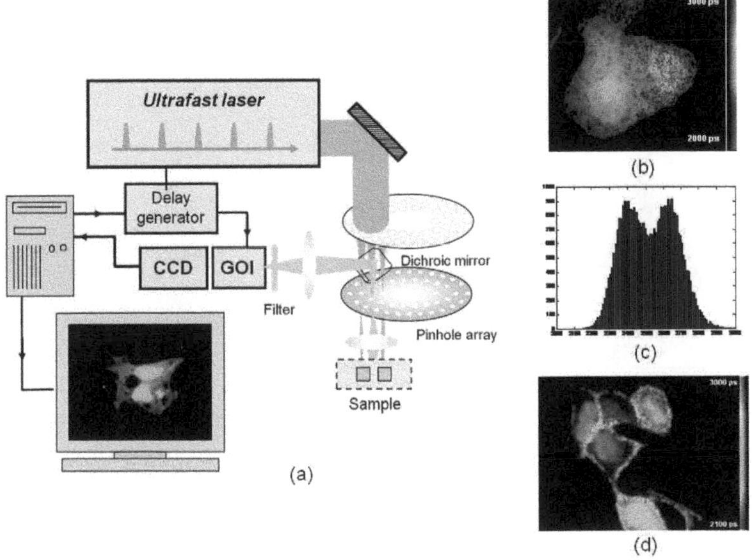

FIGURE 6.12: (a) Schematic of Nipkow disc FLIM microscope with (b) optically sectioned FLIM image of live cells expressing EGFP (left) and EGFP-mRFP FRET construct with (c) corresponding lifetime histogram and (d) FLIM-FRET image recorded with 5 s acquisition of MDCK cells expressing EGFP-Raf-RBD and H-Ras-mRFP, stimulated with EGF for 10 min. (Figure adapted from [61]).

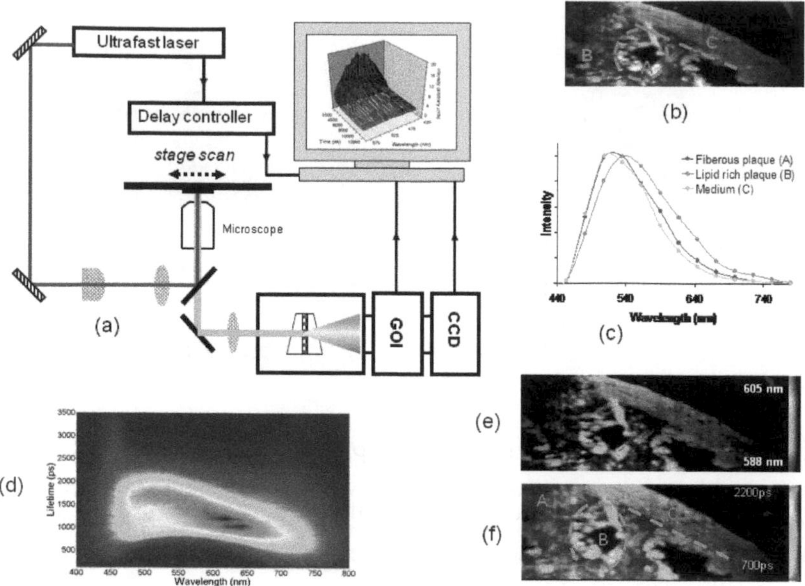

FIGURE 6.14: (a) Experimental setup for line-scanning hyperspectral FLIM, (b) integrated intensity image of sample of frozen human artery exhibiting atherosclerosis, (c) time-integrated spectra of sample regions corresponding to medium and fiberous and lipid rich plaques, (d) autofluorescence lifetime-emission matrix, (e) map of time-integrated central wavelength and (f) spectrally integrated lifetime map of sample autofluorescence. Adapted from [193].

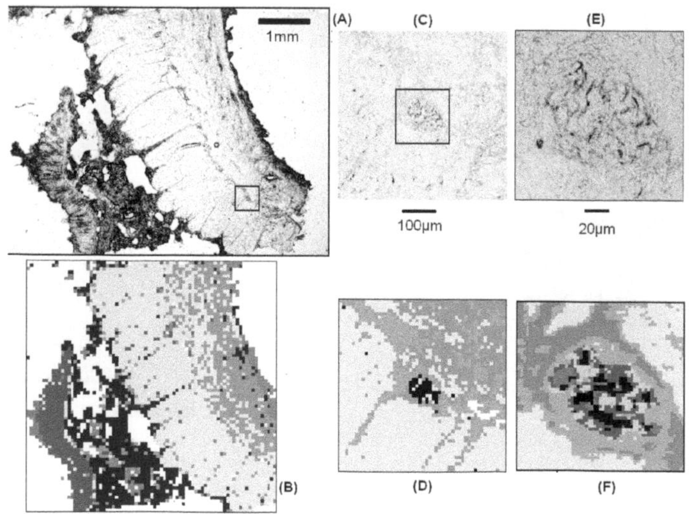

FIGURE 7.10: Photomicrographs (A, C, E) Raman images (B, D, F) of colon tissue. A step size of 60 μm gives gross overview of the tissue morphology. A step size of 10 μm of the boxed region in (A) resolves a ganglion. A step size of 2.5 μm of the boxed region in (C) displays subcellular features of the ganglion. Raman images were segmented by cluster analysis as previously described [?].

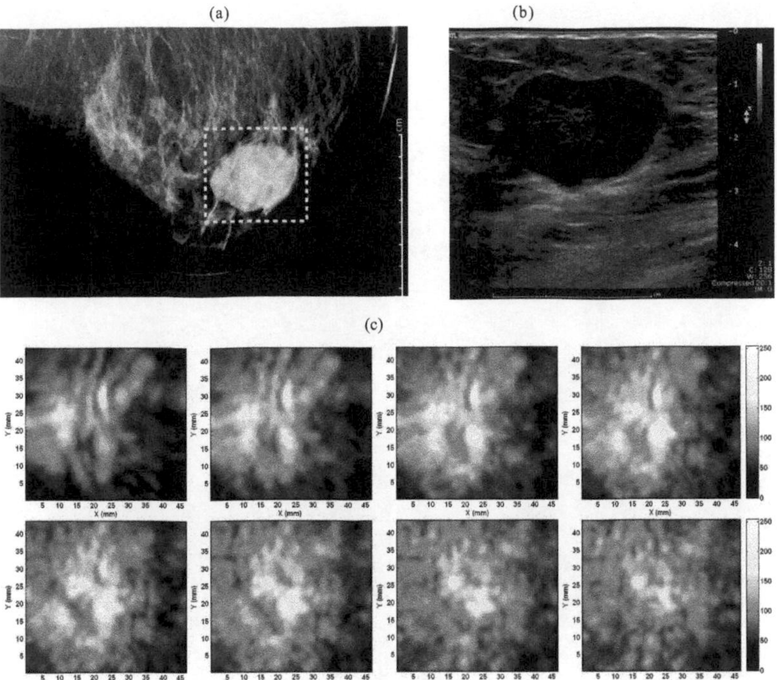

FIGURE 12.4: X-ray mammography (a), ultrasound (b), and OA slice images (c) of a 57-year-old woman with invasive ductal carcinoma exhibiting neuroendocrine differentiation in right breast (Case 2) [23]. The overall appearance of the tumor in (a) and (b) indicated benignity. The OA slice images shows high intensity distribution in a ring shape that can be attributed to higher vascular densities at tumor periphery. The interslice spacing is 1 mm with the first slice 9.5 mm below the illuminated breast surface. © OSA [23].

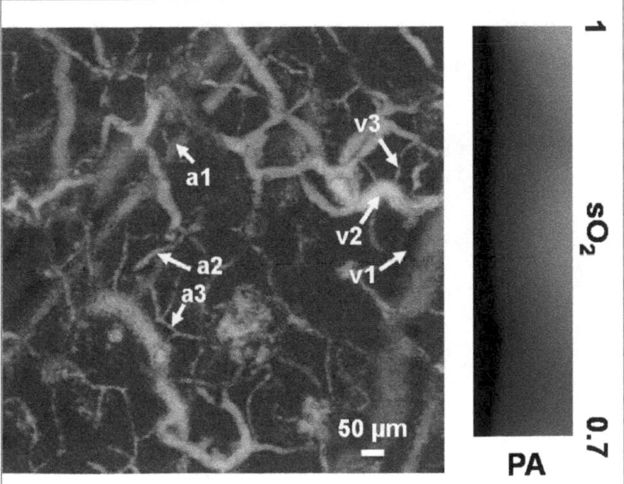

FIGURE 13.10: Pseudocolor sO_2 mapping of a microvascular network in a nude mouse ear acquired *in vivo* by OR-PAM. a1-a3; arterioles of different diameters; v1–v3: venules of different diameters. PA: photoacoustic signal amplitude.

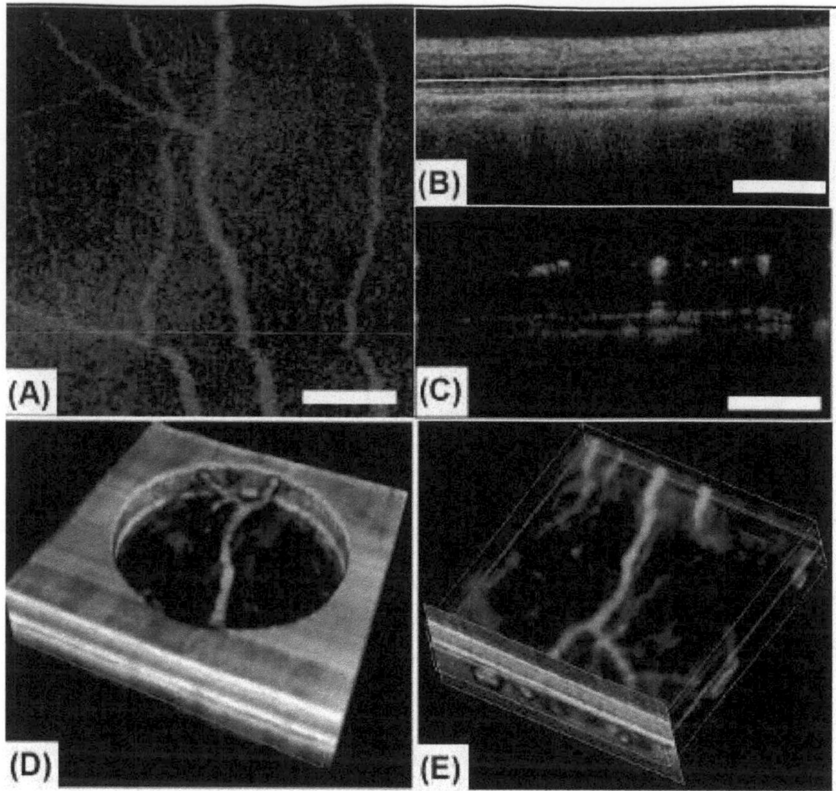

FIGURE 15.7: *In vivo* volumetric imaging of the posterior chamber of an eye from a volunteer. (A) OCT fundus image of the scanned volume. (B) OCT cross-sectional image at the position marked red in (A), in which four lines separate the blood flows in the retina and choroids. (C) Flying through a movie that represents the 2D OMAG flow images within the scanned volume. (D) Volumetric rendering of the merged structural and flow images with a cut through in the center of the structural image. (E) Volumetric rending of the blood flow image where flows in the retina are coded with green and those in choroids with red. Scale bar = 500 μm [36].

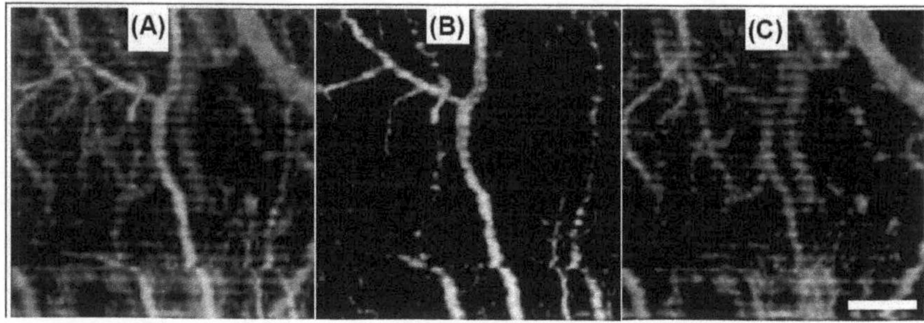

FIGURE 15.8: *x-y* projection images from 3D OMAG blood flow images. (A) Projection image from the whole scanned volume with the blood vessels in the retina are coded with green color, and those in choroids with red color. (B) *x-y* projection image from the blood vessels within the retina only. (C) *x-y* projection image from the blood vessels within the choroids only. Scale bar = 500 μm [36].

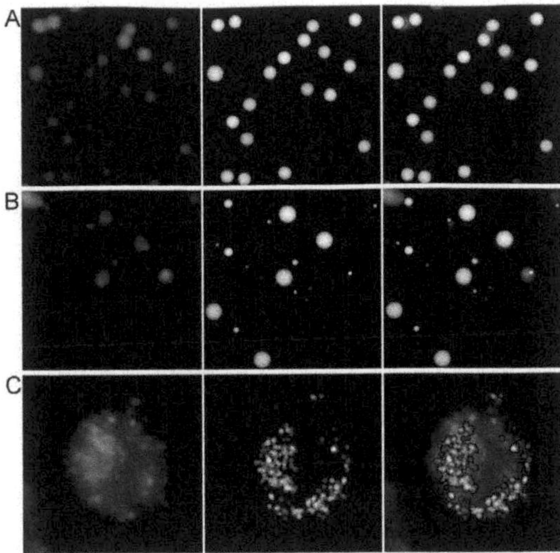

FIGURE 18.5: Fluorescence image of the suspensions of carboxylate-modified 1.9 μm diameter microspheres exhibiting red fluorescence (left side), the image reconstructed from the CLASS data (middle) and the overlay of the images (right side) (a). Same set of the images of the mixture of three sizes of fluorescent beads with sizes 0.5 μm, 1.1 μm, and 1.9 μm mixed in a ratio of 4:2:1 (a). Same set of the images of live 16HBE14o- human bronchial epithelial cells with lysosomes stained with lysosome-specific fluorescence dye (a) [1].

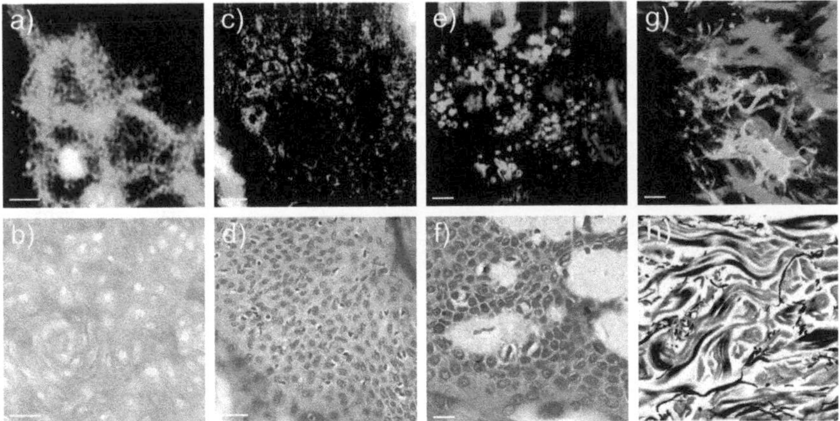

FIGURE 20.1: Comparison between *en face* optically sectioned human skin sample imaged with combined TPE intrinsic fluorescence (green) and SHG (red) microscopy and corresponding microphotographs taken with optical bright field microscopy on 5 μm thick *en face* sliced human skin samples stained with hematoxylin-and-eosin. Horny layer in human skin at 10 μm depth acquired by TPF (a) and corresponding histological image (b). Scale bars, 20 μm. Spinal layer in human skin at 45-μm depth acquired by TPF (c) and corresponding histological image (d). Scale bars, 20 μm. Dermo-epidermal junction in human skin at 70-μm depth acquired by TPF and SHG (e) and corresponding histological image (f). Scale bars, 25 μm. Elastic (green) and collagen (red) fibers in human skin at 120 μm depth acquired by TPF and SHG (g) and corresponding histological image (h) stained with Verhoeff-Van Gieson stain. Scale bars, 25 μm.

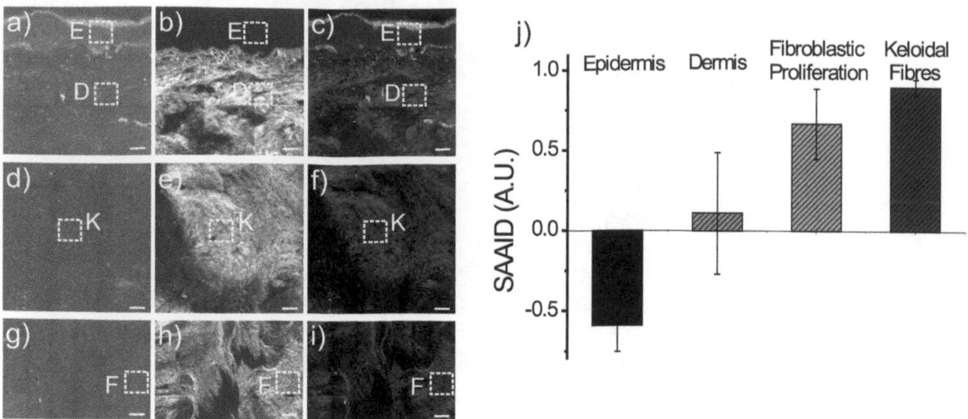

FIGURE 20.2: Transversal optically sectioned *ex vivo* skin sample, containing both epidermis (E) and healthy dermis (D) tissue, imaged using TPF microscopy (a), SHG microscopy (b), and the merge between the SHG and TPF image (c). Scale bars, 60 μm. Transversal optically sectioned *ex vivo* scar sample, containing keloidal fibers (K), imaged using TPF microscopy (d), SHG microscopy (e), and the merge between SHG and TPF image (f). Scale bars, 60 μm. Transversal optically sectioned *ex vivo* skin sample, containing fibrotic dermis tissue (F), imaged using TPF microscopy (g), SHG microscopy (h), and the merge between SHG and TPF image (i). Scale bars, 120 μm. Epidermis (E), healthy dermis (D), fibroblastic proliferation (F) and keloidal fibers (K) SAAID scores, calculated in five different zones for each of the four regions, are plotted in a bar graph (j).

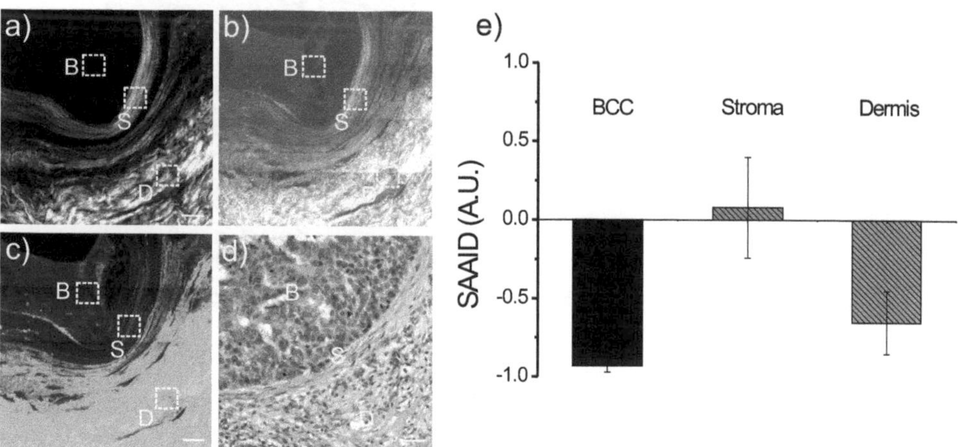

FIGURE 20.3: Transversal optically sectioned *ex vivo* BCC sample imaged using SHG microscopy (a), TPF microscopy (b), merging the SHG and the TPF image (c), and the corresponding histological image (d). Scale bars, 80 μm. BCC (B), tumor-stroma interface (S) and surrounding connective tissue (D) are highlighted by the corresponding letters. The SAAID score, calculated for the three highlighted regions (dashed line), is plotted in the bar graph in (e).

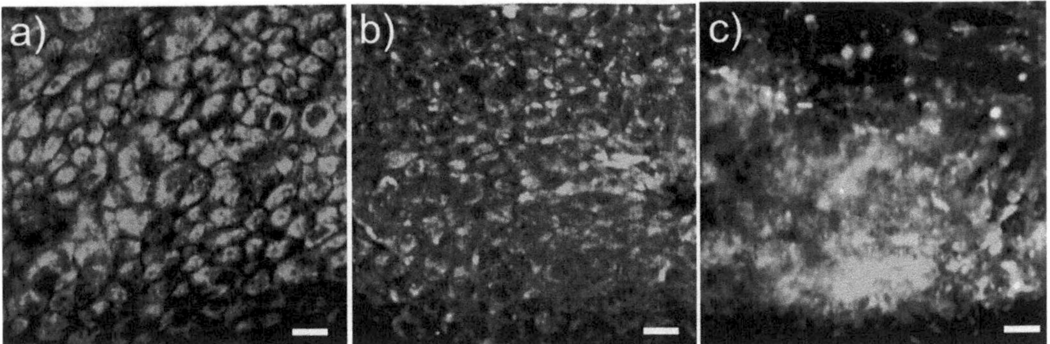

FIGURE 20.4: Two representative images of healthy bladder mucosa (a) and carcinoma *in situ* (b) acquired at 30-μm depth from tissue surface on *ex vivo* fresh biopsies. Images have been acquired using a two-photon autofluorescence of NADH with an 740 nm excitation wavelength. Scale bars, 10 μm. Tumor-connective tissue margin (c) for a carcinoma *in situ*, imaged by combined TPF microscopy (green) and SHG microscopy (blue). Scale bars, 15 μm.

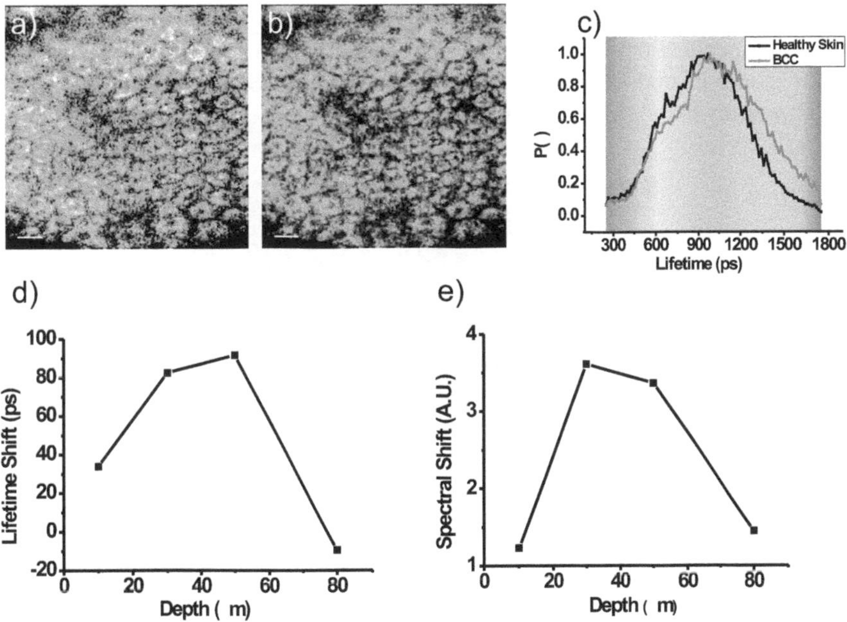

FIGURE 20.7: TPF image of human healthy epidermis spinal layer (a), acquired by photon counting method at 50-μm depth from the skin surface with an excitation wavelength of 740 nm and the corresponding FLIM image (b), obtained after system response de-convolution and single-exponential fit. Field of view, 200 μm. Scale bars, 20 μm. The color-coded scale for FLIM image is represented in (c) where normalized lifetime distributions for images acquired at the same depth for healthy skin (black line) and BCC (red line) are plotted. The lifetime shift and the spectral shift (measure of the area enclosed between healthy skin and BCC emission spectra) are plotted versus the depth of recording in (d) and (e), respectively.

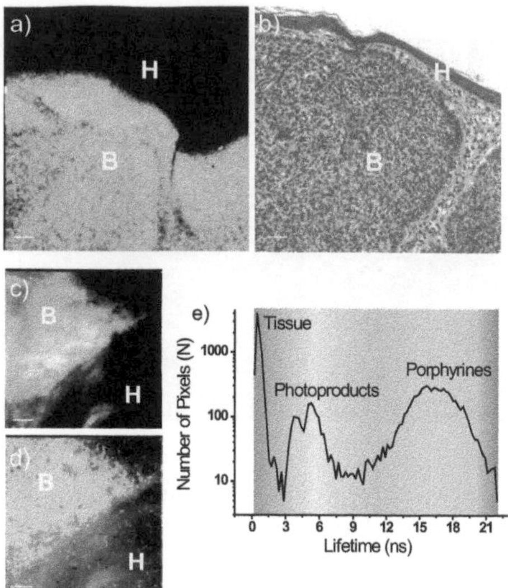

FIGURE 20.8: Transversal optically sectioned human skin tissue image acquired using TPF on a fresh ALA-treated sample (a) and the microphotograph taken after conventional histopathological examination and visualization by optical microscope on the same sample (b). Scale bars, 30 μm. *En face* optically sectioned human skin tissue image acquired using TPF on a fresh sample, excised 4 hours after ALA application (c) with the corresponding high contrast FLIM image (d), obtained after system response de-convolution and three-exponential fit. Scale bars, 20 μm. B: Basal Cell Carcinoma. H: healthy tissue. Mean lifetime distribution (e) of the image pixels of the image in (d).

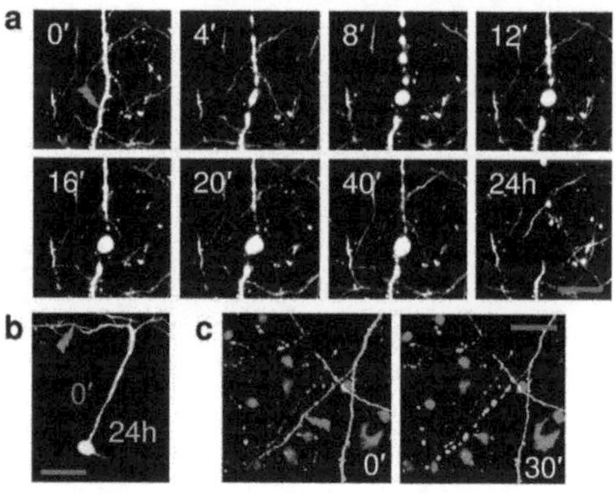

FIGURE 20.9: Laser-induced lesion of a single dendrite. (a) Time-lapse images of an irradiated dendrite (at $\approx$ 100 μm depth) in a GFP-M transgenic mouse. The dendrite has been irradiated (as indicated by the tip of the red lightning) just after the acquisition of the first image. Each image is a maximum-intensity projection of a set of optical sections acquired at a 1 μm z-step. Yellow numbers refer to the time from irradiation. Scale bar, 25 μm. (b) Overlay of maximum-intensity z-projections (from 500 to 100-μm depth) of a layer-5 pyramidal neuron before (red) and 24 h after (green) dendritic dissection (GFP-M line). This merge shows the integrity of the remaining structure after dissection. Scale bar, 60 μm. (c) The specificity of the laser dissection is demonstrated in two-color imaging experiments with labelled astrocytes. Laser nanosurgery is shown to disrupt only the GFP expressing dendrite (green), while the surrounding astrocytes (red) loaded with an intravital morphological dye (sulforhodamine 101) remain unaffected. Scale bar, 25 μm.

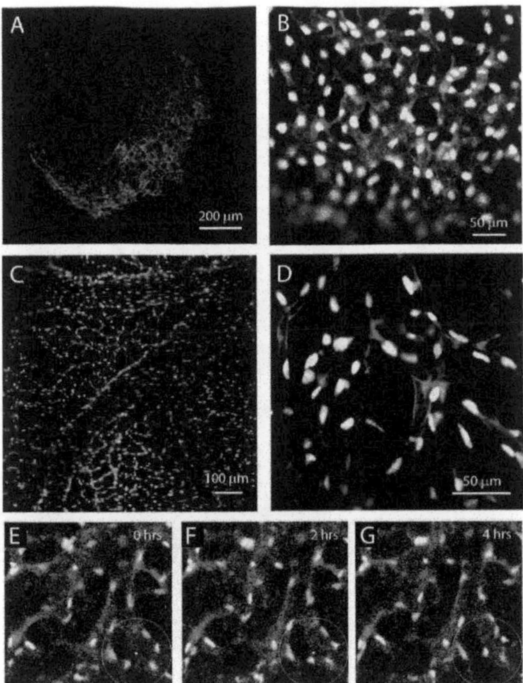

FIGURE 22.1: Imaging of embryonic vasculature in Tg(Flk1::myr-mCherry) X Tg(Flk1::H2B-EYFP) mouse line using confocal microscopy. (A, B) Vascular plexus of the yolk sac labeled by mCherry and EYFP at 8.5 dpc. (C) Remodeled vasculature of the yolk sac at 10.5 dpc. (D) Vasculature of the embryonic trunk at 12.5 dpc. (E–G) Vascular remodeling in the yolk sac captured by live imaging of embryo culture.

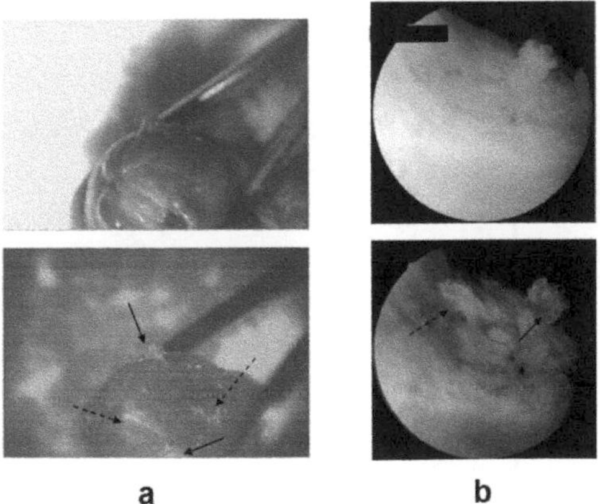

FIGURE 25.4: Examples of ALA-PpIX fluorescence imaging. a) in an oral cancer animal tumor model after intravenous administration of ALA, using green light excitation, showing the red PpIX fluorescence. b) bladder cancer detected during diagnostic cystoscopy in a patient, after instillation of ALA into the bladder and using blue light excitation (adapted from Jocham et al. [3],with permission). In each case, the corresponding white-light image is shown at the top. The solid arrows indicate frank tumor, while the dashed arrows indicate regions of early tumor tissue that are not easily seen under white light.

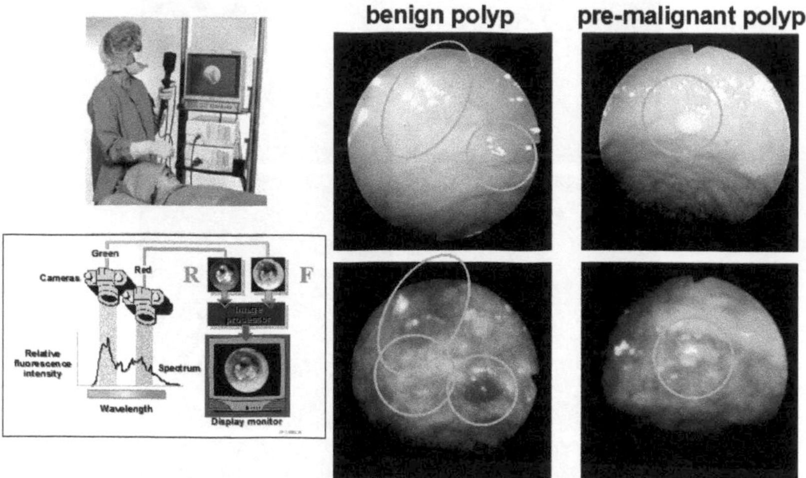

FIGURE 25.15: Autofluorescence endoscopic imaging. a) instrument and principle of operation based on detection of red and green fluorescence emission (courtesy Xillix Tech., Canada). b) example of white-light and autofluorescence images of polyps in the colon: the benign polyp shows similar fluorescence as the adjacent normal tissue, while the premalignant polyp (adenoma) has enhanced red-to-green fluorescence (courtesy Dr. N. Marcon, Toronto).

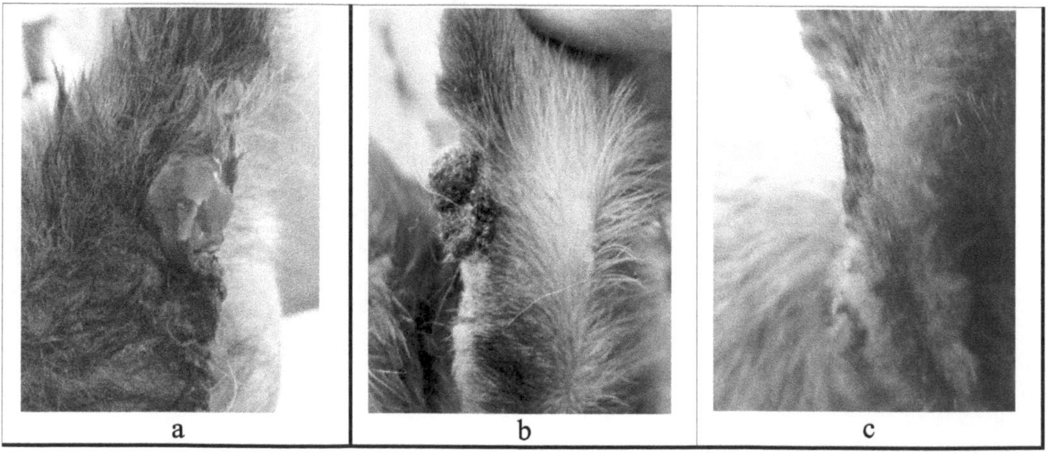

FIGURE 29.21: The cutaneous squamous cell carcinoma of the dog: a –before GNSs – PPTT, b – immediately after GNSs – PPTT, c – 5 days after GNSs – PPTT.

[103] B. Berge and J. Peseux, "Variable focal lens controlled by an external voltage: an application of electrowetting," *Eur. Phys. J. E* **3**, 159–163 (2000).

[104] S. Kwon, V. Milanovic, and L.P. Lee, "Large-displacement vertical microlens scanner with low driving voltage," *IEEE Photonics Technol. Lett.* **14**, 1572–1574 (2002).

[105] B. Qi, A.P. Himmer, L.M. Gordon, X.D.V. Yang, L.D. Dickensheets, and I.A. Vitkin, "Dynamic focus control in high-speed optical coherence tomography based on a microelectromechanical mirror," *Opt. Commun.* **232**, 123–128 (2004).

[106] A. Bjarklev, J. Broeng, and A.S. Bjarklev, *Photonic Crystal Fibers,* Kluwer Academic Publishers, Norwell, MA (2003).

[107] M.T. Myaing, J.Y. Ye, T.B. Norris, T. Thomas, J.R. Baker, W.J. Wadsworth, G. Bouwmans, J.C. Knight, and P.S.J. Russell, "Enhanced two-photon biosensing with double-clad photonic crystal fibers," *Opt. Lett.* **28**, 1224–1226 (2003).

[108] L. Fu, X.S. Gan, and M. Gu, "Nonlinear optical microscopy based on double-clad photonic crystal fibers," *Opt. Express* **13**, 5528–5534 (2005).

[109] G.P. Agrawal, *Nonlinear Fiber Optics,* Academic Press, San Diego (1995).

[110] E.B. Treacy, "Optical pulse compression with diffraction gratings," *IEEE J. Quantum Electron.* **Qe 5**, 454–458 (1969).

[111] W.H. Reeves, D.V. Skryabin, F. Biancalana, J.C. Knight, P.S. Russell, F.G. Omenetto, A. Efimov, and A.J. Taylor, "Transformation and control of ultra-short pulses in dispersion-engineered photonic crystal fibres," *Nature* **424**, 511–515 (2003).

[112] X. Li, Y. Leng, D. MacDonald, D. Wang, M.J. Cobb, W.A., and X.D. Li, "Flexible scanning microendoscope for two-photon fluorescence and SHG Imaging," in *Conference on Lasers and Electro-Optics, Optical Society America,* Baltimore, Maryland (2007).

[113] Y. Wu, Y. Leng, X. Li, D.J. MacDonald, M.J. Cobb, and X.D. Li, "Scanning fiber-optic endomicroscope system for nonlinear optical imaging of tissue," in *Topical Meeting on Biomedical Optics, Optical Society America,* St. Petersburg, Florida (2008).

[114] W.Y. Jung, S. Tang, D.T. McCormic, T.Q. Xie, Y.C. Ahn, J.P. Su, I.V. Tomov, T.B. Krasieva, B.J. Tromberg, and Z.P. Chen, "Miniaturized probe based on a microelectromechanical system mirror for multiphoton microscopy," *Opt. Lett.* **33**, 1324–1326 (2008).

[115] H. Kano and H. Hamaguchi, "In-vivo multi-nonlinear optical imaging of a living cell using a supercontinuum light source generated from a photonic crystal fiber," *Opt. Express* **14**, 2798–2804 (2006).

[116] M.J. Cobb, X.M. Liu, and X.D. Li, "Continuous focus tracking for real-time optical coherence tomography," *Opt. Lett.* **30**, 1680–1682 (2005).

[117] A.G. Podoleanu, J.A. Rogers, and D.A. Jackson, "Three dimensional OCT images from retina and skin," *Opt. Express* **7**, 292–298 (2000).

[118] M. Pircher, E. Gotzinger, and C.K. Hitzenberger, "Dynamic focus in optical coherence tomography for retinal imaging," *J. Biomed. Opt.* **11**, 054013 (2006).

[119] B.E. Bouma and G.J. Tearney, "Power-efficient nonreciprocal interferometer and linear-scanning fiber-optic catheter for optical coherence tomography," *Opt. Lett.* **24**, 531–533 (1999).

[120] X.D. Li, S.A. Boppart, J. Van Dam, H. Mashimo, M. Mutinga, W. Drexler, M. Klein, C. Pitris, M.L. Krinsky, M.E. Brezinski, and J.G. Fujimoto, "Optical coherence tomography: advanced technology for the endoscopic imaging of Barrett's esophagus," *Endoscopy* **32**, 921–930 (2000).

[121] G.J. Tearney, M.E. Brezinski, B.E. Bouma, S.A. Boppart, C. Pitris, J.F. Southern, and J.G. Fujimoto, "In vivo endoscopic optical biopsy with optical coherence tomography," *Science* **276**, 2037–2039 (1997).

[122] A. Das, M.V. Sivak, A. Chak, R.C. Wong, V. Westphal, A.M. Rollins, J. Izatt, G.A. Isenberg, and J. Willis, "Role of high resolution endoscopic imaging using optical coherence tomography (OCT) in patients with Barrett's esophagus (BE)." *Gastrointest. Endosc.* **51**, Ab93–Ab93 (2000).

[123] S. Brand, J.M. Poneros, B.E. Bouma, G.J. Tearney, C.C. Compton, and N.S. Nishioka, "Optical coherence tomography in the gastrointestinal tract," *Endoscopy* **32**, 796–803 (2000).

[124] M. Tsuboi, A. Hayashi, N. Ikeda, H. Honda, Y. Kato, S. Ichinose, and H. Kato, "Optical coherence tomography in the diagnosis of bronchial lesions," *Lung Cancer* **49**, 387–394 (2005).

[125] A.M. Rollins, R. Ung-arunyawee, A. Chak, R.C.K. Wong, K. Kobayashi, M.V. Sivak, and J.A. Izatt, "Real-time in vivo imaging of human gastrointestinal ultrastructure by use of endoscopic optical coherence tomography with a novel efficient interferometer design," *Opt. Lett.* **24**, 1358–1360 (1999).

[126] H.L. Fu, Y.X. Leng, M.J. Cobb, K. Hsu, J.H. Hwang, and X.D. Li, "Flexible miniature compound lens design for high-resolution optical coherence tomography balloon imaging catheter," *J. Biomed. Opt.* **13**, 060502 (2008).

[127] B.J. Vakoc, M. Shishko, S.H. Yun, W.Y. Oh, M.J. Suter, A.E. Desjardins, J.A. Evans, N.S. Nishioka, G.J. Tearney, and B.E. Bouma, "Comprehensive esophageal microscopy by using optical frequency-domain imaging (with video)," *Gastrointest. Endosc.* **65**, 898–905 (2007).

[128] J.H. Hwang, M.J. Cobb, M.B. Kimmey, and X.D. Li, "Optical coherence tomography imaging of the pancreas: a needle-based approach," *Clin. Gastroenterol. Hepatol.* **3**, S49–52 (2005).

22

Advanced Optical Imaging of Early Mammalian Embryonic Development

Irina V. Larina and Mary E. Dickinson

Department of Molecular Physiology and Biophysics, Baylor College of Medicine, One Baylor Plaza, Houston, TX, USA 77030

Kirill V. Larin

Biomedical Engineering Program, University of Houston, N207 Engineering Building 1, Houston, TX 77204, USA

Different dynamic aspects of early mammalian cardiovascular development can only be addressed by live embryonic imaging. Optical imaging has a clear advantage of high resolution compared to the other imaging techniques (such as micro-MRI and ultrasound bio-microscopy), while the imaging depth is acceptable due to small sizes of early embryos. Some optical imaging techniques are particularly well-suited for early developmental studies and have been used to answer many interesting questions about the mechanisms of invertebrate and vertebrate ontogeny. In this chapter we specifically focused on two approaches, confocal microscopy of vital fluorescent reporters and optical coherence tomography, discussing their advantages and limitations for live dynamic imaging of cultured mammalian embryos.

Key words: embryo, mammalian, live imaging, fluorescent proteins, optical coherence tomography, embryo culture, hemodynamics, cardiodynamics

22.1 Introduction

Very little is known about human embryonic cardiovascular dynamics because currently available imaging methods are insufficient to visualize the cardiovascular system in the first few weeks of pregnancy when the heart begins to beat. The mouse model provides a great system for studying mammalian development since its developmental processes are similar to humans and because of the wealth of available methods to manipulate the genome and create mutant animals. Many mutant mouse lines with various cardiovascular developmental defects linked to human diseases are developed and are being identified regularly using single-gene targeting approaches and large-scale screens creating a pressing need for imaging techniques allowing the study of functional effects of the genetic manipulations on cardiovascular development and dynamics.

In mice, the embryonic heart starts to form on 7.5 days postcoitum (dpc) (0.5 dpc is counted as noon of the day of the vaginal plug). At about the same time endothelial cells and blood cells appear in blood islands of the yolk sac. Within 24 hrs endothelial cells migrate, proliferate, and interconnect to form a vascular plexus, while cardiac progenitors form the heart tube consisting of an endocardial layer surrounded by cardiomyocytes. At about 8.5 days, the embryonic heart starts to beat, to circulate plasma and later blood cells through the vascular system. Shear stress created by the blood flow facilitates remodeling of the vascular plexus into more mature branching circulatory system, while the heart tube expands, loops, forms chambers and valves, and forms a four-chambered functional pump. Since the most structural rearrangements of the heart are happening during embryonic days 7.5 to 12.5, it is extremely important to focus analysis of cardiodynamics on these stages.

Even though mouse embryos develop inside the uterus, which restricts access for light microscopy, novel methods for manipulating mouse embryos *ex vivo* and protocols for growing mouse embryos on the microscopic stage have recently been established [1–4]. By controlling temperature, pH, and supplementing growth medium with a number of factors, mouse embryos dissected out of the uterus with a yolk sac intact can be grown in a static culture for more than 24 hours. This technique allows direct visualization of tissues during the embryonic growth and development at 6.5 dpc to about 10.5 dpc. Later-stage embryos (up to 13.5 dpc) can be maintained and imaged *exo utero* for shorter periods of time.

Here we are describing recent results on the live imaging of mouse embryonic cardiovascular development in static culture using two optical approaches, confocal microscopy of vital fluorescent markers and Optical Coherence Tomography (OCT).

22.2 Imaging Vascular Development Using Confocal Microscopy of Vital Fluorescent Markers

Time-lapse fluorescence imaging combined with fluorescent protein reporter lines is a powerful tool for studying cell migration and organ development as well as for researching cellular consequences of genetic manipulations [1]. Several genetic strands have been engineered that express fluorescent proteins in specific tissues of the embryo, for example, ε-globin-EGFP expressing green fluorescent protein (GFP) in primitive erythroblasts, Flk1::H2B-EYFP expressing nuclear yellow fluorescent protein (YFP) in the endothelial cells, and Oct4-GFP expressing GFP in primordial germ cells. Crossing different mutant lines to transgenic mice carrying fluorescent markers expands the potential of optical imaging in studying molecular mechanisms and signaling events regulating embryonic development.

Fluorescent imaging has been used in different models to visualize and analyze the cellular dynamics during development *in vivo* [5–7]. The use of tagged fluorescent proteins under control of specific regulatory elements has significantly increased the ability to track individual cells, characterize tissue dynamics, and understand the formation of complex cytoarchitectures [3, 5]. Using ε-globin-EGFP transgenic mouse line expressing green fluorescent protein (GFP) in embryonic blood cells we have been successful in quantifying early circulation events, in studying the initiation of blood formation, and in characterizing mutants with circulation defects [4, 8].

Recently, we developed a novel transgenic model, in which mCherry fluorescent protein is expressed in the embryonic endothelium and endocardium [9]. The mCherry fluorescent protein fused to a myristoylation motif (myr) and followed by a SV40 polyadenylation site was cloned between Flk1 regulatory elements. The Flk1 promoter was used because, as was shown previously [10], it drives expression in the embryonic vasculature in the Tg(Flk1::H2B-EYFP) line. The myristoylation motif allows for membrane localization of the fluorescent protein, and therefore, outlines the structure of the vasculature and reveals cellular morphology and cell-cell boundaries between the endothelial cells.

One benefit of the Tg(Flk1::myr-mCherry) transgenic line is the ability to cross this line with other fluorescent protein markers with minimal spectral overlap. As mCherry has an emission peak at 610 nm, it is easily separated from blue, green, or yellow fluorescent proteins. We have crossed Tg(Flk1::myr-mCherry) mice with mice from the Tg(Flk1::H2B-EYFP) line, which expresses H2B-EYFP in the nuclei of embryonic endothelial cells [10]. By crossing the new Tg(Flk1::myr-mCherry) line with mice from the Tg(Flk1::H2B-EYFP) line, we showed that both fluorescent proteins can be easily detected in different subcellular compartments. In fact, these markers provide a powerful combination since the EYFP marker allows for the analysis of cell division and the tracking of each endothelial cell whereas the membrane-targeted mCherry marker reveals cell boundaries and highlights vessel organization. This combination of markers allows identifying each endothelial cell within a vessel to reveal the distribution of individual cells within each vessel segment. In addition, because the Tg(Flk1::H2B-EYFP) line has been well characterized [10] and the expression of fluorescent proteins in both lines is driven by the same Flk1 regulatory elements, we used this cross to verify overlap of the expression patterns.

In embryos homozygous for the Flk1::myr-mCherry transgene, the level of fluorescence is notably higher than in embryos heterozygous for the insertion. Fluorescence of mCherry is first detectable at the early headfold stage (7.5 dpc) in the blood islands of the yolk sac. At this stage, the fluorescence is relatively dim and is only detectable in homozygous embryos. At 8.0 dpc, mCherry outlines the vascular plexus as it develops in the yolk sac. By 8.5 dpc the vascular plexus is brightly highlighted by the transgene (Figure 22.1A), and membrane boundaries between endothelial cells could be detected in some cells at 63× magnification (Figure 22.1B).

While both markers remain in the endothelial cells, the vascular plexus of the embryonic yolk sac is undergoing remodeling into the more mature circulatory system consisting of larger vessels and branching from them progressively smaller vessels. Figure 22.1C shows remodeled yolk sac vasculature at the embryonic day 10.5 labeled by the mCherry and EYFP. In the embryo proper, the markers are co-expressed from 8.5 dpc until late gestation throughout the vasculature of different embryonic tissues, such as trunk, skin, brain, and eye. The detailed subcellular localization can be appreciated in a high magnification views of embryonic vessels with outlined membrane boundaries and brightly labeled nuclei. As an example, Figure 22.1D shows expression of the fluorescent markers in the embryonic trunk at 12.5 dpc. During late gestation, expression of the myr-mCherry and the H2B-EYFP decreases in major blood vessels, but remained in some smaller vessels throughout postnatal life. The mCherry could be detected in the adult transgenic mice in many of the micro vessels throughout the body, for example, in the skin, the brain, and the muscle. The mCherry expression in the small capillaries is providing a quick and convenient way to genotype transgene carriers.

To visualize dynamic changes during formation of the vasculature in the embryonic yolk sac, time lapse confocal microscopy can be performed on live embryonic cultures grown at the microscopic stage. The expression of this marker from early stages, when the vascular plexus is forming, throughout the remodeling process has provided useful insights into the dynamics of vessel formation and maturation. At early stages, confocal time-lapse analysis shows many small sprouts forming to increase the number of vessel branches, whereas at later stages, after blood flow begins, many vessel branches regress. Using this technique, we were able to image formation of the vasculature in the primitive plexus of the yolk sac by starting the culture at 7.5 dpc when mCherry was detected only in the region of the blood islands [9]. The expression of mCherry and EYFP became evident throughout the yolk sac about 8 hours later; however, defined vascular structures were difficult to visualize until 12 hours after the beginning of the culture. The level of expression of the fluorescent protein increased continuously during the course of the time lapse, and by 20 hours the

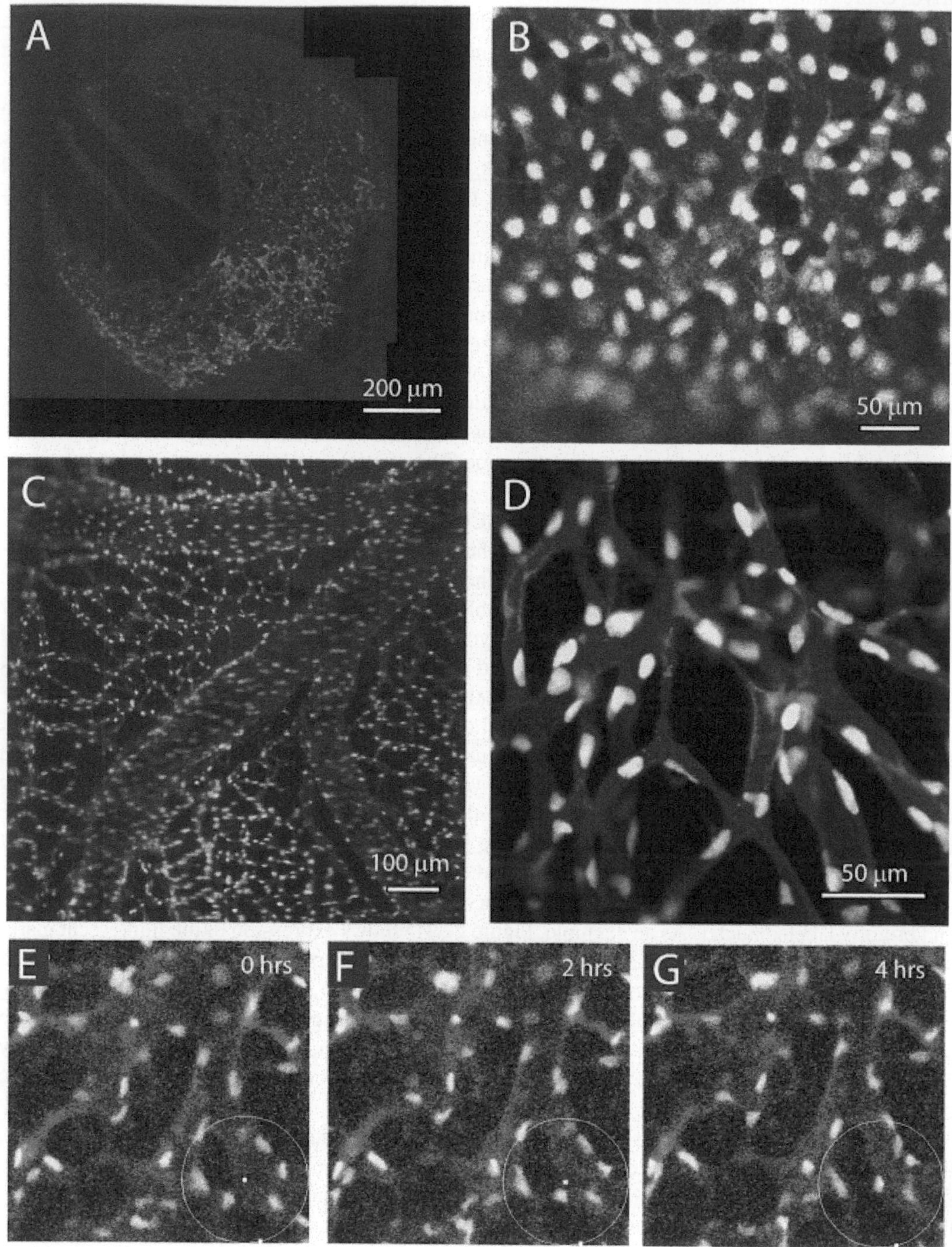

FIGURE 22.1: (See in color after page 572.) Imaging of embryonic vasculature in Tg(Flk1::myr-mCherry) X Tg(Flk1::H2B-EYFP) mouse line using confocal microscopy. (A, B) Vascular plexus of the yolk sac labeled by mCherry and EYFP at 8.5 dpc. (C) Remodeled vasculature of the yolk sac at 10.5 dpc. (D) Vasculature of the embryonic trunk at 12.5 dpc. (E–G) Vascular remodeling in the yolk sac captured by live imaging of embryo culture.

entire vascular plexus was brightly outlined to allow detection of cellular movements during the formation of the vascular plexus. We observed multiple sprouting events of endothelial cells in the same area of the yolk sac and formation of vascular branches. This is in contrast to data taken from embryos at later stages that shows the regression of many vessel segments. Figure 22.1(E–G) shows images of Tg(Flk1::myr-mCherry)XTg(Flk1::H2B-EYFP) embryonic yolk sac during live embryo culture on the imaging stage that started at about 8.5 dpc, a time when blood flow is just beginning and the fluorescent markers are labeling the immature vascular plexus. The time lapse (images taken every 10 minutes, but shown for 4 hr intervals) captured the regression of a vessel segment during vascular remodeling. The outlined vascular branch is thinning and breaks, while the endothelial cell composing the branch is joining the larger of two vessels (as evident by the movement of the EYFP labeled nucleus) with presumably higher flow. Analysis of live dynamic changes during the remodeling on the cellular level helps to understand the process of refinement of the vascular structure with some vessels regressing and those along the major flow trajectories remaining. These observations are further supporting recent studies demonstrating that vessel remodeling in the yolk sac is triggered by hemodynamic force [4, 8]. Further studies are underway to determine if sprouting persists and vessel regression is absent if normal blood flow is disrupted.

Even though the Tg(Flk1::myr-mCherry) line was designed to allow live imaging of developing vasculature, the mCherry marker can be fixed to allow immunohistochemical analysis in combination with other fluorescent markers. According to the original designers of the mCherry protein, this marker is quite stable to photobleaching [11]; mCherry is over tenfold more photostable than mRFP1 and only about 40% less photostable than EGFP. In consistence with this report, no decrease in fluorescence intensity of mCherry due to photobleaching was noticed during experiments.

One of the limitations of time-lapse imaging of live embryos is that repetitive imaging with high intensity lasers can be harmful to embryo growth. However, according to our observations, normal time-lapse sequences can be acquired for at least 24 hours, similar to other fluorescent protein transgenic lines that we have used [10].

Expression of the mCherry reporter in the embryonic endocardium provided sufficient contrast to acquire images of the mouse embryonic heart tube as it was beating. Recently, great strides have been made in understanding how the zebrafish embryonic heart functions. Rapid confocal imaging of transgenic zebrafish embryos expressing fluorescent proteins in cardiac cells has been used to define how the embryonic heart acts as a pump and how blood flow changes as the morphology of the heart becomes more complex [12–14]. Similar studies have not been undertaken in mouse embryos because reporter lines with robust expression in the early endocardium have not been available for similar analysis. Tg(Flk1::myr-mCherry) embryos are suitable for repeated imaging at 6 frames per second, opening the door for further analysis into the mechanisms of heart function for studying the aberrant heart in mutant models.

Even though confocal microscopy of vital fluorescent markers can provide valuable information about early developmental events, this imaging technique is associated with some limitations, the major of which is the imaging depth. The typical imaging depth for bright fluorescent markers (such as mCherry and EYFP) is about 200 μm. Therefore, it's perfect for visualization of the yolk sac and embryonic structures within the imaging distance, but deep embryonic tissues are not accessible for live confocal imaging. Until embryonic turning (about 16 somites on day 9.0) the developing heart is positioned on the outside of the embryo under the yolk sac, making it accessible for confocal imaging, but during embryonic turning the heart moves away from the yolk sac making confocal microscopy not applicable for analysis of cardiodynamics at these stages. Besides, the expression of the mCherry in the endocardium decreases after the beginning of circulation, which further complicates this analysis. This limitation can be addressed by Optical Coherence Tomography (as discussed below) which has an order of magnitude higher imaging depth to allow deep embryonic tissue imaging.

22.3 Live Imaging of Mammalian Embryos with OCT

Optical coherence tomography (OCT) is a 3-D imaging modality with the capability to image 2 to 3 millimeters into soft tissue and still maintains a reasonably high spatial resolution ($\sim$ 2–20 μm) [15, 16]. Implementation of Doppler measurements into the OCT facilitates the acquisition of both structural and velocity information at the same spatial and temporal resolution [17]. OCT has significant advantages over other methods such as laser scanning microscopy, which has limited imaging depth, or ultrasound that allows deep tissue imaging but with lower spatial resolution, which makes OCT an ideal imaging technique for studying early embryonic mouse development.

Several research groups have successfully applied OCT for live imaging of embryonic cardiodynamics and blood flow in Drosophila [18], *Xenopus laevis* [19, 20], quail [21], and chick [22]. Despite the critical need to understand mammalian embryonic development, studies on OCT imaging of mammalian cardiovascular dynamics are limited. Jenkins et al. applied OCT for imaging of extracted mouse embryonic hearts at 12.5 dpc and 13.5 dpc [23]. The same group reported 3-dimensional OCT images of 13.5 dpc (days postcoitum) beating, embryonic mouse hearts that were excised and externally paced [24]. Likewise, Luo et al. imaged beating 10.5 dpc hearts in embryos that were maintained outside the uterus, but with slower than normal heart beats [25]. These studies have revealed the potential of OCT for imaging embryonic mouse development, but further studies from these groups have been limited by the challenge of maintaining normal physiology. We combined mouse embryo manipulation protocols and static embryo culture previously optimized for confocal microscopy with Swept Source OCT (SS-OCT) for structural and hemodynamic analysis of the embryos [26, 27].

SS-OCT system used in these experiments (Figure 22.2) employs a broadband swept-source laser (Thorlabs, SL1325-P16) with output power $P = 12$ mW at central wavelength $\lambda_0 = 1325$ nm and spectral width $\Delta\lambda = 110$ nm. The scanning rate over the full operating wavelength range is 16 kHz. An interferogram is detected by a balanced-receiver configuration that reduces source intensity noise as well as autocorrelation noise from the sample (Thorlabs, PDB140C) and is digitized using 14-bit digitizer. Mach-Zehnder interferometer (MZI)-based optical frequency clock is used to calibrate the OCT interference signals from the optical time-delay domain in the frequency-domain before application of Fast Fourier Transform (FFT) algorithms. FFT reconstructs an OCT intensity in-depth profile (A-scan) from a single scan over the operating wavelength range, resulting in the A-scan acquisition and processing rate of 16 kHz. The scanning head of the OCT system was positioned inside the commercial 37°C, 5% CO_2 incubator. The dissection station was heated and maintained at 37°C using a custom-made heater box and conventional heater. Dissected embryos were transferred to 37°C, 5% CO_2 incubator for at least 1 hour for recovery and kept in the incubator until imaging (up to 4 hours after the dissection).

22.3.1 Structural 3-D imaging of live embryos with SS-OCT

Structural OCT is very effective as a method to produce 3-D reconstructions of entire embryos, similar to ultrasound, but with cellular resolution. Figure 22.3 shows examples of 3-D reconstructions of embryos at different developmental stages acquired with SS-OCT. The reconstructions provided sufficient details about embryonic structures (as labeled on the panels) that cannot be revealed at a similar resolution by other imaging techniques. During early stages (until about 9.5 dpc) the whole embryo is within the imaging depth of the system. As the embryo continues to grow, it's becoming more challenging to image and characterize development and dynamics of specific organs (such as heart) and requires careful positioning of the embryo on the imaging stage. Removal of the yolk sac significantly improves imaging quality to allow structural morphological analysis (Figure

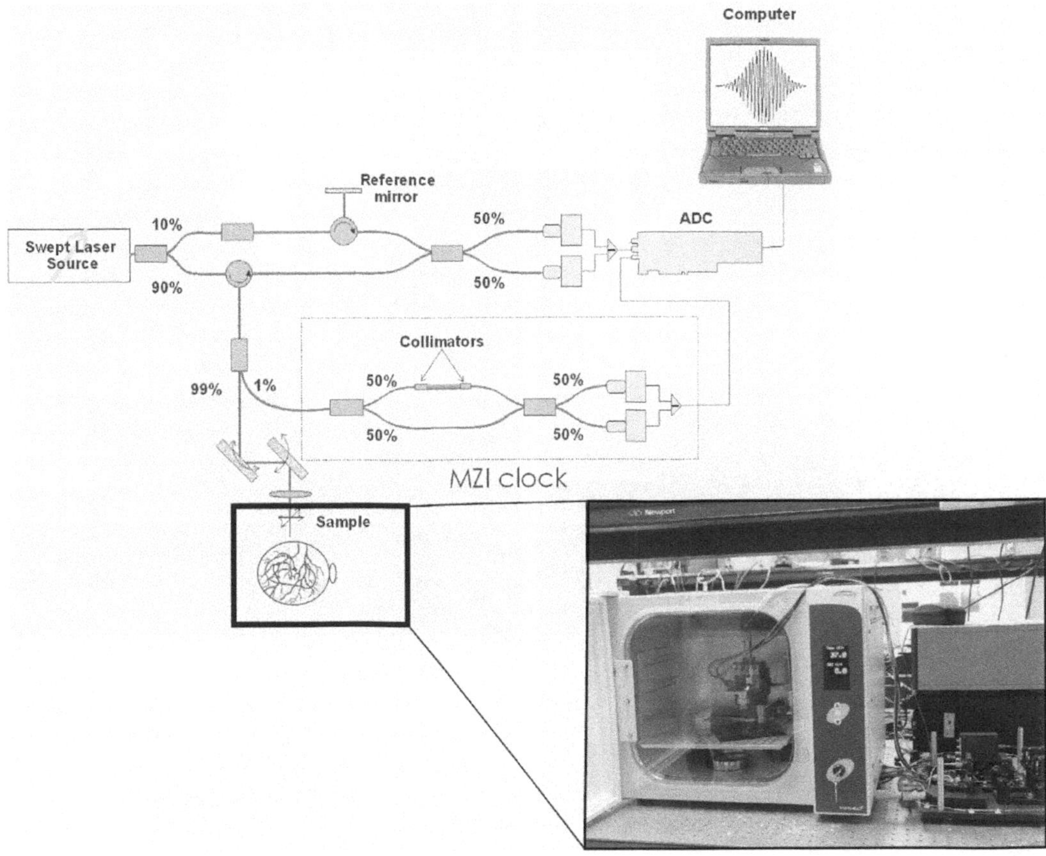

FIGURE 22.2: SS-OCT experimental setup for live embryonic imaging.

22.3F), however, this procedure disrupts circulatory network of the embryo and, therefore, cannot be performed with live imaging.

Figure 22.4 shows cross-sections through the heart region of the live embryo at 8.5 dpc. As one can see from the panels, the internal structure of the heart as well as other details of the embryo are clearly outlined demonstrating the capability of the SS-OCT for detailed structural morphological analysis of the heart as well as other embryonic organs. Figure 22.5 shows representative frames from time lapses acquired with SS-OCT from the 8.5 dpc embryo (about 8–10 somites). This stage corresponds to the beginning of blood circulation, when a majority of the blood cells is still found in blood islands with only a fraction of cells circulating. In the images, individual circulating blood cells are clearly visible in the inflow tract and the heart, the sinus venosus, and the vitelline vein. Although we can detect single blood cells and the frame rate of the system is sufficient to follow their movement, we found it technically difficult to orient the imaging plane as to capture the three-dimensional trajectory of the moving cells and to determine cell velocity by direct cell tracking. While structural SS-OCT imaging is great for visualizing blood flow at the early stages of circulation, Doppler SS-OCT can provide measurements for detailed hemodynamic analysis, as discussed below.

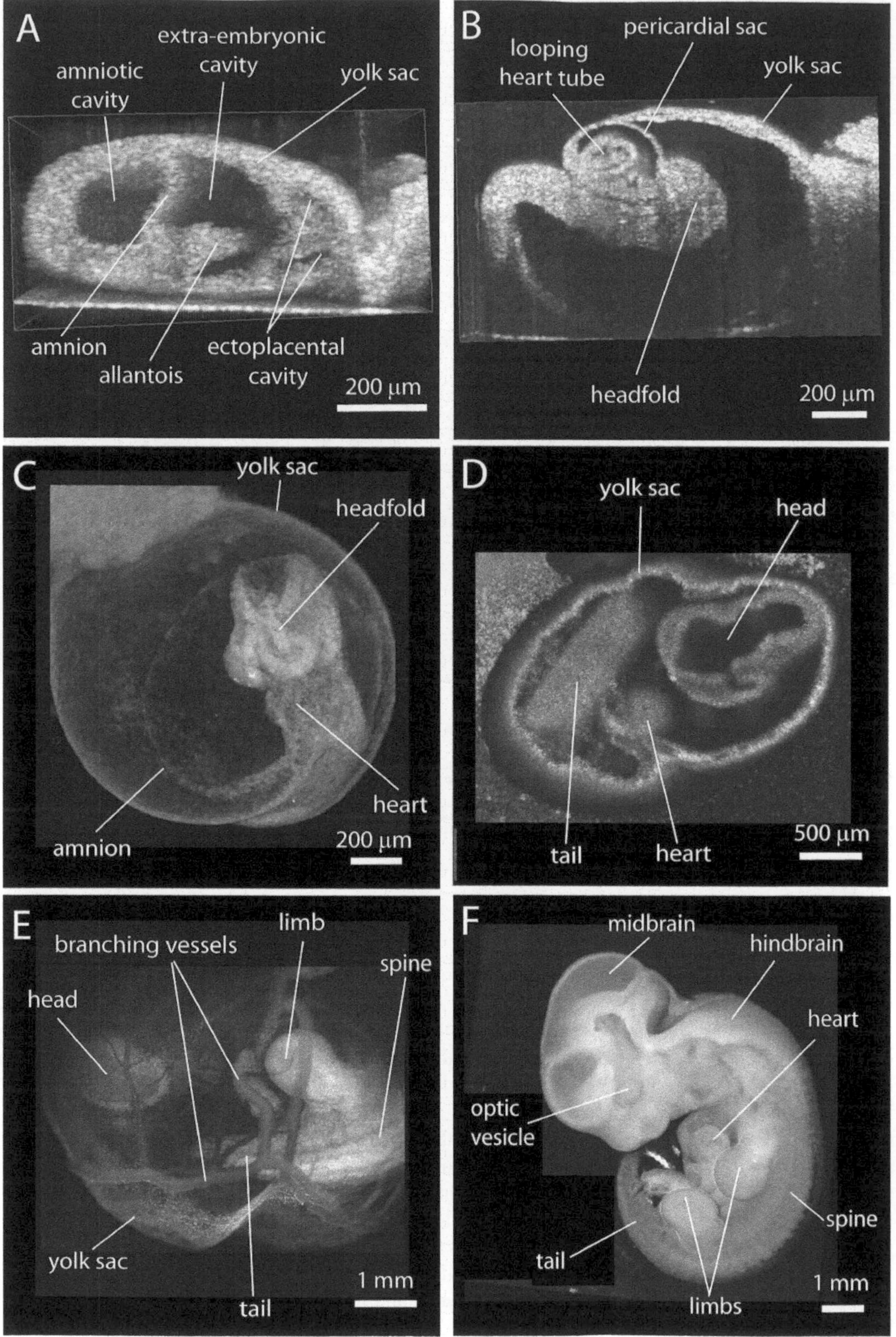

FIGURE 22.3: Structural imaging of early embryos with SS-OCT. (A–E) 3-D reconstructions of live embryos with the yolk sac at 7.5 dpc, 8.5 dpc, 8.75 dpc, 9.5 dpc, and 10.5 dpc, respectively. (F) 3-D reconstruction of 10.5 dpc embryo.

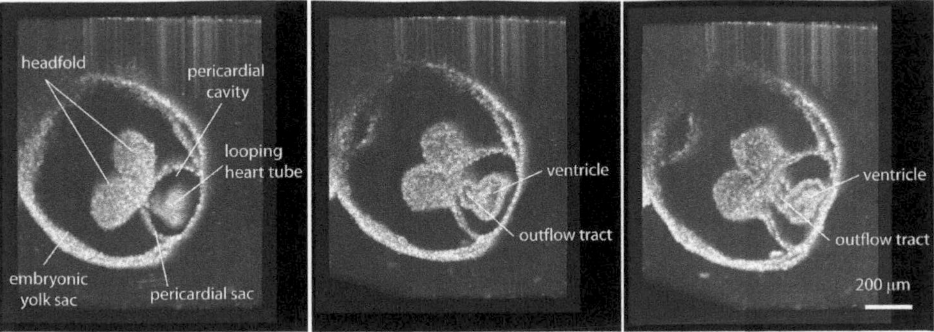

FIGURE 22.4: Structural imaging of the developing embryonic heart. Cross-sections through the heart of the 3-D reconstructions of live embryo at 8.5 dpc.

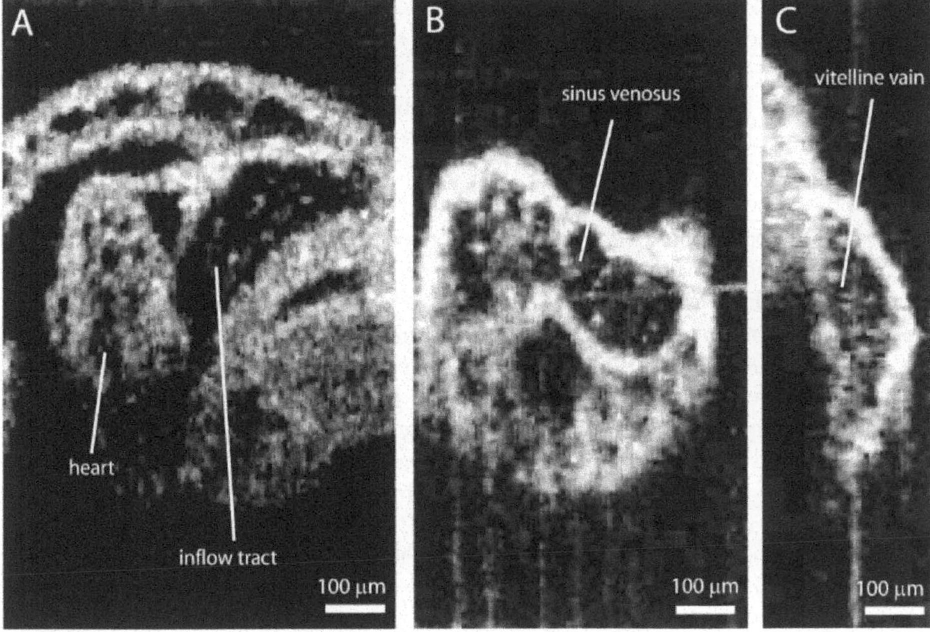

FIGURE 22.5: Structural imaging of blood flow at the early stages of circulation. Individual blood cells and small groups of cells are distinguishable in the inflow tract of the heart (A), sinus venosus (B), and the vitelline vein (C) of the 8.5 dpc embryo.

22.3.2 Doppler SS-OCT imaging of blood flow

Doppler OCT velocity imaging can be performed at the same spatial and temporal resolution as the structural OCT imaging. Figure 22.6 shows representative examples of structural and corresponding Doppler SS-OCT frames taken from the time lapses acquired from a yolk sac vessel at 9.5 dpc (Figures 22.6A and 22.6B, respectively) and an embryonic vessel at 10.5 dpc (Figures 22.6C and 22.6D, respectively). If the direction of the blood flow is known, blood flow at each pixel can be reconstructed according to the formula [28]:

$$v = \Delta\phi/(2n\langle k\rangle\tau\cos(\beta)), \tag{22.1}$$

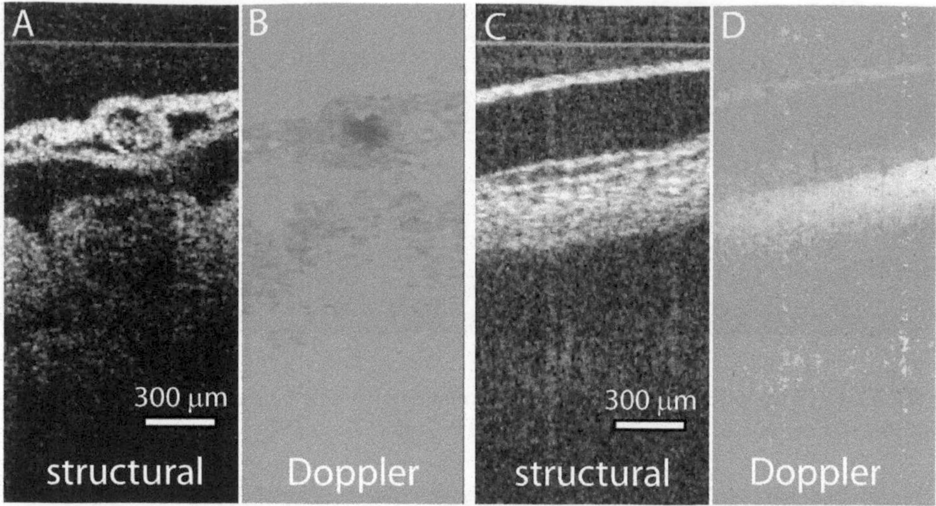

FIGURE 22.6: Doppler SS-OCT imaging of blood flow. Structural (A, C) and corresponding Doppler images (B, D) of the yolk sac vessel at the embryonic stage 9.5 dpc and embryonic vessel at 10.5 dpc, respectively.

where $\Delta\phi$ is the Doppler phase shift calculated between successive A-scans, n is the refractive index, $\langle k \rangle$ is the average wave number, τ is the time between A-scans, and β is the angle between the flow direction and the laser beam. An angle β can be calculated from structural 2-D and 3-D data sets acquired from the embryos. In the experiments presented here the refractive index was assumed to be $n = 1.4$. We recently applied Doppler SS-OCT analysis to reconstruct spatially and temporally resolved Doppler shift velocity profiles from a yolk sac vessel and the dorsal aorta deep within the 9.5 dpc mouse embryo [27], when the blood flow is well established. Yolk sac measurements made with this system were similar to values previously measured by fast scanning confocal microscopy, while the dorsal aorta flow profiles are unique for this imaging technique and demonstrate the advantage of the Doppler SS-OCT over the confocal microscopy to detect blood flow deep in tissue suggesting that Doppler SS-OCT is an effective way to make blood flow measurements in early embryos.

Hemodynamic analysis with Doppler OCT can also be performed at early stages of circulation while blood flow is establishing [26]. Figure 22.7A shows a structural image of the dorsal aorta within the 8.5 dpc embryo and corresponding Doppler velocity maps acquired at different phases of the heartbeat cycle from the same area of the embryo just a few hours after the beginning of heartbeat when blood circulation first begins. At this stage, the majority of blood cells are still found in the blood islands with limited numbers of circulating erythroblasts [8]. The Doppler velocity images were taken at 512 A-scans per frame at 25 fps. The Doppler shift signal from a small group of cells, as well as individual circulating blood cells, are clearly distinguishable in the images. Doppler velocities from all individual detectable blood cells in Figure 22.7A were measured for each acquired Doppler time frame in the time lapse. Figure 22.7B shows an average blood flow velocity plotted vs. time. Dynamics of the blood flow velocity in time reveals the pulsatile nature of the flow and allows analyzing hemodynamic changes during the heartbeat. The heart rate (about 2 beats per second) correlates well with previously reported measurements at this embryonic stage and the flow velocity values acquired in the dorsal aorta are similar to published blood flow measurements in the yolk sac at the same embryonic stage [4], supporting the physiological relevance of these data.

Blood flow velocity profiles across the vessel at different phases of the heartbeat cycle were reconstructed by analyzing individual time frames and measuring the Doppler velocity shift from each

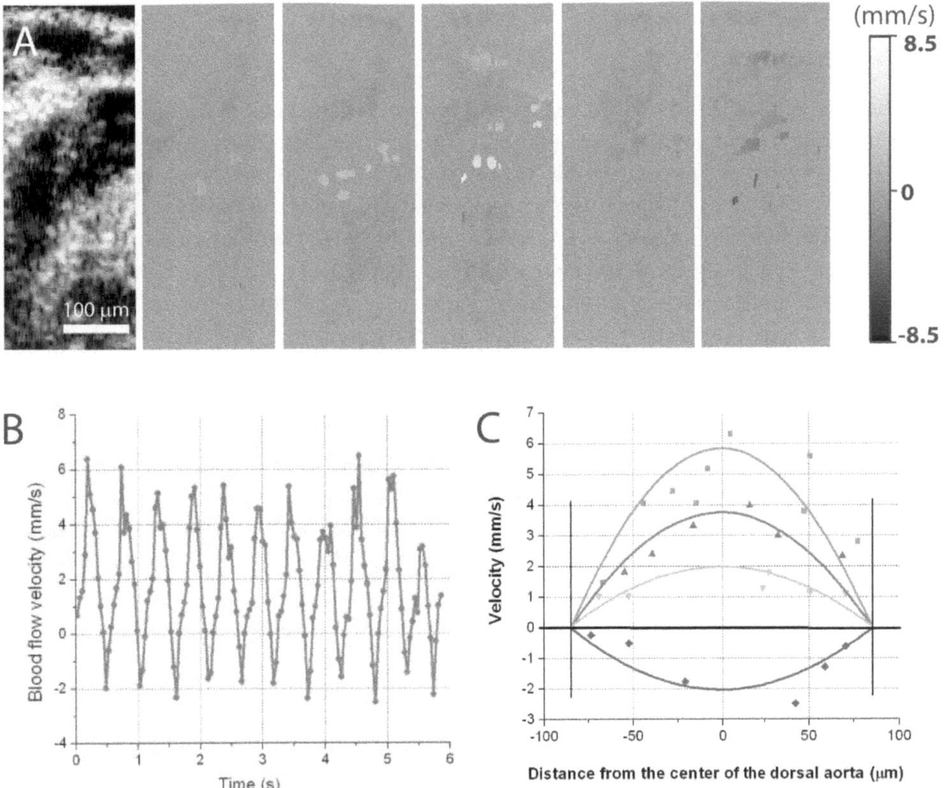

FIGURE 22.7: Doppler OCT velocity measurements from individual blood cells. (A) Structural and corresponding Doppler velocity images acquired at different phases of the heartbeat cycle at 8.5 dpc. Individual blood cells are distinguishable in the dorsal aorta. (B) Average blood flow velocity as a function of time in the corresponding area of the dorsal aorta. (C) Blood flow velocity profiles at different phases of the cardiac cycle. Each data point corresponds to the Doppler OCT velocity measurement from an individual cell. The data points were regressed using parabolic fits.

visible blood cell and the distance of the blood cell to the vessel wall. The data points for each time frame were regressed by a parabolic fit. Figure 22.7C shows the data points and the corresponding fits for different phases of the cardiac cycle. Even though the profiles were reconstructed from a very limited number of cells, it is clear that cell velocity is greater in the center of the vessel than near the vessel wall, and the measurements fit well with the parabolic profile suggesting that laminar flow is present. These measurements demonstrate that SS-OCT can provide sensitive spatially resolved hemodynamic measurements at the earliest stages of blood circulation.

Fast scanning confocal microscopy of fluorescently labeled blood cells can provide sensitive flow measurements with exceptional resolution in early mouse embryos but the imaging depth of this technique is about 200 μm, which limits its applicability for deep embryonic vessels. On the other hand, high frequency Doppler ultrasound has high imaging depth, but lower spatial resolution (30–50 μm). Doppler OCT is an excellent compromise with reasonable depth penetration and single cell resolution, which allows sensitive flow measurements even from single cells deep within the embryonic circulatory system.

22.4 Conclusion

In this chapter, we discussed two optical approaches for dynamic imaging of cultured mammalian embryos during development. Confocal microscopy combined with high-contrast fluorescent protein reporter lines enables to visualize the formation and remodeling of the embryonic vasculature with subcellular resolution, however, the imaging depth of this technique is limited to about 200 μm. Besides, fluorescence imaging is relying on availability of transgenic mouse lines with vital reporters labeling specific cells of interest, which are limited. On the other hand, OCT allows to image millimeters in the embryonic tissue at the expense of lower resolution. Additionally, optical clearing methods might potentially prove useful for enhancing the OCT signal from the internal structures [29, 30]. While confocal microscopy in combination with vital fluorescent reporters can be used for subcellular analysis of vascular development and hemodynamics in the embryonic yolk sac and superficial embryonic tissues right underneath the yolk sac, OCT can be used for dynamic 3-D structural embryonic imaging and blood flow analysis deep within the embryos. These complimentary approaches can potentially be used to characterize normal development, to study the effects of pharmaceutical agents, to analyze cardiovascular defects in mutant animals, and can provide significant insights into understanding how genetic signaling pathways and physiological inputs regulate human development.

Acknowledgments

This work has been supported, in part, by Postdoctoral Fellowship from the American Heart Association (to IVL), grants from the National Institutes of Health (HL077187 to MED), and the DOD ONR YIP (to KVL).

References

[1] E.A. Jones, D. Crotty, P.M. Kulesa, C.W. Waters, M.H. Baron, S.E. Fraser, and M.E. Dickinson, "Dynamic in vivo imaging of postimplantation mammalian embryos using whole embryo culture," *Genesis* **34**, 228–235 (2002).

[2] E.A. Jones, M.H. Baron, S.E. Fraser, and M.E. Dickinson, "Dynamic *in vivo* imaging of mammalian hematovascular development using whole embryo culture," *Methods Mol. Med.* **105**, 381–394 (2005).

[3] S.G. Megason and S.E. Fraser, "Digitizing life at the level of the cell: high-performance laser-scanning microscopy and image analysis for in toto imaging of development," *Mech. Dev.* **120**, 1407–1420 (2003).

[4] E.A.V. Jones, M.H. Baron, S.E. Fraser, and M.E. Dickinson, "Measuring hemodynamic changes during mammalian development," *Am. J. Physiol. Heart. Circ. Physiol.* **287**, H1561–1569 (2004).

[5] A.K. Hadjantonakis, M.E. Dickinson, S.E. Fraser, and V.E. Papaioannou, "Technicolour transgenics: imaging tools for functional genomics in the mouse," *Nat. Rev. Genet.* **4**, 613–625 (2003).

[6] J.W. Lichtman and S.E. Fraser, "The neuronal naturalist: watching neurons in their native habitat," *Nat. Neurosci.* **4**(Suppl), 1215–1220 (2001).

[7] P.M. Kulesa, "Developmental imaging: insights into the avian embryo," *Birth Defects Res. C: Embryo Today* **72**, 260–266 (2004).

[8] J.L. Lucitti, E.A.V. Jones, C. Huang, J. Chen, S.E. Fraser, and M.E. Dickinson, "Vascular remodeling of the mouse yolk sac requires hemodynamic force," *Development* **134**, 3317–3326 (2007).

[9] I. Larina, W. Shen, O. Kelly, A. Hadjantonakis, M. Baron, and M. Dickinson, "A membrane associated mCherry fluorescent reporter line for studying vascular remodeling and cardiac function during murine embryonic development," *Anatomical Record* **292**, 333–341 (2009).

[10] S.T. Fraser, A.-K. Hadjantonakis, K.E. Sahr, S. Willey, O.G. Kelly, E.A.V. Jones, M.E. Dickinson, and M.H. Baron, "Using a histone yellow fluorescent protein fusion for tagging and tracking endothelial cells in ES cells and mice," *Genesis* **42**, 162–171 (2005).

[11] N.C. Shaner, R.E. Campbell, P.A. Steinbach, B.N.G. Giepmans, A.E. Palmer, and R.Y. Tsien, "Improved monomeric red, orange and yellow fluorescent proteins derived from *Discosoma sp.* red fluorescent protein," *Nature Biotechnol.* **22**, 1567–1572 (2004).

[12] M. Liebling, A.S. Forouhar, R. Wolleschensky, B. Zimmermann, R. Ankerhold, S.E. Fraser, M. Gharib, and M.E. Dickinson, "Rapid three-dimensional imaging and analysis of the beating embryonic heart reveals functional changes during development," *Development. Dynam.* **235**, 2940–2948 (2006).

[13] M. Liebling, A.S. Forouhar, M. Gharib, S.E. Fraser, and M.E. Dickinson, "Four-dimensional cardiac imaging in living embryos via postacquisition synchronization of nongated slice sequences," *J. Biomed. Opt.* **10**, 054001 (2005).

[14] A.S. Forouhar, M. Liebling, A. Hickerson, A. Nasiraei-Moghaddam, H.-J. Tsai, J.R. Hove, S.E. Fraser, M.E. Dickinson, and M. Gharib, "The embryonic vertebrate heart tube is a dynamic suction pump," *Science* **312**, 751–753 (2006).

[15] D. Huang, E.A. Swanson, C.P. Lin, J.S. Schuman, W.G. Stinson, W. Chang, M.R. Hee, T. Flotte, K. Gregory, C.A. Puliafito, and et al., "Optical coherence tomography," *Science* **254**, 1178–1181 (1991).

[16] P.H. Tomlins and R.K. Wang, "Theory, developments and applications of optical coherence tomography," *J. Phys. D: Appl. Phys.* **38**, 2519–2535 (2005).

[17] Z. Chen, T. Milner, S. Srinivas, X. Wang, A. Malekafzali, M. van Gemert, and J. Nelson, "Noninvasive imaging of *in vivo* blood flow velocity using optical Doppler tomography," *Opt. Lett.* **22**, 1119–1121 (1997).

[18] M.A. Choma, S.D. Izatt, R.J. Wessells, R. Bodmer, and J.A. Izatt, "*In vivo* imaging of the adult drosophila melanogaster heart with real-time optical coherence tomography," *Circulation* **114**, 35–36 (2006).

[19] A. Mariampillai, B.A. Standish, N.R. Munce, C. Randall, G. Liu, J.Y. Jiang, A.E. Cable, I.A. Vitkin, and V.X.D. Yang, "Doppler optical cardiogram gated 2D color flow imaging at 1000 fps and 4D *in vivo* visualization of embryonic heart at 45 fps on a swept source OCT system," *Opt. Exp.* **15**, 1627–1638 (2007).

[20] S.A. Boppart, G.J. Tearney, B.E. Bouma, J.F. Southern, M.E. Brezinski, and J.G. Fujimoto, "Noninvasive assessment of the developing Xenopus cardiovascular system using optical coherence tomography," *Proc. Nat. Acad. Sci.* **94**, 4256–4261 (1997).

[21] M.W. Jenkins, O.Q. Chughtai, A.N. Basavanhally, M. Watanabe, and A.M. Rollins, "*In vivo* gated 4D imaging of the embryonic heart using optical coherence tomography," *J. Biomed. Opt.* **12**, 030505 (2007).

[22] M.W. Jenkins, D.C. Adler, M. Gargesha, R. Huber, F. Rothenberg, J. Belding, M. Watanabe, D.L. Wilson, J.G. Fujimoto, and A.M. Rollins, "Ultrahigh-speed optical coherence tomography imaging and visualization of the embryonic avian heart using a buffered Fourier Domain Mode Locked laser," *Opt. Exp.* **15**, 6251–6267 (2007).

[23] M.W. Jenkins, P. Patel, H.Y. Deng, M.M. Montano, M. Watanabe, and A.M. Rollins, "Phenotyping transgenic embryonic murine hearts using optical coherence tomography," *Appl. Opt.* **46**, 1776–1781 (2007).

[24] M.W. Jenkins, F. Rothenberg, D. Roy, V.P. Nikolski, Z. Hu, M. Watanabe, D.L. Wilson, I.R. Efimov, and A.M. Rollins, "4D embryonic cardiography using gated optical coherence tomography," *Opt. Exp.* **14**, 736–748 (2006).

[25] W. Luo, D. L. Marks, T.S. Ralston, and S.A. Boppart, "Three-dimensional optical coherence tomography of the embryonic murine cardiovascular system," *J. Biomed. Opt.* **11**, 021014 (2006).

[26] I.V. Larina, S. Ivers, S. Syed, M.E. Dickinson, and K.V. Larin, "Hemodynamic measurements from individual blood cells in early mammalian embryos with Doppler swept source OCT," *Opt. Lett.* **34**, 986–988 (2009).

[27] I.V. Larina, N. Sudheendran, M. Ghosn, J. Jiang, A. Cable, K.V. Larin, and M.E. Dickinson, "Live imaging of blood flow in mammalian embryos using Doppler swept source optical coherence tomography," *J. Biomed. Opt.* **13**, 060506-1–3 (2008).

[28] B. Vakoc, S. Yun, J. de Boer, G. Tearney, and B. Bouma, "Phase-resolved optical frequency domain imaging," *Opt. Exp.* **13**, 5483–5493 (2005).

[29] V.V. Tuchin, *Optical Clearing of Tissues and Blood,* **PM 154** SPIE Press, Bellingham, WA, 2006.

[30] I.V. Larina, E.F. Carbajal, V.V. Tuchin, M.E. Dickinson, and K.V. Larin, "Enhanced OCT imaging of embryonic tissue with optical clearing," *Laser Phys. Lett.* **5**, 476–480 (2008).

23

Terahertz Tissue Spectroscopy and Imaging

Maxim Nazarov and Alexander Shkurinov
Moscow State University, Russia

Valery V. Tuchin
Saratov State University,
Institute of Precise Mechanics and Control of RAS, Russia

X.-C. Zhang
Center for Terahertz Research, Rensselaer Polytechnic Institute, Troy, USA

Several methods of studying biological tissues with THz radiation are discussed. The specific features of radiation of this frequency range interaction with bio-objects include low scattering (due to the large wavelength), noninvasiveness (due to small photon energy), excitation of phonon modes of polycrystalline organic molecules, and high absorption and dispersion of water, the major component of biological tissues. The intensive development of this field of study became possible with the advent of the THz-time-domain technique. This method has a subpicosecond time resolution, provides direct-phase information, measures the reflection of short pulses from interfaces, and allows for broadband, low-frequency spectroscopy. All of these features may help to better visualize tissue properties and hidden tissue structures.

Key words: femtosecond laser pulses, terahertz spectroscopy and imaging, attenuated total internal reflection (ATR), biological molecules, tissues, dehydration, skin, teeth, muscle, blood, absorption coefficient, refractive index

23.1 Introduction: The Specific Properties of the THz Frequency Range for Monitoring of Tissue Properties

In light of developments over the last several decades, the notion of "THz time domain spectroscopy" (THz-TDS), implies generation and detection of THz pulses in a coherent manner using visible femtosecond laser pulses. The terahertz frequency range, located between IR and microwave ranges (1 THz $\rightarrow$ 300 μm $\rightarrow$ 33 cm^{-1} $\rightarrow$ 4.1 meV $\rightarrow$ 47.6 K), is one of the last-to-be-studied ranges in the whole scale from radio to X-rays. Effective broadband, coherent THz emitters and low-noise detectors became available only within the last few decades, mainly due to the development of lasers with ultrashort pulse duration.

THz radiation is a viable prospect for spectroscopy [1], biomedical applications [2, 3], and imaging. The big advantage of THz-based medical imaging is that THz photons are nonionizing. Many vibration transitions of small biomolecules correspond to THz frequencies. Moreover, tissue inhomogeneities less than 100 μm in diameter, which cause strong scattering of visible and IR light, do not cause strong scattering in the THz range. Biological tissues have small dispersion but large absorption in the THz range. The use of THz spectroscopy to study systems of biological significance seems reasonable, due to the high sensitivity to water properties, hydrogen bonds, and molecule conformation. Early experiments on THz imaging [4] have demonstrated rather good contrast between muscle and fat tissue layers. THz imaging also was used to reveal the contrast between regions of healthy skin and basal cell carcinoma including *in vivo* studies [5]. Studies of excised tissues have revealed that THz spectroscopy may be used to identify differences between healthy and malignant tissues [6], and to distinguish between organ types [2, 7].

Tissue response in THz frequencies is very sensitive to the presence of free and bounded water. This unique feature might allow for the differentiation between benign and malignant tumors in cancer diagnostics as the water-bond state is different in those two types of tumors types [8]. Thus, it is desirable to develop THz-techniques allowing for the observation of metabolic and pathological processes in tissues. Determination of absorption bands and transparency windows for certain tissues and substances participating in metabolic processes is necessary for the development of THz tomography. Such knowledge of tissue properties will allow researchers to detect and monitor substances with characteristically complex refraction spectra. These studies could be important for the precise marking of the margins of a pathological focus. For most far-infrared absorption spectroscopy techniques, the absorption of amino acids and many other biomolecules in the 1 to 3 THz spectral region is masked by much stronger water absorption, creating a major challenge to overcome. THz spectroscopy is also highly sensitive to the enantiomeric crystalline structures of biomolecules. The stereochemistry of many biologic and pharmaceutical species plays a significant role in whether they are functional or toxic.

Several detailed reviews concerning THz frequency application in biology [3, 9], spectroscopy [10, 11], and medicine [3, 12] already exist. In this chapter we will discuss both the recently published results and new studies on THz-spectroscopy of biologically related molecules and biological tissues. The absorption terahertz spectra of some biological molecules and tissues are presented in Figure 23.1.

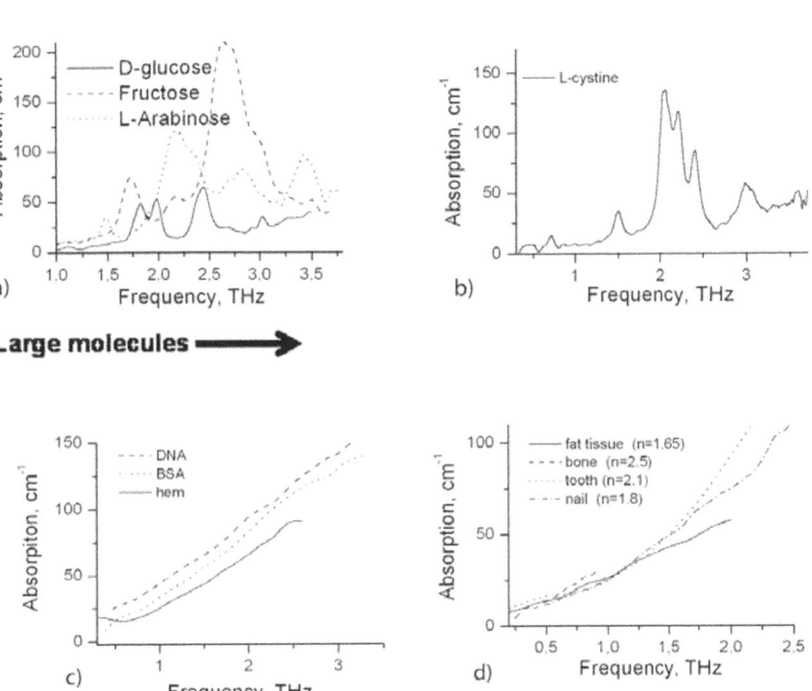

FIGURE 23.1: Typical terahertz absorption spectra of biological samples: a) sugars, b) polypeptide, c) large biomolecules, d) tissues.

23.2 Optics of THz Frequency Range: Brief Review on THz Generation and Detection Techniques

23.2.1 CW lamp and laser sources, CW detectors

In contrast to the single-cycle pulse in THz-TDS, other far-IR sources, such as arc lamps and globars, provide continuous wave (CW) radiation; free electron lasers or synchrotrons produce powerful pulses with many oscillation periods. Absorption spectra obtained by CW or THz-TDS systems are similar; the only difference is in the width of the frequency ranges. Absorption spectra of several DNA and amino acids (up to 6 THz) and liver tissue obtained with the help of parametric laser system radiation and bolometer detection are similar to THz-TDS spectra. Other sources include fixed frequency, far-IR lasers, tunable sources based on mixing a fixed frequency laser with a tunable microwave generator, or mixing together the radiation of two lasers. Highly sensitive hot electron bolometers are good detectors of CW THz radiation. Also, relatively high power CW sources are better for imaging applications [13].

23.2.2 FTIR

Fourier transform IR spectroscopy (FTIR) is an established technique. A FTIR-spectrometer consists of a broadband source of infrared radiation ("globar" – a high-pressure mercury lamp),

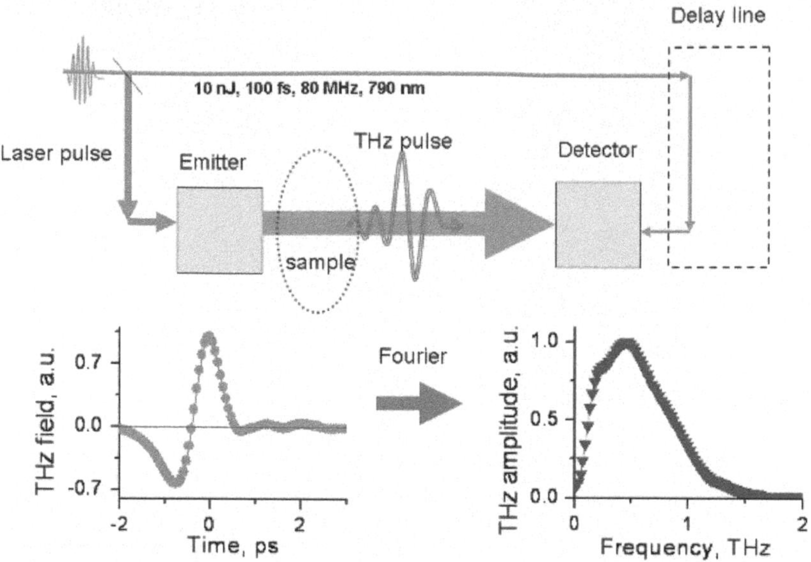

FIGURE 23.2: The principle scheme of THz time-domain spectrometer.

a Michelson interferometer, and a broadband detector (pyroelectric IR-detector or helium-cooled bolometer). The detector records an interferogram in the course of a mirror scan. Fourier transformation of the interferogram provides frequency resolution. There is no time-resolution and direct-phase information in this system.

In many experiments, identical absorption spectra information from FTIR and THz-TDS systems has been shown [14]. The difference in the two systems is in the dynamic and spectral ranges. At frequencies roughly lower than 100 cm^{-1}, THz-TDS provides better dynamic range and spectral resolution. On the other hand, intramolecular modes of biomolecules are mostly present at frequencies above 200 cm^{-1}, therefore more convenient for the FTIR technique.

23.2.3 THz-TDS, ATR

23.2.3.1 Introduction

The advantages of THz-TDS [15] are the following: it is a table-top system providing a straightforward phase (refraction) information, good signal-to-noise ratio (SNR) at low frequencies, picosecond time resolution, and a wide spectral range. The measured amplitude and phase are directly related to the absorption coefficient and index of refraction of the sample medium, and thus the complex permittivity of the sample can be obtained in a straightforward manner.

To generate THz pulses, several methods could be used. Typically photoconductive dipole antennas, a semiconductor surface, or nonlinear crystals serve as THz pulse emitters. In the first two cases, radiation is generated due to the formation of the transient photocarriers in a semiconductor irradiated by light pulses [16] and due to photocurrent pulse caused by the action of the external or internal electric field. In the case of the nonlinear crystal signal on the difference frequency, lying in THz-range (optical rectification) is generated [17].

In its turn, terahertz pulses could be detected by a similar photoconductive antenna or via electro-optical effect in the nonlinear crystal [17]. Propagating the THz pulse field induces birefringence in the crystal; at a given instant, the polarization state of a probe laser pulse changes proportionally to the THz field amplitude. Experimentally, the difference of powers of orthogonally polarized components of the probe laser pulse or a current induced in the antenna depending on the time delay between terahertz and optical pulses are measured. In general, when a semiconductor device is replaced by a nonlinear-optical device, the frequency range of a spectrometer is shifted from the lower (0.5 ± 0.45 THz) to the higher frequency region (2 ± 1.5 THz). An important part of THz-TDS system is a time delay between the optical beam and the THz beam. Often, the delay stage is placed in the pump beam, since it provides better stability than using it in the probe arm. For some delay line position, the THz field is measured in a particular moment (relative to pulse maxima in the time domain), since laser pulses coming at MHz repetition rate are very identical; from thousands of pulses, one acquires a THz pulse field value for this delay time. Then the delay line is tuned to the next point to get the next field value for the next delay time. Data acquisition is done by a lock-in amplifier with a chopper in the arm of the generator beam.

The typical pulsed terahertz spectrometer used in this study (Figure 23.2) is described in detail in [18, 19]. Depending on the absorption of the sample under study and the frequency range of interest, the type of detector and generator could be chosen [20]. For example, we used 90-fs laser pulses at 790 nm, a low-temperature grown GaAs semiconductor surface as a THz generator and a 0.3-mm-thick $\langle 110 \rangle$ ZnTe crystal as an electro-optical detector. By using the Fourier transform of the temporal pulse profile, the complex spectrum containing information on the refractive index and absorption coefficient of the medium were calculated [19]. In most experiments 50 ps time sampling of the signal with 1024 counts and 300 ms acquisition time of each count were provided. This allowed us to perform measurements with the SNR more than 10^3 in the spectral range from 0.3 to 2.5 THz with a spectral resolution of 10 GHz. The laser and terahertz pulses had a repetition period of 12 ns; with the energy of THz probing pulse being of 10^{-13}J. Such low pulse energy should not damage the tissue.

23.2.3.2 Methods for measuring the absorption coefficient and refractive index

A specific feature of pulsed terahertz spectroscopy is the possibility of direct measurements of electromagnetic field strength and direction, which contains phase information. To calculate optical properties of a tissue it is necessary to reconstruct optical parameters from the measured transmission spectra $T(\omega)$ [21]. From the experiment and subsequent Fourier transform of the temporal profiles of pulses we obtain the incident THz pulse amplitude $E_{\text{reference}}(\omega)$, and the amplitude $E_{\text{sample}}(\omega)$, of the transmitted pulse. Then we reproduce the transmission coefficient of the sample as:

$$T(\omega) = E_{\text{sample}}/E_{\text{reference}} = T_0 FP(\omega)RL(\omega), \tag{23.1}$$

where

$$T_0(\omega) = \exp\{-i(n_{\text{sample}} - n_{\text{air}})d\omega/c\}, \tag{23.2}$$

contains the basic information of the medium (absorption coefficient and refractive index);

$$RL(\omega) = 4n_{\text{sample}}n_{\text{air}}/(n_{\text{sample}} + n_{\text{air}})^2 = 1 - R(\omega)^2, \tag{23.3}$$

accounts for reflection losses from the sample boundaries;

$$FP(\omega) = \{1 - R(\omega)^2 \exp(-i2n_{\text{sample}}d\omega/c)\}^{-1}. \tag{23.4}$$

describes multiple reflected pulses in a parallel plate, i.e., Fabry-Perot modes; $n(\omega) = n'(\omega) - in''(\omega)$ is the complex refractive index for the corresponding medium; $n''(\omega) = \alpha(\omega)c/\omega$, relies on the absorption coefficient included in the imaginary part of n; $\omega = 2\pi f$ is the cyclic frequency; $R(\omega) = (n_{sample} - n_{\text{air}})/(n_{sample} + n_{\text{air}})$ is the complex reflection coefficient; c is the light speed; $n_{\text{air}} \approx 1$; and d is the sample thickness. Note that we use the absorption coefficient for the field, while in other methods the power absorption coefficient, which is twice as large, is commonly used.

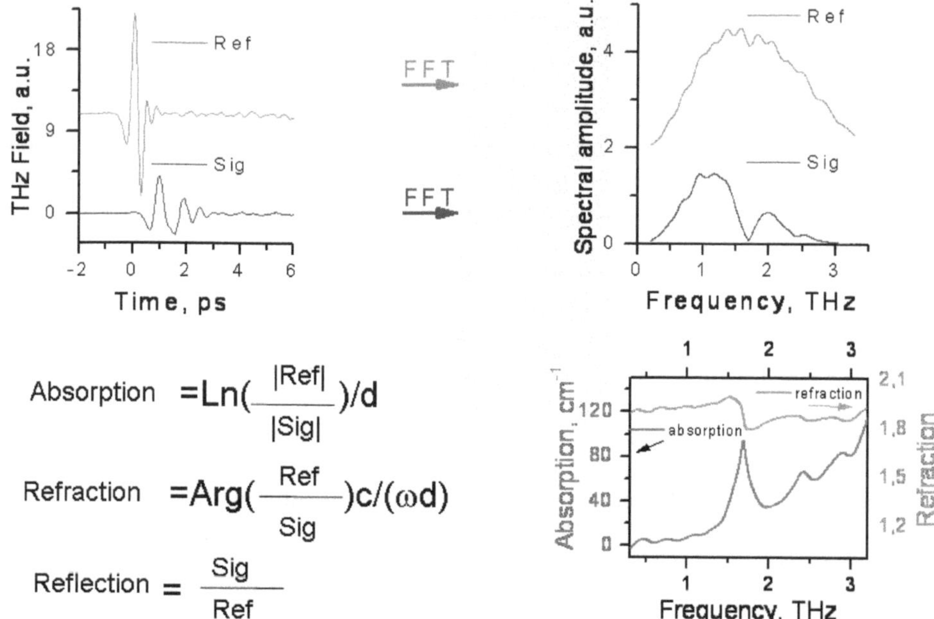

FIGURE 23.3: Schematics of data evaluation.

Equation (23.1) cannot be solved analytically, but for the most cases, the approximate solution is valid. By assuming that $RL(\omega) \approx$ const, $FP(\omega) \approx 1$, we obtain:

$$\alpha(\omega) = \ln\{|T(\omega)|\}/d + \ln\{1 - R_{av}^2\}/d, \tag{23.5}$$

where $R_{av} = (n_{av} - 1)/(n_{av} + 1)$ is the reflection coefficient, n_{av} is the real part of the refractive index of the sample averaged over the frequency range of measurements. The averaged refractive index n_{av} can be easily determined from the measurements of the delay time Δt of a pulse propagating through the sample: $n_{av} = 1 + \Delta tc/d$. The frequency dependence of the refractive index has the form:

$$n'(\omega) = \arg\{T(\omega)\}c/(\omega d) + 1. \tag{23.6}$$

Figure 23.3 presents spectra calculated using simple Eqs. (23.5) and (23.6). At studies of thin samples with a large refractive index, for example tooth slices and pressed thin pellets, it is necessary to use more rigorous equations accounting for multiple reflections from the sample edges, i.e., the Fabry-Perot modes.

To characterize the response of a medium completely we need both α and n', which comprise the dielectric function, and vice versa

$$\varepsilon(\omega) = [n'(\omega) - i\alpha(\omega)c/\omega]^2, \tag{23.7}$$

$$n'(\omega) = \mathrm{Re}[\varepsilon(\omega)^{1/2}], \quad \alpha(\omega) = (\omega/c)\mathrm{Im}[\varepsilon^*(\omega)^{1/2}]. \tag{23.8}$$

23.2.3.3 THz reflection spectroscopy

Transmission terahertz spectroscopy is not suitable for natural biological objects due to strong absorption by water or large sample sizes in *in vivo* measurements. Reflection spectroscopy is a more viable prospect in these instances, and provides distant noninvasive measurements without thickness ranging. *In vivo* skin studies are the most appropriate for this technique. The complex reflection coefficient $\tilde{R}_p(\omega)$ contains information about the refraction index and absorption coefficient of the medium:

$$\tilde{R}_p(\omega) = \frac{n^2(\omega)\cos(\theta) - \sqrt{n^2(\omega) - \sin^2(\theta)}}{n^2(\omega)\cos(\theta) + \sqrt{n^2(\omega) - \sin^2(\theta)}}. \tag{23.9}$$

It is characterized by amplitude R and phase ϕ as $\tilde{R}_p = Re^{i\varphi}$, where subscript p denotes radiation polarization in the incidence plane; θ is the incident angle. The Fresnel formula takes into account the complex part of the refractive index $n(\omega) = n'(\omega) - i\alpha(\omega)c/\omega$. The comparison of modelled reflection spectra for different incident angles θ has shown that $p-$polarization is the most informative in the sense of detection of resonance absorption, in particular the phase part of p-polarization. We chose the incident angle as $\theta_0 = \tan^{-1}(n') - 5°$, corresponding to the high sensitivity and reliability of the method to characteristic absorption lines of the media [19]. When θ_0 is close to the Brewster angle $\theta_{Br} = \tan^{-1}(n')$ resonance lines are well pronounced in reflection spectra, but at $\theta_0 = \theta_{Br}$ the amplitude R tends to zero and that complicates the experiment.

in the investigation of biological tissues, two principle questions arise: the probing depth and the issue of imaging multiple tissue layers. The method under discussion brings information only about an interface with a special resolution of tens of micrometers. For layered media with layer thickness greater than 100 μm, it will be possible to separate pulses reflected from different layers. Furthermore, expression (23.9) could be applied for each layer with its interface and bulk parameters. In the visible and NIR this method has been developed for a three-layered skin model [22], which can be easily adapted to the THz range.

23.2.3.4 Attenuated total internal reflection

To study soft tissues or aqueous solutions the attenuated total reflection (ATR) spectroscopy turned out to be a more prospective method (Figure 23.4). In this scheme it is possible to control the penetration depth δ in the matter $\delta \approx (\lambda/\pi)(n_{pr}^2 \sin\theta - n_{\text{sample}}^2)^{1/2}$ (see 23.3). With the silicon Dowe prism (1.5 × 2 cm basis, apex angle of 90°, the refractive index of silicon in THz is 3.42, the dispersion and absorption are practically negligible) a desirable sensitivity to the water concentration in soft tissue was obtained. The prism placed in a collimated THz beam conserves the beam direction. The spectrum of transmitted radiation measured for a clean prism is used as a reference. To measure the reflection spectra the sample is attached to the prism surface (or a drop of a liquid is placed on the prism surface).

The main advantage of the ATR method relative to reflection spectroscopy is in the simplicity of the reference spectra measurement, and the reflection amplitude is maximal. The main difficulty of the ATR method is an optical contact problem for hard sample studies. Simulation of a three-layered system (prism-air-matter) using the experimental parameters has shown that the existence of a 10 μm layer of air between the prism surface and matter under study is critical for getting information about studied material. For liquid samples optical contact is always good. Equation (23.9) (Fresnel formula) describing reflection is valid also for the ATR, but one must account for the refraction of the prism material n_{pr} and substitute n by n/n_{pr} in Eq. (23.9). The free-surface reflection mode and the prism scheme provide considerably different shapes for both the reflection amplitude and phase spectra.

The prism scheme is more convenient for soft tissues studies, since a sample attaching to the prism

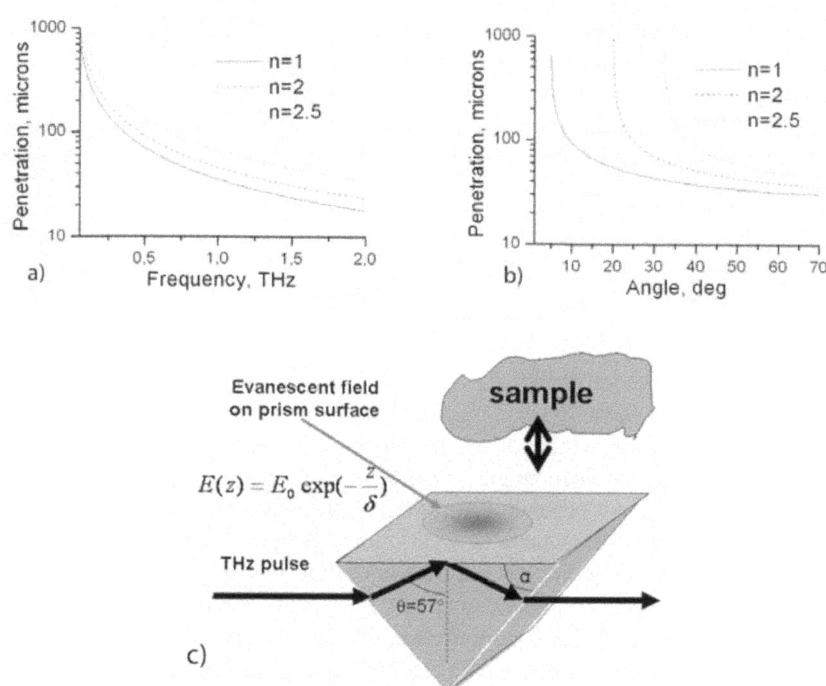

FIGURE 23.4: Probing depth of ATR spectroscopy. The legend lists the sample refraction index; penetration depth of terahertz waves for different frequencies (a) and different incident angles (b); schematics of wave interaction with Dowe prism.

basis forms a flat smooth interface with the tissue sample. We have thus measured the reflection spectra of the finger skin (Caucasian male, 32 years-old, thumb, Figure 23.13).

Since the ATR method provides measurements only for superficial tissue layers up to 10 μm, the skin was probed only within the stratum corneum. In this layer water content is minimal and is equal to 15% (by weight); the main content is formed by proteins (70%) and lipids (15%) [23]. As a result, the skin reflection spectrum is considerably different from the water reflection (see Figure 23.10). We can control evanescent field penetration depth by changing the incidence angle, thus we can measure the essential surface properties of the samples (up to $\lambda/50$ in thickness).

The other fruitful application of ATR is measurements in solutions. Crystalline sugars, amino-acids, and peptides show characteristic absorption lines in the THz frequency range. In the natural environment (water solution) they show featureless spectra with shapes similar to a solvent. The prism used in solution experiments was optimized to study small changes of parameters of water containing media, so relatively small concentrations of molecules like glucose, albumin (see Subsection 23.5.3), and DNA were detected. The sensitivity to water is due to the strong dispersion of liquid water in the 0.1–1 THz range and the angle of incidence on silicon-water interface being close to the ATR limit.

23.3 Biological Molecular Fingerprints

23.3.1 Introduction

Cell and tissue components (DNA, proteins, amino-acids, sugars, etc.) all have fingerprints from the 0.5 to 5–10 THz range, but most of those resonance absorption lines are well visible in case these molecular components are studied in an ordered solid state environment; for instance, in the molecular crystal form [24], they are related to intermolecular oscillations, sensitive to isomers, conformers, and other states of the molecules. Whereas in the crystal the well-defined periodicity and bond length distribution give rise to various lattice modes, these features merge into a broad featureless spectrum of the amorphous system (for example in sugars), where the bonds are randomly distributed. In contrast to the larger systems, a common feature of these studies is the observation of a relatively small number of sharp vibrational bands attributed to the vibrational modes clearly related to the lattice of the solid state materials. In condensed-phase biosamples, intermolecular interactions, including van der Waals forces and hydrogen bonding, modify the mode structure of intramolecular vibrations and also give rise to additional vibrational modes involving collective dynamics of several molecules. The intermolecular interactions are usually weaker than intramolecular interactions, and the characteristic signatures of the intermolecular modes often emerge in the THz region. The origin of the THz spectroscopic features can be clarified in the following way: a crystalline solid has ions arranged in a periodic structure on the microscopic level. The ions, however, are not completely static: a closer look may find that each ion wiggles in the vicinity of its lattice site, while the average positions retain the periodic arrangement. A collective oscillation of the ions with a well-defined frequency and wavelength is called a normal mode of lattice vibration, and its quantization is called a *phonon*. The phonon resonances are of great interest because, in general, the normal mode frequencies fall into the THz region. Not all normal modes, however, interact with electromagnetic radiation and also with the THz ones. Only the long wavelength optical modes in ionic crystals can be involved in such interactions. Consequently, THz spectroscopy probes the crystalline mode of the system.

The THz spectra of crystalline oligopeptides, simple sugars, pharmaceutical species, and other small biomolecules have been reported with a relatively small number of sharp vibrational bands. The spectral features of the observed modes strongly depend on temperature: the absorption lines undergo severe broadening and shift as temperature varies. Vibrational motion in biological molecules is often characterized on the picosecond time scale, and can therefore, as we indicated before, be studied by THz spectroscopy. The vibrational modes involve the intramolecular dynamics of stretch, bend, and torsion among bonded atoms. Normal mode analysis is a standard technique to identify and characterize the vibrational dynamics. In general, molecules have $3N-6$ normal modes, where N is the number of atoms. Biological molecules consisting of a large number of atoms, therefore, have complicated mode structures.

In a natural tissue-like environment (in water solution) the continuum of strongly damped intermolecular modes makes featureless spectra; only at higher (4–10 THz) frequencies are some intramolecular modes in solution observed [14]. For large biomolecules the overall dielectric response in the THz range is broad and featureless. The response, however, is highly sensitive to hydration, temperature, binding, and conformational change [9]. The THz spectra for the polynucleotides and polypeptides comes from the collective vibrational modes associated with the 3-D structure. The altering of molecular structure, either by denaturing or during cell or tissue functioning, should induce a change of density of states of the collective vibrational mode. This makes Terahertz sensitive to biomolecule conformation.

For high enough overage power of THz radiation, long exposition of proteins changes their chemical and biological properties [25]. A detailed explanation of these phenomenons is still required.

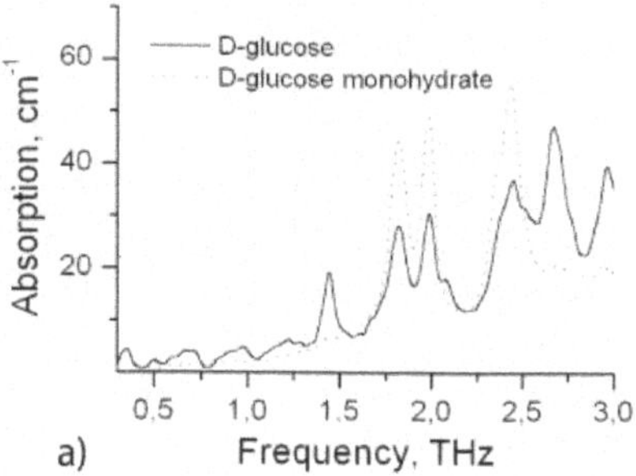

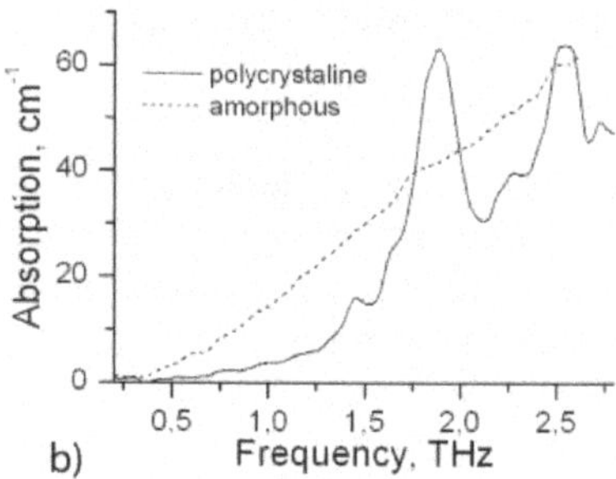

FIGURE 23.5: Saccharide spectra: a) D-glucose in different states; b) polycrystalline and amorphous sugar.

23.3.2 Sugars

It is possible to differentiate sugars with various hydrate forms by their THz spectra. For example, lactose and glucose form hydrates with distinct THz spectra. Spectroscopy of polycrystalline and amorphous saccharides in the range of 0.5 to 4 THz has shown that polycrystalline sugars have distinct absorption bands in this range, whereas absorption by the amorphous forms were featureless and increased monotonically with frequency. With a temperature increase, all absorption features of glucose and fructose broadened and red shifted. The sharp spectral features arise from intermolecular vibrational modes of the long-range order, controlled by noncovalent bonds between the sugar molecules [24].

The absorption spectra of several saccharides are presented in Figures 23.1 and 23.5. Each saccharide has its own characteristic absorption bands. Note that terahertz spectroscopy is sensitive to the presence of bounded water in molecules (glucose monohydrate and glucose have different absorption spectra).

Different saccharides exhibit both similar absorption bands (in the ranges of 1.6–1.9 THz and 2.6–2.8 THz) and characteristic bands [19]. In general, the number of bands and nonresonance background absorption increase with increasing frequency. The lowest-frequency band can be treated as the characteristic one because its mean frequency depends on the type of saccharide. It is equal to 0.54, 1.44, 1.48, 1.72, 1.82, and 1.83 THz for lactose, glucose monohydrate, arabinose, fructose, glucose, and saccharose, respectively. This means that it is possible to determine the relative content of different components in their dry mixture when their absorption spectra are known.

We have made an attempt to detect the presence of glucose in hemoglobin preliminary incubated with glucose at physiological conditions [26]. However, even in a dried sample the absorption of hemoglobin was too strong in the entire spectral range and had no specific spectral features, which did not allow us to detect the presence of glucose at small concentrations, i.e., to see polycrystalline characteristic absorption bands of glucose on the background of the hemoglobin absorption spectrum.

23.3.3 Polypeptides

To understand the THz response of proteins, we shall start with polypeptides, which are the basis of protein macromolecules. The terahertz response of small biological molecules or polypeptides

(Figure 23.1) reveal many intermolecular and intramolecular vibrational fingerprints [27]. Several short-chain solid-state dipeptides [Leu-Gly] and tripeptides [Val-Gly-Gly] reveal many distinct spectral features [11]. The terahertz spectral response was also sensitive to the sequences of amino-acid chains.

Short-chain crystalline polypeptides may be uniquely "fingerprinted" by simply acquiring the THz spectrum in the 50 to 500 cm^{-1} (1.5–15 THz) spectral range. Several mixed polypeptide sequences with up to three amino acids were also studied (e.g., Val-Gly-Gly, glutathione), but sharp spectral features were not present for species containing more than 10 amino acids [15]. This observation may be the result of overlapping spectral density as the molecular weight increases, or the spectral features are swamped by hydrogen-bonded water, which is difficult to remove from these complex biomaterials.

It is generally difficult to distinguish the origin of the resonant feature, whether it comes from intramolecular or intermolecular vibrations, unless studies on temperature or molecular environment dependence are carefully performed. In most cases, intermolecular transmissions are observed. For atomic substitution (–OH for l-serine versus –SH for l-cysteine), significant differences were noted in the THz spectrum [15], indicating the crystalline lattice interactions are strongly affected by the simple atomic substitution.

23.3.4 Proteins

Since the first three-dimensional image of a protein structure was produced by X-ray crystallography in 1958, nearly 40,000 crystal structures of biomolecules have been determined. From the intensive studies of several decades, we now have a very good understanding of the complex atomic arrangements of large biological molecules. The protein structure is organized into four levels. Primary structure refers to the linear chain of amino acids or the peptide sequence. Secondary structure contains a periodic formation of polypeptides linked by hydrogen bonds. The most common types of secondary structures are α-helices and β-sheets. The tertiary structure of protein molecules is formed as secondary structures fold into a unique three-dimensional globular structure. Quaternary structure refers to a protein complex formed by interactions among protein molecules. The three-dimensional structure of proteins is crucial information to understand how they function in biological systems. Yet it is not sufficient to gain a complete understanding of the processes, because the proteins are not entirely rigid, and the relative motions between the functional groups play a very important role. The functional rearrangements of structure are called "conformational changes," and the resulting structures are called "conformations." The conformational changes often occur on the picosecond time scale, and hence the large-amplitude vibrational modes lie in the THz region and the THz technique can be used to detect changes in biomolecules, such as proteins. For example, bovine serum albumin (BSA) and collagen were studied [28]. Mickan et al. [29] used differential THz time-domain spectroscopy for sensing of bioaffinity. Binding was observed by measuring the transmission of a thin layer of biotin bounded to the sensor protein avidin. The absorption spectrum of solvated BSA is experimentally determined in Ref. [30]. The authors have extracted the terahertz molar absorption of the solvated BSA from the much stronger water absorption and observed for the solvated protein a dense overlapping spectrum of vibrational modes that increases monotonically with increasing frequency. Experiments done with myoglobin and white hen egg lysozyme have shown that hydration induced the increased absorption above the expected for the water content [31]. No evidence of distinct and strong spectral features was found, suggesting that no specific collective vibrations dominate the protein's spectrum of motions. The shape of the observed spectrum resembles the ideal quadratic spectral density expected for a disordered ionic solid (and for large biomolecules). The dry BSA spectrum is presented in Figure 23.6.

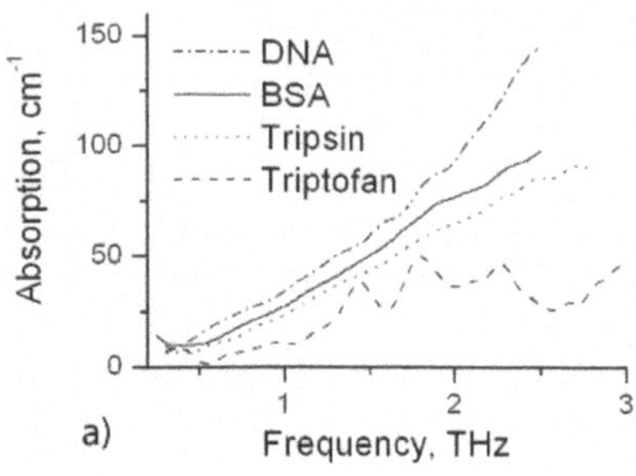

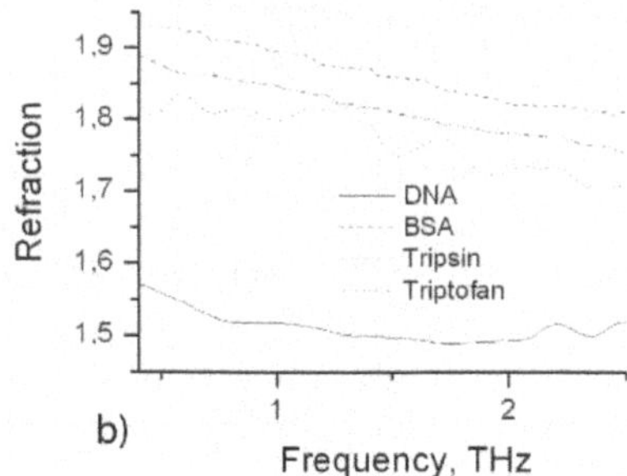

FIGURE 23.6: Different types of biomolecule spectra: absorption coefficient (a) and refraction index (b).

23.3.5 Amino-acids and nucleobases

THz transmittance spectra of the five nucleobases (adenine, guanine, cytosine, uracil, and thymine) show various distinguishable resonance peaks in a polycrystalline state [13]. For cxample, guanine showed peaks at 2.57, 3.00, 4.31, 4.84, and 5.44 THz, of which the two peaks at 3.00 and 4.84 THz were very strong. For each of the other nucleobases, two strong absorption bands with peaks at 2.85 and 3.39 THz (cytosine), 3.05 and 4.18 THz (adenine), and 3.34 and 3.84THz (uracil) were also observed. Corresponding nucleosides, dA, dG, dC, and dT (d = dioxyribose) of DNA were also measured using THz-TDS in the range from 0.5 to 4.0 THz at 10 K and 300 K [1]. The spectra above 1.5 THz of dA, dG, dC, and dT are similar to the nucleobases but a second group of resonances from 1 THz to 2 THz have narrow asymmetric line shapes attributed to the sugar groups.

Amino acids, like alanine and tyrosine [13], also have several strong characteristic absorption bands; some other aminoacids were studied by authors of Ref. [32]. The spectra of nucleic acids and nucleosides are presented in Ref. [24], and nucleobases spectra in Ref. [33].

23.3.6 DNA

In principal, an atomic level pattern of the concerted motions of polypeptide chains and DNA may be accessible through the accurate measurement of low-frequency vibrational spectra. Large biomolecules like proteins, DNA, and RNA have been predicted to show strong absorption features (twist and H-bond motions at 0–2 THz) in the far-infrared range because of the large masses involved [34]. Molecular mechanics models for proteins [35, 36] also predict vibrational modes in the THz range. However, for proteins and DNAs with about 30 kDalton (kDa) molecular weights and large numbers of constituent peptide units or bases, one would expect the density of overlapping states to be so high in terahertz frequency range that contributing absorption bands would "smear out." While the components of the DNA (the nucleobases) show distinct vibrational features [37] so far no distinct absorption peaks have been observed in DNA [38] (23.3.5).

Samples of biologically active molecules often require an environment that adversely affects far-infrared measurements. Proteins are usually wrapped in a solvatization layer of water molecules and contain counter-ions to the various polar amino-acid groups. These additional components also absorb strongly in the THz range making identification of absorption bands difficult. Thus, if the THz dielectric response is sensitive to the overall density of states, binding effects should be detectable. Using different measuring systems it was shown that biomolecular binding effects are detectable in the terahertz range. A pioneer work on the demonstration of THz sensitivity to biomolecular functioning was performed by the Kurz's group, where a significant change in both

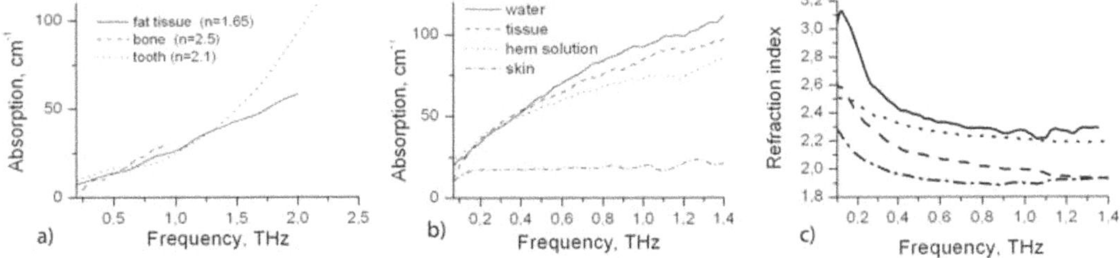

FIGURE 23.7: Typical spectral characteristics of tissue and water: a) extracted from transmission measurements, b) and c) – from ATR, legend for b) and c) is the same.

the absorption and refractive index for single-stranded versus hybridized DNA was found [39]. Again, there was an overall change in the dielectric response without strong frequency dependence.

23.4 Properties of Biological Tissues in the THz Frequency Range

Different types of tissues have featureless absorption and reflection spectra, similar to one another. The comparison of biological tissues (Figure 23.7, Table 23.1) has shown that they have high and comparable absorption but their refractive indexes are considerably different. The latter allows one to measure amplitudes of pulses reflected from different layers of the tissue. The spatial resolution of tissue terahertz probing depends on the signal-to-noise ratio provided by a pulsed terahertz spectrometer and should be considerably increased for available systems to reach desirable subcellular resolution. Sometimes sample thickness and surface profile can have significantly more influence on evaluated spectra than tissue type. At present, THz waves have some limitations to be used for *in vivo* diagnostics of tissues other than skin. However, studies of excised fresh tissue samples can be used for pathological THz diagnostic purposes [12]. Fitzgerald et al. investigated freshly excised human tissue, using a broadband THz pulsed imaging system comprising frequencies of 0.5–2.5 THz [40]. The refractive index and absorption coefficient were found to be different for skin, adipose tissue, striated muscle, vein, and nerve. The main influence on THz tissue spectra comes from water concentration within tissue. Similarity in spectra shape follow from the heavy water concentration in tissues (the absorption spectrum of water has the same shape [3, 41], but the absorption coefficient value of pure water is several times higher). The absorption spectra of tissue components (particular biomolecules) have a similar shape (linear or quadratic increase in absorption and a decrease in the refractive index with frequency). The absence of characteristic lines in the spectra of "dry" tissues can be explained by the overlap of many spectral lines.

One of the ways for careful study of soft tissue by THz-TDs is to use tissue dehydration technology [12] that conserves information about tissue properties, and makes transmission measurements easier. Another way is to use ATR.

Authors of paper [12] suggested tissue lyophilization by freeze-drying as a viable solution for overcoming hydration and freshness problems (Figure 23.8). Lyophilization removes large amounts of water while retaining sample freshness. In addition, lyophilized tissue samples are easy to handle and
their textures and dimensions do not vary over time, allowing for consistent and stable THz measurements.

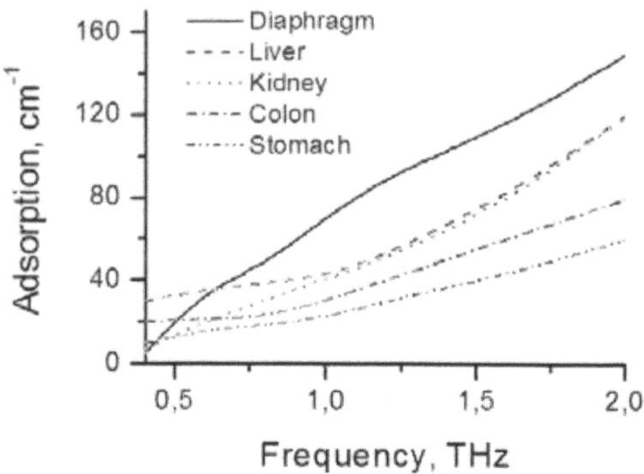

FIGURE 23.8: Nitrogen-dried fresh tissues, adapted from [12].

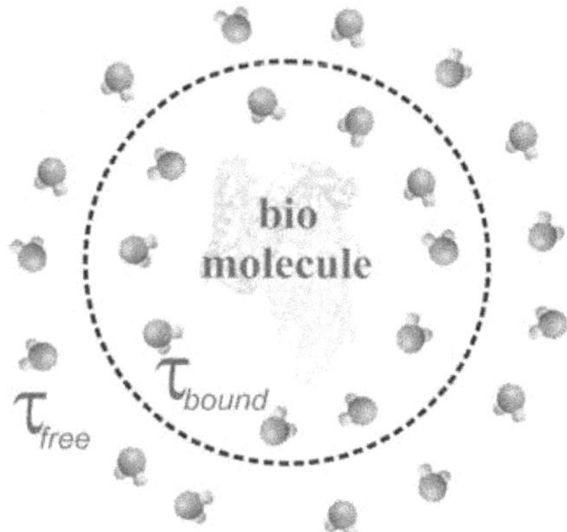

FIGURE 23.9: Water-protein interaction [42].

23.5 Water Content in Tissues and Its Interaction with Terahertz Radiation

The dynamics of water molecules interacting with biomolecules is of fundamental importance for understanding the physical and chemical processes in biological systems (see Figure 23.9). The mutual interaction influences the rotational relaxation and hydrogen-bond stretching of water molecules in an aqueous environment. Since the key mechanisms of water molecule dynamics are on the picosecond time scale, THz spectroscopy can make a direct observation of how water molecules in the vicinity of biomolecules behave differently from those in bulk water [42].

This is somewhat counterintuitive. Biological water is expected to be less responsive to THz radiation because its motions are hindered by protein surfaces. One plausible explanation is that sugar molecules may become more flexible in an aqueous environment than in a solid state. No microscopic mechanism, however, has yet been identified to explain the increase of absorption. It surely is an interesting subject for future studies.

23.5.1 Data on water content in various tissues

The most important substance contained in living systems is liquid water, with the average water content in biological tissues reaching 75% [43].Water molecular properties, dynamics, and spectral response are well described (see for example papers [10, 45]). Typical spectral features of water solutions in the THz range are broad and weak. For example, such spectra were observed for water-alcohol mixtures [10]. Water content in different types of tissues is also known [44], these data are summarized in Table 23.1 together with data on THz absorption and refraction.

TABLE 23.1: Water content, absorption coefficient, and refraction index at 1 THz of various tissues

Tissue or tissue component	Rat Water cont. [44]	Human Water cont. [44]	Power absorption at 1 THz, cm^{-1}	Refraction at 1 THz
Water	1	1	240	2.2
Lungs	0.79	0.81		
Kidneys	0.77	0.78	90^b	
Brain	0.79	0.77		
Heart	0.78	0.76		
Muscle	0.76	0.76	160^a; 120^d	2.1^a
Liver	0.71	0.75	90^b	
Skin	0.65	0.72	168^c	2.06^c
Epidermis			40^e	2.0^e
Stomach	0.75	0.71	46^b	
Bone	0.45	0.44	60^a	2.5^a, 2.49^g
Adipose	0.18	–	22^c	1.54^c; 1.58^d
Tooth				
Enamel		0.01	60^e	3.1^e; 3.09^d
Dentine		0.10	60^e	2.4^e, 2.51^d
Fat			40^a	1.65^a

^{a}pig; $^{b-g}$human (^{b} [12]; ^{c}[3]; ^{d}[40]; ^{g}[52])

23.5.2 THz spectra of water solution

Water is a polar molecule and therefore very absorbing of THz radiation. The specificity of the THz detection is hydrogen bonding involved in the creation of the biological systems under study. Hydrogen bond forms via an attractive intermolecular interaction between an electronegative atom and a hydrogen atom bonded to another electronegative atom. Usually the electronegative atom is either oxygen or nitrogen with a partial negative charge. Positively charged hydrogen is nothing more than a bare proton with little screening, hence the hydrogen bond is much stronger than nominal dipole-dipole intermolecular interactions or van der Waals forces. Hydrogen bonding energy is about a tenth of that for covalent or ionic bonds, resulting in modes in the THz frequency range. Because of the relatively strong interaction, the hydrogen bond plays a crucial role in biophysical and biochemical processes in naturally occurring biosystems.

Studying water solutions with THz time-domain spectroscopy provides information about the intermolecular dynamics. The hydration layer around a solute such as a protein, increases the ter-

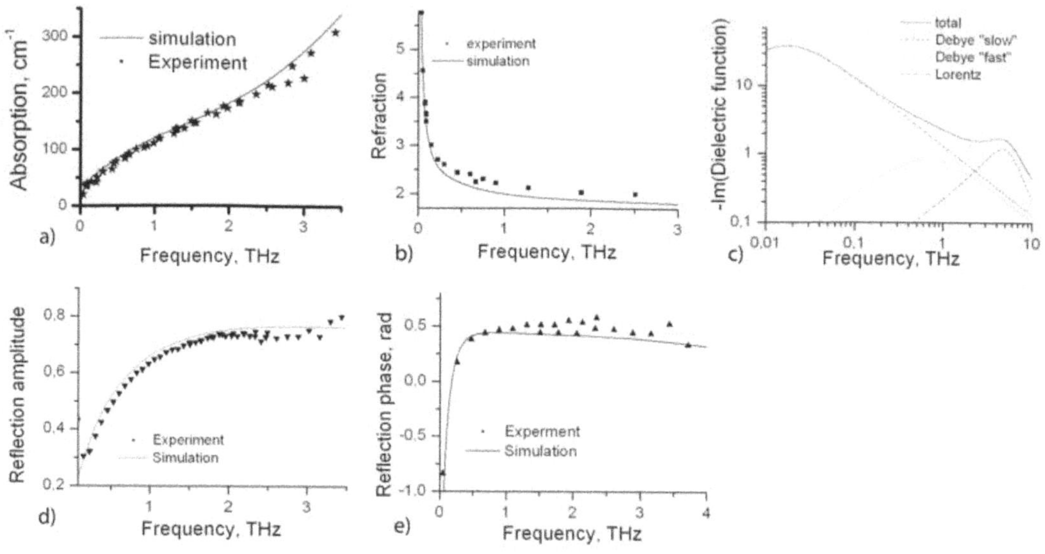

FIGURE 23.10: Water spectra.

ahertz absorption due to the coherent oscillations of the hydration water and the solute and slows down the relaxation process. At higher solute concentrations, the solvation layers overlap and the absorption may decrease.

The dielectric response of liquid water is governed by several physical processes. Major contributions are attributed to two types of relaxation dynamics characterized by fast (~10 fs) and slow (~10 ps) relaxation times. The slow relaxation is associated with rotational dynamics, yet the origin of fast relaxation is not yet clearly understood. Strong hydrogen-bond interactions between water molecules also make a sizable contribution. The intermolecular stretch mode is resonant at 5.6 THz and has a broad line width. Putting all the contributions together, we can write the dielectric constant of water as (23.10).

A 200-μm-thick water layer attenuates the radiation intensity at THz frequency by an order of magnitude, while a 1-mm-thick water layer is almost opaque in the whole terahertz range. The dielectric function of water for THz is being extensively studied [41] and it can be described by the two-component Debye model with the Lorentz term [45]:

$$\varepsilon(\omega) = \varepsilon_\infty + (\varepsilon_s - \varepsilon_1)/(1 + i\omega\tau_d) + (\varepsilon_1 - \varepsilon_\infty)/(1 + i\omega\tau_2) + A/(\omega_0^2 - \omega^2 + i\omega\gamma) + \ldots. \quad (23.10)$$

where τ_d= 9.36 ps, τ_2= 0.3 ps are the slow and fast relaxation times, ε_∞=2.5 is the susceptibility at high frequencies, ε_s= 80.2, ε_1= 5.3 are the contributions to susceptibility from first and second Debye additives, A = 38 (THz/2 $\pi)^2$, ω_0= 5.6 THz/2 π, γ = 5.9 THz are the amplitude, frequency, and line width of Lorentz additive, respectively. All constants are taken for 20°C; with temperature decrease both absorption and reflection decreases [11]. The absorption, refraction, and reflection spectra of liquid water were calculated using Eqs. (23.10), (23.8), and (23.9). The corresponding measured spectra are presented in Figure 23.10. Note that the ATR method expands the spectral range of measurements from 1 THz up to 3.5 THz [46].

Most substances in tissues are presented as aqueous solutions. In general, they could be detected by THz-TDs. However, the spectral selectivity in this case is low, because the difference is mainly in spectral amplitude, not in the shape.

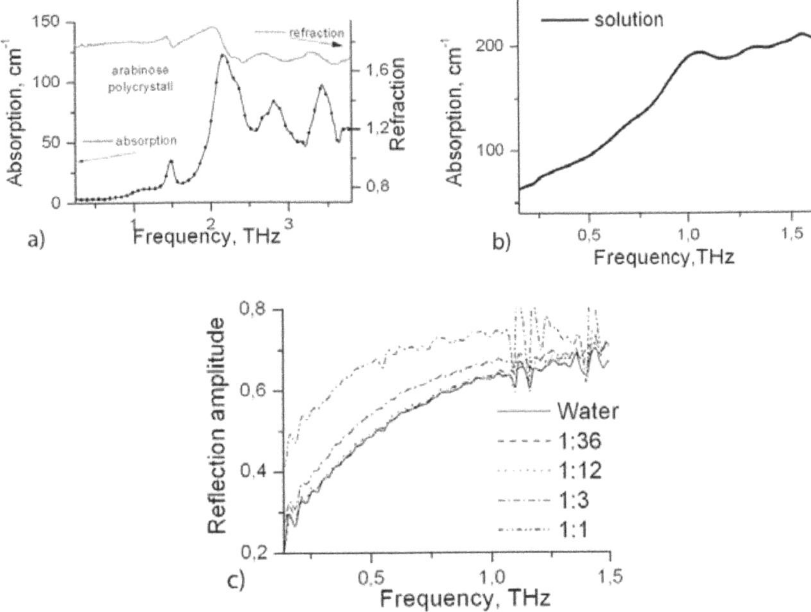

FIGURE 23.11: Arabinose spectra: a) polycrystalline state, b) saturated water solution, c) ATR reflection for different water/arabinose concentrations.

The use of THz spectroscopy to probe biomolecules in aqueous environments has shown that it is a sensitive tool to detect solute-induced changes in water's hydrogen bonding network near the vicinity of a biomolecule [47]. Highly accurate absorption measurements were performed for bulk water and lactose and compared with total absorption measured for a series of solutions with different concentrations of lactose in water. Absorption for solutions was found to be greater than the fractional sum (by volume) of the bulk absorptions. This study demonstrates that THz spectroscopy probes the coherent solvent dynamics on the subpicosecond timescale and gives quantitative measure of the extent the solute influence of biological water (i.e., water near the surface of a biomolecule).

Solutions in the THz range are easy to investigate in ATR configuration. Adding of solvable substances into water changes the ATR data, and that is demonstrated for the spectrum of saturated arabinose solution in water (at 21°C, the arabinose : water mass ratio 1:1, see Figure 23.11 c). For comparison, the absorption spectra of a dry arabinose (a 200-μm-thick pellet with a diameter of 5 mm pressed from powder at a pressure of 1 ton) and of a solution are presented in Figure 23.11 a, b. The intramolecular modes of organic crystals disappear in the solution, and molecular vibration modes are strongly broadened and cannot be resolved. The main difference from pure water absorption is in the range from 0.5 to 1 THz.

Similar behavior was observed for BSA and hemoglobin solutions. However, we have observed small changes in the low frequencies (Figure 23.12) after incubation of hemoglobin and BSA with glucose at concentration twice above physiological limit (0.72–5.4 g/L). Glucose when bounded to large molecules changes their response to the THz radiation. The nature of this influence should be further studied.

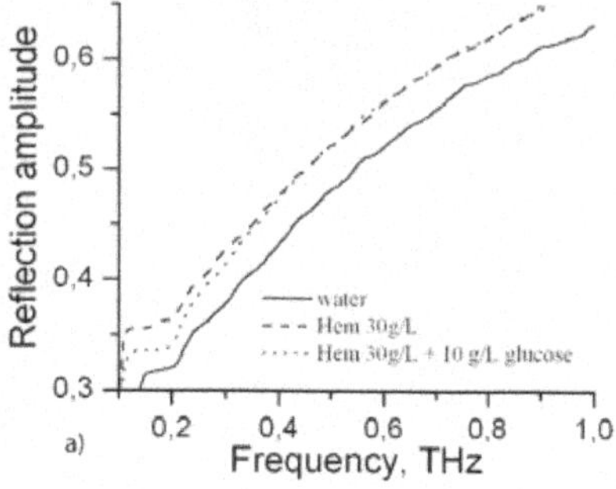

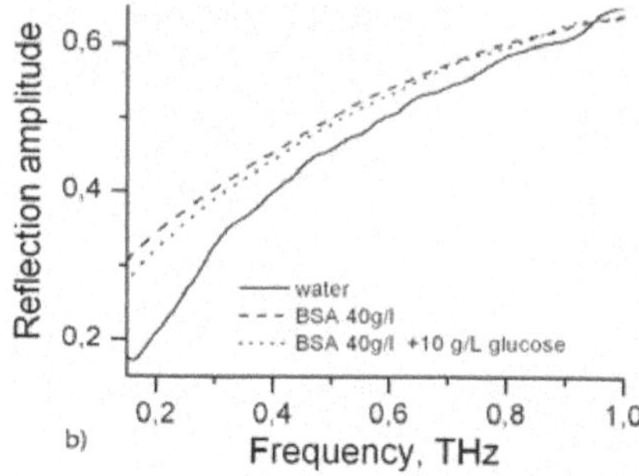

FIGURE 23.12: ATR spectra of biomolecule solutions with and without glucose: a) hemoglobin, b) albumin.

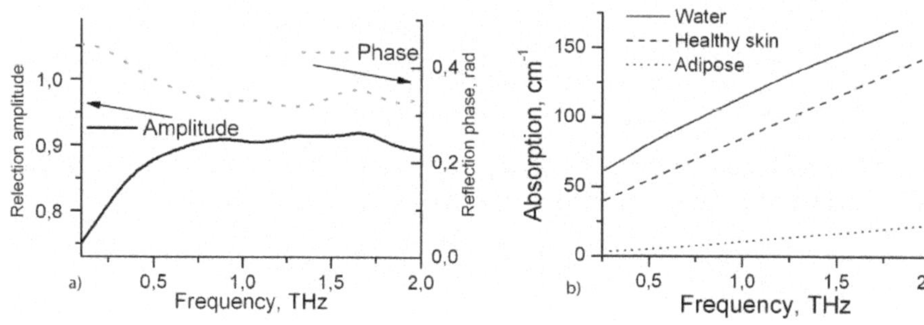

FIGURE 23.13: Skin spectra: a) total internal reflection, b) absorption, adapted from Ref. [3].

23.5.3 Skin

Skin is the most convenient object for THz-TDs studies and imaging. It is easy to measure *in vivo,* in ordinary or total internal reflection configuration. There is evidence of THz sensitivity to cancer lesions of the skin that is related to changes in water concentration or its state within the malignant tumor area. Ordinary reflection technique gives higher values of absorption coefficients than total internal reflection because of the varied penetration depth of radiation inside the skin. Less absorption for ATR is connected with the small probing depth that allows one to measure absorption only of the upper skin layer – stratum corneum, which contains a small amount of water.

Normal skin is comprised of three main layers: the stratum corneum, epidermis, and dermis. The thickness of the stratum corneum of the human palm can be up to 200 μm. The TPI system [3] was able to resolve the stratum corneum as the layered structure that gave rise to multiple reflections. The contrast seen in the images was due to the combined changes in the refractive index and absorption of the tissue.

23.5.4 Muscles

Muscle tissue (human and porcine) have been studied in a number of papers [12, 19, 40], all showing water-like THz spectra with absorption amplitudes of 1.5–2 times lower.

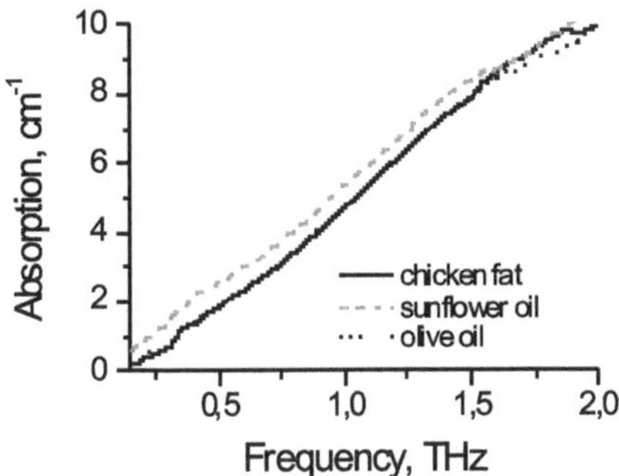

FIGURE 23.14: Spectral curves for absorption coefficient of chicken fat tissue and plant oils.

23.5.5 Liver

Some selectivity of the THz response of liver could be connected with a different content of fat and water in this tissue [58].

23.5.6 Fat

THz measurements of relevant biological materials include measurements in the THz range of two types of animal fat and plant oil from five different plants [31]. These authors found for the plant oils that the refractive index decreases slowly while the absorption coefficients increase with increasing frequency. For the animal fats, the refractive index was almost constant (1.5 for porcine fat and 1.4 for bovine fat) while the absorption also increased with frequency. In addition, they found that the refractive index increased with temperature for the animal fats but the absorption coefficient is almost unchanged. In another investigation terahertz spectra were measured for chicken fat at 21°C and vegetable oils (sunflower and olive) [19]. The spectral shape (see Figure 23.14) is almost the same in this range for all kinds of fat and oil. Note that fat tissue and oils have good transparency, weak dispersion, and a peculiarity in the range from 2 to 2.5 THz. This means that pulsed terahertz spectroscopy can be used to measure the thickness of fat layers (by the delay of THz pulse reflected from the layer boundaries).

There is a considerable difference between fat tissue (containing water) and rectified fat (see Figure 23.13). For the former, water spectral behavior dominates and the absorption coefficient is seven times higher.

23.5.7 Blood, hemoglobin, myoglobin

Terahertz blood properties are similar to that of water. In its turn dry hemoglobin, the main component of blood, has a strong featureless absorption that is increasing with frequency, like for other large biomolecules. In a THz study of myoglobin authors of paper [31] reported an increase in the absorption per protein molecule in the presence of "biological" water in contrast to a reduction expected on theoretical grounds from a decrease in the 2.5 Debye moment of bulk water to 0.8 Debye because of hindered motions experienced by "biological" water. Similarly to large biomolecules described above, a nearly continuous broadband absorption that increased linearly with frequency was observed for hemoglobin and myoglobin.

In experiments with dry hemoglobin we did not observe noticeable changes when hemoglobin was incubated with glucose in the solution before being dehydrated. Despite the fact that polycrystalline glucose has strong resonance absorption in the THz range, all hemoglobin spectra were similar to water by their shape. The refractive index for dried hemoglobin decreased with frequency; a similar decrease is present for other large biomolecules in a polycrystalline state (Figure 23.6 b).

23.5.8 Hard tissue

23.5.8.1 Tooth

Tooth tissues have been studied using THZ-TDS in a number of papers [3, 19, 48]. At present, more detailed information on these tissues is needed, including the monitoring of different types of dentine and obtaining more precise spectral characteristics. Although the water content in dentine and enamel is relatively small (see Table 23.1), it affects their optical properties, which was also demonstrated by other methods [49, 50]. Terahertz tooth spectra may be important for the development of terahertz tooth tomography, in particular, for diagnostics of some diseases related to water content in tooth tissues and for *in vivo* monitoring of tooth liquor and drug delivery. The monitoring of pathological changes in tooth tissues requires the knowledge of the main optical characteristics of various tooth tissues in the terahertz frequency range. We have measured locally (the diameter of the probe beam was 1 mm) the transmission spectra of different tooth samples (see Figure 23.15), including dentine areas with different concentrations and orientations of dentine tubules [51]. The difference between values of the refractive index of dentine and enamel is known to be considerable. In contrast, the absorption spectra of different samples without pathology are similar.

The refractive index averaged over measured frequencies is practically the same (n = 2.4) for all dentine samples and is considerably larger for enamel (n = 3.2). The refractive index dispersion varies only slightly in different tooth samples. The absorption spectrum can be approximated by the quadratic dependence $\alpha(f) = A + B_1 f + B_2 f^2$ (at least in the frequency range from 0.2 to 2.5 THz), where the coefficient B_2 plays the main role (23.5.8.2). The averaged coefficients A, B_1, B_2 are 13, −21, and 36, respectively (for f measured in THz and α in cm^{-1}). The refractive index can be assumed as a constant (Figure 23.15) for most of the applications.

In local studies of different tooth regions (dentine, enamel, and the enamel-dentinal junction), the spatial resolution was increased by using a pinhole of diameter D = 1 mm placed on the sample surface in which radiation was focused. However, a broadband ($\lambda = 1000$–$100\ \mu$m) THz pulse can be focused only down to a spot of diameter 0.5–3 mm. For $D = 1$ mm, the THz pulse amplitude is reduced insignificantly, which retains the SNR equal to 10^3, which is necessary for reliable spectral measurements. The tissue image contrast in pulsed terahertz spectroscopy cannot be increased simply by using shorter wavelengths (higher terahertz frequencies) for two reasons. First, absorption in biological tissues increases with frequency and, second, the efficiency of generation and detection in THz-TDs decreases at frequencies higher than 2−3 THz [17]. The reasonable optimal value of D, providing the required spatial resolution and spectrum quality, was 500 μm in our case in the frequency range from 0.3 to 2.5 THz.

23.5.8.2 Bone

Stringer et al. have used THz-TDS to measure parameters of the cortical bone derived from the human femur [52]. They found a linear relationship between the THz transmission and the sample density. There is evidence that the sample state of hydration is an important factor resulting in THz waves response of tissues.

23.5.9 Tissue dehydration

Authors of paper [12] have suggested lyophilization (freeze-drying) and nitrogen-drying for tissue sample preparation as an alternative to fresh tissue samples in studies with THz spectroscopy. It was shown that experiments with fresh tissues might contribute large errors at calculations in-

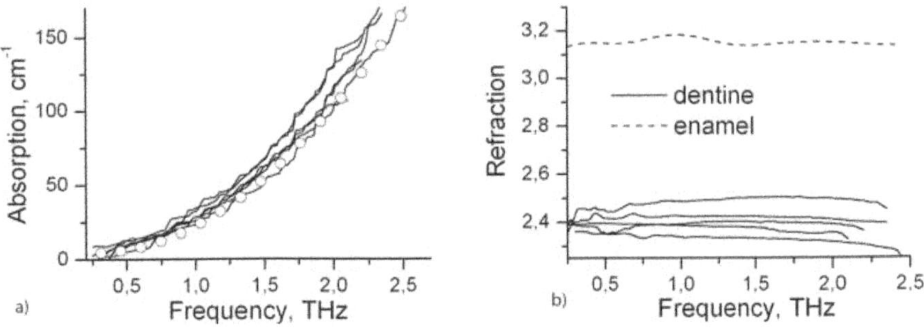

FIGURE 23.15: Tooth absorption (a) and refraction (b) spectra, each line corresponds to different local tooth areas.

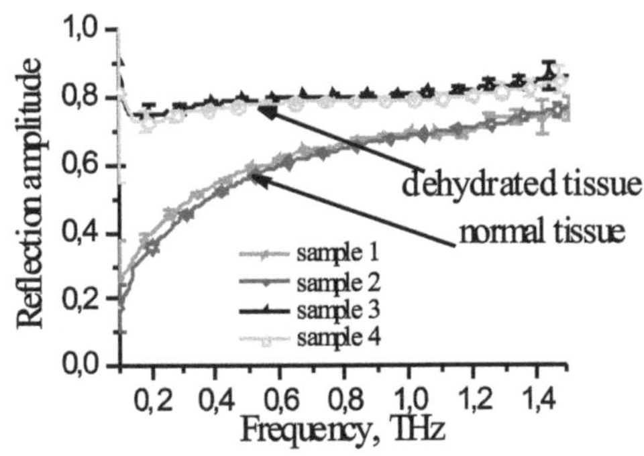

FIGURE 23.16: Muscle tissue dehydration.

volving tissue thickness. The high water content of fresh tissues also creates some variability in the THz transmission spectra. Complete water removal via nitrogen drying gets progressively difficult and residual water is inevitably included in THz spectroscopic measurements. Easy handling of lyophilized tissue, fast and effective removal of water, and structural preservation are the benefits of lyophilization techniquc. For dried tissues, noticeable differences are observed between tissue types. Further investigations to find optimal lyophilization procedure for specific tissue types and to quantify tissue dehydration are needed. Changes in weight before and after measurements may be good indicators of the extent of dehydration. Absorption spectral shape mostly conserves after dehydration and amplitude significantly decreases.

We have studied dehydration effects on THz response by the ATR method. Porcine muscle slices ($2 \times 10 \times 20$ mm) were placed in glycerol for 30 min, where free water was replaced within the tissue via osmotic action of glycerol [53]. The result was a 10% decrease of sample mass and a 5% increase of a visible light transmission.

After removing the water, the THz reflection spectra considerably increased at a low frequency range and noticeably changed their shape. This experiment demonstrates high THz sensitivity to water concentration in tissues (Figure 23.16).

23.6 THz Imaging: Techniques and Applications

23.6.1 Introduction

THz imaging of plant and animal materials debuted in 1995, and in this context it is often called T-ray imaging [54]. It has been possible to image biological samples that have a high contrast of aqueous and nonaqueous regions. Image contrast is mostly based on water/fat content ratios in different tissues. Malignant tissue contrasting on the image of a mouse liver with the developed tumor lesions has been demonstrated. Typical spatial resolution was of 250 μm laterally and 20 μm axially. Images were often obtained by raster-scanning of the sample through the focused THz beam. It is also possible to collect a two-dimensional (2D) image using a CCD camera as a detector, which allows very high data acquisition rates [55]. THz tomography, wherein the pulse reflected off an object or transmitted through it is analyzed, is now being developed [56, 57]. As a medical imaging tool, THz-TDS methods are still under development but are likely to complement existing imaging modalities [5].

A single frequency terahertz imaging system can be based on a quantum cascade laser or difference frequency system of two lasers [13] and provides a relatively large dynamic range and high spatial resolution [58].

23.6.2 Human breast

Fitzgerald et al. [5] have investigated the feasibility of THz imaging for mapping margins of tumors in freshly excised human breast tissue. The size and shape of tumor regions in the THz images were compared with the corresponding histological sections and a good correlation was found. In addition, they have also performed a spectroscopic study comparing the THz spectrum. Both the absorption coefficient and refractive index were higher for tumor tissue. These changes were consistent with higher water content and structural changes, such as increased cell and protein density.

23.6.3 Skin

A contrast has been observed between basal cell carcinoma (BCC) and normal tissue in THz images of skin cancer [5]. The study in Ref. [5] revealed that both the refractive index and the absorption coefficient of BCC were higher than that for the normal tissue.

23.6.4 Tooth

The differences in the refractive indexes of the enamel, dentine, and pulp enable the three regions to be identified [5] by the delay of the transmitted terahertz pulse.

23.6.5 Nanoparticle-enabled terahertz imaging

In [59] the principle of nanoparticle-contrast-agent-enabled terahertz imaging (COTHI) technique is demonstrated. With this technique, a significant sensitivity enhancement is achieved in THz cancer imaging with nanoparticle contrast agents that can be targeted to cancerous tumors. As high sensitivity is achieved by the surface plasmon resonance through IR laser irradiation, the resolution of the CATHI technique can be as good as a few microns.

23.7 Summary

The terahertz absorption and refraction spectra of many biological substances and tissues are studied with terahertz time-domain spectrometers and related methods developed for measurement processing. Small organic molecules have characteristic absorption bands in the terahertz frequency range. Large molecules and tissues have considerable absorption, which almost linearly increases with frequency. The absorption coefficient and refractive index of liquid water determine the applicability of THz pulses for studying biological objects. Substances that strongly absorb THz radiation are better studied by reflection spectroscopy. THz spectroscopy has been established as a technique to probe biomolecular hydration, binding, conformational change, and oxidation state change. The contrast mechanisms are not yet established, but it is underway.

Biomedical applications are also under development. The challenge of probing biomolecule spectra in a highly absorbing media is great, and the influence of hydrogen-bonding on water solutions is being studied now.

THz imaging has made significant advancements. Today, portable imaging systems and spectrometers are available commercially providing an excellent progression from the bench-top systems used in early research. Imaging times are now down from days to minutes. To progress even further, more efficient or compact emitters and detectors are required. At present, a dynamical range of 1000 for a broadband spectroscopy limits tissue and liquid studies up to 100-μm-layer samples. With this in mind, there is still much work to be done.

Acknowledgments:

Authors are grateful to O.S. Zhernovaya, O.P. Cherkasova, and E.A. Kuleshov for assistance in experiments. This work was partly supported by CRDF BRHE grant RUXO-006-SR-06 and RFBR grant 050332877. VVT was supported by grants No. 224014 PHOTONICS4LIFE-FP7-ICT-2007-2, RF President's 208.2008.2, and No. 2.1.1/4989, 2.2.1./2950 of RF Program on the Development of High School Potential.

References

[1] B.M. Fisher, H. Helm, and P.U. Jepsen "Chemical recognition with broadband THz Spectroscopy," *Proc. IEEE* **95**, 1592–1604 (2007).

[2] E. Pickwell, B.E. Cole, A.J. Fitzgerald, et al., "*In vivo* study of human skin using pulsed terahertz radiation," *Phys. Med. Biol.* **49**, 1595–1607 (2004).

[3] E. Pickwell and V.P. Wallace, "Topical review biomedical applications of terahertz technology," *J. Phys. D: Appl. Phys.* **39**, R301–R310 (2006).

[4] B.B. Hu and M.C. Nuss, "Imaging with terahertz waves," *Opt. Lett.* **20**, 1716–1718 (1995).

[5] V.P. Wallace, A.J. Fitzgerald, S. Shankar, et al., "Terahertz pulsed imaging of basal cell carcinoma ex vivo and in vivo," *B. J. Derm.* **151,** 424–432 (2004).

[6] R.M. Woodward, V.P. Wallace, D.D. Arnone, et al., "Terahertz pulsed imaging of skin cancer in the time and frequency domain," *J. Biol. Phys.* **29**, 257–261 (2003).

[7] M. He, A.K. Azad, S. Ye et.al., "Far-infrared signature of animal tissues characterized by terahertz time-domain spectroscopy," *Opt. Commun.* **259**, 389–392 (2006).

[8] S.H. Chung, A.E. Cerussi, C. Klifa, et al., "*In vivo* water state measurements in breast cancer using broadband diffuse optical spectroscopy," *Phys. Med. Biol.* **53,** 6713–6727 (2008).

[9] A.G. Markelz, "Invited Review: terahertz dielectric sensitivity to biomolecular structure and function," *IEEE J. Sel. Top. Quantum Electron.* **14**, 180–190 (2008).

[10] M.C. Beard, G.M. Turner, and C.A. Schmuttenmaer, "Feature Article: terahertz spectroscopy," *J. Phys. Chem. B* **106**, 7146–7159 (2002).

[11] J.H. Son, "Terahertz electromagnetic interactions with biological matter and their applications," *J. Appl. Phys.* **105**, 102033–102043 (2009).

[12] G.M. Png, J.W. Choi, B.W. Ng, et al., "The impact of hydration changes in fresh bio-tissue on THz spectroscopic measurements," *Phys. Med. Biol.* **53**, 3501–3517 (2008).

[13] J. Nishizawa, T. Sasaki, and T. Tanno, "Coherent terahertz-wave generation from semiconductors and its applications in biological sciences" *J. Phys. Chem. Solids*, **69**, 693–701 (2008).

[14] O. Esenturk, A. Evans, and E.J. Heilweil, "Terahertz spectroscopy of dicyanobensens: anomalus absorption intencities and spectral calculations," *Chem. Phys. Lett.* **442**, 71–77 (2007).

[15] S.L. Dexheimer (ed.), *Terahertz Spectroscopy: Principles and Applications*, Taylor & Francis Group, Boca Raton (2008).

[16] X.C. Zhang, B.B. Hu, J.T. Darrow, et al., "Generation of femtosecond electromagnetic pulses from semiconductor surfaces," *Appl. Phys. Lett.* **56**, 1011–1013 (1990).

[17] M.M. Nazarov, S.A. Makarova, A.P. Shkurinov, et al., "The use of combination of nonlinear optical materials to control THz pulse generation and detection," *Appl. Phys. Lett.*, **92**, 021114–021117 (2008).

[18] M.M. Nazarov, A.P. Shkurinov, and V.V. Tuchin, "Tooth study by terahertz time-domain spectrometer," *Proc. SPIE* **6791**, 679109-1–9 (2008).

[19] M.M. Nazarov, A.P. Shkurinov, E.A. Kuleshov, et al., "Terahertz time-domain spectroscopy of biological tissues," *Quantum Electronics* **38,** 647–654 (2008).

[20] M.M. Nazarov, A.P. Shkurinov, V.V. Tuchin et al., "Modification of terahertz pulsed spectrometer to study biological samples," *Proc. SPIE* **6535**, 65351J-1-7 (2007).

[21] W. Withayachumnankul, B.Ferguson, T.Rainsford, et al., "Material parameter extraction for terahertz time-domain spectroscopy using fixed-point iteration," *Proc. SPIE* **5840**, 221–231 (2005).

[22] L.E. Dolotov, Yu.P. Sinichkin, V.V. Tuchin, et al., "Design and evaluation of a novel portable erythema-melanin-meter," *Lasers Surg. Med.*, **34**, 127–135 (2004).

[23] H. Schaefer and T.E. Redelmeier (eds.), *Skin Barrier, Principles of Percutaneous Absorption*, Karger, Basel (1996).

[24] M. Walther, B.M. Fischer, H. Helm, et al., "Noncovalent intermolecular forces in polycrystalline and amorphous saccharides in the far infrared," *Chem. Phys.* **288**, 261–268 (2003).

[25] V.I. Fedorov and S.S. Popova, "Low THz range and reaction of biological object of different level," *Millimetre Waves in Biol. and Med.* **2**, 3–19 (2006).

[26] V.V. Tuchin (ed.), *Handbook of Optical Sensing of Glucose in Biological Fluids and Tissues*, CRC Press, Taylor & Francis Group, London (2009).

[27] N.N. Brandt, A.Yu. Chikishev, A.V. Kargovsky, et al., "Terahertz time-domain and Raman spectroscopy of the sulfur-containing peptide dimers: low-frequency markers of disulfide bridges," *Vibrational Spectroscopy* **47**, 53–58 (2008).

[28] B.M. Fischer, M. Walther, and P.U. Jepsen, "Far-infrared vibrational modes of DNA components studied by terahertz time-domain spectroscopy," *Phys. Med. Biol.* **47**, 3807–3814 (2002).

[29] S.P. Mickan, A. Menikh, H. Liu et al., " Label-free bioaffinity detection using terahertz technology," *Phys. Med. Biol.* **47**, 3789–3795 (2002).

[30] J. Xu, K.W. Plaxco, and S.J. Allen, "Probing the collective vibrational dynamics of a protein in liquid water by terahertz absorption spectroscopy," *Protein Sci.* **15**, 1175–1181 (2006).

[31] C.F. Zhang, E. Tarhan, A.K. Ramdas, et al., "Broadened far-infrared absorption spectra for hydrated and dehydrated myoglobin," *J. Phys. Chem. B.* **108**, 10077–10082 (2004).

[32] B.M. Fischer, M. Hoffmann, H. Helm, et al., "Chemical recognition in terahertz time-domain spectroscopy," *Semicond. Sci. Technol.* **20**, S246–S253 (2005).

[33] Y.C. Shen, P.C. Upadhya, E.H. Linfield et al., "Observation of far-infrared emission from excited cytosine molecules," *Appl. Phys. Lett.* **87**, 011105-1–3 (2005).

[34] L.L. Van Zandt and V.K. Saxena, "Millimeter-microwave spectrum of DNA: six predictions for spectroscopy," *Phys. Rev.* A **39**, 2672–2674 (1989).

[35] S.E. Whitmire, D. Wolpert, A.G. Markelz, et al., "Protein flexibility and conformational state: a comparison of collective vibrational modes of wild-type and D96N Bacteriorhodopsin," *Biophys. J.* **85**, 1269–1277 (2003).

[36] A. Markelz, S. Whitmire, J. Hillebrecht, et al., "THz time-domain spectroscopy of biomolecular conformational modes," *Phys. Med. Biol.* **47**, 3797–3805 (2002).

[37] B.M. Fischer, M. Walther, and P.U. Jepsen, "Far-infrared vibrational modes of DNA components studied by terahertz time-domain spectroscopy," *Phys. Med. Biol.* **47**, 3807–3814 (2002).

[38] A.G. Markelz, A. Roitberg, and E.J. Heilweil, "Terahertz applications to biomolecular sensing," *Chem. Phys. Lett.* **320**, 42–48 (2000).

[39] M. Nagel, F. Richter, P. Haring-Bolívar, et al., "A functionalized THz sensor for marker-free DNA analysis," *Phys. Med. Biol.* **48**, 3625–3636 (2003).

[40] A.J. Fitzgerald, E. Berry, N.N. Zinov'ev, et al., "Catalogue of human tissue optical properties at terahertz frequencies," *J. Bio. Phys.* **129**, 123–128 (2003).

[41] K.N. Woods and H. Wiedemann, "The influence of chain dynamics on the far-infrared spectrum of liquid methanol," *J. Chem. Phys.* **123**, 134507–134510 (2005).

[42] B. Born and M. Havenith, "Terahertz dance of proteins and sugars with water", *J. Infrared Milli. Terahz. Waves,* **30**, 1245–1254 (2009).

[43] V.V. Tuchin, *Tissue Optics: Light Scattering Methods and Instruments for Medical Diagnosis*, second edition, **PM166**, SPIE Press, Bellingham, WA (2007).

[44] R.F. Reinoso, B.A. Telfer, and M. Rowland, "Tissue water content in rats measured by desiccation," *J. Pharmacol. Toxicol. Methods*, **38**, 87–92 (1997).

[45] M. Nagai, H. Yada, T. Arikawa, et al., "Terahertz time-domain attenuated total reflection spectroscopy in water and biological solution," *Int. J. Infrared Milli. Waves* **27**, 505–515 (2006).

[46] T. Arikawa, M. Nagai, and K. Tanaka, "Characterizing hydration state in solution using terahertz time-domain attenuated total reflection spectroscopy," *Chem. Phys. Lett.* **457**, 12–17, (2008).

[47] U. Heugen, G. Schwaab, E. Bründermann, et al., "Solute-induced retardation of water dynamics probed directly by terahertz spectroscopy," *PNAS* **103**, 12301–12306 (2006).

[48] D.A. Crawley, C. Lonbottom, V.P. Wallace, et al., "Three-dimensional terahertz pulse imaging of dental tissue," *J. Biomed. Opt.* **8**, 303–307 (2003).

[49] F.W. Wehrli and M.A. Fernandez-Seara, "Nuclear magnetic resonance studies of bone water," *Annal. Biomed. Eng.* **33**, 79–86 (2005).

[50] A. Kishen and A. Rafique, "Investigations on the dynamics of water in the macrostructural dentine," *J. Biomed. Opt.* **11**, 054018-1–8 (2006).

[51] V. Imbeni, R.K. Nalla, C. Bosi, et al., "On the in vitro fracture toughness of human dentin," *J. Biomed. Mater. Res.* **66A**, 1–9 (2003).

[52] M.R. Stringer, D.N. Lund, A.P. Foulds, et al.,"The analysis of human cortical bone by terahertz time-domain spectroscopy," *Phys. Med. Biol.* **50**, 3211–3219 (2005).

[53] L. Oliveira, A. Lage, M.P. Clemente, and V. Tuchin, "Optical characterization and composition of abdominal wall muscle from rat," *Optics and Lasers in Engineering* **47**, N6, 667–672 (2009).

[54] S. Mickan, D. Abbott, J. Munch, et al., "Analysis of system trade-offs for terahertz imaging," *Microelectron. J.* **31**, 503–514 (2000).

[55] Z. Jiang and X. C. Zhang, "Terahertz imaging via electrooptic effect," *IEEE Trans. Microwave Theory Tech.* **47**, 2644–2650 (1999).

[56] R.A. Cheville, R.W. McGowan, and D. Grischkowsky, "Mechanisms responsible for terahertz glory scattering from dielectric spheres," *Phys. Rev. Lett.* **80**, 269–272 (1998).

[57] D.M. Mittleman, S. Hunsche, L. Boivin, et al., "T-ray tomography," *Opt. Lett.* **22**, 904–906 (1997).

[58] S.M. Kim, F. Hatami, J.S. Harris et al., "Biomedical terahertz imaging with a quantum cascade laser," *Appl. Phys. Lett.* **88**, 153903–153905 (2006).

[59] S.J. Oh, J. Kang, I. Maeng, J.-S. Suh, Y.-M. Huh, S. Haam, and J.-H. Son, "Nanoparticle-enabled terahertz imaging for cancer diagnosis," *Opt. Express* **17**, 3469–3475 (2009).

24

Nanoparticles as Sunscreen Compound: Risks and Benefits

Alexey P. Popov

Optoelectronics and Measurement Techniques Laboratory, Department of Electrical and Information Engineering, Faculty of Technology, University of Oulu and Infotech Oulu, Oulu, 90014, Finland, International Laser Center, M.V. Lomonosov Moscow State University, Moscow, 119991, Russia

Alexander V. Priezzhev

Physics Department and International Laser Center, M.V. Lomonosov Moscow State University, Moscow, 119991, Russia

Juergen Lademann

Center of Experimental and Applied Cutaneous Physiology, Department of Dermatology, Universitätsmedizin-Charité Berlin, Berlin, Germany

Risto Myllylä

Optoelectronics and Measurement Techniques Laboratory, Department of Electrical and Information Engineering, Faculty of Technology, University of Oulu and Infotech Oulu, Oulu, 90014, Finland

This chapter presents experimental and theoretical data on UV protection by use of nanoparticles as part of sunscreens and some side effects associated with this use.

Key words: nanoparticles, sunscreen, UV, penetration, toxicity, skin, Monte Carlo simulation, Mie theory, scattering, absorption, erythema, cancer, stratum corneum

24.1 Introduction

Nanotechnology is a term for techniques, materials, and devices that operate at the nanometer scale. It is defined as the design, characterization, production, and application of structures, devices, and systems by controlling shape and size at the nanoscale and represents one of the most promising technologies of the 21st century, and has been considered to be a new industrial revolution [1].

Nanotechnology has been penetrating deeper and deeper into everyday life and today nanomaterials are widely used in consumer products, for example in cosmetics, sunscreens, paints, coatings, food, cleaning agents, for water and air purification [2] and are expected to be increasingly used in the medical field in diagnostics, imaging, therapy, and drug delivery.

Nanoparticles are determined as particles whose dimensions do not exceed 100 nm. Such small dimensions may affect physical and chemical properties of the materials due to a larger amount of atoms being located on the particle surface, e.g., phototoxicity. For usage in sunscreens about 1000 tons of insoluble nanoparticles (titanium dioxide and zinc oxide) were produced in 2003/2004 [3]. Their extensive usage has recently raised concerns of nongovernmental organizations about their safety [4]. Such organizations insist on moratorium for nanoparticle usage until the particles are extensively explored and their harmlessness is proved by multiple independent experiments worldwide.

In this chapter, attention is mainly paid to the investigation of TiO_2 nanoparticles as the most widely used in sunscreens. Besides discussing protective properties of the nanoparticles, we consider penetration of the particles into the superficial layer of human skin after multiple applications, as well as their phototoxicity.

24.2 Nanoparticles in Sunscreens

Human beings exist surrounded by the environment, which might be potentially dangerous. The human body is covered by skin, the largest organ of the body. The function of skin is protection of inner organs from various kinds of outer hazards, such as thermal, mechanical, chemical, biological, optical, etc. One of these functions is the protection from the solar ultraviolet (UV) light. The main contributors to the UV attenuation are stratum corneum (horny layer, superficial skin layer) consisting of dead corneocytes and located in epidermis melanocytes.

The Earth is covered by the atmosphere comprising gases such as nitrogen (N_2), oxygen (O_2), ozone (O_3), carbon dioxide (CO_2), water vapors (H_2O), etc. The so called "ozone layer" absorbs the most dangerous, the shortest-wavelength (100–280 nm) fraction of the UV solar radiation. However, the long-wavelength part (280–400 nm) penetrates into the lower atmosphere and reaches the Earth's surface affecting human beings and other living creatures.

Due to the depletion of ozone because of human industrial activity, from one side, and the increasing desire of people to stay outdoors during holidays, additional protection (to what is provided by skin) is required. That is the main purpose of sunscreen use. Protective properties of sunscreens are caused by absorption and scattering of light by their ingredients, chemical and physical components [5, 6]. Chemical components are organic molecules absorbing UV light. Physical components are mineral insoluble nanoparticles having both absorbing and scattering properties.

Sunscreens have been widely used for many years already. During recent years, great concerns have been raised about the effect of nanoparticles used in sunscreens on human health. Among them are the fears about particle penetration into viable layers of skin through the stratum corneum after multiple applications of sunscreens onto the body surface. Because of small particle sizes (at a range of tens of nanometers) these concerns are not groundless. Contact of a foreign body with cells of a living organism could be potentially dangerous. In the case of nanoparticles, such a contact may, e.g., induce the generation of free radicals leading to inflammation and possibly provoking cancer.

Most sunscreens nowadays contain insoluble nano-sized particles of titanium dioxide (TiO_2) and zinc oxide (ZnO) [7], which light-attenuation efficiency depends on their size. Particles of TiO_2 absorb and scatter UV radiation the most efficiently at sizes of 60–120 nm [8], while particles of ZnO have an optimal diameter of 20–30 nm [9]. As such particles scatter UV and not visible light, the sunscreens appear to be transparent. Nanosized TiO_2 is from 14 nm in diameter. ZnO is generally used in the form of particles 30–200 nm in diameter. The surface of these mineral particles is frequently covered with inert coating materials, such as SiO_2 or Al_2O_3, to improve their dispersion in sunscreen formulations.

TiO_2 and ZnO are both semiconductor materials. Their electronic energy structure is represented by two bands, namely, the conduction band and valence band with an energy gap (band gap) between them. Being excited by a photon, an electron from the valence band goes to the conduction band, generating a positive-charged hole in the valence band. Only photons with energies larger than the band gap can be absorbed. The formed charged carriers can recombine either radiatively (by emitting a photon) or nonradiatively (releasing energy as heat). Electrons and holes can recombine on a particle surface reducing or oxidizing surrounding molecules, thus producing radicals.

Both TiO_2 and ZnO are wide-band gap semiconductors. The band gap of TiO_2 varies between 3.00 and 3.20 eV (3.00 eV for rutile, 3.11 eV for brookite and 3.20 eV for anatase forms) [10] and that of ZnO is 3.37 eV at room temperature. Another difference between the materials is the interband transitions: they are allowed in ZnO (direct) and symmetry-forbidden in TiO_2 (indirect). Despite the fact that the band gap of TiO_2 allows absorption of light with longer wavelengths than that of ZnO, the absorption efficiency of TiO_2 is shifted to the UVB wavelengths (280–315 nm).

Among the three existing crystal modifications of TiO_2 (anatase, rutile and brookite), rutile and anatase are used widely. Anatase is thermodynamically stable for diameters smaller than 11 nm; and rutile is stable for sizes larger than 35 nm [11], the latter being the most stable at high temperatures.

Nanoparticles tend to form aggregates and agglomerates of 100–200 nm in size resulting in worsening their protecting properties in the UVB range and shifting the pronounced attenuation to the longer wavelength UVA (315–400 nm) and visible regions of the solar spectrum. The typical organization of anatase particles of different diameters (25 and 400 nm) are shown in Figure 24.1. Nevertheless, novel manufacturing technologies, e.g., mechanochemical processing (MCP^{TM}), enable production of nanopowders without such disadvantages and with a narrow size distribution (e.g., 25 ± 4 nm), which can be successfully used in sunscreens [7]. The usual size distribution of TiO_2 particles is 15–20% from the mean value [12].

24.3 Penetration of Nanoparticles into Skin

24.3.1 Skin structure

Skin occupies an area of about 2 m^2 and has a weight of about 3 kg (up to 20 kg if fatty tissue is taken into account) for adult individuals [13]. The skin thickness depends on the region and varies between 1.5 and 4 mm [14].

The methods for studying skin structure are either invasive or noninvasive. The former includes, for example, biopsy, when skin samples are taken and then analyzed with a microscope. The latter are nondamaging techniques, often implying optical methods, which use nonionizing radiation and suitable spatial resolution, e.g., optical coherence tomography (OCT) [15–18], laser scanning microscopy [19–21], and fluorescence imaging [22].

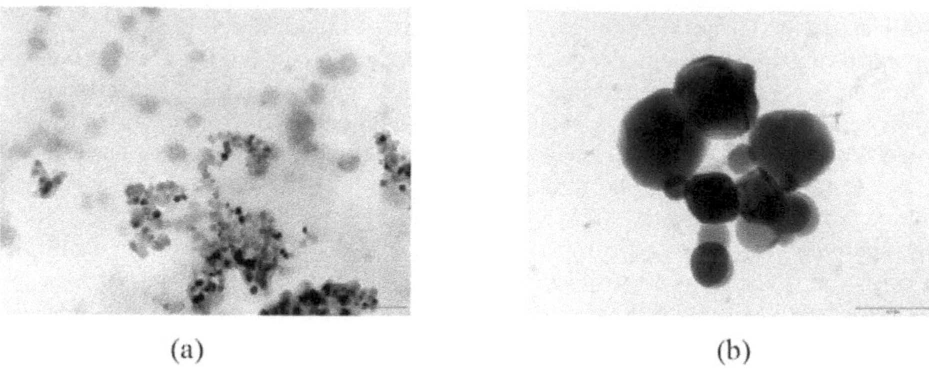

(a) (b)

FIGURE 24.1: TEM photos of 25- (a) and 400-nm (b) TiO_2 nanoparticles (magnification: ×89). Scale: the bar at the right low corner corresponds to 0.2 μm. Courtesy E.V. Zagainova.

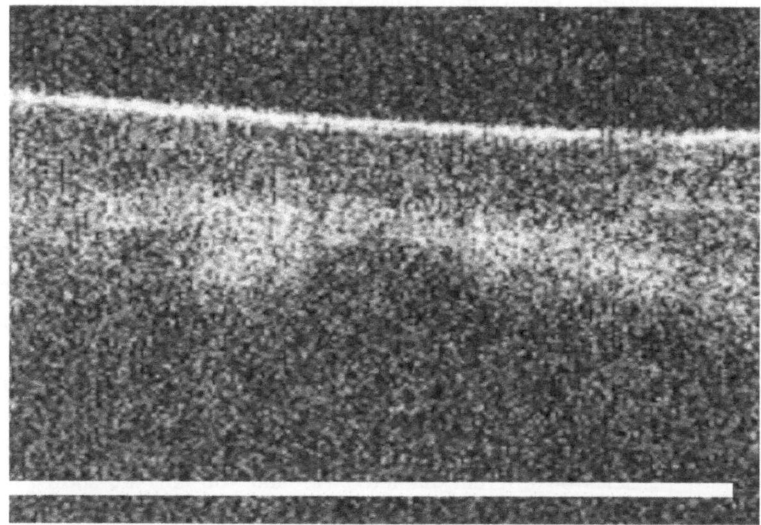

FIGURE 24.2: An OCT image of human skin *in vivo* (flexor forearm). Scale: the bar at the bottom corresponds to 1 mm. Reprinted with permission from [23]. Copyright 2008, University of Oulu.

The layers of skin (starting from the skin surface) are stratum corneum, viable epidermis, dermis, and hypodermis (a layer of subcutaneous fat). Sweat glands, sebaceous glands, and hair follicles are also considered as parts of skin. An image obtained by optical coherence tomography from a flexor forearm skin of a healthy 27-year-old male is given in Figure 24.2. The layered structure is clearly seen.

24.3.2 Stratum corneum

The outermost part of the skin is called the *stratum corneum* (horny layer). It serves as a biological cover for the whole body. Such a cover hampers the penetration of substances, organisms, or radiation thus protecting inner layers of skin from outer biological and nonbiological hazards. The published data concerning the thickness of the stratum corneum differ: 6–40 μm on such common areas as the abdomen, flexor forearm, thigh, and back [24, 25].

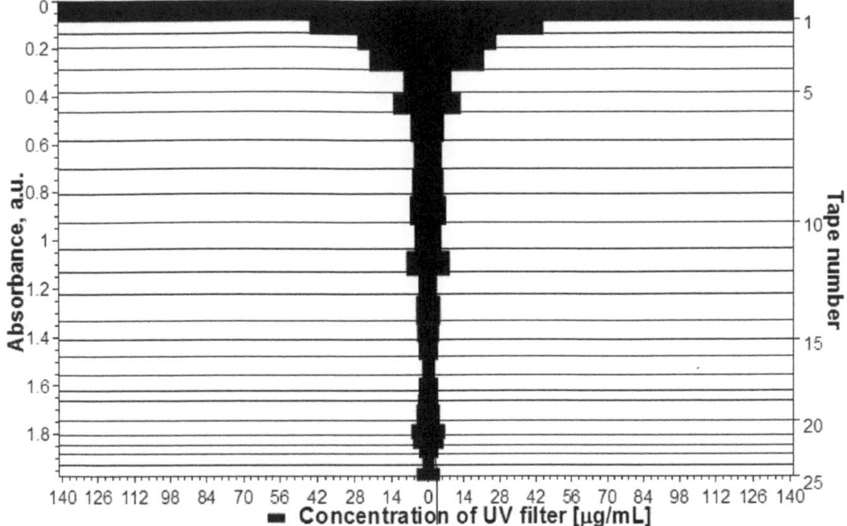

FIGURE 24.3: In-depth distribution of chemical UV filter (utilized in sunscreens), within the porcine stratum corneum, as revealed by the tape stripping technique [23]. Copyright 2008, University of Oulu.

The stratum corneum is formed by corneocytes, dead polyhedral-shaped cells without nuclei, approximately 40 μm in diameter and 0.5-μm-thick [26]. Their cellular organelles and cytoplasm disappear during the cornification process. There is an overlap between adjacent corneocytes which increases stratum corneum cohesion. The intercellular region is filled with a lipid, which represents a continuous medium and is required for barrier function.

The stratum corneum consists of about 15 layers, although exceeding this value significantly (5–10 times) in places of intensive use such as soles and palms. The upper layer (stratum disjunctum) contains about 3–5 layers and undergoes continuous desquamation. Water makes up 15% of its weight. The lower layer (stratum compactum) is thicker, more densely packed, and more regular. Additionally, it is more hydrated than the stratum disjunctum (30%). The cells of stratum corneum renew themselves continuously: desquamated cells are replaced by "fresh" ones coming from the epidermis. One cell layer takes 24 hours to form, so the whole stratum corneum renews itself completely in two weeks.

24.3.3 Permeability of stratum corneum

Although the stratum corneum is a heterogeneous structure, in cases of liquid substance penetration, it behaves as a homogeneous membrane and the diffusion law is correct [27]. No indication of a tortuous pathway for water diffusing across the stratum corneum was found; the calculated diffusion path length was equal to the physical thickness of the layer [28]. The penetration of substances can occur by intercellular (along lipid lamellae), by transcellular (through corneocytes), as well as via glands (sweat sebaceous) and along hairs. None of these ways can be treated exclusively.

The in-depth penetration profile of substances topically applied onto the skin is dependent on the vehicle used. In experiments with vanillin mixed with two kinds of emulsions, it was shown that the use of ethanol enhances penetration in comparison to w/o (water-in-oil) emulsion [29]. Application of vehicles also affects the cohesion properties of corneocytes: o/w (oil-in-water) emulsion increases it and ethanol decreases it [30].

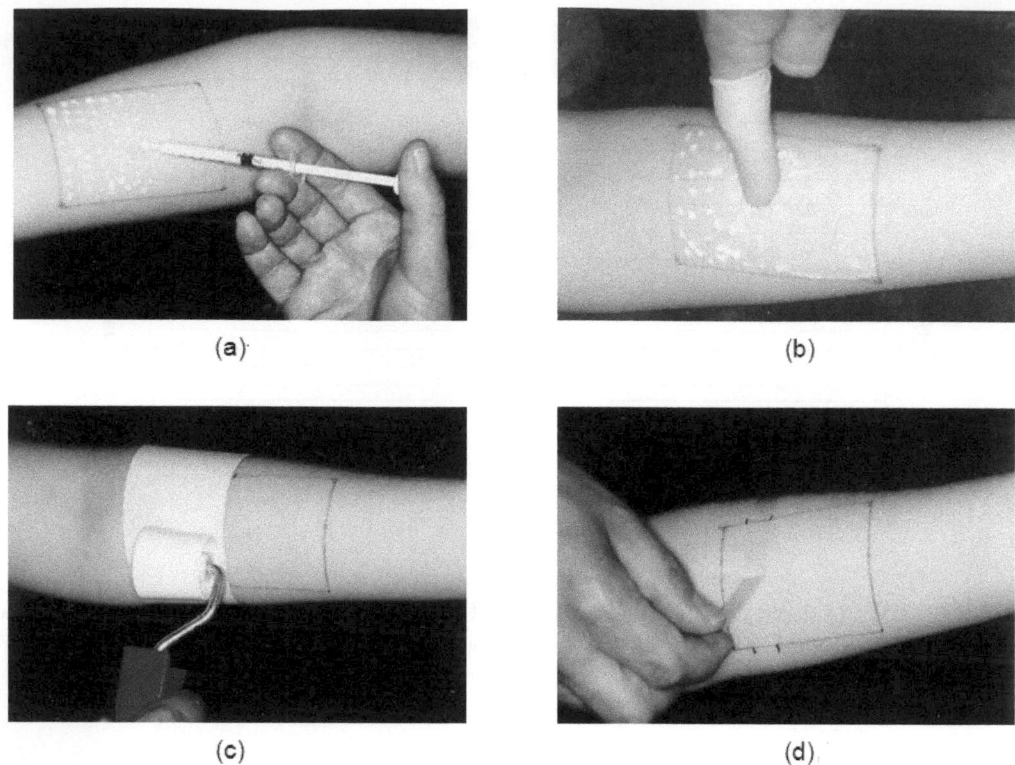

FIGURE 24.4: Application of a sunscreen and tape stripping: application of the emulsion (a), homogeneous distribution with a saturated glove finger (b), pressing of the tape by a roller (c), removing of the adhesive tape (d). Reprinted with permission from [23]. Copyright 2008, University of Oulu.

Figure 24.3 illustrates an in-depth penetration profile of a chemical UV filter (used in sunscreens) reconstructed after tape stripping using spectroscopy methods. The concentration of the UV filter substance was determined for each tape strip, as described previously [6].

24.3.4 Penetration of nanoparticles into human skin

Our experiments employed the previously-described tape stripping technique [31]. The study was performed with 6 healthy volunteers with skin types II and III. An emulsion containing coated titanium dioxide particles (mean diameter 100 nm) was studied, and 2 mg/cm^2 of this preparation, according to the COLIPA standard [32], were applied onto the flexor forearm. A skin area of 10×8 cm^2 was marked with a permanent marker. 160 mg of the selected emulsion was applied with a syringe and distributed homogeneously with a gloved finger. The procedure is illustrated by Figure 24.4.

After application of the emulsion, the volunteers rested for 1 h without sweating; covering the test area was forbidden. The emulsion with TiO_2 nanoparticles was administered 5 times over a period of 4 days. It was allowed to wash the treated skin area and wear any clothes. The process modeled a situation on a beach. The tape stripping started on the fourth day 1 h after application. Multiple applications are needed for a more homogeneous distribution of administered particles. As shown [33], the homogeneous distribution can increase the effectiveness of sunscreens by a factor of 10 in terms of SPF (sun protection factor), and inversely, an uneven distribution of the applied sunscreens

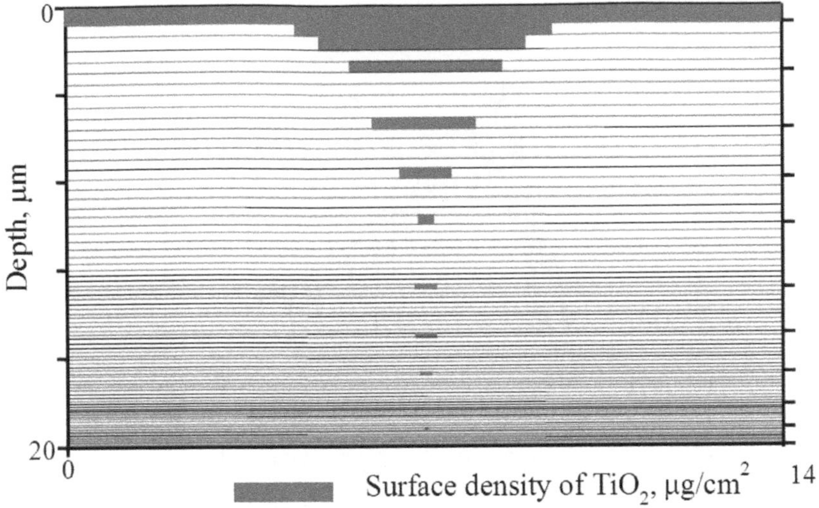

FIGURE 24.5: Experimental in-depth profile of TiO_2 particles localized within the stratum corneum, obtained by the tape-stripping technique. Reprinted with permission from [41]. Copyright 2005, IOP Publishing Ltd.

decreases SPF and leads to the necessity of using the step model of the treated skin [34].

Thin strips of stratum corneum (each of about 1–μm-thick) were removed consecutively using *tesa* adhesive tape. The penetration profile was obtained by analyzing the amount of the stratum corneum (amount of corneocytes) and that of TiO_2 removed with each tape strip. The thickness of the skin strip relative to the whole thickness of the horny layer was determined spectroscopically [29, 35]. Due to the presence of particles, the NIR light was used for this purpose [36]. The surface concentration of the particles in each strip was estimated by X-ray fluorescence measurements [37], which yielded about 14 μg/cm^2 in the first strip and almost zero in the strip taken from the depth of 15 μm. Most of the particles were located within the depth range of 0–3 μm. The results of the procedure are shown in Figure 24.5. They look similar to those from Figure 24.3. Experimental investigation of nanoparticle penetration through the skin was also studied by other research groups and similar results were obtained. Namely, using similar technique on 3 volunteers, 90% of 20-nm TiO_2 component of sunscreens was found within the first 15 tape strips after 5 hours after emulsion application; no particles were detected in the follicle, viable epidermis, or dermis [38]. Multiphoton microscopy imaging with a combination of scanning electron microscopy and an energy-dispersive X-ray technique were used to prove that 20–30 nm ZnO particles stayed in the stratum corneum and accumulated in skin folds and follicle roots of human skin *in vivo* and *in vitro* [39, 40].

Evaluation of the volume concentration C of TiO_2 particles in the superficial strip (see Figure 24.5) can be performed as follows:

$$C = \frac{N \cdot V_0}{V} = \frac{M}{\rho_0 \cdot V_0} \cdot \frac{V_0}{V} = \frac{M}{\rho_0 \cdot V}, \tag{24.1}$$

where N is the number of TiO_2 particles with volume V_0 and density ρ_0 each, within a strip of volume V. The total mass of all the TiO_2 particles inside the strip is M. The volume V equals to the strip thickness (in our case 0.75 μm) multiplied by the surface area (1 cm^2). As can be deduced from Figure 24.5, the mass M equals to 14 μg (because the area is 1 cm^2). True density of TiO_2 (rutile crystal modification) ρ_0 is 4 g cm^{-3}. Thus, the volume concentration of TiO_2 particles within the superficial strip is about 5%. In deeper parts the concentration is much smaller.

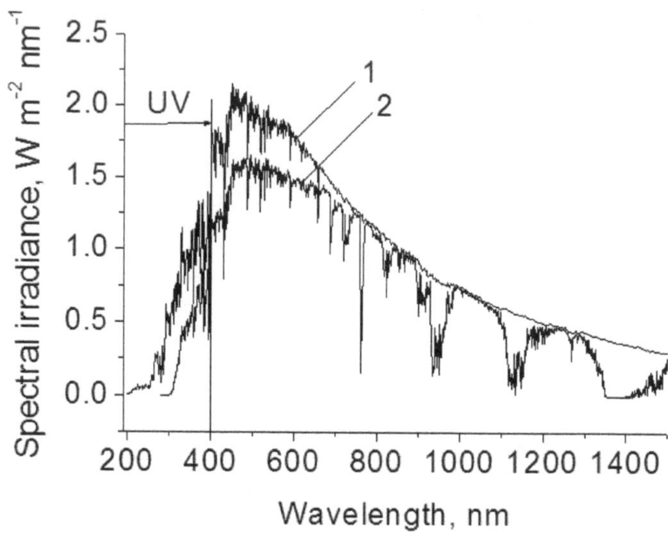

FIGURE 24.6: Solar spectrum above the atmosphere (1) and at sea level (2). Reprinted with permission from [23]. Copyright 2008, University of Oulu.

24.4 UV-Light-Blocking Efficacy of Nanoparticles

24.4.1 Solar radiation

The solar spectrum above the atmosphere and at sea level is represented in Figure 24.6. Due to absorption by certain compounds in the atmosphere (ozone O_3, oxygen O_2, carbon dioxide CO_2, water vapor H_2O), the spectrum reaching the Earth surface loses some spectral lines in its different parts.

UV radiation is a part of the solar spectrum occupying a range of 100–400 nm. It is conventionally divided into three subranges [42]: UVC (100–280 nm), UVB (280–315 nm), and UVA (315–400 nm). The former is completely absorbed by the ozone layer of the atmosphere located at a height of 18–40 km above the Earth's surface. The latter two are also attenuated: UVB – about 25 times, UVA – about 2 times, so that both fractions represent finally 5% of the solar intensity at sea level. They reach the atmosphere-ground interface and affect humans. Both types of rays could be harmful if a certain dose of radiation on the skin is exceeded.

24.4.2 Effect of UV radiation on skin

The influence of UV radiation can be either acute or chronic [43]. The former includes sunburn and sun tanning, as well as production of vitamin D. The chronic effects are skin cancer and photoaging. The UVB fraction is responsible for sunburn and increases the risk of the certain types of skin cancer (basal cell carcinoma and squamous cell carcinoma) due to direct DNA damage, which is mainly the formation of thymine-thymine dimers. The UVA fraction causes sun tanning, photoaging, and malignant melanoma by indirect DNA damage (by means of free radical formation). Although malignant melanoma is rare, it is responsible for 75% of all skin-cancer-related deaths. According to investigations, it is found that 92% of all melanoma cases are caused by indirect DNA damage and only 8% of the melanoma is caused by direct DNA damage [44]. Moderate sun

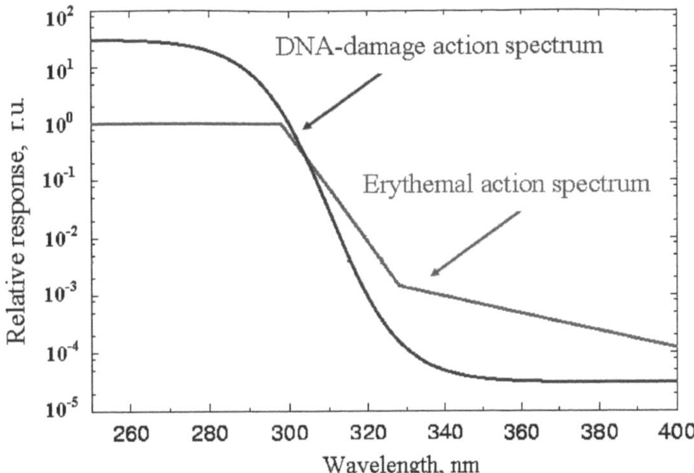

FIGURE 24.7: Erythemal and DNA-damage action spectra. Reprinted with permission from [41]. Copyright 2005, IOP Publishing Ltd.

tanning can prevent sunburn due to increased production of melanin, which is a natural protector against overexposure of human skin to UV radiation because of pronounced absorption within the UV spectral range.

24.4.3 Action spectrum and effective spectrum

The terrestrial solar spectrum contains a pronounced maximum somewhere near 500 nm (green light); the spectral intensity of UV radiation (400 nm and shorter) is considerably lower. Nevertheless, photons of the shortest fraction of the solar spectrum are the most powerful and therefore more dangerous than those of the other spectral regions. Action spectra of the susceptibility of the human skin to erythema [42] and of generalized DNA damage [45] due to UV radiation are shown in Figure 24.7.

As can be seen from the graph, erythema can more or less indicate the increasing probability of DNA damage, at least for wavelengths shorter than 310 nm. Multiplying the plots from Figures 24.6 and 24.7 (erythemal action spectrum), one can evaluate the effective spectrum taking into account the spectral intensity of wavelengths within the original solar spectrum (Figure 24.6). The result of such a procedure is shown in Figure 24.8. One can conclude from this plot that the erythema dangerous zone is 305–320 nm (mostly UVB range).

24.4.4 Mie calculations of cross-sections and anisotropy scattering factor of nanoparticles

Optical parameters, such as scattering and absorption coefficients (μ_s and μ_a respectively) for a medium partially filled with TiO_2 (or Si) particles, are needed as input data for the Monte Carlo simulations of photon migration within skin with TiO_2 particles. They can be expressed using scattering σ_s and absorption σ_a cross-sections of a particle. Thus, using the same notations as in the equation (24.1) we find that

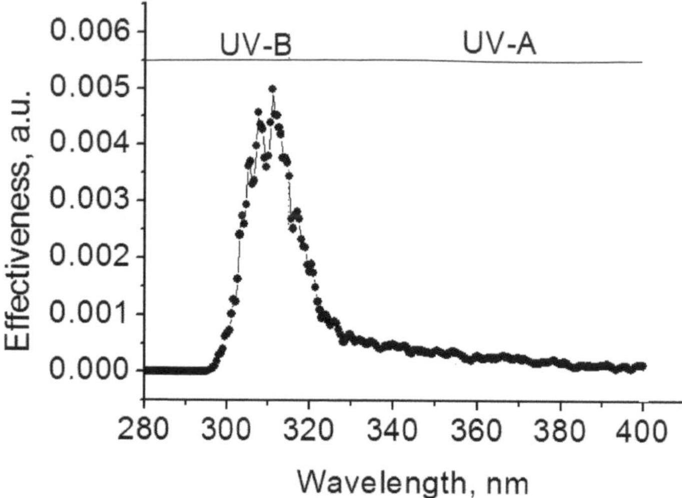

FIGURE 24.8: Effective spectrum of solar radiation within the UV spectral range. Reprinted with permission from [41]. Copyright 2005, IOP Publishing Ltd.

$$\mu_s = \frac{N \cdot \sigma_s}{V} = \frac{C}{V_0} \cdot Q_s \cdot \frac{\pi \cdot d^2}{4} = C \cdot Q_s \cdot \frac{\pi \cdot d^2/4}{\pi \cdot d^3/6} = 1.5 \cdot \frac{Q_s \cdot C}{d}, \tag{24.2}$$

$$\mu_a = \frac{N \cdot \sigma_a}{V} = \frac{C}{V_0} \cdot Q_a \cdot \frac{\pi \cdot d^2}{4} = C \cdot Q_a \cdot \frac{\pi \cdot d^2/4}{\pi \cdot d^3/6} = 1.5 \cdot \frac{Q_a \cdot C}{d}, \tag{24.3}$$

here $Q_s = \sigma_s/\sigma_g$ and $Q_a = \sigma_a/\sigma_g$ are relative (dimensionless) light scattering and absorption efficiency factors respectively, and $\sigma_g = \pi d^2/4$ is the geometrical cross-section of the particle, d is the particle diameter, C is volume concentration of particles. The efficiency factors Q_s and Q_a were determined using the Mie scattering theory with the help of MieTab 7.23 software. As reported in [46], particles 35–200 nm in diameter are spherical. Even for prolate pigment TiO_2 particles with the aspect ratio of about 1.5 the spherical approximation is acceptable [47]. Real $Re(n_p)$ and imaginary $Im(n_p)$ parts of the refractive index of TiO_2 and Si particles for the light with the wavelength λ required as input data for this software are represented in Table 24.1.

TABLE 24.1: Real and imaginary parts of refractive indices of TiO_2 (rutile) and Si for 310- and 400-nm UV light, adapted from [48].

λ, nm	$Re(n_p) - iIm(n_p)$	
	TiO_2	Si
310	$3.56 - i1.72$	$5.01 - i3.59$
400	$3.3 - i0.008$	$5.57 - i0.387$

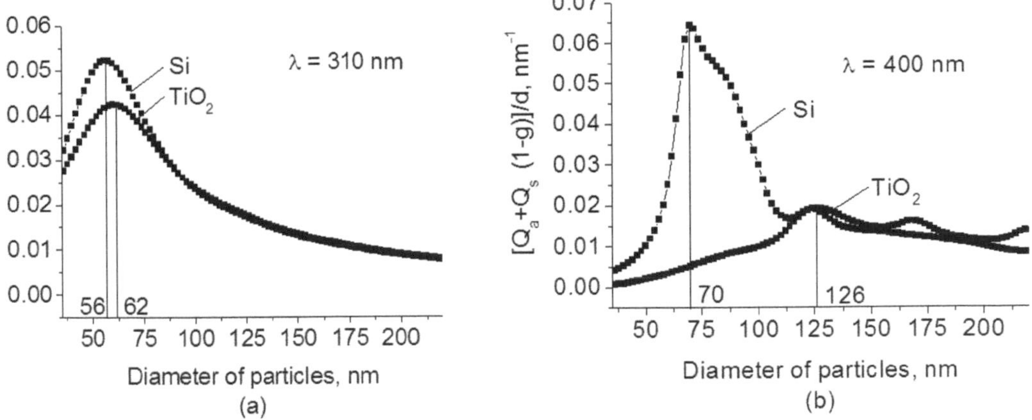

FIGURE 24.9: Dependences of quantity $[Q_a + Q_s(1-g)]/d$, on particle diameter for TiO_2 and Si nanoparticles for 310-nm (a) and 400-nm (b) light. Reprinted with permission from [52]. Copyright 2009, American Institute of Physics.

Assuming the refractive index of the horny layer $n_m = 1.53$ [49], a quantity $[Q_a + Q_s(1-g)]/d$, where g is scattering anisotropy factor, was constructed for the 310- and 400-nm radiation for the diameters of TiO_2 (and Si) particles ranging from 35 to 200 nm with a step of 2 nm. Such a quantity satisfactorily predicts the most attenuating particle sizes [50, 51]. The results of the calculations are depicted in Figure 24.9.

In both parts of the figure the curves corresponding to silicon have more pronounced peaks than those corresponding to titanium dioxide. It means that for the diameters of particles corresponding to the maxima, silicon better attenuates the UV light of the indicated wavelengths. The positions of the maxima are at 56 and 60 nm for Si and TiO_2 particles, respectively, for the 310-nm radiation. For the 400-nm light the maxima are located at 70 (for Si) and 126 (for TiO_2) nm. As can be concluded from the plots, silicon nanoparticles are better protectors almost within the whole range of the considered particle sizes. The most drastic difference between Si and TiO_2 curves is for the wavelength of 400 nm. The Si peak is shifted to the smaller diameters and is much larger than the TiO_2 peaks. This differs from the 310-nm curves, for which the difference of the maxima is not so large and the maxima are located rather close to each other.

24.4.5 Model of stratum corneum with particles

Skin surface is not plane [53] because of furrows and wrinkles located on its surface. After application of a sunscreen, nanoparticles are accumulated in such structures, resulting in an uneven distribution over the skin surface. However, uneven formations represent only 10% of the skin surface structure (age of the volunteers: up to 45 years), so the most part of the skin surface can be approximated as a flat surface. Additionally, the area covered by hair follicles is much smaller than the rest of the skin surface [54] because of their small surface density. So it is reasonable for sunscreen protection to take into consideration only skin surface free from follicles. The content of TiO_2 particles in such follicles is quite low (less than 1% of applied) as shown experimentally [37]. In our experiments described above the sunscreen with the particles was applied many times during the four days, resulting in a more homogeneous distribution than if administered only once. Taking this into account, a plane-layer model of the stratum corneum treated with a sunscreen was developed.

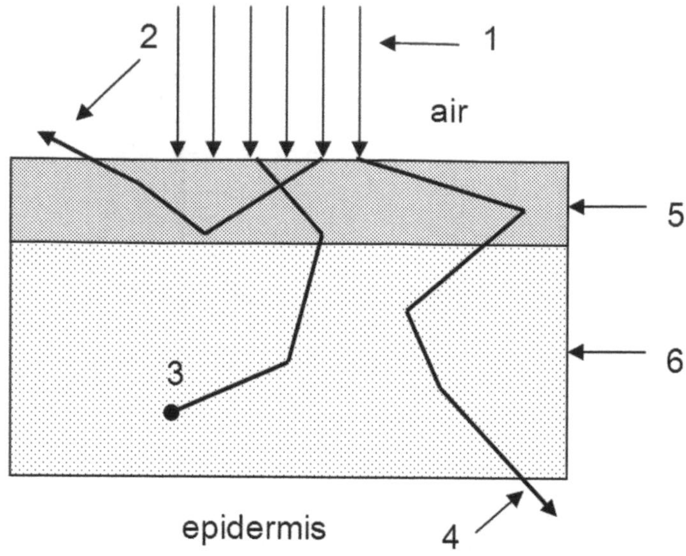

FIGURE 24.10: Model of the stratum corneum partially filled with particles used in Monte Carlo simulations: 1 – incident photon, 2 – diffusely reflected photon, 3 – absorbed photon, 4 – transmitted photon, 5 – upper part of stratum corneum (1-μm-thick, with particles), 6 – lower part of stratum corneum (without particles). Thickness of the whole stratum corneum (5 and 6) is 20 μm. Reprinted with permission from [23]. Copyright 2008, University of Oulu.

The Monte Carlo method implemented in a developed computer 3D code using Delphi® software environment, was applied to simulate the propagation of UV light within the stratum corneum with embedded TiO_2 (or Si) particles [55–57]. Figure 24.10 illustrates the geometry of the sample model. Incident light is directed perpendicularly to the air-skin interface.

We consider the integral characteristics of the registered radiation (over the whole surface or within the whole layer). The computational model of the stratum corneum consists of an infinitely wide plane layer, the upper part of which (1-μm-thick) contains TiO_2 (or Si) particles. Some authors try to take into account the real conditions representing the interfaces between the lower skin layers as quasi-random periodic structures [58, 59]. The total thickness of both parts is 20 μm, which corresponds to the real dimension of this skin layer on the back and arms [60]; only the palms and soles have thicker stratum corneum (up to 150 μm) because of more intensive use of hands and feet [61].

Optical properties of the stratum corneum (without particles) for the considered wavelengths are shown in Table 24.2. A more detailed description of the algorithm can be found elsewhere [23]. The volume concentration of the particles C used in the simulations was 1%.

TABLE 24.2: Optical properties of stratum corneum for 310- and 400-nm UV radiation, adapted from [62, 49].

λ,nm	μ_s, mm^{-1}	μ_a, mm^{-1}	g	n_m
310	240	60	0.9	1.53
400	200	23	0.9	1.53

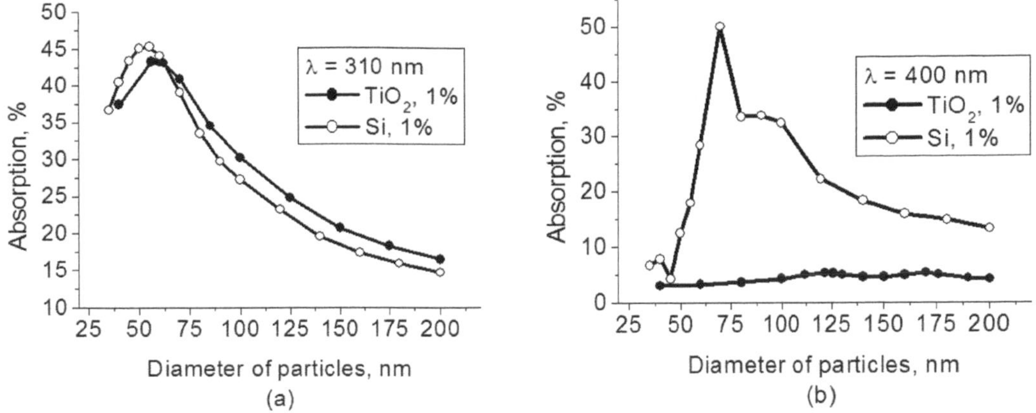

FIGURE 24.11: Absorption of 310- (a) and 400-nm (b) light within the upper part of stratum corneum (with particles). Reprinted with permission from [52]. Copyright 2009, American Institute of Physics.

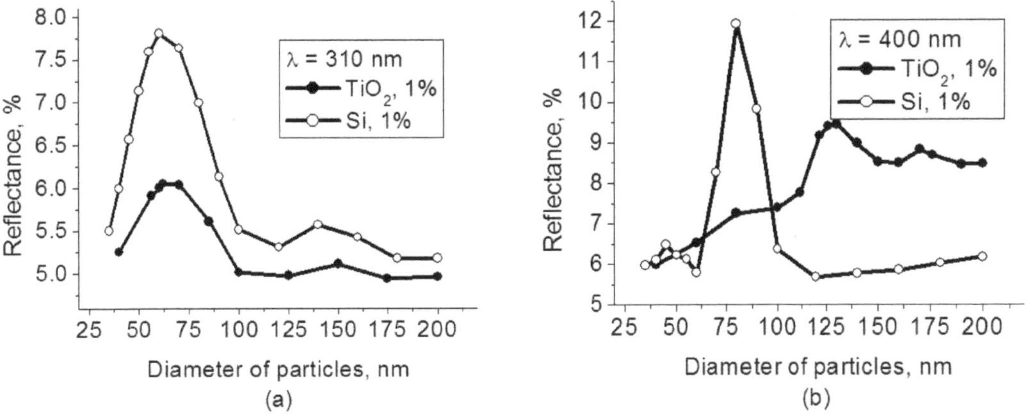

FIGURE 24.12: Reflectance of 310- (a) and 400-nm (b) light from the whole stratum corneum in the presence of TiO_2 or Si nanoparticles within its uppermost 1-μm-thick part. Reprinted with permission from [52]. Copyright 2009, American Institute of Physics.

24.4.6 Results of simulations

Figure 24.11 illustrates the dependences of absorption on particle diameters within the upper part of the stratum corneum where nanoparticles (TiO_2 or Si) are present. One can see that the curves in both parts of the figure have similar shapes as those in Figure 24.9. Positions of the highest values of absorption also coincide with those in Figure 24.9 except for the TiO_2 curve at 400-nm light: the most absorbing particles have diameters of 122 (not 126) nm. About 50% of the incident radiation is absorbed if optimal-sized particles of both types are used. It is worth noting that titanium dioxide particles absorb 400-nm light very weakly (no more than 5% of the incident radiation).

Reflectance from the whole stratum corneum (i.e., photons emerging from both parts of the layer, with and without particles, registered on the front surface of the layer) is depicted in Figure 24.12.

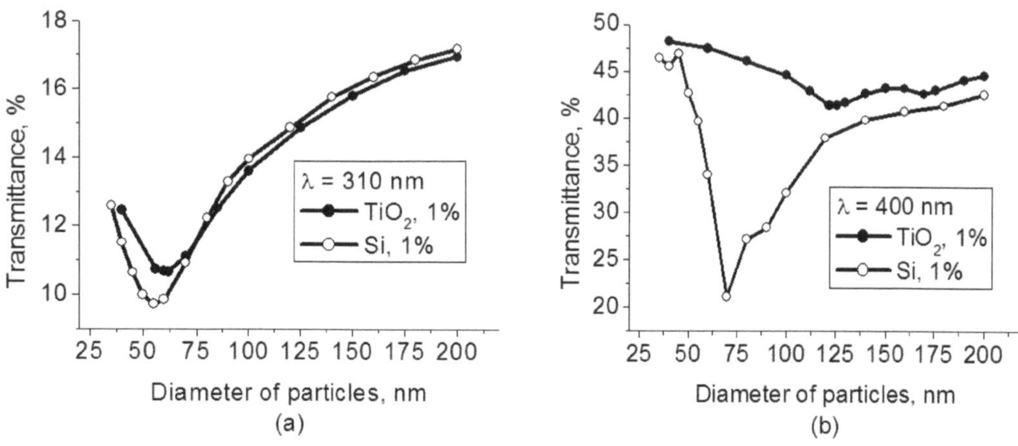

FIGURE 24.13: Transmittance of 310- (a) and 400-nm (b) light through the whole stratum corneum in the presence of TiO_2 or Si nanoparticles within its uppermost 1-μm-thick part. Reprinted with permission from [52]. Copyright 2009, American Institute of Physics.

The reflection does not exceed 12% even for the most reflecting particles. The maxima are shifted by a couple of nanometers to the larger sizes in comparison to the curves in Figure 24.9. This is explained in such a way that locations of the maxima of the extinction (Figure 24.9) and scattering curves (not shown) do not coincide. If illuminated with the 310-nm light, reflectance is higher for silicon nanoparticles than for titanium dioxide ones for all considered sizes. If illuminated with the 400-nm light, smaller (up to 90 nm in diameter) Si particles reflect light better than TiO_2 ones; for larger (90–200 nm) sizes the situation is reverse.

If the particles are used as UV light protectors, their most important characteristic is how they attenuate radiation in all ways (both absorption and reflection). Transmittance of light of the two wavelengths is shown in Figure 24.13. The curves look like inverted ones from Figure 24.9, as they should be: the higher the extinction, the lower the transmittance. Three diameters of the most attenuating particles are coincident with those corresponding to the maxima in Figure 24.9, namely 56 and 70 nm for Si (for 310- and 400-nm light, respectively) and 62 nm (for 310-nm light) for TiO_2 particles. A slight mismatch is observed for TiO_2 and 400-nm light: instead of 126 nm the optimal size is 122 nm. Nevertheless, it can be concluded that the extinction curves adequately describe the attenuation of light by particles. Simulation of UV light transmittance on emulsions with particles also showed that the optimal sizes of Si particles are smaller that those of TiO_2 [63].

In Figure 24.14 the comparison between the following three cases for 310- and 400-nm radiation is made: 1) stratum corneum without particles (0%); 2) stratum corneum with optimal-sized TiO_2 particles (1%); and 3) stratum corneum with optimal-sized Si particles (1%). The particles are located within the 1-μm-thick upper part of the stratum corneum, according to our model described above. Each column consists of the four components of the radiation (counting from the bottom): absorbed within the upper part (with particles), absorbed within the lower part (without particles), reflected from the whole stratum corneum and transmitted through the whole stratum corneum (entering the epidermis located beneath the stratum corneum).

Consider first the case of 310-nm radiation [Figure 24.14 (a)]. If particles of any of the two types are added, absorption increases dramatically in the upper part (from 5 to 40%), almost to the same level. The increase in absorption in this part causes its decrease in the lower part (from 65% to 40% of the incident radiation). Nevertheless, the total absorption increases (from 70% to 85%) in the

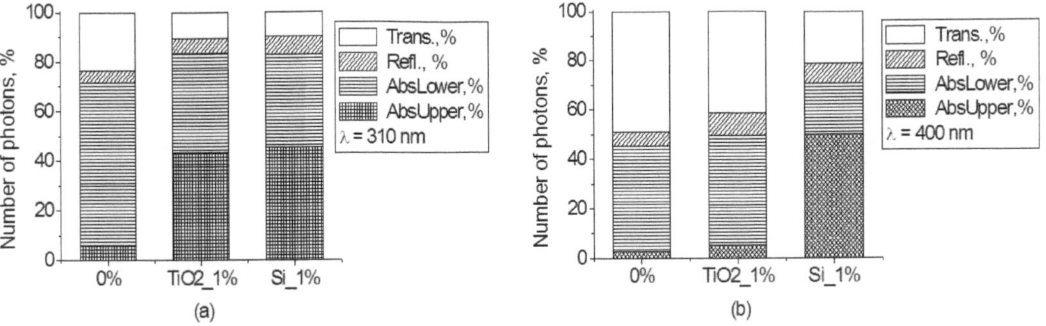

FIGURE 24.14: Effect of the optimal-sized TiO_2 and Si nanoparticles homogeneously distributed within the uppermost 1-μm-thick part of the stratum corneum for 310- (a) and 400-nm (b) light. Reprinted with permission from [52]. Copyright 2009, American Institute of Physics.

presence of the particles. Reflectance also increases. Because of these two factors, transmittance decreases from more than 20 to around 10%.

The difference between the effects of the two types of optimal-sized particles is more pronounced if the radiation with the wavelength of 400 nm is considered [Figure 24.14 (b)]. TiO_2 particles are weak absorbers in this case and the absorption in the upper part does not differ much from that of the stratum corneum without particles (still some synergy exists between the particles and the stratum corneum in increasing the absorbed radiation [64]); their main contribution to light attenuation is caused by increased reflection (from 5% to 10%). In contrast to this, Si particles are such effective absorbers at this wavelength that the absorption in the upper part exceeds the absorption in the whole stratum corneum without particles or with TiO_2 particles. The reflection is at the same level as for titanium dioxide particles. The discussed reasons result in the transmittance values: about 45% and 20% in the presence of TiO_2 and Si particles, respectively.

In the experiment carried out to compare its results with the simulations, the transmittance of UV light through a layer of o/w emulsion (L'Oréal, France) with nanoparticles UV-TITAN M 160 (Kemira, Finland) was measured with the spectrophotometer Lambda 20 (Perkin Elmer, Germany). In the corresponding simulations, 100-nm TiO_2 nanoparticles with a volume fraction of 0.2% embedded into a 20-μm-thick layer of the transparent medium with $n_m = 1.4$ were considered. Such a low concentration was chosen to be within the frames of independent scattering for all wavelengths of the spectrum (288–800 nm).

Comparison between the experimental and simulation results for TiO_2 particles of an average size of 100 nm is shown in Figure 24.15 (a). The curves represent the extinction (reciprocal to transmittance) dependencies on the wavelength. The discrepancy between the experimental and simulation curves is caused by the assumption of medium transparency and lower particle concentration (0.2%) used for simulations. The assumption of medium transparency was used because the absorption coefficient of the emulsion was not known. Some important features can be revealed from the experimental curve. As seen, the maximal extinction is achieved at the wavelength of 360 nm. By using the Mie theory, and taking into account the optical parameters of TiO_2 particles at 360-nm light ($n_p = 3.54$, $k_p = 0.16$ [48]), a quantity representing a combination of relative scattering and absorption efficiency factor, g-factor and particle diameter was calculated for the diameters of 35–200 nm and depicted in Figure 24.15 (b). The position of the maximum of the drawn curve indicates the size of the most attenuating particles [50]. In our case it is 98 nm. Using the optical parameters of the TiO_2 particles for UV and visible spectral ranges (Table 24.3), the relative scattering and absorption efficiency factors as well as g-factor, a simulation curve was calculated [Figure

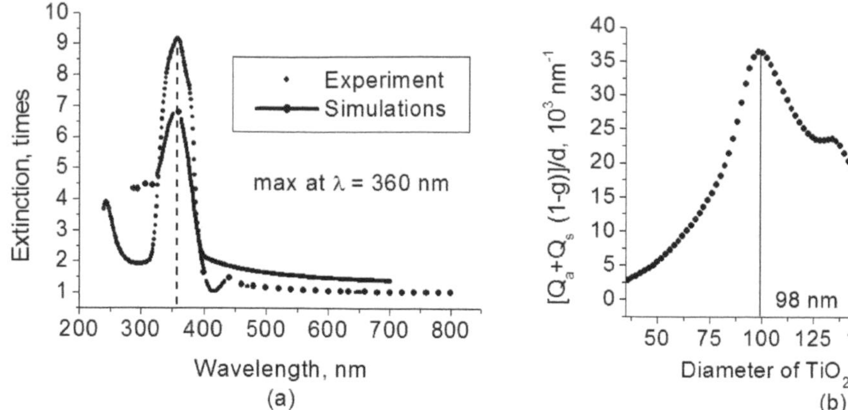

FIGURE 24.15: Experimental and simulated attenuation curves (a) and dependence of quantity $[Q_a + Q_s(1-g)]/d$ on a particle diameter for 360-nm light calculated using the Mie theory (b). Reprinted with permission from [52]. Copyright 2009, American Institute of Physics.

24.15 (a)]. The maxima of the both curves in Figure 24.15 (a) clearly coincide. This proves that the size of nanoparticles embedded in the emulsion in the experiment was about 100 nm.

TABLE 24.3: Optical parameters of TiO_2 particles in UV and visible spectrum ranges, adapted from [48].

Wavelength, nm	Refractive index		Wavelength, nm	Refractive index	
	$Re(n_p)$	$Im(n_p)$		$Re(n_p)$	$Im(n_p)$
288	3.07	2.40	580	2.72	0
294	3.42	2.47	600	2.70	0
307	3.56	1.72	620	2.69	0
318	4.46	1.92	633	2.68	0
360	3.54	0.16	660	2.66	0
400	3.13	0.008	680	2.65	0
440	2.96	0.053	700	2.64	0
460	2.90	0	720	2.63	0
480	2.86	0	740	2.63	0
500	2.82	0	760	2.62	0
520	2.79	0	780	2.61	0
540	2.76	0	800	2.61	0
560	2.74	0			

Figure 24.16 represents the dependence of optimal diameters (from the viewpoint of light attenuation) of TiO_2 particles, embedded in the stratum corneum, on the wavelengths. It was calculated according to the Mie theory. The dependence is close to being linear: the larger the light wavelength – the larger the TiO_2 particles should be to attenuate light the most effectively. This prediction is also supported by the simulations of optical coherence tomography signals from the stratum corneum partially filled with TiO_2 particles [65]: for 633-nm light, the weakest signal from the rear border of the layer was observed for the sizes of 200-nm (the largest size of all considered).

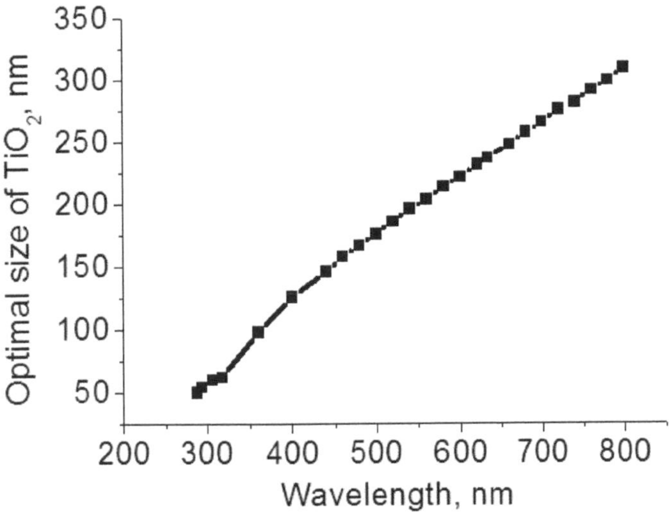

FIGURE 24.16: Optimal sizes of TiO_2 imbedded into the stratum corneum. Reprinted with permission from [23]. Copyright 2008, University of Oulu.

24.5 Toxicity of Nanoparticles

24.5.1 Free radicals

Free radicals are molecules with an unpaired electron on the external orbital and of high chemical reactivity. All radicals found in the human organism are either natural or alien. The former are inherently produced in the organism during chemical reactions: reactive oxygen species (ROS) such as superoxide, singlet oxygen, and hydroxyl radical, as well as nitric oxide, etc. They play an important role as regulatory mediators in signalling processes [66, 67]. Under normal circumstances, the amount of free radicals is balanced by enzymes and antioxidants, while, for example, pathogenesis of cancer, diabetes mellitus, atherosclerosis, neurodegenerative diseases, ischemia, etc., result in excessive production of radicals [67]. The alien free radicals appear as a consequence of the effect of ionizing radiation, UV light, xenobiotics, etc., on human tissue and are harmful.

Experimental detection of free radicals can be done directly and indirectly. The former implements the effects of either electron paramagnetic resonance (EPR) or chemoluminescence. The latter includes investigations of the end products of the reactions with free radicals involved or application of inhibitors. The EPR technique is based on the absorption of microwave radiation by an unpaired electron of a molecule located in a magnetic field.

24.5.2 EPR technique

The EPR method allows direct detecting of short-lived free radicals only at quite low temperatures (77 K, liquid nitrogen), owing to their sufficient steady-state concentrations only under such conditions [68]. In order to achieve suitable concentrations at room temperature, spin traps and spin markers are used (e.g., PCA, DPPH, Tempol, TEMPO, DMPO, 4-POBN). Spin traps are molecules, which bind to short-lived free radicals and form detectable stable forms of radicals (spin adducts). Spin markers are stabilized radicals contributing to the EPR signal; however, being in contact with short-lived free radicals the stabilized radicals lose or add an electron and become undetectable. A decrease in the EPR signal in this case quantifies the free radicals under investigation. The EPR methodology is a useful tool for the noninvasive *in vivo* measurements of skin barrier function, drug-skin interaction and cutaneous oxygen tension.

As already mentioned earlier, skin is a barrier, in particular, against UV light. Excessive doses of UV radiation can cause direct or indirect (via formation of free radicals) DNA damage leading to carcinogenesis [69]. The skin is a suitable object for EPR investigations because of its surface location and relatively small thickness. It is possible to monitor the penetration of spin traps and spin markers inside the skin [70, 71] and the in-depth appearance of generated radicals by means of EPR imaging [72]. According to investigations, the UV-induced radicals include ROS and lipid radicals [73–75] as well as melanin radicals [76]. It was shown experimentally that in human skin *in vivo* the UVA part of the UV spectrum (320-400 nm) is mainly responsible for the generation of free radicals (80-90% of total amount) because of higher penetration depth, in contrast to UVB light (280–320 nm), which contributes to radical generation only in the epidermis (down to a depth of 200 μm) [77].

24.5.3 Experiments with TiO_2 nanoparticles: Materials

In the experiments, the o/w emulsion without any UV filters (placebo) with or without coated TiO_2 nanoparticles embedded was used. The surface density of the substance was 2 mg/cm^2 corresponding to the recommendations of COLIPA [32] for sunscreen applications. One type of spin marker (PCA) was used to detect short-lived free radicals emerging under UV irradiation. It was chosen from other available spin markers owing to its satisfactory stability for our applications: after a 3 minute-long UV irradiation, the EPR signal decrease was about 1% and after a 18 minute-long irradiation – only 6% [78]. Additionally, in porcine skin the amount of antioxidants is low, thus prolonging the existence on PCA. As a source of UV radiation (280–400 nm) a solar simulator was used; the radiation intensity was 4.3 mW/cm^2, which corresponds to the solar UV intensity (4.6 mW/cm^2). In contact with free radicals, PCA loses its free electron and the signal intensity in the EPR signal decreases correspondingly.

24.5.4 Raman spectroscopy

The Raman spectroscopy system [79] used in the experiments, was based on a CW Ar^+ laser (λ = 514.5 nm, 9 mW). In order to reveal the crystal modification of the used TiO_2 particles, Raman spectra of the powders were measured. The obtained spectra are presented in Figure 24.17. The signal produced by 25-nm particles is much smaller than the signal produced by 400-nm ones, owing to features of particles-light interaction. Three obvious peaks are seen in the graph. These indicate, according to [81], that the crystal modification of the particles under investigation is anatase.

24.5.5 Mie calculations

MieTab 7.23 software (http://amiller.nmsu.edu/mietab.html) was used for Mie calculations. The radiation wavelength, refractive indices of the particles and the surrounding medium, as well as particle sizes were required as input parameters.

The interaction between the particles and UV radiation is described by the Mie theory (for spherical particles, which is true for particles larger than 25 nm). The spectrum of the UV light source is depicted in Figure 24.18. For the calculations, two wavelengths were chosen: 310 nm and 335 nm. The first one was chosen because it is close to the maximum of the source spectrum and corresponds to the erythemal maximum [41]. The second one was taken for comparison.

Titanium dioxide is a birefringent crystal, with different refractive indices for light polarized perpendicular or parallel to the optic axis. In the "average index" approximation, the particles

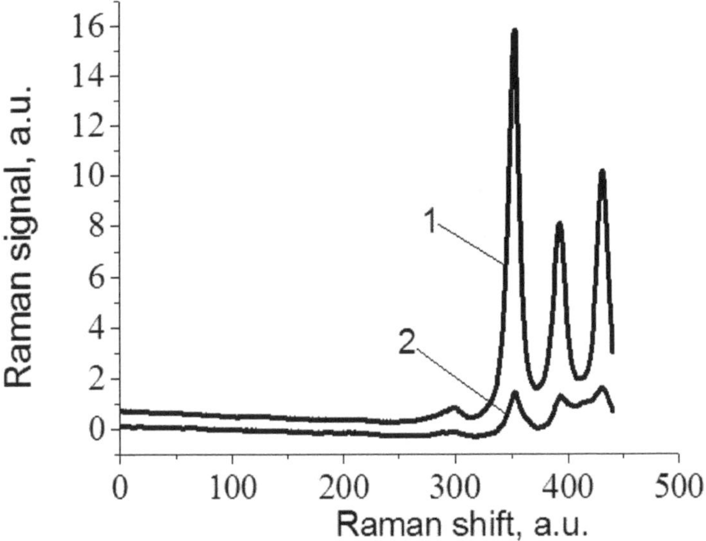

FIGURE 24.17: Signal of Raman scattering from powder TiO_2 nanoparticles with diameters of 400 nm (1) and 25 nm (2). Reprinted with permission from [80]. Copyright 2009, SPIE.

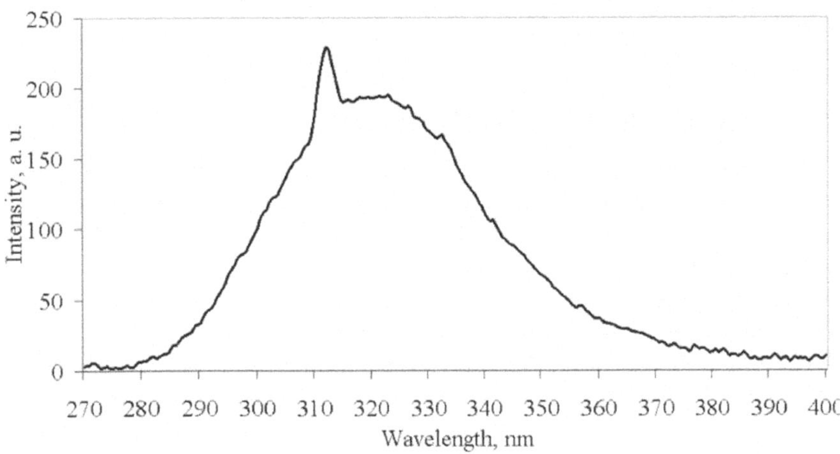

FIGURE 24.18: Spectrum of the UV lamp used in the experiment. Reprinted with permission from [80]. Copyright 2009, SPIE.

are supposed to be isotropic, with real and imaginary parts of the refractive index equal to $n_p = (2n_o + n_e)/3$ and $k_p = (2k_o + k_e)/3$, where n_o and k_o (n_e and k_e) are the ordinary (extraordinary) real and imaginary parts of the refractive index, respectively [82]. For the 310-nm and 335-nm UV radiation these constants taken from [83] result in: $n_p - ik_p = 3.48 - i0.83$ (for 310 nm) and $n_p - ik_p = 3.37 - i0.25$ (for 335 nm). The refractive index of the surrounding medium (placebo) was 1.4, the diameters of the particles were considered to be 2–200 nm with 2-nm steps. The result of the calculation is shown in Figure 24.19. The value (Q_a/d), where d is a particle diameter, is proportional to the absorption coefficient of a particle suspension [84] and therefore takes into account the presence of other particles of the same type in the sample.

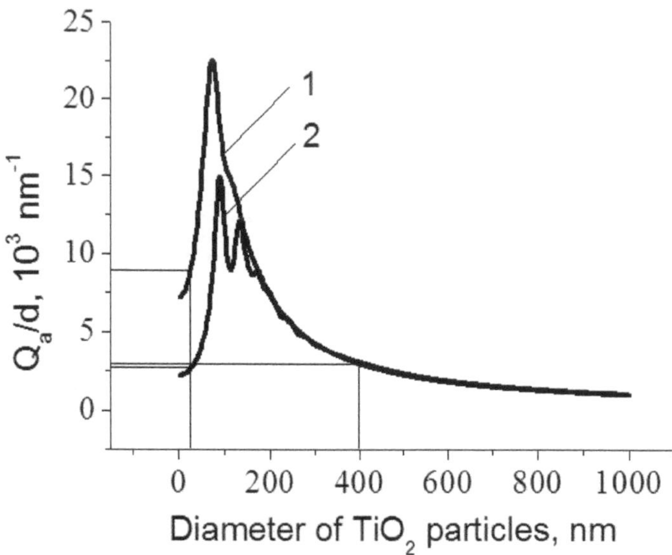

FIGURE 24.19: Relative absorption efficiency factor referred to the particle diameter (Q_{abs}/d) for 310- (1) and 335-nm (2) UV radiation calculated according to the Mie theory; maxima of the curves correspond to the particle diameters of 74 (1) and 90 (2) nm. Reprinted with permission from [80]. Copyright 2009, SPIE.

24.5.6 Experiments I: Emulsion on glass slides

Figure 24.20 shows the mean values with standard deviations of the results obtained from 5 samples on glass with or without particles. Dependences of the EPR signal amplitudes on time for the irradiated and nonirradiated samples are depicted. Statistically, there is no effect of UV radiation in the presence of large (400 nm) particles and the placebo without particles. However, the effect is distinctly seen in the case of small (25 nm) particles, although they were irradiated for a shorter time (1 min). This can be explained by the Mie theory. Considering that Figure 24.19 represents the absorption efficiency curves for 310-nm and 335-nm radiation, we can conclude that 25-nm particles absorb UV light much more efficiently than the 400-nm ones for λ = 310 nm or at the same level for λ = 335 nm at the same volume concentrations. It is known that the particles tend to form aggregates and agglomerates, although ultrasonic stirring was used during the process of embedding particles into the placebo. The formation of the above-mentioned structures causes an increase in the average size of particles resulting in the increased absorption efficacy of the 25-nm particles and decreasing that of the 400-nm particles. More pronounced absorption means more active production of free short-lived radicals. The curve corresponding to the nonirradiated large particles looks very similar to that of the placebo and different from that of small particles.

24.5.7 Experiments II: Emulsion on porcine skin

Ears of freshly slaughtered domestic pigs were used for these experiments. 20 tape strips were taken from the same area to remove the stratum corneum in part in order to ease the penetration of the PCA into the skin [71]. Corresponding to [85], with this number of strips about 50% of stratum corneum was removed. The PCA served as a spin marker for revealing the production of free radicals. In the investigation of free radicals with the EPR technique, the deeper the PCA penetrated, the higher the amount of skin volume was involved.

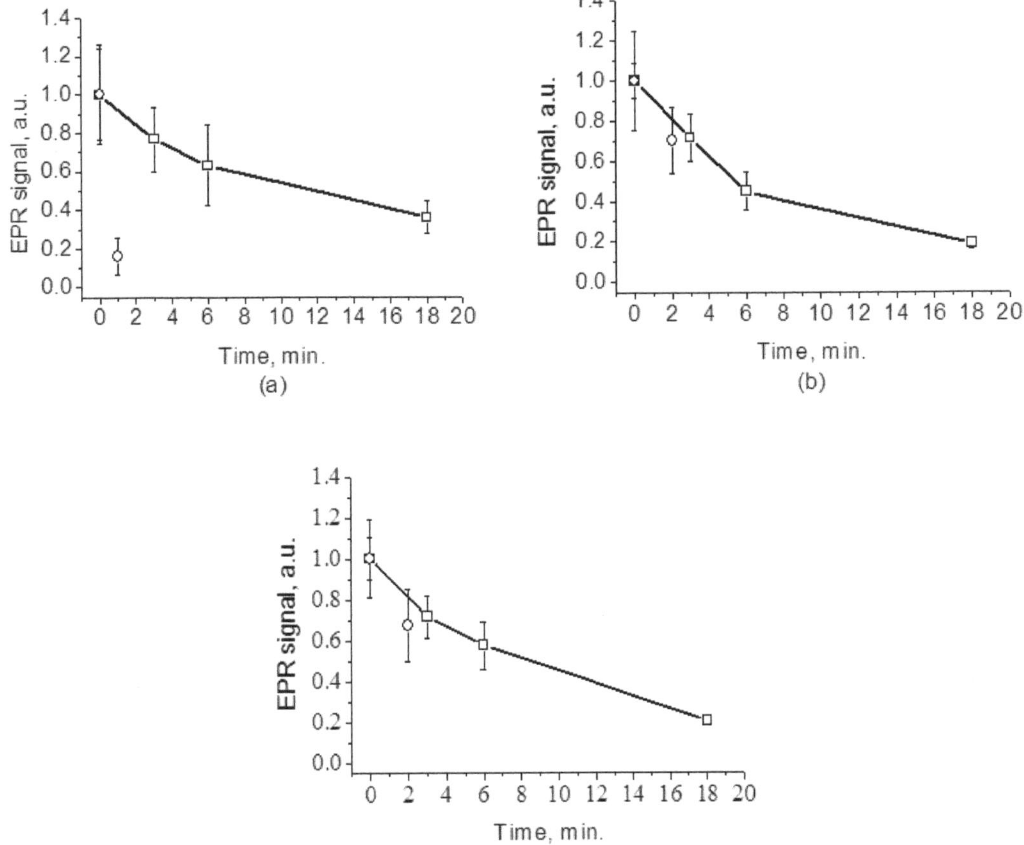

FIGURE 24.20: Temporal dependences of amplitudes of the EPR signals obtained from the samples on glass slides: the placebo with 25-nm particles (a), the placebo with 400-nm particles (b), and the placebo without particles (c) UV-irradiated (open circles) during 1 min (a) or 2 min (b, c) and not irradiated (open squares). Zero-time corresponds to the beginning of UV irradiation (for the irradiated samples). Reprinted with permission from [80]. Copyright 2009, SPIE.

As with the experiments that applied particles onto glass slides, three different substances were investigated: the placebo with 25-nm particles, the placebo with 400-nm particles and the placebo without particles. The substances were topically applied onto the skin, with a surface density of 2 mg/cm^2.

The samples were irradiated immediately after preparation. The results of the measurements with standard deviations are presented in Figure 24.21. Even with statistical errors taken into account, the effect of UV irradiation is clearly seen in all cases (the points corresponding to such samples are located below the nonirradiated), the magnitude of the effect is almost the same for the samples with the particles in the placebo, with the placebo only and for the skin samples without the placebo and particles. This means that the amounts of short-lived free radicals appearing under UV irradiation are comparable and do not depend on the presence of the particles on the skin surface. In other words, the contribution of skin to free-radical generation under UV irradiation exceeds that of the particles.

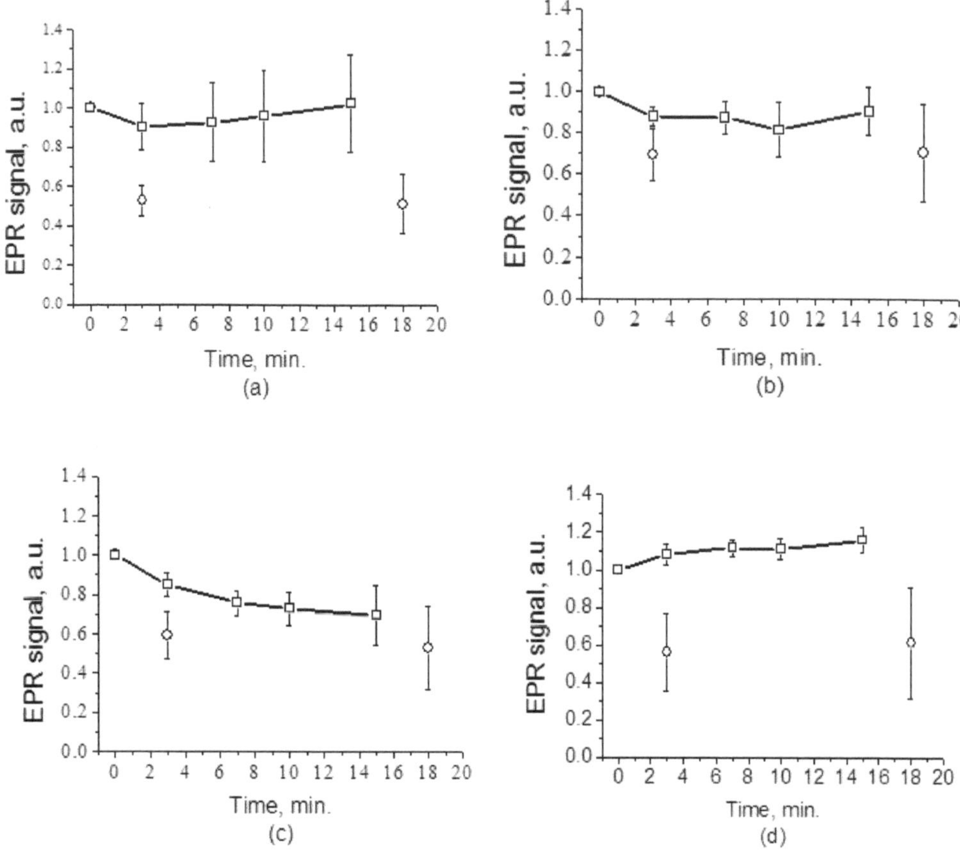

FIGURE 24.21: Temporal dependences of amplitudes of the EPR signals obtained from the samples on porcine skin *in vitro*: the placebo with 25- (a) or 400-nm (b) particles, the placebo without particles (c), and skin without the placebo and particles (d) UV-irradiated during 3 min (open circles) and not irradiated (open squares). For averaging, 4 skin samples with the placebo and each type of the particles, 6 skin samples with the placebo only and 6 skin samples without the placebo and particles were measured. Zero-time corresponds to the beginning of UV irradiation (for the irradiated samples). Reprinted with permission from [80]. Copyright 2009, SPIE.

24.6 Conclusion

In this chapter, we discussed some issues concerning the advantages and possible disadvantages of nanoparticle use for sunscreen applications. Our experimental results (as well as those of other groups) indicate that nanoparticles of TiO_2 and ZnO do not penetrate into viable epidermis, thus decreasing public concerns about the use of sunscreens.

We showed theoretically, how skin optical properties could be modified by means of TiO_2 (rutile form) and Si nanoparticles. For the two wavelengths of light considered, the sizes of the most attenuating particles are the following: 56 and 70 nm for Si (for 310- and 400-nm light, respectively) and 62 and 122 nm for TiO_2 particles (for 310- and 400-nm light, respectively). The mechanisms behind these effects are absorption of the UV radiation of the shorter wavelength for both types of particles; scattering for TiO_2 and absorption (and to a lesser extent scattering) for Si nanoparticles of the longer wavelength. Although the possible existence of SiO_2 shells on the particles was not taken into account in the calculations, the effect of extinction would be even more pronounced in that case because the real part of the refractive index of SiO_2 is 1.47–1.49 for the radiation of 290–400 nm and hence is lower than that of skin (1.53).

We proved experimentally by means of EPR spectroscopy, small (25 nm in diameter) coated nanoparticles of titanium dioxide (anatase form) are more photoactive (phototoxic) than large (400 nm in diameter) particles. This effect is clearly seen if the particles mixed with the placebo are applied on glass slides. The Mie theory explains the difference in light absorption by particles of the mentioned sizes and, as a consequence, the difference in amount of generated free radicals. However, if the particles are applied onto the porcine skin *in vitro*, no distinct difference is observed. This is caused by high skin contribution to the formation of radicals. In comparison to the skin's ability to produce radicals, the nanoparticles do not play a significant role in the concentrations used (2 mg/cm^2).

Acknowledgments

Financial support from Infotech Oulu Graduate School, Tauno Tönningin Foundation, Tekniikan edistämissäätiö, Academy of Finland (all Finland), and German Student Exchange Service (Germany) is highly appreciated.

References

[1] G.J. Nohynek, J. Lademann, C. Ribaud, and M. Roberts, "Grey Goo on the skin? Nanotechnology, cosmetic and sunscreen safety," *Crit. Rev. Toxicol.* **37**, 251–277 (2007).

[2] U. Diebold, "The surface science of titanium dioxide," *Surf. Sci. Rep.* **48**, 53–229 (2003).

[3] P.J.A. Borm, D. Robbins, S. Haubold, T. Kuhbusch, H. Fissan, K. Donaldson, R.P.F. Schins, V. Stone, W. Kreyling, J. Lademann, J. Krutmann, D. Warheit, and E. Oberdörster, "The potential risks of nanomaterials: a review conducted out for ECETOC," *Particle Fibre Toxicol.* **3**, 1–36 (2006).

[4] ETC group, "No small matter II: the case for a global moratorium. Size matters!," Occasional Paper Series **7** (1), 1–14 (2003).

[5] R.F. Edlich, K.L. Winter, H.W. Lim, M.J. Cox, D.G. Becker, J.H. Horovitz, L.S. Nichter, L.D. Britt, and W.B. Long, "Photoprotection by sunscreens with topical antioxidants and systematic antioxidants to reduce sun exposure," *J. Long-Term Effects Med. Implants* **14**, 317–340 (2004).

[6] U. Jacobi, H.-J. Weigmann, M. Baumann, A.-I. Reiche, W. Sterry, and J. Lademann, "Lateral spreading of topically applied UV filter substances investigated by tape stripping," *Skin Pharmacol. Physiol.* **17**, 17–22 (2004).

[7] B. Innes, T. Tsuzuki, H. Dawkins, J. Dunlop, G. Trotter, M.R. Nearn, and P.G. McCormick, "Nanotechnology and the cosmetic chemist," *Cosmetics, Aerosols and Toiletries in Australia* **15,** 10–12, 21–24 (2002).

[8] A.P. Popov, J. Lademann, A.V. Priezzhev, and R. Myllylä, "Effect of size of TiO_2 nanoparticles embedded into stratum corneum on ultraviolet-A and ultraviolet-B sun-blocking properties of the skin," *J. Biomed. Opt.* **10**, 064037-1–9 (2005).

[9] S.E. Cross, B. Innes, M.S. Roberts, T. Tsuzuki, T.A. Robertson, and P. McCormick, "Human skin penetration of sunscreen nanoparticles: in vitro assessment of a novel micronized zinc oxide formulation," *Skin Pharmacol. Physiol.* **20**, 148–154 (2007).

[10] J.-G. Li, T. Ishigaki, and X. Sun, "Anatase, brookite, and rutile nanocrystals via redox reactions under mild hydrothermal conditions: phase-selective synthesis and physicochemical properties," *J. Phys. Chem. C* **111**, 4969–4976 (2007).

[11] X.H. Zhang, S.J. Chua, A.M. Yong, and S.Y. Chow, "Exciton radiative lifetime in ZnO quantum dots embedded in SiO_x matrix," *Appl. Phys. Lett.* **88**, 221903-1–3 (2006).

[12] P.P. Ahonen, J. Joutensaari, O. Richard, U. Tapper, D. Brown, J. Jokiniemi, and E. Kauppinen, "Mobility size development and the crystallization path during aerosol decomposition synthesis of TiO_2 particles," *J. Aerosol. Sci.* **32**, 615–630 (2001).

[13] R.D. Mosteller, "Simplified calculation of body surface area," *N. Engl. Med.* **317**, 1098 (1987).

[14] P. Fritsch, *Dermatologie und Venerologie*, Springer-Verlag, Berlin-Heidelberg (1998).

[15] E.V. Zagainova, M.V. Shirmanova, M.Yu. Kirillin, B.N. Khlebtsov, A.G. Orlova, I.V. Balalaeva, M.A. Sirotkina, M.L. Bugrova, P.D. Agrba, and V.A. Kamensky, "Contrasting properties of gold nanoparticles for optical coherence tomography: phantom, *in vivo* studies and Monte Carlo simulation," *Phys. Med. Biol.* **53**, 4995–5009 (2008).

[16] I. Meglinski, M. Kirillin, V. Kuzmin, and R. Myllylä, "Simulation of polarization-sensitive optical coherence tomography images by a Monte Carlo method," *Opt. Lett.* **33**, 1581–1583 (2004).

[17] U. Jacobi, E. Waibler, W. Sterry, and J. Lademann, "*In vivo* determination of the long-term reservoir of the horny layer using laser scanning microscopy," *Laser Phys.* **15**, 565–569 (2005).

[18] J. Lademann, A. Knüttel, H. Richter, N. Otberg, R. von Pelchrzim, H. Audring, H. Meffert, W. Sterry, and K. Hoffmann, "Application of optical coherent tomography for skin diagnostics," *Laser Phys.* **15**, 288–294 (2005).

[19] L.E. Meyer, N. Otberg, H. Richter, W. Sterry, and J. Lademann, "New prospects in dermatology: fiber-based confocal scanning laser microscopy," *Laser Phys.* **16**, 758–764 (2006).

[20] A. Teichmann, H. Sadeyh Pour Soleh, S. Schanzer, H. Richter, A. Schwarz, and J. Lademann, "Evaluation of the efficacy of skin care products by laser scanning microscopy," *Laser Phys. Lett.* **3**, 507–509 (2006).

[21] T. Rieger, A. Teichmann, H. Richter, W. Sterry, and J. Lademann, "Application of *in-vivo* laser scanning microscopy for evaluation of barrier creams," *Laser Phys. Lett.* **4**, 72–76 (2007).

[22] I. Riemann, E. Dimitrow, P. Fischer, A. Reif, M. Kaatz, P. Elsner, and K. König, "High resolution multiphoton tomography of human skin *in vivo* and *in vitro*," *Proc. SPIE* **5463**, 21–28 (2004).

[23] A.P. Popov, *TiO_2 Nanoparticles as UV Protectors in Skin*, Acta Universitatis Ouluensis, Oulu (2008).

[24] R.L. Anderson and J.M. Cassidy, "Variations in physical dimensions and chemical composition of human stratum corneum," *J. Invest. Dermatol.* **61**, 30–32 (1973).

[25] K.A. Holbrook and G.F. Odland, "Regional differences in the thickness (cell layers) of the human stratum corneum: an ultrastructural analysis," *J. Invest. Dermatol.* **62**, 415–422 (1974).

[26] H. Schaefer and T. Redelmeier, *Skin Barrier: Principles of Percutaneous Absorption*, Karger, Basel (1996).

[27] Y.N. Kalia, F. Pirot, and R.H. Guy, "Homogeneous transport in a heterogeneous membrane: water diffusion across human stratum corneum *in vivo*," *Biophys. J.* **71**, 2692–2700 (1996).

[28] D.A. Schwindt, K.P. Wilhelm, and H.I. Maibach, "Water diffusion characteristics of human stratum corneum at different anatomical sites *in vivo*," *J. Invest. Dermatol.* **111**, 385–389 (1998).

[29] U. Jacobi, N. Meykadeh, W. Sterry, and J. Lademann, "Effect of the vehicle on the amount of stratum corneum removed by tape stripping," *J. Dt. Dermatol. Gesell.* **1**, 884–889 (2003).

[30] J. Lademann, A. Ingevicius, O. Zurbau, H.D. Liess, S. Schanzer, H.-J. Weigmann, C. Antoniou, R. von Pelchrzim, and W. Sterry, "Penetration studies of topically applied substances: optical determination of the amount of stratum corneum removed by tape stripping," *J. Biomed. Opt.* **11**, 054026-1–6 (2006).

[31] H.-J. Weigmann, J. Lademann, S. Schanzer, U. Lindemann, R. von Pelchrzim, H. Schaefer, W. Sterry, and V. Shah, "Correlation of the local distribution of topically applied substances inside the stratum corneum determined by tape-stripping to differences in bioavailability," *Skin Pharmacol. Appl. Skin Physiol.* **14**, 98–102 (2001).

[32] COLIPA, European Cosmetic, Toiletry and Perfumery Association, "COLIPA sun protection factor test method 94/289" (1994).

[33] J. Lademann, A. Rudolph, U. Jacobi, H.-J. Weigmann, H. Schafer, W. Sterry, and M. Meinke, "Influence of nonhomogeneous distribution of topically applied UV filters on sun protection factors," *J. Biomed. Opt.* **9**, 1358–1362 (2004).

[34] L. Ferrero, M. Pissavini, and L. Zastrow, "Spectroscopy of sunscreen products," *Proc. European UV Sunfilters Conference*, Paris, France, 52–64 (1999).

[35] H.-J. Weigmann, U. Lindemann, C. Antoniou, G.N. Tsikrikas, A.I. Stratigos, A. Katsambas, W. Sterry, and J. Lademann, "UV/VIS absorbance allows rapid, accurate, and reproducible mass determination of corneocytes removed by tape stripping," *Skin Pharmacol. Appl. Skin Physiol.* **16**, 217–227 (2003).

[36] A.P. Popov, A.V. Priezzhev, J. Lademann, and R. Myllylä, "Advantages of NIR radiation use for optical determination of skin horny layer thickness with embedded TiO_2 nanoparticles during tape stripping procedure," *Laser Phys.* **16**, 751–757 (2006).

[37] J. Lademann, H.-J. Weigmann, C. Rickmeyer, H. Barthelmes, H. Schaefer, G. Mueller, and W. Sterry, "Penetration of titanium dioxide microparticles in a sunscreen formulation into the horny layer and the follicular orifice," *Skin Pharmacol. Appl. Skin Physiol.* **12**, 247–256 (1999).

[38] A. Mavon, C. Miquel, O. Lejeune, B. Payre, and P. Moretto, "*In vitro* percutaneous absorption and *in vivo* stratum corneum distribution of an organic and a mineral sunscreen," *Skin Pharmacol. Physiol.* **20**, 10–20 (2007).

[39] M.S. Roberts, M.J. Roberts, T.A. Robertson, W. Sanchez, C. Thörling, Y. Zou, X. Zhao, W. Becker, and A. Zvyagin, "*In vitro* and *in vivo* imagin of xenobiotics transport in human skin and in the rat liver," *J. Biophoton.* **1**, 478–493 (2008).

[40] A.V. Zvyagin, X. Zhao, A. Gierden, W. Sanchez, J.A. Ross, and M.S. Roberts, "Imaging of zinc oxide nanoparticle penetration in human skin *in vitro* and *in vivo*," *J. Biomed. Opt.* **13**, 064031-1–9 (2008).

[41] A.P. Popov, A.V. Priezzhev, J. Lademann, and R. Myllylä, "TiO_2 nanoparticles as an effective UV-B radiation skin-protective compound in sunscreens," *J. Phys. D: Appl. Phys.* **38**, 2564–2570 (2005).

[42] A.F. McKinlay and B.L. Diffey, "A reference action spectrum for ultraviolet induced erythema in human skin," *CIE J.* **6**, 17–22 (1987).

[43] B.L. Diffey, "Solar ultraviolet radiation effects on biological systems," *Phys. Med. Biol.* **36**, 299–328 (1991).

[44] H. Davies, G.R. Bignell, C. Cox, P. Stephens, S. Edkins et al., "Mutations of the BRAF gene in human cancer," *Nature* **417**, 949–954 (2002).

[45] R.B. Setlow, E. Grist, K. Thompson, and A.D. Woodhead, "Wavelengths effective in induction of malignant melanoma," *Proc. Nat. Acad. Sci.* **90**, 6666–6670 (1993).

[46] Y. Shao and D. Slossmann, "Effect of particle size on performance of physical sunscreen formulas," *Proc. Personal Care Ingredients Asia Conference*, Shanghai, P.R. China, 1–9 (1999).

[47] L.E. McNeil and R.H. French, "Multiple scattering from rutile TiO_2 particles," *Acta Mater.* **48**, 4571–4576 (2000).

[48] E.D. Palik, *Handbook of Optical Constants of Solids*, Academic Press, Orlando (1985).

[49] V.V. Tuchin, *Tissue Optics: Light Scattering Methods and Instruments for Medical Diagnosis*, SPIE Tutorial Texts in Optical Engineering **TT38**, SPIE Press, Bellingham, WA (2000).

[50] A.P. Popov, A.V. Priezzhev, J. Lademann, and R. Myllylä, "Effect of multiple scattering of light by titanium dioxide nanoparticles implanted into a superficial skin layer on radiation transmission in different wavelength ranges," *Quantum Electron.* **37**, 17–21 (2007).

[51] A.P. Popov, A.V. Priezzhev, J. Lademann, and R. Myllylä, "Influence of multiple light scattering on TiO_2 nanoparticles imbedded into stratum corneum, on light transmittance in UV and visible wavelength regions," *Proc. SPIE* **6535**, 65351E-1–6 (2007).

[52] A.P. Popov, A.V. Priezzhev, J. Lademann, and R. Myllylä, "Biophysical mechanisms of modification of skin optical properties in the UV wavelength range with nanoparticles," *J. Appl. Phys.* **105**, 101901-1–2 (2009).

[53] U. Jacobi, M. Chen, G. Frankowski, R. Sinkgraven, M. Hund, B. Rzany, W. Sterry, and J. Lademann, "*In vivo* determination of skin surface topography using an optical 3D device," *Skin Res. Technol.* **10**, 207–214 (2004).

[54] N. Otberg, H. Richter, H. Schaefer, U. Blume-Peytavi, W. Sterry, and J. Lademann, "Variations of hair follicle size and distribution in different body sites," *J. Invest. Dermatol.* **122**, 14–19 (2004).

[55] A.P. Popov, A.V. Priezzhev, J. Lademann, and R. Myllylä, "Manipulation of optical properties of human skin by light scattering nanoparticles of titanium dioxide," *Proc. SPIE* **5578**, 269–277 (2004).

[56] A.P. Popov, A.V. Priezzhev, and J. Lademann, "Control of optical properties of human skin by embedding light scattering nanoparticles," *Proc. SPIE* **5850**, 286–293 (2005).

[57] A.P. Popov, A.V. Priezzhev, J. Lademann, and R. Myllylä, "Efficiency of TiO_2 nanoparticles of different sizes as UVB light skin-protective fraction in sunscreens," *Proc. SPIE* **5771**, 336–343 (2005).

[58] M.Yu. Kirillin, A.V. Priezzhev, and R. Myllylä, "Effect of coherence length and numerical aperture on the formation of OCT signals from model biotissues," *Proc. SPIE* **6534**, 6534H-1–10 (2007).

[59] M.Yu. Kirillin, A.V. Priezzhev, and R. Myllylä, "Contribution of various scattering orders to OCT images of skin," *Proc. SPIE* **6627**, 66270Q-1–7 (2007).

[60] E. Bordenave, E. Abraham, G. Jonusauskas, N. Tsurumach, J. Oberlé, C. Rullière, P.E. Minot, M. Lassègues, and J.E. Surlève Bazeille, "Wide-field optical coherence tomography: imaging of biological tissues," *Appl. Opt.* **41**, 2059–2064 (2002).

[61] P.J. Caspers, G.W. Lucassen, and G.J. Pupplels, "Combined *in vivo* confocal Raman spectroscopy and confocal microscopy of human skin," *Biophys. J.* **85**, 572–580 (2003).

[62] M.J.C. van Gemert, S.L. Jacques, H.J.C.M. Sterenborg, and V.M. Star, "Skin Optics," *IEEE Tranc. Biomed. Eng.* **36**, 1146–1154 (1989).

[63] A.P. Popov, A.V. Priezzhev, J. Lademann, and R. Myllylä, "Monte Carlo calculations of UV protective properties of emulsions containing TiO_2, Si and SiO_2 nanoparticles," *Proc. SPIE* **7022**, 702211-1–7 (2008).

[64] J. Lademann, S. Schanzer, U. Jacobi, H. Schaefer, F. Pflücker, H. Driller, J. Beck, M. Meinke, A. Roggan, and W. Sterry, "Synergy effects between organic and inorganic UV filters in sunscreens," *J. Biomed. Opt.* **10,** 014008-1–7 (2005).

[65] A.P. Popov, M.Yu. Kirillin, A.V. Priezzhev, J. Lademann, J. Hast, and R. Myllylä, "Optical sensing of titanium dioxide nanoparticles within horny layer of human skin and their protecting effect against solar UV radiation," *Proc. SPIE* **5702,** 113–122 (2005).

[66] D. Darr and I. Fridovich, "Free radicals in cutaneous biology," *J. Invest. Dermatol.* **102**, 671–675 (1994).

[67] W. Dröge, "Free radicals in the physiological control of cell function," *Physiol. Rev.* **82**, 47–95 (2002).

[68] J. Thiele and P. Elsner (eds.), *Oxidants and Antioxidants in Cutaneous Biology* **29**, Karger, Basel (2001).

[69] H.N. Ananthaswamy and W.E. Pierceall, "Molecular mechanisms of ultraviolet radiation carcinogenesis," *Photochem. Photobiol.* **52**, 1119–1136 (1990).

[70] T. Herrling, J. Fuchs, and N. Groth, "Kinetic measurements using EPR imaging with a modulated field gradient," *J. Magn. Reson.* **154**, 6–14 (2002).

[71] M. Meinke, S. Haag, N. Groth, F. Klein, R. Lauster, W. Sterry, and J. Lademann, "Method for detection of free radicals in skin by EPR spectroscopy after UV irradiation," *SÖFW-Journal* **134**, 2–10 (2008).

[72] T. Herrling, K. Jung, and J. Fuchs, "Measurements of UV-generated free radicals/reactive oxygen species (ROS) in skin," *Spectrochim. Acta A* **63**, 840–845 (2005).

[73] J. Nishi, R. Ogura, M. Sugiyama, T. Hidaka, and M. Kohno, "Involvement of active oxygen in lipid peroxide radical reaction of epidermal homogenate following ultraviolet light exposure," *J. Invest. Dermatol.* **97**, 115–119 (1991).

[74] R. Ogura, M. Sugiyama, J. Nishi, and N. Haramaki, "Mechanism of lipid formation following exposure of epidermal homogenate to ultraviolet light," *J. Invest. Dermatol.* **97**, 1044–1047 (1991).

[75] T. Herrling, J. Fuchs, J. Rehberg, and N. Groth, "UV-induced free radical in the skin detected by ESR spectroscopy and imaging using nitroxides," *Free Radic. Biol. Med.* **35**, 59–67 (2003).

[76] B. Collins, T.O. Poehler, W.A. Bryden, "EPR persistence measurements of UV-induced melanin free radicals in whole skin," *Photochem. Photobiol.* **62**, 557–560 (1995).

[77] T. Herrling, L. Zastrow, J. Fuchs, and N. Groth, "Electron spin resonance detection of UVA-induced free radicals," *Skin Pharmacol. Appl. Skin Physiol.* **15**, 381–383 (2002).

[78] S. Haag, *Nachweis von Freien Radikalen in Biologischen Proben mittels ESR-Spektroskopie*, Diploma thesis, Technical University Berlin, Berlin (2007).

[79] M.E. Darvin, I. Gersonde, H. Albrecht, W. Sterry, and J. Lademann, "*In vivo* Raman spectroscopic analysis of the influence of UV radiation on caratenoid antioxidant substance degradation of the human skin," *Laser Phys.* **16**, 833–837 (2006).

[80] A.P. Popov, S. Haag, M. Meinke, J. Lademann, A.V. Priezzhev, and R. Myllylä, "Effect of size of TiO_2 nanoparticles applied to glass slide and porcine skin on generation of free radicals under ultraviolet irradiation," *J. Biomed. Opt.* **14**, 021011-1–7 (2009).

[81] K.D.O. Jackson, "A guide to identifying common inorganic fillers and activators using vibrational spectroscopy," *Internet J. Vibr. Spectr.* **3** (1998). http://www.ijvs.com/volume2/edition3/section3.html#jackson

[82] B.R. Palmer, P. Stamatakis, C.F. Bohren, and G.C. Salzman, "A multiple scattering model for opacifying particles in polymer films," *J. Coat. Technol.* **61**, 41–47 (1989).

[83] G.E. Jellison Jr., L.A. Boatner, and J.D. Budai, "Spectroscopic ellipsometry of thin film and bulk anatase (TiO_2)," *J. Appl. Phys.* **93**, 9537–9541 (2003).

[84] A.P. Popov, A.V. Priezzhev, J. Lademann, and R. Myllylä, "The effect of nanometer particles of titanium oxide on the protective properties of skin in the UV region," *J. Opt. Technol.* **73**, 208–211 (2006).

[85] A.P. Popov, J. Lademann, A.V. Priezzhev, and R. Myllylä, "Reconstruction of stratum corneum profile of porcine ear skin after tape stripping using UV/VIS spectroscopy," *Proc. SPIE* **6628**, 66281S-1-6 (2007).

25

Photodynamic Therapy/Diagnostics: Principles, Practice, and Advances

Brian C. Wilson
Department of Medical Biophysics, University of Toronto/Ontario Cancer Institute, Toronto, ON, Canada

The term photodynamic therapy (PDT) as applied to biomedical science and, more particularly, to clinical medicine is generally defined as the use of a compound or drug (photosensitizer) that has no or minimal effect alone but which, when activated by light, generates one or more reactive chemical species that are able to modify or kill cells and tissues. As historically defined, the PDT reaction should be mediated by oxygen, through the generation of reactive oxygen, most commonly singlet-state oxygen ($^1O_2$). However, there are photosensitizers under development that may use oxygen-independent photophysical pathways and we will include them as *de facto* photodynamic agents also.

The clinical status of PDT is complex, since there are numerous different photosensitizers that have been used in patients to treat a wide variety of diseases, from cancer to infection to abnormal blood vessel proliferation. Some photosensitizers have been approved in different countries for different applications, while others are still in the clinical-trial or preclinical evaluation stage. There are also several "hardware" technologies required to apply PDT in patients, including different light sources (lasers, light emitting diodes and filtered lamps), light delivery devices (based particularly on optical fibers to reach deep within the body), and "dosimetry" instruments to measure the local light intensity or photosensitizer concentration or oxygen concentration, or combinations of these. Correspondingly, biophysical and biological models of PDT have been developed to systematize the complex interactions of light, photosensitizer and oxygen, and the resulting effects on cells and tissues. At the cellular level, there is a good understanding of the basic mechanisms of PDT-induced damage, with one or more different mechanisms of cell death being involved (necrosis, apoptosis, autophagy), depending on the PDT parameters and cellular characteristics. Likewise, tissue may be modified or destroyed by PDT either directly by the cell death or indirectly by disruption of the vasculature, and both local and systemic immune responses can be involved in some cases.

The multicomponent nature of the photophysical, photochemical, and photobiological processes in PDT make it a challenging modality to optimize, and much of the clinical progress to date has been largely empirically based. On the other hand, this same complexity provides the opportunity for PDT to have many potential applications in medicine and, increasingly also, as a tool in biological research.

Since many PDT photosensitizers are also fluorescent, the development of PDT as a treatment technique has often been in step with the development of photodynamic diagnosis (PDD), leading to the concept of "seek and destroy," for example as applied to early-stage cancers. Photosensitizer-based fluorescence imaging/spectroscopy is emerging also for guiding surgical resections, particularly of tumor tissue.

While much of the more than 100-year history of PDT/PDD has been dominated by the use of molecular photosensitizers and activation by CW (continuous wave) light, in the past decade a number of radically-different approaches have emerged. These include: photodynamic molecular beacons in which the photoactivation is quenched until the molecules reach specific targets; the use of nanoparticles, either as photosensitizers or as carriers thereof; and the use of ultrashort pulsed laser activation in so-called 2-photon PDT. Thus, traditional photodynamic therapy is converging with advances from other fields to further exploit the basic concept of light-activated drugs.

This chapter will focus on the science and technology underlying both established and emerging methods in PDT and PDD. Clinical applications will be used to illustrate the challenges and opportunities in translating these principles into practical and cost-effective techniques.

Key words: photodynamic, photosensitizers, photophysics, photochemistry, photobiology, lasers, LEDs, dosimetry

25.1 Historical Introduction

The history of PDT/PDD has been reviewed in several texts [1, 2]. The first recorded observation of the photodynamic effect was made in the last decade of the 19th century by Oscar Raab (a medical student in the laboratory of H. von Tappenier in Germany), who noted that acridine dye could kill microorganisms (paramecia) in the presence of light. Interestingly, this was immediately linked to the fluorescent properties of the dye, i.e. the emission of longer wavelength light when exposed to short wavelength light, and led von Tappenier to speculate on the potential uses of fluorescence in medicine. Thus, PDT and PDD have been conceptually strongly linked from the beginning. The medical applications were subsequently pursued using the dye eosin (as in the H&E tissue stain used commonly in histopathology) to treat skin tumors, applying the dye applied topically. The need for oxygen to be present in these reactions was also demonstrated, and the term "photodynamic" was coined in 1907.

The therapeutic use of PDT then all but disappeared until the early 1970s. During the intervening 60 years, investigations continued sporadically on the fluorescent properties of dyes for the purpose of tumor localization, in particular with porphyrin compounds, including hematoporphyrin (HP) that was first prepared in the mid 19th century by acid treatment of blood. Porphyrins are an important class of compounds that are involved in particular in the synthesis of hemoglobin. As illustrated in Figure 25.1, these are large flat molecules comprising 4 pyrrol rings, with various side-chains that distinguish the different species. From the photophysical perspective, they are important in that they a) have a broad spectrum of light absorption, from the violet/blue through to the red part of the visible wavelength range, b) often are fluorescent, with a characteristic red emission, and c) in some cases can generate singlet oxygen upon light absorption. HP was found to have

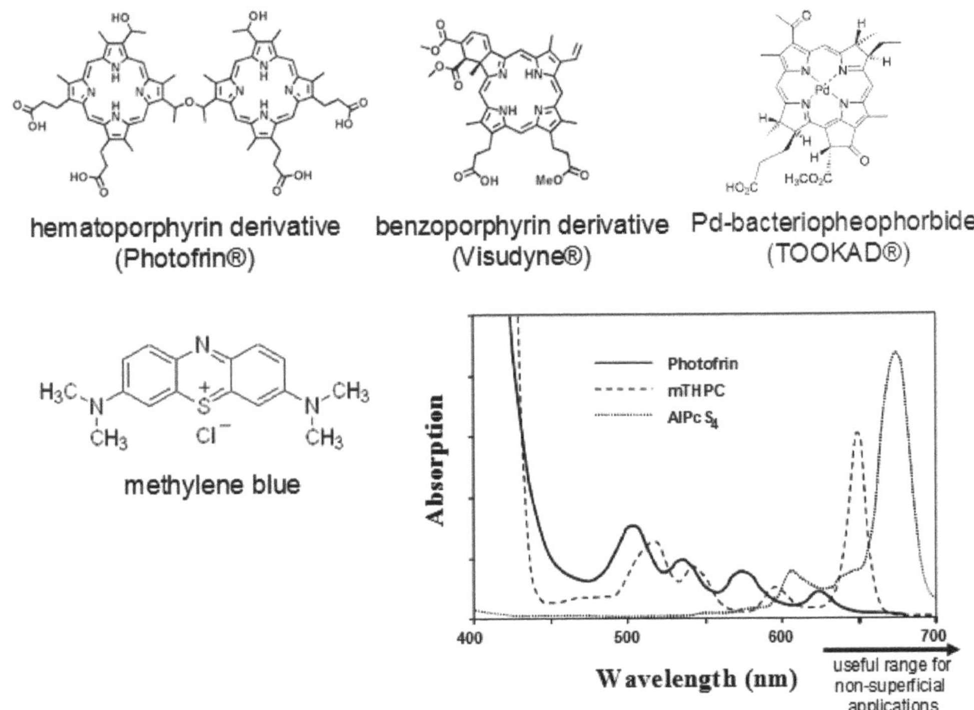

FIGURE 25.1: Molecular structures and absorption spectra of some representative PDT photosensitizers.

photodynamic effects and several animal studies were reported in the 1940s, but more attention was focused on attempts to use its fluorescence to localize tumors. For this, several clinical studies were carried out in the 1950s that had overall mixed success, since the results were generally inconsistent. This situation was improved by the discovery by Schwartz of HpD, a derivative of HP that had superior tumor-localizing and also photodynamic properties to HP. In the 1970s, HpD-based fluorescence spectroscopy/imaging was used during endoscopic examination of the bronchus. The ability to detecting early lesions that were occult to white-light endoscopy lead to many studies of HpD fluorescence diagnostics and localization, and subsequently to the development of new fluorescence imaging systems and investigation of other fluorescent "contrast agents" [3], including PDT photosensitizers (at subtherapeutic doses).

Returning to the subject of photodynamic *therapy*, in 1978 Dougherty and colleagues in Buffalo, NY carried out the first major clinical study of HpD-based PDT patients with primary or recurrent skin tumors of a variety of histopathologic types. They used red light from a filtered xenon arc lamp, applied 1–7 days after systemic injection of HpD. Despite the limitations of HpD, particularly the generalized skin photosensitivity lasting for up to several weeks, the results of these and subsequent studies in a variety of different types of cancer in different locations (deep-seated as well as superficial) were very encouraging. This, combined with the emergence of medical laser sources and optical fibers that followed the invention of the first laser in 1960 and then of optical fibers, enabled the first government approval for PDT, in Canada, in 1993. This was for the treatment of patients with superficial recurrent cancer of the wall of the bladder, and the photosensitizer used was a commercial version of HpD (Photofrin®). Photofrin-PDT has been approved subsequently for a variety of different types, locations, and stages of solid tumors or premalignant lesions (dysplasias) by various governments worldwide. Renewed interest in PDD PDT with other photosensitizers has also entered clinical practice, the most notable being that of a different porphyrin compound (benzoporphyrin derivative, BPD) in treatment of wet-form age-related macular degeneration, in which abnormal growth of blood vessels in retinal choroid can lead to blindness [4]. This has been used in over 2 million patients worldwide to slow or halt progression of the disease. Most recently, PDT has been approved, also initially in Canada, for treatment of periodontitis, i.e. gum infection, the intent being to kill the microorganisms (in this case bacteria), bringing PDT back full circle to the original observations over 100 years ago.

Perhaps ironically, however, considering its historical connection to PDT, PDD using photosensitizers has not had the same impact to date. Rather, the use of tissue autofluorescence-based endoscopic diagnosis, which does not involve the application of an exogenous compound, has been more fully developed and is approved and commercially available in several countries [5]. However, there is significant renewed interest in PDD, with the development of novel targeted fluorescent contrast agents, not only for early cancer detection but also for fluorescence image-guided surgery [6].

Before considering the details of PDT and its applications, it is worth noting that the scientific, technological and clinical challenges posed in developing PDT into a true clinical modality have stimulated research and development that underlies many aspects of biophotonics in general. First, it was a major motivation in the early 1980s for the development of techniques to measure the optical absorption and elastic scattering properties of tissues and of corresponding quantitative models of light propagation in tissues (diffusion theory, Monte Carlo simulations) to understand and optimize light delivery in PDT of solid tumors. This was discussed in one of the earliest overview papers on the physics of PDT, recently updated and extended [7]. Second, it spurred the development of many different fiberoptic-based light delivery devices, in order to "shape" the light distribution to the target tissue to be treated by PDT [7, 8], and was one of the first modalities to utilize novel light sources in medicine, such as the KTP-pumped dye laser, high-power diode lasers and LED arrays [7, 9]. Third, it expanded greatly the field of photobiology, particularly in its medical applications, and has led to many new insights into how cells respond to photochemical damage at the molecular level [10]. Thus, PDT may be considered one of the "pivotal challenges" in biophotonics [11].

25.2 Photophysics of PDT/PDD

We start with a discussion of the photophysical processes in PDT, i.e., what are the various energy states of the photosensitizer, transitions between these states and interactions with other molecules, particularly oxygen? Since this drives all of the subsequent chemistry and biology of PDT, it is important to understand this in some detail [12]. For all the photosensitizers in current clinical use and with many of those under development, the primary photophysical process involved is believed to be the generation of singlet oxygen, $^1O_2$, by the process illustrated in Figure 25.2.

In this so-called Type II reaction the singlet oxygen is produced from ground-state oxygen ($^2O_3$) that is present in the target cells or tissues being raised to an excited energy state by energy exchange from the triplet state (T_1) of the photosensitizer. This triplet state is generated from the singlet state (S_1) that is formed by absorption of a photon of light by a ground-state photosensitizer molecule (S_0): typical absorption spectra are also shown in Figure 25.1. In this process, the S_0 state of the photosensitizer is regenerated, so that the cycle can repeat: typically, this may occur many thousands of times for a given molecule during a PDT treatment, so that in effect the photosensitizer serves as a catalyst for the conversion of light energy into energy of the excited $^1O_2$ state. The $^1O_2$ molecule then is responsible for chemical alterations in target biomolecules in the cells/tissue, resulting in one or more of the various biological effects that will be discussed below. For maximum efficacy,

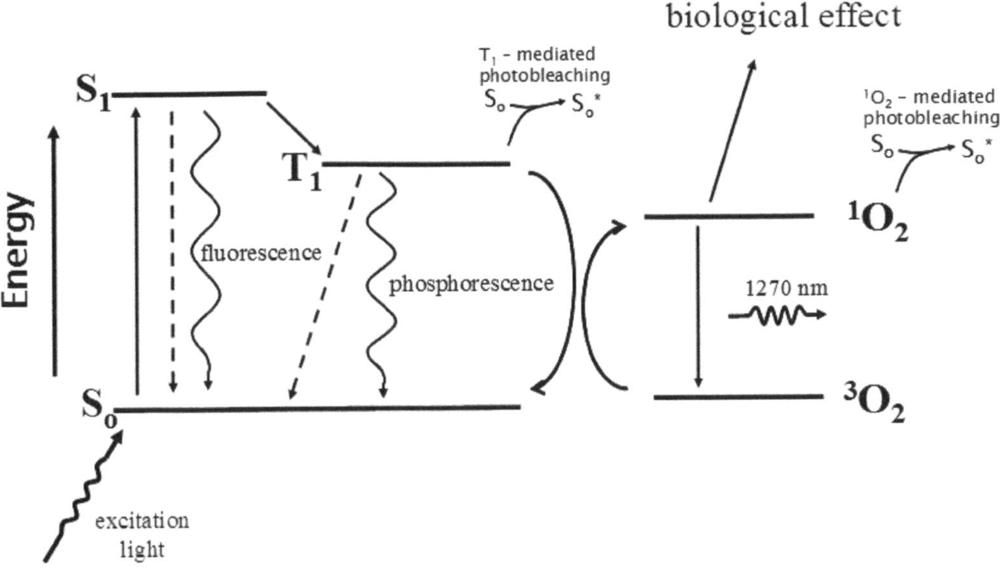

FIGURE 25.2: Jablonski energy diagram for Type II photophysics, showing absorption of a photon by the molecular ground state (S_0), raising it to an excited singlet state (S_1), from where it may i) undergo intersystem crossing to the triplet state (T_1), ii) decay radiatively, emitting a fluorescence photon or iii) de-excite nonradiatively (dashed line). The T_1 state can then either decay back to S_0 by phosphorescence emission, or can exchange energy with ground-state oxygen ($^2O_3$) to generate the excited singlet oxygen state ($^1O_2$). $^1O_2$ may then cause biochemical changes with nearby molecules in the cells or tissue or may decay radiatively to $^2O_3$, emitting a near-infrared photon. Two photobleaching pathways are also indicated.

the quantum efficiency of the photosensitizer should be high, i.e., the probability that a molecule of $^1O_2$ will be generated following absorption of a photon by the photosensitizer ground state, should close to 100%. It is the fact that the ground state of oxygen ($^3O_2$) is unusual in being a triplet state, making the $T_1 \rightarrow ^3O_2$ transition quantum-mechanically allowed (no change of spin), that allows photosensitizers to have usefully high $^1O_2$ quantum yields.

The energy gap between the excited and ground-state oxygen molecules is close to 1 eV, so that this would be the minimum photon energy required to trigger the Type-II reaction. In practice, the efficiency of $^1O_2$ production is limited to photon energies above about 1.5 eV, corresponding to wavelengths below about 800 nm: the longest wavelength of activation used to date in any clinical studies of PDT is 763 nm using a bacteriopheophorbide photosensitizer (TOOKAD) [13]. The shortest wavelength used clinically is blue light from a filtered fluorescent lamp to treat a very superficial skin condition (actinic keratosis caused by overexposure to sunlight) [14].

In the alternative, Type I, reaction the chemical changes in the biomolecules (which are generally different from those in Type II reactions, although the "downstream" biological effects may be similar) are caused by interactions of either the excited singlet (S_1) or triplet (T_1) states themselves, without the oxygen intermediary. Unless otherwise stated, we will assume that Type II processes are dominant.

The various excited states in Figure 25.2 may de-excite either radiatively, i.e., with the emission of visible or near-infrared (NIR) light, or nonradiatively (internally by energy transfer as in $T_1 \rightarrow ^2O_3$, or by collisional de-excitation with other molecules resulting in heat). The $S_1 \rightarrow S_0$ transition corresponds to fluorescence emission, the basis of PDD. The higher the $^1O_2$ quantum efficiency, the lower the fluorescence yield, but typically a fluorescence quantum efficiency of a few percent is adequate for fluorescence spectroscopy or imaging, and several clinical photosensitizers fall into this category. This is useful, not only for fluorescence diagnostics but for measuring the photosensitizer uptake prior to light activation for PDT and also, through photobleaching, for PDT dosimetry, as discussed later. The emitted fluorescence photons from $S_1 \rightarrow S_0$ de-excitation have a spread in energy (wavelength), since S_0 can be in a range of vibrational energy states. Thus, the fluorescence emission spectrum has a finite width. (The corresponding S_1 vibrational states can be ignored, since they are very short-lived compared with the fluorescence lifetime, so that effectively all the fluorescence occurs from the lowest S_1 state.) The fluorescence lifetime of PDT photosensitizers is typically in the ns range.

Since it is quantum-mechanically forbidden, the triplet-to-ground state transition ($T_1 \rightarrow S_0$), corresponding to phosphorescence emission has low probability in most cases. To date this emission has not been exploited in PDT, although other molecules (e.g., palladium porphines) designed to have high phosphorescence quantum yield are used as intravital probes of cell/tissue oxygenation [15]: since the oxygen serves to quench the triplet state, the measured phosphorescence lifetime decreases as the local oxygen concentration increases. Because of the low phosphorescence probability of most PDT compounds, the T_1 state is long lived (typically $\sim \mu_s$), which increases the probability of energy transfer to ground-state oxygen and, hence, generation of $^1O_2$. The 3rd radiative transition is that of $^1O_2 \rightarrow ^3O_2$, i.e., decay of singlet oxygen back to the ground (triplet) state. The energy of this transition corresponds to a wavelength of around 1270 nm, which can be detected by photomultiplier tubes having an extended-wavelength response. This is the basis for so-called SOLD (singlet oxygen luminescence dosimetry) in PDT [16], which will be discussed later. The $^1O_2$ lifetime in water is about 3 μs. In biological media such as cells and tissues, the lifetime is believed to be an order-of-magnitude shorter, although there is some controversy about this. Due to the high reactivity of $^1O_2$ with biomolecules, the probability of luminescence decay is very low ($\sim$1 in 10^6-10^8), which makes it technically challenging to detect and quantify accurately. Assuming that the lifetime is short, the diffusion distance, i.e., the average distance travelled by a molecule of $^1O_2$ before it interacts, is very short: possibly as low as 10 nm by some estimates, which is much smaller than typical micron-sized mammalian cells. Hence, the chemical damage and consequent biological effects of $^1O_2$ are extremely dependent on the microlocalization of the photosensitizer molecule at the time of excitation. This is supported by, for example, correlation between the localization of many porphyrin-like photosensitizers to mitochondrial membranes and apoptotic ("programmed") cell death associated with mitochondria damage (see below).

It is important to note that the quantum efficiency of these various processes is independent of the wavelength of the excitation light, as long as the photon energy is high enough to generate the S_1 state. However, the quantum yields and the spectra of the S_0 absorption and of the fluorescence emission depend strongly on the photosensitizer molecular structure and whether or not the photosensitizer is in a monomeric state or shows some degree of aggregation. They can also be significantly affected by the local microenvironment, which can vary with the photosensitizer microlocalization in cells or tissues. These effects are important in understanding and optimizing the factors that influence the efficacy of PDT in particular applications.

A final aspect of PDT photophysics that also has significant practical consequences is photosensitizer photobleaching [7,17], illustrated in Figure 25.3. Although the exact mechanism(s) for this are occasionally obscure, a useful operational definition for photobleaching in PDT is the loss of fluorescence as the PDT light application proceeds. This can be monitored either by point fluorescence spectroscopy at a location of interest or by fluorescence imaging. If this is also a measure of the loss of photodynamic effect, resulting in decreasing $^1O_2$ production as treatment proceeds, then in principle the photobleaching could be used to monitor the treatment, as discussed below. If photobleaching does occur, then in some cases there is a corresponding appearance of a "photoproduct," typically seen as an additional signature in the absorption and/or fluorescence spectrum of

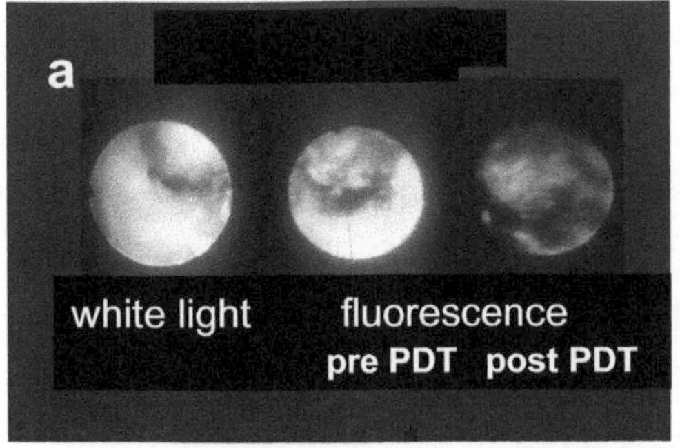

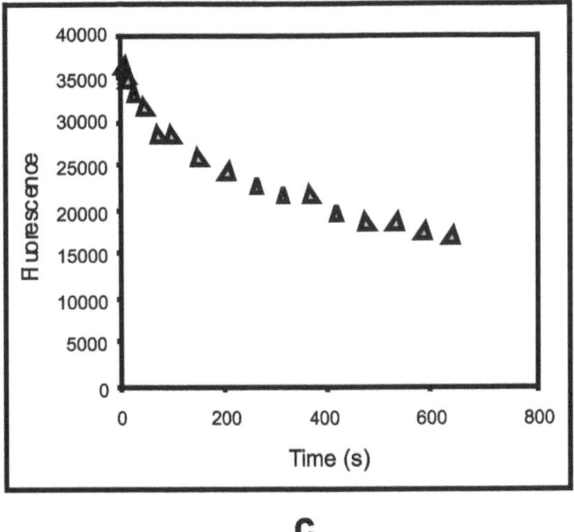

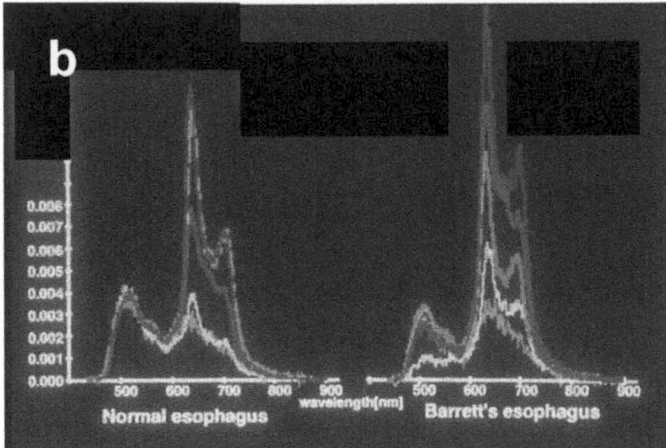

FIGURE 25.3: Photosensitizer photobleaching. a) loss of Photofrin fluorescence seen during PDT irradiation of a tumor in the esophagus (courtesy Dr. N. Marcon, Toronto). Reduced fluorescence brightness is seen in the post-PDT fluorescence image compared to the pre-PDT image. b) fluorescence spectra of Photofrin measured during PDT for Barret's esophagus, showing decreasing signal as treatment proceeds. c) fluorescence photobleaching in cells *in vitro*: the photobleaching constant, β [cm^2J^{-1}] can be derived by fitting to an exponential curve $F=F_o e^{-\beta\Phi}$, where Φ is the light fluence.

the photosensitizer [18]. This could also be used as a dosimeter. The question arises as to whether or not the photoproduct is itself a photosensitizer and contributes to the PDT effect (if its absorption spectrum overlaps with that of the light source being used). This is probably highly dependent on the particular photosensitizer.

It may appear that the photophysics of PDT is inordinately complicated and that there is only partial, conditional, or qualitative understanding of this first and critical stage in the PDT process. While this is true, and the situation is compounded rather than resolved by the continued appearance of new photosensitizers (some involving also Type I interactions), significant progress has been possible in practice by semi-empirical approaches to the practice of PDT. Nevertheless, quantitative techniques for PDT dosimetry are used frequently in preclinical research, and increasingly for clinical treatments, especially where the target tissue is of complex geometry and close to critical normal organs [7].

25.3 Photochemistry of PDT/PDD

Since the discovery of HpD, there has been a considerable effort to develop new photosensitizers for PDT with improved properties, including:

- *chemistry:* single, inexpensive, synthetic compounds rather than a blood-derived mixture of many different porphyrins,
- *photophysics:* higher singlet oxygen quantum yield, larger extinction coefficient (optical absorbance per unit photosensitizer concentration) at longer (far-red or near-infrared) wavelengths where the light can penetrate deeper in tissues,
- *low skin photosensitization:* either through rapid clearance or low uptake in skin,
- *good target to nontarget uptake ratio:* typically < 5 for Photofrin in tumors,
- *fluorescent:* for ease of localization, potential photobleaching dosimetry, and PDD,
- *low systemic (dark) toxicity.*

Many of these so-called second-generation photosensitizers are porphyrin-based, while others utilize a different core chemical structure. As illustrated in Figure 25.1, most are based on large, ring-like molecules, since these give favorable photophysical properties in the visible/NIR range. Important points to note are that 1) the particular metal ion, if any, in the macromolecular ring structures strongly affects the photophysical properties, 2) the pharmacokinetics and localization of the photosensitizer are often controlled by the side chains, since these affect such properties as the water/lipid solubility and charge, and 3) the longer wavelength absorption generally involves a larger ring structure, which ultimately limits the (photo)stability of the molecule. The properties of PDT agents have been reviewed in some detail (see [12] and refs. therein), so here we will highlight only examples of photosensitizers that have some particular features and that are either in clinical use or under active investigation, and also some of the newer concepts that are under development.

In the early 1980's, Kennedy and Pottier reported the use of ALA as a PDT "prodrug." Since then, this idea has been extensively investigated and several ALA-based agents are approved, primarily to date for dermatological applications [19]. ALA stands for (5 or delta)-aminolevulinic acid. This is a naturally-occurring small molecule that is key in the control of heme biosynthesis, leading to production of hemoglobin in blood. When cells are supplied with excess ALA, heme synthesis is stepped up, following a sequence of steps involving various porphyrin species, the penultimate of which is a fluorescent photosensitizing molecule, protoporphyrin IX (PpIX). PpIX itself can be chemically synthesized and administered as a PDT compound, but then the specific advantages of ALA-generated drug are lost. In particular, endogenous ALA-PpIX synthesis is highly tissue-specific, with tumors usually showing higher levels of PpIX than many normal tissues, possibly because of low levels of the enzyme ferrochelatase, that is required to convert PpIX to heme, and/or to low levels of available iron. ALA can be administered topically to skin lesions, and variants of ALA with higher lipid solubility have been developed with the objective of achieving better penetration into tissue. ALA-PpIX has also proven to be very useful in PDD, because of the high tumor specificity. This has been applied in many different clinical trials of ALA fluorescence endoscopy, in particular in the bronchus, bladder, and gastrointestinal tract (esophagus, stomach, and colon), as well as in imaging skin tumors [20]. Figure 25.4 shows examples of these applications. ALA-PpIX also provides fluorescence image contrast in tumor surgery, so that it can be used to guide resection of malignant tissue. Recent clinical trials, where fluorescence excitation (with blue light) and detection (in the red region of the spectrum) has been incorporated into an intraoperative microscope,

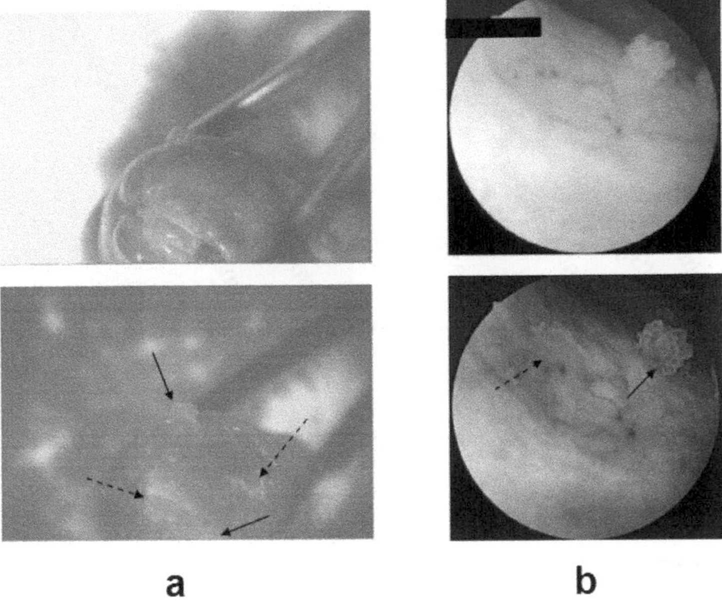

FIGURE 25.4: (See in color after page 572.) Examples of ALA-PpIX fluorescence imaging. a) in an oral cancer animal tumor model after intravenous administration of ALA, using green light excitation, showing the red PpIX fluorescence. b) bladder cancer detected during diagnostic cystoscopy in a patient, after instillation of ALA into the bladder and using blue light excitation (adapted from Jocham et al. [3],with permission). In each case, the corresponding white-light image is shown at the top. The solid arrows indicate frank tumor, while the dashed arrows indicate regions of early tumor tissue that are not easily seen under white light.

have shown great promise in neurosurgery to increase survival in patients with malignant brain tumors [6] and this "fluorescence-guided resection" approach is now being extended to other types of tumor, such as the oral cavity and the prostate. The technique will be discussed further below.

As mentioned in the introduction, PDT is emerging as a useful anti-infective modality [21]. In the 1980s/90s, it was investigated as a means to purge blood of viral agents, such as HIV and hepatitis prior to blood transfusion, but this is not currently in use, because of a combination of negative factors (cost, some damage to normal blood cells and concern over residual material in the blood). More recently, PDT using small cationic photosensitizers, like methylene blue and derivatives thereof, have been investigated for treatment or prevention of localized bacterial infection. This approach has become particularly important with the emergence of bacterial strains that are resistant to multiple antibiotics. As has been observed also in mammalian cells [22], the induction of resistance to PDT appears to be very low, due to the fact that cell kill does not involve direct damage to the genetic material but rather to membranes: in the case of bacteria, to the cell wall. Note that the structure of photosensitizers that are useful for anti-infective PDT is completely different than for PDT targeted at mammalian cells, although ALA-PpIX has also been explored for the former.

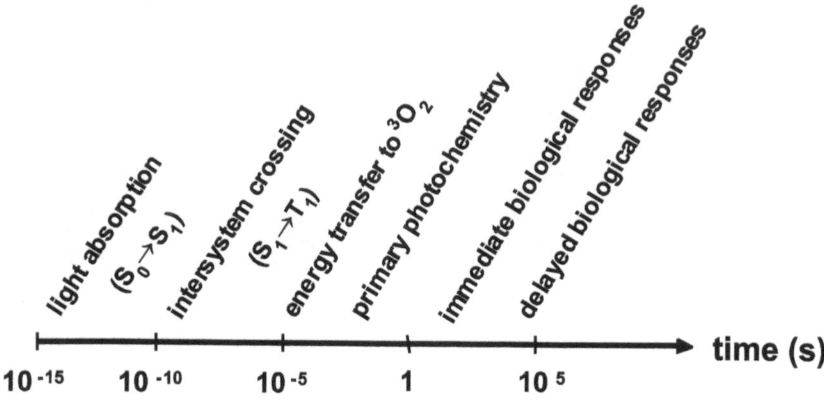

FIGURE 25.5: Order-of-magnitude time scales for the sequence of events in PDT.

25.4 Photobiology of PDT

The biological effects of singlet oxygen are initiated by the immediate direct damage to cell structures. Since the $^1O_2$ diffusion distance is small, this occurs primarily at the site of localization of the photosensitizer molecules. Most photosensitizers localize in cell membranes, commonly either the plasma membrane in the case of short incubation time or to the mitochondrial membrane, since this is a favored site for photosensitizer binding, although localization in other organelles is also possible for some photosensitizers. Unless a specific targeting strategy is used, as discussed below, PDT agents do not usually localize significantly to nuclear DNA, so that the main molecules that are chemically modified by the singlet oxygen are membrane lipids and proteins. Multiple chemical changes have been described [12], which then cause membrane disruption and/or trigger cell signaling processes, leading either to cell death or to repair. Subsequently, the dead cells can either be cleared by the body or, if the number of cells killed is large enough, can lead to either local tissue and/or systemic responses.

It is important to appreciate the huge range of time scales over which the photophysical, photochemical, and photobiological effects occur, which range from $\sim 10^{-14}$ s for the initial photon absorption event to days or months for the final biological effects to be complete, as summarized in Figure 25.5.

When mammalian cells and tissues are exposed to PDT, there are 3 main classes of biological effect that may occur, depending on the photosensitizer, its site of localization (at the organ/tissue, cellular and intracellular levels) and the local PDT dose (as represented, for example, by the amount of singlet oxygen generated). The first effect is direct cell kill due to the photochemical damage caused in the target cells containing the photosensitizer. This can be seen when cells growing in culture (*in vitro*) are incubated with the photosensitizer and then exposed to an appropriate dose of light of the correct wavelength to activate it (Figure 25.6a-top). While there are differences in such cell survival curves for different cell types exposed to PDT, most mammalian cells show qualitatively similar overall responses to PDT. Cell death following PDT *in vitro* may be mediated through different mechanisms [23]. The first is necrosis, in which the cells die in an uncontrolled, energy-independent manner, typically due to damaged plasma membrane.

An alternative death pathway, which is commonly found in PDT with many photosensitizers and typically requires lower PDT doses, is that of apoptosis or programmed cell death (Fig 6a-middle), which is a controlled, energy-dependent process that is usually due to damage to the mitochondria.

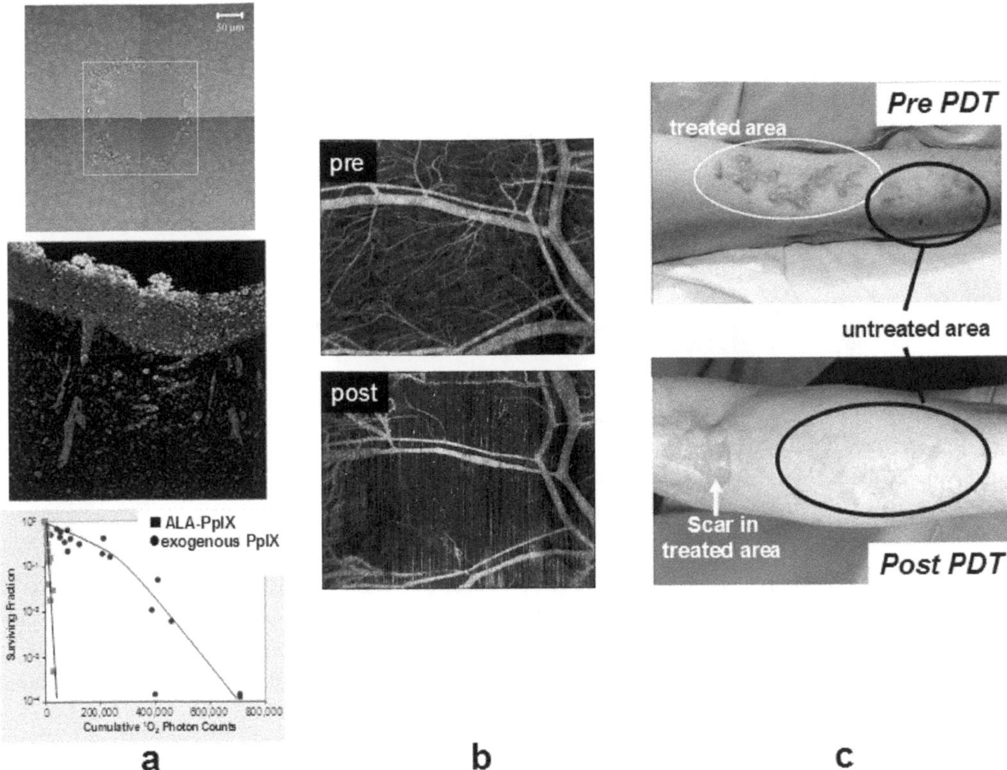

FIGURE 25.6: Examples of the main biological effects in PDT. a) *in vitro* cell kill: necrosis in a cell monolayer with loss of adherence in the treated area and uptake of stain in the adjacent cells (top), apoptosis in brain tumor cells indicated by the bright regions corresponding to an apoptosis-specific fluorescent stain (middle), and (bottom) cell survival as a function of cumulative $^1O_2$ in tumor cells showing higher kill when the photosensitizer is located in the mitochrondria (ALA-PpIX) than in the cell membrane (exogenous PpIX). b) effects on vasculature, showing the shut-down of blood flow in the microvessels following PDT (bottom), imaged using speckle-variance Doppler optical coherence tomography [27]. c) immune effect of PDT, where treatment of skin tumor in one area resulted in disappearance of tumors outside the treatment field (adapted from Thong et al. [28], with permission).

Many studies have been carried out to illuminate the molecular pathways involved in this form of cell death [24]. It is important to reiterate that the intracellular site of photosensitizer localization is a critical factor in determining the specific pathways. Recently, a third mechanism has been identified in PDT cell death, namely autophagy [25], which is also a controlled death process. The importance of this in the PDT response of normal and pathological tissues is not yet well understood. An advantage of apoptosis, and possibly also autophagy, is that it does not trigger an inflammatory response as can happen with tissue necrosis and which can result in secondary collateral damage to adjacent normal tissue.

For the reason indicated above, there is usually little direct involvement of DNA damage. This is in contrast to other treatments such as radiation therapy or many forms of chemotherapy. Recent work (see Section 25.7) has aimed deliberately to target the photosensitizer to the nucleus (or other organelles): while this may increase the efficiency of PDT kill, care will be needed not to lose the advantages of nonnuclear targeting, namely lack of mutagenic effects, low induction of resistance,

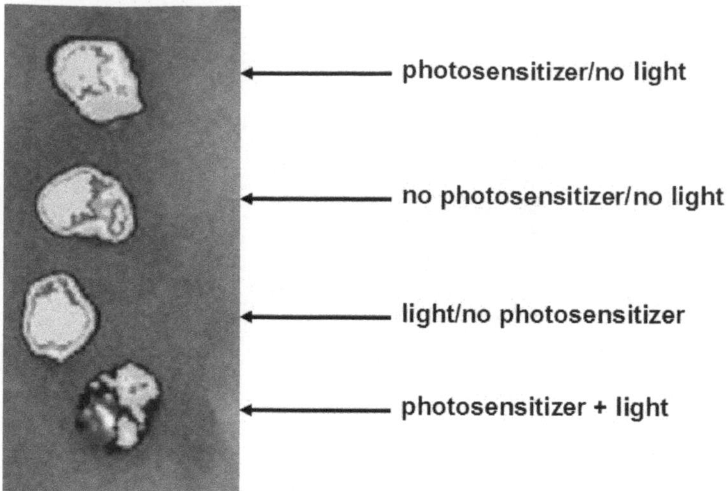

FIGURE 25.7: Example of PDT killing of bacteria (*E. coli*) in infected skin wounds, treated using a conjugate of the photosensitizer chlorine e6. The bacteria carry the luficerase gene for bioluminescence. The bioluminescence images show PDT effect plus 3 control conditions. The bioluminescence signal is clearly reduced following PDT treatment, indicating bacterial killing (adapted from Demidova et al. [26], with permission).

and lack of cross-resistance to treatments that do target DNA. It is also notable that cell death with PDT usually occurs without requiring cell division, again unlike other treatments where the damage is not manifest until the cells attempt to proliferate.

Figure 25.7 shows PDT killing of bacteria in infected skin wounds [26]. For anti-infective PDT, while $^1O_2$ is still the primary cytotoxic agent for Type II photosensitizers, the cell kill is believed to be mainly due to damage to the cell wall, which is structurally and chemically unlike the mammalian cell wall. Because Gram-positive and Gram-negative bacteria have very different cell wall thickness and structure, their PDT sensitivities can vary: in general Gm^+ bacteria such as *Staphylococcus* are easier to kill (i.e., require a lower photosensitizer and/or light dose) than Gm^- strains such as *Pseudomonas*. Highly charge cationic photosensitizers are usually much more effective than anionic compounds, especially for Gm^+ bacteria since they are better able to penetrate into the cell wall. Other microorganisms such as yeasts, fungi, and spores also show quite variable PDT sensitivity. Selective killing of microorganisms, for example, in locally infected tissues is often achieved by exploiting the much faster drug uptake than in mammalian cells, by applying the light within a few minutes of applying the photosensitizer.

The second class of action of PDT in tissue *in vivo* is that of vascular damage, as illustrated in Figure 25.6b. This is seen most easily if the photosensitizer is activated by light exposure while it is still in the circulation, which is the basis for Visudyne-PDT treatment of age-related macular degeneration (AMD), in which the light is given within a few minutes after intravenous injection of the photosensitizer [4]. The abnormal blood vessels in the back of the eye that cause loss of central vision in this disease are shut down by the treatment. A simplified explanation of the processes involves direct PDT damage to the endothelial cells that line the vessels, which then lift off the basement membrane (analogous to the cells in Figure 25.6a-top). This causes circulating platelets to aggregate, triggering formation of thrombus that occludes the blood flow. Clearly, this vascular mechanism of action is of prime importance in treating a vascular pathology such as AMD, but it also may play a role in other treatments, such as PDT of solid tumors. Photosensitizers have been developed that are designed to have very rapid clearance from the circulation and require that the light be applied shortly after, or even during, photosensitizer administration. An example is that of the bacteriopheophorbide agent TOOKAD (WST09) that has been used in clinical trials in prostate cancer [29] and can shut down the vasculature of the entire prostate gland, thereby depriving tumor cells of oxygen and nutrients.

The third class of action that may occur *in vivo* is that of immune modulation [28, 30]. This is a much less well-understood mechanism. Both local and distant (systemic) immune responses have been seen following PDT in both animal models and patients, and these effects can be either inhibitory or the immune system can be stimulated. This has not been much exploited deliberately in the clinic to date: rather, most observations in patients have been unintentional or at least not part of the deliberate therapeutic stratagem. One example of a very positive immune response is the work of M. Olivo and colleagues in Singapore [28], illustrated in Figure 25.6c, who observed disappearance of tumors that lay outside the light treatment field following PDT of cutaneous lesions. Some of the important questions in this are: how do the immune effects depend on the PDT parameters (photosensitizer and doses), what are the underlying biological mechanisms, can these be exploited in a reproducible and enhanced way, and how important are these effects in clinical PDT treatments?

25.5 PDT Instrumentation

Instrumentation plays a central role in making PDT a practical end effective treatment modality, and may be considered at several different levels, which we will discuss in turn:

- light sources
- light delivery devices
- dose monitoring techniques and instruments
- techniques and instruments for monitoring the biological responses to PDT

25.5.1 Light sources

As with other light-tissue mechanisms of therapeutic action (photothermal or photomechanical), the amount of light required in PDT is substantial compared to ionizing radiation. Thus, a total absorbed energy dose in radiotherapy of solid tumors is ~50 Gy, i.e., 0.05 J/g of tissue, whereas PDT typically requires $\sim 10^2$–10^3 Joules of absorbed energy per gram of tissue. This is due to the fact that relatively little damage to DNA is required to disrupt the ability of cells to replicate: in principle, a single lesion is sufficient, whereas much more biochemical damage is required to cell membranes before cell death is induced. For the simplest case of surface illumination, the incident energy density is then typically a few hundred Joules per cm^2. For a light treatment time of, say, 15 min, this requires a delivered optical power density of a few hundred $mWcm^{-2}$, so that a 10 cm^2 area requires a light source capable of delivering $\sim$ 1 W total power. The requirements can easily be several times higher so that, taking into account losses in delivering the light to the target tissue, the light source output is typically in the 1–10 Watt range at the appropriate wavelength for the particular photosensitizer used.

There are 3 main light source technologies [7,31], with different advantages and limitations: examples of clinical systems are shown in Figure 25.8. Lasers found in clinical PDT are now almost exclusively solid-state diode lasers, although prior to about the mid-1990s dye lasers pumped by argon-ion or by frequency-doubled Nd:YAG (KTP) lasers were used and other technologies such

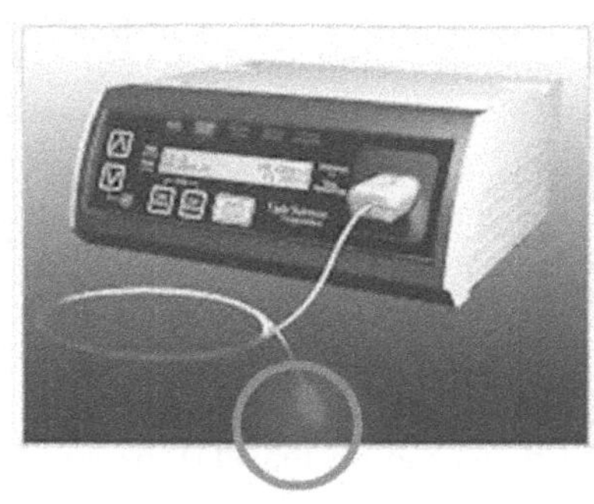
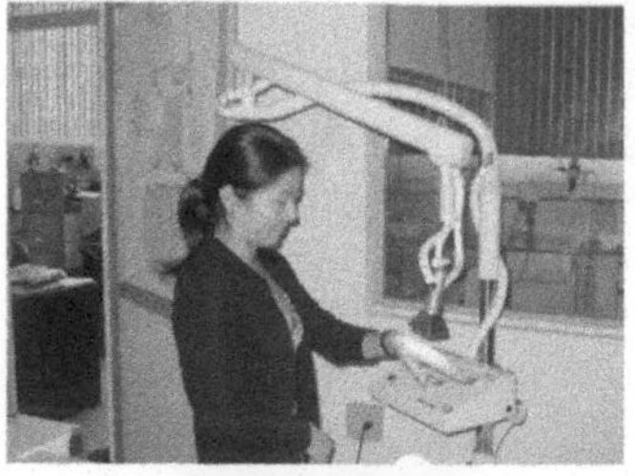
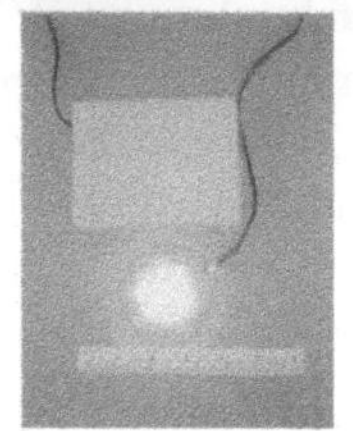
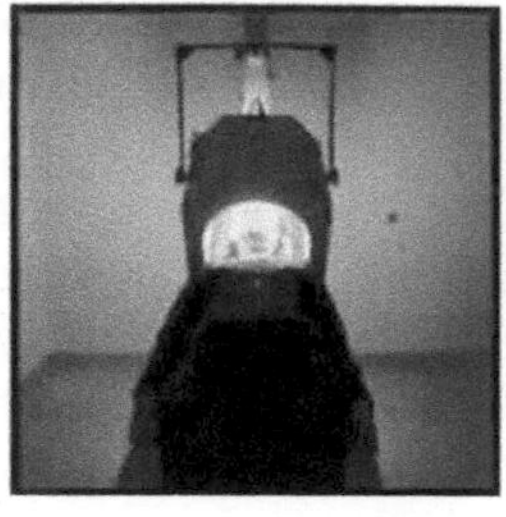

FIGURE 25.8: Examples of clinical light sources for PDT. a) diode laser, with fiberoptic delivery. b) LED arrays, including semiconductor array source (top: courtesy Dr. C. Kurachi, Saõ Carlos, Brazil, pictured) and a prototype OLED "patch" (bottom: from Moseley et al. [32], with permission). c) filtered fluorescent-lamp array used for blue-light PDT of superficial skin disease.

as metal-vapor and excimer lasers were also found. The diode lasers are typically coupled into multimodal optical fibers for delivery to the tissues, with coupling losses being $< 20\%$ for fiber core diameters of 200–600 μm, and this is a major advantage for interstitial or endoscopic delivery, as will be seen in the next section. The cost per Watt is typically about $10K: this is an order of magnitude higher than the cost of a diode laser of equivalent power found in the research laboratory or industrial applications, due in part to the need to have the laser device incorporated into a clinical "system," with appropriate safety features and user controls, and to have the system approved for human use (e.g., by the FDA). Given this high cost, a drawback of diode lasers is that different systems are needed for each wavelength required to match different photosensitizers.

Semiconductor light emitting diodes (LEDs) are finding increasing use as PDT light sources, having the advantages of high power at relatively lower cost and the capability to tailor the individual devices into linear or 2D arrays of different configurations. The former have been used for interstitial applications in which a string of LEDs a few cm in length is mounted in a catheter-type sheath, which is then inserted directly into the target tissue. In open-beam configurations for surface use, the wide divergence of LEDs can be a problem in capturing and directing the light with low losses. Removing the heat, given the relatively low electrical-to-optical conversion efficiency of LEDs (less than $\sim$15%), is also a significant technical challenge. Both active (air and fluid flow) and passive (heat sinking, phase-change materials) cooling schemes are used, with the former generally increasing the cost and complexity of the device. Unlike lasers, LEDs are not monochromatic, having typical bandwidths of $\sim$25 nm. Thus, depending on the spectral overlap between the LED emission and the photosensitizer absorption spectra, the effective light intensity should be reduced by about 50% compared to using a laser at the absorption maximum. This needs to be taken into account in determining the light dose used. The same point applies also to filtered lamps. Recently, organic LEDs have been introduced for PDT [32]. These are inexpensive, flexible, and available in the large-area format, all of which should be advantageous. The main limitation at present is the low output (typically $\sim$ mWcm^{-2}), so that treatment times in the order of hours may be needed. This matches well with their use as "light patches."

High brightness lamps are also used clinically in PDT, as in the case of the U-shaped bank of blue (wavelength-filtered) fluorescent lamps used in ALA-PpIX treatment of the face and scalp. While lamps have the convenience of being able (within limits) to select the wavelength band by the filter used and the total output power can be high, the usable power within the activation spectral band of the photosensitizer is much lower and it is difficult to couple the light efficiently into even large-diameter fiberoptic bundles or liquid light guides. Their utility to date has been primarily situations where large surface areas of tissue are to be treated. So-called intense pulsed lights (IPL), with pulse lengths in the 1–100 ms range and based on flashlamp technology are found in dermatology for various applications and have been used in clinical trials for ALA-PDT treatments in the skin [33]. They are reported to cause less pain than CW activation for the same clinical efficacy, perhaps due to the optical pulse length being shorter than the time required for nerve stimulation.

Note that all the light sources above, except for IPLs, operate either in CW mode or, e.g., in the case of the KTP-pumped dye laser or pulsed flash lamp, quasi-CW mode. Although pulsed, the pulse length is long enough and the peak pulse power density is low enough that there is no significant nonlinear absorption by the photosensitizer. Two-photon PDT, which will be discussed below, requires short pulse lasers, e.g., Ti:sapphire operating in the 100 fs/80 MHz regime, that are still large-frame, bench-top devices.

There are other possible ways to generate the light for PDT treatments that are more speculative. For example, chemiluminescence may be able to generate enough light of the relevant wavelength for some purposes: most probably this will be limited to situations in which very large surface areas are to be treated, such that the volume-to-surface ratio of the chemiluminescent material is low, since the total number of photons generated is proportional to the volume. Nevertheless, chemiluminescence "patches" have been reported for treating skin lesions on the face in patients. An extension of this concept is to use bioluminescence as the photoactivation source. *Prima facie*, this would appear to generate an impossibly low light flux, considering the total energy densities usually needed in PDT. Nevertheless, A. MacRobert and colleagues [34] have been able to show cell kill *in vitro* using this approach, but it is important to note that activation of the photosensitizer is believed to be through direct resonant energy transfer from the bioluminescent interaction (BRET), rather than by bioluminescence emission of light photons followed by their re-absorption. This requires that the bioluminescence occurs in very close proximity to the photosensitizer molecule, given the $1/r^6$ dependence of resonant energy transfer.

25.5.2 Light delivery and distribution

The transfer of the light from the source to required body site and "shaping" this to conform to the target volume is an important challenge in effective PDT treatment. There are several classes of treatment "geometry," as illustrated in Figure 25.9 [7]. The first of these, surface irradiation, is used in treating lesions on the skin. The surface of other accessible tissues, such as the oral cavity may also be illuminated in this way, as can hollow organs if the light is delivered endoscopically (see below). The light "dose" or dose rates are typically cited in terms of $\mathrm{J\,cm^{-2}}$ or $\mathrm{mW\,cm^{-2}}$, respectively, these being measures of the incident delivered light.

The actual fluence rate at the surface may be higher than the incident value, by up to about 5-fold, due to light that is backscattered from the deeper regions of the tissue. This depends on the absorption and elastic scattering properties of the tissue (primarily the transport albedo, $a' = \mu_s'/(\mu_a + \mu_s')$, where μ_a is the absorption coefficient and μ_s' is the transport scattering coefficient of the tissue at the treatment wavelength). Beyond an initial "build-up" region, the exponential fall-off in fluence (rate) with depth in the tissue is determined by μ_a and μ_s', and the effective penetration depth ($\delta_{\text{eff}} = 1/\mu_{\text{eff}}$, where the effective attenuation coefficient $\mu_{\text{eff}} \sim \sqrt{3\mu_a\mu_s'}$) is a useful measure of this. δ_{eff} is strongly wavelength and tissue dependent, generally increasing in the red and near-infrared regions. There is a large literature on tissue optical properties and their measurement and on the corresponding spatial distribution of light in tissue, both theoretical and experimental [7, 35]. Suffice it to say here that in practical terms and for most applications to date,

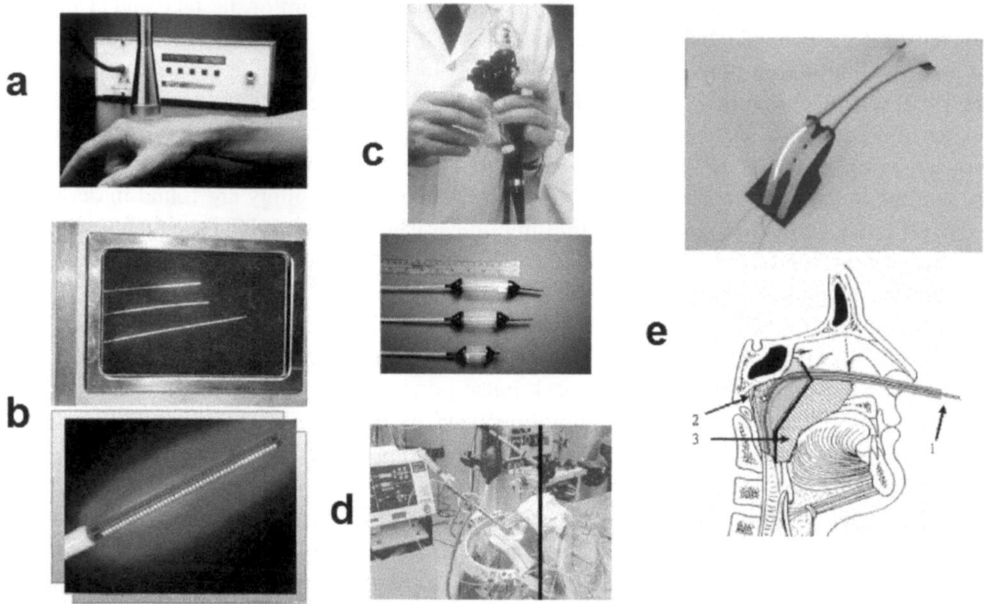

FIGURE 25.9: Methods of light delivery/tailoring for PDT. a) surface irradiation with lamp source. b) diffusing fibers (top) and LED linear array (bottom) for interstitial irradiation. c) endoscopic irradiation, with diffusing fiber (top) and balloon irradiators (bottom). D) intracavitary irradiation postsurgery in the brain. e) applicator for irradiation of the nasopharynx (from Nyst et al. [39], with permission).

effective penetration depths of a few mm are typical in soft tissues. The resulting effective treatment depth, e.g., the depth of necrosis induced by PDT, d_n, is usually a few times the δ_{eff} value at the treatment wavelength, depending on the photosensitizer, its concentration, and the tissue sensitivity to PDT damage. Achieving greater treatment depth by increasing the light is ultimately limited by the maximum incident intensity that can be used without causing tissue heating (less than $\sim$200 mWcm^{-2}) and by the practical treatment time.

Surface irradiation of the retinal layers of the eye for PDT of macular degeneration is a special case of surface illumination, in which the light is applied through the pupil, usually by coupling a diode laser into a fundus camera that images the retina and allows the light beam to be directed by the operator to the area of abnormal vasculature.

The second type of treatment geometry is when the light is delivered to below the tissue surface (interstitial treatment), most commonly by inserting one or more optical fibers designed specifically for this purpose. The fiber tips may be modified to scatter the light along some length (up to about 10 cm), so that effectively a cylinder of tissue is irradiated. This is achieved by stripping the fiber cladding and coating it with a light-scattering material [36] or, in a new approach, by writing a Bragg grating into the fiber [37]. The linear LED arrays referred to above serve the same function as these line diffusing fibers, although the power per cm that can be used is much smaller. Typically, interstitial treatments are specified in terms of mW$\cdot$cm^{-1} or J$\cdot$cm^{-1} of fiber length (or, in the case of very short or point diffusers, mW and J). Heating of the tissue due the light absorption limits the power delivery to about 200 mW or 200 mW per cm of diffusing tip. For both point and cylindrical diffusers, the fall-off in fluence (rate) with radial distance from the fiber is faster than with wide-area surface irradiation for the same tissue optical properties: e.g., for a point source in tissue the fluence falls off approximately as $(1/r)\exp(-r/\delta_{\text{eff}})$, where the additional $(1/r)$ term arises due to diffusion of the multiply-scattered photons. Again, the spherical or cylindrical radius of necrosis is typically a few times larger than the effective penetration depth. The total treatment volume may be increased substantially by placing multiple interstitial fibers spaced throughout the target tissue, as is done for example in treating the whole prostate gland [29]. Interstitial treatments may also be used within hollow organs by placing the fibers through the instrument channel of an endoscope (normally used for taking tissue biopsies). This represents the third geometry, where either surface or interstitial treatment can be carried out within hollow organs Clearly, this is best achieved with a diode laser as the light source. (Arrays of long diffusing fibers have been proposed for the purpose of irradiating tissue surfaces, in the form of a light "mesh" or "blanket," although these have not yet been reported clinically.)

Finally, within either hollow organs such as the esophagus or bladder or within tissue cavities following surgical resection (e.g., of a tumor mass), the light may be spread over the whole inner surface of the cavity. This may be done either by filling the cavity with a liquid that is highly light scattering (usually Intralipid®, a lipoprotein colloidal suspension used as a nutritional fluid) and/or by using a balloon. An optical fiber (or LED source) is then placed near the center of the volume to give uniform illumination. Since the cavity acts as a (low efficiency) integrating sphere, the total (primary + scattered) light fluence rate at the surface can be several times the primary intensity: factors of up to 7-fold have been reported, for example, in red-light PDT of the whole bladder [38], and this can significantly alter the treatment dosimetry.

Finally, a number of light distributors have been developed for specialized applications, such as treatment of tumors that have a complex shape and are difficult to reach: one example for treating in the nasopharynx is shown in Figure 25.9e [39].

25.5.3 Dose monitoring

Most clinical procedures for PDT define the "doses" simply in terms of the administered quantity of photosensitizer (e.g., mg per kg body weight) and the delivered light (J, $J \cdot cm^{-1}$ of fiber length or $J \cdot cm^{-2}$ of tissue surface area). The values used for any particular condition are usually derived from dose-ranging clinical trials, in which the optimum parameters are obtained for some best benefit-to-risk ratio averaged over the patient cohorts. While this can be satisfactory, it does not take into account the patient-to-patient or target lesion-to-lesion variability due differences in i) photosensitizer pharmacokinetics and uptake/microdistribution in the target tissue, ii) tissue optical properties that affect the spatial distribution of the light, iii) tissue oxygenation (for Type II photochemistry), and iv) different intrinsic photosensitivity of either the target tissue or of adjacent tissues that may also be in the light treatment field. Factors (i)–(iii) impinge directly on the local generation of a singlet oxygen, while the last factor may result in a different therapeutic outcome even for the same photophysical dose. The variation in these factors can be substantial, and this is the primary motivation for the development of PDT dosimetry techniques and technologies. Failure to measure some or all of these factors can confound the interpretation of preclinical research findings, especially for *in vivo* animal models, and can result in empirical clinical trials requiring large numbers of patients to reach statistical significance. Furthermore, these factors can interact in a spatially- and temporally-dependent manner, even during the treatment: e.g., the concentration of photosensitizer in the tissue can increase the optical absorption and so reduce the light penetration and, in turn, the photosensitizer can photobleach, while a high drug-light product can deplete the tissue oxygen as it is converted to $^1O_2$.

The challenges and potential approaches to PDT dosimetry have recently been discussed in depth [7], so that here we will simply summarize the main approaches and the current status. In so-called explicit dosimetry, the individual physical dose factors – light, photosensitizer, tissue oxygen – are measured separately. This usually involves sampling these factors at a limited number of positions in the tissue, or a volume-averaged value is used. In the simplest case, one or more photodetectors (e.g., photodiodes) may be placed on the tissue surface to monitor the incident light intensity. This

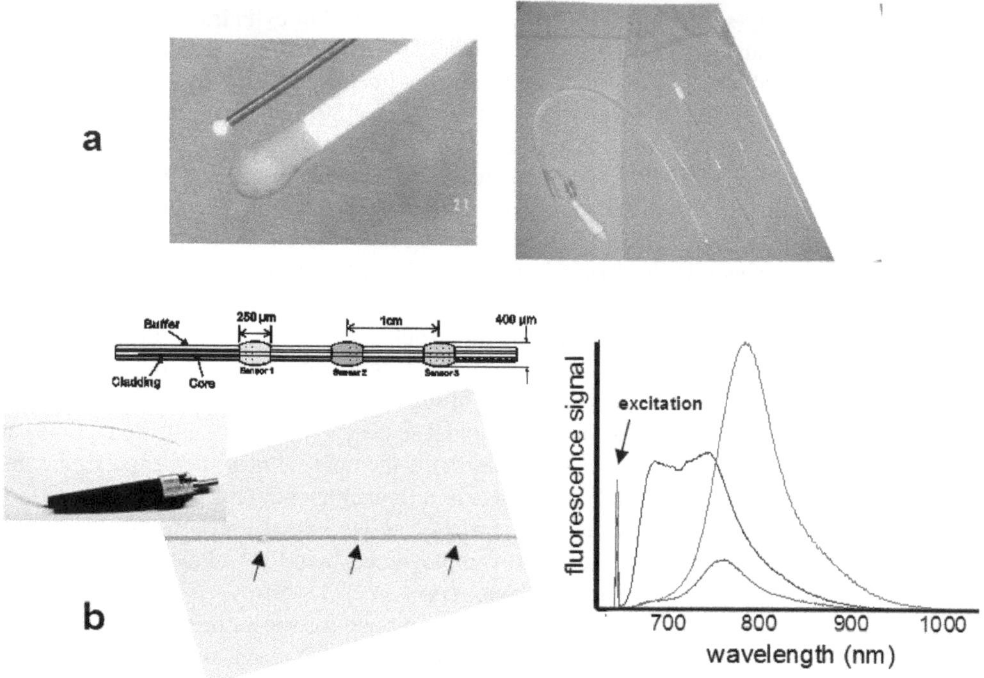

FIGURE 25.10: Optical fiber detector probes for localized measurements of the light fluence, Φ. a) based on light-scattering coating. b) based on multiple point fluorophores [41]: the emission spectra, $F_i(\lambda)$, from each fluorophore shown on the right are used as the basis spectra to deconvolve the fluence values from the measured fluorescence spectrum, $F_{meas}(\lambda) = C.\Sigma\Phi_i.F_i(\lambda)$, where C is a calibration constant (courtesy Dr. L. Lilge, Toronto).

has been used particularly in irradiating a large complex surface such as the abdominal or thoracic cavity, where the light source needs to be moved around in order to minimize shadowing effects [40].

More commonly, local light measurements can be made using an optical fiber probe placed on or inserted into the tissue and connected to a calibrated photodetector such as a photodiode. Since the local light fluence (Jcm^{-2}) is the most relevant metric, these detectors should have an isotropic response, i.e., the signal should be independent of the direction of the light. This can be achieved by making a small diffusing tip on the end of the fiber that randomizes the light, a known fraction of which is then detected. A probe of this type is illustrated in Figure 25.10a. In an alternative device that is still under development [41], a material that fluoresces upon excitation at the PDT treatment wavelength is incorporated into a small length of the fiber after local stripping of the cladding. This allows measurements at several points simultaneously by using materials of different fluorescence emission spectra: the measured spectrum is then de-convolved to separate the signals from each point (Figure 25.10b). With either device, point interstitial fluence-rate measurements can be made continuously during treatment, and the integrated light fluence calculated to account for changes due to the applied light or the tissue optical properties, e.g., from altered blood flow or photosensitizer photobleaching.

If these measurements are made at known distances from the light source (e.g., below the irradiated surface or at a radial distance from an interstitial source), then it is possible to estimate the tissue optical properties (absorption and transport scattering coefficients) and, thereby, reconstruct the distribution of the light throughout the tissue volume [42]. This can be useful to validate treatment planning (see below) and in the retrospective analysis of the tissue response.

For the photosensitizer, fluorescence and/or diffuse reflectance/transmittance spectroscopy can be used, either at the tissue surface or by interstitial fiber probes. Fluorescence is usually the more sensitive measure, as long as the photosensitizer has reasonable fluorescence quantum yield. The fluorescence can also be imaged from the tissue surface. The major challenge is to convert the fluorescence signal into a quantitative estimate of the photosensitizer concentration (local or volume averaged), because of the attenuation of the excitation and emission light by the tissue absorption and scattering. Two main approaches have been explored to address this problem: either the measurement is made such that the effects of attenuation are minimized, or the optical properties are also measured and used to correct for the attenuation. An example of the former is the use of a very small separation between the optical fiber that delivers the excitation light and the fiber that collects the resulting fluorescence emission. Many such fiber pairs may be combined to give an estimate of the photosensitizer concentration averaged over a tissue volume: analogous interstitial probes have also been reported [43]. For the second quantification technique, measuring the diffuse (multiply scattered) light at the excitation and emission wavelengths at several source-detector separations can yield the optical properties, and Figure 25.11 shows a recently developed device that combines this with the measurement of the fluorescence itself. Fluorescence differential path length spectroscopy, in which in which the fluorescence signal is corrected only for the tissue light absorption, represents an alternative approach to quantifying the photosensitizer concentration in tissue [45].

Measuring the tissue oxygenation is challenging, not only for PDT but also in other applications such as radiotherapy where tumor hypoxia is an important factor influencing the treatment outcome. Point measurements of the oxygen tension (pO_2) can be made using polarigraphic microelectrodes or using fiberoptic probes that incorporate an oxygen-dependent fluorescent material, both of which measure the pO_2 over a very local volume. Alternatively, an indirect estimate of tissue oxygenation can be made using optical diffuse reflectance spectroscopy [44] and detecting the spectral signatures that are characteristic of hemoglobin and oxyhemoglobin, from which the oxygen saturation, SO_2, can be estimated.

Recently, the group of Andersson-Engels and colleagues in Lund, Sweden have reported a single instrument that combines all 3 dose measurements (light, photosensitizer, oxygen) [46], which should facilitate the development of more fully integrated PDT dosimetry measurements in clinical trials and, eventually, in routine clinical use. It will then be possible to ascertain the added value of such individualized dosimetry, which will likely vary with the application: the greatest value will be in situations where a) the target volume is substantial and complex, b) the target is in close proximity to critical normal structures, and c) the PDT treatment is aggressive. The archetypal example to date has been in PDT of prostate cancer, where the whole prostate is the target and damage to the normal rectal wall, urinary sphincter, and erectile nerve bundles must be avoided [29]. The other case where dosimetry has been important is in treatment of early cancers of hollow organs, such as the bladder and esophagus, because of the need to carefully balance tumor destruction and damage to the underlying muscle layer of the organ. Of course, if photosensitizers with very high target-to-non target specificity become available, then, in principle, one could simply "overdose" with light, assuming that oxygen is in excess. This is not the case at present (see Section 25.7).

In the second ("implicit") approach to PDT dosimetry, rather than measuring the separate physical factors, the photobleaching of the photosensitizer may serve as a surrogate combined dose metric. Many photosensitizers are photolabile, showing decreasing fluorescence (and/or absorbance) with increasing light exposure (Figure 25.3). The advantage of this approach is that, in principle, a single measurement can be made (as a function of time and/or position during treatment), and this can be technically quite straightforward utilizing either point fluorescence spectroscopy or fluorescence imaging (as in PDD). The main challenge lies in the interpretation of the photobleaching data. Recent studies have shown that there can be a strong correlation between, say, PDT cell kill *in vitro* and the degree of photobeaching [47]. However, other work [48] suggests that, depending

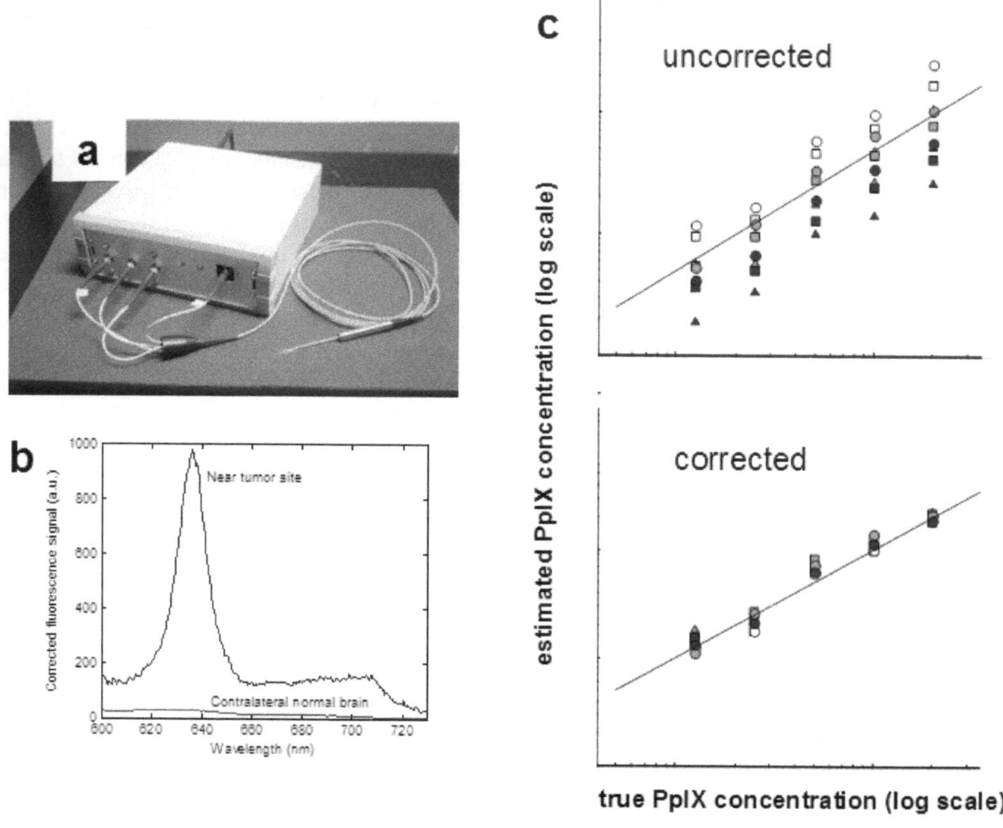

FIGURE 25.11: Fiber-optic probe used for localized measurements of photosensitizer concentration in tissue. a) prototype instrument. b) example of measured fluorescence spectra from ALA-PpIX in brain tumor and normal brain. c) estimated concentration of PpIX in tissue-simulating phantoms of different optical absorption and scattering (denoted by different symbols), before and after correction for light attenuation. Note the large reduction in the variance on the calculated concentration after applying the attenuation correction (courtesy Anthony Kim, Toronto).

on the photosensitizer and the mechanism of photobleaching (e.g., whether or not it is oxygen dependent), the relationship may break down, particularly under low oxygen conditions where other photophysical pathways over and above those shown in Figure 25.2 may occur. More work is needed to understand the details of photobleaching of different photosensitizers before this can be used as a stand-alone dosimetry technique. Nevertheless, it may be very useful to show whether or not there are areas of the target tissues that are relatively undertreated. An alternative approach that derives from the same underlying concept is to measure the appearance of optical signals (absorption or fluorescence) due to generation of photoproducts during light irradiation [18]. While this likely occurs with only a few of the current photosensitizers, the detection scheme can also be quite simple and inexpensive. Again, as with photobleaching, a challenge is to relate the photoproduct signal to the effective PDT dose to the tissue.

Consider again the Jablonski diagram of Figure 25.2. Since singlet oxygen is the putative "active agent" in PDT, it would be an advantage to measure this directly and this is the approach used in singlet oxygen luminescence dosimetry (SOLD), based on the detection of the near-infrared ($\sim$1270 nm) radiative decay of $^1O_2$ back to ground-state oxygen: see reference [49] for an extensive review of this technique. Unfortunately, because $^1O_2$ reacts so readily with biomolecules (leading to the therapeutic effects), the probability of radiative decay in cells and tissues is very low ($\sim$1:10^6-10^8) and the lifetime is less than 1 μs in most studies. This results in a very low photon budget in any detection scheme. Nevertheless, the technique has been successfully demonstrated in several studies, both in cells *in vitro* and in animal models *in vivo*, using a pulsed laser source for excitation and a cooled photomultiplier tube with extended-near infrared sensitivity operated in time-correlated single photon counting mode. Good correlations between the integrated $^1O_2$ level during treatment with different PDT responses have confirmed the central role of $^1O_2$ in PDT and the potential for SOLD to be used either a) for PDT dosimetry, possibly with predictive capability, and/or b) as a gold standard against which other explicit or implicit dosimetry techniques may be tested/calibrated. Since this instrumentation is fairly expensive and cumbersome, there has been work to use a simpler and cheaper pulsed diode laser as the SOLD activation source [50], which also has the advantage that more wavelengths are available to match the PDT treatment wavelengths so that the $^1O_2$ measurement is integrated over the corresponding treatment depth. Studies of the relative sensitivity of the two forms of instrument have not been reported. With either approach, it is critical to make measurements at other wavelengths around the 1270 $^1O_2$ emission peak in order to subtract the background fluorescence/phosphorescence from the photosensitizer and tissue. Examples of SOLD are shown in Figure 25.12 to illustrate the technologies and the type of results they produce: this includes an example of singlet oxygen imaging or mapping in which the spatial distribution of the luminescence signal is measured by scanning the laser beam. The recent development of near-infrared doped InGaAs cameras with high quantum yield at 1270 nm may be a breakthrough for $^1O_2$ imaging.

Other groups have tried to circumvent the SOLD technique by using $^1O_2$-sensitive chemiluminescence reporter molecules that give high optical signal [59]: the limitation is that the delivery to the target may be a challenge *in vivo*. Finally, it should be noted that there are reports of modeling and experimental studies in tumors that the $^1O_2$ signal, at least as averaged over a tumor tissue volume, may not correlate with the final biological outcome [51]. This may be attributable to secondary effects, in particular the immune response that, while triggered by the $^1O_2$-induced damage, is not directly dependent on the overall level of singlet oxygen generated throughout the tumor volume.

25.5.4 PDT response modeling

Having knowledge of the separate physical dose parameters from one or more of the above explicit dosimetry methods allows the estimation of the effective PDT dose that, ideally, is predictive of the treatment outcome. Most work on this to date has been based on the so-called PDT Threshold model [52] which comes from the common observation that the boundary of PDT effect, e.g., necrosis, is often sharp and distinct, even although the light distribution decreases continuously with distance from the light source. The simplest explanation is that a minimum local concentration of singlet oxygen is required to produce a given biological effect. The threshold dose can then be rigorously defined in terms of this $^1O_2$ level which, in turn, can be calculated from the measured local light fluence, photosensitizer concentration and oxygen levels and is a measure of the intrinsic photodynamic sensitivity of the tissue. The threshold model has been validated for a number of different photosensitizers in different tissues, mainly using tissue necrosis as the biological endpoint. There is some evidence that it applies also to vascularly-targeted PDT. However, other endpoints may not show this threshold behavior: for example, studies of the distribution of PDT-induced apoptosis in tissue are best explained by a stochastic model, whereby the probability of apoptosis falls off continuously and proportionally with the singlet oxygen concentration rather than being constant over some range and then falling sharply to zero [53].

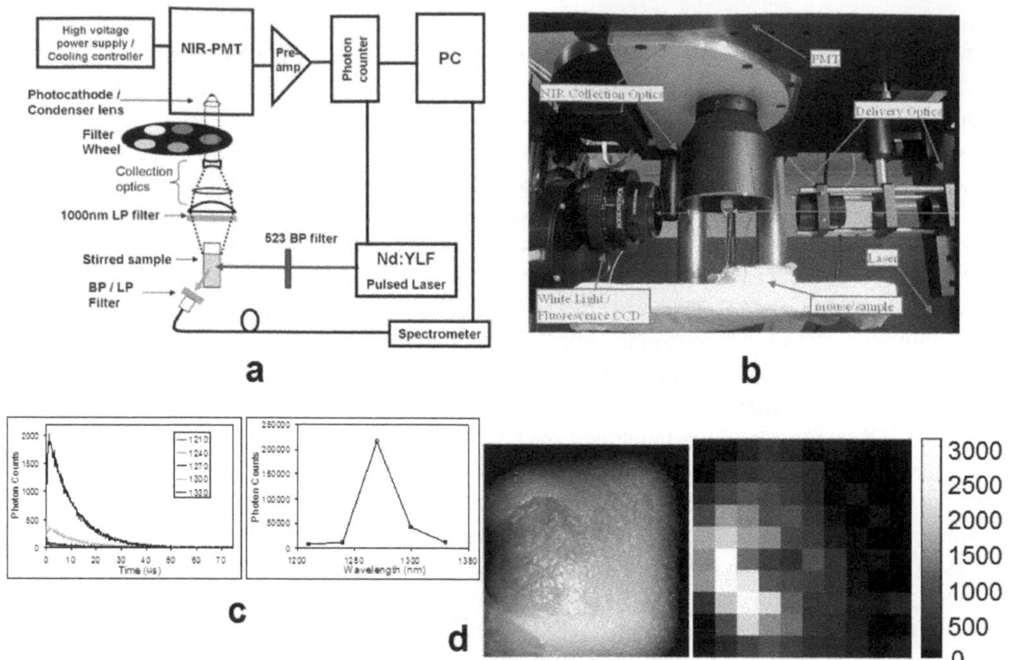

FIGURE 25.12: (See in color after page 572.) Singlet oxygen luminescence dosimetry. a) system diagram, as set up for measurements in a cuvette. b) photograph of the system, as set up for combined $^1O_2$ luminescence mapping and fluorescence imaging. c) time-resolved luminescence signal following a short laser pulse, with detection at different wavelengths (left) and resulting time-integrated signal as a function of wavelength (right). d) $^1O_2$ luminescence map of tumor in an animal model: the field-of-view is approximately the same in the white-light (left) and luminescence (right) images.

25.5.5 PDT biological response monitoring

Independent of whether or not some form of physical dosimetry is used to monitor treatments and possibly optimize this for individual patients, the assessment of the biological response to PDT is important. This may be considered from two different perspectives: a) to assess the effectiveness of treatment as part of patient follow-up, which may then be used to decide on further treatment options, including repeat PDT, and b) to make adjustments during the PDT treatment itself, also for optimization. These two strategies pose different scientific and practical challenges and may involve very different monitoring techniques and instrumentation. Here, we will concentrate on options for the latter, where optical techniques may play a particularly central role. Prior to addressing this, it is worth mentioning that posttreatment monitoring has mainly relied on established medical techniques and procedures, in particular various forms of radiological imaging: computed tomography, magnetic resonance imaging, ultrasound, and radionuclide imaging (including positron emission tomography). Some of these techniques provide more information on the cellular responses of the tissues, while others report changes in vascularity or blood flow. The choice of technique clearly then depends on the anticipated form of PDT damage. Some clinical applications are, at least currently, not amenable to imaging: for example the effectiveness of anti-infective PDT treatment is usually assessed by swabbing the area and performing microbiological assay.

Intratreatment monitoring of response is a very interesting challenge: for maximum utility, one would like to be able to assess the effects of treatment throughout the target tissue (for efficacy) and

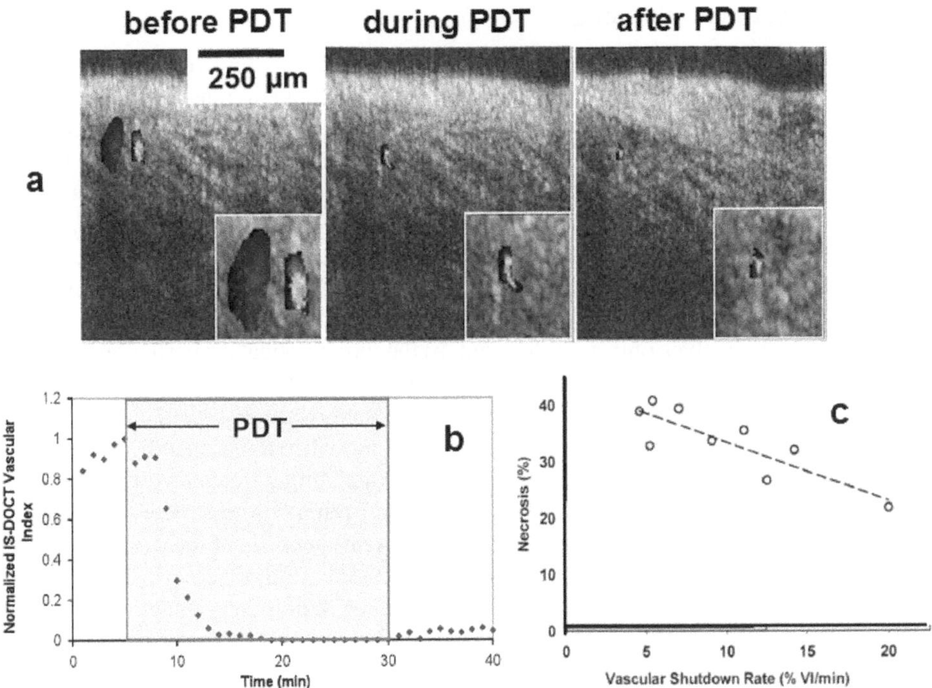

FIGURE 25.13: (See in color after page 572.) Example of tissue response monitoring during PDT, using DOCT. a) gray-scale (tissue structure) and superimposed Doppler blood flow (color scale) images in a prostate tumor model, showing shut down of an artery-vein pair, with the latter also shown on an expanded scale in each image. b) representative response curve, showing a measure of the microvascular volume plotted as a function of time. c) inverse correlation between PDT-induced tumor tissue necrosis and the rate of vascular shut-down during treatment. (Courtesy Dr. B. Standish, Toronto).

adjacent tissues (for safety) as the light is being applied, so that adjustments could be made "on line." This requires that the relevant biological response(s) occur on this short time scale. Evidence that his may be possible, at least in terms of the vascular response of tissues, comes a) indirectly from changes that are seen in the light distribution in tissue that are likely due to changing blood content and/or oxygenation, and b) directly from monitoring blood flow. An example of (a) is the use of diffuse correlation spectroscopy by Yodh and colleagues [54], while direct vascular responses have been demonstrated using Doppler optical tomography [55], as shown in Figure 25.13. In both cases, it is important to note that significant changes were seen while only part way through the treatment and that these could be quantified and were correlated with an independent, posttreatment biological endpoint (e.g., tissue necrosis). This gives hope that a predictive algorithm could be established that would allow feedback to adjust the light, or possibly also the photosensitizer delivery and/or tissue oxygenation. Optical techniques such as DCS or DOCT are suitable for clinical use. There are a variety of other optical methods, both imaging and spectroscopic, used in preclinical studies to facilitate direct measurement of cellular, vascular, or molecular changes due to PDT, including bioluminescence imaging [26,56], intravital confocal fluorescence microscopy [57] and hyperspectral imaging [58]. Photoacoustic imaging [59] could also be used for PDT response monitoring.

25.5.6 PDT treatment planning

Complementary to dosimetric measurements in PDT, in the last few years several groups have developed approaches to pretreatment planning applied to PDT, analogous to that used routinely in radiation therapy, particularly brachytherapy with implanted radioactive sources: see reference [60] and references therein. The concept is most applicable to treatment of complex 3D tissue targets situated where complete 1'coverage" is required while minimizing collateral (i.e., unintended) damage to adjacent tissues. To date this has been applied mainly to whole-prostate PDT, although the concept is widely applicable, particularly for tumors in solid organs such as the head & neck and pancreas. PDT treatment planning is intrinsically more challenging than for radiotherapy, since a) the photosensitizer uptake/distribution in the tissue plays a role in addition to the light distribution, whereas radiotherapy involves only a single agent, b) the tissue optical properties and resulting light distribution are much more heterogeneous and variable from tissue to tissue than for ionizing radiation, c) PDT is a much more dynamic process, as evidenced by the observed changes in the tissue during treatment, and d) PDT is usually a single-session modality rather than involving many fractions as in radiotherapy. Figure 25.14 shows an example in planning treatment of the whole prostate gland. This has proven useful not only for planning and optimizing treatment prospectively (pretreatment planning) but also for retrospective (posttreatment) analysis of the outcome of treatment, for example correlating the pretreatment plans with the distribution of necrotic or avascular tissues assessed at different times after treatment by MRI or other radiological imaging techniques [29]. This allows continuous refinement of the treatment planning, e.g., by updating the tissue optical properties or PDT threshold doses [42].

25.6 PDD Technologies

Photodynamic diagnostics refers primarily to the use of fluorescence imaging of PDT photosensitizers for the purpose of disease detection or localization. As indicated in Section 1, the early development of photosensitizers was in fact driven in part by the need for better fluorescence diagnostic agents. Since, for most PDT compounds the fluorescence yield is reasonably high (a few percent), the fluorescence signal is strong and relatively easy to detect spectroscopically or by imaging (Figures 25.3 and 25.4). In the 1980s, technology became available that allowed fluorescence images to be made during endoscopy, by adding short-wavelength light source and a sensitive camera (e.g., intensified CCD) to detect the fluorescent emission from the tissue after filtering the diffusely reflected excitation light. This was initially used mainly in the bronchus, with the primary motivation being to detect the levels of hematoporphyrin derivative in the tumor tissue to be targeted for PDT treatment. It became clear, however, that this also allowed detection of tumors and premalignant lesions in the lung, i.e., it could be used as a diagnostic technique per se. Studies were performed by assess the lowest dose of photosensitizer that could be used for this purpose (so as to minimize skin photosensitivity) and it was found that, at least in the case of premalignant lesions in the lung, better diagnostic accuracy was obtained without any photosensitizer, i.e., using the natural endogenous (auto)fluorescence of the tissue. This drove the development of autofluorescence bronchoscopy as a technology in its own right, independent of PDT, leading to its approval as a clinical technique. Subsequently, the approach was used in other organs, such as the gastrointestinal tract, oral cavity, cervix, and bladder [3,5,61].

Figure 25.15 shows examples of autofluorescence endoscopic equipment and images. Autofluorescence endoscopy is now combined with standard white-light endoscopy and other methods such a narrow-band imaging (in which the separate red-green-blue diffuse reflectance images are captured) in multimodality instruments [62]. Many studies have been done to understand the changes

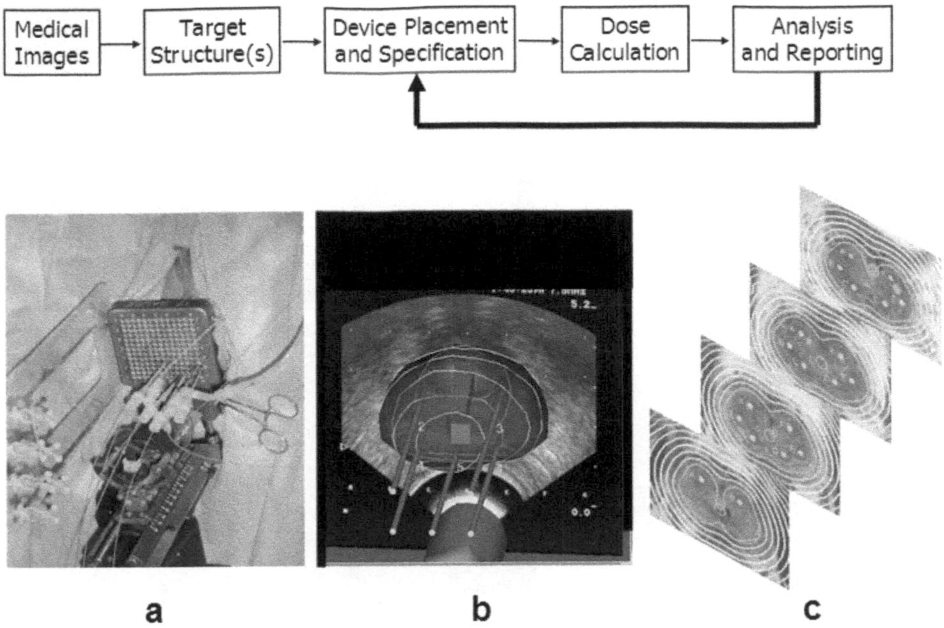

FIGURE 25.14: Example of PDT treatment of the prostate, using multiple interstitial cyilindrical diffusing fibers [60]. a) set up, with the source (and detector) fibers placed through a template. b) 3D representation of the whole-prostate target superimposed on a transrectal ultrasound image. c) treatment plan showing several transverse sections through the prostate and the iso-fluence contour lines: 2 cm long fibers, 360 J/cm; tissue optical properties μ_a= 0.4 cm^{-1}, μ_s' = 3.25 cm^{-1}; contours 512, 256, 128, 64, 32, 16, 8, 4, 2, 1, 0.5, 0.25, 0.125, and 0.0625 J/cm^{-2}. The flow chart above indicates the typical sequence of steps in PDT treatment planning.

in tissue that underlie the contrast obtained in autofluorescence imaging of different organs [63].

With the development of ALA-PpIX as a photosensitizer, PpIX fluorescence has been widely investigated for tumor detection/localization, for example in bladder cancer as illustrated in Figure 25.4b. Whether autofluorescence or photosensitizer-contrast imaging provides higher diagnostic accuracy probably depends on the organ being examined and on the pathological stage.

A recent advance in fluorescence imaging *in vivo* is the development of molecular beacons (MBs) [64]. The principle, shown in Figure 25.16, is to link the fluorescent molecule to a second molecule that quenches (suppresses) the fluorescence, for example due to Föster resonance energy transfer (FRET), in which the energy of the excited singlet state is transferred to the quencher (which needs to have spectral overlap with the fluorophore), before it can de-excite by fluorescence emission. The linker can be of several types, including enzyme-cleavable sequences or "hairpin" nucleic acid loops that open upon hybridization with target-specific RNA. Thus, the fluorophore is "silent" until it reaches and interacts with the target cells. MBs are commercially available that target a variety of tissue characteristics/pathologies, both cancer and nononcologic, and are now entering clinical trials. As will be discussed below, the same principle can be applied to quench the photodynamic action to increase the target specificity.

A different major application of PDD is the use of fluorescence image-guided surgery. This has been established, for example, using the near-infrared dye indocyanine green (which is a weak photosensitizer) to assess the spread of a tumor through the lymphatic system [65]. The other development, which arises directly from the use of PDT for brain tumor treatment, is the use of

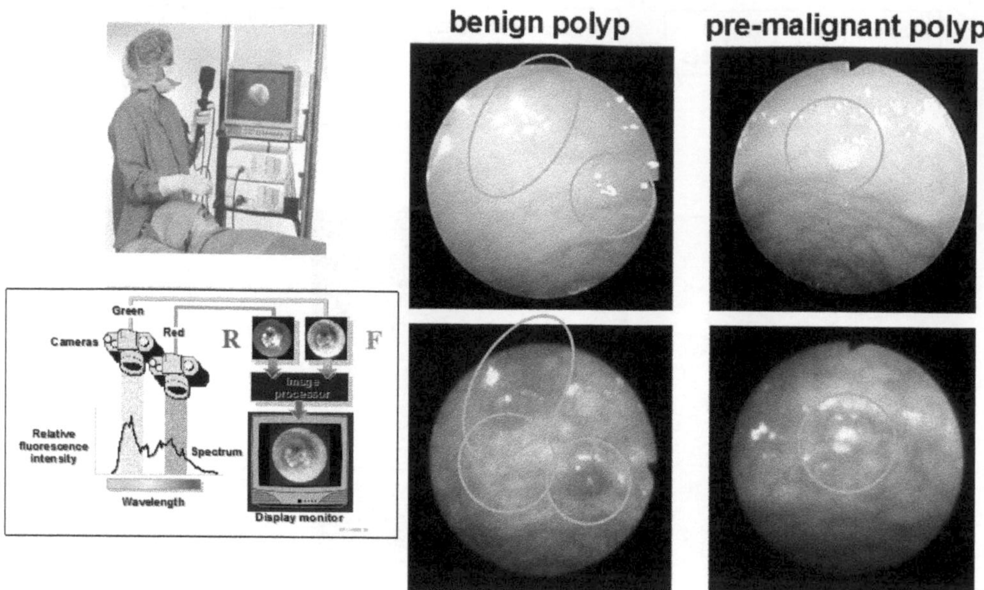

FIGURE 25.15: (See in color after page 572.) Autofluorescence endoscopic imaging. a) instrument and principle of operation based on detection of red and green fluorescence emission (courtesy Xillix Tech., Canada). b) example of white-light and autofluorescence images of polyps in the colon: the benign polyp shows similar fluorescence as the adjacent normal tissue, while the premalignant polyp (adenoma) has enhanced red-to-green fluorescence (courtesy Dr. N. Marcon, Toronto).

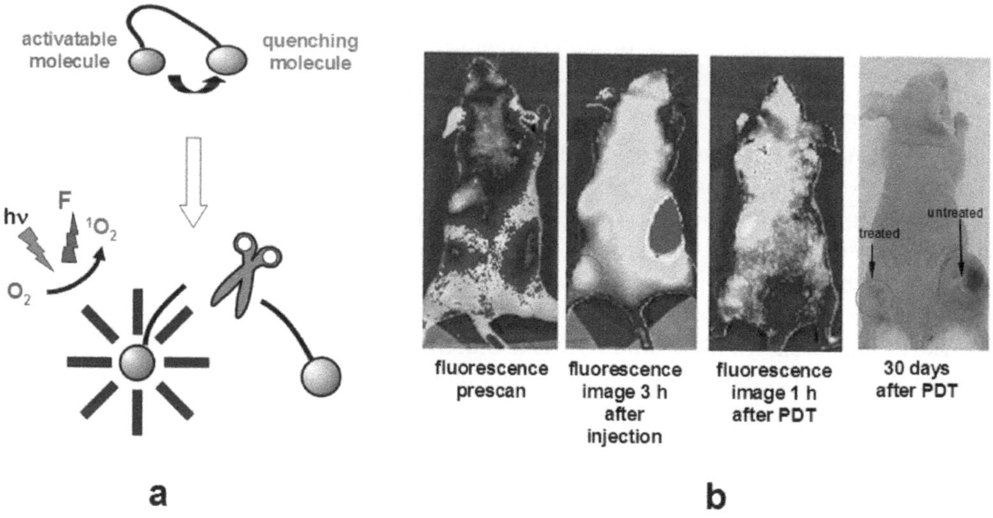

FIGURE 25.16: Molecular beacons. a) the principle of fluorescence detection or singlet oxygen generation after unquenching (in this case by enzyme cleavage). b) an example of a tumor-specific enzyme-cleavable beacon in a mouse model bearing tumors on each hind leg, with one treated by PDT. Uptake and quenching of the beacon is visible in the fluorescence images (note that there is also a high background signal from some of the other organs here). The untreated tumor continues to grow, while the treated tumor is not visible (courtesy Dr. G. Zheng, Toronto).

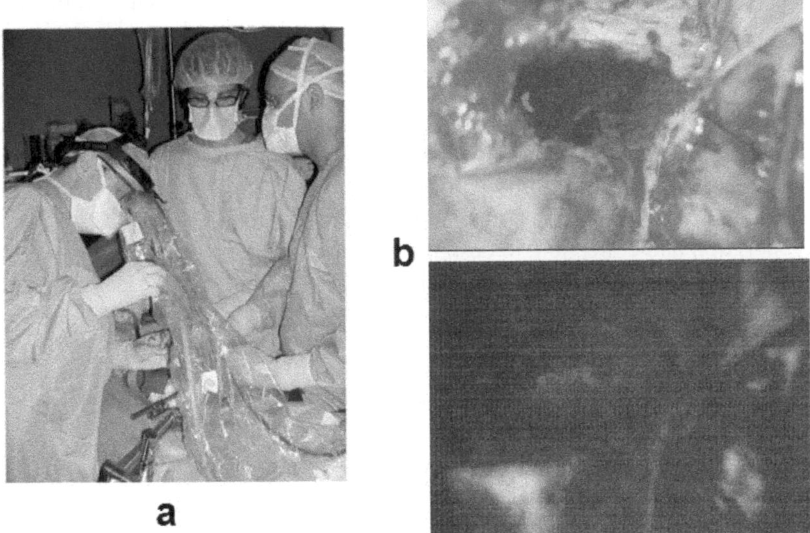

FIGURE 25.17: Fluorescence image-guided tumor resection. a) prototype camera system developed in the author's lab, in use during prostate surgery [67], b) image of tumor and surrounding brain tissue during malignant brain tumor resection using Photofrin to provide the fluorescence contrast, taken using an early prototype camera [66]. In (b) the upper white-light image was taken after completion of standard surgical resection, and no tumor is visible: fluorescence is seen in the lower image and was confirmed by biopsy to be residual tumor (courtesy Dr. P. Muller, Toronto).

fluorescence imaging to identify residual tumor at the end of standard resection [66]. An example of a camera system designed for this purpose is shown in Figure 25.17a. As seen in Figure 25.17b, following apparent complete resection of malignant gliomas, i.e., where no tumor is seen by normal white-light visualization, there is nevertheless a high probability of a residual tumor, since most of these patients go on to have local recurrences. The residual tumor tissue can, however, be visualized by the photosensitizer fluorescence. This was initially demonstrated in a rat glioma model and subsequently in patients undergoing Photofrin-PDT [66]. Most work was subsequently done with ALA-PpIX, and randomized clinical trials by W. Stummer and colleagues in Germany has shown that a significant increase in survival can be achieved compared to standard resection procedures [6]. In terms of hardware for the imaging, 2 main approaches have been taken. Yang and coworkers have developed "free-standing" systems that incorporate an excitation light and sensitive camera ($\pm$ point fluorescence spectroscopy) as a single device: several generations have been reported, becoming more compact over time [66,67], although with some trade-off in performance. The German group has instead integrated fluorescence imaging into existing intra-operative microscopes [6]. The advantage of the free-standing system is that it can be used independently of whether or not an intra-operative microscope system is available, since this restricts the integrated system to "high end" applications such as neurosurgery. On the other hand, the integrated systems provide maximum user convenience.

There are several strategies under investigation to improve fluorescence image-guided resection, including: algorithms and ancillary devices to make the fluorescent signal less dependent on the tissue optical properties, background autofluorescence, and imaging geometry [68]; techniques to make the images quantitative and so measure the concentration of the fluorophore in the tissues, which may improve the reliability and reproducibility of the technique, and; coregistering the fluorescence images with pre- and intraoperative volumetric MRI data to help localize the tumor target and critical normal brain structures. Further advances in photodynamic diagnostics, such as novel targeted fluorescence contrast agents, may apply also to surgical guidance. It should be noted that there are several alternative, or possibly complementary, techniques for optical guidance in surgery, including: point spectroscopies based on tissue autofluorescence (CW and time-resolved), near-infrared Raman or nonlinear light-tissue interactions (coherent anti-Stokes Raman and second-harmonic generation), and high-resolution imaging by optical coherence tomography or intravital confocal microscopy/ microendoscopy. Each of these involves major challenges to optimize the optical signal generation and collection, to design and fabricate ergonomic and cost-effective clinical instruments, and to perform preclinical and clinical validation studies.

25.7 Novel Directions in PDT

There are numerous developments underway or conceived that depart in fundamental ways from the established current practice, in terms of the photophysics or photochemistry that are exploited, the photobiological mechanisms and objectives, or the clinical and nonclinical applications. This is a rapidly growing subject, so here we will summarize only some major topics: for more extensive information, the reader is referred to the recent book of Hamblin and Mroz [69] and references therein.

25.7.1 Photophysics-based developments

Referring to Figure 25.2, the standard PDT effect is based on the absorption by the photosensitizer molecule of a single photon of sufficient energy (short enough wavelength) to raise the molecule to the S_1 excited state. Alternatively, 2 photons of half the energy (twice the wavelength) can be absorbed simultaneously. However, the probability of this 2-γ absorption is very low, so that the instantaneous power density must be very high ($\sim 10^{10}$ Wcm^{-2}), but at the same time the total energy must be limited in order to avoid photothermal effects. Hence, the pulse length must be very short ($<\sim$500 fs). This is now technologically feasible in the laboratory with Ti:sapphire lasers and will likely become clinically practical with the rapid development of fiber lasers. Even with such pulses, the 2-γ cross-section of conventional PDT compounds is low, so that molecules that are specifically designed to have high cross-sections are required [70]. (It is of interest to note that quantum dot nanoparticles have extremely high 2γ cross sections, although toxicity remains a concern for clinical applications.)

The motivation for this approach is twofold. First, the use of long wavelength light (typically ~800–950 nm) gives increased penetration in tissue. This could translate into higher effective treatment depth, although this may be countered by the fact that the probability of 2-γ absorption is proportional to the square of the light intensity (I^2) or fluence rate at any point, and this falls off with depth more rapidly than does the fluence rate itself. Nevertheless, positive findings have been reported of this increased treatment depth *in vivo* [71]. The second potential advantage comes from the I^2 dependence itself: if the beam is tightly focused, the PDT effect can be confined to a diffraction-limited spot at depth in the tissue, thereby avoiding activation of photosensitizer above or below this depth (see Figure 25.18b), as long as the 1-γ absorption coefficient at the longer wavelength is negligible. This has been demonstrated both in cells and to target single blood vessels *in vivo* [72,73], as illustrated in Figure 25.18c. Potential applications of this point-by-point PDT targeting include various vascular lesions such as AMD and port-wine stain, and melanoma (which is difficult to treat by conventional PDT because of the poor light penetration due to melanin ab-

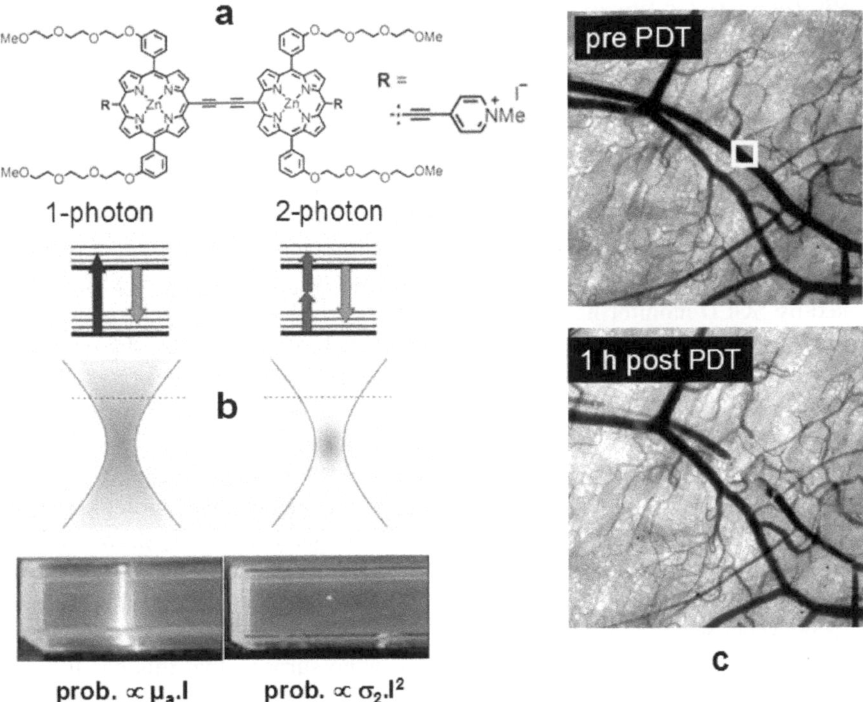

FIGURE 25.18: 2-photon PDT. a) photosensitizer based on a conjugated porphyrin dimer [70] designed to have a very high 2-photon cross-section ($\sigma_2 \sim$ 17,000 GM units at 916 nm). b) the principle of spatial confinement of the PDT effect using 2-γ activation: for 1-γ PDT, the probability of absorption by the photosensitizer is proportional to the absorption coefficient times the light intensity, I, while for 2-γ activation, the probability is proportional to I^2. c) example of single-vessel closure by 2-photon PDT, imaged by Doppler optical coherence tomography: the treated area is indicated by the white square (approx 80 x 80 μm).

sorption). While 2-γ is conceptually compelling and the principle has been demonstrated, there are significant scientific and technical challenges to translate this into the clinic, including developing the required technologies (fs laser source and delivery to the tissue, beam scanning and control, and the imaging and monitoring of the target as treatment proceeds) and understanding the biological responses (e.g., of focal treatment of blood vessels or cells/tissues).

A second form of 2-γ PDT is so-called 2-photon/2-color activation in which two independent short laser pulses are used sequentially in order i) to generate the S_1 and hence the T_1 state, and ii) to raise the T_1 state to a higher T_n state, which then interacts with the biological substrate. As demonstrated by R. Redmond and colleagues [74], this has the advantage that cell killing can be achieved even in the absence of oxygen (by non-$^1O_2$ mediated pathways), which would allow effective treatment even under hypoxic conditions such as found in many solid tumors. The challenges again are in the chemistry (to find molecules with good optical absorption characteristics in both the S_0 and T_1 states) and in the laser sources needed to generate successive pulses of the right wavelengths, pulse lengths, and pulse-pulse time intervals: at present such sources are not clinically practicable.

25.7.2 Photosensitizer-based

New photosensitizer-based developments include both novel forms of the active material and techniques for targeted delivery. With respect to the former, there is of course continuous development of "classical" PDT molecules, with variations on the structure to improve the performance. These are generally incremental advances, but are likely to lead to the next generation of clinical compounds. In the medium term, a likely candidate new class of PDT agents is that of photodynamic molecular beacons. As discussed in the context of PDD, just as fluorophores can be quenched and unquenched, so photosensitizers can be attached to quenching moieties, thereby increasing their target specificity, since they cannot be activated until unquenching takes place. This has been demonstrated by SOLD monitoring of the $^1O_2$ generation before and after targeting and also in a number of *in vitro* and *in vivo* model systems (Figure 25.16b), with both enzyme cleavage-based and hairpin nucleic acid sequence unfolding-based beacons [75,76]. However, as with conventional photosensitizers, photodynamic molecular beacons must still be delivered to the target efficiently in order to achieve adequate concentration and potency.

As indicated earlier, most photosensitizers localize in the mitochrondria. There is interest in targeting other organelles, such as the nucleus, in order to trigger other biological pathways and responses, possibly with higher efficiency. For example, Bisland et al. [77] showed that a branched peptide (loligomer) structure could deliver photosensitizers to the DNA, increasing the cell kill for the same total concentration of photosensitizer by 100-1000 fold. Other molecular entities are being explored for the same purpose [78].

With the recent explosion of nanotechnologies, it is no surprise that this has impacted significantly on PDT and PDD, although not yet at the clinical level. The main interest has been to use nanoparticles for molecularly-targeted delivery of the photosensitizer or fluorophore [79]. It is noted that some existing photosensitizers already use nanoparticle delivery, mainly to overcome their poor water solubility: e.g., benzoporphyrin derivative is delivered in liposomal nanoparticles in the clinical formulation for Visudyne-PDT of macular degeneration. In general, the advantages of nanoparticles is that a) they can carry a large "payload" of photosensitizer, b) their surface can be coated to provide good systemic circulation (to reduce sequestration in the liver), and c) they can be decorated with targeting moieties such as antibodies or peptides to bind to specific receptor on the target cells. A recent further development the use of "nanocells." These are compound nanoparticles that comprise an inner core of a polymer nanoparticle conjugated to the photosensitizer molecule and an outer liposome-like structure that encapsulates both this nanoparticle core and a second active agent that is synergistic with the PDT effect [80]. The converse arrangement can also be used, with the photosensitizer and complementary agent switched.

A second possibility for NP-based PDT is to use nanoparticles that themselves have photodynamic properties: this includes quantum dots and porous silicon, both of which are capable of generating $^1O_2$ upon light absorption. Qdots may also be conjugated to a molecular photosensitizer to serve as a FRET pair, with the Qdot absorbing the light and transferring the energy to the photosensitizer. There is no doubt that many variations-on-a-theme for NP-based PDT will be appear in the next few years, as well as completely novel ways to exploit the unique properties of different classes of nanoparticle. These will be both monofunctional, i.e., acting only as PDT materials, and bi- or multifunctional, combining PDT properties and other optical and nonoptical features (e.g., PDT + MRI contrast).

25.7.3 Photobiology-based

PDT has traditionally been delivered primarily as an "acute" modality, with the highest acceptable photosensitizer and light doses given in a single treatment. There is increasing interest in the possibility of fractionating the photosensitizer and/or light delivery. This should be distinguished from light delivery using rapid on-off cycles, which has been investigated as a way to increase the

PDT effect through reducing the effects of photochemical depletion of tissue oxygen, by allowing reperfusion in the dark intervals (typically up to 30 s) [81]. In the extreme version of fractionation, known as metronomic PDT, *m*PDT, both drug and light are delivered continuously at very low rates over an extended period (hours or days). In some cases this is driven by the characteristics of the light sources, as with interstitial LEDs where the power is limited by the self-heating, and with OLEDs where at present the light output is low. However, the real motivation for *m*PDT is to select or enhance specific mechanisms of action. This has been most investigated using ALA-PpIX in brain tumors, where delivering ALA and light very slowly has been shown to produce tumor cell-specific apoptosis without any necrosis of either the tumor or normal brain [82]. This has the advantage of avoiding an induced inflammatory response, thereby minimizing collateral damage. It remains to be seen if this can produce adequate tumor control (cell kill > cell proliferation), and there are technology challenges in both continuous drug administration and in safely delivering light to deep-seated tumor tissues over an extended period. However, preclinical studies have been encouraging [83].

Other strategies to achieve different biological endpoints include, for example, PDT to modulate bone growth [84], where the effects are likely mediated not by cell kill but by stimulating angiogenesis. There is also evidence from preclinical studies that ALA-PDT may be effective against the hyperexcited neuronal cells in epilepsy, which may or may not involve actual cell death [85]. A third example is the use of sublethal doses of PDT to transiently open the blood brain barrier, thereby improving delivery of therapeutic drugs (including PDT sensitizers) or biologics [86]: this is due to a reversible response of the endothelial cells lining the blood vessels, which form very tight junctions in the normal brain tissue.

25.7.4 Applications-based

There are many new applications of PDT being explored to address specific clinical or nonclinical problems, of which the following are examples to illustrate the wide range. In oncology, preclinical work has shown the technical feasibility and potential of PDT to de-bulk metastatic tumors in the vertebrae in order to facilitate injection of plastic into the bone to provide mechanical stability (vertebroplasty): this involves introduction of new mechanical devices to integrate PDT light delivery with the orthopedic vertebroplasty instrumentation [87], and clinical trials are in progress. In infectious diseases, PDT of osteomyelitis (bone infection) [88] and tropical diseases such as leishmaniasis and tuberculosis [89] shows promising efficacy and may provide a very cost-effective solution to these widespread conditions. For surgical applications, photochemical bonding of tissues (e.g., nerves, blood vessels) has seen a recent revival of interest, with studies using the photosensitizer Rose Bengal showing that collagen cross-linking can be achieved that gives very high mechanical strength without the problems associated with thermal bonding techniques (such as laser) [90]. Finally, as a generic approach to improved delivery of drugs or biologics (e.g., viruses) into cells, the group of K. Berg and coworkers have pioneered photochemical internalization (PCI). In this techniques, the photosensitzer is first located in the cell plasma membrane, the active agent of interest is then invaginated into endocytic vesicles that enter the cell, whereupon light irradiation disrupts the vesicle membrane, releasing the active agent [91]. Also generically, there is increasing investigation of PDT used in combination with other treatment modalities, including surgery, chemotherapy, anti-angiogenics, and hyperthermia.

In the preclinical domain, the possibilities to use photochemistry to modify artificial tissue constructs, either structurally and/or functionally, for different applications in tissue engineering have had some initial exploration [92]. One example is that of M. Stoichet and colleagues, where 2-γ PDT was used to create channels for preferential cell migration and growth in a 3D hydrogel matrix [93]. One could envisage using this to create quite complex 3D structures such as artificial blood vessels for implantation and tissue regeneration.

25.8 Conclusions

For the many different applications, PDT offers distinct potential advantages compared with other treatment modalities, including low systemic toxicity, repeatability of treatments, relatively low cost, and lack of induced resistance. The challenges lie in the greater complexity that arises from a) its being a multicomponent technique, involving both drugs, and optical radiation, b) the dynamic interdependence of these factors (and with tissue oxygenation), and c) the multiplicity of photobiological mechanisms. This last is a tremendous advantage for optimizing PDT for different clinical targets, but also complicates the interpretation of PDT biological responses. It has also proven easier for PDT to be widely accepted into clinical practice where the method is technically similar to existing treatments: this is the case, for example, in treating macular degeneration in the eye and in skin lesions, since ophthalmologists and dermatologists, respectively, are very familiar with laser/optical technologies. There is much higher education, training, and acceptance barrier with other medical specialties. Increasingly, as with all new modalities, PDT also faces the challenge of cost containment.

An exciting aspect of PFT/PDD is that the research and development in are intrinsically highly multidisciplinary, involving chemists, optical physicists and engineers, molecular and cell biologists, and different medical specialists. For each discipline there are many existing and emerging challenges to be addressed, both in the individual domains but also in the integration of the different technologies.

In terms of the emerging and future directions, PDT/PDD, as with other modalities, is likely to be accelerated by convergence with other sciences and technologies. Thus: advances in knowledge and techniques in molecular biology are enabling molecular targeting of PDT agents in a disease-specific way; nanotechnologies (particularly nanoparticle constructs) will drive new approaches to photosensitizer delivery and photoactivation processes; changing demographics on a global scale are posing new health problems in which PDT may play an important role, as in the case of infectious diseases and, conversely, these new applications will required new technological solutions, particularly as the use of PDT moves beyond the expert hospital environment into the community clinic or even the home-care setting; and, finally, combining PDT/PDD with other modalities, including but not restricted to those based on photonics, will require levels of scientific and technological integration that have not been used to date in this field.

Acknowledgments

The author wishes to acknowledge the following agencies for support of PDT research: Canadian Cancer Society Research Institute (PDT dosimetry), National Institutes of Health (CA43892: PDT in brain and prostate cancer), Canadian Institute for Photonic Innovations (2-γ PDT), Canadian Institutes of Health Research (PDT molecular beacons and PDT in spinal metastasis), Ontario Institute for Cancer Research and Canadian Breast Cancer Foundation (PDT for spinal metastases).

References

[1] J. Moan and Q. Peng, "An outline of the hundred-year history of PDT," *Anticancer Res.* **23**, 3591–3600 (2003).

[2] T.J. Dougherty, C.J. Gomer, B.W. Henderson, et al., "Photodynamic therapy," *J. Nat. Cancer Inst.* **90**, 889–905 (1998).

[3] D. Jocham, H. Stepp, and R. Waidelich, "Photodynamic diagnosis in urology: state-of-the-art," *Eur. Urol.* **53**, 1138–1150 (2008).

[4] G. Donati, "Emerging therapies for neovascular age-related macular degeneration: state of the art," *Opthalmol.* **221**, 366–377 (2007).

[5] G.W. Falk, "Autofluorescence endoscopy," *Gastrointest. Clin. N. Am.* **19**, 209–220 (2009).

[6] W. Stummer, U. Pichlmeier, T. Meinel, et al., "Fluorescence-guided surgery with 5-aminolevulinic acid for resection of malignant glioma: a randomised controlled multicentre phase III trial," *Lancet Oncol.* **7**, 392–401 (2006).

[7] B.C. Wilson and M.S. Patterson, "The physics, biophysics and technology of photodynamic therapy," *Phys. Med. Biol.* **53**, 61–109 (2008).

[8] L.H. Murrer, J.P. Marijnissen, and W.M. Star, "Light distribution by linear diffusing sources for photodyamic therapy," *Phys. Med. Biol.* **41**, 951–961 (1996).

[9] L. Brancaleon and H. Moseley, "Laser and non-laser light sources for photodynamic therapy," *Lasers Med. Sci.* **17**, 173–186 (2002).

[10] E. Buytaert, M. Dewaele, and P. Agostinis, "Molecular effectors of multiple cell death pathways initiated by photodynamic therapy," *Biochim. Biophys. Acta* **1776**, 86–107 (2007).

[11] B.C. Wilson, "Detection and treatment of dysplasia in Barrett's esophagus: a pivotal challenge in translating biophotonics from bench to bedside," *J. Biomed. Opt.* **12**, 051401 (2007).

[12] K. Plaetzer, B. Krammer, J. Berlanda, et al., "Photophysics and photochemistry of photodynamic therapy: fundamental aspects," *Lasers Med. Sci.* **24**, 259–268 (2009).

[13] A. Brandis, O. Mazor, E. Neumark, et al., "Novel water-soluble bacteriochlorophyll derivatives for vascular-targeted photodynamic therapy: synthesis, solubility, phototoxicity and the effect of serum proteins," *Photochem. Photobiol.* **81**, 983–993 (2005).

[14] A. Stritt, H. F. Merk, L. R. Braathen, et al., "Photodynamic therapy in the treatment of actinic keratosis," *Photochem. Photobiol.* **84**, 388–398 (2008).

[15] S.A. Vinogradov, P.P. Grosul, V.V. Rozhkov, et al., "Oxygen distributions in tissue measured by phosphorescence quenching," *Adv. Exp. Med. Biol.* **510**, 181–185 (2003).

[16] M.T. Jarvi, M.J. Niedre, and B.C. Wilson, "Singlet oxygen luminescence dosimetry (SOLD). for photodynamic therapy: current status and future prospects," *Photochem. Photobiol.* **82**, 1198–1210 (2006).

[17] B.C. Wilson, R.A. Weersink, and L. Lilge, "Fluorescence in photodynamic therapy dosimetry," in *Fluorescence in Biomedicine*, B. Pogue and M. Mycek (eds.), Marcel Dekker, NY, 529–562 (2003).

[18] H.S. Zeng, M. Korbelik, D.I. Mclean, et al., "Monitoring photoproduct formation and photobleaching by fluorescence spectroscopy has the potential to improve PDT dosimetry with a verteporfin-like photosensitizer," *Photochem. Photobiol.* **75**, 398–405 (2002).

[19] B. Krammer and B. Plaetzer, "ALA and its clinical impact, from bench to bedside," *Photochem. Photobiol. Sci.* **7**, 283–289 (2008).

[20] J. de Leeuw, N. van der Beek, W.D. Neugebauer, et al., "Fluorescence detection and diagnosis of non-melanoma skin cancer at an early stage," *Lasers Surg. Med.* **41**, 96–103 (2009).

[21] G. Jori, "Photodynamic therapy of microbial infections: state of the art and perspectives," *J. Environ. Pathol. Toxicol. Oncol.* **25**, 505–519 (2006).

[22] B.C. Wilson, M. Olivo, and G. Singh, "Subcellular localization of Photofrin and aminolevulinic acid and photodynamic cross-resistance *in vitro* in radiation induced fibrosarcoma cells sensitive or resistant to Photofrin-mediated photodynamic therapy," *Photochem. Photobiol.* **65**, 166–176 (1997).

[23] B.W. Henderson and T.J. Dougherty, "How does photodynamic therapy work?" *Photochem. Photobiol.* **55**, 145–157 (1992).

[24] E. Buytaert, M. Dewaele, and P. Agositinis, "Molecular effectors of multiple cell death pathways initiated by photodynamic therapy," *Biochem. Biophys. Acta* **1776**, 87–106 (2007).

[25] D. Kessel and N.L. Oleinick, "Initiation of autophagy by photodynamic therapy," *Meth. Enzymol.* **453**, 1–16, (2009).

[26] T.N. Demidova, F. Gad, T. Zahra, et al., "Monitoring photodynamic therapy of localized infections by bioluminescence imaging of genetically engineered bacteria," *J. Photochem. Photobiol.* **B3**, 15–25 (2005).

[27] A. Miriampillai, B.A. Standish, E.H. Moriyama, et al., "Speckle variance detection of mirovasculature using swept-source optical coherence tomography," *Opt. Lett.* **33**, 1530–1532 (2008).

[28] P.S. Thong, K.W. Ong, N.S. Goh, K.W. Kho, V. Manivasager, R. Bhuvaneswari, and M. Olivo, "Photodynamic-therapy-activated immune response against distant untreated tumours in recurrent angiosarcoma," *Lancet Oncol.* **8**, 950–952 (2007).

[29] J. Trachtenberg, R.A. Weersink, S.R.H. Davidson, et al., "Vascular-targeted photodynamic therapy (padoporfin, WST09) for recurrent prostate cancer after failure of external beam radiotherapy: a study of escalating light doses," *BJU International* **102**, 556–562 (2008).

[30] M. Korbelik, "PDT host response and its role in the therapy outcome," *Lasers Surg. Med.* **38**, 500–508 (2006).

[31] L. Brancaleon and H. Moseley, "Laser and non-laser light sources for photodynamic therapy," *Lasers Med. Sci.* **17**, 173–186 (2002).

[32] H. Moseley, J.W. Allen, S. Ibbotson, et al., "Ambulatory photodynamic therapy: a new concept in delivering photodynamic therapy," *Br. J. Dermatol.* **154**, 747–750 (2006).

[33] M.H. Gold, "Acne and PDT: new techniques with lasers and light sources," *Lasers Med. Sci.* **22**, 67–72 (2007).

[34] T. Theodossiou, J.S. Hothersall, E.A. Woods, et al., "Firefly luciferin- activated rose bengal: *in vitro* photodynamic therapy by intracellular chemiluminescence in transgenic NIH 3T3 cells," *Cancer Res.* **63**, 1818–1821 (2003).

[35] V.V. Tuchin, *Tissue Optics: Light Scattering Methods and Instruments for Medical Diagnosis*, 2nd edition, **PM 166**, SPIE Press, Bellingham, WA (2007).

[36] L.H. Murrer, J.P. Marijnissen, and W.M. Star, "Light distribution by linear diffusing sources for photodynamic therapy," *Phys. Med. Biol.* **41**, 951–961 (1996).

[37] A. Rendon, R. Weersink, and L. Lilge, "Towards conformal light delivery using tailored cylindrical diffusers: attainable light dose distributions," *Phys. Med. Biol.* **51**, 5967–5975 (2006).

[38] M. Star, H.P.A. Marijnissen, H. Jansen, et al., "Light dosimetry for photodynamic therapy by whole bladder wall irradiation," *Photochem. Photobiol.* **46**, 619–624 (1987).

[39] J. Nyst, R.L. van Veen, I.B. Tan, et al., "Performance of a dedicated light delivery and dosimetry device for photodynamic therapy of nasopharyngeal carcinoma: phantom and volunteer experiments," *Lasers Surg. Med.* **39**, 647–653 (2007).

[40] A. Cengel, E. Glatstein, and S.M. Hahn, "Intraperitoneal photodynamic therapy," *Cancer Treat. Res.* **134**, 493–514 (2007).

[41] B. Lai, M. Loshchenov, A. Douplik, et al., "Three-dimensional fluence rate measurement and data acquisition system for minimally invasive light therapies," *Rev. Sci. Instr.* **80**, 043104 (2009).

[42] R. Weersink, A. Bogaards, M. Gertner, et al., "Techniques for delivery and monitoring of TOOKAD (WST09)-mediated photodynamic therapy of the prostate: clinical experience and practicalities," *J. Photochem. Photobiol.* **79**, 211–222 (2005).

[43] W. Pogue and G. Burke, "Fiber-optic bundle design for quantitative fluorescence measurement from tissue," *Appl. Opt.* **37**, 7429–7436 (1998).

[44] G. Yu, T. Durduran, C. Zhou, et al., "Real-time in situ monitoring of human prostate photodynamic therapy with diffuse light," *Photochem. Photobiol.* **82**, 1279–1284 (2006).

[45] B. Kruijt, S. Kascakova, H.S. de Bruijn, et al., "*In vivo* quantification of chromophore concentration using fluorescence differential path length spectroscopy," *J. Biomed. Opt.* **14**, 034022 (2009).

[46] M.S. Thompson, A. Johansson, T. Johansson, S. Andersson-Engels, and N. Bendsoe, "Clinical system for interstitial photodynamic therapy with combined on-line dosimetry measurements," *Appl. Opt.* **44**, 4023–4031 (2005).

[47] J.S. Dysart, G. Singh, and M.S. Patterson, "Calculation of singlet oxygen dose from photosensitizer fluorescence and photobleaching during mTHPC photodynamic therapy of MLL cells," *Photochem. Photobiol.* **81**, 196–205 (2005).

[48] S. Pratavieira, P.F.C. Menezes, C. Kurachi, et al., "Photodegradation of hematoporphyrin in solution: anomalous behavior at low oxygen concentration," *Laser Phys.: Laser Meth. Chem. Biol. Med.* **19**, 1263–1271 (2009).

[49] M. Jarvi, M.J. Niedre, and B.C. Wilson, "Singlet oxygen luminescence dosimetry (SOLD). for photodynamic therapy: current status and future prospects," *Photochem. Photobiol.* **82**, 1198–1210 (2006).

[50] L. Lee, A. Zhu, M.F. Minhaj, et al., "Pulsed diode laser-based monitor for singlet molecular oxygen," *J. Biomed. Opt.* **13**, 034010 (2008).

[51] K.K. Wang, S. Mitra, and T.H. Foster, "Photodynamic dose does not correlate with long-term tumor response to mTHPC-PDT performed at several drug-light intervals," *Med. Phys.* **35**, 3518–3526 (2008).

[52] T.J. Farrell, B.C. Wilson, M.S. Patterson, et al., "Comparison of the in vivo photodynamic threshold dose for Photofrin, mono- and tetrasulfonated aluminum phthalocyanine using a rat liver model," *Photochem. Photobiol.* **68,** 394–399 (1998).

[53] L. Lilge, M. Portnoy, and B.C. Wilson, "Apoptosis induced *in vivo* by photodynamic therapy in normal brain and intracranial tumour tissue," *Br. J. Cancer* **83**, 1110–1117 (2000).

[54] G.Q. Yu, T. Durduran, C. Zhou, et al., "Noninvasive monitoring of murine tumor blood flow during and after photodynamic therapy provides early assessment of therapeutic efficacy," *Clin. Cancer Res.* **11,** 3543–3552 (2005).

[55] A. Standish, K.K.C. Lee, X. Jin, et al., "Interstitial Doppler optical coherence tomography as a local tumor necrosis predictor in photodynamic therapy of prostatic carcinoma: an *in vivo* study," *Cancer Res.* **68**, 9987–9995 (2008).

[56] H. Moriyama, S.K. Bisland, L. Lilge, et al., "Bioluminescence imaging of the response of rat gliosarcoma to ALA-PpIX-mediated photodynamic therapy," *Photochem. Photobiol.* **80**, 242–249 (2004).

[57] M. Khurana, E.H. Moriyama, A. Mariampillai, et al., "Intravital high-resolution optical imaging of individual vessel response to photodynamic treatment, *J. Biomed. Optics* **13**, 040502 (2008).

[58] K. Chang, I. Rizvi, N. Solban, et al., "*In vivo* optical molecular imaging of vascular endothelial growth factor for monitoring cancer treatment," *Clin. Cancer Res.* **14**, 4146–4153 (2008).

[59] E.W. Stein, K. Maslov, and L.V. Wang, "Noninvasive, *in vivo* imaging of blood-oxygenation dynamics within the mouse brain using photoacoustic microscopy," *J. Biomed. Opt.* **14**, 020502 (2009).

[60] S.R.H. Davidson, R.A. Weersink, M.A. Haider, et al., "Treatment planning and dose analysis for interstitial photodynamic therapy of prostate cancer," *Phys. Med. Biol.* **54**, 2293–2313 (2009).

[61] G.A. Wagnieres, W.M. Star, and B.C. Wilson, "*In vivo* fluorescence spectroscopy and imaging for oncological applications," *Photochem. Photobiol.* **68**, 603–632 (1998).

[62] W.L. Curvers, R. Singh, L.-M. Wong-Kee Song, et al., "Endoscopic tri-modal imaging for detection of early neoplasia in Barrett's oesophagus: a multi-centre feasibility study using high-resolution endoscopy, autofluorescence imaging and narrow band imaging incorporated in one endoscopy system," *Gut* **57**, 167–172 (2008).

[63] R.S. DaCosta, H. Andersson, M. Cirocco, et al., "Autofluorescence characterization of isolated whole crypts and primary cultured human epithelial cells from normal, hyperplastic and adenomatous colonic mucosa," *J. Clin. Pathol.* **58,** 766–774 (2005).

[64] C.H. Tung, U. Mahmood, S. Bredow, et al., "*In vivo* imaging of proteolytic enzyme activity using a novel molecular reporter," *Cancer Res.* **60**, 4953–4958 (2000).

[65] L. Troyan, V. Kianzad, S.L. Gibbs-Strauss, et al., "The FLARE intraoperative near-infrared fluorescence imaging system: a first-in-human clinical trial in breast cancer sentinel lymph node mapping," *Ann. Surg. Oncol.* **16** (10), 2943–2952 (2009)

[66] V. Yang, P.J. Muller, P. Herman, et al., "A multispectral fluorescence imaging system: design and initial clinical tests in intra-operative Photofrin-photodynamic therapy of brain tumors," *Lasers Surg. Med.* **32**, 224–232 (2003).

[67] A. Bogaards, A. Varma, K. Zhang, et al., "Fluorescence image-guided brain tumour resection with adjuvant metronomic photodynamic therapy: pre-clinical model and technology development," *Photochem. Photobiol. Sci.* **4**, 438–442 (2005).

[68] A. Bogaards, H.J. Sterenborg, J. Trachtenberg, et al., "*In vivo* quantification of fluorescent molecular markers in real-time by ratio imaging for diagnostic screening and image-guided surgery," *Lasers Surg. Med.* **39**, 605–613 (2007).

[69] M. Hamblin and P. Mroz (eds.), *Advances in Photodynamic Therapy: Basic, Translational and Clinical*, Artec, Boston/London (2008).

[70] E. Dahlstedt, H.A. Collins, M. Balaz, et al., "One- and two-photon activated phototoxicity of conjugated porphyrin dimers with high two-photon absorption cross sections," *Org. Biomol. Chem.* **7**, 897–904 (2009).

[71] R. Starkey, A.K. Rebane, M.A. Drobizhev, et al., "New two-photon activated photodynamic therapy sensitizers induce xenograft tumor regressions after near-IR laser treatment through the body of the host mouse," *Clin. Cancer Res.* **14**, 6564–6573 (2008).

[72] H.A. Collins, M. Khurana, E.H. Moriyama, et al., "Blood vessel closure using photosensitisers engineered for two-photon excitation," *Nature Photonics* **2**, 420–424 (2008).

[73] M. Khurana, E.H. Moriyama, A. Mariampillai, et al., "Intravital high-resolution optical imaging of individual vessel response to photodynamic treatment," *J. Biomed. Opt.* **13**, 040502 (2008).

[74] G. Smith, W.G. McGimpsey, M.C. Lynch, I.E. Kochevar, and R.W. Redmond, "An efficient oxygen independent 2-photon photosensitization mechanism," *Photochem. Photobiol.* **59**, 135–139 (1994).

[75] J. Chen, K. Stefflova, M.J. Niedre, et al., "Protease-triggered photosensitizing beacon based on singlet oxygen quenching and activation," *J. Am. Chem. Soc.* **126**, 11450–11451 (2004).

[76] J. Chen, J.F. Lovell, P.C. Lo, et al., A "Tumor-mRNA triggered photodynamic molecular beacon based on oliogonucleotide hairpin control of singlet oxygen production," *Photochem. Photobiol. Sci.* 7, 675–680 (2008).

[77] S.K. Bisland, D. Singh, and J. Gariépy, "Potentiation of chlorin e6 photodynamic activity *in vitro* with peptide-based intracellular vehicles," *Bioconj. Chem.* **10**, 982–992 (1999).

[78] K.P. Mahon, T.B. Potocky, D. Blair, et al., "Deconvolution of the cellular oxidative stress response with organelle-specific Peptide conjugates," *Chem. Biol.* **14**, 923–930 (2007).

[79] B.C. Wilson, "Photonic and non-photonic based nanoparticles in cancer imaging and therapeutics," in: *Photon-Based Nanoscience and Nanobiotechnology*, J. Dubowski and S. Tanev (eds.), Springer, Dordrecht, 121–151 (2006).

[80] P. Rai, S. Chang, Z. Mai, et al., "Nanotechnology-based combination therapy improves treatment response in cancer models," *Proc. SPIE* **7380**, 73800W-1–11 (2009).

[81] H. Foster and L. Gao, "Dosimetry in photodynamic therapy – oxygen and the critical importance of capillary density," *Radiat. Res.* **130,** 379–383 (1992).

[82] S.K. Bisland, L. Lilge, A. Lin, et al., "Metronomic photodynamic therapy as a new paradigm for photodynamic therapy: rationale and pre-clinical evaluation of technical feasibility for treating malignant brain tumors," *Photochem. Photobiol.* **80**, 2–30 (2004).

[83] N. Davies and B.C. Wilson, "Interstitial *in vivo* ALA-PpIX mediated metronomic photodynamic therapy (mPDT). using the CNS-1 astrocytoma with bioluminescence monitoring," *Photodiag. Photodyn. Ther.* **4**, 202–212 (2007).

[84] S.K. Bisland, C. Johnson, M. Diab, et al., "A new technique for physiodesis using photodynamic therapy," *Clin. Orthoped. Rel. Res.* **461**, 153–161 (2007).

[85] E. Zusman, M. Sidu, V. Coon, et al., "Photodynamic therapy for epilepsy," *Proc. SPIE* **6139**, 210–218 (2006).

[86] H. Hirschberg, M.J. Zhang, H.M. Gach, et al., "Targeted delivery of bleomycin to the brain using photo-chemical internalization of *Clostridium* perfringens epsilon prototoxin," *J. Neurooncol.* **95**, 317–329 (2009).

[87] S. Burch, A. Bogaards, J. Siewerdsen, et al., "Photodynamic therapy for the treatment of metastatic lesions in bone: studies in rat and porcine models," *J. Biomed. Opt.* **10**, 1–13 (2005).

[88] S.K. Bisland, C. Chien, B.C. Wilson, et al., "Pre-clinical *in vitro* and *in vivo* studies to examine the potential use of photodynamic therapy in the treatment of osteomyelitis," *Photochem. Photobiol. Sci.* **5**, 31–38 (2006).

[89] O.E. Akilov, W. Yousaf, S.X. Lukjan, et al., "Optimization of topical photodynamic therapy with 3,7-bis(di-n-butylamino) phenothiazin-5-ium bromide for cutaneous leishmaniasis," *Lasers Surg. Med.* **41**, 358–365 (2009).

[90] C. O'Neill, J.M. Winograd, J.L. Zeballos, et al., "Microvascular anastomosis using a photochemical tissue bonding technique," *Lasers Surg. Med.* **39**, 716–722 (2007).

[91] A. Høgset, L. Prasmickaite, K.P. Selbo, M. Hellum, B.Ø. Engesæter, A. Bonsted, and K. Berg, "Photochemical internalisation in drug and gene delivery," *Adv. Drug Deliv. Rev.* **56**, 95–115 (2004).

[92] B.C. Wilson, "The potential of photodynamic therapy in regenerative medicine," *J. Craniofac. Surg.* **14**, 278–283 (2003).

[93] Y. Luo and M.S. Shoichet, "A photolabile hydrogel for guided three-dimensional cell growth and migration," *Nat. Mater.* **3**, 249–253 (2004).

26

Advances in Low-Intensity Laser and Phototherapy

Ying-Ying Huang
Wellman Center for Photomedicine, Massachusetts General Hospital, Boston MA; Department of Dermatology, Harvard Medical School, Boston MA; Aesthetic and Plastic Center, Guangxi Medical University, Nanning, Guangxi, P.R. China

Aaron C.-H. Chen
Wellman Center for Photomedicine, Massachusetts General Hospital, Boston MA; Boston University School of Medicine, Graduate Medical Sciences, Boston MA

Michael R. Hamblin
Wellman Center for Photomedicine, Massachusetts General Hospital, Boston MA Department of Dermatology, Harvard Medical School, Boston MA Harvard-MIT Division of Health Sciences and Technology, Cambridge MA

The use of low levels of visible or near infrared light for reducing pain, inflammation and edema, promoting healing of wounds, deeper tissues and nerves, and preventing tissue damage and death has been known for almost forty years since the invention of lasers. Originally thought to be a peculiar property of laser light (soft or cold lasers), the subject has now broadened to include photobiomodulation and photobiostimulation using noncoherent light. Despite many reports of positive findings from experiments conducted *in vitro*, in animal models and in randomized controlled clinical trials, low level light therapy (LLLT) remains controversial. This controversy is likely due to two main reasons; first, the biochemical mechanisms underlying the positive effects are incompletely understood, and second, the complexity of rationally choosing among a large number of illumination parameters such as wavelength, fluence, power density, pulse structure, and treatment timing has led to the publication of a number of negative studies as well as many positive ones. In particular, a biphasic dose response has been frequently observed where low levels of light have a much better effect than higher levels. This chapter will cover some of the proposed cellular chromophores responsible for the effect of visible light on mammalian cells, including cytochrome c oxidase (with absorption peaks in the near infrared region of the spectrum) and photoactive porphyrins and flavins. Mitochondria are thought to be a likely site for the initial effects of light and the involvement of nitric oxide had been postulated. Activation of the respiratory chain leads to increased ATP production, modulation of reactive oxygen species, and activation of transcription factors. These effects in turn lead to increased cell proliferation and migration (particularly by fibroblasts), modulation in levels of cytokines, growth factors and inflammatory mediators, prevention of cell death by anti-apoptotic signaling, and increased tissue oxygenation. The results of these biochemical and cellular changes in animals and patients include such benefits as increased healing in chronic wounds, improvements in sports injuries and carpal tunnel syndrome, pain reduction in arthritis and neuropathies, and amelioration of damage after heart attacks, stroke, nerve injury, and retinal toxicity.

Key words: photobiomodulation, low-level laser therapy, ROS, mitochondria, cytochrome c oxidase, NF-kB, gene transcription, AP1, cells, blood, healing, pain relief

26.1 Historical Introductions

In 1967 Endre Mester wanted to test if laser radiation might cause cancer in mice [1]. He applied laser treatment with a low-powered ruby laser (694 nm) to one mice group, which had shaved dorsal hair. And to his surprise, instead of getting cancer, the treated group hair grew back more quickly than the untreated group. This was the first demonstration of "laser biostimulation." Currently, low-level laser (or light) therapy (LLLT), also known as "cold laser," "soft laser," "biostimulation," or "photobiomodulation" is practiced as part of physical therapy in many parts of the world.

In fact, light therapy is one of the oldest therapeutic methods used by humans (historically as solar therapy by Egyptians, later as UV therapy for which Nils Finsen won the Nobel prize in 1904 [2]. The use of lasers and LEDs as light sources was the next step in the technological development of light therapy, which is now applied to many thousands of people worldwide each day. In LLLT the question is no longer whether light has biological effects but rather how energy from therapeutic lasers and LEDs works at the cellular and organism levels and what are the optimal light parameters for different uses of these light sources.

26.2 Cellular Chromophores

The first law of photobiology states that for low-power visible light to have any effect on a living biological system, the photons must be absorbed by electronic absorption bands belonging to some molecular chromophore or photoacceptor [3]. A chromophore is a molecule (or part of a molecule) where the energy difference between electrons in two different molecular orbitals falls within the energy possesed by photons in the visible spectrum. One approach to finding the identity of this chromophore is to carry out action spectra. An action spectrum is the rate of a physiological activity plotted against the wavelength of light. It shows which wavelength of light is most effectively used in a specific chemical reaction. It was suggested in 1989 that the mechanism of LLLT at the cellular level was based on the absorption of monochromatic visible and NIR radiation by components of the cellular respiratory chain [4].

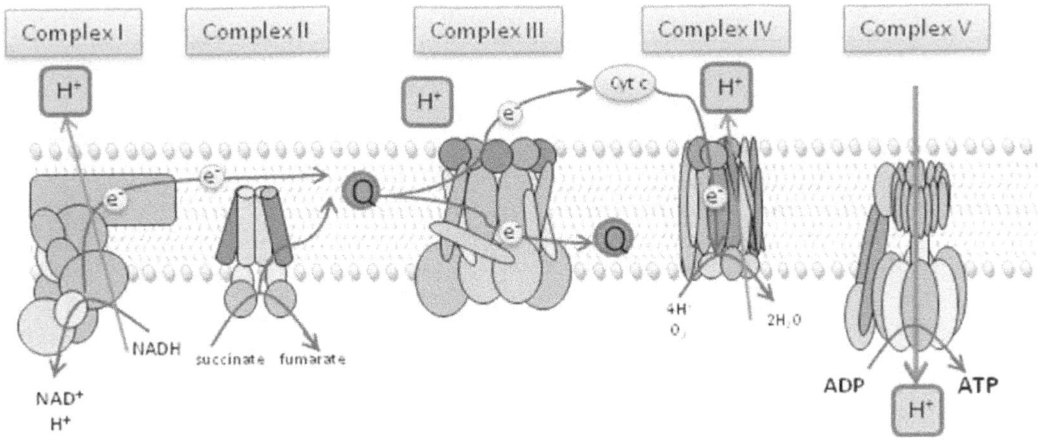

FIGURE 26.1: Mitochondrial respiratory chain.

26.2.1 Mitochondria

Mitochondria play an important role in energy generation and metabolism. Current research about the mechanism of LLLT effects inevitably involves mitochondria. Mitochondria are distinct organelles with two membranes and are usually rod-shaped, ranging from 1 to 10 μm in size. Mitochondria are sometimes described as "cellular power plants," because they convert food molecules into energy in the form of ATP via the process of oxidative phosphorylation. Mitochondria continue the process of catabolism using metabolic pathways including the Krebs cycle, fatty acid oxidation, and amino acid oxidation.

26.2.2 Mitochondrial Respiratory Chain

The mitochondrial electron transport chain consists of a series of metalloproteins bound to the inner membrane of the mitochondria. The inner mitochondrial membrane contains five complexes of integral membrane proteins: NADH dehydrogenase (Complex I), succinate dehydrogenase (Complex II), cytochrome c reductase (Complex III), cytochrome c oxidase (Complex IV), and ATP synthase and two freely diffusible molecules ubiquinone and cytochrome c that shuttle electrons from one complex to the next (Figure 26.1). The respiratory chain accomplishes the stepwise transfer of electrons from NADH and FADH2 (produced in the citric acid or Krebs cycle) to oxygen molecules to form (with the aid of protons) water molecules harnessing the energy released by this transfer to the pumping of protons (H+) from the matrix to the intermembrane space. The gradient of protons formed across the inner membrane by this process of active transport forms a miniature battery. The protons can flow back down this gradient, reentering the matrix, only through another complex of integral proteins in the inner membrane, the ATP synthase complex [5].

26.2.3 Tissue photobiology

An important consideration involves the optical properties of tissue. Both the absorption and scattering of light in tissue are wavelength dependent (both much higher in the blue region of the spectrum than the red) and the principle tissue chromophores (hemoglobin and melanin) have high absorption bands at wavelengths shorter than 600 nm. Water begins to absorb significantly at wavelengths greater than 1150 nm. For these reasons there is a so-called "optical window" in tissue

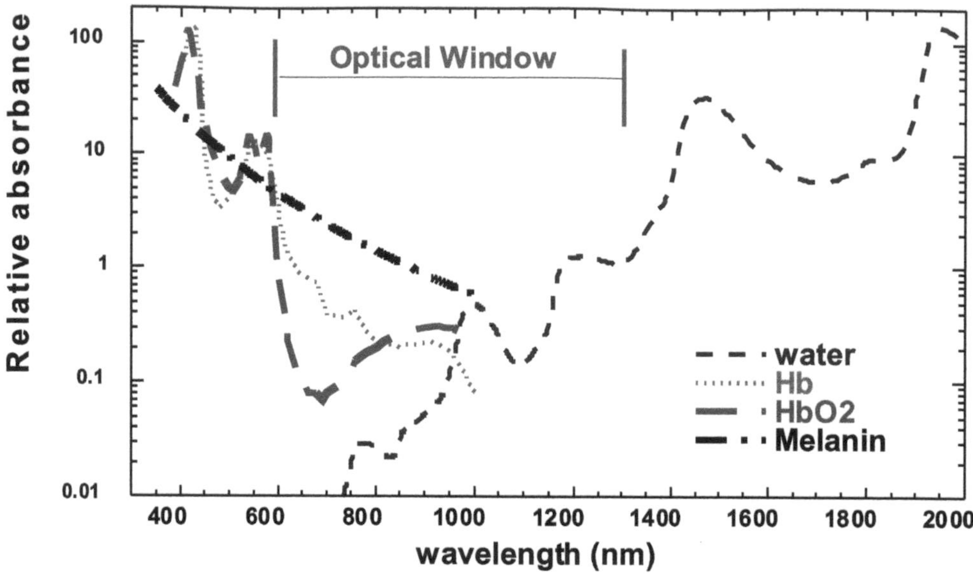

FIGURE 26.2: Tissue optical window.

covering the red and near-infrared wavelengths, where the effective tissue penetration of light is maximized (Figure 26.2). Therefore, although blue, green, and yellow light may have significant effects on cells growing in optically transparent culture medium, the use of LLLT in animals and patients almost exclusively involves red and near-infrared light (600–950 nm) [6].

26.2.4 Cytochrome c oxidase is a photoacceptor

Absorption spectra obtained for cytochrome c oxidase in different oxidation states were recorded and found to be very similar to the action spectra for biological responses to light. Therefore, it was proposed that cytochrome c oxidase (Cox) is the primary photoacceptor for the red-NIR range in mammalian cells [7]. Whelan's group has also suggested that Cox is the critical chromophore responsible for stimulatory effects of irradiation with infrared light [8-10]. *In vitro*, Wong-Riley et al. demonstrated that infrared irradiation reversed the reduction in Cox activity produced by the blockade of voltage dependent sodium channels with tetrodotoxin and up-regulated Cox activity in primary neuronal cells [11]. *In vivo*, Eells et al. demonstrated that rat retinal neurons are protected from damage induced by methanol intoxication. The actual toxic metabolite formed from methanol is formic acid, which inhibits Cox [12]. Moreover, increased activity of cytochrome c oxidase and an increase in polarographically measured oxygen uptake were observed in illuminated mitochondria [13].

26.2.5 Photoactive porphyrins

Another class of cellular molecule that might act as a chromophore is that of porphyrin. This is the "singlet-oxygen hypothesis." Porphyrins are formed in mitochondria as part of the heme biosynthesis cycle. The absorption spectrum of porphyrins like protoporphyrin IX (PPIX) has five bands spanning the region from 400 nm to 630 nm, and decreasing in size as they move toward the red. Molecules with visible absorption bands like PPIX lacking transition metal coordination

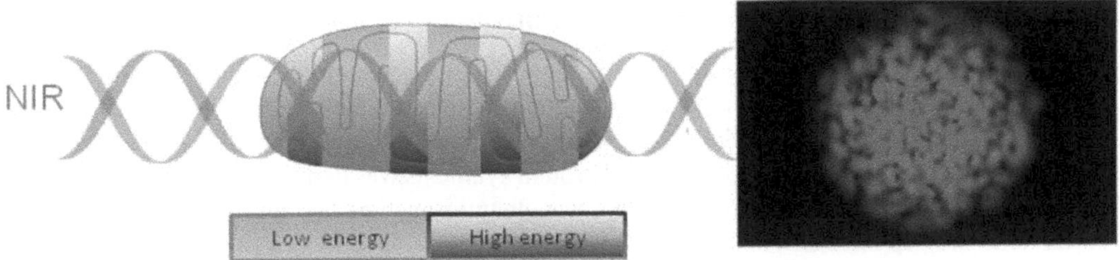

FIGURE 26.3: Laser speckle pattern.

centers [14] can be converted into a long-lived triplet state after photon absorption. This triplet state can interact with ground-state oxygen with energy transfer leading to production of a reactive species, singlet oxygen. This is the same molecule utilized in photodynamic therapy (PDT) to kill cancer cells, destroy blood vessels, and kill microbes. Researchers in PDT have known for a long time that very low doses of PDT can cause cell proliferation and tissue stimulation instead of the killing observed at high doses [15].

26.2.6 Flavoproteins

Another class of molecule that might act as a photoacceptor, is that of flavoproteins that has been mainly proposed by Lubart's group [16]. Flavoproteins are proteins that contain a derivative of riboflavin: flavin adenine dinucleotide (FAD) or flavin mononucleotide (FMN). Flavoproteins are involved in a wide array of biological processes, including, but by no means limited to, bioluminescence, removal of radicals contributing to oxidative stress, photosynthesis, DNA repair, and apoptosis. The spectroscopic properties of the flavin cofactor make it a natural reporter for changes occurring within the active site [17].

26.2.7 Laser speckle effects in mitochondria

It should be mentioned that there is another mechanism that has been proposed to account for low-level laser effects on tissue. This explanation relies on the phenomenon of laser speckle, which is peculiar to laser light. The speckle effect is a result of the interference of many waves, having different phases, which add together to give a resultant wave whose amplitude, and therefore intensity, varies randomly. Each point on illuminated tissue acts as a source of secondary spherical waves. The light at any point in the scattered light field is made up of waves, which have been scattered from each point on the illuminated surface. If the surface is rough enough to create path-length differences exceeding one wavelength, giving rise to phase changes greater than 2π, the amplitude, and hence the intensity, of the resultant light varies randomly (Figure 26.3). It is proposed that the variation in intensity between speckle spots that are about 1 micron apart can give rise to small but steep temperature gradients within subcellular organelles such as mitochondria without causing photochemistry [18]. These temperature gradients are proposed to cause some unspecified changes in mitochondrial metabolism. This hypothesis would explain reports that some LLLT effects in cells and tissues are more pronounced when coherent laser light is used than comparable noncoherent light from LED or filtered lamp sources that are of similar wavelength ranges although not monochromatic.

26.2.8 LLLT enhances ATP synthesis in mitochondria

Several pieces of evidence suggest that mitochondria are responsible for the cellular response to red visible and NIR light. The most popular system to study is the effects of HeNe laser illumination of mitochondria isolated from rat liver, in which increased proton electrochemical potential and ATP synthesis was found [19]. Increased RNA and protein synthesis was demonstrated after 5 J/cm^2 [20]. Irradiation of mitochondria with light at other wavelengths such as 660 nm [21], 650 nm, and 725 nm [22] also showed increases in oxygen consumption, membrane potential, and enhanced synthesis of NADH and ATP. Irradiation with light at 633 nm increased the mitochondrial membrane potential and proton gradient, and increased the rate of ADP/ATP exchange [23], as well as RNA and protein synthesis in the mitochondria. It is also believed that mitochondria are the primary targets when the whole cells are irradiated with light at 630, 632.8, or 820 nm [24–26].

26.3 LLLT and Signaling Pathways

It is clear that signal transduction pathways must operate in cells in order to transduce the signal from the cellular chromophores or photoacceptors that absorb photon energy to the biochemical machinery that operates in gene transcription. As described above it is thought that LLLT is based on the ability of light to alter cell metabolism as a result of its being absorbed by mitochondria and cytochrome c oxidase in particular [27]. However, relatively little is known as to how mitochondria send signals to the nucleus and how the nucleus controls the expression of individual genes.

26.3.1 Redox sensitive pathway

Several classes of molecules such as reactive oxygen species (ROS) and reactive nitrogen species (RNS) are involved in the signaling pathways from mitochondria to nuclei. ROS first reported by Harman [28] are ions or very small molecules and encompass a variety of partially reduced metabolites of oxygen (e.g., superoxide anions, hydrogen peroxide, and hydroxyl radicals) possessing higher reactivities than molecular oxygen due to the presence of unpaired valence shell electrons. The combination of the products of the reduction potential and reducing capacity of the linked redox couples present in cells and tissues represent the redox environment (redox state) of the cell. Redox couples present in the cell include: nicotinamide adenine dinucleotide (oxidized/ reduced forms), nicotinamide adenine dinucleotide phosphate NADP/NADPH, glutathione/glutathione disulfide couple, and thioredoxin/thioredoxin disulfide couple [29]. The term redox signaling is widely used to describe a regulatory process in which the signal is delivered through redox chemistry, including bacteria, to induce protective responses against oxidative damage and to reset the original state of "redox homeostasis" after temporary exposure to ROS [30]. The primary ROS produced in mitochondria is superoxide anion (O2-), which is converted to H_2O_2 either by spontaneous dismutation or by the enzyme superoxide dismutase (SOD). It has been reported that several molecules could be the sensors of ROS [31]. It is proposed that LLLT produces a shift in overall cell redox potential in the direction of greater oxidation [32]. Lubart group found red light illumination increased ROS generation and cell redox activity [33]. They also reported the ROS generation can be detected by electron spin resonance (ESR) techniques [34, 35].

Because of the ubiquitous nature of oxidative stress and the damaging effects of ROS, cells respond to these adverse conditions by modulation of their antioxidant levels, induction of new gene expression, and protein modification [36]. The homeostatic modulation of oxidant levels is a highly efficient mechanism that appeared early in evolution, allowing all cells to tightly control their redox status within a very narrow range. There are numerous molecular sensors within cells that can

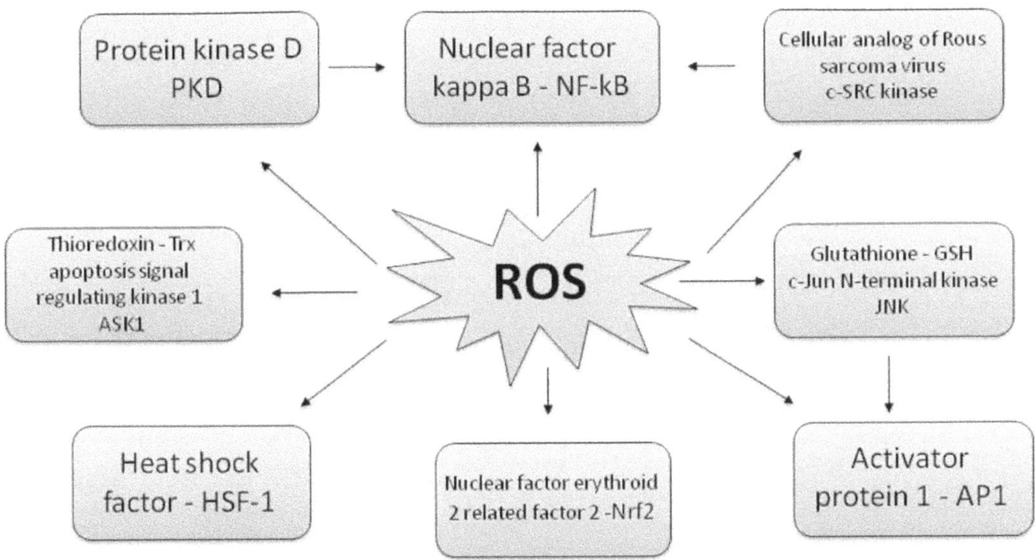

FIGURE 26.4: Schematic diagram of cellular sensors, signal transduction pathways, and transcription factors that govern the response of cells to ROS.

detect ROS by becoming oxidized and the new chemical structures formed initiate signal transduction pathways that can lead to a cascade of reactions [37]. As a consequence, a cell may change its rate of proliferation, or die, depending on the signal networks that it operates. An intracellular oscillation of oxidant levels has been previously experimentally linked to maintenance of the rate of cell proliferation [38]. Figure 26.4 graphically illustrates the range of ROS sensors, signal transduction intermediates and transcription factors that have been reported to govern cellular response to oxidative stress.

26.3.2 Cyclic AMP-dependent signaling pathway

Cyclic AMP (cAMP) is produced by adenylate cyclase, which is stimulated by G-protein in response to binding of an extracellular signaling molecule (hormone, neuromediator) to its cognate G-protein coupled receptor. Then cAMP activates protein kinase A, which regulates numerous cellular processes. This pathway is commonly involved in cell protection. Phosphorylation of the signaling intermediate CREB activates the cAMP response element (CRE)-mediated gene expression. Many inflammation-induced genes, such as c-fos, prodynorphin, cyclooxygenase-2, neurokinin-1, and tyrosine receptor kinase B contain CRE sites in their promoter regions and c-fos is regulated by the extracellular signal-regulated protein kinase pathway in cortical neurons; noxious stimulation and inflammation induce CREB phosphorylation in dorsal horn neurons.

26.3.3 Nitric oxide signaling

Light mediated vasodilation was first described in 1968 by R.F. Furchgott, in his nitric oxide research that lead to his receipt of a Nobel Prize thirty years later in 1998 [39]. Later studies conducted by other researchers confirmed and extended Furchgott's early work and demonstrate the ability of light to influence the localized production or release of NO and stimulate vasodilation through the effect NO on cGMP. This finding suggests that properly designed illumination devices may be effective, noninvasive therapeutic agents for patients who would benefit from increased

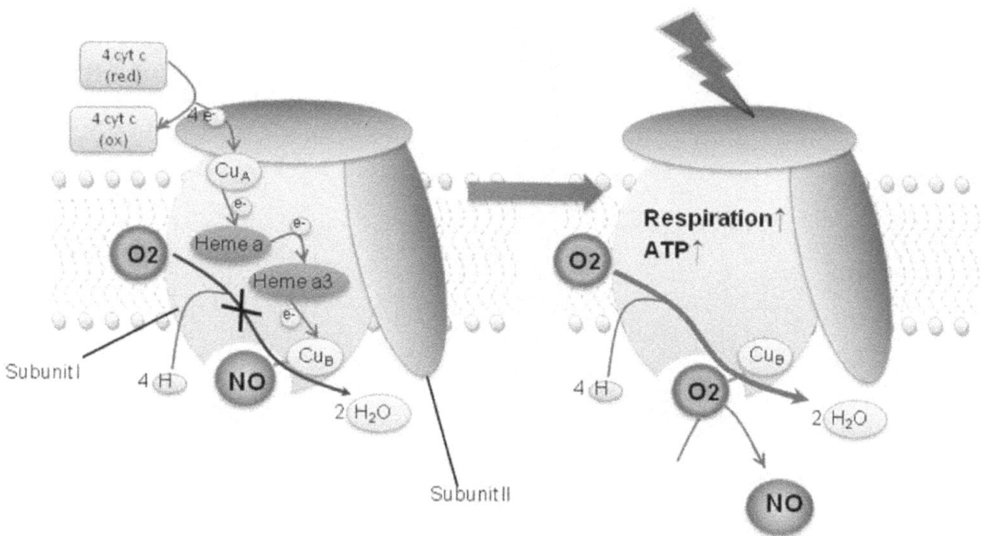

FIGURE 26.5: Cytochrome c oxidase and nitric oxide.

localized NO availability. However, the wavelengths that are most effective on this light-mediated release of NO are different from those used in LLT being in the UVA and blue range [40] that are absorbed by hemoglobin and that illumination can release the NO from hemoglobin (specifically from the nitrosothiols in the beta chain of the hemoglobin molecule) in red blood cells [41–43].

As mentioned in 26.2.4 cytochrome C oxidase (Cox) is situated on the inner membrane of the mitochondrion, where it catalyzes the oxidation of cytochrome C and the reduction of O_2 to water in a process linked to the pumping of protons out of the mitochondrial matrix. The enzyme contains 2 heme (aa and a3) and 2 copper centers (CuA and CuB), of which the heme iron of cytochrome a3 together with CuB, in their reduced form, form the O_2-binding site. NO closely resembles O_2 and therefore can also bind to this site. In the mid 1990s it was demonstrated that NO inhibits the activity of Cox [44–46]. It has been proposed that LLLT might work by photodissociating NO from the Cox, thereby reversing the signaling consequences of excessive NO binding (Figure 26.5) [47].

Light can indeed reverse the inhibition caused by NO binding to Cox, both in isolated mitochondria and in whole cells [48]. Light can also protect cells against NO-induced cell death. These experiments used light in the visible spectrum, with wavelengths from 600 to 630 nm. NIR also seems to have effects on Cox in conditions where NO is unlikely to be present.

Other *in vivo* studies on use of 780 nm for stimulating bone healing in rats [49], the use of 804-nm laser to decrease damage inflicted in rat hearts after creation of heart attacks [50], have shown significant increases of nitric oxide in illuminated tissues after LLLT. On the other hand, studies have been reported on the use of red and NIR LLLT to treat mice with arthritis caused by intra-articular injection of zymosan [51], and studies with a 660-nm laser for strokes created in rats [52] have both shown reduction of NO in the tissues. These authors explained this observation by proposing that LLLT inhibited inducible nitric oxide synthase.

26.3.4 G-protein pathway

G protein-coupled receptors (GPCRs) are the largest superfamily of proteins in the body. One thousand different GPCRs have been identified, and some are activated by exogenous stimuli such as light, odors, and taste.

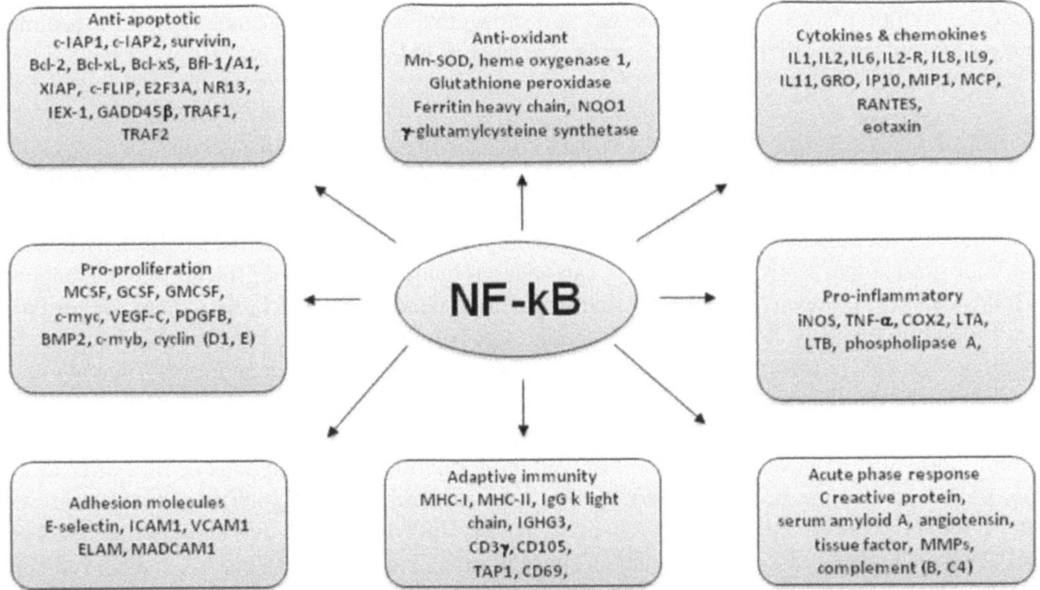

FIGURE 26.6: NF-κB target genes.

P2Y receptors are ubiquitous GPCRs that respond to adenine and/or uridine nucleotides. They are coupled to specific cellular functions including angiogenesis, neurotransmission, wound healing, morphogenesis, cell proliferation, and apoptosis. Light-induced neurite extension may be due to light interaction with cellular organelles leading to an increase in adenine or uridine nucleotides that stimulate the P2Y receptors [53].

26.4 Gene Transcription after LLLT

Transcription factors are proteins that can translocate from the cytosol to the nucleus where they can bind to specific sequences of DNA (consensus binding site or response element) and thereby controls the transfer (or transcription) of genetic information from DNA to RNA. Transcription factors perform this function alone, or with other proteins in a complex, by promoting (as an activator), or blocking (as a repressor) the recruitment of RNA polymerase (the enzyme which activates the transcription of genetic information from DNA to RNA) to specific genes. There are approximately 2600 proteins in the human genome that contain DNA-binding domains and most of these are presumed to function as transcription factors. Figure 26.6 shows many of the NF-kB responsive (target) genes divided into groups. Several important regulation pathways are mediated through the cellular redox state. Changes in redox state induce the activation of numerous intracellular signaling pathways, regulate nucleic acid synthesis, protein synthesis, enzyme activation, and cell cycle progression [36]. These cytosolic responses in turn induce transcriptional changes. Several transcription factors are regulated by changes in the cellular redox state. Among them redox factor –1 (Ref-1)-dependent activator protein-1 (AP-1) (Fos and Jun), nuclear factor κB (NF-κB), p.53, activating transcription factor/cAMP-response element–binding protein (ATF/ CREB), hypoxia-inducible factor (HIF)-1α, and HIF-like factor. As a rule, the oxidized forms of redox-dependent transcription factors have low DNA-binding activity. Ref-1 is an important factor for the specific reduction of these transcription factors. However, it was also shown that low levels of oxidants appear to stimulate proliferation and differentiation of some type of cells [54–56].

26.4.1 NF-κB

Nuclear factor kappa B (NF-κB) is a transcription factor regulating multiple gene expression [57], and has been shown to govern various cellular functions, including inflammatory and stress-induced responses and survival [58]. NF-κB activation is governed by negative feedback by IκB, an inhibitor protein that binds to NF-κB, but can undergo ubiquitination and proteasomal degradation (Figure 26.7) [59], thus freeing NF-κB to translocate to the nucleus and initiate transcription [60]. Understanding the activation mechanisms that govern NF-κB may be important in studying tissue repair or even cancer progression. NF-κB is a redox-sensitive transcription factor [61], that has been proposed to be the sensor for oxidative stress [62]. Reactive oxygen species (ROS) can both activate NF-κB [63], and have been shown to be involved in NF-κB activation by other pro-inflammatory stimuli [64]. Several laboratories have observed the formation of ROS in cells *in vitro* after LLLT [35, 65–67], and it has been proposed that ROS are involved in the signaling pathways initiated after photons are absorbed by the mitochondria in cells [68]. Rizzi et al. has showed that histological abnormalities with increase in collagen concentration, and oxidative stress were observed after trauma. The associated reduction of inducible nitric oxide synthase overexpression and collagen production suggest that the NF-κB pathway may be a signaling route involved in the pathogenesis of muscle trauma [69].

26.4.2 AP-1

The transcription factor AP-1 was one of the first mammalian transcription factors to be identified and is involved in cellular proliferation, transformation, and death [70]. AP-1 is not a single protein, but a complex array of heterodimers composed of proteins that belong to the Jun, Fos, Maf, and ATF subfamilies, which recognize specific nuclear target sequences (Figure 26.8) [71]. Different dimeric combinations can stimulate a variety of gene expression patterns. AP-1 can be activated by growth factors, cytokines, hypoxia, ionizing, and UV radiation. It has been also shown that AP1 activity is under the control of the redox state of the cell [72]. AP-1 mediated gene transcription can also be initiated by oxidative stress [73]. It was initially suggested that asbestos-induced alterations in cellular thiol stats caused subsequent AP-1 activation in rat pleural mesothelial cells [74]. These data were supported by work showing ROS generation during the interaction of silica and asbestos with epidermal cells of AP-1 luciferase reporter mice [75]. Addition of superoxide dismutase and catalase inhibited this activation.

26.4.3 HIF-1

The hypoxia-inducible factor (HIF-1) is a ubiquitous transcription factor involved in the control of cell and tissue responses to hypoxia, specifically in angiogenesis, hematopoiesis, and anaerobic energy metabolism. It may serve as a cellular oxygen sensor. HIF-1 consists of an oxygen-regulated HIF-1α subunit and a constitutively expressed HIF-1β subunit. HIF-1 activates the transcription of target genes by binding to a core DNA sequence. Hundreds of human genes are regulated by HIF-1 [76]. There are over 70 of these genes have been established as direct targets by identification of critical HIF-1 binding sites [77].

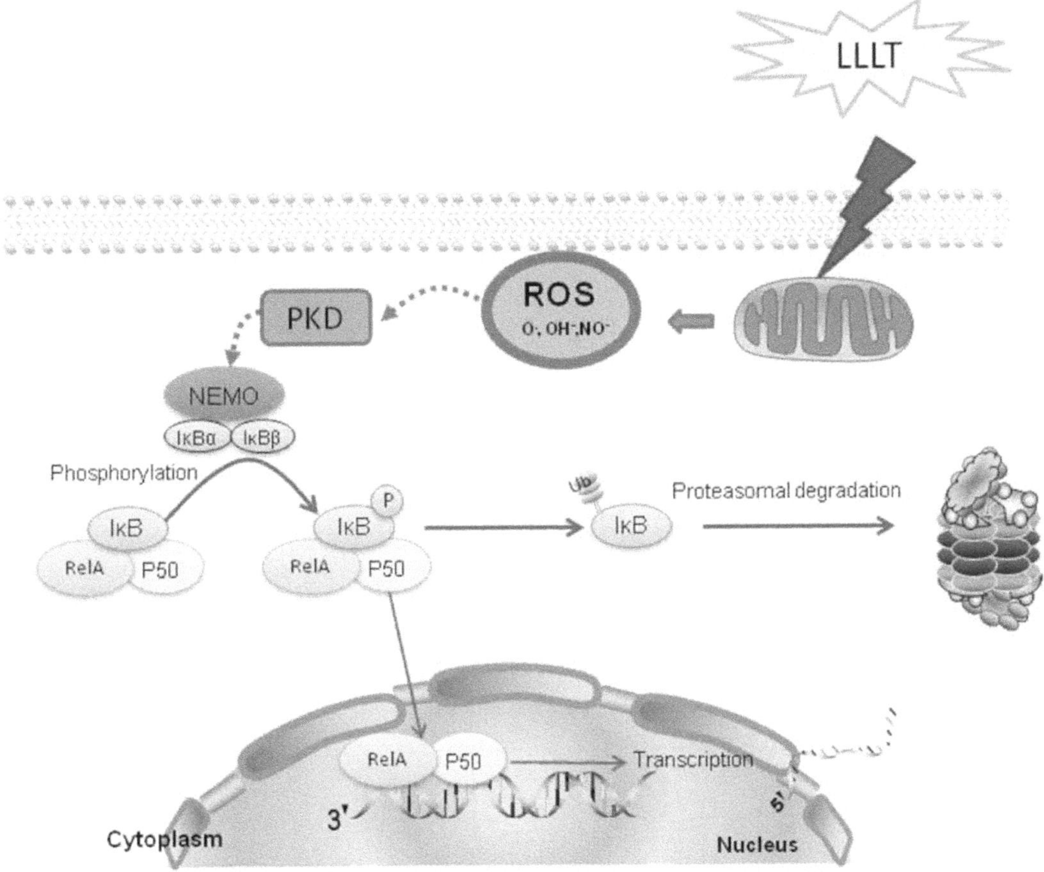

FIGURE 26.7: NF-κB activation.

26.4.4 Ref-1

The human DNA repair enzyme Redox factor –1 (Ref-1) is a dual function protein that plays an important role in both DNA base excision repair and in transcriptional responses to oxidative stress [78]. Ref-1 is one of the key enzymes involved in the repair process. Moreover, in addition to its DNA repair activity, Ref-1 has also been found to facilitate the DNA-binding activity of several transcription factors through both redox-dependent and redox-independent mechanisms [79].

26.5 Cellular Effects

Although underlying mechanism of LLLT are still not clearly understood, *in vitro* and clinical experiments indicate LLLT can prevent cell apoptosis and improve cell proliferation, migration and adhesion, and support the clinical application of LLLT (Figure 26.9).

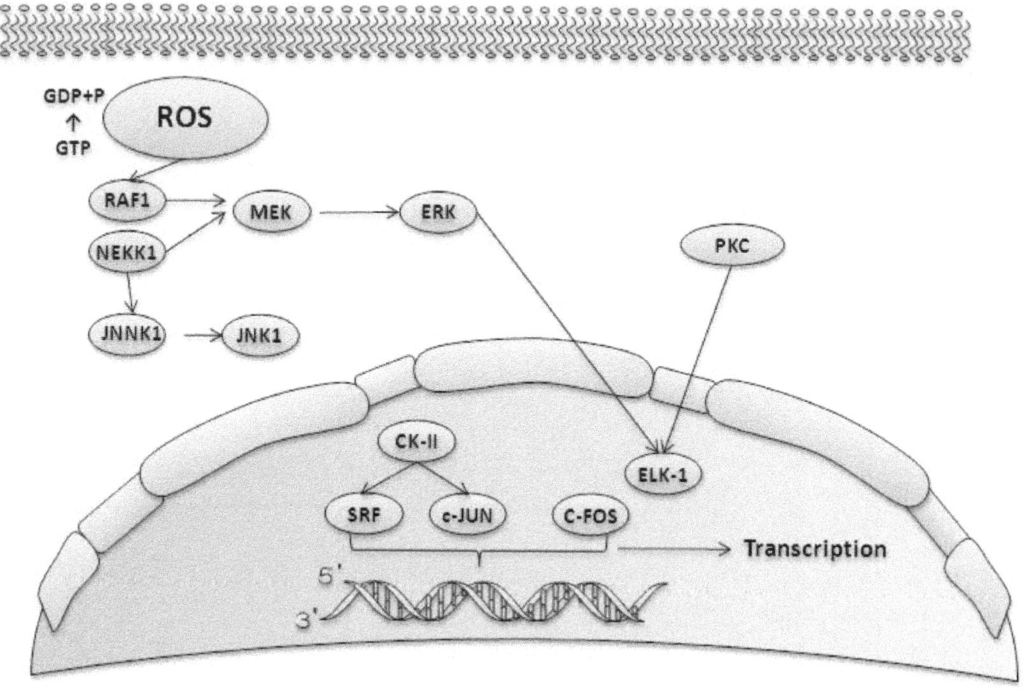

FIGURE 26.8: AP1 pathway.

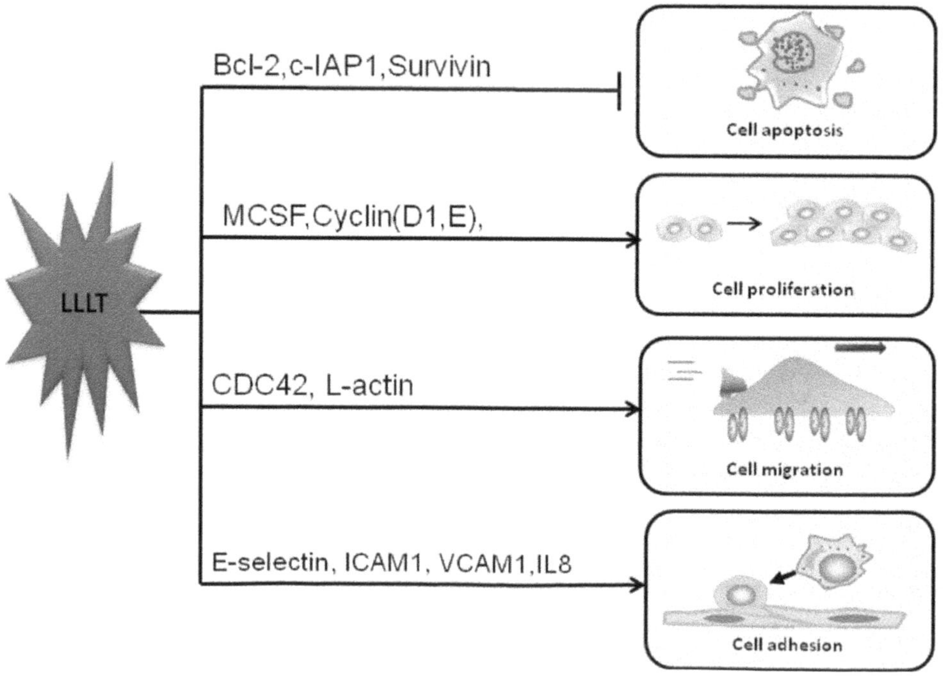

FIGURE 26.9: Cellular effects of LLLT.

26.5.1 Prevention of apoptosis

As discussed above one effect of LLLT is to relieve a blockade of the cytochrome c oxidase whether by NO or any other inhibitory molecules, and consequently lowers the likelihood of apoptosis in many conditions. Eells found that NIR phototherapy counters methanol poisoning, which injures the retina and optic nerve, often causing blindness. The toxic metabolite is formic acid, which inhibits cytochrome oxidase. Wong-Riley and her colleagues at the Medical College of Wisconsin in Milwaukee have shown that NIR phototherapy can also reverse the effects of cyanide on cell cultures. Cyanide poisons cells by binding to cytochrome c oxidase. LLLT could halve the rate of apoptosis in cultured neurons, even when given before cyanide treatment. The results of the study demonstrated that LLLT, in addition to providing positive biomodulation, acts in the re-establishment of cellular homeostasis when the cells are maintained under the condition of nutritional stress; it also prevented apoptosis in cells [8].

NIR LLLT is also proposed to be effective in chronic inflammatory conditions where spreading apoptosis can magnify the amount of cell death and increase the consequential loss of function of the organ affected. As examples it has been proposed to use NIR LLLT on other conditions in which cells die by apoptosis, including acute ischemic stroke and myocardial infarction. The release of cytochrome c from the mitochondria into the cytoplasm is a potent apoptotic signal. Cytochrome c release results in the activation of caspase-3 and activation of apoptotic pathways. The apoptotic cells appear as soon as a few hours after stroke, but the cell numbers peak at 24–48 h after reperfusion [80]. In rat models of stroke, cytoplasmic cytochrome c can be detected for 24 h after the occlusion. *In vitro* 810-nm light can prevent the TTX induced decrease in cytochrome oxidase activity. LLLT may also be able to maintain cytochrome c oxidase activity *in vivo* by preventing release of cytochrome c into the cytoplasm [12]. This would result in the prevention of apoptosis. The release of cytochrome c is regulated by the Bcl/Bax system. Bax promotes release and Bcl decreases release. In myofiber cultures *in vitro*, NIR light promotes Bcl-2 expression and inhibits Bax expression. This fits with the prevention of cytochrome C. Eells and her colleagues have shown that NIR phototherapy could cut the rate of apoptosis by 50% in a rat model of retinitis pigmentosa, in which photoreceptors die by apoptosis during postnatal development causing retinal degeneration and blindness.

26.5.2 Proliferation

LLLT has been shown to enhance cell proliferation *in vitro* in several cell systems: fibroblasts [81–83], keratinocytes [84], endothelial cells [85], and lymphocytes [86, 87]. *In vitro* study by Luciana show that LLLT with administrated fluence of 2 J/cm^2 resulted in increased proliferation rate on fibroblasts grown in a nutritional deficit setting (5% serum), while LLLT had low or no effect on fibroblasts grown under an ideal growth condition (10% serum). Using the same fluence, a shorter application time of LLLT gave a higher increase in cellular proliferation rate [88], however, the mechanism by which LLLT effects cellular proliferation is still under discussion.

26.5.3 Migration

There is an increasing interest in the LLLT effect on cell migration. Hawkins' experiment suggested that 5J/cm^2 LLLT stimulates migration, proliferation, and metabolism of wounded fibroblasts to accelerate wound closure [89]. Recently results from Mvula indicate that 5 J/cm^2 of laser irradiation can positively affect human adipose stem cells by increasing cellular viability, proliferation, migration, and expression of beta1-integrin [90]. However, the research findings are still ambiguous and need further investigation.

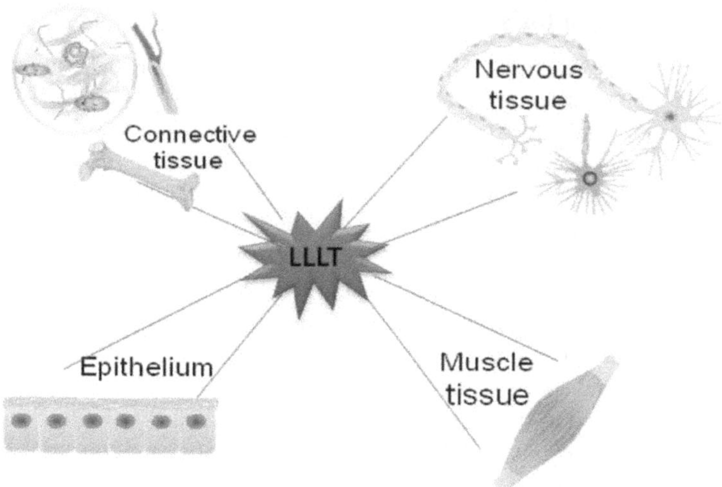

FIGURE 26.10: LLLT and tissue effects.

26.5.4 Adhesion

As we discuss before, LLLT is widely used for the treatment of wound healing. One of the research interests about the mechanism of wound healing is adhesion interaction between two cells and cells with extracellular matrices, which play an important role in wound healing. Karu's experiment [91] indicated that pulsed IR radiation at 820 nm increases the cell-matrix attachment.

26.6 Tissue Effects

Based on morphology, animal tissues can be grouped into four basic types: epithelium, connective tissue, muscle tissue, and nervous tissue. (Figure 26.10)

26.6.1 Epithelium

Epithelial tissues are formed by layers of cells that cover organ surfaces such as the surface of the skin, the airways, and the inner lining of the digestive tract. LLLT on epithelial tissues may cause epithelial cells proliferation [85], migration, and adhesion as mentioned before. LLLT can also activate cell secretion of different growth factors, such as vascular endothelial growth factor (VEGF) [92], and transforming growth factor-β (TGF-β) [93].

26.6.2 Connective tissue

Connective tissue is comprised of cells separated by nonliving material, which is called the extracellular matrix. Connective tissue proper includes the following five types: loose connective, dense connective, elastic, reticular, and adipose. The category "specialized connective tissue" consists of bone, cartilage, and blood. Collagen is the main protein of connective tissue in animals and the most abundant protein in mammals. LLLT can increase collagen synthesis by fibroblasts [93]. LLLT was found to be an effective lymphatic drainage in postmastectomy lymphedema patients indicated that after 1 or 3 months of 2 cycles of active laser treatment reducing the volume of the affected arm, extracellular fluid, and tissue hardness [94]. Finally, LLLT may enhance neovascularisation, promote angiogenesis [95], and increase blood flow [96].

26.6.3 Muscle tissue

Muscle cells form the active contractile tissue of the body known as muscle tissue. Muscle tissue is separated into three distinct categories: smooth muscle, which is found in the inner linings of organs; skeletal muscle, which is found attached to bone providing for gross movement; and cardiac muscle that is found in the heart, allowing it to contract and pump blood throughout an organism. LLLT increased cGMP synthesis in penile smooth muscle cells *in vitro* [97]. LLLT also modulated matrix metalloproteinase activity and gene expression in porcine aortic smooth muscle cells [98]. Shefer at el. reported LLLT can stimulate reactivation and proliferation in isolated myofibers. They show that LLLT promoted the survival of fibers and their adjacent cells, as well as cultured myogenic cells, under serum-free conditions that normally lead to apoptosis [99]. They also discovered LLLT on activation skeletal muscle satellite cells, enhancing their proliferation and inhibit differentiation in vitro [100] and LLLT regulation of protein synthesis of skeletal muscle cells [101]. Oron reported a beneficial effect of LLLT on repair processes after injury or ischemia in skeletal and heart muscles [102].

26.7 Animal and Clinical Studies of LLLT

In this section we will cover applications of LLLT to the following broad areas of medicine: inflammatory disorders, stimulation of healing, relief of pain, and improvement of appearance. The very important application of LLLT to the central nervous system, that includes repair of spinal cord injuries, reduction of brain damage after stroke and traumatic brain injury, and the treatment of degenerative brain conditions such as Alzheimer's and Parkinson's diseases will be covered in the next chapter.

26.7.1 LLLT in inflammatory disorders

LLLT was conducted into different animal model of inflammatory disorders such as burns and peritonitis [103]. The 904 nm GaAs laser was found to reduce inflammatory cell migration in mice with lipopolysaccharide-induced peritonitis [103]. Anti-inflammatory effects of low-level laser therapy (660 nm) were reported in the early phase in carrageenan-induced pleurisy in the rat [104] and the inflammatory response induced by snake venom [105]. Albertini's group reported in several papers LLLT effects on different red wavelengths (660 nm and 684 nm) in carrageenan-induced rat paw edema model of acute inflammatory process [106–108].

26.7.1.1 Carpal tunnel syndrome

Carpal tunnel syndrome (CTS) is an entrapment neuropathy of the median nerve at the wrist, leading to paresthesia, numbness, and muscle weakness in the hand. It is one of the most common peripheral nerve disorders. Most cases of CTS are idiopathic (without known cause), genetic factors determine most of the risk, and the role of arm use and other environmental factors is disputed.

Naeser [109] found LLLT with continuous wave, 15 mW, 632.8 nm, and infrared laser (pulsed, 9.4 W, 904 nm) have a significant effect on CTS treatment in 2002. Recently, Elwakil's [110] study was conducted to evaluate the effectiveness of LLLT by Helium Neon (He-Ne) laser (632.8 nm) for CTS in comparison to the standard open carpal tunnel release surgery. LLLT showed significant outcomes in all parameters of subjective complaints. Ekim [111] reported LLLT with a power output of 50 mW and wavelength of 780 nm, dosage 1.5 J/cm^2, is an effective treatment in patients with rheumatoid arthritis (RA) in carpal tunnel syndrome (CTS).

26.7.1.2 Mucositis

Mucositis is the painful inflammation and ulceration of the mucous membranes lining the digestive tract, usually as an adverse effect of chemotherapy and radiotherapy treatment for cancer. The potential for lesion and pain control using laser (He/Ne) treatment was initially studied in France [112]. Wong reported that a 830-nm laser with 0.7–0.8 J/cm^2 significantly reduced the incidence and the severity of mucositis in chemotherapy patients [113]. Nes [114] reported a significant reduction of mucositis pain, by using an GaAlAs laser, with a wavelength of 830 nm, a potency of 250 mW, and an energy given of 35 J/cm^2. Recently, Jaguar reported patients received GaAlAs and diode laser therapy with 660 nm wavelength, power 10 mW and the energy density delivered to the oral mucosa was 2,5 J/cm^2. LLLT significantly reduced the incidence and the severity of mucositis in chemotherapy patients [115].

26.7.1.3 Joint disorders

There are different forms of joint disorders; each has a different cause. The most common form of arthritis, osteoarthritis (degenerative joint disease) is a result of trauma to the joint, infection of the joint, or age. Emerging evidence suggests that abnormal anatomy might contribute to the early development of osteoarthritis. Other arthritis forms are rheumatoid arthritis and psoriatic arthritis, autoimmune diseases in which the body attacks itself. Septic arthritis is caused by joint infection. Gouty arthritis is caused by deposition of uric acid crystals in the joint, causing inflammation.

Animal studies: Hamblin group tested LLLT on rats that had zymosan injected into their knee joints to induce inflammatory arthritis. Illumination with an 810-nm laser was highly effective (almost as good as dexamethasone) at reducing swelling and a longer illumination time (10 or 100 minutes compared to 1 minute) was more important in determining effectiveness than either the total fluence delivered or the irradiance. LLLT induced reduction of joint swelling correlated with reduction in the inflammatory marker serum prostaglandin E2 (PGE2) [116, 117].

Clinical studies: Brosseau [118] recently reported a meta-analysis. Thirteen trials were included, with 212 patients randomized to laser and 174 patients to placebo laser, and 68 patients received active laser on one hand and placebo on the opposite hand, relative to a separate control group, LLLT reduced pain by 70% relative to placebo and reduced morning stiffness. There were no significant differences between subgroups based on LLLT dosage, wavelength, site of application, or treatment length. There are still more clinical trials required to establish the LLLT dosage and the treatment length. Jamtvedt also concluded that there is moderate-quality evidence that low-level laser therapy reduces pain and improves function by Mata assay [119].

Recently, Bjordal et al. [120] conducted a systematic review of laser therapy with location-specific doses for chronic joint pain and found that this therapy significantly reduces pain and improves health status of patients with joint disorders.

26.7.1.4 Musculoskeletal disorders

These reports indicate that low-level laser therapy is effective in relieving pain from several musculoskeletal conditions including tendonitis [121], shoulder injuries, muscle spasm, myofasciitis, and fibromyalgia [122], While studying the effect of laser therapy on painful conditions, McNeely et al. reported that lasers achieved the premier overall ranking for pain relief compared with the other electrophysical modalities [123]. A rapid relief from neck pain [124], as well as back pain [125], has been reported by several investigators following laser application.

26.7.2 LLLT in healing

26.7.2.1 Wound healing

The literature on LLLT applied to a stimulation of wound healing in a variety of animal models contains both positive and negative studies. The reasons for the conflicting reports, sometimes in very similar wound models, are probably diverse. It is probable that applications of LLLT in animal models will be more effective if carried out on models that have some intrinsic disease state. LLLT significantly improves wound healing in both diabetic rats and diabetic mice. LLLT was also effective in X-radiation impaired wound healing in mice. Furthermore, the total collagen content was significantly increased at 2 months when compared with control wounds. The beneficial effect of LLLT on wound healing can be explained by considering several basic biological mechanisms including the induction of expression cytokines and growth factors known to be responsible for the many phases of wound healing. First, there is a report [121] that HeNe laser (632.7 nm) increased both protein and mRNA levels of interleukins 1 and 8 in keratinocytes. These are cytokines responsible for the initial inflammatory phase of wound healing. Second, there are reports [122] that LLLT can up-regulate cytokines responsible for fibroblast proliferation and migration. Third, it has been reported [92] that LLLT can increase growth factors such as VEGF responsible for the neovascularization necessary for wound healing. Fourth, TGF-β is a growth factor responsible for inducing collagen synthesis from fibroblasts and has been reported to be upregulated by LLLT [93]. Fifth, there are reports [123, 124] that LLLT can induce fibroblasts to undergo the transformation into myofibloblasts, a cell type that expresses smooth muscle α-actin and desmin and has the phenotype of contractile cells that hasten wound contraction. Our group found that wound healing was significantly stimulated in some strains of mice but not in other strains. Illuminated wounds started to contract while control wounds initially expanded for the first 24 hours. We found a biphasic dose-response curve for fluence of 635-nm light with a maximum positive effect at 2 J/cm^2. 820 nm was found to be the best wavelength tested compared to 635, 670, and 720 nm. We found no difference between noncoherent 635+/-15-nm light from a lamp and coherent 633-nm light from a He/Ne laser. LLLT increased the number of alpha-smooth muscle actin-positive cells at the wound edge [125].

26.7.2.2 Bone

LLLT was reported to affect the proliferation and maturation of human osteoblasts cells. In animal studies Silva et al. [126] investigated that low-intensity laser (GaAsAl, 830 nm) irradiation on improved bone repair in the femurs of mice. The laser of low power promotes the expression of bFGF in the periodontal tissue and alveolar bone remodeling was reported [127].

26.7.2.3 Tendon

In clinical studies LLLT administered with optimal doses of 904 nm and possibly 632 nm wavelengths directly to the lateral elbow tendon insertions, seem to offer short-term pain relief and less disability in LET, both alone and in conjunction with an exercise regimen. This finding contradicts the conclusions of previous reviews that failed to assess treatment procedures, wavelengths, and optimal doses.

26.7.2.4 Cartilage

A recent study investigated low-level HeNe laser therapy can increase histological parameters of immobilized articular cartilage in rabbits. It was observed that LLLT had significantly increased the depth of the chondrocytes in four-week immobilized femoral articular cartilage and the depth of articular cartilage in seven-week immobilized knees [128].

26.7.2.5 Muscle

LLLT has been shown to promote proliferation of different types of muscle cells including skeletal muscle, smooth muscle, and cardiac muscle. The effect of phototherapy on the process of muscle regeneration following injury was studied in Oron's lab in mammals [129]. Laser irradiation (Ga-As, 810 nm) was reported markedly protected skeletal muscles from degeneration following acute injury in rats [130]. Lopes-Martins reported that LLLT doses of 0.5 and 1.0 J/cm2 can prevent development of muscular fatigue in rats during repeated tetanic contractions [131]. Leal Jr. and Lopes-Martins found Ga-Al-As 655-nm and 830-nm laser can both delay the onset of skeletal muscle fatigue in high-intensity exercises [132, 133].

26.7.2.6 Nerve

Cells comprising the central nervous system (see next chapter) and peripheral nervous system are classified as neural tissue. Neuroregeneration refers to the regrowth or repair of the nervous tissues, cells, or cell products. As a result of the high incidence of neurological injuries, nerve regeneration, and repair is becoming a rapidly growing field. Animal models have been employed to study LLLT effects in nerve repair [134, 135]. Anders et al. [136] studied LLLT for regenerating crushed rat facial nerves; by comparing 361, 457, 514, 633, 720, and 1064 nm and found best response with 162.4 J/cm^2 of 633-nm HeNe laser. Rochkind carried out a clinical study on LLLT in patients with long-term peripheral nerve injury noninvasive 780-nm laser. Phototherapy can progressively improve nerve function, which leads to significant functional recovery [137].

26.7.3 LLLT in pain relief

In recent years, there is growing interest in the use of laser biostimulation as a therapeutic modality for pain management. Alterations in neuronal activity have been suggested to play a role in pain relief by laser therapy.

Recently, another novel LLLT mechanism was observed in an isolated nerve model study, which focused on the effect of a single 30 s, 830 nm (cw) laser exposure in a rat dorsal root ganglion (DRG) culture model. In this model LLLT blocked the axonal flow of small-diameter nerves, this resulted in a decrease in mitochondrial membrane potential (MMP) with a concurrent decrease in available ATP required for nerve function. Laser-induced neural blockade is a consequence of such changes and provide a mechanism for a neural basis of laser-induced pain relief [138].

Many published reports documented the positive findings of laser biostimulation in pain management. This level of evidence relates to chronic neck pain [139], tendonitis [140], chronic joint disorders [120], and chronic pain. Randomized controlled trials (RCTs) provide evidence for the efficacy of laser therapy in chronic low back pain. Yaksich et al. reported a good improvement rate in approximately 60% of cases of postherpetic neuralgia [141]. Pinheiro confirmed LLLT is an effective tool for disorders including temporomandibular joint (TMJ) pain, trigeminal neuralgia, with treatment of 632.8, 670, and 830 nm diode lasers [142].

For clinical use, wavelength is generally recognized as one of the most important parameters. In the treatment of painful conditions, both visible (e.g., λ= 632.8, 670 nm) and infrared (e.g., λ= 780, 810–830, 904 nm) wavelengths have been used. In support of inhibitory mechanisms are a number of studies that demonstrate that 830 nm, continuous wave (cw) slows nerve conduction velocity and increases latencies in median [143] and sural nerves [144]. Such variability in outcomes may be due to the multiplicity of parameters used, including wavelengths, energy, and power densities, with differing frequencies of application.

Diabetic neuropathy is a painful condition frequently affecting the feet of diabetics and leading to development of chronic ulcers and serious falls. Monochromatic infrared energy (Anodyne, 930 nm) has been tested with mixed results [145–148]. Reversal of diabetic neuropathy was associated with an immediate reduction in the absolute number of falls, a reduced fear of falling, and improved activities of daily living [149].

26.7.4 LLLT in aesthetic applications

26.7.4.1 LLLT for hair regrowth

Since the first pioneering publication of Mester [1] reported stimulation of hair growth in mice, there have been virtually no follow-up studies on LLLT stimulation for hair growth in animal models.

Despite the fact that LLLT devices are widely marketed and used for hair regrowth, there have been only a few literature reports containing some observations of LLLT-induced hair growth in patients, and amelioration or treatment of any type of alopecia. A Japanese group reported [150] on the use of Super Lizer (a linear polarized light source providing 1.8 W of 600–1600-nm light) to treat alopecia areata. It produced significant hair growth compared to nontreated lesions in 47% of patients. A report from Finland [151] compared three different light sources used for male-pattern baldness (HeNe laser, In GaAl diode laser at 670-nm, and noncoherent 635-nm LED) and measured blood flow in the scalp.

Schwartz et al. [152] reported in 2002 that helium/neon laser irradiation (3 J/cm^2) augmented the level of nerve growth factor (NGF) mRNA fivefold and increased NGF release to the medium of myotubes cultured *in vitro*. This correlated with a transient elevation of intracellular calcium in the myotubes. Yu and coworkers found a significant increase in NGF release from cultured human keratinocytes [153]. Therefore, it is postulated that LLLT may influence hair regrowth via the NGF signaling system.

26.7.4.2 Acne therapy

Acne vulgaris is one of the most common dermatologic disorders encountered in everyday practice. Treatment options for this, such as antibiotic resistance and slow onset of action from many topical therapies, have led researchers to seek out alternative therapies, especially for those suffering from moderate to severe inflammatory acne vulgaris. Lasers and light sources are finding increased usage in the treatment of inflammatory acne vulgaris. Light sources including blue lights and intense pulsed lights are becoming regular additions to routine medical management to enhance the therapeutic response. Recently, there have been some reports that blue light is effective in treating acne, because the porphyrin produced by *Propionibacterium acnes* absorbs blue light [154] and kills the bacteria via PDT action. Papageorgiou [155] reported that phototherapy with mixed blue-red light, probably by combining antibacterial (blue light) and anti-inflammatory action (red light), is a more effective means of treating acne vulgaris of mild to moderate severity.

26.7.4.3 Wrinkle reduction

Zelickson et al. [156] have shown that new collagen was formed 12 weeks after treatment with a 585-nm pulsed dye laser. Several reports showed improvement in skin collagen and objectively measured data showed significant reductions of wrinkles (maximum: 36%) and increases of skin elasticity (maximum: 19%) compared to baseline on the treated face in treatment groups after low-level laser irradiation [157]. Omi reported the efficacy in collagen replenishment after irradiation of 3 J/cm^2, 585-nm laser (Chromogenex V3), with a spot size of 5-mm diameter in patients [158]. Trelles got good results with combined wavelengths of 595 plus 1450 nm for skin rejuvenation in clinical studies [159, 160]. Lee et al. investigated the clinical efficacy of LED phototherapy for skin rejuvenation through the comparison 830 and 633 nm LEDs. They showed significant reductions of wrinkles (maximum: 36%) and increases of skin elasticity (maximum: 19%) compared to baseline on the treated face in the three treatment groups. A marked increase in the amount of collagen and elastic fibers in all treatment groups was observed. Ultrastructural examination demonstrated highly activated fibroblasts, surrounded by abundant elastic and collagen fibers. Immunohistochemistry showed increases in several markers of healthy skin after LED phototherapy. They concluded that 830 and 633-nm LED phototherapy is an effective approach for skin rejuvenation [161].

26.8 Conclusion

Advances in molecular and cellular biology are being made at an increasing rate worldwide, and these advances far surpass whatever advances are being made in the field of LLLT. We believe that ongoing scientific discoveries and the steadily increased rational understanding of the basic cellular chromophores, signal transduction pathways, transcription factors activation, and the subsequent changes in cell and tissue growth, survival and healing observed both in animals and in patients, will allow LLLT to gain more widespread acceptance in the scientific and medical community.

Acknowledgments

Work in the authors' laboratory is supported by the US NIH (R01AI050875) and U.S. Air Force MFEL program (contract FA9550-04-1-0079).

References

[1] E. Mester, B. Szende, and P. Gartner, "The effect of laser beams on the growth of hair in mice," *Radiobiol. Radiother. (Berl.)* **9**, 621–626 (1968).

[2] R. Roelandts, "The history of phototherapy: something new under the sun?," *J. Am. Acad. Dermatol.* **46**, 926–930 (2002).

[3] J.C. Sutherland, "Biological effects of polychromatic light," *Photochem. Photobiol.* **76**, 164–170 (2002).

[4] T. Karu, "Laser biostimulation: a photobiological phenomenon," *J. Photochem. Photobiol. B.* **3**, 638–640 (1989).

[5] J.F. Turrens, "Mitochondrial formation of reactive oxygen species," *J. Physiol.* **552**, 335–344 (2003).

[6] T.I. Karu and N.I. Afanasyeva, "Cytochrome c oxidase as the primary photoacceptor upon laser exposure of cultured cells to visible and near IR-range light," *Dokl. Akad. Nauk.* **342**, 693–695 (1995).

[7] T.I. Karu and S.F. Kolyakov, "Exact action spectra for cellular responses relevant to phototherapy," *Photomed. Laser Surg.* **23**, 355–361 (2005).

[8] C.M. Carnevalli, C.P. Soares, R.A. Zangaro, et al., "Laser light prevents apoptosis in Cho K-1 cell line," *J. Clin. Laser Med. Surg.* **21**, 193–196 (2003).

[9] H.T. Whelan, E.V. Buchmann, A. Dhokalia, et al., "Effect of NASA light-emitting diode irradiation on molecular changes for wound healing in diabetic mice," *J. Clin. Laser Med. Surg.* **21**, 67–74 (2003).

[10] H.T. Whelan, R.L. Smits, Jr., E.V. Buchman, et al., "Effect of NASA light-emitting diode irradiation on wound healing," *J. Clin. Laser Med. Surg.* **19**, 305–314 (2001).

[11] M.T. Wong-Riley, H.L. Liang, J.T. Eells, et al., "Photobiomodulation directly benefits primary neurons functionally inactivated by toxins: role of cytochrome c oxidase," *J. Biol. Chem.* **280**, 4761–4771 (2005).

[12] J.T. Eells, M.M. Henry, P. Summerfelt, et al., "Therapeutic photobiomodulation for methanol-induced retinal toxicity," *Proc. Natl. Acad. Sci. U. S. A.* **100**, 3439–3444 (2003).

[13] D. Pastore, M. Greco, V.A. Petragallo, et al., "Increase in <–H+/e- ratio of the cytochrome c oxidase reaction in mitochondria irradiated with helium-neon laser," *Biochem. Mol. Biol. Int.* **34**, 817–826 (1994).

[14] H. Friedmann, R. Lubart, I. Laulicht, et al., "A possible explanation of laser-induced stimulation and damage of cell cultures," *J. Photochem. Photobiol. B.* **11**, 87–91 (1991).

[15] K. Plaetzer, T. Kiesslich, B. Krammer, et al., "Characterization of the cell death modes and the associated changes in cellular energy supply in response to AlPcS4-PDT," *Photochem. Photobiol. Sci.* **1**, 172–177 (2002).

[16] M. Eichler, R. Lavi, A. Shainberg, et al., "Flavins are source of visible-light-induced free radical formation in cells," *Lasers Surg. Med.* **37**, 314–319 (2005).

[17] V. Massey, "The chemical and biological versatility of riboflavin," *Biochem. Soc. Trans.* **28**, 283–296 (2000).

[18] A.N. Rubinov and A.A. Afanas'ev, "Nonresonance mechanisms of biological effects of coherent and incoherent light," *Opt. Spectr.* **98**, 943–948 (2005).

[19] S. Passarella, E. Casamassima, S. Molinari, et al., "Increase of proton electrochemical potential and ATP synthesis in rat liver mitochondria irradiated in vitro by helium-neon laser," *FEBS Lett.* **175**, 95–99 (1984).

[20] M. Greco, G. Guida, E. Perlino, et al., "Increase in RNA and protein synthesis by mitochondria irradiated with helium-neon laser," *Biochem. Biophys. Res. Commun.* **163**, 1428–1434 (1989).

[21] W. Yu, J.O. Naim, M. McGowan, et al., "Photomodulation of oxidative metabolism and electron chain enzymes in rat liver mitochondria," *Photochem. Photobiol.* **66**, 866–871 (1997).

[22] M.W. Gordon, "The correlation between in vivo mitochondrial changes and tryptophan pyrrolase activity," *Arch. Biochem. Biophys.* **91**, 75–82 (1960).

[23] S. Passarella, A. Ostuni, A. Atlante, et al., "Increase in the ADP/ATP exchange in rat liver mitochondria irradiated in vitro by helium-neon laser," *Biochem. Biophys. Res. Commun.* **156**, 978–986 (1988).

[24] T. Karu, L. Pyatibrat and G. Kalendo, "Irradiation with He-Ne laser increases ATP level in cells cultivated in vitro," *J. Photochem. Photobiol. B.* **27**, 219–223 (1995).

[25] L.E. Bakeeva, V.M. Manteifel, E.B. Rodichev, et al., "Formation of gigantic mitochondria in human blood lymphocytes under the effect of an He-Ne laser," *Mol. Biol. (Mosk.)* **27**, 608–617 (1993).

[26] V.M. Manteifel, T.N. Andreichuk, T.I. Karu, et al., "Activation of transcription function in lymphocytes after irradiation with He-Ne-laser," *Mol. Biol. (Mosk.)* **24**, 1067–1075 (1990).

[27] T.I. Karu, L.V. Pyatibrat and N.I. Afanasyeva, "A novel mitochondrial signaling pathway activated by visible-to-near infrared radiation," *Photochem. Photobiol.* **80**, 366–372 (2004).

[28] D. Harman, "Aging: a theory based on free radical and radiation chemistry," *J. Gerontol.* **11**, 298–300 (1956).

[29] F.Q. Schafer and G.R. Buettner, "Redox environment of the cell as viewed through the redox state of the glutathione disulfide/glutathione couple," *Free Radic. Biol. Med.* **30**, 1191–1212 (2001).

[30] W. Droge, "Free radicals in the physiological control of cell function," *Physiol. Rev.* **82**, 47–95 (2002).

[31] P. Storz, "Mitochondrial ROS–radical detoxification, mediated by protein kinase D," *Trends Cell Biol.* **17**, 13–18 (2007).

[32] T. Karu, "Primary and secondary mechanisms of action of visible to near-IR radiation on cells," *J. Photochem. Photobiol. B.* **49**, 1–17 (1999).

[33] R. Lavi, A. Shainberg, H. Friedmann, et al., "Low energy visible light induces reactive oxygen species generation and stimulates an increase of intracellular calcium concentration in cardiac cells," *J. Biol. Chem.* **278**, 40917–40922 (2003).

[34] R. Lavi, M. Sinyakov, A. Samuni, et al., "ESR detection of 1O2 reveals enhanced redox activity in illuminated cell cultures," *Free Radic. Res.* **38**, 893–902 (2004).

[35] M. Eichler, R. Lavi, H. Friedmann, et al., "Red light-induced redox reactions in cells observed with TEMPO," *Photomed. Laser Surg.* **25**, 170–174 (2007).

[36] H. Liu, R. Colavitti, I.I. Rovira, et al., "Redox-dependent transcriptional regulation," *Circ. Res.* **97**, 967–974 (2005).

[37] R. Schreck and P.A. Baeuerle, "A role for oxygen radicals as second messengers," *Trends Cell Biol.* **1**, 39–42 (1991).

[38] K. Irani, Y. Xia, J.L. Zweier, et al., "Mitogenic signaling mediated by oxidants in Ras-transformed fibroblasts," *Science* **275**, 1649–1652 (1997).

[39] S.J. Ehrreich and R.F. Furchgott, "Relaxation of mammalian smooth muscles by visible and ultraviolet radiation," *Nature* **218**, 682–684 (1968).

[40] H. Chaudhry, M. Lynch, K. Schomacker, et al., "Relaxation of vascular smooth muscle induced by low-power laser radiation," *Photochem. Photobiol.* **58**, 661–669 (1993).

[41] R. Mittermayr, A. Osipov, C. Piskernik, et al., "Blue laser light increases perfusion of a skin flap via release of nitric oxide from hemoglobin," *Mol. Med.* **13**, 22–29 (2007).

[42] Yu.A. Vladimirov, G.I. Klebanov, G.G. Borisenko, et al., "Molecular and cellular mechanisms of the low intensity laser radiation effect," *Biofizika* **49**, 339–350 (2004).

[43] Yu.A. Vladimirov, A.N. Osipov, and G.I. Klebanov, "Photobiological principles of therapeutic applications of laser radiation," *Biochemistry* (Moscow) **69**, 81–90 (2004).

[44] M.W. Cleeter, J.M. Cooper, V.M. Darley-Usmar, et al., "Reversible inhibition of cytochrome c oxidase, the terminal enzyme of the mitochondrial respiratory chain, by nitric oxide: implications for neurodegenerative diseases," *FEBS Lett.* **345**, 50–54 (1994).

[45] G.C. Brown and C.E. Cooper, "Nanomolar concentrations of nitric oxide reversibly inhibit synaptosomal respiration by competing with oxygen at cytochrome oxidase," *FEBS Lett.* **356**, 295–298 (1994).

[46] M. Schweizer and C. Richter, "Nitric oxide potently and reversibly deenergizes mitochondria at low oxygen tension," *Biochem. Biophys. Res. Commun.* **204**, 169–175 (1994).

[47] T.I. Karu, L.V. Pyatibrat, and N.I. Afanasyeva, "Cellular effects of low power laser therapy can be mediated by nitric oxide," *Lasers Surg. Med.* **36**, 307–314 (2005).

[48] V. Borutaite, A. Budriunaite, and G.C. Brown, "Reversal of nitric oxide-, peroxynitrite- and S-nitrosothiol-induced inhibition of mitochondrial respiration or complex I activity by light and thiols," *Biochim. Biophys. Acta* **1459**, 405–412 (2000).

[49] G.A. Guzzardella, M. Fini, P. Torricelli, et al., "Laser stimulation on bone defect healing: an in vitro study," *Lasers Med. Sci.* **17**, 216–220 (2002).

[50] H. Tuby, L. Maltz, and U. Oron, "Modulations of VEGF and iNOS in the rat heart by low level laser therapy are associated with cardioprotection and enhanced angiogenesis," *Lasers Surg. Med.* **38**, 682–688 (2006).

[51] Y. Moriyama, E.H. Moriyama, K. Blackmore, et al., "In vivo study of the inflammatory modulating effects of low-level laser therapy on iNOS expression using bioluminescence imaging," *Photochem. Photobiol.* **81**, 1351–1355 (2005).

[52] M.C. Leung, S.C. Lo, F.K. Siu, et al., "Treatment of experimentally induced transient cerebral ischemia with low energy laser inhibits nitric oxide synthase activity and up-regulates the expression of transforming growth factor-beta 1," *Lasers Surg. Med.* **31**, 283–288 (2002).

[53] V.E. Klepeis, I. Weinger, E. Kaczmarek, et al., "P2Y receptors play a critical role in epithelial cell communication and migration," *J. Cell. Biochem.* **93**, 1115–1133 (2004).

[54] M. Yang, N.B. Nazhat, X. Jiang, et al., "Adriamycin stimulates proliferation of human lymphoblastic leukaemic cells via a mechanism of hydrogen peroxide (H_2O_2) production," *Br. J. Haematol.* **95**, 339–344 (1996).

[55] W.G. Kirlin, J. Cai, S.A. Thompson, et al., "Glutathione redox potential in response to differentiation and enzyme inducers," *Free Radic. Biol. Med.* **27**, 1208–1218 (1999).

[56] S. Alaluf, H. Muir-Howie, H.L. Hu, et al., "Atmospheric oxygen accelerates the induction of a post-mitotic phenotype in human dermal fibroblasts: the key protective role of glutathione," *Differentiation* **66**, 147–155 (2000).

[57] T. Wang, X. Zhang, and J.J. Li, "The role of NF-kappaB in the regulation of cell stress responses," *Int. Immunopharmacol.* **2**, 1509–1520 (2002).

[58] V.R. Baichwal and P.A. Baeuerle, "Activate NF-kappa B or die?," *Curr. Biol.* **7**, R94–R96 (1997).

[59] T. Henkel, T. Machleidt, I. Alkalay, et al., "Rapid proteolysis of I kappa B-alpha is necessary for activation of transcription factor NF-kappa B," *Nature* **365**, 182–185 (1993).

[60] A.Hoffmann, A. Levchenko, M.L. Scott, et al., "The IkappaB-NF-kappaB signaling module: temporal control and selective gene activation," *Science* **298**, 1241–1245 (2002).

[61] C.T. D'Angio and J.N. Finkelstein, "Oxygen regulation of gene expression: a study in opposites," *Mol. Genet. Metab.* **71**, 371–380 (2000).

[62] N. Li and M. Karin, "Is NF-kappaB the sensor of oxidative stress?," *FASEB J.* **13**, 1137–1143 (1999).

[63] R. Schreck, P. Rieber and P.A. Baeuerle, "Reactive oxygen intermediates as apparently widely used messengers in the activation of the NF-kappa B transcription factor and HIV-1," *Embo J.* **10**, 2247–2258 (1991).

[64] R. Schreck, R. Grassmann, B. Fleckenstein, et al., "Antioxidants selectively suppress activation of NF-kappa B by human T-cell leukemia virus type I Tax protein," *J. Virol.* **66**, 6288–6293 (1992).

[65] R. Lubart, M. Eichler, R. Lavi, et al., "Low-energy laser irradiation promotes cellular redox activity," *Photomed. Laser Surg.* **23**, 3–9 (2005).

[66] E. Alexandratou, D. Yova, P. Handris, et al., "Human fibroblast alterations induced by low power laser irradiation at the single cell level using confocal microscopy," *Photochem. Photobiol. Sci.* **1**, 547–552 (2002).

[67] G. Pal, A. Dutta, K. Mitra, et al., "Effect of low intensity laser interaction with human skin fibroblast cells using fiber-optic nano-probes," *J. Photochem. Photobiol. B.* **86**, 252–261 (2007).

[68] J. Tafur and P.J. Mills, "Low-intensity light therapy: exploring the role of redox mechanisms," *Photomed. Laser Surg.* epub ahead of print (2008).

[69] C.F. Rizzi, J.L. Mauriz, D.S. Freitas Correa, et al., "Effects of low-level laser therapy (LLLT) on the nuclear factor (NF)-kappaB signaling pathway in traumatized muscle," *Lasers Surg. Med.* **38**, 704–713 (2006).

[70] E. Shaulian and M. Karin, "AP-1 as a regulator of cell life and death," *Nat. Cell Biol.* **4**, E131–E136 (2002).

[71] Y. Chinenov and T.K. Kerppola, "Close encounters of many kinds: Fos-Jun interactions that mediate transcription regulatory specificity," *Oncogene* **20**, 2438–2452 (2001).

[72] S. Bergelson, R. Pinkus and V. Daniel, "Intracellular glutathione levels regulate Fos/Jun induction and activation of glutathione S-transferase gene expression," *Cancer Res.* **54**, 36–40 (1994).

[73] D.M. Flaherty, M.M. Monick, A.B. Carter, et al., "Oxidant-mediated increases in redox factor-1 nuclear protein and activator protein-1 DNA binding in asbestos-treated macrophages," *J. Immunol.* **168**, 5675–5681 (2002).

[74] Y.M. Janssen, N.H. Heintz, and B.T. Mossman, "Induction of c-fos and c-jun proto-oncogene expression by asbestos is ameliorated by N-acetyl-L-cysteine in mesothelial cells," *Cancer Res.* **55**, 2085–2089 (1995).

[75] M. Ding, Z. Dong, F. Chen, et al., "Asbestos induces activator protein-1 transactivation in transgenic mice," *Cancer Res.* **59**, 1884–1889 (1999).

[76] D.J. Manalo, A. Rowan, T. Lavoie, et al., "Transcriptional regulation of vascular endothelial cell responses to hypoxia by HIF-1," *Blood* **105**, 659–669 (2005).

[77] C.T. Taylor, "Mitochondria and cellular oxygen sensing in the HIF pathway," *Biochem. J.* **409**, 19–26 (2008).

[78] A.R. Evans, M. Limp-Foster, and M.R. Kelley, "Going APE over ref-1," *Mutat. Res.* **461**, 83–108 (2000).

[79] G. Fritz, "Human APE/Ref-1 protein," *Int. J. Biochem. Cell Biol.* **32**, 925–929 (2000).

[80] M. Asahi and H. Yamamura, "Physiological roles of cyclic nucleotide dependent protein kinases in advanced brain functions," *Tanpakushitsu Kakusan Koso.* **42**, 403–410 (1997).

[81] H.H. van Breugel and P.R. Bar, "Power density and exposure time of He-Ne laser irradiation are more important than total energy dose in photo-biomodulation of human fibroblasts in vitro," *Lasers Surg. Med.* **12**, 528–537 (1992).

[82] W. Yu, J.O. Naim, and R.J. Lanzafame, "The effect of laser irradiation on the release of bFGF from 3T3 fibroblasts," *Photochem. Photobiol.* **59**, 167–170 (1994).

[83] R. Lubart, Y. Wollman, H. Friedmann, et al., "Effects of visible and near-infrared lasers on cell cultures," *J. Photochem. Photobiol. B.* **12**, 305–310 (1992).

[84] N. Grossman, N. Schneid, H. Reuveni, et al., "780 nm low power diode laser irradiation stimulates proliferation of keratinocyte cultures: involvement of reactive oxygen species," *Lasers Surg. Med.* **22**, 212–218 (1998).

[85] P. Moore, T.D. Ridgway, R.G. Higbee, et al., "Effect of wavelength on low-intensity laser irradiation-stimulated cell proliferation in vitro," *Lasers Surg. Med.* **36**, 8–12 (2005).

[86] I. Stadler, R. Evans, B. Kolb, et al., "In vitro effects of low-level laser irradiation at 660 nm on peripheral blood lymphocytes," *Lasers Surg. Med.* **27**, 255–261 (2000).

[87] A.D. Agaiby, L.R. Ghali, R. Wilson, et al., "Laser modulation of angiogenic factor production by T-lymphocytes," *Lasers Surg. Med.* **26**, 357–363 (2000).

[88] L. Almeida-Lopes, J. Rigau, R.A. Zangaro, et al., "Comparison of the low level laser therapy effects on cultured human gingival fibroblasts proliferation using different irradiance and same fluence," *Lasers Surg. Med.* **29**, 179–184 (2001).

[89] D. Hawkins, N. Houreld, and H. Abrahamse, "Low level laser therapy (LLLT) as an effective therapeutic modality for delayed wound healing," *Ann. NY Acad. Sci.* **1056**, 486–493 (2005).

[90] B. Mvula, T. Mathope, T. Moore, et al., "The effect of low level laser irradiation on adult human adipose derived stem cells," *Lasers Med. Sci.* **23**, 277–282 (2008).

[91] T.I. Karu, L.V. Pyatibrat and G.S. Kalendo, "Cell attachment to extracellular matrices is modulated by pulsed radiation at 820 nm and chemicals that modify the activity of enzymes in the plasma membrane," *Lasers Surg. Med.* **29**, 274–281 (2001).

[92] N. Kipshidze, V. Nikolaychik, M.H. Keelan, et al., "Low-power helium: neon laser irradiation enhances production of vascular endothelial growth factor and promotes growth of endothelial cells in vitro," *Lasers Surg. Med.* **28**, 355–364 (2001).

[93] A. Khanna, L.R. Shankar, M.H. Keelan, et al., "Augmentation of the expression of proangiogenic genes in cardiomyocytes with low dose laser irradiation in vitro," *Cardiovasc. Radiat. Med.* **1**, 265–269 (1999).

[94] C.J. Carati, S.N. Anderson, B.J. Gannon, et al., "Treatment of postmastectomy lymphedema with low-level laser therapy: a double blind, placebo-controlled trial," *Cancer.* **98**, 1114–1122 (2003).

[95] T. Hagen, C.T. Taylor, F. Lam, et al., "Redistribution of intracellular oxygen in hypoxia by nitric oxide: effect on HIF1alpha," *Science* **302**, 1975–1978 (2003).

[96] N.M. Burduli and A.A. Gazdanova, "Laser Doppler fluometry in assessment of endothelium state in patients with coronary heart disease and its correction by intravenous laser irradiation of blood," *Klin. Med. (Mosk.)* **86**, 44–47 (2008).

[97] N. Kipshidze, J.R. Petersen, J. Vossoughi, et al., "Low-power laser irradiation increases cyclic GMP synthesis in penile smooth muscle cells in vitro," *J. Clin. Laser Med. Surg.* **18**, 291–294 (2000).

[98] L. Gavish, L. Perez and S.D. Gertz, "Low-level laser irradiation modulates matrix metalloproteinase activity and gene expression in porcine aortic smooth muscle cells," *Lasers Surg. Med.* **38**, 779–786 (2006).

[99] G. Shefer, T.A. Partridge, L. Heslop, et al., "Low-energy laser irradiation promotes the survival and cell cycle entry of skeletal muscle satellite cells," *J. Cell Sci.* **115**, 1461–1469 (2002).

[100] G. Shefer, U. Oron, A. Irintchev, et al., "Skeletal muscle cell activation by low-energy laser irradiation: a role for the MAPK/ERK pathway," *J. Cell Physiol.* **187**, 73–80 (2001).

[101] G. Shefer, I. Barash, U. Oron, et al., "Low-energy laser irradiation enhances de novo protein synthesis via its effects on translation-regulatory proteins in skeletal muscle myoblasts," *Biochim. Biophys. Acta.* **1593**, 131–139 (2003).

[102] U. Oron, "Photoengineering of tissue repair in skeletal and cardiac muscles," *Photomed. Laser Surg.* **24**, 111–120 (2006).

[103] F. Correa, R.A. Lopes-Martins, J.C. Correa, et al., "Low-level laser therapy (GaAs lambda = 904 nm) reduces inflammatory cell migration in mice with lipopolysaccharide-induced peritonitis," *Photomed. Laser Surg.* **25**, 245–249 (2007).

[104] E.S. Boschi, C.E. Leite, V.C. Saciura, et al., "Anti-inflammatory effects of low-level laser therapy (660 nm) in the early phase in carrageenan-induced pleurisy in rat," *Lasers Surg. Med.* **40**, 500–508 (2008).

[105] A.M. Barbosa, A.B. Villaverde, L. Guimaraes-Souza, et al., "Effect of low-level laser therapy in the inflammatory response induced by Bothrops jararacussu snake venom," *Toxicon.* **51**, 1236–1244 (2008).

[106] R. Albertini, A.B. Villaverde, F. Aimbire, et al., "Anti-inflammatory effects of low-level laser therapy (LLLT) with two different red wavelengths (660 nm and 684 nm) in carrageenan-induced rat paw edema," *J. Photochem. Photobiol. B*. **89**, 50–55 (2007).

[107] R. Albertini, F.S. Aimbire, F.I. Correa, et al., "Effects of different protocol doses of low power gallium-aluminum-arsenate (Ga-Al-As) laser radiation (650 nm) on carrageenan induced rat paw ooedema," *J. Photochem. Photobiol. B*. **74**, 101–107 (2004).

[108] F. Aimbire, R. Albertini, M.T. Pacheco, et al., "Low-level laser therapy induces dose-dependent reduction of TNFalpha levels in acute inflammation," *Photomed. Laser Surg*. **24**, 33–37 (2006).

[109] M.A. Naeser, K.A. Hahn, B.E. Lieberman, et al., "Carpal tunnel syndrome pain treated with low-level laser and microamperes transcutaneous electric nerve stimulation: a controlled study," *Arch. Phys. Med. Rehabil*. **83**, 978–988 (2002).

[110] T.F. Elwakil, A. Elazzazi, and H. Shokeir, "Treatment of carpal tunnel syndrome by low-level laser versus open carpal tunnel release," *Lasers Med. Sci*. **22**, 265–270 (2007).

[111] A. Ekim, O. Armagan, F. Tascioglu, et al., "Effect of low level laser therapy in rheumatoid arthritis patients with carpal tunnel syndrome," *Swiss Med. Wkly*. **137**, 347–352 (2007).

[112] G. Ciais, M. Namer, M. Schneider, et al., "Laser therapy in the prevention and treatment of mucositis caused by anticancer chemotherapy," *Bull. Cancer*. **79**, 183–191 (1992).

[113] S.F. Wong and P. Wilder-Smith, "Pilot study of laser effects on oral mucositis in patients receiving chemotherapy," *Cancer J*. **8**, 247–254 (2002).

[114] A.G. Nes and M.B. Posso, "Patients with moderate chemotherapy-induced mucositis: pain therapy using low intensity lasers," *Int. Nurs. Rev*. **52**, 68–72 (2005).

[115] G.C. Jaguar, J.D. Prado, I.N. Nishimoto, et al., "Low-energy laser therapy for prevention of oral mucositis in hematopoietic stem cell transplantation," *Oral Dis*. **13**, 538–543 (2007).

[116] D.M. Ferreira, R.A. Zangaro, A.B. Villaverde, et al., "Analgesic effect of He-Ne (632.8 nm) low-level laser therapy on acute inflammatory pain," *Photomed. Laser Surg*. **23**, 177–181 (2005).

[117] A.P. Castano, T. Dai, I. Yaroslavsky, et al., "Low-level laser therapy for zymosan-induced arthritis in rats: importance of illumination time," *Lasers Surg. Med*. **39**, 543–550 (2007).

[118] L. Brosseau, V. Robinson, G. Wells, et al., "WITHDRAWN: Low level laser therapy (classes III) for treating osteoarthritis," *Cochrane Database Syst. Rev*. CD002046 (2007).

[119] G. Jamtvedt, K.T. Dahm, A. Christie, et al., "Physical therapy interventions for patients with osteoarthritis of the knee: an overview of systematic reviews," *Phys. Ther*. **88**, 123–136 (2008).

[120] J.M. Bjordal, C. Couppe, R.T. Chow, et al., "A systematic review of low level laser therapy with location-specific doses for pain from chronic joint disorders," *Aust. J. Physiother*. **49**, 107–116 (2003).

[121] H.S. Yu, K.L. Chang, C.L. Yu, et al., "Low-energy helium-neon laser irradiation stimulates interleukin-1 alpha and interleukin-8 release from cultured human keratinocytes," *J. Invest. Dermatol*. **107**, 593–596 (1996).

[122] V.K. Poon, L. Huang and A. Burd, "Biostimulation of dermal fibroblast by sublethal Q-switched Nd:YAG 532 nm laser: collagen remodeling and pigmentation," *J. Photochem. Photobiol. B*. **81**, 1–8 (2005).

[123] M.L. McNeely, S. Armijo Olivo, and D.J. Magee, "A systematic review of the effectiveness of physical therapy interventions for temporomandibular disorders," *Phys. Ther.* **86**, 710–725 (2006).

[124] E.J. Neiburger, "Rapid healing of gingival incisions by the helium-neon diode laser," *J. Mass. Dent. Soc.* **48**, 8–13, 40 (1999).

[125] T.N. Demidova-Rice, E.V. Salomatina, A.N. Yaroslavsky, et al., "Low-level light stimulates excisional wound healing in mice," *Lasers Surg. Med.* **39**, 706–715 (2007).

[126] A.R. Medrado, L.S. Pugliese, S.R. Reis, et al., "Influence of low level laser therapy on wound healing and its biological action upon myofibroblasts," *Lasers Surg. Med.* **32**, 239–244 (2003).

[127] X. Zhu, Y. Chen and X. Sun, "A study on expression of basic fibroblast growth factors in periodontal tissue following orthodontic tooth movement associated with low power laser irradiation," *Hua Xi Kou Qiang Yi Xue Za Zhi* **20**, 166–168 (2002).

[128] M. Bayat, E. Ansari, N. Gholami, et al., "Effect of low-level helium-neon laser therapy on histological and ultrastructural features of immobilized rabbit articular cartilage," *J. Photochem. Photobiol. B.* **87**, 81–87 (2007).

[129] N. Weiss and U. Oron, "Enhancement of muscle regeneration in the rat gastrocnemius muscle by low energy laser irradiation," *Anat. Embryol. (Berl.)* **186**, 497–503 (1992).

[130] D. Avni, S. Levkovitz, L. Maltz, et al., "Protection of skeletal muscles from ischemic injury: low-level laser therapy increases antioxidant activity," *Photomed. Laser Surg.* **23**, 273–277 (2005).

[131] R.A. Lopes-Martins, R.L. Marcos, P.S. Leonardo, et al., "Effect of low-level laser (Ga-Al-As 655 nm) on skeletal muscle fatigue induced by electrical stimulation in rats," *J. Appl. Physiol.* **101**, 283–288 (2006).

[132] E.C. Leal Junior, R.A. Lopes-Martins, F. Dalan, et al., "Effect of 655-nm low-level laser therapy on exercise-induced skeletal muscle fatigue in humans," *Photomed. Laser Surg.* **26**, 419–424 (2008).

[133] E.C. Leal Junior, R.A. Lopes-Martins, A.A. Vanin, et al., "Effect of 830 nm low-level laser therapy in exercise-induced skeletal muscle fatigue in humans," *Lasers Med. Sci.* (2008).

[134] D. Gigo-Benato, S. Geuna, and S. Rochkind, "Phototherapy for enhancing peripheral nerve repair: a review of the literature," *Muscle Nerve* **31**, 694–701 (2005).

[135] J.J. Anders, S. Geuna, and S. Rochkind, "Phototherapy promotes regeneration and functional recovery of injured peripheral nerve," *Neurol. Res.* **26**, 233–239 (2004).

[136] J.J. Anders, R.C. Borke, S.K. Woolery, et al., "Low power laser irradiation alters the rate of regeneration of the rat facial nerve," *Lasers Surg. Med.* **13**, 72–82 (1993).

[137] S. Rochkind, V. Drory, M. Alon, et al., "Laser phototherapy (780 nm), a new modality in treatment of long-term incomplete peripheral nerve injury: a randomized double-blind placebo-controlled study," *Photomed. Laser Surg.* **25**, 436–442 (2007).

[138] R.T. Chow, M.A. David, and P.J. Armati, "830 nm laser irradiation induces varicosity formation, reduces mitochondrial membrane potential and blocks fast axonal flow in small and medium diameter rat dorsal root ganglion neurons: implications for the analgesic effects of 830 nm laser," *J. Peripher. Nerv. Syst.* **12**, 28–39 (2007).

[139] R.T. Chow and L. Barnsley, "Systematic review of the literature of low-level laser therapy (LLLT) in the management of neck pain," *Lasers Surg. Med.* **37**, 46–52 (2005).

[140] J.M. Bjordal, R.A. Lopes-Martins, and V.V. Iversen, "A randomised, placebo controlled trial of low level laser therapy for activated Achilles tendinitis with microdialysis measurement of peritendinous prostaglandin E2 concentrations," *Br. J. Sports Med.* **40**, 76–80; discussion 76–80 (2006).

[141] I. Yaksich, L.C. Tan, and V. Previn, "Low energy laser therapy for treatment of post-herpetic neuralgia," *Ann. Acad. Med. Singapore.* **22**, 441–442 (1993).

[142] A.L. Pinheiro, E.T. Cavalcanti, T.I. Pinheiro, et al., "Low-level laser therapy is an important tool to treat disorders of the maxillofacial region," *J. Clin. Laser Med. Surg.* **16**, 223–226 (1998).

[143] G.D. Baxter, D.M. Walsh, J.M. Allen, et al., "Effects of low intensity infrared laser irradiation upon conduction in the human median nerve in vivo," *Exp. Physiol.* **79**, 227–234 (1994).

[144] E. Vinck, P. Coorevits, B. Cagnie, et al., "Evidence of changes in sural nerve conduction mediated by light emitting diode irradiation," *Lasers Med. Sci.* **20**, 35–40 (2005).

[145] L.A. Lavery, D.P. Murdoch, J. Williams, et al., "Does anodyne light therapy improve peripheral neuropathy in diabetes? A double-blind, sham-controlled, randomized trial to evaluate monochromatic infrared photoenergy," *Diabetes Care* **31**, 316–321 (2008).

[146] L.B. Harkless, S. DeLellis, D.H. Carnegie, et al., "Improved foot sensitivity and pain reduction in patients with peripheral neuropathy after treatment with monochromatic infrared photo energy–MIRE," *J. Diabetes Complications.* **20**, 81–87 (2006).

[147] J.K. Clifft, R.J. Kasser, T.S. Newton, et al., "The effect of monochromatic infrared energy on sensation in patients with diabetic peripheral neuropathy: a double-blind, placebo-controlled study," *Diabetes Care* **28**, 2896–2900 (2005).

[148] D.R. Leonard, M.H. Farooqi and S. Myers, "Restoration of sensation, reduced pain, and improved balance in subjects with diabetic peripheral neuropathy: a double-blind, randomized, placebo-controlled study with monochromatic near-infrared treatment," *Diabetes Care* **27**, 168–172 (2004).

[149] M.W. Powell, D.H. Carnegie, and T.J. Burke, "Reversal of diabetic peripheral neuropathy with phototherapy (MIRE) decreases falls and the fear of falling and improves activities of daily living in seniors," *Age Ageing* **35**, 11–16 (2006).

[150] M. Yamazaki, Y. Miura, R. Tsuboi, et al., "Linear polarized infrared irradiation using Super Lizer is an effective treatment for multiple-type alopecia areata," *Int. J. Dermatol.* **42**, 738–740 (2003).

[151] P.J. Pontinen, T. Aaltokallio, and P.J. Kolari, "Comparative effects of exposure to different light sources (He-Ne laser, InGaAl diode laser, a specific type of noncoherent LED) on skin blood flow for the head," *Acupunct. Electrother. Res.* **21**, 105–118 (1996).

[152] F. Schwartz, C. Brodie, E. Appel, et al., "Effect of helium/neon laser irradiation on nerve growth factor synthesis and secretion in skeletal muscle cultures," *J. Photochem. Photobiol. B.* **66**, 195–200 (2002).

[153] H.S. Yu, C.S. Wu, C.L. Yu, et al., "Helium-neon laser irradiation stimulates migration and proliferation in melanocytes and induces repigmentation in segmental-type vitiligo," *J. Invest. Dermatol.* **120**, 56–64 (2003).

[154] T. Omi, P. Bjerring, S. Sato, et al., "420 nm intense continuous light therapy for acne," *J. Cosmet. Laser Ther.* **6**, 156–162 (2004).

[155] P. Papageorgiou, A. Katsambas, and A. Chu, "Phototherapy with blue (415 nm) and red (660 nm) light in the treatment of acne vulgaris," *Br. J. Dermatol.* **142**, 973–978 (2000).

[156] B.D. Zelickson, S.L. Kilmer, E. Bernstein, et al., "Pulsed dye laser therapy for sun damaged skin," *Lasers. Surg. Med.* **25**, 229–236 (1999).

[157] C.H. Chen, J.L. Tsai, Y.H. Wang, et al., "Low-level laser irradiation promotes cell proliferation and mRNA expression of type I collagen and decorin in porcine achilles tendon fibroblasts in vitro," *J. Orthop. Res.* (2008).

[158] T. Omi, S. Kawana, S. Sato, et al., "Cutaneous immunological activation elicited by a low-fluence pulsed dye laser," *Br. J. Dermatol.* **153** Suppl 2, 57–62 (2005).

[159] M.A. Trelles, I. Allones, J.L. Levy, et al., "Combined nonablative skin rejuvenation with the 595- and 1450-nm lasers," *Dermatol. Surg.* **30**, 1292–1298 (2004).

[160] M.A. Trelles, S. Mordon, and R.G. Calderhead, "Facial rejuvenation and light: our personal experience," *Lasers Med. Sci.* **22**, 93–99 (2007).

[161] S.Y. Lee, K.H. Park, J.W. Choi, et al., "A prospective, randomized, placebo-controlled, double-blinded, and split-face clinical study on LED phototherapy for skin rejuvenation: clinical, profilometric, histologic, ultrastructural, and biochemical evaluations and comparison of three different treatment settings," *J. Photochem. Photobiol. B.* **88**, 51–67 (2007).

27

Low-Level Laser Therapy in Stroke and Central Nervous System

Ying-Ying Huang
Wellman Center for Photomedicine, Massachusetts General Hospital, Boston MA; Department of Dermatology, Harvard Medical School, Boston MA Aesthetic Plastic Laser Center, Guangxi Medical University, Nanning, Guangxi, P. R. China

Michael R Hamblin
Wellman Center for Photomedicine, Massachusetts General Hospital, Boston MA Department of Dermatology, Harvard Medical School, Boston MA Harvard-MIT Division of Health Sciences and Technology, Cambridge MA

Luis De Taboada
PhotoThera Inc, Carlsbad, CA

Low-level laser/light therapy (LLLT) for neurological disorders in the central nervous system (CNS) is currently an experimental concept. The broad goals for clinical utilization are the prevention and/or repair of damage, relief of symptoms, slowing of disease progression, and correction of genetic abnormalities. Experimental studies have tested and continue to test these goals by investigating LLLT in animal models of diseases and injuries that affect the brain and spinal cord. Successful clinical trials have been carried out of transcranial laser therapy for stroke. Discoveries concerning the molecular basis of various neurological diseases, combined with advances that have been made in understanding the molecular and cellular mechanisms in LLLT, both *in vitro* and *in vivo*, have allowed rational light-based therapeutic approaches for a wide variety of CNS disorders to be investigated. Limitations in knowledge are still apparent, such as the optimal wavelength, light source, doses, pulsed or CW, polarization state, treatment timing, and repetition frequency. Collaborative efforts between clinicians and basic researchers will likely increase the usage and understanding of effective laser-based therapies in the CNS.

Key words: low-level laser therapy, neurological disorders, stroke, ROS, mitochondria, photobiomodulation.

27.1 Introduction

Low-level laser (or light) therapy (LLLT) has been clinically applied for many indications in medicine that require the following processes: protection from cell and tissue death, stimulation of healing and repair of injuries, and reduction of pain, swelling, and inflammation. One area that is attracting growing interest is the use of LLLT to treat stroke, traumatic brain injury, neurodegenerative diseases, and spinal cord injuries. The notable lack of any effective drug-based therapies for most of these diseases has motivated researchers to consider the use of light as a real approach to mitigating what is considered to be a group of serious diseases; the fact that near-infrared light can penetrate into the brain and spinal cord allows noninvasive treatment to be carried out with a low likelihood of treatment-related adverse events. Although in the past it was generally accepted that the central nervous system could not repair itself, recent discoveries in the field of neuronal stem cells have brought this dogma into question. LLLT may have beneficial effects in the acute treatment of brain damage after stroke or injury, but may also favorably impact the more chronic degenerative brain diseases.

27.2 Photobiology of Low-Level Laser Therapy

The first law of photobiology states that for low-power light to have any effect on a living biological system, the photons must be absorbed by electronic absorption bands belonging to some molecular chromophore or photoacceptor [1]. One approach to finding the identity of this chromophore(s) relevant to LLLT is to carry out action spectra that show which wavelength(s) of light is most effectively used in a specific chemical reaction. It was suggested in 1989 by Karu that the mechanism of LLLT at the cellular level was based on the absorption of monochromatic visible and near infrared light (NIRL) by components of the cellular respiratory chain [2].

There are perhaps three main areas of medicine and veterinary practice where LLT has a major role to play. These are wound healing, tissue repair, and prevention of tissue death; relief of inflammation in chronic diseases and injuries with its associated pain and edema; and relief of neurogenic pain and some neurological problems. Figure 27.1 shows the first law of photobiology and the major roles of LLLT.

Biological responses of living cells to photon irradiation are initiated by photon absorption in intracellular chromophores or photoacceptors. The absorbed photon's energy excites the photoacceptor molecule into a more energetic (higher) electronic state, resulting in a physical and/or chemical molecular change ultimately leading to the cell's biological response. Potential intracellular photoacceptors have been suggested in the previous chapter (Chapter 28). However, in all cases the photon/photoacceptor interaction depends on the absorption spectrum of the photoacceptor and the photon wavelength; matching the wavelength to the peak in the photoacceptor absorption spectrum maximizes the probability of absorption and very likely the cell's biological response.

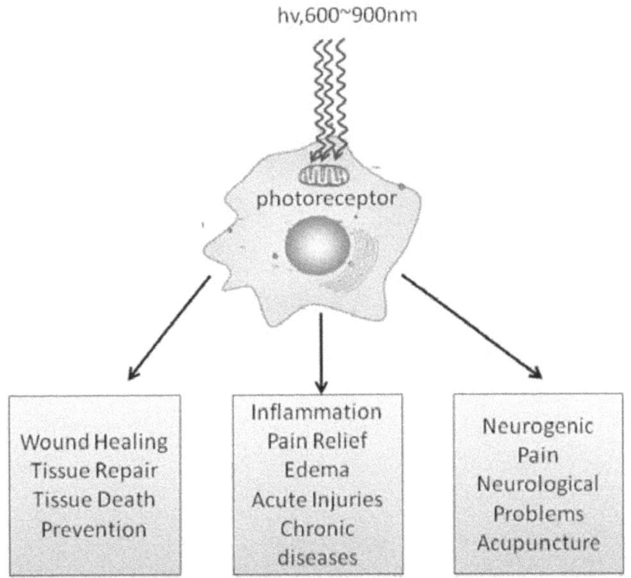

FIGURE 27.1: Schematic representation of the main areas of application of LLLT.

27.3 LLLT Effects on Nerves

The main effects of LLLT on mammalian cells in general have been discussed in detail in the preceding chapter and are graphically illustrated in Figure 27.2. To summarize: there is good evidence that LLLT increases mitochondrial respiration probably by activating cytochrome c oxidase [3], increases the production of adenosine triphosphate (ATP) [4], and increases the amount of intracellular reactive oxygen species [5]. In many cases increases in cell proliferation, migration, and resistance to death have been observed [6]. Recently, the Hamblin group also found LLLT using 810-nm increased activation of the transcription factor NF-kB [7].

27.3.1 LLLT on neuronal cells

The nervous system is divided into two parts: the central nervous system (CNS), which consists of the brain and spinal cord, and the peripheral nervous system (PNS), which consists of cranial and spinal nerves along with their associated ganglia. The nervous system is, on a small scale, primarily made up of neurons. Neurons exist in a number of different shapes and sizes and can be classified by their morphology and function. The anatomist Camillo Golgi grouped neurons into two types; type I with long axons used to move signals over long distances and type II without axons. Type I cells can be further divided by where the cell body or soma is located. The basic morphology of type I neurons, represented by spinal motor neurons, consists of a cell body called the soma and a long thin axon that is covered by the myelin sheath. Around the cell body is a branching dendritic tree that receives signals from other neurons. The end of the axon has branching terminals (axon terminal) that release transmitter substances into a gap called the synaptic cleft between the terminals and the dendrites of the next neuron [8]. Figure 27.3 shows LLLT interacting with neuronal cells.

In cultured human neuronal cells, LLLT resulted in doubling of the ATP content. LLLT also increased heat shock proteins and preserved mitochondrial function [3]. Dr. Oron showed that

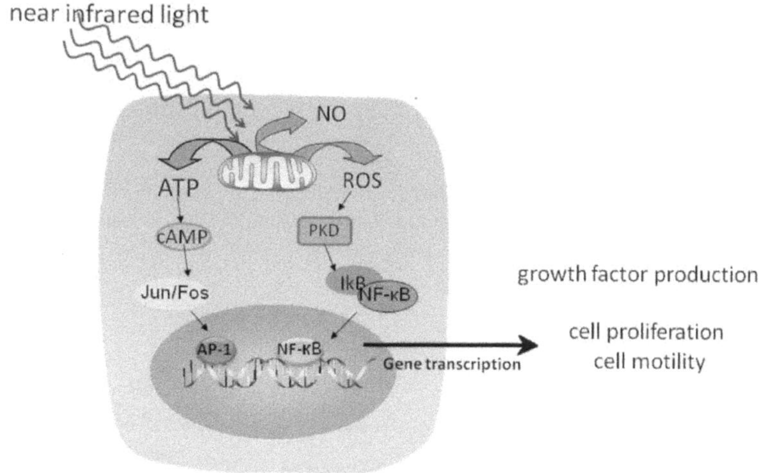

FIGURE 27.2: Mechanism of low-level light therapy (LLLT).

Ga-As laser (808 nm, 50 mW/cm^2, 0.05 J/cm^2) irradiation can enhance ATP production in normal human neural progenitor cells in culture, ATP concentrations were measured 10 min after laser application [9].

Ignatov et al. [10] used intracellular dialysis and membrane voltage clamping were used to show that He-Ne laser irradiation of a pond snail neuron at a dose of 0.07 mJ increases the amplitude of the potential-dependent slow potassium current, while a dose of 7 mJ decreased this current. Bupivacaine can also suppress the potassium current, and it was found that the combined application of laser irradiation at a dose of 0.7 mJ increased the blocking effect of 10 μM bupivacaine on the slow potassium current, while an irradiation dose of 0.07 mJ weakened the effect of bupivacaine.

Recently, Ying et al. reported that pretreatment with NIR light via light-emitting diode (LED) significantly suppressed rotenone or MPP(+)-induced apoptosis in both striatal and visual cortical neurons from newborn rats, and that pretreatment plus LED treatment during neurotoxin exposure was significantly better than LED treatment alone during exposure to neurotoxins [11].

These *in vitro* findings suggest that low-level laser therapy (LLLT) could have beneficial effects in animal models of neurological disorders and could be a treatment for stroke.

27.3.2 LLLT on nerves *in vivo*

In 1986 Nissan first used laser therapy for nerves on a sciatic nerve of rats with He-Ne Laser (4 J/cm^2) [12]. Animal models have been employed to study LLLT effects in enhancing peripheral nerve repair [13, 14]. Anders et al. [15] studied LLLT for regenerating crushed rat facial nerves; by comparing 361, 457, 514, 633, 720, and 1064 nm and found the best response with 162.4 J/cm^2 of 633 nm He-Ne laser.

27.4 Human Skull Transmission Measurements

Wan et al. in 1981 reported the transmittance of nonionizing radiation in human tissues [16]. Spectral transmittance of 400–865 nm radiation through various human structures, including the

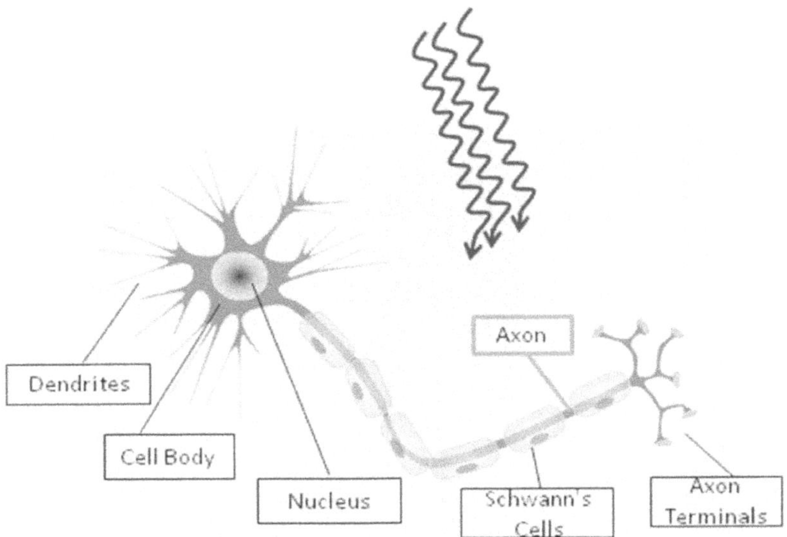

FIGURE 27.3: LLLT on neuronal cells.

skull with scalp, the chest wall, abdominal wall and scrotum, is presented. There is essentially no visible light of wavelengths shorter than 500 nm transmitted through the chest or the abdominal wall. Transmittance of all tissues increases progressively with wavelengths from 600 to 814 nm. Tissue thickness, optical absorption, and scattering are major influencing factors [17]. Lychagov et al. presented the recent results of measurements of transmittance of high-power laser irradiation through skull bones and scalp [18]. Figure 27.4 shows the *ex vivo* human skull transmission measurement experimental setup. Character of transmittance was investigated and characteristics of heterogeneity of the scattering structure of the skull bones were shown. Besides that, temperature variations of skull and scalp surfaces under exposure of high-power laser irradiation during experiments was controlled. Experimental results were verified by Monte-Carlo simulations. Figure 27.5 shows the transmission of 808-nm NIR light through fresh human cadaver brain as measured by an infrared camera.

27.5 The Problem of Stroke

27.5.1 Epidemic of stroke

Stroke is the third leading cause of death in the United States after heart disease and cancer [19]. Each year, approximately 780,000 people experience a new or recurrent stroke. Approximately 600,000 of these are first attacks, and 180,000 are recurrent attacks [20]. The 3-month mortality rate from ischemic stroke is approximately 12%. Internationally, millions of people have a new or recurrent stroke each year, and nearly a quarter of these people die. Globally, stroke death rates vary widely; the highest rates are in Portugal, China, Korea, and most of Eastern Europe and the lowest rates are in Switzerland, Canada, and the United States.

Strokes can be classified into two major categories, ischemic and hemorrhagic. Ischemic strokes account for over 80% of all strokes. The most common cause of ischemic stroke is the blockage of an artery in the brain by a clot, thrombosis, embolism, or stenosis. Hemorrhagic strokes are

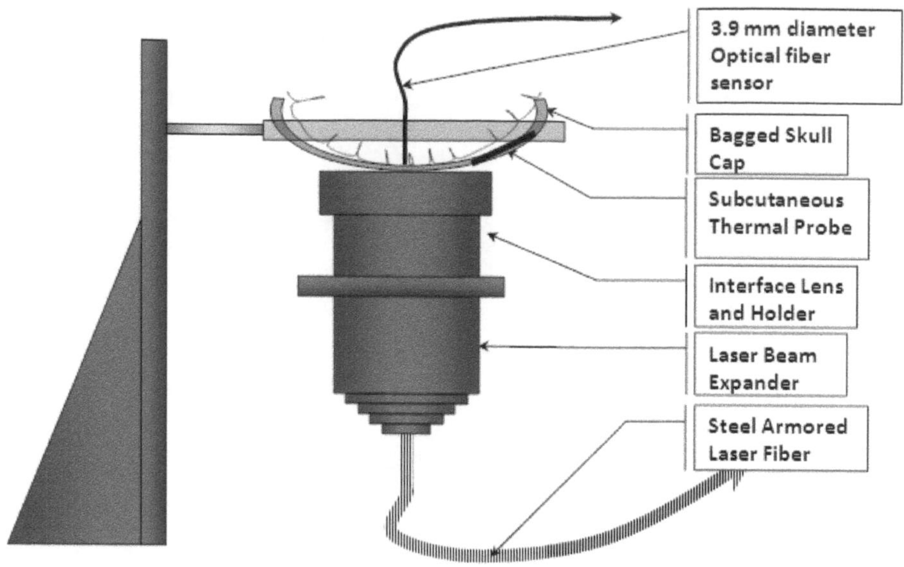

FIGURE 27.4: *Ex vivo* human skull transmission measurement.

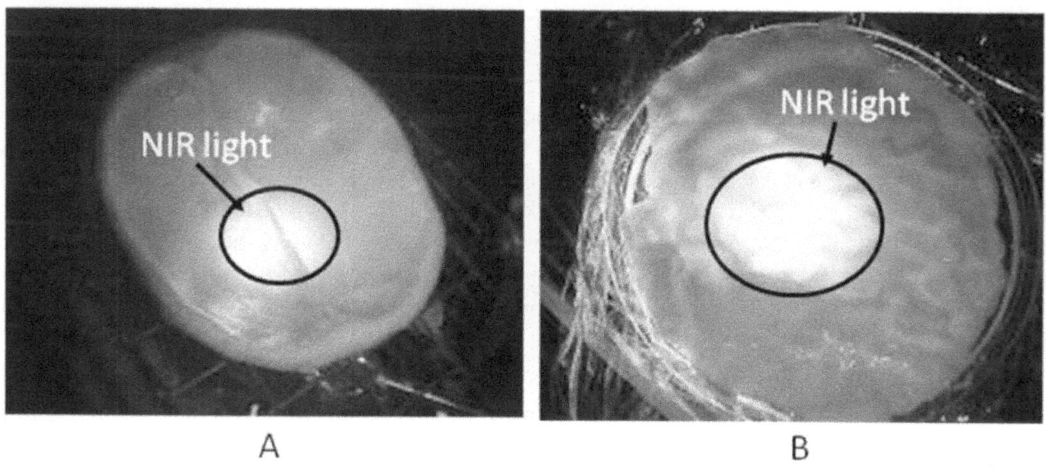

FIGURE 27.5: IR camera photos – fresh human cadaver brain. 808-nm infrared wavelength transmits through skin, bone, and dura to reach the cortex (A – skull cap; B – 15-mm-thick slice of cerebrum).

the result of the rupture of a cerebral artery, which can lead to spasm of the artery and various degrees of bleeding. Until recently, the care of subjects with ischemic stroke was largely supportive, focusing on prevention and treatment of respiratory and cardiovascular complications. Common acute complications of stroke include pneumonia, urinary tract infection, and pulmonary embolism. Long-term morbidity in survivors of stroke is common, with ambulation difficulty in 20%, need for assistance in activities of daily living in 30%, and vocational disability in 50–70% of patients.

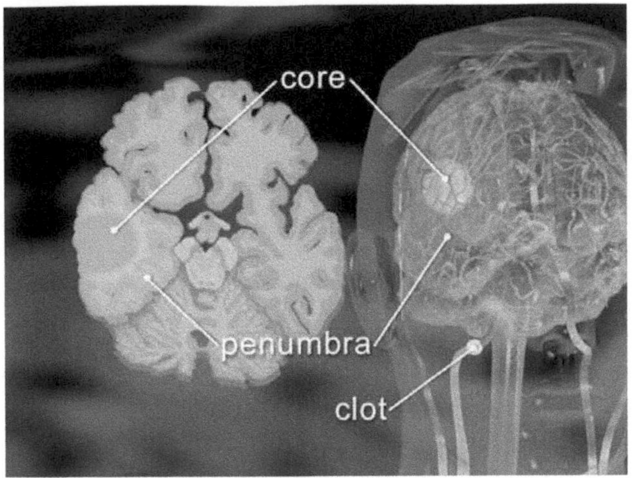

FIGURE 27.6: Acute ischemic stroke physiology.

27.5.2 Mechanisms of brain injury after stroke

Ischemic stroke is a complex entity with multiple etiologies and variable clinical manifestations. Thrombosis can form in the extracranial and intracranial arteries when the intima is roughened and plaque forms along the injured vessel. The endothelial injury permits platelets to adhere and aggregate, then coagulation is activated and thrombus develops at the site of the plaque. When the compensatory mechanism of collateral circulation fails, perfusion is compromised, leading to decreased perfusion and cell death. During an embolic stroke, a clot travels to the brain through an artery. Cells in the core ischemic zone die within minutes. Ischemia impairs the patient's neurologic function. Cells in the penumbra of the injury remain potentially salvageable for hours [21]. Figure 27.6 shows the physiology of acute ischemic stroke.

When an ischemic stroke occurs, the blood supply to the brain is interrupted, and brain cells are deprived of the glucose and oxygen they need to function. The human brain comprises 2% of body weight but requires 20% of total oxygen consumption [22]. The brain requires this large amount of oxygen to generate sufficient ATP by oxidative phosphorylation to maintain and restore ionic gradients. One estimate suggests that the Na+/K+ATPase found on the plasma membrane of neurons, consumes 70% of the energy supplied to the brain. This ion pump maintains the high intracellular K+ concentration and the low intracellular Na+ concentrate ion necessary for the propagation of action potentials. After global ischemia, mitochondrial inhibition of ATP synthesis leads to the ATP being consumed within 2 min, this causes neuronal plasma membrane depolarization, release of potassium into the extracellular space and entry of sodium into cells [23]. Energy failure also prevents the plasma membrane Ca^{2+} ATPase from maintaining the very low concentrations of calcium that are normally present within each cell.

Stroke can manifest as focal injuries (contusions, lacerations, hematomas) or diffuse axonal injury (DAI) or a combination thereof. Diffuse axonal injury has been recognized as one of the main consequences of blunt head trauma and blast injuries [24]. Secondary injuries are attributable to further cellular damage that results from follow-up effects of the primary injuries and develop over a period of hours or days following initial traumatic assault. Secondary brain injury is mediated through excitotoxic cell death in which injured neurons depolarize and release glutamate [25]. Neighboring cells are in turn depolarized by the excessive glutamate concentrations, and results in a vicious spiral of increasing glutamate concentration. Depolarized neurons suffer a huge influx of sodium and calcium but could survive if they had sufficient ATP to power the Na+/K+-ATPase pumps and handle this osmotic load. However in the setting of metabolic failure they exhibit cytotoxic edema, with eventual loss of viability [26]. An influx of calcium linked to delayed damage, a decrease in mitochondrial membrane potential, and increased production of reactive oxygen species are observed. Other biochemical processes exacerbating stroke include the activation of astrocytes and microglia leading to increased cytokines contributing to inflammation and characterized by increased prostaglandin E2 [27]. Excitotoxic cell death mediated by glutamate also affects glial cells as well as neurons, in particular oligodendrocytes [28]. Reactive oxygen species (ROS) have been implicated as key participants of other acute CNS injuries, such as TBI, spinal cord injury, and ischemia, as well as chronic neurodegenerative diseases [29]. Stroke causes physical disruption of neuronal tissue that sets into motion secondary damage resulting in the death of additional tissue. Hypoperfusion and ischemia induced increases in the formation of superoxide and nitric oxide (NO) have been reported after stroke, predicting a role for oxidant stress during damage [30]. Cellular protection against these ROS involves an elaborate antioxidant defense system, including that associated with manganese superoxide dismutase (MnSOD).

27.5.3 Thrombolysis therapy of stroke

Thrombolytic therapy is the only intervention of proven and substantial benefit for select patients with acute cerebral ischemia [31]. The evidence based for thrombolysis therapy includes 21 completed randomized controlled clinical trials enrolling 7152 patients, using various agents, doses, time windows, and intravenous or intra-arterial modes of administration [32].

The main agent that has been employed is recombinant tissue plasminogen activator (t-PA) [33], t-PA is produced endogenously by endothelial cells and is relatively fibrin specific. T-PA works by converting the proenzyme, plasminogen to the activated enzyme plasmin. Activated plasmin in turn dissolves fibrin clots into low molecular weight fibrin degradation products. Other thrombolytics that have been used include streptokinase [34, 35].

Time lost is brain lost in acute cerebral ischemia. In a typical middle cerebral artery ischemic stroke, two million nerve cells are lost each minute in which reperfusion has not been achieved [36]. A pooled analysis of all 2775 patients enrolled in the first 6 intravenous tPA trials provided clear and convincing evidence of a time-dependent benefit of thrombolytic therapy [37].

27.5.4 Investigational neuroprotectants and pharmacological intervention

Neuroprotection is defined as any strategy, or combination of strategies, that antagonizes, interrupts, or slows the sequence of injurious biochemical and molecular events that, if left unchecked, would eventuate in irreversible ischemic injury to the brain. There have been an enormous variety of agents and strategies that have received clinical scrutiny, each justified by a pathophysiological rationale. In all, approximately 165 ongoing or completed clinical trials have been published [38] and are summarized in Figure 27.7. There has been an almost universal outcome of failure for all of these trials, with exceptions of some small hint of efficacy in only a few cases.

27.6 TLT for Stroke

The beneficial effects of NIR light on cells and neurons *in vitro*, together with the demonstrated ability of NIR light to penetrate into the brain, strongly suggested that transcranial LLLT should be studied as a therapy for stroke.

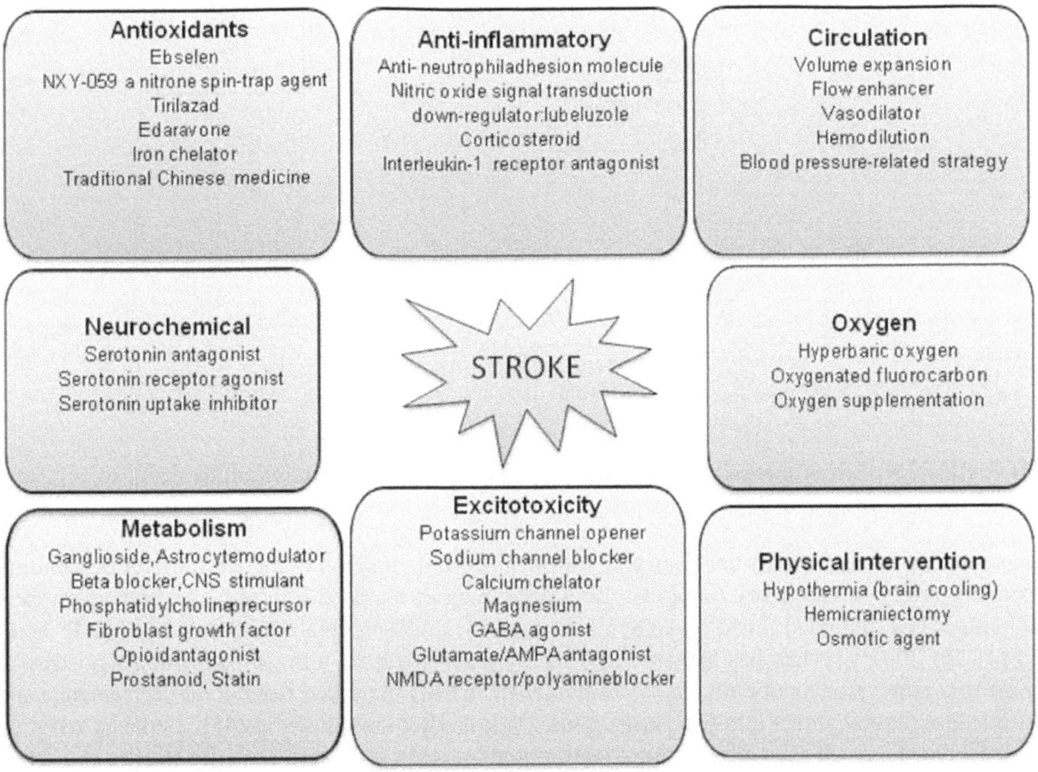

Antioxidants
Ebselen
NXY-059 a nitrone spin-trap agent
Tirilazad
Edaravone
Iron chelator
Traditional Chinese medicine

Anti-inflammatory
Anti-neutrophiladhesion molecule
Nitric oxide signal transduction
down-regulator:lubeluzole
Corticosteroid
Interleukin-1 receptor antagonist

Circulation
Volume expansion
Flow enhancer
Vasodilator
Hemodilution
Blood pressure-related strategy

Neurochemical
Serotonin antagonist
Serotonin receptor agonist
Serotonin uptake inhibitor

STROKE

Oxygen
Hyperbaric oxygen
Oxygenated fluorocarbon
Oxygen supplementation

Metabolism
Ganglioside,Astrocytemodulator
Beta blocker,CNS stimulant
Phosphatidylcholineprecursor
Fibroblast growth factor
Opioidantagonist
Prostanoid, Statin

Excitotoxicity
Potassium channel opener
Sodium channel blocker
Calcium chelator
Magnesium
GABA agonist
Glutamate/AMPAantagonist
NMDA receptor/polyamineblocker

Physical intervention
Hypothermia (brain cooling)
Hemicraniectomy
Osmotic agent

FIGURE 27.7: Clinical trials of pharmacological and physical therapies for stroke.

Transcranial laser therapy (TLT) at 808–810 nm can penetrate the brain and was shown to cause the enhanced production of ATP in the rat cerebral cortex [3]. Findings of increased neurogenesis in the subventricular zone (SVZ) were reported in an ischemic stroke animal model treated with TLT [39]. Based on these findings, it is thought that TLT may have multiple mechanisms of action and could be beneficial in acute ischemic stroke [40].

27.6.1 TLT in animal models for stroke

Although there is no one animal model that identically mimics stroke in humans, the use of animal models is essential for the development of therapeutic interventions for stroke. Outcome measures for animal models involve functional measures and evaluation of the infarct size. Recommendations from the Stroke Therapy Academic Industry Roundtable (STAIR) for preclinical animal models are for initial studies to be conducted in the rat followed by a second species, specifically primates. Ischemia is typically induced by occluding the middle cerebral artery (MCA) in the animal. The MCA is most often used to simulate human stroke as most human strokes are due to the occlusion of this vessel or one of its branches [41]. Animals that have been used in ischemia models for various specific reasons include rats, mice, gerbils, cats, rabbits, dogs, pigs, and nonhuman primates. Figure 27.8 shows noninvasive parameter measurement of animal models' skull.

The first animal model for TLT on the brain was reported by Azbel, using a rat model, for findings of synaptic conductance of rat hippocampal neurons in 1993 [42]. *In vivo* studies have suggested that infrared laser therapy could be beneficial for the treatment of acute myocardial infarction as shown by Ad and Oron who showed in 2001 that TLT reduced the loss of myocardial tissue and

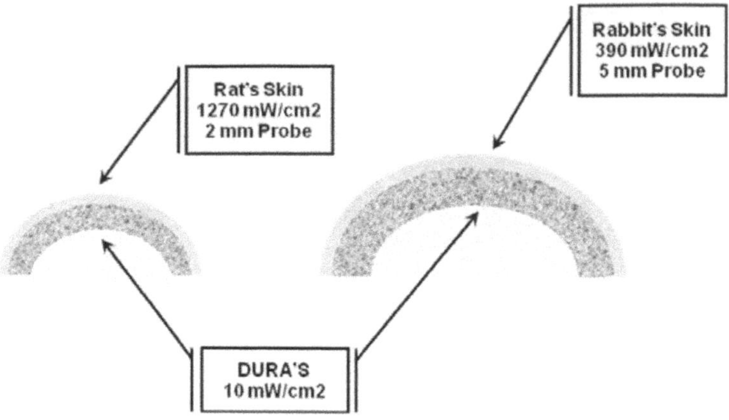

FIGURE 27.8: Transmission measurements in the skull of different animal models.

the severity of acute myocardial infarction following chronic ligation of the left anterior descending coronary artery in laboratory rats [43]. In 2002, Leung et al. used a model of transient cerebral ischemia, TLT inhibited nitric oxide synthase activity, and unregulated expression of TGF beta-1 [44]. In 2004 Lapchak has shown that laser treatment with 7.5 mw/cm^2 at 6 hours poststroke onset in a rabbit small clot embolic stroke model (RSCEM) improved behavioral performance and produced a durable effect that was measurable 21 days after embolization [45]. Lapchak reported their further research showed TLT improved motor function following embolic strokes in rabbits [46, 47]. In 2006, De Taboada et al. [48] showed in 2 different animal models a positive impact of infrared laser therapy on the experimental, ischemic stroke treatment outcomes in New Zealand rabbits subjected to a RSCEM and also in Sprague-Dawley rats (permanent middle cerebral artery occlusion). De Taboada et al. have also shown that laser treatment up to 24 hours poststroke onset in permanent middle cerebral artery occlusion showed significant improvement in neurological deficits that was evident at 14, 21, and 28 days poststroke when compared with the sham control group [48]. Meanwhile, Oron reported that a noninvasive intervention of TLT issued 24 hours after acute stroke may provide a significant functional benefit with an underlying mechanism possibly being induction of neurogenesis [39]. Currently, the putative mechanism for infrared laser therapy in stroke involves the stimulation of mitochondria, which then leads to preservation of tissue in the ischemic penumbra and enhanced neuron recovery. The exact mechanistic pathways remain to be elucidated. Figure 27.9 shows how the treatment is carried out in a mouse model of stroke.

27.6.2 TLT in clinical trials for stroke

The NeuroThera Effectiveness and Safety Trial–1 (NEST-1) evaluated the safety and preliminary effectiveness of the NeuroThera Laser System in the ability to improve 90-day outcomes in ischemic stroke patients treated within 24 hours from stroke onset [49]. The NEST-1 was a prospective, intention-to-treat, multicenter, international, double-blind trial involving 120 ischemic stroke patients treated, randomized 2:1 ratio, with 79 patients in the active treatment group and 41 in the sham (placebo) control group. Figure 27.10 shows TLT on stroke in clinical. Only patients with baseline stroke severity measured by the National Institutes of Health Stroke Scale (NIHSS) scores of 7 to 22 were included. Treatment consisted of the application of a hand-held 808-nm laser probe to twenty predetermined locations on the shaved scalp for 2 minutes in each location. The NEST-1 study indicates that infrared laser therapy has shown initial safety and effectiveness for the treatment of ischemic stroke in humans when initiated within 24 hours of stroke onset and a second larger trial was warranted.

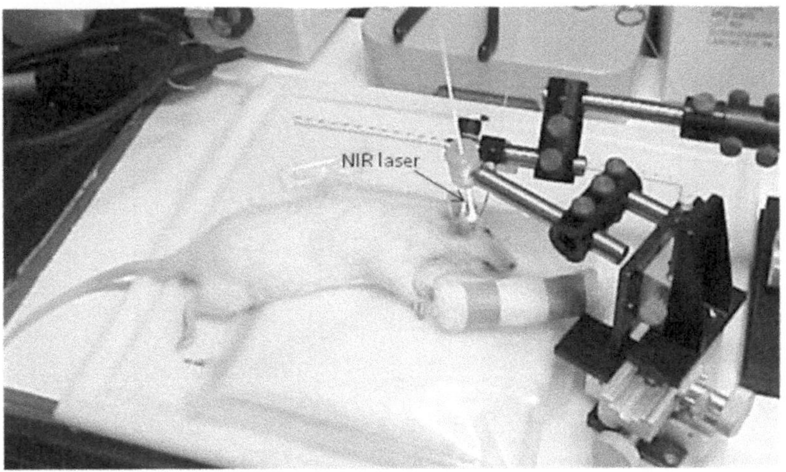

FIGURE 27.9: Animal model of TLT in the brain and CNS.

Recently, this second clinical trial of the effectiveness and safety trial of NIR laser treatment within 24 hours from stroke onset (NEST-2) was finished [50]. This study was a prospective, double-blind, randomized, sham-controlled, parallel group, multicenter study that included sites in Sweden, Germany, Peru, and the United States, and enrolled 660 subjects. Subjects were followed for 90 days poststroke onset. The primary effectiveness endpoint for this study was the binary endpoint that defined success as a modified Rankin Scale (mRS) score of 0–2 and failure as a mRS score of 3–6 at 90 days or the last rating. TLT was applied within 24 hours from stroke onset. The study demonstrated safety but did not meet formal statistical significance for efficacy. Of 658 randomized patients, 331 received TLT and 327 received sham; 120 (36.3%) in the TLT group achieved favorable outcome versus 101 (30.9%), in the sham group ($P = 0.094$), odds ratio 1.38 (95% CI, 0.95 to 2.00). Comparable results were seen for the other outcome measures. Although no prespecified test achieved significance, a post hoc analysis of patients with a baseline National Institutes of Health Stroke Scale (NIHSS) score of < 16 showed a favorable outcome at 90 days on the primary end point ($p < 0.044$). Mortality rates and serious adverse events did not differ between groups with 17.5% and 17.4% mortality, 37.8% and 41.8% serious adverse events for TLT and sham, respectively. A third pivotal trial of TLT with refined baseline NIHSS exclusion criteria (< 16) is now planned.

27.7 LLLT for CNS Damage

The central nervous system (CNS) is the largest part of the nervous system, and includes the brain and spinal cord. The spinal cavity holds and protects the spinal cord, while the head contains and protects the brain. The CNS is covered by the meninges, a three-layered protective coat. The brain is also protected by the skull, and the spinal cord is also protected by the vertebrae. Different reports show LLLT has a significant effect on a wide range of CNS disorders. Major conditions include: epilepsy; traumatic brain injury; neurodegenerative disorders, including Alzheimer's disease; headache, motion sickness and vertigo, movement disorders, such as Parkinson's disease; multiple sclerosis; neuromuscular disorders; neuropathic pain; sleep disorders; infections of the brain; and cerebrovascular disease, such as stroke and transient ischemic attack. Figure 27.11 shows some CNS disorders that have been treated with LLLT.

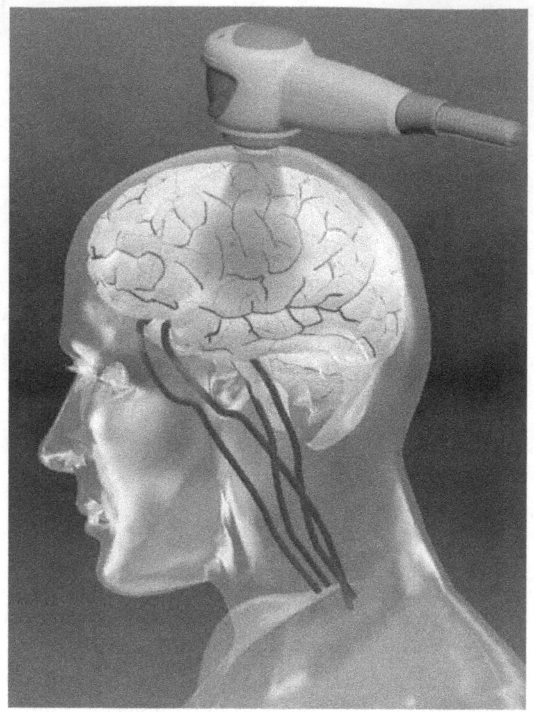

FIGURE 27.10: TLT for stroke.

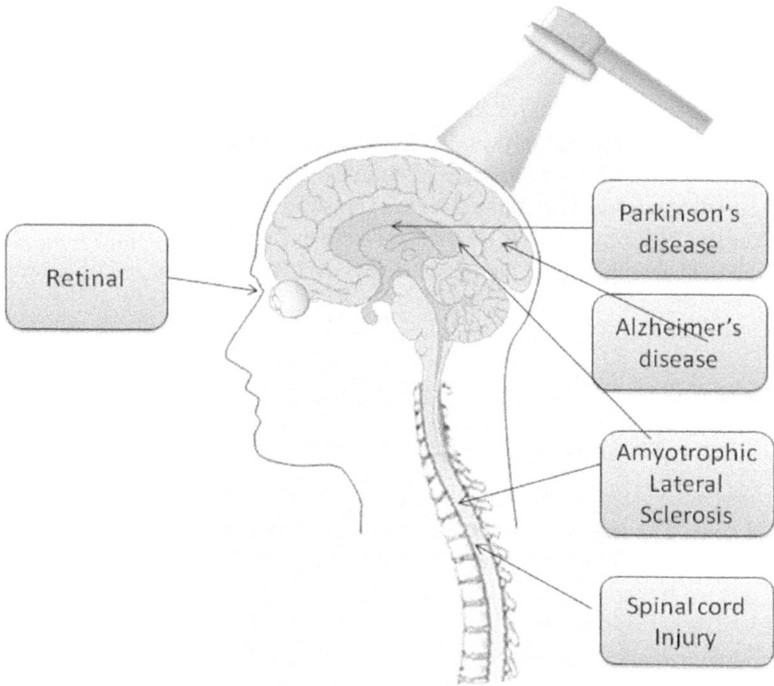

FIGURE 27.11: LLLT for central nervous system disorders.

27.7.1 Traumatic brain injury (TBI)

Traumatic brain injury (TBI) occurs when an outside force traumatically injures the brain. TBI can result in neurological impairment because of immediate CNS tissue disruption (primary injury), but, additionally, surviving cells may be secondarily damaged by complex mechanisms triggered by the primary event, leading to further damage and disability [51]. LLLT has been found to modulate various biological processes. Oron et al. [52] showed that TLT given 4 h following TBI provides a significant long-term functional neurological benefit. Further confirmatory trials are warranted. TBI was induced by a weight-drop device, and motor function was assessed 1 h posttrauma using a neurological severity score (NSS). Mice were then divided into three groups of eight mice each: one control group that received a sham LLLT procedure and was not irradiated; and two groups that received LLLT at two different doses (10 and 20 mW/cm^2) transcranially. An 808-nm Ga-As diode laser was employed transcranially 4 h posttrauma to illuminate the entire cortex of the brain. Motor function was assessed up to 4 weeks, and lesion volume was measured.

27.7.2 Spinal cord injury (SCI)

The spinal cord is a long, thin, tubular bundle of nerves that is an extension of the central nervous system from the brain and is enclosed in and protected by the bony vertebral column. Spinal cord injuries (SCI) can be caused by trauma to the spinal column. The vertebral bones or intervertebral disks can shatter, causing the spinal cord to be punctured by a sharp fragment of bone. Usually victims of spinal cord injuries will suffer loss of feeling in certain parts of their body.

As yet LLLT has only been studied in animal models of spinal cord injury: A pilot study by Rochkind [53] examined the effects of composite implants of cultured embryonal nerve cells, and laser irradiation on the regeneration and repair of the completely transected spinal cord on rats. Three months after transection, implantation and laser irradiation (with 780 nm, 250 mW, 30 min daily), intensive axonal sprouting occurred in the group with implantation and laser. The study suggests *in vitro* composite implants are a regenerative and reparative source for reconstructing the transected spinal cord. Postoperative low-power laser irradiation enhances axonal sprouting and spinal cord repair. Rochkind's recent study showed postoperative 780-nm laser phototherapy enhanced the regenerative process of the peripheral nerve after reconnection of the nerve defect using a polyglycolic acid neurotube [54]. Byrnes et al. used 1.600 J/cm^2 of 810-nm diode laser to improve healing and functionality in a T9 dorsal hemisection of the spinal cord in rats [55]. Light (810 nm) was applied transcutaneously at the lesion site immediately after injury and daily for 14 consecutive days. A laser diode with an output power of 150 mW was used for the treatment. The daily dosage at the surface of the skin overlying the lesion site was 1.589 J/cm^2. The average length of axonal re-growth in the rats in the light treatment groups with the hemisection and contusion injuries was significantly longer than the comparable untreated control groups ($3.66/pm0.26$ mm, hemisection; 2.89 ± 0.84 mm, contusion). The total axon number in the light therapy groups was significantly higher compared to the untreated groups for both injury models ($p < 0.05$). Light therapy applied noninvasively promotes axonal regeneration and functional recovery in acute SCI caused by different types of trauma [56].

27.7.3 Reversal of neurotoxicity

Neurotoxicity occurs when the exposure to natural or artificial toxic substances, alters the normal activity of the nervous system in such a way as to cause damage to nervous tissue.

Studies from Whelan's group [57] have explored the use of 670-nm LEDs in combating neuronal damage caused by neurotoxins on rat models. Methanol intoxication is caused by the metabolic conversion to formic acid that produces injury to the retina and optic nerve, resulting in blindness. Using a rat model and the electroretinogram as a sensitive indicator of retinal function, they found a significant recovery of rod- and cone-mediated functions in 670-nm LED treatments (4J/cm^2), methanol-intoxicated rats, and histopathologic evidence of retinal protection.

A subsequent study [58] explored the effects of an irreversible inhibitor of cytochrome c oxidase, potassium cyanide in primary cultured neurons. LED treatment partially restored enzyme activity blocked by 10–100 μM KCN. It significantly reduced neuronal cell death induced by 300 μM KCN from 83.6 to 43.5%. LED significantly restored neuronal ATP content only at 10 μM KCN but not at higher concentrations of KCN tested. In contrast, LED was able to completely reverse the detrimental effect of tetrodotoxin, which only indirectly down-regulated enzyme levels. Among the wavelengths tested (670, 728, 770, 830, and 880 nm), the most effective ones (830 nm, 670 nm) paralleled the NIR absorption spectrum of oxidized cytochrome c oxidase.

27.8 LLLT for Neurodegenerative Diseases

27.8.1 Neurodegenerative disease

A neurodegenerative disease is a disorder caused by the deterioration of certain nerve cells (neurons). Changes in these cells cause them to function abnormally, eventually bringing about their death. The diseases, Alzheimer's disease (AD), Parkinson's disease (PD), and Amyotrophic Lateral Sclerosis (ALS), as well as multiple sclerosis (MS), are all due to neuronal degeneration in the central nervous system [59]. The chronic, unrelenting, progressive nature of these devastating degenerative diseases has motivated the search for therapies that could slow down or arrest the downward course experienced by most patients, and even more desirable would be a therapy that could actually reverse the neuronal damage. Transcranial light therapy is considered to have the potential to accomplish these goals.

27.8.2 Parkinson's disease

Parkinson's disease (PD) is a degenerative disorder of the central nervous system that often impairs the sufferer's motor skills, speech, and other functions [60].

Trimmer et al. showed near infrared light therapy (810-nm diode laser, 50 mW/cm^2) can restore axonal transport to normal levels in sporadic PD human transmitochondrial cybrid "cytoplastic hybrid" neuronal cells [61].

The influence of laser therapy on the course of PD was studied *in vivo* and *in vitro*. This influence appeared adaptogenic both in the group with elevated and low MAO B and Cu/Zn SOD activity. Laser therapy resulted in reduction of the neurological deficit, normalization of the activity of MAO B, Cu/Zn-SOD, and immune indices [62].

27.8.3 Alzheimer's disease

Alzheimer's disease (AD), also called Alzheimer disease, Senile Dementia of the Alzheimer Type (SDAT) or simply Alzheimer's, is the most common form of dementia [63]. Though the cause of AD is still unknown, impaired oxidative balance, mitochondrial dysfunction, and disordered cytochrome c oxidase play important roles in the pathogenesis of Alzheimer's disease [64, 65]. The British Broadcasting Corporation (BBC) news reported an experimental helmet made of 700 LEDs that delivered near infrared light (1070 nm) was being tested by scientists as a treatment for Alzheimer's disease. Current treatments for Alzheimer's delay progression of the disease but cannot reverse memory loss [66].

27.8.4 Amyotrophic lateral sclerosis (ALS)

Amyotrophic Lateral Sclerosis (ALS) is a progressive, usually fatal, neurodegenerative disease caused by the degeneration of motor neurons, the nerve cells in the central nervous system that control voluntary muscle movement. As a motor neuron disease, the disorder causes muscle weakness and atrophy throughout the body as both the upper and lower motor neurons degenerate, ceasing to send messages to the muscles. Unable to function, the muscles gradually weaken, develop fasciculations (twitches) because of denervation, and eventually atrophy because of that denervation. Mitochondrial dysfunction and oxidative stress play an important role in motor neuron loss in ALS. Light therapy has biomodulatory effects on mitochondria. A study examined the synergistic effect of LLLT and riboflavin on the survival of motor neurons in a mouse model of ALS. They used 810-nm diode laser, 140-mW output power, 1.4-cm^2 spot area, 120-seconds treatment duration, and 12 J/ cm^2-energy density. The results came out ineffective in altering disease progression in the G93A SOD1 mice using these laser parameters [67].

27.9 LLLT for Psychiatric Disorders

Light therapy has been widely used for some psychiatric disorders for nearly 30 years [68]. Light therapy has been tested in the following psychiatric disorders: seasonal affective disorder (SAD) [69], premenstrual dysphoric disorder [70], antepartum and postpartum major depressive disorder [71, 72], bulimia nervosa [73], and adult attention-deficit disorder [74]. These studies used full sunlight and products (such as light boxes) using very intense artificial illumination that are effective. Newer research indicates that using a lower intensity of certain wavelengths of light, i.e., the "blue" wavelengths, may be at least as efficacious as using 2.500 lux [75]. Another approach to treating depression and other psychiatric disorders is the use of transcranial magnetic stimulation (TMS). TMS is a noninvasive method to excite neurons in the brain in which weak electric currents are induced in the tissue by rapidly changing magnetic fields (electromagnetic induction). This way, brain activity can be triggered with minimal discomfort, and the functionality of the circuitry and connectivity of the brain can be studied. Repetitive TMS can produce longer lasting changes. Numerous small-scale pilot studies have shown it could be a treatment tool for various neurological conditions (e.g., migraine, stroke, Parkinsons Disease, dystonia, tinnitus) and psychiatric conditions (e.g., major depression, auditory hallucinations) [76]. The success of TMS has suggested that similar stimulation could be applied to brain neurons using transcranial NIR light therapy, but as yet there are no authoritative publications and only anecdotal reports of success.

LLLT in various forms has been applied in patients for treating psychiatric disorders such as depression [77, 78], smoking cessation [79], and alcoholism [80].

27.10 Conclusions and Future Outlook

The remarkable effects of TLT in remedying CNS damage and slowing degenerative neurological disease in a noninvasive manner with little evidence of any adverse side effects, suggest that its application will only increase. Advances that are being made in understanding the molecular and cellular basis for the action of red and near-infrared light on cells and tissues, will only serve to increase the acceptance of TLT by the medical profession at large. The clinical trials for stroke described in section 27.6.2 will provide additional credibility to the use of TLT in the brain. The wider availability of diode lasers and light-emitting diode arrays with sufficient power to deliver therapeutically effective wavelengths and fluences of light to deep-lying structures in the brain and spinal cord will go further to encourage more widespread clinical studies. Stroke, traumatic brain injury, depression, Alzheimer's, AND Parkinson's diseases together with spinal cord injury are responsible for much mortality and morbidity in the modern age, and this will only increase as the Western population ages and lives longer. If LLLT can make even a small contribution to mitigating the loss of life, suffering, disability, and financial burden caused by CNS disorders, it will make the efforts of researchers in the LLLT field worthwhile.

References

[1] J.C. Sutherland, "Biological effects of polychromatic light," *Photochem. Photobiol.* **76,** 164–170 (2002).

[2] T. Karu, "Laser biostimulation: a photobiological phenomenon," *J. Photochem. Photobiol. B* **3,** 638–640 (1989).

[3] J. Streeter, L. De Taboada, and U. Oron, "Mechanisms of action of light therapy for stroke and acute myocardial infarction," *Mitochondrion* **4,** 569–576 (2004).

[4] T.I. Karu, L.V. Piatibrat, and G.S. Kalendo, "Suppression of the intracellular concentration of ATP by irradiating with a laser pulse of wavelength 820 nm," *Dokl. Akad. Nauk.* **364**, 399–401 (1999).

[5] R. Lavi, A. Shainberg, H. Friedmann, et al., "Low energy visible light induces reactive oxygen species generation and stimulates an increase of intracellular calcium concentration in cardiac cells," *J. Biol. Chem.* **278,** 40917–40922 (2003).

[6] P. Moore, T.D. Ridgway, R.G. Higbee, et al., "Effect of wavelength on low-intensity laser irradiation-stimulated cell proliferation *in vitro*," *Lasers Surg. Med.* **36**, 8–12 (2005).

[7] A.C.-H. Chen, P.R. Arany, Y.-Y. Huang, et al., "Low level laser therapy activates NF-kB via generation of reactive oxygen species in mouse embryonic fibroblasts," *Proc. SPIE* **7165**, 10.1117/12.809605, eds. M.R. Hamblin, R.W. Waynant, and J.J. Anders (2009).

[8] L.R. Squire, F. Bloom, and N. Spitzer, *Fundamental Neuroscience*, Academic Press (2008).

[9] U. Oron, S. Ilic, L. De Taboada, et al., "Ga-As (808 nm) laser irradiation enhances ATP production in human neuronal cells in culture," *Photomed. Laser Surg.* **25**, 180–182 (2007).

[10] Y.D. Ignatov, A.I. Vislobokov, T.D. Vlasov, et al., "Effects of helium-neon laser irradiation and local anesthetics on potassium channels in pond snail neurons" *Neurosci. Behav. Physiol.* **35**, 871–875 (2005).

[11] R. Ying, H.L. Liang, H.T. Whelan, et al., "Pretreatment with near-infrared light via light-emitting diode provides added benefit against rotenone- and MPP+-induced neurotoxicity," *Brain Res.* **1243**, 167–173 (2008).

[12] M. Nissan, S. Rochkind, N. Razon, et al., "HeNe laser irradiation delivered transcutaneously: its effect on the sciatic nerve of rats," *Lasers Surg. Med.* **6**, 435–438 (1986).

[13] D. Gigo-Benato, S. Geuna, and S. Rochkind, "Phototherapy for enhancing peripheral nerve repair: a review of the literature," *Muscle Nerve* **31**, 694–701 (2005).

[14] J.J. Anders, S. Geuna, and S. Rochkin, "Phototherapy promotes regeneration and functional recovery of injured peripheral nerve," *Neurol. Res.* **26**, 233–239 (2004).

[15] J.J. Anders, R.C. Borke, S.K. Woolery, et al., "Low power laser irradiation alters the rate of regeneration of the rat facial nerve," *Lasers Surg. Med.* **13**, 72–82 (1993).

[16] S. Wan, J.A. Parrish, R.R. Anderson, et al., "Transmittance of nonionizing radiation in human tissues," *Photochem. Photobiol.* **34**, 679–681 (1981).

[17] E. Salomatina, B. Jiang, J. Novak, et al., "Optical properties of normal and cancerous human skin in the visible and near-infrared spectral range," *J. Biomed. Opt.* **11**, 064026 (2006).

[18] V.V. Lychagov, V.V. Tuchin, M.A. Vilensky, et al., "Experimental study of NIR transmittance of the human skull," *Proc. SPIE* **6085**, 10.1117/12.650116, ed. V.V. Tuchin (2006).

[19] P.A. Wolf, "An overview of the epidemiology of stroke," *Stroke* **21**, Suppl. 11, 4–6 (1990).

[20] W. Rosamond, K. Flegal, K. Furie, et al., "Heart disease and stroke statistics-2008 update: a report from the American Heart Association Statistics Committee and Stroke Statistics Subcommittee," *Circulation*, **117**, E25–E146 (2008).

[21] A. Durukan and T. Tatlisumak, "Acute ischemic stroke: overview of major experimental rodent models, pathophysiology, and therapy of focal cerebral ischemia," *Pharmacol. Biochem. Behav*. **87**, 179–197 (2007).

[22] L. Edvinsson and D.N. Krause, eds. *Cerebral Blood Flow and Metabolism*, 2nd ed., Lippincott, Williams, and Wilkins, Philadelphia, PA (2002).

[23] L.R. Caplan (ed.), *Caplan's Stroke: A Clinical Approach*, 3rd ed., Butterworth-Heinemann, Boston (2000).

[24] K.H. Taber, D.L. Warden, and R.A. Hurley, "Blast-related traumatic brain injury: what is known?" *J. Neuropsychiatry Clin. Neurosci*. **18**, 141–145 (2006).

[25] A.M. Palmer, D.W. Marion, M.L. Botscheller, et al., "Traumatic brain injury-induced excitotoxicity assessed in a controlled cortical impact model," *J. Neurochem.* **61**, 2015–2024 (1993).

[26] H. Pasantes-Morales and K. Tuz, "Volume changes in neurons: hyperexcitability and neuronal death," *Contrib. Nephrol.* **152**, 221–240 (2006).

[27] M. Ahmad, A.S. Ahmad, H. Zhuang, et al., "Stimulation of prostaglandin E2-EP3 receptors exacerbates stroke and excitotoxic injury," *J. Neuroimmunol.* **184**, 172–179 (2007).

[28] C. Matute, E. Alberdi, M. Domercq, et al., "Excitotoxic damage to white matter," *J. Anat.* **210**, 693–702 (2007).

[29] Y. Ikeda and D.M. Long, "The molecular basis of brain injury and brain edema: the role of oxygen free radicals," *Neurosurgery*, **27**, 1–11 (1990).

[30] L. Cherian and C.S. Robertson, "L-arginine and free radical scavengers increase cerebral blood flow and brain tissue nitric oxide concentrations after controlled cortical impact injury in rats," *J. Neurotrauma* **20**, 77–85 (2003).

[31] H. Adams, R. Adams, G. Del Zoppo, et al., "Guidelines for the early management of patients with ischemic stroke – 2005 guidelines update – a scientific statement from the Stroke Council of the American Heart Association/American Stroke Association," *Stroke* **36**, 916–923 (2005).

[32] P. Sandercock, R. Lindley, J. Wardlaw, et al., "The third international stroke trial (IST-3) of thrombolysis for acute ischaemic stroke," *Trials* **9**, 37 (2008).

[33] O. Barushka, T. Yaakobi, and U. Oron, "Effect of low-energy laser (He-Ne) irradiation on the process of bone repair in the rat tibia," *Bone* **16**, 47–55 (1995).

[34] A. Furlan, R. Higashida, L. Wechsler, et al., "Intra-arterial prourokinase for acute ischemic stroke: the PROACT II study: a randomized controlled trial; prolyse in acute cerebral thromboembolism," *JAMA* **282**, 2003–2011 (1999).

[35] M. Hommel, C. Cornu, F. Boutitie, et al., "Thrombolytic therapy with streptokinase in acute ischemic stroke: the Multicenter Acute Stroke Trial-Europe Study Group," *N. Engl. J. Med.* **335**, 145–150 (1996).

[36] J.L. Saver, "Time is brain-quantified," *Stroke* **37**, 263–266 (2006).

[37] W. Hacke, G. Donnan, C. Fieschi, et al., "Association of outcome with early stroke treatment: pooled analysis of ATLANTIS, ECASS, and NINDS rt-PA stroke trials," *Lancet* **363**, 768–774 (2004).

[38] M.D. Ginsberg, "Neuroprotection for ischemic stroke: past, present and future" *Neuropharmacology*, **55**, 363–389 (2008).

[39] A. Oron, U. Oron, J. Chen, et al., "Low-level laser therapy applied transcranially to rats after induction of stroke significantly reduces long-term neurological deficits," *Stroke* **37**, 2620–2624 (2006).

[40] Y. Lampl, "Laser treatment for stroke," *Expert. Rev. Neurother.* **7**, 961–965 (2007).

[41] S.P. Finklestein, M. Fisher, A.J. Furlan, et al., "Recommendations for standards regarding preclinical neuroprotective and restorative drug development," *Stroke* **30**, 2752–2758 (1999).

[42] D.I. Azbel, N.V. Egorushkina, I. Kuznetsova, et al., "The effect of the blood serum from patients subjected to intravenous laser therapy on the parameters of synaptic transmission," *Biull. Eksp. Biol. Med.* **116**, 149–151 (1993).

[43] N. Ad and U. Oron, "Impact of low level laser irradiation on infarct size in the rat following myocardial infarction," *Int. J. Cardiol.* **80**, 109–116 (2001).

[44] M.C. Leung, S.C. Lo, F.K. Siu, et al., "Treatment of experimentally induced transient cerebral ischemia with low energy laser inhibits nitric oxide synthase activity and up-regulates the expression of transforming growth factor-beta 1," *Lasers Surg. Med.* **31**, 283–288 (2002).

[45] P.A. Lapchak, J. Wei, and J.A. Zivin, "Transcranial infrared laser therapy improves clinical rating scores after embolic strokes in rabbits," *Stroke* **35**, 1985–1988 (2004).

[46] P.A. Lapchak, K.F. Salgado, C.H. Chao, et al., "Transcranial near-infrared light therapy improves motor function following embolic strokes in rabbits: an extended therapeutic window study using continuous and pulse frequency delivery modes," *Neuroscience* **148**, 907–914 (2007).

[47] P.A. Lapchak and D.M. Araujo, "Advances in ischemic stroke treatment: neuroprotective and combination therapies," *Expert. Opin. Emerg. Drugs* **12**, 97–112 (2007).

[48] L. De Taboada, S. Ilic, S. Leichliter-Martha, et al., "Transcranial application of low-energy laser irradiation improves neurological deficits in rats following acute stroke," *Lasers Surg. Med.* **38**, 70–73 (2006).

[49] Y. Lampl, J.A. Zivin, M. Fisher, et al., "Infrared laser therapy for ischemic stroke: a new treatment strategy: results of the NeuroThera Effectiveness and Safety Trial-1 (NEST-1)," *Stroke* **38**, 1843–1849 (2007).

[50] J.A. Zivin, G.W. Albers, N. Bornstein, et al., "Effectiveness and safety of transcranial laser therapy for acute ischemic stroke," *Stroke* (2009).

[51] G.M. Teasdale and D.I. Graham, "Craniocerebral trauma: protection and retrieval of the neuronal population after injury," *Neurosurgery* **43**, 723–737; discussion 737–728 (1998).

[52] A. Oron, U. Oron, J. Streeter, et al., "Low-level laser therapy applied transcranially to mice following traumatic brain injury significantly reduces long-term neurological deficits," *J. Neurotrauma* **24**, 651–656 (2007).

[53] S. Rochkind, A. Shahar, M. Amon, et al., "Transplantation of embryonal spinal cord nerve cells cultured on biodegradable microcarriers followed by low power laser irradiation for the treatment of traumatic paraplegia in rats," *Neurol. Res.* **24**, 355–360 (2002).

[54] S. Rochkind, L. Leider-Trejo, M. Nissan, et al., "Efficacy of 780-nm laser phototherapy on peripheral nerve regeneration after neurotube reconstruction procedure (double-blind randomized study)," *Photomed. Laser Surg.* **25**, 137–143 (2007).

[55] K.R. Byrnes, R.W. Waynant, I.K. Ilev, et al., "Light promotes regeneration and functional recovery and alters the immune response after spinal cord injury," *Lasers Surg. Med.* **36**, 171–185 (2005).

[56] X. Wu, A.E. Dmitriev, M.J. Cardoso, et al., "810 nm wavelength light: an effective therapy for transected or contused rat spinal cord," *Lasers Surg. Med.* **41**, 36–41 (2009).

[57] J.T. Eells, M.M. Henry, P. Summerfelt, M.T.T. Wong-Riley, E.V. Buchmann, M. Kane, N.T. Whelan, and H.T. Whelan, "Therapeutic photobiomodulation for methanol-induced retinal toxicity," *Proc. Natl. Acad. Sci. USA* **100**, 3439–3444 (2003).

[58] M.T. Wong-Riley, H.L. Liang, J.T. Eells, et al., "Photobiomodulation directly benefits primary neurons functionally inactivated by toxins: role of cytochrome c oxidase," *J. Biol. Chem.* **280**, 4761–4771 (2005).

[59] R.M. Friedlander, "Apoptosis and caspases in neurodegenerative diseases," *N. Engl. J. Med.* **348**, 1365–1375 (2003).

[60] J. Jankovic, "Parkinson's disease: clinical features and diagnosis," *J. Neurol. Neurosurg. Psychiatry* **79**, 368–376 (2008).

[61] P.A. Trimmer, K.M. Schwartz, M.K. Borland, et al., "Reduced axonal transport in Parkinson's disease cybrid neurites is restored by light therapy," *Mol. Neurodegener.* (2009).

[62] T.V. Vitreshchak, V.V. Mikhailov, M.A. Piradov, et al., "Laser modification of the blood in vitro and in vivo in patients with Parkinson's disease," *Bull. Exp. Biol. Med.* **135**, 430–432 (2003).

[63] D.J. Selkoe, "Alzheimer's disease: genes, proteins, and therapy," *Physiol. Rev.* **81**, 741–766 (2001).

[64] M.A. Smith, C.A. Rottkamp, A. Nunomura, et al., "Oxidative stress in Alzheimer's disease," *Biochim. Biophys. Acta* **1502**, 139–144 (2000).

[65] P.G. Sullivan and M.R. Brown, "Mitochondrial aging and dysfunction in Alzheimer's disease," *Prog. Neuropsychopharmacol. Biol. Psychiatry* **29**, 407–410 (2005).

[66] *Alzheimer's helmet therapy hope* [BBC News] 1/25/2008 [cited 2009 2/25]; Health: Available from: http://news.bbc.co.uk/go/pr/fr/–/2/hi/health/7208768.stm.

[67] H. Moges, O.M. Vasconcelos, W.W. Campbell, et al., "Light therapy and supplementary Riboflavin in the SOD1 transgenic mouse model of familial amyotrophic lateral sclerosis (FALS)," *Lasers Surg. Med.* **41**, 52–59 (2009).

[68] D.F. Kripke, S.C. Risch, and D. Janowsky, "Bright white light alleviates depression," *Psychiatry Res.* **10**, 105–112 (1983).

[69] J.S. Terman, M. Terman, D. Schlager, et al., "Efficacy of brief, intense light exposure for treatment of winter depression," *Psychopharmacol. Bull.* **26**, 3–11 (1990).

[70] B.L. Parry, S.L. Berga, N. Mostofi, et al., "Morning versus evening bright light treatment of late luteal phase dysphoric disorder," *Am. J. Psychiatry* **146**, 1215–1217 (1989).

[71] C.N. Epperson, M. Terman, J.S. Terman, et al., "Randomized clinical trial of bright light therapy for antepartum depression: preliminary findings," *J. Clin. Psychiatry* **65**, 421–425 (2004).

[72] M. Corral, A. Kuan, and D. Kostaras, "Bright light therapy's effect on postpartum depression," *Am. J. Psychiatry* **157**, 303–304 (2000).

[73] R.W. Lam, E.M. Goldner, L. Solyom, et al., "A controlled study of light therapy for bulimia nervosa," *Am. J. Psychiatry* **151**, 744–750 (1994).

[74] Y.E. Rybak, H.E. McNeely, B.E. Mackenzie, et al., "An open trial of light therapy in adult attention-deficit/hyperactivity disorder," *J. Clin. Psychiatry* **67**, 1527–1535 (2006).

[75] A. Wirz-Justice, P. Graw, K. Krauchi, et al., "Light therapy in seasonal affective disorder is independent of time of day or circadian phase," *Arch. Gen. Psychiatry* **50**, 929–937 (1993).

[76] A. Pascual-Leone, N. Davey, J. Rothwell, et al. (eds), *Handbook of Transcranial Magnetic Stimulation*, Oxford University Press, London (2002).

[77] J.I. Quah-Smith, W.M. Tang, and J. Russell, "Laser acupuncture for mild to moderate depression in a primary care setting–a randomised controlled trial," *Acupunct. Med.* **23**, 103–111 (2005).

[78] A. Gur, M. Karakoc, K. Nas, et al., "Effects of low power laser and low dose amitriptyline therapy on clinical symptoms and quality of life in fibromyalgia: a single-blind, placebo-controlled trial," *Rheumatol. Int.* **22**, 188–193 (2002).

[79] C. Yiming, Z. Changxin, W.S. Ung, et al., "Laser acupuncture for adolescent smokers – a randomized double-blind controlled trial," *Am. J. Chin. Med.* **28**, 443–449 (2000).

[80] J. Zalewska-Kaszubska and D. Obzejta, "Use of low-energy laser as adjunct treatment of alcohol addiction," *Lasers Med. Sci.* **19**, 100–104 (2004).

28

Advances in Cancer Photothermal Therapy

Wei R. Chen
Biomedical Engineering Program, Department of Engineering and Physics, College of Mathematics and Science, University of Central Oklahoma, Edmond, OK

Xiaosong Li
Department of Oncology, the First Affiliated Hospital of Chinese PLA General Hospital, Beijing, China

Mark F. Naylor
Department of Dermatology, University of Oklahoma College of Medicine at Tulsa, Tulsa, OK

Hong Liu
Center for Bioengineering and School of Electrical and Computer Engineering, University of Oklahoma, Norman, OK

Robert E. Nordquist
Wound Healing of Oklahoma, Inc., 14 NE 48th Street, Oklahoma City, OK

Photothermal interactions have a direct impact on biological tissues. Due to the high sensitivity of tumor tissue to temperature increase, photothermal therapy has attracted increased attention from researchers and clinicians. Thermal therapy started with whole-body hyperthermia. Recently, the thermal therapy has employed modern devices and methodologies, such as lasers, radio frequency radiation, and ultrasound. Selective phototherapy has been used for cancer treatment using endogenous and exogenous agents to enhance the energy absorption on target tissue to induce desired photophysical reactions. To achieve long-term effects, immunotherapy has been used in combination with phototherapies, particularly with photothermal therapy to induce tumor-specific immune responses.

This article reviews the basic biological responses induced by photothermal therapy, particularly in cancer treatment. Selective photothermal therapy using exogenous dyes will be introduced. The combination of phototherapy and immunotherapy in cancer treatment will be covered in detail in terms of preclinical and preliminary clinical results. Nanotechnology has become a fast-growing field for selective photothermal therapy. The use of nanoparticles, nanoclusters, and nanotubes in photothermal and immunological cancer treatment will also be discussed.

Key words: Selective photothermal therapy, photoimmunological responses, photothermal immunotherapy, nanotechnology in cancer treatment.

28.1 Introduction

Light has been an integral part of our lives since the beginning of time. It not only has long fascinated poets, writers, and philosophers, it also has attracted the attention of scientists, engineers, and medical professionals. Long before its true nature as an electromagnetic wave was unveiled by James C. Maxwell in 1865, light had been used for medicinal purposes in both diagnosis and treatment of illness.

Light became a true medical tool after the laser was invented. Coherent light with controllable wavelengths revolutionized the medical use of light. The intensity of light and its versatile methods of delivery made the laser not only a tool of convenience, but also in many cases a tool of necessity. Lasers have become the major source of light for optical diagnostic modalities, such as light spectroscopy, optical coherent tomography, light endoscopy, and fluorescence spectroscopy.

In therapeutics, lasers have been used to facilitate photophysical interactions [1–4] for biomedical interventions. When laser energy is absorbed by organized tissue, it results in three photophysical interactions: photomechanical, photochemical, and photothermal. Photomechanical reaction induces tissue stress and may lead to surface tissue break up and ejection. Photochemical reaction causes the change of chemical bonds and forms toxic radicals, such as the release of singlet oxygen, leading to the death of organized tissues, as in the case of photodynamic therapy [5–12]. Photothermal reaction produces hyperthermia and coagulation of tissue; it can be an effective means in tumor tissue destruction due to the sensitivity of tumor cell to temperature elevation [1–4, 13]. Among the three interactions, photothermal is the most commonly encountered phenomenon in almost all the laser applications. It causes direct and controllable thermal damage to tissue. Because of its effect on tissue, thermal damage has been an important factor when a laser is used in clinical treatment. While its enhancement and control are needed when the lesion removal and tumor destruction are concerned, efforts are always required to avoid or to reduce its impact on normal tissue.

Because tumor cells are more sensitive to temperature increase, thermal energy has been used for tumor destruction. However, tissue optical properties can be changed when energy is absorbed to cause temperature increase. Furthermore, energy intensity in tissue follows an exponential decline due to attenuation. As a consequence, photothermal effect is often only "skin" deep. Often the denatured tissue significantly increases the energy absorption and prevents the photothermal destruction from occurring in the target tissue deeply buried in normal tissue. This has provided a challenge for noninvasive photothermal treatment of deep lesions.

It is well known, however, that systemic diseases such as metastatic tumors cannot be effectively eradicated by local treatment modalities. In fact, according to the principles of angiogenesis, the treatment of primary tumors using local interventions such as surgery and radiation may stimulate the growth of metastases at remote sites [14–16]. Therefore, a more holistic, systemic approach is needed. The active enhancement of host immunological systems using immunostimulants, in conjunction with phototherapies, particularly with selective photothermal interactions, has proven to be synergistic for the systemic control of cancers in preclinical and preliminary clinical studies. Ultimately, this approach may prove to be effective for the treatment of metastatic tumors.

28.2 Thermal Effects on Biological Tissues

28.2.1 Tissue responses to temperature increase

It has been well established that organized tissue responds strongly to temperature increase. When the tissue temperature approaches and exceeds 40°C, there is an increased blood flow in both tumor and normal tissues [17–18]. As tumor temperature reaches 41.5°C, cellular cytotoxicity occurs [19]. Temperatures above 42.5°C can result in vascular destruction within tumor tissue [20]. The thermal impact on tissue changes drastically when temperature exceeds 43°C; the rate of "cell kill" doubles for every 1°C increase beyond 43°C and decreases by a factor of 4 to 6 for every 1°C drop below 43°C [21–22]. It has been observed that tumor tissue is more sensitive to temperature increase than normal tissue [13].

28.2.2 Tumor tissue responses to thermal therapy

An important reason for the observed cytotoxicity associated with temperature increase is that thermal effects are more profound in the acidotic (low pH) conditions present in poorly oxygenated tumors. Therefore, heat has a greater cytotoxic effect in tumors than in normal tissue [23]. However, heat alone has only a limited effect in cancer treatment; in the early whole-body hyperthermia treatment, only a small portion of complete responses and short response duration were observed [24–30]. Furthermore, when conventional hyperthermia is used in cancer treatment, damage to surrounding normal tissue is often unavoidable.

28.2.3 Immune responses induced by photothermal therapy

Photothermal therapy can be considered as a form of local hyperthermia, in many ways similar to radiofrequency (RF) [31], microwave [32], and extracorporeal high-intensity focused ultrasound (HIFU) [33]. Since evidence shows that raising the body temperature can induce immune responses against tumors, photothermal therapy can play an important role in cancer treatment.

The immunomodulatory function of thermal therapy has been found to be sensitively regulated by temperature, as different levels of heating can bring different modulatory effects on different sensitive targets [19]. Temperature between 39–40°C, called fever-range temperature, can modulate the activities of immune cells, including antigen presenting cells (APC), T cells, and natural killing cells (NK cells). Heat shock temperature is in the range of 41–43°C, which can increase the immunogenicity of tumor cells. Cytotoxic temperature is above 43°C, which can create an antigen source for induction of an antitumor immune response [34].

Photothermal therapy can induce high temperature increase in target tissue, which can ablate tumor cells directly and release a large load of tumor antigen for the generation of antitumor immunity. Tumor cells swell and break into pieces allowing antigen release with the increase of temperature. These antigens include tumor-associated antigens, thermally induced heat shock proteins (HSP), and a large amount of self-antigens. APCs, particularly dendric cells (DC), can capture these antigens and migrate to lymph nodes. They present the antigens to T cells to induce an immune response that can be effective against tumor cells.

The *in situ* hyperthermia-based vaccination approach can provide a tumor vaccine, whereupon after local hyperthermia treatment tumor tissue could serve as an antigen depot for inducing tumor-specific immune response. In combination with traditional tumor vaccines, hyperthermia-induced antigen release can "boost" immunological memory and amplify immune response by antigen spreading, thereby enhancing the vaccine effect.

However, as the immune systems of cancer patients are often compromised, tumor debris may not be sufficient in inducing a potent antitumor response [35]. Additional immunological interventions may be required to invoke the immune system to achieve an effective and protective immune response against residual tumor cells.

Since DCs are rare cells in the blood, the addition of cytokines, such as granulocyte monocyte colony stimulating factor (GM-CSF), can increase the number of circulating DCs that can infiltrate heat-treated tumors, thereby, amplifying the antitumor immune response [36]. Similar results were also obtained in a murine metastatic lung cancer, where irradiation of primary tumor in combination with systemic administration of Fms-like tyrosine kinase 3 ligand (Flt3L), another cytokine that stimulates DC proliferation, induced a strong tumor-specific immunity that eradicated microscopic lung metastases and provided a long-term cure [37–38]. Intratumoral injection of autologous immature DCs into heat-treated tumors can also induce a strong antitumor immunity [39–40]. It has been demonstrated that cytotoxic T-lymphocyte antigen 4 (CTLA-4) blocking antibodies [34] and TLR9 ligand [41] could further amplify the tumor-specific immunity.

Efforts have been made to optimize the treatment method and to improve the therapeutic effect and minimize adverse effects. Kah et al. [42] reported that the combination of photodynamic therapy (PDT) and PTT treatment can further reduce the cell viability compared to PDT or PTT alone. Researchers are monitoring photothermal therapy using photoacoustic imaging, and temperature measurement [43]. To optimize the treatment of photothermal therapy, Shah et al. investigated the feasibility of ultrasound imaging to monitor temperature changes during photothermal treatment [44]. The experiments performed on tissue-mimicking phantoms and *ex vivo* animal tissue samples showed its advantages to remotely guide photothermal therapy.

In the path to clinical application, many details remain to be developed, including technique development, preclinical evaluation, and clinical protocol design. As noted above, the emerging evidence demonstrates the exciting prospect of photothermal therapy. With further research, this promising method will be accepted by an oncologist. It eventually could benefit millions of cancer patients.

The relationship between photothermal therapy and immunological responses remains an exciting topic. However, to optimize the photothermal therapy, specific targeting of tumor cells and active stimulation of host immune systems in addition to the thermally induced immunological responses are needed. These approaches will be discussed in the following sections.

28.3 Selective Photothermal Interaction in Cancer Treatment

28.3.1 Near-infrared laser for tissue irradiation

To effectively utilize the thermal interaction for cancer treatment, selective targeting is crucial. The organized tissue has a window for light penetration in the near-infrared region. Laser energy with such wavelengths will be able to penetrate tissue with little damage to the tissue on the path of the laser beam. The high penetrating power of the near-infrared lasers also means less laser-tissue interaction of any sort, hence not causing desirable tissue destruction, unless the irradiation fluence is high enough for the surface action.

28.3.2 Selective photothermal interaction using light absorbers

To utilize the penetrating laser beam for enhancement of laser-tissue interaction, *in situ* light absorbers with appropriate absorption spectra are needed. While endogenous light absorbers can be used for this purpose, it is most common to use exogenous absorbers to achieve the selective light

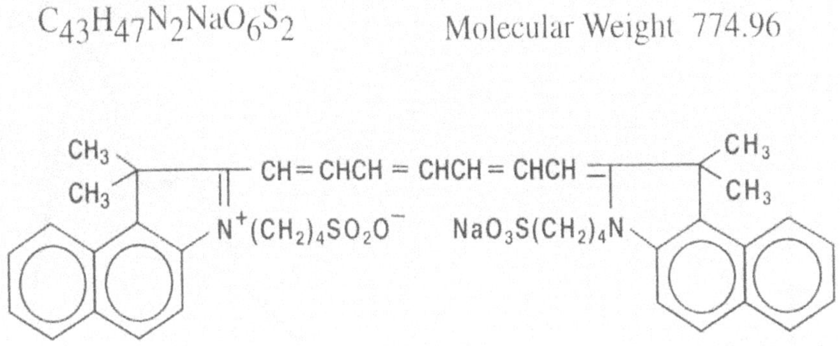

$C_{43}H_{47}N_2NaO_6S_2$ Molecular Weight 774.96

FIGURE 28.1: Molecular structure of indocyanine green (ICG).

absorption for appropriate wavelengths and appropriate levels of tissue absorption.

The absorber molecules in the photothermal interaction can absorb light to reach an excited high-energy state. When the molecules return to their ground states, the energy released is in the form of photons with a wavelength outside the tissue "transparent window." These photons will be readily absorbed by the surrounding tissue. If the activation of the absorber molecules continues with a sufficiently high rate, the rate of heat production in the tissue can exceed the rate of heat dissipation. As a consequence, the tissue temperature increases steadily. The dose of the dye and the laser fluence can be adjusted in order to achieve a desirable temperature elevation in target tissue, resulting in selective photothermal tissue destruction.

The selective photothermal laser-tissue interaction using a near-infrared laser and an *in situ* light-absorbing dye was proposed by Chen et al. [45]. Specifically, an 805-nm diode laser and indocyanine green (ICG) were first used for thermal treatment of rat tumors, both *in vitro* and *in vivo* [46–48].

28.3.3 Indocyanine green

Indocyanine green, a water-soluble compound, has a high absorption peak around 790 nm with a broad absorption shoulder. The molecular structure and its absorption spectra are given in Figures 28.1 and 28.2. It has been used in both lesion detection [49–50] and lesion treatment [51–55]. ICG has been approved by FDA to be used in cardiovascular, ophthalmic, and hepatic studies in humans [56–68]. The almost perfect match between the 805-nm laser and the absorption peak of ICG makes the selective photothermal reaction possible.

28.3.4 *In vivo* selective laser-photothermal tissue interaction

The combination of an 805-nm diode laser and ICG was used to treat mammary tumors in rats. When the primary tumor reached a size of 0.2 cm^3 to 1 cm^3, the ICG solution was injected into the center of the tumors and followed by irradiation of the laser. Tissue histology in Figures 28.3 to 28.5 shows clear coupling of the 805-nm laser and ICG. Even as deep as 2 cm from the treatment surface, the photothermal damage was apparent (Figure 28.3). When ICG was not present, the skin over the tumor could be totally spared, but the tumor containing ICG underneath could sustain damage, as evidenced in Figure 28.4.

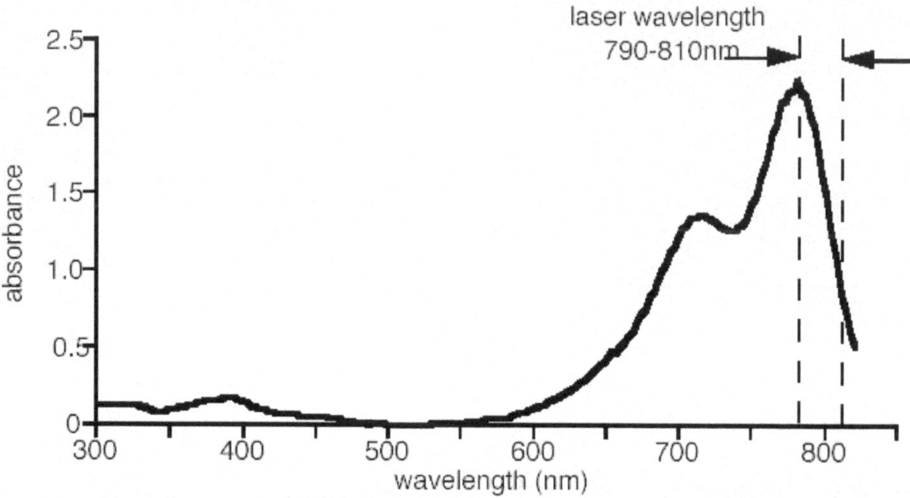

FIGURE 28.2: Absorption spectrum of indocyanine green (ICG).

FIGURE 28.3: Photomicrograph of a tumor sample immediately after treatment with 805-nm laser and ICG. The center field of the sample, about 2 cm from the treatment surface, suffered apparent photothermal damage.

Posttreatment observation, however, revealed that this photothermal interaction, although being at a large and controllable scale, still left many survival tumor cells. These residual tumor cells could still be seen three days after treatment and many were capable of dividing and multiplying, as shown in Figure 28.5.

28.3.5 Laser-ICG photothermal effect on survival of tumor-bearing rats

Rats bearing metastatic breast tumor DMBA-4 were treated with the laser after the injection of ICG, with different combinations of ICG dose and laser settings. All the treated rats eventually died with multiple tumors. The postoperation data of rat survival are summarized in Table 28.1.

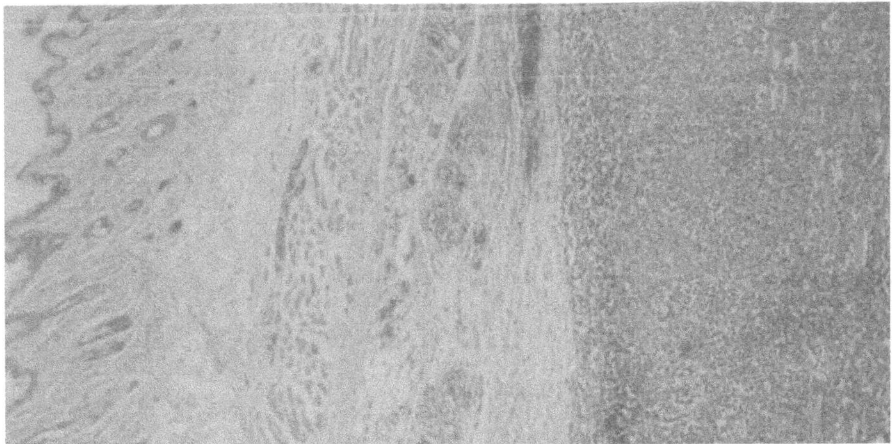

FIGURE 28.4: Photomicrograph of a tumor sample immediately after treatment with 805-nm laser and ICG. The epidermis and subcutaneous tissue appeared to have all the normal characteristics, but the tumor beneath the skin apparently suffered severe injury, as evidenced by large clefts and loss of cytoplasmic elements.

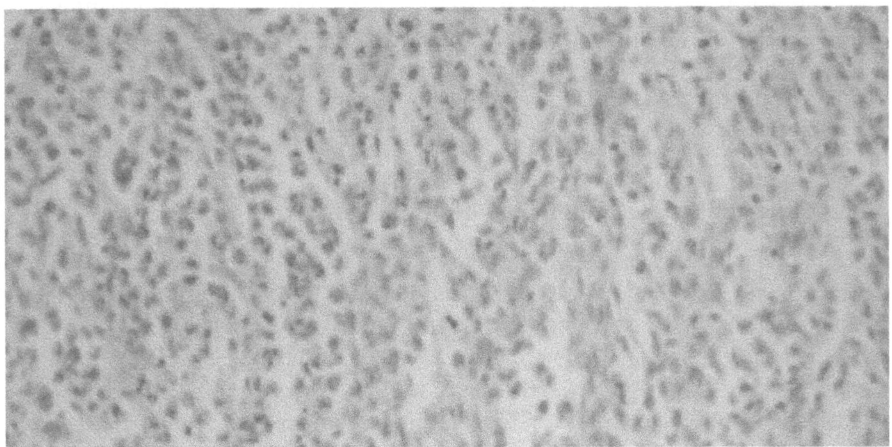

FIGURE 28.5: Photomicrograph of a tumor sample 3 days after treatment with 805-nm laser and ICG. The field, about 1 cm from the treatment surface, shows survival tumor cells and many are in the dividing stage.

The results of photothermal laser tissue interactions and the survival data revealed the limitation of thermal therapy for tumor treatment.

TABLE 28.1: Survival time of tumor-bearing rats after tumor inoculation

Treatment	Number of Rats	Survival Days after Tumor Inoculation
Control	33	31.5±3.7
Laser + ICG	16	29.9±3.8

28.4 Selective Photothermal Therapy Using Nanotechnology

28.4.1 Nanotechnology in biomedical fields

Nanostructures have been used in biomedical areas since their discovery. Nanomaterials have been mainly used in disease diagnosis, targeted drug delivery, and disease treatment [69,70]. The major efforts in biomedical applications of nanotechnology related to the photothermal treatment of cancer have been mainly in the areas of immunological enhancement and selective photothermal interaction.

28.4.2 Nanotechnology for immunological enhancement

Nanoparticles have been used as carriers for vaccines. To enhance the immune response, nanovaccine was used, which is constructed by co-encapsulating a tumor specific antigen and an adjuvant using nanoemulsion [71]. Nanoparticles conjugated with immune cells were used to form a vaccine, to resist tumor growth in mice [72]. Efforts were also made by others to use nanoparticles as carriers for vaccination [73]. Antiepidermal growth factor receptor antibody conjugated gold nanoparticles were used to enhance laser-induced tumor cell death [74]. Magnetic nanoparticles were used to induce immune responses [75–76]. Others used antibody-conjugated nanoparticles to locate tumors [77–80].

28.4.3 Nanotechnology for enhancement of photothermal interactions

Current research in the area of selective photothermal therapy focuses on the use of new agents to absorb laser energy to generate heat in targeted tissue [81]. Many groups have shown strong interests in nanotechnology, in combination with near-infrared laser light.

To further enhance the photothermal effect, many methods have been applied in controlling the cellular uptake of the gold nanoparticles. The early work by Hirsch et al. [82] used an 805-nm laser and directly injected gold nanoparticles into target tissue to treat animal liver cancer. Others also used nanomaterials for selective photothermal destruction of tumor cells [83–84]. Recent study using single-welled carbon nanotubes with a 980-nm laser also showed promising results in treatment of animal tumors [85].

28.4.4 Antibody-conjugated nanomaterials for enhancement of photothermal destruction of tumors

Photothermal therapy using the absorption properties of antibody-conjugated nanomaterials [73, 86–88] has demonstrated selective killing of cancer cells, while leaving healthy cells unaffected. The antibody conjugation provided accumulation of nanoparticles in target tissue and irradiation of laser light of appropriate wavelength provided the selective photothermal destruction.

Melancon et al. described a new class of molecular-specific photothermal coupling agents based on hollow gold nanoshells covalently attached to the monoclonal antibody directed at epidermal growth factor receptor (EGFR) [89]. Bernardi et al. used the antibody against HER2 to target gold-silica nanoshells to medulloblastoma cells, and the antibody against interleukin-13 receptor-alpha 2 (IL13Ralpha2) to glioma cells [90]. They believed that the use of antibody-tagged gold-silica nanoshells to selectively target cancer cells presented a promising new strategy for the treatment of central nervous system tumors that would minimize the damage and toxicity to the surrounding normal brain tissue.

Gobin reported selective targeting of prostate cancer cells with nanoshells conjugated to ephrinA I. EphrinA I is a ligand for EphA2 receptor that is overexpressed on PC-3 cells [91].

Li et al. demonstrated that the cellular uptake of gold nanoparticles conjugated with transferrin molecules was six times of that in the absence of this interaction [92]. Hence, the laser irradiation power was reduced by two orders of magnitude while achieving the same efficiency. The cellular uptake by the normal cells was only one fourth of that by the cancerous cells. The different concentration between normal cells and cancerous cells makes it possible to increase the targeting effect.

The work of Dickerson demonstrates the feasibility of *in vivo* treatment of deep-tissue malignancies using plasmonic gold nanorods [93]. Dramatic size decreases in squamous cell carcinoma xenografts were observed for direct and intravenous administration of pegylated gold nanorods in nu/nu mice after near-infrared laser irradiation.

Recently, dendrimer-encapsulated gold nanoparticles have been prepared and identified for their potential use towards the photothermal treatment of malignant tissue [94]. In one study, amine-terminated G5-PAMAM dendrimer-entrapped gold nanoparticles were prepared and covalently conjugated to flourescein and folic acid for targeted delivery to tumor cells overexpressing folic acid receptors, as reported by the Baker group [95]. The dendrimers were shown to specifically bind to KB cells *in vitro* and were internalized into lysosomes within 2 h. The applicability of these particles for targeted hyperthermia treatment and as electron-dense contrast agents was recognized and *in vivo* performance studies are currently underway. Some scholars even predicted that the field of oncology will soon be revolutionized by novel strategies for diagnosis and therapy employing dendrimer-based nanodevices [96].

Some other newly invented materials like quantum dots (QDs) have primarily been developed as fluorescent probes with unique optical properties. Research has been done to investigate its utilities in photoacoustic (PA) and photothermal (PT) effect [97].

In summary, nanotechnology has been demonstrated its great potential in thermal therapy for cancers.

28.5 Photothermal Immunotherapy

As discussed in the previous sections, while photothermal therapy can be used for a large scale, controlled destruction of cancer cells, it has limitations in completely eradicating cancers, particularly metastatic tumors, either due to the incomplete destruction of target tumors cells, or due to metastasis of the tumor to other sites. Therefore, a holistic approach is needed. Furthermore, thermal therapy itself can induce various immune responses. However, such responses often are not enough to systemically control the tumors. To effectively augment photothermal therapy, active immunological stimulation can be used. Photothermal immunotherapy, a combination of local selective photothermal treatment and local immunological stimulation using adjuvant, was proposed in 1997 [98]. The method of photothermal immunotherapy and its preclinical and preliminary clinical outcomes are discussed in this section.

28.5.1 Procedures of photothermal immunotherapy

28.5.1.1 Components of photothermal immunotherapy

Photothermal immunotherapy consists of three major components:

- A near-infrared laser;
- A light-absorbing dye with an absorption peak corresponding to the laser light;
- An immunological stimulant.

In the early study of photothermal immunotherapy, an 805-nm diode laser was used, due to its ability to penetrate biological tissues without significant thermal damage by itself to normal tissue. To achieve the desirable thermal effect, indocyanine green (ICG) was used due to its high absorption peak around 800 nm, a perfect match to the infrared laser wavelength (see Section 28.3). The immunoadjuvant was glycated chitosan (GC), a specially synthesized compound [98–100]. The synthesis process of GC made chitosan soluble while enhancing the immunological stimulation ability of chitosan.

28.5.1.2 Treatment procedures of photothermal immunotherapy

Photothermal immunotherapy consists of two major treatment procedures:

- Local administration of the light-absorbing dye and/or immunoadjuvant to the target tumors;
- Noninvasive, local irradiation of the target tumors.

The standard procedure of administration of dye and immunoadjuvant is local injection. Based on the burden of the primary tumor, appropriate doses of ICG and GC were used. After the injection of ICG and GC, laser light was directed to the treatment site through optical fibers. A microdiffuse lens attached to the end of the fiber was usually used to ensure the uniform delivery of laser energy to the treatment surface. The protocol in animal studies required only a local treatment. The detailed treatment procedures were described in previous publications [98–100].

28.5.2 Effects of photothermal immunotherapy in preclinical studies

Animal tumor models, including breast tumors in rats and mice, melanoma tumors in mice, and prostate tumors in rats, were used in the development in photothermal immunotherapy. The animal studies showed great potential of photothermal immunotherapy [100–103]. The outcomes of the treatment include:

- Regression of treated primary tumors;
- Eradication of untreated metastases;
- Long-term survival of treated animals;
- Induced tumor resistance;
- Induced immunological effect.

28.5.2.1 Regression of treated primary tumors and untreated metastases

After the photothermal immunotherapy, all the treated primary tumors usually continued to grow, only at a slower rate compared with the untreated control group. All the tumor-bearing rats in both treated and control groups developed metastases in the remote inguinal and axillary areas two to

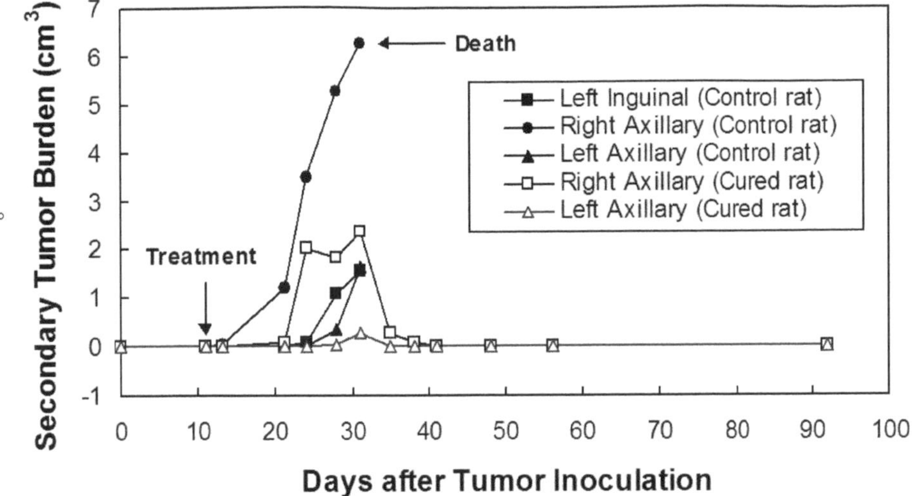

FIGURE 28.6: The size of metastases in the unimplanted inguinal and axillary areas of a successfully treated tumor-bearing rat (open squares and triangles) and of an untreated control tumor-bearing rat (solid squares, circles, and triangles). For the laser immunotherapy treated rat, the metastatic tumor burdens continued increasing until day 30 and then regression began; the metastatic tumors disappeared around day 40 and no recurrence was observed. For the untreated rat, the metastases continued to grow until the time of death (the upper arrow). The laser immunotherapy treatment took place 11 days after the tumor inoculation (the lower arrow).

three weeks after the inoculation of the primary tumor. Success of photothermal immunotherapy manifested only about 4 to 6 weeks after treatment, when the metastases began to regress, as shown in Figure 28.6. Primary tumors in successfully treated rats also followed the same regression pattern, as shown in Figure 28.7.

The successfully treated tumor-bearing animals usually had a long-term survival time. Furthermore, these animals could resist repeated challenges with the tumors of the same origin and also escalated tumor doses.

28.5.2.2 Immunological responses induced by photothermal immunotherapy

To study the immune responses induced by photothermal immunotherapy at the cellular level, histochemical assays were performed. Two assays, fluorescent labeling of living tumor cells and tissue immunoperoxidase of preserved tumor cells, were used to probe the tumor-specific antibodies in the successfully treated tumor-bearing rats [100]. In both cases, the sera of rats were used as the sources of primary antibodies.

Using immunoperoxidase assay to detect antibody binding to the preserved tumor tissue sections also showed that the sera from an untreated control tumor showed very little peroxidase activity, indicating the lack of tumor-specific antibodies, while photothermal immunotherapy treatment resulted in induction of specific antibodies in the rat sera, as evidenced by the intense staining at the plasma membranes of preserved tumor tissue.

Using fluorescent assay to detect antibody binding to the live tumors cells, the histochemical assay showed strong antibody binding using sera from successfully treated rats, when compared with sera from the untreated control tumor-bearing rats.

Adoptive immunity transfer was also carried out using the spleen cells from successfully treated animals as immune cells [101]. These immune cells were mixed with tumor cells and injected in

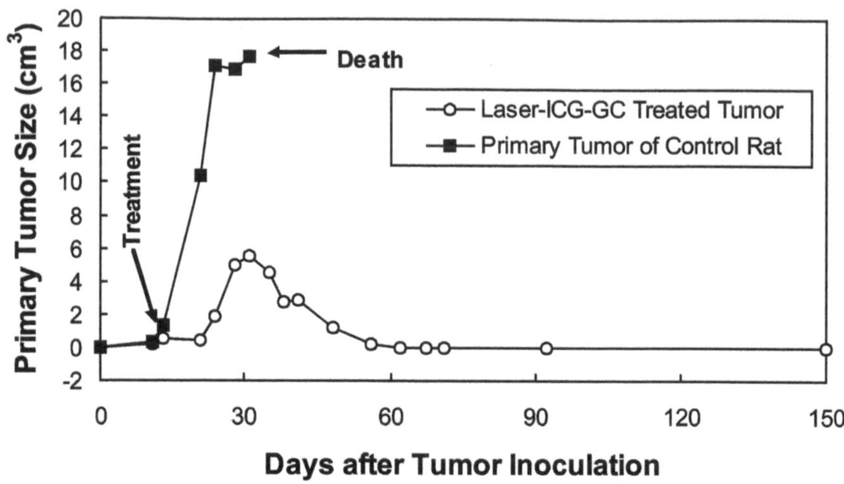

FIGURE 28.7: The size of the primary tumor of a successfully treated tumor-bearing rat (open circles) by laser immunotherapy and an untreated control tumor rat (solid squares). The treatment occurred eleven days after tumor inoculation (the lower arrow). The tumor burden continued increasing and around day 30 the regression began. The treated tumor became dead tissue about six weeks after treatment and was gradually absorbed. The treated rat was free of tumor 180 days after treatment, and tumor recurrence was not observed. The untreated rat's tumor continued to grow until the time of death (the upper arrow) at day 31 with multiple metastases.

naive animals. The resistance of the animals to the tumor cells was observed, as shown in Figure 28.8.

The spleen cells from a photothermal immunotherapy cured rat provided 100% protection to the recipients; neither primary nor metastatic tumors were observed among this group of rats. The control rats all died with multiple metastases within 35 days of tumor inoculation. The spleen cells from a healthy rat did not provide any protection to the recipients.

28.5.3 Possible immunological mechanism of photothermal immunotherapy

It has been well established that an immunological response is crucial in cancer treatment. In addition to traditional immunotherapy, such a systemic response in the host is also the goal of many other treatment modalities.

The immune response is an extremely sophisticated process. It can involve humoral and/or cellular immune activities. The results of photothermal immunotherapy showed the possibility of both cellular and humoral immunological responses, although the precise mechanism was not clear.

It is hypothesized that the induction of tumor immunity observed in the photothermal immunotherapy is the result of the combined photothermal and photoimmunological interactions. The photothermal interaction using the 805-nm laser and indocyanine green created a selective tissue destruction zone in the target tumor mass, while sparing the surrounding normal tissue. It reduced tumor burden in a controlled, large scale. Although this thermal reaction did not result in a complete removal or total acute tumor eradication, it served an additional purpose: exposure of the tumor antigen(s). Since the thermal interaction occurred inside the tissue, due to the lack of oxygen

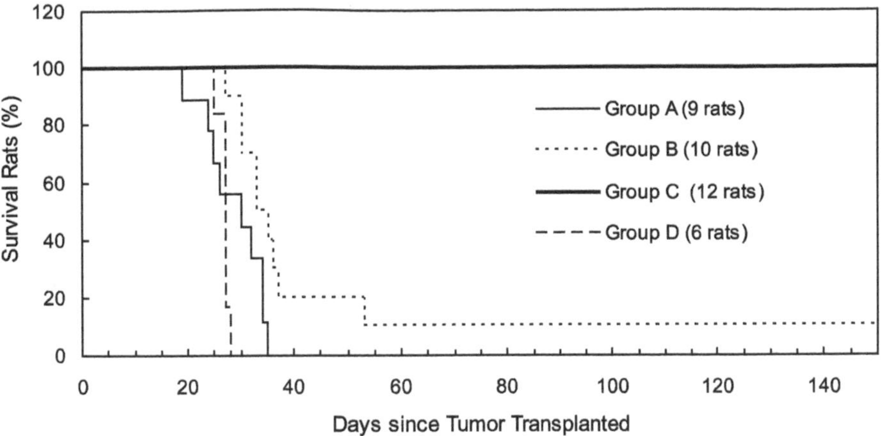

FIGURE 28.8: Rat survival curves in the adoptive immunity experiments using rat splenocytes as immune cells. Data were collected from two separate experiments. Viable tumor cells were admixed with spleen cells from different rats, then injected to naive rats. The ratio is 40 million spleen cells to 100,000 tumor cells per rat. The thin solid line represents the rats in Group A, the tumor control rats. The dotted line represents the rats in Group B, injected with tumor cells admixed with spleen cells from an untreated tumor-bearing rat. The thick solid line represents the rats in Group C, injected with tumor cells admixed with spleen cells from photothermal immunotherapy cured rats. The thin dash line represents the result using spleen cells from naive rats (Group D). The spleen cells from photothermal immunotherapy cured rats totally inhibited the tumor growth; all the rats survived and none developed tumors in Group C. In comparison, only the spleen cells from tumor-bearing rats had a certain impact on rat survival with a 10% survival (Group B). Rats in all three groups except for rats in Group C developed metastases.

the tumor tissue would not be totally denatured. The tumor cells, broken up by the laser energy while not removed, still contained tumor antigens to be recognized by the host immune system. The administration of glycated chitosan augmented the photothermal effect in two unique ways. First, it served as a carrier for ICG so that the ICG molecules could stay *in situ* for an extended period during which laser treatment could take place. Second, after the laser treatment, the host immune system, with the help of the adjuvant, could be directed to recognize the specific tumor antigens. Eventually, the immune system would produce tumor-specific antibodies and initiate a systemic assault against the residual tumor cells and metastases. This action is expected to last until all the tumor cells in the host are eradicated. Furthermore, the general immune response can lead to a tumor-specific immunity against future tumor exposure. The observed experimental results strongly support this mechanism.

Although the understanding of the working mechanism is very important, the crucial issue has always been how to induce the appropriate immune response in combating metastatic tumors.

28.5.4 Photothermal immunotherapy in clinical studies

The concept of photothermal immunotherapy has been applied to treat late-stage melanoma patients. To facilitate immunological stimulation, a unique toll-like receptor agonist, imiquimod, (Aldara^{TM}, imiquimod 5% cream, 3M Pharmaceuticals, St. Paul, Minnesota) was used as the immunoadjuvant. Imiquimod, a drug for topical application, has been approved by the FDA as a monotherapy for the treatment of warts, actinic keratoses, and superficial basal cell carcinoma. Its

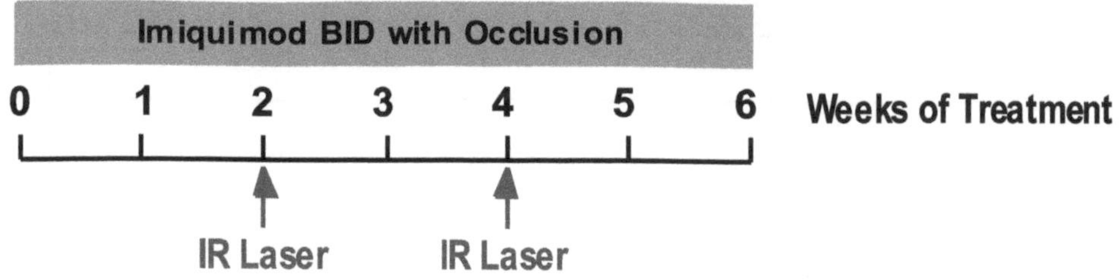

FIGURE 28.9: Treatment cycle of photothermal immunotherapy for late-state melanoma patients.

strong immunological stimulating effect made imiquimod a good candidate for use in combination with photothermal therapy.

The procedure for each cycle of photothermal immunotherapy for the melanoma treatment is given in Figure 28.9. The six-week treatment cycle was carried out on a designated 200-cm^2 area of skin, and consisted of six weeks of topical imiquimod treatment started two weeks before the first laser treatment session. Two laser treatment sessions were performed for each treatment cycle at week 2 and 4. Additional treatment cycles were carried out in the same treatment area or in different areas, if the response to the treatment was not complete.

The preliminary clinical trials showed promising results [104]. The responses of the locally treated tumors were immediate. In many cases, the untreated metastases surrounding the treated tumors were also affected, a clear sign of immunological effects.

Shown in Figure 28.10 are the treatment effects of a stage IV melanoma patient. The original tumor and local metastases were on the left arm of the patient (Figure 28.10, Panel A). As shown in the figure, imiquimod induced all of the laser-treated areas to ulcerate. Almost all the visible metastases showed marked inflammation and were shrinking (Figure 28.10, Panel B). The degree of inflammation seen by the end of the treatment cycle was impressive (Figure 28.10, Panel C). Although the thermal injury was severe after the laser treatment, the skin lesion recovered nicely, and eight weeks after the treatment, all the tumors, treated and untreated, were all eradicated (Figure 28.10, Panel D).

Furthermore, chest X-rays of this patient were taken during and after the photothermal immunotherapy. Lung X-rays at the beginning of the treatment cycle showed two lung metastases (Figure 28.11, Panel A). Three weeks after the second laser treatment, these nodules had reached maximum size (Figure 28.11, Panel B). By seven weeks following completion of the treatment cycle, the nodules had begun to shrink (Figure 28.11, Panel C). Seven months after completing the treatment cycle, the lung nodules had been cleared completely (Figure 28.11, Panel D).

This specific case and other clinical results strongly indicated the feasibility of the combination of thermal and immunological stimulation for the treatment of metastatic tumors.

28.6 Conclusion

Photothermal phenomenon is an exciting topic in research and in clinical applications. The photothermal-tissue interaction alone can cause direct tissue damage. With appropriate laser parameters, this interaction can be used in a variety of clinical applications such as lesion ablation, resurfacing, and tissue welding. The thermal damage can also be used to treat tumors. However, direct use of the laser limits the effective treatment area only to the surface tissue, or to the region

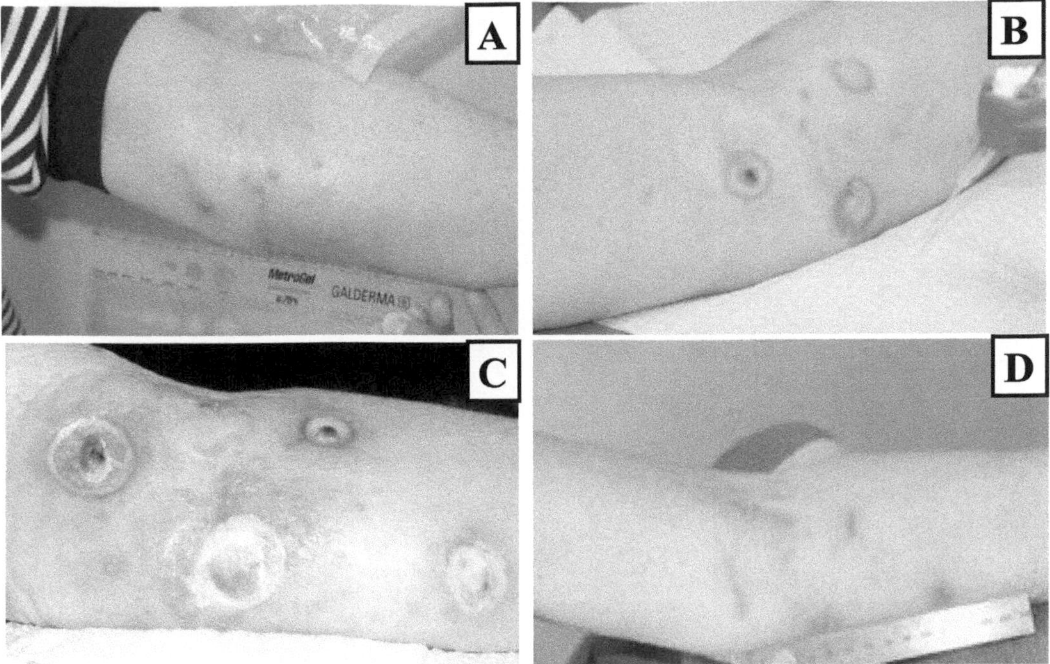

FIGURE 28.10: Treatment area of melanoma on the arm of a patient using photothermal immunotherapy. Panel A shows the treatment area after application of imiquimod twice a day for two weeks. Panel B shows the treatment area immediately after the first laser session. Panel C shows the treatment area one week after the second laser treatment. Panel D shows the treatment area six weeks after completion of the first treatment cycle.

in the close vicinity of the laser source, due to the absorption of the laser energy by the tissue.

To overcome the major shortcoming of laser thermal interaction, a near-infrared laser and a laser-absorbing dye can be employed to produce a selective photothermal tissue interaction. However, the photothermal therapy itself has its limitations, particularly in dealing with metastatic tumors.

The ultimate cure of cancer, particularly metastatic cancers, should be an immunological one. Only when the host immune system is fully stimulated to fight a specific target can the effect be systemic and long-term. This long-term effect can be achieved by using combined photothermal and photoimmunological interactions in a novel treatment modality. The photothermal immunotherapy results from animal experiments showed a great potential in treating metastatic tumors. The total eradication of tumors, both treated primary tumors and untreated secondary tumors, showed the acute efficacy of the method. The resistance to tumor rechallenge showed the establishment of a long-term immunity induced by photothermal immunotherapy. The histochemical and bioimmunological assays indicated an induced immune response at both cellular and molecular levels. The protective power of the immune spleen cells in adoptive immune transfer showed the long-lasting immune effect induced by photothermal immunotherapy.

In conclusion, laser photothermal interaction can be a precursor for immunotherapy. The combined photothermal and immunological effects could prove to be an efficacious means in treating metastatic, late-stage tumors.

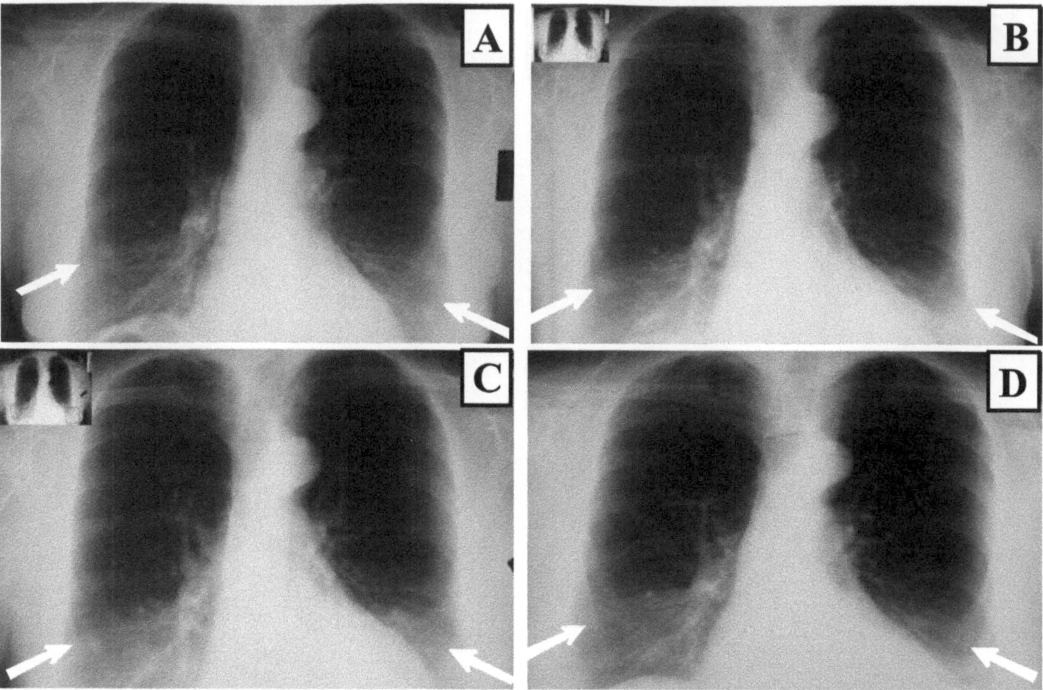

FIGURE 28.11: Posterior-anterior views of the chest X-ray of a patient treated using photothermal immunotherapy. Panel A: a chest X-ray taken three days after the first laser treatment demonstrates a definite ($\approx$ 1.2 cm) metastasis in the right lateral lung base and one ($\approx$ 1.5 cm) near the left costophrenic angle (arrows indicate locations of the metastases). Panel B: three weeks after the second laser treatment, metastases have actually increased in size although their outlines are less distinct; there is a small pleural effusion in the left costophrenic angle. Panel C: eight weeks after the second laser treatment, the metastases are clearly shrinking although still present both on plain films and in CT images. Panel D: seven months after the second laser treatment, the plain films are back to baseline (no evidence of pulmonary metastases).

Acknowledgment

This research is supported in part by the College of Graduate Studies and Research, University of Central Oklahoma, and by a grant from the U.S. National Institute of Health (P20 RR016478 from the INBRE Program of the National Center for Research Resources).

References

[1] J.L. Boulnois, "Photophysical processes in recent medical laser developments: a review," *Lasers Med. Sci.* **1**, 47–66 (1986).

[2] S.L. Jacques, "Laser-tissue interactions," *ancer Bull.* **41**, 211–218 (1989).

[3] S. Thomsen, "Pathologic analysis of photothermal and photomechanical effects of laser-tissue interactions," *Photochem. Photobiol.* **53**, 825–835 (1991).

[4] S.L. Jacques, "Laser-tissue interactions. Photochemical, photothermal, and photomechanical," *Surg. Clin. North Am.* **72**, 531–558 (1992).

[5] T.J. Dougherty, J.E. Kaufman, A. Goldfarb, K.R. Weishaupt, D.G. Boyle, and A. Mlttleman, "Photoradiation therapy for the treatment of malignant tumors," *Cancer Res.* **38**, 2628–2635 (1978).

[6] M.J. Manyak, A. Russo, P.D. Smith, and E. Glatsein. "Photodynamic therapy," *J. Clin. Oncol.* **6**, 380–391 (1988).

[7] T.J. Dougherty, "Photoradiation therapy for cutaneous and subcutaneous malignancies," *J. Invest. Dermatol.* **77**, 122–124 (1981).

[8] A. Dahlman, "Laser photoradiation therapy of cancer," *Cancer Res.* **43,** 430–434 (1983).

[9] T.J. Dougherty, "Photodynamic therapy: status and potential," *Oncology* **3,** 67–78 (1989).

[10] S.L. Marcus, "Photodynamic therapy of human cancer: clinical status, potential and needs," *Future Directions and Applications in Photodynamic Therapy* (C.J. Gomer, eds.), 5–56, SPIE Press, Bellingham, WA, 1990.

[11] N.A. Buskard, "The use of photodynamic therapy in cancer," *Sem. Oncol.* **21**, 1–27 (1994).

[12] A.M. Fisher, A.L. Murphee, and C.J. Gomer, "Clinical and pre-clinical photodynamic therapy," *Lasers Surg. Med.* **17**, 2–31 (1995).

[13] L.J. Anghileri and J. Robert, *Hyperthermia in Cancer Treatment*, CRC Press, Boca Raton, FL (1986).

[14] K. Camphausen, M.A. Moses, W.D. Beecken, M.K. Khan, J. Folkman, and M.S. O'Reilly, "Radiation therapy to a primary tumor accelerates metastatic growth in mice," *Cancer Res.* **61**, 2207–2211 (2001).

[15] W.H. Clark Jr., D.E. Elder, D.I.V. Guerry, L.E. Braitman, B.J. Trock, D. Schultz, M. Synnestevdt, and A.C. Halpern, "Model predicting survival in stage I melanoma based on tumor progression," *J. Natl. Cancer Inst.* **81**, 1893–1904 (1989).

[16] M.S. O'Reilly, R. Rosenthal, E. Sage, S. Smith, L. Holmgren, M.A. Moses, Y. Shing, and J. Folkman, "The suppression of tumor metastases by a primary tumor," *Surg. Forum.* **44**, 474–476 (1993).

[17] C.W. Song, A. Lokshina, J.G. Rhee, M. Patten, and S.H. Levitt, "Implication of blood flow in hyperthermic treatment of tumors," *IEEE Trans. Biomed. Eng.* **31**, 9–16 (1984).

[18] H.S. Reinhold and B. Endrich, "Tumor microcirculation as a target for hyperthermia: a review," *Int. J. Hyperthermia* **2**, 111–137 (1986).

[19] H.G. Zhang, K. Mehta, P. Cohen, and C. Guha, "Hyperthermia on immune regulation: a temperature's story," *Cancer Lett.* **2**, 191–204 (2008).

[20] W.C. Dewey, L.E. Hopwood, S.A. Sapareto, and L.E. Gerweck, "Cellular responses to combinations of hyperthermia and radiation," *Radiology* **123**, 463–474 (1977).

[21] S.A. Sapareto and W.C. Dewey, "Thermal dose determination in cancer therapy," *Int. J. Radiat. Oncol. Biol. Phys.* **10**, 787–800 (1984).

[22] S.B. Field and C.C. Morris, "The relationship between heating time and temperature: its relevance to clinical hyperthermia," *Radiother. Oncol.* **1**, 179–186 (1983).

[23] A.J. Thistlethwaite, D.B. Leeper, D.J. Moylan, and R.E. Nerlinger, "pH distribution in human tumors," *Int. J. Radiat. Oncol. Biol. Phys.* **11**, 1647–1652 (1986).

[24] M.W. Dewhirst, W.G. Conner, T.E. Moon, and H.B. Roth, "Response of spontaneous animal tumors to heat and/or radiation: preliminary results of a phase III trial," *J. Natl. Cancer Inst.* **61**, 395–397 (1982).

[25] M.W. Dewhirst, W.G. Conner, and D.A. Sim, "Preliminary results of a phase III trial of spontaneous animal tumors to heat and/or radiation: early normal tissue response and tumor volume influence on initial response," *Int. J. Radiat. Oncol. Biol. Phys.* **8**, 1951–1962 (1982).

[26] J.H. Kim, E.W. Hahn, and N. Tokita, "Combination hyperthermia and radiation therapy for cutaneous malignant melanoma," *Cancer* **41**, 2143–2148 (1978).

[27] H.H. Leveen, S. Wapnick, V. Riccone, G. Falk, and N. Ahmed, "Tumor eradication by radiofrequency therapy," *J. Am. Med. Assoc.* **235**, 2198–2200 (1976).

[28] M.R. Manning, T.C. Cetas, R.C. Miller, J.R. Oleson, W.G. Conner, and E.W. Gerner, "Results of a phase I trial employing hyperthermia alone or in combination with external beam or interstitial radiotherapy," *Cancer* **49**, 205–216 (1982).

[29] J.B. Marmor, D. Pounds, N. Hahn, and G.M. Hahn, "Treating spontaneous tumors in dogs and cats by ultrasound induced hyperthermia," *Int. J. Radiat. Oncol. Biol. Phys.* **4**, 967–973 (1978).

[30] J.B. Marmor, D. Pounds, D.B. Postic, and G.M. Hahn, "Treatment of superficial human neoplasms by local hyperthermia induced by ultrasound," *Cancer* **43**, 196–205 (1979).

[31] P. Mertyna, M.W. Dewhirst, E. Halpern, W. Goldberg, and S.N. Goldberg, "Radiofrequency ablation: the effect of distance and baseline temperature on thermal dose required for coagulation," *Int. J. Hyperthermia* **7**, 550–559 (2008).

[32] G. Gravante, S.L. Ong, M.S. Metcalfe, A. Strickland, A.R. Dennison, and D.M. Lloyd, "Hepatic microwave ablation: a review of the histological changes following thermal damage," *Liver Int.* **7**, 911–921 (2008).

[33] F. Feng, A. Mal, M. Kabo, J.C. Wang, and Y. Bar-Cohen, "The mechanical and thermal effects of focused ultrasound in a model biological material," *J. Acoust. Soc. Am.* **4 Pt 1**, 2347–2355 (2005).

[34] M.H. den Brok, R.P. Sutmuller, R. van der Voort, E.J. Bennink, C.G. Figdor, T.J. Ruers, and G.J. Adema, "*In situ* tumor ablation creates an antigen source for the generation of antitumor immunity," *Cancer Res.* **64**, 4024–4029 (2004).

[35] E. Jager, D. Jager, and A. Knuth, "Antigen-specific immunotherapy and cancer vaccines," *Int. J. Cancer* **106**, 820–817 (2003).

[36] A. Mukhopadhaya, J. Mendecki, X. Dong, L. Liu, S. Kalnicki, M. Garg, A. Alfieri and C. Guha, "Localized hyperthermia combined with intratumoral dendritic cells induces systemic antitumor immunity," *Cancer Res.* **67**, 7798–7806 (2007).

[37] P.K. Chakravarty, A. Alfieri, E.K. Thomas, V. Beri, K.E. Tanaka, B. Vikram, and C. Guha, "Flt3-ligand administration after radiation therapy prolongs survival in a murine model of metastatic lung cancer," *Cancer Res.* **59**, 6028–6032 (1999).

[38] P.K. Chakravarty, C. Guha, A. Alfieri, V. Beri, Z. Niazova, N.J. Deb, Z. Fan, E.K. Thomas, and B. Vikram, "Flt3L therapy following localized tumor irradiation generates long-term protective immune response in metastatic lung cancer: its implication in designing a vaccination strategy," *Oncology* **70**, 245–254 (2006).

[39] J. Guo, J. Zhu, X. Sheng, X. Wang, L. Qu, Y. Han, Y. Liu, H. Zhang, L. Huo, S. Zhang, B. Lin, and Z. Yang, "Intratumoral injection of dendritic cells in combination with local hyperthermia induces systemic antitumor effect in patients with advanced melanoma," *Int. J. Cancer* **120**, 2418–2425 (2007).

[40] K. Tanaka, A. Ito, T. Kobayashi, T. Kawamura, S. Shimada, K. Matsumoto, T. Saida, and H. Honda, "Intratumoral injection of immature dendritic cells enhances antitumor effect of hyperthermia using magnetic nanoparticles," *Int. J. Cancer* **116**, 624–633 (2005).

[41] M.H. den Brok, R.P. Sutmuller, S. Nierkens, E.J. Bennink, L.W. Toonen, C.G. Figdor, T.J. Ruers, and G.J. Adema, "Synergy between in situ cryoablation and TLR9 stimulation results in a highly effective *in vivo* dendritic cell vaccine," *Cancer Res.* **66**, 7285–7292 (2006).

[42] J.C. Kah, R.C. Wan, K.Y. Wong, S. Mhaisalkar, C.J. Sheppard, and M. Olivo, "Combinatorial treatment of photothermal therapy using gold nanoshells with conventional photodynamic therapy to improve treatment efficacy: an in vitro study," *Lasers Surg. Med.* **40**, 584–589 (2008).

[43] J. Shah, S. Park, S. Aglyamov, T. Larson, L. Ma, K. Sokolov, K. Johnston, T. Milner, and S.Y. Emelianov, "Photoacoustic imaging and temperature measurement for photothermal cancer therapy," *J. Biomed. Opt.* **13**, 034024 (2008).

[44] J. Shah, S.R. Aglyamov, K. Sokolov, T.E. Milner, and S.Y. Emelianov, "Ultrasound imaging to monitor photothermal therapy – feasibility study," *Opt. Express.* **16**, 3776–3785 (2008).

[45] W.R. Chen, R.L. Adams, S. Heaton, D.T. Dickey, K.E. Bartels, and R.E. Nordquist, "Chromophore-enhanced laser-tumor tissue photothermal interaction using 808 nm diode laser," *Cancer Lett.* **88**, 15–19 (1995).

[46] W.R. Chen, R.L. Adams, K.E. Bartels, and R.E. Nordquist, "Chromophore-enhanced *in vivo* tumor cell destruction using an 808-nm diode laser," *Cancer Lett.* **94**, 125–131 (1995).

[47] W.R. Chen, R.L. Adams, C.L. Phillips, and R.E. Nordquist, "Indocyanine green *in situ* administration and photothermal destruction of tumor tissue using an 808-nm diode laser," *Proc SPIE* **2681**, 94–101 (1996).

[48] W.R. Chen, R.L. Adams, A.K. Higgins, K.E. Bartels, and R.E. Nordquist, "Photothermal effects on murine mammary tumors using indocyanine green and an 808-nm diode laser: an in vivo efficacy study," *Cancer Lett.* **98**, 169–173 (1996).

[49] X. Li, B. Beauviot, R. White, S. Nioka, B. Chance, and A. Yodh, "Tumor localization using fluorescence of indocyanine green (ICG) in rat models," *Proc SPIE* **2389**, 789–797 (1995).

[50] S. Zhao, M.A. O'Leary, S. Nioka, and B. Chance "Breast tumor detection using continuous wave light source," *Proc SPIE* **2389**, 809–817 (1995).

[51] R.R. Anderson, C.A. Puliafito, E.S. Gragoudas, and R.F. Steinert, "Dye enhanced laser photocoagulation of choroidal vessels," *Invest. Ophthalmol. Vis. Sci.* **25**, 89 (1984).

[52] B. Jean, J. Maier, M. Braun, C.S. Kischkel, M. Baumann, and H.-J. Thiel, "Target dyes in ophthalmology, Part I–principles and indications," *Lasers Light Ophthalmol.* **3**, 39–45 (1990).

[53] C.A. Puliafito, T.K. Destro, and E. Dobi, "Dye enhanced photocoagulation of choroidal neovascularization," *Invest. Ophthalmol. Vis. Sci.* **29**, 414 (1988).

[54] J.H. Suh, T. Miki, A. Obana, K. Shiraki, M. Matsumoto, and T. Sakamoto, "Indocyanine green enhancement of diode laser photocoagulation," *J. Eye* **7**, 1697–1699 (1990).

[55] J.H. Suh, T. Miki, A. Obana, K. Shiraki, and M. Matsumoto, "Effects of indocyanine green dye enhanced diode laser photocoagulation in non-pigmented rabbit eyes," *Osaka City Med. J.* **37**, 89–106 (1991).

[56] L.A. Bauer, J.R. Horn, and K.E. Opheim, "Variability of indocyanine green pharmacokinetics in healthy adults," *Clin. Pharm.* **8**, 54–55 (1989).

[57] J.T. DiPiro, K.D. Hooker, J.C. Sherman, M.G. Gaines, and J.J. Wynn, "Effect of experimental hemorrhagic shock on hepatic drug elimination," *Crit. Care Med.* **20**, 810–815 (1992).

[58] L.A. Bauer, K. Murray, J.R. Horn, K. Opheim, and J. Olsen, "Influence of nifedipine therapy on indocyanine green and oral propranolol pharmacokinetics," *Eur. J. Clin. Pharmacol.* **37**, 257–260 (1989).

[59] P.A. Soons, A. de Boer, A.F. Cohen, and D.D. Breimer, "Assessment of hepatic blood flow in healthy subjects by continuous infusion of indocyanine green," *Br. J. Clin. Pharmacol.* **32**, 697–704 (1991).

[60] A. Kubota, A. Okada, Y. Fukui, H. Kawahara, K. Imura, and S. Kamata, "Indocyanine green test is a reliable indicator of postoperative liver function in biliary atresia," *J. Pediatr. Gastroenterol. Nutr.* **16**, 61–65 (1993).

[61] D. Clements, E. Elias, and P. McMaster, "Preliminary study of indocyanine green clearance in primary biliary cirrhosis," *Scand. J. Gastroenterol.* **26**, 119–123 (1991).

[62] M. Vaubourdolle, V. Gufflet, O. Chazouilleres, J. Giboudeau, and R. Poupon, "Indocyanine green-sulfobromophthalein pharmacokinetics for diagnosing primary biliary cirrhosis and assessing histological severity," *Clin. Chem.* **37**, 1688–1690 (1991).

[63] H.A. Wynne, J. Goudevenos, M.D. Rawlins, O.F.W. James, P.C. Adams, and K.W. Woodhouse, "Hepatic drug clearance: the effect of age using indocyanine green as a model compound," *Br. J. Clin. Pharmac.* **30**, 634–637 (1990).

[64] P.A. Soons, J.M. Kroon, and D.D. Breimer, "Effects of single-dose and short-term oral nifedipine on indocyanine green clearance as assessed by spectrophotometry and high performance liquid chromatography," *J. Clin. Pharmacol.* **30**, 693–698 (1990).

[65] W.E. Evans, M.V. Relling, S. de Graaf, J.H. Rodman, J.A. Pieper, M.L. Christensen, and W.R. Crom, "Hepatic drug clearance in children: studies with indocyanine green as a model substrate," *J. Pharm. Sci.* **78**, 452–456 (1989).

[66] G.L. Kearns, G.B. Mallory, W.R. Crom, and W.E. Evans, "Enhanced hepatic drug clearance in patients with cystic fibrosis," *J. Pediatrics* **117**, 972–979 (1990).

[67] M.J. Avram, T.C. Krejcie, and T.K. Henthorn, "The relationship of age to the pharmacokinetics of early drug distribution: the concurrent disposition of thiopental and indocyanine green," *Anesthesiology* **72**, 403–411 (1990).

[68] A. Scheider, A. Voeth, A. Kaboth, and L. Neuhauser, "Fluorescence characteristics of indocyanine green in the normal choroid and in subretinal neovascular membranes," *German J. Ophthalmal.* **1**, 7–11 (1992).

[69] G.S. Terentyuk, G.N. Maslyakova, L.V. Suleymanova, B.N. Khlebtsov, B.Ya. Kogan, G.G. Akchurin, A.V. Shantrocha, I.L. Maksimova, N.G. Khlebtsov, and V.V. Tuchin, "Circulation and distribution of gold nanoparticles and induced alterations of tissue morphology at intravenous particle delivery," *J. Biophoton.* **2** (5), 292–302 (2009).

[70] G. Akchurin, B. Khlebtsov, G. Akchurin, V. Tuchin, V. Zharov, and N. Khlebtsov, "Gold nanoshell photomodification under a single-nanosecond laser pulse accompanied by color-shifting and bubble formation phenomena," *Nanotechnology* **19**, 015701-1–8 (2008).

[71] R. Shi, L. Hong, D. Wu, X. Ning, Y. Chen, T. Lin, D. Fan, and K. Wu, "Enhanced immune response to gastric cancer specific antigen Peptide by coencapsulation with CpG oligodeoxynucleotides in nanoemulsion," *Cancer Biol. Ther.* **4**, 218–224 (2005).

[72] T. Fifis, A. Gamvrellis, B. Crimeen-Irwin, G.A. Pietersz, J. Li, P.L. Mottram, I.F.C. McKenzie, and M. Plebanski, "Size-dependent immunogenicity: therapeutic and protective properties of nano-vaccines against tumors," *J. Immunol.* **173**, 3148–3154 (2004).

[73] N. Dinauer, S. Balthasar, C. Weber, J. Kreuter, K. Langer, and H. Von Briesen, "Selective targeting of antibody-conjugated nanoparticles to leukemic cells and primary T-lymphocytes," *Biomaterials* **26**, 5898–5906 (2005).

[74] X. Huang, P.K. Jain, I.H. El-Sayed, and M.A. El-Sayed, "Determination of the minimum temperature required for selective photothermal destruction of cancer cells with the use of immunotargeted gold nanoparticles," *Photochem Photobiol.* **82**, 412–417 (2006).

[75] A. Ito, H. Honda, and T. Kobayashi, "Cancer immunotherapy based on intracellular hyperthermia using magnetite nanoparticles: a novel concept of 'heat-controlled necrosis' with heat shock protein expression," *Cancer Immunol. Immunother.* **55**, 320–328 (2006).

[76] A. Ito, Y. Kuga, H. Honda, H. Kikkawa, A. Horiuchi Y. Watanabe, and T. Kobayashi, "Magnetite nanoparticle-loaded anti-HER2 immunoliposomes for combination of antibody therapy with hyperthermia," *Cancer Lett.* **212**, 167–175 (2004).

[77] L. Li, C.A. Wartchow, S.N. Danthi, Z. Shen, N. Dechene, J. Pease, H.S. Choi, T. Doede, P. Chu, S. Ning, D.Y. Yee, M.D. Bednarski, and S.J. Knox, "A novel antiangiogenesis therapy using an integrin antagonist or anti–Flk-1 antibody coated 90Y-labeled nanoparticles," *Int. J. Radiat. Oncol. Biol. Phys.* **58**, 1215–1227 (2004).

[78] S. De Jong, G. Chikh, L. Sekirov, S. Raney, S. Semple, S. Klimuk, N. Yuan, M. Hope, P. Cullis, and Y. Tam "Encapsulation in liposomal nanoparticles enhances the immunostimulatory, adjuvant and anti-tumor activity of subcutaneously administered CpG ODN," *Cancer Immunol. Immunother.* **56**, 1251–1264 (2007).

[79] I.H. El-Sayed, X. Huang, and M.A. El-Sayed, "Selective laser photo-thermal therapy of epithelial carcinoma using anti-EGFR antibody conjugated gold nanoparticles," *Cancer Lett.* **239**, 129–135 (2006).

[80] M.R. McDevitt, D. Chattopadhyay, B.J. Kappel, J.S. Jaggi, S.R. Schiffman, C. Antczak, J.T. Njardarson, R. Brentjens, and D.A. Scheinberg, "Tumor targeting with antibody-functionalized, radiolabeled carbon nanotubes," *J. Nucl. Med.* **48**, 1180–1189 (2007).

[81] I.L. Maksimova, G.G. Akchurin, B.N. Khlebtsov, G.S. Terentyuk, G.G. Akchurin, I.A. Ermolaev, A.A. Skaptsov, E.P. Soboleva, N.G. Khlebtsov, and V.V. Tuchin, "Near-infrared laser

photothermal therapy of cancer by using gold nanoparticles: computer simulations and experiment," *Med. Laser. Appl.* **22**, 199–206 (2007).

[82] L.R. Hirsch, R.J. Stafford, J.A. Bankson, S.R. Sershen, B. Rivera, R.E. Price, J.D. Hazle, N.J. Halas, and J.L. West, "Nanoshell-mediated near-infrared thermal therapy of tumors under magnetic resonance guidance," *PNAS* **100**, 13549–13554 (2003).

[83] D.P. O'Neal, L.R. Hirsch, N.J. Halas, J.D. Paynea, and J.L. West, "Photo-thermal tumor ablation in mice using near infrared-absorbing nanoparticles," *Cancer Lett.* **209**, 171–176 (2004).

[84] N.W.S. Kam, M.J. O'Connell, J.A. Wisdom, and H. Dai, "Carbon nanotubes as multifunctional biological transporters and near-infrared agents for selective cancer cell destruction," *Proc. Natl. Acad. Sci. USA* **102**, 11600–11605 (2005).

[85] F. Zhou, D. Xing, Z. Ou, B. Wu, D.E. Resasco, and W.R. Chen, "Cancer photothermal therapy in the near-infrared region by using single-walled carbon nanotubes," *J. Biomed. Opt.* **14**, 021009 (2009).

[86] C. Loo, A. Lowery, N. Halas, J. West, and R. Drezek, "Immunotargeted nanoshells for integrated cancer imaging and therapy," *Nano Lett.* **5**, 709–711 (2005).

[87] X. Huang, I.H. El-Sayed, W. Qian, and M.A. El-Sayed, "Cancer cell imaging and photothermal therapy in the near-infrared region by using gold nanorods," *J. Am. Chem. Soc.* **128**, 2115–2120 (2006).

[88] Z. Liu, W. Cai, L. He, N. Nakayama, K. Chen, X. Sun, X. Chen, and H. Dai, "*In vivo* biodistribution and highly efficient tumour targeting of carbon nanotubes in mice," *Nat. Nanotech.* **2**, 47–52 (2007).

[89] M.P. Melancon, W. Lu, Z. Yang, R. Zhang, Z. Cheng, A.M. Elliot, J. Stafford, T. Olson, J.Z. Zhang, and C. Li, "*In vitro* and *in vivo* targeting of hollow gold nanoshells directed at epidermal growth factor receptor for photothermal ablation therapy," *Mol. Cancer Ther.* **7**, 1730–1739 (2008).

[90] R.J. Bernardi, A.R. Lowery, P.A. Thompson, S.M. Blaney, and J.L. West, "Immunonanoshells for targeted photothermal ablation in medulloblastoma and glioma: an *in vitro* evaluation using human cell lines," *J. Neurooncol.* **86**, 165–172 (2008).

[91] A.M. Gobin, J.J. Moon, and J.L. West, "EphrinA I-targeted nanoshells for photothermal ablation of prostate cancer cells," *Int. J. Nanomedicine* **3**, 351–358 (2008).

[92] J.L. Li, L. Wang, X.Y. Liu, Z.P. Zhang, H.C. Guo, W.M. Liu, and S.H. Tang, "In vitro cancer cell imaging and therapy using transferrin-conjugated gold nanoparticles," *Cancer Lett.* **274**, 319–316 (2009).

[93] E.B. Dickerson, E.C. Dreaden, X. Huang, I.H. El-Sayed, H. Chu, S. Pushpanketh, J.F. McDonald, and M.A. El-Sayed, "Gold nanorod assisted near-infrared plasmonic photothermal therapy (PPTT) of squamous cell carcinoma in mice," *Cancer Lett.* **269**, 57–66 (2008).

[94] J.B. Wolinsky and M.W. Grinstaff, "Therapeutic and diagnostic applications of dendrimers for cancer treatment," *Adv. Drug Deliv. Rev.* **10**, 1037–1055 (2008).

[95] X. Shi, S. Wang, S. Meshinchi, M.E. Van Antwerp, X. Bi, I. Lee, and J.R. Baker, Jr., "Dendrimer-entrapped gold nanoparticles as a platform for cancer-cell targeting and imaging," *Small* **3,** 1245–1252 (2007).

[96] I.J. Majoros, C.R. Williams, and J.R. Baker Jr., "Current dendrimer applications in cancer diagnosis and therapy," *Curr. Top. Med. Chem.* **8**, 1169–1179 (2008).

[97] E.V. Shashkov, M. Everts, E.I. Galanzha, and V.P. Zharov, "Quantum dots as multimodal photoacoustic and photothermal contrast agents," *Nano Lett.* **8**, 3953–3958 (2008).

[98] W.R. Chen, R.L. Adams, R. Carubelli, and R.E. Nordquist, "Laser-photosensitizer assisted immunotherapy: a novel modality in cancer treatment," *Cancer Lett.* **115**, 25–30 (1997).

[99] W.R. Chen, R. Carubelli, H. Liu, and R.E. Nordquist, "Laser immunotherapy: a novel treatment modality for metastatic tumors," *Mol. Biotechnol.* **25**, 37–43 (2003).

[100] W.R. Chen, W.G. Zhu, J.R. Dynlacht, H. Liu, and R.E. Nordquist, "Long-term tumor resistance induced by laser photo-immunotherapy," *Int. J. Cancer* **81**, 808–812 (1999).

[101] W.R. Chen, A.K. Singhal, H. Liu, and R.E. Nordquist, "Laser immunotherapy induced antitumor immunity and its adoptive transfer," (Advances in Brief) *Cancer Res.* **61**, 459–461 (2001).

[102] W.R. Chen, J.W. Ritchey, K.E. Bartels, H. Liu, and R.E. Nordquist, "Effect of different components of laser immunotherapy in treatment of metastatic tumors in rats," *Cancer Res.* **62**, 4295–4299 (2002).

[103] W.R. Chen, S.W. Jeong, M.D. Lucroy, R.F. Wolf, E.W. Howard, H. Liu, and R.E. Nordquist, "Induced anti-tumor immunity against DMBA-4 metastatic mammary tumors in rats using a novel approach," *Int. J. Cancer* **107**, 1053–1057 (2003).

[104] M.F. Naylor, W.R. Chen, T.K. Teague, L. Perry, and R.E. Nordquist, "*In situ* photo immunotherapy: a tumor-directed treatment modality for melanoma," *Br. J. Dermatol.* **155**, 1287–1292 (2006).

29

Cancer Laser Thermotherapy Mediated by Plasmonic Nanoparticles

Georgy S. Terentyuk

Saratov State University, The First Veterinary Clinic, Saratov, Russia

Garif G. Akchurin and Irina L. Maksimova

Saratov State University, Saratov, Russia

Galina N. Maslyakova

Saratov Medical State University, Saratov, Russia

Nikolai G. Khlebtsov

Institute of Biochemistry and Physiology of Plants and Microorganisms RAS, Saratov, Russia

Valery V. Tuchin

Saratov State University, Institute of Precise Mechanics and Control RAS, Saratov, Russia

We describe an application of plasmonic silica/gold nanoshells to produce a controllable local laser hyperthermia or thermolysis of tissues with the aim of thc enhancement of cancer photothermal therapy. Laser irradiation parameters are optimized on the basis of preliminary experimental studies using a test-tube phantom and laboratory rats. Temperature distributions on the animal skin surface at hypodermic and intramuscular injection of gold nanoparticle suspensions and affected by the laser radiation are measured *in vivo* with a thermal imaging system. It is shown that the temperature in the volume region of nanoparticle location can substantially exceed the surface temperature recorded by the thermal imaging system. The results of temperature measurements are compared with tissue histology results. Nanoshell-based photothermal therapy in several animal models of human tumors gave highly promising results; the dosage information, thermal response, and tumor outcomes for these experiments are presented.

Key words: laser photothermal cancer therapy, gold nanoparticles, pharmacokinetics, particle biodistribution

29.1 Introduction

The available hyperthermic techniques for cancer tumor therapy [1–4] possess low spatial selectivity in the heating of tumors and surrounding healthy tissues. One of the ways to improve the laser heating spatial selectivity is tumor tissue photothermal labeling by gold nanoparticles with different shapes and structures, such as nanoshells [5–7], nanorods [8–10], nanocages [11], and others [11–14]. The spectral tuning to desirable wavelength range of the single-particle plasmon resonance and control of the ratio of absorption and scattering efficiencies for a particle are achieved by producing particles with a different size, shape, and structure [15–18]. By exposing nanoparticles to laser radiation on the wavelength close to their plasmon-resonant absorption band [19], it is possible to produce a localized heating of cells labeled by nanoparticle without any harmful action on surrounding healthy tissues [20, 21]. Such an approach has been developed over the last five years and it has been called plasmonic photothermal therapy (PPTT) [22]. The spectral tuning of nanoparticle resonance to the "therapeutic optical window" (from 750 to 1100 nm) and achieving the reasonable ratio between the absorption and scattering efficiencies are reported [23–27]. In practice, almost all papers on PPTT used diode lasers with the wavelength around 805–810 nm. Wei and coworkers [9] demonstrated that gold nanorods conjugated to folate ligands can be used for hyperthermic therapy of oral cancer cells with a CW Ti:Sapphire laser. Severe destruction of cell membranes was observed at laser irradiation with energy density after 30 s irradiation as low as 30 J/cm^2.

Addressed accumulation of particles to a biological target can be achieved by passive or active delivery. In the first case, the nonfunctionalized particles are accumulated within a desired biological target according for the size-dependent permeability. For example, 130-nm silica/gold nanoshells can pass through the vessel walls leading to their accumulation in the surrounding tumor tissue [28]. In addition, the surface PEGylation of GNSs ensures their stability against agglomeration at physiological conditions during systemic circulation [35, 36]. The second ("active") approach is based on the usage of particles with the surface attached biospecific probing molecules that can be bonded to the target molecular sites on the tumor cell membrane. The surface molecular attachment procedure is called "functionalization" [29] and the term "conjugate" is often used to designate functionalized particles [30, 31]. Prior to laser treatment, the refractive index matching agents (e.g., PEG diacrylate) may be applied over the tumor surface [32] to reduce intensity of backreflected light from the skin superficial layers.

Gold nanoshells (GNSs) belong to a prospective class of optical adjustable nanoparticles with dielectric (silica) core included in a thin metallic (gold) shell [33]. By an appropriate choice of the ratio of core diameter to shell thickness and surface functionalization, the so-called immunotargeted GNSs can be engineered that effectively absorb and scatter light at a desired wavelength from visible to infrared, and provide selective labeling of cancer cells [34]. Specific surface chemistries are necessary to provide uptake of nanoshells by tumor cells and for targeting specific cell types by bioconjugate strategies. The absorption cross-section of a single GNS is high enough to provide a competition of nanoparticle technology with applications of indocyanine green dye—a typical photothermal sensitizer used in laser cancer therapy [37]. Enhanced efficiency of absorption and scattering of GNSs have been used to develop a combined approach, allowing one to provide selective destruction of GNS-labeled carcinoma cells via PPTT and monitor photothermal action by concurrent imaging of tumor tissues under action by optical coherence tomography [38].

Pharmacokinetic, toxicity, and organ/tissue distribution properties of functionalized NPs are of great interest from the clinical point of view because of the potential of this nanotechnology for human cancer treatment. After the initial injection of NPs into animal blood stream, the systemic circulation distributes the NPs towards all organs in the body. To evaluate the distribution of administrated NPs inside the body over the various organ systems and within the particular organs, one needs systematic studies on various animal models. Several publications have shown distributions of NPs between different animal organs including spleen, heart, liver, and brain. For instance, Hillyer and Albrecht [39] after oral administration of colloidal gold (CG) NPs to mice observed an increased concentration of gold in mouse organs as particle size decreased from 58 to 4 nm. For 13-nm CG NPs, the highest amount of gold was detected in liver and spleen after intraperitoneal administration [40]. Niidome et al. [41] showed that after intravenously injection of gold nanorods (NRs) these particles accumulated within 30 min predominantly in the liver. The coating of gold NRs by PEG-thiols resulted in their prolonged circulation. De Jong et al. [42] performed a kinetic study to evaluate the gold NPs size-dependent organ distributions for 10-, 50-, 100-, and 250-nm particles in the rat. Rats were intravenously injected in the tail vein with suspension of gold NPs and their distribution was measured quantitatively in 24 h after injection. The authors also found that the tissue type distribution of gold NPs is size dependent with the smallest 10 nm NPs possessing the most widespread organ distribution.

Katti and coworkers [43] published detailed *in vitro* analysis and *in vivo* pharmacokinetics studies for gold NPs within the nontoxic phytochemical gumarabic matrix in pigs to gain insight into the organ specific localization. Recently, Kogan and coworkers reported a preliminary data on biodistribution of PEG-coated 15–50-nm gold NRs and 130-nm silica/gold NSs in tumor-bearing mice [36]. The kinetics of gold distribution was evaluated for blood, liver, tumor, and muscles over 1 hr (NRs) and 24 hrs (NSs) time intervals. It was found that the gold NRs were nonspecifically accumulated in tumors with significant contrast in comparison with healthy tissues, while silica/gold NSs demonstrated a smoother distribution. Terentyuk and et al. [44] were evaluated the particle-size effects on the distribution of PEG-coated CG NPs of 15 and 50 nm in diameter and 160 nm-silica/gold NSs in rats and rabbits.

There are several papers on *in vitro* experiments related to application of NSs to PPTT of cancer cells [5, 6, 45–48], however the number of *ex vivo* and *in vivo* studies is quite limited. Moreover, although several preclinical studies on the utility of GNSs for PPTT applications are available [5, 28, 32, 49], there remain several areas still to be studied more thoroughly. First of all the question is about controlled and localized hyperthermia without significant overheating of both tumor and surrounding normal tissues. The distribution of elevated temperature under PPTT treatment is determined by absorption of light by nanoparticles acting as point-wise local heat sources, and by thermal diffusion over surrounding tissues. Practically at PPTT one needs to provide the temperature elevation ΔT in the range from 10 to 20°C. To achieve such a temperature increase, the GNS concentration of $(1 \div 5) \times 10^9$ particles/mL, the laser power density of $(1 \div 5)$ W/cm^2, and the treatment exposures between 1 and 5 min are used. It should be noted that biological effects have nonlinear dependence on changes in the particle concentration and the delivered laser power density, which is defined by the type of tissue and thermoregulation ability of a living organism. A possible method for intelligent PPTT may be adaptive heating in real-time control mode, as described in Ref. [49]. Such a technology implies a precise control of local temperature within tumor and surrounding tissues. Thermocouples provide a direct but invasive method of temperature monitoring [32], which can be used for laboratory experiments only. In addition, the thermocouple response may be affected by unwanted direct laser heating rather than by the thermal equilibrium with the bulk tissue. More sophisticated and nondirect techniques for measurements of bulk tissue temperature such as temperature-dependent phase transition of hydrogels [50] or ultrasound imaging [51] have been recently described. The *in vivo* temperature distributions within the bulk tissue could be also measured by magnetic resonance temperature imaging (MRTI) technique based on the proton-resonance frequency-shift method [32, 49]. The temperature resolution of MRTI measurements is less than 1°C, which is close to the resolution of IR thermography. The IR thermography is a simpler and less expensive technique but is restricted to retrieving surface-temperature data only. To reconstruct in-depth temperature profiles of tissue basing on surface temperature measurements, the designing of special algorithms is needed [4]. Precise control over the local temperature distribution is the key factor to be considered in the context of enhanced PPTT efficacy. Depending on the accuracy of heating protocol and on the rise of tumor temperature laser heating can result in both: 1) tumor cell apoptosis and/or necrosis, and 2) accelerated tumor growth. Specifically, heating up to 39 to 45°C may lead to the acceleration of biological reactions accompanied by the production of heat shock proteins and by intensive growth of the tumor [52]. This is a generally accepted concept that one needs to ensure a steep rise in tumor temperature followed up by a constant temperature conditions over a given time period [21]. The use of plasmon-resonant particles in general meets the basic goals of PPTT related to controllable tumor heating. It is significant that gold nanoparticle light-mediated therapy is on the way to clinical trials in humans; however, extensive animal studies should be done prior. To characterize tissue heating effects, the morphological investigations in the framework of therapeutic pathomorphology should be provided. Although such an approach has been used for the estimation of hyperthermic effects of gold nanorods in tumor cells, studies of morphological changes under *in vivo* conditions are quite rare [5, 32] and they still need additional efforts to be completely understood.

This chapter is focused on the investigation of heating kinetics, spatial temperature distribution, and morphological alterations of tissues on the dependence of laser irradiation parameters and nanoparticle concentration. We describe experimental results for phantoms and laboratory animals at several temperature regimes, corresponding morphological tissue patterns are presented and discussed. Physics of plasmonic nanoparticles, fabrication techniques, description of their optical properties, and some biomedical applications can be found in Chapter 2.

29.2 Characteristics of Gold Nanoparticles

The protocol of manufacturing nanoshells having a silica core of 140 nm in diameter with a 20-nm-thick gold shell is described in paper [53]. An analogous protocol was used to prepare PEG-coated 15-nm and 50-nm CG NPs that had been fabricated as described in Ref. [54]. The gold concentration of all NPs solutions corresponds to 0.01% $HAuCl_4$ (the gold concentration is about 57 μg/ml). The poly(ethylene glycol) molecules have been attached (using thiol end-groups) to the surface of nanoparticles to provide their protection from reticuloendothelial systems of the organisms, increase the biocompatibility and biostability of nanoparticles, and contribute prolongation of their circulation in blood. Parameters of fabricated gold nanoparticles were tested by optical spectral analysis of particle suspensions (measuring of optical density in the spectral range corresponding to plasmonic resonance) to determine the wavelength of plasmonic resonance and its spectral width and by transmission electron microscopy (TEM) to identify their form and diameter and size-distribution function. Quasi-elastic or angular elastic light scattering techniques also were used to estimate hydrodynamic or actual mean diameter of particles in suspensions.

Spectral characteristics of gold nanoparticles calculated using the Mie theory are presented in Figure 29.1a. Water suspensions of gold bulk nanospheres of 15 and 50 nm in diameter have extinction maxima at the wavelengths near 520 nm. They are characterized by strong absorption and weak scattering. Layered nanoparticles, such as gold nanoshells with the ratio of core diameter to shell thickness as 140/20 nm, are the best for providing tissue hyperthermia by 810-nm laser diode due to extinction maximum at this particular wavelength. The measured spectrum of optical density of silica/gold nanoshells 140/20 nm is presented in Figure 29.1b. There is a good correlation between

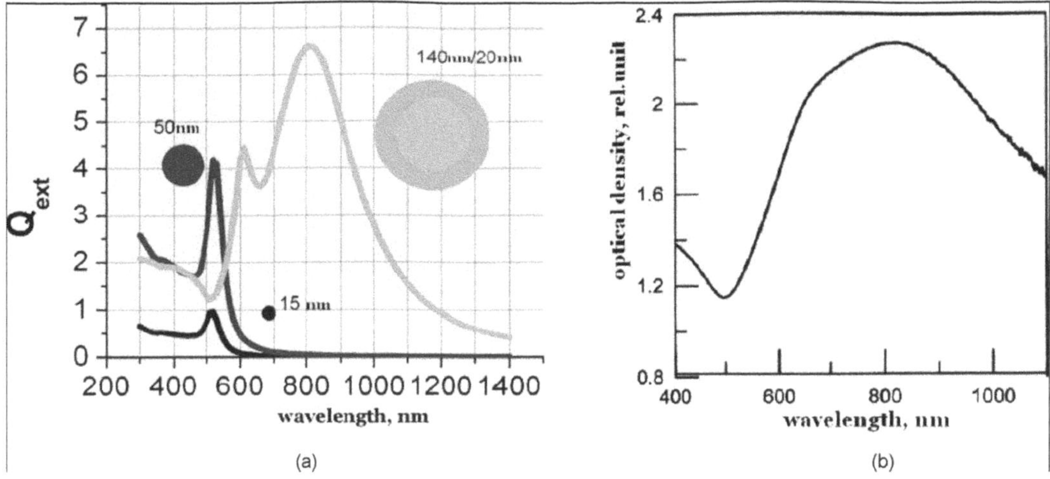

FIGURE 29.1: (a) Theoretical spectral curves of extinction cross-section for gold bulk nanoparticles of 15 and 50 nm in diameter, and silica/gold nanoshells (140/20 nm). (b) Experimental spectral curve of optical density for silica/gold nanoshells (140/20 nm), fabricated in the Laboratory of Nanobiotechnology at the Institute of Biochemistry and Physiology of Plants and Microorganisms RAS, Saratov, Russia (Specord BS-250 spectrophotometer (Analytik Jena, Germany), 1 cm cuvettes).

calculated and measured spectra; however, the experimental curve is smoother due to particle size heterogeneity.

29.3 Calculation of the Temperature Fields and Model Experiments

To simulate the spatial distribution of light absorbance related to the process of electromagnetic wave propagation through a system of discrete scattering particles with consideration of multiple scattering effects, the computer Monte Carlo codes were used. The diffuse scattering of laser radiation ($\lambda = 808$ nm) by the systems of gold nanoshells was calculated accounting for actual absorption and scattering spectra of plasmon-resonant particles that strongly depend on particle size and shape. By changing the diameter of spherical gold particles, it is possible to obtain spectral characteristics with the center of absorption peak in different parts of visible spectrum. Additional possibilities for obtaining the desired spectral characteristics come from using the suspension of nonspherical or layered particles. In these cases, the plasmon resonance shifts to the NIR region, where the biological tissue transparency window is located. Figure 29.2 shows the spatial distribution of absorption intensity for silica/gold nanoshells in water. The spectral dependences of the gold optical constants were taken from a spline corresponding to the experimental data [55, 56]. The extinction and scattering efficiencies are defined as the ratio of the integral optical cross-sections to the geometrical cross-section. For the silica core diameters and the shell thicknesses about 140 and 15 nm, respectively, the extinction peak is localized near 810 nm. The spatial distribution of the absorption is controlled by change of particle concentration and accordingly by the multiplicity of scattering.

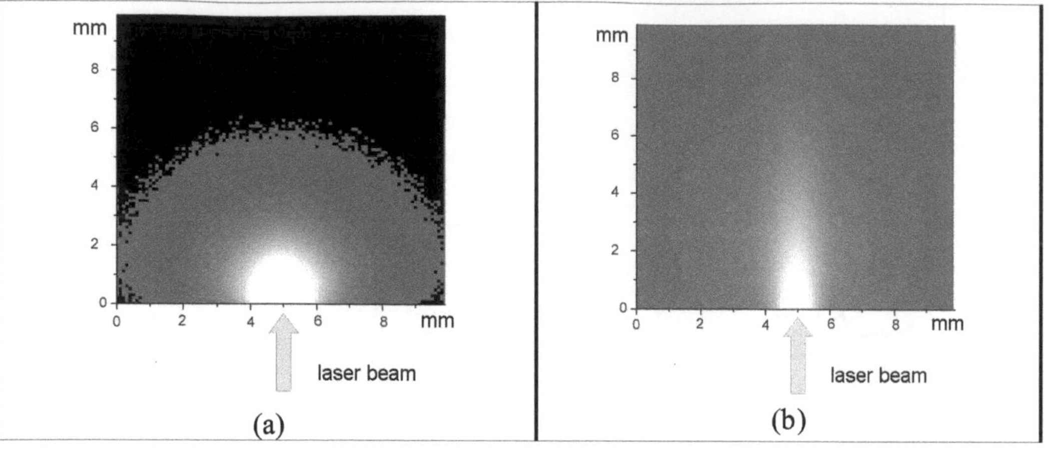

FIGURE 29.2: Computer simulation of the spatial distribution of the absorbed energy of laser beam (808 nm) redistribution with the water suspensions silica/gold nanoshells (140/20 nm): (a) particle concentration of 5×10^9 ml^{-1} and (b) particle concentration of 1×10^9 ml^{-1}. The incident light propagates from the bottom to the top (positive z-direction), the scattered layer is supposed to be infinite for x- and y-axes [20].

At high particle concentration the laser radiation is mainly absorbed near the sample surface and it is also strongly scattered (Figure 29.2a). With the small concentrations of particles, the radiation penetrates more deeply (Figure 29.2b) due to less scattering and absorption.

It is also interesting to consider situations when the laser beam diameter is comparable with the diameter of the region of high concentration of nanoparticles or larger [57]. In calculations, the region containing nanoparticles was assumed as spherical. Figure 29.3 shows that upon irradiation by a broad collimated beam, a greater part of photons are absorbed in the front half of the region (region 2), whereas upon focusing the beam to the center of the region, absorbed photons are localized near the beam waist and on the front surface of this spherical region. The grey scale (at the right) corresponds to the density of absorbed photons within the area of 10 mm^2 in the range 0–0.01% of the total number of incident photons.

To calculate the spatial distribution of the temperature, we employed the two-dimensional Poisson equation as a mathematical model. The axial symmetry of the problem allows one to restrict the solution by the two-dimensional Poisson equation written in cylindrical coordinates:

$$\mathrm{div}\{\lambda_T \mathrm{grad}[T(r,z)]\} = Q_h(r,z), \tag{29.1}$$

where λ_T is the coefficient of thermal conductivity, $T(r,z)$ is the temperature distribution, $Q_h(r,z)$ is the distribution of the power of the internal heat sources. The thermal boundary conditions have the following forms:

$$\lambda_T \frac{\partial T}{\partial r}\,|_{r=R} = \alpha_1\,(T\,|_{r=R} - T_0)\,, \quad \lambda_T \frac{\partial T}{\partial z}\,|_{z=0} = \alpha_2\,(T\,|_{z=0} - T_0) \tag{29.2}$$

Here, α_1 and α_2 are the coefficients of heat transfer at the boundaries of the computational domain and T_0 is the external temperature of surrounding medium ($T_0 = 25°$C).

The Poisson Eq. (29.1) was solved by the Galerkin finite-element method. The temperature was approximated by a linear combination of the basis (shape) functions of the linear triangular finite elements. To solve the system of algebraic equations, we used the Gauss method as applied to band

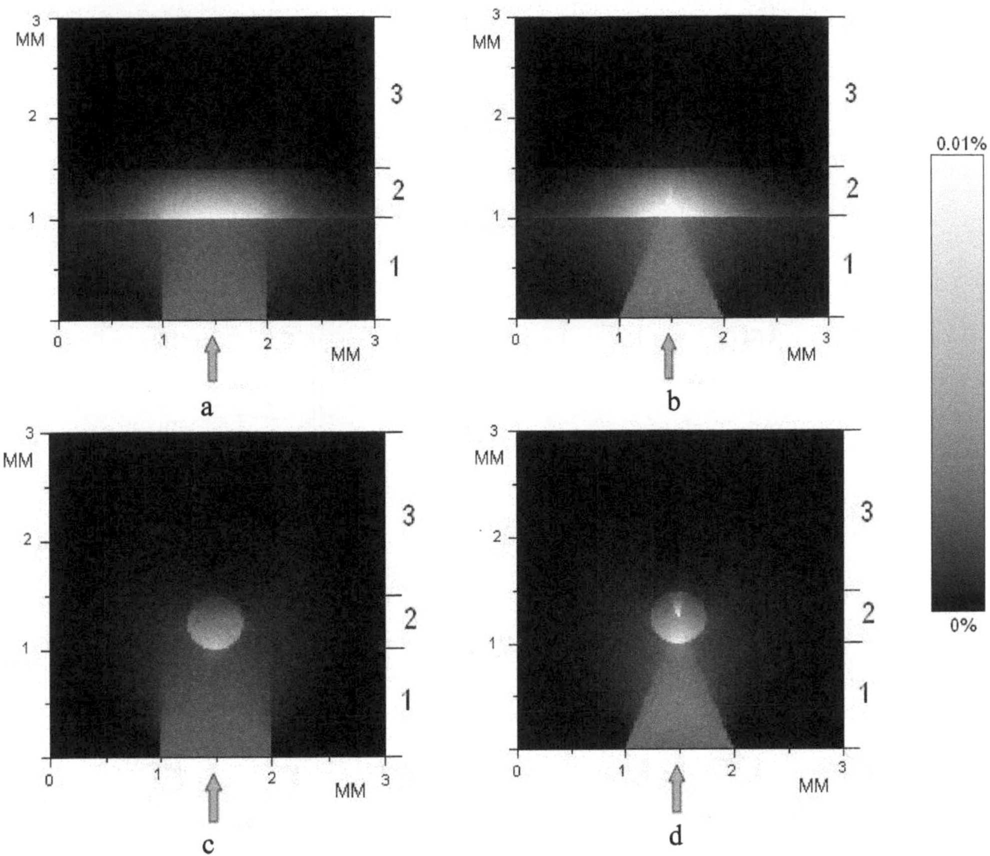

FIGURE 29.3: Computer simulations of the spatial distribution of laser energy in structured systems (tissue models) containing nanoparticles: (a, b) layered structure: 1-mm-thick layer I contains scattering particles of 0.5 μm in diameter with weak absorption and volume fraction of 0.1 (tissue layer model); 0.5-mm-thick layer 2 contains, along with the tissue modeling particles, 15-nm-thick gold nanoshells with a silicon core of diameter 140 nm; layer 3, as layer 1, contains only tissue particles; (c, d) structure as a sphere of diameter 0.5 mm, which contains, along with tissue particles (c), plasmon-resonance gold nanoshells (d) with the same parameters as for (a) and (b) [57].

matrices. The calculations were carried out with a nonuniform finite element grid possessing 1800 nodes. Figure 29.4 shows isotherms of the stationary temperature fields (a and b).

Experimental studies included the determination of spatial distribution of temperature with different in-depth location and concentrations of nanoparticles in samples of biological tissues. An IR thermograph IRISYS 4010 was used for noninvasive monitoring of the surface temperature. The thermal imaging system is based on the uncooled bolometric matrix providing temperature recording in the range from −10 to +250°C, with a temperature resolution of 0.15°C. The system is sufficiently fast (8 frames per second) to provide animal studies *in vivo*. The wavelength range of the thermal imager comprises 8 to 14 μm, thus the scattered laser radiation (808 nm) does not affect thermal-vision measurements. Figure 29.5 shows the measured thermograms for colloidal solution of silica/gold nanoshells in a cylindrical test tube.

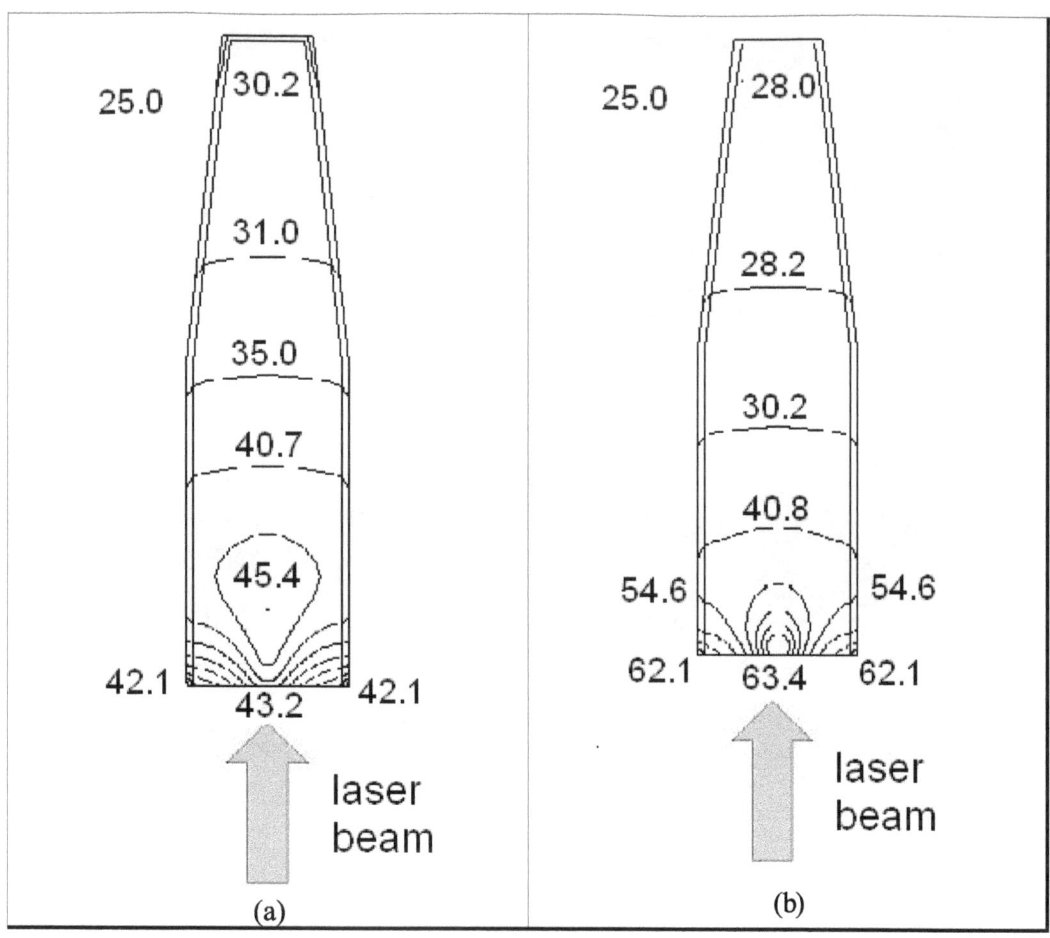

FIGURE 29.4: The stationary isotherms calculated using the Poisson equation for a two-dimensional axial symmetrical model of an Eppendorf vial. The spatial distribution of the thermal sources $Q_h(r, z)$ was obtained from the computer simulations shown in Figure 29.2. [20].

The laser radiation propagates to the end of a test tube parallel to its axes. The thermogram was recorded in a lateral direction of the test tube. Such a configuration allows one to obtain a depth profile of temperature. All samples were irradiated by a semiconductor laser PhotoThera Laser System operating at 808 nm and output power 2W for a continuous wave (CW) mode. The Gaussian beam diameter at a level $1/e^2$ was about 5 mm and the power density was about 10 W/cm^2.

Tissue phantom studies and corresponding computer simulations are necessary to solve the inverse problem for reconstruction of in-depth temperature profiles using surface thermal imaging data for the specified experimental conditions. For optimization of laser heating technology, in-depth temperature distributions were investigated using 2-ml test tubes (1 cm in diameter) containing solutions of silica/gold nanoshells (140/20 nm) with different concentrations (see Figure 29.6). Heating was provided by irradiating of a CW NIR diode laser (808 nm) during 2 min (total dose of 120 J, power density 4 W/cm^2). From the measurements, it follows that at high concentrations of particles the whole phantom volume is not heated "directly" by laser light, but only the part that is closer to the incident laser beam. The remaining volume of nanoparticle solution is heated up

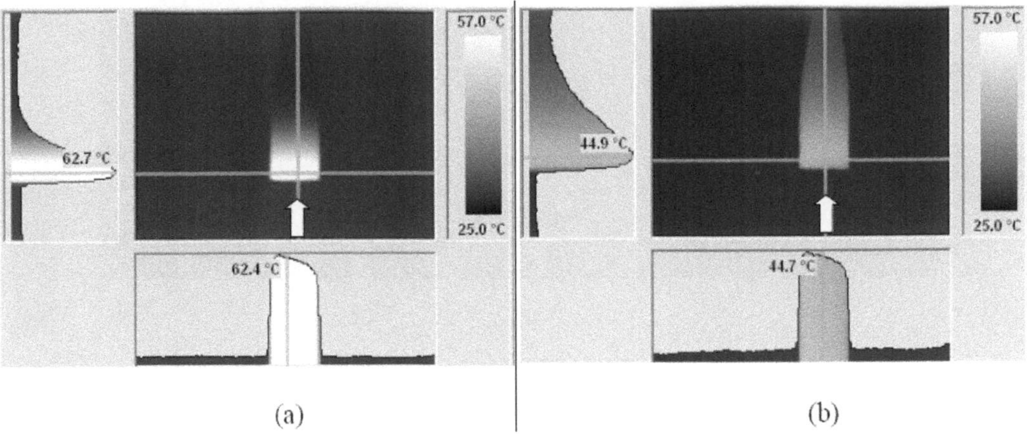

FIGURE 29.5: Thermogram of plasmon-resonant nanoshells in a standard Eppendorf vial (1.5 ml) after 2 min irradiation with a 2W-power laser beam: (a) particle concentration is 5×10^9 ml^{-1}and (b) 0.61×10^9 ml^{-1}. The incident light propagates from the bottom to the top.

indirectly due to a long-distant thermal diffusion process. Thus, some inhomogeneity of heating could be expected. Indeed, heating of all tissue components is due to the thermal diffusion phenomenon, however, in the regions where the laser beam penetrates and its intensity is sufficient, the short-distant thermal diffusion from localized heat sources (nanoparticles) is much faster. Thus, "directly" means only fast heat transfer to surrounding tissue from actual directly heated nanoparticles.

With lesser concentration of particles, the whole volume is heated "directly" by laser light, but this heating could be insufficient because of low concentrations of heated particles. Evident inhomogeneity of in-depth heating of nanoparticle solutions is followed from data presented in Figure 29.6. The effective heating depth for highly concentrated solutions ($N = 5 \times 10^9$ cm^{-3}) does not exceed 5–7 mm. This puts restrictions on the nanoparticle concentration for clinical applications for tumors with sizes bigger than 1 cm^3. If the concentration decreases four times, the effective heating depth is increased two times. Uniform temperature distribution in the test tube is recorded only at the thinning of the nanoparticle solution 16 times. In this case, maximal temperature alterations do not exceed 18°C, while at a concentration of 5×10^9 cm^{-3} the temperature rise is 40°C. It is necessary to note that 808-nm laser radiation is also absorbed by water, thus in control measurements with the test tube filled up by a physiological solution, background heating was about 6°C. The temperature measurements using a test tube allowed us to determine the upper limit of the temperature increase in tissues.

The concentration of gold nanoparticles in tissues is conceivably lower than in the initial injected solution. To estimate the temperature rise in tissues, experimental dependence of a temperature rise on concentration of gold nanoparticles in a phantom can be used (see Figure 29.7). Temperature kinetics for the solution of gold nanoparticles with a concentration of $N = 5 \times 10^9$ cm^{-3} is shown in Figure 29.8. The observed temporal dependence of a temperature rise is nonlinear. The most heating occurs over the first 100 sec of laser irradiation, then a temperature increase is saturated with time. After laser switch off, the temperature decreases more slowly than it increased at laser heating. Even after 5 min elapsed since the laser was switched off, the temperature of the nanoparticle solution was about 10°C above the reference one. Evidently, the cooling rate depends on the thermal diffusivity of material and a sample volume.

The characteristic thermal time response of an object is defined by its dimension R_0 (the radius for a cylinder form) and the thermal diffusivity of its material a_T [4]:

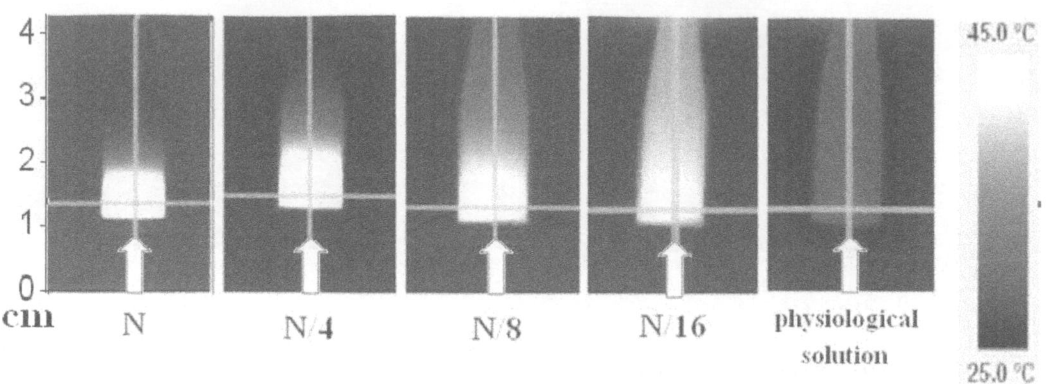

FIGURE 29.6: Thermograms of the test tube containing solutions of silica/gold nanoshells (140/20 nm) and physiological solution heated by diode laser (808 nm). Maximal particle concentration is $N = 5 \times 10^9$ cm^{-3} [21].

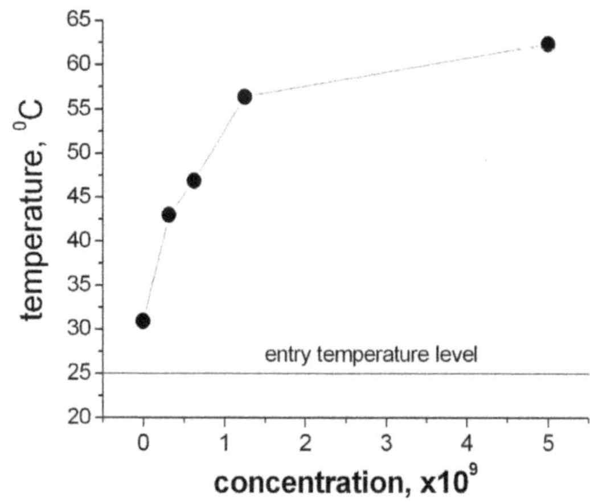

FIGURE 29.7: Maximal temperature of gold nanoparticle solution with different concentrations measured at the end of CW laser irradiation within 2 min (power 2 W, total energy 120 J, power density 4 W/cm^2) [21].

$$\tau_T \approx (R_0)^2 / a_T \tag{29.3}$$

Experimental values for the thermal diffusivity a_T of many soft tissues lie within the rather narrow range from 1.03×10^{-7} m^2/s (hydrated collagen containing 50% of water) to 1.46×10^{-7} m^2/s (pure water). Therefore, the characteristic thermal time response for our experiments with phantoms and tissues according to expression (29.3) is around 200 to 300 sec. This value is in good agreement with our experimental results presented in Figure 29.8. Therefore, the real time of an object overheating exceeds the laser exposure time. This finding is necessary to take into account assigning a duration of local laser hyperthermia mediated by gold nanoparticles.

Irradiation with short laser pulses has been shown to lead to the rapid heating of the particles and vaporization of a thin layer of fluid surrounding each particle, producing a microscopic version of underwater explosion, and cavitation bubble formation [22]. Zharov et al. [58] also found that nano-

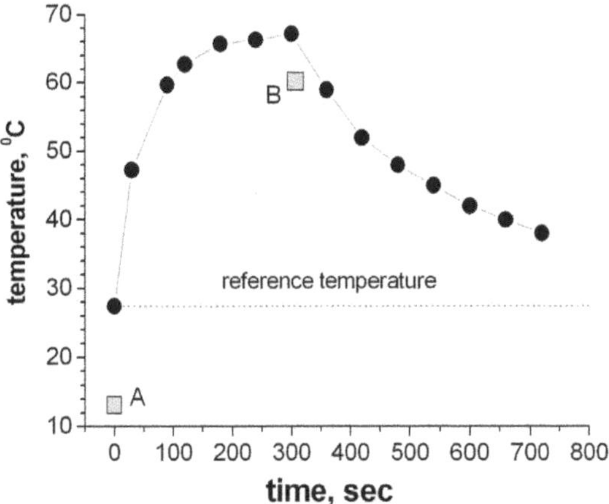

FIGURE 29.8: Temporal dependence of temperature in the center of the laser spot of the CW (4 W/cm^2) diode laser (808 nm) used for heating the solution of gold nanoparticles with a concentration of $N = 5 \times 10^9$ cm^{-3} in the test tube; A – denotes laser switch on and B – switch off [21].

clusters formed by the assembly of gold nanoparticles on the human breast cancer cell membrane significantly enhance the bubble formation causing more efficient cancer cell killing.

Laser-induced generation of vapor bubbles in water around plasmonic nanoparticles was experimentally studied in papers [59, 60]. Nanoparticle-generated bubbles spatially localize a laser-induced thermal field and also amplify the optical scattering relative to that by gold nanoparticles. To optimize the conditions of bubble generation around plasmonic nanoparticles, the bubble lifetimes and threshold fluencies were determined as functions of the parameters of the laser (pulse duration, fluence, interpulse interval), nanoparticles (size, shape, aggregation state), and the sample chamber. Nanoparticle-generated bubbles were suggested to be considered as nanosized optical sensors and sources of localized thermal and mechanical impact. Laser-induced photothermal (PT) response of the system includes the initial NP laser heating, which in turn causes fast heat transfer to the surrounding medium due to thermal diffusion; vaporization of surrounding medium if the temperature exceeds the vaporization threshold, bubble formation, and corresponding generation of shock and acoustic waves [59]. Local vaporization around a NP still remains the most underrecognized PT phenomenon because its nanoscale nature complicates the analysis of bubble generation. However, the bubble mode of the PT interaction may localize in space and time laser-induced thermal and mechanical impacts around plasmonic NPs. Such localization cannot be achieved either with long or ultrashort laser pulses. Too long pulses (or CW optical activation) cause the thermal field to spread over a large space (many orders of magnitude larger than the NP size) due to thermal diffusion (an estimation shows that during approximately 1 msec heat is redistributed smoothly within the tissue for typical concentrations of gold nanoparticles with typical sizes). Ultrashort laser pulses concentrate the thermal field within the NP, but generate pressure waves that also spread over a large volume. In contrast, a bubble may potentially concentrate the laser-induced thermal field and mechanical (pressure) impact around a NP with their characteristic size and duration determined by the bubble diameter (in the range of 50–1000 nm) and lifetime (in the range of 5–500 ns), respectively. These two parameters may be considered as the measures of the PT process in space and time and can be potentially controlled through the parameters of NPs and laser radiation. NP-generated vapor bubbles have already improved the selectivity and efficacy of PT therapy [61–63].

29.4 Circulation and Distribution of Gold Nanoparticles and Induced Alterations of Tissue Morphology at Intravenous Particle Delivery

The specific physicochemical properties at the nanoscale are expected to result in increased reactivity of the biological system. Thus, in addition to their beneficial effects, engineered nanoparticles of different types may also represent a potential hazard to human health. Several studies propose that nanoparticles have a different toxicity profile compared with the larger particles [64, 65]. Kinetic properties are considered to be an important descriptor for potential human toxicity and thus for human health risk. It is important to know the total amount of NPs delivered and absorbed within the body. The distribution of absorbed nanoparticles inside the body over the various organ systems and within the organs needs to be determined. After the initial delivery of nanoparticles the systemic circulation can distribute the particles towards all organs and tissues in the body. Several studies have shown distribution of particles to many organs including the liver, spleen, heart, and brain [39, 42].

The lungs, liver, and the spleen are the most critical organs in terms of particle trapping and sequestration [66]. The liver has the highest microvasculature number and density, with a size of 10–13 μm in diameter. The endothelial cells of the sinusoidal walls, where liver Kupffer cells are attached, have numerous small pores ranging in size from 100–300 nm [67, 68]. The Kupffer cells constitute approximately 30% of liver sinusoidal cells. Therefore, nanoparticles are likely to be sequestered in the liver sinusoid and phagocytosed by Kupffer cells. Finally, the spleen is most likely the site where intravenously injected particles are trapped because the microcirculation of the spleen is quite complex. The major role of the spleen is to remove damaged or old erythrocytes, pathogens, and particles from the circulation. Everyday, approximately 10^{11} erythrocytes are phagocytized by macrophages in the red pulp cord. The venous sinuses (sinusoids) are enveloped by a framework of reticular fibers that lie between the splenic cords. These venous sinuses are 100–150 μm wide and are lined with discontinuous endothelium that allows blood cells to re-enter to the circulation. There are small slits between the endothelial cells, referred to as interendothelial slits, that are approximately 4-μm wide, depending on the species [69]. Normal erythrocytes, which are 7–12 μm in diameter, are able to squeeze through the interendothelial slits to re-enter the circulation, while damaged rigid erythrocytes are unable to pass through these narrow slits because of their loss of flexibility [70]. Similarly, it is likely that rigid particles larger than the slits size would be trapped in the red pulp due to the limited size of the splenic interendothelial slits and, as a result, would be phagocytized by splenic macrophages. Aside from the geometric trap and phagocytosis mediated by tissue macrophages, intravenously administered particles may encounter additional circulating phagocytic cells, such as monocytes. For example, the half-life of systemically circulating aminomodified small particles with sizes between 100 nm and 1 μm is only 80–300 s because of monocyte particle uptake [70]. Thus, developing a drug delivery strategy to minimize the contact and recognition of the delivery carrier by phagocytes and to maximize the time remaining in the circulation is critical. Polyethylenglycol (PEG) cover provides a shielding effect, by delaying recognition and sequestration by circulatory monocytes and tissue macrophages of NPs covered by PEG.

As a result of abnormal organization and structure of the tumor vessels, blood flow in tumor vessels is, in general, slower and is associated with a characteristic transcapillary "leaking" phenomenon. Most of the blood vessels in the internal region of a tumor are venules, while tissue within the periphery of the tumor is more viable and contain arteries or arterioles. Therefore, the pressure differences between arterioles and venules in the necrotic core are extremely low, but are larger in viable rims of the tumor. This heterogeneity in blood flow within the tumor partially explains the uneven drug distribution pattern observed within some tumors. Intratumoral injection of therapeutics may be one way of bypassing endothelial barriers, since it is associated with an increase in the levels and retention of therapeutic molecules near the tumor mass while preventing systemic side effects [71].

Interstitial fluid pressure (IFP) is increased in most solid tumors, including breast [72, 73], melanoma, head and neck carcinoma, and colorectal carcinoma [74]. Increased IFP contributes to decreased transcapillary transport in tumors and drug retention time in the tumor.

The use of gold nanoparticles as a therapeutic modality for remotely controlled thermal ablation will hold promise in the development of novel therapy over conventional chemotherapy. Systemic injection of gold nanoshell followed by near infrared treatments effectively inhibited tumor growth and prolonged tumor-free survival in mice bearing xenograft tumors [5, 38].

In the case of *in vivo* applications, PEGylated nanoparticles are preferentially accumulated within tumor tissues due to the enhanced permeability and retention effect, known as the "golden standard" for drug design [22]. Compared to normal tissue, the blood vessels in tumor tissue are more leaky, and thus, macromolecular or polymeric molecules preferentially extravasate into tumor tissue. Due to the decreased lymphatics, the tumor tissue retains large molecules for a longer time, whereas normal tissue quickly clears out the external particles. In the course of tumor genesis, blood vessels around the tumor undergo dramatic morphological changes and the endothelial cells create a large number of fenestrations, with sizes about 200–300 nm and sometimes up to 1200 nm [75].

The cell membrane acts as a regulator and defensive unit to protect the cell from the outside environment by controlling the influx and outflow of chemicals, proteins, and other biologically significant compounds permitting the cell's functionality and survival. However, membranes could be additional barriers for drug delivery. Many types of cells including endothelial cells, fibroblasts, osteoclasts, and pericytes have some phagocytic or pinocytic activity [76]. Pinocytosis refers to the uptake of fluids and solutes and is closely related to receptor-mediated endocytosis. For example, one of the roles of endothelial cells is to transport nutrients from the blood to adjacent tissue, and therefore, possesses a high phagocytic nature. Pinocytosis and receptor-mediated endocytosis share a clathrin-based mechanism and usually occur independently of actin polymerization. By contrast, phagocytosis, which is uptake of large particles by the cells, occurs by an actin-dependent mechanism and is usually independent on clathrin. Both nonspecific binding and surface receptor binding events could trigger further receptor recruitment and surface migration events, to possibly strengthen the binding.

For *in vivo* experiments [44], the white laboratory rats (weight 180–200 g) and rabbits (weight 3 kg) were used. To evaluate the circulation kinetics of smallest particles in blood, 5 ml of freshly prepared PEG-coated 15-nm CG NPs was injected in the vein *saphena medialis* of rabbits ($n = 8$). Before injection nanoparticles were dissolved in 0.9%-NaCl. For blank stress control, 5 ml of 0.9%-NaCl solution was injected in the vein *saphena medialis* of rabbits ($n = 2$). To evaluate the natural level of gold, 2 ml of blood was collected before NPs injection. In 15, 30, 60, and 120 min after injection, blood was collected for analyses by atomic absorptive spectrometer (AAS-3, Carl Zeiss). In 72 hrs after injection, the following organs were collected for histological examination: brain, kidney, liver, lung, and spleen.

We also investigated the distribution of NPs over different organs of white laboratory rats ($n = 30$, blank control group $n = 4$). Solutions of 15-nm, 50-nm gold NPs, and 160-nm Au/SiO_2 nanoshells were used. One milliliter of each freshly prepared solution was injected in the tail vein. Gold concentration of all injected solutions was about 57 mg/ml, which corresponds to the standard synthetic conditions with 0.01% Trichloroisocyanuric acid. In 24 hrs after injection, blood, brain, kidney, liver, lung, and spleen were collected. The organs were weighed, the tissue samples were homogenized and frozen for determination of gold content by AAS-3. EDTA blood was stored in the refrigerator (4–7°C).

To prepare the samples of blood and visceral organs for AAS analysis, the method of mineralization (i.e., acid decomposition) at 600–630°C was used. For mineralization in a muffle oven, the sulfuric acid was added to samples during ashing. A mixture of two acids, HNO_3 and HCl, was used for complete mineralization of the sample organic components. To ensure the formation of a

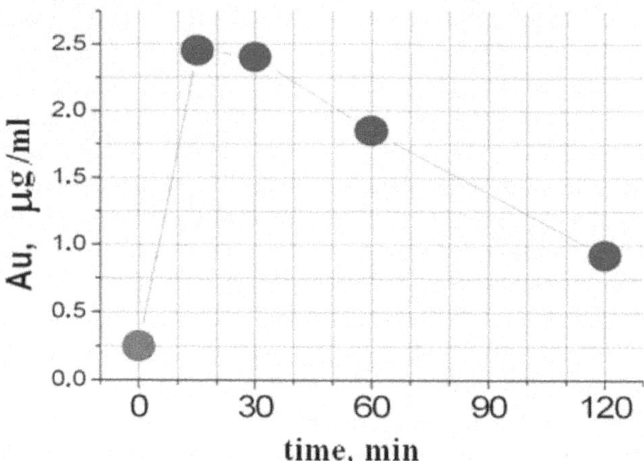

FIGURE 29.9: Dependence of the gold concentration in rabbit blood on the time after intravenous injection of 15 nm PEG-coated gold nanoparticles.

strong gold complex in the solutions under study, we used a 0.5 N HCl solution. The gold content in sample solutions was evaluated at a resonant emission line 242.8 nm with the spectral slit width of 35 nm. The oxyacetylene torch was used for measurements with the sampling mass of about of 1 g. The measurement protocol allows for the minimum detectable gold concentration of about 0.02 μg/ml. For calibration purposes, we used a stock solution (gold concentration of 1000 μg/ml) and its corresponding dilutions. The linearity range of concentration measurements was 0.2–20 μg/ml.

Results of circulation kinetics of PEG-coated 15-nm gold nanoparticles are presented in Figure 29.9. The highest concentration of gold (about of 2.5 μg/ml) as a characteristic of NPs concentration in the rabbit blood was found 10–30 min after administration. Then, the concentration was gradually decreased. In 2 hrs after injection, the gold concentration remained within the notable level (about 1 μg/ml), which is higher than the blank control level concentration of gold in blood (about 0.25 μg/ml). The decrease in the blood gold concentration can be explained by distribution of NPs over inner organs, because the corresponding increase in the organ's gold content is observed significantly later (several hours after injection).

In paper [44] the biodistribution of gold NPs over the rat brain, kidney, lung, liver, spleen, and blood in terms of the gold concentration measured 24 hrs after intravenous injection was investigated. For all particle sizes, the maximal concentration of gold was observed in the liver and spleen. It may be explained by the fenestrated structure of capillaries in these organs. On the other hand, the endothelium of the kidney *glomeruli* is also known to be fenestrated significantly, but our data shows rather low level of NPs accumulation in the kidney. Perhaps, the basal membrane of the *glomeruli* presents a barrier for gold NPs accumulation. Further, if we compare the data for 15, 50, and 160-nm NPs, the latter show highest accumulation in liver, spleen, and lung. It follows from results of [44] that the size-dependent NPs accumulation is observed for liver and spleen only, where the gold concentration is decreased with the decrease in the NPs size. In contrast to this finding, the gold content in the brain was negligible – very close to control value, however somewhat of an increase is seen for smaller 15-nm NPs. The gold concentration in the kidney and lung are almost constant and also equal to the control level for all NPs sizes. In general, we did not find significant differences in the gold concentration of NPs accumulated in the brain, kidney, and lung in comparison with the control level. For the spleen, the gold content was almost equal to that for 50- and 160-nm NPs. Furthermore, the gold content of these NPs in blood was close to the control level. By contrast, 15-nm NPs demonstrated a notable ability for recirculation and their concentration in blood was higher than for 50- and 160-nm NPs in 24 hrs after injection. In agreement with previous observations [36], the silica/gold NSs showed a quick appearance in the blood (within the first 10–20 min) followed by a significant decrease in 24 hrs after injection. Note that the gold NRs demonstrated a slower time-dependent decrease [36] in the blood flow. For rabbits, analogous biodistribution results have been found in 72 hrs after NPs injection. For instance, the insert in Figure 2.21 (Chapter 2) shows data for 15-nm gold NPs. Unfortunately, the whole set of biodistribution data for rabbits turned out to be not statistically reliable and they are not shown here. Our results are in satisfactory agreement with the data published quite recently by De Jong and coworkers [42] for 10-, 50-, 100-, and 250-nm CG NPs. In particular, they also observed maximal accumulation of gold in the rat liver followed by spleen. Furthermore, the maximal gold content was detected for 100 nm NPs, followed by 250-nm and 50-nm NPs. However, two differences between our data and those from Ref. [42] should be noted here. First, for 10 nm NPs, the gold concentration in liver, spleen, kidney, and brain were significantly higher than analogous gold concentration for 50-nm particles [42]. Perhaps, these differences are related to the differences in sizes of our (15 nm) and their (10 nm) smallest NPs. Second, De Jong and coworkers [42] reported quite untypical accumulation of 50 nm NPs in the lung. Specifically, the gold concentration of 50-nm NPs in lung was 20–30 times higher than the average (almost constant) level for 10-, 100-, and 250-nm NPs. Our data do not confirm such an anomaly.

In the paper [77] tissue/organ distribution of 15-, 50-, 100- and 200-nm size gold NP in mice was demonstrated (see Table 29.1). The size-dependent particle distribution of gold NP was found. In contrast to other results [42, 44] in this study smaller-sized 15 nm-gold NPs showed higher concentration in tissues compared to larger ones. Gold nanoparticles were mainly accumulated in the liver followed by the lung, spleen, and kidney. Interestingly, 15- and 50-nm gold NP can able to pass the blood–brain barrier as evident from the gold concentration in brain.

TABLE 29.1: Biodistribution of Au NPs over white mouse organs in terms of Au per g of tissue /and percentage of a given dose. Adapted from Ref. [77].

Organs	Top number: μg Au per g of tissue Bottom number: % of a given dose			
	Au 15 -	Au 50 -	Au 100 -	Au 200 -
Lung	32.3	18.7	15.2	19.4
	0.24	0.073	0.059	0.047
Liver	52.3	21.3	27.1	58.8
	0.40	0.08	0.11	0.14
Kidney	25.5	3.75	1.29	9.35
	0.193	0.014	0.005	0.022
Spleen	5.5	11.5	12.9	28.9
	0.041	0.045	0.050	0.070
Heart	1.05	0.97	3.24	2.86
	0.007	0.003	0.012	0.006
Brain	10.0	9.1	6.0	0.15
	0.075	0.036	0.023	0.0003
Blood	0.56	0.59	ND	0.11
	0.004	0.002		0.0002
Stomach	3.6	0.25	0.80	0.15
	0.027	0.0009	0.003	0.0003
Pancreas	ND	1.73	7.52	6.77
		0.006	0.029	0.016

The first histological examination of animal tissues after intravenous injection of NPs are presented in [44]. For histological examination, the fixation of samples was carried out with 10%-solution of formaldehyde. After fixation, the histological slides were prepared and stained by routine techniques. As a rule, the histological description is based on the examination of a series of slides rather than on a particular sample. Before consideration of results for experimental samples, we provide a set of typical histological images for normal control tissues (see Figure 29.10). It follows from Figure 2.21 (Chapter 2) that 160-nm silica/gold NSs are strongly accumulated in the rat liver and spleen [44]. This is why we performed histological studies for such NPs first, followed by examination of samples for other NPs sizes. In 24 hrs after the intravenous injection of 160-nm NSs, we observed the following changes in the rat spleen (see Figure 29.11a): first, although the general view of the tissue structure is close to that for control samples, we note some vacuolar degeneration of the endothelial cells in the artery walls; further, the white pulp is normal but no germinal centers are observed; in addition, one can observe notable congestion of the blood in the red pulp, significant amounts of hemolized erythrocytes, and accumulation of the granular pigment of a black-brown color (Figure 29.11a). In the rat liver, the general view of the tissue structure is kept unchangeable. However, the intralobular sinusoidal capillaries are dilated and contain erythrocytes. Some granular pigmented structures of a black-brown color are located in the lumen of blood vessels. In some parts of the samples, a colored black-brown pigment is located irregularly (inside hepatocytes and between them), while in the other parts it presents in the cytoplasm of the Kupffer cells (Figure 29.11b). In the lung samples, we observe that a moderate hyperemia, the foci of acute emphysema are characteristic. Interalveolar *septae* are thin or dilated due to edema. One can note a separation of the blood cells and plasma as well as a vacuolar degeneration of the endothelial cells of blood vessels. In some vessels, we found a desquamation of endothelial cells in the lumen accompanied by an absence of a typical black-brown staining (Figure 29.11c). In spite of the small accumulation of NPs in kidney, we can identify some kind of ischemia in these organs. Specifically, the *glomeruli* are moderately hyperemic and one can see a proliferation of *mesangial* cells in some *glomeruli.* Additionally, a granular degeneration of the epithelium of the proximal convoluted tubules occurs. The blood vessel walls are thickened and a vacuolar degeneration of the endothelial cells can be identified (Figure 29.11d). In the brain samples, we did not observe significant structural changes except for a moderate hyperemia, pericellular edema, and weak degeneration of the nerve cells (Figure 29.12). In general, one can say that these images are very close to those for control samples. This conclusion holds for particles with all sizes used in our experiments.

In some blood vessels, we observed unusual separation of the blood cells and plasma and moderate degeneration of the parenchymatous cells. Figure 29.13 shows an example of intravenous administration of 50-nm nanoparticles. Near the border between the separated plasma and the erythrocyte deposits, one can observe a black pigment (see Figure 29.13). The exact mechanism of blood separation is not clear.

Figure 29.14 shows histological examination of rat organs after administration of 50-nm gold NPs. The following changes in the organs can be identified. In the spleen, a general view of structure remains unchangeable but the white pulp is damaged, the peripheral zone of lymphoid nodules is indistinct, and the nodules have no germinal centers. The majority of lymphocytes have indistinct outlines and a light-blue cytoplasm. In the red pulp, we can see severe congestion of the blood. There is a large amount of black-brown staining in both the white and red pulp. Multiple bodies (they look like apoptotic bodies) are visible between the normal lymphocytes (Figure 29.14a).

The structure of the liver is normal. In the central veins we can see the separation of blood cells and plasma and a black-brown pigment. Severe degeneration of hepatocytes and moderate congestion of the blood are found (Figure 29.14b). In the lung we noted mild hyperemia, focal hemorrhages, and foci of acute emphysema. In the foci of hemorrhages, one can see notable accumulation of the black-brown pigment. The same pigment is located in the lumen of the large blood vessels (Figure 29.14c). The changes of kidney correspond to a moderate hyperemia. In the lumen of blood vessels, black staining is accumulated. The structural changes in the *glomeruli* are quite

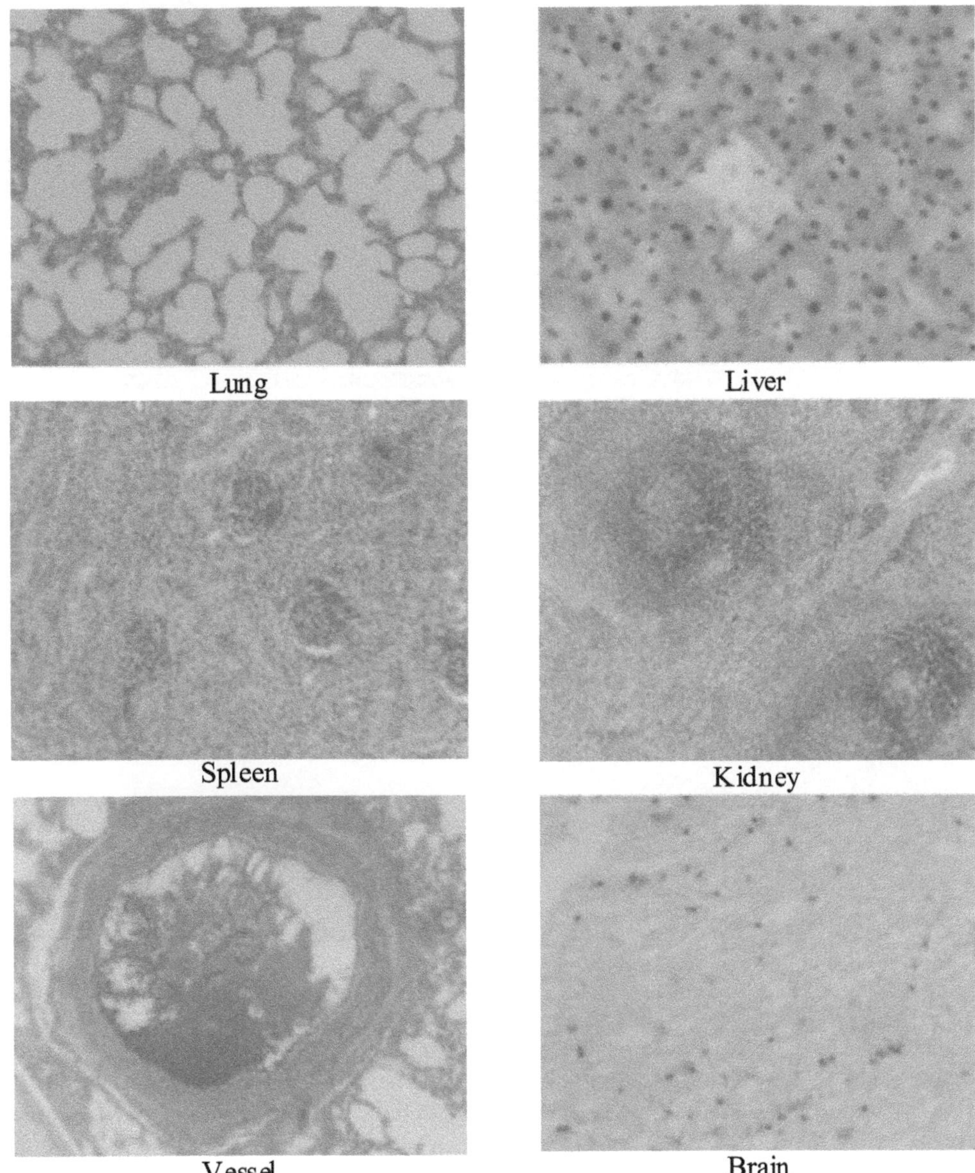

FIGURE 29.10: Typical histological images of normal rat tissues. Staining by hematoxylin – eosin (×150).

different. It is possible to observe the collapse of capillaries in some *glomeruli*, whereas in the other *glomeruli* we observe some kind of hyperemia with accumulation of single erythrocytes and fluid in the Bowman's capsule. There is a granular degeneration of the epithelium of proximal convoluted tubules (Figure 29.14d).

The final set of histological images was obtained for rat organs after intravenous administration of 15-nm gold NPs. First, a general view of the spleen structure is kept unchanged. White pulp looks like normal tissue although large macrophages with pale cytoplasm are presented. In the red

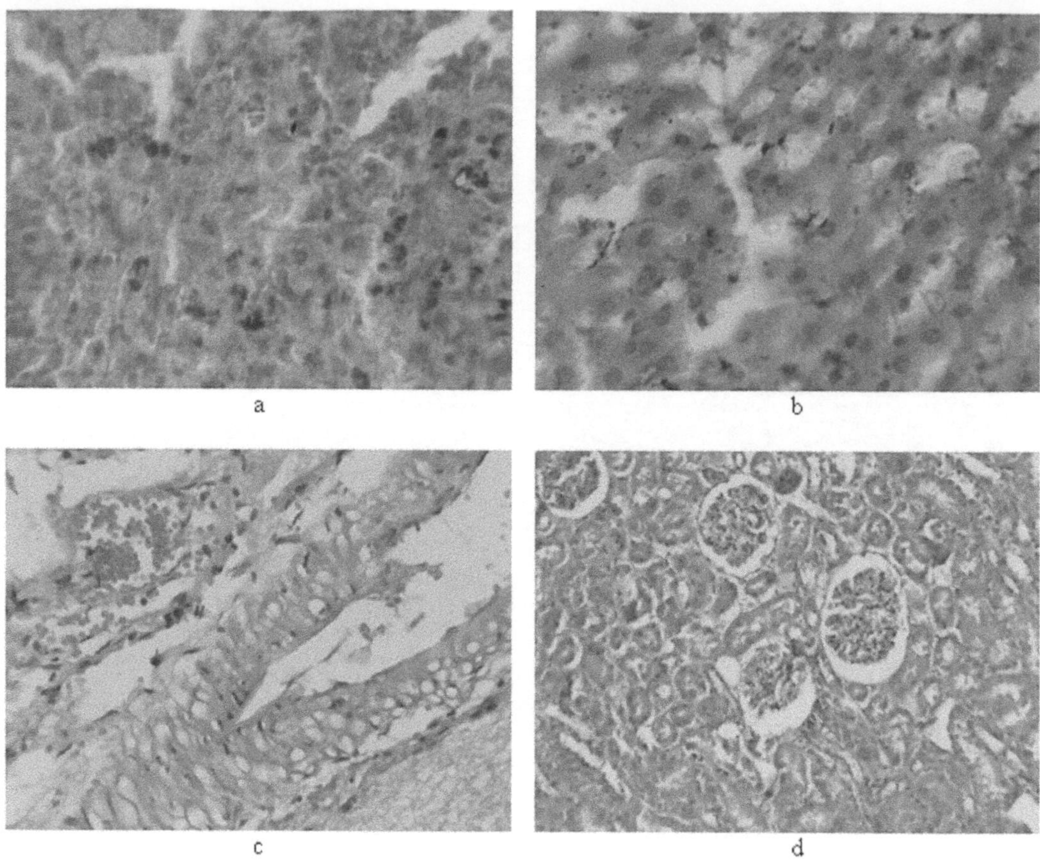

FIGURE 29.11: Histological images of rat organs after intravenous administration of 160-nm gold nanoshells. Staining by hematoxylin – eosin (×150). a – spleen, b – liver, c – lung, d – kidney.

pulp, we can see severe congestion of the blood. A large amount of pigment is located in the red pulp, while less of this pigment is observed near the lymphoid nodules. The general liver structure is unchanged. We observed severe degeneration of hepatocytes and a moderate congestion of the blood. In addition, rather small granules of brown-yellow staining are located in some parts of the liver between hepatocytes and inside them. In the lung samples, one can observe focal hemorrhages with depositions of brown-yellow staining. Mild hyperemia and the blood separation are less expressed. There is a small amount of brown-yellow staining. Finally, for the kidney samples, one can observe a moderate hyperemia of the cortex and an ischemia of the *glomeruli.* A granular and vacuolar degeneration of the epithelium of tubules are common. One can note some black-colored fields in the lumen of some blood vessels. Our AAS measurements and direct observation with the dark-field light microscopy (not shown here) revealed significant number of 160-nm NSs in the liver and spleen as compared with the smaller (15- and 50-nm) NPs accumulation. However, it should be emphasized that the morphological changes under 160-nm NPs treatment were less expressed than those for 15- and 50-nm particles.

In addition to light microscopy of histological preparations, we performed a TEM examination of rat liver tissues taken in 72 hrs after intravenous injection of 15-nm gold NPs (Figure 29.15). For TEM examinations, the 35-mm strips of a tissue were divided by a blade to the pieces of about 1×1 mm. Further fixation, dehydration, impregnation, and filling into the resin were carried out

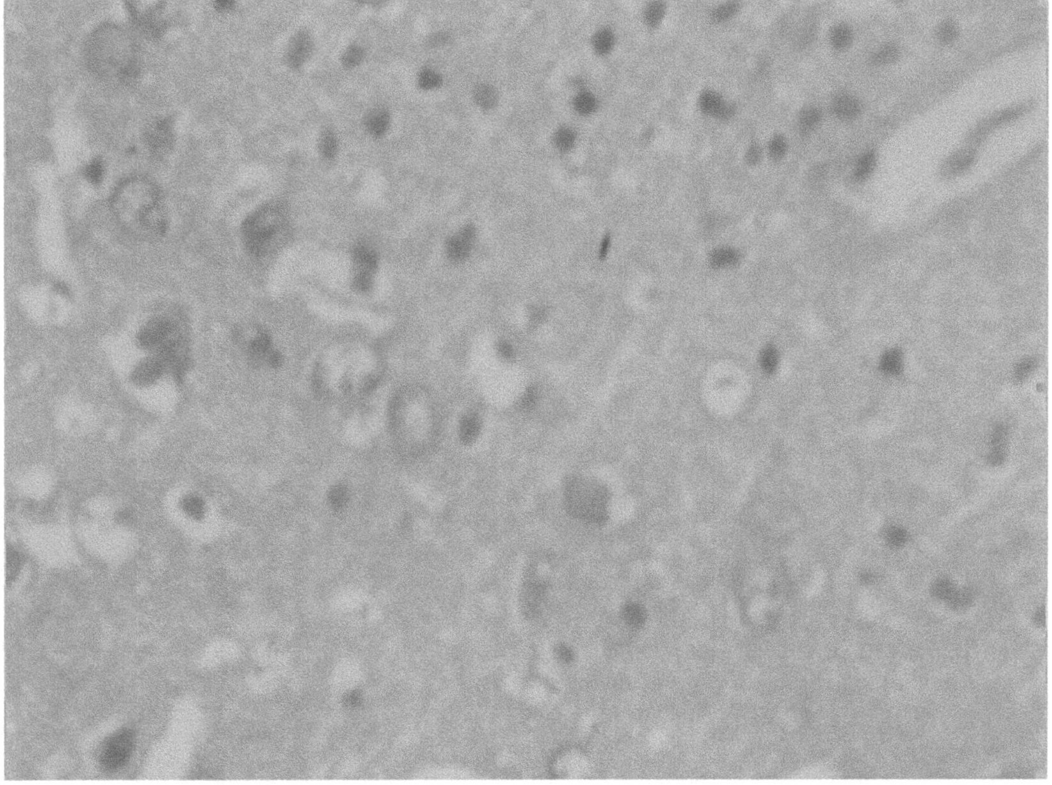

FIGURE 29.12: Histological image of rat brain after intravenous administration of gold NPs. Staining by hematoxylin – eosin (×150). This image is typical for particles of all sizes (15, 50, and 160 nm) and it is very similar to the control samples.

by standard procedures. Thin (40–80 μm) slices of tissue samples were obtained by the ultratome "Reichert." The slices were mounted to the grids without the base layers and contrasted by the saturated alcoholic solutions of uranyl acetate at a temperature of 56°C during 10 min. The specimens were investigated using electron microscope Hitachi Hu-1a in TEM mode. Recall that for the liver, we observed maximal accumulation of all NPs. During investigation no significant distortions were observed in the cell microstructure except for the vacuolar degeneration in hepatocytes [78].

29.5 Local Laser Hyperthermia and Thermolysis of Normal Tissues, Transplanted and Spontaneous Tumors

Investigation of thermal effects and alterations of tissue morphology induced by laser irradiation at subcutaneous and/or intramuscular injection of 0.1-ml silica/gold nanoshell suspensions was carried out using white laboratory rats. In the solution of designed silica/gold nanoshells (140/20 nm), the gold content was 150 μg/ml. A similar concentration expressed in the number of particles, $N = 5 \times 10^9$ cm^{-3}, was used in *in vitro* experiments with tissue phantoms. In *in vivo* studies, 48 animals were divided into two groups, 24 rats in each; rats from the first and the second groups were

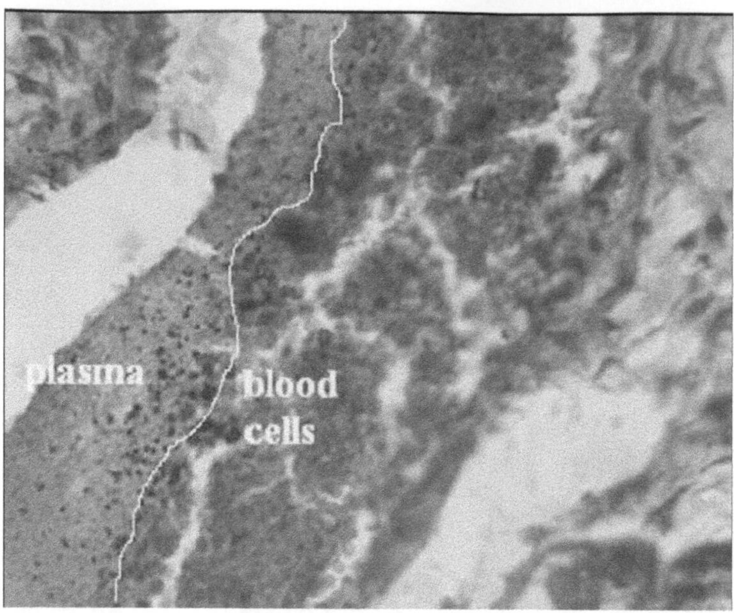

FIGURE 29.13: Rat blood vessel containing separated blood cells and deposits of black-brown pigment. Staining by hematoxylin – eosin (×150).

irradiated by continuous-wave and pulsed laser light, accordingly. Each of the 24 animals in each group was injected with a solution of gold nanoparticle suspension once subcutaneously and then in 20 min intramuscularly. The first injection was done subcutaneously in front of the abdominal wall, the second intramuscular one was done into the hip far enough from the first injection area. After two minutes elapsed after each injection, the skin area around the injection puncture was exposure to laser light to produce photothermal effects mediated by the inserted nanoparticles within the skin or within the muscle tissue. A 20-min delay and different locations of treated areas allowed us to consider tissue photothermal responses at subcutaneous and intramuscular gold nanoparticle delivery for the particular animal as well as independent ones. We used the diode laser, providing a mean wavelength of 808 nm, optical-fiber output, and CW and pulse modes of operation. The CW laser output power was 2 W. In the pulse mode, the peak power was 8 W, with an on–off time ratio of 0.25, and pulse duration of 1 msec. The preliminary estimation of heating kinetics and size of the heated area was carried out at different concentrations of nanoparticles in a test tube. For the noninvasive measurement of spatial distribution of the surface temperature of objects under investigation, the IR Imager IRISYS 4010 was used. For providing experiments *in vivo*, general anesthesia was applied. Tissue samples for histological examination were collected after the injection of nanoparticles before, after, and 24-hrs past laser heating. Heating duration was from 10 to 360 sec. To prove the evidence of nanoparticle delivery into the tissue and to evaluate their spatial distribution within the tissue qualitatively, TEM and dark-field optical microscopy were used (Figure 29.16).

The experimental animals were fixed at the horizontal position supine with hair shaved off. Every animal was labeled by a symmetrical centerline of stomach as shown in Figure 29.17. Three areas to the left from the center line were used to control the heating under laser radiation without nanoparticles. The control areas are marked by dark dots (•, ••, and •••). About 0.1 ml of the nanoshell suspension was injected in each of three areas to the right from the center line. In the areas marked as X, XX, and XXX, the injections were intradermal, subcutaneous, and intramuscular, respectively. The intramuscular injection depth was about 5 mm, the laser power was of 2.5 W, and the distance between the fiber and skin surface was about 15 mm. The coagulation of tissue within

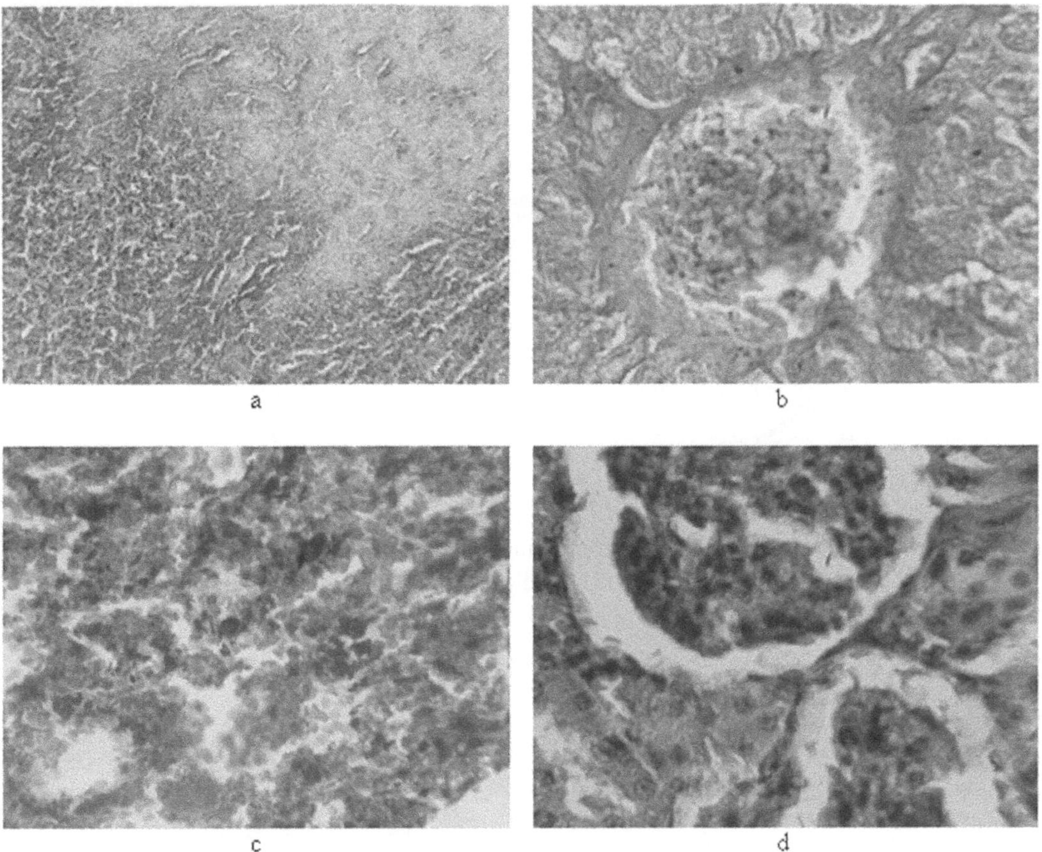

FIGURE 29.14: Histological changes in rat organs after intravenous administration of 50-nm gold NPs (hematoxylin – eosin, ×150). a – spleen, b – liver, c – hemorrhage in the lung with pigment, d – kidney.

the area marked by the X symbol was observed visually already after 10 s of irradiation.

The control area (without nanoparticles) of rat tissues was heated up to 46°C. This temperature cannot lead to irreversible injuries of tissues, although such heating can lead to the damage of cells for sufficiently prolonged treatment. In the case of the intramuscular GNPs administration (area XXX) we did not observe significant changes in the color and structure of the tissue surface during irradiation time. However, the surface temperature was approximately 7°' higher than in the control measurements without nanoparticles. Computer simulation showed that the temperature in the region of nanoparticle localization could substantially exceed the surface temperature recorded by the thermal imaging system. The local temperature can reach or slightly exceed 60°' with the hypodermic injection of nanoparticles. The denaturation of proteins should occur rapidly at this temperature. It can be complete or partial, reversible or irreversible. Clearly, the degree and speed of such protein denaturation depends strongly on the value and duration of temperature excess, as well as on the protein nature. The critical denaturation temperature for the majority of tissue components is about 57°'. After 15–20 s we observed the surface albication of skin followed by the thermal damage of the skin.

With the intracutaneous administration of nanoparticles, the visual changes in the biological tissue were observed for irradiation times less than 10 s. The noticeable dehydration of biological

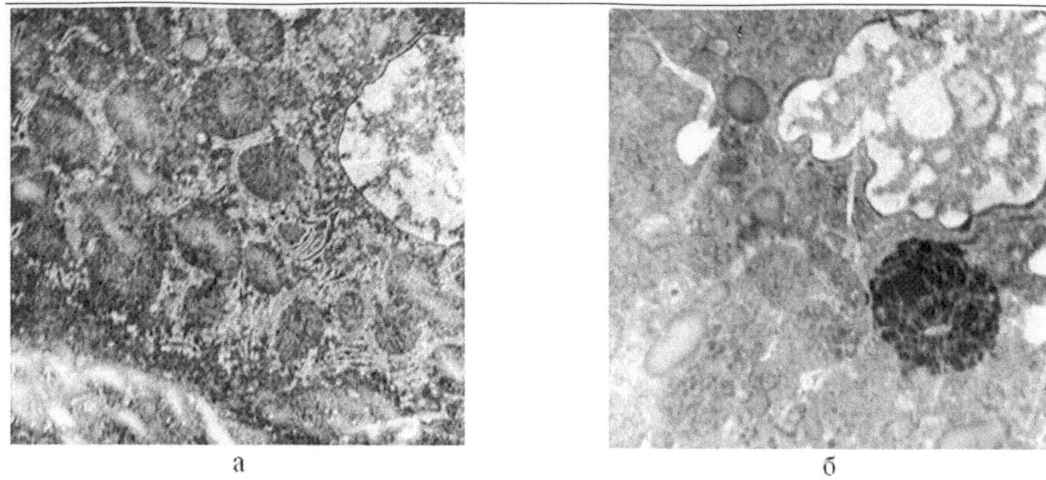

FIGURE 29.15: TEM images of rat liver after intravenous administration of 15-nm gold nanoparticles (×8000): a – fragment of hepatocyte and bile capillary, b – Kupffer cell.

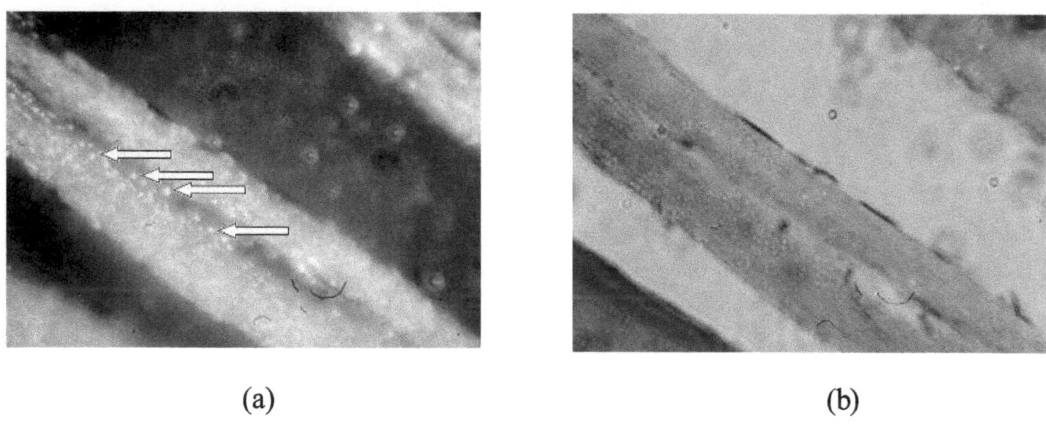

FIGURE 29.16: Histological study of muscle tissue (×200). Staining by hematoxylin/eosin: (a) dark-field image and (b) optical transmission image. The white arrows show the gold particle clusters.

tissues begins at a temperature near 70°'. After the water withdrawal, the dried tissue is heated rapidly to a temperature of 150°', at which the carbonization process begins. In short, hydrogen leaves the organic molecules and a fine dispersed carbon (soot) is formed, i.e., the carbonization occurs. The maximum value of the local temperature in this case can exceed 180°'. This value exceeds the temperature substantially, which was observed in the experiments with the large volumes of the aqueous suspension of gold nanoparticles. We can explain this difference by rapid expulsion of water from the local section of biological tissue. The process of evaporation was also observed in the test tube. Note that the evaporation takes rather long time periods because of the large volume of liquid, and only a small part of the liquid volume was actually evaporated after irradiation

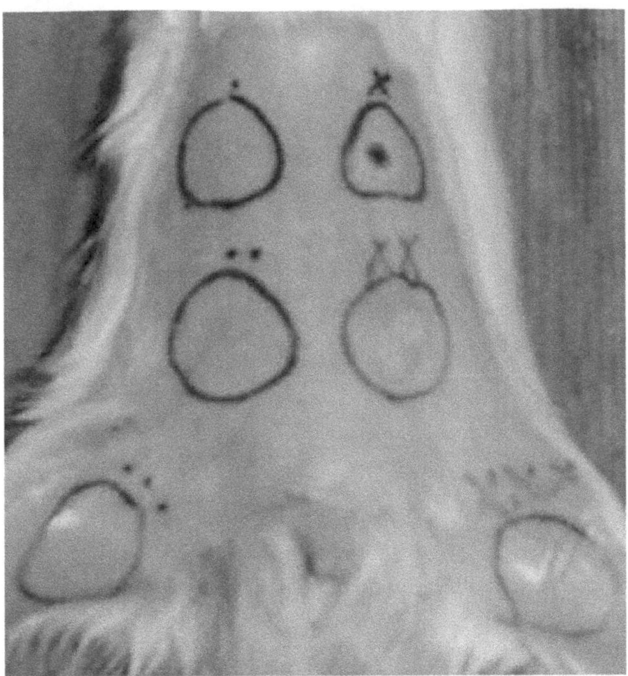

FIGURE 29.17: Laser photothermolysis (808 nm) of superficial rat tissue with intradermal injection of gold nanoparticles. Symbols •, ••, and •••stand for control (without nanoparticles); symbols X and XX designate the intradermal and subcutaneous injection of 0.1 ml nanoparticle solution; symbol XXX designate the intramuscular injection of 0.1-ml nanoparticles to depth of about 5 mm.

during 2–3 min. In these experiments and in a number of rare cases, we observed the formation of nanoparticle aggregates.

With the aim of *in vivo* determination of the spatial distribution of temperature corresponding to different location depths and concentration of nanoparticles in tissues, the solution of silica/gold nanoshells (140/20 nm, 150 μg/ml of gold) was injected (0.1 ml) into white laboratory rats subcutaneously or intramuscularly.

In vivo temperature spatial distributions were measured for CW and pulsed laser irradiation and different injection depths of nanoparticles into tissues (Figure 29.18). It is observed that heating happens more rapidly and temperature rise is substantially higher at laser irradiation of the skin with inserted nanoparticles than for skin free of particles. The rate of heating is very important for the photothermal therapy of cancer. Temperature rising to 46–50°C is accepted as optimal for this purpose [79]. Such temperatures admittedly stimulate apoptosis in tumor tissue, in contrast to heating in the range from 37 to 43°C, which is very undesirable [80]. In our experiments, by using nanoshells the time interval for tissue heating up to 46°C was 15 to 20 sec, and in their absence it was more than tenfold prolonged, up to 3 to 4 min.

It is interesting to note that both temperature kinetic curves for CW and pulsed irradiation of tissue with nanoparticles have a local maximum at 60 sec (Figure 29.18). The origin of this maximum may be explained by the manifestation of the compensatory reaction of the organism to temperature rise, which has a characteristic time of about 100 sec. With the fast heating caused by nanoparticles, the inertial mechanism of heat regulation could not provide any immediate temperature compensation. In contrast, essentially good compensation is realized for laser heating without inserted particles, and some compensation is seen in a larger time scale for heating with inserted nanoparticles. Also it

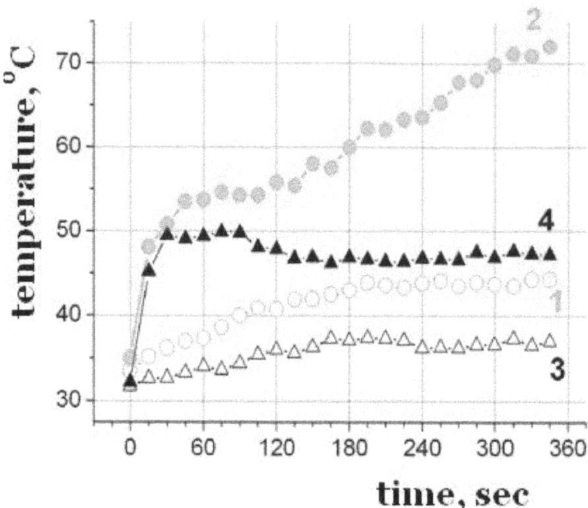

FIGURE 29.18: *In vivo* measured temporal dependences of rat skin temperature at CW (circles) and pulse (triangles) laser heating, 1, 3—without nanoparticles, 2, 4—with silica/gold nanoshells (140/20 nm), subcutaneous injection.

is important to note that laser pulse heating is more controllable in a large time scale and has similar short time kinetics as CW. In that sense, pulsed irradiation could be preferable because it gives fast and self-limited temperature rise.

The comparison of curves 2 and 4 (Figure 29.18) shows that the increase of effectiveness of hyperthermia is possible by a combination of different heating modes. It is reasonable to use the prior CW mode with a higher power density of laser radiation, and switch over to pulse mode or CW mode with a lower power.

At 24 hrs after laser action, biopsy samples of irradiated tissues were taken for histological investigation. Dark-field microscopy was used for imaging of nanoparticles in the muscle tissue (Figure 29.16). *In vivo* thermal measurements and histological studies of the tissue before and after laser irradiation were compared.

Figure 29.19 shows the thermograms and structure of the skin and muscle tissue before and after laser action on tissue with the administrated nanoshells. In the control samples, we did not observe any pathological changes (Figures 29.19a and 29.19d). For subcutaneously administrated nanoshells after laser irradiation, the epidermis of the skin is partly absent with the formation of vesicles filled up by serous fluid. In the derma, we found severe edema with disorientation of collagen fibers (Figure 29.19c). In the muscle tissue we observed edema, hyperemia, and inflammatory infiltration by leukocytes. The control of the spatial distribution of the temperature of the skin surface using the IR imager showed that after 30 sec of laser action, the maximal temperature in the center of the laser spot is about 65°C (Figure 29.19b). At intramuscular administration of the nanoshells, we revealed moderate edema with defibrillation of collagen fibers in the muscle tissue (Figure 29.19f). In the connective tissue, we observed inflammatory leukocyte infiltration and edema. In the case of intramuscular particle administration, we did not observe significant changes in the color or structure of the tissue surface during the light exposure time. However, the surface temperature was approximately 7°C higher than in the control measurements without nanoparticles. Thus, the laser thermolysis can be achieved in the region of nanoparticle localization without damage of the surface tissues. This conclusion is confirmed by histological examinations.

Nanoshell-based photothermal therapy in several animal models of human tumors have produced

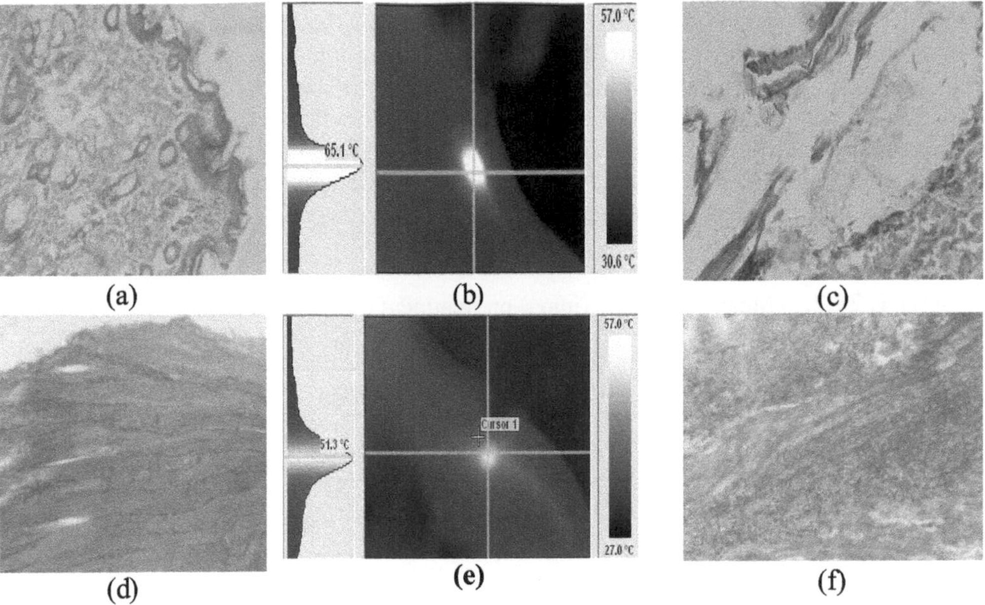

FIGURE 29.19: Comparison of local temperature rise and corresponding alterations in tissue morphology: (a) histology of rat skin before laser action, (b) thermogram at laser heating with subcutaneous administration of silica/gold nanoshells (140/20 nm), (c) histology of rat skin after 30 sec of laser action, (d) histology of rat muscle tissue before laser action, (e) thermogram at laser heating with intramuscular administration of silica/gold nanoshells (140/20 nm), and (f) histology of rat muscle tissue after 30 sec of laser action [21].

highly promising results; here we are presenting brief information about nanoparticle dosage, thermal response, and tumor outcomes for these experiments [81–83]. Nanoshell-based therapy was first demonstrated in tumors grown in mice [5]. Subcutaneous tumors were grown in mice to a size of $\approx$1.0 cm in diameter. PEGylated nanoshells were directly injected into the tumors under magnetic resonance imaging (MRI) guidance. Control tumors received saline injections. The tumors were subsequently exposed to NIR light, and the tumor temperature, as well as the temperature of the adjacent tissue, was monitored during and after laser irradiation. The mice were euthanized, and tumors were excised for histological evaluation. Analysis of the nanoshell-based photothermal treatment reveals tissue damage in an area of a similar extent as that exposed to laser irradiation. Magnetic resonance thermal imaging (MRTI) was used to monitor the temperature profile of the tumor during and after irradiation. Analysis of these temperature maps reveals an average temperature increase of 37.4 $\pm$ 6.6°C after 4–6 min of irradiation. These temperatures were sufficient to induce irreversible tissue damage. Nanoshell-free control samples show an average temperature rise of 9.1 $\pm$ 4.7°C, considered to be safe for cell viability.

Subsequent experiments were then conducted to determine the therapeutic efficacy and animal survival times by monitoring the tumor growth and regression over a period of 90 days [28]. In these studies, tumors were grown subcutaneously in mice, and PEGylated nanoshells were injected systemically via the tail vein, accumulating in the tumor over 6 hrs. The tumors were then irradiated with a diode NIR laser at a wavelength of 808 nm at a power density of 4 W/cm^2 for 3 min. A sham treatment group received the same laser treatment following saline injection, and control group received no treatment. The follow up of this treatment is the change in tumor size over the first 10 days, which indicates a dramatic difference in tumor size for the three groups under study (Table 29.2). In the nanoshell-treated group there was 100% resorption of the tumor to the 10th day. In the rest of the study, this result persisted, whereas in the sham and control groups, the tumor size became large enough (tumor was doubled in size and had more than 5% of body weight) and these mice were euthanized. Table 29.3 presents the plot of the survival statistics for the three investigated groups of mice. Average survival time for the nanoshell-treated group was > 60 days, the control group was 10.1 days, and the sham treatment group was 12.5 days [28].

TABLE 29.2: Mean tumor size (mm^2) on treatment day [28].

	Treatment group	Control group	Sham treatment
Day 0	20±2	19±3	28
Day 10	0	76±39	85

TABLE 29.3: Survival for the first 25 days [28].

% Surviving	Day 0	Day 5	Day 10	Day 15	Day 20	Day 25
Nanoshell-treated group ($n = 7$)	100	100	100	100	100	100
Control group ($n = 9$)	100	100	89	0	0	0
Sham treatment group ($n = 8$)	100	100	88	25	0	0

The passive accumulation of nanoshells in tumors could not provide effective conditions for tumor ablation, because of low selectivity of nanoparticle accumulation. To optimize the procedure Stern et al. have evaluated the effect of nanoshell concentration on tumor ablation in a human prostate cancer model in mice [84]. Tumors were grown subcutaneously. PEGylated nanoshells of two different dosages (7.0 μL/g (low dose) and 8.5 μL/g of body weight (high dose)) were delivered into the mice via tail vein injection. The nanoshells were allowed to accumulate for 18 hrs at which time the tumors were irradiated using a Diomed NIR laser at 810 nm for 3 min. Tumor size was measured during 21 days. In the low-dose group, only partial tumor ablation was achieved. Nine of the ten tumors showed arrested growth (mean volume 49.2 mm^3 from a baseline of 41.6 mm^3), as opposed to the control sample where the tumor burden tripled in 21 days (126.4 mm^3 from a baseline of 43.5 mm^3). Histologically, the tumors showed partial ablation with patchy areas of survived tumor cells. In the high-dose treatment group, to the 21st day a complete tumor deletion was observed (Figure 29.20). Histological evaluation also confirmed tumor necrosis for this dosage. A well-circumscribed eschar was formed over the laser treated region by day 1. This eschar fell off by day 21 revealing normal healthy skin. The control samples that did not receive any nanoshells did not form an eschar over the laser-treated areas. In the high-dose group, the average achieved temperature was 65.4°C, which is known to be effective in thermal ablation therapy.

Paper [85] describes some preliminary clinical results of silica/gold nanoshells (140/20 nm)-mediated laser thermal ablation of spontaneous tumors of skin and oral mucous of cats and dogs. For preoperation estimation of the tumor size and control of local heating and temperature rise under the laser radiation, the infrared thermography was used. It was also effective for monitoring the healing process in the post-operative period and for early detection of the suppurative inflammation.

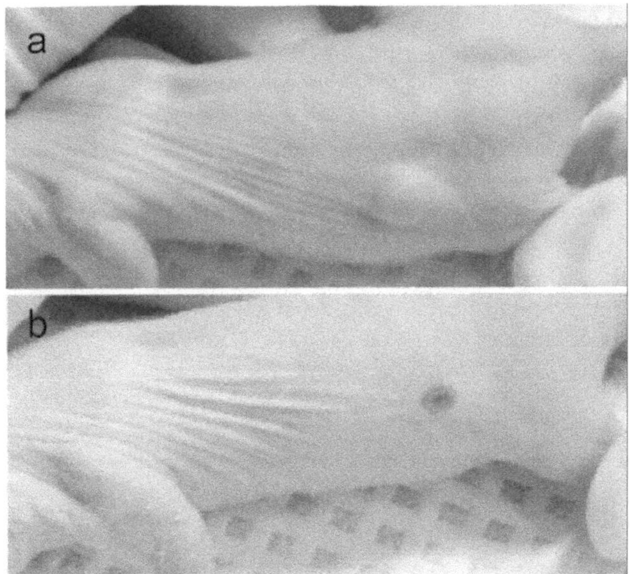

FIGURE 29.20: Photothermal tumor ablation using 3 min exposure of NIR laser (810 nm) in 18 hrs after intravein injection of PEGylated nanoshells of concentration 8.5 μL/g of body weight (mean temperature rise of 65.4°C): (a) tumor before treatment; (b) complete ablation of tumor. The eschar formed over the laser-treated region fell off by day 21 exposing healthy skin [84].

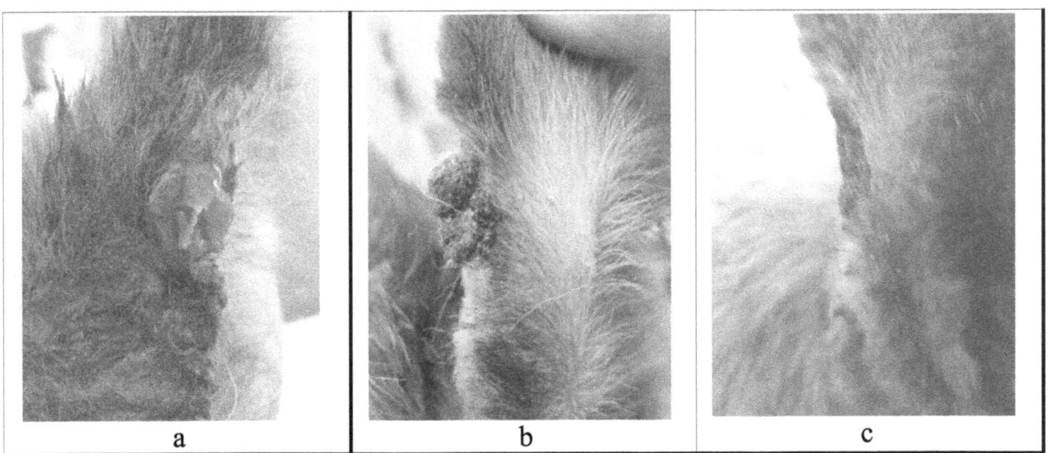

FIGURE 29.21: (See in color after page 572.) The cutaneous squamous cell carcinoma of the dog: a –before GNSs – PPTT, b – immediately after GNSs – PPTT, c – 5 days after GNSs – PPTT.

The informative ability of the IR thermography was proved for diagnostics of the skin melanoma, the cancer of tongue, larynx, the squamous cell carcinoma, and basal cell carcinoma of the pinna [86]. Authors of paper [85] have presented the following experimental examples of successful phototreatment: the melanoma of mucous of dog mouth (two cases), oral squamous cell carcinoma (two cases), the cutaneous and the external ear canal squamous cell carcinoma.

Nanoparticles were injected directly into the region of the tumor. In the first two cases, the treatment was provided in the late stage of pathologic process and the recovery could not be achieved. Nevertheless, we observed during the period of several months the stabilization of the pathological process and even a stable improvement of the clinical state during the complex application of the laser PPTT and immunotherapy. In the remaining cases, after laser hyperthermia, a mild inflammatory reaction was observed, and the epithelization with the formation of scar occurred in 7–12 days (Figure 29.21). Additionally, some remission process was observed within 3–6 months; it was accomplished by the absence of metastasis during 3 months. Experiments showed effective destruction of cancer cells of the ear, mouth, and skin by the local injection of plasmonic gold nanoshells followed by diode laser (808 nm) irradiation. For the destruction of such tumor cells, the pulse duration was about 1 msec (on–off time ratio was about 1/4), the average power densities were about 1–3 W/cm^2, and the laser fluences were about 100–200 J/cm^2.

29.6 Conclusions

To ensure an effective nonspecific (i.e., nontargeted) accumulation of NPs in the tumor, one has to increase the time of NPs circulation. One possible approach is the covering of gold NPs by tiolated PEG molecules. Smaller 15-nm gold NPS are able to circulate in the rabbit blood much longer than the larger NPs. In contrast to published data for circulation kinetics of silica/gold NSs in mice [36], the gold concentration of 15-nm NPs (2 hrs after injection in rabbit) was only two times lower than the maximal concentration at 20–30 min. In agreement with published data [36, 40, 42], paper [44] shows that the gold NPs are mainly accumulated in the spleen and liver in 24 hrs after injection. Furthermore, the biodistribution of NPs over different rat organs was shown to be size-dependent, especially for the spleen, liver, and blood. The maximal NP accumulation was found for 160-nm silica/gold NSs. Analogous biodistribution results have been found for the rabbit organs. Note that significant statistical variations in the detected gold concentrations within one group of rats have been pointed out in Ref. [42]. In 24 hrs after injection of NPs, authors [44] demonstrated histological changes not only in the liver and spleen, but also in other organs with a small level of NP accumulation. Histological examination of different samples allows one to assume the size-dependent character of morphological tissue changes caused, possibly, by NP treatment. Specifically, 160-nm NPs affected the vessel wall and caused the vacuolar degeneration of the endothelial cells. In fact, an unusual separation of the blood cells and plasma and a moderate degeneration of the parenchymatous cells were observed. In general, both 50-and 15-nm NPs caused more expressed changes in the inner organs as revealed from histological images, except for the brain, where the structural changes were not significant and not size-dependent. This finding could be related to the defense barrier function of cerebral membranes. On the other hand, slightly enhanced content of 15-nm gold NPs in the brain correlate with an analogous observation by De Jong and coworkers [42] for 10-nm particles, whereas no gold was detected for larger (50–250 nm) NPs.

By experimental investigations using a test tube it is found that at CW laser heating, the time needed to reach the steady-state temperature conditions is 100 to 150 sec. For instance, both CW and pulse laser heating have similar short time kinetics (temperature rise up to 46–50°C over time no more than 40 sec); however, due to the compensatory response of the organism, pulsed laser heating is more controllable than CW on a large time scale. In that sense, pulsed irradiation (on–off time ratio of 0.25, and pulse duration of 1 msec) is preferable because it gives fast and prolonged self-limited temperature rise. Gold nanoshells mediated laser destruction of muscle tissue is achieved *in vivo* in the region of nanoparticle localization (at a depth of 4 mm) without any damage to superficial tissues. In spite of similar power energy densities used in the pulse and CW mode, the nanoparticle mediated laser heating of tissue is different in these two modes. It is found that the elevated temperature of the object under study is kept for some time after laser action. This finding is necessary to account for assigning the duration of local laser/gold nanoparticle hyperthermia in clinical studies. In particular, optimal nanoparticle concentrations that provides smooth heating of tissue volume allows for the treatment of tumors with sizes no bigger than 1 cm^3. Thus, to provide smooth heating of big tumors at a high concentration of nanoparticles, multibeam irradiation could be effective.

Most of the current *in vivo* studies have been provided for subcutaneous tumors, which are easily accessible to NIR light applied over the skin surface. NIR light penetration in the tissue is a few centimeters. For treatment of deeper tissues, fiber optic deliver systems could be effective. Nanotherapy strategy may function as a stand-alone therapy but is also likely to be adjunctive to current systemic therapies. Since nanoshell-based photothermal therapy is functioning in ways completely different from conventional therapies, its resistance mechanisms are unlikely to be similar. Nanoshell-based photothermal therapy in several animal models of human tumors have produced highly promising results [81].

PPTT was effectively applied to treatment of spontaneous tumors (cutaneous squamous cell and basal cell carcinoma of the external ear canal, squamous cell carcinoma, and malignant melanoma of oral mucous) of a dog and cats at local injection of the nanoparticle into the tumor tissue. Some positive therapeutic effect was also observed in the course of complex therapy.

Acknowledgments

This research was supported by grants of RFBR (N07-02-01434a and N08-02-00399a), RF Program on the Development of High School Potential (N2.1.1/4989 and N2.2.1.1/2950), the President of the RF for the support of Leading Scientific Schools (N208.2008.2), FP7-ICT-2007-2 PHOTONICS4LIFE (N 224014), and the Special Program of RF "Scientific and Pedagogical Personnel of Innovative Russia," Governmental contract 02.740.11.0484.

Authors are thankful to I.A. Ermolaev and A.A. Skaptsov for their help in computer modeling of thermal effects.

References

[1] G. Müller and A. Roggan (Eds.), *Laser-Induced Interstitial Thermotherapy*, **PM25**, SPIE Press, Bellingham, WA (1995).

[2] P. Wust, B. Hildebrandt, G. Sreenivasa, B. Rau, J. Gellermann, H. Riess, R. Felix, and P.M. Schlag, "Hyperthermia in combined treatment of cancer," *Lancet Oncol.* **3**, 487–497 (2002).

[3] H.P. Berlien and G.J. Mueller, *Applied Laser Medicine*, Springer-Verlag, Berlin (2003).

[4] V.V. Tuchin, *Tissue Optics: Light Scattering Methods and Instruments for Medical Diagnosis*, 2nd ed., **PM 166**, SPIE Press, Bellingham, WA (2007).

[5] L.R. Hirsch, R.J. Stafford, J.A. Bankson, S.R. Sershen, B. Rivera, R.E. Price, J.D. Hazle, N.J. Halas, and J.L. West, "Nanoshell-mediated near-infrared thermal therapy of tumors under magnetic resonance guidance," *Proc. Natl. Acad. Sci. U.S.A.* **100**, 13549–13554 (2003).

[6] C. Loo, A. Lin, L. Hirsch, M. Lee, J. Barton, N. Halas, J. West, and R. Drezek, "Nanoshell-enabled photonics-based imaging and therapy of cancer," *Technol. Cancer Res. Treat.* **3**, 33–40 (2004).

[7] I.L. Maksimova, G.G. Akchurin, B.N. Khlebtsov, G.S. Terentyuk, G.G. Akchurin Jr., I.A. Ermolaev, A.A. Skaptsov, E.P. Soboleva, N.G. Khlebtsov, and V.V. Tuchin, "Near-infrared laser photothermal therapy of cancer by using gold nanoparticles: computer simulations and experiment," *Med. Laser Appl.* **22**, 199–206 (2007).

[8] X. Huang, I.H. El-Sayed, W. Qian, and M.A. El-Sayed, "Cancer cell imaging and photothermal therapy in the near-infrared region by using gold nanorods," *J. Am. Chem. Soc.* **128**, 2115–2120 (2006).

[9] T.B. Huff, L. Tong, Y. Zhao, M.N. Hansen, J.X. Cheng, and A. Wei, "Hyperthermic effects of gold nanorods on tumor cells," *Nanomedicine* **2**, 125–132 (2007).

[10] E.B. Dickerson, E.C. Dreaden, X. Huang, I.H. El-Sayed, H. Chu, S. Pushpanketh, J.F. McDonald, and M.A. El-Sayed, "Gold nanorod assisted near-infrared plasmonic photothermal therapy (PPTT) of squamous cell carcinoma in mice," *Cancer Lett.* **269**, 57–66 (2008).

[11] J. Chen, D. Wang, J.Xi, L. Au, A. Siekkinen, A. Warsen, Z.Y. Li, H. Zhang, Y. Xia, and X. Li, "Immuno gold nanocages with tailored optical properties for targeted photothermal destruction of cancer cells," *Nano Lett.* **7**, 1318–1322 (2007).

[12] P.K. Jain, I.H. El-Sayed, and M.A. El-Sayed, "Au nanoparticles target cancer," *Nanotoday* **2**, 18–29 (2007).

[13] D. Pissuwan, S.M. Valenzuela, and M.B. Cortie, "Therapeutic possibilities of plasmonically heated gold nanoparticles," *Trends Biotechnol.* **24**, 62–67 (2006).

[14] R. Visaria, J.C. Bischof, M. Loren, B. Williams, E. Ebbini, G. Paciotti, and R. Griffin, "Nanotherapeutics for enhancing thermal therapy of cancer," *Int. J. Hyperthermia* **23**, 501–511 (2007).

[15] K.-S. Lee and M.A. El-Sayed, "Dependence of the enhanced optical scattering efficiency relative to that of absorption for gold metal nanorods on aspect ratio, size, end-cap shape, and medium refractive index," *J. Phys. Chem.* **B 109**, 20331–20338 (2005).

[16] B.N. Khlebtsov, V.P. Zharov, A.G. Melnikov, V.V. Tuchin, and N.G. Khlebtsov, "Optical amplification of photothermal therapy with gold nanoparticles and nanoclusters," *Nanotechnology* **17**, 5167–5179 (2006).

[17] P.K. Jain, K.S. Lee, I.H. El-Sayed, and M.A. El- Sayed, "Calculated absorption and scattering properties of gold nanoparticles of different size, shape, and composition: applications in biological imaging and biomedicine," *J. Phys. Chem. B* **110**, 7238–7248 (2006).

[18] C. J. Noguez, "Surface plasmons on metal nanoparticles: the influence of shape and physical environment," *J. Phys. Chem. C* **111**, 3806–3819 (2007).

[19] U. Kreibig and M. Vollmer, *Optical Properties of Metal Clusters*, Springer-Verlag, Berlin (1995).

[20] G.S. Terentyuk, I.L. Maksimova, V.V. Tuchin, V.P. Zharov, B.N. Khlebtsov, V.A. Bogatyrev, L.A. Dykman, and N.G. Khlebtsov, "Application of gold nanoparticles to X-ray diagnostics and photothermal therapy of cancer," *Proc. SPIE* **6536**, 65360B-1–12 (2007).

[21] G.S. Terentyuk, G.N. Maslyakova, L.V. Suleymanova, N.G. Khlebtsov, B.N. Khlebtsov, G.G. Akchurin, I.L. Maksimova, and V.V. Tuchin, "Laser-induced tissue hyperthermia mediated by gold nanoparticles: toward cancer phototherapy," *J. Biomed. Opt.* 1**4**(2), 021016-1–8 (2009).

[22] X. Huang, P.K. Jain, I.H. El-Sayed, and M.A. El-Sayed, "Plasmonic photothermal therapy (PPTT) using gold nanoparticles," *Lasers Med. Sci.* **23**, 217–228 (2008).

[23] S. Oldenburg, R.D. Averitt, S. Westcott, and N.J. Halas, "Nanoengineering of optical resonances," *Chem. Phys. Lett.* **288**, 243–247 (1998).

[24] L. Tong and J.-X. Cheng, "Gold nanorod-mediated photothermolysis induces apoptosis of macrophages via damage of mitochondria," *Nanomedicine*, **4**, 265–276 (2009).

[25] N. Harris, M.J. Ford, and M.B. Cortie, "Optimization of plasmonic heating by gold nanospheres and nanoshells," *J. Phys. Chem. B* **110**, 10701–10707 (2006).

[26] N.G. Khlebtsov, "Optics and biophotonics of nanoparticles with a plasmon resonance," *Quantum Electron.* **38**, 504–529 (2008).

[27] V. Myroshnychenko, J. Rodriguez-Fernandez, I. Pastoriza-Santos, A.M. Funston, C. Novo, P. Mulvaney, L.M. Liz-Marzan, and F.J. Garcia de Abajo, "Modelling the optical response of gold nanoparticles," *Chem. Soc. Rev.* **37**, 1792–1805 (2008).

[28] D.P. O'Neal, L.R. Hirsch, N.J. Halas, J.D. Payne, and J.L. West, "Photo-thermal tumor ablation in mice using near infrared-absorbing nanoparticles," *Cancer Lett.* **209**, 171–176 (2004).

[29] W.R. Glomm "Functionalized gold nanoparticles for application in biotechnology," *J. Dispers. Sci. Technol.* **26**, 389–414 (2005).

[30] V.A. Bogatyrev and L.A. Dykman "Gold nanoparticles: preparation, functionalization, and applications in biochemistry and immunochemistry," *Russ. Chem. Rev.* **76**, 181–194 (2007).

[31] N.G. Khlebtsov, A.G. Melnikov, V.A. Bogatyrev, and L.A. Dykman, "Optical properties and biomedical applications of nanostructures based on gold and silver bioconjugates," in *Photopolarimetry in Remote Sensing, NATO Science Series, ii: Mathematics, Physics, and Chemistry* **161**, G. Videen, Y.S. Yatskiv, M.I. Mishchenko (Eds.), Dordrecht, Kluwer, 265–308 (2004).

[32] P. Diagaradjane, A. Shetty, J.C. Wang, A.M. Elliott, J. Schwartz, S. Shentu, H.C. Park, A. Deorukhkar, R.J. Stafford, S.H. Cho, J.W. Tunnell, J.D. Hazle, and S. Krishnan, "Modulation of

in vivo tumor radiation response via gold nanoshell-mediated vascular-focused hyperthermia: characterizing an integrated antihypoxic and localized vascular disrupting targeting strategy," *Nano Lett.* **8**, 1492–1500 (2008).

[33] L.R. Hirsch, A.M. Gobin, A.R. Lowery, F. Tam, R.A. Drezek, N.J. Halas, and J.L. West, "Metal nanoshells," *Ann. Biomed. Eng.* **34**, 15–22 (2006).

[34] P.K. Jain, K.S. Lee, I.H. El-Sayed, and M.A. El-Sayed, "Calculated absorption and scattering properties of gold nanoparticles of different size, shape, and composition: applications in biological imaging and biomedicine," *J. Phys. Chem. B* **110**, 7238–7248 (2006).

[35] R.T. Zaman, P. Diagaradjane, J.C. Wang, J. Schwartz, N. Rajaram, K.L. Gill-Sharp, S.H. Cho, H.G. Rylander III, J.D. Payne, S. Krishnan, and J.W. Tunnell, "*In vivo* detection of gold nanoshells in tumors using diffuse optical spectroscopy," *IEEE J. Sel. Top. Quant. Electron.* **13**, 1715–1720 (2007).

[36] B. Kogan, N. Andronova, N. Khlebtsov, B. Khlebtsov, V. Rudoy, O. Dementieva, E. Sedykh, and L. Bannykh, "Pharmacokinetic study of PEGylated plasmon resonant gold nanoparticles in tumor-bearing mice," *Tech. Proc. 2008 NSTI Nanotechnol. Conf. Trade Show, NSTINanotech, Nanotechnol.* **2**, 65–68 (2008).

[37] V.G. Liu, T.M. Cowan, S.W. Jeong, S.L. Jacques, E.C. Lemley, and W.R. Chen, "Selective photothermal interaction using an 805-nm diode laser and indocyanine green in gel phantom and chicken breast tissue," *Lasers Med. Sci.* **17**, 272–279 (2002).

[38] A.M. Gobin, M.H. Lee, N.J. Halas, W.D. James, R.A. Drezek, and J.L. West, "Near-infrared resonant nanoshells for combined optical imaging and photothermal cancer therapy," *Nano Lett.* **7**, 1929–1934 (2007).

[39] J.F. Hillyer and R.M. Albrecht, "Gastrointestinal persorption and tissue distribution of differently sized colloidal gold nanoparticles," *J. Pharm. Sci.* **90,** 1927–1936 (2001).

[40] J.F. Hillyer and R.M. Albrecht, "Correlative instrumental neutron activation analysis, light microscopy, transmission electron microscopy, and X-ray microanalysis for qualitative and quantitative detection of colloidal gold spheres in biological specimens," *Microsc. Microanal.* **4,** 481–490 (1999).

[41] T. Niidome, M. Yamagata, Y. Okamoto, Y. Akiyama, H. Takahashi, T. Kawano, Y. Katayama, and Y. Niidome, "PEG-modified gold nanorods with a stealth character for in vivo application," *J. Control. Release* **114**, 343–347 (2006).

[42] W.H. De Jong, W.I. Hagens, P. Krystek, M.C. Burger, A.J.A.M. Sips, and R.E. Geertsma, "Particle size-dependent organ distribution of gold nanoparticles after intravenous administration," *Biomaterials* **29,** 1912–1919 (2008).

[43] V. Kattumuri, K. Katti, S. Bhaskaran, E.J. Boote, S.W. Casteel, G.M. Fent, D.J. Robertson, M. Chandrasekhar, R. Kannan, and K.V. Katti, "Gum arabic as a phytochemical construct for the stabilization of gold nanoparticles: in vivo pharmacokinetics and X-ray-contrast-imaging studies," *Small* **3**, 333–341 (2007).

[44] G.S. Terentyuk, G.N. Maslyakova, L.V. Suleymanova, B.N. Khlebtsov, B.Ya. Kogan, G.G. Akchurin, A.V. Shantrocha, I.L. Maksimova, N.G. Khlebtsov, and V.V. Tuchin, "Circulation and distribution of gold nanoparticles and induced alterations of tissue morphology at intravenous particle delivery," *J. Biophoton.* **2**(5), 292–302 (2009).

[45] C. Loo, A. Lowery, N. Halas, J. West, and R. Drezek, "Immunotargeted nanoshells for integrated cancer imaging and therapy," *Nano Lett.* **5**, 709–711 (2005).

[46] R.J. Bernardi, A.R. Lowery, P.A. Thompson, S.M. Blaney, and J.L. West, "Immunonanoshells for targeted photothermal ablation in medulloblastoma and glioma: an *in vitro* evaluation using human cell lines," *J. Neuro-Oncol.* **86**, 165–172 (2008).

[47] T.S. Hauck and W.C.W. Chan, "Gold nanoshells in cancer imaging and therapy: towards clinical application," *Nanomedicine* **2**, 735–738 (2007).

[48] A.R. Lowery, A.M. Gobin, E.S. Day, N.J. Halas, and J.L. West, "Immunonanoshells for targeted photothermal ablation of tumor cells," *Int. J. Nanomed.* **1**(20), 149–154 (2006).

[49] Y. Feng, D. Fuentes, A. Hawkins, J. Bass, M.N. Rylander, A. Elliott, A. Shetty, R. J. Stafford, and J. T. Oden, "Nanoshell-mediated laser surgery simulation for prostate cancer treatment," *Eng. Comput.* **25**(1), 3–13 (2009).

[50] M. Bikram, A.M. Gobin, R.E. Whitmire, and J.L. West, "Temperature-sensitive hydrogels with SiO_2–Au nanoshells for controlled drug delivery," *J. Controlled Release* **123** 219–227 (2007).

[51] J. Shah, S.R. Aglyamov, K. Sokolov, T.E. Milner, and S.Y. Emelianov, "Ultrasound imaging to monitor photothermal therapy—feasibility study," *Opt. Express* **16**, 3776–3785 (2008).

[52] W.L. Yang, D.G. Nair, R. Makizumi, G. Gallos, X. Ye, R.R. Sharma, and T.S. Ravikumar, "Heat shock protein 70 is induced in mouse human colon tumor xenografts after sublethal radiofrequency ablation," *Ann. Surg. Oncol.* **11**, 399–406 (2004).

[53] S.L. Westcott, S.J. Oldenburg, T.R. Lee, and N.J. Halas, "Formation and adsorption of gold nanoparticle-cluster on functionalized silica nanoparticle surfaces," *Langmuir* **14**, 5396–5401 (1998).

[54] N.G. Khlebtsov, V.A. Bogatyrev, L.A. Dykman, and A.G. Melnikov, "Spectral extinction of colloidal gold and its biospecific conjugates," *J. Colloid Interface Sci.* **180**, 436–445 (1996).

[55] G.B. Irani, T. Huen, and F. Wooten, "Optical constants of silver and gold in the visible and vacuum ultraviolet," *J. Opt. Soc. Am.* **61,** 128–129 (1971).

[56] P.B. Johnson and R.W. Christy, "Optical constants of noble metals," *Phys Rev B* **12**, 4370–4379 (1973).

[57] I.L. Maksimova, G.G. Akchurin, G.S. Terentyuk, B.N. Khlebtsov, G.G. Akchurin Jr., I.A. Ermolaev, A.A. Skaptsov, E.M. Revzina, V.V. Tuchin, and N.G. Khlebtsov "Laser photothermolysis of biological tissues by using plasmon-resonsnce particles," *Quantum Electronics* **38**(6), 536–542 (2008).

[58] V.P. Zharov, E.N. Galitovskaya, C. Johnson, and T. Kelly, "Synergistic enhancement of selective nanophotothermolysis with gold nanoclusters: potential for cancer therapy," *Lasers Surg. Med.* **37**, 219–226 (2005).

[59] D. Lapotko, "Optical excitation and detection of vapor bubbles around plasmonic nanoparticles," *Opt. Express*. **17**, 4, 2538–2556 (2009).

[60] E. Faraggi, B.S. Gerstman, and J. Sun, "Biophysical effects of pulsed lasers in the retina and other tissues containing strongly absorbing particles: shockwave and explosive bubble generation," *J. Biomed. Opt.* **10**, 064029 (2005).

[61] D. Lapotko, E. Lukianova, M. Potapnev, O. Aleinikova, and A. Oraevsky, "Method of laser activated nanothermolysis for elimination of tumor cells," *Cancer Lett.* **239**, 36–45 (2006).

[62] D. Lapotko, E. Lukianova, and A. raevsky, "Selective laser nano-thermolysis of human leukemia cells with microbubbles generated around clusters of gold nanoparticles," *Lasers Surg. Med.* **38**, 631–642 (2006).

[63] Y. Hleb, J.H. Hafner, J.N. Myers, E.Y. Hanna, and D.O. Lapotko, "LANTCET: elimination of solid tumor cells with photothermal bubbles generated around clusters of gold nanoparticles," *Nanomedicine* **3**, 648–667 (2008).

[64] K. Donaldson, V. Stone, A. Clouter, L. Renwick, and W. MacNee, "Ultrafine particles," *Occup. Environ. Med.* **58**, 211–216 (2001).

[65] G. Oberdorster, E. Oberdorster, and J. Oberdorster, "Nanotoxicology: an emerging discipline evolving from studies of ultrafine particles," *Environ. Health Perspect.* **113**, 823–839 (2005).

[66] T. Tanaka, P. Decuzzi, M. Cristofanilli, J.H. Sakamoto, E. Tasciotti, F.M. Robertson, and M. Ferrari, "Nanotechnology for breast cancer therapy," *Biomed. Microdevices* **11**, 49–63 (2009).

[67] E. Wisse, F. Braet, D. Luo, et al., "Structure and function of sinusoidal lining cells in the liver," *Toxicol. Pathol.* **24**, 100–111 (1996).

[68] D.C. Bibby, J.E. Talmadge, M.K. Dalal, et al., "Pharmacokinetics and biodistribution of RGDtargeted doxorubicin-loaded nanoparticles in tumor-bearing mice," *Int. J. Pharm.* **293**, 281–290 (2005).

[69] T. Fujita, "A scanning electron microscope study of the human spleen," *Arch. Histol. Jpn.* **37**, 187–216 (1974).

[70] T. Murakami, T. Fujita, and M. Miyoshi, "Closed circulation in the rat spleen as evidenced by scanning electron microscopy of vascular casts," *Experientia* **29**, 1374–1375 (1973).

[71] M. Azemar, S. Djahansouzi, E. Jager, et al., "Regression of cutaneous tumor lesions in patients intratumorally injected with a recombinant single-chain antibodytoxin targeted to ErbB2/HER2" *Breast Cancer Res. Treat.* **82**, 155–164 (2003).

[72] J.R. Less, M.C. Posner, Y. Boucher, D. Borochovitz, N. Wolmark, and R.K. Jain, "Interstitial hypertension in human breast and colorectal tumors," *Cancer Res.* **52**, 6371–6374 (1992).

[73] S.D. Nathanson and L. Nelson, "Interstitial fluid pressure in breast cancer, benign breast conditions, and breast parenchyma," *Ann. Surg. Oncol.* 1, 333–338 (1994).

[74] C.H. Heldin, K. Rubin, K. Pietras, and A. Ostman, "High interstitial fluid pressure—an obstacle in cancer therapy," *Nat. Rev. Cancer* **4**, 806–813 (2004).

[75] H. Hashizume, P. Baluk, S. Morikawa, J.W. McLean, G. Thurston, S. Roberge, et al., "Openings between defective endothelial cells explain tumor vessel leakiness," *Am. J. Pathol.* **156**, 1363–1380 (2000).

[76] P. Henneke and D.T. Golenbock, "Phagocytosis, innate immunity, and host-pathogen specificity," *J. Exp. Med.* **199**, 1–4 (2004).

[77] G. Sonavane, K. Tomoda, and K. Makino, "Biodistribution of colloidal gold nanoparticles after intravenous administration: effect of particle size," *Coll. Surf. B: Biointerf.* **66**, 274–280 (2008).

[78] C. Murphy, A. Gole, J. Stone, P. Sisco, A Alkilany, E. Goldsmith, and S.C. Baxter, "Gold nanoparticles in biology: beyond toxicity to cellular imaging," *Acc. Chem. Res.* **41**, 1721–1730 (2008).

[79] K. Ivarsson, L. Myllymäki, K. Jansner, U. Stenram, and K.G. Tranberg, "Resistance to tumour challenge after tumour laser thermotherapy is associated with a cellular immune response," *Br. J. Cancer* **93**, 435–440 (2005).

[80] S. Basu, R.J. Binder, R. Suto, K.M. Anderson, and P.K. Srivastava, "Necrotic but not apoptotic cell death releases heat shock proteins, which deliver a partial maturation signal to dendritic cells and activate the NF-kappaB pathway," *Int. Immunol.* **12**, 1539–1546 (2000).

[81] S. Lal, S. E. Clare, and N.J. Halas "Nanoshell-enabled photothermal cancer therapy: impending clinical impact," *Accounts Chem. Res.* **41**, 12, 1842–1851 (2008).

[82] I. El-Sayed, X. Huang, and M. El-Sayed "Selective laser photo-thermal therapy of epithelial carcinoma using anti-EGFR antibody conjugated gold nanoparticles," *Cancer Lett.* **239**, 129–35 (2006).

[83] D. Lapotko, E. Lukianova, M. Potapnev, O. Aleinikova, and A. Oraevsky, "Elimination of leukemic cells from human transplants by laser nano-thermolysis," *Proc. SPIE* **6086**, 135–142 (2006).

[84] J.M. Stern, J. Stanfield, W. Kabbani, J.-T. Hsieh, and J.A. Cadeddu, "Selective prostate cancer thermal ablation with laser activated gold nanoshells," *J. Urol.* **179**, 748–753 (2008).

[85] G.S. Terentyuk, G.G. Akchurin., I.L. Maksimova, G.N. Maslyakova, L.V. Suleymanova, and V.V. Tuchin, "Optimization of laser heating with the treatment of spontaneous tumors of domestic animals by the use of the thermography," *Proc. SPIE* **6791**, 67910Q-1–10 (2008).

[86] L.G. Rosenfeld and N.N. Kolotilov, "Remote infrared thermography in oncology," *Onkologiya* **3**, 103–106 (2001) [in Russian].

30

"All Laser" Corneal Surgery by Combination of Femtosecond Laser Ablation and Laser Tissue Welding

Francesca Rossi, Paolo Matteini, Fulvio Ratto

Istituto di Fisica Applicata "Nello Carrara," Consiglio Nazionale delle Ricerche, Sesto Fiorentino, Italy

Luca Menabuoni, Ivo Lenzetti

Ospedale Misericordia e Dolce, Unità Operativa Oculistica, Azienda USL 4, Prato, Italy

Roberto Pini

Istituto di Fisica Applicata "Nello Carrara," Consiglio Nazionale delle Ricerche, Sesto Fiorentino, Italy

In the last decade, laser welding of corneal tissue has been clinically performed in corneal transplant operations for suturing the transplanted corneal flap. The technique is based on the photothermal activation of the stromal tissue, previously stained with a proper chromophore (e.g., indocyanine green) in order to produce absorption of the radiation emitted by a low-power diode laser (at 810 nm) only in the stained tissue. The resulting temperature rise induces modifications in the stromal collagen, creating intramolecular bridges between the cut walls. The results are the sealing of the wound, with negligible thermal damage to the surroundings, and an improved healing process. Recently, this laser welding technique has been proposed to be used in combination with femtosecond laser intratissue cutting of corneal tissue: a flap of preset and constant thickness is sculptured both in the donor and in the recipient eye. The transplanted corneal button thus perfectly matches the recipient bed. The combination of femtosecond laser cutting and diode laser welding thus provides an innovative "all laser" surgery, in which conventional surgical tools, like blades and stitches, can be avoided and replaced by more precise and effective laser tools.

Key words: femtosecond laser, diode laser, indocyanine green, laser welding, laser cutting, corneal transplant

30.1 Basic Principles of Femtosecond Laser Ablation

Femtosecond (Fs) lasers represent new revolutionary tools for intratissue surgery and micromanipulation. These lasers emit femtosecond pulse trains in the near infrared, with repetition rates from hundreds of MHz to kHz, depending on the laser scheme. They are used for inducing nonlinear phenomena in biotissues, which are commonly transparent to these near infrared wavelengths, such as corneal tissue. When Fs laser pulses are focused in a tiny volume (typical focus spot in the order of a few microns), very high power densities are reached ($\sim 10^{13}$ W/cm^2), thus inducing nonlinear multiphoton absorption [1, 2], which generates free electrons within the focal volume. The free electron density produced by multiphoton ionization may overcome a threshold value for generating an optical breakdown: in this condition the free electrons absorb photons through a process called "inverse Bremsstrahlung absorption," thus increasing their energy. After a certain number of these absorption events (depending on the laser wavelength), a free electron has acquired sufficient energy to produce another free electron by impact ionization, i.e., by collision with a heavy particle. Recurring Bremsstrahlung absorption and impact ionization give rise to avalanche ionization and plasma formation. As a consequence of the very fast plasma heating and expansion, shock wave and cavitation bubble generation may be observed in biological tissues and water. The induced bubbles have a particular dynamics: they grow to a maximum diameter (which can be controlled and limited up to within 5–15 microns) inducing ruptures inside the tissue, thus enabling intratissue surgery with a micrometrical precision.

The potentials of femtosecond laser applications for intratissue cutting have been recognized through the years together with the technological improvement of the devices. In some application cases Fs lasers have reached the clinical phase to produce intracellular and intratissue cutting, especially for use in eye surgery [3–5].

30.2 Femtosecond Laser Preparation of Ocular Flaps

Here we describe clinical applications of Fs laser in corneal surgery in order to perform precise cutting of corneal flaps. Clinical trials have been carried out with a commercial Fs laser (IntraLase^{TM}60 FS, Abbott Medical Optics, Santa Ana – CA, USA). This is a Neodymium: Glass laser, emitting near infrared pulses of 600–800 fs duration, with a repetition rate of 60 kHz. The system employs a quartz window that is placed in contact with the external surface of the cornea in order to provide applanation of the cornea, enabling a more controllable and stable laser cutting procedure. The advantages of using a Fs laser for the resection of corneal flaps in comparison with conventional mechanical tools, such as microkeratomes, are several. The Fs laser offers a very high cutting precision (in the order of 5 μm) and repeatability: this means that donor and recipient eyes may be suitably cut in order to achieve an almost perfect matching of the transplanted tissue with the acceptor bed, which results in an effective postoperative healing phase and better vision recovery. Beside this, the flap edges may be designed according to various shapes in order to match different surgical needs.

Commercially available Fs laser systems for corneal trephination can perform both lamellar and vertical cuts, at variable depths in the corneal stroma (see Figure 30.1, where the different corneal structures are evidenced) and at different angles. More specifically, the lamellar cut is obtained in a plane parallel to a lamellar plane, i.e., orthogonal to the propagation of light impinging the eye in normal vision: to obtain such an intrastromal resection plane, the focus of the Fs laser is scanned in a raster pattern, delivering contiguous spots. The diameter of the flat resection area

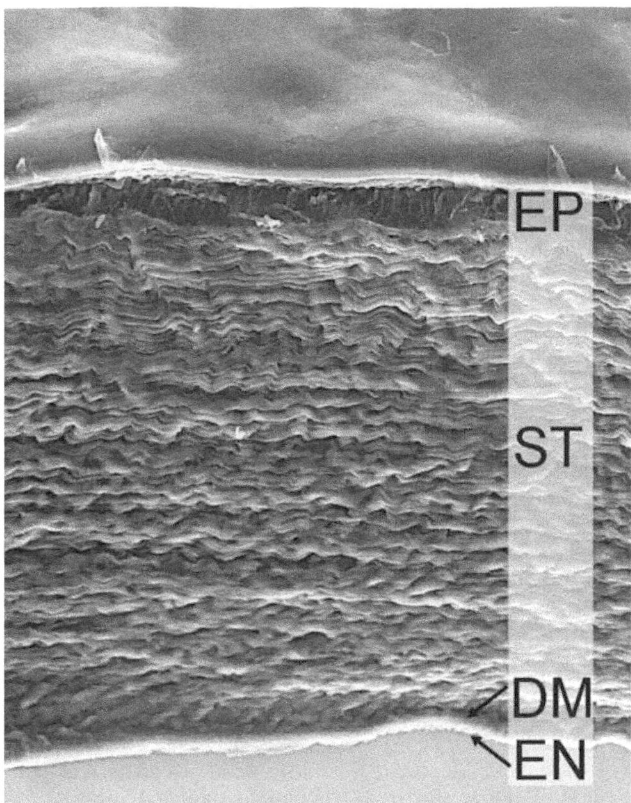

FIGURE 30.1: Image of a cross-section of a porcine cornea taken with a scanning electron microscope (400 X, pressure 5.00 Torr). The different corneal layers are (from top to bottom): epithelium (EP), stroma (ST), Descemet's membrane (DM), and endothelium. Lamellar planes are evident in the stromal layer, running parallel to the external surface of the cornea.

is typically in the range 6–9 mm in corneal transplant procedures. The side cut of the corneal flap is performed by means of a circular pattern of continuous spots, sequentially moved in the postero-anterior direction, i.e., through the stroma from the inner toward the external surface of the cornea. The deposited energy per pulse may be varied (typically in the range 0.5–2.5 μJ), in order to control the efficiency of the cutting effect in the various resection phases: for example, the energy required for the trephination phase to produce a vertical cut across the stromal lamellae is higher then that required for the lamellar resection, performed in parallel to the lamellar layers. This cutting procedure enables to vary also the angle of the side cut of the corneal flap. The result is a manifold-shaped corneal button, properly designed for the required corneal surgery. Commonly used shapes in penetrating keratoplasty (i.e., full-thickness corneal transplant) are for example the "mushroom" one, with a larger diameter cut anteriorly, or the "top-hat" shape, with a larger diameter cut posteriorly, as well as more complicated profiles like the "Christmas Tree" shape [6].

Fs laser resection can thus offer improved cutting precision and flexibility in comparison with mechanical cutting tools, as well as a lower risk to produce incomplete and decentered flaps, perforations, and surgical traumas to the epithelium [6–9]. On the other hand, rare but possible risks associated with the Fs cutting have to be taken into account, such as: 1) the transient light sensitivity syndrome, probably due to a biomechanical response of the corneal cellular components to near infrared light; 2) diffuse lamellar keratitis, eventually related to microscopic collateral effects of photodisruption; and 3) opaque bubble-layer formation, due to the encapsulation of gas bubbles in the intralamellar spaces [6, 7]. The first two problems have been solved improving the femtosecond laser efficiency, using a higher repetition rate (new models with 80–150 kHz repetition rates are now available) and a lower energy per pulse, while the opaque bubble layer may be avoided by selecting a proper side-cut profile, enabling the gas produced after photodisruption to be dispersed in the anterior chamber or in the outside [6]. Performance improvements are expected by the next generation of Fs laser in order to enhance safety and efficiency of cutting operations in eye surgery. On the technological side, the new systems will be more compact, less dependent on the environment conditions (such as temperature and humidity), and less expensive, thus providing a more diffuse and easier utilization of these laser-assisted surgical techniques.

30.3 Low-Power Diode Laser Welding of Ocular Tissues

Laser welding of biological tissues is a technique used to produce immediate closure of wounds. For over 20 years several experiments have been carried out using a variety of lasers for sealing many tissue types, including that of blood vessels, urethra, nerves, dura mater, skin, stomach, and colon [10]. Laser welding enables a reduction in foreign-body reaction, bleeding, suture and needle trauma, as well as in surgical times, and skill requirements. The achievement of a watertight closure of the weld is also another important advantage. However, one possible limitation of laser tissue welding is the damage to tissue by excessive laser heating, which can cause irreversible damage to structural proteins.

Although the precise molecular mechanism of laser tissue welding is still unknown, it is widely considered to be a thermal phenomenon. Laser irradiation induces thermal changes in connective-tissue proteins within cut tissues, resulting in a bond between the two adjoining edges. The control of the dosimetry of the laser irradiation and of the induced temperature rise are crucial in order to minimize the risk of heat damage to the tissue. To overcome this problem and to improve the localization of laser light absorption into tissue, the application of photo-enhancing chromophores has been proposed as a safer technique. The use of wavelength-specific chromophores makes a differential absorption possible between the stained region and the surrounding tissue. The advantage is primarily a selective absorption of laser radiation by the target without the need for a precise focusing of the laser beam. Moreover, lower laser irradiances can be used because of the increased absorption of stained tissue. Various chromophores have been employed as laser absorbers, including indocyanine green (ICG) [11], fluorescein, [12] basic fuchsin, and fen 6 [13]. The addition of solders has also been proposed in order to strengthen the wound during and after laser application. The most useful surgical adhesives are blood, plasma, fibrinogen, and albumin. Following coagulation, these materials act as glues that form an interdigitated matrix among the collagen fibers.

In ophthalmology, experimental studies of laser-induced suturing of corneal tissue on animal models have been reported since 1992 by various authors [14–17]. Major potentialities in using lasers for the welding of corneal tissue are a reduction of postoperative astigmatism, surgical time, and foreign body reaction, when used in substitution of conventional suturing. Most of the laser welding procedures that have been proposed up to now were based on the use of water as an endogenous chromophore for absorbing near- and far-infrared laser light. However, they did not demonstrate it to be safe enough to reach the clinical phase, since the achievement of successful welding of corneal cuts was frequently accompanied by heat side effects that caused partial stromal coagulation and affected corneal transparency. Improved results were obtained using ICG as an exogenous chromophore topically applied to the corneal wound to enhance the absorption of low-power near-infrared diode laser radiation [18–20]. This dye is characterized by high optical absorption around 800 nm (see Figure 30.2), as the stroma is almost transparent at this wavelength

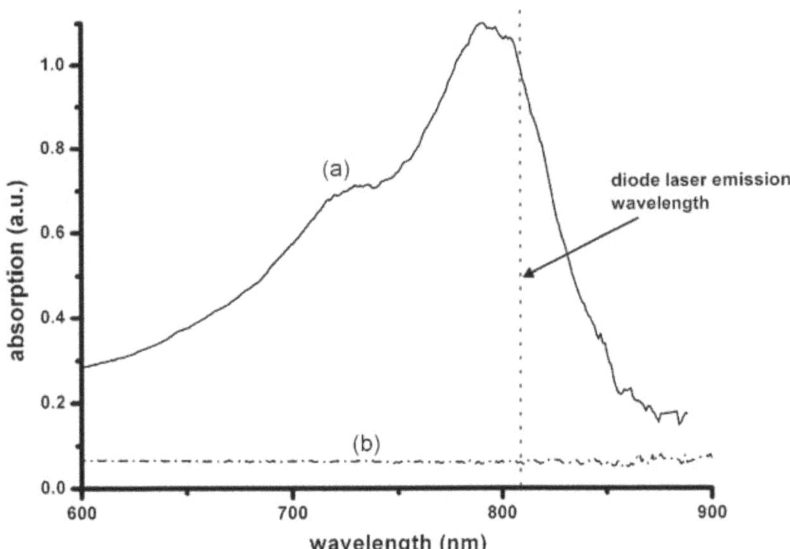

FIGURE 30.2: Absorption curve of an ICG-stained porcine cornea (a) compared with the one of a native cornea (b). Diode laser peak wavelength at 810 nm is also shown.

[21, 22]. The welding of the corneal wound is achieved after staining the cut walls with a water solution of ICG and irradiating with a diode laser, emitting at 810 nm that operates at low power (12.5 to 16.7 W/cm^2 at the corneal surface). The result is a localized heating of the cut that induces a mild and controlled welding of the stromal collagen, thereby minimizing the risk of thermal injury.

These features enabled us to receive the approval from the Italian Ministry of Health for the clinical use of this technique in the transplant of the cornea for performing laser-induced suturing of the donor button in substitution of the conventional continuous suture [20]. In parallel, we have continued to perform experimental studies on animal models aimed at a better comprehension of the corneal welding processes: 1) the healing process of laser-welded corneal tissue was studied in New Zealand albino rabbits during a follow-up period of 90 days, demonstrating shorter recovery times and better restoration of the corneal architecture, in comparison with conventional suturing procedures [23,24]; also, 2) the temperature dynamics during laser-induced corneal welding was experimentally studied by thermometric analysis and compared with a theoretical model describing the temperature distribution at the welded site during laser irradiation; the study indicated a lower temperature increase inside stromal tissue in comparison with that induced by other laser welding procedures not employing photo-enhancing chromophores [25,26].

In the optimized laser welding procedure employed in clinical applications, a very dense preparation of ICG (10% w/w) is used to stain the selected target, e.g., the corneal cut walls. After a few minutes the stained tissue is washed with abundant water, in order to remove the ICG solution in excess. Then the stained wound is irradiated by near infrared diode laser radiation, which induces the laser welding effect.

Laser light irradiation may be performed in continuous wave emission or in the pulsed mode. The continuous wave modality is typically used to close a corneal wound, such as in cataract surgery and in penetrating or lamellar keratoplasty [27, 28]: very low laser power densities (10–20 W/cm^2) are delivered in the noncontact mode, thus inducing a moderate temperature rise within the tissue (in the range of 55–60°C), which enables the sealing of the apposed margins, with moderate strength of the weld. On the other hand, the pulsed irradiation modality provides a stronger welding effect, since it induces local collagen denaturation at temperatures above 65°C. It is used in the contact mode (i.e., with the fiber tip in contact with the tissue) when apposition of tissue flaps is required, such as in the transplant of the endothelium and in the closure of capsulorhexes, i.e., openings on the surface of the lens capsule. In these cases welding operations can be performed in the anterior chamber of the eye, where conventional suturing is not applicable [28–32].

The laser used in clinical welding procedures is an AlGaAs diode laser (Mod. WELD 800, El.En., Florence, Italy), which can be operated in both pulsed and continuous wave emissions, with a power output adjustable in the range 0.5–10.0 W. Laser light is coupled to a 300 μm core fiber, whose tip is mounted in a hand piece, enabling easy handling under a surgical microscope.

Clinical practice carried out in more than 100 patients indicated that inflammatory and foreign body reactions, as typically observed when using standard suturing techniques, are significantly reduced; surgical induced astigmatism is limited (typically below 3.5 Diopters) and maintained at a constant value during the postoperative period.

30.4 Combining Femtosecond Laser Cutting and Diode Laser Suturing

The advantages of laser welding procedures in corneal surgery can be exploited when used in combination with Fs laser cutting of the cornea. The result is a unique "all laser" surgery, which employs only laser tools to enable precise preparation of the donor flap and recipient corneal bed, and to secure the transplanted tissue to the patient's eye in its final position. This combined technique may be used in the transplant of an entire corneal button (penetrating keratoplasty), as well as for the substitution of corneal layers of various thicknesses, thus implementing corneal surgery techniques, while minimizing collateral effects and surgical traumas. Fs laser cutting is first used for the preparation of the donor tissue: the corneal-scleral button, provided by the Eye Bank, is cut with the Fs laser while placed on an artificial anterior chamber, i.e., a support that confers the correct internal pressure and curvature to the excised eye portion, as in a real eye. Then the recipient bed in the patient's eye is prepared with the same Fs laser, according to the particular shape that is required by the specific surgical need. The donor flap is then positioned on the recipient eye after removal of the ill layer. In some cases, the application of a limited number of interrupted stitches is needed to favor the apposition of the edges prior to welding operations. Then the laser welding procedure is performed along the edges of the corneal wound. Typical procedures based on the "all laser" surgery technique are penetrating keratoplasty, anterior lamellar keratoplasty (i.e., the transplant of the upper portion of the corneal stroma, see Figure 30.4 (a) and (b)), and endothelial transplant (i.e., the transplant of the inner corneal layers including the endothelium and the Descemet's membrane, see Figure 30.4 (c)) [28–30, 32,33].

30.4.1 Penetrating keratoplasty

In penetrating keratoplasty (PK), the flexibility of the Fs laser cutting operations is exploited to obtain a complex-pattern trephination (side cut), both in the donor and recipient eye. In the "all laser" PK performed clinically by our team, a corneal cut with a diameter of 8.75 mm and a side-cut profile according to the "Christmas Tree" shape is created by Fs sculpturing in the donor cornea and in the recipient patient eye. Then the transplanted corneal button is secured by eight or sixteen interrupted stitches to ensure its correct positioning. The laser welding procedure is then performed by using the continuous wave irradiation modality, as described before (see Figure 30.3).

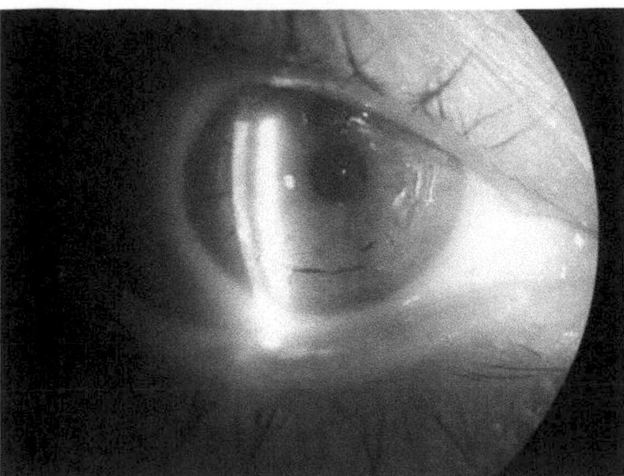

FIGURE 30.3: A transplanted eye 3 days after penetrating keratoplasty. The side cut profile was performed according to the "Christmas Tree" shape, and the transplanted button was secured by 16 interrupted stitches and laser welding.

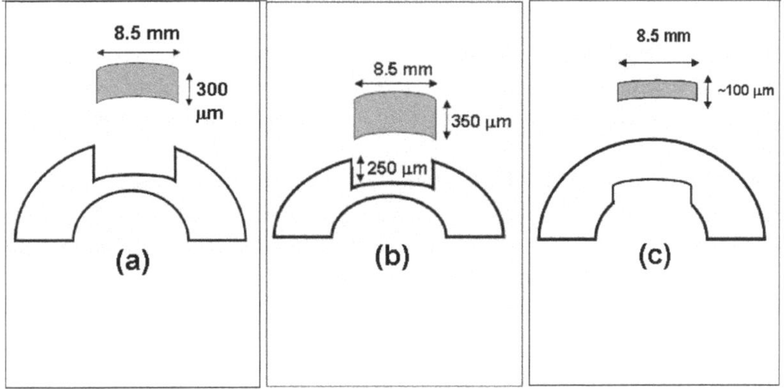

FIGURE 30.4: Schematic sketch of the flap preparation in the "all laser" lamellar surgery (a), anterior lamellar keratoplasty (b), and endothelial (or "deep lamellar") keratoplasty (c).

30.4.2 Anterior lamellar keratoplasty

Anterior lamellar keratoplasty is a new technique for corneal transplant enabling less traumatic surgical operations, with preservation of the inner layers of the patient's cornea. It is suitable in cases of particular pathologies, such as leukoma or keratoconus [29]. The flap preparation is performed by the use of the Fs laser (see Figure 30.4). The Fs cutting procedure is designed to perform in the patient eye a flat raster pattern at a selected lamellar depth in the corneal stroma, so as to remove only the diseased anterior portion of the cornea. The trephination phase is then performed by producing a side cut in between the previously created resection plane and the external surface of the cornea. The donor cornea is treated in a similar way, after it has been positioned on the artificial anterior chamber, as described before. The thickness of the donor flap may be slightly different from the recipient bed depth, depending on the pathology to be treated. For instance, in the cases of central keratoconus, in which the thickness of the central region of cornea may by pathologically reduced, a thicker corneal flap is transplanted in order to restore the original corneal thickness. Laser welding procedure is then performed along the button edges.

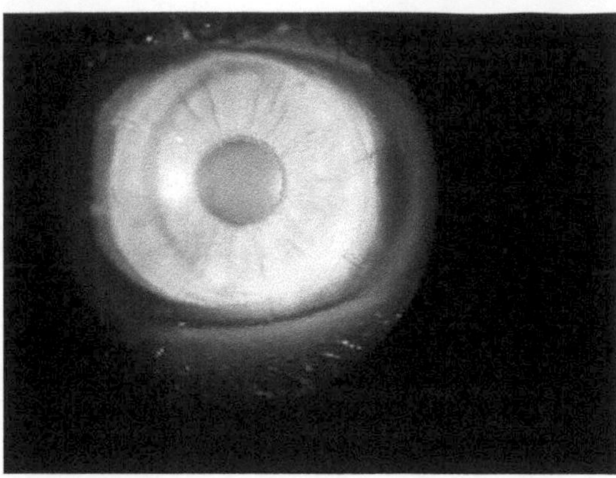

FIGURE 30.5: Anterior lamellar keratoplasty in keratoconus. The image is taken two months after surgery: no inflammatory reaction was detected during the postoperative period, and the visual recovery was satisfactory.

In our experience, about 20 cases of both leukoma and keratoconus have been treated with this technique. In the case of leukoma, donor and host buttons are prepared with exactly the same shape, i.e., 8.5-mm diameter and 300-μm thickness ((Fig 30.4 (a)); no suture is applied and the button is secured by laser welding alone. In the case of keratoconus, a 8.5 mm in diameter, 250-μm deep recipient corneal bed is trephined, while the donor button is shaped to be 100–150 μm thicker (Fig 30.4 (b)). After implantation, 8 supporting stitches are applied and then continuous diode laser welding of the cut edge is performed. After surgery, objective observations typically indicate a good morphological aspect of the cornea and a very good control of the postoperative astigmatism (see Figure 30.5).

30.4.3 Endothelial transplantation (deep lamellar keratoplasty)

A new surgery technique is now under development: the deep lamellar keratoplasty, used to obtain the endothelial transplant. The endothelium is the inner layer of the cornea (see Figure 30.1). In this case the trephination, vertical cut is first performed, with a defined circular perimeter; then a lamellar plane is performed 150–200 μm anteriorly in respect to the endothelium plane. This horizontal plane intentionally has a larger diameter in respect to the circular cut performed vertically, in order to ensure that the two cuts meet. The endothelial flap is cut in a similar shape both in the donor and host eye. In the host eye a small incision is performed at the external periphery of the cornea: through this the donor flap is inserted and assured in the correct position. This very thin corneal flap cannot be sutured with a conventional stitching technique: this is the reason why in conventional endothelium transplant (without laser welding), displacement of the donor graft is observed in about 30% of the cases. The laser welding procedure is in this case the key factor of the surgery success: by employing the pulsed technique it is possible to block the endothelium in its final position. In the treated eyes, 100-μm deep donor button (after epithelium removal) was performed within the Fs cutting phase. In our experience, diode laser spot welding of the periphery of the donor endothelium flap was performed by irradiating the deep layer from the external surface of the cornea [29, 30] (Figure 30.6).

Observations during the healing process indicated that the detachment of the transplanted flap never occurred and a good vision recovery was observed within 7 postoperative days. Optical coherence tomography (OCT) examinations of the treated corneas showed a perfect adhesion of the endothelium to the recipient eye (see Figure 30.7).

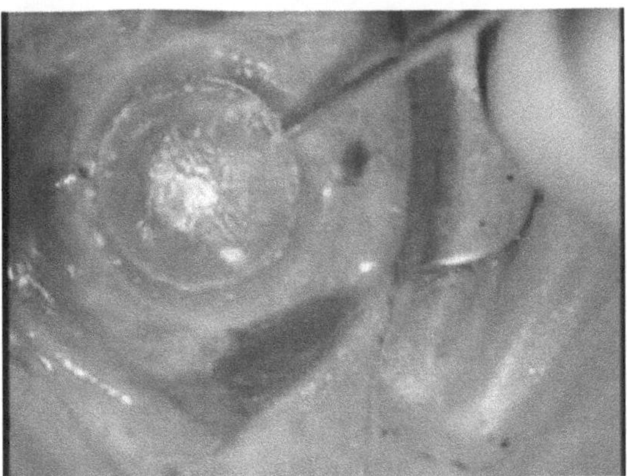

FIGURE 30.6: Laser welding procedure in an endothelial transplant: the peripherally ICG-stained donor flap is secured in the correct final position by a few laser welding spots at the interface of the host and donor corneal layers. Laser irradiation can be performed from the outside of the eye, thus avoiding the risk of mechanical traumas to the endothelial tissue.

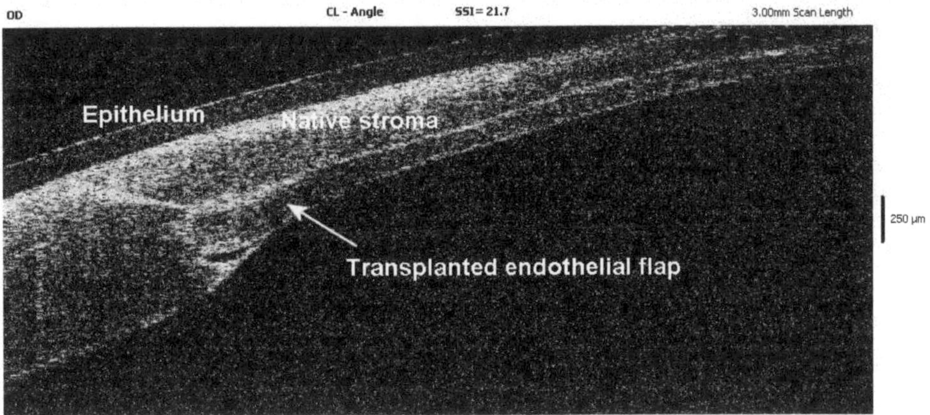

FIGURE 30.7: OCT image of a transplanted endothelial flap, 1 week postoperative. The perfect continuity between the transplanted flap and the native stroma is evident (arrow).

This is a clear example of the advantages offered by an "all laser" surgery: Fs ablation provides a perfect sculpturing of both donor and recipient corneas, not achievable with mechanical cutting tools, while laser welding represents the key factor of the postoperative success, since it makes it possible to block the transplanted endothelium in its final position, avoiding the risk of dislocation.

30.5 Conclusions

"All laser" surgery represents a combination of new laser techniques that are proposed to provide less traumatic and minimally invasive surgical procedures. Present clinical applications are in penetrating, lamellar, and endothelial keratoplasty. A femtosecond laser is used to provide a cutting phase characterized by very high precision and flexibility, enabling complex trephination patterns reproducible at the micrometrical scale. Exactly the same pattern can be obtained in both donor and recipient eyes, thus providing a perfect matching of the wound edges, lesser surgical-induced trauma, and a shorter recovery time. The laser welding procedure enables an immediate sealing of the wound, which may prevent postoperative inflammations and infections, or, in the case of endothelial transplant, avoid the risk of postoperative dislocation of the transplanted flap.

Even if presently implemented only in ophthalmic surgery, "all laser" surgery could be extended to other clinical fields, such as microvascular and reconstructive surgeries. Future improvements are reasonably expected also on the technological side: femtosecond laser systems are nowadays expensive and complex devices, but more compact and lower-cost schemes are under study, which will probably exploit highly precise laser-induced cutting procedures. Even the laser welding technique requires some further developments in order to provide more reliable and stable laser-absorbing dyes; in this respect, studies on the use of new and more efficient and stable nanostructured chromophores are underway [34].

Acknowledgment

The authors wish to thank the financial support offered by the project "SALTO" of the Health Board of Tuscany.

References

[1] A. Vogel and V. Venugopalan, "Mechanisms of pulsed laser ablation of biological tissues," *Chem. Rev.* **103**, 577–644 (2003).

[2] A. Vogel, J. Noack, G. Hüttman, and G. Paltauf, "Mechanisms of femtosecond laser nanosurgery of cells and tissues," *Appl. Phys. B* **81**, 1015–1047 (2005).

[3] K. König, O. Krauss, and I. Riemann, "Intratissue surgery with 80 MHz nanojoule femtosecond laser pulses in the near infrared," *Opt. Expr.* **10**, 171–176 (2002).

[4] H. Lubatschowski, "Overview of commercially available femtosecond lasers in refractive surgery," *J. Refract. Surg.* **24**, 102–107 (2008).

[5] T. Ripken, U. Oberheide, C. Ziltz, W. Ertmer, G. Gerten, and H. Lubatschowski, "Fs-laser induced elasticity changes to improve presbyopic lens accommodation," *Proc. SPIE* **5688**, 278–287 (2005).

[6] H.K. Soong and J.B. Malta, "Femtosecond lasers in ophthalmology," *Am. J. Ophthalm.* **147**, 189–197 (2009).

[7] I. Kaiserman, H.S. Maresky, I. Bahar, and D.S. Rootman, "Incidence, possible risk factors, and potential effects of an opaque bubble layer created by a femtosecond laser," *J. Cataract Refract. Surg.* **34**, 417–423 (2008).

[8] M.P. Holzer, T.M. Rabsilber, and G.U. Auffarth, "Femtosecond laser-assisted corneal flap cuts: morphology, accuracy, and histopathology," *Invest. Ophthalmol. Vis. Sci.* **47**, 2828–2831 (2006).

[9] J.S. Mehat, R. Shilbayeh, Y.M. Por, et al., "Femtosecond laser creation of donor cornea buttons for Descemet-stripping endothelial keratoplasty," *J. Cataract. Refract. Surg.* **34**, 1970–1975 (2008).

[10] K.M. McNally, "Laser tissue welding," Chap. 39, in *Biomedical Photonics Handbook*, T. Vo-Dihn, ed., CRC Press, Boca Raton, 1–45 (2003).

[11] S.D. DeCoste, W. Farinelli, T. Flotte, et al., "Dye-enhanced laser welding for skin closure," *Lasers Surg. Med.* **12**, 25–32 (1992).

[12] D.P. Poppas, P. Sutaria, R.E. Sosa, et al., "Chromophore enhanced laser welding of canine ureters *in vitro* using a human protein solder: a preliminary step for laparoscopic tissue welding," *J. Urol.* **150**, 1052–1055 (1993).

[13] S.G. Brooks, S. Ashley, H. Wright, et al., "The histological measurement of laser-induced thermal damage in vascular tissue using the stain picrosirius red F3BA," *Lasers Med. Sci.*, **6**, 399–405 (1991).

[14] N.L. Burstein, J.M. Williams, M.J. Nowicki, et al., "Corneal welding using hydrogen fluoride laser," *Arch. Ophthalmol.* **110**, 12–13 (1992).

[15] T.J. Desmettre, S.R. Mordon, and V. Mitchell, "Tissue welding for corneal wound suture with CW 1.9 micro diode laser: an in vivo preliminary study," *Proc SPIE* **2623**, 372–379 (1996).

[16] G. Trabucchi, P.G. Gobbi, R. Brancato, et al., "Laser welding of corneal tissue: preliminary experiences using 810 nm and 1950 nm diode lasers," *Proc. SPIE* **2623**, 380–387 (1996).

[17] E. Strassmann, N. Loya, D.D. Gaton, et al., "Laser soldering of the cornea in a rabbit model using a controlled-temperature CO2 laser system," *Proc. SPIE* **4244**, 253–265 (2001).

[18] L. Menabuoni, B. Dragoni, and R. Pini, "Preliminary experiences on diode laser welding in corneal transplantation," *Proc. SPIE* **2922**, 449–452 (1996).

[19] L. Menabuoni, F. Mincione, B. Dragoni, G.P. Mincione, and R. Pini, "Laser welding to assist penetrating keratoplasty: *in vivo* studies," *Proc. SPIE* **3195**, 25–28 (1998).

[20] R. Pini, L. Menabuoni, and L. Starnotti, "First application of laser welding in clinical transplantation of the cornea," *Proc. SPIE* **4244**, 266–271 (2001).

[21] D. Stanescu-Segall and T.L. Jackson, "Vital staining with indocyanine green: a review of the clinical and experimental studies relating to safety," *Eye* **23**, 504–508 (2009).

[22] M.L.J. Landsman, G. Kwant, G.A. Mook, and W.G. Zijlstra, "Light-absorbing properties, stability, and spectral stabilization of indocyanine green," *J. Appl. Physiol.* **40**, 575–583 (1976).

[23] F. Rossi, R. Pini, L. Menabuoni, R. Mencucci, U. Menchini, S. Ambrosini, and G. Vannelli, "Experimental study on the healing process following laser welding of the cornea," *J. Biomed. Opt.* **10**, 024004 (2005).

[24] P. Matteini, F. Rossi, L. Menabuoni, and R. Pini, "Microscopic characterization of collagen modifications induced by low-temperature diode-laser welding of corneal tissue," *Lasers Surg. Med.* **39**, 597–604 (2007).

[25] F. Rossi, R. Pini, and L. Menabuoni, "Experimental and model analysis on the temperature dynamics during diode laser welding of the cornea," *J. Biomed. Opt.* **12**, 014031 (2007).

[26] F. Rossi, P. Matteini, R. Pini, and L. Menabuoni, "Temperature control during diode laser welding in a human cornea," *Proc. SPIE* **6632**, 663215 (2007).

[27] L. Menabuoni, R. Pini, F. Rossi, et al., "Laser-assisted corneal welding in cataract surgery: retrospective study," *J. Cataract Refract. Surg.* **33**, 1608–1612 (2007).

[28] L. Menabuoni, R. Pini, M. Fantozzi, et al., "All-laser sutureless lamellar keratoplasty (ALSL-LK): a first case report," *Invest. Ophthalmol. Vis. Sci.* **47**, E-Abstract 2356 (2006).

[29] F. Rossi, P. Matteini, F. Ratto, L. Menabuoni, I. Lenzetti, and R. Pini, "Laser tissue welding in ophthalmic surgery," *J. Biophotonics* **1**, 331–342 (2008).

[30] L. Menabuoni, I. Lenzetti, T. Rutili, et al., "Combining femtosecond and diode lasers to improve endothelial keratoplasty outcome: a preliminary study," *Invest. Ophthalmol. Vis. Sci.* **48**, E-Abstract 4711 (2007).

[31] R. Pini, F. Rossi, L. Menabuoni, I. Lenzetti, S. Yoo, and J.-M. Parel, "A new technique for the closure of the lens capsule by laser welding," *Ophthalmic Surg. Lasers Imaging* **39**, 260–261 (2008).

[32] L. Menabuoni, I. Lenzetti, L. Cortesini, et al., "Technical improvements in DSAEK performed with the combined use of femtosecond and diode lasers," *Invest. Ophthalmol. Vis. Sci.* **49**: E-Abstract 2319 (2008).

[33] R. Pini, F. Rossi, P. Matteini, F. Ratto, L. Menabuoni, I. Lenzetti, S. H. Yoo, and J.-M. Parel, "Combining femtosecond laser ablation and diode laser welding in lamellar and endothelial corneal transplants," *Proc. SPIE* **6844**, 684411 (2008).

[34] F. Ratto, P. Matteini, F. Rossi, et al., "Photothermal effects in connective tissues mediated by laser-activated gold nanorods," *Nanomedicine* **5** (2), 143–151 (2009).

Index

Milton Keynes UK
Ingram Content Group UK Ltd.
UKHW050306111024
449327UK00043B/2008